AN INTRO... ION TO MATH... ISTICS A... ATIONS

Fifth Edition

...J. LARSEN
...rbilt University

...S L. MARX
...f West Florida

... Columbus Indianapolis New York San Francisco
Upper Saddle River Amsterdam Cape Town Dubai
London Madrid Milan Munich Paris Montréal
Toronto Delhi Mexico City São Paulo Sydney
Hong Kong Seoul Singapore Taipei Tokyo

Editor in Chief: Deirdre Lynch
Acquisitions Editor: Christopher Cummings
Associate Editor: Christina Lepre
Assistant Editor: Dana Jones
Senior Managing Editor: Karen Wernholm
Associate Managing Editor: Tamela Ambush
Senior Production Project Manager: Peggy McMahon
Senior Design Supervisor: Andrea Nix
Cover Design: Tina Gleason
Interior Design: Tamara Newnam
Marketing Manager: Alex Gay
Marketing Assistant: Kathleen DeChavez
Senior Author Support/Technology Specialist: Joe Vetere
Manufacturing Manager: Evelyn Beaton
Senior Manufacturing Buyer: Carol Melville
Production Coordination, Technical Illustrations, and Composition: Integra Software Services, Inc.

Prentice Hall
is an imprint of

1 2 3 4 5 6 7 8 9 10—EB—14 13 12 11 10

ISBN-13: 978-0-321-76656-4
ISBN-10: 0-321-76656-3

Table of Contents

4 Special Distributions 221

5 Estimation 281

6 Hypothesis Testing 350

7 Inferences Based on the Normal Distribution 385

8 Types of Data: A Brief Overview 430

9 Two-Sample Inferences 457

10 Goodness-of-Fit Tests 493

PREFACE

The first edition of this text was published in 1981. Each subsequent revision since then has undergone more than a few changes. Topics have been added, computer software and simulations introduced, and examples redone. What has *not* changed over the years is our pedagogical focus. As the title indicates, this book is an introduction to mathematical statistics *and its applications*. Those last three words are not an afterthought. We continue to believe that mathematical statistics is best learned and most effectively motivated when presented against a backdrop of real-world examples and all the issues that those examples necessarily raise.

We recognize that college students today have more mathematics courses to choose from than ever before because of the new specialties and interdisciplinary areas that continue to emerge. For students wanting a broad educational experience, an introduction to a given topic may be all that their schedules can reasonably accommodate. Our response to that reality has been to ensure that each edition of this text provides a more comprehensive and more usable treatment of statistics than did its predecessors.

Traditionally, the focus of mathematical statistics has been fairly narrow—the subject's objective has been to provide the theoretical foundation for all of the various procedures that are used for describing and analyzing data. What it has not spoken to at much length are the important questions of *which* procedure to use in a given situation, and *why*. But those are precisely the concerns that every user of statistics must inevitably confront. To that end, adding features that can create a path from the theory of statistics to its practice has become an increasingly high priority.

New to This Edition

- Beginning with the third edition, Chapter 8, titled "Data Models," was added. It discussed some of the basic principles of experimental design, as well as some guidelines for knowing how to begin a statistical analysis. In this fifth edition, the Data Models ("Types of Data: A Brief Overview") chapter has been substantially rewritten to make its main points more accessible.
- Beginning with the fourth edition, the end of each chapter except the first featured a section titled "Taking a Second Look at Statistics." Many of these sections describe the ways that statistical terminology is often misinterpreted in what we see, hear, and read in our modern media. Continuing in this vein of interpretation, we have added in this fifth edition comments called "About the Data." These sections are scattered throughout the text and are intended to encourage the reader to think critically about a data set's assumptions, interpretations, and implications.
- Many examples and case studies have been updated, while some have been deleted and others added.
- Section 3.8, "Transforming and Combining Random Variables," has been rewritten.

- Section 3.9, "Further Properties of the Mean and Variance," now includes a discussion of covariances so that sums of random variables can be dealt with in more generality.
- Chapter 5, "Estimation," now has an introduction to bootstrapping.
- Chapter 7, "Inferences Based on the Normal Distribution," has new material on the noncentral t distribution and its role in calculating Type II error probabilities.
- Chapter 9, "Two-Sample Inferences," has a derivation of Welch's approximation for testing the differences of two means in the case of unequal variances.

We hope that the changes in this edition will not undo the best features of the first four. What made the task of creating the fifth edition an enjoyable experience was the nature of the subject itself and the way that it can be beautifully elegant and down-to-earth practical, all at the same time. Ultimately, our goal is to share with the reader at least some small measure of the affection we feel for mathematical statistics and its applications.

Supplements

Instructor's Solutions Manual. This resource contains worked-out solutions to all text exercises and is available for download from the Pearson Education Instructor Resource Center.

Student Solutions Manual ISBN-10: 0-321-69402-3; ISBN-13: 978-0-321-69402-7. Featuring complete solutions to selected exercises, this is a great tool for students as they study and work through the problem material.

Acknowledgments

We would like to thank the following reviewers for their detailed and valuable comments, criticisms, and suggestions:

Dr. Abera Abay, *Rowan University*
Kyle Siegrist, *University of Alabama in Huntsville*
Ditlev Monrad, *University of Illinois at Urbana-Champaign*
Vidhu S. Prasad, *University of Massachusetts, Lowell*
Wen-Qing Xu, *California State University, Long Beach*
Katherine St. Clair, *Colby College*
Yimin Xiao, *Michigan State University*
Nicolas Christou, *University of California, Los Angeles*
Daming Xu, *University of Oregon*
Maria Rizzo, *Ohio University*
Dimitris Politis, *University of California at San Diego*

Finally, we convey our gratitude and appreciation to Pearson Arts & Sciences Associate Editor for Statistics Christina Lepre; Acquisitions Editor Christopher Cummings; and Senior Production Project Manager Peggy McMahon, as well as to Project Manager Amanda Zagnoli of Elm Street Publishing Services, for their excellent teamwork in the production of this book.

Richard J. Larsen
Nashville, Tennessee
Morris L. Marx
Pensacola, Florida

Chapter 1

INTRODUCTION

"Until the phenomena of any branch of knowledge have been submitted to measurement and number it cannot assume the status and dignity of a science."

—Francis Galton

1.1 An Overview

Sir Francis Galton was a preeminent biologist of the nineteenth century. A passionate advocate for the theory of evolution (his nickname was "Darwin's bulldog"), Galton was also an early crusader for the study of statistics and believed the subject would play a key role in the advancement of science:

> Some people hate the very name of statistics, but I find them full of beauty and interest. Whenever they are not brutalized, but delicately handled by the higher methods, and are warily interpreted, their power of dealing with complicated phenomena is extraordinary. They are the only tools by which an opening can be cut through the formidable thicket of difficulties that bars the path of those who pursue the Science of man.

Did Galton's prediction come to pass? Absolutely—try reading a biology journal or the analysis of a psychology experiment before taking your first statistics course. Science and statistics have become inseparable, two peas in the same pod. What the good gentleman from London failed to anticipate, though, is the extent to which *all* of us—not just scientists—have become enamored (some would say obsessed) with numerical information. The stock market is awash in averages, indicators, trends, and exchange rates; federal education initiatives have taken standardized testing to new levels of specificity; Hollywood uses sophisticated demographics to see who's watching what, and why; and pollsters regularly tally and track our every opinion, regardless of how irrelevant or uninformed. In short, we have come to expect everything to be measured, evaluated, compared, scaled, ranked, and rated—and if the results are deemed unacceptable for whatever reason, we demand that someone or something be held accountable (in some appropriately quantifiable way).

To be sure, many of these efforts are carefully carried out and make perfectly good sense; unfortunately, others are seriously flawed, and some are just plain nonsense. What they all speak to, though, is the clear and compelling need to know something about the subject of statistics, its uses and its misuses.

This book addresses two broad topics—the *mathematics of statistics* and the *practice of statistics*. The two are quite different. The former refers to the probability theory that supports and justifies the various methods used to analyze data. For the most part, this background material is covered in Chapters 2 through 7. The key result is the *central limit theorem*, which is one of the most elegant and far-reaching results in all of mathematics. (Galton believed the ancient Greeks would have personified and deified the central limit theorem had they known of its existence.) Also included in these chapters is a thorough introduction to combinatorics, the mathematics of systematic counting. Historically, this was the very topic that launched the development of probability in the first place, back in the seventeenth century. In addition to its connection to a variety of statistical procedures, combinatorics is also the basis for every state lottery and every game of chance played with a roulette wheel, a pair of dice, or a deck of cards.

The practice of statistics refers to all the issues (and there are many!) that arise in the design, analysis, and interpretation of data. Discussions of these topics appear in several different formats. Following most of the case studies throughout the text is a feature entitled "About the Data." These are additional comments about either the particular data in the case study or some related topic suggested by those data. Then near the end of most chapters is a Taking a Second Look at Statistics section. Several of these deal with the *misuses* of statistics—specifically, inferences drawn incorrectly and terminology used inappropriately. The most comprehensive data-related discussion comes in Chapter 8, which is devoted entirely to the critical problem of knowing how to *start* a statistical analysis—that is, knowing which procedure should be used, and why.

More than a century ago, Galton described what he thought a knowledge of statistics should entail. Understanding "the higher methods," he said, was the key to ensuring that data would be "delicately handled" and "warily interpreted." The goal of this book is to make that happen.

1.2 Some Examples

Statistical methods are often grouped into two broad categories—descriptive statistics and inferential statistics. The former refers to all the various techniques for summarizing and displaying data. These are the familiar bar graphs, pie charts, scatterplots, means, medians, and the like, that we see so often in the print media. The much more mathematical inferential statistics are procedures that make generalizations and draw conclusions of various kinds based on the information contained in a set of data; moreover, they calculate the probability of the generalizations being correct.

Described in this section are three case studies. The first illustrates a very effective use of several descriptive techniques. The latter two illustrate the sorts of questions that inferential procedures can help answer.

Case Study 1.2.1

Pictured at the top of Figure 1.2.1 is the kind of information routinely recorded by a seismograph—listed chronologically are the occurrence times and Richter magnitudes for a series of earthquakes. As raw data, the numbers are largely

(Continued on next page)

meaningless: No patterns are evident, nor is there any obvious connection between the frequencies of tremors and their severities.

Episode number	Date	Time	Severity (Richter scale)
⋮	⋮	⋮	⋮
217	6/19	4:53 P.M.	2.7
218	7/2	6:07 A.M.	3.1
219	7/4	8:19 A.M.	2.0
220	8/7	1:10 A.M.	4.1
221	8/7	10:46 P.M.	3.6
⋮	⋮	⋮	⋮

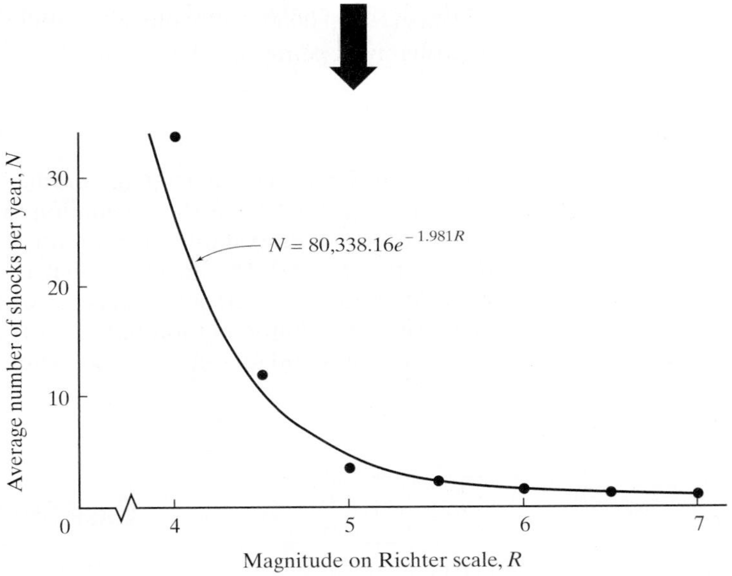

Average number of shocks per year, N

$N = 80{,}338.16e^{-1.981R}$

Magnitude on Richter scale, R

Figure 1.2.1

Shown at the bottom of the figure is the result of applying several descriptive techniques to an actual set of seismograph data recorded over a period of several years in southern California (67). Plotted above the Richter (R) value of 4.0, for example, is the average number (N) of earthquakes occurring per year in that region having magnitudes in the range 3.75 to 4.25. Similar points are included for R-values centered at 4.5, 5.0, 5.5, 6.0, 6.5, and 7.0. Now we can see that earthquake frequencies and severities are clearly related: Describing the (N, R)'s exceptionally well is the equation

$$N = 80{,}338.16e^{-1.981R} \tag{1.2.1}$$

which is found using a procedure described in Chapter 9. (*Note*: Geologists have shown that the model $N = \beta_0 e^{\beta_1 R}$ describes the (N, R) relationship all over the world. All that changes from region to region are the numerical values for β_0 and β_1.)

(Continued on next page)

(Case Study 1.2.1 continued)

Notice that Equation 1.2.1 is more than just an elegant summary of the observed (N, R) relationship. Rather, it allows us to estimate the likelihood of future earthquake catastrophes for large values of R that have never been recorded. For example, many Californians worry about the "Big One," a monster tremor—say, $R = 10.0$—that breaks off chunks of tourist-covered beaches and sends them floating toward Hawaii. How often might we expect that to happen? Setting $R = 10.0$ in Equation 1.2.1 gives

$$N = 80{,}338.16e^{-1.98(10.0)}$$
$$= 0.0002 \text{ earthquake per year}$$

which translates to a prediction of one such megaquake every five thousand years $(= 1/0.0002)$. (Of course, whether that estimate is alarming or reassuring probably depends on whether you live in San Diego or Topeka. . . .)

About the Data The megaquake prediction prompted by Equation 1.2.1 raises an obvious question: Why is the calculation that led to the model $N = 80{,}338.16e^{-1.981R}$ not considered an example of inferential statistics even though it did yield a prediction for $R = 10$? The answer is that Equation 1.2.1—by itself—does not tell us anything about the "error" associated with its predictions. In Chapter 11, a more elaborate probability method based on Equation 1.2.1 is described that does yield error estimates and qualifies as a bona fide inference procedure.

Case Study 1.2.2

Claims of disputed authorship can be very difficult to resolve. Speculation has persisted for several hundred years that some of William Shakespeare's works were written by Sir Francis Bacon (or maybe Christopher Marlowe). And whether it was Alexander Hamilton or James Madison who wrote certain of the Federalist Papers is still an open question. Less well known is a controversy surrounding Mark Twain and the Civil War.

One of the most revered of all American writers, Twain was born in 1835, which means he was twenty-six years old when hostilities between the North and South broke out. At issue is whether he was ever a participant in the war—and, if he was, on which side. Twain always dodged the question and took the answer to his grave. Even had he made a full disclosure of his military record, though, his role in the Civil War would probably still be a mystery because of his self-proclaimed predisposition to be less than truthful. Reflecting on his life, Twain made a confession that would give any would-be biographer pause: "I am an old man," he said, "and have known a great many troubles, but most of them never happened."

What some historians think might be the clue that solves the mystery is a set of ten essays that appeared in 1861 in the *New Orleans Daily Crescent*. Signed

(Continued on next page)

"Quintus Curtius Snodgrass," the essays purported to chronicle the author's adventures as a member of the Louisiana militia. Many experts believe that the exploits described actually did happen, but Louisiana field commanders had no record of anyone named Quintus Curtius Snodgrass. More significantly, the pieces display the irony and humor for which Twain was so famous.

Table 1.2.1 summarizes data collected in an attempt (16) to use statistical inference to resolve the debate over the authorship of the Snodgrass letters. Listed are the proportions of three-letter words (1) in eight essays known to have been written by Mark Twain and (2) in the ten Snodgrass letters.

Researchers have found that authors tend to have characteristic word-length profiles, regardless of what the topic might be. It follows, then, that if Twain and Snodgrass were the same person, the proportion of, say, three-letter words that they used should be roughly the same. The bottom of Table 1.2.1 shows that, on the average, 23.2% of the words in a Twain essay were three letters long; the corresponding average for the Snodgrass letters was 21.0%.

If Twain and Snodgrass were the same person, the difference between these average three-letter proportions should be close to 0: for these two sets of essays, the difference in the averages was 0.022 (=0.232 − 0.210). How should we interpret the difference 0.022 in this context? Two explanations need to be considered:

1. The difference, 0.022, is sufficiently small (i.e., close to 0) that it does not rule out the possibility that Twain and Snodgrass were the same person.

or

2. The difference, 0.022, is so large that the only reasonable conclusion is that Twain and Snodgrass were not the same person.

Choosing between explanations 1 and 2 is an example of hypothesis testing, which is a very frequently encountered form of statistical inference.

The principles of hypothesis testing are introduced in Chapter 6, and the particular procedure that applies to Table 1.2.1 first appears in Chapter 9. So as not to spoil the ending of a good mystery, we will defer unmasking Mr. Snodgrass until then.

Table 1.2.1

Twain	Proportion	QCS	Proportion
Sergeant Fathom letter	0.225	Letter I	0.209
Madame Caprell letter	0.262	Letter II	0.205
Mark Twain letters in		Letter III	0.196
Territorial Enterprise		Letter IV	0.210
First letter	0.217	Letter V	0.202
Second letter	0.240	Letter VI	0.207
Third letter	0.230	Letter VII	0.224
Fourth letter	0.229	Letter VIII	0.223
First *Innocents Abroad* letter		Letter IX	0.220
First half	0.235	Letter X	0.201
Second half	0.217		
Average:	**0.232**		**0.210**

Case Study 1.2.3

It may not be made into a movie anytime soon, but the way that statistical inference was used to spy on the Nazis in World War II is a pretty good tale. And it certainly did have a surprise ending!

The story began in the early 1940s. Fighting in the European theatre was intensifying, and Allied commanders were amassing a sizeable collection of abandoned and surrendered German weapons. When they inspected those weapons, the Allies noticed that each one bore a different number. Aware of the Nazis' reputation for detailed record keeping, the Allies surmised that each number represented the chronological order in which the piece had been manufactured. But if that was true, might it be possible to use the "captured" serial numbers to estimate the total number of weapons the Germans had produced?

That was precisely the question posed to a group of government statisticians working out of Washington, D.C. Wanting to estimate an adversary's manufacturing capability was, of course, nothing new. Up to that point, though, the only sources of that information had been spies and traitors; using serial numbers was something entirely new.

The answer turned out to be a fairly straightforward application of the principles that will be introduced in Chapter 5. If n is the total number of captured serial numbers and x_{max} is the largest captured serial number, then the estimate for the total number of items produced is given by the formula

$$\text{estimated output} = [(n+1)/n]x_{max} - 1 \tag{1.2.2}$$

Suppose, for example, that $n = 5$ tanks were captured and they bore the serial numbers 92, 14, 28, 300, and 146, respectively. Then $x_{max} = 300$ and the estimated total number of tanks manufactured is 359:

$$\begin{aligned}\text{estimated output} &= [(5+1)/5]300 - 1 \\ &= 359\end{aligned}$$

Did Equation 1.2.2 work? Better than anyone could have expected (probably even the statisticians). When the war ended and the Third Reich's "true" production figures were revealed, it was found that serial number estimates were far more accurate in every instance than all the information gleaned from traditional espionage operations, spies, and informants. The serial number estimate for German tank production in 1942, for example, was 3400, a figure very close to the actual output. The "official" estimate, on the other hand, based on intelligence gathered in the usual ways, was a grossly inflated 18,000 (64).

About the Data Large discrepancies, like 3400 versus 18,000 for the tank estimates, were not uncommon. The espionage-based estimates were consistently erring on the high side because of the sophisticated Nazi propaganda machine that deliberately exaggerated the country's industrial prowess. On spies and would-be adversaries, the Third Reich's carefully orchestrated dissembling worked exactly as planned; on Equation 1.2.2, though, it had no effect whatsoever!

1.3 A Brief History

For those interested in how we managed to get to where we are (or who just want to procrastinate a bit longer), Section 1.3 offers a brief history of probability and statistics. The two subjects were not mathematical littermates—they began at different times in different places for different reasons. How and why they eventually came together makes for an interesting story and reacquaints us with some towering figures from the past.

Probability: The Early Years

No one knows where or when the notion of chance first arose; it fades into our prehistory. Nevertheless, evidence linking early humans with devices for generating random events is plentiful: Archaeological digs, for example, throughout the ancient world consistently turn up a curious overabundance of *astragali*, the heel bones of sheep and other vertebrates. Why should the frequencies of these bones be so disproportionately high? One could hypothesize that our forebears were fanatical foot fetishists, but two other explanations seem more plausible: The bones were used for religious ceremonies *and for gambling*.

Astragali have six sides but are not symmetrical (see Figure 1.3.1). Those found in excavations typically have their sides numbered or engraved. For many ancient civilizations, astragali were the primary mechanism through which oracles solicited the opinions of their gods. In Asia Minor, for example, it was customary in divination rites to roll, or *cast*, five astragali. Each possible configuration was associated with the name of a god and carried with it the sought-after advice. An outcome of (1, 3, 3, 4, 4), for instance, was said to be the throw of the savior Zeus, and its appearance was taken as a sign of encouragement (34):

> One one, two threes, two fours
> The deed which thou meditatest, go do it boldly.
> Put thy hand to it. The gods have given thee
> favorable omens
> Shrink not from it in thy mind, for no evil
> shall befall thee.

Figure 1.3.1

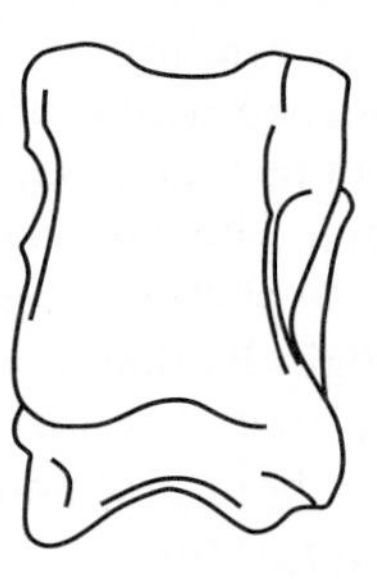

Sheep astragalus

A (4, 4, 4, 6, 6), on the other hand, the throw of the child-eating Cronos, would send everyone scurrying for cover:

> Three fours and two sixes. God speaks as follows.
> Abide in thy house, nor go elsewhere,

> Lest a ravening and destroying beast come nigh thee.
> For I see not that this business is safe. But bide
> thy time.

Gradually, over thousands of years, astragali were replaced by dice, and the latter became the most common means for generating random events. Pottery dice have been found in Egyptian tombs built before 2000 B.C.; by the time the Greek civilization was in full flower, dice were everywhere. (*Loaded* dice have also been found. Mastering the mathematics of probability would prove to be a formidable task for our ancestors, but they quickly learned how to cheat!)

The lack of historical records blurs the distinction initially drawn between divination ceremonies and recreational gaming. Among more recent societies, though, gambling emerged as a distinct entity, and its popularity was irrefutable. The Greeks and Romans were consummate gamblers, as were the early Christians (91).

Rules for many of the Greek and Roman games have been lost, but we can recognize the lineage of certain modern diversions in what was played during the Middle Ages. The most popular dice game of that period was called *hazard*, the name deriving from the Arabic *al zhar*, which means "a die." Hazard is thought to have been brought to Europe by soldiers returning from the Crusades; its rules are much like those of our modern-day craps. Cards were first introduced in the fourteenth century and immediately gave rise to a game known as *Primero*, an early form of poker. Board games such as backgammon were also popular during this period.

Given this rich tapestry of games and the obsession with gambling that characterized so much of the Western world, it may seem more than a little puzzling that a formal study of probability was not undertaken sooner than it was. As we will see shortly, the first instance of anyone *conceptualizing* probability in terms of a mathematical model occurred in the sixteenth century. That means that more than 2000 years of dice games, card games, and board games passed by before someone finally had the insight to write down even the simplest of probabilistic abstractions.

Historians generally agree that, as a subject, probability got off to a rocky start because of its incompatibility with two of the most dominant forces in the evolution of our Western culture, Greek philosophy and early Christian theology. The Greeks were comfortable with the notion of chance (something the Christians were not), but it went against their nature to suppose that random events could be quantified in any useful fashion. They believed that any attempt to reconcile mathematically what *did* happen with what *should have* happened was, in their phraseology, an improper juxtaposition of the "earthly plane" with the "heavenly plane."

Making matters worse was the antiempiricism that permeated Greek thinking. Knowledge, to them, was not something that should be derived by experimentation. It was better to reason out a question logically than to search for its explanation in a set of numerical observations. Together, these two attitudes had a deadening effect: The Greeks had no motivation to think about probability in any abstract sense, nor were they faced with the problems of interpreting data that might have pointed them in the direction of a probability calculus.

If the prospects for the study of probability were dim under the Greeks, they became even worse when Christianity broadened its sphere of influence. The Greeks and Romans at least accepted the *existence* of chance. However, they believed their gods to be either unable or unwilling to get involved in matters so mundane as the outcome of the roll of a die. Cicero writes:

> Nothing is so uncertain as a cast of dice, and yet there is no one who plays often who does not make a Venus-throw[1] and occasionally twice and thrice in succession. Then are we, like fools, to prefer to say that it happened by the direction of Venus rather than by chance?

For the early Christians, though, there was no such thing as chance: Every event that happened, no matter how trivial, was perceived to be a direct manifestation of God's deliberate intervention. In the words of St. Augustine:

> Nos eas causas quae dicuntur fortuitae ... non dicimus
> nullas, sed latentes; easque tribuimus vel veri Dei ...
> (We say that those causes that are said to be by chance
> are not non-existent but are hidden, and we attribute
> them to the will of the true God ...)

Taking Augustine's position makes the study of probability moot, and it makes a probabilist a heretic. Not surprisingly, nothing of significance was accomplished in the subject for the next fifteen hundred years.

It was in the sixteenth century that probability, like a mathematical Lazarus, arose from the dead. Orchestrating its resurrection was one of the most eccentric figures in the entire history of mathematics, Gerolamo Cardano. By his own admission, Cardano personified the best and the worst—the Jekyll and the Hyde—of the Renaissance man. He was born in 1501 in Pavia. Facts about his personal life are difficult to verify. He wrote an autobiography, but his penchant for lying raises doubts about much of what he says. Whether true or not, though, his "one-sentence" self-assessment paints an interesting portrait (127):

> Nature has made me capable in all manual work, it has given me the spirit of a philosopher and ability in the sciences, taste and good manners, voluptuousness, gaiety, it has made me pious, faithful, fond of wisdom, meditative, inventive, courageous, fond of learning and teaching, eager to equal the best, to discover new things and make independent progress, of modest character, a student of medicine, interested in curiosities and discoveries, cunning, crafty, sarcastic, an initiate in the mysterious lore, industrious, diligent, ingenious, living only from day to day, impertinent, contemptuous of religion, grudging, envious, sad, treacherous, magician and sorcerer, miserable, hateful, lascivious, obscene, lying, obsequious, fond of the prattle of old men, changeable, irresolute, indecent, fond of women, quarrelsome, and because of the conflicts between my nature and soul I am not understood even by those with whom I associate most frequently.

Formally trained in medicine, Cardano's interest in probability derived from his addiction to gambling. His love of dice and cards was so all-consuming that he is said to have once sold all his wife's possessions just to get table stakes! Fortunately, something positive came out of Cardano's obsession. He began looking for a mathematical model that would describe, in some abstract way, the outcome of a random event. What he eventually formalized is now called the *classical definition of probability*: If the total number of possible outcomes, all equally likely, associated with some action is n, and if m of those n result in the occurrence of some given event, then the probability of that event is m/n. If a fair die is rolled, there are $n = 6$ possible outcomes. If the event "Outcome is greater than or equal to 5" is the one in

---

[1] When rolling four astragali, each of which is numbered on *four* sides, a Venus-throw was having each of the four numbers appear.

Figure 1.3.2

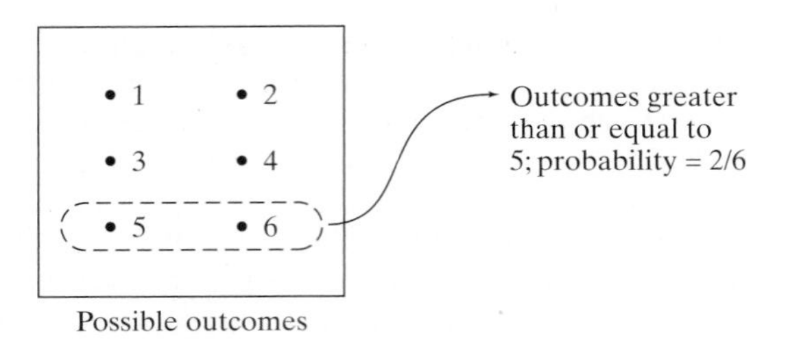

which we are interested, then $m = 2$ (the outcomes 5 and 6) and the probability of the event is $\frac{2}{6}$, or $\frac{1}{3}$ (see Figure 1.3.2).

Cardano had tapped into the most basic principle in probability. The model he discovered may seem trivial in retrospect, but it represented a giant step forward: His was the first recorded instance of anyone computing a *theoretical*, as opposed to an empirical, probability. Still, the actual impact of Cardano's work was minimal. He wrote a book in 1525, but its publication was delayed until 1663. By then, the focus of the Renaissance, as well as interest in probability, had shifted from Italy to France.

The date cited by many historians (those who are not Cardano supporters) as the "beginning" of probability is 1654. In Paris a well-to-do gambler, the Chevalier de Méré, asked several prominent mathematicians, including Blaise Pascal, a series of questions, the best known of which is the *problem of points*:

> Two people, A and B, agree to play a series of fair games until one person has won six games. They each have wagered the same amount of money, the intention being that the winner will be awarded the entire pot. But suppose, for whatever reason, the series is prematurely terminated, at which point A has won five games and B three. How should the stakes be divided?

[The correct answer is that A should receive seven-eighths of the total amount wagered. (*Hint*: Suppose the contest were resumed. What scenarios would lead to A's being the first person to win six games?)]

Pascal was intrigued by de Méré's questions and shared his thoughts with Pierre Fermat, a Toulouse civil servant and probably the most brilliant mathematician in Europe. Fermat graciously replied, and from the now-famous Pascal-Fermat correspondence came not only the solution to the problem of points but the foundation for more general results. More significantly, news of what Pascal and Fermat were working on spread quickly. Others got involved, of whom the best known was the Dutch scientist and mathematician Christiaan Huygens. The delays and the indifference that had plagued Cardano a century earlier were not going to happen again.

Best remembered for his work in optics and astronomy, Huygens, early in his career, was intrigued by the problem of points. In 1657 he published *De Ratiociniis in Aleae Ludo* (*Calculations in Games of Chance*), a very significant work, far more comprehensive than anything Pascal and Fermat had done. For almost fifty years it was the standard "textbook" in the theory of probability. Not surprisingly, Huygens has supporters who feel that *he* should be credited as the founder of probability.

Almost all the mathematics of probability was still waiting to be discovered. What Huygens wrote was only the humblest of beginnings, a set of fourteen propositions bearing little resemblance to the topics we teach today. But the foundation was there. The mathematics of probability was finally on firm footing.

Statistics: From Aristotle to Quetelet

Historians generally agree that the basic principles of statistical reasoning began to coalesce in the middle of the nineteenth century. What triggered this emergence was the union of three different "sciences," each of which had been developing along more or less independent lines (195).

The first of these sciences, what the Germans called *Staatenkunde*, involved the collection of comparative information on the history, resources, and military prowess of nations. Although efforts in this direction peaked in the seventeenth and eighteenth centuries, the concept was hardly new: Aristotle had done something similar in the fourth century B.C. Of the three movements, this one had the least influence on the development of modern statistics, but it did contribute some terminology: The word *statistics*, itself, first arose in connection with studies of this type.

The second movement, known as *political arithmetic*, was defined by one of its early proponents as "the art of reasoning by figures, upon things relating to government." Of more recent vintage than Staatenkunde, political arithmetic's roots were in seventeenth-century England. Making population estimates and constructing mortality tables were two of the problems it frequently dealt with. In spirit, political arithmetic was similar to what is now called *demography*.

The third component was the development of a *calculus of probability*. As we saw earlier, this was a movement that essentially started in seventeenth-century France in response to certain gambling questions, but it quickly became the "engine" for analyzing all kinds of data.

Staatenkunde: The Comparative Description of States

The need for gathering information on the customs and resources of nations has been obvious since antiquity. Aristotle is credited with the first major effort toward that objective: His *Politeiai*, written in the fourth century B.C., contained detailed descriptions of some 158 different city-states. Unfortunately, the thirst for knowledge that led to the *Politeiai* fell victim to the intellectual drought of the Dark Ages, and almost two thousand years elapsed before any similar projects of like magnitude were undertaken.

The subject resurfaced during the Renaissance, and the Germans showed the most interest. They not only gave it a name, *Staatenkunde*, meaning "the comparative description of states," but they were also the first (in 1660) to incorporate the subject into a university curriculum. A leading figure in the German movement was Gottfried Achenwall, who taught at the University of Göttingen during the middle of the eighteenth century. Among Achenwall's claims to fame is that he was the first to use the word *statistics* in print. It appeared in the preface of his 1749 book *Abriss der Statswissenschaft der heutigen vornehmsten europaishen Reiche und Republiken*. (The word *statistics* comes from the Italian root *stato*, meaning "state," implying that a statistician is someone concerned with government affairs.) As terminology, it seems to have been well-received: For almost one hundred years the word *statistics* continued to be associated with the comparative description of states. In the middle of the nineteenth century, though, the term was redefined, and statistics became the new name for what had previously been called political arithmetic.

How important was the work of Achenwall and his predecessors to the development of statistics? That would be difficult to say. To be sure, their contributions were more indirect than direct. They left no methodology and no general theory. But they did point out the need for collecting accurate data and, perhaps more importantly, reinforced the notion that something complex—even as complex as an entire nation—can be effectively studied by gathering information on its component parts. Thus, they were lending important support to the then-growing belief that *induction*, rather than *deduction*, was a more sure-footed path to scientific truth.

Political Arithmetic

In the sixteenth century the English government began to compile records, called *bills of mortality*, on a parish-to-parish basis, showing numbers of deaths and their underlying causes. Their motivation largely stemmed from the plague epidemics that had periodically ravaged Europe in the not-too-distant past and were threatening to become a problem in England. Certain government officials, including the very influential Thomas Cromwell, felt that these bills would prove invaluable in helping to control the spread of an epidemic. At first, the bills were published only occasionally, but by the early seventeenth century they had become a weekly institution.[2]

Figure 1.3.3 (on the next page) shows a portion of a bill that appeared in London in 1665. The gravity of the plague epidemic is strikingly apparent when we look at the numbers at the top: Out of 97,306 deaths, 68,596 (over 70%) were caused by the plague. The breakdown of certain other afflictions, though they caused fewer deaths, raises some interesting questions. What happened, for example, to the 23 people who were "frighted" or to the 397 who suffered from "rising of the lights"?

Among the faithful subscribers to the bills was John Graunt, a London merchant. Graunt not only read the bills, he studied them intently. He looked for patterns, computed death rates, devised ways of estimating population sizes, and even set up a primitive life table. His results were published in the 1662 treatise *Natural and Political Observations upon the Bills of Mortality*. This work was a landmark: Graunt had launched the twin sciences of vital statistics and demography, and, although the name came later, it also signaled the beginning of political arithmetic. (Graunt did not have to wait long for accolades; in the year his book was published, he was elected to the prestigious Royal Society of London.)

High on the list of innovations that made Graunt's work unique were his objectives. Not content simply to describe a situation, although he was adept at doing so, Graunt often sought to go beyond his data and make generalizations (or, in current statistical terminology, draw *inferences*). Having been blessed with this particular turn of mind, he almost certainly qualifies as the world's first statistician. All Graunt really lacked was the probability theory that would have enabled him to frame his inferences more mathematically. That theory, though, was just beginning to unfold several hundred miles away in France (151).

Other seventeenth-century writers were quick to follow through on Graunt's ideas. William Petty's *Political Arithmetick* was published in 1690, although it had probably been written some fifteen years earlier. (It was Petty who gave the movement its name.) Perhaps even more significant were the contributions of Edmund Halley (of "Halley's comet" fame). Principally an astronomer, he also dabbled in political arithmetic, and in 1693 wrote *An Estimate of the Degrees of the Mortality of Mankind, drawn from Curious Tables of the Births and Funerals at the city of Breslaw; with an attempt to ascertain the Price of Annuities upon Lives*. (Book titles

[2] An interesting account of the bills of mortality is given in Daniel Defoe's *A Journal of the Plague Year*, which purportedly chronicles the London plague outbreak of 1665.

The bill for the year—A General Bill for this present year, ending the 19 of December, 1665, according to the Report made to the King's most excellent Majesty, by the Co. of Parish Clerks of Lond., & c.—gives the following summary of the results; the details of the several parishes we omit, they being made as in 1625, except that the out-parishes were now 12:—

Buried in the 27 Parishes within the walls	15,207
Whereof of the plague	9,887
Buried in the 16 Parishes without the walls	41,351
Whereof of the plague.	28,838
At the Pesthouse, total buried	159
Of the plague	156
Buried in the 12 out-Parishes in Middlesex and surrey	18,554
Whereof of the plague	21,420
Buried in the 5 Parishes in the City and Liberties of Westminster	12,194
Whereof the plague	8,403
The total of all the christenings	9,967
The total of all the burials this year	97,306
Whereof of the plague	68,596

Abortive and Stillborne	617	Griping in the Guts	1,288	Palsie	30
Aged	1,545	Hang'd & made away themselved	7	Plague	68,596
Ague & Feaver	5,257	Headmould shot and mould fallen	14	Plannet	6
Appolex and Suddenly	116	Jaundice	110	Plurisie	15
Bedrid	10	Impostume	227	Poysoned	1
Blasted	5	Kill by several accidents	46	Quinsie	35
Bleeding	16	King's Evill	86	Rickets	535
Cold & Cough	68	Leprosie	2	Rising of the Lights	397
Collick & Winde	134	Lethargy	14	Rupture	34
Comsumption & Tissick	4,808	Livergrown	20	Scurry	105
Convulsion & Mother	2,036	Bloody Flux, Scowring & Flux	18	Shingles & Swine Pox	2
Distracted	5	Burnt and Scalded	8	Sores, Ulcers, Broken and	
Dropsie & Timpany	1,478	Calenture	3	Bruised Limbs	82
Drowned	50	Cancer, Cangrene & Fistula	56	Spleen	14
Executed	21	Canker and Thrush	111	Spotted Feaver & Purples	1,929
Flox & Smallpox	655	Childbed	625	Stopping of the Stomach	332
Found Dead in streets, fields, &c.	20	Chrisomes and Infants	1,258	Stone and Stranguary	98
French Pox	86	Meagrom and Headach	12	Surfe	1,251
Frighted	23	Measles	7	Teeth & Worms	2,614
Gout & Sciatica	27	Murthered & Shot	9	Vomiting	51
Grief	46	Overlaid & Starved	45	Wenn	8

Christened-Males	5,114	Females	4,853	In all	9,967
Buried-Males	58,569	Females	48,737	In all	97,306

Of the Plague	68,596
Increase in the Burials in the 130 Parishes and the Pesthouse this year	79,009
Increase of the Plague in the 130 Parishes and the Pesthouse this year	68,590

Figure 1.3.3

were longer then!) Halley shored up, mathematically, the efforts of Graunt and others to construct an accurate mortality table. In doing so, he laid the foundation for the important theory of annuities. Today, all life insurance companies base their premium schedules on methods similar to Halley's. (The first company to follow his lead was The Equitable, founded in 1765.)

For all its initial flurry of activity, political arithmetic did not fare particularly well in the eighteenth century, at least in terms of having its methodology fine-tuned. Still, the second half of the century did see some notable achievements in improving the quality of the databases: Several countries, including the United States in 1790, established a periodic census. To some extent, answers to the questions that interested Graunt and his followers had to be deferred until the theory of probability could develop just a little bit more.

Quetelet: The Catalyst

With political arithmetic furnishing the data and many of the questions, and the theory of probability holding out the promise of rigorous answers, the birth of statistics was at hand. All that was needed was a catalyst—someone to bring the two together. Several individuals served with distinction in that capacity. Carl Friedrich Gauss, the superb German mathematician and astronomer, was especially helpful in showing how statistical concepts could be useful in the physical sciences. Similar efforts in France were made by Laplace. But the man who perhaps best deserves the title of "matchmaker" was a Belgian, Adolphe Quetelet.

Quetelet was a mathematician, astronomer, physicist, sociologist, anthropologist, and poet. One of his passions was collecting data, and he was fascinated by the regularity of social phenomena. In commenting on the nature of criminal tendencies, he once wrote (70):

> Thus we pass from one year to another with the sad perspective of seeing the same crimes reproduced in the same order and calling down the same punishments in the same proportions. Sad condition of humanity! ... We might enumerate in advance how many individuals will stain their hands in the blood of their fellows, how many will be forgers, how many will be poisoners, almost we can enumerate in advance the births and deaths that should occur. There is a budget which we pay with a frightful regularity; it is that of prisons, chains and the scaffold.

Given such an orientation, it was not surprising that Quetelet would see in probability theory an elegant means for expressing human behavior. For much of the nineteenth century he vigorously championed the cause of statistics, and as a member of more than one hundred learned societies, his influence was enormous. When he died in 1874, statistics had been brought to the brink of its modern era.

1.4 A Chapter Summary

The concepts of probability lie at the very heart of all statistical problems. Acknowledging that fact, the next two chapters take a close look at some of those concepts. Chapter 2 states the axioms of probability and investigates their consequences. It also covers the basic skills for algebraically manipulating probabilities and gives an introduction to combinatorics, the mathematics of counting. Chapter 3 reformulates much of the material in Chapter 2 in terms of *random variables*, the latter being a concept of great convenience in applying probability to statistics. Over the years, particular measures of probability have emerged as being especially useful: The most prominent of these are profiled in Chapter 4.

Our study of statistics proper begins with Chapter 5, which is a first look at the theory of parameter estimation. Chapter 6 introduces the notion of hypothesis testing, a procedure that, in one form or another, commands a major share of the remainder of the book. From a conceptual standpoint, these are very important chapters: Most formal applications of statistical methodology will involve either parameter estimation or hypothesis testing, or both.

Among the probability functions featured in Chapter 4, the *normal distribution*—more familiarly known as the *bell-shaped curve*—is sufficiently important to merit even further scrutiny. Chapter 7 derives in some detail many of the properties and applications of the normal distribution as well as those of several related probability functions. Much of the theory that supports the methodology appearing in Chapters 9 through 13 comes from Chapter 7.

Chapter 8 describes some of the basic principles of experimental "design." Its purpose is to provide a framework for comparing and contrasting the various statistical procedures profiled in Chapters 9 through 14.

Chapters 9, 12, and 13 continue the work of Chapter 7, but with the emphasis on the comparison of several populations, similar to what was done in Case Study 1.2.2. Chapter 10 looks at the important problem of assessing the level of agreement between a set of data and the values predicted by the probability model from which those data presumably came. Linear relationships are examined in Chapter 11.

Chapter 14 is an introduction to nonparametric statistics. The objective there is to develop procedures for answering some of the same sorts of questions raised in Chapters 7, 9, 12, and 13, but with fewer initial assumptions.

As a general format, each chapter contains numerous examples and case studies, the latter including actual experimental data taken from a variety of sources, primarily newspapers, magazines, and technical journals. We hope that these applications will make it abundantly clear that, while the general orientation of this text is theoretical, the consequences of that theory are never too far from having direct relevance to the "real world."

Chapter 2

PROBABILITY

One of the most influential of seventeenth-century mathematicians, Fermat earned his living as a lawyer and administrator in Toulouse. He shares credit with Descartes for the invention of analytic geometry, but his most important work may have been in number theory. Fermat did not write for publication, preferring instead to send letters and papers to friends. His correspondence with Pascal was the starting point for the development of a mathematical theory of probability.

—Pierre de Fermat (1601–1665)

Pascal was the son of a nobleman. A prodigy of sorts, he had already published a treatise on conic sections by the age of sixteen. He also invented one of the early calculating machines to help his father with accounting work. Pascal's contributions to probability were stimulated by his correspondence, in 1654, with Fermat. Later that year he retired to a life of religious meditation.

—Blaise Pascal (1623–1662)

2.1 Introduction

Experts have estimated that the likelihood of any given UFO sighting being genuine is on the order of one in one hundred thousand. Since the early 1950s, some ten thousand sightings have been reported to civil authorities. What is the probability that at least one of those objects was, in fact, an alien spacecraft? In 1978, Pete Rose of the Cincinnati Reds set a National League record by batting safely in forty-four consecutive games. How unlikely was that event, given that Rose was a lifetime .303 hitter? By definition, the *mean free path* is the average distance a molecule in a gas travels before colliding with another molecule. How likely is it that the distance a molecule travels between collisions will be at least twice its mean free path? Suppose a boy's mother and father both have genetic markers for sickle cell anemia, but neither parent exhibits any of the disease's symptoms. What are the chances that their son will also be asymptomatic? What are the odds that a poker player is dealt a full house or that a craps-shooter makes his "point"? If a woman has lived to age seventy, how likely is it that she will die before her ninetieth birthday? In 1994, Tom Foley was Speaker of the House and running for re-election. The day after the election, his race had still not been "called" by any of the networks: he trailed his Republican challenger by 2174 votes, but 14,000 absentee ballots remained to be counted. Foley, however, conceded. Should he have waited for the absentee ballots to be counted, or was his defeat at that point a virtual certainty?

As the nature and variety of these questions would suggest, probability is a subject with an extraordinary range of real-world, everyday applications. What began as an exercise in understanding games of chance has proven to be useful everywhere. Maybe even more remarkable is the fact that the solutions to all of these diverse questions are rooted in just a handful of definitions and theorems. Those results, together with the problem-solving techniques they empower, are the sum and substance of Chapter 2. We begin, though, with a bit of history.

The Evolution of the Definition of Probability

Over the years, the definition of probability has undergone several revisions. There is nothing contradictory in the multiple definitions—the changes primarily reflected the need for greater generality and more mathematical rigor. The first formulation (often referred to as the *classical* definition of probability) is credited to Gerolamo Cardano (recall Section 1.3). It applies only to situations where (1) the number of possible outcomes is finite and (2) all outcomes are equally likely. Under those conditions, the probability of an event comprised of m outcomes is the ratio m/n, where n is the total number of (equally likely) outcomes. Tossing a fair, six-sided die, for example, gives $m/n = \frac{3}{6}$ as the probability of rolling an even number (that is, either 2, 4, or 6).

While Cardano's model was well-suited to gambling scenarios (for which it was intended), it was obviously inadequate for more general problems, where outcomes are not equally likely and/or the number of outcomes is not finite. Richard von Mises, a twentieth-century German mathematician, is often credited with avoiding the weaknesses in Cardano's model by defining "empirical" probabilities. In the von Mises approach, we imagine an experiment being repeated over and over again *under presumably identical conditions*. Theoretically, a running tally could be kept of the number of times (m) the outcome belongs to a given event divided by n, the total number of times the experiment is performed. According to von Mises, the probability of the given event is the limit (as n goes to infinity) of the ratio m/n. Figure 2.1.1 illustrates the empirical probability of getting a head by tossing a fair coin: as the number of tosses continues to increase, the ratio m/n converges to $\frac{1}{2}$.

Figure 2.1.1

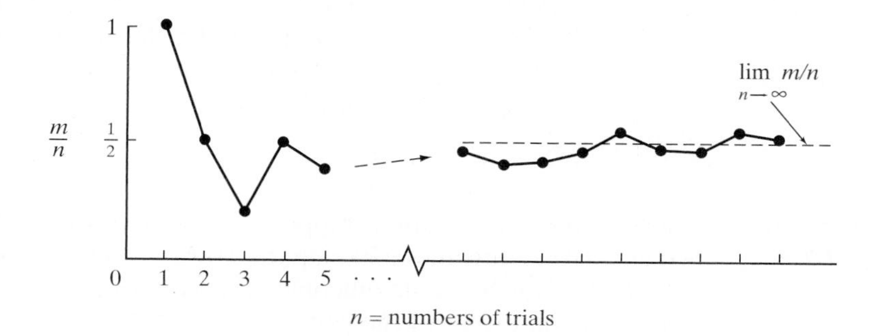

The von Mises approach definitely shores up some of the inadequacies seen in the Cardano model, but it is not without shortcomings of its own. There is some conceptual inconsistency, for example, in extolling the limit of m/n as a way of defining a probability *empirically*, when the very act of repeating an experiment under identical conditions an infinite number of times is physically impossible. And left unanswered is the question of how large n must be in order for m/n to be a good approximation for $\lim m/n$.

Andrei Kolmogorov, the great Russian probabilist, took a different approach. Aware that many twentieth-century mathematicians were having success developing subjects axiomatically, Kolmogorov wondered whether probability might similarly be defined operationally, rather than as a ratio (like the Cardano model) or as a limit (like the von Mises model). His efforts culminated in a masterpiece of mathematical elegance when he published *Grundbegriffe der Wahrscheinlichkeitsrechnung (Foundations of the Theory of Probability)* in 1933. In essence, Kolmogorov was able to show that a maximum of four simple axioms is necessary and sufficient to define the way any and all probabilities must behave. (These will be our starting point in Section 2.3.)

We begin Chapter 2 with some basic (and, presumably, familiar) definitions from set theory. These are important because probability will eventually be defined as a *set function*—that is, a mapping from a set to a number. Then, with the help of Kolmogorov's axioms in Section 2.3, we will learn how to calculate and manipulate probabilities. The chapter concludes with an introduction to *combinatorics*—the mathematics of systematic counting—and its application to probability.

2.2 Sample Spaces and the Algebra of Sets

The starting point for studying probability is the definition of four key terms: *experiment, sample outcome, sample space*, and *event*. The latter three, all carryovers from classical set theory, give us a familiar mathematical framework within which to work; the former is what provides the conceptual mechanism for casting real-world phenomena into probabilistic terms.

By an *experiment* we will mean any procedure that (1) can be repeated, theoretically, an infinite number of times; and (2) has a well-defined set of possible outcomes. Thus, rolling a pair of dice qualifies as an experiment; so does measuring a hypertensive's blood pressure or doing a spectrographic analysis to determine the carbon content of moon rocks. Asking a would-be psychic to draw a picture of an image presumably transmitted by another would-be psychic does *not* qualify as an experiment, because the set of possible outcomes cannot be listed, characterized, or otherwise defined.

Each of the potential eventualities of an experiment is referred to as a *sample outcome*, s, and their totality is called the *sample space*, S. To signify the membership of s in S, we write $s \in S$. Any designated collection of sample outcomes, including individual outcomes, the entire sample space, and the null set, constitutes an *event*. The latter is said to *occur* if the outcome of the experiment is one of the members of the event.

Example 2.2.1 Consider the experiment of flipping a coin three times. What is the sample space? Which sample outcomes make up the event A: Majority of coins show heads?

Think of each sample outcome here as an ordered triple, its components representing the outcomes of the first, second, and third tosses, respectively. Altogether, there are eight different triples, so those eight comprise the sample space:

$$S = \{\text{HHH, HHT, HTH, THH, HTT, THT, TTH, TTT}\}$$

By inspection, we see that four of the sample outcomes in S constitute the event A:

$$A = \{\text{HHH, HHT, HTH, THH}\}$$

Example 2.2.2

Imagine rolling two dice, the first one red, the second one green. Each sample outcome is an ordered pair (face showing on red die, face showing on green die), and the entire sample space can be represented as a 6×6 matrix (see Figure 2.2.1).

Face showing on green die

Face showing on red die	1	2	3	4	5	6
1	(1, 1)	(1, 2)	(1, 3)	(1, 4)	(1, 5)	(1, 6)
2	(2, 1)	(2, 2)	(2, 3)	(2, 4)	(2, 5)	(2, 6)
3	(3, 1)	(3, 2)	(3, 3)	(3, 4)	(3, 5)	(3, 6)
4	(4, 1)	(4, 2)	(4, 3)	(4, 4)	(4, 5)	(4, 6)
5	(5, 1)	(5, 2)	(5, 3)	(5, 4)	(5, 5)	(5, 6)
6	(6, 1)	(6, 2)	(6, 3)	(6, 4)	(6, 5)	(6, 6)

A

Figure 2.2.1

Gamblers are often interested in the event A that the sum of the faces showing is a 7. Notice in Figure 2.2.1 that the sample outcomes contained in A are the six diagonal entries, $(1, 6)$, $(2, 5)$, $(3, 4)$, $(4, 3)$, $(5, 2)$, and $(6, 1)$.

Example 2.2.3

A local TV station advertises two newscasting positions. If three women (W_1, W_2, W_3) and two men (M_1, M_2) apply, the "experiment" of hiring two coanchors generates a sample space of ten outcomes:

$$S = \{(W_1, W_2), (W_1, W_3), (W_2, W_3), (W_1, M_1), (W_1, M_2), (W_2, M_1), (W_2, M_2), (W_3, M_1), (W_3, M_2), (M_1, M_2)\}$$

Does it matter here that the two positions being filled are equivalent? Yes. If the station were seeking to hire, say, a sports announcer and a weather forecaster, the number of possible outcomes would be twenty: (W_2, M_1), for example, would represent a different staffing assignment than (M_1, W_2).

Example 2.2.4

The number of sample outcomes associated with an experiment need not be finite. Suppose that a coin is tossed until the first tail appears. If the first toss is itself a tail, the outcome of the experiment is T; if the first tail occurs on the second toss, the outcome is HT; and so on. Theoretically, of course, the first tail may *never* occur, and the infinite nature of S is readily apparent:

$$S = \{\text{T, HT, HHT, HHHT}, \ldots\}$$

Example 2.2.5

There are three ways to indicate an experiment's sample space. If the number of possible outcomes is small, we can simply list them, as we did in Examples 2.2.1 through 2.2.3. In some cases it may be possible to *characterize* a sample space by showing the structure its outcomes necessarily possess. This is what we did in Example 2.2.4.

A third option is to state a mathematical formula that the sample outcomes must satisfy.

A computer programmer is running a subroutine that solves a general quadratic equation, $ax^2 + bx + c = 0$. Her "experiment" consists of choosing values for the three coefficients $a, b,$ and c. Define (1) S and (2) the event A: Equation has two equal roots.

First, we must determine the sample space. Since presumably no combinations of finite a, b, and c are inadmissible, we can characterize S by writing a series of inequalities:

$$S = \{(a, b, c) : -\infty < a < \infty, -\infty < b < \infty, -\infty < c < \infty\}$$

Defining A requires the well-known result from algebra that a quadratic equation has equal roots if and only if its discriminant, $b^2 - 4ac$, vanishes. Membership in A, then, is contingent on $a, b,$ and c satisfying an equation:

$$A = \{(a, b, c) : b^2 - 4ac = 0\}$$ ■

Questions

2.2.1. A graduating engineer has signed up for three job interviews. She intends to categorize each one as being either a "success" or a "failure" depending on whether it leads to a plant trip. Write out the appropriate sample space. What outcomes are in the event A: Second success occurs on third interview? In B: First success never occurs? (*Hint*: Notice the similarity between this situation and the coin-tossing experiment described in Example 2.2.1.)

2.2.2. Three dice are tossed, one red, one blue, and one green. What outcomes make up the event A that the sum of the three faces showing equals 5?

2.2.3. An urn contains six chips numbered 1 through 6. Three are drawn out. What outcomes are in the event "Second smallest chip is a 3"? Assume that the order of the chips is irrelevant.

2.2.4. Suppose that two cards are dealt from a standard 52-card poker deck. Let A be the event that the sum of the two cards is 8 (assume that aces have a numerical value of 1). How many outcomes are in A?

2.2.5. In the lingo of craps-shooters (where two dice are tossed and the underlying sample space is the matrix pictured in Figure 2.2.1) is the phrase "making a hard eight." What might that mean?

2.2.6. A poker deck consists of fifty-two cards, representing thirteen denominations (2 through ace) and four suits (diamonds, hearts, clubs, and spades). A five-card hand is called a *flush* if all five cards are in the same suit but not all five denominations are consecutive. Pictured in the next column is a flush in hearts. Let N be the set of five cards in hearts that are *not* flushes. How many outcomes are in N? [*Note*: In poker, the denominations (A, 2, 3, 4, 5) are considered to be consecutive (in addition to sequences such as (8, 9, 10, J, Q)).]

		Denominations												
		2	3	4	5	6	7	8	9	10	J	Q	K	A
Suits	D													
	H	X	X				X				X	X		
	C													
	S													

2.2.7. Let P be the set of right triangles with a 5″ hypotenuse and whose height and length are a and b, respectively. Characterize the outcomes in P.

2.2.8. Suppose a baseball player steps to the plate with the intention of trying to "coax" a base on balls by never swinging at a pitch. The umpire, of course, will necessarily call each pitch either a ball (B) or a strike (S). What outcomes make up the event A, that a batter walks on the sixth pitch? (*Note*: A batter "walks" if the fourth ball is called before the third strike.)

2.2.9. A telemarketer is planning to set up a phone bank to bilk widows with a Ponzi scheme. His past experience (prior to his most recent incarceration) suggests that each phone will be in use half the time. For a given phone at a given time, let 0 indicate that the phone is available and let 1 indicate that a caller is on the line. Suppose that the telemarketer's "bank" is comprised of four telephones.

(a) Write out the outcomes in the sample space.
(b) What outcomes would make up the event that exactly two phones are being used?
(c) Suppose the telemarketer had k phones. How many outcomes would allow for the possibility that at most one more call could be received? (*Hint*: How many lines would have to be busy?)

2.2.10. Two darts are thrown at the following target:

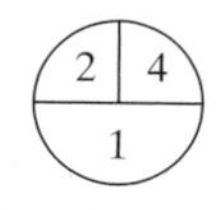

(a) Let (u, v) denote the outcome that the first dart lands in region u and the second dart, in region v. List the sample space of (u, v)'s.
(b) List the outcomes in the sample space of *sums*, $u + v$.

2.2.11. A woman has her purse snatched by two teenagers. She is subsequently shown a police lineup consisting of five suspects, including the two perpetrators. What is the sample space associated with the experiment "Woman picks two suspects out of lineup"? Which outcomes are in the event A: She makes at least one incorrect identification?

2.2.12. Consider the experiment of choosing coefficients for the quadratic equation $ax^2 + bx + c = 0$. Characterize the values of a, b, and c associated with the event A: Equation has complex roots.

2.2.13. In the game of craps, the person rolling the dice (the *shooter*) wins outright if his first toss is a 7 or an 11. If his first toss is a 2, 3, or 12, he loses outright. If his first roll is something else, say, a 9, that number becomes his "point" and he keeps rolling the dice until he either rolls another 9, in which case he wins, or a 7, in which case he loses. Characterize the sample outcomes contained in the event "Shooter wins with a point of 9."

2.2.14. A probability-minded despot offers a convicted murderer a final chance to gain his release. The prisoner is given twenty chips, ten white and ten black. All twenty are to be placed into two urns, according to any allocation scheme the prisoner wishes, with the one proviso being that each urn contain at least one chip. The executioner will then pick one of the two urns at random and from that urn, one chip at random. If the chip selected is white, the prisoner will be set free; if it is black, he "buys the farm." Characterize the sample space describing the prisoner's possible allocation options. (Intuitively, which allocation affords the prisoner the greatest chance of survival?)

2.2.15. Suppose that ten chips, numbered 1 through 10, are put into an urn at one minute to midnight, and chip number 1 is quickly removed. At one-half minute to midnight, chips numbered 11 through 20 are added to the urn, and chip number 2 is quickly removed. Then at one-fourth minute to midnight, chips numbered 21 to 30 are added to the urn, and chip number 3 is quickly removed. If that procedure for adding chips to the urn continues, how many chips will be in the urn at midnight (148)?

Unions, Intersections, and Complements

Associated with events defined on a sample space are several operations collectively referred to as the *algebra of sets*. These are the rules that govern the ways in which one event can be combined with another. Consider, for example, the game of craps described in Question 2.2.13. The shooter wins on his initial roll if he throws either a 7 or an 11. In the language of the algebra of sets, the event "Shooter rolls a 7 *or* an 11" is the *union* of two simpler events, "Shooter rolls a 7" and "Shooter rolls an 11." If E denotes the union and if A and B denote the two events making up the union, we write $E = A \cup B$. The next several definitions and examples illustrate those portions of the algebra of sets that we will find particularly useful in the chapters ahead.

Definition 2.2.1. Let A and B be any two events defined over the same sample space S. Then

a. The *intersection* of A and B, written $A \cap B$, is the event whose outcomes belong to both A and B.
b. The *union* of A and B, written $A \cup B$, is the event whose outcomes belong to either A or B or both.

Example 2.2.6

A single card is drawn from a poker deck. Let A be the event that an ace is selected:

$$A = \{\text{ace of hearts, ace of diamonds, ace of clubs, ace of spades}\}$$

Let B be the event "Heart is drawn":

$$B = \{\text{2 of hearts, 3 of hearts}, \ldots, \text{ ace of hearts}\}$$

Then

$$A \cap B = \{\text{ace of hearts}\}$$

and

$$A \cup B = \{\text{2 of hearts, 3 of hearts}, \ldots, \text{ ace of hearts, ace of diamonds, ace of clubs, ace of spades}\}$$

(Let C be the event "Club is drawn." Which cards are in $B \cup C$? In $B \cap C$?) ■

Example 2.2.7

Let A be the set of x's for which $x^2 + 2x = 8$; let B be the set for which $x^2 + x = 6$. Find $A \cap B$ and $A \cup B$.

Since the first equation factors into $(x + 4)(x - 2) = 0$, its solution set is $A = \{-4, 2\}$. Similarly, the second equation can be written $(x + 3)(x - 2) = 0$, making $B = \{-3, 2\}$. Therefore,

$$A \cap B = \{2\}$$

and

$$A \cup B = \{-4, -3, 2\}$$ ■

Example 2.2.8

Consider the electrical circuit pictured in Figure 2.2.2. Let A_i denote the event that switch i fails to close, $i = 1, 2, 3, 4$. Let A be the event "Circuit is not completed." Express A in terms of the A_i's.

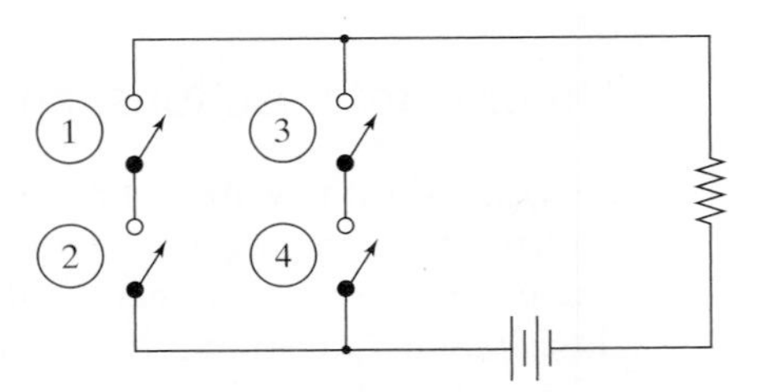

Figure 2.2.2

Call the ① and ② switches line a; call the ③ and ④ switches line b. By inspection, the circuit fails only if *both* line a and line b fail. But line a fails only if *either* ① *or* ② (or both) fail. That is, the event that line a fails is the union $A_1 \cup A_2$. Similarly, the failure of line b is the union $A_3 \cup A_4$. The event that the circuit fails, then, is an intersection:

$$A = (A_1 \cup A_2) \cap (A_3 \cup A_4)$$ ■

Definition 2.2.2. Events A and B defined over the same sample space are said to be *mutually exclusive* if they have no outcomes in common—that is, if $A \cap B = \emptyset$, where $\emptyset$ is the null set.

Example 2.2.9

Consider a single throw of two dice. Define A to be the event that the *sum* of the faces showing is odd. Let B be the event that the two faces themselves are odd. Then clearly, the intersection is empty, the sum of two odd numbers necessarily being even. In symbols, $A \cap B = \emptyset$. (Recall the event $B \cap C$ asked for in Example 2.2.6.) ■

> **Definition 2.2.3.** Let A be any event defined on a sample space S. The *complement* of A, written A^C, is the event consisting of all the outcomes in S other than those contained in A.

Example 2.2.10

Let A be the set of (x, y)'s for which $x^2 + y^2 < 1$. Sketch the region in the xy-plane corresponding to A^C.

From analytic geometry, we recognize that $x^2 + y^2 < 1$ describes the interior of a circle of radius 1 centered at the origin. Figure 2.2.3 shows the complement—the points on the circumference of the circle and the points outside the circle.

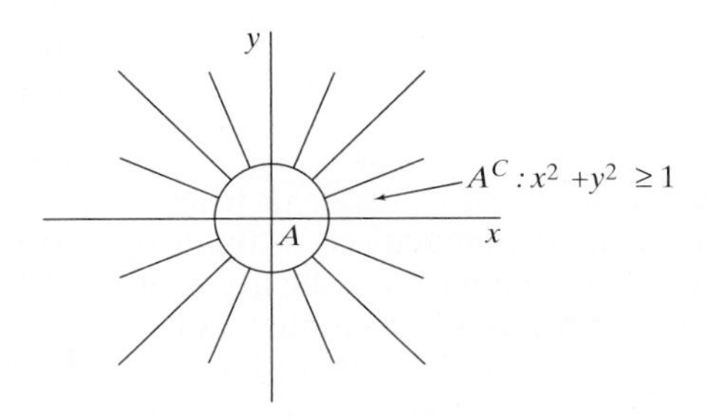

Figure 2.2.3 ■

The notions of union and intersection can easily be extended to more than two events. For example, the expression $A_1 \cup A_2 \cup \cdots \cup A_k$ defines the set of outcomes belonging to *any* of the A_i's (or to any combination of the A_i's). Similarly, $A_1 \cap A_2 \cap \cdots \cap A_k$ is the set of outcomes belonging to *all* of the A_i's.

Example 2.2.11

Suppose the events $A_1, A_2, \ldots, A_k$ are intervals of real numbers such that

$$A_i = \{x : 0 \leq x < 1/i\}, \quad i = 1, 2, \ldots, k$$

Describe the sets $A_1 \cup A_2 \cup \cdots \cup A_k = \cup_{i=1}^{k} A_i$ and $A_1 \cap A_2 \cap \cdots \cap A_k = \cap_{i=1}^{k} A_i$.

Notice that the A_i's are telescoping sets. That is, A_1 is the interval $0 \leq x < 1$, A_2 is the interval $0 \leq x < \frac{1}{2}$, and so on. It follows, then, that the *union* of the k A_i's is simply A_1 while the *intersection* of the A_i's (that is, their overlap) is A_k. ■

Questions

2.2.16. Sketch the regions in the xy-plane corresponding to $A \cup B$ and $A \cap B$ if

$$A = \{(x, y): 0 < x < 3, 0 < y < 3\}$$

and

$$B = \{(x, y): 2 < x < 4, 2 < y < 4\}$$

2.2.17. Referring to Example 2.2.7, find $A \cap B$ and $A \cup B$ if the two equations were replaced by inequalities: $x^2 + 2x \leq 8$ and $x^2 + x \leq 6$.

2.2.18. Find $A \cap B \cap C$ if $A = \{x: 0 \leq x \leq 4\}$, $B = \{x: 2 \leq x \leq 6\}$, and $C = \{x: x = 0, 1, 2, \ldots\}$.

2.2.19. An electronic system has four components divided into two pairs. The two components of each pair are wired in parallel; the two pairs are wired in series. Let A_{ij} denote the event "ith component in jth pair fails," $i = 1, 2$; $j = 1, 2$. Let A be the event "System fails." Write A in terms of the A_{ij}'s.

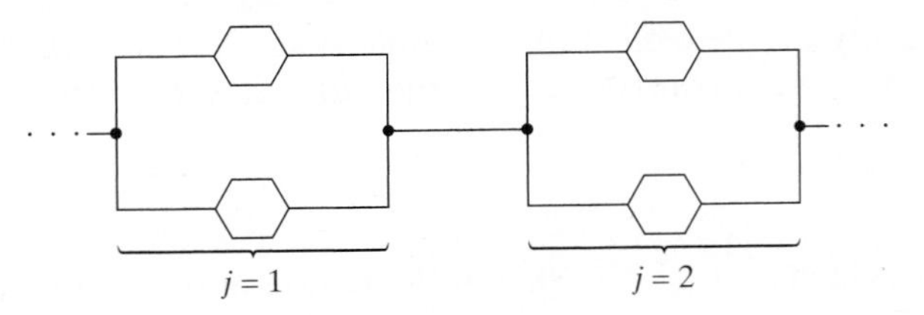

2.2.20. Define $A = \{x : 0 \le x \le 1\}$, $B = \{x : 0 \le x \le 3\}$, and $C = \{x : -1 \le x \le 2\}$. Draw diagrams showing each of the following sets of points:

(a) $A^C \cap B \cap C$
(b) $A^C \cup (B \cap C)$
(c) $A \cap B \cap C^C$
(d) $[(A \cup B) \cap C^C]^C$

2.2.21. Let A be the set of five-card hands dealt from a 52-card poker deck, where the denominations of the five cards are all consecutive—for example, (7 of hearts, 8 of spades, 9 of spades, 10 of hearts, jack of diamonds). Let B be the set of five-card hands where the suits of the five cards are all the same. How many outcomes are in the event $A \cap B$?

2.2.22. Suppose that each of the twelve letters in the word

$$T \quad E \quad S \quad S \quad E \quad L \quad L \quad A \quad T \quad I \quad O \quad N$$

is written on a chip. Define the events F, R, and C as follows:

F: letters in first half of alphabet
R: letters that are repeated
V: letters that are vowels

Which chips make up the following events?

(a) $F \cap R \cap V$
(b) $F^C \cap R \cap V^C$
(c) $F \cap R^C \cap V$

2.2.23. Let A, B, and C be any three events defined on a sample space S. Show that

(a) the outcomes in $A \cup (B \cap C)$ are the same as the outcomes in $(A \cup B) \cap (A \cup C)$.
(b) the outcomes in $A \cap (B \cup C)$ are the same as the outcomes in $(A \cap B) \cup (A \cap C)$.

2.2.24. Let $A_1, A_2, \ldots, A_k$ be any set of events defined on a sample space S. What outcomes belong to the event

$$(A_1 \cup A_2 \cup \cdots \cup A_k) \cup \left(A_1^C \cap A_2^C \cap \cdots \cap A_k^C\right)$$

2.2.25. Let A, B, and C be any three events defined on a sample space S. Show that the operations of union and intersection are *associative* by proving that

(a) $A \cup (B \cup C) = (A \cup B) \cup C = A \cup B \cup C$
(b) $A \cap (B \cap C) = (A \cap B) \cap C = A \cap B \cap C$

2.2.26. Suppose that three events—A, B, and C—are defined on a sample space S. Use the union, intersection, and complement operations to represent each of the following events:

(a) none of the three events occurs
(b) all three of the events occur
(c) only event A occurs
(d) exactly one event occurs
(e) exactly two events occur

2.2.27. What must be true of events A and B if

(a) $A \cup B = B$
(b) $A \cap B = A$

2.2.28. Let events A and B and sample space S be defined as the following intervals:

$$S = \{x : 0 \le x \le 10\}$$
$$A = \{x : 0 < x < 5\}$$
$$B = \{x : 3 \le x \le 7\}$$

Characterize the following events:

(a) A^C
(b) $A \cap B$
(c) $A \cup B$
(d) $A \cap B^C$
(e) $A^C \cup B$
(f) $A^C \cap B^C$

2.2.29. A coin is tossed four times and the resulting sequence of heads and/or tails is recorded. Define the events A, B, and C as follows:

A: exactly two heads appear
B: heads and tails alternate
C: first two tosses are heads

(a) Which events, if any, are mutually exclusive?
(b) Which events, if any, are subsets of other sets?

2.2.30. Pictured on the next page are two organizational charts describing the way upper management vets new proposals. For both models, three vice presidents—1, 2, and 3—each voice an opinion.

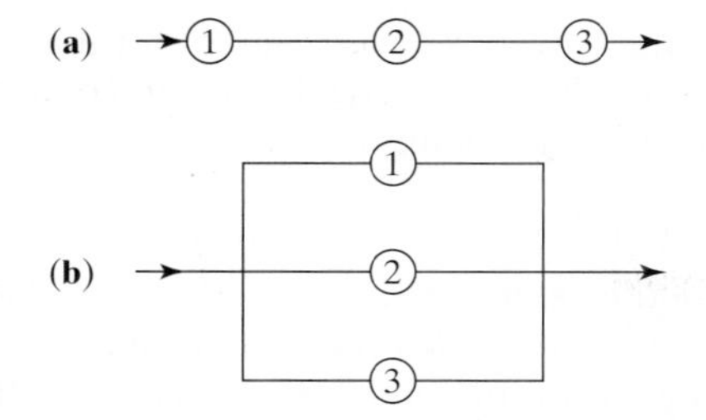

For (a), all three must concur if the proposal is to pass; if any one of the three favors the proposal in (b), it passes. Let A_i denote the event that vice president i favors the proposal, $i = 1, 2, 3$, and let A denote the event that the proposal passes. Express A in terms of the A_i's for the two office protocols. Under what sorts of situations might one system be preferable to the other?

Expressing Events Graphically: Venn Diagrams

Relationships based on two or more events can sometimes be difficult to express using only equations or verbal descriptions. An alternative approach that can be highly effective is to represent the underlying events graphically in a format known as a *Venn diagram*. Figure 2.2.4 shows Venn diagrams for an intersection, a union, a complement, and two events that are mutually exclusive. In each case, the shaded interior of a region corresponds to the desired event.

Figure 2.2.4

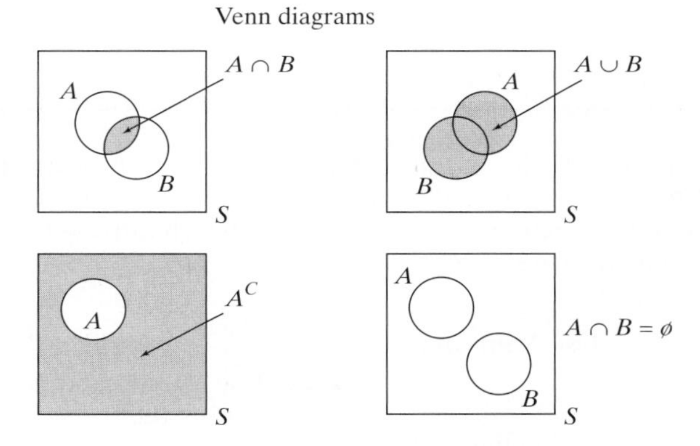

Example 2.2.12

When two events A and B are defined on a sample space, we will frequently need to consider

a. the event that *exactly one* (of the two) occurs.
b. the event that *at most one* (of the two) occurs.

Getting expressions for each of these is easy if we visualize the corresponding Venn diagrams.

The shaded area in Figure 2.2.5 represents the event E that either A or B, but not both, occurs (that is, *exactly one* occurs).

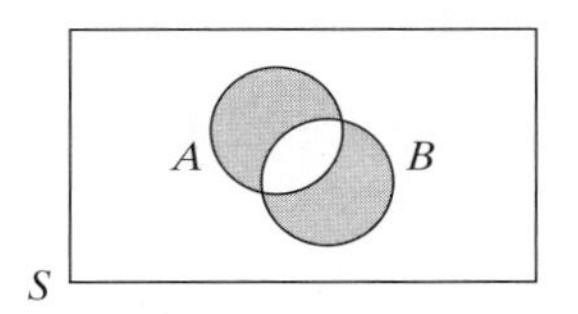

Figure 2.2.5

Just by looking at the diagram we can formulate an expression for E. The portion of A, for example, included in E is $A \cap B^C$. Similarly, the portion of B included in E is $B \cap A^C$. It follows that E can be written as a union:

$$E = (A \cap B^C) \cup (B \cap A^C)$$

(Convince yourself that an equivalent expression for E is $(A \cap B)^C \cap (A \cup B)$.)

Figure 2.2.6 shows the event F that *at most one* (of the two events) occurs. Since the latter includes every outcome except those belonging to *both* A and B, we can write

$$F = (A \cap B)^C$$

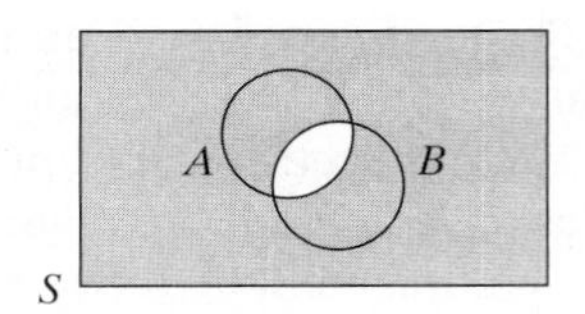

Figure 2.2.6

Questions

2.2.31. During orientation week, the latest Spiderman movie was shown twice at State University. Among the entering class of 6000 freshmen, 850 went to see it the first time, 690 the second time, while 4700 failed to see it either time. How many saw it twice?

2.2.32. Let A and B be any two events. Use Venn diagrams to show that

(a) the complement of their intersection is the union of their complements:

$$(A \cap B)^C = A^C \cup B^C$$

(b) the complement of their union is the intersection of their complements:

$$(A \cup B)^C = A^C \cap B^C$$

(These two results are known as *DeMorgan's laws*.)

2.2.33. Let A, B, and C be any three events. Use Venn diagrams to show that

(a) $A \cap (B \cup C) = (A \cap B) \cup (A \cap C)$
(b) $A \cup (B \cap C) = (A \cup B) \cap (A \cup C)$

2.2.34. Let A, B, and C be any three events. Use Venn diagrams to show that

(a) $A \cup (B \cup C) = (A \cup B) \cup C$
(b) $A \cap (B \cap C) = (A \cap B) \cap C$

2.2.35. Let A and B be any two events defined on a sample space S. Which of the following sets are necessarily subsets of which other sets?

$$A \quad B \quad A \cup B \quad A \cap B \quad A^C \cap B$$

$$A \cap B^C \quad (A^C \cup B^C)^C$$

2.2.36. Use Venn diagrams to suggest an equivalent way of representing the following events:

(a) $(A \cap B^C)^C$
(b) $B \cup (A \cup B)^C$
(c) $A \cap (A \cap B)^C$

2.2.37. A total of twelve hundred graduates of State Tech have gotten into medical school in the past several years. Of that number, one thousand earned scores of twenty-seven or higher on the MCAT and four hundred had GPAs that were 3.5 or higher. Moreover, three hundred had MCATs that were twenty-seven or higher *and* GPAs that were 3.5 or higher. What proportion of those twelve hundred graduates got into medical school with an MCAT lower than twenty-seven and a GPA below 3.5?

2.2.38. Let A, B, and C be any three events defined on a sample space S. Let $N(A)$, $N(B)$, $N(C)$, $N(A \cap B)$, $N(A \cap C)$, $N(B \cap C)$, and $N(A \cap B \cap C)$ denote the numbers of outcomes in all the different intersections in which A, B, and C are involved. Use a Venn diagram to suggest a formula for $N(A \cup B \cup C)$. [*Hint:* Start with the sum $N(A) + N(B) + N(C)$ and use the Venn diagram to identify the "adjustments" that need to be made to that sum before it can equal $N(A \cup B \cup C)$.] As a precedent, note that $N(A \cup B) = N(A) + N(B) - N(A \cap B)$. There, in the case of *two* events, subtracting $N(A \cap B)$ is the "adjustment."

2.2.39. A poll conducted by a potential presidential candidate asked two questions: (1) Do you support the candidate's position on taxes? and (2) Do you support the candidate's position on homeland security? A total of twelve hundred responses were received; six hundred said "yes" to the first question and four hundred said "yes" to the second. If three hundred respondents said "no" to the taxes question and "yes" to the homeland security question, how many said "yes" to the taxes question but "no" to the homeland security question?

2.2.40. For two events A and B defined on a sample space S, $N(A \cap B^C) = 15$, $N(A^C \cap B) = 50$, and $N(A \cap B) = 2$. Given that $N(S) = 120$, how many outcomes belong to neither A nor B?

2.3 The Probability Function

Having introduced in Section 2.2 the twin concepts of "experiment" and "sample space," we are now ready to pursue in a formal way the all-important problem of assigning a *probability* to an experiment's outcome—and, more generally, to an event. Specifically, if A is any event defined on a sample space S, the symbol $P(A)$ will denote the *probability of* A, and we will refer to P as the *probability function*. It is, in effect, a mapping from a set (i.e., an event) to a number. The backdrop for our discussion will be the unions, intersections, and complements of set theory; the starting point will be the axioms referred to in Section 2.1 that were originally set forth by Kolmogorov.

If S has a finite number of members, Kolmogorov showed that as few as three axioms are necessary and sufficient for characterizing the probability function P:

Axiom 1. *Let A be any event defined over S. Then $P(A) \geq 0$.*

Axiom 2. $P(S) = 1$.

Axiom 3. *Let A and B be any two mutually exclusive events defined over S. Then*

$$P(A \cup B) = P(A) + P(B)$$

When S has an infinite number of members, a fourth axiom is needed:

Axiom 4. *Let $A_1, A_2, \ldots,$ be events defined over S. If $A_i \cap A_j = \emptyset$ for each $i \neq j$, then*

$$P\left(\bigcup_{i=1}^{\infty} A_i\right) = \sum_{i=1}^{\infty} P(A_i)$$

From these simple statements come the general rules for manipulating the probability function that apply no matter what specific mathematical form the function may take in a particular context.

Some Basic Properties of P

Some of the immediate consequences of Kolmogorov's axioms are the results given in Theorems 2.3.1 through 2.3.6. Despite their simplicity, several of these properties—as we will soon see—prove to be immensely useful in solving all sorts of problems.

Theorem 2.3.1 $P(A^C) = 1 - P(A)$.

Proof By Axiom 2 and Definition 2.2.3,

$$P(S) = 1 = P(A \cup A^C)$$

But A and A^C are mutually exclusive, so

$$P(A \cup A^C) = P(A) + P(A^C)$$

and the result follows. □

Theorem 2.3.2 $P(\emptyset) = 0$.

Proof Since $\emptyset = S^C$, $P(\emptyset) = P(S^C) = 1 - P(S) = 0$. □

Theorem 2.3.3 *If $A \subset B$, then $P(A) \leq P(B)$.*

Proof Note that the event B may be written in the form

$$B = A \cup (B \cap A^C)$$

where A and $(B \cap A^C)$ are mutually exclusive. Therefore,

$$P(B) = P(A) + P(B \cap A^C)$$

which implies that $P(B) \geq P(A)$ since $P(B \cap A^C) \geq 0$. □

Theorem 2.3.4 *For any event A, $P(A) \leq 1$.*

Proof The proof follows immediately from Theorem 2.3.3 because $A \subset S$ and $P(S) = 1$. □

Theorem 2.3.5 *Let $A_1, A_2, \ldots, A_n$ be events defined over S. If $A_i \cap A_j = \emptyset$ for $i \neq j$, then*

$$P\left(\bigcup_{i=1}^{n} A_i\right) = \sum_{i=1}^{n} P(A_i)$$

Proof The proof is a straightforward induction argument with Axiom 3 being the starting point. □

Theorem 2.3.6 $P(A \cup B) = P(A) + P(B) - P(A \cap B)$.

Proof The Venn diagram for $A \cup B$ certainly suggests that the statement of the theorem is true (recall Figure 2.2.4). More formally, we have from Axiom 3 that

$$P(A) = P(A \cap B^C) + P(A \cap B)$$

and

$$P(B) = P(B \cap A^C) + P(A \cap B)$$

Adding these two equations gives

$$P(A) + P(B) = [P(A \cap B^C) + P(B \cap A^C) + P(A \cap B)] + P(A \cap B)$$

By Theorem 2.3.5, the sum in the brackets is $P(A \cup B)$. If we subtract $P(A \cap B)$ from both sides of the equation, the result follows. □

Example 2.3.1 Let A and B be two events defined on a sample space S such that $P(A)=0.3$, $P(B)=0.5$, and $P(A\cup B)=0.7$. Find (a) $P(A\cap B)$, (b) $P(A^C\cup B^C)$, and (c) $P(A^C\cap B)$.

a. Transposing the terms in Theorem 2.3.6 yields a general formula for the probability of an intersection:

$$P(A\cap B)=P(A)+P(B)-P(A\cup B)$$

Here

$$\begin{aligned}P(A\cap B)&=0.3+0.5-0.7\\&=0.1\end{aligned}$$

b. The two cross-hatched regions in Figure 2.3.1 correspond to A^C and B^C. The union of A^C and B^C consists of those regions that have cross-hatching in either or both directions. By inspection, the only portion of S *not* included in $A^C\cup B^C$ is the intersection, $A\cap B$. By Theorem 2.3.1, then,

$$\begin{aligned}P(A^C\cup B^C)&=1-P(A\cap B)\\&=1-0.1\\&=0.9\end{aligned}$$

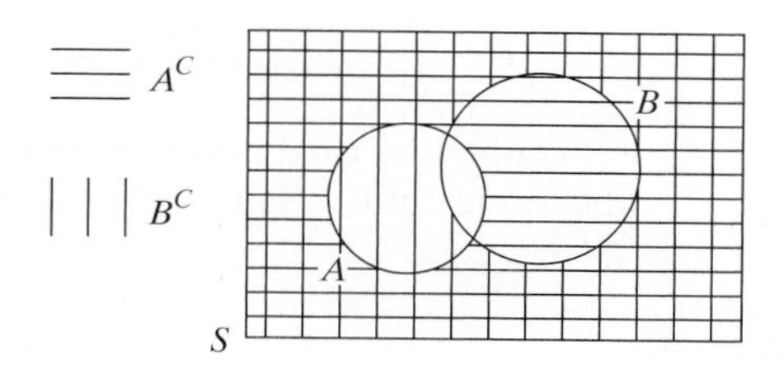

Figure 2.3.1

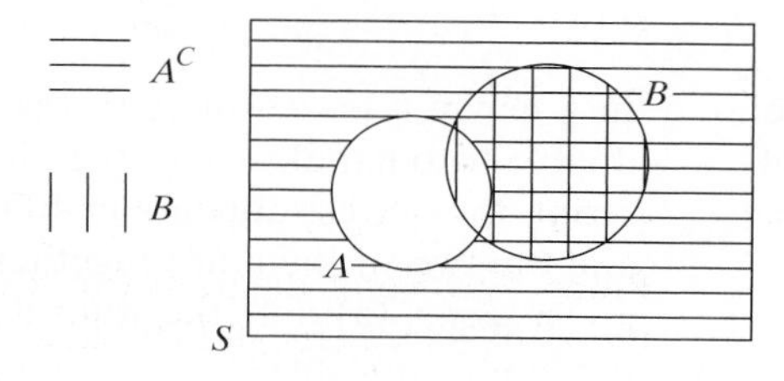

Figure 2.3.2

c. The event $A^C\cap B$ corresponds to the region in Figure 2.3.2 where the cross-hatching extends in *both* directions—that is, everywhere in B except the intersection with A. Therefore,

$$\begin{aligned}P(A^C\cap B)&=P(B)-P(A\cap B)\\&=0.5-0.1\\&=0.4\end{aligned}$$

■

Example 2.3.2

Show that

$$P(A \cap B) \geq 1 - P(A^C) - P(B^C)$$

for any two events A and B defined on a sample space S.

From Example 2.3.1a and Theorem 2.3.1,

$$P(A \cap B) = P(A) + P(B) - P(A \cup B)$$
$$= 1 - P(A^C) + 1 - P(B^C) - P(A \cup B)$$

But $P(A \cup B) \leq 1$ from Theorem 2.3.4, so

$$P(A \cap B) \geq 1 - P(A^C) - P(B^C)$$

■

Example 2.3.3

Two cards are drawn from a poker deck without replacement. What is the probability that the second is higher in rank than the first?

Let A_1, A_2, and A_3 be the events "First card is lower in rank," "First card is higher in rank," and "Both cards have same rank," respectively. Clearly, the three A_i's are mutually exclusive and they account for all possible outcomes, so from Theorem 2.3.5,

$$P(A_1 \cup A_2 \cup A_3) = P(A_1) + P(A_2) + P(A_3) = P(S) = 1$$

Once the first card is drawn, there are three choices for the second that would have the same rank—that is, $P(A_3) = \frac{3}{51}$. Moreover, symmetry demands that $P(A_1) = P(A_2)$, so

$$2P(A_2) + \frac{3}{51} = 1$$

implying that $P(A_2) = \frac{8}{17}$.

■

Example 2.3.4

In a newly released martial arts film, the actress playing the lead role has a stunt double who handles all of the physically dangerous action scenes. According to the script, the actress appears in 40% of the film's scenes, her double appears in 30%, and the two of them are together 5% of the time. What is the probability that in a given scene, (a) only the stunt double appears and (b) neither the lead actress nor the double appears?

a. If L is the event "Lead actress appears in scene" and D is the event "Double appears in scene," we are given that $P(L) = 0.40$, $P(D) = 0.30$, and $P(L \cap D) = 0.05$. It follows that

$$P(\text{Only double appears}) = P(D) - P(L \cap D)$$
$$= 0.30 - 0.05$$
$$= 0.25$$

(recall Example 2.3.1c).

b. The event "Neither appears" is the complement of the event "At least one appears." But $P(\text{At least one appears}) = P(L \cup D)$. From Theorems 2.3.1 and 2.3.6, then,

$$\begin{aligned} P(\text{Neither appears}) &= 1 - P(L \cup D) \\ &= 1 - [P(L) + P(D) - P(L \cap D)] \\ &= 1 - [0.40 + 0.30 - 0.05] \\ &= 0.35 \end{aligned}$$

■

Example 2.3.5

Having endured (and survived) the mental trauma that comes from taking two years of chemistry, a year of physics, and a year of biology, Biff decides to test the medical school waters and sends his MCATs to two colleges, X and Y. Based on how his friends have fared, he estimates that his probability of being accepted at X is 0.7, and at Y is 0.4. He also suspects there is a 75% chance that at least one of his applications will be rejected. What is the probability that he gets at least one acceptance?

Let A be the event "School X accepts him" and B the event "School Y accepts him." We are given that $P(A) = 0.7$, $P(B) = 0.4$, and $P(A^C \cup B^C) = 0.75$. The question is asking for $P(A \cup B)$.

From Theorem 2.3.6,

$$P(A \cup B) = P(A) + P(B) - P(A \cap B)$$

Recall from Question 2.2.32 that $A^C \cup B^C = (A \cap B)^C$, so

$$P(A \cap B) = 1 - P[(A \cap B)^C] = 1 - 0.75 = 0.25$$

It follows that Biff's prospects are not all that bleak—he has an 85% chance of getting in somewhere:

$$\begin{aligned} P(A \cup B) &= 0.7 + 0.4 - 0.25 \\ &= 0.85 \end{aligned}$$

■

Comment Notice that $P(A \cup B)$ varies directly with $P(A^C \cup B^C)$:

$$\begin{aligned} P(A \cup B) &= P(A) + P(B) - [1 - P(A^C \cup B^C)] \\ &= P(A) + P(B) - 1 + P(A^C \cup B^C) \end{aligned}$$

If $P(A)$ and $P(B)$, then, are fixed, we get the curious result that Biff's chances of getting at least one acceptance increase if his chances of at least one rejection increase.

Questions

2.3.1. According to a family-oriented lobbying group, there is too much crude language and violence on television. Forty-two percent of the programs they screened had language they found offensive, 27% were too violent, and 10% were considered excessive in both language and violence. What percentage of programs did comply with the group's standards?

2.3.2. Let A and B be any two events defined on S. Suppose that $P(A) = 0.4$, $P(B) = 0.5$, and $P(A \cap B) = 0.1$. What is the probability that A or B but not both occur?

2.3.3. Express the following probabilities in terms of $P(A)$, $P(B)$, and $P(A \cap B)$.

(a) $P(A^C \cup B^C)$
(b) $P(A^C \cap (A \cup B))$

2.3.4. Let A and B be two events defined on S. If the probability that at least one of them occurs is 0.3 and the probability that A occurs but B does not occur is 0.1, what is $P(B)$?

2.3.5. Suppose that three fair dice are tossed. Let A_i be the event that a 6 shows on the ith die, $i = 1, 2, 3$. Does $P(A_1 \cup A_2 \cup A_3) = \frac{1}{2}$? Explain.

2.3.6. Events A and B are defined on a sample space S such that $P((A \cup B)^C) = 0.5$ and $P(A \cap B) = 0.2$. What is the probability that either A or B but not both will occur?

2.3.7. Let $A_1, A_2, \ldots, A_n$ be a series of events for which $A_i \cap A_j = \emptyset$ if $i \neq j$ and $A_1 \cup A_2 \cup \cdots \cup A_n = S$. Let B be any event defined on S. Express B as a union of intersections.

2.3.8. Draw the Venn diagrams that would correspond to the equations (a) $P(A \cap B) = P(B)$ and (b) $P(A \cup B) = P(B)$.

2.3.9. In the game of "odd man out" each player tosses a fair coin. If all the coins turn up the same except for one, the player tossing the different coin is declared the odd man out and is eliminated from the contest. Suppose that three people are playing. What is the probability that someone will be eliminated on the first toss? (*Hint:* Use Theorem 2.3.1.)

2.3.10. An urn contains twenty-four chips, numbered 1 through 24. One is drawn at random. Let A be the event that the number is divisible by 2 and let B be the event that the number is divisible by 3. Find $P(A \cup B)$.

2.3.11. If State's football team has a 10% chance of winning Saturday's game, a 30% chance of winning two weeks from now, and a 65% chance of losing both games, what are their chances of winning exactly once?

2.3.12. Events A_1 and A_2 are such that $A_1 \cup A_2 = S$ and $A_1 \cap A_2 = \emptyset$. Find p_2 if $P(A_1) = p_1$, $P(A_2) = p_2$, and $3p_1 - p_2 = \frac{1}{2}$.

2.3.13. Consolidated Industries has come under considerable pressure to eliminate its seemingly discriminatory hiring practices. Company officials have agreed that during the next five years, 60% of their new employees will be females and 30% will be minorities. One out of four new employees, though, will be a white male. What percentage of their new hires will be minority females?

2.3.14. Three events—A, B, and C—are defined on a sample space, S. Given that $P(A) = 0.2$, $P(B) = 0.1$, and $P(C) = 0.3$, what is the smallest possible value for $P[(A \cup B \cup C)^C]$?

2.3.15. A coin is to be tossed four times. Define events X and Y such that

X: first and last coins have opposite faces
Y: exactly two heads appear

Assume that each of the sixteen head/tail sequences has the same probability. Evaluate

(a) $P(X^C \cap Y)$
(b) $P(X \cap Y^C)$

2.3.16. Two dice are tossed. Assume that each possible outcome has a $\frac{1}{36}$ probability. Let A be the event that the sum of the faces showing is 6, and let B be the event that the face showing on one die is twice the face showing on the other. Calculate $P(A \cap B^C)$.

2.3.17. Let A, B, and C be three events defined on a sample space, S. Arrange the probabilities of the following events from smallest to largest:

(a) $A \cup B$
(b) $A \cap B$
(c) A
(d) S
(e) $(A \cap B) \cup (A \cap C)$

2.3.18. Lucy is currently running two dot-com scams out of a bogus chatroom. She estimates that the chances of the first one leading to her arrest are one in ten; the "risk" associated with the second is more on the order of one in thirty. She considers the likelihood that she gets busted for both to be 0.0025. What are Lucy's chances of avoiding incarceration?

2.4 Conditional Probability

In Section 2.3, we calculated probabilities of certain events by manipulating other probabilities whose values we were given. Knowing $P(A)$, $P(B)$, and $P(A \cap B)$, for example, allows us to calculate $P(A \cup B)$ (recall Theorem 2.3.6). For many real-world situations, though, the "given" in a probability problem goes beyond simply knowing a set of other probabilities. Sometimes, we know *for a fact* that certain events *have already occurred*, and those occurrences may have a bearing on the probability we are trying to find. In short, the probability of an event A may have to be "adjusted" if we know for certain that some related event B has already occurred. Any probability that is revised to take into account the (known) occurrence of other events is said to be a ***conditional probability***.

Consider a fair die being tossed, with A defined as the event "6 appears." Clearly, $P(A) = \frac{1}{6}$. But suppose that the die has already been tossed—by someone who refuses to tell us whether or not A occurred but does enlighten us to the extent of confirming that B occurred, where B is the event "Even number appears." What are the chances of A now? Here, common sense can help us: There are three equally likely even numbers making up the event B—one of which satisfies the event A, so the "updated" probability is $\frac{1}{3}$.

Notice that the effect of additional information, such as the knowledge that B has occurred, is to revise—indeed, to *shrink*—the original sample space S to a new set of outcomes S'. In this example, the original S contained six outcomes, the conditional sample space, three (see Figure 2.4.1).

Figure 2.4.1

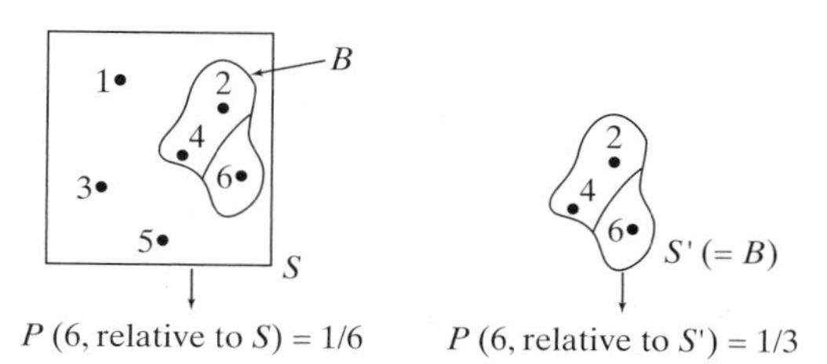

The symbol $P(A|B)$—read "the probability of A given B"—is used to denote a conditional probability. Specifically, $P(A|B)$ refers to the probability that A *will occur* given that B *has already occurred*.

It will be convenient to have a formula for $P(A|B)$ that can be evaluated in terms of the original S, rather than the revised S'. Suppose that S is a finite sample space with n outcomes, all equally likely. Assume that A and B are two events containing a and b outcomes, respectively, and let c denote the number of outcomes in the intersection of A and B (see Figure 2.4.2). Based on the argument suggested in Figure 2.4.1, the *conditional probability of A given B is the ratio of c to b*. But c/b can be written as the quotient of two other ratios,

Figure 2.4.2

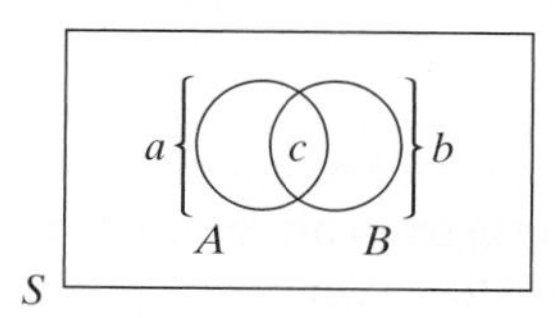

$$\frac{c}{b} = \frac{c/n}{b/n}$$

so, for this particular case,

$$P(A|B) = \frac{P(A \cap B)}{P(B)} \tag{2.4.1}$$

The same underlying reasoning that leads to Equation 2.4.1, though, holds true even when the outcomes are not equally likely or when S is uncountably infinite.

Definition 2.4.1. Let A and B be any two events defined on S such that $P(B) > 0$. The conditional probability of A, assuming that B has already occurred, is written $P(A|B)$ and is given by

$$P(A|B) = \frac{P(A \cap B)}{P(B)}$$

Comment Definition 2.4.1 can be cross-multiplied to give a frequently useful expression for the probability of an intersection. If $P(A|B) = P(A \cap B)/P(B)$, then

$$P(A \cap B) = P(A|B)P(B) \tag{2.4.2}$$

Example 2.4.1

A card is drawn from a poker deck. What is the probability that the card is a club, given that the card is a king?

Intuitively, the answer is $\frac{1}{4}$: The king is equally likely to be a heart, diamond, club, or spade. More formally, let C be the event "Card is a club"; let K be the event "Card is a king." By Definition 2.4.1,

$$P(C|K) = \frac{P(C \cap K)}{P(K)}$$

But $P(K) = \frac{4}{52}$ and $P(C \cap K) = P(\text{Card is a king of clubs}) = \frac{1}{52}$. Therefore, confirming our intuition,

$$P(C|K) = \frac{1/52}{4/52} = \frac{1}{4}$$

[Notice in this example that the conditional probability $P(C|K)$ is numerically the same as the unconditional probability $P(C)$—they both equal $\frac{1}{4}$. This means that our knowledge that K has occurred gives us no additional insight about the chances of C occurring. Two events having this property are said to be *independent*. We will examine the notion of independence and its consequences in detail in Section 2.5.] ■

Example 2.4.2

Our intuitions can often be fooled by probability problems, even ones that appear to be simple and straightforward. The "two boys" problem described here is an often-cited case in point.

Consider the set of families having two children. Assume that the four possible birth sequences—(younger child is a boy, older child is a boy), (younger child is a boy, older child is a girl), and so on—are equally likely. What is the probability that both children are boys given that at least one is a boy?

The answer is *not* $\frac{1}{2}$. The correct answer can be deduced from Definition 2.4.1. By assumption, each of the four possible birth sequences—(b, b), (b, g), (g, b), and (g, g)—has a $\frac{1}{4}$ probability of occurring. Let A be the event that both children are boys, and let B be the event that at least one child is a boy. Then

$$P(A|B) = P(A \cap B)/P(B) = P(A)/P(B)$$

since A is a subset of B (so the overlap between A and B is just A). But A has one outcome $\{(b, b)\}$ and B has three outcomes $\{(b, g), (g, b), (b, b)\}$. Applying Definition 2.4.1, then, gives

$$P(A|B) = (1/4)/(3/4) = \frac{1}{3}$$

Another correct approach is to go back to the sample space and deduce the value of $P(A|B)$ from first principles. Figure 2.4.3 shows events A and B defined on the four family types that comprise the sample space S. Knowing that B has occurred redefines the sample space to include *three* outcomes, each now having a $\frac{1}{3}$ probability. Of those three possible outcomes, one—namely, (b, b)—satisfies the event A. It follows that $P(A|B) = \frac{1}{3}$.

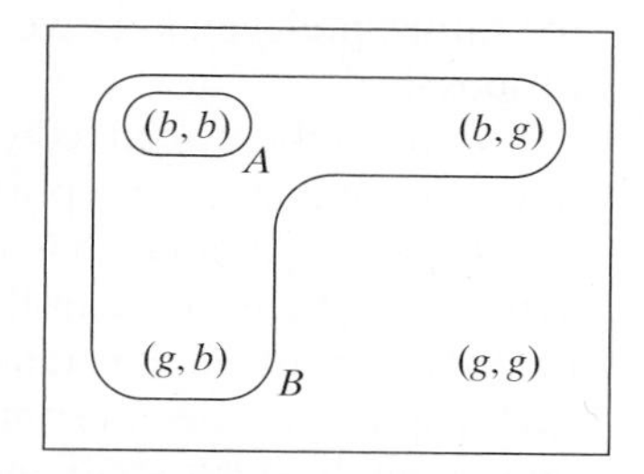

S = sample space of two-child families
[outcomes written as (first born, second born)]

Figure 2.4.3

■

Example 2.4.3

Two events A and B are defined such that (1) the probability that A occurs but B does not occur is 0.2, (2) the probability that B occurs but A does not occur is 0.1, and (3) the probability that neither occurs is 0.6. What is $P(A|B)$?

The three events whose probabilities are given are indicated on the Venn diagram shown in Figure 2.4.4. Since

$$P(\text{Neither occurs}) = 0.6 = P((A \cup B)^C)$$

it follows that

$$P(A \cup B) = 1 - 0.6 = 0.4 = P(A \cap B^C) + P(A \cap B) + P(B \cap A^C)$$

so

$$\begin{aligned} P(A \cap B) &= 0.4 - 0.2 - 0.1 \\ &= 0.1 \end{aligned}$$

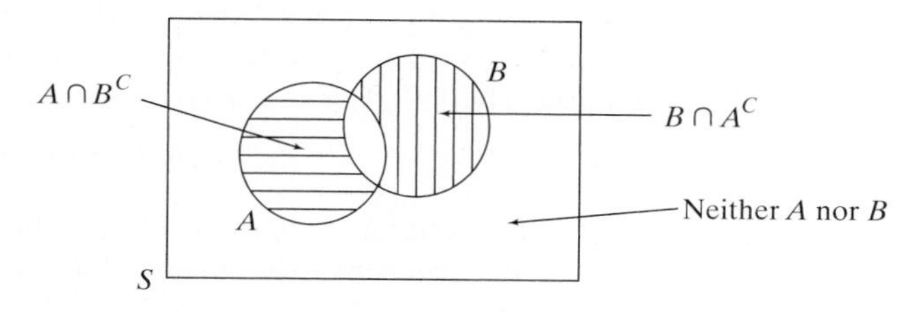

Figure 2.4.4

From Definition 2.4.1, then,

$$P(A|B) = \frac{P(A \cap B)}{P(B)} = \frac{P(A \cap B)}{P(A \cap B) + P(B \cap A^C)}$$
$$= \frac{0.1}{0.1 + 0.1}$$
$$= 0.5$$

■

Example 2.4.4

The possibility of importing liquified natural gas (LNG) from Algeria has been suggested as one way of coping with a future energy crunch. Complicating matters, though, is the fact that LNG is highly volatile and poses an enormous safety hazard. Any major spill occurring near a U.S. port could result in a fire of catastrophic proportions. The question, therefore, of the *likelihood* of a spill becomes critical input for future policymakers who may have to decide whether or not to implement the proposal.

Two numbers need to be taken into account: (1) the probability that a tanker will have an accident near a port, and (2) the probability that a major spill will develop *given* that an accident has happened. Although no significant spills of LNG have yet occurred anywhere in the world, these probabilities can be approximated from records kept on similar tankers transporting less dangerous cargo. On the basis of such data, it has been estimated (42) that the probability is $8/50{,}000$ that an LNG tanker will have an accident on any one trip. Given that an accident *has* occurred, it is suspected that only three times in fifteen thousand will the damage be sufficiently severe that a major spill would develop. What are the chances that a given LNG shipment would precipitate a catastrophic disaster?

Let A denote the event "Spill develops" and let B denote the event "Accident occurs." Past experience is suggesting that $P(B) = 8/50{,}000$ and $P(A|B) = 3/15{,}000$. Of primary concern is the probability that an accident will occur *and* a spill will ensue—that is, $P(A \cap B)$. Using Equation 2.4.2, we find that the chances of a catastrophic accident are on the order of three in one hundred million:

$$P(\text{Accident occurs and spill develops}) = P(A \cap B)$$
$$= P(A|B)P(B)$$
$$= \frac{3}{15{,}000} \cdot \frac{8}{50{,}000}$$
$$= 0.000000032$$

■

Example 2.4.5

Max and Muffy are two myopic deer hunters who shoot simultaneously at a nearby sheepdog that they have mistaken for a 10-point buck. Based on years of well-documented ineptitude, it can be assumed that Max has a 20% chance of hitting a stationary target at close range, Muffy has a 30% chance, and the probability is 0.06 that they will both be on target. Suppose that the sheepdog is hit and killed by exactly one bullet. What is the probability that Muffy fired the fatal shot?

Let A be the event that Max hit the dog, and let B be the event that Muffy hit the dog. Then $P(A) = 0.2$, $P(B) = 0.3$, and $P(A \cap B) = 0.06$. We are trying to find

$$P(B|(A^C \cap B) \cup (A \cap B^C))$$

where the event $(A^C \cap B) \cup (A \cap B^C)$ is the union of A and B minus the intersection—that is, it represents the event that either A or B but not both occur (recall Figure 2.4.4).

Notice, also, from Figure 2.4.4 that the intersection of B and $(A^C \cap B) \cup (A \cap B^C)$ is the event $A^C \cap B$. Therefore, from Definition 2.4.1,

$$\begin{aligned} P(B|(A^C \cap B) \cup (A \cap B^C)) &= [P(A^C \cap B)]/[P\{(A^C \cap B) \cup (A \cap B^C)\}] \\ &= [P(B) - P(A \cap B)]/[P(A \cup B) - P(A \cap B)] \\ &= [0.3 - 0.06]/[0.2 + 0.3 - 0.06 - 0.06] \\ &= 0.63 \end{aligned}$$

■

Example 2.4.6

The highways connecting two resort areas at A and B are shown in Figure 2.4.5. There is a direct route through the mountains and a more circuitous route going through a third resort area at C in the foothills. Travel between A and B during the winter months is not always possible, the roads sometimes being closed due to snow and ice. Suppose we let E_1, E_2, and E_3 denote the events that highways AB, AC, and BC are passable, respectively, and we know from past years that on a typical winter day,

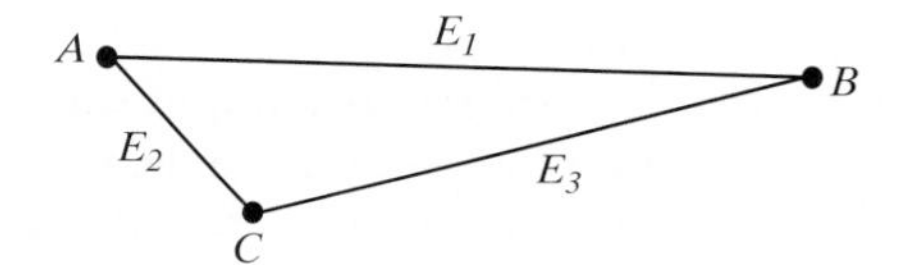

Figure 2.4.5

$$P(E_1) = \frac{2}{5}, \quad P(E_2) = \frac{3}{4}, \quad P(E_3) = \frac{2}{3}$$

and

$$P(E_3|E_2) = \frac{4}{5}, \quad P(E_1|E_2 \cap E_3) = \frac{1}{2}$$

What is the probability that a traveler will be able to get from A to B?

If E denotes the event that we *can* get from A to B, then

$$E = E_1 \cup (E_2 \cap E_3)$$

It follows that

$$P(E) = P(E_1) + P(E_2 \cap E_3) - P[E_1 \cap (E_2 \cap E_3)]$$

Applying Equation 2.4.2 three times gives

$$\begin{aligned} P(E) &= P(E_1) + P(E_3|E_2)P(E_2) - P[E_1|(E_2 \cap E_3)]P(E_2 \cap E_3) \\ &= P(E_1) + P(E_3|E_2)P(E_2) - P[E_1|(E_2 \cap E_3)]P(E_3|E_2)P(E_2) \\ &= \frac{2}{5} + \left(\frac{4}{5}\right)\left(\frac{3}{4}\right) - \left(\frac{1}{2}\right)\left(\frac{4}{5}\right)\left(\frac{3}{4}\right) = 0.7 \end{aligned}$$

(Which route should a traveler starting from A try first to maximize the chances of getting to B?)

■

Case Study 2.4.1

Several years ago, a television program (inadvertently) spawned a conditional probability problem that led to more than a few heated discussions, even in the national media. The show was *Let's Make a Deal*, and the question involved the strategy that contestants should take to maximize their chances of winning prizes.

On the program, a contestant would be presented with three doors, behind one of which was the prize. After the contestant had selected a door, the host, Monty Hall, would open one of the other two doors, showing that the prize was not there. Then he would give the contestant a choice—either stay with the door initially selected or switch to the "third" door, which had not been opened.

For many viewers, common sense seemed to suggest that switching doors would make no difference. By assumption, the prize had a one-third chance of being behind each of the doors when the game began. Once a door was opened, it was argued that each of the remaining doors now had a one-half probability of hiding the prize, so contestants gained nothing by switching their bets.

Not so. An application of Definition 2.4.1 shows that it *did* make a difference—contestants, in fact, *doubled* their chances of winning by switching doors. To see why, consider a specific (but typical) case: the contestant has bet on Door #2 and Monty Hall has opened Door #3. Given that sequence of events, we need to calculate and compare the conditional probability of the prize being behind Door #1 and Door #2, respectively. If the former is larger (and we will prove that it is), the contestant should switch doors.

Table 2.4.1 shows the sample space associated with the scenario just described. If the prize is actually behind Door #1, the host has no choice but to open Door #3; similarly, if the prize is behind Door #3, the host has no choice but to open Door #1. In the event that the prize is behind Door #2, though, the host would (theoretically) open Door #1 half the time and Door #3 half the time.

Table 2.4.1

(Prize Location, Door Opened)	Probability
(1, 3)	1/3
(2, 1)	1/6
(2, 3)	1/6
(3, 1)	1/3

Notice that the four outcomes in S are not equally likely. There is necessarily a one-third probability that the prize is behind each of the three doors. However, the two choices that the host has when the prize is behind Door #2 necessitate that the two outcomes (2, 1) and (2, 3) share the one-third probability that represents the chances of the prize being behind Door #2. Each, then, has the one-sixth probability listed in Table 2.4.1.

(Continued on next page)

Let A be the event that the prize is behind Door #2, and let B be the event that the host opened Door #3. Then

$$P(A|B) = P(\text{Contestant wins by not switching}) = [P(A \cap B)]/P(B)$$
$$= [\tfrac{1}{6}]/[\tfrac{1}{3} + \tfrac{1}{6}]$$
$$= \tfrac{1}{3}$$

Now, let A^* be the event that the prize is behind Door #1, and let B (as before) be the event that the host opens Door #3. In this case,

$$P(A^*|B) = P(\text{Contestant wins by switching}) = [P(A^* \cap B)]/P(B)$$
$$= [\tfrac{1}{3}]/[\tfrac{1}{3} + \tfrac{1}{6}]$$
$$= \tfrac{2}{3}$$

Common sense would have led us astray again! If given the choice, contestants should have *always* switched doors. Doing so upped their chances of winning from one-third to two-thirds.

Questions

2.4.1. Suppose that two fair dice are tossed. What is the probability that the sum equals 10 given that it exceeds 8?

2.4.2. Find $P(A \cap B)$ if $P(A) = 0.2$, $P(B) = 0.4$, and $P(A|B) + P(B|A) = 0.75$.

2.4.3. If $P(A|B) < P(A)$, show that $P(B|A) < P(B)$.

2.4.4. Let A and B be two events such that $P((A \cup B)^C) = 0.6$ and $P(A \cap B) = 0.1$. Let E be the event that either A or B but not both will occur. Find $P(E|A \cup B)$.

2.4.5. Suppose that in Example 2.4.2 we ignored the ages of the children and distinguished only *three* family types: (boy, boy), (girl, boy), and (girl, girl). Would the conditional probability of both children being boys given that at least one is a boy be different from the answer found on p. 35? Explain.

2.4.6. Two events, A and B, are defined on a sample space S such that $P(A|B) = 0.6$, P(At least one of the events occurs) $= 0.8$, and P(Exactly one of the events occurs) $= 0.6$. Find $P(A)$ and $P(B)$.

2.4.7. An urn contains one red chip and one white chip. One chip is drawn at random. If the chip selected is red, that chip together with two additional red chips are put back into the urn. If a white chip is drawn, the chip is returned to the urn. Then a second chip is drawn. What is the probability that both selections are red?

2.4.8. Given that $P(A) = a$ and $P(B) = b$, show that

$$P(A|B) \geq \frac{a+b-1}{b}$$

2.4.9. An urn contains one white chip and a second chip that is equally likely to be white or black. A chip is drawn at random and returned to the urn. Then a second chip is drawn. What is the probability that a white appears on the second draw given that a white appeared on the first draw? (*Hint:* Let W_i be the event that a white chip is selected on the ith draw, $i = 1, 2$. Then $P(W_2|W_1) = \frac{P(W_1 \cap W_2)}{P(W_1)}$. If both chips in the urn are white, $P(W_1) = 1$; otherwise, $P(W_1) = \frac{1}{2}$.)

2.4.10. Suppose events A and B are such that $P(A \cap B) = 0.1$ and $P((A \cup B)^C) = 0.3$. If $P(A) = 0.2$, what does $P[(A \cap B)|(A \cup B)^C]$ equal? (*Hint*: Draw the Venn diagram.)

2.4.11. One hundred voters were asked their opinions of two candidates, A and B, running for mayor. Their responses to three questions are summarized below:

	Number Saying "Yes"
Do you like A?	65
Do you like B?	55
Do you like both?	25

(a) What is the probability that someone likes neither?
(b) What is the probability that someone likes exactly one?
(c) What is the probability that someone likes at least one?
(d) What is the probability that someone likes at most one?

(e) What is the probability that someone likes exactly one given that he or she likes at least one?
(f) Of those who like at least one, what proportion like both?
(g) Of those who do not like A, what proportion like B?

2.4.12. A fair coin is tossed three times. What is the probability that at least two heads will occur given that at most two heads have occurred?

2.4.13. Two fair dice are rolled. What is the probability that the number on the first die was at least as large as 4 given that the sum of the two dice was 8?

2.4.14. Four cards are dealt from a standard 52-card poker deck. What is the probability that all four are aces given that at least three are aces? (*Note*: There are 270,725 different sets of four cards that can be dealt. Assume that the probability associated with each of those hands is 1/270,725.)

2.4.15. Given that $P(A \cap B^C) = 0.3$, $P((A \cup B)^C) = 0.2$, and $P(A \cap B) = 0.1$, find $P(A|B)$.

2.4.16. Given that $P(A) + P(B) = 0.9$, $P(A|B) = 0.5$, and $P(B|A) = 0.4$, find $P(A)$.

2.4.17. Let A and B be two events defined on a sample space S such that $P(A \cap B^C) = 0.1$, $P(A^C \cap B) = 0.3$, and $P((A \cup B)^C) = 0.2$. Find the probability that at least one of the two events occurs given that at most one occurs.

2.4.18. Suppose two dice are rolled. Assume that each possible outcome has probability 1/36. Let A be the event that the sum of the two dice is greater than or equal to 8, and let B be the event that at least one of the dice shows a 5. Find $P(A|B)$.

2.4.19. According to your neighborhood bookie, five horses are scheduled to run in the third race at the local track, and handicappers have assigned them the following probabilities of winning:

Horse	Probability of Winning
Scorpion	0.10
Starry Avenger	0.25
Australian Doll	0.15
Dusty Stake	0.30
Outandout	0.20

Suppose that Australian Doll and Dusty Stake are scratched from the race at the last minute. What are the chances that Outandout will prevail over the reduced field?

2.4.20. Andy, Bob, and Charley have all been serving time for grand theft auto. According to prison scuttlebutt, the warden plans to release two of the three next week. They all have identical records, so the two to be released will be chosen at random, meaning that each has a two-thirds probability of being included in the two to be set free. Andy, however, is friends with a guard who will know ahead of time which two will leave. He offers to tell Andy the name of one prisoner *other than himself* who will be released. Andy, however, declines the offer, believing that if he learns the name of one prisoner scheduled to be released, then *his* chances of being the other person set free will drop to one-half (since only two prisoners will be left at that point). Is his concern justified?

Applying Conditional Probability to Higher-Order Intersections

We have seen that conditional probabilities can be useful in evaluating intersection probabilities—that is, $P(A \cap B) = P(A|B)P(B) = P(B|A)P(A)$. A similar result holds for higher-order intersections. Consider $P(A \cap B \cap C)$. By thinking of $A \cap B$ as a single event—say, D—we can write

$$\begin{aligned} P(A \cap B \cap C) &= P(D \cap C) \\ &= P(C|D)P(D) \\ &= P(C|A \cap B)P(A \cap B) \\ &= P(C|A \cap B)P(B|A)P(A) \end{aligned}$$

Repeating this same argument for n events, $A_1, A_2, \ldots, A_n$, gives a formula for the general case:

$$\begin{aligned} P(A_1 \cap A_2 \cap \cdots \cap A_n) = P(A_n|A_1 \cap A_2 \cap \cdots \cap A_{n-1}) \\ \cdot P(A_{n-1}|A_1 \cap A_2 \cap \cdots \cap A_{n-2}) \cdot \cdots \cdot P(A_2|A_1) \cdot P(A_1) \end{aligned} \tag{2.4.3}$$

Example 2.4.7 An urn contains five white chips, four black chips, and three red chips. Four chips are drawn sequentially and without replacement. What is the probability of obtaining the sequence (white, red, white, black)?

Figure 2.4.6. shows the evolution of the urn's composition as the desired sequence is assembled. Define the following four events:

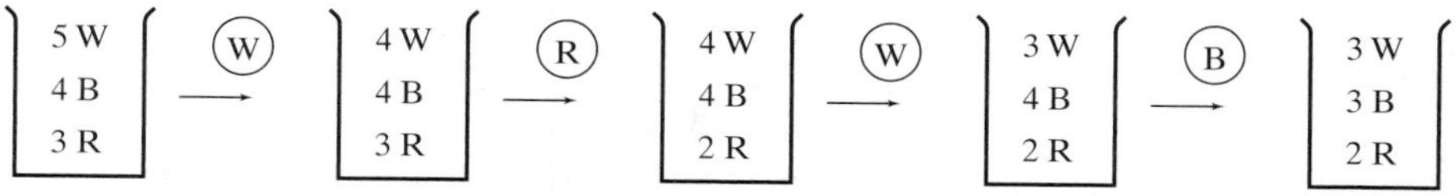

5 W, 4 B, 3 R → (W) → 4 W, 4 B, 3 R → (R) → 4 W, 4 B, 2 R → (W) → 3 W, 4 B, 2 R → (B) → 3 W, 3 B, 2 R

Figure 2.4.6

A: white chip is drawn on first selection
B: red chip is drawn on second selection
C: white chip is drawn on third selection
D: black chip is drawn on fourth selection

Our objective is to find $P(A \cap B \cap C \cap D)$.

From Equation 2.4.3,

$$P(A \cap B \cap C \cap D) = P(D|A \cap B \cap C) \cdot P(C|A \cap B) \cdot P(B|A) \cdot P(A)$$

Each of the probabilities on the right-hand side of the equation here can be gotten by just looking at the urns pictured in Figure 2.4.6: $P(D|A \cap B \cap C) = \frac{4}{9}$, $P(C|A \cap B) = \frac{4}{10}$, $P(B|A) = \frac{3}{11}$, and $P(A) = \frac{5}{12}$. Therefore, the probability of drawing a (white, red, white, black) sequence is 0.02:

$$\begin{aligned} P(A \cap B \cap C \cap D) &= \frac{4}{9} \cdot \frac{4}{10} \cdot \frac{3}{11} \cdot \frac{5}{12} \\ &= \frac{240}{11{,}880} \\ &= 0.02 \end{aligned}$$

■

Case Study 2.4.2

Since the late 1940s, tens of thousands of eyewitness accounts of strange lights in the skies, unidentified flying objects, and even alleged abductions by little green men have made headlines. None of these incidents, though, has produced any hard evidence, any irrefutable *proof* that Earth has been visited by a race of extraterrestrials. Still, the haunting question remains—are we alone in the universe? Or are there other civilizations, more advanced than ours, making the occasional flyby?

Until, or unless, a flying saucer plops down on the White House lawn and a strange-looking creature emerges with the proverbial "Take me to your leader" demand, we may never know whether we have any cosmic neighbors. Equation 2.4.3, though, can help us speculate on the *probability* of our not being alone.

(Continued on next page)

(Case Study 2.4.2 continued)

Recent discoveries suggest that planetary systems much like our own may be quite common. If so, there are likely to be many planets whose chemical makeups, temperatures, pressures, and so on, are suitable for life. Let those planets be the points in our sample space. Relative to them, we can define three events:

A: life arises
B: technical civilization arises (one capable of interstellar communication)
C: technical civilization is flourishing *now*

In terms of A, B, and C, the probability that a habitable planet is presently supporting a technical civilization is the probability of an intersection—specifically, $P(A \cap B \cap C)$. Associating a number with $P(A \cap B \cap C)$ is highly problematic, but the task is simplified considerably if we work instead with the equivalent conditional formula, $P(C|B \cap A) \cdot P(B|A) \cdot P(A)$.

Scientists speculate (153) that life of some kind may arise on one-third of all planets having a suitable environment and that life on maybe 1% of all those planets will evolve into a technical civilization. In our notation, $P(A) = \frac{1}{3}$ and $P(B|A) = \frac{1}{100}$.

More difficult to estimate is $P(C|A \cap B)$. On Earth, we have had the capability of interstellar communication (that is, radio astronomy) for only a few decades, so $P(C|A \cap B)$, *empirically*, is on the order of 1×10^{-8}. But that may be an overly pessimistic estimate of a technical civilization's ability to endure. It may be true that if a civilization can avoid annihilating itself when it first develops nuclear weapons, its prospects for longevity are fairly good. If that were the case, $P(C|A \cap B)$ might be as large as 1×10^{-2}.

Putting these estimates into the computing formula for $P(A \cap B \cap C)$ yields a range for the probability of a habitable planet currently supporting a technical civilization. The chances may be as small as 3.3×10^{-11} or as "large" as 3.3×10^{-5}:

$$(1 \times 10^{-8})\left(\frac{1}{100}\right)\left(\frac{1}{3}\right) < P(A \cap B \cap C) < (1 \times 10^{-2})\left(\frac{1}{100}\right)\left(\frac{1}{3}\right)$$

or

$$0.000000000033 < P(A \cap B \cap C) < 0.000033$$

A better way to put these figures in some kind of perspective is to think in terms of *numbers* rather than probabilities. Astronomers estimate there are 3×10^{11} habitable planets in our Milky Way galaxy. Multiplying that total by the two limits for $P(A \cap B \cap C)$ gives an indication of *how many* cosmic neighbors we are likely to have. Specifically, $3 \times 10^{11} \cdot 0.000000000033 \doteq 10$, while $3 \times 10^{11} \cdot 0.000033 \doteq 10{,}000{,}000$. So, on the one hand, we may be a galactic rarity. At the same time, the probabilities do not preclude the very real possibility that the Milky Way is abuzz with activity and that our neighbors number in the millions.

Questions

2.4.21. An urn contains six white chips, four black chips, and five red chips. Five chips are drawn out, one at a time and without replacement. What is the probability of getting the sequence (black, black, red, white, white)? Suppose that the chips are numbered 1 through 15. What is the probability of getting a specific sequence—say, (2, 6, 4, 9, 13)?

2.4.22. A man has n keys on a key ring, one of which opens the door to his apartment. Having celebrated a bit too much one evening, he returns home only to find himself unable to distinguish one key from another. Resourceful, he works out a fiendishly clever plan: He will choose a key at random and try it. If it fails to open the door, he will discard it and choose at random one of the remaining $n-1$ keys, and so on. Clearly, the probability that he gains entrance with the first key he selects is $1/n$. Show that the probability the door opens with the *third* key he tries is also $1/n$. (*Hint:* What has to happen before he even gets to the third key?)

2.4.23. Suppose that four cards are drawn from a standard 52-card poker deck. What is the probability of drawing, in order, a 7 of diamonds, a jack of spades, a 10 of diamonds, and a 5 of hearts?

2.4.24. One chip is drawn at random from an urn that contains one white chip and one black chip. If the white chip is selected, we simply return it to the urn; if the black chip is drawn, that chip—together with another black—are returned to the urn. Then a second chip is drawn, with the same rules for returning it to the urn. Calculate the probability of drawing two whites followed by three blacks.

Calculating "Unconditional" and "Inverse" Probabilities

We conclude this section with two very useful theorems that apply to *partitioned* sample spaces. By definition, a set of events $A_1, A_2, \ldots, A_n$ "partition" S if every outcome in the sample space belongs to one and only one of the A_i's—that is, the A_i's are mutually exclusive and their union is S (see Figure 2.4.7).

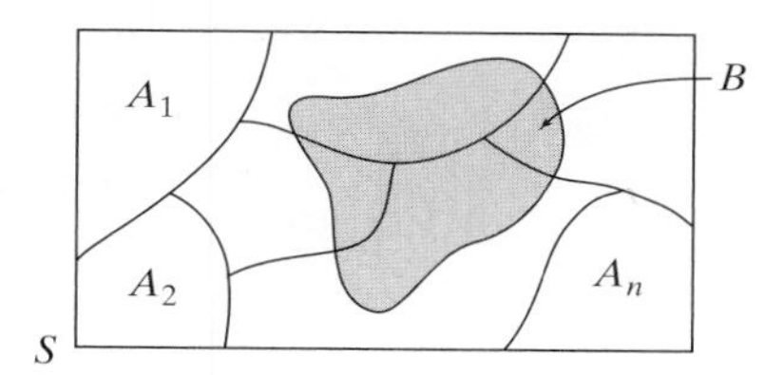

Figure 2.4.7

Let B, as pictured, denote any event defined on S. The first result, Theorem 2.4.1, gives a formula for the "unconditional" probability of B (in terms of the A_i's). Then Theorem 2.4.2 calculates the set of conditional probabilities, $P(A_j|B)$, $j = 1, 2, \ldots, n$.

Theorem 2.4.1 *Let $\{A_i\}_{i=1}^n$ be a set of events defined over S such that $S = \bigcup_{i=1}^n A_i$, $A_i \cap A_j = \emptyset$ for $i \neq j$, and $P(A_i) > 0$ for $i = 1, 2, \ldots, n$. For any event B,*

$$P(B) = \sum_{i=1}^{n} P(B|A_i)P(A_i)$$

Proof By the conditions imposed on the A_i's,

$$B = (B \cap A_1) \cup (B \cap A_2) \cup \cdots \cup (B \cap A_n)$$

and

$$P(B) = P(B \cap A_1) + P(B \cap A_2) + \cdots + P(B \cap A_n)$$

But each $P(B \cap A_i)$ can be written as the product $P(B|A_i)P(A_i)$, and the result follows. □

Example 2.4.8

Urn I contains two red chips and four white chips; urn II, three red and one white. A chip is drawn at random from urn I and transferred to urn II. Then a chip is drawn from urn II. What is the probability that the chip drawn from urn II is red?

Let B be the event "Chip drawn from urn II is red"; let A_1 and A_2 be the events "Chip transferred from urn I is red" and "Chip transferred from urn I is white," respectively. By inspection (see Figure 2.4.8), we can deduce all the probabilities appearing in the right-hand side of the formula in Theorem 2.4.1:

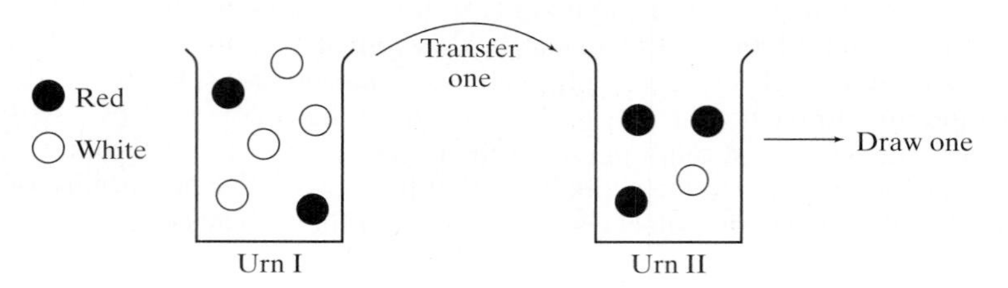

Figure 2.4.8

$$P(B|A_1) = \frac{4}{5} \qquad P(B|A_2) = \frac{3}{5}$$
$$P(A_1) = \frac{2}{6} \qquad P(A_2) = \frac{4}{6}$$

Putting all this information together, we see that the chances are two out of three that a red chip will be drawn from urn II:

$$\begin{aligned} P(B) &= P(B|A_1)P(A_1) + P(B|A_2)P(A_2) \\ &= \frac{4}{5} \cdot \frac{2}{6} + \frac{3}{5} \cdot \frac{4}{6} \\ &= \frac{2}{3} \end{aligned}$$

■

Example 2.4.9

A standard poker deck is shuffled and the card on top is removed. What is the probability that the *second* card is an ace?

Define the following events:

B: second card is an ace
A_1: top card was an ace
A_2: top card was not an ace

Then $P(B|A_1) = \frac{3}{51}$, $P(B|A_2) = \frac{4}{51}$, $P(A_1) = \frac{4}{52}$, and $P(A_2) = \frac{48}{52}$. Since the A_i's partition the sample space of two-card selections, Theorem 2.4.1 applies. Substituting into the expression for $P(B)$ shows that $\frac{4}{52}$ is the probability that the second card is an ace:

$$\begin{aligned} P(B) &= P(B|A_1)P(A_1) + P(B|A_2)P(A_2) \\ &= \frac{3}{51} \cdot \frac{4}{52} + \frac{4}{51} \cdot \frac{48}{52} \\ &= \frac{4}{52} \end{aligned}$$

Comment Notice that $P(B) = P(\text{2nd card is an ace})$ is numerically the same as $P(A_1) = P(\text{first card is an ace})$. The analysis in Example 2.4.9 illustrates a basic principle in probability that says, in effect, "What you don't know, doesn't matter." Here, removal of the top card is irrelevant to any subsequent probability calculations *if the identity of that card remains unknown.* ■

Example 2.4.10

Ashley is hoping to land a summer internship with a public relations firm. If her interview goes well, she has a 70% chance of getting an offer. If the interview is a bust, though, her chances of getting the position drop to 20%. Unfortunately, Ashley tends to babble incoherently when she is under stress, so the likelihood of the interview going well is only 0.10. What is the probability that Ashley gets the internship?

Let B be the event "Ashley is offered internship," let A_1 be the event "Interview goes well," and let A_2 be the event "Interview does not go well." By assumption,

$$P(B|A_1) = 0.70 \qquad P(B|A_2) = 0.20$$
$$P(A_1) = 0.10 \qquad P(A_2) = 1 - P(A_1) = 1 - 0.10 = 0.90$$

According to Theorem 2.4.1, Ashley has a *25%* chance of landing the internship:

$$\begin{aligned} P(B) &= P(B|A_1)P(A_1) + P(B|A_2)P(A_2) \\ &= (0.70)(0.10) + (0.20)(0.90) \\ &= 0.25 \end{aligned}$$

■

Example 2.4.11

In an upstate congressional race, the incumbent Republican (R) is running against a field of three Democrats (D_1, D_2, and D_3) seeking the nomination. Political pundits estimate that the probabilities of D_1, D_2, or D_3 winning the primary are 0.35, 0.40, and 0.25, respectively. Furthermore, results from a variety of polls are suggesting that R would have a 40% chance of defeating D_1 in the general election, a 35% chance of defeating D_2, and a 60% chance of defeating D_3. Assuming all these estimates to be accurate, what are the chances that the Republican will retain his seat?

Let B denote the event that "R wins general election," and let A_i denote the event "D_i wins Democratic primary," $i = 1, 2, 3$. Then

$$P(A_1) = 0.35 \qquad P(A_2) = 0.40 \qquad P(A_3) = 0.25$$

and

$$P(B|A_1) = 0.40 \qquad P(B|A_2) = 0.35 \qquad P(B|A_3) = 0.60$$

so

$$\begin{aligned} P(B) &= P(\text{Republican wins general election}) \\ &= P(B|A_1)P(A_1) + P(B|A_2)P(A_2) + P(B|A_3)P(A_3) \\ &= (0.40)(0.35) + (0.35)(0.40) + (0.60)(0.25) \\ &= 0.43 \end{aligned}$$

■

Example 2.4.12 Three chips are placed in an urn. One is red on both sides, a second is blue on both sides, and the third is red on one side and blue on the other. One chip is selected at random and placed on a table. Suppose that the color showing on that chip is red. What is the probability that the color underneath is also red (see Figure 2.4.9)?

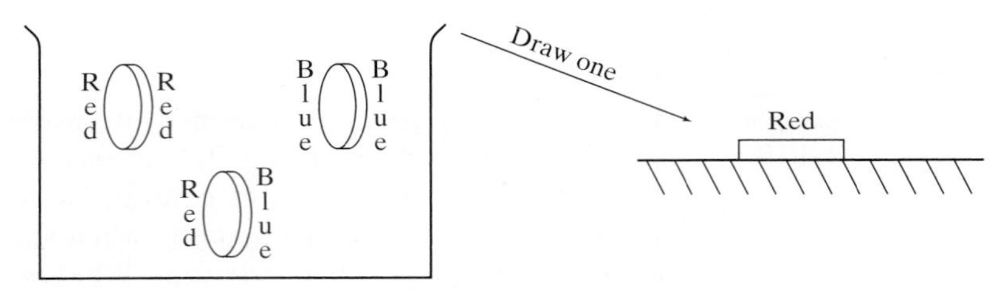

Figure 2.4.9

At first glance, it may seem that the answer is one-half: We know that the blue/blue chip has not been drawn, and only one of the remaining two—the red/red chip—satisfies the event that the color underneath is red. If this game were played over and over, though, and records were kept of the outcomes, it would be found that the proportion of times that a red top has a red bottom is two-thirds, not the one-half that our intuition might suggest. The correct answer follows from an application of Theorem 2.4.1.

Define the following events:

A: bottom side of chip drawn is red
B: top side of chip drawn is red
A_1: red/red chip is drawn
A_2: blue/blue chip is drawn
A_3: red/blue chip is drawn

From the definition of conditional probability,

$$P(A|B) = \frac{P(A \cap B)}{P(B)}$$

But $P(A \cap B) = P(\text{Both sides are red}) = P(\text{red/red chip}) = \frac{1}{3}$. Theorem 2.4.1 can be used to find the denominator, $P(B)$:

$$\begin{aligned} P(B) &= P(B|A_1)P(A_1) + P(B|A_2)P(A_2) + P(B|A_3)P(A_3) \\ &= 1 \cdot \frac{1}{3} + 0 \cdot \frac{1}{3} + \frac{1}{2} \cdot \frac{1}{3} \\ &= \frac{1}{2} \end{aligned}$$

Therefore,

$$P(A|B) = \frac{1/3}{1/2} = \frac{2}{3}$$

Comment The question posed in Example 2.4.12 gives rise to a simple but effective con game. The trick is to convince a "mark" that the initial analysis given above is correct, meaning that the bottom has a fifty-fifty chance of being the same color as the top. Under that incorrect presumption that the game is "fair," both participants put up the same amount of money, but the gambler (knowing the correct analysis) always bets that the bottom is the same color as the top. In the long run, then, the con artist will be winning an even-money bet two-thirds of the time! ■

Questions

2.4.25. A toy manufacturer buys ball bearings from three different suppliers—50% of her total order comes from supplier 1, 30% from supplier 2, and the rest from supplier 3. Past experience has shown that the quality-control standards of the three suppliers are not all the same. Two percent of the ball bearings produced by supplier 1 are defective, while suppliers 2 and 3 produce defective bearings 3% and 4% of the time, respectively. What proportion of the ball bearings in the toy manufacturer's inventory are defective?

2.4.26. A fair coin is tossed. If a head turns up, a fair die is tossed; if a tail turns up, two fair dice are tossed. What is the probability that the face (or the sum of the faces) showing on the die (or the dice) is equal to 6?

2.4.27. Foreign policy experts estimate that the probability is 0.65 that war will break out next year between two Middle East countries if either side significantly escalates its terrorist activities. Otherwise, the likelihood of war is estimated to be 0.05. Based on what has happened this year, the chances of terrorism reaching a critical level in the next twelve months are thought to be three in ten. What is the probability that the two countries will go to war?

2.4.28. A telephone solicitor is responsible for canvassing three suburbs. In the past, 60% of the completed calls to Belle Meade have resulted in contributions, compared to 55% for Oak Hill and 35% for Antioch. Her list of telephone numbers includes one thousand households from Belle Meade, one thousand from Oak Hill, and two thousand from Antioch. Suppose that she picks a number at random from the list and places the call. What is the probability that she gets a donation?

2.4.29. If men constitute 47% of the population and tell the truth 78% of the time, while women tell the truth 63% of the time, what is the probability that a person selected at random will answer a question truthfully?

2.4.30. Urn I contains three red chips and one white chip. Urn II contains two red chips and two white chips. One chip is drawn from each urn and transferred to the other urn. Then a chip is drawn from the first urn. What is the probability that the chip ultimately drawn from urn I is red?

2.4.31. Medical records show that 0.01% of the general adult population not belonging to a high-risk group (for example, intravenous drug users) are HIV-positive. Blood tests for the virus are 99.9% accurate when given to someone infected and 99.99% accurate when given to someone not infected. What is the probability that a random adult not in a high-risk group will test positive for the HIV virus?

2.4.32. Recall the "survival" lottery described in Question 2.2.14. What is the probability of release associated with the prisoner's optimal strategy?

2.4.33. State College is playing Backwater A&M for the conference football championship. If Backwater's first-string quarterback is healthy, A&M has a 75% chance of winning. If they have to start their backup quarterback, their chances of winning drop to 40%. The team physician says that there is a 70% chance that the first-string quarterback will play. What is the probability that Backwater wins the game?

2.4.34. An urn contains forty red chips and sixty white chips. Six chips are drawn out and discarded, and a seventh chip is drawn. What is the probability that the seventh chip is red?

2.4.35. A study has shown that seven out of ten people will say "heads" if asked to call a coin toss. Given that the coin is fair, though, a head occurs, on the average, only five times out of ten. Does it follow that you have the advantage if you let the other person call the toss? Explain.

2.4.36. Based on pretrial speculation, the probability that a jury returns a guilty verdict in a certain high-profile murder case is thought to be 15% if the defense can discredit the police department and 80% if they cannot. Veteran court observers believe that the skilled defense attorneys have a 70% chance of convincing the jury that the police either contaminated or planted some of the key evidence. What is the probability that the jury returns a guilty verdict?

2.4.37. As an incoming freshman, Marcus believes that he has a 25% chance of earning a GPA in the 3.5 to 4.0 range, a 35% chance of graduating with a 3.0 to 3.5 GPA, and a 40% chance of finishing with a GPA less than 3.0. From what the pre-med advisor has told him, Marcus has an 8 in 10 chance of getting into medical school if his GPA is above 3.5, a 5 in 10 chance if his GPA is in the 3.0 to 3.5 range, and only a 1 in 10 chance if his GPA falls below

3.0. Based on those estimates, what is the probability that Marcus gets into medical school?

2.4.38. The governor of a certain state has decided to come out strongly for prison reform and is preparing a new early release program. Its guidelines are simple: prisoners related to members of the governor's staff would have a 90% chance of being released early; the probability of early release for inmates not related to the governor's staff would be 0.01. Suppose that 40% of all inmates are related to someone on the governor's staff. What is the probability that a prisoner selected at random would be eligible for early release?

2.4.39. Following are the percentages of students of State College enrolled in each of the school's main divisions. Also listed are the proportions of students in each division who are women.

Division	%	% Women
Humanities	40	60
Natural science	10	15
History	30	45
Social science	20	75
	100	

Suppose the registrar selects one person at random. What is the probability that the student selected will be a male?

Bayes' Theorem

The second result in this section that is set against the backdrop of a partitioned sample space has a curious history. The first explicit statement of Theorem 2.4.2, coming in 1812, was due to Laplace, but it was named after the Reverend Thomas Bayes, whose 1763 paper (published posthumously) had already outlined the result. On one level, the theorem is a relatively minor extension of the definition of conditional probability. When viewed from a loftier perspective, though, it takes on some rather profound philosophical implications. The latter, in fact, have precipitated a schism among practicing statisticians: "Bayesians" analyze data one way; "non-Bayesians" often take a fundamentally different approach (see Section 5.8).

Our use of the result here will have nothing to do with its statistical interpretation. We will apply it simply as the Reverend Bayes originally intended, as a formula for evaluating a certain kind of "inverse" probability. If we know $P(B|A_i)$ for all i, the theorem enables us to compute conditional probabilities "in the other direction"—that is, we can deduce $P(A_j|B)$ from the $P(B|A_i)$'s.

Theorem 2.4.2 *(Bayes') Let $\{A_i\}_{i=1}^n$ be a set of n events, each with positive probability, that partition S in such a way that $\cup_{i=1}^n A_i = S$ and $A_i \cap A_j = \emptyset$ for $i \neq j$. For any event B (also defined on S), where $P(B) > 0$,*

$$P(A_j|B) = \frac{P(B|A_j)P(A_j)}{\sum_{i=1}^{n} P(B|A_i)P(A_i)}$$

for any $1 \leq j \leq n$.

Proof From Definition 2.4.1,

$$P(A_j|B) = \frac{P(A_j \cap B)}{P(B)} = \frac{P(B|A_j)P(A_j)}{P(B)}$$

But Theorem 2.4.1 allows the denominator to be written as $\sum_{i=1}^{n} P(B|A_i)P(A_i)$, and the result follows. □

Problem-Solving Hints

(Working with Partitioned Sample Spaces)

Students sometimes have difficulty setting up problems that involve partitioned sample spaces—in particular, ones whose solution requires an application of either Theorem 2.4.1 or 2.4.2—because of the nature and amount of information that need to be incorporated into the answers. The "trick" is learning to identify which part of the "given" corresponds to B and which parts correspond to the A_i's. The following hints may help.

1. As you read the question, pay particular attention to the last one or two sentences. Is the problem asking for an *unconditional probability* (in which case Theorem 2.4.1 applies) or a *conditional probability* (in which case Theorem 2.4.2 applies)?
2. If the question is asking for an unconditional probability, let B denote the event whose probability you are trying to find; if the question is asking for a conditional probability, let B denote the event that has *already happened*.
3. Once event B has been identified, reread the beginning of the question and assign the A_i's.

Example 2.4.13

A biased coin, twice as likely to come up heads as tails, is tossed once. If it shows heads, a chip is drawn from urn I, which contains three white chips and four red chips; if it shows tails, a chip is drawn from urn II, which contains six white chips and three red chips. Given that a white chip was drawn, what is the probability that the coin came up tails (see Figure 2.4.10)?

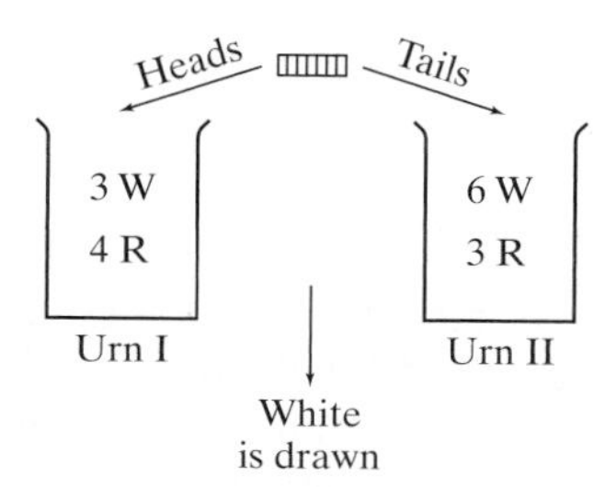

Figure 2.4.10

Since $P(\text{heads}) = 2P(\text{tails})$, it must be true that $P(\text{heads}) = \frac{2}{3}$ and $P(\text{tails}) = \frac{1}{3}$. Define the events

B: white chip is drawn
A_1: coin came up heads (i.e., chip came from urn I)
A_2: coin came up tails (i.e., chip came from urn II)

Our objective is to find $P(A_2|B)$. From Figure 2.4.10,

$$P(B|A_1) = \frac{3}{7} \qquad P(B|A_2) = \frac{6}{9}$$

$$P(A_1) = \frac{2}{3} \qquad P(A_2) = \frac{1}{3}$$

so

$$P(A_2|B) = \frac{P(B|A_2)P(A_2)}{P(B|A_1)P(A_1) + P(B|A_2)P(A_2)}$$
$$= \frac{(6/9)(1/3)}{(3/7)(2/3) + (6/9)(1/3)}$$
$$= \frac{7}{16}$$

■

Example 2.4.14

During a power blackout, one hundred persons are arrested on suspicion of looting. Each is given a polygraph test. From past experience it is known that the polygraph is 90% reliable when administered to a guilty suspect and 98% reliable when given to someone who is innocent. Suppose that of the one hundred persons taken into custody, only twelve were actually involved in any wrongdoing. What is the probability that a given suspect is innocent given that the polygraph says he is guilty?

Let B be the event "Polygraph says suspect is guilty," and let A_1 and A_2 be the events "Suspect is guilty" and "Suspect is not guilty," respectively. To say that the polygraph is "90% reliable when administered to a guilty suspect" means that $P(B|A_1) = 0.90$. Similarly, the 98% reliability for innocent suspects implies that $P(B^C|A_2) = 0.98$, or, equivalently, $P(B|A_2) = 0.02$.

We also know that $P(A_1) = \frac{12}{100}$ and $P(A_2) = \frac{88}{100}$. Substituting into Theorem 2.4.2, then, shows that the probability a suspect is innocent given that the polygraph says he is guilty is 0.14:

$$P(A_2|B) = \frac{P(B|A_2)P(A_2)}{P(B|A_1)P(A_1) + P(B|A_2)P(A_2)}$$
$$= \frac{(0.02)(88/100)}{(0.90)(12/100) + (0.02)(88/100)}$$
$$= 0.14$$

■

Example 2.4.15

As medical technology advances and adults become more health conscious, the demand for diagnostic screening tests inevitably increases. Looking for problems, though, when no symptoms are present can have undesirable consequences that may outweigh the intended benefits.

Suppose, for example, a woman has a medical procedure performed to see whether she has a certain type of cancer. Let B denote the event that the test *says* she has cancer, and let A_1 denote the event that she actually *does* (and A_2, the event that she *does not*). Furthermore, suppose the prevalence of the disease and the precision of the diagnostic test are such that

$$P(A_1) = 0.0001 \qquad [\text{and} \quad P(A_2) = 0.9999]$$
$$P(B|A_1) = 0.90 = P(\text{Test } says \text{ woman has cancer when, in fact, she does})$$
$$P(B|A_2) = P\left(B|A_1^C\right) = 0.001 = P(\text{false positive}) = P(\text{Test } says \text{ woman has cancer when, in fact, she does not})$$

What is the probability that she *does* have cancer, given that the diagnostic procedure says she does? That is, calculate $P(A_1|B)$.

Although the method of solution here is straightforward, the actual numerical answer is not what we would expect. From Theorem 2.4.2,

$$P(A_1|B) = \frac{P(B|A_1)P(A_1)}{P(B|A_1)P(A_1) + P(B|A_1^C)P(A_1^C)}$$

$$= \frac{(0.9)(0.0001)}{(0.9)(0.0001) + (0.001)(0.9999)}$$

$$= 0.08$$

So, only 8% of those women identified as having cancer actually do! Table 2.4.2 shows the strong dependence of $P(A_1|B)$ on $P(A_1)$ and $P(B|A_1^C)$.

Table 2.4.2

$P(A_1)$	$P(B\|A_1^C)$	$P(A_1\|B)$
0.0001	0.001	0.08
	0.0001	0.47
0.001	0.001	0.47
	0.0001	0.90
0.01	0.001	0.90
	0.0001	0.99

In light of these probabilities, the practicality of screening programs directed at diseases having a low prevalence is open to question, especially when the diagnostic procedure, itself, poses a nontrivial health risk. (For precisely those two reasons, the use of chest X-rays to screen for tuberculosis is no longer advocated by the medical community.) ■

Example 2.4.16

According to the manufacturer's specifications, your home burglar alarm has a 95% chance of going off if someone breaks into your house. During the two years you have lived there, the alarm has gone off on five different nights, each time for no apparent reason. Suppose the alarm goes off tomorrow night. What is the probability that someone is trying to break into your house? (*Note*: Police statistics show that the chances of any particular house in your neighborhood being burglarized on any given night are two in ten thousand.)

Let B be the event "Alarm goes off tomorrow night," and let A_1 and A_2 be the events "House is being burglarized" and "House is not being burglarized," respectively. Then

$$P(B|A_1) = 0.95$$

$$P(B|A_2) = 5/730 \quad \text{(i.e., five nights in two years)}$$

$$P(A_1) = 2/10,000$$

$$P(A_2) = 1 - P(A_1) = 9998/10,000$$

The probability in question is $P(A_1|B)$.

Intuitively, it might seem that $P(A_1|B)$ should be close to 1 because the alarm's "performance" probabilities look good—$P(B|A_1)$ is close to 1 (as it should be) and $P(B|A_2)$ is close to 0 (as it should be). Nevertheless, $P(A_1|B)$ turns out to be surprisingly small:

$$\begin{aligned}P(A_1|B) &= \frac{P(B|A_1)P(A_1)}{P(B|A_1)P(A_1) + P(B|A_2)P(A_2)} \\ &= \frac{(0.95)(2/10{,}000)}{(0.95)(2/10{,}000) + (5/730)(9998/10{,}000)} \\ &= 0.027\end{aligned}$$

That is, if you hear the alarm going off, the probability is only 0.027 that your house is being burglarized.

Computationally, the reason $P(A_1|B)$ is so small is that $P(A_2)$ is so large. The latter makes the denominator of $P(A_1|B)$ large and, in effect, "washes out" the numerator. Even if $P(B|A_1)$ were substantially increased (by installing a more expensive alarm), $P(A_1|B)$ would remain largely unchanged (see Table 2.4.3).

Table 2.4.3

$P(B\|A_1)$	0.95	0.97	0.99	0.999
$P(A_1\|B)$	0.027	0.028	0.028	0.028

■

Questions

2.4.40. Urn I contains two white chips and one red chip; urn II has one white chip and two red chips. One chip is drawn at random from urn I and transferred to urn II. Then one chip is drawn from urn II. Suppose that a red chip is selected from urn II. What is the probability that the chip transferred was white?

2.4.41. Urn I contains three red chips and five white chips; urn II contains four reds and four whites; urn III contains five reds and three whites. One urn is chosen at random and one chip is drawn from that urn. Given that the chip drawn was red, what is the probability that III was the urn sampled?

2.4.42. A dashboard warning light is supposed to flash red if a car's oil pressure is too low. On a certain model, the probability of the light flashing when it should is 0.99; 2% of the time, though, it flashes for no apparent reason. If there is a 10% chance that the oil pressure really is low, what is the probability that a driver needs to be concerned if the warning light goes on?

2.4.43. Building permits were issued last year to three contractors starting up a new subdivision: Tara Construction built two houses; Westview, three houses; and Hearthstone, six houses. Tara's houses have a 60% probability of developing leaky basements; homes built by Westview and Hearthstone have that same problem 50% of the time and 40% of the time, respectively. Yesterday, the Better Business Bureau received a complaint from one of the new homeowners that his basement is leaking. Who is most likely to have been the contractor?

2.4.44. Two sections of a senior probability course are being taught. From what she has heard about the two instructors listed, Francesca estimates that her chances of passing the course are 0.85 if she gets Professor X and 0.60 if she gets Professor Y. The section into which she is put is determined by the registrar. Suppose that her chances of being assigned to Professor X are four out of ten. Fifteen weeks later we learn that Francesca did, indeed, pass the course. What is the probability she was enrolled in Professor X's section?

2.4.45. A liquor store owner is willing to cash personal checks for amounts up to \$50, but she has become wary of customers who wear sunglasses. Fifty percent of checks written by persons wearing sunglasses bounce. In contrast, 98% of the checks written by persons not wearing sunglasses clear the bank. She estimates that 10% of her customers wear sunglasses. If the bank returns a check and marks it "insufficient funds," what is the probability it was written by someone wearing sunglasses?

2.4.46. Brett and Margo have each thought about murdering their rich Uncle Basil in hopes of claiming their inheritance a bit early. Hoping to take advantage of Basil's predilection for immoderate desserts, Brett has put rat poison into the cherries flambé; Margo, unaware of Brett's activities, has laced the chocolate mousse with cyanide. Given the amounts likely to be eaten, the probability of the rat poison being fatal is 0.60; the cyanide, 0.90. Based on other dinners where Basil was presented with the same dessert options, we can assume that he has a 50% chance of asking for the cherries flambé, a 40% chance of ordering the chocolate mousse, and a 10% chance of skipping dessert altogether. No sooner are the dishes cleared away than Basil drops dead. In the absence of any other evidence, who should be considered the prime suspect?

2.4.47. Josh takes a twenty-question multiple-choice exam where each question has five possible answers. Some of the answers he knows, while others he gets right just by making lucky guesses. Suppose that the conditional probability of his knowing the answer to a randomly selected question given that he got it right is 0.92. How many of the twenty questions was he prepared for?

2.4.48. Recently the U.S. Senate Committee on Labor and Public Welfare investigated the feasibility of setting up a national screening program to detect child abuse. A team of consultants estimated the following probabilities: (1) one child in ninety is abused, (2) a screening program can detect an abused child 90% of the time, and (3) a screening program would incorrectly label 3% of all nonabused children as abused. What is the probability that a child is actually abused given that the screening program makes that diagnosis? How does the probability change if the incidence of abuse is one in one thousand? Or one in fifty?

2.4.49. At State University, 30% of the students are majoring in humanities, 50% in history and culture, and 20% in science. Moreover, according to figures released by the registrar, the percentages of women majoring in humanities, history and culture, and science are 75%, 45%, and 30%, respectively. Suppose Justin meets Anna at a fraternity party. What is the probability that Anna is a history and culture major?

2.4.50. An "eyes-only" diplomatic message is to be transmitted as a binary code of 0's and 1's. Past experience with the equipment being used suggests that if a 0 is sent, it will be (correctly) received as a 0 90% of the time (and mistakenly decoded as a 1 10% of the time). If a 1 is sent, it will be received as a 1 95% of the time (and as a 0 5% of the time). The text being sent is thought to be 70% 1's and 30% 0's. Suppose the next signal sent is received as a 1. What is the probability that it was sent as a 0?

2.4.51. When Zach wants to contact his girlfriend and he knows she is not at home, he is twice as likely to send her an e-mail as he is to leave a message on her answering machine. The probability that she responds to his e-mail within three hours is 80%; her chances of being similarly prompt in answering a phone message increase to 90%. Suppose she responded within two hours to the message he left this morning. What is the probability that Zach was communicating with her via e-mail?

2.4.52. A dot-com company ships products from three different warehouses (A, B, and C). Based on customer complaints, it appears that 3% of the shipments coming from A are somehow faulty, as are 5% of the shipments coming from B, and 2% coming from C. Suppose a customer is mailed an order and calls in a complaint the next day. What is the probability the item came from Warehouse C? Assume that Warehouses A, B, and C ship 30%, 20%, and 50% of the dot-com's sales, respectively.

2.4.53. A desk has three drawers. The first contains two gold coins, the second has two silver coins, and the third has one gold coin and one silver coin. A coin is drawn from a drawer selected at random. Suppose the coin selected was silver. What is the probability that the other coin in that drawer is gold?

2.5 Independence

Section 2.4 dealt with the problem of reevaluating the probability of a given event in light of the additional information that some other event has already occurred. It often is the case, though, that the probability of the given event remains unchanged, regardless of the outcome of the second event—that is, $P(A|B) = P(A) = P(A|B^C)$. Events sharing this property are said to be *independent*. Definition 2.5.1 gives a necessary and sufficient condition for two events to be independent.

Definition 2.5.1. Two events A and B are said to be *independent* if $P(A \cap B) = P(A) \cdot P(B)$.

Comment The fact that the probability of the intersection of two independent events is equal to the product of their individual probabilities follows immediately from our first definition of independence, that $P(A|B) = P(A)$. Recall that the definition of conditional probability holds true for *any* two events A and B [provided that $P(B > 0)$]:

$$P(A|B) = \frac{P(A \cap B)}{P(B)}$$

But $P(A|B)$ can equal $P(A)$ *only* if $P(A \cap B)$ factors into $P(A)$ times $P(B)$.

Example 2.5.1

Let A be the event of drawing a king from a standard poker deck and B, the event of drawing a diamond. Then, by Definition 2.5.1, A and B are independent because the probability of their intersection—drawing a king of diamonds—is equal to $P(A) \cdot P(B)$:

$$P(A \cap B) = \frac{1}{52} = \frac{1}{4} \cdot \frac{1}{13} = P(A) \cdot P(B)$$

■

Example 2.5.2

Suppose that A and B are independent events. Does it follow that A^C and B^C are also independent? That is, does $P(A \cap B) = P(A) \cdot P(B)$ guarantee that $P(A^C \cap B^C) = P(A^C) \cdot P(B^C)$?

Yes. The proof is accomplished by equating two different expressions for $P(A^C \cup B^C)$. First, by Theorem 2.3.6,

$$P(A^C \cup B^C) = P(A^C) + P(B^C) - P(A^C \cap B^C) \tag{2.5.1}$$

But the union of two complements is the complement of their intersection (recall Question 2.2.32). Therefore,

$$P(A^C \cup B^C) = 1 - P(A \cap B) \tag{2.5.2}$$

Combining Equations 2.5.1 and 2.5.2, we get

$$1 - P(A \cap B) = 1 - P(A) + 1 - P(B) - P(A^C \cap B^C)$$

Since A and B are independent, $P(A \cap B) = P(A) \cdot P(B)$, so

$$\begin{aligned} P(A^C \cap B^C) &= 1 - P(A) + 1 - P(B) - [1 - P(A) \cdot P(B)] \\ &= [1 - P(A)][1 - P(B)] \\ &= P(A^C) \cdot P(B^C) \end{aligned}$$

the latter factorization implying that A^C and B^C are, themselves, independent. (If A and B are independent, are A and B^C independent?)

■

Example 2.5.3

Electronics Warehouse is responding to affirmative-action litigation by establishing hiring goals by race and sex for its office staff. So far they have agreed to employ the 120 people characterized in Table 2.5.1. How many black women do they need in order for the events A: Employee is female and B: Employee is black to be independent?

Let x denote the number of black women necessary for A and B to be independent. Then

$$P(A \cap B) = P(\text{black female}) = x/(120 + x)$$

must equal

$$P(A)P(B) = P(\text{female})P(\text{black}) = [(40+x)/(120+x)] \cdot [(30+x)/(120+x)]$$

Setting $x/(120+x) = [(40+x)/(120+x)] \cdot [(30+x)/(120+x)]$ implies that $x = 24$ black women need to be on the staff in order for A and B to be independent.

Table 2.5.1

	White	Black
Male	50	30
Female	40	

■

Comment Having shown that "Employee is female" and "Employee is black" are independent, does it follow that, say, "Employee is male" and "Employee is white" are independent? Yes. By virtue of the derivation in Example 2.5.2, the independence of events A and B implies the independence of events A^C and B^C (as well as A and B^C and A^C and B). It follows, then, that the $x = 24$ black women not only makes A and B independent, it also implies, more generally, that "race" and "sex" are independent.

Example 2.5.4

Suppose that two events, A and B, each having nonzero probability, are mutually exclusive. Are they also independent?

No. If A and B are mutually exclusive, then $P(A \cap B) = 0$. But $P(A) \cdot P(B) > 0$ (by assumption), so the equality spelled out in Definition 2.5.1 that characterizes independence is not met. ■

Deducing Independence

Sometimes the physical circumstances surrounding two events make it obvious that the occurrence (or nonoccurrence) of one has absolutely no influence or effect on the occurrence (or nonoccurrence) of the other. If that should be the case, then the two events will necessarily be *independent* in the sense of Definition 2.5.1.

Suppose a coin is tossed twice. Clearly, whatever happens on the first toss has no physical connection or influence on the outcome of the second. If A and B, then, are events defined on the second and first tosses, respectively, it would have to be the case that $P(A|B) = P(A|B^C) = P(A)$. For example, let A be the event that the second toss of a fair coin is a head, and let B be the event that the first toss of that coin is a tail. Then

$$\begin{aligned} P(A|B) &= P(\text{head on second toss} \mid \text{tail on first toss}) \\ &= P(\text{head on second toss}) = \frac{1}{2} \end{aligned}$$

Being able to infer that certain events are independent proves to be of enormous help in solving certain problems. The reason is that many events of interest are, in fact, intersections. If those events are independent, then the probability of that intersection reduces to a simple product (because of Definition 2.5.1)—that is, $P(A \cap B) = P(A) \cdot P(B)$. For the coin tosses just described,

$$\begin{aligned}P(A \cap B) &= P(\text{head on second toss} \cap \text{tail on first toss})\\ &= P(A) \cdot P(B)\\ &= P(\text{head on second toss}) \cdot P(\text{tail on first toss})\\ &= \frac{1}{2} \cdot \frac{1}{2}\\ &= \frac{1}{4}\end{aligned}$$

Example 2.5.5

Myra and Carlos are summer interns working as proofreaders for a local newspaper. Based on aptitude tests, Myra has a 50% chance of spotting a hyphenation error, while Carlos picks up on that same kind of mistake 80% of the time. Suppose the copy they are proofing contains a hyphenation error. What is the probability it goes undetected?

Let A and B be the events that Myra and Carlos, respectively, catch the mistake. By assumption, $P(A) = 0.50$ and $P(B) = 0.80$. What we are looking for is the probability of the complement of a union. That is,

$$\begin{aligned}P(\text{Error goes undetected}) &= 1 - P(\text{Error } \textit{is} \text{ detected})\\ &= 1 - P(\text{Myra or Carlos or both see the mistake})\\ &= 1 - P(A \cup B)\\ &= 1 - \{P(A) + P(B) - P(A \cap B)\} \quad \text{(from Theorem 2.3.6)}\end{aligned}$$

Since proofreaders invariably work by themselves, events A and B are necessarily independent, so $P(A \cap B)$ would reduce to the product $P(A) \cdot P(B)$. It follows that such an error would go unnoticed *10%* of the time:

$$\begin{aligned}P(\text{Error goes undetected}) &= 1 - \{0.50 + 0.80 - (0.50)(0.80)\} = 1 - 0.90\\ &= 0.10\end{aligned}$$

■

Example 2.5.6

Suppose that one of the genes associated with the control of carbohydrate metabolism exhibits two alleles—a dominant W and a recessive w. If the probabilities of the WW, Ww, and ww genotypes in the present generation are p, q, and r, respectively, for both males and females, what are the chances that an individual in the *next* generation will be a ww?

Let A denote the event that an offspring receives a w allele from her father; let B denote the event that she receives the recessive allele from her mother. What we are looking for is $P(A \cap B)$.

According to the information given,

$$\begin{aligned}p &= P(\text{Parent has genotype WW}) = P(\text{WW})\\ q &= P(\text{Parent has genotype Ww}) = P(\text{Ww})\\ r &= P(\text{Parent has genotype ww}) = P(\text{ww})\end{aligned}$$

If an offspring is equally likely to receive either of her parent's alleles, the probabilities of A and B can be computed using Theorem 2.4.1:

$$P(A) = P(A \mid WW)P(WW) + P(A \mid Ww)P(Ww) + P(A \mid ww)P(ww)$$
$$= 0 \cdot p + \frac{1}{2} \cdot q + 1 \cdot r$$
$$= r + \frac{q}{2} = P(B)$$

Lacking any evidence to the contrary, there is every reason here to assume that A and B are independent events, in which case

$$P(A \cap B) = P(\text{Offspring has genotype ww})$$
$$= P(A) \cdot P(B)$$
$$= \left(r + \frac{q}{2}\right)^2$$

(This particular model for allele segregation, together with the independence assumption, is called *random Mendelian mating.*) ■

Example 2.5.7

Emma and Josh have just gotten engaged. What is the probability that they have different blood types? Assume that blood types for both men and women are distributed in the general population according to the following proportions:

Blood Type	Proportion
A	40%
B	10%
AB	5%
O	45%

First, note that the event "Emma and Josh have *different* blood types" includes more possibilities than does the event "Emma and Josh have the *same* blood type." That being the case, the complement will be easier to work with than the question originally posed. We can start, then, by writing

$$P(\text{Emma and Josh have different blood types})$$
$$= 1 - P(\text{Emma and Josh have the same blood type})$$

Now, if we let E_X and J_X represent the events that Emma and Josh, respectively, have blood type X, then the event "Emma and Josh have the same blood type" is a union of intersections, and we can write

$$P(\text{Emma and Josh have the same blood type}) = P\{(E_A \cap J_A) \cup (E_B \cap J_B) \cup (E_{AB} \cap J_{AB}) \cup (E_O \cap J_O)\}$$

Since the four intersections here are mutually exclusive, the probability of their union becomes the sum of their probabilities. Moreover, "blood type" is not a factor in the selection of a spouse, so E_X and J_X are independent events and $P(E_X \cap J_X) = P(E_X)P(J_X)$. It follows, then, that Emma and Josh have a *62.5%* chance of having different blood types:

$$\begin{aligned} P(\text{Emma and Josh have different blood types}) &= 1 - \{P(E_A)P(J_A) + P(E_B)P(J_B) \\ &\quad + P(E_{AB})P(J_{AB}) + P(E_O)P(J_O)\} \\ &= 1 - \{(0.40)(0.40) + (0.10)(0.10) \\ &\quad + (0.05)(0.05) + (0.45)(0.45)\} \\ &= 0.625 \end{aligned}$$

Questions

2.5.1. Suppose that $P(A \cap B) = 0.2$, $P(A) = 0.6$, and $P(B) = 0.5$.

(a) Are A and B mutually exclusive?
(b) Are A and B independent?
(c) Find $P(A^C \cup B^C)$.

2.5.2. Spike is not a terribly bright student. His chances of passing chemistry are 0.35; mathematics, 0.40; and both, 0.12. Are the events "Spike passes chemistry" and "Spike passes mathematics" independent? What is the probability that he fails both subjects?

2.5.3. Two fair dice are rolled. What is the probability that the number showing on one will be twice the number appearing on the other?

2.5.4. Urn I has three red chips, two black chips, and five white chips; urn II has two red, four black, and three white. One chip is drawn at random from each urn. What is the probability that both chips are the same color?

2.5.5. Dana and Cathy are playing tennis. The probability that Dana wins at least one out of two games is 0.3. What is the probability that Dana wins at least one out of four?

2.5.6. Three points, X_1, X_2, and X_3, are chosen at random in the interval $(0, a)$. A second set of three points, Y_1, Y_2, and Y_3, are chosen at random in the interval $(0, b)$. Let A be the event that X_2 is between X_1 and X_3. Let B be the event that $Y_1 < Y_2 < Y_3$. Find $P(A \cap B)$.

2.5.7. Suppose that $P(A) = \frac{1}{4}$ and $P(B) = \frac{1}{8}$.

(a) What does $P(A \cup B)$ equal if
  1. A and B are mutually exclusive?
  2. A and B are independent?
(b) What does $P(A \mid B)$ equal if
  1. A and B are mutually exclusive?
  2. A and B are independent?

2.5.8. Suppose that events A, B, and C are independent.

(a) Use a Venn diagram to find an expression for $P(A \cup B \cup C)$ that does *not* make use of a complement.
(b) Find an expression for $P(A \cup B \cup C)$ that *does* make use of a complement.

2.5.9. A fair coin is tossed four times. What is the probability that the number of heads appearing on the first two tosses is equal to the number of heads appearing on the second two tosses?

2.5.10. Suppose that two cards are drawn simultaneously from a standard 52-card poker deck. Let A be the event that both are either a jack, queen, king, or ace of hearts, and let B be the event that both are aces. Are A and B independent? (*Note*: There are 1326 equally likely ways to draw two cards from a poker deck.)

Defining the Independence of More Than Two Events

It is not immediately obvious how to extend Definition 2.5.1 to, say, *three* events. To call A, B, and C independent, should we require that the probability of the three-way intersection factors into the product of the three original probabilities,

$$P(A \cap B \cap C) = P(A) \cdot P(B) \cdot P(C) \tag{2.5.3}$$

or should we impose the definition we already have on the three *pairs* of events:

$$P(A \cap B) = P(A) \cdot P(B)$$
$$P(B \cap C) = P(B) \cdot P(C) \qquad (2.5.4)$$
$$P(A \cap C) = P(A) \cdot P(C)$$

Actually, neither condition by itself is sufficient. If three events satisfy Equations 2.5.3 *and* 2.5.4, we will call them independent (or *mutually independent*), but Equation 2.5.3 does not imply Equation 2.5.4, nor does Equation 2.5.4 imply Equation 2.5.3 (see Questions 2.5.11 and 2.5.12).

More generally, the independence of n events requires that the probabilities of all possible intersections equal the products of all the corresponding individual probabilities. Definition 2.5.2 states the result formally. Analogous to what was true in the case of *two* events, the practical applications of Definition 2.5.2 arise when n events are mutually independent, and we can calculate $P(A_1 \cap A_2 \cap \cdots \cap A_n)$ by computing the product $P(A_1) \cdot P(A_2) \cdots P(A_n)$.

Definition 2.5.2. Events $A_1, A_2, \ldots, A_n$ are said to be *independent* if for every set of indices $i_1, i_2, \ldots, i_k$ between 1 and n, inclusive,

$$P(A_{i_1} \cap A_{i_2} \cap \cdots \cap A_{i_k}) = P(A_{i_1}) \cdot P(A_{i_2}) \cdot \cdots \cdot P(A_{i_k})$$

Example 2.5.8

An insurance company plans to assess its future liabilities by sampling the records of its current policyholders. A pilot study has turned up three clients—one living in Alaska, one in Missouri, and one in Vermont—whose estimated chances of surviving to the year 2015 are 0.7, 0.9, and 0.3, respectively. What is the probability that by the end of 2014 the company will have had to pay death benefits to exactly one of the three?

Let A_1 be the event "Alaska client survives through 2014." Define A_2 and A_3 analogously for the Missouri client and Vermont client, respectively. Then the event E: "Exactly one dies" can be written as the union of three intersections:

$$E = \left(A_1 \cap A_2 \cap A_3^C\right) \cup \left(A_1 \cap A_2^C \cap A_3\right) \cup \left(A_1^C \cap A_2 \cap A_3\right)$$

Since each of the intersections is mutually exclusive of the other two,

$$P(E) = P\left(A_1 \cap A_2 \cap A_3^C\right) + P\left(A_1 \cap A_2^C \cap A_3\right) + P\left(A_1^C \cap A_2 \cap A_3\right)$$

Furthermore, there is no reason to believe that for all practical purposes the fates of the three are not independent. That being the case, each of the intersection probabilities reduces to a product, and we can write

$$\begin{aligned} P(E) &= P(A_1) \cdot P(A_2) \cdot P\left(A_3^C\right) + P(A_1) \cdot P\left(A_2^C\right) \cdot P(A_3) + P\left(A_1^C\right) \cdot P(A_2) \cdot P(A_3) \\ &= (0.7)(0.9)(0.7) + (0.7)(0.1)(0.3) + (0.3)(0.9)(0.3) \\ &= 0.543 \end{aligned}$$

■

Comment "Declaring" events independent for reasons other than those prescribed in Definition 2.5.2 is a necessarily subjective endeavor. Here we might feel fairly certain that a "random" person dying in Alaska will not affect the survival chances of a "random" person residing in Missouri (or Vermont). But there may be special circumstances that invalidate that sort of argument. For example, what if the three individuals in question were mercenaries fighting in an African border war and were all crew members assigned to the same helicopter? In practice, all we can do is look at each situation on an individual basis and try to make a reasonable judgment as to whether the occurrence of one event is likely to influence the outcome of another event.

Example 2.5.9

Protocol for making financial decisions in a certain corporation follows the "circuit" pictured in Figure 2.5.1. Any budget is first screened by 1. If he approves it, the plan is forwarded to 2, 3, and 5. If either 2 or 3 concurs, it goes to 4. If either 4 or 5 says "yes," it moves on to 6 for a final reading. Only if 6 is also in agreement does the proposal pass. Suppose that 1, 5, and 6 each has a 50% chance of saying "yes," whereas 2, 3, and 4 will each concur with a probability of 0.70. If everyone comes to a decision independently, what is the probability that a budget will pass?

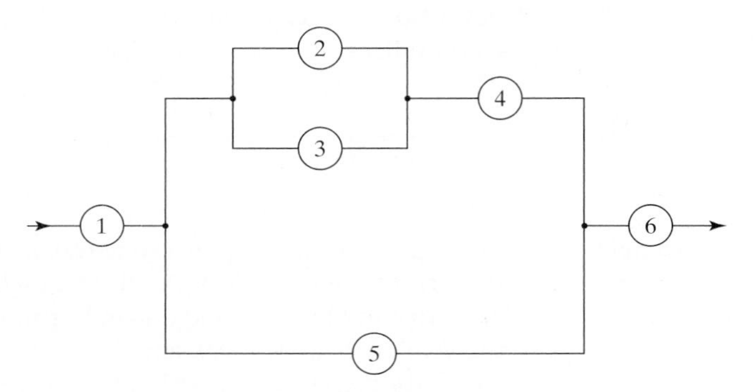

Figure 2.5.1

Probabilities of this sort are calculated by reducing the circuit to its component unions and intersections. Moreover, if all decisions are made independently, which is the case here, then every intersection becomes a product.

Let A_i be the event that person i approves the budget, $i = 1, 2, \ldots, 6$. Looking at Figure 2.5.1, we see that

$$\begin{aligned} P(\text{Budget passes}) &= P(A_1 \cap \{[(A_2 \cup A_3) \cap A_4] \cup A_5\} \cap A_6) \\ &= P(A_1)P\{[(A_2 \cup A_3) \cap A_4] \cup A_5\}P(A_6) \end{aligned}$$

By assumption, $P(A_1) = 0.5$, $P(A_2) = 0.7$, $P(A_3) = 0.7$, $P(A_4) = 0.7$, $P(A_5) = 0.5$, and $P(A_6) = 0.5$, so

$$\begin{aligned} P\{[(A_2 \cup A_3) \cap A_4]\} &= [P(A_2) + P(A_3) - P(A_2)P(A_3)]P(A_4) \\ &= [0.7 + 0.7 - (0.7)(0.7)](0.7) \\ &= 0.637 \end{aligned}$$

Therefore,

$$\begin{aligned} P(\text{Budget passes}) &= (0.5)\{0.637 + 0.5 - (0.637)(0.5)\}(0.5) \\ &= 0.205 \end{aligned}$$

■

Repeated Independent Events

We have already seen several examples where the event of interest was actually an intersection of independent simpler events (in which case the probability of the intersection reduced to a product). There is a special case of that basic scenario that deserves special mention because it applies to numerous real-world situations. If the events making up the intersection all arise from the same physical circumstances and assumptions (i.e., they represent repetitions of the same experiment), they are referred to as *repeated independent trials*. The number of such trials may be finite or infinite.

Example 2.5.10

Suppose the string of Christmas tree lights you just bought has twenty-four bulbs wired in series. If each bulb has a 99.9% chance of "working" the first time current is applied, what is the probability that the string itself will *not* work?

Let A_i be the event that the ith bulb fails, $i = 1, 2, \ldots, 24$. Then

$$\begin{aligned} P(\text{String fails}) &= P(\text{At least one bulb fails}) \\ &= P(A_1 \cup A_2 \cup \cdots \cup A_{24}) \\ &= 1 - P(\text{String works}) \\ &= 1 - P(\text{All twenty-four bulbs work}) \\ &= 1 - P(A_1^C \cap A_2^C \cap \cdots \cap A_{24}^C) \end{aligned}$$

If we assume that bulb failures are independent events,

$$P(\text{String fails}) = 1 - P(A_1^C)P(A_2^C)\cdots P(A_{24}^C)$$

Moreover, since all the bulbs are presumably manufactured the same way, $P(A_i^C)$ is the same for all i, so

$$\begin{aligned} P(\text{String fails}) &= 1 - \{P(A_i^C)\}^{24} \\ &= 1 - (0.999)^{24} \\ &= 1 - 0.98 \\ &= 0.02 \end{aligned}$$

The chances are one in fifty, in other words, that the string will not work the first time current is applied. ■

Example 2.5.11

During the 1978 baseball season, Pete Rose of the Cincinnati Reds set a National League record by hitting safely in forty-four consecutive games. Assume that Rose was a .300 hitter and that he came to bat four times each game. If each at-bat is assumed to be an independent event, what probability might reasonably be associated with a hitting streak of that length?

For this problem we need to invoke the repeated independent trials model *twice*—once for the four at-bats making up a game and a second time for the forty-four games making up the streak. Let A_i denote the event "Rose hit safely in ith game," $i = 1, 2, \ldots, 44$. Then

$$\begin{aligned} P(\text{Rose hit safely in forty-four consecutive games}) &= P(A_1 \cap A_2 \cap \cdots \cap A_{44}) \\ &= P(A_1) \cdot P(A_2) \cdot \cdots \cdot P(A_{44}) \end{aligned} \qquad (2.5.5)$$

Since all the $P(A_i)$'s are equal, we can further simplify Equation 2.5.5 by writing

$$P(\text{Rose hit safely in forty-four consecutive games}) = [P(A_1)]^{44}$$

To calculate $P(A_1)$ we should focus on the *complement* of A_1. Specifically,

$$\begin{aligned} P(A_1) &= 1 - P\left(A_1^C\right) \\ &= 1 - P(\text{Rose did } not \text{ hit safely in Game 1}) \\ &= 1 - P(\text{Rose made four outs}) \\ &= 1 - (0.700)^4 \quad (\text{Why?}) \\ &= 0.76 \end{aligned}$$

Therefore, the probability of a .300 hitter putting together a forty-four-game streak (during a given set of forty-four games) is 0.0000057:

$$\begin{aligned} P(\text{Rose hit safely in forty-four consecutive games}) &= (0.76)^{44} \\ &= 0.0000057 \end{aligned}$$

■

Comment The analysis described here has the basic "structure" of a repeated independent trials problem, but the assumptions that the latter makes are not entirely satisfied by the data. Each at-bat, for example, is not really a repetition of the same experiment, nor is $P(A_i)$ the same for all i. Rose would obviously have had different probabilities of getting a hit against different pitchers. Moreover, although "four" was probably the typical number of official at-bats that he had during a game, there would certainly have been many instances where he had either fewer or more. Modest deviations from game to game, though, would not have had a major effect on the probability associated with Rose's forty-four-game streak.

Example 2.5.12

In the game of craps, one of the ways a player can win is by rolling (with two dice) one of the sums 4, 5, 6, 8, 9, or 10, and then rolling that sum again before rolling a sum of 7. For example, the sequence of sums 6, 5, 8, 8, 6 would result in the player winning on his fifth roll. In gambling parlance, "6" is the player's "point," and he "made his point." On the other hand, the sequence of sums 8, 4, 10, 7 would result in the player losing on his fourth roll: his point was an 8, but he rolled a sum of 7 before he rolled a second 8. What is the probability that a player wins with a point of 10?

Table 2.5.2

Sequence of Rolls	Probability
(10, 10)	(3/36)(3/36)
(10, no 10 or 7, 10)	(3/36)(27/36)(3/36)
(10, no 10 or 7, no 10 or 7, 10)	(3/36)(27/36)(27/36)(3/36)
⋮	⋮

Table 2.5.2 shows some of the ways a player can make a point of 10. Each sequence, of course, is an intersection of independent events, so its probability becomes a product. The event "Player wins with a point of 10" is then the union

of all the sequences that could have been listed in the first column. Since all those sequences are mutually exclusive, the probability of winning with a point of 10 reduces to the sum of an infinite number of products:

$$\begin{aligned} P(\text{Player wins with a point of } 10) &= \frac{3}{36} \cdot \frac{3}{36} + \frac{3}{36} \cdot \frac{27}{36} \cdot \frac{3}{36} \\ &\quad + \frac{3}{36} \cdot \frac{27}{36} \cdot \frac{27}{36} \cdot \frac{3}{36} + \cdots \\ &= \frac{3}{36} \cdot \frac{3}{36} \sum_{k=0}^{\infty} \left(\frac{27}{36}\right)^k \end{aligned} \tag{2.5.6}$$

Recall from algebra that if $0 < r < 1$,

$$\sum_{k=0}^{\infty} r^k = 1/(1-r)$$

Applying the formula for the sum of a geometric series to Equation 2.5.6 shows that the probability of winning at craps with a point of 10 is $\frac{1}{36}$:

$$\begin{aligned} P(\text{Player wins with a point of } 10) &= \frac{3}{36} \cdot \frac{3}{36} \cdot \frac{1}{\left(1 - \frac{27}{36}\right)} \\ &= \frac{1}{36} \end{aligned}$$

Table 2.5.3

Point	P (makes point)
4	1/36
5	16/360
6	25/396
8	25/396
9	16/360
10	1/36

Table 2.5.3 shows the probabilities of a person "making" each of the possible six points—4, 5, 6, 8, 9, and 10. According to the rules of craps, a player wins by either (1) getting a sum of 7 or 11 on the first roll or (2) getting a 4, 5, 6 , 8 , 9, or 10 on the first roll and making the point. But $P(\text{sum} = 7) = 6/36$ and $P(\text{sum} = 11) = 2/36$, so

$$\begin{aligned} P(\text{Player wins}) &= \frac{6}{36} + \frac{2}{36} + \frac{1}{36} + \frac{16}{360} + \frac{25}{396} + \frac{25}{396} + \frac{16}{360} + \frac{1}{36} \\ &= 0.493 \end{aligned}$$

As even-money games go, craps is relatively fair—the probability of the shooter winning is not much less than 0.500. ■

Example 2.5.13

A transmitter is sending a binary code (+ and − signals) that must pass through three relay signals before being sent on to the receiver (see Figure 2.5.2). At each relay station, there is a 25% chance that the signal will be reversed—that is

$$P(+ \text{ is sent by relay } i \mid - \text{ is received by relay } i)$$
$$= P(- \text{ is sent by relay } i \mid + \text{ is received by relay } i)$$
$$= 1/4, i = 1, 2, 3$$

Suppose + symbols make up 60% of the message being sent. If the signal + is received, what is the probability a + was sent?

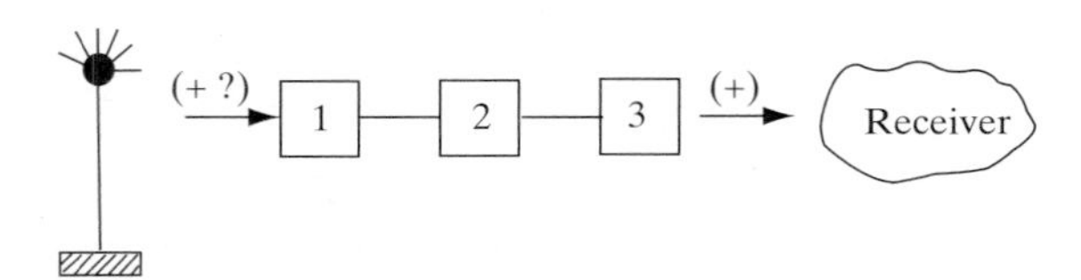

Figure 2.5.2

This is basically a Bayes' Theorem (Theorem 2.4.2) problem, but the three relay stations introduce a more complex mechanism for transmission error. Let A be the event "+ is transmitted from tower" and B be the event "+ is received from relay 3." Then

$$P(A|B) = \frac{P(B|A)P(A)}{P(B|A)P(A) + P(B|A^C)P(A^C)}$$

Notice that a + can be received from relay 3 given that a + was initially sent from the tower if either (1) all relay stations function properly or (2) any *two* of the stations make transmission errors. Table 2.5.4 shows the four mutually exclusive ways (1) and (2) can happen. The probabilities associated with the message transmissions at each relay station are shown in parentheses. Assuming the relay station outputs are independent events, the probability of an entire transmission sequence is simply the product of the probabilities in parentheses in any given row. These overall probabilities are listed in the last column; their sum, 36/64, is $P(B|A)$. By a similar analysis, we can show that

$$P(B|A^C) = P(+ \text{ is received from relay } 3 \mid - \text{ is transmitted from tower}) = 28/64$$

Finally, since $P(A) = 0.6$ and then $P(A^C) = 0.4$, the conditional probability we are looking for is 0.66:

$$P(A|B) = \frac{\left(\frac{36}{64}\right)(0.6)}{\left(\frac{36}{64}\right)(0.6) + \left(\frac{28}{64}\right)(0.4)} = 0.66$$

■

Table 2.5.4

	Signal transmitted by			
Tower	Relay 1	Relay 2	Relay 3	Probability
+	+(3/4)	−(1/4)	+(1/4)	3/64
+	−(1/4)	−(3/4)	+(1/4)	3/64
+	−(1/4)	+(1/4)	+(3/4)	3/64
+	+(3/4)	+(3/4)	+(3/4)	27/64
				36/64

Example 2.5.14

Andy, Bob, and Charley have gotten into a disagreement over a female acquaintance, Donna, and decide to settle their dispute with a three-cornered pistol duel. Of the three, Andy is the worst shot, hitting his target only 30% of the time. Charley, a little better, is on-target 50% of the time, while Bob never misses (see Figure 2.5.3). The rules they agree to are simple: They are to fire at the targets of their choice in succession, and cyclically, in the order Andy, Bob, Charley, and so on, until only one of them is left standing. On each "turn," they get only one shot. If a combatant is hit, he no longer participates, either as a target or as a shooter.

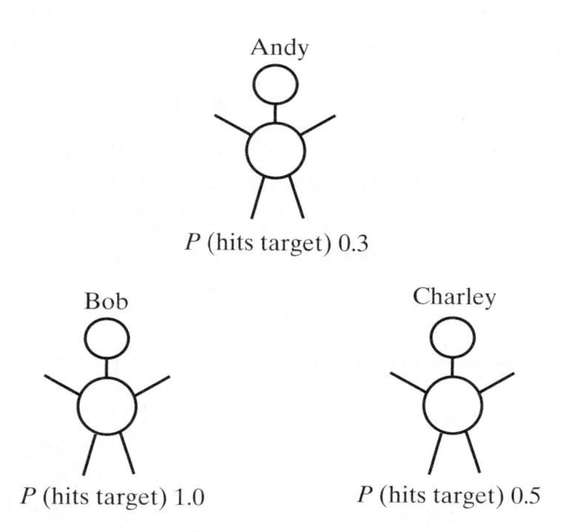

Figure 2.5.3

As Andy loads his revolver, he mulls over his options (his objective is clear—to maximize his probability of survival). According to the rule he can shoot either Bob or Charley, but he quickly rules out shooting at the latter because it would be counterproductive to his future well-being: If he shot at Charley and had the misfortune of hitting him, it would be Bob's turn, and Bob would have no recourse but to shoot at Andy. From Andy's point of view, this would be a decidedly grim turn of events, since Bob never misses. Clearly, Andy's only option is to shoot at Bob. This leaves two scenarios: (1) He shoots at Bob and hits him, or (2) he shoots at Bob and misses.

Consider the first possibility. If Andy hits Bob, Charley will proceed to shoot at Andy, Andy will shoot back at Charley, and so on, until one of them hits the other. Let CH_i and CM_i denote the events "Charley hits Andy with the ith shot" and "Charley misses Andy with the ith shot," respectively. Define AH_i and AM_i analogously. Then Andy's chances of survival (given that he has killed Bob) reduce to a countably infinite union of intersections:

$$P(\text{Andy survives}) = P[(CM_1 \cap AH_1) \cup (CM_1 \cap AM_1 \cap CM_2 \cap AH_2) \cup (CM_1 \cap AM_1 \cap CM_2 \cap AM_2 \cap CM_3 \cap AH_3) \cup \cdots]$$

Note that each intersection is mutually exclusive of all of the others and that its component events are independent. Therefore,

$$P(\text{Andy survives}) = P(CM_1)P(AH_1) + P(CM_1)P(AM_1)P(CM_2)P(AH_2) + P(CM_1)P(AM_1)P(CM_2)P(AM_2)P(CM_3)P(AH_3) + \cdots$$

$$= (0.5)(0.3) + (0.5)(0.7)(0.5)(0.3) + (0.5)(0.7)(0.5)(0.7)(0.5)(0.3) + \cdots$$

$$= (0.5)(0.3) \sum_{k=0}^{\infty} (0.35)^k$$

$$= (0.15)\left(\frac{1}{1-0.35}\right) = \frac{3}{13}$$

Now consider the second scenario. If Andy shoots at Bob and misses, Bob will undoubtedly shoot and hit Charley, since Charley is the more dangerous adversary. Then it will be Andy's turn again. Whether he would see another tomorrow would depend on his ability to make that very next shot count. Specifically,

$$P(\text{Andy survives}) = P(\text{Andy hits Bob on second turn}) = 3/10$$

But $\frac{3}{10} > \frac{3}{13}$, so Andy is better off *not* hitting Bob with his first shot. And because we have already argued that it would be foolhardy for Andy to shoot at Charley, Andy's optimal strategy is clear—deliberately miss both Bob and Charley with the first shot. ■

Questions

2.5.11. Suppose that two fair dice (one red and one green) are rolled. Define the events

A: a 1 or a 2 shows on the red die
B: a 3, 4, or 5 shows on the green die
C: the dice total is 4, 11, or 12

Show that these events satisfy Equation 2.5.3 but not Equation 2.5.4.

2.5.12. A roulette wheel has thirty-six numbers colored red or black according to the pattern indicated below:

Roulette wheel pattern																	
1	2	3	4	5	6	7	8	9	10	11	12	13	14	15	16	17	18
R	R	R	R	R	B	B	B	B	R	R	R	R	B	B	B	B	B
36	35	34	33	32	31	30	29	28	27	26	25	24	23	22	21	20	19

Define the events

A: red number appears
B: even number appears
C: number is less than or equal to 18

Show that these events satisfy Equation 2.5.4 but not Equation 2.5.3.

2.5.13. How many probability equations need to be verified to establish the mutual independence of *four* events?

2.5.14. In a roll of a pair of fair dice (one red and one green), let A be the event the red die shows a 3, 4, or 5; let B be the event the green die shows a 1 or a 2; and let C be the event the dice total is 7. Show that A, B, and C are independent.

2.5.15. In a roll of a pair of fair dice (one red and one green), let A be the event of an odd number on the red die, let B be the event of an odd number on the green die, and let C be the event that the sum is odd. Show that any pair of these events is independent but that A, B, and C are not mutually independent.

2.5.16. On her way to work, a commuter encounters four traffic signals. Assume that the distance between each of the four is sufficiently great that her probability of getting a green light at any intersection is independent of what happened at any previous intersection. The first two lights are green for forty seconds of each minute; the last two, for thirty seconds of each minute. What is the probability that the commuter has to stop at least three times?

2.5.17. School board officials are debating whether to require all high school seniors to take a proficiency exam before graduating. A student passing all three parts (mathematics, language skills, and general knowledge) would be awarded a diploma; otherwise, he or she would receive only a certificate of attendance. A practice test given to this year's ninety-five hundred seniors resulted in the following numbers of failures:

Subject Area	Number of Students Failing
Mathematics	3325
Language skills	1900
General knowledge	1425

If "Student fails mathematics," "Student fails language skills," and "Student fails general knowledge" are independent events, what proportion of next year's seniors can be expected to fail to qualify for a diploma? Does independence seem a reasonable assumption in this situation?

2.5.18. Consider the following four-switch circuit:

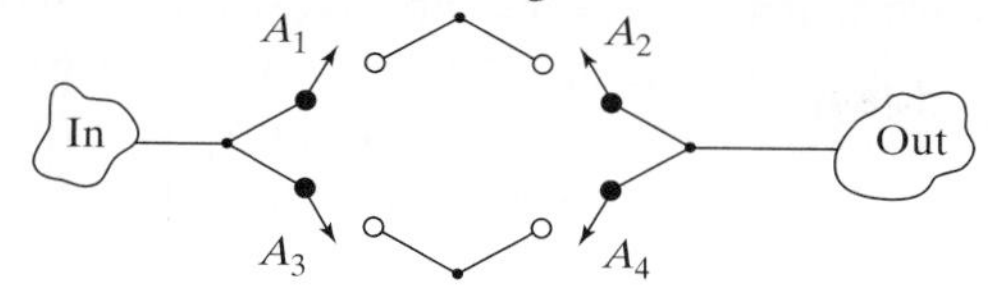

If all switches operate independently and $P(\text{Switch closes}) = p$, what is the probability the circuit is completed?

2.5.19. A fast-food chain is running a new promotion. For each purchase, a customer is given a game card that may win \$10. The company claims that the probability of a person winning at least once in five tries is 0.32. What is the probability that a customer wins \$10 on his or her first purchase?

2.5.20. Players A, B, and C toss a fair coin in order. The first to throw a head wins. What are their respective chances of winning?

2.5.21. In a certain third world nation, statistics show that only two out of ten children born in the early 1980s reached the age of twenty-one. If the same mortality rate is operative over the next generation, how many children does a woman need to bear if she wants to have at least a 75% probability that at least one of her offspring survives to adulthood?

2.5.22. According to an advertising study, 15% of television viewers who have seen a certain automobile commercial can correctly identify the actor who does the voice-over. Suppose that ten such people are watching TV and the commercial comes on. What is the probability that at least one of them will be able to name the actor? What is the probability that exactly one will be able to name the actor?

2.5.23. A fair die is rolled and then n fair coins are tossed, where n is the number showing on the die. What is the probability that no heads appear?

2.5.24. Each of m urns contains three red chips and four white chips. A total of r samples with replacement are taken from each urn. What is the probability that at least one red chip is drawn from at least one urn?

2.5.25. If two fair dice are tossed, what is the smallest number of throws, n, for which the probability of getting at least one double 6 exceeds 0.5? (*Note:* This was one of the first problems that de Méré communicated to Pascal in 1654.)

2.5.26. A pair of fair dice are rolled until the first sum of 8 appears. What is the probability that a sum of 7 does not precede that first sum of 8?

2.5.27. An urn contains w white chips, b black chips, and r red chips. The chips are drawn out at random, one at a time, with replacement. What is the probability that a white appears before a red?

2.5.28. A Coast Guard dispatcher receives an SOS from a ship that has run aground off the shore of a small island. Before the captain can relay her exact position, though, her radio goes dead. The dispatcher has n helicopter crews he can send out to conduct a search. He suspects the ship is somewhere either south in area I (with probability p) or north in area II (with probability $1 - p$). Each of the n rescue parties is equally competent and has probability r of locating the ship given it has run aground in the sector being searched. How should the dispatcher deploy the helicopter crews to maximize the probability that one of them will find the missing ship? (*Hint:* Assume that m search crews are sent to area I and $n - m$ are sent to area II. Let B denote the event that the ship is found, let A_1 be the event that the ship is in area I, and let A_2 be the event that the ship is in area II. Use Theorem 2.4.1 to get an expression for $P(B)$; then differentiate with respect to m.)

2.5.29. A computer is instructed to generate a random sequence using the digits 0 through 9; repetitions are permissible. What is the shortest length the sequence can be and still have at least a 70% probability of containing at least one 4?

2.5.30. A box contains a two-headed coin and eight fair coins. One coin is drawn at random and tossed n times. Suppose all n tosses come up heads. Show that the limit of the probability that the coin is fair is 0 as n goes to infinity.

2.6 Combinatorics

Combinatorics is a time-honored branch of mathematics concerned with counting, arranging, and ordering. While blessed with a wealth of early contributors (there are references to combinatorial problems in the Old Testament), its emergence as a separate discipline is often credited to the German mathematician and philosopher Gottfried Wilhelm Leibniz (1646–1716), whose 1666 treatise, *Dissertatio de arte combinatoria*, was perhaps the first monograph written on the subject (107).

Applications of combinatorics are rich in both diversity and number. Users range from the molecular biologist trying to determine how many ways genes can be positioned along a chromosome, to a computer scientist studying queuing priorities, to a psychologist modeling the way we learn, to a weekend poker player wondering whether he should draw to a straight, or a flush, or a full house. Surprisingly enough, despite the considerable differences that seem to distinguish one question from another, solutions to all of these questions are rooted in the same set of four basic theorems and rules.

Counting Ordered Sequences: The Multiplication Rule

More often than not, the relevant "outcomes" in a combinatorial problem are ordered sequences. If two dice are rolled, for example, the outcome (4, 5)—that is, the first die comes up 4 and the second die comes up 5—is an ordered sequence of length two. The number of such sequences is calculated by using the most fundamental result in combinatorics, the *multiplication rule*.

Multiplication Rule *If operation A can be performed in m different ways and operation B in n different ways, the sequence (operation A, operation B) can be performed in $m \cdot n$ different ways.*

Proof At the risk of belaboring the obvious, we can verify the multiplication rule by considering a *tree* diagram (see Figure 2.6.1). Since each version of A can be followed by any of n versions of B, and there are m of the former, the total number of "A, B" sequences that can be pieced together is obviously the product $m \cdot n$. □

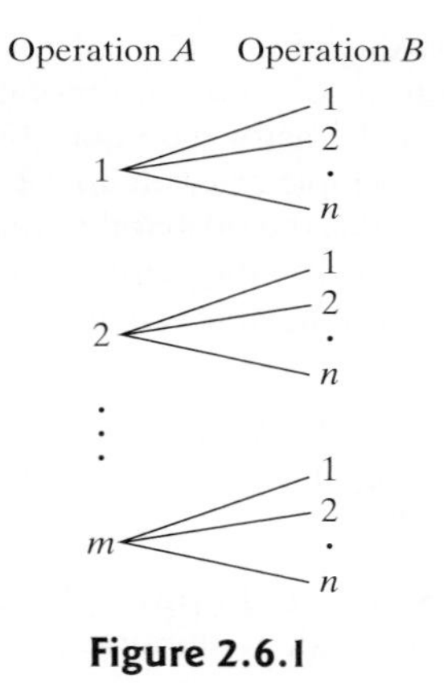

Figure 2.6.1

Corollary 2.6.1 *If operation A_i, $i = 1, 2, \ldots, k$, can be performed in n_i ways, $i = 1, 2, \ldots, k$, respectively, then the ordered sequence (operation A_1, operation $A_2, \ldots,$ operation A_k) can be performed in $n_1 \cdot n_2 \cdot \cdots \cdot n_k$ ways.* ◀

Example 2.6.1 The combination lock on a briefcase has two dials, each marked off with sixteen notches (see Figure 2.6.2). To open the case, a person first turns the left dial in a certain direction for two revolutions and then stops on a particular mark. The right dial is set in a similar fashion, after having been turned in a certain direction for two revolutions. How many different settings are possible?

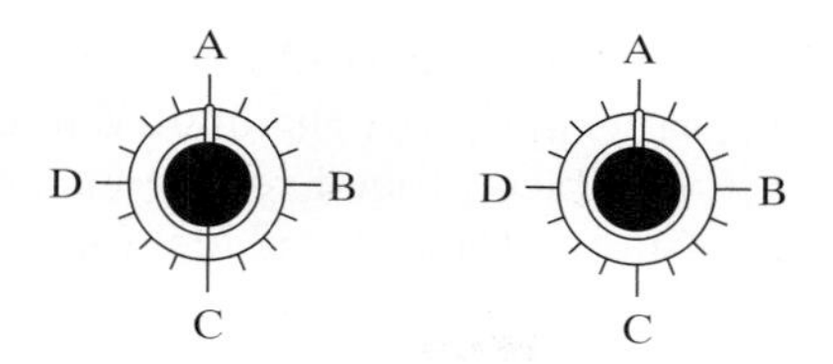

Figure 2.6.2

In the terminology of the multiplication rule, opening the briefcase corresponds to the four-step sequence (A_1, A_2, A_3, A_4) detailed in Table 2.6.1. Applying the previous corollary, we see that 1024 different settings are possible:

$$\begin{aligned}\text{number of different settings} &= n_1 \cdot n_2 \cdot n_3 \cdot n_4 \\ &= 2 \cdot 16 \cdot 2 \cdot 16 \\ &= 1024\end{aligned}$$

Table 2.6.1

Operation	Purpose	Number of Options
A_1	Rotating the left dial in a particular direction	2
A_2	Choosing an endpoint for the left dial	16
A_3	Rotating the right dial in a particular direction	2
A_4	Choosing an endpoint for the right dial	16

Comment Designers of locks should be aware that the number of dials, as opposed to the number of notches on each dial, is the critical factor in determining how many different settings are possible. A two-dial lock, for example, where each dial has twenty notches, gives rise to only $2 \cdot 20 \cdot 2 \cdot 20 = 1600$ settings. If those forty notches, though, are distributed among *four* dials (ten to each dial), the number of different settings increases a hundredfold to *160,000* $(= 2 \cdot 10 \cdot 2 \cdot 10 \cdot 2 \cdot 10 \cdot 2 \cdot 10)$. ■

Example 2.6.2 Alphonse Bertillon, a nineteenth-century French criminologist, developed an identification system based on eleven anatomical variables (height, head width, ear length, etc.) that presumably remain essentially unchanged during an individual's adult life. The range of each variable was divided into three subintervals: small, medium, and large. A person's *Bertillon configuration* is an ordered sequence of eleven letters, say,

$$s, s, m, m, l, s, l, s, s, m, s$$

where a letter indicates the individual's "size" relative to a particular variable. How populated does a city have to be before it can be guaranteed that at least two citizens will have the same Bertillon configuration?

Viewed as an ordered sequence, a Bertillon configuration is an eleven-step classification system, where three options are available at each step. By the multiplication rule, a total of 3^{11}, or 177,147, distinct sequences are possible. Therefore, any city with at least 177,148 adults would necessarily have at least two residents with the same pattern. (The limited number of possibilities generated by the configuration's variables proved to be one of its major weaknesses. Still, it was widely used in Europe for criminal identification before the development of fingerprinting.) ■

Example 2.6.3

In 1824 Louis Braille invented what would eventually become the standard alphabet for the blind. Based on an earlier form of "night writing" used by the French army for reading battlefield communiqués in the dark, Braille's system replaced each written character with a six-dot matrix:

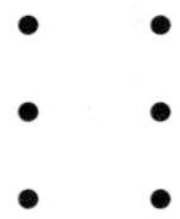

• •
• •
• •

where certain dots were raised, the choice depending on the character being transcribed. The letter *e*, for example, has two raised dots and is written

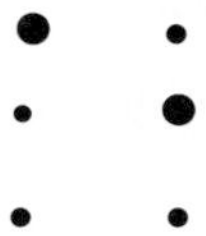

● •
• ●
• •

Punctuation marks, common words, suffixes, and so on, also have specified dot patterns. In all, how many different characters can be enciphered in Braille?

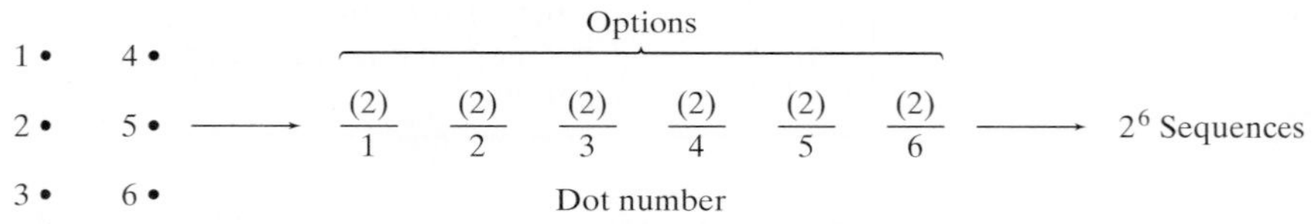

Figure 2.6.3

Think of the dots as six distinct operations, numbered 1 to 6 (see Figure 2.6.3). In forming a Braille letter, we have two options for each dot: We can raise it or *not* raise it. The letter *e*, for example, corresponds to the six-step sequence (raise, do not raise, do not raise, do not raise, raise, do not raise). The number of such sequences, with $k = 6$ and $n_1 = n_2 = \cdots = n_6 = 2$, is 2^6, or *64*. One of those sixty-four configurations, though, has *no* raised dots, making it of no use to a blind person. Figure 2.6.4 shows the entire sixty-three-character Braille alphabet.

■

Example 2.6.4

The annual NCAA ("March Madness") basketball tournament starts with a field of sixty-four teams. After six rounds of play, the squad that remains unbeaten is declared the national champion. How many different configurations of winners and losers are possible, starting with the first round? Assume that the initial pairing of the sixty-four invited teams into thirty-two first-round matches has already been done.

Counting the number of ways a tournament of this sort can play out is an exercise in applying the multiplication rule twice. Notice, first, that the thirty-two first-round games can be decided in 2^{32} ways. Similarly, the resulting sixteen second-round games can generate 2^{16} different winners, and so on. Overall, the tournament can be pictured as a six-step sequence, where the number of possible outcomes at

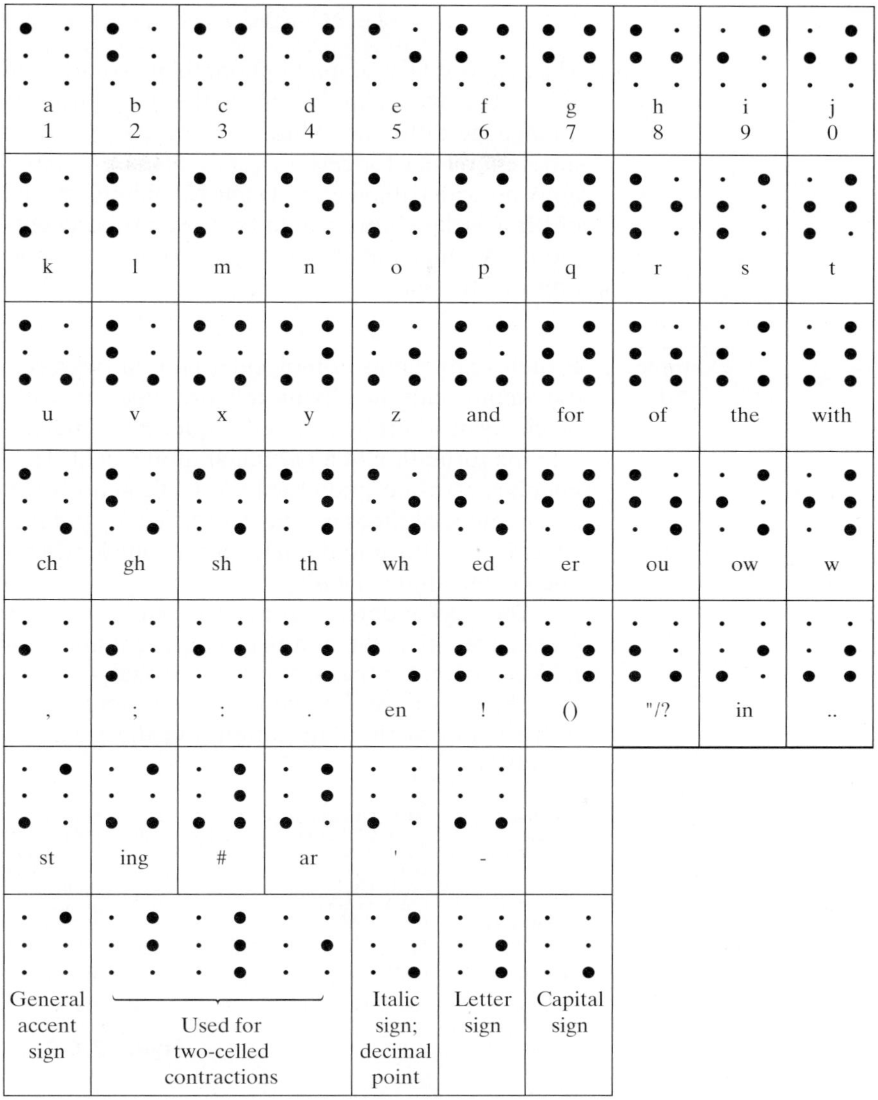

a 1	b 2	c 3	d 4	e 5	f 6	g 7	h 8	i 9	j 0
k	l	m	n	o	p	q	r	s	t
u	v	x	y	z	and	for	of	the	with
ch	gh	sh	th	wh	ed	er	ou	ow	w
,	;	:	.	en	!	()	"/?	in	..
st	ing	#	ar	'	-				
General accent sign	Used for two-celled contractions			Italic sign; decimal point	Letter sign	Capital sign			

Figure 2.6.4

the six steps are 2^{32}, 2^{16}, 2^8, 2^4, 2^2, and 2^1, respectively. It follows that the number of possible tournaments (not all of which, of course, would be equally likely!) is the product $2^{32} \cdot 2^{16} \cdot 2^8 \cdot 2^4 \cdot 2^2 \cdot 2^1$, or 2^{63}. ■

Example 2.6.5

In a famous science fiction story by Arthur C. Clarke, "The Nine Billion Names of God," a computer firm is hired by the lamas in a Tibetan monastery to write a program to generate all possible names of God. For reasons never divulged, the lamas believe that all such names can be written using no more than nine letters. If no letter combinations are ruled inadmissible, is the "nine billion" in the story's title a large enough number to accommodate all possibilities?

No. The lamas are in for a fleecing. The total number of names, N, would be the sum of all one-letter names, two-letter names, and so on. By the multiplication rule, the number of k-letter names is 26^k, so

$$N = 26^1 + 26^2 + \cdots + 26^9 \quad = \quad 5,646,683,826,134$$

The proposed list of nine billion, then, would be more than 5.6 trillion names short! (*Note*: The discrepancy between the story's title and the N we just computed is more a language difference than anything else. Clarke was British, and the British have different names for certain numbers than we have in the United States. Specifically, an American trillion is the English's billion, which means that the American editions of Mr. Clarke's story would be more properly entitled "The Nine Trillion Names of God." A more puzzling question, of course, is why "nine" appears in the title as opposed to "six.") ■

Example 2.6.6

Proteins are chains of molecules chosen (with repetition) from some twenty different amino acids. In a living cell, proteins are synthesized through the *genetic code*, a mechanism whereby ordered sequences of nucleotides in the messenger RNA dictate the formation of a particular amino acid. The four key nucleotides are adenine, guanine, cytosine, and uracil (A, G, C, and U). Assuming A, G, C, or U can appear any number of times in a nucleotide chain and that all sequences are physically possible, what is the minimum length the nucleotides must have if they are to be able to encode the amino acids?

The answer derives from a trial-and-error application of the multiplication rule. Given a length r, the number of different nucleotide sequences would be 4^r. We are looking, then, for the smallest r such that $4^r \geq 20$. Clearly, $r = 3$.

The entire genetic code for the amino acids is shown in Figure 2.6.5. For a discussion of the duplication and the significance of the three missing triplets, see (194).

Alanine	GCU, GCC, GCA, GCG	Leucine	UUA, UUG, CUU, CUC, CUA, CUG
Arginine	CGU, CGC, CGA,CGG,AGA, AGG	Lysine	AAA, AAG
Asparagine	AAU, AAC	Methionine	AUG
Aspartic acid	GAU, GAC	Phynylalanine	UUU, UUC
Cysteine	UGU, UGC	Proline	CCU,CCC, CCA, CCG
Glutamic acid	GAA, GAG	Serine	UCU, UCC, UCA, UCG, AGU, AGC
Glutamine	CAA, CAG	Threonine	ACU, ACC, ACA, ACG
Glycine	GGU, GGC, GGA, GGG	Tryptophan	UGG
Histidine	CAU, CAC	Tyrosine	UAU, UAC
Isoleucine	AUU, AUC, AUA	Valine	GUU, GUC,GUA,GUG

Figure 2.6.5 ■

Problem-Solving Hints

(Doing combinatorial problems)

Combinatorial questions sometimes call for problem-solving techniques that are not routinely used in other areas of mathematics. The three listed below are especially helpful.

1. Draw a diagram that shows the structure of the outcomes that are being counted. Be sure to include (or indicate) all relevant variations. A case in point is Figure 2.6.3. Almost invariably, diagrams such as these will suggest the formula, or combination of formulas, that should be applied.
2. Use enumerations to "test" the appropriateness of a formula. Typically, the answer to a combinatorial problem—that is, the number of ways to do something—will be so large that listing all possible outcomes is not

feasible. It often *is* feasible, though, to construct a simple, but analogous, problem for which the entire set of outcomes can be identified (and counted). If the proposed formula does not agree with the simple-case enumeration, we know that our analysis of the original question is incorrect.

3. If the outcomes to be counted fall into structurally different categories, the total number of outcomes will be the *sum* (not the product) of the number of outcomes in each category. Recall Example 2.6.5. The categories there are the nine different name lengths.

Questions

2.6.1. A chemical engineer wishes to observe the effects of temperature, pressure, and catalyst concentration on the yield resulting from a certain reaction. If she intends to include two different temperatures, three pressures, and two levels of catalyst, how many different runs must she make in order to observe each temperature-pressure-catalyst combination exactly twice?

2.6.2. A coded message from a CIA operative to his Russian KGB counterpart is to be sent in the form Q4ET, where the first and last entries must be consonants; the second, an integer 1 through 9; and the third, one of the six vowels. How many different ciphers can be transmitted?

2.6.3. How many terms will be included in the expansion of

$$(a+b+c)(d+e+f)(x+y+u+v+w)$$

Which of the following will be included in that number: *aeu*, *cdx*, *bef*, *xvw*?

2.6.4. Suppose that the format for license plates in a certain state is two letters followed by four numbers.

(a) How many different plates can be made?
(b) How many different plates are there if the letters can be repeated but no two numbers can be the same?
(c) How many different plates can be made if repetitions of numbers and letters are allowed except that no plate can have four zeros?

2.6.5. How many integers between 100 and 999 have distinct digits, and how many of those are odd numbers?

2.6.6. A fast-food restaurant offers customers a choice of eight toppings that can be added to a hamburger. How many different hamburgers can be ordered?

2.6.7. In baseball there are twenty-four different "base-out" configurations (runner on first—two outs, bases loaded—none out, and so on). Suppose that a new game, sleazeball, is played where there are seven bases (excluding home plate) and each team gets five outs an inning. How many base-out configurations would be possible in sleazeball?

2.6.8. When they were first introduced, postal zip codes were five-digit numbers, theoretically ranging from 00000 to 99999. (In reality, the lowest zip code was 00601 for San Juan, Puerto Rico; the highest was 99950 for Ketchikan, Alaska.) An additional four digits have been added, so each zip code is now a nine-digit number. How many zip codes are at least as large as 60000–0000, are even numbers, and have a 7 as their third digit?

2.6.9. A restaurant offers a choice of four appetizers, fourteen entrees, six desserts, and five beverages. How many different meals are possible if a diner intends to order only three courses? (Consider the beverage to be a "course.")

2.6.10. An octave contains twelve distinct notes (on a piano, five black keys and seven white keys). How many different eight-note melodies within a single octave can be written if the black keys and white keys need to alternate?

2.6.11. Residents of a condominium have an automatic garage door opener that has a row of eight buttons. Each garage door has been programmed to respond to a particular set of buttons being pushed. If the condominium houses 250 families, can residents be assured that no two garage doors will open on the same signal? If so, how many additional families can be added before the eight-button code becomes inadequate? (*Note:* The order in which the buttons are pushed is irrelevant.)

2.6.12. In international Morse code, each letter in the alphabet is symbolized by a series of dots and dashes: the letter *a*, for example, is encoded as "· –". What is the minimum number of dots and/or dashes needed to represent any letter in the English alphabet?

2.6.13. The decimal number corresponding to a sequence of n binary digits $a_0, a_1, \ldots, a_{n-1}$, where each a_i is either 0 or 1, is defined to be

$$a_0 2^0 + a_1 2^1 + \cdots + a_{n-1} 2^{n-1}$$

For example, the sequence 0 1 1 0 is equal to 6 ($= 0 \cdot 2^0 + 1 \cdot 2^1 + 1 \cdot 2^2 + 0 \cdot 2^3$). Suppose a fair coin is tossed nine times. Replace the resulting sequence of H's and

T's with a binary sequence of 1's and 0's (1 for H, 0 for T). For how many sequences of tosses will the decimal corresponding to the observed set of heads and tails exceed 256?

2.6.14. Given the letters in the word

$$Z\ O\ M\ B\ I\ E\ S$$

in how many ways can two of the letters be arranged such that one is a vowel and one is a consonant?

2.6.15. Suppose that two cards are drawn—in order—from a standard 52-card poker deck. In how many ways can the first card be a club and the second card be an ace?

2.6.16. Monica's vacation plans require that she fly from Nashville to Chicago to Seattle to Anchorage. According to her travel agent, there are three available flights from Nashville to Chicago, five from Chicago to Seattle, and two from Seattle to Anchorage. Assume that the numbers of options she has for return flights are the same. How many round-trip itineraries can she schedule?

Counting Permutations (when the objects are all distinct)

Ordered sequences arise in two fundamentally different ways. The first is the scenario addressed by the multiplication rule—a process is comprised of k operations, each allowing n_i options, $i = 1, 2, \ldots, k$; choosing one version of each operation leads to $n_1 n_2 \ldots n_k$ possibilities.

The second occurs when an ordered arrangement of some specified length k is formed from a finite collection of objects. Any such arrangement is referred to as a *permutation of length k*. For example, given the three objects A, B, and C, there are six different permutations of length two that can be formed if the objects cannot be repeated: AB, AC, BC, BA, CA, and CB.

Theorem 2.6.1 *The number of permutations of length k that can be formed from a set of n distinct elements, repetitions not allowed, is denoted by the symbol $_nP_k$, where*

$$_nP_k = n(n-1)(n-2)\cdots(n-k+1) = \frac{n!}{(n-k)!}$$

Proof Any of the n objects may occupy the first position in the arrangement, any of $n-1$ the second, and so on—the number of choices available for filling the kth position will be $n-k+1$ (see Figure 2.6.6). The theorem follows, then, from the multiplication rule: There will be $n(n-1)\cdots(n-k+1)$ ordered arrangements. □

$$\text{Choices:}\quad \frac{n}{1}\quad \frac{n-1}{2}\quad \cdots \quad \frac{n-(k-2)}{k-1}\quad \frac{n-(k-1)}{k}$$

Position in sequence

Figure 2.6.6

Corollary 2.6.2 *The number of ways to permute an entire set of n distinct objects is $_nP_n = n(n-1)(n-2)\cdots 1 = n!$.* ◄

Example 2.6.7 How many permutations of length $k = 3$ can be formed from the set of $n = 4$ distinct elements, A, B, C, and D?

According to Theorem 2.6.1, the number should be 24:

$$\frac{n!}{(n-k)!} = \frac{4!}{(4-3)!} = \frac{4\cdot 3\cdot 2\cdot 1}{1} = 24$$

Confirming that figure, Table 2.6.2 lists the entire set of 24 permutations and illustrates the argument used in the proof of the theorem.

Table 2.6.2

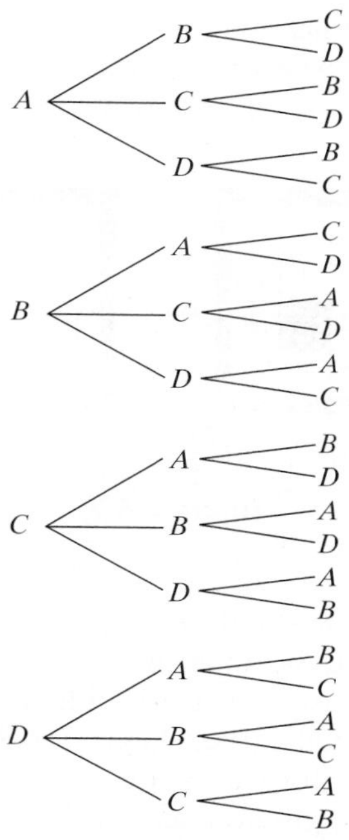

First	Second	Third	No.	Permutation
A	B	C	1.	(ABC)
		D	2.	(ABD)
	C	B	3.	(ACB)
		D	4.	(ACD)
	D	B	5.	(ADB)
		C	6.	(ADC)
B	A	C	7.	(BAC)
		D	8.	(BAD)
	C	A	9.	(BCA)
		D	10.	(BCD)
	D	A	11.	(BDA)
		C	12.	(BDC)
C	A	B	13.	(CAB)
		D	14.	(CAD)
	B	A	15.	(CBA)
		D	16.	(CBD)
	D	A	17.	(CDA)
		B	18.	(CDB)
D	A	B	19.	(DAB)
		C	20.	(DAC)
	B	A	21.	(DBA)
		C	22.	(DBC)
	C	A	23.	(DCA)
		B	24.	(DCB)

■

Example 2.6.8

In her sonnet with the famous first line, "How do I love thee? Let me count the ways," Elizabeth Barrett Browning listed eight ways. Suppose Ms. Browning had decided that writing greeting cards afforded her a better format for expressing her feelings. For how many years could she have corresponded with her favorite beau on a daily basis and never sent the same card twice? Assume that each card contains exactly four of the eight "ways" and that order matters.

In selecting the verse for a card, Ms. Browning would be creating a permutation of length $k = 4$ from a set of $n = 8$ distinct objects. According to Theorem 2.6.1,

$$\text{number of different cards} = {}_8P_4 = \frac{8!}{(8-4)!} = 8 \cdot 7 \cdot 6 \cdot 5$$

$$= 1680$$

At the rate of a card a day, she could have kept the correspondence going for more than four and one-half years. ■

Example 2.6.9

Years ago—long before Rubik's Cubes and electronic games had become epidemic—puzzles were much simpler. One of the more popular combinatorial-related diversions was a four-by-four grid consisting of fifteen movable squares and one empty space. The object was to maneuver as quickly as possible an arbitrary configuration (Figure 2.6.7a) into a specific pattern (Figure 2.6.7b). How many different ways could the puzzle be arranged?

Take the empty space to be square number 16 and imagine the four rows of the grid laid end to end to make a sixteen-digit sequence. Each permutation of that sequence corresponds to a different pattern for the grid. By the corollary to Theorem 2.6.1, the number of ways to position the tiles is 16!, or more than twenty trillion (20,922,789,888,000, to be exact). *That total is more than fifty times the number of stars in the entire Milky Way galaxy.* (*Note:* Not all of the 16! permutations can be generated without physically removing some of the tiles. Think of the two-by-two version of Figure 2.6.7 with tiles numbered 1 through 3. How many of the 4! theoretical configurations can actually be formed?)

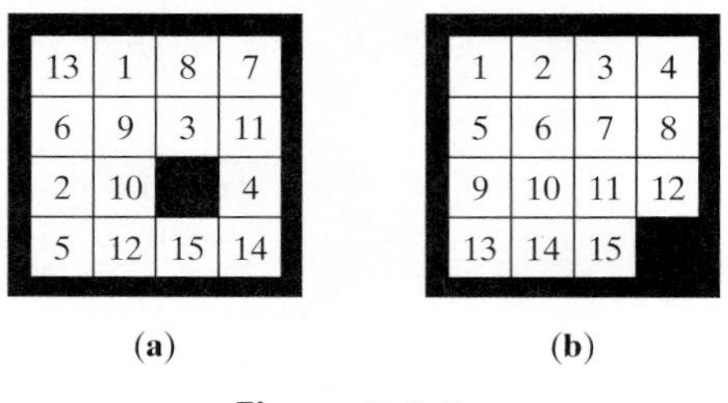

Figure 2.6.7

Example 2.6.10 A deck of fifty-two cards is shuffled and dealt face up in a row. For how many arrangements will the four aces be adjacent?

This is a good example illustrating the problem-solving benefits that come from drawing diagrams, as mentioned earlier. Figure 2.6.8 shows the basic structure that needs to be considered: The four aces are positioned as a "clump" somewhere between or around the forty-eight non-aces.

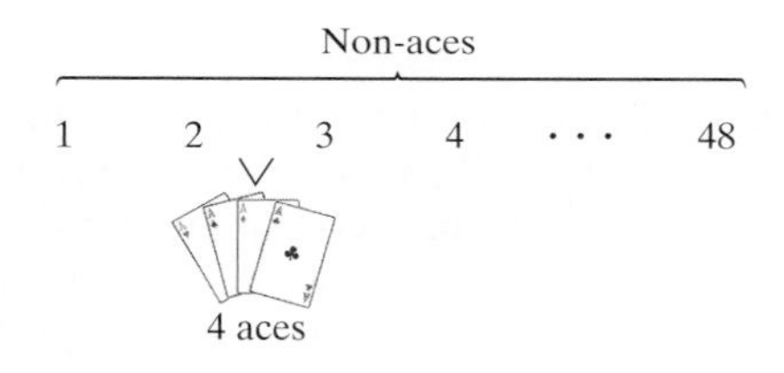

Figure 2.6.8

Clearly, there are forty-nine "spaces" that could be occupied by the four aces (in front of the first non-ace, between the first and second non-aces, and so on). Furthermore, by the corollary to Theorem 2.6.1, once the four aces are assigned to one of those forty-nine positions, they can still be permuted in ${}_4P_4 = 4!$ ways. Similarly, the forty-eight non-aces can be arranged in ${}_{48}P_{48} = 48!$ ways. It follows from the multiplication rule, then, that the number of arrangements having consecutive aces is the product $49 \cdot 4! \cdot 48!$, or, approximately, 1.46×10^{64}.

Comment Computing $n!$ can be quite cumbersome, even for n's that are fairly small: We saw in Example 2.6.9, for instance, that 16! is already in the trillions. Fortunately, an easy-to-use approximation is available. According to *Stirling's formula*,

$$n! \doteq \sqrt{2\pi}\, n^{n+1/2} e^{-n}$$

In practice, we apply Stirling's formula by writing

$$\log_{10}(n!) \doteq \log_{10}\left(\sqrt{2\pi}\right) + \left(n + \frac{1}{2}\right)\log_{10}(n) - n\log_{10}(e)$$

and then exponentiating the right-hand side.

In Example 2.6.10, the number of arrangements was calculated to be $49 \cdot 4! \cdot 48!$, or $24 \cdot 49!$. Substituting into Stirling's formula, we can write

$$\begin{aligned}\log_{10}(49!) &\doteq \log_{10}\left(\sqrt{2\pi}\right) + \left(49 + \frac{1}{2}\right)\log_{10}(49) - 49\log_{10}(e)\\ &\approx 62.783366\end{aligned}$$

Therefore,

$$\begin{aligned}24 \cdot 49! &\doteq 24 \cdot 10^{62.78337}\\ &= 1.46 \times 10^{64}\end{aligned}$$

Example 2.6.11 In chess a rook can move vertically and horizontally (see Figure 2.6.9). It can capture any unobstructed piece located anywhere in its own row or column. In how many ways can eight distinct rooks be placed on a chessboard (having eight rows and eight columns) so that no two can capture one another?

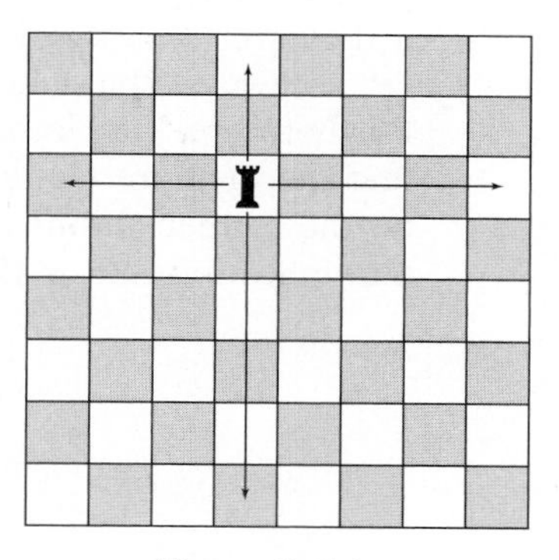

Figure 2.6.9

■

To start with a simpler problem, suppose that the eight rooks are all identical. Since no two rooks can be in the same row or same column (why?), it follows that each row must contain exactly one. The rook in the first row, however, can be in any of eight columns; the rook in the second row is then limited to being in one of seven columns, and so on. By the multiplication rule, then, the number of noncapturing configurations for eight identical rooks is ${}_8P_8$, or 8! (see Figure 2.6.10).

Now imagine the eight rooks to be distinct—they might be numbered, for example, 1 through 8. The rook in the first row could be marked with any of eight numbers; the rook in the second row with any of the remaining seven numbers; and so on. Altogether, there would be 8! numbering patterns *for each configuration*. The total number of ways to position eight distinct, noncapturing rooks, then, is $8! \cdot 8!$, or 1,625,702,400.

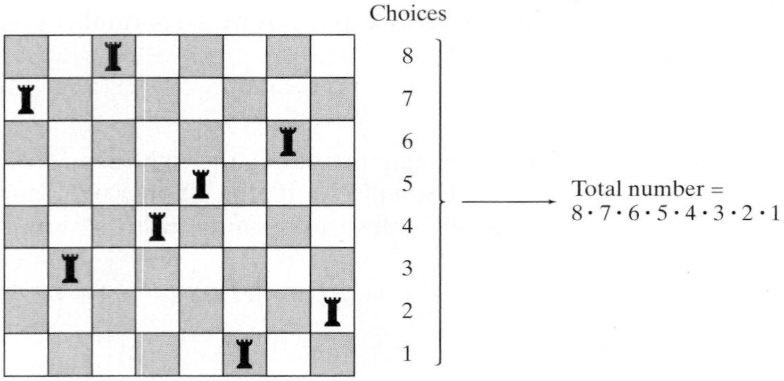

Choices

8
7
6
5
4
3
2
1

Total number = $8 \cdot 7 \cdot 6 \cdot 5 \cdot 4 \cdot 3 \cdot 2 \cdot 1$

Figure 2.6.10

Example 2.6.12 A new horror movie, *Friday the 13th, Part X*, will star Jason's great-grandson (also named Jason) as a psychotic trying to dispatch (as gruesomely as possible) eight camp counselors, four men and four women. (a) How many scenarios (i.e., victim orders) can the screenwriters devise, assuming they want Jason to do away with all the men before going after any of the women? (b) How many scripts are possible if the only restriction imposed on Jason is that he save Muffy for last?

a. Suppose the male counselors are denoted A, B, C, and D and the female counselors, W, X, Y, and Z. Among the admissible plots would be the sequence pictured in Figure 2.6.11, where B is done in first, then D, and so on. The men, if they are to be restricted to the first four positions, can still be permuted in ${}_4P_4 = 4!$ ways. The same number of arrangements can be found for the women. Furthermore, the plot in its entirety can be thought of as a two-step sequence: first the men are eliminated, then the women. Since 4! ways are available to do the former and 4! the latter, the total number of different scripts, by the multiplication rule, is 4!4!, or *576*.

Men				Women			
B	D	A	C	Y	Z	W	X
1	2	3	4	5	6	7	8

Order of killing

Figure 2.6.11

b. If the only condition to be met is that Muffy be dealt with last, the number of admissible scripts is simply ${}_7P_7 = 7!$, that being the number of ways to permute the other seven counselors (see Figure 2.6.12).

B	W	Z	C	Y	A	D	*Muffy*
1	2	3	4	5	6	7	8

Order of killing

Figure 2.6.12 ■

Example 2.6.13 Consider the set of nine-digit numbers that can be formed by rearranging without repetition the integers 1 through 9. For how many of those permutations will the 1 and the 2 precede the 3 and the 4? That is, we want to count sequences like 7 2 5 *1* *3* 6 9 *4* 8 but not like 6 8 *1* 5 *4* 2 7 *3* 9.

At first glance, this seems to be a problem well beyond the scope of Theorem 2.6.1. With the help of a symmetry argument, though, its solution is surprisingly simple.

Think of just the digits 1 through 4. By the corollary on p. 74, those four numbers give rise to $4!(=24)$ permutations. Of those twenty-four, only four—(1, 2, 3, 4), (2, 1, 3, 4), (1, 2, 4, 3), and (2, 1, 4, 3)—have the property that the 1 and the 2 come before the 3 and the 4. It follows that $\frac{4}{24}$ of the total number of nine-digit permutations should satisfy the condition being imposed on 1, 2, 3, and 4. Therefore,

$$\begin{aligned}\text{number of permutations where 1 and 2 precede 3 and 4} &= \frac{4}{24} \cdot 9! \\ &= 60{,}480\end{aligned}$$

■

Questions

2.6.17. The board of a large corporation has six members willing to be nominated for office. How many different "president/vice president/treasurer" slates could be submitted to the stockholders?

2.6.18. How many ways can a set of four tires be put on a car if all the tires are interchangeable? How many ways are possible if two of the four are snow tires?

2.6.19. Use Stirling's formula to approximate 30!. (*Note:* The exact answer is 265,252,859,812,268,935,315,188, 480,000,000.)

2.6.20. The nine members of the music faculty baseball team, the Mahler Maulers, are all incompetent, and each can play any position equally poorly. In how many different ways can the Maulers take the field?

2.6.21. A three-digit number is to be formed from the digits 1 through 7, with no digit being used more than once. How many such numbers would be less than 289?

2.6.22. Four men and four women are to be seated in a row of chairs numbered 1 through 8.

(a) How many total arrangements are possible?
(b) How many arrangements are possible if the men are required to sit in alternate chairs?

2.6.23. An engineer needs to take three technical electives sometime during his final four semesters. The three are to be selected from a list of ten. In how many ways can he schedule those classes, assuming that he never wants to take more than one technical elective in any given term?

2.6.24. How many ways can a twelve-member cheerleading squad (six men and six women) pair up to form six male-female teams? How many ways can six male-female teams be positioned along a sideline? What might the number $6!6!2^6$ represent? What might the number $6!6!2^6 2^{12}$ represent?

2.6.25. Suppose that a seemingly interminable German opera is recorded on all six sides of a three-record album. In how many ways can the six sides be played so that at least one is out of order?

2.6.26. A group of n families, each with m members, are to be lined up for a photograph. In how many ways can the nm people be arranged if members of a family must stay together?

2.6.27. Suppose that ten people, including you and a friend, line up for a group picture. How many ways can the photographer rearrange the line if she wants to keep exactly three people between you and your friend?

2.6.28. Use an induction argument to prove Theorem 2.6.1. (*Note*: This was the first mathematical result known to have been proved by induction. It was done in 1321 by Levi ben Gerson.)

2.6.29. In how many ways can a pack of fifty-two cards be dealt to thirteen players, four to each, so that every player has one card of each suit?

2.6.30. If the definition of $n!$ is to hold for all nonnegative integers n, show that it follows that $0!$ must equal 1.

2.6.31. The crew of Apollo 17 consisted of a pilot, a copilot, and a geologist. Suppose that NASA had actually trained nine aviators and four geologists as candidates for the flight. How many different crews could they have assembled?

2.6.32. Uncle Harry and Aunt Minnie will both be attending your next family reunion. Unfortunately, they hate each other. Unless they are seated with at least two people between them, they are likely to get into a shouting match. The side of the table at which they will be seated has seven chairs. How many seating arrangements are available for those seven people if a safe distance is to be maintained between your aunt and your uncle?

2.6.33. In how many ways can the digits 1 through 9 be arranged such that

(a) all the even digits precede all the odd digits?
(b) all the even digits are adjacent to each other?
(c) two even digits begin the sequence and two even digits end the sequence?
(d) the even digits appear in either ascending or descending order?

Counting Permutations (when the objects are not all distinct)

The corollary to Theorem 2.6.1 gives a formula for the number of ways an entire set of n objects can be permuted *if the objects are all distinct*. Fewer than $n!$ permutations are possible, though, if some of the objects are identical. For example, there are $3! = 6$ ways to permute the three distinct objects A, B, and C:

ABC
ACB
BAC
BCA
CAB
CBA

If the three objects to permute, though, are A, A, and B—that is, if two of the three are identical—the number of permutations decreases to three:

AAB
ABA
BAA

As we will see, there are many real-world applications where the n objects to be permuted belong to r different categories, each category containing one or more identical objects.

Theorem 2.6.2 *The number of ways to arrange n objects, n_1 being of one kind, n_2 of a second kind, ..., and n_r of an rth kind, is*

$$\frac{n!}{n_1!\,n_2!\cdots n_r!}$$

where $\sum_{i=1}^{r} n_i = n$.

Proof Let N denote the total number of such arrangements. For any one of those N, the similar objects (if they were actually different) could be arranged in $n_1!n_2!\cdots n_r!$ ways. (Why?) It follows that $N \cdot n_1!n_2!\cdots n_r!$ is the total number of ways to arrange n (distinct) objects. But $n!$ equals that same number. Setting $N \cdot n_1!n_2!\cdots n_r!$ equal to $n!$ gives the result. □

Comment Ratios like $n!/(n_1!n_2!\cdots n_r!)$ are called *multinomial coefficients* because the general term in the expansion of

$$(x_1 + x_2 + \cdots + x_r)^n$$

is

$$\frac{n!}{n_1!n_2!\cdots n_r!}x_1^{n_1}x_2^{n_2}\cdots x_r^{n_r}$$

Example 2.6.14

A pastry in a vending machine costs 85¢. In how many ways can a customer put in two quarters, three dimes, and one nickel?

Figure 2.6.13

If all coins of a given value are considered identical, then a typical deposit sequence, say, *QDDQND* (see Figure 2.6.13), can be thought of as a permutation of $n = 6$ objects belonging to $r = 3$ categories, where

$$n_1 = \text{number of nickels} = 1$$

$$n_2 = \text{number of dimes} = 3$$

$$n_3 = \text{number of quarters} = 2$$

By Theorem 2.6.2, there are sixty such sequences:

$$\frac{n!}{n_1!n_2!n_3!} = \frac{6!}{1!3!2!} = 60$$

Of course, had we assumed the coins were distinct (having been minted at different places and different times), the number of distinct permutations would have been 6!, or 720. ■

Example 2.6.15

Prior to the seventeenth century there were no scientific journals, a state of affairs that made it difficult for researchers to document discoveries. If a scientist sent a copy of his work to a colleague, there was always a risk that the colleague might claim it as his own. The obvious alternative—wait to get enough material to publish a book—invariably resulted in lengthy delays. So, as a sort of interim documentation, scientists would sometimes send each other anagrams—letter puzzles that, when properly unscrambled, summarized in a sentence or two what had been discovered.

When Christiaan Huygens (1629–1695) looked through his telescope and saw the ring around Saturn, he composed the following anagram (191):

$$aaaaaaa, ccccc, d, eeeee, g, h, iiiiiii, llll, mm,$$

$$nnnnnnnnn, oooo, pp, q, rr, s, ttttt, uuuuu$$

How many ways can the sixty-two letters in Huygens's anagram be arranged?

Let $n_1(=7)$ denote the number of a's, $n_2(=5)$ the number of c's, and so on. Substituting into the appropriate multinomial coefficient, we find

$$N = \frac{62!}{7!5!1!5!1!1!7!4!2!9!4!2!1!2!1!5!5!}$$

as the total number of arrangements. To get a feeling for the magnitude of N, we need to apply Stirling's formula to the numerator. Since

$$62! \doteq \sqrt{2\pi}e^{-62}62^{62.5}$$

then

$$\begin{aligned}\log(62!) &\doteq \log\left(\sqrt{2\pi}\right) - 62 \cdot \log(e) + 62.5 \cdot \log(62) \\ &\doteq 85.49731\end{aligned}$$

The antilog of 85.49731 is 3.143×10^{85}, so

$$N \doteq \frac{3.143 \times 10^{85}}{7!5!1!5!1!1!7!4!2!9!4!2!1!2!1!5!5!}$$

is a number on the order of 3.6×10^{60}. Huygens was clearly taking no chances! (*Note:* When appropriately rearranged, the anagram becomes "Annulo cingitur tenui, plano, nusquam cohaerente, ad eclipticam inclinato," which translates to "Surrounded by a thin ring, flat, suspended nowhere, inclined to the ecliptic.") ■

Example 2.6.16

What is the coefficient of x^{23} in the expansion of $(1+x^5+x^9)^{100}$?

To understand how this question relates to permutations, consider the simpler problem of expanding $(a+b)^2$:

$$\begin{aligned}(a+b)^2 &= (a+b)(a+b) \\ &= a \cdot a + a \cdot b + b \cdot a + b \cdot b \\ &= a^2 + 2ab + b^2\end{aligned}$$

Notice that each term in the first $(a+b)$ is multiplied by each term in the second $(a+b)$. Moreover, the coefficient that appears in front of each term in the expansion corresponds to the number of ways that that term can be formed. For example, the 2 in the term $2ab$ reflects the fact that the product ab can result from two different multiplications:

$$\underbrace{(a+b)(a+b)}_{ab} \qquad \text{or} \qquad (a+\underbrace{b)\,(a}_{ab}+b)$$

By analogy, the coefficient of x^{23} in the expansion of $(1+x^5+x^9)^{100}$ will be the number of ways that one term from each of the one hundred factors $(1+x^5+x^9)$ can be multiplied together to form x^{23}. The only factors that will produce x^{23}, though, are the set of two x^9's, one x^5, and ninety-seven 1's:

$$x^{23} = x^9 \cdot x^9 \cdot x^5 \cdot 1 \cdot 1 \cdots 1$$

It follows that the *coefficient* of x^{23} is the number of ways to permute two x^9's, one x^5, and ninety-seven 1's. So, from Theorem 2.6.2,

$$\text{coefficient of } x^{23} = \frac{100!}{2!1!97!}$$
$$= 485,100$$

■

Example 2.6.17

A palindrome is a phrase whose letters are in the same order whether they are read backward or forward, such as Napoleon's lament

Able was I ere I saw Elba.

or the often-cited

Madam, I'm Adam.

Words themselves can become the units in a palindrome, as in the sentence

Girl, bathing on Bikini, eyeing boy,
finds boy eyeing bikini on bathing girl.

Suppose the members of a set consisting of four objects of one type, six of a second type, and two of a third type are to be lined up in a row. How many of those permutations are palindromes?

Think of the twelve objects to arrange as being four A's, six B's, and two C's. If the arrangement is to be a palindrome, then half of the A's, half of the B's, and half of the C's must occupy the first six positions in the permutation. Moreover, the final six members of the sequence must be in the reverse order of the first six. For example, if the objects comprising the first half of the permutation were

$$C \quad A \quad B \quad A \quad B \quad B$$

then the last six would need to be in the order

$$B \quad B \quad A \quad B \quad A \quad C$$

It follows that the number of palindromes is the number of ways to permute the first six objects in the sequence, because once the first six are positioned, there is only one arrangement of the last six that will complete the palindrome. By Theorem 2.6.2, then,

$$\text{number of palindromes} = 6!/(2!3!1!) = 60$$

■

Example 2.6.18

A deliveryman is currently at Point X and needs to stop at Point 0 before driving through to Point Y (see Figure 2.6.14). How many different routes can he take without ever going out of his way?

Notice that any admissible path from, say, X to 0 is an ordered sequence of 11 "moves"—nine east and two north. Pictured in Figure 2.6.14, for example, is the particular X to 0 route

$$E \quad E \quad N \quad E \quad E \quad E \quad E \quad N \quad E \quad E \quad E$$

Similarly, any acceptable path from 0 to Y will necessarily consist of five moves east and three moves north (the one indicated is $E\ E\ N\ N\ E\ N\ E\ E$).

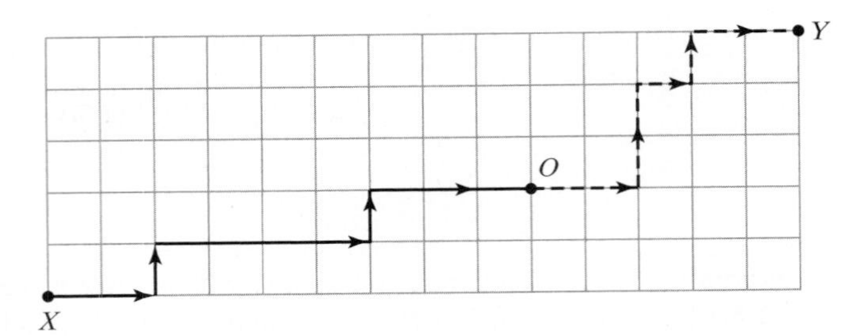

Figure 2.6.14

Since each path from X to 0 corresponds to a unique permutation of nine E's and two N's, the *number* of such paths (from Theorem 2.6.2) is the quotient

$$11!/(9!2!) = 55$$

For the same reasons, the number of different paths from 0 to Y is

$$8!/(5!3!) = 56$$

By the multiplication rule, then, the total number of admissible routes from X to Y that pass through 0 is the product of 55 and *56*, or *3080*. ■

Example 2.6.19

A burglar is trying to deactivate an alarm system that has a six-digit entry code. He notices that three of the keyboard buttons—the 3, the 4, and the 9—are more polished than the other seven, suggesting that only those three numbers appear in the correct entry code. Trial and error may be a feasible strategy, but earlier misadventures have convinced him that if his probability of guessing the correct code in the first thirty minutes is not at least 70%, the risk of getting caught is too great. Given that he can try a different permutation every five seconds, what should he do? He could look for an unlocked window to crawl through (or, here's a thought, get an honest job!). Deactivating the alarm, though, is not a good option.

Table 2.6.3 shows that 570 six-digit permutations can be made from the numbers 3, 4, and 9.

Table 2.6.3

Form of Permutations	Example	Number
One digit appears four times; other digits appear once	449434	$6!/(4!1!1!) \times 3 = 90$
One digit appears three times; another appears twice; and a third appears once	944334	$6!/(3!2!1!) \times 3! = 360$
Each digit appears twice	439934	$6!/(2!2!2!) \times 1 = 120$
		TOTAL: 570

Guessing at the rate of one permutation every five seconds would allow 360 permutations to be tested in thirty minutes, but 360 is only 63% of 570, so the burglar's 70% probability criteria of success would not be met. (*Question*: The first factors in Column 3 of Table 2.6.3 are applications of Theorem 2.6.2 to the sample permutations shown in Column 2. What do the second factors in Column 3 represent?) ■

Questions

2.6.34. Which state name can generate more permutations, TENNESSEE or FLORIDA?

2.6.35. How many numbers greater than four million can be formed from the digits 2, 3, 4, 4, 5, 5, 5?

2.6.36. An interior decorator is trying to arrange a shelf containing eight books, three with red covers, three with blue covers, and two with brown covers.

(a) Assuming the titles and the sizes of the books are irrelevant, in how many ways can she arrange the eight books?
(b) In how many ways could the books be arranged if they were all considered distinct?
(c) In how many ways could the books be arranged if the red books were considered indistinguishable, but the other five were considered distinct?

2.6.37. Four Nigerians (A, B, C, D), three Chinese (#, *, &), and three Greeks (α, β, γ) are lined up at the box office, waiting to buy tickets for the World's Fair.

(a) How many ways can they position themselves if the Nigerians are to hold the first four places in line; the Chinese, the next three; and the Greeks, the last three?
(b) How many arrangements are possible if members of the same nationality must stay together?
(c) How many different queues can be formed?
(d) Suppose a vacationing Martian strolls by and wants to photograph the ten for her scrapbook. A bit myopic, the Martian is quite capable of discerning the more obvious differences in human anatomy but is unable to distinguish one Nigerian (N) from another, one Chinese (C) from another, or one Greek (G) from another. Instead of perceiving a line to be $B*\beta AD\#\&C\alpha\gamma$, for example, she would see *NCGNNCCNGG*. From the Martian's perspective, in how many different ways can the ten funny-looking Earthlings line themselves up?

2.6.38. How many ways can the letters in the word

SLUMGULLION

be arranged so that the three L's precede all the other consonants?

2.6.39. A tennis tournament has a field of $2n$ entrants, all of whom need to be scheduled to play in the first round. How many different pairings are possible?

2.6.40. What is the coefficient of x^{12} in the expansion of $(1+x^3+x^6)^{18}$?

2.6.41. In how many ways can the letters of the word

ELEEMOSYNARY

be arranged so that the S is always immediately followed by a Y?

2.6.42. In how many ways can the word *ABRACADABRA* be formed in the array pictured below? Assume that the word must begin with the top A and progress diagonally downward to the bottom A.

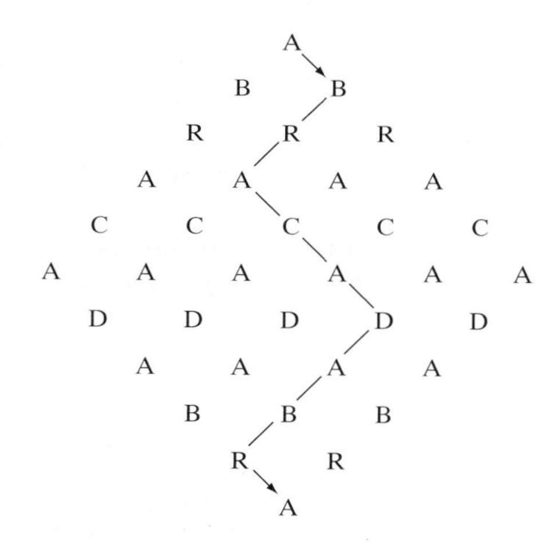

2.6.43. Suppose a pitcher faces a batter who never swings. For how many different ball/strike sequences will the batter be called out on the fifth pitch?

2.6.44. What is the coefficient of $w^2x^3yz^3$ in the expansion of $(w+x+y+z)^9$?

2.6.45. Imagine six points in a plane, no three of which lie on a straight line. In how many ways can the six points be used as vertices to form two triangles? (*Hint:* Number the points 1 through 6. Call one of the triangles A and the other B. What does the permutation

A	A	B	B	A	B
1	2	3	4	5	6

represent?)

2.6.46. Show that $(k!)!$ is divisible by $k!^{(k-1)!}$. (*Hint:* Think of a related permutation problem whose solution would require Theorem 2.6.2.)

2.6.47. In how many ways can the letters of the word

$$BROBDINGNAGIAN$$

be arranged without changing the order of the vowels?

2.6.48. Make an anagram out of the familiar expression STATISTICS IS FUN. In how many ways can the letters in the anagram be permuted?

2.6.49. Linda is taking a five-course load her first semester: English, math, French, psychology, and history. In how many different ways can she earn three A's and two B's? Enumerate the entire set of possibilities. Use Theorem 2.6.2 to verify your answer.

Counting Combinations

Order is not always a meaningful characteristic of a collection of elements. Consider a poker player being dealt a five-card hand. Whether he receives a 2 of hearts, 4 of clubs, 9 of clubs, jack of hearts, and ace of diamonds *in that order*, or in any one of the other $5! - 1$ permutations of those particular five cards is irrelevant—the hand is still the same. As the last set of examples in this section bears out, there are many such situations—problems where our only legitimate concern is with the composition of a set of elements, not with any particular arrangement of them.

We call a collection of k *unordered* elements a *combination of size* k. For example, given a set of $n = 4$ distinct elements—A, B, C, and D—there are *six* ways to form combinations of size 2:

A and B	B and C
A and C	B and D
A and D	C and D

A general formula for counting combinations can be derived quite easily from what we already know about counting permutations.

Theorem 2.6.3 *The number of ways to form combinations of size k from a set of n distinct objects, repetitions not allowed, is denoted by the symbols $\binom{n}{k}$ or ${}_nC_k$, where*

$$\binom{n}{k} = {}_nC_k = \frac{n!}{k!(n-k)!}$$

Proof Let the symbol $\binom{n}{k}$ denote the number of combinations satisfying the conditions of the theorem. Since each of those combinations can be ordered in $k!$ ways, the product $k!\binom{n}{k}$ must equal the number of *permutations* of length k that can be formed from n distinct elements. But n distinct elements can be formed into permutations of length k in $n(n-1)\cdots(n-k+1) = n!/(n-k)!$ ways. Therefore,

$$k!\binom{n}{k} = \frac{n!}{(n-k)!}$$

Solving for $\binom{n}{k}$ gives the result. □

Comment It often helps to think of combinations in the context of drawing objects out of an urn. If an urn contains n chips labeled 1 through n, the number of ways we can reach in and draw out different samples of size k is $\binom{n}{k}$. In deference to this sampling interpretation for the formation of combinations, $\binom{n}{k}$ is usually read "n things taken k at a time" or "n choose k."

Comment The symbol $\binom{n}{k}$ appears in the statement of a familiar theorem from algebra,

$$(x+y)^n = \sum_{k=0}^{n} \binom{n}{k} x^k y^{n-k}$$

Since the expression being raised to a power involves two terms, x and y, the constants $\binom{n}{k}$, $k = 0, 1, \ldots, n$, are commonly referred to as *binomial coefficients*.

Example 2.6.20

Eight politicians meet at a fund-raising dinner. How many greetings can be exchanged if each politician shakes hands with every other politician exactly once?

Imagine the politicians to be eight chips—1 through 8—in an urn. A handshake corresponds to an unordered sample of size 2 chosen from that urn. Since repetitions are not allowed (even the most obsequious and overzealous of campaigners would not shake hands with himself!), Theorem 2.6.3 applies, and the total number of handshakes is

$$\binom{8}{2} = \frac{8!}{2!6!}$$

or 28. ■

Example 2.6.21

A chemist is trying to synthesize a part of a straight-chain aliphatic hydrocarbon polymer that consists of twenty-one radicals—ten ethyls (E), six methyls (M), and five propyls (P). Assuming all arrangements of radicals are physically possible, how many different polymers can be formed if no two of the methyl radicals are to be adjacent?

Imagine arranging the E's and the P's without the M's. Figure 2.6.15 shows one such possibility. Consider the sixteen "spaces" between and outside the E's and P's as indicated by the arrows in Figure 2.6.15. In order for the M's to be nonadjacent, they must occupy any six of these locations. But those six spaces can be chosen in $\binom{16}{6}$ ways. And for each of the $\binom{16}{6}$ positionings of the M's, the E's and P's can be permuted in $\frac{15!}{10!5!}$ ways (Theorem 2.6.2).

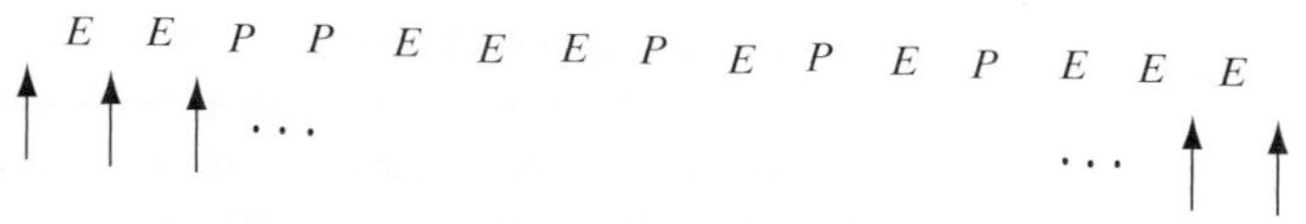

E E P P E E E P E P E P E E E

↑ ↑ ↑ . . . . . . ↑ ↑

Figure 2.6.15

So, by the multiplication rule, the total number of polymers having nonadjacent methyl radicals is 24,048,024:

$$\binom{16}{6} \cdot \frac{15!}{10!5!} = \frac{16!}{10!6!} \frac{15!}{10!5!} = (8008)(3003) = 24,048,024$$

■

Example 2.6.22

Binomial coefficients have many interesting properties. Perhaps the most familiar is Pascal's triangle,[1] a numerical array where each entry is equal to the sum of the two numbers appearing diagonally above it (see Figure 2.6.16). Notice that each entry in Pascal's triangle can be expressed as a binomial coefficient, and the relationship just described appears to reduce to a simple equation involving those coefficients:

$$\binom{n+1}{k} = \binom{n}{k} + \binom{n}{k-1} \tag{2.6.1}$$

Prove that Equation 2.6.1 holds for all positive integers n and k.

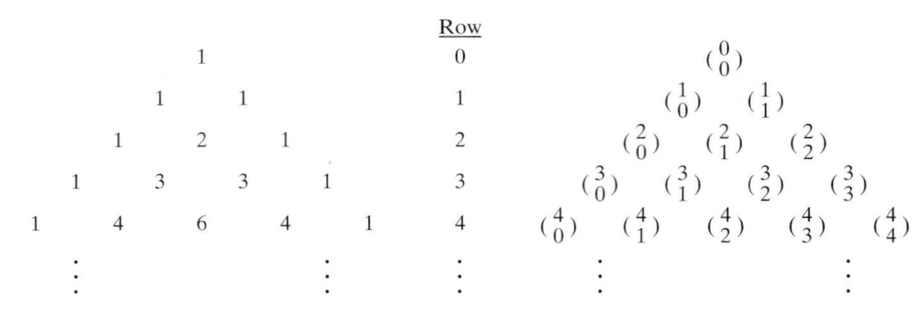

Figure 2.6.16

Consider a set of $n+1$ distinct objects $A_1, A_2, \ldots, A_{n+1}$. We can obviously draw samples of size k from that set in $\binom{n+1}{k}$ different ways. Now, consider any particular object—for example, A_1. Relative to A_1, each of those $\binom{n+1}{k}$ samples belongs to one of two categories: those containing A_1 and those not containing A_1. To form samples containing A_1, we need to select $k-1$ additional objects from the remaining n. This can be done in $\binom{n}{k-1}$ ways. Similarly, there are $\binom{n}{k}$ ways to form samples not containing A_1. Therefore, $\binom{n+1}{k}$ must equal $\binom{n}{k} + \binom{n}{k-1}$. ■

Example 2.6.23

The answers to combinatorial questions can sometimes be obtained using quite different approaches. What invariably distinguishes one solution from another is the way in which outcomes are characterized.

For example, suppose you have just ordered a roast beef sub at a sandwich shop, and now you need to decide which, if any, of the available toppings (lettuce, tomato, onions, etc.) to add. If the shop has eight "extras" to choose from, how many different subs can you order?

One way to answer this question is to think of each sub as an ordered sequence of length eight, where each position in the sequence corresponds to one of the toppings. At each of those positions, you have two choices—"add" or "do not add" that particular topping. Pictured in Figure 2.6.17 is the sequence corresponding to the sub that has lettuce, tomato, and onion but no other toppings. Since two choices ("add" or "do not add") are available for each of the eight toppings, the multiplication rule

[1] Despite its name, Pascal's triangle was not discovered by Pascal. Its basic structure had been known hundreds of years before the French mathematician was born. It was Pascal, though, who first made extensive use of its properties.

Add?

Y	Y	Y	N	N	N	N	N
Lettuce	Tomato	Onion	Mustard	Relish	Mayo	Pickles	Peppers

Figure 2.6.17

tells us that the number of different roast beef subs that could be requested is 2^8, or *256*.

An ordered sequence of length eight, though, is not the only model capable of characterizing a roast beef sandwich. We can also distinguish one roast beef sub from another by the particular *combination* of toppings that each one has. For example, there are $\binom{8}{4} = 70$ different subs having exactly four toppings. It follows that the total number of different sandwiches is the total number of different combinations of size k, where k ranges from 0 to 8. Reassuringly, that sum agrees with the ordered sequence answer:

$$\begin{aligned}\text{total number of different roast beef subs} &= \binom{8}{0} + \binom{8}{1} + \binom{8}{2} + \cdots + \binom{8}{8} \\ &= 1 + 8 + 28 + \cdots + 1 \\ &= 256\end{aligned}$$

What we have just illustrated here is another property of binomial coefficients—namely, that

$$\sum_{k=0}^{n} \binom{n}{k} = 2^n \tag{2.6.2}$$

The proof of Equation 2.6.2 is a direct consequence of Newton's binomial expansion (see the second comment following Theorem 2.6.3). ■

Questions

2.6.50. How many straight lines can be drawn between five points (A, B, C, D, and E), no three of which are collinear?

2.6.51. The Alpha Beta Zeta sorority is trying to fill a pledge class of nine new members during fall rush. Among the twenty-five available candidates, fifteen have been judged marginally acceptable and ten highly desirable. How many ways can the pledge class be chosen to give a two-to-one ratio of highly desirable to marginally acceptable candidates?

2.6.52. A boat has a crew of eight: Two of those eight can row only on the stroke side, while three can row only on the bow side. In how many ways can the two sides of the boat be manned?

2.6.53. Nine students, five men and four women, interview for four summer internships sponsored by a city newspaper.

(a) In how many ways can the newspaper choose a set of four interns?

(b) In how many ways can the newspaper choose a set of four interns if it must include two men and two women in each set?

(c) How many sets of four can be picked such that not everyone in a set is of the same sex?

2.6.54. The final exam in History 101 consists of five essay questions that the professor chooses from a pool of seven that are given to the students a week in advance. For how many possible sets of questions does a student need to be prepared? In this situation, does order matter?

2.6.55. Ten basketball players meet in the school gym for a pickup game. How many ways can they form two teams of five each?

2.6.56. Your statistics teacher announces a twenty-page reading assignment on Monday that is to be finished by Thursday morning. You intend to read the first x_1 pages

Monday, the next x_2 pages Tuesday, and the final x_3 pages Wednesday, where $x_1 + x_2 + x_3 = 20$, and each $x_i \geq 1$. In how many ways can you complete the assignment? That is, how many different sets of values can be chosen for x_1, x_2, and x_3?

2.6.57. In how many ways can the letters in

$$MISSISSIPPI$$

be arranged so that no two I's are adjacent?

2.6.58. Prove that $\sum_{k=0}^{n} \binom{n}{k} = 2^n$. (*Hint:* Use the binomial expansion mentioned on p. 87.)

2.6.59. Prove that

$$\binom{n}{0}^2 + \binom{n}{1}^2 + \cdots + \binom{n}{n}^2 = \binom{2n}{n}$$

(*Hint:* Rewrite the left-hand side as

$$\binom{n}{0}\binom{n}{n} + \binom{n}{1}\binom{n}{n-1} + \binom{n}{2}\binom{n}{n-2} + \cdots$$

and consider the problem of selecting a sample of n objects from an original set of $2n$ objects.)

2.6.60. Show that

$$\binom{n}{1} + \binom{n}{3} + \cdots = \binom{n}{0} + \binom{n}{2} + \cdots$$

(*Hint:* Consider the expansion of $(x - y)^n$.)

2.6.61. Prove that successive terms in the sequence $\binom{n}{0}$, $\binom{n}{1}, \ldots, \binom{n}{n}$ first increase and then decrease. [*Hint:* Examine the ratio of two successive terms, $\binom{n}{j+1} \Big/ \binom{n}{j}$.]

2.6.62. Mitch is trying to add a little zing to his cabaret act by telling four jokes at the beginning of each show. His current engagement is booked to run four months. If he gives one performance a night and never wants to repeat the same set of jokes on any two nights, what is the minimum number of jokes he needs in his repertoire?

2.6.63. Compare the coefficients of t^k in $(1+t)^d(1+t)^e = (1+t)^{d+e}$ to prove that

$$\sum_{j=0}^{k} \binom{d}{j}\binom{e}{k-j} = \binom{d+e}{k}$$

2.7 Combinatorial Probability

In Section 2.6 our concern focused on counting the number of ways a given operation, or sequence of operations, could be performed. In Section 2.7 we want to couple those enumeration results with the notion of probability. Putting the two together makes a lot of sense—there are many combinatorial problems where an enumeration, by itself, is not particularly relevant. A poker player, for example, is not interested in knowing the total *number* of ways he can draw to a straight; he *is* interested, though, in his *probability* of drawing to a straight.

In a combinatorial setting, making the transition from an enumeration to a probability is easy. If there are n ways to perform a certain operation and a total of m of those satisfy some stated condition—call it A—then $P(A)$ is defined to be the ratio m/n. This assumes, of course, that all possible outcomes are equally likely.

Historically, the "m over n" idea is what motivated the early work of Pascal, Fermat, and Huygens (recall Section 1.3). Today we recognize that not all probabilities are so easily characterized. Nevertheless, the m/n model—the so-called *classical* definition of probability—is entirely appropriate for describing a wide variety of phenomena.

Example 2.7.1

An urn contains eight chips, numbered 1 through 8. A sample of three is drawn without replacement. What is the probability that the largest chip in the sample is a 5?

Let A be the event "Largest chip in sample is a 5." Figure 2.7.1 shows what must happen in order for A to occur: (1) the 5 chip must be selected, and (2) two chips must be drawn from the subpopulation of chips numbered 1 through 4. By the multiplication rule, the number of samples satisfying event A is the product $\binom{1}{1} \cdot \binom{4}{2}$.

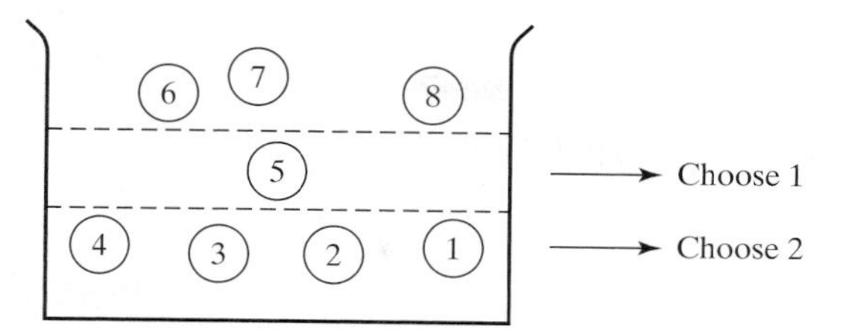

Figure 2.7.1

The sample space S for the experiment of drawing three chips from the urn contains $\binom{8}{3}$ outcomes, all equally likely. In this situation, then, $m = \binom{1}{1} \cdot \binom{4}{2}$, $n = \binom{8}{3}$, and

$$P(A) = \frac{\binom{1}{1} \cdot \binom{4}{2}}{\binom{8}{3}}$$
$$= 0.11$$

Example 2.7.2

An urn contains n red chips numbered 1 through n, n white chips numbered 1 through n, and n blue chips numbered 1 through n (see Figure 2.7.2). Two chips are drawn at random and without replacement. What is the probability that the two drawn are either the same color or the same number?

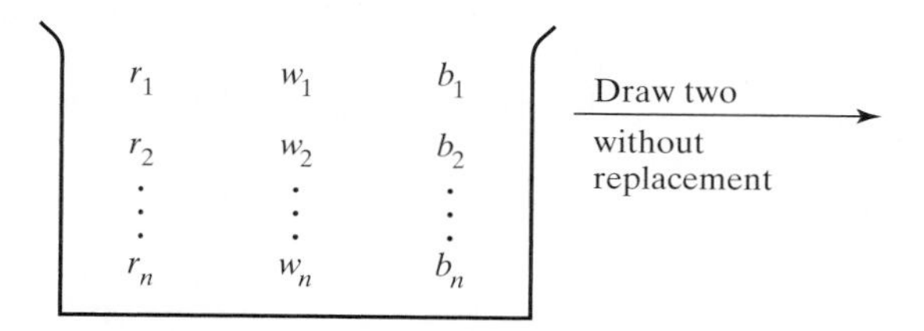

Figure 2.7.2

Let A be the event that the two chips drawn are the same color; let B be the event that they have the same number. We are looking for $P(A \cup B)$.

Since A and B here are mutually exclusive,

$$P(A \cup B) = P(A) + P(B)$$

With $3n$ chips in the urn, the total number of ways to draw an unordered sample of size 2 is $\binom{3n}{2}$. Moreover,

$$\begin{aligned} P(A) &= P(2\,\text{reds} \cup 2\,\text{whites} \cup 2\,\text{blues}) \\ &= P(2\,\text{reds}) + P(2\,\text{whites}) + P(2\,\text{blues}) \\ &= 3\binom{n}{2} \Big/ \binom{3n}{2} \end{aligned}$$

and

$$P(B) = P(\text{two 1's} \cup \text{two 2's} \cup \cdots \cup \text{two } n\text{'s})$$
$$= n\binom{3}{2} \Big/ \binom{3n}{2}$$

Therefore,

$$P(A \cup B) = \frac{3\binom{n}{2} + n\binom{3}{2}}{\binom{3n}{2}}$$
$$= \frac{n+1}{3n-1}$$

■

Example 2.7.3 Twelve fair dice are rolled. What is the probability that

a. the first six dice all show one face and the last six dice all show a second face?
b. not all the faces are the same?
c. each face appears exactly twice?

a. The sample space that corresponds to the "experiment" of rolling twelve dice is the set of ordered sequences of length twelve, where the outcome at every position in the sequence is one of the integers 1 through 6. If the dice are fair, all 6^{12} such sequences are equally likely.

Let A be the set of rolls where the first six dice show one face and the second six show another face. Figure 2.7.3 shows one of the sequences in the event A. Clearly, the face that appears for the first half of the sequence could be any of the six integers from 1 through 6.

Faces

2	2	2	2	2	2	4	4	4	4	4	4
1	2	3	4	5	6	7	8	9	10	11	12

Position in sequence

Figure 2.7.3

Five choices would be available for the last half of the sequence (since the two faces cannot be the same). The number of sequences in the event A, then, is ${}_6P_2 = 6 \cdot 5 = 30$. Applying the "*m*/*n*" rule gives

$$P(A) = 30/6^{12} = 1.4 \times 10^{-8}$$

b. Let B be the event that not all the faces are the same. Then

$$P(B) = 1 - P(B^C)$$
$$= 1 - 6/12^6$$

since there are six sequences—(1, 1, 1, 1, 1, 1, 1, 1, 1, 1, 1, 1,), . . ., (6, 6, 6, 6, 6, 6, 6, 6, 6, 6, 6, 6,)—where the twelve faces *are* all the same.

c. Let C be the event that each face appears exactly twice. From Theorem 2.6.2, the *number* of ways each face can appear exactly twice is $12!/(2!\cdot 2!\cdot 2!\cdot 2!\cdot 2!\cdot 2!)$. Therefore,

$$\begin{aligned} P(C) &= \frac{12!/(2!\cdot 2!\cdot 2!\cdot 2!\cdot 2!\cdot 2!)}{6^{12}} \\ &= 0.0034 \end{aligned}$$

■

Example 2.7.4

A fair die is tossed n times. What is the probability that the sum of the faces showing is $n+2$?

The sample space associated with rolling a die n times has 6^n outcomes, all of which in this case are equally likely because the die is presumed fair. There are two "types" of outcomes that will produce a sum of $n+2$: (a) $n-1$ 1's and one 3 and (b) $n-2$ 1's and two 2's (see Figure 2.7.4). By Theorem 2.6.2 the number of sequences having $n-1$ 1's and one 3 is $\frac{n!}{1!(n-1)!}=n$; likewise, there are $\frac{n!}{2!(n-2)!}=\binom{n}{2}$ outcomes having $n-2$ 1's and two 2's. Therefore,

$$P(\text{sum}=n+2)=\frac{n+\binom{n}{2}}{6^n}$$

■

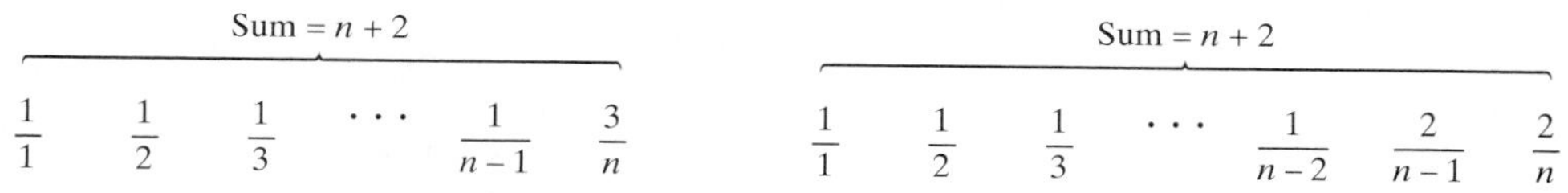

Sum $= n+2$: $\frac{1}{1}\ \frac{1}{2}\ \frac{1}{3}\ \cdots\ \frac{1}{n-1}\ \frac{3}{n}$

Sum $= n+2$: $\frac{1}{1}\ \frac{1}{2}\ \frac{1}{3}\ \cdots\ \frac{1}{n-2}\ \frac{2}{n-1}\ \frac{2}{n}$

Figure 2.7.4

Example 2.7.5

Two monkeys, Mickey and Marian, are strolling along a moonlit beach when Mickey sees an abandoned Scrabble set. Investigating, he notices that some of the letters are missing, and what remain are the following fifty-nine:

A	B	C	D	E	F	G	H	I	J	K	L	M
4	1	2	2	7	1	1	3	5	0	3	5	1

N	O	P	Q	R	S	T	U	V	W	X	Y	Z
3	2	0	0	2	8	4	2	0	1	0	2	0

Mickey, being of a romantic bent, would like to impress Marian, so he rearranges the letters in hopes of spelling out something clever. (*Note*: The rearranging is random because Mickey can't spell; fortunately, Marian can't read, so it really doesn't matter.) What is the probability that Mickey gets lucky and spells out

She walks in beauty, like the night
Of cloudless climes and starry skies

As we might imagine, Mickey would have to get *very* lucky. The total number of ways to permute fifty-nine letters—four A's, one B, two C's, and so on—is a direct application of Theorem 2.6.2:

$$\frac{59!}{4!1!2!\ldots 2!0!}$$

But of that number of ways, only one is the couplet he is hoping for. So, since he is arranging the letters randomly, making all permutations equally likely, the probability of his spelling out Byron's lines is

$$\frac{1}{\dfrac{59!}{4!1!2!\ldots 2!0!}}$$

or, using Stirling's formula, about 1.7×10^{-61}. Love may conquer all, but it won't beat those odds: Mickey would be well advised to start working on Plan B. ■

Example 2.7.6

Suppose that k people are selected at random from the general population. What are the chances that at least two of those k were born on the same day? Known as the *birthday problem*, this is a particularly intriguing example of combinatorial probability because its statement is so simple, its analysis is straightforward, yet its solution, as we will see, is strongly contrary to our intuition.

Picture the k individuals lined up in a row to form an ordered sequence. If leap year is omitted, each person might have any of 365 birthdays. By the multiplication rule, the group as a whole generates a sample space of 365^k birthday sequences (see Figure 2.7.5).

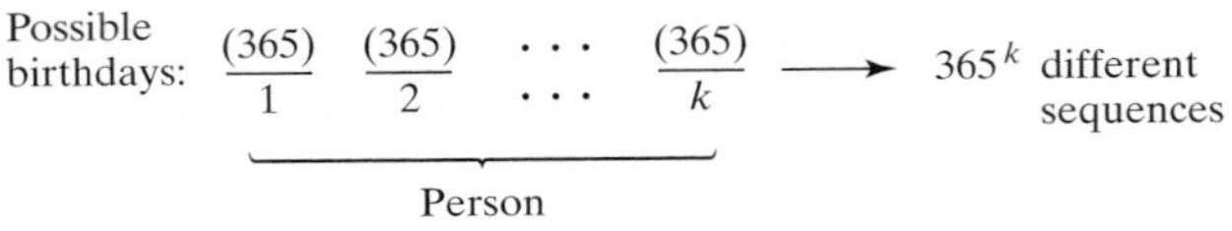

$$\text{Possible birthdays:}\quad \underbrace{\frac{(365)}{1}\quad \frac{(365)}{2}\quad \cdots \quad \frac{(365)}{k}}_{\text{Person}} \longrightarrow 365^k \text{ different sequences}$$

Figure 2.7.5

Define A to be the event "At least two people have the same birthday." If each person is assumed to have the same chance of being born on any given day, the 365^k sequences in Figure 2.7.5 are equally likely, and

$$P(A) = \frac{\text{number of sequences in } A}{365^k}$$

Counting the number of sequences in the numerator here is prohibitively difficult because of the complexity of the event A; fortunately, counting the number of sequences in A^c is quite easy. Notice that each birthday sequence in the sample space belongs to exactly one of two categories (see Figure 2.7.6):

1. At least two people have the same birthday.
2. All k people have different birthdays.

It follows that

$$\text{number of sequences in } A = 365^k - \text{number of sequences where all } k \text{ people have different birthdays}$$

The number of ways to form birthday sequences for k people subject to the restriction that all k birthdays must be different is simply the number of ways to form permutations of length k from a set of 365 distinct objects:

$${}_{365}P_k = 365(364)\cdots(365-k+1)$$

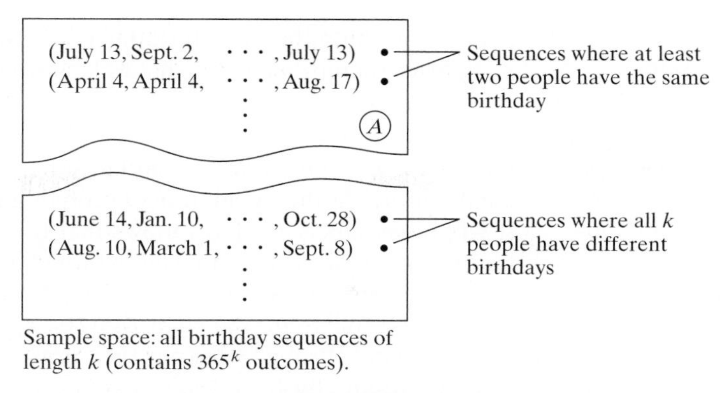

Figure 2.7.6

Therefore,

$$P(A) = P(\text{At least two people have the same birthday})$$
$$= \frac{365^k - 365(364)\cdots(365-k+1)}{365^k}$$

Table 2.7.1 shows $P(A)$ for k values of 15, 22, 23, 40, 50, and 70. Notice how the $P(A)$'s greatly exceed what our intuition would suggest.

Comment Presidential biographies offer one opportunity to "confirm" the unexpectedly large values that Table 2.7.1 gives for $P(A)$. Among our first $k = 40$ presidents, two did have the same birthday: Harding and Polk were both born on November 2. More surprising, though, are the death dates of the presidents: John Adams, Jefferson, and Monroe all died on July 4, and Fillmore and Taft both died on March 8.

Table 2.7.1

k	$P(A) = P$ (at least two have same birthday)
15	0.253
22	0.476
23	0.507
40	0.891
50	0.970
70	0.999

■

Comment The values for $P(A)$ in Table 2.7.1 are actually slight *underestimates* for the true probabilities that at least two of k people will be born on the same day. The assumption made earlier that all 365^k birthday sequences are equally likely is not entirely correct: Births are somewhat more common during the summer than they are during the winter. It has been proven, though, that any sort of deviation from the equally likely model will serve only to *increase* the chances that two or more people will share the same birthday (117). So, if $k = 40$, for example, the probability is slightly greater than 0.891 that at least two were born on the same day.

Example 2.7.7

One of the more instructive—and to some, one of the more useful—applications of combinatorics is the calculation of probabilities associated with various poker hands. It will be assumed in what follows that five cards are dealt from a poker deck and that no other cards are showing, although some may already have been dealt. The sample space is the set of $\binom{52}{5} = 2{,}598{,}960$ different hands, each having probability $1/2{,}598{,}960$. What are the chances of being dealt (a) a *full house*, (b) *one pair*, and (c) a *straight*? [Probabilities for the various other kinds of poker hands (two pairs, three-of-a-kind, flush, and so on) are gotten in much the same way.]

a. *Full house.* A full house consists of three cards of one denomination and two of another. Figure 2.7.7 shows a full house consisting of three 7's and two queens. Denominations for the three-of-a-kind can be chosen in $\binom{13}{1}$ ways. Then, given that a denomination has been decided on, the three requisite suits can be selected in $\binom{4}{3}$ ways. Applying the same reasoning to the pair gives $\binom{12}{1}$ available denominations, each having $\binom{4}{2}$ possible choices of suits. Thus, by the multiplication rule,

$$P(\text{full house}) = \frac{\binom{13}{1}\binom{4}{3}\binom{12}{1}\binom{4}{2}}{\binom{52}{5}} = 0.00144$$

	2	3	4	5	6	7	8	9	10	J	Q	K	A
D													
H						×					×		
C						×							
S						×					×		

Figure 2.7.7

b. *One pair.* To qualify as a one-pair hand, the five cards must include two of the same denomination and three "single" cards—cards whose denominations match neither the pair nor each other. Figure 2.7.8 shows a pair of 6's. For the pair, there are $\binom{13}{1}$ possible denominations and, once selected, $\binom{4}{2}$ possible suits. Denominations for the three single cards can be chosen $\binom{12}{3}$ ways (see Question 2.7.16), and each card can have any of $\binom{4}{1}$ suits. Multiplying these factors together and dividing by $\binom{52}{2}$ gives a probability of 0.42:

$$P(\text{one pair}) = \frac{\binom{13}{1}\binom{4}{2}\binom{12}{3}\binom{4}{1}\binom{4}{1}\binom{4}{1}}{\binom{52}{5}} = 0.42$$

	2	3	4	5	6	7	8	9	10	J	Q	K	A
D			×										×
H					×		×						
C					×								
S													

Figure 2.7.8

c. *Straight.* A straight is five cards having consecutive denominations *but not all in the same suit*—for example, a 4 of diamonds, 5 of hearts, 6 of hearts, 7 of clubs, and 8 of diamonds (see Figure 2.7.9). An ace may be counted "high" or "low," which means that (10, jack, queen, king, ace) is a straight and so is (ace, 2, 3, 4, 5). (If five consecutive cards are all in the same suit, the hand is called a *straight flush*. The latter is considered a fundamentally different type of hand in the sense that a straight flush "beats" a straight.) To get the numerator for $P(\text{straight})$, we will first ignore the condition that all five cards not be in the same suit and simply count the number of hands having consecutive denominations. Note there are ten sets of consecutive denominations of length five: (ace, 2, 3, 4, 5), (2, 3, 4, 5, 6), ..., (10, jack, queen, king, ace). With no restrictions on the suits, each card can be either a diamond, heart, club, or spade. It follows, then, that the number of five-card hands having consecutive denominations is $10 \cdot \binom{4}{1}^5$. But forty ($= 10 \cdot 4$) of those hands are straight flushes. Therefore,

$$P(\text{straight}) = \frac{10 \cdot \binom{4}{1}^5 - 40}{\binom{52}{5}} = 0.00392$$

Table 2.7.2 shows the probabilities associated with all the different poker hands. Hand i beats hand j if $P(\text{hand } i) < P(\text{hand } j)$.

	2	3	4	5	6	7	8	9	10	J	Q	K	A
D			×				×						
H				×	×								
C						×							
S													

Figure 2.7.9

Table 2.7.2

Hand	Probability
One pair	0.42
Two pairs	0.048
Three-of-a-kind	0.021
Straight	0.0039
Flush	0.0020
Full house	0.0014
Four-of-a-kind	0.00024
Straight flush	0.000014
Royal flush	0.0000015

■

Problem-Solving Hints

(Doing combinatorial probability problems)

Listed on p. 72 are several hints that can be helpful in counting the number of ways to do something. Those same hints apply to the solution of combinatorial *probability* problems, but a few others should be kept in mind as well.

1. The solution to a combinatorial probability problem should be set up as a quotient of numerator and denominator *enumerations*. Avoid the temptation to multiply probabilities associated with each position in the sequence. The latter approach will always "sound" reasonable, but it will frequently oversimplify the problem and give the wrong answer.
2. Keep the numerator and denominator consistent with respect to *order*—if permutations are being counted in the numerator, be sure that permutations are being counted in the denominator; likewise, if the outcomes in the numerator are combinations, the outcomes in the denominator must also be combinations.
3. The number of outcomes associated with any problem involving the rolling of n six-sided dice is 6^n; similarly, the number of outcomes associated with tossing a coin n times is 2^n. The number of outcomes associated with dealing a hand of n cards from a standard 52-card poker deck is ${}_{52}C_n$.

Questions

2.7.1. Ten equally qualified marketing assistants are candidates for promotion to associate buyer; seven are men and three are women. If the company intends to promote four of the ten at random, what is the probability that exactly two of the four are women?

2.7.2. An urn contains six chips, numbered 1 through 6. Two are chosen at random and their numbers are added together. What is the probability that the resulting sum is equal to 5?

2.7.3. An urn contains twenty chips, numbered 1 through 20. Two are drawn simultaneously. What is the probability that the numbers on the two chips will differ by more than 2?

2.7.4. A bridge hand (thirteen cards) is dealt from a standard 52-card deck. Let A be the event that the hand contains four aces; let B be the event that the hand contains four kings. Find $P(A \cup B)$.

2.7.5. Consider a set of ten urns, nine of which contain three white chips and three red chips each. The tenth contains five white chips and one red chip. An urn is picked at random. Then a sample of size 3 is drawn without replacement from that urn. If all three chips drawn are white, what is the probability that the urn being sampled is the one with five white chips?

2.7.6. A committee of fifty politicians is to be chosen from among our one hundred U.S. senators. If the selection is done at random, what is the probability that each state will be represented?

2.7.7. Suppose that n fair dice are rolled. What are the chances that all n faces will be the same?

2.7.8. Five fair dice are rolled. What is the probability that the faces showing constitute a "full house"—that is, three faces show one number and two faces show a second number?

2.7.9. Imagine that the test tube pictured contains $2n$ grains of sand, n white and n black. Suppose the tube is vigorously shaken. What is the probability that the two colors of sand will completely separate; that is, all of one color fall to the bottom, and all of the other color lie on top? (*Hint:* Consider the $2n$ grains to be aligned in a row. In how many ways can the n white and the n black grains be permuted?)

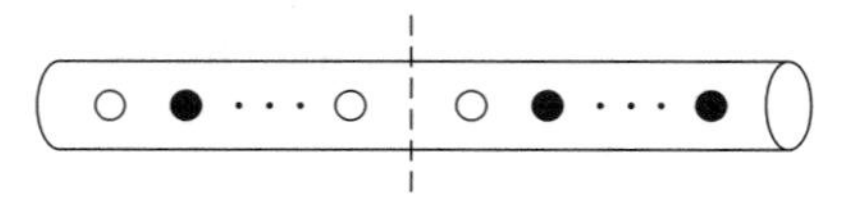

2.7.10. Does a monkey have a better chance of rearranging

$$ACCLLUUS \quad \text{to spell} \quad CALCULUS$$

or

$$AABEGLR \quad \text{to spell} \quad ALGEBRA?$$

2.7.11. An apartment building has eight floors. If seven people get on the elevator on the first floor, what is the probability they all want to get off on different floors? On the same floor? What assumption are you making? Does it seem reasonable? Explain.

2.7.12. If the letters in the phrase

$$A\ ROLLING\ STONE\ GATHERS\ NO\ MOSS$$

are arranged at random, what are the chances that not all the S's will be adjacent?

2.7.13. Suppose each of ten sticks is broken into a long part and a short part. The twenty parts are arranged into ten pairs and glued back together so that again there are ten sticks. What is the probability that each long part will be paired with a short part? (*Note:* This problem is a model for the effects of radiation on a living cell. Each chromosome, as a result of being struck by ionizing radiation, breaks into two parts, one part containing the centromere. The cell will die unless the fragment containing the centromere recombines with a fragment not containing a centromere.)

2.7.14. Six dice are rolled one time. What is the probability that each of the six faces appears?

2.7.15. Suppose that a randomly selected group of k people are brought together. What is the probability that exactly one pair has the same birthday?

2.7.16. For one-pair poker hands, why is the number of denominations for the three single cards $\binom{12}{3}$ rather than $\binom{12}{1}\binom{11}{1}\binom{10}{1}$?

2.7.17. Dana is not the world's best poker player. Dealt a 2 of diamonds, an 8 of diamonds, an ace of hearts, an ace of clubs, and an ace of spades, she discards the three aces. What are her chances of drawing to a flush?

2.7.18. A poker player is dealt a 7 of diamonds, a queen of diamonds, a queen of hearts, a queen of clubs, and an ace of hearts. He discards the 7. What is his probability of drawing to either a full house or four-of-a-kind?

2.7.19. Tim is dealt a 4 of clubs, a 6 of hearts, an 8 of hearts, a 9 of hearts, and a king of diamonds. He discards the 4 and the king. What are his chances of drawing to a straight flush? To a flush?

2.7.20. Five cards are dealt from a standard 52-card deck. What is the probability that the sum of the faces on the five cards is 48 or more?

2.7.21. Nine cards are dealt from a 52-card deck. Write a formula for the probability that three of the five even numerical denominations are represented twice, one of the three face cards appears twice, and a second face card appears once. (*Note:* Face cards are the jacks, queens, and kings; 2, 4, 6, 8, and 10 are the even numerical denominations.)

2.7.22. A coke hand in bridge is one where none of the thirteen cards is an ace or is higher than a 9. What is the probability of being dealt such a hand?

2.7.23. A pinochle deck has forty-eight cards, two of each of six denominations (9, J, Q, K, 10, A) and the usual four suits. Among the many hands that count for meld is a *roundhouse*, which occurs when a player has a king and queen of each suit. In a hand of twelve cards, what is the probability of getting a "bare" roundhouse (a king and queen of each suit and no other kings or queens)?

2.7.24. A somewhat inebriated conventioneer finds himself in the embarrassing predicament of being unable to predetermine whether his next step will be forward or backward. What is the probability that after hazarding n such maneuvers he will have stumbled forward a distance of r steps? (*Hint:* Let x denote the number of steps he takes forward and y, the number backward. Then $x + y = n$ and $x - y = r$.)

2.8 Taking a Second Look at Statistics (Monte Carlo Techniques)

Recall the von Mises definition of probability given on p. 17. If an experiment is repeated n times under identical conditions, and if the event E occurs on m of those repetitions, then

$$P(E) = \lim_{n \to \infty} \frac{m}{n} \tag{2.8.1}$$

To be sure, Equation 2.8.1 is an asymptotic result, but it suggests an obvious (and *very* useful) approximation—if n is finite,

$$P(E) \doteq \frac{m}{n}$$

In general, efforts to estimate probabilities by simulating repetitions of an experiment (usually with a computer) are referred to as *Monte Carlo* studies. Usually the technique is used in situations where an exact probability is difficult to calculate. It can also be used, though, as an empirical justification for choosing one proposed solution over another.

For example, consider the game described in Example 2.4.12 An urn contains a red chip, a blue chip, and a two-color chip (red on one side, blue on the other). One chip is drawn at random and placed on a table. The question is, if *blue* is showing, what is the probability that the color underneath is also *blue*?

Pictured in Figure 2.8.1 are two ways of conceptualizing the question just posed. The outcomes in (a) are assuming that a *chip* was drawn. Starting with that premise, the answer to the question is $\frac{1}{2}$—the red chip is obviously eliminated and only one of the two remaining chips is blue on both sides.

Figure 2.8.1

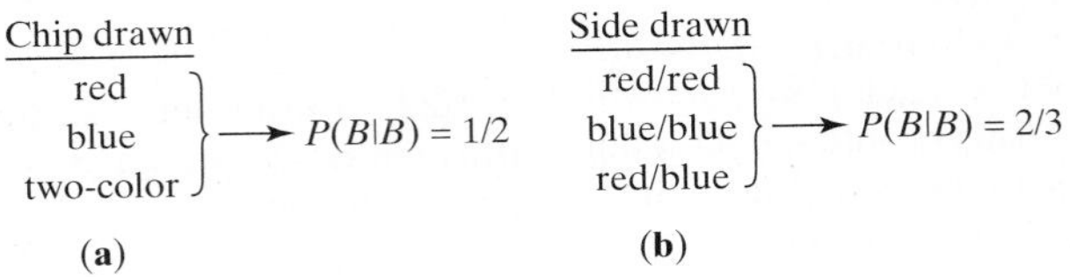

(a) Chip drawn: red, blue, two-color → $P(B|B) = 1/2$

(b) Side drawn: red/red, blue/blue, red/blue → $P(B|B) = 2/3$

Table 2.8.1

Trial #	S	U	Trial #	S	U	Trial #	S	U	Trial #	S	U
1	R	B	26	B	R	51	B	R	76	B	B*
2	B	B*	27	R	R	52	R	B	77	B	B*
3	B	R	28	R	B	53	B	B*	78	R	R
4	R	R	29	R	B	54	R	B	79	B	B*
5	R	B	30	R	R	55	R	R	80	R	R
6	R	B	31	R	B	56	R	B	81	R	B
7	R	R	32	B	B*	57	R	R	82	R	B
8	R	R	33	R	B	58	B	B*	83	R	R
9	B	B*	34	B	B*	59	B	R	84	B	R
10	B	R	35	B	B*	60	B	B*	85	B	R
11	R	R	36	R	R	61	B	R	86	R	R
12	B	B*	37	B	R	62	R	B	87	B	B*
13	R	R	38	B	B*	63	R	R	88	R	B
14	B	R	39	R	R	64	R	R	89	B	R
15	B	B*	40	B	B*	65	B	B*	90	R	R
16	B	B*	41	B	B*	66	B	R	91	R	B
17	R	B	42	B	R	67	R	R	92	R	R
18	B	R	43	B	B*	68	B	B*	93	R	R
19	B	B*	44	B	B*	69	B	B*	94	R	B
20	B	B*	45	B	B*	70	R	R	95	B	B*
21	R	R	46	R	R	71	R	R	96	B	B*
22	R	R	47	B	B*	72	B	B*	97	B	R
23	B	B*	48	B	B*	73	R	B	98	R	R
24	B	R	49	R	R	74	R	R	99	B	B*
25	B	B*	50	R	R	75	B	B*	100	B	B*

By way of contrast, the outcomes in (b) are assuming that the *side* of a chip was drawn. If so, the blue color showing could be any of three blue sides, two of which are blue underneath. According to model (b), then, the probability of both sides being blue is $\frac{2}{3}$.

The formal analysis on p. 46, of course, resolves the debate—the correct answer is $\frac{2}{3}$. But suppose that such a derivation was unavailable. How might we assess the relative plausibilities of $\frac{1}{2}$ and $\frac{2}{3}$? The answer is simple—just play the game a number of times and see what proportion of outcomes that show blue on top have blue underneath.

To that end, Table 2.8.1 summarizes the results of one hundred random drawings. For a total of fifty-two, blue was showing (S) when the chip was placed on a table; for thirty-six of the trials (those marked with an asterisk), the color underneath (U) was also blue. Using the approximation suggested by Equation 2.8.1,

$$P(\text{Blue is underneath} \mid \text{Blue is on top}) = P(B \mid B) \doteq \frac{36}{52} = 0.69$$

a figure much more consistent with $\frac{2}{3}$ than with $\frac{1}{2}$.

The point of this example is not to downgrade the importance of rigorous derivations and exact answers. Far from it. The application of Theorem 2.4.1 to solve the problem posed in Example 2.4.12 is obviously superior to the Monte Carlo approximation illustrated in Table 2.8.1. Still, replications of experiments can often provide valuable insights and call attention to nuances that might otherwise go unnoticed. As a problem-solving technique in probability and combinatorics, they are extremely important.

Chapter 3

RANDOM VARIABLES

One of a Swiss family producing eight distinguished scientists, Jakob was forced by his father to pursue theological studies, but his love of mathematics eventually led him to a university career. He and his brother, Johann, were the most prominent champions of Leibniz's calculus on continental Europe, the two using the new theory to solve numerous problems in physics and mathematics. Bernoulli's main work in probability, Ars Conjectandi, *was published after his death by his nephew, Nikolaus, in 1713.*

—Jakob (Jacques) Bernoulli (1654–1705)

3.1 Introduction

Throughout Chapter 2, probabilities were assigned to *events*—that is, to sets of sample outcomes. The events we dealt with were composed of either a finite or a countably infinite number of sample outcomes, in which case the event's probability was simply the sum of the probabilities assigned to its outcomes. One particular probability function that came up over and over again in Chapter 2 was the assignment of $\frac{1}{n}$ as the probability associated with each of the n points in a finite sample space. This is the model that typically describes games of chance (and all of our combinatorial probability problems in Chapter 2).

The first objective of this chapter is to look at several other useful ways for assigning probabilities to sample outcomes. In so doing, we confront the desirability of "redefining" sample spaces using functions known as *random variables*. How and why these are used—and what their mathematical properties are—become the focus of virtually everything covered in Chapter 3.

As a case in point, suppose a medical researcher is testing eight elderly adults for their allergic reaction (yes or no) to a new drug for controlling blood pressure. One of the $2^8 = 256$ possible sample points would be the sequence (yes, no, no, yes, no, no, yes, no), signifying that the first subject had an allergic reaction, the second did not, the third did not, and so on. Typically, in studies of this sort, the particular subjects experiencing reactions is of little interest: what does matter is the *number* who show a reaction. If that were true here, the outcome's relevant information (i.e., the number of allergic reactions) could be summarized by the number *3*.[1]

Suppose X denotes the number of allergic reactions among a set of eight adults. Then X is said to be a *random variable* and the number 3 is the *value* of the random variable for the outcome (yes, no, no, yes, no, no, yes, no).

In general, random variables are functions that associate numbers with some attribute of a sample outcome that is deemed to be especially important. If X denotes the random variable and s denotes a sample outcome, then $X(s) = t$, where t is a real number. For the allergy example, $s =$ (yes, no, no, yes, no, no, yes, no) and $t = 3$.

Random variables can often create a dramatically simpler sample space. That certainly is the case here—the original sample space has *256* $(= 2^8)$ outcomes, each being an ordered sequence of length eight. The random variable X, on the other hand, has only *nine* possible values, the integers from 0 to 8, inclusive.

In terms of their fundamental structure, all random variables fall into one of two broad categories, the distinction resting on the number of possible values the random variable can equal. If the latter is finite or countably infinite (which would be the case with the allergic reaction example), the random variable is said to be *discrete*; if the outcomes can be any real number in a given interval, the number of possibilities is uncountably infinite, and the random variable is said to be *continuous*. The difference between the two is critically important, as we will learn in the next several sections.

The purpose of Chapter 3 is to introduce the important definitions, concepts, and computational techniques associated with random variables, both discrete and continuous. Taken together, these ideas form the bedrock of modern probability and statistics.

3.2 Binomial and Hypergeometric Probabilities

This section looks at two specific probability scenarios that are especially important, both for their theoretical implications as well as for their ability to describe real-world problems. What we learn in developing these two models will help us understand random variables in general, the formal discussion of which begins in Section 3.3.

The Binomial Probability Distribution

Binomial probabilities apply to situations involving a series of independent and identical trials, where each trial can have only one of two possible outcomes. Imagine three distinguishable coins being tossed, each having a probability p of coming up heads. The set of possible outcomes are the eight listed in Table 3.2.1. If the probability of any of the coins coming up heads is p, then the probability of the *sequence* (H, H, H) is p^3, since the coin tosses qualify as independent trials. Similarly, the

[1] By Theorem 2.6.2, of course, there would be a total of *fifty-six* $(= 8!/3!5!)$ outcomes having exactly three yeses. All fifty-six would be equivalent in terms of what they imply about the drug's likelihood of causing allergic reactions.

probability of (T, H, H) is $(1-p)p^2$. The fourth column of Table 3.2.1 shows the probabilities associated with each of the three-coin sequences.

Table 3.2.1

1st Coin	2nd Coin	3rd Coin	Probability	Number of Heads
H	H	H	p^3	3
H	H	T	$p^2(1-p)$	2
H	T	H	$p^2(1-p)$	2
T	H	H	$p^2(1-p)$	2
H	T	T	$p(1-p)^2$	1
T	H	T	$p(1-p)^2$	1
T	T	H	$p(1-p)^2$	1
T	T	T	$(1-p)^3$	0

Suppose our main interest in the coin tosses is the *number* of heads that occur. Whether the actual sequence is, say, (H, H, T) or (H, T, H) is immaterial, since each outcome contains exactly two heads. The last column of Table 3.2.1 shows the number of heads in each of the eight possible outcomes. Notice that there are *three* outcomes with exactly two heads, each having an individual probability of $p^2(1-p)$. The probability, then, of the event "two heads" is the sum of those three individual probabilities—that is, $3p^2(1-p)$. Table 3.2.2 lists the probabilities of tossing k heads, where $k = 0, 1, 2$, or 3.

Table 3.2.2

Number of Heads	Probability
0	$(1-p)^3$
1	$3p(1-p)^2$
2	$3p^2(1-p)$
3	p^3

Now, more generally, suppose that n coins are tossed, in which case the number of heads can equal any integer from 0 through n. By analogy,

$$P(k \text{ heads}) = \begin{pmatrix} \text{number of} \\ \text{ways to arrange } k \\ \text{heads and } n-k \text{ tails} \end{pmatrix} \cdot \begin{pmatrix} \text{probability of} \\ \text{any particular sequence} \\ \text{having } k \text{ heads} \\ \text{and } n-k \text{ tails} \end{pmatrix}$$

$$= \begin{pmatrix} \text{number of ways} \\ \text{to arrange } k \\ \text{heads and } n-k \text{ tails} \end{pmatrix} \cdot p^k(1-p)^{n-k}$$

The number of ways to arrange k H's and $n-k$ T's, though, is $\frac{n!}{k!(n-k)!}$, or $\binom{n}{k}$ (recall Theorem 2.6.2).

Theorem 3.2.1 *Consider a series of n independent trials, each resulting in one of two possible outcomes, "success" or "failure." Let $p = P$ (success occurs at any given trial) and assume that p remains constant from trial to trial. Then*

$$P(k \text{ successes}) = \binom{n}{k} p^k(1-p)^{n-k}, \; k = 0, 1, \ldots, n$$

Comment The probability assignment given by the equation in Theorem 3.2.1 is known as the *binomial distribution*.

Example 3.2.1

An information technology center uses nine aging disk drives for storage. The probability that any one of them is out of service is 0.06. For the center to function properly, at least seven of the drives must be available. What is the probability that the computing center can get its work done?

The probability that a drive is available is $p = 1 - 0.06 = 0.94$. Assuming the devices operate independently, the number of disk drives available has a binomial distribution with $n = 9$ and $p = 0.94$. The probability that at least seven disk drives work is a reassuring 0.986:

$$\binom{9}{7}(0.94)^7(0.06)^2 + \binom{9}{8}(0.94)^8(0.06)^1 + \binom{9}{7}(0.94)^9(0.06)^0 = 0.986$$

■

Example 3.2.2

Kingwest Pharmaceuticals is experimenting with a new affordable AIDS medication, PM-17, that may have the ability to strengthen a victim's immune system. Thirty monkeys infected with the HIV complex have been given the drug. Researchers intend to wait six weeks and then count the number of animals whose immunological responses show a marked improvement. Any inexpensive drug capable of being effective 60% of the time would be considered a major breakthrough; medications whose chances of success are 50% or less are not likely to have any commercial potential.

Yet to be finalized are guidelines for interpreting results. Kingwest hopes to avoid making either of two errors: (1) rejecting a drug that would ultimately prove to be marketable and (2) spending additional development dollars on a drug whose effectiveness, in the long run, would be 50% or less. As a tentative "decision rule," the project manager suggests that unless *sixteen or more* of the monkeys show improvement, research on PM-17 should be discontinued.

a. What are the chances that the "sixteen or more" rule will cause the company to reject PM-17, *even if the drug is 60% effective*?

b. How often will the "sixteen or more" rule allow a 50%-effective drug to be perceived as a major breakthrough?

(a) Each of the monkeys is one of $n = 30$ independent trials, where the outcome is either a "success" (Monkey's immune system is strengthened) or a "failure" (Monkey's immune system is not strengthened). By assumption, the probability that PM-17 produces an immunological improvement in any given monkey is $p = P$ (success) $= 0.60$.

By Theorem 3.2.1, the probability that exactly k monkeys (out of thirty) will show improvement after six weeks is $\binom{30}{k}(0.60)^k(0.40)^{30-k}$. The probability, then, that the "sixteen or more" rule will cause a 60%-effective drug to be discarded is the sum of "binomial" probabilities for k values ranging from 0 to 15:

$$P(60\%\text{-effective drug fails "sixteen or more" rule}) = \sum_{k=0}^{15}\binom{30}{k}(0.60)^k(0.40)^{30-k}$$

$$= 0.1754$$

Roughly 18% of the time, in other words, a "breakthrough" drug such as PM-17 will produce test results so mediocre (as measured by the "sixteen or more" rule) that the company will be misled into thinking it has no potential.

(b) The other error Kingwest can make is to conclude that PM-17 warrants further study when, in fact, its value for p is below a marketable level. The chance that particular incorrect inference will be drawn here is the probability that the number of successes will be greater than or equal to sixteen when $p = 0.5$. That is,

$$\begin{aligned}
P(50\%\text{-effective PM-17 appears to be marketable}) \\
&= P(\text{Sixteen or more successes occur}) \\
&= \sum_{k=16}^{30} \binom{30}{k} (0.5)^k (0.5)^{30-k} \\
&= 0.43
\end{aligned}$$

Thus, even if PM-17's success rate is an unacceptably low 50%, it has a 43% chance of performing sufficiently well in thirty trials to satisfy the "sixteen or more" criterion.

Comment Evaluating binomial summations can be tedious, even with a calculator. Statistical software packages offer a convenient alternative. Appendix 3.A.1 describes how one such program, Minitab, can be used to answer the sorts of questions posed in Example 3.2.2. ■

Example 3.2.3

The Stanley Cup playoff in professional hockey is a seven-game series, where the first team to win four games is declared the champion. The series, then, can last anywhere from four to seven games (just like the World Series in baseball). Calculate the likelihoods that the series will last four, five, six, or seven games. Assume that (1) each game is an independent event and (2) the two teams are evenly matched.

Consider the case where Team A wins the series in *six* games. For that to happen, they must win exactly three of the first five games *and* they must win the sixth game. Because of the independence assumption, we can write

$$\begin{aligned}
P(\text{Team A wins in six games}) &= P(\text{Team A wins three of first five}) \cdot \\
&\quad P(\text{Team A wins sixth}) \\
&= \left[\binom{5}{3}(0.5)^3(0.5)^2\right] \cdot (0.5) = 0.15625
\end{aligned}$$

Since the probability that Team B wins the series in six games is the same (why?),

$$\begin{aligned}
P(\text{Series ends in six games}) &= P(\text{Team A wins in six games} \cup \\
&\quad \text{Team B wins in six games}) \\
&= P(\text{A wins in six}) + P(\text{B wins in six}) \qquad \text{(why?)} \\
&= 0.15625 + 0.15625 \\
&= 0.3125
\end{aligned}$$

A similar argument allows us to calculate the probabilties of four-, five-, and seven-game series:

$$P(\text{four-game series}) = 2(0.5)^4 = 0.125$$

$$P(\text{five-game series}) = 2\left[\binom{4}{3}(0.5)^3(0.5)\right](0.5) = 0.25$$

$$P(\text{seven-game series}) = 2\left[\binom{6}{3}(0.5)^3(0.5)^3\right](0.5) = 0.3125$$

Having calculated the "theoretical" probabilities associated with the possible lengths of a Stanley Cup playoff raises an obvious question: How do those likelihoods compare with the actual distribution of playoff lengths? Between 1947 and 2006 there were sixty playoffs (the 2004–05 season was cancelled). Column 2 in Table 3.2.3 shows the proportion of playoffs that have lasted four, five, six, and seven games, respectively.

Table 3.2.3

Series Length	Observed Proportion	Theoretical Probability
4	$17/60 = 0.283$	0.125
5	$15/60 = 0.250$	0.250
6	$16/60 = 0.267$	0.3125
7	$12/60 = 0.200$	0.3125

Source: statshockey.homestead.com/stanleycup.html

Clearly, the agreement between the entries in Columns 2 and 3 is not very good: Particularly noticeable is the excess of short playoffs (four games) and the deficit of long playoffs (seven games). What this "lack of fit" suggests is that one or more of the binomial distribution assumptions is not satisfied. Consider, for example, the parameter p, which we assumed to equal $\frac{1}{2}$. In reality, its value might be something quite different—just because the teams playing for the championship won their respective divisions, it does not necessarily follow that the two are equally good. Indeed, if the two contending teams were frequently mismatched, the consequence would be an increase in the number of short playoffs and a decrease in the number of long playoffs. It may also be the case that momentum is a factor in a team's chances of winning a given game. If so, the independence assumption implicit in the binomial model is rendered invalid. ■

Example 3.2.4

The junior mathematics class at Superior High School knows that the probability of making a 600 or greater on the SAT Reasoning Test in Mathematics is 0.231, while the similar probability for the Critical Reading Test is 0.191. The math students issue a challenge to their math-averse classmates. Each group will select four students and have them take the respective test. The mathematics students will win the challenge if more of their members exceed 600 on the mathematics test than do the other students on the Critical Reading Test. What is the probability that the mathematics students win the challenge?

Let M denote the number of mathematics scores of 600 or more and CR denote the similar number for the critical reading testees. In this notation, a typical combination in which the mathematics class wins is $CR = 2$, $M = 3$. The probability of this combination is

$$P(CR = 2, M = 3) = P(CR = 2)P(M = 3)$$

because events involving CR and M are independent. But

$$P(CR = 2) \cdot P(M = 3) = \left[\binom{4}{2}(0.191)^2(0.809)^2\right] \cdot \left[\binom{4}{3}(0.231)^3(0.769)^1\right]$$
$$= (0.143)(0.038) = 0.0054$$

Table 3.2.4 below lists all of these *joint probabilities* to four decimal places for the various values of CR and M. The shaded cells are those where mathematics wins the challenge.

Table 3.2.4

CR \ M	0	1	2	3	4
0	0.1498	0.1800	0.0811	0.0162	0.0012
1	0.1415	0.1700	0.0766	0.0153	0.0012
2	0.0501	0.0602	0.0271	0.0054	0.0004
3	0.0079	0.0095	0.0043	0.0009	0.0001
4	0.0005	0.0006	0.0003	0.0001	0.0000

The sum of the probabilities in the cells is 0.3775.

The moral of the story is that the mathematics students need to study more probability. ■

Questions

3.2.1. An investment analyst has tracked a certain blue-chip stock for the past six months and found that on any given day, it either goes up a point or goes down a point. Furthermore, it went up on 25% of the days and down on 75%. What is the probability that at the close of trading four days from now, the price of the stock will be the same as it is today? Assume that the daily fluctuations are independent events.

3.2.2. In a nuclear reactor, the fission process is controlled by inserting special rods into the radioactive core to absorb neutrons and slow down the nuclear chain reaction. When functioning properly, these rods serve as a first-line defense against a core meltdown. Suppose a reactor has ten control rods, each operating independently and each having an 0.80 probability of being properly inserted in the event of an "incident." Furthermore, suppose that a meltdown will be prevented if at least half the rods perform satisfactorily. What is the probability that, upon demand, the system will fail?

3.2.3. In 2009 a donor who insisted on anonymity gave seven-figure donations to twelve universities. A media report of this generous but somewhat mysterious act identified that all of the universities awarded had female presidents. It went on to say that with about 23% of U.S. college presidents being women, the probability of a dozen randomly selected institutions having female presidents is about 1/50,000,000. Is this probability approximately correct?

3.2.4. An entrepreneur owns six corporations, each with more than \$10 million in assets. The entrepreneur consults the *U.S. Internal Revenue Data Book* and discovers that the IRS audits 15.3% of businesses of that size. What is the probability that two or more of these businesses will be audited?

3.2.5. The probability is 0.10 that ball bearings in a machine component will fail under certain adverse conditions of load and temperature. If a component containing eleven ball bearings must have a least eight of them functioning to operate under the adverse conditions, what is the probability that it will break down?

3.2.6. Suppose that since the early 1950s some ten-thousand independent UFO sightings have been reported to civil authorities. If the probability that any sighting is genuine is on the order of one in one hundred thousand, what is the probability that at least one of the ten-thousand was genuine?

3.2.7. Doomsday Airlines ("Come Take the Flight of Your Life") has two dilapidated airplanes, one with two engines, and the other with four. Each plane will land safely only if at least half of its engines are working. Each engine on each aircraft operates independently and each has probability $p = 0.4$ of failing. Assuming you wish to maximize your survival probability, which plane should you fly on?

3.2.8. Two lighting systems are being proposed for an employee work area. One requires fifty bulbs, each having a probability of 0.05 of burning out within a month's time. The second has one hundred bulbs, each with a 0.02 burnout probability. Whichever system is installed will be inspected once a month for the purpose of replacing burned-out bulbs. Which system is likely to require less maintenance? Answer the question by comparing the probabilities that each will require at least one bulb to be replaced at the end of thirty days.

3.2.9. The great English diarist Samuel Pepys asked his friend Sir Isaac Newton the following question: Is it more likely to get at least one 6 when six dice are rolled, at least two 6's when twelve dice are rolled, or at least three 6's when eighteen dice are rolled? After considerable correspondence [see (158)]. Newton convinced the skeptical Pepys that the first event is the most likely. Compute the three probabilities.

3.2.10. The gunner on a small assault boat fires six missiles at an attacking plane. Each has a 20% chance of being on-target. If two or more of the shells find their mark, the plane will crash. At the same time, the pilot of the plane fires ten air-to-surface rockets, each of which has a 0.05 chance of critically disabling the boat. Would you rather be on the plane or the boat?

3.2.11. If a family has four children, is it more likely they will have two boys and two girls or three of one sex and one of the other? Assume that the probability of a child being a boy is $\frac{1}{2}$ and that the births are independent events.

3.2.12. Experience has shown that only $\frac{1}{3}$ of all patients having a certain disease will recover if given the standard treatment. A new drug is to be tested on a group of twelve volunteers. If the FDA requires that at least seven of these patients recover before it will license the new drug, what is the probability that the treatment will be discredited even if it has the potential to increase an individual's recovery rate to $\frac{1}{2}$?

3.2.13. Transportation to school for a rural county's seventy-six children is provided by a fleet of four buses. Drivers are chosen on a day-to-day basis and come from a pool of local farmers who have agreed to be "on call." What is the smallest number of drivers who need to be in the pool if the county wants to have at least a 95% probability on any given day that all the buses will run? Assume that each driver has an 80% chance of being available if contacted.

3.2.14. The captain of a Navy gunboat orders a volley of twenty-five missiles to be fired at random along a five-hundred-foot stretch of shoreline that he hopes to establish as a beachhead. Dug into the beach is a thirty-foot-long bunker serving as the enemy's first line of defense. The captain has reason to believe that the bunker will be destroyed if at least three of the missiles are on-target. What is the probability of that happening?

3.2.15. A computer has generated seven random numbers over the interval 0 to 1. Is it more likely that (a) exactly three will be in the interval $\frac{1}{2}$ to 1 or (b) fewer than three will be greater than $\frac{3}{4}$?

3.2.16. Listed in the following table is the length distribution of World Series competition for the 58 series from 1950 to 2008 (there was no series in 1994).

World Series Lengths

Number of Games, X	Number of Years
4	12
5	10
6	12
7	24
	58

Source: espn.go.com/mlb/worldseries/history/winners

Assuming that each World Series game is an independent event and that the probability of either team's winning any particular contest is 0.5, find the probability of each series length. How well does the model fit the data? (Compute the "expected" frequencies, that is, multiply the probability of a given-length series times 58).

3.2.17. Use the expansion of $(x+y)^n$ (recall the comment in Section 2.6 on p. 67) to verify that the binomial probabilities sum to 1; that is, $\sum_{k=0}^{n} \binom{n}{k} p^k (1-p)^{n-k} = 1$.

3.2.18. Suppose a series of n independent trials can end in one of *three* possible outcomes. Let k_1 and k_2 denote the number of trials that result in outcomes 1 and 2, respectively. Let p_1 and p_2 denote the probabilities associated with outcomes 1 and 2. Generalize Theorem 3.2.1 to deduce a formula for the probability of getting k_1 and k_2 occurrences of outcomes 1 and 2, respectively.

3.2.19. Repair calls for central air conditioners fall into three general categories: coolant leakage, compressor failure, and electrical malfunction. Experience has shown that the probabilities associated with the three are 0.5, 0.3, and 0.2, respectively. Suppose that a dispatcher has logged in ten service requests for tomorrow morning. Use the answer to Question 3.2.18 to calculate the probability that three of those ten will involve coolant leakage and five will be compressor failures.

The Hypergeometric Distribution

The second "special" distribution that we want to look at formalizes the urn problems that frequented Chapter 2. Our solutions to those earlier problems tended to be enumerations in which we listed the entire set of possible samples, and then counted the ones that satisfied the event in question. The inefficiency and redundancy of that approach should now be painfully obvious. What we are seeking here is a general formula that can be applied to any and all such problems, much like the expression in Theorem 3.2.1 can handle the full range of questions arising from the binomial model.

Suppose an urn contains r red chips and w white chips, where $r+w=N$. Imagine drawing n chips from the urn one at a time without replacing any of the chips selected. At each drawing we record the color of the chip removed. The question is, what is the probability that exactly k red chips are included among the n that are removed?

Notice that the experiment just described is similar in some respects to the binomial model, but the method of sampling creates a critical distinction. *If* each chip drawn was replaced prior to making another selection, then each drawing would be an independent trial, the chances of drawing a red in any given trial would be a constant r/N, and the probability that exactly k red chips would ultimately be included in the n selections would be a direct application of Theorem 3.2.1:

$$P(k \text{ reds drawn}) = \binom{n}{k} (r/N)^k (1-r/N)^{n-k}, \quad k = 0, 1, 2, \ldots, n$$

However, if the chips drawn are *not* replaced, then the probability of drawing a red on any given attempt is not necessarily r/N: Its value would depend on the colors of the chips selected earlier. Since $p = P(\text{Red is drawn}) = P(\text{success})$ does not remain constant from drawing to drawing, the binomial model of Theorem 3.2.1 does not apply. Instead, probabilities that arise from the "no replacement" scenario just described are said to follow the *hypergeometric distribution*.

Theorem 3.2.2 *Suppose an urn contains r red chips and w white chips, where $r+w=N$. If n chips are drawn out at random, without replacement, and if k denotes the number of red chips selected, then*

$$P(k \text{ red chips are chosen}) = \frac{\binom{r}{k}\binom{w}{n-k}}{\binom{N}{n}} \qquad (3.2.1)$$

where k varies over all the integers for which $\binom{r}{k}$ *and* $\binom{w}{n-k}$ *are defined. The probabilities appearing on the right-hand side of Equation 3.2.1 are known as the* hypergeometric distribution.

Proof Assume the chips are distinguishable. We need to count the number of elements making up the event of getting k red chips and $n-k$ white chips. The number of ways to select the red chips, regardless of the order in which they are chosen, is ${}_rP_k$. Similarly, the number of ways to select the $n-k$ white chips is ${}_wP_{n-k}$. However, the order in which the white chips are selected does matter. Each outcome is an n-long ordered sequence of red and white. There are $\binom{n}{k}$ ways to choose where in the sequence the red chips go. Thus, the number of elements in the event of interest is $\binom{n}{k}{}_rP_k\,{}_wP_{n-k}$. Now, the total number of ways to choose n elements from N, in order, without replacement is ${}_NP_n$, so

$$P(k \text{ red chips are chosen}) = \frac{\binom{n}{k}{}_rP_k\,{}_wP_{n-k}}{{}_NP_n}$$

This quantity, while correct, is not in the form of the statement of the theorem. To make that conversion, we have to change all of the terms in the expression to factorials:

$$\begin{aligned} P(k \text{ red chips are chosen}) &= \frac{\binom{n}{k}{}_rP_k\,{}_wP_{n-k}}{{}_NP_n} \\ &= \frac{\dfrac{n!}{k!(n-k)!}\,\dfrac{r!}{(r-k)!}\,\dfrac{w!}{(w-n+k)!}}{\dfrac{N!}{(N-n)!}} \\ &= \frac{\dfrac{r!}{k!(r-k)!}\,\dfrac{w!}{(n-k)!(w-n+k)!}}{\dfrac{N!}{n!(N-n)!}} = \frac{\binom{r}{k}\binom{w}{n-k}}{\binom{N}{n}} \end{aligned}$$

□

Comment The appearance of binomial coefficients suggests a model of selecting unordered subsets. Indeed, one can consider the model of selecting a subset of size n simultaneously, where order doesn't matter. In that case, the question remains: What is the probability of getting k red chips and $n-k$ white chips? A moment's reflection will show that the hypergeometric probabilities given in the statement of the theorem also answer that question. So, if our interest is simply counting the number of red and white chips in the sample, the probabilities are the same whether the drawing of the sample is simultaneous or the chips are drawn in order without repetition.

Comment The name *hypergeometric* derives from a series introduced by the Swiss mathematician and physicist Leonhard Euler, in 1769:

$$1 + \frac{ab}{c}x + \frac{a(a+1)b(b+1)}{2!c(c+1)}x^2 + \frac{a(a+1)(a+2)b(b+1)(b+2)}{3!c(c+1)(c+2)}x^3 + \cdots$$

This is an expansion of considerable flexibility: Given appropriate values for a, b, and c, it reduces to many of the standard infinite series used in analysis. In particular, if a is set equal to 1, and b and c are set equal to each other, it reduces to the familiar *geometric* series,

$$1 + x + x^2 + x^3 + \cdots$$

hence the name *hypergeometric*. The relationship of the probability function in Theorem 3.2.2 to Euler's series becomes apparent if we set $a=-n, b=-r, c=w-n+1$, and multiply the series by $\binom{w}{n}/\binom{N}{n}$. Then the coefficient of x^k will be

$$\frac{\binom{r}{k}\binom{w}{n-k}}{\binom{N}{n}}$$

the value the theorem gives for $P(k$ red chips are chosen).

Example 3.2.5

A hung jury is one that is unable to reach a unanimous decision. Suppose that a pool of twenty-five potential jurors is assigned to a murder case where the evidence is so overwhelming against the defendant that twenty-three of the twenty-five would return a guilty verdict. The other two potential jurors would vote to acquit regardless of the facts. What is the probability that a twelve-member panel chosen at random from the pool of twenty-five will be unable to reach a unanimous decision?

Think of the jury pool as an urn containing twenty-five chips, twenty-three of which correspond to jurors who would vote "guilty" and two of which correspond to jurors who would vote "not guilty." If either or both of the jurors who would vote "not guilty" are included in the panel of twelve, the result would be a hung jury. Applying Theorem 3.2.2 (twice) gives *0.74* as the probability that the jury impanelled would not reach a unanimous decision:

$$\begin{aligned} P(\text{Hung jury}) &= P(\text{Decision is not unanimous}) \\ &= \binom{2}{1}\binom{23}{11}\bigg/\binom{25}{12} + \binom{2}{2}\binom{23}{10}\bigg/\binom{25}{12} \\ &= 0.74 \end{aligned}$$

■

Example 3.2.6

The Florida Lottery features a number of games of chance, one of which is called Fantasy Five. The player chooses five numbers from a card containing the numbers 1 through 36. Each day five numbers are chosen at random, and if the player matches all five, the winnings can be as much as $200,000 for a $1 bet.

Lottery games like this one have spawned a mini-industry looking for biases in the selection of the winning numbers. Websites post various "analyses" claiming certain numbers are "hot" and should be played. One such examination focused on the frequency of winning numbers between 1 and 12. The probability of such occurrences fits the hypergeometric distribution, where $r=12$, $w=24$, $n=5$, and $N=36$. For example, the probability that *three* of the five numbers are 12 or less is

$$\frac{\binom{12}{3}\binom{24}{2}}{\binom{36}{5}} = \frac{60{,}720}{376{,}992} = 0.161$$

Notice how that compares to the *observed* proportion of drawings with exactly three numbers between 1 and 12. Of the 2008 daily drawings—366 of them—there were sixty-five with three numbers 12 or less, giving a relative frequency of $65/366 = 0.178$.

The full breakdown of observed and expected probabilities for winning numbers between 1 and 12 is given in Table 3.2.5.

The naive or dishonest commentator might claim that the lottery "likes" numbers ≤ 12 since the proportion of tickets drawn with three, four, or five numbers $\leq$ 12 is

$$0.178 + 0.038 + 0.005 = 0.221$$

Table 3.2.5

No. Drawn ≤ 12	Observed Proportion	Hypergeometric Probability
0	0.128	0.113
1	0.372	0.338
2	0.279	0.354
3	0.178	0.161
4	0.038	0.032
5	0.005	0.002

Source: www.flalottery.com/exptkt/ff.html

This figure is in excess of the sum of the hypergeometric probabilities for $k = 3$, 4, and 5:

$$0.161 + 0.032 + 0.002 = 0.195$$

However, we shall see in Chapter 10 that such variation is well within the random fluctuations expected for truly random drawings. No bias can be inferred from these results. ■

Example 3.2.7

When a bullet is fired it becomes scored with minute striations produced by imperfections in the gun barrel. Appearing as a series of parallel lines, these striations have long been recognized as a basis for matching a bullet with a gun, since repeated firings of the same weapon will produce bullets having substantially the same configuration of markings. Until recently, deciding how close two patterns had to be before it could be concluded the bullets came from the same weapon was largely subjective. A ballistics expert would simply look at the two bullets under a microscope and make an informed judgment based on past experience. Today, however, criminologists are beginning to address the problem more quantitatively, partly with the help of the hypergeometric distribution.

Suppose a bullet is recovered from the scene of a crime, along with the suspect's gun. Under a microscope, a grid of m cells, numbered 1 to m, is superimposed over the bullet. If m is chosen large enough that the width of the cells is sufficiently small, each of that evidence bullet's n_e striations will fall into a different cell (see Figure 3.2.1a). Then the suspect's gun is fired, yielding a test bullet, which will have a total of n_t striations located in a possibly different set of cells (see Figure 3.2.1b). How might we assess the similarities in cell locations for the two striation patterns?

As a model for the striation pattern on the evidence bullet, imagine an urn containing m chips, with n_e corresponding to the striation locations. Now, think of the striation pattern on the *test* bullet as representing a sample of size n_t from the evidence urn. By Theorem 3.2.2, the probability that k of the cell locations will be shared by the two striation patterns is

$$\frac{\binom{n_e}{k}\binom{m-n_e}{n_t-k}}{\binom{m}{n_t}}$$

Suppose the bullet found at a murder scene is superimposed with a grid having $m = 25$ cells, n_e of which contain striations. The suspect's gun is fired and the bullet is found to have $n_t = 3$ striations, one of which matches the location of one of the striations on the evidence bullet. What do you think a ballistics expert would conclude?

Striations (total of n_e)

Evidence bullet

1 2 3 4 5 . . . m

(a)

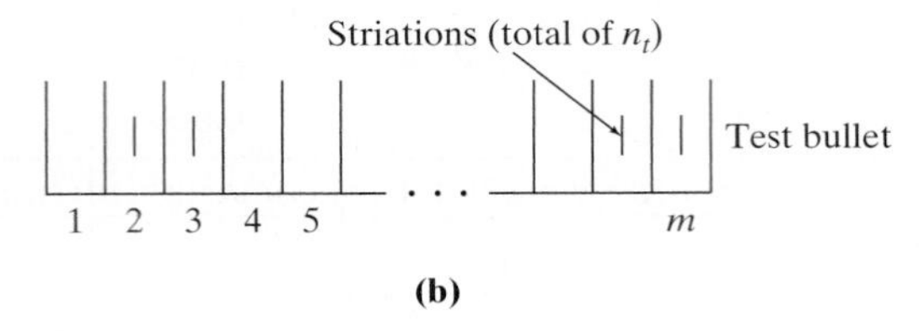

(b)

Figure 3.2.1

Intuitively, the similarity between the two bullets would be reflected in the probability that *one or more* striations in the suspect's bullet match the evidence bullet. The smaller that probability is, the stronger would be our belief that the two bullets were fired by the same gun. Based on the values given for m, n_e, and n_t,

$$\begin{aligned} P(\text{one or more matches}) &= \frac{\binom{4}{1}\binom{21}{2}}{\binom{25}{3}} + \frac{\binom{4}{2}\binom{21}{1}}{\binom{25}{3}} + \frac{\binom{4}{3}\binom{21}{0}}{\binom{25}{3}} \\ &= 0.42 \end{aligned}$$

If P(one or more matches) had been a very small number—say, 0.001—the inference would have been clear-cut: The same gun fired both bullets. But, here with the probability of one or more matches being so large, we cannot rule out the possibility that the bullets were fired by two different guns (and, presumably, by two different people). ■

Example 3.2.8

A tax collector, finding himself short of funds, delayed depositing a large property tax payment ten different times. The money was subsequently repaid, and the whole amount deposited in the proper account. The tip-off to this behavior was the delay of the deposit. During the period of these irregularities, there was a total of 470 tax collections.

An auditing firm was preparing to do a routine annual audit of these transactions. They decided to randomly sample nineteen of the collections (approximately 4%) of the payments. The auditors would assume a pattern of malfeasance only if they saw three or more irregularities. What is the probability that three or more of the delayed deposits would be chosen in this sample?

This kind of audit sampling can be considered a hypergeometric experiment. Here, $N = 470$, $n = 19$, $r = 10$, and $w = 460$. In this case it is better to calculate the desired probability via the complement—that is,

$$1 - \frac{\binom{10}{0}\binom{460}{19}}{\binom{470}{19}} - \frac{\binom{10}{1}\binom{460}{18}}{\binom{470}{19}} - \frac{\binom{10}{2}\binom{460}{17}}{\binom{470}{19}}$$

The calculation of the first hypergeometric term is

$$\frac{\binom{10}{0}\binom{460}{19}}{\binom{470}{19}} = 1 \cdot \frac{460!}{19!441!} \cdot \frac{19!451}{470!} = \frac{451}{470} \cdot \frac{450}{469} \cdot \cdots \cdot \frac{442}{461} = 0.6592$$

To compute hypergeometric probabilities where the numbers are large, a useful device is a *recursion formula*. To that end, note that the ratio of the $k+1$ term to the k term is

$$\frac{\binom{r}{k+1}\binom{w}{n-k-1}}{\binom{N}{n}} \div \frac{\binom{r}{k}\binom{w}{n-k}}{\binom{N}{n}} = \frac{n-k}{k+1} \cdot \frac{r-k}{w-n+k+1}$$

(See Question 3.2.30.)

Therefore,

$$\frac{\binom{10}{1}\binom{460}{18}}{\binom{470}{19}} = 0.6592 \cdot \frac{19+0}{1+0} \cdot \frac{10-0}{460-19+0+1} = 0.2834$$

and

$$\frac{\binom{10}{2}\binom{460}{17}}{\binom{470}{19}} = 0.2834 \cdot \frac{19-1}{1+1} \cdot \frac{10-1}{460-19+1+1} = 0.0518$$

The desired probability, then, is $1 - 0.6592 - 0.2834 - 0.0518 = 0.0056$, which shows that a larger audit sample would be necessary to have a reasonable chance of detecting this sort of impropriety. ■

Case Study 3.2.1

Biting into a plump, juicy apple is one of the innocent pleasures of autumn. Critical to that enjoyment is the *firmness* of the apple, a property that growers and shippers monitor closely. The apple industry goes so far as to set a lowest acceptable limit for firmness, which is measured (in lbs) by inserting a probe into the apple. For the Red Delicious variety, for example, firmness is supposed to be at least 12 lbs; in the state of Washington, wholesalers are not allowed to sell apples if more than 10% of their shipment falls below that 12-lb limit.

All of this raises an obvious question: How can shippers demonstrate that their apples meet the 10% standard? Testing each one is not an option—the probe that measures firmness renders an apple unfit for sale. That leaves *sampling* as the only viable strategy.

Suppose, for example, a shipper has a supply of 144 apples. She decides to select 15 at random and measure each one's firmness, with the intention of selling the remaining apples if 2 or fewer in the sample are substandard. What are the consequences of her plan? More specifically, does it have a good chance of "accepting" a shipment that meets the 10% rule and "rejecting" one that does not? (If either or both of those objectives are not met, the plan is inappropriate.)

For example, suppose there are actually *10* defective apples among the original 144. Since $\frac{10}{144} \times 100 = 6.9\%$, that shipment would be suitable for sale because fewer than 10% failed to meet the firmness standard. The question is,

(Continued on next page)

(Case Study 3.2.1 continued)

how likely is it that a sample of 15 chosen at random from that shipment will pass inspection?

Notice, here, that the number of substandard apples in the sample has a hypergeometric distribution with $r = 10$, $w = 134$, $n = 15$, and $N = 144$. Therefore,

$$\begin{aligned} P(\text{Sample passes inspection}) &= P(2 \text{ or fewer substandard apples are found}) \\ &= \frac{\binom{10}{0}\binom{134}{15}}{\binom{144}{15}} + \frac{\binom{10}{1}\binom{134}{14}}{\binom{144}{15}} + \frac{\binom{10}{2}\binom{134}{13}}{\binom{144}{15}} \\ &= 0.320 + 0.401 + 0.208 = 0.929 \end{aligned}$$

So, the probability is reassuringly high that a supply of apples this good would, in fact, be judged acceptable to ship. Of course, it also follows from this calculation that roughly *7%* of the time, the number of substandard apples found will be *greater* than 2, in which case the apples would be (incorrectly) assumed to be unsuitable for sale (earning them an undeserved one-way ticket to the applesauce factory . . .).

How good is the proposed sampling plan at recognizing apples that would, in fact, be inappropriate to ship? Suppose, for example, that *30*, or *21%*, of the 144 apples would fall below the 12-lb limit. Ideally, the probability here that a sample passes inspection should be small. The number of substandard apples found in this case would be hypergeometric with $r = 30$, $w = 114$, $n = 15$, and $N = 144$, so

$$\begin{aligned} P(\text{Sample passes inspection}) &= \frac{\binom{30}{0}\binom{114}{15}}{\binom{144}{15}} + \frac{\binom{30}{1}\binom{114}{14}}{\binom{144}{15}} + \frac{\binom{30}{2}\binom{114}{13}}{\binom{144}{15}} \\ &= 0.024 + 0.110 + 0.221 = 0.355 \end{aligned}$$

Here the bad news is that the sampling plan will allow a 21% defective supply to be shipped *36%* of the time. The good news is that *64%* of the time, the number of substandard apples in the sample will exceed 2, meaning that the correct decision "not to ship" will be made.

Figure 3.2.2 shows P(Sample passes) plotted against the percentage of defectives in the entire supply. Graphs of this sort are called *operating characteristic* (or *OC*) *curves*: They summarize how a sampling plan will respond to all possible levels of quality.

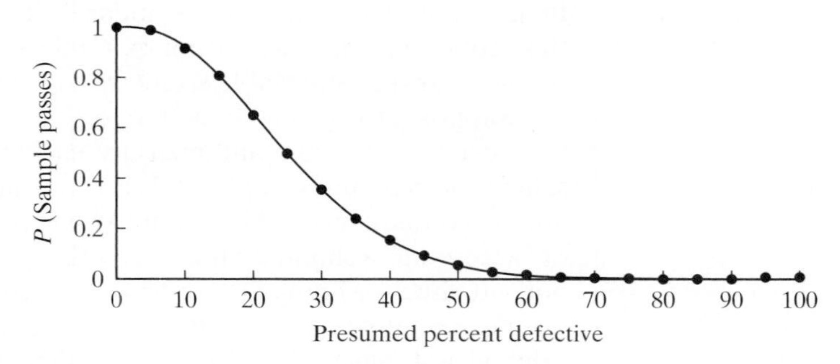

Figure 3.2.2

(Continued on next page)

> **Comment** Every sampling plan invariably allows for two kinds of errors—rejecting shipments that should be accepted and accepting shipments that should be rejected. In practice, the probabilities of committing these errors can be manipulated by redefining the decision rule and/or changing the sample size. Some of these options will be explored in Chapter 6.

Questions

3.2.20. A corporate board contains twelve members. The board decides to create a five-person Committee to Hide Corporation Debt. Suppose four members of the board are accountants. What is the probability that the Committee will contain two accountants and three nonaccountants?

3.2.21. One of the popular tourist attractions in Alaska is watching black bears catch salmon swimming upstream to spawn. Not all "black" bears are black, though—some are tan-colored. Suppose that six black bears and three tan-colored bears are working the rapids of a salmon stream. Over the course of an hour, six different bears are sighted. What is the probability that those six include at least twice as many black bears as tan-colored bears?

3.2.22. A city has 4050 children under the age of ten, including 514 who have not been vaccinated for measles. Sixty-five of the city's children are enrolled in the ABC Day Care Center. Suppose the municipal health department sends a doctor and a nurse to ABC to immunize any child who has not already been vaccinated. Find a formula for the probability that exactly k of the children at ABC have not been vaccinated.

3.2.23. Country A inadvertently launches ten guided missiles—six armed with nuclear warheads—at Country B. In response, Country B fires seven antiballistic missiles, each of which will destroy exactly one of the incoming rockets. The antiballistic missiles have no way of detecting, though, which of the ten rockets are carrying nuclear warheads. What are the chances that Country B will be hit by at least one nuclear missile?

3.2.24. Anne is studying for a history exam covering the French Revolution that will consist of five essay questions selected at random from a list of ten the professor has handed out to the class in advance. Not exactly a Napoleon buff, Anne would like to avoid researching all ten questions but still be reasonably assured of getting a fairly good grade. Specifically, she wants to have at least an 85% chance of getting at least four of the five questions right. Will it be sufficient if she studies eight of the ten questions?

3.2.25. Each year a college awards five merit-based scholarships to members of the entering freshman class who have exceptional high school records. The initial pool of applicants for the upcoming academic year has been reduced to a "short list" of eight men and ten women, all of whom seem equally deserving. If the awards are made at random from among the eighteen finalists, what are the chances that both men and women will be represented?

3.2.26. Keno is a casino game in which the player has a card with the numbers 1 through 80 on it. The player selects a set of k numbers from the card, where k can range from one to fifteen. The "caller" announces twenty winning numbers, chosen at random from the eighty. The amount won depends on how many of the called numbers match those the player chose. Suppose the player picks ten numbers. What is the probability that among those ten are six winning numbers?

3.2.27. A display case contains thirty-five gems, of which ten are real diamonds and twenty-five are fake diamonds. A burglar removes four gems at random, one at a time and without replacement. What is the probability that the last gem she steals is the second real diamond in the set of four?

3.2.28. A bleary-eyed student awakens one morning, late for an 8:00 class, and pulls two socks out of a drawer that contains two black, six brown, and two blue socks, all randomly arranged. What is the probability that the two he draws are a matched pair?

3.2.29. Show directly that the set of probabilities associated with the hypergeometric distribution sum to 1. (*Hint:* Expand the identity

$$(1+\mu)^N = (1+\mu)^r(1+\mu)^{N-r}$$

and equate coefficients.)

3.2.30. Show that the ratio of two successive hypergeometric probability terms satisfies the following equation,

$$\frac{\binom{r}{k+1}\binom{w}{n-k-1}}{\binom{N}{n}} \div \frac{\binom{r}{k}\binom{w}{n-k}}{\binom{N}{n}} = \frac{n-k}{k+1} \cdot \frac{r-k}{w-n+k+1}$$

for any k where both numerators are defined.

3.2.31. Urn I contains five red chips and four white chips; urn II contains four red and five white chips. Two chips are drawn simultaneously from urn I and placed into urn II. Then a single chip is drawn from urn II. What is the probability that the chip drawn from urn II is white? (*Hint:* Use Theorem 2.4.1.)

3.2.32. As the owner of a chain of sporting goods stores, you have just been offered a "deal" on a shipment of one hundred robot table tennis machines. The price is right, but the prospect of picking up the merchandise at midnight from an unmarked van parked on the side of the New Jersey Turnpike is a bit disconcerting. Being of low repute yourself, you do not consider the legality of the transaction to be an issue, but you do have concerns about being cheated. If too many of the machines are in poor working order, the offer ceases to be a bargain. Suppose you decide to close the deal only if a sample of ten machines contains no more than one defective. Construct the corresponding operating characteristic curve. For approximately what incoming quality will you accept a shipment 50% of the time?

3.2.33. Suppose that r of N chips are red. Divide the chips into three groups of sizes n_1, n_2, and n_3, where $n_1 + n_2 + n_3 = N$. Generalize the hypergeometric distribution to find the probability that the first group contains r_1 red chips, the second group r_2 red chips, and the third group r_3 red chips, where $r_1 + r_2 + r_3 = r$.

3.2.34. Some nomadic tribes, when faced with a life-threatening, contagious disease, try to improve their chances of survival by dispersing into smaller groups. Suppose a tribe of twenty-one people, of whom four are carriers of the disease, split into three groups of seven each. What is the probability that at least one group is free of the disease? (*Hint:* Find the probability of the complement.)

3.2.35. Suppose a population contains n_1 objects of one kind, n_2 objects of a second kind, . . . , and n_t objects of a tth kind, where $n_1 + n_2 + \cdots + n_t = N$. A sample of size n is drawn at random and without replacement. Deduce an expression for the probability of drawing k_1 objects of the first kind, k_2 objects of the second kind, . . . , and k_t objects of the tth kind by generalizing Theorem 3.2.2.

3.2.36. Sixteen students—five freshmen, four sophomores, four juniors, and three seniors—have applied for membership in their school's Communications Board, a group that oversees the college's newspaper, literary magazine, and radio show. Eight positions are open. If the selection is done at random, what is the probability that each class gets two representatives? (*Hint:* Use the generalized hypergeometric model asked for in Question 3.2.35.)

3.3 Discrete Random Variables

The binomial and hypergeometric distributions described in Section 3.2 are special cases of some important general concepts that we want to explore more fully in this section. Previously in Chapter 2, we studied in depth the situation where every point in a sample space is equally likely to occur (recall Section 2.6). The sample space of independent trials that ultimately led to the binomial distribution presented a quite different scenario: specifically, individual points in S had different probabilities. For example, if $n = 4$ and $p = \frac{1}{3}$, the probabilities assigned to the sample points (s, f, s, f) and (f, f, f, f) are $(1/3)^2(2/3)^2 = \frac{4}{81}$ and $(2/3)^4 = \frac{16}{81}$, respectively. Allowing for the possibility that different outcomes may have different probabilities will obviously broaden enormously the range of real-world problems that probability models can address.

How to assign probabilities to outcomes that are not binomial or hypergeometric is one of the major questions investigated in this chapter. A second critical issue is the nature of the sample space itself and whether it makes sense to redefine the outcomes and create, in effect, an alternative sample space. Why we would want to do that has already come up in our discussion of independent trials. The "original" sample space in such cases is a set of ordered sequences, where the ith member of a sequence is either an "s" or an "f," depending on whether the ith trial ended in success or failure, respectively. However, knowing which particular trials ended in success is typically less important than knowing the *number* that did (recall the medical researcher discussion on p. 102). That being the case, it often makes sense to replace each ordered sequence with the number of successes that sequence contains. Doing so collapses the original set of 2^n ordered sequences (i.e., outcomes) in S to the set of $n+1$ integers ranging from 0 to n. The probabilities assigned to those integers, of course, are given by the binomial formula in Theorem 3.2.1.

In general, a function that assigns numbers to outcomes is called a *random variable*. The purpose of such functions in practice is to define a new sample space whose outcomes speak more directly to the objectives of the experiment. That was the rationale that ultimately motivated both the binomial and hypergeometric distributions.

The purpose of this section is to (1) outline the general conditions under which probabilities can be assigned to sample spaces and (2) explore the ways and means of redefining sample spaces through the use of random variables. The notation introduced in this section is especially important and will be used throughout the remainder of the book.

Assigning Probabilities: The Discrete Case

We begin with the general problem of assigning probabilities to sample outcomes, the simplest version of which occurs when the number of points in S is either finite or countably infinite. The probability functions, $p(s)$, that we are looking for in those cases satisfy the conditions in Definition 3.3.1.

Definition 3.3.1. Suppose that S is a finite or countably infinite sample space. Let p be a real-valued function defined for each element of S such that

a. $0 \leq p(s)$ for each $s \in S$
b. $\sum_{s \in S} p(s) = 1$

Then p is said to be a *discrete probability function*.

Comment Once $p(s)$ is defined for all s, it follows that the probability of any event A—that is, $P(A)$—is the sum of the probabilities of the outcomes comprising A:

$$P(A) = \sum_{s \in A} p(s) \tag{3.3.1}$$

Defined in this way, the function $P(A)$ satisfies the probability axioms given in Section 2.3. The next several examples illustrate some of the specific forms that $p(s)$ can have and how $P(A)$ is calculated.

Example 3.3.1

Ace-six flats are a type of crooked dice where the cube is foreshortened in the one-six direction, the effect being that 1's and 6's are more likely to occur than any of the other four faces. Let $p(s)$ denote the probability that the face showing is s. For many ace-six flats, the "cube" is asymmetric to the extent that $p(1) = p(6) = \frac{1}{4}$, while $p(2) = p(3) = p(4) = p(5) = \frac{1}{8}$. Notice that $p(s)$ here qualifies as a discrete probability function because each $p(s)$ is greater than or equal to 0 and the sum of $p(s)$, over all s, is $1\left[= 2\left(\frac{1}{4}\right) + 4\left(\frac{1}{8}\right)\right]$.

Suppose A is the event that an even number occurs. It follows from Equation 3.3.1 that $P(A) = P(2) + P(4) + P(6) = \frac{1}{8} + \frac{1}{8} + \frac{1}{4} = \frac{1}{2}$.

Comment If two ace-six flats are rolled, the probability of getting a sum equal to 7 is equal to $2p(1)p(6) + 2p(2)p(5) + 2p(3)p(4) = 2\left(\frac{1}{4}\right)^2 + 4\left(\frac{1}{8}\right)^2 = \frac{3}{16}$. If two *fair* dice are rolled, the probability of getting a sum equal to 7 is $2p(1)p(6) + 2p(2)p(5) + 2p(3)p(4) = 6\left(\frac{1}{6}\right)^2 = \frac{1}{6}$, which is less than $\frac{3}{16}$. Gamblers cheat with ace-six flats by switching back and forth between fair dice and ace-six flats, depending on whether or not they want a sum of 7 to be rolled. ■

Example 3.3.2

Suppose a fair coin is tossed until a head comes up for the first time. What are the chances of that happening on an odd-numbered toss?

Note that the sample space here is countably infinite and so is the set of outcomes making up the event whose probability we are trying to find. The $P(A)$ that we are looking for, then, will be the sum of an infinite number of terms.

Let $p(s)$ be the probability that the first head appears on the sth toss. Since the coin is presumed to be fair, $p(1) = \frac{1}{2}$. Furthermore, we would expect that half the time, when a tail appears, the next toss would be a head, so $p(2) = \frac{1}{2} \cdot \frac{1}{2} = \frac{1}{4}$. In general, $p(s) = \left(\frac{1}{2}\right)^s, s = 1, 2, \ldots$.

Does $p(s)$ satisfy the conditions stated in Definition 3.3.1? Yes. Clearly, $p(s) \geq 0$ for all s. To see that the sum of the probabilities is 1, recall the formula for the sum of a geometric series: If $0 < r < 1$,

$$\sum_{s=0}^{\infty} r^s = \frac{1}{1-r} \tag{3.3.2}$$

Applying Equation 3.3.2 to the sample space here confirms that $P(S) = 1$:

$$P(S) = \sum_{s=1}^{\infty} p(s) = \sum_{s=1}^{\infty} \left(\frac{1}{2}\right)^s = \sum_{s=0}^{\infty} \left(\frac{1}{2}\right)^s - \left(\frac{1}{2}\right)^0 = 1 \Big/ \left(1 - \frac{1}{2}\right) - 1 = 1$$

Now, let A be the event that the first head appears on an odd-numbered toss. Then $P(A) = p(1) + p(3) + p(5) + \cdots$ But

$$p(1) + p(3) + p(5) + \cdots = \sum_{s=0}^{\infty} p(2s+1) = \sum_{s=0}^{\infty} \left(\frac{1}{2}\right)^{2s+1} = \left(\frac{1}{2}\right) \sum_{s=0}^{\infty} \left(\frac{1}{4}\right)^s$$
$$= \left(\frac{1}{2}\right)\left[1 \Big/ \left(1 - \frac{1}{4}\right)\right] = \frac{2}{3}$$

■

Case Study 3.3.1

For good pedagogical reasons, the principles of probability are always introduced by considering events defined on familiar sample spaces generated by simple experiments. To that end, we toss coins, deal cards, roll dice, and draw chips from urns. It would be a serious error, though, to infer that the importance of probability extends no further than the nearest casino. In its infancy,

(Continued on next page)

gambling and probability were, indeed, intimately related: Questions arising from games of chance were often the catalyst that motivated mathematicians to study random phenomena in earnest. But more than 340 years have passed since Huygens published *De Ratiociniis*. Today, the application of probability to gambling is relatively insignificant (the NCAA March basketball tournament notwithstanding) compared to the depth and breadth of uses the subject finds in business, medicine, engineering, and science.

Probability functions—properly chosen—can "model" complex real-world phenomena every bit as well as $P(\text{heads}) = \frac{1}{2}$ describes the behavior of a fair coin. The following set of actuarial data is a case in point. Over a period of three years (= 1096 days) in London, records showed that a total of 903 deaths occurred among males eighty-five years of age and older (180). Columns 1 and 2 of Table 3.3.1 give the breakdown of those 903 deaths according to the number occurring on a given day. Column 3 gives the *proportion* of days for which exactly s elderly men died.

Table 3.3.1

(1) Number of Deaths, s	(2) Number of Days	(3) Proportion [= Col.(2)/1096]	(4) $p(s)$
0	484	0.442	0.440
1	391	0.357	0.361
2	164	0.150	0.148
3	45	0.041	0.040
4	11	0.010	0.008
5	1	0.001	0.003
6+	0	0.000	0.000
	1096	1	1

For reasons that we will go into at length in Chapter 4, the probability function that describes the behavior of this particular phenomenon is

$$p(s) = P(s \text{ elderly men die on a given day})$$
$$= \frac{e^{-0.82}(0.82)^s}{s!}, \qquad s = 0, 1, 2, \ldots \tag{3.3.3}$$

How do we know that the $p(s)$ in Equation 3.3.3 is an appropriate way to assign probabilities to the "experiment" of elderly men dying? Because it accurately predicts what happened. Column 4 of Table 3.3.1 shows $p(s)$ evaluated for $s = 0, 1, 2, \ldots$. To two decimal places, the agreement between the entries in Column 3 and Column 4 is perfect.

Example 3.3.3 Consider the following experiment: Every day for the next month you copy down each number that appears in the stories on the front pages of your hometown newspaper. Those numbers would necessarily be extremely diverse: One might be the age of a celebrity who had just died, another might report the interest rate currently paid on government Treasury bills, and still another might give the number of square feet of retail space recently added to a local shopping mall.

Suppose you then calculated the proportion of those numbers whose leading digit was a 1, the proportion whose leading digit was a 2, and so on. What relationship would you expect those proportions to have? Would numbers starting with a 2, for example, occur as often as numbers starting with a 6?

Let $p(s)$ denote the probability that the first significant digit of a "newspaper number" is s, $s = 1, 2, \ldots, 9$. Our intuition is likely to tell us that the nine first digits should be equally probable—that is, $p(1) = p(2) = \cdots = p(9) = \frac{1}{9}$. Given the diversity and the randomness of the numbers, there is no obvious reason why one digit should be more common than another. Our intuition, though, would be wrong—first digits are *not* equally likely. Indeed, they are not even close to being equally likely!

Credit for making this remarkable discovery goes to Simon Newcomb, a mathematician who observed more than a hundred years ago that some portions of logarithm tables are used more than others (78). Specifically, pages at the beginning of such tables are more dog-eared than pages at the end, suggesting that users have more occasion to look up logs of numbers starting with small digits than they do numbers starting with large digits.

Almost fifty years later, a physicist, Frank Benford, reexamined Newcomb's claim in more detail and looked for a mathematical explanation. What is now known as *Benford's law* asserts that the first digits of many different types of measurements, or combinations of measurements, often follow the discrete probability model:

$$p(s) = P(\text{1st significant digit is } s) = \log\left(1 + \frac{1}{s}\right), \quad s = 1, 2, \ldots, 9$$

Table 3.3.2 compares Benford's law to the uniform assumption that $p(s) = \frac{1}{9}$, for all s. The differences are striking. According to Benford's law, for example, 1's are the most frequently occurring first digit, appearing 6.5 times ($= 0.301/0.046$) as often as 9's.

Table 3.3.2

s	"Uniform" Law	Benford's Law
1	0.111	0.301
2	0.111	0.176
3	0.111	0.125
4	0.111	0.097
5	0.111	0.079
6	0.111	0.067
7	0.111	0.058
8	0.111	0.051
9	0.111	0.046

Comment A key to *why* Benford's law is true is the differences in proportional changes associated with each leading digit. To go from one thousand to two thousand, for example, represents a 100% increase; to go from eight thousand to nine thousand, on the other hand, is only a 12.5% increase. That would suggest that evolutionary phenomena such as stock prices would be more likely to start with 1's and 2's than with 8's and 9's—and they are. Still, the precise conditions under which $p(s) = \log\left(1 + \frac{1}{s}\right)$, $s = 1, 2, \ldots, 9$, are not fully understood and remain a topic of research. ■

Example 3.3.4 Is

$$p(s) = \frac{1}{1+\lambda}\left(\frac{\lambda}{1+\lambda}\right)^s, \quad s = 0, 1, 2, \ldots; \quad \lambda > 0$$

a discrete probability function? Why or why not?

To qualify as a discrete probability function, a given $p(s)$ needs to satisfy parts (a) and (b) of Definition 3.3.1. A simple inspection shows that part (a) is satisfied. Since $\lambda > 0$, $p(s)$ is, in fact, greater than or equal to 0 for all $s = 0, 1, 2, \ldots$. Part (b) is satisfied if the sum of all the probabilities defined on the outcomes in S is 1. But

$$\begin{aligned}\sum_{\text{all } s \in S} p(s) &= \sum_{s=0}^{\infty} \frac{1}{1+\lambda}\left(\frac{\lambda}{1+\lambda}\right)^s \\ &= \frac{1}{1+\lambda}\left(\frac{1}{1-\frac{\lambda}{1+\lambda}}\right) \qquad \text{(why?)} \\ &= \frac{1}{1+\lambda} \cdot \frac{1+\lambda}{1} \\ &= 1\end{aligned}$$

The answer, then, is "yes"—$p(s) = \frac{1}{1+\lambda}\left(\frac{\lambda}{1+\lambda}\right)^s, s = 0, 1, 2, \ldots; \lambda > 0$ *does* qualify as a discrete probability function. Of course, whether it has any practical value depends on whether the set of values for $p(s)$ actually do describe the behavior of real-world phenomena. ■

Defining "New" Sample Spaces

We have seen how the function $p(s)$ associates a probability with each outcome, s, in a sample space. Related is the key idea that outcomes can often be grouped or reconfigured in ways that may facilitate problem solving. Recall the sample space associated with a series of n independent trials, where each s is an ordered sequence of successes and failures. The most relevant information in such outcomes is often the *number* of successes that occur, not a detailed listing of which trials ended in success and which ended in failure. That being the case, it makes sense to define a "new" sample space by grouping the original outcomes according to the number of successes they contained. The outcome $(f, f, \ldots, f)$, for example, had 0 successes. On the other hand, there were n outcomes that yielded 1 success—$(s, f, f, \ldots, f)$, $(f, s, f, \ldots, f)$, $\ldots$, and $(f, f, \ldots, s)$. As we saw earlier in this chapter, that particular regrouping of outcomes ultimately led to the binomial distribution.

The function that replaces the outcome $(s, f, f, \ldots, f)$ with the numerical value 1 is called a *random variable*. We conclude this section with a discussion of some of the concepts, terminology, and applications associated with random variables.

Definition 3.3.2. A function whose domain is a sample space S and whose values form a finite or countably infinite set of real numbers is called a *discrete random variable*. We denote random variables by uppercase letters, often X or Y.

Example 3.3.5 Consider tossing two dice, an experiment for which the sample space is a set of ordered pairs, $S = \{(i, j) \mid i = 1, 2, \ldots, 6;\ j = 1, 2, \ldots, 6\}$. For a variety of games ranging from Monopoly to craps, the *sum* of the numbers showing is what matters on a given turn. That being the case, the original sample space S of thirty-six ordered pairs would not provide a particularly convenient backdrop for discussing the rules of those games. It would be better to work directly with the sums. Of course, the eleven possible sums (from 2 to 12) are simply the different values of the random variable X, where $X(i, j) = i + j$.

Comment In the above example, suppose we define a random variable X_1 that gives the result on the first die and a random variable X_2 that gives the result on the second die. Then $X = X_1 + X_2$. Note how easily we could extend this idea to the toss of *three* dice, or *ten* dice. The ability to conveniently express complex events in terms of simpler ones is an advantage of the random variable concept that we will see playing out over and over again. ■

The Probability Density Function

We began this section discussing the function $p(s)$, which assigns a probability to each outcome s in S. Now, having introduced the notion of a random variable X as a real-valued function defined on S—that is, $X(s) = k$—we need to find a mapping analogous to $p(s)$ that assigns probabilities to the different values of k.

Definition 3.3.3. Associated with every discrete random variable X is a *probability density function* (or *pdf*), denoted $p_X(k)$, where

$$p_X(k) = P(\{s \in S \mid X(s) = k\})$$

Note that $p_X(k) = 0$ for any k not in the range of X. For notational simplicity, we will usually delete all references to s and S and write $p_X(k) = P(X = k)$.

Comment We have already discussed at length two examples of the function $p_X(k)$. Recall the binomial distribution derived in Section 3.2. If we let the random variable X denote the number of successes in n independent trials, then Theorem 3.2.1 states that

$$P(X = k) = p_X(k) = \binom{n}{k} p^k (1 - p)^{n-k}, \quad k = 0, 1, \ldots, n$$

A similar result was given in that same section in connection with the hypergeometric distribution. If a sample of size n is drawn without replacement from an urn containing r red chips and w white chips, and if we let the random variable X denote the number of red chips included in the sample, then (according to Theorem 3.2.2),

$$P(X = k) = p_X(k) = \binom{r}{k}\binom{w}{n-k} \Big/ \binom{r+w}{n}$$

Example 3.3.6

Consider again the rolling of two dice as described in Example 3.3.5. Let i and j denote the faces showing on the first and second die, respectively, and define the random variable X to be the sum of the two faces: $X(i, j) = i + j$. Find $p_X(k)$.

According to Definition 3.3.3, each value of $p_X(k)$ is the sum of the probabilities of the outcomes that get mapped by X onto the value k. For example,

$$\begin{aligned} P(X=5) = p_X(5) &= P(\{s \in S \mid X(s) = 5\}) \\ &= P[(1,4), (4,1), (2,3), (3,2)] \\ &= P(1,4) + P(4,1) + P(2,3) + P(3,2) \\ &= \frac{1}{36} + \frac{1}{36} + \frac{1}{36} + \frac{1}{36} \\ &= \frac{4}{36} \end{aligned}$$

assuming the dice are fair. Values of $p_X(k)$ for other k are calculated similarly. Table 3.3.3 shows the random variable's entire pdf.

Table 3.3.3

k	$p_X(k)$	k	$p_X(k)$
2	1/36	8	5/36
3	2/36	9	4/36
4	3/36	10	3/36
5	4/36	11	2/36
6	5/36	12	1/36
7	6/36		

■

Example 3.3.7

Acme Industries typically produces three electric power generators a day; some pass the company's quality-control inspection on their first try and are ready to be shipped; others need to be retooled. The probability of a generator needing further work is 0.05. If a generator is ready to ship, the firm earns a profit of \$10,000. If it needs to be retooled, it ultimately costs the firm \$2,000. Let X be the random variable quantifying the company's daily profit. Find $p_X(k)$.

The underlying sample space here is a set of $n = 3$ independent trials, where $p = P(\text{Generator passes inspection}) = 0.95$. If the random variable X is to measure the company's daily profit, then

$$\begin{aligned} X = \$10{,}000 &\times (\text{no. of generators passing inspection}) \\ &- \$2{,}000 \times (\text{no. of generators needing retooling}) \end{aligned}$$

For instance, $X(s, f, s) = 2(\$10{,}000) - 1(\$2{,}000) = \$18{,}000$. Moreover, the random variable X equals \$18,000 whenever the day's output consists of two successes and one failure. That is, $X(s, f, s) = X(s, s, f) = X(f, s, s)$. It follows that

$$P(X = \$18{,}000) = p_X(18{,}000) = \binom{3}{2}(0.95)^2(0.05)^1 = 0.135375$$

Table 3.3.4 shows $p_X(k)$ for the four possible values of k (\$30,000, \$18,000, \$6,000, and −\$6,000).

Table 3.3.4

No. Defectives	k = Profit	$p_X(k)$
0	\$30,000	0.857375
1	\$18,000	0.135375
2	\$6,000	0.007125
3	−\$6,000	0.000125

■

Example 3.3.8

As part of her warm-up drill, each player on State's basketball team is required to shoot free throws until two baskets are made. If Rhonda has a 65% success rate at the foul line, what is the pdf of the random variable X that describes the number of throws it takes her to complete the drill? Assume that individual throws constitute independent events.

Figure 3.3.1 illustrates what must occur if the drill is to end on the kth toss, $k = 2, 3, 4, \ldots$: First, Rhonda needs to make exactly one basket sometime during the first $k-1$ attempts, and, second, she needs to make a basket on the kth toss. Written formally,

$$\begin{aligned} p_X(k) &= P(X = k) = P(\text{Drill ends on } k\text{th throw}) \\ &= P[(1 \text{ basket and } k-2 \text{ misses in first } k-1 \text{ throws}) \cap (\text{basket on } k\text{th throw})] \\ &= P(1 \text{ basket and } k-2 \text{ misses}) \cdot P(\text{basket}) \end{aligned}$$

Exactly one basket

Miss	Basket	Miss	...	Miss	Basket
1	2	3		$k-1$	k

Attempts

Figure 3.3.1

Notice that $k-1$ different sequences have the property that exactly one of the first $k-1$ throws results in a basket:

$$k-1 \text{ sequences} \begin{cases} \frac{B}{1} \ \frac{M}{2} \ \frac{M}{3} \ \frac{M}{4} \ \cdots \ \frac{M}{k-1} \\ \frac{M}{1} \ \frac{B}{2} \ \frac{M}{3} \ \frac{M}{4} \ \cdots \ \frac{M}{k-1} \\ \quad\quad \vdots \\ \frac{M}{1} \ \frac{M}{2} \ \frac{M}{3} \ \frac{M}{4} \ \cdots \ \frac{B}{k-1} \end{cases}$$

Since each sequence has probability $(0.35)^{k-2}(0.65)$,

$$P(1 \text{ basket and } k-2 \text{ misses}) = (k-1)(0.35)^{k-2}(0.65)$$

Therefore,

$$\begin{aligned} p_X(k) &= (k-1)(0.35)^{k-2}(0.65) \cdot (0.65) \\ &= (k-1)(0.35)^{k-2}(0.65)^2, \quad k = 2, 3, 4, \ldots \end{aligned} \tag{3.3.4}$$

Table 3.3.5 shows the pdf evaluated for specific values of k. Although the range of k is infinite, the bulk of the probability associated with X is concentrated in the values 2 through 7: It is highly unlikely, for example, that Rhonda would need more than seven shots to complete the drill.

Table 3.3.5

k	$p_X(k)$
2	0.4225
3	0.2958
4	0.1553
5	0.0725
6	0.0317
7	0.0133
8+	0.0089

■

The Cumulative Distribution Function

In working with random variables, we frequently need to calculate the probability that the value of a random variable is somewhere between two numbers. For example, suppose we have an integer-valued random variable. We might want to calculate an expression like $P(s \leq X \leq t)$. If we know the pdf for X, then

$$P(s \leq X \leq t) = \sum_{k=s}^{t} p_X(k)$$

But depending on the nature of $p_X(k)$ and the number of terms that need to be added, calculating the sum of $p_X(k)$ from $k = s$ to $k = t$ may be quite difficult. An alternate strategy is to use the fact that

$$P(s \leq X \leq t) = P(X \leq t) - P(X \leq s - 1)$$

where the two probabilities on the right represent *cumulative* probabilities of the random variable X. If the latter were available (and they often are), then evaluating $P(s \leq X \leq t)$ by one simple subtraction would clearly be easier than doing all the calculations implicit in $\sum_{k=s}^{t} p_X(k)$.

Definition 3.3.4. Let X be a discrete random variable. For any real number t, the probability that X takes on a value $\leq t$ is the *cumulative distribution function* (*cdf*) of X [written $F_X(t)$]. In formal notation, $F_X(t) = P(\{s \in S \mid X(s) \leq t\})$. As was the case with pdfs, references to s and S are typically deleted, and the cdf is written $F_X(t) = P(X \leq t)$.

Example 3.3.9

Suppose we wish to compute $P(21 \leq X \leq 40)$ for a binomial random variable X with $n = 50$ and $p = 0.6$. From Theorem 3.2.1, we know the formula for $p_X(k)$, so $P(21 \leq X \leq 40)$ can be written as a simple, although computationally cumbersome, sum:

$$P(21 \leq X \leq 40) = \sum_{k=21}^{40} \binom{50}{k} (0.6)^k (0.4)^{50-k}$$

Equivalently, the probability we are looking for can be expressed as the difference between two cdfs:

$$P(21 \leq X \leq 40) = P(X \leq 40) - P(X \leq 20) = F_X(40) - F_X(20)$$

As it turns out, values of the cdf for a binomial random variable are widely available, both in books and in computer software. Here, for example, $F_X(40) = 0.9992$ and $F_X(20) = 0.0034$, so

$$P(21 \le X \le 40) = 0.9992 - 0.0034$$
$$= 0.9958$$

■

Example 3.3.10 Suppose that two fair dice are rolled. Let the random variable X denote the larger of the two faces showing: (a) Find $F_X(t)$ for $t = 1, 2, \ldots, 6$ and (b) Find $F_X(2.5)$.

a. The sample space associated with the experiment of rolling two fair dice is the set of ordered pairs $s = (i, j)$, where the face showing on the first die is i and the face showing on the second die is j. By assumption, all thirty-six possible outcomes are equally likely. Now, suppose t is some integer from 1 to 6, inclusive. Then

$$\begin{aligned} F_X(t) &= P(X \le t) \\ &= P[\text{Max}\,(i, j) \le t] \\ &= P(i \le t \quad \text{and} \quad j \le t) \qquad \text{(why?)} \\ &= P(i \le t) \cdot P(j \le t) \qquad \text{(why?)} \\ &= \frac{t}{6} \cdot \frac{t}{6} \\ &= \frac{t^2}{36}, \quad t = 1, 2, 3, 4, 5, 6 \end{aligned}$$

b. Even though the random variable X has nonzero probability only for the integers 1 through 6, the cdf is defined for *any* real number from $-\infty$ to $+\infty$. By definition, $F_X(2.5) = P(X \le 2.5)$. But

$$\begin{aligned} P(X \le 2.5) &= P(X \le 2) + P(2 < X \le 2.5) \\ &= F_X(2) + 0 \end{aligned}$$

so

$$F_X(2.5) = F_X(2) = \frac{2^2}{36} = \frac{1}{9}$$

What would the graph of $F_X(t)$ as a function of t look like?

■

Questions

3.3.1. An urn contains five balls numbered 1 to 5. Two balls are drawn simultaneously.

(a) Let X be the larger of the two numbers drawn. Find $p_X(k)$.
(b) Let V be the sum of the two numbers drawn. Find $p_V(k)$.

3.3.2. Repeat Question 3.3.1 for the case where the two balls are drawn *with replacement.*

3.3.3. Suppose a fair die is tossed three times. Let X be the largest of the three faces that appear. Find $p_X(k)$.

3.3.4. Suppose a fair die is tossed three times. Let X be the number of different faces that appear (so $X = 1, 2$, or 3). Find $p_X(k)$.

3.3.5. A fair coin is tossed three times. Let X be the number of heads in the tosses minus the number of tails. Find $p_X(k)$.

3.3.6. Suppose die one has spots $1, 2, 2, 3, 3, 4$ and die two has spots $1, 3, 4, 5, 6, 8$. If both dice are rolled, what is the sample space? Let $X =$ total spots showing. Show that the pdf for X is the same as for normal dice.

3.3.7. Suppose a particle moves along the x-axis beginning at 0. It moves one integer step to the left or right with equal probability. What is the pdf of its position after four steps?

3.3.8. How would the pdf asked for in Question 3.3.7 be affected if the particle was twice as likely to move to the right as to the left?

3.3.9. Suppose that five people, including you and a friend, line up at random. Let the random variable X denote the number of people standing between you and your friend. What is $p_X(k)$?

3.3.10. Urn I and urn II each have two red chips and two white chips. Two chips are drawn simultaneously from each urn. Let X_1 be the number of red chips in the first sample and X_2 the number of red chips in the second sample. Find the pdf of $X_1 + X_2$.

3.3.11. Suppose X is a binomial random variable with $n = 4$ and $p = \frac{2}{3}$. What is the pdf of $2X + 1$?

3.3.12. Find the cdf for the random variable X in Question 3.3.3.

3.3.13. A fair die is rolled four times. Let the random variable X denote the number of 6's that appear. Find and graph the cdf for X.

3.3.14. At the points $x = 0, 1, \ldots, 6$, the cdf for the discrete random variable X has the value $F_X(x) = x(x+1)/42$. Find the pdf for X.

3.3.15. Find the pdf for the discrete random variable X whose cdf at the points $x = 0, 1, \ldots, 6$ is given by $F_X(x) = x^3/216$.

3.4 Continuous Random Variables

The statement was made in Chapter 2 that all sample spaces belong to one of two generic types—*discrete* sample spaces are ones that contain a finite or a countably infinite number of outcomes and *continuous* sample spaces are those that contain an uncountably infinite number of outcomes. Rolling a pair of dice and recording the faces that appear is an experiment with a discrete sample space; choosing a number at random from the interval [0, 1] would have a continuous sample space.

How we assign probabilities to these two types of sample spaces is different. Section 3.3 focused on discrete sample spaces. Each outcome s is assigned a probability by the discrete probability function $p(s)$. If a random variable X is defined on the sample space, the probabilities associated with its outcomes are assigned by the probability density function $p_X(k)$. Applying those same definitions, though, to the outcomes in a continuous sample space will not work. The fact that a continuous sample space has an uncountably infinite number of outcomes eliminates the option of assigning a probability to each point as we did in the discrete case with the function $p(s)$. We begin this section with a particular pdf defined on a discrete sample space that suggests how we might define probabilities, in general, on a continuous sample space.

Suppose an electronic surveillance monitor is turned on briefly at the beginning of every hour and has a 0.905 probability of working properly, regardless of how long it has remained in service. If we let the random variable X denote the hour at which the monitor first fails, then $p_X(k)$ is the product of k individual probabilities:

$$\begin{aligned} p_X(k) &= P(X = k) = P(\text{Monitor fails for the first time at the } k\text{th hour}) \\ &= P(\text{Monitor functions properly for first } k-1 \text{ hours} \cap \text{Monitor fails at the } k\text{th hour}) \\ &= (0.905)^{k-1}(0.095), \quad k = 1, 2, 3, \ldots \end{aligned}$$

Figure 3.4.1 shows a probability histogram of $p_X(k)$ for k values ranging from 1 to 21. Here the height of the kth bar is $p_X(k)$, and since the width of each bar is 1, the *area* of the kth bar is also $p_X(k)$.

Now, look at Figure 3.4.2, where the exponential curve $y = 0.1e^{-0.1x}$ is superimposed on the graph of $p_X(k)$. Notice how closely the area under the curve approximates the area of the bars. It follows that the probability that X lies in some given interval will be numerically similar to the integral of the exponential curve above that same interval.

Figure 3.4.1

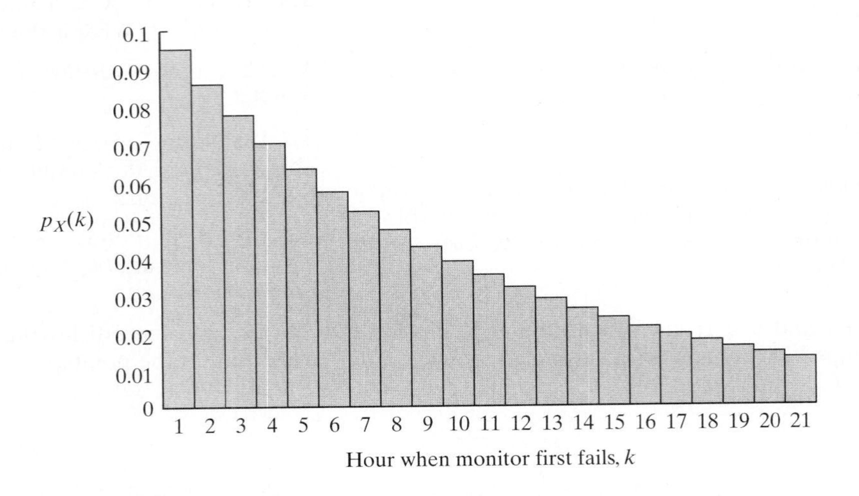

Figure 3.4.2

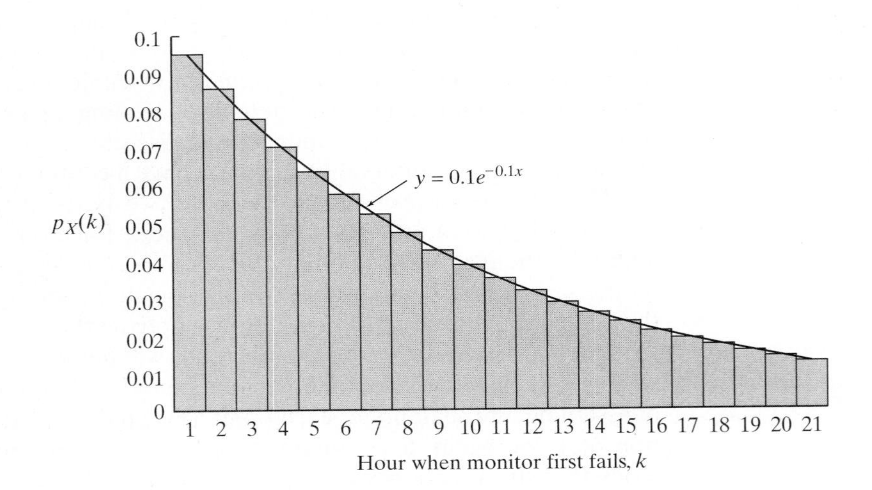

For example, the probability that the monitor fails sometime during the first four hours would be the sum

$$
\begin{aligned}
P(0 \leq X \leq 4) &= \sum_{k=0}^{4} p_X(k) \\
&= \sum_{k=0}^{4} (0.905)^{k-1}(0.095) \\
&= 0.3297
\end{aligned}
$$

To four decimal places, the corresponding area under the exponential curve is the same:

$$
\int_0^4 0.1e^{-0.1x}\, dx = 0.3297
$$

Implicit in the similarity here between $p_X(k)$ and the exponential curve $y = 0.1e^{-0.1x}$ is our sought-after alternative to $p(s)$ for continuous sample spaces. Instead of defining probabilities for individual points, we will define probabilities for *intervals* of points, and those probabilities will be areas under the graph of some function (such as $y = 0.1e^{-0.1x}$), where the shape of the function will reflect the desired probability "measure" to be associated with the sample space.

Definition 3.4.1. A probability function P on a set of real numbers S is called *continuous* if there exists a function $f(t)$ such that for any closed interval $[a, b] \subset S$, $P([a, b]) = \int_a^b f(t)\,dt$.

Comment If a probability function P satisfies Definition 3.4.1, then $P(A) = \int_A f(t)\,dt$ for any set A where the integral is defined.

Conversely, suppose a function $f(t)$ has the two properties

1. $f(t) \geq 0$ for all t.
2. $\int_{-\infty}^{\infty} f(t)\,dt = 1$.

If $P(A) = \int_A f(t)\,dt$ for all A, then P will satisfy the probability axioms given in Section 2.3.

Choosing the Function $f(t)$

We have seen that the probability structure of any sample space with a finite or countably infinite number of outcomes is defined by the function $p(s) = P(\text{Outcome is } s)$. For sample spaces having an uncountably infinite number of possible outcomes, the function $f(t)$ serves an analogous purpose. Specifically, $f(t)$ defines the probability structure of S in the sense that the probability of any *interval* in the sample space is the *integral* of $f(t)$. The next set of examples illustrate several different choices for $f(t)$.

Example 3.4.1

The continuous equivalent of the equiprobable probability model on a discrete sample space is the function $f(t)$ defined by $f(t) = 1/(b-a)$ for all t in the interval $[a, b]$ (and $f(t) = 0$, otherwise). This particular $f(t)$ places equal probability weighting on every closed interval of the same length contained in the interval $[a, b]$. For example, suppose $a = 0$ and $b = 10$, and let $A = [1, 3]$ and $B = [6, 8]$. Then $f(t) = \frac{1}{10}$, and

$$P(A) = \int_1^3 \left(\frac{1}{10}\right) dt = \frac{2}{10} = P(B) = \int_6^8 \left(\frac{1}{10}\right) dt$$

(See Figure 3.4.3.)

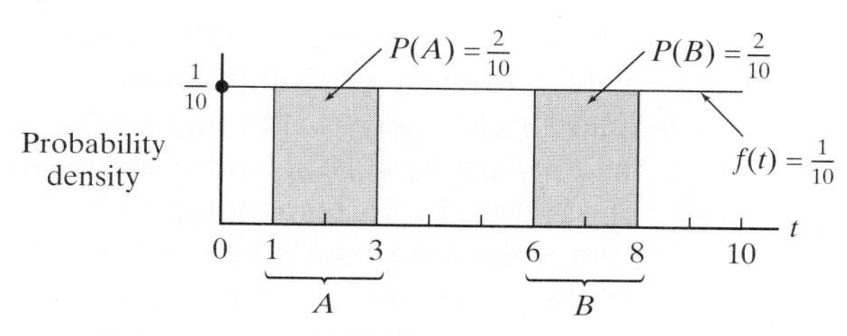

Figure 3.4.3

■

Example 3.4.2

Could $f(t) = 3t^2, 0 \le t \le 1$, be used to define the probability function for a continuous sample space whose outcomes consist of all the real numbers in the interval [0, 1]? Yes, because (1) $f(t) \ge 0$ for all t, and (2) $\int_0^1 f(t)\,dt = \int_0^1 3t^2\,dt = t^3\big|_0^1 = 1$.

Notice that the shape of $f(t)$ (see Figure 3.4.4) implies that outcomes close to 1 are more likely to occur than are outcomes close to 0. For example, $P\left(\left[0, \frac{1}{3}\right]\right) = \int_0^{1/3} 3t^2\,dt = t^3\big|_0^{1/3} = \frac{1}{27}$, while $P\left(\left[\frac{2}{3}, 1\right]\right) = \int_{2/3}^1 3t^2\,dt = t^3 \int_{2/3}^1 = 1 - \frac{8}{27} = \frac{19}{27}$.

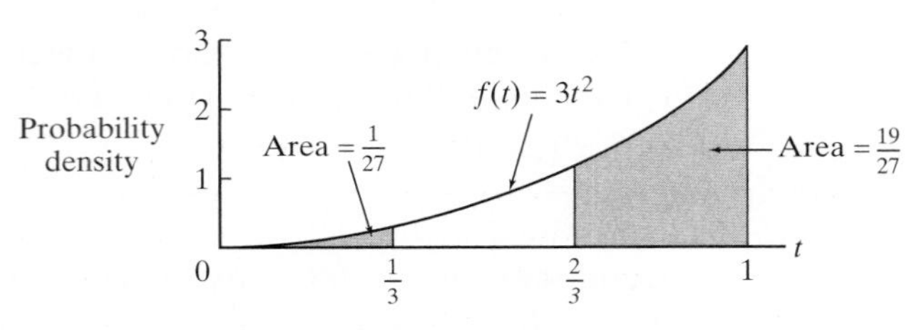

Figure 3.4.4

Example 3.4.3

By far the most important of all continuous probability functions is the "bell-shaped" curve, known more formally as the *normal* (or *Gaussian*) *distribution*. The sample space for the normal distribution is the entire real line; its probability function is given by

$$f(t) = \frac{1}{\sqrt{2\pi}\sigma} \exp\left[-\frac{1}{2}\left(\frac{t-\mu}{\sigma}\right)^2\right], \quad -\infty < t < \infty, \quad -\infty < \mu < \infty, \quad \sigma > 0$$

Depending on the values assigned to the parameters μ and σ, $f(t)$ can take on a variety of shapes and locations; three are illustrated in Figure 3.4.5.

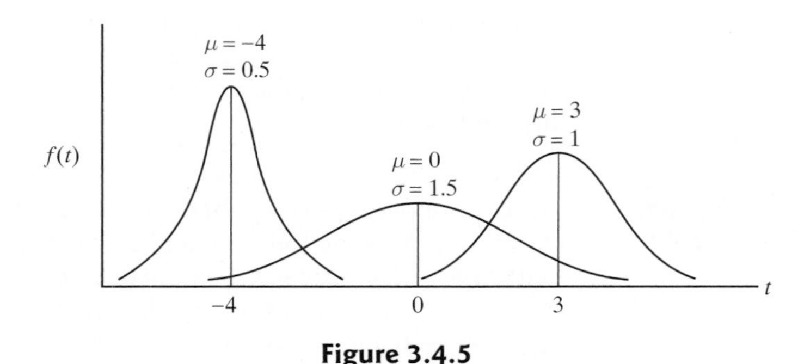

Figure 3.4.5

Fitting $f(t)$ to Data: The Density-Scaled Histogram

The notion of using a continuous probability function to approximate an integer-valued discrete probability model has already been discussed (recall Figure 3.4.2). The "trick" there was to replace the spikes that define $p_X(k)$ with rectangles whose heights are $p_X(k)$ and whose widths are 1. Doing that makes the sum of the areas of the rectangles corresponding to $p_X(k)$ equal to 1, which is the same as the total area under the approximating continuous probability function. Because of the equality of those two areas, it makes sense to superimpose (and compare) the "histogram" of $p_X(k)$ and the continuous probability function on the same set of axes.

Now, consider the related, but slightly more general, problem of using a continuous probability function to model the distribution of a set of n measurements,

$y_1, y_2, \ldots, y_n$. Following the approach taken in Figure 3.4.2, we would start by making a histogram of the n observations. The problem, though, is that the sum of the areas of the bars comprising that histogram would not necessarily equal 1.

As a case in point, Table 3.4.1 shows a set of forty observations. Grouping those y_i's into five classes, each of width 10, produces the distribution and histogram pictured in Figure 3.4.6. Furthermore, suppose we have reason to believe that these forty y_i's may be a random sample from a uniform probability function defined over the interval [20, 70]—that is,

$$f(t) = \frac{1}{70-20} = \frac{1}{50}, \quad 20 \le t \le 70$$

Table 3.4.1

33.8	62.6	42.3	62.9	32.9	58.9	60.8	49.1	42.6	59.8
41.6	54.5	40.5	30.3	22.4	25.0	59.2	67.5	64.1	59.3
24.9	22.3	69.7	41.2	64.5	33.4	39.0	53.1	21.6	46.0
28.1	68.7	27.6	57.6	54.8	48.9	68.4	38.4	69.0	46.6

(recall Example 3.4.1). How can we appropriately draw the distribution of the y_i's and the uniform probability model on the same graph?

Figure 3.4.6

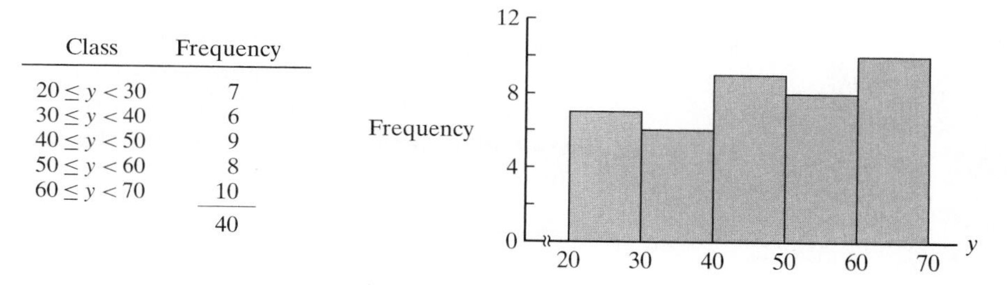

Class	Frequency
$20 \le y < 30$	7
$30 \le y < 40$	6
$40 \le y < 50$	9
$50 \le y < 60$	8
$60 \le y < 70$	10
	40

Note, first, that $f(t)$ and the histogram are not compatible in the sense that the area under $f(t)$ is (necessarily) 1 $(= 50 \times \frac{1}{50})$, but the sum of the areas of the bars making up the histogram is 400:

$$\text{histogram area} = 10(7) + 10(6) + 10(9) + 10(8) + 10(10)$$
$$= 400$$

Nevertheless, we can "force" the total area of the five bars to match the area under $f(t)$ by redefining the scale of the vertical axis on the histogram. Specifically, *frequency* needs to be replaced with the analog of *probability density,* which would be the scale used on the vertical axis of any graph of $f(t)$. Intuitively, the density associated with, say, the interval [20, 30) would be defined as the quotient

$$\frac{7}{40 \times 10}$$

because integrating that constant over the interval [20, 30) would give $\frac{7}{40}$, and the latter does represent the estimated probability that an observation belongs to the interval [20, 30).

Figure 3.4.7 shows a histogram of the data in Table 3.4.1, where the height of each bar has been converted to a *density,* according to the formula

$$\text{density (of a class)} = \frac{\text{class frequency}}{\text{total no. of observations} \times \text{class width}}$$

Superimposed is the uniform probability model, $f(t) = \frac{1}{50}$, $20 \leq t \leq 70$. Scaled in this fashion, areas under both $f(t)$ and the histogram are 1.

Figure 3.4.7

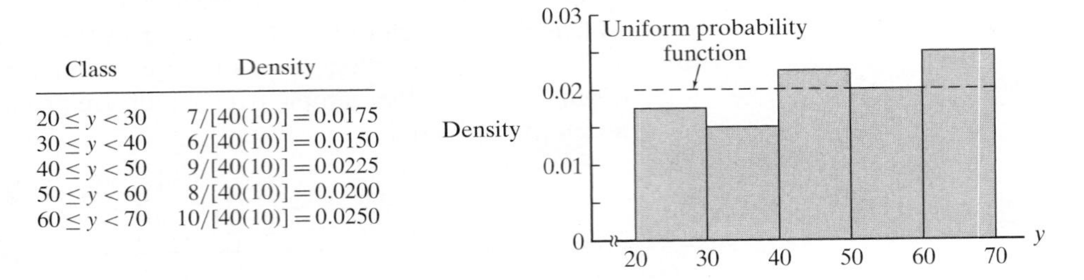

Class	Density
$20 \leq y < 30$	$7/[40(10)] = 0.0175$
$30 \leq y < 40$	$6/[40(10)] = 0.0150$
$40 \leq y < 50$	$9/[40(10)] = 0.0225$
$50 \leq y < 60$	$8/[40(10)] = 0.0200$
$60 \leq y < 70$	$10/[40(10)] = 0.0250$

In practice, density-scaled histograms offer a simple, but effective, format for examining the "fit" between a set of data and a presumed continuous model. We will use it often in the chapters ahead. Applied statisticians have especially embraced this particular graphical technique. Indeed, computer software packages that include *Histograms* on their menus routinely give users the choice of putting either *frequency* or *density* on the vertical axis.

Case Study 3.4.1

Years ago, the V805 transmitter tube was standard equipment on many aircraft radar systems. Table 3.4.2 summarizes part of a reliability study done on the V805; listed are the lifetimes (in hrs) recorded for 903 tubes (35). Grouped into intervals of width 80, the densities for the nine classes are shown in the last column.

Table 3.4.2

Lifetime (hrs)	Number of Tubes	Density
0–80	317	0.0044
80–160	230	0.0032
160–240	118	0.0016
240–320	93	0.0013
320–400	49	0.0007
400–480	33	0.0005
480–560	17	0.0002
560–700	26	0.0002
700+	20	0.0002
	903	

Experience has shown that lifetimes of electrical equipment can often be nicely modeled by the exponential probability function,

$$f(t) = \lambda e^{-\lambda t}, \quad t > 0$$

where the value of λ (for reasons explained in Chapter 5) is set equal to the reciprocal of the average lifetime of the tubes in the sample. Can the distribution of these data also be described by the exponential model?

(Continued on next page)

One way to answer such a question is to superimpose the proposed model on a graph of the density-scaled histogram. The extent to which the two graphs are similar then becomes an obvious measure of the appropriateness of the model.

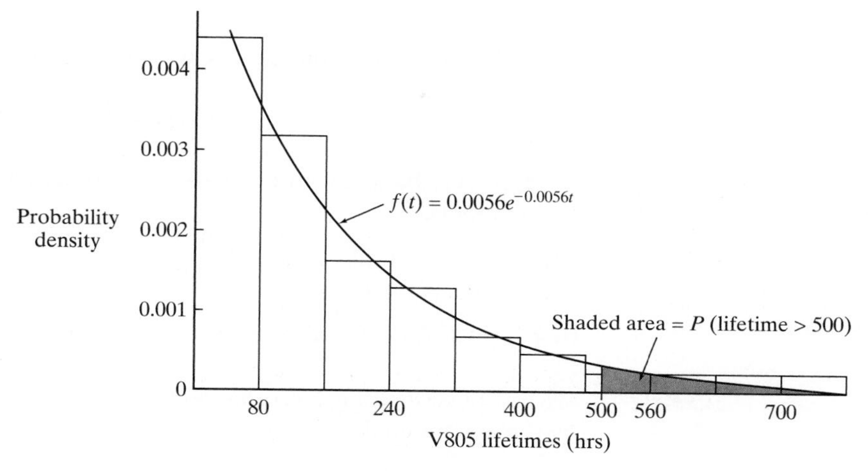

Figure 3.4.8

For these data, λ would be 0.0056. Figure 3.4.8 shows the function

$$f(t) = 0.0056e^{-0.0056t}$$

plotted on the same axes as the density-scaled histogram. Clearly, the agreement is excellent, and we would have no reservations about using areas under $f(t)$ to estimate lifetime probabilities. How likely is it, for example, that a V805 tube will last longer than five hundred hrs? Based on the exponential model, that probability would be 0.0608:

$$\begin{aligned} P(\text{V805 lifetime exceeds 500 hrs}) &= \int_{500}^{\infty} 0.0056e^{-0.0056y}\,dy \\ &= -e^{-0.0056y}\Big|_{500}^{\infty} = e^{-0.0056(500)} = e^{-2.8} = 0.0608 \end{aligned}$$

Continuous Probability Density Functions

We saw in Section 3.3 how the introduction of discrete random variables facilitates the solution of certain problems. The same sort of function can also be defined on sample spaces with an uncountably infinite number of outcomes. Usually, the sample space is an interval of real numbers—finite or infinite. The notation and techniques for this type of random variable replace sums with integrals.

Definition 3.4.2. Let Y be a function from a sample space S to the real numbers. The function Y is a called a *continuous random variable* if there exists a function $f_Y(y)$ such that for any real numbers a and b with $a < b$

$$P(a \le Y \le b) = \int_a^b f_Y(y)dy$$

The function $f_Y(y)$ is the *probability density function (pdf)* for Y.

As in the discrete case, the *cumulative distribution function (cdf)* is defined by

$$F_Y(y) = P(Y \leq y)$$

The cdf in the continuous case is just an integral of $f_Y(y)$, that is,

$$F_Y(y) = \int_{-\infty}^{y} f_Y(t)dt$$

Let $f(y)$ be an arbitrary real-valued function defined on some subset S of the real numbers. If

1. $f(y) \geq 0$ for all y in S and
2. $\int_s f_Y(y)dy = 1$

then $f(y) = f_Y(y)$ for all y, where the random variable Y is the identity mapping.

Example 3.4.4

We saw in Case Study 3.4.1 that lifetimes of V805 radar tubes can be nicely modeled by the exponential probability function

$$f(t) = 0.0056e^{-0.0056t}, \quad t > 0$$

To couch that statement in random variable notation would simply require that we define Y to be the life of a V805 radar tube. Then Y would be the identity mapping, and the pdf for the random variable Y would be the same as the probability function, $f(t)$. That is, we would write

$$f_Y(y) = 0.0056e^{-0.0056y}, \quad y \geq 0$$

Similarly, when we work with the bell-shaped normal distribution in later chapters, we will write the model in random variable notation as

$$f_Y(y) = \frac{1}{\sqrt{2\pi}\sigma}e^{-\frac{1}{2}\left(\frac{y-\mu}{\sigma}\right)^2}, \quad -\infty < y < \infty$$

■

Example 3.4.5

Suppose we would like a continuous random variable Y to "select" a number between 0 and 1 in such a way that intervals near the middle of the range would be more likely to be represented than intervals near either 0 or 1. One pdf having that property is the function $f_Y(y) = 6y(1-y)$, $0 \leq y \leq 1$ (see Figure 3.4.9). Do we know for certain that the function pictured in Figure 3.4.9 is a "legitimate" pdf? Yes, because $f_Y(y) \geq 0$ for all y, and $\int_0^1 6y(1-y)\,dy = 6[y^2/2 - y^3/3]\big|_0^1 = 1$.

Comment To simplify the way pdfs are written, it will be assumed that $f_Y(y) = 0$ for all y outside the range actually specified in the funtion's definition. In Example 3.4.5,

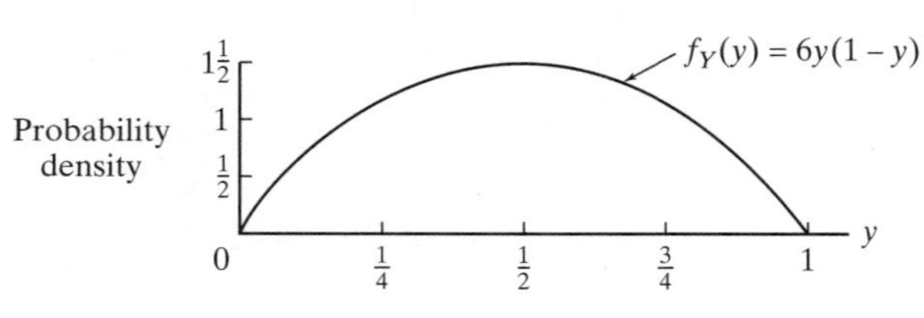

Figure 3.4.9

for instance, the statement $f_Y(y) = 6y(1-y)$, $0 \le y \le 1$, is to be interpreted as an abbreviation for

$$f_Y(y) = \begin{cases} 0, & y < 0 \\ 6y(1-y), & 0 \le y \le 1 \\ 0, & y > 1 \end{cases}$$

■

Continuous Cumulative Distribution Functions

Associated with every random variable, discrete or continuous, is a cumulative distribution function. For discrete random variables (recall Definition 3.3.4), the cdf is a nondecreasing step function, where the "jumps" occur at the values of t for which the pdf has positive probability. For continuous random variables, the cdf is a monotonically nondecreasing continuous function. In both cases, the cdf can be helpful in calculating the probability that a random variable takes on a value in a given interval. As we will see in later chapters, there are also several important relationships that hold for continuous cdfs and pdfs. One such relationship is cited in Theorem 3.4.1.

Definition 3.4.3. The cdf for a continuous random variable Y is an indefinite integral of its pdf:

$$F_Y(y) = \int_{-\infty}^{y} f_Y(r)\, dr = P(\{s \in S \mid Y(s) \le y\}) = P(Y \le y)$$

Theorem 3.4.1 *Let $F_Y(y)$ be the cdf of a continuous random variable Y. Then*

$$\frac{d}{dy} F_Y(y) = f_Y(y)$$

Proof The statement of Theorem 3.4.1 follows immediately from the Fundamental Theorem of Calculus. □

Theorem 3.4.2 *Let Y be a continuous random variable with cdf $F_Y(y)$. Then*

a. $P(Y > s) = 1 - F_Y(s)$

b. $P(r < Y \le s) = F_Y(s) - F_Y(r)$

c. $\lim_{y \to \infty} F_Y(y) = 1$

d. $\lim_{y \to -\infty} F_Y(y) = 0$

Proof

a. $P(Y > s) = 1 - P(Y \le s)$ since $(Y > s)$ and $(Y \le s)$ are complementary events. But $P(Y \le s) = F_Y(s)$, and the conclusion follows.

b. Since the set $(r < Y \le s) = (Y \le s) - (Y \le r)$, $P(r < Y \le s) = P(Y \le s) - P(Y \le r) = F_Y(s) - F_Y(r)$.

c. Let $\{y_n\}$ be a set of values of Y, $n = 1, 2, 3, \ldots$, where $y_n < y_{n+1}$ for all n, and $\lim_{n \to \infty} y_n = \infty$. If $\lim_{n \to \infty} F_Y(y_n) = 1$ for every such sequence $\{y_n\}$, then $\lim_{y \to \infty} F_Y(y) = 1$. To that end, set $A_1 = (Y \le y_1)$, and let $A_n = (y_{n-1} < Y \le y_n)$ for $n = 2, 3, \ldots$. Then

$F_Y(y_n) = P(\cup_{k=1}^{n} A_k) = \sum_{k=1}^{n} P(A_k)$, since the A_k's are disjoint. Also, the sample space $S = \cup_{k=1}^{\infty} A_k$, and by Axiom 4, $1 = P(S) = P(\cup_{k=1}^{\infty} A_k) = \sum_{k=1}^{\infty} P(A_k)$. Putting these equalities together gives $1 = \sum_{k=0}^{\infty} P(A_k) = \lim_{n\to\infty} \sum_{k=0}^{n} P(A_k) = \lim_{n\to\infty} F_Y(y_n)$.

d.
$$\lim_{y\to-\infty} F_Y(y) = \lim_{y\to-\infty} P(Y \le y) = \lim_{y\to-\infty} P(-Y \ge -y) = \lim_{y\to-\infty} [1 - P(-Y \le -y)]$$
$$= 1 - \lim_{y\to-\infty} P(-Y \le -y) = 1 - \lim_{y\to\infty} P(-Y \le y)$$
$$= 1 - \lim_{y\to\infty} F_{-Y}(y) = 0$$ □

Questions

3.4.1. Suppose $f_Y(y) = 4y^3, 0 \le y \le 1$. Find $P(0 \le Y \le \frac{1}{2})$.

3.4.2. For the random variable Y with pdf $f_Y(y) = \frac{2}{3} + \frac{2}{3}y, 0 \le y \le 1$, find $P(\frac{3}{4} \le Y \le 1)$.

3.4.3. Let $f_Y(y) = \frac{3}{2}y^2, -1 \le y \le 1$. Find $P(|Y - \frac{1}{2}| < \frac{1}{4})$. Draw a graph of $f_Y(y)$ and show the area representing the desired probability.

3.4.4. For persons infected with a certain form of malaria, the length of time spent in remission is described by the continuous pdf $f_Y(y) = \frac{1}{9}y^2, 0 \le y \le 3$, where Y is measured in years. What is the probability that a malaria patient's remission lasts longer than one year?

3.4.5. The length of time, Y, that a customer spends in line at a bank teller's window before being served is described by the exponential pdf $f_Y(y) = 0.2e^{-0.2y}, y \ge 0$.

(a) What is the probability that a customer will wait more than ten minutes?

(b) Suppose the customer will leave if the wait is more than ten minutes. Assume that the customer goes to the bank twice next month. Let the random variable X be the number of times the customer leaves without being served. Calculate $p_X(1)$.

3.4.6. Let n be a positive integer. Show that $f_Y(y) = (n+2)(n+1)y^n(1-y), 0 \le y \le 1$, is a pdf.

3.4.7. Find the cdf for the random variable Y given in Question 3.4.1. Calculate $P(0 \le Y \le \frac{1}{2})$ using $F_Y(y)$.

3.4.8. If Y is an exponential random variable, $f_Y(y) = \lambda e^{-\lambda y}, y \ge 0$, find $F_Y(y)$.

3.4.9. If the pdf for Y is

$$f_Y(y) = \begin{cases} 0, & |y| > 1 \\ 1 - |y|, & |y| \le 1 \end{cases}$$

find and graph $F_Y(y)$.

3.4.10. A continuous random variable Y has a cdf given by

$$F_Y(y) = \begin{cases} 0 & y < 0 \\ y^2 & 0 \le y < 1 \\ 1 & y \ge 1 \end{cases}$$

Find $P(\frac{1}{2} < Y \le \frac{3}{4})$ two ways—first, by using the cdf and second, by using the pdf.

3.4.11. A random variable Y has cdf

$$F_Y(y) = \begin{cases} 0 & y < 1 \\ \ln y & 1 \le y \le e \\ 1 & e < y \end{cases}$$

Find

(a) $P(Y < 2)$
(b) $P(2 < Y \le 2\frac{1}{2})$
(c) $P(2 < Y < 2\frac{1}{2})$
(d) $f_Y(y)$

3.4.12. The cdf for a random variable Y is defined by $F_Y(y) = 0$ for $y < 0$; $F_Y(y) = 4y^3 - 3y^4$ for $0 \le y \le 1$; and $F_Y(y) = 1$ for $y > 1$. Find $P(\frac{1}{4} \le Y \le \frac{3}{4})$ by integrating $f_Y(y)$.

3.4.13. Suppose $F_Y(y) = \frac{1}{12}(y^2 + y^3), 0 \le y \le 2$. Find $f_Y(y)$.

3.4.14. In a certain country, the distribution of a family's disposable income, Y, is described by the pdf $f_Y(y) = ye^{-y}, y \ge 0$. Find $F_Y(y)$.

3.4.15. The logistic curve $F(y) = \frac{1}{1+e^{-y}}, -\infty < y < \infty$, can represent a cdf since it is increasing, $\lim_{y\to-\infty} \frac{1}{1+e^{-y}} = 0$, and $\lim_{y\to+\infty} \frac{1}{1+e^{-y}} = 1$. Verify these three assertions and also find the associated pdf.

3.4.16. Let Y be the random variable described in Question 3.4.1. Define $W = 2Y$. Find $f_W(w)$. For which values of w is $f_W(w) \ne 0$?

3.4.17. Suppose that $f_Y(y)$ is a continuous and symmetric pdf, where *symmetry* is the property that $f_Y(y) = f_Y(-y)$ for all y. Show that $P(-a \le Y \le a) = 2F_Y(a) - 1$.

3.4.18. Let Y be a random variable denoting the age at which a piece of equipment fails. In reliability theory, the probability that an item fails at time y given that it has survived until time y is called the *hazard rate*, $h(y)$. In terms of the pdf and cdf,

$$h(y) = \frac{f_Y(y)}{1 - F_Y(y)}$$

Find $h(y)$ if Y has an exponential pdf (see Question 3.4.8).

3.5 Expected Values

Probability density functions, as we have already seen, provide a global overview of a random variable's behavior. If X is discrete, $p_X(k)$ gives $P(X = k)$ for all k; if Y is continuous, and A is any interval or a countable union of intervals, $P(Y \in A) = \int_A f_Y(y)\,dy$. Detail that explicit, though, is not always necessary—or even helpful. There are times when a more prudent strategy is to focus the information contained in a pdf by summarizing certain of its features with single numbers.

The first such feature that we will examine is *central tendency*, a term referring to the "average" value of a random variable. Consider the pdfs $p_X(k)$ and $f_Y(y)$ pictured in Figure 3.5.1. Although we obviously cannot predict with certainty what values any future X's and Y's will take on, it seems clear that X values will tend to lie somewhere near μ_X, and Y values somewhere near μ_Y. In some sense, then, we can characterize $p_X(k)$ by μ_X, and $f_Y(y)$ by μ_Y.

Figure 3.5.1

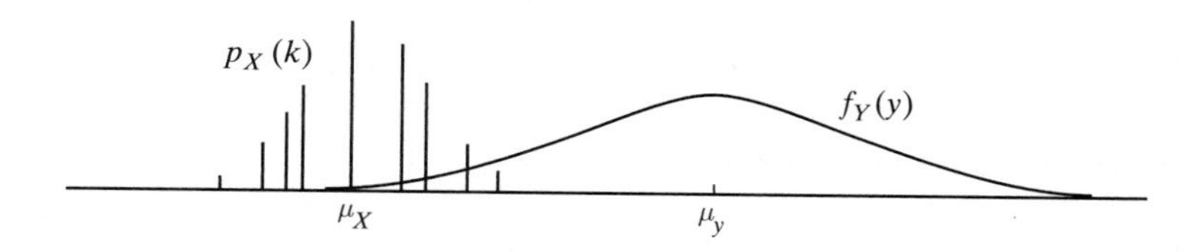

The most frequently used measure for describing central tendency—that is, for quantifying μ_X and μ_Y—is the *expected value*. Discussed at some length in this section and in Section 3.9, the expected value of a random variable is a slightly more abstract formulation of what we are already familiar with in simple discrete settings as the arithmetic average. Here, though, the values included in the average are "weighted" by the pdf.

Gambling affords a familiar illustration of the notion of an expected value. Consider the game of roulette. After bets are placed, the croupier spins the wheel and declares one of thirty-eight numbers, 00, 0, 1, 2, ..., 36, to be the winner. Disregarding what seems to be a perverse tendency of many roulette wheels to land on numbers for which no money has been wagered, we will assume that each of these thirty-eight numbers is equally likely (although only the eighteen numbers 1, 3, 5, ..., 35 are considered to be odd and only the eighteen numbers 2, 6, 4, ..., 36 are considered to be even). Suppose that our particular bet (at "even money") is \$1 on odds. If the random variable X denotes our winnings, then X takes on the value *1* if an odd number occurs, and -1 otherwise. Therefore,

$$p_X(1) = P(X = 1) = \frac{18}{38} = \frac{9}{19}$$

and

$$p_X(-1) = P(X = -1) = \frac{20}{38} = \frac{10}{19}$$

Then $\frac{9}{19}$ of the time we will win \$1 and $\frac{10}{19}$ of the time we will lose \$1. Intuitively, then, if we persist in this foolishness, we stand to *lose*, on the average, a little more than 5¢ each time we play the game:

$$\text{“expected” winnings} = \$1 \cdot \frac{9}{19} + (-\$1) \cdot \frac{10}{19}$$
$$= -\$0.053 \doteq -5¢$$

The number -0.053 is called the *expected value of X*.

Physically, an expected value can be thought of as a center of gravity. Here, for example, imagine two bars of height $\frac{10}{19}$ and $\frac{9}{19}$ positioned along a weightless X-axis at the points -1 and $+1$, respectively (see Figure 3.5.2). If a fulcrum were placed at the point -0.053, the system would be in balance, implying that we can think of that point as marking off the center of the random variable's distribution.

Figure 3.5.2

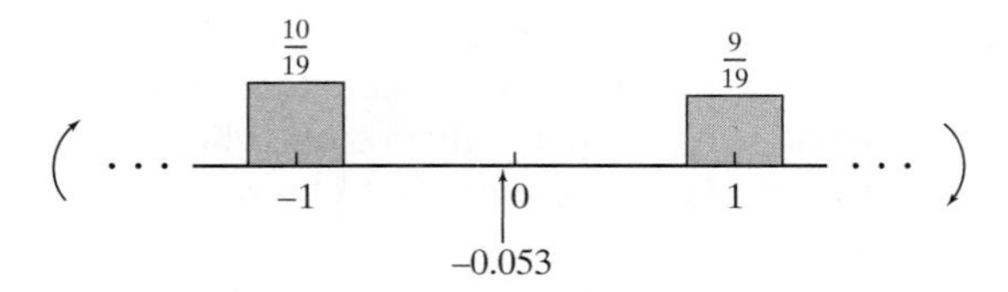

If X is a discrete random variable taking on each of its values with the same probability, the expected value of X is simply the everyday notion of an arithmetic average or mean:

$$\text{expected value of } X = \sum_{\text{all } k} k \cdot \frac{1}{n} = \frac{1}{n} \sum_{\text{all } k} k$$

Extending this idea to a discrete X described by an arbitrary pdf, $p_X(k)$, gives

$$\text{expected value of } X = \sum_{\text{all } k} k \cdot p_X(k) \tag{3.5.1}$$

For a continuous random variable Y, the summation in Equation 3.5.1 is replaced by an integration and $k \cdot p_X(k)$ becomes $y \cdot f_Y(y)$.

Definition 3.5.1. Let X be a discrete random variable with probability function $p_X(k)$. The *expected value of X* is denoted $E(X)$ (or sometimes μ or μ_X) and is given by

$$E(X) = \mu = \mu_X = \sum_{\text{all } k} k \cdot p_X(k)$$

Similarly, if Y is a continuous random variable with pdf $f_Y(y)$,

$$E(Y) = \mu = \mu_Y = \int_{-\infty}^{\infty} y \cdot f_Y(y)\, dy$$

Comment We assume that both the sum and the integral in Definition 3.5.1 converge absolutely:

$$\sum_{\text{all } k} |k| p_X(k) < \infty \qquad \int_{-\infty}^{\infty} |y| f_Y(y)\, dy < \infty$$

If not, we say that the random variable has no finite expected value. One immediate reason for requiring *absolute* convergence is that a convergent sum that is not absolutely convergent depends on the order in which the terms are added, and order should obviously not be a consideration when defining an average.

Example 3.5.1

Suppose X is a binomial random variable with $p = \frac{5}{9}$ and $n = 3$. Then $p_X(k) = P(X = k) = \binom{3}{k}\left(\frac{5}{9}\right)^k\left(\frac{4}{9}\right)^{3-k}$, $k = 0, 1, 2, 3$. What is the expected value of X?

Applying Definition 3.5.1 gives

$$E(X) = \sum_{k=0}^{3} k \cdot \binom{3}{k}\left(\frac{5}{9}\right)^k\left(\frac{4}{9}\right)^{3-k}$$

$$= (0)\left(\frac{64}{729}\right) + (1)\left(\frac{240}{729}\right) + (2)\left(\frac{300}{729}\right) + (3)\left(\frac{125}{729}\right) = \frac{1215}{729} = \frac{5}{3} = 3\left(\frac{5}{9}\right)$$

Comment Notice that the expected value here reduces to five-thirds, which can be written as three times five-ninths, the latter two factors being n and p, respectively. As the next theorem proves, that relationship is not a coincidence. ■

Theorem 3.5.1

Suppose X is a binomial random variable with parameters n and p. Then $E(X) = np$.

Proof According to Definition 3.5.1, $E(X)$ for a binomial random variable is the sum

$$E(X) = \sum_{k=0}^{n} k \cdot p_X(k) = \sum_{k=0}^{n} k\binom{n}{k} p^k (1-p)^{n-k}$$

$$= \sum_{k=0}^{n} \frac{k \cdot n!}{k!(n-k)!} p^k (1-p)^{n-k}$$

$$= \sum_{k=1}^{n} \frac{n!}{(k-1)!(n-k)!} p^k (1-p)^{n-k} \tag{3.5.2}$$

At this point, a trick is called for. If $E(X) = \sum_{\text{all } k} g(k)$ can be factored in such a way that $E(X) = h \sum_{\text{all } k} p_{X^*}(k)$, where $p_{X^*}(k)$ is the pdf for some random variable X^*, then $E(X) = h$, since the sum of a pdf over its entire range is 1. Here, suppose that np is factored out of Equation 3.5.2. Then

$$E(X) = np \sum_{k=1}^{n} \frac{(n-1)!}{(k-1)!(n-k)!} p^{k-1} (1-p)^{n-k}$$

$$= np \sum_{k=1}^{n} \binom{n-1}{k-1} p^{k-1} (1-p)^{n-k}$$

Now, let $j = k - 1$. It follows that

$$E(X) = np \sum_{j=0}^{n-1} \binom{n-1}{j} p^j (1-p)^{n-j-1}$$

Finally, letting $m = n - 1$ gives

$$E(X) = np \sum_{j=0}^{m} \binom{m}{j} p^j (1-p)^{m-j}$$

and, since the value of the sum is 1 (why?),

$$E(X) = np \tag{3.5.3}$$

□

Comment The statement of Theorem 3.5.1 should come as no surprise. If a multiple-choice test, for example, has one hundred questions, each with five possible answers, we would "expect" to get twenty correct, just by guessing. But if the random variable X denotes the number of correct answers (out of one hundred), $20 = E(X) = 100\left(\frac{1}{5}\right) = np$.

Example 3.5.2

An urn contains nine chips, five red and four white. Three are drawn out at random without replacement. Let X denote the number of red chips in the sample. Find $E(X)$.

From Section 3.2, we recognize X to be a hypergeometric random variable, where

$$P(X = k) = p_X(k) = \frac{\binom{5}{k}\binom{4}{3-k}}{\binom{9}{3}}, \quad k = 0, 1, 2, 3$$

Therefore,

$$E(X) = \sum_{k=0}^{3} k \cdot \frac{\binom{5}{k}\binom{4}{3-k}}{\binom{9}{3}}$$

$$= (0)\left(\frac{4}{84}\right) + (1)\left(\frac{30}{84}\right) + (2)\left(\frac{40}{84}\right) + (3)\left(\frac{10}{84}\right)$$

$$= \frac{5}{3}$$

■

Comment As was true in Example 3.5.1, the value found here for $E(X)$ suggests a general formula—in this case, for the expected value of a hypergeometric random variable.

Theorem 3.5.2

Suppose X is a hypergeometric random variable with parameters r, w, and n. That is, suppose an urn contains r red balls and w white balls. A sample of size n is drawn simultaneously from the urn. Let X be the number of red balls in the sample. Then $E(X) = \frac{rn}{r+w}$.

Proof See Question 3.5.25. □

Comment Let p represent the proportion of red balls in an urn—that is, $p = \frac{r}{r+w}$. The formula, then, for the expected value of a hypergeometric random variable has the same structure as the formula for the expected value of a binomial random variable:

$$E(X) = \frac{rn}{r+w} = n\frac{r}{r+w} = np$$

Example 3.5.3

Among the more common versions of the "numbers" racket is a game called D.J., its name deriving from the fact that the winning ticket is determined from Dow Jones averages. Three sets of stocks are used: Industrials, Transportations, and Utilities. Traditionally, the three are quoted at two different times, 11 A.M. and noon. The last digits of the earlier quotation are arranged to form a three-digit number; the noon quotation generates a second three-digit number, formed the same way. Those two numbers are then added together and the last three digits of that sum become the winning pick. Figure 3.5.3 shows a set of quotations for which *906* would be declared the winner.

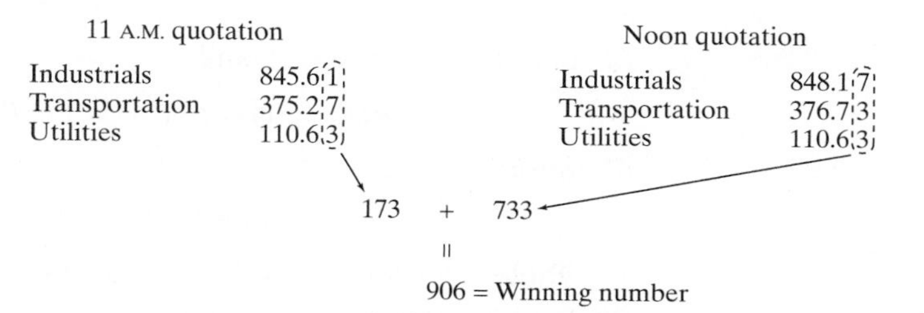

Figure 3.5.3

The payoff in D.J. is 700 to 1. Suppose that we bet \$5. How much do we stand to win, or lose, *on the average*?

Let p denote the probability of our number being the winner and let X denote our earnings. Then

$$X = \begin{cases} \$3500 & \text{with probability } p \\ -\$5 & \text{with probability } 1-p \end{cases}$$

and

$$E(X) = \$3500 \cdot p - \$5 \cdot (1-p)$$

Our intuition would suggest (and this time it would be correct!) that each of the possible winning numbers, 000 through 999, is equally likely. That being the case, $p = 1/1000$ and

$$E(X) = \$3500 \cdot \left(\frac{1}{1000}\right) - \$5 \cdot \left(\frac{999}{1000}\right) = -\$1.50$$

On the average, then, we lose \$1.50 on a \$5.00 bet. ■

Example 3.5.4

Suppose that fifty people are to be given a blood test to see who has a certain disease. The obvious laboratory procedure is to examine each person's blood individually, meaning that fifty tests would eventually be run. An alternative strategy is to divide each person's blood sample into two parts—say, A and B. All of the A's would then be mixed together and treated as one sample. If that "pooled" sample proved to be negative for the disease, all fifty individuals must necessarily be free of the infection, and no further testing would need to be done. If the pooled sample gave a positive reading, of course, all fifty B samples would have to be analyzed separately. Under what conditions would it make sense for a laboratory to consider pooling the fifty samples?

In principle, the pooling strategy is preferable (i.e., more economical) if it can substantially reduce the number of tests that need to be performed. Whether or not it can do so depends ultimately on the probability p that a person is infected with the disease.

Let the random variable X denote the number of tests that will have to be performed if the samples are pooled. Clearly,

$$X = \begin{cases} 1 & \text{if none of the fifty is infected} \\ 51 & \text{if at least one of the fifty is infected} \end{cases}$$

But

$$P(X = 1) = p_X(1) = P(\text{None of the fifty is infected})$$
$$= (1 - p)^{50}$$

(assuming independence), and

$$P(X = 51) = p_X(51) = 1 - P(X = 1) = 1 - (1 - p)^{50}$$

Therefore,

$$E(X) = 1 \cdot (1 - p)^{50} + 51 \cdot [1 - (1 - p)^{50}]$$

Table 3.5.1 shows $E(X)$ as a function of p. As our intuition would suggest, the pooling strategy becomes increasingly feasible as the prevalence of the disease diminishes. If the chance of a person being infected is 1 in 1000, for example, the pooling strategy requires an average of only 3.4 tests, a dramatic improvement over the fifty tests that would be needed if the samples were tested one by one. On the other hand, if 1 in 10 individuals is infected, pooling would be clearly inappropriate, requiring *more* than fifty tests [$E(X) = 50.7$].

Table 3.5.1

p	$E(X)$
0.5	51.0
0.1	50.7
0.01	20.8
0.001	3.4
0.0001	1.2

■

Example 3.5.5

Consider the following game. A fair coin is flipped until the first tail appears; we win \$2 if it appears on the first toss, \$4 if it appears on the second toss, and, in general, \$$2^k$ if it first occurs on the kth toss. Let the random variable X denote our winnings. How much should we have to pay in order for this to be a fair game? [*Note:* A fair game is one where the difference between the ante and $E(X)$ is 0.]

Known as the St. Petersburg paradox, this problem has a rather unusual answer. First, note that

$$p_X(2^k) = P(X = 2^k) = \frac{1}{2^k}, \quad k = 1, 2, \ldots$$

Therefore,

$$E(X) = \sum_{\text{all } k} 2^k p_X(2^k) = \sum_{k=1}^{\infty} 2^k \cdot \frac{1}{2^k} = 1 + 1 + 1 + \cdots$$

which is a divergent sum. That is, X does not have a finite expected value, so in order for this game to be fair, our ante would have to be an infinite amount of money! ■

Comment Mathematicians have been trying to "explain" the St. Petersburg paradox for almost two hundred years (56). The answer seems clearly absurd—no gambler would consider paying even $25 to play such a game, much less an infinite amount—yet the computations involved in showing that X has no finite expected value are unassailably correct. Where the difficulty lies, according to one common theory, is with our inability to put in perspective the very small probabilities of winning very large payoffs. Furthermore, the problem assumes that our opponent has infinite capital, which is an impossible state of affairs. We get a much more reasonable answer for $E(X)$ if the stipulation is added that our winnings can be at most, say, $1000 (see Question 3.5.19) or if the payoffs are assigned according to some formula other than 2^k (see Question 3.5.20).

Comment There are two important lessons to be learned from the St. Petersburg paradox. First is the realization that $E(X)$ is not necessarily a meaningful characterization of the "location" of a distribution. Question 3.5.24 shows another situation where the formal computation of $E(X)$ gives a similarly inappropriate answer. Second, we need to be aware that the notion of expected value is not necessarily synonymous with the concept of *worth*. Just because a game, for example, has a positive expected value—even a very *large* positive expected value—does not imply that someone would want to play it. Suppose, for example, that you had the opportunity to spend your last $10,000 on a sweepstakes ticket where the prize was $1 billion but the probability of winning was only one in ten thousand. The expected value of such a bet would be over $90,000,

$$\begin{aligned} E(X) &= \$1{,}000{,}000{,}000\left(\frac{1}{10{,}000}\right) + (-\$10{,}000)\left(\frac{9{,}999}{10{,}000}\right) \\ &= \$90{,}001 \end{aligned}$$

but it is doubtful that many people would rush out to buy a ticket. (Economists have long recognized the distinction between a payoff's numerical value and its perceived desirability. They refer to the latter as *utility*.)

Example 3.5.6

The distance, Y, that a molecule in a gas travels before colliding with another molecule can be modeled by the exponential pdf

$$f_Y(y) = \frac{1}{\mu} e^{-y/\mu}, \quad y \geq 0$$

where μ is a positive constant known as the *mean free path*. Find $E(Y)$.

Since the random variable here is continuous, its expected value is an integral:

$$E(Y) = \int_0^{\infty} y \frac{1}{\mu} e^{-y/\mu}\, dy$$

Let $w = y/\mu$, so that $dw = 1/\mu\ dy$. Then $E(Y) = \mu \int_0^\infty we^{-w}dw$. Setting $u = w$ and $dv = e^{-w}dw$ and integrating by parts gives

$$E(Y) = \mu[-we^{-w} - e^{-w}]\Big|_0^\infty = \mu \qquad (3.5.4)$$

Equation 3.5.4 shows that μ is aptly named—it does, in fact, represent the average distance a molecule travels, free of any collisions. Nitrogen (N_2), for example, at room temperature and standard atmospheric pressure has $\mu = 0.00005$ cm. An N_2 molecule, then, travels that far before colliding with another N_2 molecule, *on the average.* ■

Example 3.5.7

One continuous pdf that has a number of interesting applications in physics is the *Rayleigh distribution,* where the pdf is given by

$$f_Y(y) = \frac{y}{a^2}e^{-y^2/2a^2}, \quad a > 0; \quad 0 \le y < \infty \qquad (3.5.5)$$

Calculate the expected value for a random variable having a Rayleigh distribution.

From Definition 3.5.1,

$$E(Y) = \int_0^\infty y \cdot \frac{y}{a^2} e^{-y^2/2a^2} dy$$

Let $v = y/(\sqrt{2}a)$. Then

$$E(Y) = 2\sqrt{2}a \int_0^\infty v^2 e^{-v^2} dv$$

The integrand here is a special case of the general form $v^{2k}e^{-v^2}$. For $k = 1$,

$$\int_0^\infty v^{2k}e^{-v^2}dv = \int_0^\infty v^2 e^{-v^2} dv = \frac{1}{4}\sqrt{\pi}$$

Therefore,

$$E(Y) = 2\sqrt{2}a \cdot \frac{1}{4}\sqrt{\pi}$$
$$= a\sqrt{\pi/2}$$

■

Comment The pdf here is named for John William Strutt, Baron Rayleigh, the nineteenth- and twentieth-century British physicist who showed that Equation 3.5.5 is the solution to a problem arising in the study of wave motion. If two waves are superimposed, it is well known that the height of the resultant at any time t is simply the algebraic sum of the corresponding heights of the waves being added (see Figure 3.5.4). Seeking to extend that notion, Rayleigh posed the following question: If n waves, each having the same amplitude h and the same wavelength, are superimposed randomly with respect to phase, what can we say about the amplitude R of the resultant? Clearly, R is a random variable, its value depending on the particular collection of phase angles represented by the sample. What Rayleigh was able to show in his 1880 paper (166) is that when n is large, the probabilistic behavior of R is described by the pdf

$$f_R(r) = \frac{2r}{nh^2} \cdot e^{-r^2/nh^2}, \quad r > 0$$

which is just a special case of Equation 3.5.5 with $a = \sqrt{2/nh^2}$.

Figure 3.5.4

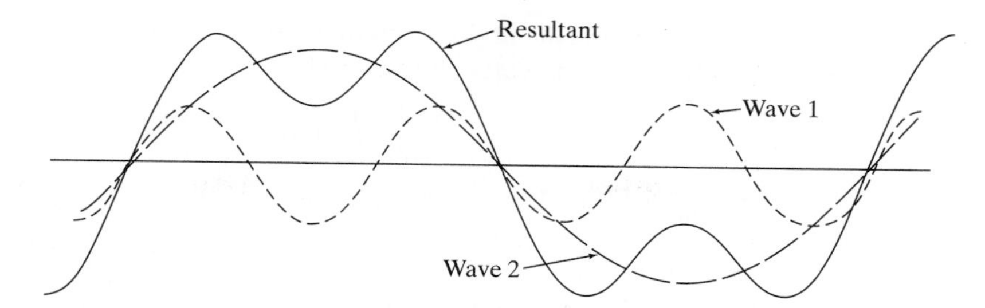

A Second Measure of Central Tendency: The Median

While the expected value is the most frequently used measure of a random variable's central tendency, it does have a weakness that sometimes makes it misleading and inappropriate. Specifically, if one or several possible values of a random variable are either much smaller or much larger than all the others, the value of μ can be distorted in the sense that it no longer reflects the center of the distribution in any meaningful way. For example, suppose a small community consists of a homogeneous group of middle-range salary earners, and then Bill Gates moves to town. Obviously, the town's average salary before and after the multibillionaire arrives will be quite different, even though he represents only one new value of the "salary" random variable.

It would be helpful to have a measure of central tendency that is not so sensitive to "outliers" or to probability distributions that are markedly skewed. One such measure is the *median*, which, in effect, divides the area under a pdf into two equal areas.

Definition 3.5.2. If X is a discrete random variable, the median, m, is that point for which $P(X < m) = P(X > m)$. In the event that $P(X \leq m) = 0.5$ and $P(X \geq m') = 0.5$, the median is defined to be the arithmetic average, $(m + m')/2$.

If Y is a continuous random variable, its median is the solution to the integral equation $\int_{-\infty}^{m} f_Y(y)\,dy = 0.5$.

Example 3.5.8

If a random variable's pdf is symmetric, both μ and m will be equal. Should $p_X(k)$ or $f_Y(y)$ not be symmetric, though, the difference between the expected value and the median can be considerable, especially if the asymmetry takes the form of extreme skewness. The situation described here is a case in point.

Soft-glow makes a 60-watt light bulb that is advertised to have an average life of one thousand hours. Assuming that the performance claim is valid, is it reasonable for consumers to conclude that the Soft-glow bulbs they buy will last for approximately one thousand hours?

No! If the average life of a bulb is one thousand hours, the (continuous) pdf, $f_Y(y)$, modeling the length of time, Y, that it remains lit before burning out is likely to have the form

$$f_Y(y) = 0.001e^{-0.001y}, \quad y > 0 \tag{3.5.6}$$

(for reasons explained in Chapter 4). But Equation 3.5.6 is a very skewed pdf, having a shape much like the curve drawn in Figure 3.4.8. The median for such a distribution will lie considerably to the left of the mean.

More specifically, the median lifetime for these bulbs—according to Definition 3.5.2—is the value m for which

$$\int_0^m 0.001e^{-0.001y}\,dy = 0.5$$

But $\int_0^m 0.001e^{-0.001y}\,dy = 1 - e^{-0.001m}$. Setting the latter equal to 0.5 implies that

$$m = (1/-0.001)\ln(0.5) = 693$$

So, even though the *average* life of one of these bulbs is 1000 hours, there is a 50% chance that the one you buy will last less than 693 hours. ■

Questions

3.5.1. Recall the game of Keno described in Question 3.2.26. The following are all the payoffs on a \$1 wager where the player has bet on ten numbers. Calculate $E(X)$, where the random variable X denotes the amount of money won.

Number of Correct Guesses	Payoff	Probability
< 5	−\$ 1	.935
5	2	.0514
6	18	.0115
7	180	.0016
8	1,300	1.35×10^{-4}
9	2,600	6.12×10^{-6}
10	10,000	1.12×10^{-7}

3.5.2. The roulette wheels in Monte Carlo typically have a 0 but not a 00. What is the expected value of betting on red in this case? If a trip to Monte Carlo costs \$3000, how much would a player have to bet to justify gambling there rather than Las Vegas?

3.5.3. The pdf describing the daily profit, X, earned by Acme Industries was derived in Example 3.3.7. Find the company's *average* daily profit.

3.5.4. In the game of redball, two drawings are made without replacement from a bowl that has four white ping-pong balls and two red ping-pong balls. The amount won is determined by how many of the red balls are selected. For a \$5 bet, a player can opt to be paid under either Rule A or Rule B, as shown. If you were playing the game, which would you choose? Why?

A		*B*	
No. of Red Balls Drawn	Payoff	No. of Red Balls Drawn	Payoff
0	0	0	0
1	\$2	1	\$1
2	\$10	2	\$20

3.5.5. Suppose a life insurance company sells a \$50,000, five-year term policy to a twenty-five-year-old woman. At the beginning of each year the woman is alive, the company collects a premium of \$$P$. The probability that the woman dies and the company pays the \$50,000 is given in the table below. So, for example, in Year 3, the company loses \$50,000 – \$$P$ with probability 0.00054 and gains \$$P$ with probability 1 – 0.00054 = 0.99946. If the company expects to make \$1000 on this policy, what should P be?

Year	Probability of Payoff
1	0.00051
2	0.00052
3	0.00054
4	0.00056
5	0.00059

3.5.6. A manufacturer has one hundred memory chips in stock, 4% of which are likely to be defective (based on past experience). A random sample of twenty chips is selected and shipped to a factory that assembles laptops. Let X denote the number of computers that receive faulty memory chips. Find $E(X)$.

3.5.7. Records show that 642 new students have just entered a certain Florida school district. Of those 642, a total of 125 are not adequately vaccinated. The district's physician has scheduled a day for students to receive whatever shots they might need. On any given day, though, 12% of the district's students are likely to be absent. How many new students, then, can be expected to remain inadequately vaccinated?

3.5.8. Calculate $E(Y)$ for the following pdfs:

(a) $f_Y(y) = 3(1-y)^2, 0 \le y \le 1$
(b) $f_Y(y) = 4ye^{-2y}, y \ge 0$
(c) $f_Y(y) = \begin{cases} \frac{3}{4}, & 0 \le y \le 1 \\ \frac{1}{4}, & 2 \le y \le 3 \\ 0, & \text{elsewhere} \end{cases}$
(d) $f_Y(y) = \sin y, \quad 0 \le y \le \frac{\pi}{2}$

3.5.9. Recall Question 3.4.4, where the length of time Y (in years) that a malaria patient spends in remission has pdf $f_Y(y) = \frac{1}{9}y^2, 0 \le y \le 3$. What is the average length of time that such a patient spends in remission?

3.5.10. Let the random variable Y have the uniform distribution over $[a, b]$; that is, $f_Y(y) = \frac{1}{b-a}$ for $a \le y \le b$. Find $E(Y)$ using Definition 3.5.1. Also, deduce the value of $E(Y)$, knowing that the expected value is the center of gravity of $f_Y(y)$.

3.5.11. Show that the expected value associated with the exponential distribution, $f_Y(y) = \lambda e^{-\lambda y}$, $y > 0$, is $1/\lambda$, where λ is a positive constant.

3.5.12. Show that

$$f_Y(y) = \frac{1}{y^2}, \quad y \ge 1$$

is a valid pdf but that Y does not have a finite expected value.

3.5.13. Based on recent experience, ten-year-old passenger cars going through a motor vehicle inspection station have an 80% chance of passing the emissions test. Suppose that two hundred such cars will be checked out next week. Write two formulas that show the number of cars that are expected to pass.

3.5.14. Suppose that fifteen observations are chosen at random from the pdf $f_Y(y) = 3y^2, 0 \le y \le 1$. Let X denote the number that lie in the interval $(\frac{1}{2}, 1)$. Find $E(X)$.

3.5.15. A city has 74,806 registered automobiles. Each is required to display a bumper decal showing that the owner paid an annual wheel tax of $50. By law, new decals need to be purchased during the month of the owner's birthday. How much wheel tax revenue can the city expect to receive in November?

3.5.16. Regulators have found that twenty-three of the sixty-eight investment companies that filed for bankruptcy in the past five years failed because of fraud, not for reasons related to the economy. Suppose that nine additional firms will be added to the bankruptcy rolls during the next quarter. How many of those failures are likely to be attributed to fraud?

3.5.17. An urn contains four chips numbered 1 through 4. Two are drawn without replacement. Let the random variable X denote the larger of the two. Find $E(X)$.

3.5.18. A fair coin is tossed three times. Let the random variable X denote the total number of heads that appear times the number of heads that appear on the first and third tosses. Find $E(X)$.

3.5.19. How much would you have to ante to make the St. Petersburg game "fair" (recall Example 3.5.5) if the most you could win was $1000? That is, the payoffs are $\$2^k$ for $1 \le k \le 9$, and $1000 for $k \ge 10$.

3.5.20. For the St. Petersburg problem (Example 3.5.5), find the expected payoff if

(a) the amounts won are c^k instead of 2^k, where $0 < c < 2$.
(b) the amounts won are $\log 2^k$. [This was a modification suggested by D. Bernoulli (a nephew of James Bernoulli) to take into account the decreasing marginal utility of money—the more you have, the less useful a bit more is.]

3.5.21. A fair die is rolled three times. Let X denote the number of different faces showing, $X = 1, 2, 3$. Find $E(X)$.

3.5.22. Two distinct integers are chosen at random from the first five positive integers. Compute the expected value of the absolute value of the difference of the two numbers.

3.5.23. Suppose that two evenly matched teams are playing in the World Series. On the average, how many games will be played? (The winner is the first team to get four victories.) Assume that each game is an independent event.

3.5.24. An urn contains one white chip and one black chip. A chip is drawn at random. If it is white, the "game" is over; if it is black, that chip and another black one are put into the urn. Then another chip is drawn at random from the "new" urn and the same rules for ending or continuing the game are followed (i.e., if the chip is white, the game is over; if the chip is black, it is placed back in the urn, together with another chip of the same color). The drawings continue until a white chip is selected. Show that the expected number of drawings necessary to get a white chip is not finite.

3.5.25. A random sample of size n is drawn without replacement from an urn containing r red chips and w white chips. Define the random variable X to be the number of red chips in the sample. Use the summation technique described in Theorem 3.5.1 to prove that $E(X) = rn/(r + w)$.

3.5.26. Given that X is a nonnegative, integer-valued random variable, show that

$$E(X) = \sum_{k=1}^{\infty} P(X \ge k)$$

3.5.27. Find the median for each of the following pdfs:

(a) $f_Y(y) = (\theta + 1)y^\theta, \ 0 \le y \le 1$, where $\theta > 0$
(b) $f_Y(y) = y + \frac{1}{2}, \ 0 \le y \le 1$

The Expected Value of a Function of a Random Variable

There are many situations that call for finding the expected value of a *function* of a random variable—say, $Y = g(X)$. One common example would be change of scale problems, where $g(X) = aX + b$ for constants a and b. Sometimes the pdf of the new random variable Y can be easily determined, in which case $E(Y)$ can be calculated by simply applying Definition 3.5.1. Often, though, $f_Y(y)$ can be difficult to derive, depending on the complexity of $g(X)$. Fortunately, Theorem 3.5.3 allows us to calculate the expected value of Y without knowing the pdf for Y.

Theorem 3.5.3 *Suppose X is a discrete random variable with pdf $p_X(k)$. Let $g(X)$ be a function of X. Then the expected value of the random variable $g(X)$ is given by*

$$E[g(X)] = \sum_{\text{all } k} g(k) \cdot p_X(k)$$

provided that $\sum_{\text{all } k} |g(k)| p_X(k) < \infty$.

If Y is a continuous random variable with pdf $f_Y(y)$, and if $g(Y)$ is a continuous function, then the expected value of the random variable $g(Y)$ is

$$E[g(Y)] = \int_{-\infty}^{\infty} g(y) \cdot f_Y(y)\, dy$$

provided that $\int_{-\infty}^{\infty} |g(y)| f_Y(y)\, dy < \infty$.

Proof We will prove the result for the discrete case. See (146) for details showing how the argument is modified when the pdf is continuous. Let $W = g(X)$. The set of all possible k values, $k_1, k_2, \ldots$, will give rise to a set of w values, $w_1, w_2, \ldots$, where, in general, more than one k may be associated with a given w. Let S_j be the set of k's for which $g(k) = w_j$ [so $\cup_j S_j$ is the entire set of k values for which $p_X(k)$ is defined]. We obviously have that $P(W = w_j) = P(X \in S_j)$, and we can write

$$\begin{aligned} E(W) = \sum_j w_j \cdot P(W = w_j) &= \sum_j w_j \cdot P(X \in S_j) \\ &= \sum_j w_j \sum_{k \in S_j} p_X(k) \\ &= \sum_j \sum_{k \in S_j} w_j \cdot p_X(k) \\ &= \sum_j \sum_{k \in S_j} g(k) p_X(k) \quad \text{(why?)} \\ &= \sum_{\text{all } k} g(k) p_X(k) \end{aligned}$$

Since it is being assumed that $\sum_{\text{all } k} |g(k)| p_X(k) < \infty$, the statement of the theorem holds. □

Corollary 3.5.1 *For any random variable W, $E(aW + b) = aE(W) + b$, where a and b are constants.* ◄

Proof Suppose W is continuous; the proof for the discrete case is similar. By Theorem 3.5.3, $E(aW + b) = \int_{-\infty}^{\infty} (aw + b) f_W(w)\, dw$, but the latter can be written $a\int_{-\infty}^{\infty} w \cdot f_W(w)\, dw + b\int_{-\infty}^{\infty} f_W(w)\, dw = aE(W) + b \cdot 1 = aE(W) + b$. □

Example 3.5.9

Suppose that X is a random variable whose pdf is nonzero only for the three values -2, 1, and $+2$:

k	$p_X(k)$
-2	$\frac{5}{8}$
1	$\frac{1}{8}$
2	$\frac{2}{8}$
	1

Let $W = g(X) = X^2$. Verify the statement of Theorem 3.5.3 by computing $E(W)$ two ways—first, by finding $p_W(w)$ and summing $w \cdot p_W(w)$ over w and, second, by summing $g(k) \cdot p_X(k)$ over k.

By inspection, the pdf for W is defined for only two values, 1 and 4:

$w\ (=k^2)$	$p_W(w)$
1	$\frac{1}{8}$
4	$\frac{7}{8}$
	1

Taking the first approach to find $E(W)$ gives

$$E(W) = \sum_w w \cdot p_W(w) = 1 \cdot \left(\frac{1}{8}\right) + 4 \cdot \left(\frac{7}{8}\right)$$

$$= \frac{29}{8}$$

To find the expected value via Theorem 3.5.3, we take

$$E[g(X)] = \sum_k k^2 \cdot p_X(k) = (-2)^2 \cdot \frac{5}{8} + (1)^2 \cdot \frac{1}{8} + (2)^2 \cdot \frac{2}{8}$$

with the sum here reducing to the answer we already found, $\frac{29}{8}$.

For this particular situation, neither approach was easier than the other. In general, that will not be the case. Finding $p_W(w)$ is often quite difficult, and on those occasions Theorem 3.5.3 can be of great benefit. ■

Example 3.5.10

Suppose the amount of propellant, Y, put into a can of spray paint is a random variable with pdf

$$f_Y(y) = 3y^2, \quad 0 < y < 1$$

Experience has shown that the largest surface area that can be painted by a can having Y amount of propellant is twenty times the area of a circle generated by a radius of Y ft. If the Purple Dominoes, a newly formed urban gang, have just stolen their first can of spray paint, can they expect to have enough to cover a $5' \times 8'$ subway panel with grafitti?

No. By assumption, the maximum area (in ft^2) that can be covered by a can of paint is described by the function

$$g(Y) = 20\pi Y^2$$

According to the second statement in Theorem 3.5.3, though, the average value for $g(Y)$ is slightly less than the desired 40 ft^2:

$$\begin{aligned}
E[g(Y)] &= \int_0^1 20\pi y^2 \cdot 3y^2\,dy \\
&= \left.\frac{60\pi y^5}{5}\right|_0^1 \\
&= 12\pi \\
&= 37.7 \text{ ft}^2
\end{aligned}$$

■

Example 3.5.11

A fair coin is tossed until a head appears. You will be given $\left(\frac{1}{2}\right)^k$ dollars if that first head occurs on the kth toss. How much money can you expect to be paid?

Let the random variable X denote the toss at which the first head appears. Then

$$\begin{aligned}
p_X(k) = P(X = k) &= P(\text{1st } k-1 \text{ tosses are tails and } k\text{th toss is a head}) \\
&= \left(\frac{1}{2}\right)^{k-1} \cdot \frac{1}{2} \\
&= \left(\frac{1}{2}\right)^k, \quad k = 1, 2, \ldots
\end{aligned}$$

Moreover,

$$\begin{aligned}
E(\text{amount won}) = E\left[\left(\frac{1}{2}\right)^X\right] &= E[g(X)] = \sum_{\text{all } k} g(k) \cdot p_X(k) \\
&= \sum_{k=1}^{\infty} \left(\frac{1}{2}\right)^k \cdot \left(\frac{1}{2}\right)^k \\
&= \sum_{k=1}^{\infty} \left(\frac{1}{2}\right)^{2k} = \sum_{k=1}^{\infty} \left(\frac{1}{4}\right)^k \\
&= \sum_{k=0}^{\infty} \left(\frac{1}{4}\right)^k - \left(\frac{1}{4}\right)^0 \\
&= \frac{1}{1-\frac{1}{4}} - 1 \\
&= \$0.33
\end{aligned}$$

■

Example 3.5.12

In one of the early applications of probability to physics, James Clerk Maxwell (1831–1879) showed that the speed S of a molecule in a perfect gas has a density function given by

$$f_S(s) = 4\sqrt{\frac{a^3}{\pi}}\, s^2 e^{-as^2}, \quad s > 0$$

where a is a constant depending on the temperature of the gas and the mass of the particle. What is the average *energy* of a molecule in a perfect gas?

Let m denote the molecule's mass. Recall from physics that energy (W), mass (m), and speed (S) are related through the equation

$$W = \frac{1}{2}mS^2 = g(S)$$

To find $E(W)$ we appeal to the second part of Theorem 3.5.3:

$$\begin{aligned} E(W) &= \int_0^{\infty} g(s) f_S(s)\, ds \\ &= \int_0^{\infty} \frac{1}{2}ms^2 \cdot 4\sqrt{\frac{a^3}{\pi}} s^2 e^{-as^2}\, ds \\ &= 2m\sqrt{\frac{a^3}{\pi}} \int_0^{\infty} s^4 e^{-as^2}\, ds \end{aligned}$$

We make the substitution $t = as^2$. Then

$$E(W) = \frac{m}{a\sqrt{\pi}} \int_0^{\infty} t^{3/2} e^{-t}\, dt$$

But

$$\int_0^{\infty} t^{3/2} e^{-t}\, dt = \left(\frac{3}{2}\right)\left(\frac{1}{2}\right)\sqrt{\pi} \quad \text{(see Section 4.4.6)}$$

so

$$\begin{aligned} E(\text{energy}) = E(W) &= \frac{m}{a\sqrt{\pi}} \left(\frac{3}{2}\right)\left(\frac{1}{2}\right)\sqrt{\pi} \\ &= \frac{3m}{4a} \end{aligned}$$

■

Example 3.5.13 Consolidated Industries is planning to market a new product and they are trying to decide how many to manufacture. They estimate that each item sold will return a profit of m dollars; each one not sold represents an n-dollar loss. Furthermore, they suspect the demand for the product, V, will have an exponential distribution,

$$f_V(v) = \left(\frac{1}{\lambda}\right) e^{-v/\lambda}, \quad v > 0$$

How many items should the company produce if they want to maximize their expected profit? (Assume that n, m, and λ are known.)

If a total of x items are made, the company's profit can be expressed as a function $Q(v)$, where

$$Q(v) = \begin{cases} mv - n(x - v) & \text{if } v < x \\ mx & \text{if } v \geq x \end{cases}$$

and v is the number of items sold. It follows that their *expected* profit is

$$\begin{aligned} E[Q(V)] &= \int_0^{\infty} Q(v) \cdot f_V(v)\, dv \\ &= \int_0^{x} [(m+n)v - nx]\left(\frac{1}{\lambda}\right) e^{-v/\lambda}\, dv + \int_x^{\infty} mx \cdot \left(\frac{1}{\lambda}\right) e^{-v/\lambda}\, dv \qquad (3.5.7) \end{aligned}$$

The integration here is straightforward, though a bit tedious. Equation 3.5.7 eventually simplifies to

$$E[Q(V)] = \lambda \cdot (m+n) - \lambda \cdot (m+n)e^{-x/\lambda} - nx$$

To find the optimal production level, we need to solve $dE[Q(V)]/dx = 0$ for x. But

$$\frac{dE[Q(V)]}{dx} = (m+n)e^{-x/\lambda} - n$$

and the latter equals zero when

$$x = -\lambda \cdot \ln\left(\frac{n}{m+n}\right)$$

■

Example 3.5.14

A point, y, is selected at random from the interval [0, 1], dividing the line into two segments (see Figure 3.5.5). What is the expected value of the ratio of the shorter segment to the longer segment?

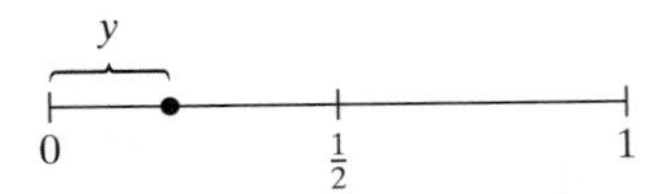

Figure 3.5.5

Notice, first, that the function

$$g(Y) = \frac{\text{shorter segment}}{\text{longer segment}}$$

has two expressions, depending on the location of the chosen point:

$$g(Y) = \begin{cases} y/(1-y), & 0 \leq y \leq \frac{1}{2} \\ (1-y)/y, & \frac{1}{2} < y \leq 1 \end{cases}$$

By assumption, $f_Y(y) = 1, 0 \leq y \leq 1$, so

$$E[g(Y)] = \int_0^{\frac{1}{2}} \frac{y}{1-y} \cdot 1\, dy + \int_{\frac{1}{2}}^1 \frac{1-y}{y} \cdot 1\, dy$$

Writing the second integrand as $(1/y - 1)$ gives

$$\int_{\frac{1}{2}}^1 \frac{1-y}{y} \cdot 1\, dy = \int_{\frac{1}{2}}^1 \left(\frac{1}{y} - 1\right) dy = (\ln y - y)\Big|_{\frac{1}{2}}^1$$

$$= \ln 2 - \frac{1}{2}$$

By symmetry, though, the two integrals are the same, so

$$E\left(\frac{\text{shorter segment}}{\text{longer segment}}\right) = 2\ln 2 - 1$$

$$= 0.39$$

On the average, then, the longer segment will be a little more than $2\frac{1}{2}$ times the length of the shorter segment. ■

Questions

3.5.28. Suppose X is a binomial random variable with $n=10$ and $p=\frac{2}{5}$. What is the expected value of $3X-4$?

3.5.29. A typical day's production of a certain electronic component is twelve. The probability that one of these components needs rework is 0.11. Each component needing rework costs \$100. What is the average daily cost for defective components?

3.5.30. Let Y have probability density function

$$f_Y(y)=2(1-y),\ 0\le y\le 1$$

Suppose that $W=Y^2$, in which case

$$f_W(w)=\frac{1}{\sqrt{w}}-1,\ 0\le w\le 1$$

Find $E(W)$ in two different ways.

3.5.31. A tool and die company makes castings for steel stress-monitoring gauges. Their annual profit, Q, in hundreds of thousands of dollars, can be expressed as a function of product demand, y:

$$Q(y)=2(1-e^{-2y})$$

Suppose that the demand (in thousands) for their castings follows an exponential pdf, $f_Y(y)=6e^{-6y}$, $y>0$. Find the company's expected profit.

3.5.32. A box is to be constructed so that its height is five inches and its base is Y inches by Y inches, where Y is a random variable described by the pdf, $f_Y(y)=6y(1-y)$, $0<y<1$. Find the expected volume of the box.

3.5.33. Grades on the last Economics 301 exam were not very good. Graphed, their distribution had a shape similar to the pdf

$$f_Y(y)=\frac{1}{5000}(100-y),\quad 0\le y\le 100$$

As a way of "curving" the results, the professor announces that he will replace each person's grade, Y, with a new grade, $g(Y)$, where $g(Y)=10\sqrt{Y}$. Will the professor's strategy be successful in raising the class average above 60?

3.5.34. If Y has probability density function

$$f_Y(y)=2y,\ 0\le y\le 1$$

then $E(Y)=\frac{2}{3}$. Define the random variable W to be the squared deviation of Y from its mean, that is, $W=\left(Y-\frac{2}{3}\right)^2$. Find $E(W)$.

3.5.35. The hypotenuse, Y, of the isosceles right triangle shown is a random variable having a uniform pdf over the interval [6, 10]. Calculate the expected value of the triangle's area. Do not leave the answer as a function of a.

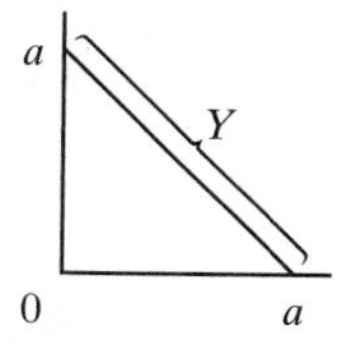

3.5.36. An urn contains n chips numbered 1 through n. Assume that the probability of choosing chip i is equal to ki, $i=1,2,\ldots,n$. If one chip is drawn, calculate $E\left(\frac{1}{X}\right)$, where the random variable X denotes the number showing on the chip selected. [*Hint*: Recall that the sum of the first n integers is $n(n+1)/2$.]

3.6 The Variance

We saw in Section 3.5 that the location of a distribution is an important characteristic and that it can be effectively measured by calculating either the mean or the median. A second feature of a distribution that warrants further scrutiny is its *dispersion*—that is, the extent to which its values are spread out. The two properties are totally different: Knowing a pdf's location tells us absolutely nothing about its dispersion. Table 3.6.1, for example, shows two simple discrete pdfs with the same expected value (equal to zero), but with vastly different dispersions.

Table 3.6.1

k	$p_{X_1}(k)$	k	$p_{X_2}(k)$
-1	$\frac{1}{2}$	$-1{,}000{,}000$	$\frac{1}{2}$
1	$\frac{1}{2}$	$1{,}000{,}000$	$\frac{1}{2}$

It is not immediately obvious how the dispersion in a pdf should be quantified. Suppose that X is any discrete random variable. One seemingly reasonable approach would be to average the deviations of X from their mean—that is, calculate the expected value of $X - \mu$. As it happens, that strategy will not work because the negative deviations will exactly cancel the positive deviations, making the numerical value of such an average always zero, regardless of the amount of spread present in $p_X(k)$:

$$E(X - \mu) = E(X) - \mu = \mu - \mu = 0 \tag{3.6.1}$$

Another possibility would be to modify Equation 3.6.1 by making all the deviations positive—that is, to replace $E(X - \mu)$ with $E(|X - \mu|)$. This does work, and it *is* sometimes used to measure dispersion, but the absolute value is somewhat troublesome mathematically: It does not have a simple arithmetic formula, nor is it a differentiable function. *Squaring* the deviations proves to be a much better approach.

Definition 3.6.1. The *variance* of a random variable is the expected value of its squared deviations from μ. If X is discrete, with pdf $p_X(k)$,

$$\text{Var}(X) = \sigma^2 = E[(X - \mu)^2] = \sum_{\text{all } k} (k - \mu)^2 \cdot p_X(k)$$

If Y is continuous, with pdf $f_Y(y)$,

$$\text{Var}(Y) = \sigma^2 = E[(Y - \mu)^2] = \int_{-\infty}^{\infty} (y - \mu)^2 \cdot f_Y(y)\, dy$$

[If $E(X^2)$ or $E(Y^2)$ is not finite, the variance is not defined.]

Comment One unfortunate consequence of Definition 3.6.1 is that the units for the variance are the square of the units for the random variable: If Y is measured in inches, for example, the units for $\text{Var}(Y)$ are inches squared. This causes obvious problems in relating the variance back to the sample values. For that reason, in applied statistics, where unit compatibility is especially important, dispersion is measured not by the variance but by the *standard deviation*, which is defined to be the square root of the variance. That is,

$$\sigma = \text{standard deviation} = \begin{cases} \sqrt{\sum_{\text{all } k} (k - \mu)^2 \cdot p_X(k)} & \text{if } X \text{ is discrete} \\ \sqrt{\int_{-\infty}^{\infty} (y - \mu)^2 \cdot f_Y(y)\, dy} & \text{if } Y \text{ is continuous} \end{cases}$$

Comment The analogy between the expected value of a random variable and the center of gravity of a physical system was pointed out in Section 3.5. A similar equivalency holds between the variance and what engineers call a *moment of inertia*. If a set of weights having masses $m_1, m_2, \ldots$ are positioned along a (weightless) rigid bar at distances $r_1, r_2, \ldots$ from an axis of rotation (see Figure 3.6.1), the moment of inertia of the system is defined to be value $\sum_i m_i r_i^2$. Notice, though, that if the masses were the probabilities associated with a discrete random variable and if the axis of rotation were actually μ, then $r_1, r_2, \ldots$ could be written $(k_1 - \mu), (k_2 - \mu), \ldots$ and $\sum_i m_i r_i^2$ would be the same as the variance, $\sum_{\text{all } k} (k - \mu)^2 \cdot p_X(k)$.

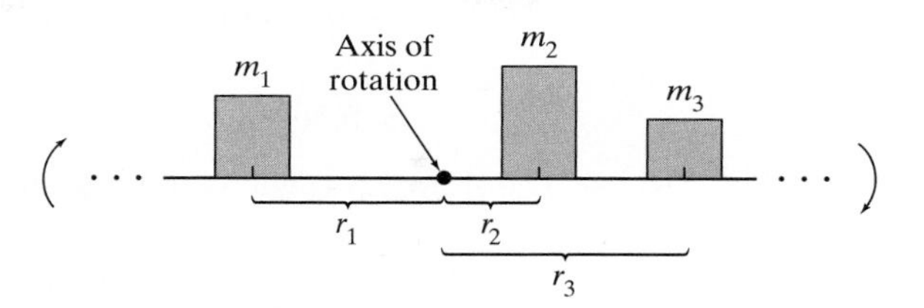

Figure 3.6.1

Definition 3.6.1 gives a formula for calculating σ^2 in both the discrete and the continuous cases. An equivalent—but easier-to-use—formula is given in Theorem 3.6.1.

Theorem 3.6.1 *Let W be any random variable, discrete or continuous, having mean μ and for which $E(W^2)$ is finite. Then*

$$\text{Var}(W) = \sigma^2 = E(W^2) - \mu^2$$

Proof We will prove the theorem for the continuous case. The argument for discrete W is similar. In Theorem 3.5.3, let $g(W) = (W - \mu)^2$. Then

$$\text{Var}(W) = E[(W - \mu)^2] = \int_{-\infty}^{\infty} g(w) f_W(w)\, dw = \int_{-\infty}^{\infty} (w - \mu)^2 f_W(w)\, dw$$

Squaring out the term $(w - \mu)^2$ that appears in the integrand and using the additive property of integrals gives

$$\begin{aligned}
\int_{-\infty}^{\infty} (w-\mu)^2 f_W(w)\, dw &= \int_{-\infty}^{\infty} (w^2 - 2\mu w + \mu^2) f_W(w)\, dw \\
&= \int_{-\infty}^{\infty} w^2 f_W(w)\, dw - 2\mu \int_{-\infty}^{\infty} w f_W(w)\, dw + \int_{-\infty}^{\infty} \mu^2 f_W(w)\, dw \\
&= E(W^2) - 2\mu^2 + \mu^2 = E(W^2) - \mu^2
\end{aligned}$$

Note that the equality $\int_{-\infty}^{\infty} w^2 f_W(w)\, dw = E(W^2)$ also follows from Theorem 3.5.3. □

Example 3.6.1 An urn contains five chips, two red and three white. Suppose that two are drawn out at random, *without replacement*. Let X denote the number of red chips in the sample. Find Var(X).

Note, first, that since the chips are not being replaced from drawing to drawing, X is a hypergeometric random variable. Moreover, we need to find μ, regardless of which formula is used to calculate σ^2. In the notation of Theorem 3.5.2, $r = 2$, $w = 3$, and $n = 2$, so

$$\mu = rn/(r + w) = 2 \cdot 2/(2 + 3) = 0.8$$

To find Var(X) using Definition 3.6.1, we write

$$\text{Var}(X) = E[(X-\mu)^2] = \sum_{\text{all } x} (x-\mu)^2 \cdot f_X(x)$$

$$= (0-0.8)^2 \cdot \frac{\binom{2}{0}\binom{3}{2}}{\binom{5}{2}} + (1-0.8)^2 \cdot \frac{\binom{2}{1}\binom{3}{1}}{\binom{5}{2}} + (2-0.8)^2 \cdot \frac{\binom{2}{2}\binom{3}{0}}{\binom{5}{2}}$$

$$= 0.36$$

To use Theorem 3.6.1, we would first find $E(X^2)$. From Theorem 3.5.3,

$$E(X^2) = \sum_{\text{all } x} x^2 \cdot f_X(x) = 0^2 \cdot \frac{\binom{2}{0}\binom{3}{2}}{\binom{5}{2}} + 1^2 \cdot \frac{\binom{2}{1}\binom{3}{1}}{\binom{5}{2}} + 2^2 \cdot \frac{\binom{2}{2}\binom{3}{0}}{\binom{5}{2}}$$

$$= 1.00$$

Then

$$\text{Var}(X) = E(X^2) - \mu^2 = 1.00 - (0.8)^2$$

$$= 0.36$$

confirming what we calculated earlier. ■

In Section 3.5 we encountered a change of scale formula that applied to expected values. For any constants a and b and any random variable W, $E(aW+b) = aE(W)+b$. A similar issue arises in connection with the *variance* of a linear transformation: If Var$(W) = \sigma^2$, what is the variance of $aW+b$?

Theorem 3.6.2 *Let W be any random variable having mean μ and where $E(W^2)$ is finite. Then $Var(aW+b) = a^2 Var(W)$.*

Proof Using the same approach taken in the proof of Theorem 3.6.1, it can be shown that $E[(aW+b)^2] = a^2E(W^2) + 2ab\mu + b^2$. We also know from the corollary to Theorem 3.5.3 that $E(aW+b) = a\mu + b$. Using Theorem 3.6.1, then, we can write

$$\text{Var}(aW+b) = E[(aW+b)^2] - [E(aW+b)]^2$$

$$= [a^2E(W^2) + 2ab\mu + b^2] - [a\mu + b]^2$$

$$= [a^2E(W^2) + 2ab\mu + b^2] - [a^2\mu^2 + 2ab\mu + b^2]$$

$$= a^2[E(W^2) - \mu^2] = a^2\text{Var}(W)$$ □

Example 3.6.2 A random variable Y is described by the pdf

$$f_Y(y) = 2y, \quad 0 \le y \le 1$$

What is the standard deviation of $3Y+2$?

First, we need to find the variance of Y. But

$$E(Y) = \int_0^1 y \cdot 2y\, dy = \frac{2}{3}$$

and

$$E(Y^2) = \int_0^1 y^2 \cdot 2y\, dy = \frac{1}{2}$$

so

$$\text{Var}(Y) = E(Y^2) - \mu^2 = \frac{1}{2} - \left(\frac{2}{3}\right)^2$$
$$= \frac{1}{18}$$

Then, by Theorem 3.6.2,

$$\text{Var}(3Y + 2) = (3)^2 \cdot \text{Var}(Y) = 9 \cdot \frac{1}{18}$$
$$= \frac{1}{2}$$

which makes the standard deviation of $3Y + 2$ equal to $\sqrt{\frac{1}{2}}$ or *0.71*. ■

Questions

3.6.1. Find Var(X) for the urn problem of Example 3.6.1 if the sampling is done *with* replacement.

3.6.2. Find the variance of Y if

$$f_Y(y) = \begin{cases} \frac{3}{4}, & 0 \le y \le 1 \\ \frac{1}{4}, & 2 \le y \le 3 \\ 0, & \text{elsewhere} \end{cases}$$

3.6.3. Ten equally qualified applicants, six men and four women, apply for three lab technician positions. Unable to justify choosing any of the applicants over all the others, the personnel director decides to select the three at random. Let X denote the number of men hired. Compute the standard deviation of X.

3.6.4. Compute the variance for a uniform random variable defined on the unit interval.

3.6.5. Use Theorem 3.6.1 to find the variance of the random variable Y, where

$$f_Y(y) = 3(1 - y)^2, \quad 0 < y < 1$$

3.6.6. If

$$f_Y(y) = \frac{2y}{k^2}, \quad 0 \le y \le k$$

for what value of k does Var(Y) = 2?

3.6.7. Calculate the standard deviation, σ, for the random variable Y whose pdf has the graph shown below:

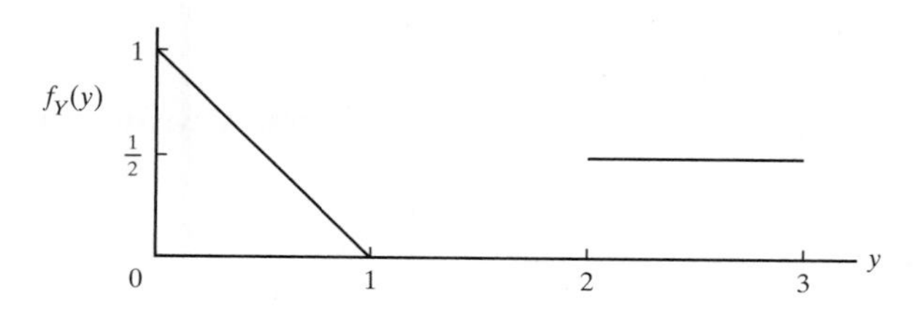

3.6.8. Consider the pdf defined by

$$f_Y(y) = \frac{2}{y^3}, \quad y \ge 1$$

Show that (a) $\int_1^\infty f_Y(y)\,dy = 1$, (b) $E(Y) = 2$, and (c) Var(Y) is not finite.

3.6.9. Frankie and Johnny play the following game. Frankie selects a number at random from the interval $[a, b]$. Johnny, not knowing Frankie's number, is to pick a second number from that same inverval and pay Frankie an amount, W, equal to the squared difference between the two [so $0 \le W \le (b - a)^2$]. What should be Johnny's strategy if he wants to minimize his expected loss?

3.6.10. Let Y be a random variable whose pdf is given by $f_Y(y) = 5y^4, 0 \le y \le 1$. Use Theorem 3.6.1 to find Var(Y).

3.6.11. Suppose that Y is an exponential random variable, so $f_Y(y) = \lambda e^{-\lambda y}$, $y \ge 0$. Show that the variance of Y is $1/\lambda^2$.

3.6.12. Suppose that Y is an exponential random variable with $\lambda = 2$ (recall Question 3.6.11). Find $P[Y > E(Y) + 2\sqrt{\text{Var}(Y)}]$.

3.6.13. Let X be a random variable with finite mean μ. Define for every real number a, $g(a) = E[(X - a)^2]$. Show that

$$g(a) = E[(X - \mu)^2] + (\mu - a)^2.$$

What is another name for $\min g(a)$?

3.6.14. Let Y have the pdf given in Question 3.6.5. Find the variance of W, where $W = -5Y + 12$.

3.6.15. If Y denotes a temperature recorded in degrees Fahrenheit, then $\frac{5}{9}(Y - 32)$ is the corresponding temperature in degrees Celsius. If the standard deviation for a set of temperatures is 15.7°F, what is the standard deviation of the equivalent Celsius temperatures?

3.6.16. If $E(W) = \mu$ and $\text{Var}(W) = \sigma^2$, show that

$$E\left(\frac{W-\mu}{\sigma}\right) = 0 \quad \text{and} \quad \text{Var}\left(\frac{W-\mu}{\sigma}\right) = 1$$

3.6.17. Suppose U is a uniform random variable over $[0, 1]$.

(a) Show that $Y = (b-a)U + a$ is uniform over $[a, b]$.
(b) Use part (a) and Question 3.6.4 to find the variance of Y.

3.6.18. Recovering small quantities of calcium in the presence of magnesium can be a difficult problem for an analytical chemist. Suppose the amount of calcium Y to be recovered is uniformly distributed between 4 and 7 mg. The amount of calcium recovered by one method is the random variable

$$W_1 = 0.2281 + (0.9948)Y + E_1$$

where the error term E_1 has mean 0 and variance 0.0427 and is independent of Y.

A second procedure has random variable

$$W_2 = -0.0748 + (1.0024)Y + E_2$$

where the error term E_2 has mean 0 and variance 0.0159 and is independent of Y.

The better technique should have a mean as close as possible to the mean of $Y (= 5.5)$, and a variance as small as possible. Compare the two methods on the basis of mean and variance.

Higher Moments

The quantities we have identified as the mean and the variance are actually special cases of what are referred to more generally as the *moments* of a random variable. More precisely, $E(W)$ is the *first moment about the origin* and σ^2 is the *second moment about the mean*. As the terminology suggests, we will have occasion to define higher moments of W. Just as $E(W)$ and σ^2 reflect a random variable's location and dispersion, so it is possible to characterize other aspects of a distribution in terms of other moments. We will see, for example, that the skewness of a distribution—that is, the extent to which it is not symmetric around μ—can be effectively measured in terms of a *third* moment. Likewise, there are issues that arise in certain applied statistics problems that require a knowledge of the flatness of a pdf, a property that can be quantified by the *fourth* moment.

Definition 3.6.2. Let W be any random variable with pdf $f_W(w)$. For any positive integer r,

1. The rth *moment of W about the origin*, μ_r, is given by

$$\mu_r = E(W^r)$$

provided $\int_{-\infty}^{\infty} |w|^r \cdot f_W(w)\, dw < \infty$ (or provided the analogous condition on the *summation* of $|w|^r$ holds, if W is discrete). When $r = 1$, we usually delete the subscript and write $E(W)$ as μ rather than μ_1.

2. The rth *moment of W about the mean*, μ'_r, is given by

$$\mu'_r = E[(W-\mu)^r]$$

provided the finiteness conditions of part 1 hold.

Comment We can express μ'_r in terms of μ_j, $j = 1, 2, \ldots, r$, by simply writing out the binomial expansion of $(W-\mu)^r$:

$$\mu'_r = E[(W-\mu)^r] = \sum_{j=0}^{r} \binom{r}{j} E(W^j)(-\mu)^{r-j}$$

Thus,

$$\mu_2' = E[(W-\mu)^2] = \sigma^2 = \mu_2 - \mu_1^2$$
$$\mu_3' = E[(W-\mu)^3] = \mu_3 - 3\mu_1\mu_2 + 2\mu_1^3$$
$$\mu_4' = E[(W-\mu)^4] = \mu_4 - 4\mu_1\mu_3 + 6\mu_1^2\mu_2 - 3\mu_1^4$$

and so on.

Example 3.6.3

The *skewness* of a pdf can be measured in terms of its third moment about the mean. If a pdf is symmetric, $E[(W-\mu)^3]$ will obviously be zero; for pdfs not symmetric, $E[(W-\mu)^3]$ will not be zero. In practice, the symmetry (or lack of symmetry) of a pdf is often measured by the *coefficient of skewness*, γ_1, where

$$\gamma_1 = \frac{E[(W-\mu)^3]}{\sigma^3}$$

Dividing μ_3' by σ^3 makes γ_1 dimensionless.

A second "shape" parameter in common use is the *coefficient of kurtosis*, γ_2, which involves the *fourth* moment about the mean. Specifically,

$$\gamma_2 = \frac{E[(W-\mu)^4]}{\sigma^4} - 3$$

For certain pdfs, γ_2 is a useful measure of peakedness: relatively flat pdfs are said to be *platykurtic*; more peaked pdfs are called *leptokurtic*. ■

Earlier in this chapter we encountered random variables whose means do not exist—recall, for example, the St. Petersburg paradox. More generally, there are random variables having certain of their higher moments finite and certain others, not finite. Addressing the question of whether or not a given $E(W^j)$ is finite is the following existence theorem.

Theorem 3.6.3

If the kth moment of a random variable exists, all moments of order less than k exist.

Proof Let $f_Y(y)$ be the pdf of a continuous random variable Y. By Definition 3.6.2, $E(Y^k)$ exists if and only if

$$\int_{-\infty}^{\infty} |y|^k \cdot f_Y(y)\,dy < \infty \tag{3.6.2}$$

Let $1 \le j < k$. To prove the theorem we must show that

$$\int_{-\infty}^{\infty} |y|^j \cdot f_Y(y)\,dy < \infty$$

is implied by Inequality 3.6.2. But

$$\begin{aligned}
\int_{-\infty}^{\infty} |y|^j \cdot f_Y(y)\,dy &= \int_{|y|\le 1} |y|^j \cdot f_Y(y)\,dy + \int_{|y|>1} |y|^j \cdot f_Y(y)\,dy \\
&\le \int_{|y|\le 1} f_Y(y)\,dy + \int_{|y|>1} |y|^j \cdot f_Y(y)\,dy \\
&\le 1 + \int_{|y|>1} |y|^j \cdot f_Y(y)\,dy \\
&\le 1 + \int_{|y|>1} |y|^k \cdot f_Y(y)\,dy < \infty
\end{aligned}$$

Therefore, $E(Y^j)$ exists, $j = 1, 2, \ldots, k-1$. The proof for discrete random variables is similar. □

Questions

3.6.19. Let Y be a uniform random variable defined over the interval $(0, 2)$. Find an expression for the rth moment of Y about the origin. Also, use the binomial expansion as described in the Comment to find $E[(Y-\mu)^6]$.

3.6.20. Find the coefficient of skewness for an exponential random variable having the pdf

$$f_Y(y) = e^{-y}, \quad y > 0$$

3.6.21. Calculate the coefficient of kurtosis for a uniform random variable defined over the unit interval, $f_Y(y) = 1$, for $0 \le y \le 1$.

3.6.22. Suppose that W is a random variable for which $E[(W-\mu)^3] = 10$ and $E(W^3) = 4$. Is it possible that $\mu = 2$?

3.6.23. If $Y = aX + b$, $a > 0$, show that Y has the same coefficients of skewness and kurtosis as X.

3.6.24. Let Y be the random variable of Question 3.4.6, where for a positive integer n, $f_Y(y) = (n+2)(n+1)y^n(1-y)$, $0 \le y \le 1$.

(a) Find $\text{Var}(Y)$.

(b) For any positive integer k, find the kth moment around the origin.

3.6.25. Suppose that the random variable Y is described by the pdf

$$f_Y(y) = c \cdot y^{-6}, \quad y > 1$$

(a) Find c.

(b) What is the highest moment of Y that exists?

3.7 Joint Densities

Sections 3.3 and 3.4 introduced the basic terminology for describing the probabilistic behavior of a *single* random variable. Such information, while adequate for many problems, is insufficient when more than one variable are of interest to the experimenter. Medical researchers, for example, continue to explore the relationship between blood cholesterol and heart disease, and, more recently, between "good" cholesterol and "bad" cholesterol. And more than a little attention—both political and pedagogical—is given to the role played by K–12 funding in the performance of would-be high school graduates on exit exams. On a smaller scale, electronic equipment and systems are often designed to have built-in redundancy: Whether or not that equipment functions properly ultimately depends on the reliability of two different components.

The point is, there are many situations where two relevant random variables, say, X and Y,[2] are defined on the same sample space. Knowing only $f_X(x)$ and $f_Y(y)$, though, does not necessarily provide enough information to characterize the all-important *simultaneous* behavior of X and Y. The purpose of this section is to introduce the concepts, definitions, and mathematical techniques associated with distributions based on two (or more) random variables.

Discrete Joint Pdfs

As we saw in the single-variable case, the pdf is defined differently depending on whether the random variable is discrete or continuous. The same distinction applies

---

[2] For the next several sections we will suspend our earlier practice of using X to denote a discrete random variable and Y to denote a continuous random variable. The category of the random variables will need to be determined from the context of the problem. Typically, though, X and Y will either be both discrete or both continuous.

to joint pdfs. We begin with a discussion of joint pdfs as they apply to two discrete random variables.

Definition 3.7.1. Suppose S is a discrete sample space on which two random variables, X and Y, are defined. The *joint probability density function of X and Y (or joint pdf)* is denoted $p_{X,Y}(x, y)$, where

$$p_{X,Y}(x, y) = P(\{s|X(s) = x \quad \text{and} \quad Y(s) = y\})$$

Comment A convenient shorthand notation for the meaning of $p_{X,Y}(x, y)$, consistent with what we used earlier for pdfs of single discrete random variables, is to write $p_{X,Y}(x, y) = P(X = x, Y = y)$.

Example 3.7.1 A supermarket has two express lines. Let X and Y denote the number of customers in the first and in the second, respectively, at any given time. During nonrush hours, the joint pdf of X and Y is summarized by the following table:

		X			
		0	1	2	3
Y	0	0.1	0.2	0	0
	1	0.2	0.25	0.05	0
	2	0	0.05	0.05	0.025
	3	0	0	0.025	0.05

Find $P(|X - Y| = 1)$, the probability that X and Y differ by exactly 1.

By definition,

$$\begin{aligned} P(|X - Y| = 1) &= \sum_{|x-y|=1} \sum p_{X,Y}(x, y) \\ &= p_{X,Y}(0, 1) + p_{X,Y}(1, 0) + p_{X,Y}(1, 2) \\ &\quad + p_{X,Y}(2, 1) + p_{X,Y}(2, 3) + p_{X,Y}(3, 2) \\ &= 0.2 + 0.2 + 0.05 + 0.05 + 0.025 + 0.025 \\ &= 0.55 \end{aligned}$$

[Would you expect $p_{X,Y}(x, y)$ to be symmetric? Would you expect the event $|X - Y| \geq 2$ to have zero probability?] ■

Example 3.7.2 Suppose two fair dice are rolled. Let X be the sum of the numbers showing, and let Y be the larger of the two. So, for example,

$$p_{X,Y}(2, 3) = P(X = 2, Y = 3) = P(\emptyset) = 0$$

$$p_{X,Y}(4, 3) = P(X = 4, Y = 3) = P(\{(1, 3)(3, 1)\}) = \frac{2}{36}$$

and

$$p_{X,Y}(6, 3) = P(X = 6, Y = 3) = P(\{(3, 3)\} = \frac{1}{36}$$

The entire joint pdf is given in Table 3.7.1.

Table 3.7.1

x \ y	1	2	3	4	5	6	Row totals
2	1/36	0	0	0	0	0	1/36
3	0	2/36	0	0	0	0	2/36
4	0	1/36	2/36	0	0	0	3/36
5	0	0	2/36	2/36	0	0	4/36
6	0	0	1/36	2/36	2/36	0	5/36
7	0	0	0	2/36	2/36	2/36	6/36
8	0	0	0	1/36	2/36	2/36	5/36
9	0	0	0	0	2/36	2/36	4/36
10	0	0	0	0	1/36	2/36	3/36
11	0	0	0	0	0	2/36	2/36
12	0	0	0	0	0	1/36	1/36
Col. totals	1/36	3/36	5/36	7/36	9/36	11/36	

Notice that the row totals in the right-hand margin of the table give the pdf for X. Similarly, the column totals along the bottom detail the pdf for Y. Those are not coincidences. Theorem 3.7.1 gives a formal statement of the relationship between the joint pdf and the individual pdfs. ■

Theorem 3.7.1 *Suppose that $p_{X,Y}(x, y)$ is the joint pdf of the discrete random variables X and Y. Then*

$$p_X(x) = \sum_{\text{all } y} p_{X,Y}(x, y) \quad \textit{and} \quad p_Y(y) = \sum_{\text{all } x} p_{X,Y}(x, y)$$

Proof We will prove the first statement. Note that the collection of sets $(Y = y)$ for all y forms a partition of S; that is, they are disjoint and $\bigcup_{\text{all } y}(Y = y) = S$. The set $(X = x) = (X = x) \cap S = (X = x) \cap \bigcup_{\text{all } y}(Y = y) = \bigcup_{\text{all } y}[(X = x) \cap (Y = y)]$, so

$$p_X(x) = P(X = x) = P\left(\bigcup_{\text{all } y}[(X = x) \cap (Y = y)]\right)$$

$$= \sum_{\text{all } y} P(X = x, Y = y) = \sum_{\text{all } y} p_{X,Y}(x, y)$$

□

Definition 3.7.2. An individual pdf obtained by summing a joint pdf over all values of the other random variable is called a *marginal pdf*.

Continuous Joint Pdfs

If X and Y are both continuous random variables, Definition 3.7.1 does not apply because $P(X = x, Y = y)$ will be identically 0 for all (x, y). As was the case in single-variable situations, the joint pdf for two continuous random variables will be defined as a function that when integrated yields the probability that (X, Y) lies in a specified region of the xy-plane.

Definition 3.7.3. Two random variables defined on the same set of real numbers are *jointly continuous* if there exists a function $f_{X,Y}(x, y)$ such that for any region R in the xy-plane, $P[(X, Y) \in R] = \int\int_R f_{X,Y}(x, y)\, dx\, dy$. The function $f_{X,Y}(x, y)$ is the *joint pdf of X and Y*.

Comment Any function $f_{X,Y}(x, y)$ for which

1. $f_{X,Y}(x, y) \geq 0$ for all x and y
2. $\int_{-\infty}^{\infty}\int_{-\infty}^{\infty} f_{X,Y}(x, y)\, dx\, dy = 1$

qualifies as a joint pdf. We shall employ the convention of naming the domain only where the joint pdf is nonzero; everywhere else it will be assumed to be zero. This is analogous, of course, to the notation used earlier in describing the domain of single random variables.

Example 3.7.3 Suppose that the variation in two continuous random variables, X and Y, can be modeled by the joint pdf $f_{X,Y}(x, y) = cxy$, for $0 < y < x < 1$. Find c.

By inspection, $f_{X,Y}(x, y)$ will be nonnegative as long as $c \geq 0$. The particular c that qualifies $f_{X,Y}(x, y)$ as a joint pdf, though, is the one that makes the volume under $f_{X,Y}(x, y)$ equal to 1. But

$$\int\int_S cxy\, dy\, dx = 1 = c\int_0^1\left[\int_0^x (xy)\, dy\right]dx = c\int_0^1 x\left(\frac{y^2}{2}\Big|_0^x\right)dx$$

$$= c\int_0^1\left(\frac{x^3}{2}\right)dx = c\frac{x^4}{8}\Big|_0^1 = \left(\frac{1}{8}\right)c$$

Therefore, $c = 8$. ■

Example 3.7.4 A study claims that the daily number of hours, X, a teenager watches television and the daily number of hours, Y, he works on his homework are approximated by the joint pdf

$$f_{X,Y}(x, y) = xye^{-(x+y)}, \quad x > 0, \quad y > 0$$

What is the probability that a teenager chosen at random spends at least twice as much time watching television as he does working on his homework?

The region, R, in the xy-plane corresponding to the event "$X \geq 2Y$" is shown in Figure 3.7.1. It follows that $P(X \geq 2Y)$ is the volume under $f_{X,Y}(x, y)$ above the region R:

$$P(X \geq 2Y) = \int_0^{\infty}\int_0^{x/2} xye^{-(x+y)}\, dy\, dx$$

Separating variables, we can write

$$P(X \geq 2Y) = \int_0^{\infty} xe^{-x}\left[\int_0^{x/2} ye^{-y} dy\right]dx$$

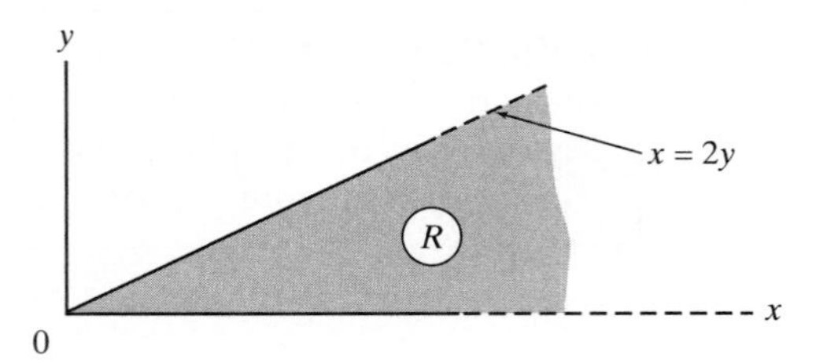

Figure 3.7.1

and the double integral reduces to $\frac{7}{27}$:

$$\begin{aligned}
P(X \geq 2Y) &= \int_0^{\infty} xe^{-x}\left[1-\left(\frac{x}{2}+1\right)e^{-x/2}\right]dx \\
&= \int_0^{\infty} xe^{-x}dx - \int_0^{\infty} \frac{x^2}{2}e^{-3x/2}\,dx - \int_0^{\infty} xe^{-3x/2}dx \\
&= 1 - \frac{16}{54} - \frac{4}{9} \\
&= \frac{7}{27}
\end{aligned}$$

■

Geometric Probability

One particularly important special case of Definition 3.7.3 is the *joint uniform pdf*, which is represented by a surface having a constant height everywhere above a specified rectangle in the xy-plane. That is,

$$f_{X,Y}(x, y) = \frac{1}{(b-a)(d-c)}, \quad a \leq x \leq b, c \leq y \leq d$$

If R is some region in the rectangle where X and Y are defined, $P((X, Y) \in R)$ reduces to a simple ratio of areas:

$$P((X, Y) \in R) = \frac{\text{area of } R}{(b-a)(d-c)} \tag{3.7.1}$$

Calculations based on Equation 3.7.1 are referred to as *geometric probabilities*.

Example 3.7.5

Two friends agree to meet on the University Commons "sometime around 12:30." But neither of them is particularly punctual—or patient. What will actually happen is that each will arrive at random sometime in the interval from 12:00 to 1:00. If one arrives and the other is not there, the first person will wait fifteen minutes or until 1:00, whichever comes first, and then leave. What is the probability that the two will get together?

To simplify notation, we can represent the time period from 12:00 to 1:00 as the interval from zero to sixty minutes. Then if x and y denote the two arrival times, the sample space is the 60×60 square shown in Figure 3.7.2. Furthermore, the event M, "The two friends meet," will occur if and only if $|x - y| \leq 15$ or, equivalently,

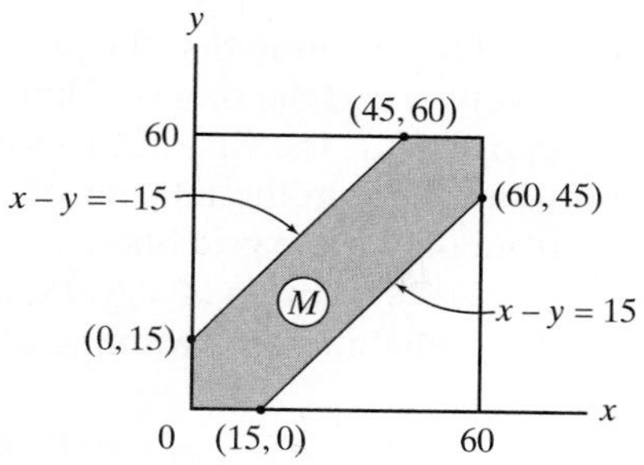

Figure 3.7.2

if and only if $-15 \le x - y \le 15$. These inequalities appear as the shaded region in Figure 3.7.2.

Notice that the areas of the triangles above and below M are each equal to $\frac{1}{2}(45)(45)$. It follows that the two friends have a 44% chance of meeting:

$$
\begin{aligned}
P(M) &= \frac{\text{area of } M}{\text{area of } S} \\
&= \frac{(60)^2 - 2\left[\frac{1}{2}(45)(45)\right]}{(60)^2} \\
&= 0.44
\end{aligned}
$$

■

Example 3.7.6

A carnival operator wants to set up a ringtoss game. Players will throw a ring of diameter d onto a grid of squares, the side of each square being of length s (see Figure 3.7.3). If the ring lands entirely inside a square, the player wins a prize. To ensure a profit, the operator must keep the player's chances of winning down to something less than one in five. How small can the operator make the ratio d/s?

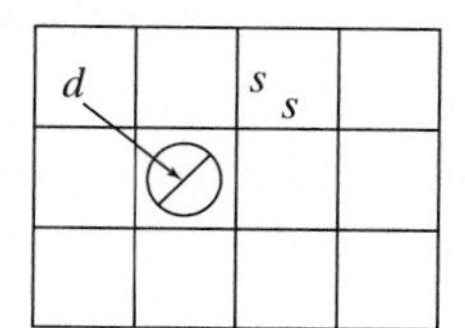

Figure 3.7.3

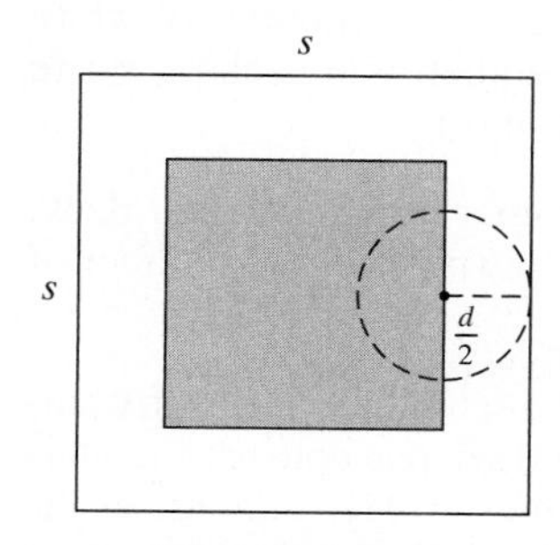

Figure 3.7.4

First, assume that the player is required to stand far enough away that no skill is involved and the ring is falling at random on the grid. From Figure 3.7.4, we see that in order for the ring not to touch any side of the square, the ring's center must be somewhere in the interior of a smaller square, each side of which is a distance $d/2$ from one of the grid lines.

Since the area of a grid square is s^2 and the area of an interior square is $(s-d)^2$, the probability of a winning toss can be written as the ratio:

$$P(\text{Ring touches no lines}) = \frac{(s-d)^2}{s^2}$$

But the operator requires that

$$\frac{(s-d)^2}{s^2} \leq 0.20$$

Solving for d/s gives

$$\frac{d}{s} \geq 1 - \sqrt{0.20} = 0.55$$

That is, if the diameter of the ring is at least 55% as long as the side of one of the squares, the player will have no more than a 20% chance of winning. ■

Questions

3.7.1. If $p_{X,Y}(x, y) = cxy$ at the points $(1, 1)$, $(2, 1)$, $(2, 2)$, and $(3, 1)$, and equals 0 elsewhere, find c.

3.7.2. Let X and Y be two continuous random variables defined over the unit square. What does c equal if $f_{X,Y}(x, y) = c(x^2 + y^2)$?

3.7.3. Suppose that random variables X and Y vary in accordance with the joint pdf, $f_{X,Y}(x, y) = c(x + y)$, $0 < x < y < 1$. Find c.

3.7.4. Find c if $f_{X,Y}(x, y) = cxy$ for X and Y defined over the triangle whose vertices are the points $(0, 0)$, $(0, 1)$, and $(1, 1)$.

3.7.5. An urn contains four red chips, three white chips, and two blue chips. A random sample of size 3 is drawn without replacement. Let X denote the number of white chips in the sample and Y the number of blue chips. Write a formula for the joint pdf of X and Y.

3.7.6. Four cards are drawn from a standard poker deck. Let X be the number of kings drawn and Y the number of queens. Find $p_{X,Y}(x, y)$.

3.7.7. An advisor looks over the schedules of his fifty students to see how many math and science courses each has registered for in the coming semester. He summarizes his results in a table. What is the probability that a student selected at random will have signed up for more math courses than science courses?

		Number of math courses, X		
		0	1	2
Number of science courses, Y	0	11	6	4
	1	9	10	3
	2	5	0	2

3.7.8. Consider the experiment of tossing a fair coin three times. Let X denote the number of heads on the last flip, and let Y denote the total number of heads on the three flips. Find $p_{X,Y}(x, y)$.

3.7.9. Suppose that two fair dice are tossed one time. Let X denote the number of 2's that appear, and Y the number of 3's. Write the matrix giving the joint probability density function for X and Y. Suppose a third random variable, Z, is defined, where $Z = X + Y$. Use $p_{X,Y}(x, y)$ to find $p_Z(z)$.

3.7.10. Suppose that X and Y have a bivariate uniform density over the unit square:

$$f_{X,Y}(x, y) = \begin{cases} c, & 0 < x < 1, \quad 0 < y < 1 \\ 0, & \text{elsewhere} \end{cases}$$

(a) Find c.
(b) Find $P\left(0 < X < \frac{1}{2}, 0 < Y < \frac{1}{4}\right)$.

3.7.11. Let X and Y have the joint pdf

$$f_{X,Y}(x, y) = 2e^{-(x+y)}, \quad 0 < x < y, \quad 0 < y$$

Find $P(Y < 3X)$.

3.7.12. A point is chosen at random from the interior of a circle whose equation is $x^2 + y^2 \leq 4$. Let the random variables X and Y denote the x- and y-coordinates of the sampled point. Find $f_{X,Y}(x, y)$.

3.7.13. Find $P(X < 2Y)$ if $f_{X,Y}(x, y) = x + y$ for X and Y each defined over the unit interval.

3.7.14. Suppose that five independent observations are drawn from the continuous pdf $f_T(t) = 2t, 0 \leq t \leq 1$. Let X denote the number of t's that fall in the interval $0 \leq t < \frac{1}{3}$ and let Y denote the number of t's that fall in the interval $\frac{1}{3} \leq t < \frac{2}{3}$. Find $p_{X,Y}(1, 2)$.

3.7.15. A point is chosen at random from the interior of a right triangle with base b and height h. What is the probability that the y value is between 0 and $h/2$?

Marginal Pdfs for Continuous Random Variables

The notion of marginal pdfs in connection with discrete random variables was introduced in Theorem 3.7.1 and Definition 3.7.2. An analogous relationship holds in the continuous case—*integration*, though, replaces the summation that appears in Theorem 3.7.1.

Theorem 3.7.2 *Suppose X and Y are jointly continuous with joint pdf $f_{X,Y}(x, y)$. Then the* marginal pdfs, *$f_X(x)$ and $f_Y(y)$, are given by*

$$f_X(x) = \int_{-\infty}^{\infty} f_{X,Y}(x, y)\, dy \quad \text{and} \quad f_Y(y) = \int_{-\infty}^{\infty} f_{X,Y}(x, y)\, dx$$

Proof It suffices to verify the first of the theorem's two equalities. As is often the case with proofs for continuous random variables, we begin with the cdf:

$$F_X(x) = P(X \leq x) = \int_{-\infty}^{\infty} \int_{-\infty}^{x} f_{X,Y}(t, y)\, dt\, dy = \int_{-\infty}^{x} \int_{-\infty}^{\infty} f_{X,Y}(x, y)\, dy\, dt$$

Differentiating both ends of the equation above gives

$$f_X(x) = \int_{-\infty}^{\infty} f_{X,Y}(x, y)\, dy$$

(recall Theorem 3.4.1). □

Example 3.7.7 Suppose that two continuous random variables, X and Y, have the joint uniform pdf

$$f_{X,Y}(x, y) = \frac{1}{6}, \quad 0 \leq x \leq 3, \quad 0 \leq y \leq 2$$

Find $f_X(x)$.

Applying Theorem 3.7.2 gives

$$f_X(x) = \int_0^2 f_{X,Y}(x, y)\, dy = \int_0^2 \frac{1}{6}\, dy = \frac{1}{3}, \quad 0 \leq x \leq 3$$

Notice that X, by itself, is a uniform random variable defined over the interval [0, 3]; similarly, we would find that $f_Y(y)$ has a uniform pdf over the interval [0, 2]. ■

Example 3.7.8 Consider the case where X and Y are two continuous random variables, jointly distributed over the first quadrant of the xy-plane according to the joint pdf,

$$f_{X,Y}(x, y) = \begin{cases} y^2 e^{-y(x+1)}, & x \geq 0, \quad y \geq 0 \\ 0, & \text{elsewhere} \end{cases}$$

Find the two marginal pdfs.

First, consider $f_X(x)$. By Theorem 3.7.2,

$$f_X(x) = \int_{-\infty}^{\infty} f_{X,Y}(x, y)\, dy = \int_0^{\infty} y^2 e^{-y(x+1)}\, dy$$

In the integrand, substitute

$$u = y(x+1)$$

making $du = (x+1)\, dy$. This gives

$$f_X(x) = \frac{1}{x+1} \int_0^{\infty} \frac{u^2}{(x+1)^2} e^{-u}\, du = \frac{1}{(x+1)^3} \int_0^{\infty} u^2 e^{-u}\, du$$

After applying integration by parts (twice) to $\int_0^{\infty} u^2 e^{-u}\, du$, we get

$$\begin{aligned} f_X(x) &= \frac{1}{(x+1)^3} \left[-u^2 e^{-u} - 2ue^{-u} - 2e^{-u} \right] \Big|_0^{\infty} \\ &= \frac{1}{(x+1)^3} \left[2 - \lim_{u \to \infty} \left(\frac{u^2}{e^u} + \frac{2u}{e^u} + \frac{2}{e^u} \right) \right] \\ &= \frac{2}{(x+1)^3}, \quad x \geq 0 \end{aligned}$$

Finding $f_Y(y)$ is a bit easier:

$$\begin{aligned} f_Y(y) &= \int_{-\infty}^{\infty} f_{X,Y}(x, y)\, dx = \int_0^{\infty} y^2 e^{-y(x+1)}\, dx \\ &= y^2 e^{-y} \int_0^{\infty} e^{-yx}\, dx = y^2 e^{-y} \left(\frac{1}{y} \right) \left(-e^{-yx} \Big|_0^{\infty} \right) \\ &= y e^{-y}, \quad y \geq 0 \end{aligned}$$

■

Questions

3.7.16. Find the marginal pdf of X for the joint pdf derived in Question 3.7.5.

3.7.17. Find the marginal pdfs of X and Y for the joint pdf derived in Question 3.7.8.

3.7.18. The campus recruiter for an international conglomerate classifies the large number of students she interviews into three categories—the lower quarter, the middle half, and the upper quarter. If she meets six students on a given morning, what is the probability that they will be evenly divided among the three categories? What is the marginal probability that exactly two will belong to the middle half?

3.7.19. For each of the following joint pdfs, find $f_X(x)$ and $f_Y(y)$.

(a) $f_{X,Y}(x, y) = \frac{1}{2}, 0 \leq x \leq 2, 0 \leq y \leq 1$
(b) $f_{X,Y}(x, y) = \frac{3}{2} y^2, 0 \leq x \leq 2, 0 \leq y \leq 1$
(c) $f_{X,Y}(x, y) = \frac{2}{3}(x + 2y), 0 \leq x \leq 1, 0 \leq y \leq 1$
(d) $f_{X,Y}(x, y) = c(x + y), 0 \leq x \leq 1, 0 \leq y \leq 1$
(e) $f_{X,Y}(x, y) = 4xy, 0 \leq x \leq 1, 0 \leq y \leq 1$
(f) $f_{X,Y}(x, y) = xye^{-(x+y)}, 0 \leq x, 0 \leq y$
(g) $f_{X,Y}(x, y) = ye^{-xy-y}, 0 \leq x, 0 \leq y$

3.7.20. For each of the following joint pdfs, find $f_X(x)$ and $f_Y(y)$.

(a) $f_{X,Y}(x, y) = \frac{1}{2}, 0 \le x \le y \le 2$
(b) $f_{X,Y}(x, y) = \frac{1}{x}, 0 \le y \le x \le 1$
(c) $f_{X,Y}(x, y) = 6x, 0 \le x \le 1, 0 \le y \le 1 - x$

3.7.21. Suppose that $f_{X,Y}(x, y) = 6(1 - x - y)$ for x and y defined over the unit square, subject to the restriction that $0 \le x + y \le 1$. Find the marginal pdf for X.

3.7.22. Find $f_Y(y)$ if $f_{X,Y}(x, y) = 2e^{-x}e^{-y}$ for x and y defined over the shaded region pictured.

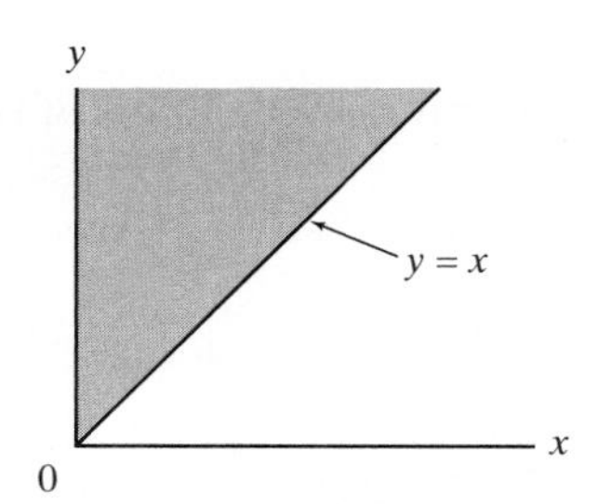

3.7.23. Suppose that X and Y are discrete random variables with

$$p_{X,Y}(x, y) = \frac{4!}{x!y!(4-x-y)!}\left(\frac{1}{2}\right)^x\left(\frac{1}{3}\right)^y\left(\frac{1}{6}\right)^{4-x-y},$$
$$0 \le x + y \le 4$$

Find $p_X(x)$ and $p_Y(x)$.

3.7.24. A generalization of the binomial model occurs when there is a sequence of n independent trials with *three* outcomes, where $p_1 = P(\text{outcome 1})$ and $p_2 = P(\text{outcome 2})$. Let X and Y denote the number of trials (out of n) resulting in outcome 1 and outcome 2, respectively.

(a) Show that $p_{X,Y}(x, y) = \frac{n!}{x!y!(n-x-y)!}p_1^x p_2^y (1 - p_1 - p_2)^{n-x-y}, 0 \le x + y \le n$
(b) Find $p_X(x)$ and $p_Y(x)$.

(*Hint:* See Question 3.7.23.)

Joint Cdfs

For a single random variable X, the cdf of X evaluated at some point x—that is, $F_X(x)$—is the probability that the random variable X takes on a value less than or equal to x. Extended to two variables, a *joint cdf* [evaluated at the point (u, v)] is the probability that $X \le u$ and, simultaneously, that $Y \le v$.

Definition 3.7.4. Let X and Y be any two random variables. The *joint cumulative distribution function of X and Y* (or *joint cdf*) is denoted $F_{X,Y}(u, v)$, where

$$F_{X,Y}(u, v) = P(X \le u \quad \text{and} \quad Y \le v)$$

Example 3.7.9

Find the joint cdf, $F_{X,Y}(u, v)$, for the two random variables X and Y whose joint pdf is given by $f_{X,Y}(x, y) = \frac{4}{3}(x + xy), 0 \le x \le 1, 0 \le y \le 1$.

If Definition 3.7.4 is applied, the probability that $X \le u$ and $Y \le v$ becomes a double integral of $f_{X,Y}(x, y)$:

$$F_{X,Y}(u, v) = \frac{4}{3}\int_0^v \int_0^u (x + xy)\,dx\,dy = \frac{4}{3}\int_0^v \left[\int_0^u (x + xy)\,dx\right] dy$$
$$= \frac{4}{3}\int_0^v \left[\frac{x^2}{2}(1 + y)\Big|_0^u\right] dy = \frac{4}{3}\int_0^v \frac{u^2}{2}(1 + y)\,dy$$
$$= \frac{4}{3}\frac{u^2}{2}\left(y + \frac{y^2}{2}\right)\Big|_0^v = \frac{4}{3}\frac{u^2}{2}\left(v + \frac{v^2}{2}\right)$$

which simplifies to

$$F_{X,Y}(u,v) = \frac{1}{3}u^2(2v+v^2)$$

[For what values of u and v is $F_{X,Y}(u,v)$ defined?] ■

Theorem 3.7.3 *Let $F_{X,Y}(u,v)$ be the joint cdf associated with the continuous random variables X and Y. Then the joint pdf of X and Y, $f_{X,Y}(x,y)$, is a second partial derivative of the joint cdf—that is, $f_{X,Y}(x,y) = \frac{\partial^2}{\partial x \partial y}F_{X,Y}(x,y)$, provided $F_{X,Y}(x,y)$ has continuous second partial derivatives.*

Example 3.7.10 What is the joint pdf of the random variables X and Y whose joint cdf is $F_{X,Y}(x,y) = \frac{1}{3}x^2(2y+y^2)$?

By Theorem 3.7.3,

$$f_{X,Y}(x,y) = \frac{\partial^2}{\partial x\, \partial y}F_{X,Y}(x,y) = \frac{\partial^2}{\partial x\, \partial y}\frac{1}{3}x^2(2y+y^2)$$

$$= \frac{\partial}{\partial y}\frac{2}{3}x(2y+y^2) = \frac{2}{3}x(2+2y) = \frac{4}{3}(x+xy)$$

Notice the similarity between Examples 3.7.9 and 3.7.10—$f_{X,Y}(x,y)$ is the same in both examples; so is $F_{X,Y}(x,y)$. ■

Multivariate Densities

The definitions and theorems in this section extend in a very straightforward way to situations involving more than two variables. The joint pdf for n discrete random variables, for example, is denoted $p_{X_1,\ldots,X_n}(x_1,\ldots,x_n)$ where

$$p_{X_1,\ldots,X_n}(x_1,\ldots,x_n) = P(X_1 = x_1,\ldots,X_n = x_n)$$

For n continuous random variables, the joint pdf is that function $f_{X_1,\ldots,X_n}(x_1,\ldots,x_n)$ having the property that for any region R in n-space,

$$P[(X_1,\ldots,X_n)\in R] = \iint_R \cdots \int f_{X_1,\ldots,X_n}(x_1,\ldots,x_n)\, dx_1 \cdots dx_n$$

And if $F_{X_1,\ldots,X_n}(x_1,\ldots,x_n)$ is the joint *cdf* of continuous random variables $X_1,\ldots,X_n$—that is, $F_{X_1,\ldots,X_n}(x_1,\ldots,x_n) = P(X_1 \le x_1,\ldots,X_n \le x_n)$—then

$$f_{X_1,\ldots,X_n}(x_1,\ldots,x_n) = \frac{\partial^n}{\partial x_1 \cdots \partial x_n}F_{X_1,\ldots,X_n}(x_1,\ldots,x_n)$$

The notion of a marginal pdf also extends readily, although in the n-variate case, a marginal pdf can, itself, be a joint pdf. Given $X_1,\ldots,X_n$, the marginal pdf of any subset of r of those variables $(X_{i_1}, X_{i_2},\ldots,X_{i_r})$ is derived by integrating (or summing) the joint pdf with respect to the remaining $n-r$ variables $(X_{j_1}, X_{j_2},\ldots,X_{j_{n-r}})$. If the X_i's are all continuous, for example,

$$f_{X_{i_1},\ldots,X_{i_r}}(x_{i_1},\ldots,x_{i_r}) = \int_{-\infty}^{\infty}\int_{-\infty}^{\infty}\cdots\int_{-\infty}^{\infty} f_{X_1,\ldots,X_n}(x_1,\ldots,x_n)\, dx_{j_1}\cdots dx_{j_{n-r}}$$

Questions

3.7.25. Consider the experiment of simultaneously tossing a fair coin and rolling a fair die. Let X denote the number of heads showing on the coin and Y the number of spots showing on the die.

(a) List the outcomes in S.
(b) Find $F_{X,Y}(1, 2)$.

3.7.26. An urn contains twelve chips—four red, three black, and five white. A sample of size 4 is to be drawn without replacement. Let X denote the number of white chips in the sample, Y the number of red. Find $F_{X,Y}(1, 2)$.

3.7.27. For each of the following joint pdfs, find $F_{X,Y}(u, v)$.

(a) $f_{X,Y}(x, y) = \frac{3}{2}y^2, 0 \le x \le 2, 0 \le y \le 1$
(b) $f_{X,Y}(x, y) = \frac{2}{3}(x + 2y), 0 \le x \le 1, 0 \le y \le 1$
(c) $f_{X,Y}(x, y) = 4xy, 0 \le x \le 1, 0 \le y \le 1$

3.7.28. For each of the following joint pdfs, find $F_{X,Y}(u, v)$.

(a) $f_{X,Y}(x, y) = \frac{1}{2}, 0 \le x \le y \le 2$
(b) $f_{X,Y}(x, y) = \frac{1}{x}, 0 \le y \le x \le 1$
(c) $f_{X,Y}(x, y) = 6x, 0 \le x \le 1, 0 \le y \le 1 - x$

3.7.29. Find and graph $f_{X,Y}(x, y)$ if the joint cdf for random variables X and Y is

$$F_{X,Y}(x, y) = xy, \qquad 0 < x < 1, \quad 0 < y < 1$$

3.7.30. Find the joint pdf associated with two random variables X and Y whose joint cdf is

$$F_{X,Y}(x, y) = (1 - e^{-\lambda y})(1 - e^{-\lambda x}), \qquad x > 0, \quad y > 0$$

3.7.31. Given that $F_{X,Y}(x, y) = k(4x^2y^2 + 5xy^4), 0 < x < 1, 0 < y < 1$, find the corresponding pdf and use it to calculate $P(0 < X < \frac{1}{2}, \frac{1}{2} < Y < 1)$.

3.7.32. Prove that

$$P(a < X \le b, c < Y \le d) = F_{X,Y}(b, d) - F_{X,Y}(a, d) - F_{X,Y}(b, c) + F_{X,Y}(a, c)$$

3.7.33. A certain brand of fluorescent bulbs will last, on the average, 1000 hours. Suppose that four of these bulbs are installed in an office. What is probability that all four are still functioning after 1050 hours? If X_i denotes the ith bulb's life, assume that

$$f_{X_1,X_2,X_3,X_4}(x_1, x_2, x_3, x_4) = \prod_{i=1}^{4} \left(\frac{1}{1000}\right) e^{-x/1000}$$

for $x_i > 0, i = 1, 2, 3, 4$.

3.7.34. A hand of six cards is dealt from a standard poker deck. Let X denote the number of aces, Y the number of kings, and Z the number of queens.

(a) Write a formula for $p_{X,Y,Z}(x, y, z)$.
(b) Find $p_{X,Y}(x, y)$ and $p_{X,Z}(x, z)$.

3.7.35. Calculate $p_{X,Y}(0, 1)$ if $p_{X,Y,Z}(x, y, z) = \frac{3!}{x!y!z!(3-x-y-z)!} \left(\frac{1}{2}\right)^x \left(\frac{1}{12}\right)^y \left(\frac{1}{6}\right)^z \cdot \left(\frac{1}{4}\right)^{3-x-y-z}$ for $x, y, z = 0, 1, 2, 3$ and $0 \le x + y + z \le 3$.

3.7.36. Suppose that the random variables X, Y, and Z have the multivariate pdf

$$f_{X,Y,Z}(x, y, z) = (x + y)e^{-z}$$

for $0 < x < 1, 0 < y < 1$, and $z > 0$. Find (a) $f_{X,Y}(x, y)$, (b) $f_{Y,Z}(y, z)$, and (c) $f_Z(z)$.

3.7.37. The four random variables W, X, Y, and Z have the multivariate pdf

$$f_{W,X,Y,Z}(w, x, y, z) = 16wxyz$$

for $0 < w < 1, 0 < x < 1, 0 < y < 1$, and $0 < z < 1$. Find the marginal pdf, $f_{W,X}(w, x)$, and use it to compute $P(0 < W < \frac{1}{2}, \frac{1}{2} < X < 1)$.

Independence of Two Random Variables

The concept of independent events that was introduced in Section 2.5 leads quite naturally to a similar definition for independent random variables.

> **Definition 3.7.5.** Two random variables X and Y are said to be *independent* if for every interval A and every interval B, $P(X \in A$ and $Y \in B) = P(X \in A)P(Y \in B)$.

Theorem 3.7.4 *The continuous random variables X and Y are independent if and only if there are functions $g(x)$ and $h(y)$ such that*

$$f_{X,Y}(x, y) = g(x)h(y) \tag{3.7.2}$$

If Equation 3.7.2 holds, there is a constant k such that $f_X(x) = kg(x)$ and $f_Y(y) = (1/k)h(y)$.

Proof First, suppose that X and Y are independent. Then $F_{X,Y}(x, y) = P(X \leq x$ and $Y \leq y) = P(X \leq x)P(Y \leq y) = F_X(x)F_Y(y)$, and we can write

$$f_{X,Y}(x, y) = \frac{\partial^2}{\partial x\, \partial y} F_{X,Y}(x, y) = \frac{\partial^2}{\partial x\, \partial y} F_X(x)F_Y(y) = \frac{d}{dx} F_X(x) \frac{d}{dy} F_Y(y) = f_X(x)f_Y(y)$$

Next we need to show that Equation 3.7.2 implies that X and Y are independent. To begin, note that

$$f_X(x) = \int_{-\infty}^{\infty} f_{X,Y}(x, y)\, dy = \int_{-\infty}^{\infty} g(x)h(y)\, dy = g(x) \int_{-\infty}^{\infty} h(y)\, dy$$

Set $k = \int_{-\infty}^{\infty} h(y)\, dy$, so $f_X(x) = kg(x)$. Similarly, it can be shown that $f_Y(y) = (1/k)h(y)$. Therefore,

$$\begin{aligned}
P(X \in A \text{ and } Y \in B) &= \int_A \int_B f_{X,Y}(x, y)\, dx\, dy = \int_A \int_B g(x)h(y)\, dx\, dy \\
&= \int_A \int_B kg(x)(1/k)h(y)\, dx\, dy = \int_A f_X(x)\, dx \int_B f_Y(y)\, dy \\
&= P(X \in A)P(Y \in B)
\end{aligned}$$

and the theorem is proved. □

Comment Theorem 3.7.4 can be adapted to the case that X and Y are discrete.

Example 3.7.11 Suppose that the probabilistic behavior of two random variables X and Y is described by the joint pdf $f_{X,Y}(x, y) = 12xy(1 - y)$, $0 \leq x \leq 1$, $0 \leq y \leq 1$. Are X and Y independent? If they are, find $f_X(x)$ and $f_Y(y)$.

According to Theorem 3.7.4, the answer to the independence question is "yes" if $f_{X,Y}(x, y)$ can be factored into a function of x times a function of y. There are such functions. Let $g(x) = 12x$ and $h(y) = y(1 - y)$.

To find $f_X(x)$ and $f_Y(y)$ requires that the "12" appearing in $f_{X,Y}(x, y)$ be factored in such a way that $g(x) \cdot h(y) = f_X(x) \cdot f_Y(y)$. Let

$$k = \int_{-\infty}^{\infty} h(y)\, dy = \int_0^1 y(1 - y)\, dy = [y^2/2 - y^3/3]\Big|_0^1 = \frac{1}{6}$$

Therefore, $f_X(x) = kg(x) = \frac{1}{6}(12x) = 2x$, $0 \leq x \leq 1$ and $f_Y(y) = (1/k)h(y) = 6y(1 - y)$, $0 \leq y \leq 1$. ■

Independence of n (>2) Random Variables

In Chapter 2, extending the notion of independence from *two* events to n events proved to be something of a problem. The independence of each subset of the n events had to be checked separately (recall Definition 2.5.2). This is not necessary in the case of n random variables. We simply use the extension of Theorem 3.7.4 to n random variables as the definition of independence in the multidimensional case. The theorem that independence is equivalent to the factorization of the joint pdf holds in the multidimensional case.

Definition 3.7.6. The n random variables $X_1, X_2, \ldots, X_n$ are said to be *independent* if there are functions $g_1(x_1), g_2(x_2), \ldots, g_n(x_n)$ such that for every $x_1, x_2, \ldots, x_n$

$$f_{X_1,X_2,\ldots,X_n}(x_1, x_2, \ldots, x_n) = g_1(x_1)g_2(x_2)\cdots g_n(x_n)$$

A similar statement holds for discrete random variables, in which case f is replaced with p.

Comment Analogous to the result for $n = 2$ random variables, the expression on the right-hand side of the equation in Definition 3.7.6 can also be written as the product of the marginal pdfs of $X_1, X_2, \ldots,$ and X_n.

Example 3.7.12

Consider k urns, each holding n chips numbered 1 through n. A chip is to be drawn at random from each urn. What is the probability that all k chips will bear the same number?

If $X_1, X_2, \ldots, X_k$ denote the numbers on the 1st, 2nd, ..., and kth chips, respectively, we are looking for the probability that $X_1 = X_2 = \cdots = X_k$. In terms of the joint pdf,

$$P(X_1 = X_2 = \cdots = X_k) = \sum_{x_1 = x_2 = \cdots = x_k} p_{X_1,X_2,\ldots,X_k}(x_1, x_2, \ldots, x_k)$$

Each of the selections here is obviously independent of all the others, so the joint pdf factors according to Definition 3.7.6, and we can write

$$\begin{aligned}
P(X_1 = X_2 = \cdots = X_k) &= \sum_{i=1}^{n} p_{X_1}(x_i) \cdot p_{X_2}(x_i) \cdots p_{X_k}(x_i) \\
&= n \cdot \left(\frac{1}{n} \cdot \frac{1}{n} \cdot \cdots \cdot \frac{1}{n}\right) \\
&= \frac{1}{n^{k-1}}
\end{aligned}$$

■

Random Samples

Definition 3.7.6 addresses the question of independence as it applies to n random variables having marginal pdfs—say, $f_{X_1}(x_1), f_{X_2}(x_2), \ldots, f_{X_n}(x_n)$—that might be quite different. A special case of that definition occurs for virtually every set of data collected for statistical analysis. Suppose an experimenter takes a set of n measurements, $x_1, x_2, \ldots, x_n$, under the same conditions. Those X_i's, then, qualify as a set of independent random variables—moreover, each represents the *same* pdf. The special—but familiar—notation for that scenario is given in Definition 3.7.7. We will encounter it often in the chapters ahead.

Definition 3.7.7. Let $X_1, X_2, \ldots, X_n$ be a set of n independent random variables, all having the same pdf. Then $X_1, X_2, \ldots, X_n$ are said to be a *random sample of size n*.

Questions

3.7.38. Two fair dice are tossed. Let X denote the number appearing on the first die and Y the number on the second. Show that X and Y are independent.

3.7.39. Let $f_{X,Y}(x, y) = \lambda^2 e^{-\lambda(x+y)}$, $0 \le x$, $0 \le y$. Show that X and Y are independent. What are the marginal pdfs in this case?

3.7.40. Suppose that each of two urns has four chips, numbered 1 through 4. A chip is drawn from the first urn and bears the number X. That chip is added to the second urn. A chip is then drawn from the second urn. Call its number Y.

(a) Find $p_{X,Y}(x, y)$.
(b) Show that $p_X(k) = p_Y(k) = \frac{1}{4}$, $k = 1, 2, 3, 4$.
(c) Show that X and Y are not independent.

3.7.41. Let X and Y be random variables with joint pdf

$$f_{X,Y}(x, y) = k, \qquad 0 \le x \le 1, \quad 0 \le y \le 1, \quad 0 \le x + y \le 1$$

Give a geometric argument to show that X and Y are not independent.

3.7.42. Are the random variables X and Y independent if $f_{X,Y}(x, y) = \frac{2}{3}(x + 2y)$, $0 \le x \le 1$, $0 \le y \le 1$?

3.7.43. Suppose that random variables X and Y are independent with marginal pdfs $f_X(x) = 2x$, $0 \le x \le 1$, and $f_Y(y) = 3y^2$, $0 \le y \le 1$. Find $P(Y < X)$.

3.7.44. Find the joint cdf of the independent random variables X and Y, where $f_X(x) = \dfrac{x}{2}$, $0 \le x \le 2$, and $f_Y(y) = 2y$, $0 \le y \le 1$.

3.7.45. If two random variables X and Y are independent with marginal pdfs $f_X(x) = 2x$, $0 \le x \le 1$, and $f_Y(y) = 1$, $0 \le y \le 1$, calculate $P\left(\frac{Y}{X} > 2\right)$.

3.7.46. Suppose $f_{X,Y}(x, y) = xye^{-(x+y)}$, $x > 0$, $y > 0$. Prove for any real numbers a, b, c, and d that

$$P(a < X < b, c < Y < d) = P(a < X < b) \cdot P(c < Y < d)$$

thereby establishing the independence of X and Y.

3.7.47. Given the joint pdf $f_{X,Y}(x, y) = 2x + y - 2xy$, $0 < x < 1$, $0 < y < 1$, find numbers a, b, c, and d such that

$$P(a < X < b, c < Y < d) \neq P(a < X < b) \cdot P(c < Y < d)$$

thus demonstrating that X and Y are not independent.

3.7.48. Prove that if X and Y are two independent random variables, then $U = g(X)$ and $V = h(Y)$ are also independent.

3.7.49. If two random variables X and Y are defined over a region in the XY-plane that is *not* a rectangle (possibly infinite) with sides parallel to the coordinate axes, can X and Y be independent?

3.7.50. Write down the joint probability density function for a random sample of size n drawn from the exponential pdf, $f_X(x) = (1/\lambda)e^{-x/\lambda}$, $x \ge 0$.

3.7.51. Suppose that X_1, X_2, X_3, and X_4 are independent random variables, each with pdf $f_{X_i}(x_i) = 4x_i^3$, $0 \le x_i \le 1$. Find

(a) $P\left(X_1 < \frac{1}{2}\right)$.
(b) $P\left(\text{exactly one } X_i < \frac{1}{2}\right)$.
(c) $f_{X_1,X_2,X_3,X_4}(x_1, x_2, x_3, x_4)$.
(d) $F_{X_2,X_3}(x_2, x_3)$.

3.7.52. A random sample of size $n = 2k$ is taken from a uniform pdf defined over the unit interval. Calculate $P\left(X_1 < \frac{1}{2}, X_2 > \frac{1}{2}, X_3 < \frac{1}{2}, X_4 > \frac{1}{2}, \ldots, X_{2k} > \frac{1}{2}\right)$.

3.8 Transforming and Combining Random Variables

Transformations

Transforming a variable from one scale to another is a problem that is comfortably familiar. If a thermometer says the temperature outside is 83°F, we know that the temperature *in degrees Celsius* is *28*:

$$°\text{C} = \left(\frac{5}{9}\right)(°\text{F} - 32) = \left(\frac{5}{9}\right)(83 - 32) = 28$$

An analogous question arises in connection with random variables. Suppose that X is a discrete random variable with pdf $p_X(k)$. If a second random variable, Y, is defined to be $aX + b$, where a and b are constants, what can be said about the pdf for Y?

Theorem 3.8.1 *Suppose X is a discrete random variable. Let $Y = aX + b$, where a and b are constants. Then $p_Y(y) = p_X\left(\frac{y-b}{a}\right)$.*

Proof $p_Y(y) = P(Y = y) = P(aX + b = y) = P\left(X = \frac{y-b}{a}\right) = p_X\left(\frac{y-b}{a}\right)$ □

Example 3.8.1 Let X be a random variable for which $p_X(k) = \frac{1}{10}$, for $k = 1, 2, \ldots, 10$. What is the probability distribution associated with the random variable Y, where $Y = 4X - 1$? That is, find $p_Y(y)$.

From Theorem 3.8.1, $P(Y = y) = P(4X - 1 = y) = P[X = (y+1)/4] = p_X\left(\frac{y+1}{4}\right)$, which implies that $p_Y(y) = \frac{1}{10}$ for the ten values of $(y+1)/4$ that equal 1, 2, ..., 10. But $(y+1)/4 = 1$ when $y = 3$, $(y+1)/4 = 2$ when $y = 7, \ldots, (y+1)/4 = 10$ when $y = 39$. Therefore, $p_Y(y) = \frac{1}{10}$, for $y = 3, 7, \ldots, 39$. ■

Next we give the analogous result for a linear transformation of a *continuous* random variable.

Theorem 3.8.2 *Suppose X is a continuous random variable. Let $Y = aX + b$, where $a \neq 0$ and b is a constant. Then*

$$f_Y(y) = \frac{1}{|a|} f_X\left(\frac{y-b}{a}\right)$$

Proof We begin by writing an expression for the cdf of Y:

$$F_Y(y) = P(Y \leq y) = P(aX + b \leq y) = P(aX \leq y - b)$$

At this point we need to consider two cases, the distinction being the sign of a. Suppose, first, that $a > 0$. Then

$$F_Y(y) = P(aX \leq y - b) = P\left(X \leq \frac{y-b}{a}\right)$$

and differentiating $F_Y(y)$ yields $f_Y(y)$:

$$f_Y(y) = \frac{d}{dy} F_Y(y) = \frac{d}{dy} F_X\left(\frac{y-b}{a}\right) = \frac{1}{a} f_X\left(\frac{y-b}{a}\right) = \frac{1}{|a|} f_X\left(\frac{y-b}{a}\right)$$

If $a < 0$,

$$F_Y(y) = P(aX \leq y - b) = P\left(X > \frac{y-b}{a}\right) = 1 - P\left(X \leq \frac{y-b}{a}\right)$$

Differentiation in this case gives

$$f_Y(y) = \frac{d}{dy} F_Y(y) = \frac{d}{dy}\left[1 - F_X\left(\frac{y-b}{a}\right)\right] = -\frac{1}{a} f_X\left(\frac{y-b}{a}\right) = \frac{1}{|a|} f_X\left(\frac{y-b}{a}\right)$$

and the theorem is proved. □

Now, armed with the multivariable concepts and techniques covered in Section 3.7, we can extend the investigation of transformations to functions defined on sets of random variables. In statistics, the most important combination of a set of random variables is often their sum, so we continue this section with the problem of finding the pdf of $X + Y$.

Finding the Pdf of a Sum

Theorem 3.8.3 *Suppose that X and Y are independent random variables. Let $W = X + Y$. Then*

1. *If X and Y are discrete random variables with pdfs $p_X(x)$ and $p_Y(y)$, respectively,*

$$p_W(w) = \sum_{\text{all } x} p_X(x) p_Y(w - x)$$

2. *If X and Y are continuous random variables with pdfs $f_X(x)$ and $f_Y(y)$, respectively,*

$$f_W(w) = \int_{-\infty}^{\infty} f_X(x) f_Y(w - x)\, dx$$

Proof

1. $$\begin{aligned} p_W(w) &= P(W = w) = P(X + Y = w) \\ &= P\left(\bigcup_{\text{all } x}(X = x, Y = w - x)\right) = \sum_{\text{all } x} P(X = x, Y = w - x) \\ &= \sum_{\text{all } x} P(X = x) P(Y = w - x) \\ &= \sum_{\text{all } x} p_X(x) p_Y(w - x) \end{aligned}$$

   where the next-to-last equality derives from the independence of X and Y.

2. Since X and Y are continuous random variables, we can find $f_W(w)$ by differentiating the corresponding cdf, $F_W(w)$. Here, $F_W(w) = P(X + Y \leq w)$ is found by integrating $f_{X,Y}(x, y) = f_X(x) \cdot f_Y(y)$ over the shaded region R, as pictured in Figure 3.8.1.

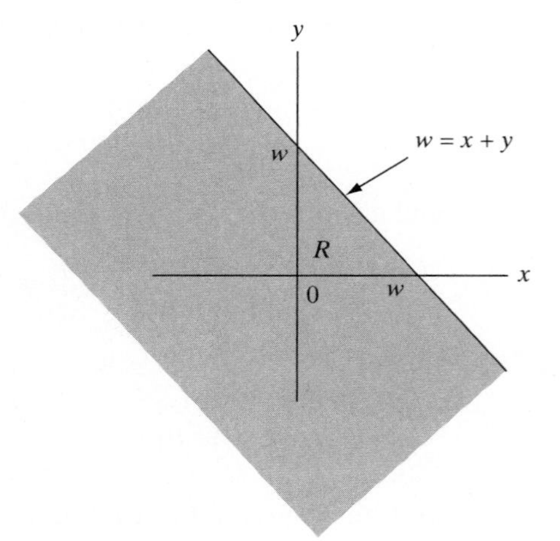

Figure 3.8.1

By inspection,

$$F_w(w) = \int_{-\infty}^{\infty} \int_{-\infty}^{w-x} f_X(x) f_Y(y)\, dy\, dx = \int_{-\infty}^{\infty} f_X(x) \left[\int_{-\infty}^{w-x} f_Y(y)\, dy\right] dx$$

$$= \int_{-\infty}^{\infty} f_X(x) F_Y(w - x)\, dx$$

Assume that the integrand in the above equation is sufficiently smooth so that differentiation and integration can be interchanged. Then we can write

$$f_W(w) = \frac{d}{dw} F_W(w) = \frac{d}{dw} \int_{-\infty}^{\infty} f_X(x) F_Y(w-x)\, dx = \int_{-\infty}^{\infty} f_X(x) \left[\frac{d}{dw} F_Y(w-x) \right] dx$$
$$= \int_{-\infty}^{\infty} f_X(x) f_Y(w-x)\, dx$$

and the theorem is proved. □

Comment The integral in part (2) above is referred to as the *convolution* of the functions f_X and f_Y. Besides their frequent appearances in random variable problems, convolutions turn up in many areas of mathematics and engineering.

Example 3.8.2

Suppose that X and Y are two independent binomial random variables, each with the same success probability but defined on m and n trials, respectively. Specifically,

$$p_X(k) = \binom{m}{k} p^k (1-p)^{m-k}, \quad k = 0, 1, \ldots, m$$

and

$$p_Y(k) = \binom{n}{k} p^k (1-p)^{n-k}, \quad k = 0, 1, \ldots, n$$

Find $p_W(w)$, where $W = X + Y$.

By Theorem 3.8.3, $p_W(w) = \sum_{\text{all } x} p_X(x) p_Y(w-x)$, but the summation over "all x" needs to be interpreted as the set of values for x and $w - x$ such that $p_X(x)$ and $p_Y(w-x)$, respectively, are both nonzero. But that will be true for all integers x from 0 to w. Therefore,

$$p_W(w) = \sum_{x=0}^{w} p_X(x) p_Y(w-x) = \sum_{x=0}^{w} \binom{m}{x} p^x (1-p)^{m-x} \binom{n}{w-x} p^{w-x} (1-p)^{n-(w-x)}$$
$$= \sum_{x=0}^{w} \binom{m}{x} \binom{n}{w-x} p^w (1-p)^{n+m-w}$$

Now, consider an urn having m red chips and n white chips. If w chips are drawn out—without replacement—the probability that exactly x red chips are in the sample is given by the hypergeometric distribution,

$$P(x \text{ reds in sample}) = \frac{\binom{m}{x}\binom{n}{w-x}}{\binom{m+n}{w}} \tag{3.8.1}$$

Summing Equation 3.8.1 from $x = 0$ to $x = w$ must equal 1 (why?), in which case

$$\sum_{x=0}^{w} \binom{m}{x} \binom{n}{w-x} = \binom{m+n}{w}$$

so

$$p_W(w) = \binom{m+n}{w} p^w (1-p)^{n+m-w}, \quad w = 0, 1, \ldots, n+m$$

Should we recognize $p_W(w)$? Definitely. Compare the structure of $p_W(w)$ to the statement of Theorem 3.2.1: The random variable W has a binomial distribution where the probability of success at any given trial is p and the total number of trials is $n + m$. ■

Comment Example 3.8.2 shows that the binomial distribution "reproduces" itself—that is, if X and Y are independent binomial random variables with the same value for p, their sum is also a binomial random variable. Not all random variables share that property. The sum of two independent uniform random variables, for example, is not a uniform random variable (see Question 3.8.3).

Example 3.8.3

Suppose a radiation monitor relies on an electronic sensor, whose lifetime X is modeled by the exponential pdf, $f_X(x) = \lambda e^{-\lambda x}, x > 0$. To improve the reliability of the monitor, the manufacturer has included an identical second sensor that is activated only in the event the first sensor malfunctions. (This is called *cold redundancy*.) Let the random variable Y denote the operating lifetime of the second sensor, in which case the lifetime of the monitor can be written as the sum $W = X + Y$. Find $f_W(w)$.

Since X and Y are both continuous random variables,

$$f_W(w) = \int_{-\infty}^{\infty} f_X(x) f_Y(w - x)\, dx \tag{3.8.2}$$

Notice that $f_X(x) > 0$ only if $x > 0$ and that $f_Y(w - x) > 0$ only if $x < w$. Therefore, the integral in Equation 3.8.2 that goes from $-\infty$ to ∞ reduces to an integral from 0 to w, and we can write

$$f_W(w) = \int_0^w f_X(x) f_Y(w - x)\, dx = \int_0^w \lambda e^{-\lambda x} \lambda e^{-\lambda(w-x)}\, dx = \lambda^2 \int_0^w e^{-\lambda x} e^{-\lambda(w-x)}\, dx$$

$$= \lambda^2 e^{-\lambda w} \int_0^w dx = \lambda^2 w e^{-\lambda w}, \quad w \geq 0$$

■

Comment By integrating $f_X(x)$ and $f_W(w)$, we can assess the improvement in the monitor's reliability afforded by the cold redundancy. Since X is an exponential random variable, $E(X) = 1/\lambda$ (recall Question 3.5.11). How different, for example, are $P(X \geq 1/\lambda)$ and $P(W \geq 1/\lambda)$? A simple calculation shows that the latter is actually *twice* the magnitude of the former:

$$P(X \geq 1/\lambda) = \int_{1/\lambda}^{\infty} \lambda e^{-\lambda x}\, dx = -e^{-u}\Big|_1^{\infty} = e^{-1} = 0.37$$

$$P(W \geq 1/\lambda) = \int_{1/\lambda}^{\infty} \lambda^2 w e^{-\lambda w}\, dw = e^{-u}(-u - 1)\Big|_1^{\infty} = 2e^{-1} = 0.74$$

Finding the Pdfs of Quotients and Products

We conclude this section by considering the pdfs for the quotient and product of two independent random variables. That is, given X and Y, we are looking for $f_W(w)$, where (1) $W = Y/X$ and (2) $W = XY$. Neither of the resulting formulas is as important as the pdf for the *sum* of two random variables, but both formulas will play key roles in several derivations in Chapter 7.

Theorem 3.8.4 *Let X and Y be independent continuous random variables, with pdfs $f_X(x)$ and $f_Y(y)$, respectively. Assume that X is zero for at most a set of isolated points. Let $W = Y/X$. Then*

$$f_W(w) = \int_{-\infty}^{\infty} |x| f_X(x) f_Y(wx)\,dx$$

Proof

$$\begin{aligned}
F_W(w) &= P(Y/X \le w)\\
&= P(Y/X \le w \quad \text{and} \quad X \ge 0) + P(Y/X \ge w \quad \text{and} \quad X < 0)\\
&= P(Y \le wX \quad \text{and} \quad X \ge 0) + P(Y \ge wX \quad \text{and} \quad X < 0)\\
&= P(Y \le wX \quad \text{and} \quad X \ge 0) + 1 - P(Y \le wX \quad \text{and} \quad X < 0)\\
&= \int_0^{\infty} \int_{-\infty}^{wx} f_X(x) f_Y(y)\,dy\,dx + 1 - \int_{-\infty}^{0} \int_{-\infty}^{wx} f_X(x) f_Y(y)\,dy\,dx
\end{aligned}$$

Then we differentiate $F_W(w)$ to obtain

$$\begin{aligned}
f_W(w) = \frac{d}{dw} F_W(w) &= \frac{d}{dw} \int_0^{\infty} \int_{-\infty}^{wx} f_X(x) f_Y(y)\,dy\,dx - \frac{d}{dw} \int_{-\infty}^{0} \int_{-\infty}^{wx} f_X(x) f_Y(y)\,dy\,dx\\
&= \int_0^{\infty} f_X(x) \left(\frac{d}{dw} \int_{-\infty}^{wx} f_Y(y)\,dy \right) dx - \int_{-\infty}^{0} f_X(x) \left(\frac{d}{dw} \int_{-\infty}^{wx} f_Y(y)\,dy \right) dx
\end{aligned} \tag{3.8.3}$$

(Note that we are assuming sufficient regularity of the functions to permit interchange of integration and differentiation.)

To proceed, we need to differentiate the function $G(w) = \int_{-\infty}^{wx} f_Y(y)\,dy$ with respect to w. By the Fundamental Theorem of Calculus and the chain rule, we find

$$\frac{d}{dw} G(w) = \frac{d}{dw} \int_{-\infty}^{wx} f_Y(y)\,dy = f_Y(wx) \frac{d}{dw} wx = x f_Y(wx)$$

Putting this result into Equation 3.8.3 gives

$$\begin{aligned}
f_W(w) &= \int_0^{\infty} x f_X(x) f_Y(wx)\,dx - \int_{-\infty}^{0} x f_X(x) f_Y(wx)\,dx\\
&= \int_0^{\infty} x f_X(x) f_Y(wx)\,dx + \int_{-\infty}^{0} (-x) f_X(x) f_Y(wx)\,dx\\
&= \int_0^{\infty} |x| f_X(x) f_Y(wx)\,dx + \int_{-\infty}^{0} |x| f_X(x) f_Y(wx)\,dx\\
&= \int_{-\infty}^{\infty} |x| f_X(x) f_Y(wx)\,dx
\end{aligned}$$

which completes the proof. □

Example 3.8.4 Let X and Y be independent random variables with pdfs $f_X(x) = \lambda e^{-\lambda x}$, $x > 0$, and $f_Y(y) = \lambda e^{-\lambda y}$, $y > 0$, respectively. Define $W = Y/X$. Find $f_W(w)$.

Substituting into the formula given in Theorem 3.8.4, we can write

$$\begin{aligned}
f_W(w) &= \int_0^{\infty} x(\lambda e^{-\lambda x})(\lambda e^{-\lambda x w})\,dx = \lambda^2 \int_0^{\infty} x e^{-\lambda(1+w)x}\,dx\\
&= \frac{\lambda^2}{\lambda(1+w)} \int_0^{\infty} x \lambda(1+w) e^{-\lambda(1+w)x}\,dx
\end{aligned}$$

Notice that the integral is the expected value of an exponential random variable with parameter $\lambda(1+w)$, so it equals $1/\lambda(1+w)$ (recall Example 3.5.6). Therefore,

$$f_W(w) = \frac{\lambda^2}{\lambda(1+w)} \frac{1}{\lambda(1+w)} = \frac{1}{(1+w)^2}, \quad w \geq 0$$

■

Theorem 3.8.5 *Let X and Y be independent continuous random variables with pdfs $f_X(x)$ and $f_Y(y)$, respectively. Let $W = XY$. Then*

$$f_W(w) = \int_{-\infty}^{\infty} \frac{1}{|x|} f_X(x) f_Y(w/x)\, dx = \int_{-\infty}^{\infty} \frac{1}{|x|} f_X(w/x) f_Y(x)\, dx$$

Proof A line-by-line, straightforward modification of the proof of Theorem 3.8.4 will provide a proof of Theorem 3.8.5. The details are left to the reader. □

Example 3.8.5 Suppose that X and Y are independent random variables with pdfs $f_X(x) = 1$, $0 \leq x \leq 1$, and $f_Y(y) = 2y$, $0 \leq y \leq 1$, respectively. Find $f_W(w)$, where $W = XY$.

According to Theorem 3.8.5,

$$f_W(w) = \int_{-\infty}^{\infty} \frac{1}{|x|} f_X(x) f_Y(w/x)\, dx$$

The region of integration, though, needs to be restricted to values of x for which the integrand is positive. But $f_Y(w/x)$ is positive only if $0 \leq w/x \leq 1$, which implies that $x \geq w$. Moreover, for $f_X(x)$ to be positive requires that $0 \leq x \leq 1$. Any x, then, from w to 1 will yield a positive integrand. Therefore,

$$f_W(w) = \int_w^1 \frac{1}{x}(1)(2w/x)\, dx = 2w \int_w^1 \frac{1}{x^2}\, dx = 2 - 2w, \quad 0 \leq w \leq 1$$

Comment Theorems 3.8.3, 3.8.4, and 3.8.5 can be adapted to situations where X and Y are not independent by replacing the product of the marginal pdfs with the joint pdf. ■

Questions

3.8.1. Let X and Y be two independent random variables. Given the marginal pdfs shown below, find the pdf of $X + Y$. In each case, check to see if $X + Y$ belongs to the same family of pdfs as do X and Y.

(a) $p_X(k) = e^{-\lambda}\frac{\lambda^k}{k!}$ and $p_Y(k) = e^{-\mu}\frac{\mu^k}{k!}$, $k = 0, 1, 2, \ldots$

(b) $p_X(k) = p_Y(k) = (1-p)^{k-1}p$, $k = 1, 2, \ldots$

3.8.2. Suppose $f_X(x) = xe^{-x}$, $x \geq 0$, and $f_Y(y) = e^{-y}$, $y \geq 0$, where X and Y are independent. Find the pdf of $X + Y$.

3.8.3. Let X and Y be two independent random variables, whose marginal pdfs are given below. Find the pdf of $X + Y$. (*Hint:* Consider two cases, $0 \leq w < 1$ and $1 \leq w \leq 2$.)

$$f_X(x) = 1, \ 0 \leq x \leq 1, \text{ and } f_Y(y) = 1, \ 0 \leq y \leq 1$$

3.8.4. If a random variable V is independent of two independent random variables X and Y, prove that V is independent of $X + Y$.

3.8.5. Let Y be a continuous nonnegative random variable. Show that $W = Y^2$ has pdf $f_W(w) = \frac{1}{2\sqrt{w}} f_Y(\sqrt{w})$. [*Hint:* First find $F_W(w)$.]

3.8.6. Let Y be a uniform random variable over the interval [0, 1]. Find the pdf of $W = Y^2$.

3.8.7. Let Y be a random variable with $f_Y(y) = 6y(1-y)$, $0 \leq y \leq 1$. Find the pdf of $W = Y^2$.

3.8.8. Suppose the velocity of a gas molecule of mass m is a random variable with pdf $f_Y(y) = ay^2e^{-by^2}$, $y \geq 0$, where a and b are positive constants depending on the gas. Find the pdf of the kinetic energy, $W = (m/2)Y^2$, of such a molecule.

3.8.9. Given that X and Y are independent random variables, find the pdf of XY for the following two sets of marginal pdfs:

(a) $f_X(x) = 1, 0 \le x \le 1$, and $f_Y(y) = 1, 0 \le y \le 1$
(b) $f_X(x) = 2x, 0 \le x \le 1$, and $f_Y(y) = 2y, 0 \le y \le 1$

3.8.10. Let X and Y be two independent random variables. Given the marginal pdfs indicated below, find the cdf of Y/X. (*Hint:* Consider two cases, $0 \le w \le 1$ and $1 < w$.)

(a) $f_X(x) = 1, 0 \le x \le 1$, and $f_Y(y) = 1, 0 \le y \le 1$
(b) $f_X(x) = 2x, 0 \le x \le 1$, and $f_Y(y) = 2y, 0 \le y \le 1$

3.8.11. Suppose that X and Y are two independent random variables, where $f_X(x) = xe^{-x}, x \ge 0$, and $f_Y(y) = e^{-y}$, $y \ge 0$. Find the pdf of Y/X.

3.9 Further Properties of the Mean and Variance

Sections 3.5 and 3.6 introduced the basic definitions related to the expected value and variance of *single* random variables. We learned how to calculate $E(W)$, $E[g(W)]$, $E(aW + b)$, $\text{Var}(W)$, and $\text{Var}(aW + b)$, where a and b are any constants and W could be either a discrete or a continuous random variable. The purpose of this section is to examine certain multivariable extensions of those results, based on the joint pdf material covered in Section 3.7.

We begin with a theorem that generalizes $E[g(W)]$. While it is stated here for the case of *two* random variables, it extends in a very straightforward way to include functions of n random variables.

Theorem 3.9.1

1. Suppose X and Y are discrete random variables with joint pdf $p_{X,Y}(x, y)$, and let $g(X, Y)$ be a function of X and Y. Then the expected value of the random variable $g(X, Y)$ is given by

$$E[g(X, Y)] = \sum_{\text{all } x}\sum_{\text{all } y} g(x, y) \cdot p_{X,Y}(x, y)$$

provided $\sum_{\text{all } x}\sum_{\text{all } y} |g(x, y)| \cdot p_{X,Y}(x, y) < \infty$.

2. Suppose X and Y are continuous random variables with joint pdf $f_{X,Y}(x, y)$, and let $g(X, Y)$ be a continuous function. Then the expected value of the random variable $g(X, Y)$ is given by

$$E[g(X, Y)] = \int_{-\infty}^{\infty}\int_{-\infty}^{\infty} g(x, y) \cdot f_{X,Y}(x, y)\, dx\, dy$$

provided $\int_{-\infty}^{\infty}\int_{-\infty}^{\infty} |g(x, y)| \cdot f_{X,Y}(x, y)\, dx\, dy < \infty$.

Proof The basic approach taken in deriving this result is similar to the method followed in the proof of Theorem 3.5.3. See (128) for details. □

Example 3.9.1

Consider the two random variables X and Y whose joint pdf is detailed in the 2×4 matrix shown in Table 3.9.1. Let

$$g(X, Y) = 3X - 2XY + Y$$

Find $E[g(X, Y)]$ two ways—first, by using the basic definition of an expected value, and second, by using Theorem 3.9.1.

Let $Z = 3X - 2XY + Y$. By inspection, Z takes on the values 0, 1, 2, and 3 according to the pdf $f_Z(z)$ shown in Table 3.9.2. Then from the basic definition

Table 3.9.1

		Y			
		0	1	2	3
X	0	$\frac{1}{8}$	$\frac{1}{4}$	$\frac{1}{8}$	0
	1	0	$\frac{1}{8}$	$\frac{1}{4}$	$\frac{1}{8}$

Table 3.9.2

z	0	1	2	3
$f_Z(z)$	$\frac{1}{4}$	$\frac{1}{2}$	$\frac{1}{4}$	0

that an expected value is a weighted average, we see that $E[g(X, Y)]$ is equal to 1:

$$\begin{aligned} E[g(X, Y)] = E(Z) &= \sum_{\text{all } z} z \cdot f_Z(z) \\ &= 0 \cdot \frac{1}{4} + 1 \cdot \frac{1}{2} + 2 \cdot \frac{1}{4} + 3 \cdot 0 \\ &= 1 \end{aligned}$$

The same answer is obtained by applying Theorem 3.9.1 to the joint pdf given in Figure 3.9.1:

$$\begin{aligned} E[g(X, Y)] &= 0 \cdot \frac{1}{8} + 1 \cdot \frac{1}{4} + 2 \cdot \frac{1}{8} + 3 \cdot 0 + 3 \cdot 0 + 2 \cdot \frac{1}{8} + 1 \cdot \frac{1}{4} + 0 \cdot \frac{1}{8} \\ &= 1 \end{aligned}$$

The advantage, of course, enjoyed by the latter solution is that we avoid the intermediate step of having to determine $f_Z(z)$. ■

Example 3.9.2 An electrical circuit has three resistors, R_X, R_Y, and R_Z, wired in parallel (see Figure 3.9.1). The nominal resistance of each is fifteen ohms, but their *actual* resistances, X, Y, and Z, vary between ten and twenty according to the joint pdf

$$f_{X,Y,Z}(x, y, z) = \frac{1}{675{,}000}(xy + xz + yz), \quad \begin{array}{l} 10 \le x \le 20 \\ 10 \le y \le 20 \\ 10 \le z \le 20 \end{array}$$

What is the expected resistance for the circuit?

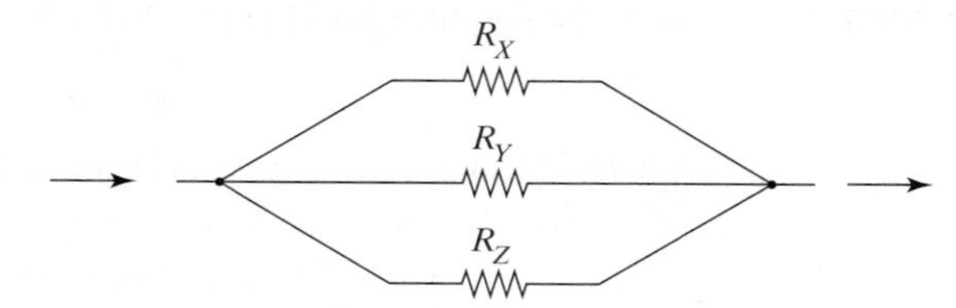

Figure 3.9.1

Let R denote the circuit's resistance. A well-known result in physics holds that

$$\frac{1}{R} = \frac{1}{X} + \frac{1}{Y} + \frac{1}{Z}$$

or, equivalently,

$$R = \frac{XYZ}{XY + XZ + YZ} = R(X, Y, Z)$$

Integrating $R(x, y, z) \cdot f_{X,Y,Z}(x, y, z)$ shows that the expected resistance is five:

$$\begin{aligned}
E(R) &= \int_{10}^{20}\int_{10}^{20}\int_{10}^{20} \frac{xyz}{xy + xz + yz} \cdot \frac{1}{675{,}000}(xy + xz + yz)\,dx\,dy\,dz \\
&= \frac{1}{675{,}000}\int_{10}^{20}\int_{10}^{20}\int_{10}^{20} xyz\,dx\,dy\,dz \\
&= 5.0
\end{aligned}$$

■

Theorem 3.9.2 *Let X and Y be any two random variables (discrete or continuous, dependent or independent), and let a and b be any two constants. Then*

$$E(aX + bY) = aE(X) + bE(Y)$$

provided $E(X)$ and $E(Y)$ are both finite.

Proof Consider the continuous case (the discrete case is proved much the same way). Let $f_{X,Y}(x, y)$ be the joint pdf of X and Y, and define $g(X, Y) = aX + bY$. By Theorem 3.9.1,

$$\begin{aligned}
E(aX + bY) &= \int_{-\infty}^{\infty}\int_{-\infty}^{\infty} (ax + by) f_{X,Y}(x, y)\,dx\,dy \\
&= \int_{-\infty}^{\infty}\int_{-\infty}^{\infty} (ax) f_{X,Y}(x, y)\,dx\,dy + \int_{-\infty}^{\infty}\int_{-\infty}^{\infty} (by) f_{X,Y}(x, y)\,dx\,dy \\
&= a\int_{-\infty}^{\infty} x\left[\int_{-\infty}^{\infty} f_{X,Y}(x, y)\,dy\right]dx + b\int_{-\infty}^{\infty} y\left[\int_{-\infty}^{\infty} f_{X,Y}(x, y)\,dx\right]dy \\
&= a\int_{-\infty}^{\infty} x f_X(x)\,dx + b\int_{-\infty}^{\infty} y f_Y(y)\,dy \\
&= aE(X) + bE(Y)
\end{aligned}$$

□

Corollary 3.9.1 *Let $W_1, W_2, \ldots, W_n$ be any random variables for which $E(W_i) < \infty$, $i = 1, 2, \ldots, n$, and let $a_1, a_2, \ldots, a_n$ be any set of constants. Then*

$$E(a_1W_1 + a_2W_2 + \cdots + a_nW_n) = a_1E(W_1) + a_2E(W_2) + \cdots + a_nE(W_n)$$

◄

Example 3.9.3 Let X be a binomial random variable defined on n independent trials, each trial resulting in success with probability p. Find $E(X)$.

Note, first, that X can be thought of as a sum, $X = X_1 + X_2 + \cdots + X_n$, where X_i represents the number of successes occurring at the ith trial:

$$X_i = \begin{cases} 1 & \text{if the } i\text{th trial produces a success} \\ 0 & \text{if the } i\text{th trial produces a failure} \end{cases}$$

(Any X_i defined in this way on an individual trial is called a *Bernoulli* random variable. Every binomial random variable, then, can be thought of as the sum of n independent Bernoullis.) By assumption, $p_{X_i}(1) = p$ and $p_{X_i}(0) = 1 - p$, $i = 1, 2, \ldots, n$. Using the corollary,

$$\begin{aligned} E(X) &= E(X_1) + E(X_2) + \cdots + E(X_n) \\ &= n \cdot E(X_1) \end{aligned}$$

the last step being a consequence of the X_i's having identical distributions. But

$$E(X_1) = 1 \cdot p + 0 \cdot (1 - p) = p$$

so $E(X) = np$, which is what we found before (recall Theorem 3.5.1). ■

Comment The problem-solving implications of Theorem 3.9.2 and its corollary should not be underestimated. There are many real-world events that can be modeled as a linear combination $a_1 W_1 + a_2 W_2 + \cdots + a_n W_n$, where the W_i's are relatively simple random variables. Finding $E(a_1 W_1 + a_2 W_2 + \cdots + a_n W_n)$ *directly* may be prohibitively difficult because of the inherent complexity of the linear combination. It may very well be the case, though, that calculating the individual $E(W_i)$'s is easy. Compare, for instance, Example 3.9.3 with Theorem 3.5.1. Both derive the formula that $E(X) = np$ when X is a binomial random variable. However, the approach taken in Example 3.9.3 (i.e., using Theorem 3.9.2) is *much* easier. The next several examples further explore the technique of using linear combinations to facilitate the calculation of expected values.

Example 3.9.4

A disgruntled secretary is upset about having to stuff envelopes. Handed a box of n letters and n envelopes, she vents her frustration by putting the letters into the envelopes *at random*. How many people, on the average, will receive their correct mail?

If X denotes the number of envelopes properly stuffed, what we want is $E(X)$. However, applying Definition 3.5.1 here would prove formidable because of the difficulty in getting a workable expression for $p_X(k)$ [see (95)]. By using the corollary to Theorem 3.9.2, though, we can solve the problem quite easily.

Let X_i denote a random variable equal to the number of correct letters put into the ith envelope, $i = 1, 2, \ldots, n$. Then X_i equals 0 or 1, and

$$p_{X_i}(k) = P(X_i = k) = \begin{cases} \dfrac{1}{n} & \text{for } k = 1 \\ \dfrac{n-1}{n} & \text{for } k = 0 \end{cases}$$

But $X = X_1 + X_2 + \cdots + X_n$ and $E(X) = E(X_1) + E(X_2) + \cdots + E(X_n)$. Furthermore, each of the X_i's has the same expected value, $1/n$:

$$E(X_i) = \sum_{k=0}^{1} k \cdot P(X_i = k) = 0 \cdot \frac{n-1}{n} + 1 \cdot \frac{1}{n} = \frac{1}{n}$$

It follows that

$$\begin{aligned} E(X) &= \sum_{i=1}^{n} E(X_i) = n \cdot \left(\frac{1}{n}\right) \\ &= 1 \end{aligned}$$

showing that, *regardless of n,* the expected number of properly stuffed envelopes is one. (Are the X_i's independent? Does it matter?) ■

Example 3.9.5

Ten fair dice are rolled. Calculate the expected value of the sum of the faces showing.

If the random variable X denotes the sum of the faces showing on the ten dice, then

$$X = X_1 + X_2 + \cdots + X_{10}$$

where X_i is the number showing on the ith die, $i = 1, 2, \ldots, 10$. By assumption, $p_{X_i}(k) = \frac{1}{6}$ for $k = 1, 2, 3, 4, 5, 6$, so $E(X_i) = \sum_{k=1}^{6} k \cdot \frac{1}{6} = \frac{1}{6} \sum_{k=1}^{6} k = \frac{1}{6} \cdot \frac{6(7)}{2} = 3.5$. By the corollary to Theorem 3.9.2,

$$\begin{aligned} E(X) &= E(X_1) + E(X_2) + \cdots + E(X_{10}) \\ &= 10(3.5) \\ &= 35 \end{aligned}$$

Notice that $E(X)$ can also be deduced here by appealing to the notion that expected values are centers of gravity. It should be clear from our work with combinatorics that $P(X=10) = P(X=60)$, $P(X=11) = P(X=59)$, $P(X=12) = P(X=58)$, and so on. In other words, the probability function $p_X(k)$ is symmetric, which implies that its center of gravity is the midpoint of the range of its X-values. It must be the case, then, that $E(X)$ equals $\frac{10+60}{2}$ or *35*. ■

Example 3.9.6

The honor count in a (thirteen-card) bridge hand can vary from zero to thirty-seven according to the formula:

$$\begin{aligned} \text{honor count} = 4 \cdot (\text{number of aces}) + 3 \cdot (\text{number of kings}) + 2 \cdot (\text{number of queens}) \\ + 1 \cdot (\text{number of jacks}) \end{aligned}$$

What is the expected honor count of North's hand?

The solution here is a bit unusual in that we use the corollary to Theorem 3.9.2 *backward*. If $X_i, i = 1, 2, 3, 4$, denotes the honor count for players North, South, East, and West, respectively, and if X denotes the analogous sum for the entire deck, we can write

$$X = X_1 + X_2 + X_3 + X_4$$

But

$$X = E(X) = 4 \cdot 4 + 3 \cdot 4 + 2 \cdot 4 + 1 \cdot 4 = 40$$

By symmetry, $E(X_i) = E(X_j)$, $i \neq j$, so it follows that $40 = 4 \cdot E(X_1)$, which implies that *ten* is the expected honor count of North's hand. (Try doing this problem directly, without making use of the fact that the deck's honor count is forty.) ■

Expected Values of Products: A Special Case

We know from Theorem 3.9.1 that for any two random variables X and Y,

$$E(XY) = \begin{cases} \sum_{\text{all } x} \sum_{\text{all } y} xy\,p_{X,Y}(x, y) & \text{if } X \text{ and } Y \text{ are discrete} \\ \int_{-\infty}^{\infty} \int_{-\infty}^{\infty} xy\,f_{X,Y}(x, y)\,dx\,dy & \text{if } X \text{ and } Y \text{ are continuous} \end{cases}$$

If, however, X and Y are independent, there is an easier way to calculate $E(XY)$.

Theorem 3.9.3 *If X and Y are independent random variables,*

$$E(XY) = E(X) \cdot E(Y)$$

provided $E(X)$ and $E(Y)$ both exist.

Proof Suppose X and Y are both discrete random variables. Then their joint pdf, $p_{X,Y}(x, y)$, can be replaced by the product of their marginal pdfs, $p_X(x) \cdot p_Y(y)$, and the double summation required by Theorem 3.9.1 can be written as the product of two single summations:

$$\begin{aligned} E(XY) &= \sum_{\text{all } x} \sum_{\text{all } y} xy \cdot p_{X,Y}(x, y) \\ &= \sum_{\text{all } x} \sum_{\text{all } y} xy \cdot p_X(x) \cdot p_Y(y) \\ &= \sum_{\text{all } x} x \cdot p_X(x) \cdot \left[\sum_{\text{all } y} y \cdot p_Y(y) \right] \\ &= E(X) \cdot E(Y) \end{aligned}$$

The proof when X and Y are both continuous random variables is left as an exercise. □

Questions

3.9.1. Suppose that r chips are drawn with replacement from an urn containing n chips, numbered 1 through n. Let V denote the sum of the numbers drawn. Find $E(V)$.

3.9.2. Suppose that $f_{X,Y}(x, y) = \lambda^2 e^{-\lambda(x+y)}$, $0 \le x$, $0 \le y$. Find $E(X + Y)$.

3.9.3. Suppose that $f_{X,Y}(x, y) = \frac{2}{3}(x + 2y)$, $0 \le x \le 1$, $0 \le y \le 1$ [recall Question 3.7.19(c)]. Find $E(X + Y)$.

3.9.4. Marksmanship competition at a certain level requires each contestant to take ten shots with each of two different handguns. Final scores are computed by taking a weighted average of 4 times the number of bull's-eyes made with the first gun plus 6 times the number gotten with the second. If Cathie has a 30% chance of hitting the bull's-eye with each shot from the first gun and a 40% chance with each shot from the second gun, what is her expected score?

3.9.5. Suppose that X_i is a random variable for which $E(X_i) = \mu$, $i = 1, 2, \ldots, n$. Under what conditions will the following be true?

$$E\left(\sum_{i=1}^{n} a_i X_i\right) = \mu$$

3.9.6. Suppose that the daily closing price of stock goes up an eighth of a point with probability p and down an eighth of a point with probability q, where $p > q$. After n days how much gain can we expect the stock to have achieved? Assume that the daily price fluctuations are independent events.

3.9.7. An urn contains r red balls and w white balls. A sample of n balls is drawn *in order* and *without* replacement. Let X_i be 1 if the ith draw is red and 0 otherwise, $i = 1, 2, \ldots, n$.

(a) Show that $E(X_i) = E(X_1)$, $i = 2, 3, \ldots, n$.
(b) Use the corollary to Theorem 3.9.2 to show that the expected number of red balls is $nr/(r + w)$.

3.9.8. Suppose two fair dice are tossed. Find the expected value of the product of the faces showing.

3.9.9. Find $E(R)$ for a two-resistor circuit similar to the one described in Example 3.9.2, where $f_{X,Y}(x, y) = k(x + y)$, $10 \le x \le 20$, $10 \le y \le 20$.

3.9.10. Suppose that X and Y are both uniformly distributed over the interval [0, 1]. Calculate the expected value of the square of the distance of the random point (X, Y) from the origin; that is, find $E(X^2 + Y^2)$. (*Hint:* See Question 3.8.6.)

3.9.11. Suppose X represents a point picked at random from the interval $[0, 1]$ on the x-axis, and Y is a point picked at random from the interval $[0, 1]$ on the y-axis. Assume that X and Y are independent. What is the expected value of the area of the triangle formed by the points $(X, 0)$, $(0, Y)$, and $(0, 0)$?

3.9.12. Suppose $Y_1, Y_2, \ldots, Y_n$ is a random sample from the uniform pdf over $[0, 1]$. The geometric mean of the numbers is the random variable $\sqrt[n]{Y_1 Y_2 \cdot \cdots \cdot Y_n}$. Compare the expected value of the geometric mean to that of the arithmetic mean $\bar{Y}$.

Calculating the Variance of a Sum of Random Variables

When random variables are not independent, a measure of the relationship between them, their *covariance*, enters into the picture.

Definition 3.9.1. Given random variables X and Y with finite variances, define the *covariance* of X and Y, written $\text{Cov}(X, Y)$, as

$$\text{Cov}(X, Y) = E(XY) - E(X)E(Y)$$

Theorem 3.9.4 *If X and Y are independent, then* $\text{Cov}(X, Y) = 0$.

Proof If X and Y are independent, by Theorem 3.9.3, $E(XY) = E(X)E(Y)$. Then

$$\text{Cov}(X, Y) = E(XY) - E(X)E(Y) = E(X)E(Y) - E(X)E(Y) = 0$$

The converse of Theorem 3.9.4 is *not* true. Just because $\text{Cov}(X, Y) = 0$, we cannot conclude that X and Y are independent. Example 3.9.7 is a case in point. □

Example 3.9.7 Consider the sample space $S = \{(-2, 4), (-1, 1), (0, 0), (1, 1), (2, 4)\}$, where each point is assumed to be equally likely. Define the random variable X to be the first component of a sample point and Y, the second. Then $X(-2, 4) = -2$, $Y(-2, 4) = 4$, and so on.

Notice that X and Y are dependent:

$$\frac{1}{5} = P(X = 1, Y = 1) \neq P(X = 1) \cdot P(Y = 1) = \frac{1}{5} \cdot \frac{2}{5} = \frac{2}{25}$$

However, the convariance of X and Y is zero:

$$E(XY) = [(-8) + (-1) + 0 + 1 + 8] \cdot \frac{1}{5} = 0$$

$$E(X) = [(-2) + (-1) + 0 + 1 + 2] \cdot \frac{1}{5} = 0$$

and

$$E(Y) = (4 + 1 + 0 + 1 + 4) \cdot \frac{1}{5} = 2$$

so

$$\text{Cov}(X, Y) = E(XY) - E(X) \cdot E(Y) = 0 - 0 \cdot 2 = 0$$ ■

Theorem 3.9.5 demonstrates the role of the covariance in finding the variance of a sum of random variables that are not necessarily independent.

Theorem 3.9.5 *Suppose X and Y are random variables with finite variances, and a and b are constants. Then*

$$\text{Var}(aX + bY) = a^2\text{Var}(X) + b^2\text{Var}(Y) + 2ab\,\text{Cov}(X, Y)$$

Proof For convenience, denote $E(X)$ by μ_X and $E(Y)$ by μ_Y. Then $E(aX+bY) = a\mu_X + b\mu_Y$ and

$$\begin{aligned}\text{Var}(aX+bY) &= E[(aX+bY)^2] - (a\mu_X + b\mu_Y)^2 \\ &= E(a^2X^2 + b^2Y^2 + 2abXY) - (a^2\mu_X^2 + b^2\mu_Y^2 + 2ab\mu_X\mu_Y) \\ &= [E(a^2X^2) - a^2\mu_X^2] + [E(b^2Y^2) - b^2\mu_Y^2] + [2abE(XY) - 2ab\mu_X\mu_Y] \\ &= a^2[E(X^2) - \mu_X^2] + b^2[E(Y^2) - \mu_Y^2] + 2ab[E(XY) - \mu_X\mu_Y] \\ &= a^2\,\text{Var}(X) + b^2\,\text{Var}(Y) + 2ab\text{Cov}(X,Y)\end{aligned}$$

□

Example 3.9.8

For the joint pdf $f_{X,Y}(x,y) = x+y,\ 0 \le x \le 1,\ 0 \le y \le 1$, find the variance of $X+Y$.

Since X and Y are not independent,

$$\text{Var}(X+Y) = \text{Var}(X) + \text{Var}(Y) + 2\text{Cov}(X,Y)$$

The pdf is symmetric in X and Y, so $\text{Var}(X) = \text{Var}(Y)$, and we can write $\text{Var}(X+Y) = 2[\text{Var}(X) + \text{Cov}(X,Y)]$.

To calculate $\text{Var}(X)$, the marginal pdf of X is needed. But

$$f_X(x) = \int_0^1 (x+y)dy = x + \frac{1}{2}$$

$$\mu_X = \int_0^1 x(x+\frac{1}{2})dx = \int_0^1 (x^2 + \frac{x}{2})dx = \frac{7}{12}$$

$$E(X^2) = \int_0^1 x^2(x+\frac{1}{2})dx = \int_0^1 (x^3 + \frac{x^2}{2})dx = \frac{5}{12}$$

$$\text{Var}(X) = E(X^2) - \mu_X^2 = \frac{5}{12} - \left(\frac{7}{12}\right)^2 = \frac{11}{144}$$

Then

$$E(XY) = \int_0^1\int_0^1 xy(x+y)dydx = \int_0^1 \left(\frac{x^2}{2} + \frac{x}{3}\right)dx = \frac{x^3}{6} + \frac{x^2}{6}\bigg|_0^1 = \frac{1}{3}$$

so, putting all of the pieces together,

$$\text{Cov}(X,Y) = 1/3 - (7/12)(7/12) = -1/144$$

and, finally, $\text{Var}(X+Y) = 2[11/144 + (-1/144)] = 5/36$

■

The two corollaries that follow are straightforward extensions of Theorem 3.9.5 to n variables. The details of the proof will be left as an exercise.

Corollary

Suppose that $W_1, W_2, \ldots, W_n$ are random variables with finite variances. Then

$$\text{Var}\left(\sum_{i=1}^{a} a_i W_i\right) = \sum_{i=1}^{n} a_i^2\text{Var}(W_i) + 2\sum_{i<j} a_i a_j \text{Cov}(W_i, W_j)$$

◄

Corollary

Suppose that $W_1, W_2, \ldots, W_n$ are independent random variables with finite variances. Then

$$\text{Var}(W_1 + W_2 + \cdots + W_n) = \text{Var}(W_1) + \text{Var}(W_2) + \cdots + \text{Var}(W_n)$$

More discussion of the covariance and its role in measuring the relationship between random variables occurs in Section 11.4.

◄

Example 3.9.9

The binomial random variable, being a sum of n independent Bernoullis, is an obvious candidate for the corollary to Theorem 3.9.5 on the sum of independent random variables. Let X_i denote the number of successes occurring on the ith trial. Then

$$X_i = \begin{cases} 1 & \text{with probability } p \\ 0 & \text{with probability } 1-p \end{cases}$$

and

$$X = X_1 + X_2 + \cdots + X_n = \text{total number of successes in } n \text{ trials}$$

Find Var(X).

Note that

$$E(X_i) = 1 \cdot p + 0 \cdot (1-p) = p$$

and

$$E\left(X_i^2\right) = (1)^2 \cdot p + (0)^2 \cdot (1-p) = p$$

so

$$\begin{aligned} \text{Var}(X_i) &= E\left(X_i^2\right) - [E(X_i)]^2 = p - p^2 \\ &= p(1-p) \end{aligned}$$

It follows, then, that the *variance of a binomial random variable is* $np(1-p)$:

$$\text{Var}(X) = \sum_{i=1}^{n} \text{Var}(X_i) = np(1-p)$$

■

Example 3.9.10

Recall the hypergeometric model—an urn contains N chips, r red and w white ($r + w = N$); a random sample of size n is selected without replacement and the random variable X is defined to be the number of red chips in the sample. As in the previous example, write X as a sum of simple random variables.

$$X_i = \begin{cases} 1 & \text{if the } i\text{th chip drawn is red} \\ 0 & \text{otherwise} \end{cases}$$

Then $X = X_1 + X_2 + \cdots + X_n$. Clearly,

$$E(X_i) = 1 \cdot \frac{r}{N} + 0 \cdot \frac{w}{N} = \frac{r}{N}$$

and $E(X) = n\left(\frac{r}{N}\right) = np$, where $p = \frac{r}{N}$.

Since $X_i^2 = X_i$, $E(X_i^2) = E(X_i) = \frac{r}{N}$ and

$$\text{Var}(X_i) = E(X_i^2) - [E(X_i)]^2 = \frac{r}{N} - \left(\frac{r}{N}\right)^2 = p(1-p)$$

Also, for any $j \neq k$,

$$\begin{aligned} \text{Cov}(X_j, X_k) &= E(X_j X_k) - E(X_j)E(X_k) \\ &= 1 \cdot P(X_j X_k = 1) - \left(\frac{r}{N}\right)^2 \\ &= \frac{r}{N} \cdot \frac{r-1}{N-1} - \frac{r^2}{N^2} = -\frac{r}{N} \cdot \frac{N-r}{N} \cdot \frac{1}{N-1} \end{aligned}$$

From the first corollary to Theorem 3.9.5, then,

$$\begin{aligned}\text{Var}(X) &= \sum_{i=1}^{n} \text{Var}(X_i) + 2\sum_{j<k} \text{Cov}(X_j, X_k) \\ &= np(1-p) - 2\binom{n}{2} p(1-p) \cdot \frac{1}{N-1} \\ &= p(1-p)\left[n - \frac{n(n-1)}{N-1}\right] \\ &= np(1-p) \cdot \frac{N-n}{N-1}\end{aligned}$$

■

Example 3.9.11

In statistics, it is often necessary to draw inferences based on $\overline{W}$, the average computed from a random sample of n observations. Two properties of $\overline{W}$ are especially important. First, if the W_i's come from a population where the mean is μ, the corollary to Theorem 3.9.2 implies that $E(\overline{W}) = \mu$. Second, if the W_i's come from a population whose variance is σ^2, then $\text{Var}(\overline{W}) = \sigma^2/n$. To verify the latter, we can appeal again to Theorem 3.9.5. Write

$$\overline{W} = \frac{1}{n}\sum_{i=1}^{n} W_i = \frac{1}{n} \cdot W_1 + \frac{1}{n} \cdot W_2 + \cdots + \frac{1}{n} \cdot W_n$$

Then

$$\begin{aligned}\text{Var}(\overline{W}) &= \left(\frac{1}{n}\right)^2 \cdot \text{Var}(W_1) + \left(\frac{1}{n}\right)^2 \cdot \text{Var}(W_2) + \cdots + \left(\frac{1}{n}\right)^2 \cdot \text{Var}(W_n) \\ &= \left(\frac{1}{n}\right)^2 \sigma^2 + \left(\frac{1}{n}\right)^2 \sigma^2 + \cdots + \left(\frac{1}{n}\right)^2 \sigma^2 \\ &= \frac{\sigma^2}{n}\end{aligned}$$

■

Questions

3.9.13. Suppose that two dice are thrown. Let X be the number showing on the first die and let Y be the larger of the two numbers showing. Find $\text{Cov}(X, Y)$.

3.9.14. Show that

$$\text{Cov}(aX + b, cY + d) = ac\text{Cov}(X, Y)$$

for any constants $a, b, c,$ and d.

3.9.15. Let U be a random variable uniformly distributed over $[0, 2\pi]$. Define $X = \cos U$ and $Y = \sin U$. Show that X and Y are dependent but that $\text{Cov}(X, Y) = 0$.

3.9.16. Let X and Y be random variables with

$$f_{X,Y}(x, y) = \begin{cases} 1, & -y < x < y, \quad 0 < y < 1 \\ 0, & \text{elsewhere} \end{cases}$$

Show that $\text{Cov}(X, Y) = 0$ but that X and Y are dependent.

3.9.17. Suppose that $f_{X,Y}(x, y) = \lambda^2 e^{-\lambda(x+y)}$, $0 \le x$, $0 \le y$. Find $\text{Var}(X + Y)$. (*Hint:* See Questions 3.6.11 and 3.9.2.)

3.9.18. Suppose that $f_{X,Y}(x, y) = \frac{2}{3}(x + 2y)$, $0 \le x \le 1$, $0 \le y \le 1$. Find $\text{Var}(X + Y)$. (*Hint:* See Question 3.9.3.)

3.9.19. For the uniform pdf defined over [0, 1], find the variance of the geometric mean when $n = 2$ (see Question 3.9.12).

3.9.20. Let X be a binomial random variable based on n trials and a success probability of p_x; let Y be an independent binomial random variable based on m trials and a success probability of p_Y. Find $E(W)$ and $\text{Var}(W)$, where $W = 4X + 6Y$.

3.9.21. Let the Poisson random variable U (see p. 227) be the number of calls for technical assistance received by a computer company during the firm's nine normal workday hours. Suppose the average number of calls per hour is 7.0 and that each call costs the company \$50. Let V be a Poisson random variable representing the number of calls for technical assistance received during a day's remaining fifteen hours. Suppose the average number of calls per hour is 4.0 for that time period and that each such call costs the company \$60. Find the expected cost and the variance of the cost associated with the calls received during a twenty-four-hour day.

3.9.22. A mason is contracted to build a patio retaining wall. Plans call for the base of the wall to be a row of fifty 10-inch bricks, each separated by $\frac{1}{2}$-inch-thick mortar. Suppose that the bricks used are randomly chosen from a population of bricks whose mean length is 10 inches and whose standard deviation is $\frac{1}{32}$ inch. Also, suppose that the mason, on the average, will make the mortar $\frac{1}{2}$ inch thick, but that the actual dimension will vary from brick to brick, the standard deviation of the thicknesses being $\frac{1}{16}$ inch. What is the standard deviation of L, the length of the first row of the wall? What assumption are you making?

3.9.23. An electric circuit has six resistors wired in series, each nominally being five ohms. What is the maximum standard deviation that can be allowed in the manufacture of these resistors if the combined circuit resistance is to have a standard deviation no greater than 0.4 ohm?

3.9.24. A gambler plays n hands of poker. If he wins the kth hand, he collects k dollars; if he loses the kth hand, he collects nothing. Let T denote his total winnings in n hands. Assuming that his chances of winning each hand are constant and independent of his success or failure at any other hand, find $E(T)$ and $\text{Var}(T)$.

3.10 Order Statistics

The single-variable transformation taken up in Section 3.4 involved a standard linear operation, $Y = aX + b$. The bivariate transformations in Section 3.8 were similarly arithmetic, typically being concerned with either sums or products. In this section we will consider a different sort of transformation, one involving the *ordering* of an entire *set* of random variables. This particular transformation has wide applicability in many areas of statistics, and we will see some of its consequences in later chapters.

Definition 3.10.1. Let Y be a continuous random variable for which $y_1, y_2, \ldots, y_n$ are the values of a random sample of size n. Reorder the y_i's from smallest to largest:

$$y'_1 < y'_2 < \cdots < y'_n$$

(No two of the y_i's are equal, except with probability zero, since Y is continuous.) Define the random variable Y'_i to have the value y'_i, $1 \leq i \leq n$. Then Y'_i is called the ith *order statistic*. Sometimes Y'_n and Y'_1 are denoted $Y_{\max}$ and $Y_{\min}$, respectively.

Example 3.10.1

Suppose that four measurements are made on the random variable Y: $y_1 = 3.4$, $y_2 = 4.6$, $y_3 = 2.6$, and $y_4 = 3.2$. The corresponding ordered sample would be

$$2.6 < 3.2 < 3.4 < 4.6$$

The random variable representing the smallest observation would be denoted Y'_1, with its value for this particular sample being 2.6. Similarly, the value for the second order statistic, Y'_2, is 3.2, and so on. ■

The Distribution of Extreme Order Statistics

By definition, every observation in a random sample has the same pdf. For example, if a set of four measurements is taken from a normal distribution with $\mu = 80$ and $\sigma = 15$, then $f_{Y_1}(y)$, $f_{Y_2}(y)$, $f_{Y_3}(y)$, and $f_{Y_4}(y)$ are all the same—each is a normal pdf with $\mu = 80$ and $\sigma = 15$. The pdf describing an *ordered* observation, though, is *not* the same as the pdf describing a *random* observation. Intuitively, that makes sense. If a single observation is drawn from a normal distribution with $\mu = 80$ and $\sigma = 15$, it would not be surprising if that observation were to take on a value near 80. On the other hand, if a random sample of $n = 100$ observations is drawn from that same distribution, we would not expect the smallest observation—that is, Y_{min}—to be anywhere near 80. Common sense tells us that that smallest observation is likely to be much smaller than 80, just as the largest observation, Y_{max}, is likely to be much larger than 80.

It follows, then, that before we can do any probability calculations—or any applications whatsoever—involving order statistics, we need to know the pdf of Y_i' for $i = 1, 2, \ldots, n$. We begin by investigating the pdfs of the "extreme" order statistics, $f_{Y_{\text{max}}}(y)$ and $f_{Y_{\text{min}}}(y)$. These are the simplest to work with. At the end of the section we return to the more general problems of finding (1) the pdf of Y_i' for any i and (2) the joint pdf of Y_i' and Y_j', where $i < j$.

Theorem 3.10.1 *Suppose that $Y_1, Y_2, \ldots, Y_n$ is a random sample of continuous random variables, each having pdf $f_Y(y)$ and cdf $F_Y(y)$. Then*

a. *The pdf of the largest order statistic is*

$$f_{Y_{\text{max}}}(y) = f_{Y_n'}(y) = n[F_Y(y)]^{n-1} f_Y(y)$$

b. *The pdf of the smallest order statistic is*

$$f_{Y_{\text{min}}}(y) = f_{Y_1'}(y) = n[1 - F_Y(y)]^{n-1} f_Y(y)$$

Proof Finding the pdfs of Y_{max} and Y_{min} is accomplished by using the now-familiar technique of differentiating a random variable's cdf. Consider, for example, the case of the largest order statistic, Y_n':

$$\begin{aligned} F_{Y_n'}(y) = F_{Y_{\text{max}}}(y) &= P(Y_{\text{max}} \leq y) \\ &= P(Y_1 \leq y, Y_2 \leq y, \cdots, Y_n \leq y) \\ &= P(Y_1 \leq y) \cdot P(Y_2 \leq y) \cdots P(Y_n \leq y) \qquad \text{(why?)} \\ &= [F_Y(y)]^n \end{aligned}$$

Therefore,

$$f_{Y_n'}(y) = d/dy[[F_Y(y)]^n] = n[F_Y(y)]^{n-1} f_Y(y)$$

Similarly, for the smallest order statistic ($i = 1$),

$$\begin{aligned} F_{Y_1'}(y) = F_{Y_{\text{min}}}(y) &= P(Y_{\text{min}} \leq y) \\ &= 1 - P(Y_{\text{min}} > y) = 1 - P(Y_1 > y) \cdot P(Y_2 > y) \cdots P(Y_n > y) \\ &= 1 - [1 - F_Y(y)]^n \end{aligned}$$

Therefore,

$$f_{Y_1'}(y) = d/dy[1 - [1 - F_Y(y)]^n] = n[1 - F_Y(y)]^{n-1} f_Y(y) \qquad \square$$

Example 3.10.2 Suppose a random sample of $n = 3$ observations—Y_1, Y_2, and Y_3—is taken from the exponential pdf, $f_Y(y) = e^{-y}$, $y \geq 0$. Compare $f_{Y_1}(y)$ with $f_{Y_1'}(y)$. Intuitively, which will be larger, $P(Y_1 < 1)$ or $P(Y_1' < 1)$?

The pdf for Y_1, of course, is just the pdf of the distribution being sampled—that is,

$$f_{Y_1}(y) = f_Y(y) = e^{-y}, \quad y \geq 0$$

To find the pdf for Y_1' requires that we apply the formula given in the proof of Theorem 3.10.1 for $f_{Y_{\min}}(y)$. Note, first of all, that

$$F_Y(y) = \int_0^y e^{-t} dt = -e^{-t}\Big|_0^y = 1 - e^{-y}$$

Then, since $n = 3$ (and $i = 1$), we can write

$$\begin{aligned} f_{Y_1'}(y) &= 3[1 - (1 - e^{-y})]^2 e^{-y} \\ &= 3e^{-3y}, \quad y \geq 0 \end{aligned}$$

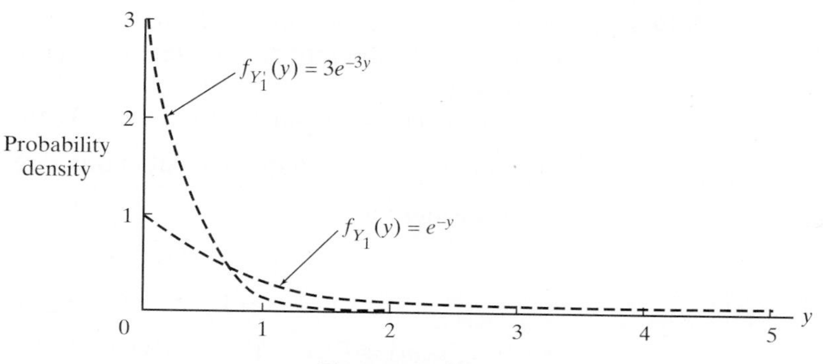

Figure 3.10.1

Figure 3.10.1 shows the two pdfs plotted on the same set of axes. Compared to $f_{Y_1}(y)$, the pdf for Y_1' has more of its area located above the smaller values of y (where Y_1' is more likely to lie). For example, the probability that the smallest observation (out of three) is less than 1 is 95%, while the probability that a random observation is less than 1 is only 63%:

$$\begin{aligned} P(Y_1' < 1) &= \int_0^1 3e^{-3y}\, dy = \int_0^3 e^{-u}\, du = -e^{-u}\Big|_0^3 = 1 - e^{-3} \\ &= 0.95 \end{aligned}$$

$$\begin{aligned} P(Y_1 < 1) &= \int_0^1 e^{-y}\, dy = -e^{-y}\Big|_0^1 = 1 - e^{-1} \\ &= 0.63 \end{aligned}$$

■

Example 3.10.3

Suppose a random sample of size 10 is drawn from a continuous pdf $f_Y(y)$. What is the probability that the largest observation, Y_{10}', is less than the pdf's median, m?

Using the formula for $f_{Y_{10}'}(y) = f_{Y_{\max}}(y)$ given in the proof of Theorem 3.10.1, it is certainly true that

$$P(Y_{10}' < m) = \int_{-\infty}^{m} 10 f_Y(y)[F_Y(y)]^9 dy \tag{3.10.1}$$

but the problem does not specify $f_Y(y)$, so Equation 3.10.1 is of no help.

Fortunately, a much simpler solution is available, even if $f_Y(y)$ were specified: The event "$Y'_{10} < m$" is equivalent to the event "$Y_1 < m \cap Y_2 < m \cap \cdots \cap Y_{10} < m$." Therefore,

$$P(Y'_{10} < m) = P(Y_1 < m, Y_2 < m, \ldots, Y_{10} < m) \quad (3.10.2)$$

But the ten observations here are independent, so the intersection probability implicit on the right-hand side of Equation 3.10.2 factors into a product of ten terms. Moreover, each of those terms equals $\frac{1}{2}$ (by definition of the median), so

$$\begin{aligned} P(Y'_{10} < m) &= P(Y_1 < m) \cdot P(Y_2 < m) \cdots P(Y_{10} < m) \\ &= \left(\tfrac{1}{2}\right)^{10} \\ &= 0.00098 \end{aligned}$$

Example 3.10.4

To find order statistics for discrete pdfs, the probability arguments of the type used in the proof of Theorem 3.10.1 can be be employed. The example of finding the pdf of $X_{\min}$ for the discrete density function $p_X(k)$, $k = 0, 1, 2, \ldots$ suffices to demonstrate this point.

Given a random sample $X_1, X_2, \ldots, X_n$ from $p_X(k)$, choose an arbitrary nonnegative integer m. Recall that the cdf in this case is given by $F_X(m) = \sum\limits_{k=0}^{m} p_k$.

Consider the events

$$A = (m \le X_1 \cap m \le X_2 \cap \cdots \cap m \le X_n) \text{ and}$$
$$B = (m + 1 \le X_1 \cap m + 1 \le X_2 \cap \cdots \cap m + 1 \le X_n)$$

Then $p_{X_{\min}}(m) = P(A \cap B^C) = P(A) - P(A \cap B) = P(A) - P(B)$, where $A \cap B = B$, since $B \subset A$.

Now $P(A) = P(m \le X_1) \cdot P(m \le X_2) \cdot \ldots \cdot P(m \le X_n) = [1 - F_X(m-1)]^n$ by the independence of the X_i. Similarly $P(B) = [1 - F_X(m)]^n$, so

$$p_{Y_{\min}}(m) = [1 - F_X(m-1)]^n - [1 - F_X(m)]^n$$

A General Formula for $f_{Y'_i}(y)$

Having discussed two special cases of order statistics, $Y_{\min}$ and $Y_{\max}$, we now turn to the more general problem of finding the pdf for the ith order statistic, where i can be any integer from 1 through n.

Theorem 3.10.2

Let $Y_1, Y_2, \ldots, Y_n$ be a random sample of continuous random variables drawn from a distribution having pdf $f_Y(y)$ and cdf $F_Y(y)$. The pdf of the ith order statistic is given by

$$f_{Y'_i}(y) = \frac{n!}{(i-1)!(n-i)!}[F_Y(y)]^{i-1}[1 - F_Y(y)]^{n-i} f_Y(y)$$

for $1 \le i \le n$.

Proof We will give a heuristic argument that draws on the similarity between the statement of Theorem 3.10.2 and the binomial distribution. For a formal induction proof verifying the expression given for $f_{Y'_i}(y)$, see (97).

Recall the derivation of the binomial probability function, $p_X(k) = P(X = k) = \binom{n}{k} p^k (1-p)^{n-k}$, where X is the number of successes in n independent trials, and p

is the probability that any given trial ends in success. Central to that derivation was the recognition that the event "$X = k$" is actually a union of all the different (mutually exclusive) sequences having exactly k successes and $n - k$ failures. Because the trials are independent, the probability of any such sequence is $p^k(1-p)^{n-k}$ and the number of such sequences (by Theorem 2.6.2) is $n!/[k!(n-k)!]$ (or $\binom{n}{k}$), so the probability that $X = k$ is the product $\binom{n}{k} p^k(1-p)^{n-k}$.

Here we are looking for the pdf of the ith order statistic at some point y—that is, $f_{Y_i'}(y)$. As was the case with the binomial, that pdf will reduce to a combinatorial term times the probability associated with an intersection of independent events. The only fundamental difference is that Y_i' is a continuous random variable, whereas the binomial X is discrete, which means that what we find here will be a probability *density* function.

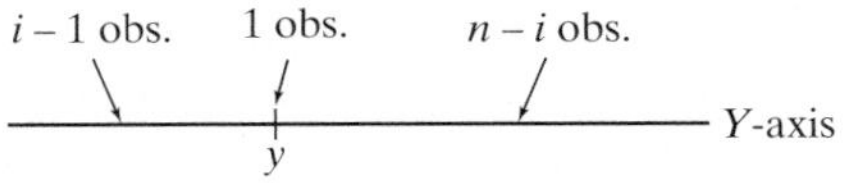

Figure 3.10.2

By Theorem 2.6.2, there are $n!/[(i-1)!1!(n-i)!]$ ways that n observations can be parceled into three groups such that the ith largest is at the point y (see Figure 3.10.2). Moreover, the likelihood associated with any particular set of points having the configuration pictured in Figure 3.10.2 will be the probability that $i - 1$ (independent) observations are all less than y, $n - i$ observations are greater than y, and one observation is at y. The probability density associated with those constraints for a given set of points would be $[F_Y(y)]^{i-1}[1 - F_Y(y)]^{n-i} f_Y(y)$. The probability density, then, that the ith order statistic is located at the point y is the product

$$f_{Y_i'}(y) = \frac{n!}{(i-1)!(n-i)!}[F_Y(y)]^{i-1}[1 - F_Y(y)]^{n-i} f_Y(y)$$

□

Example 3.10.5

Suppose that many years of observation have confirmed that the annual maximum flood tide Y (in feet) for a certain river can be modeled by the pdf

$$f_Y(y) = \frac{1}{20}, \quad 20 < y < 40$$

(*Note:* It is unlikely that flood tides would be described by anything as simple as a uniform pdf. We are making that choice here solely to facilitate the mathematics.) The Army Corps of Engineers is planning to build a levee along a certain portion of the river, and they want to make it high enough so that there is only a 30% chance that the second worst flood in the next thirty-three years will overflow the embankment. How high should the levee be? (We assume that there will be only one potential flood per year.)

Let h be the desired height. If $Y_1, Y_2, \ldots, Y_{33}$ denote the flood tides for the next $n = 33$ years, what we require of h is that

$$P(Y_{32}' > h) = 0.30$$

As a starting point, notice that for $20 < y < 40$,

$$F_Y(y) = \int_{20}^{y} \frac{1}{20}\, dy = \frac{y}{20} - 1$$

Therefore,

$$f_{Y'_{32}}(y) = \frac{33!}{31!1!}\left(\frac{y}{20}-1\right)^{31}\left(2-\frac{y}{20}\right)^{1}\cdot\frac{1}{20}$$

and h is the solution of the integral equation

$$\int_h^{40}(33)(32)\left(\frac{y}{20}-1\right)^{31}\left(2-\frac{y}{20}\right)^{1}\cdot\frac{dy}{20} = 0.30 \tag{3.10.3}$$

If we make the substitution

$$u = \frac{y}{20} - 1$$

Equation 3.10.3 simplifies to

$$P(Y'_{32} > h) = 33(32)\int_{(h/20)-1}^{1} u^{31}(1-u)\,du$$

$$= 1 - 33\left(\frac{h}{20}-1\right)^{32} + 32\left(\frac{h}{20}-1\right)^{33} \tag{3.10.4}$$

Setting the right-hand side of Equation 3.10.4 equal to 0.30 and solving for h by trial and error gives

$$h = 39.3 \text{ feet}$$

■

Joint Pdfs of Order Statistics

Finding the joint pdf of two or more order statistics is easily accomplished by generalizing the argument that derived from Figure 3.10.2. Suppose, for example, that each of n observations in a random sample has pdf $f_Y(y)$ and cdf $F_Y(y)$. The joint pdf for order statistics Y'_i and Y'_j at points u and v, where $i < j$ and $u < v$, can be deduced from Figure 3.10.3, which shows how the n points must be distributed if the ith and jth order statistics are to be located at points u and v, respectively.

Figure 3.10.3

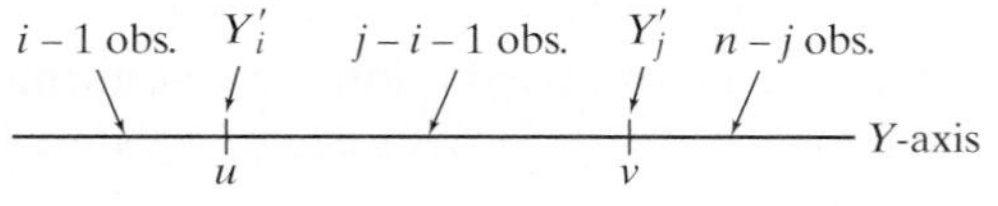

By Theorem 2.6.2, the number of ways to divide a set of n observations into groups of sizes $i-1$, 1, $j-i-1$, 1, and $n-j$ is the quotient

$$\frac{n!}{(i-1)!1!(j-i-1)!1!(n-j)!}$$

Also, given the independence of the n observations, the probability that $i-1$ are less than u is $[F_Y(u)]^{i-1}$, the probability that $j-i-1$ are between u and v is $[F_Y(v) - F_Y(u)]^{j-i-1}$, and the probability that $n-j$ are greater than v is $[1-F_Y(v)]^{n-j}$. Multiplying, then, by the pdfs describing the likelihoods that Y'_i and Y'_j would be at points u and v, respectively, gives the joint pdf of the two order statistics:

$$f_{Y'_i,Y'_j}(u,v) = \frac{n!}{(i-1)!(j-i-1)!(n-j)!}[F_Y(u)]^{i-1}[F_Y(v)-F_Y(u)]^{j-i-1}\cdot$$
$$[1-F_Y(v)]^{n-j}\,f_Y(u)\,f_Y(v) \tag{3.10.5}$$

for $i < j$ and $u < v$.

Example 3.10.6

Let Y_1, Y_2, and Y_3 be a random sample of size $n=3$ from the uniform pdf defined over the unit interval, $f_Y(y)=1, 0\le y\le 1$. By definition, the *range, R*, of a sample is the difference between the largest and smallest order statistics—in this case,

$$R = \text{range} = Y_{\max} - Y_{\min} = Y_3' - Y_1'$$

Find $f_R(r)$, the pdf for the range.

We will begin by finding the joint pdf of Y_1' and Y_3'. Then $f_{Y_1',Y_3'}(u,v)$ is integrated over the region $Y_3' - Y_1' \le r$ to find the cdf, $F_R(r) = P(R \le r)$. The final step is to differentiate the cdf and make use of the fact that $f_R(r) = F_R'(r)$.

If $f_Y(y)=1, 0\le y\le 1$, it follows that

$$F_Y(y) = P(Y\le y) = \begin{cases} 0, & y<0 \\ y, & 0\le y\le 1 \\ 1. & y>1 \end{cases}$$

Applying Equation 3.10.5, then, with $n=3$, $i=1$, and $j=3$, gives the joint pdf of Y_1' and Y_3'. Specifically,

$$\begin{aligned} f_{Y_1',Y_3'}(u,v) &= \frac{3!}{0!1!0!}u^0(v-u)^1(1-v)^0\cdot 1\cdot 1 \\ &= 6(v-u), \quad 0\le u<v\le 1 \end{aligned}$$

Moreover, we can write the cdf for R in terms of Y_1' and Y_3':

$$F_R(r) = P(R\le r) = P(Y_3' - Y_1' \le r) = P(Y_3' \le Y_1' + r)$$

Figure 3.10.4 shows the region in the $Y_1'Y_3'$-plane corresponding to the event that $R\le r$. Integrating the joint pdf of Y_1' and Y_3' over the shaded region gives

$$F_R(r) = P(R\le r) = \int_0^{1-r}\int_u^{u+r} 6(v-u)\,dv\,du + \int_{1-r}^1\int_u^1 6(v-u)\,dv\,du$$

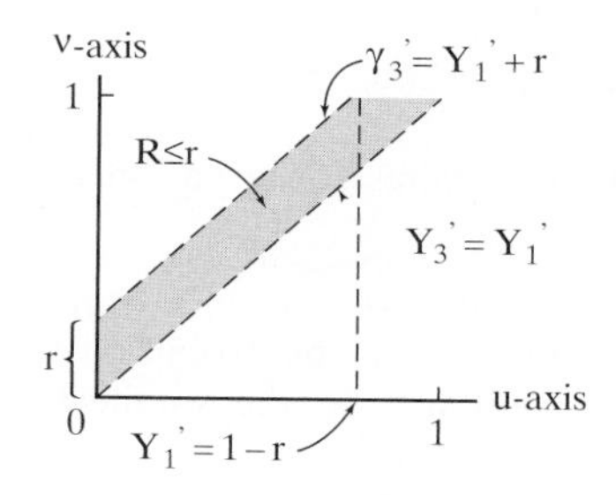

Figure 3.10.4

The first double integral equals $3r^2 - 3r^3$; the second equals r^3. Therefore,

$$F_R(r) = 3r^2 - 3r^3 + r^3 = 3r^2 - 2r^3$$

which implies that

$$f_R(r) = F_R'(r) = 6r - 6r^2, \quad 0\le r\le 1$$

■

Questions

3.10.1. Suppose the length of time, in minutes, that you have to wait at a bank teller's window is uniformly distributed over the interval (0, 10). If you go to the bank four times during the next month, what is the probability that your second longest wait will be less than five minutes?

3.10.2. A random sample of size $n = 6$ is taken from the pdf $f_Y(y) = 3y^2$, $0 \le y \le 1$. Find $P(Y_5' > 0.75)$.

3.10.3. What is the probability that the larger of two random observations drawn from any continuous pdf will exceed the sixtieth percentile?

3.10.4. A random sample of size 5 is drawn from the pdf $f_Y(y) = 2y$, $0 \le y \le 1$. Calculate $P(Y_1' < 0.6 < Y_5')$. (*Hint:* Consider the complement.)

3.10.5. Suppose that $Y_1, Y_2, \ldots, Y_n$ is a random sample of size n drawn from a continuous pdf, $f_Y(y)$, whose median is m. Is $P(Y_1' > m)$ less than, equal to, or greater than $P(Y_n' > m)$?

3.10.6. Let $Y_1, Y_2, \ldots, Y_n$ be a random sample from the exponential pdf $f_y(y) = e^{-y}$, $y \ge 0$. What is the smallest n for which $P(Y_{\text{min}} < 0.2) > 0.9$?

3.10.7. Calculate $P(0.6 < Y_4' < 0.7)$ if a random sample of size 6 is drawn from the uniform pdf defined over the interval [0, 1].

3.10.8. A random sample of size $n = 5$ is drawn from the pdf $f_Y(y) = 2y$, $0 \le y \le 1$. On the same set of axes, graph the pdfs for Y_2, Y_1', and Y_5'.

3.10.9. Suppose that n observations are taken at random from the pdf

$$f_Y(y) = \frac{1}{\sqrt{2\pi}(6)} e^{-\frac{1}{2}\left(\frac{y-20}{6}\right)^2}, \quad -\infty < y < \infty$$

What is the probability that the smallest observation is larger than twenty?

3.10.10. Suppose that n observations are chosen at random from a continuous pdf $f_Y(y)$. What is the probability that the last observation recorded will be the smallest number in the entire sample?

3.10.11. In a certain large metropolitan area, the proportion, Y, of students bused varies widely from school to school. The distribution of proportions is roughly described by the following pdf:

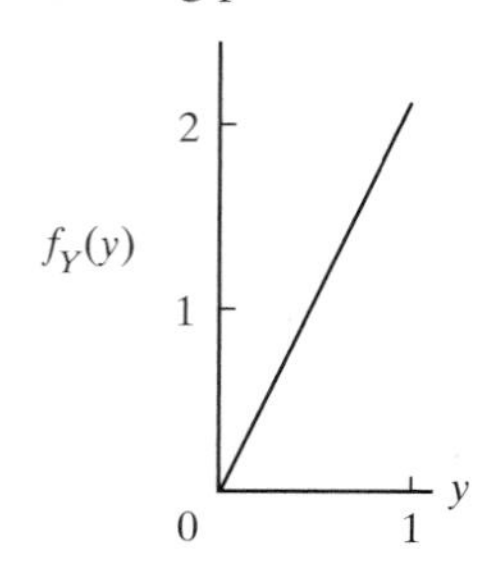

Suppose the enrollment figures for five schools selected at random are examined. What is the probability that the school with the fourth highest proportion of bused children will have a Y value in excess of 0.75? What is the probability that none of the schools will have fewer than 10% of their students bused?

3.10.12. Consider a system containing n components, where the lifetimes of the components are independent random variables and each has pdf $f_Y(y) = \lambda e^{-\lambda y}$, $y > 0$. Show that the average time elapsing before the first component failure occurs is $1/n\lambda$.

3.10.13. Let $Y_1, Y_2, \ldots, Y_n$ be a random sample from a uniform pdf over [0, 1]. Use Theorem 3.10.2 to show that $\int_0^1 y^{i-1}(1-y)^{n-i}\,dy = \dfrac{(i-1)!(n-i)!}{n!}$.

3.10.14. Use Question 3.10.13 to find the expected value of Y_i', where $Y_1, Y_2, \ldots, Y_n$ is a random sample from a uniform pdf defined over the interval [0, 1].

3.10.15. Suppose three points are picked randomly from the unit interval. What is the probability that the three are within a half unit of one another?

3.10.16. Suppose a device has three independent components, all of whose lifetimes (in months) are modeled by the exponential pdf, $f_Y(y) = e^{-y}$, $y > 0$. What is the probability that all three components will fail within two months of one another?

3.11 Conditional Densities

We have already seen that many of the concepts defined in Chapter 2 relating to the probabilities of *events*—for example, independence—have random variable counterparts. Another of these carryovers is the notion of a conditional probability, or, in what will be our present terminology, a *conditional probability density function.* Applications of conditional pdfs are not uncommon. The height and girth of a tree, for instance, can be considered a pair of random variables. While it is easy to measure girth, it can be difficult to determine height; thus it might be of interest to a lumberman to know the probabilities of a ponderosa pine's attaining certain heights given a known value for its girth. Or consider the plight of a school board member agonizing over which way to vote on a proposed budget increase. Her task would be that much easier if she knew the conditional probability that x additional tax dollars would stimulate an average increase of y points among twelfth graders taking a standardized proficiency exam.

Finding Conditional Pdfs for Discrete Random Variables

In the case of discrete random variables, a conditional pdf can be treated in the same way as a conditional probability. Note the similarity between Definitions 3.11.1 and 2.4.1.

Definition 3.11.1. Let X and Y be discrete random variables. The *conditional probability density function of Y given x*—that is, the probability that Y takes on the value y given that X is equal to x—is denoted $p_{Y|x}(y)$ and given by

$$p_{Y|x}(y) = P(Y = y \mid X = x) = \frac{p_{X,Y}(x, y)}{p_X(x)}$$

for $p_X(x) \neq 0$.

Example 3.11.1

A fair coin is tossed five times. Let the random variable Y denote the total number of heads that occur, and let X denote the number of heads occurring on the last two tosses. Find the conditional pdf $p_{Y|x}(y)$ for all x and y.

Clearly, there will be three different conditional pdfs, one for each possible value of X ($x = 0$, $x = 1$, and $x = 2$). Moreover, for each value of x there will be four possible values of Y, based on whether the first three tosses yield zero, one, two, or three heads.

For example, suppose no heads occur on the last two tosses. Then $X = 0$, and

$$\begin{aligned} p_{Y|0}(y) &= P(Y = y \mid X = 0) = P(y \text{ heads occur on first three tosses}) \\ &= \binom{3}{y}\left(\frac{1}{2}\right)^y\left(1 - \frac{1}{2}\right)^{3-y} \\ &= \binom{3}{y}\left(\frac{1}{2}\right)^3, \quad y = 0, 1, 2, 3 \end{aligned}$$

Now, suppose that $X = 1$. The corresponding conditional pdf in that case becomes

$$p_{Y|x}(y) = P(Y = y \mid X = 1)$$

Notice that $Y = 1$ if zero heads occur in the first three tosses, $Y = 2$ if one head occurs in the first three trials, and so on. Therefore,

$$\begin{aligned} p_{Y|1}(y) &= \binom{3}{y-1}\left(\frac{1}{2}\right)^{y-1}\left(1 - \frac{1}{2}\right)^{3-(y-1)} \\ &= \binom{3}{y-1}\left(\frac{1}{2}\right)^3, \quad y = 1, 2, 3, 4 \end{aligned}$$

Similarly,

$$p_{Y|2}(y) = P(Y = y \mid X = 2) = \binom{3}{y-2}\left(\frac{1}{2}\right)^3, \quad y = 2, 3, 4, 5$$

Figure 3.11.1 shows the three conditional pdfs. Each has the same shape, but the possible values of Y are different for each value of X.

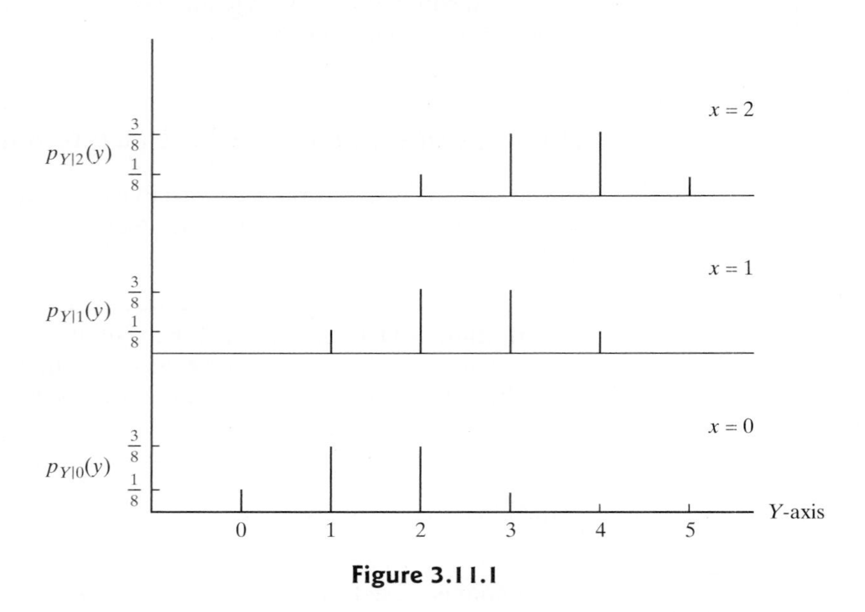

Figure 3.11.1

Example 3.11.2 Assume that the probabilistic behavior of a pair of discrete random variables X and Y is described by the joint pdf

$$p_{X,Y}(x, y) = xy^2/39$$

defined over the four points (1, 2), (1, 3), (2, 2), and (2, 3). Find the conditional probability that $X = 1$ given that $Y = 2$.

By definition,

$$\begin{aligned} p_{X|2}(1) &= P(X = 1 \text{ given that } Y = 2) \\ &= \frac{p_{X,Y}(1, 2)}{p_Y(2)} \\ &= \frac{1 \cdot 2^2/39}{1 \cdot 2^2/39 + 2 \cdot 2^2/39} \\ &= 1/3 \end{aligned}$$

Example 3.11.3 Suppose that X and Y are two independent binomial random variables, each defined on n trials and each having the same success probability p. Let $Z = X + Y$. Show that the conditional pdf $p_{X|z}(x)$ is a hypergeometric distribution.

We know from Example 3.8.2 that Z has a binomial distribution with parameters $2n$ and p. That is,

$$p_Z(z) = P(Z = z) = \binom{2n}{z} p^z (1-p)^{2n-z}, \quad z = 0, 1, \ldots, 2n.$$

By Definition 3.11.1,

$$\begin{aligned}
p_{X|z}(x) = P(X=x|Z=z) &= \frac{p_{X,Z}(x,z)}{p_Z(z)} \\
&= \frac{P(X=x \text{ and } Z=z)}{P(Z=z)} \\
&= \frac{P(X=x \text{ and } Y=z-x)}{P(Z=z)} \\
&= \frac{P(X=x)\cdot P(Y=z-x)}{P(Z=z)} \quad \text{(because } X \text{ and } Y \text{ are independent)} \\
&= \frac{\binom{n}{x}p^x(1-p)^{n-x}\cdot\binom{n}{z-x}p^{z-x}(1-p)^{n-(z-x)}}{\binom{2n}{z}p^z(1-p)^{2n-z}} \\
&= \frac{\binom{n}{x}\binom{n}{z-x}}{\binom{2n}{z}}
\end{aligned}$$

which we recognize as being the hypergeometric distribution.

Comment The notion of a conditional pdf generalizes easily to situations involving more than two discrete random variables. For example, if X, Y, and Z have the joint pdf $p_{X,Y,Z}(x,y,z)$, the *joint conditional pdf* of, say, X and Y given that $Z=z$ is the ratio

$$p_{X,Y|z}(x,y) = \frac{p_{X,Y,Z}(x,y,z)}{p_Z(z)}$$

■

Example 3.11.4 Suppose that random variables X, Y, and Z have the joint pdf

$$p_{X,Y,Z}(x,y,z) = xy/9z$$

for points (1, 1, 1), (2, 1, 2), (1, 2, 2), (2, 2, 2), and (2, 2, 1). Find $p_{X,Y|z}(x,y)$ for all values of z.

To begin, we see from the points for which $p_{X,Y,Z}(x,y,z)$ is defined that Z has two possible values, 1 and 2. Suppose $z=1$. Then

$$p_{X,Y|1}(x,y) = \frac{p_{X,Y,Z}(x,y,1)}{p_Z(1)}$$

But

$$\begin{aligned}
p_Z(1) = P(Z=1) &= P[(1,1,1)\cup(2,2,1)] \\
&= 1\cdot\frac{1}{9}\cdot 1 + 2\cdot\frac{2}{9}\cdot 1 \\
&= \frac{5}{9}
\end{aligned}$$

Therefore,

$$p_{X,Y|1}(x,y) = \frac{xy/9}{\frac{5}{9}} = xy/5 \quad \text{for} \quad (x,y)=(1,1) \quad \text{and} \quad (2,2)$$

Suppose $z = 2$. Then

$$p_Z(2) = P(Z = 2) = P[(2, 1, 2) \cup (1, 2, 2) \cup (2, 2, 2)]$$
$$= 2 \cdot \frac{1}{18} + 1 \cdot \frac{2}{18} + 2 \cdot \frac{2}{18}$$
$$= \frac{8}{18}$$

so

$$p_{X,Y|2}(x, y) = \frac{p_{X,Y,Z}(x, y, 2)}{p_Z(2)}$$
$$= \frac{x \cdot y/18}{\frac{8}{18}}$$
$$= \frac{xy}{8} \quad \text{for} \quad (x, y) = (2, 1), (1, 2), \text{ and } (2, 2)$$

Questions

3.11.1. Suppose X and Y have the joint pdf $p_{X,Y}(x, y) = \frac{x+y+xy}{21}$ for the points (1, 1), (1, 2), (2, 1), (2, 2), where X denotes a "message" sent (either $x = 1$ or $x = 2$) and Y denotes a "message" received. Find the probability that the message sent was the message received—that is, find $p_{Y|x}(y)$.

3.11.2. Suppose a die is rolled six times. Let X be the total number of 4's that occur and let Y be the number of 4's in the first two tosses. Find $p_{Y|x}(y)$.

3.11.3. An urn contains eight red chips, six white chips, and four blue chips. A sample of size 3 is drawn without replacement. Let X denote the number of red chips in the sample and Y, the number of white chips. Find an expression for $p_{Y|x}(y)$.

3.11.4. Five cards are dealt from a standard poker deck. Let X be the number of aces received, and Y the number of kings. Compute $P(X = 2|Y = 2)$.

3.11.5. Given that two discrete random variables X and Y follow the joint pdf $p_{X,Y}(x, y) = k(x + y)$, for $x = 1, 2, 3$ and $y = 1, 2, 3$,

(a) Find k.
(b) Evaluate $p_{Y|x}(1)$ for all values of x for which $p_x(x) > 0$.

3.11.6. Let X denote the number on a chip drawn at random from an urn containing three chips, numbered 1, 2, and 3. Let Y be the number of heads that occur when a fair coin is tossed X times.

(a) Find $p_{X,Y}(x, y)$.
(b) Find the marginal pdf of Y by summing out the x values.

3.11.7. Suppose X, Y, and Z have a trivariate distribution described by the joint pdf

$$p_{X,Y,Z}(x, y, z) = \frac{xy + xz + yz}{54}$$

where x, y, and z can be 1 or 2. Tabulate the joint conditional pdf of X and Y given each of the two values of z.

3.11.8. In Question 3.11.7 define the random variable W to be the "majority" of x, y, and z. For example, $W(2, 2, 1) = 2$ and $W(1, 1, 1) = 1$. Find the pdf of $W|x$.

3.11.9. Let X and Y be independent random variables where $p_x(k) = e^{-\lambda}\frac{\lambda^k}{k!}$ and $p_Y(k) = e^{-\mu}\frac{\mu^k}{k!}$ for $k = 0, 1, \ldots$. Show that the conditional pdf of X given that $X + Y = n$ is binomial with parameters n and $\frac{\lambda}{\lambda+\mu}$. (*Hint:* See Question 3.8.1.)

3.11.10. Suppose Compositor A is preparing a manuscript to be published. Assume that she makes X errors on a given page, where X has the Poisson pdf, $p_X(k) = e^{-2}2^k/k!$, $k = 0, 1, 2, \ldots$. A second compositor, B, is also working on the book. He makes Y errors on a page, where $p_Y(k) = e^{-3}3^k/k!$, $k = 0, 1, 2, \ldots$. Assume that Compositor A prepares the first one hundred pages of the text and Compositor B, the last one hundred pages. After the book is completed, reviewers (with too much time on their hands!) find that the text contains a total of 520 errors. Write a formula for the exact probability that fewer than half of the errors are due to Compositor A.

Finding Conditional Pdfs for Continuous Random Variables

If the variables X and Y are continuous, we can still appeal to the quotient $f_{X,Y}(x, y)/f_X(x)$ as the definition of $f_{Y|x}(y)$ and argue its propriety by analogy. A more satisfying approach, though, is to arrive at the same conclusion by taking the limit of Y's "conditional" *cdf*.

If X is continuous, a direct evaluation of $F_{Y|x}(y) = P(Y \leq y|X = x)$, via Definition 2.4.1, is impossible, since the denominator would be zero. Alternatively, we can think of $P(Y \leq y|X = x)$ as a limit:

$$P(Y \leq y|X = x) = \lim_{h\to 0} P(Y \leq y|x \leq X \leq x + h)$$

$$= \lim_{h\to\infty} \frac{\int_x^{x+h} \int_{-\infty}^{y} f_{X,Y}(t, u)\, du\, dt}{\int_x^{x+h} f_X(t)\, dt}$$

Evaluating the quotient of the limits gives $\frac{0}{0}$, so l'Hôpital's rule is indicated:

$$P(Y \leq y|X = x) = \lim_{h\to 0} \frac{\frac{d}{dh}\int_x^{x+h} \int_{-\infty}^{y} f_{X,Y}(t, u)\, du\, dt}{\frac{d}{dh}\int_x^{x+h} f_X(t)\, dt} \tag{3.11.1}$$

By the Fundamental Theorem of Calculus,

$$\frac{d}{dh}\int_x^{x+h} g(t)\, dt = g(x + h)$$

which simplifies Equation 3.11.1 to

$$P(Y \leq y|X = x) = \lim_{h\to 0} \frac{\int_{-\infty}^{y} f_{X,Y}[(x + h), u]\, du}{f_X(x + h)}$$

$$= \frac{\int_{-\infty}^{y} \lim_{h\to 0} f_{X,Y}(x + h, u)\, du}{\lim_{h\to 0} f_X(x + h)} = \int_{-\infty}^{y} \frac{f_{X,Y}(x, u)}{f_X(x)}\, du$$

provided that the limit operation and the integration can be interchanged [see (8) for a discussion of when such an interchange is valid]. It follows from this last expression that $f_{X,Y}(x, y)/f_X(x)$ behaves as a conditional probability density function should, and we are justified in extending Definition 3.11.1 to the continuous case.

Example 3.11.5 Let X and Y be continuous random variables with joint pdf

$$f_{X,Y}(x, y) = \begin{cases} \left(\frac{1}{8}\right)(6 - x - y), & 0 \leq x \leq 2, \quad 2 \leq y \leq 4 \\ 0, & \text{elsewhere} \end{cases}$$

Find (a) $f_X(x)$, (b) $f_{Y|x}(y)$, and (c) $P(2 < Y < 3|x = 1)$.

a. From Theorem 3.7.2,

$$f_X(x) = \int_{-\infty}^{\infty} f_{X,Y}(x, y)\, dy = \int_2^4 \left(\frac{1}{8}\right)(6 - x - y)\, dy$$

$$= \left(\frac{1}{8}\right)(6 - 2x), \quad 0 \le x \le 2$$

b. Substituting into the "continuous" statement of Definition 3.11.1, we can write

$$f_{Y|x}(y) = \frac{f_{X,Y}(x, y)}{f_X(x)} = \frac{\left(\frac{1}{8}\right)(6 - x - y)}{\left(\frac{1}{8}\right)(6 - 2x)}$$

$$= \frac{6 - x - y}{6 - 2x}, \quad 0 \le x \le 2, \quad 2 \le y \le 4$$

c. To find $P(2 < Y < 3|x = 1)$, we simply integrate $f_{Y|1}(y)$ over the interval $2 < Y < 3$:

$$P(2 < Y < 3|x = 1) = \int_2^3 f_{Y|1}(y)\, dy$$

$$= \int_2^3 \frac{5 - y}{4}\, dy$$

$$= \frac{5}{8}$$

[A partial check that the derivation of a conditional pdf is correct can be performed by integrating $f_{Y|x}(y)$ over the entire range of Y. That integral should be 1. Here, for example, when $x = 1$, $\int_{-\infty}^{\infty} f_{Y|1}(y)\, dy = \int_2^4 [(5 - y)/4]\, dy$ *does* equal 1.] ■

Questions

3.11.11. Let X be a nonnegative random variable. We say that X is *memoryless* if

$$P(X > s + t|X > t) = P(X > s) \quad \text{for all } s, t \ge 0$$

Show that a random variable with pdf $f_X(x) = (1/\lambda)e^{-x/\lambda}$, $x > 0$, is memoryless.

3.11.12. Given the joint pdf

$$f_{X,Y}(x, y) = 2e^{-(x+y)}, \quad 0 \le x \le y, \quad y \ge 0$$

find

(a) $P(Y < 1|X < 1)$.
(b) $P(Y < 1|X = 1)$.
(c) $f_{Y|x}(y)$.
(d) $E(Y|x)$.

3.11.13. Find the conditional pdf of Y given x if

$$f_{X,Y}(x, y) = x + y$$

for $0 \le x \le 1$ and $0 \le y \le 1$.

3.11.14. If

$$f_{X,Y}(x, y) = 2, \quad x \ge 0, \quad y \ge 0, \quad x + y \le 1$$

show that the conditional pdf of Y given x is uniform.

3.11.15. Suppose that

$$f_{Y|x}(y) = \frac{2y + 4x}{1 + 4x} \quad \text{and} \quad f_X(x) = \frac{1}{3} \cdot (1 + 4x)$$

for $0 < x < 1$ and $0 < y < 1$. Find the marginal pdf for Y.

3.11.16. Suppose that X and Y are distributed according to the joint pdf

$$f_{X,Y}(x, y) = \frac{2}{5} \cdot (2x + 3y), \quad 0 \le x \le 1, \quad 0 \le y \le 1$$

Find

(a) $f_X(x)$.
(b) $f_{Y|x}(y)$.
(c) $P\left(\frac{1}{4} \le Y \le \frac{3}{4}|X = \frac{1}{2}\right)$.
(d) $E(Y|x)$.

3.11.17. If X and Y have the joint pdf

$$f_{X,Y}(x, y) = 2, \quad 0 \le x < y \le 1$$

find $P\left(0 < X < \frac{1}{2} | Y = \frac{3}{4}\right)$.

3.11.18. Find $P\left(X < 1 | Y = 1\frac{1}{2}\right)$ if X and Y have the joint Pdf

$$f_{X,Y}(x, y) = xy/2, \quad 0 \le x < y \le 2$$

3.11.19. Suppose that X_1, X_2, X_3, X_4, and X_5 have the joint pdf

$$f_{X_1,X_2,X_3,X_4,X_5}(x_1, x_2, x_3, x_4, x_5) = 32x_1x_2x_3x_4x_5$$

for $0 < x_i < 1$, $i = 1, 2, \ldots, 5$. Find the joint conditional pdf of X_1, X_2, and X_3 given that $X_4 = x_4$ and $X_5 = x_5$.

3.11.20. Suppose the random variables X and Y are jointly distributed according to the Pdf

$$f_{X,Y}(x, y) = \frac{6}{7}\left(x^2 + \frac{xy}{2}\right), \quad 0 \le x \le 1, \quad 0 \le y \le 2$$

Find

(a) $f_X(x)$.
(b) $P(X > 2Y)$.
(c) $P\left(Y > 1 | X > \frac{1}{2}\right)$.

3.12 Moment-Generating Functions

Finding moments of random variables directly, particularly the higher moments defined in Section 3.6, is conceptually straightforward but can be quite problematic: Depending on the nature of the pdf, integrals and sums of the form $\int_{-\infty}^{\infty} y^r f_Y(y)\,dy$ and $\sum_{\text{all } k} k^r p_X(k)$ can be very difficult to evaluate. Fortunately, an alternative method is available. For many pdfs, we can find a *moment-generating function* (or *mgf*), $M_W(t)$, one of whose properties is that the rth derivative of $M_W(t)$ evaluated at zero is equal to $E(W^r)$.

Calculating a Random Variable's Moment-Generating Function

In principle, what we call a moment-generating function is a direct application of Theorem 3.5.3.

Definition 3.12.1. Let W be a random variable. The *moment-generating function (mgf) for W* is denoted $M_W(t)$ and given by

$$M_W(t) = E(e^{tW}) = \begin{cases} \sum_{\text{all } k} e^{tk} p_W(k) & \text{if } W \text{ is discrete} \\ \int_{-\infty}^{\infty} e^{tw} f_W(w)\,dw & \text{if } W \text{ is continuous} \end{cases}$$

at all values of t for which the expected value exists.

Example 3.12.1

Suppose the random variable X has a *geometric pdf*,

$$p_X(k) = (1 - p)^{k-1} p, \quad k = 1, 2, \ldots$$

[In practice, this is the pdf that models the occurrence of the first success in a series of independent trials, where each trial has a probability p of ending in success (recall Example 3.3.2)]. Find $M_X(t)$, the moment-generating function for X.

Since X is discrete, the first part of Definition 3.12.1 applies, so

$$M_X(t) = E(e^{tX}) = \sum_{k=1}^{\infty} e^{tk}(1 - p)^{k-1} p$$

$$= \frac{p}{1 - p} \sum_{k=1}^{\infty} e^{tk}(1 - p)^k = \frac{p}{1 - p} \sum_{k=1}^{\infty} [(1 - p)e^t]^k \tag{3.12.1}$$

The t in $M_X(t)$ can be any number in a neighborhood of zero, as long as $M_X(t) < \infty$. Here, $M_X(t)$ is an infinite sum of the terms $[(1-p)e^t]^k$, and that sum will be finite only if $(1-p)e^t < 1$, or, equivalently, if $t < \ln[1/(1-p)]$. It will be assumed, then, in what follows that $0 < t < \ln[1/(1-p)]$.

Recall that

$$\sum_{k=0}^{\infty} r^k = \frac{1}{1-r}$$

provided $0 < r < 1$. This formula can be used on Equation 3.12.1, where $r = (1-p)e^t$ and $0 < t < \ln\left[\frac{1}{(1-p)}\right]$. Specifically,

$$\begin{aligned} M_X(t) &= \frac{p}{1-p}\left[\sum_{k=0}^{\infty}[(1-p)e^t]^k - [(1-p)e^t]^0\right] \\ &= \frac{p}{1-p}\left[\frac{1}{1-(1-p)e^t} - 1\right] \\ &= \frac{pe^t}{1-(1-p)e^t} \end{aligned}$$

■

Example 3.12.2

Suppose that X is a binomial random variable with pdf

$$p_X(k) = \binom{n}{k} p^k (1-p)^{n-k}, \quad k = 0, 1, \ldots, n$$

Find $M_X(t)$.

By Definition 3.12.1,

$$\begin{aligned} M_X(t) = E(e^{tX}) &= \sum_{k=0}^{n} e^{tk}\binom{n}{k} p^k (1-p)^{n-k} \\ &= \sum_{k=0}^{n}\binom{n}{k}(pe^t)^k (1-p)^{n-k} \end{aligned} \tag{3.12.2}$$

To get a closed-form expression for $M_X(t)$—that is, to evaluate the sum indicated in Equation 3.12.2—requires a (hopefully) familiar formula from algebra: According to *Newton's binomial expansion*,

$$(x+y)^n = \sum_{k=0}^{n}\binom{n}{k} x^k y^{n-k} \tag{3.12.3}$$

for any x and y. Suppose we let $x = pe^t$ and $y = 1-p$. It follows from Equations 3.12.2 and 3.12.3, then, that

$$M_X(t) = (1 - p + pe^t)^n$$

[Notice in this case that $M_X(t)$ is defined for all values of t.] ■

Example 3.12.3

Suppose that Y has an exponential pdf, where $f_Y(y) = \lambda e^{-\lambda y}$, $y > 0$. Find $M_Y(t)$.

Since the exponential pdf describes a continuous random variable, $M_Y(t)$ is an integral:

$$M_Y(t) = E(e^{tY}) = \int_0^\infty e^{ty} \cdot \lambda e^{-\lambda y}\, dy$$
$$= \int_0^\infty \lambda e^{-(\lambda - t)y}\, dy$$

After making the substitution $u = (\lambda - t)y$, we can write

$$M_Y(t) = \int_{u=0}^\infty \lambda e^{-u} \frac{du}{\lambda - t}$$
$$= \frac{\lambda}{\lambda - t}\left[-e^{-u}\Big|_{u=0}^{\infty}\right]$$
$$= \frac{\lambda}{\lambda - t}\left[1 - \lim_{u\to\infty} e^{-u}\right] = \frac{\lambda}{\lambda - t}$$

Here, $M_Y(t)$ is finite and nonzero only when $u = (\lambda - t)y > 0$, which implies that t must be less than λ. For $t \geq \lambda$, $M_Y(t)$ fails to exist. ■

Example 3.12.4

The normal (or bell-shaped) curve was introduced in Example 3.4.3. Its pdf is the rather cumbersome function

$$f_Y(y) = (1/\sqrt{2\pi}\sigma)\exp\left[-\frac{1}{2}\left(\frac{y-\mu}{\sigma}\right)^2\right], \quad -\infty < y < \infty$$

where $\mu = E(Y)$ and $\sigma^2 = \text{Var}(Y)$. Derive the moment-generating function for this most important of all probability models.

Since Y is a continuous random variable,

$$M_Y(t) = E(e^{tY}) = (1/\sqrt{2\pi}\sigma)\int_{-\infty}^{\infty} \exp(ty)\exp\left[-\frac{1}{2}\left(\frac{y-\mu}{\sigma}\right)^2\right] dy$$

$$= (1/\sqrt{2\pi}\sigma)\int_{-\infty}^{\infty} \exp\left[-\frac{y^2 - 2\mu y - 2\sigma^2 t y + \mu^2}{2\sigma^2}\right] dy \tag{3.12.4}$$

Evaluating the integral in Equation 3.12.4 is best accomplished by completing the square of the numerator of the exponent (which means that the square of half the coefficient of y is added and subtracted). That is, we can write

$$y^2 - (2\mu + 2\sigma^2 t)y + (\mu + \sigma^2 t)^2 - (\mu + \sigma^2 t)^2 + \mu^2$$
$$= [y - (\mu + \sigma^2 t)]^2 - \sigma^4 t^2 + 2\mu t\sigma^2 \tag{3.12.5}$$

The last two terms on the right-hand side of Equation 3.12.5, though, do not involve y, so they can be factored out of the integral, and Equation 3.12.4 reduces to

$$M_Y(t) = \exp\left(\mu t + \frac{\sigma^2 t^2}{2}\right)(1/\sqrt{2\pi}\sigma)\int_{-\infty}^{\infty} \exp\left[-\frac{1}{2}\left[\frac{y - (\mu + t\sigma^2)}{\sigma}\right]^2\right] dy$$

But, together, the latter two factors equal 1 (why?), implying that the moment-generating function for a normally distributed random variable is given by

$$M_Y(t) = e^{\mu t + \sigma^2 t^2/2}$$ ■

Questions

3.12.1. Let X be a random variable with pdf $p_X(k) = 1/n$, for $k = 0, 1, 2, \ldots, n-1$ and 0 otherwise. Show that $M_X(t) = \frac{1-e^{nt}}{n(1-e^t)}$.

3.12.2. Two chips are drawn at random and without replacement from an urn that contains five chips, numbered 1 through 5. If the sum of the chips drawn is even, the random variable X equals 5; if the sum of the chips drawn is odd, $X = -3$. Find the moment-generating function for X.

3.12.3. Find the expected value of e^{3X} if X is a binominal random variable with $n = 10$ and $p = \frac{1}{3}$.

3.12.4. Find the moment-generating function for the discrete random variable X whose probability function is given by

$$p_X(k) = \left(\frac{3}{4}\right)^k \left(\frac{1}{4}\right), \quad k = 0, 1, 2, \ldots$$

3.12.5. Which pdfs would have the following moment-generating functions?

(a) $M_Y(t) = e^{6t^2}$
(b) $M_Y(t) = 2/(2-t)$
(c) $M_X(t) = \left(\frac{1}{2} + \frac{1}{2}e^t\right)^4$
(d) $M_X(t) = 0.3e^t/(1 - 0.7e^t)$

3.12.6. Let Y have pdf

$$f_Y(y) = \begin{cases} y, & 0 \le y \le 1 \\ 2-y, & 1 \le y \le 2 \\ 0, & \text{elsewhere} \end{cases}$$

Find $M_Y(t)$.

3.12.7. A random variable X is said to have a *Poisson distribution* if $p_X(k) = P(X = k) = e^{-\lambda}\lambda^k/k!$, $k = 0, 1, 2, \ldots$. Find the moment-generating function for a Poisson random variable. Recall that

$$e^r = \sum_{k=0}^{\infty} \frac{r^k}{k!}$$

3.12.8. Let Y be a continuous random variable with $f_Y(y) = ye^{-y}$, $0 \le y$. Show that $M_Y(t) = \frac{1}{(1-t)^2}$.

Using Moment-Generating Functions to Find Moments

Having practiced *finding* the functions $M_X(t)$ and $M_Y(t)$, we now turn to the theorem that spells out their relationship to X^r and Y^r.

Theorem 3.12.1 *Let W be a random variable with probability density function $f_W(w)$. [If W is continuous, $f_W(w)$ must be sufficiently smooth to allow the order of differentiation and integration to be interchanged.] Let $M_W(t)$ be the moment-generating function for W. Then, provided the rth moment exists,*

$$M_W^{(r)}(0) = E(W^r)$$

Proof We will verify the theorem for the continuous case where r is either 1 or 2. The extensions to discrete random variables and to an arbitrary positive integer r are straightforward.

For $r = 1$,

$$\begin{aligned} M_Y^{(1)}(0) &= \frac{d}{dt}\int_{-\infty}^{\infty} e^{ty} f_Y(y)\,dy \bigg|_{t=0} = \int_{-\infty}^{\infty} \frac{d}{dt} e^{ty} f_Y(y)\,dy \bigg|_{t=0} \\ &= \int_{-\infty}^{\infty} y e^{ty} f_Y(y)\,dy \bigg|_{t=0} = \int_{-\infty}^{\infty} y e^{0 \cdot y} f_Y(y)\,dy \\ &= \int_{-\infty}^{\infty} y f_Y(y)\,dy = E(Y) \end{aligned}$$

For $r = 2$,

$$M_Y^{(2)}(0) = \frac{d^2}{dt^2} \int_{-\infty}^{\infty} e^{ty} f_Y(y)\,dy \bigg|_{t=0} = \int_{-\infty}^{\infty} \frac{d^2}{dt^2} e^{ty} f_Y(y)\,dy \bigg|_{t=0}$$

$$= \int_{-\infty}^{\infty} y^2 e^{ty} f_Y(y)\,dy \bigg|_{t=0} = \int_{-\infty}^{\infty} y^2 e^{0\cdot y} f_Y(y)\,dy$$

$$= \int_{-\infty}^{\infty} y^2 f_Y(y)\,dy = E(Y^2)$$

□

Example 3.12.5

For a geometric random variable X with pdf

$$p_X(k) = (1-p)^{k-1}p, \quad k = 1, 2, \ldots$$

we saw in Example 3.12.1 that

$$M_X(t) = pe^t[1-(1-p)e^t]^{-1}$$

Find the expected value of X by differentiating its moment-generating function.

Using the product rule, we can write the first derivative of $M_X(t)$ as

$$M_X^{(1)}(t) = pe^t(-1)[1-(1-p)e^t]^{-2}(-1)(1-p)e^t + [1-(1-p)e^t]^{-1}pe^t$$

$$= \frac{p(1-p)e^{2t}}{[1-(1-p)e^t]^2} + \frac{pe^t}{1-(1-p)e^t}$$

Setting $t = 0$ shows that $E(X) = \frac{1}{p}$:

$$M_X^{(1)}(0) = E(X) = \frac{p(1-p)e^{2\cdot 0}}{[1-(1-p)e^0]^2} + \frac{pe^0}{1-(1-p)e^0}$$

$$= \frac{p(1-p)}{p^2} + \frac{p}{p}$$

$$= \frac{1}{p}$$

■

Example 3.12.6

Find the expected value of an exponential random variable with pdf

$$f_Y(y) = \lambda e^{-\lambda y}, \quad y > 0$$

Use the fact that

$$M_Y(t) = \lambda(\lambda - t)^{-1}$$

(as shown in Example 3.12.3).

Differentiating $M_Y(t)$ gives

$$M_Y^{(1)}(t) = \lambda(-1)(\lambda - t)^{-2}(-1)$$

$$= \frac{\lambda}{(\lambda - t)^2}$$

Set $t = 0$. Then

$$M_Y^{(1)}(0) = \frac{\lambda}{(\lambda - 0)^2}$$

implying that

$$E(Y) = \frac{1}{\lambda}$$

■

Example 3.12.7

Find an expression for $E(X^k)$ if the moment-generating function for X is given by

$$M_X(t) = (1 - p_1 - p_2) + p_1 e^t + p_2 e^{2t}$$

The only way to deduce a formula for an arbitrary moment such as $E(X^k)$ is to calculate the first couple moments and look for a pattern that can be generalized. Here,

$$M_X^{(1)}(t) = p_1 e^t + 2p_2 e^{2t}$$

so

$$E(X) = M_X^{(1)}(0) = p_1 e^0 + 2p_2 e^{2\cdot 0}$$
$$= p_1 + 2p_2$$

Taking the second derivative, we see that

$$M_X^{(2)}(t) = p_1 e^t + 2^2 p_2 e^{2t}$$

implying that

$$E(X^2) = M_X^{(2)}(0) = p_1 e^0 + 2^2 p_2 e^{2\cdot 0}$$
$$= p_1 + 2^2 p_2$$

Clearly, each successive differentiation will leave the p_1 term unaffected but will multiply the p_2 term by 2. Therefore,

$$E(X^k) = M_X^{(k)}(0) = p_1 + 2^k p_2$$ ■

Using Moment-Generating Functions to Find Variances

In addition to providing a useful technique for calculating $E(W^r)$, moment-generating functions can also find variances, because

$$\text{Var}(W) = E(W^2) - [E(W)]^2 \tag{3.12.6}$$

for any random variable W (recall Theorem 3.6.1). Other useful "descriptors" of pdfs can also be reduced to combinations of moments. The *skewness* of a distribution, for example, is a function of $E[(W-\mu)^3]$, where $\mu = E(W)$. But

$$E[(W-\mu)^3] = E(W^3) - 3E(W^2)E(W) + 2[E(W)]^3$$

In many cases, finding $E[(W-\mu)^2]$ or $E[(W-\mu)^3]$ could be quite difficult if moment-generating functions were not available.

Example 3.12.8

We know from Example 3.12.2 that if X is a binomial random variable with parameters n and p, then

$$M_X(t) = (1 - p + pe^t)^n$$

Use $M_X(t)$ to find the variance of X.

The first two derivatives of $M_X(t)$ are

$$M_X^{(1)}(t) = n(1 - p + pe^t)^{n-1} \cdot pe^t$$

and

$$M_X^{(2)}(t) = pe^t \cdot n(n-1)(1-p+pe^t)^{n-2} \cdot pe^t + n(1-p+pe^t)^{n-1} \cdot pe^t$$

Setting $t = 0$ gives

$$M_X^{(1)}(0) = np = E(X)$$

and

$$M_X^{(2)}(0) = n(n-1)p^2 + np = E(X^2)$$

From Equation 3.12.6, then,

$$\begin{aligned}\text{Var}(X) &= n(n-1)p^2 + np - (np)^2 \\ &= np(1-p)\end{aligned}$$

(the same answer we found in Example 3.9.9). ■

Example 3.12.9

A discrete random variable X is said to have a *Poisson* distribution if

$$p_X(k) = P(X = k) = \frac{e^{-\lambda}\lambda^k}{k!}, \quad k = 0, 1, 2, \ldots$$

(An example of such a distribution is the mortality data described in Case Study 3.3.1.) It can be shown (see Question 3.12.7) that the moment-generating function for a Poisson random variable is given by

$$M_X(t) = e^{-\lambda + \lambda e^t}$$

Use $M_X(t)$ to find $E(X)$ and $\text{Var}(X)$.

Taking the first derivative of $M_X(t)$ gives

$$M_X^{(1)}(t) = e^{-\lambda + \lambda e^t} \cdot \lambda e^t$$

so

$$\begin{aligned}E(X) = M_X^{(1)}(0) &= e^{-\lambda + \lambda e^0} \cdot \lambda e^0 \\ &= \lambda\end{aligned}$$

Applying the product rule to $M_X^{(1)}(t)$ yields the second derivative,

$$M_X^{(2)}(t) = e^{-\lambda + \lambda e^t} \cdot \lambda e^t + \lambda e^t e^{-\lambda + \lambda e^t} \cdot \lambda e^t$$

For $t = 0$,

$$\begin{aligned}M_X^{(2)}(0) = E(X^2) &= e^{-\lambda + \lambda e^0} \cdot \lambda e^0 + \lambda e^0 \cdot e^{-\lambda + \lambda e^0} \cdot \lambda e^0 \\ &= \lambda + \lambda^2\end{aligned}$$

The variance of a Poisson random variable, then, proves to be the same as its mean:

$$\begin{aligned}\text{Var}(X) &= E(X^2) - [E(X)]^2 \\ &= M_X^{(2)}(0) - \left[M_X^{(1)}(0)\right]^2 \\ &= \lambda^2 + \lambda - \lambda^2 \\ &= \lambda\end{aligned}$$

■

Questions

3.12.9. Calculate $E(Y^3)$ for a random variable whose moment-generating function is $M_Y(t) = e^{t^2/2}$.

3.12.10. Find $E(Y^4)$ if Y is an exponential random variable with $f_Y(y) = \lambda e^{-\lambda y}$, $y > 0$.

3.12.11. The form of the moment-generating function for a normal random variable is $M_Y(t) = e^{at+b^2t^2/2}$ (recall Example 3.12.4). Differentiate $M_Y(t)$ to verify that $a = E(Y)$ and $b^2 = \text{Var}(Y)$.

3.12.12. What is $E(Y^4)$ if the random variable Y has moment-generating function $M_Y(t) = (1 - \alpha t)^{-k}$?

3.12.13. Find $E(Y^2)$ if the moment-generating function for Y is given by $M_Y(t) = e^{-t+4t^2}$. Use Example 3.12.4 to find $E(Y^2)$ without taking any derivatives. (*Hint*: Recall Theorem 3.6.1.)

3.12.14. Find an expression for $E(Y^k)$ if $M_Y(t) = (1 - t/\lambda)^{-r}$, where λ is any positive real number and r is a positive integer.

3.12.15. Use $M_Y(t)$ to find the expected value of the uniform random variable described in Question 3.12.1.

3.12.16. Find the variance of Y if $M_Y(t) = e^{2t}/(1 - t^2)$.

Using Moment-Generating Functions to Identify Pdfs

Finding moments is not the only application of moment-generating functions. They are also used to identify the pdf of *sums* of random variables—that is, finding $f_W(w)$, where $W = W_1 + W_2 + \cdots + W_n$. Their assistance in the latter is particularly important for two reasons: (1) Many statistical procedures are defined in terms of sums, and (2) alternative methods for deriving $f_{W_1+W_2+\cdots+W_n}(w)$ are extremely cumbersome.

The next two theorems give the background results necessary for deriving $f_W(w)$. Theorem 3.12.2 states a key uniqueness property of moment-generating functions: If W_1 and W_2 are random variables with the same mgfs, they must necessarily have the same pdfs. In practice, applications of Theorem 3.12.2 typically rely on one or both of the algebraic properties cited in Theorem 3.12.3.

Theorem 3.12.2 *Suppose that W_1 and W_2 are random variables for which $M_{W_1}(t) = M_{W_2}(t)$ for some interval of t's containing 0. Then $f_{W_1}(w) = f_{W_2}(w)$.*

Proof See (95). □

Theorem 3.12.3

a. Let W be a random variable with moment-generating function $M_W(t)$. Let $V = aW + b$. Then

$$M_V(t) = e^{bt} M_W(at)$$

b. Let $W_1, W_2, \ldots, W_n$ be independent random variables with moment-generating functions $M_{W_1}(t), M_{W_2}(t), \ldots,$ and $M_{W_n}(t)$, respectively. Let $W = W_1 + W_2 + \cdots + W_n$. Then

$$M_W(t) = M_{W_1}(t) \cdot M_{W_2}(t) \cdots M_{W_n}(t)$$

Proof The proof is left as an exercise. □

Example 3.12.10 Suppose that X_1 and X_2 are two independent Poisson random variables with parameters λ_1 and λ_2, respectively. That is,

$$p_{X_1}(k) = P(X_1 = k) = \frac{e^{-\lambda_1}\lambda_1 k}{k!}, \quad k = 0, 1, 2, \ldots$$

and

$$p_{X_2}(k) = P(X_2 = k) = \frac{e^{-\lambda_2}\lambda_2 k}{k!}, \quad k = 0, 1, 2 \ldots$$

Let $X = X_1 + X_2$. What is the pdf for X?

According to Example 3.12.9, the moment-generating functions for X_1 and X_2 are

$$M_{X_1}(t) = e^{-\lambda_1 + \lambda_1 e^t}$$

and

$$M_{X_2}(t) = e^{-\lambda_2 + \lambda_2 e^t}$$

Moreover, if $X = X_1 + X_2$, then by part (b) of Theorem 3.12.3,

$$\begin{aligned} M_X(t) &= M_{X_1}(t) \cdot M_{X_2}(t) \\ &= e^{-\lambda_1 + \lambda_1 e^t} \cdot e^{-\lambda_2 + \lambda_2 e^t} \\ &= e^{-(\lambda_1 + \lambda_2) + (\lambda_1 + \lambda_2)e^t} \end{aligned} \tag{3.12.7}$$

But, by inspection, Equation 3.12.7 is the moment-generating function that a Poisson random variable with $\lambda = \lambda_1 + \lambda_2$ would have. It follows, then, by Theorem 3.12.2 that

$$p_X(k) = \frac{e^{-(\lambda_1 + \lambda_2)}(\lambda_1 + \lambda_2)^k}{k!}, \quad k = 0, 1, 2, \ldots$$

Comment The Poisson random variable reproduces itself in the sense that the sum of independent Poissons is also a Poisson. A similar property holds for independent normal random variables (see Question 3.12.19) and, under certain conditions, for independent binomial random variables (recall Example 3.8.2). ■

Example 3.12.11 We saw in Example 3.12.4 that a normal random variable, Y, with mean μ and variance σ^2 has pdf

$$f_Y(y) = (1/\sqrt{2\pi}\sigma)\exp\left[-\frac{1}{2}\left(\frac{y-\mu}{\sigma}\right)^2\right], \quad -\infty < y < \infty$$

and mgf

$$M_Y(t) = e^{\mu t + \sigma^2 t^2/2}$$

By definition, a *standard normal random variable* is a normal random variable for which $\mu = 0$ and $\sigma = 1$. Denoted Z, the pdf and mgf for a standard normal random variable are $f_Z(z) = (1/\sqrt{2\pi})e^{-z^2/2}$, $-\infty < z < \infty$, and $M_Z(t) = e^{t^2/2}$, respectively. Show that the ratio

$$\frac{Y - \mu}{\sigma}$$

is a standard normal random variable, Z.

Write $\frac{Y-\mu}{\sigma}$ as $\frac{1}{\sigma}Y - \frac{\mu}{\sigma}$. By part (a) of Theorem 3.12.3,

$$\begin{aligned} M_{(Y-\mu)/\sigma}(t) &= e^{-\mu t/\sigma} M_Y\left(\frac{t}{\sigma}\right) \\ &= e^{-\mu t/\sigma} e^{[\mu t/\sigma + \sigma^2 (t/\sigma)^2/2]} \\ &= e^{t^2/2} \end{aligned}$$

But $M_Z(t) = e^{t^2/2}$ so it follows from Theorem 3.12.2 that the pdf for $\frac{Y-\mu}{\sigma}$ is the same as the pdf for $f_z(z)$. (We call $\frac{Y-\mu}{\sigma}$ a Z *transformation*. Its importance will become evident in Chapter 4.) ■

Questions

3.12.17. Use Theorem 3.12.3(a) and Question 3.12.8 to find the moment-generating function of the random variable Y, where $f_Y(y) = \lambda y e^{-\lambda y}$, $y \geq 0$.

3.12.18. Let Y_1, Y_2, and Y_3 be independent random variables, each having the pdf of Question 3.12.17. Use Theorem 3.12.3(b) to find the moment-generating function of $Y_1 + Y_2 + Y_3$. Compare your answer to the moment-generating function in Question 3.12.14.

3.12.19. Use Theorems 3.12.2 and 3.12.3 to determine which of the following statements is true:

(a) The sum of two independent Poisson random variables has a Poisson distribution.
(b) The sum of two independent exponential random variables has an exponential distribution.
(c) The sum of two independent normal random variables has a normal distribution.

3.12.20. Calculate $P(X \leq 2)$ if $M_X(t) = \left(\frac{1}{4} + \frac{3}{4}e^t\right)^5$.

3.12.21. Suppose that $Y_1, Y_2, \ldots, Y_n$ is a random sample of size n from a normal distribution with mean μ and standard deviation σ. Use moment-generating functions to deduce the pdf of $\bar{Y} = \frac{1}{n}\sum_{i=1}^{n} Y_i$.

3.12.22. Suppose the moment-generating function for a random variable W is given by

$$M_W(t) = e^{-3+3e^t} \cdot \left(\frac{2}{3} + \frac{1}{3}e^t\right)^4$$

Calculate $P(W \leq 1)$. (*Hint:* Write W as a sum.)

3.12.23. Suppose that X is a Poisson random variable, where $p_X(k) = e^{-\lambda}\lambda^k/k!$, $k = 0, 1, \ldots$.

(a) Does the random variable $W = 3X$ have a Poisson distribution?
(b) Does the random variable $W = 3X + 1$ have a Poisson distribution?

3.12.24. Suppose that Y is a normal variable, where $f_Y(y) = (1/\sqrt{2\pi}\sigma)\exp\left[-\frac{1}{2}\left(\frac{y-\mu}{\sigma}\right)^2\right]$, $-\infty < y < \infty$.

(a) Does the random variable $W = 3Y$ have a normal distribution?
(b) Does the random variable $W = 3Y + 1$ have a normal distribution?

3.13 Taking a Second Look at Statistics (Interpreting Means)

One of the most important ideas coming out of Chapter 3 is the notion of the *expected value* (or *mean*) of a random variable. Defined in Section 3.5 as a number that reflects the "center" of a pdf, the expected value (μ) was originally introduced for the benefit of gamblers. It spoke directly to one of their most fundamental questions—How much will I win or lose, *on the average,* if I play a certain game? (Actually, the real question they probably had in mind was "How much are *you* going to *lose,* on the average?") Despite having had such a selfish, materialistic, gambling-oriented *raison d'etre,* the expected value was quickly embraced by (respectable) scientists and researchers of all persuasions as a preeminently useful descriptor of a distribution. Today, it would not be an exaggeration to claim that the majority of *all* statistical analyses focus on either (1) the expected value of a single random variable or (2) comparing the expected values of two or more random variables.

In the lingo of applied statistics, there are actually two fundamentally different types of "means"—*population means* and *sample means*. The term "population mean" is a synonym for what mathematical statisticians would call an expected value—that is, a population mean (μ) is a weighted average of the possible values associated with a theoretical probability model, either $p_X(k)$ or $f_Y(y)$, depending on whether the underlying random variable is discrete or continuous. A *sample mean* is the arithmetic average of a set of measurements. If, for example, n observations—$y_1, y_2, \ldots, y_n$—are taken on a continuous random variable Y, the sample mean is denoted $\bar{y}$, where

$$\bar{y} = \frac{1}{n}\sum_{i=1}^{n} y_i$$

Conceptually, sample means are *estimates* of population means, where the "quality" of the estimation is a function of (1) the sample size and (2) the standard deviation (σ) associated with the individual measurements. Intuitively, as the sample size gets larger and/or the standard deviation gets smaller, the approximation will tend to get better.

Interpreting means (either $\bar{y}$ or μ) is not always easy. To be sure, what they imply *in principle* is clear enough—both $\bar{y}$ and μ are measuring the centers of their respective distributions. Still, many a wrong conclusion can be traced directly to researchers misunderstanding the value of a mean. Why? Because the distributions that $\bar{y}$ and/or μ are *actually* representing may be dramatically different from the distributions we *think* they are representing.

An interesting case in point arises in connection with SAT scores. Each fall the average SATs earned by students in each of the fifty states and the District of Columbia are released by the Educational Testing Service (ETS). With "accountability" being one of the new paradigms and buzzwords associated with K–12 education, SAT scores have become highly politicized. At the national level, Democrats and Republicans each campaign on their own versions of education reform, fueled in no small measure by scores on standardized exams, SATs included; at the state level, legislatures often modify education budgets in response to how well or how poorly their students performed the year before. Does it make sense, though, to use SAT averages to characterize the quality of a state's education system? Absolutely not! Averages of this sort refer to very different distributions from state to state. Any attempt to interpret them at face value will necessarily be misleading.

One such state-by-state SAT comparison that appeared in the mid-90s is reproduced in Table 3.13.1. Notice that Tennessee's entry is 1023, which is the tenth highest average listed. Does it follow that Tennessee's educational system is among the best in the nation? Probably not. Most independent assessments of K–12 education rank Tennessee's schools among the weakest in the nation, not among the best. If those opinions are accurate, why do Tennessee's students do so well on the SAT?

The answer to that question lies in the academic profiles of the students who take the SAT in Tennessee. Most college-bound students in that state apply exlusively to schools in the South and the Midwest, where admissions are based on the ACT, not the SAT. The SAT is primarily used by private schools, where admissions tend to be more competitive. As a result, the students in Tennessee who take the SAT are not representative of the entire population of students in that state. A disproportionate number are exceptionally strong academically, those being the students who feel that they have the ability to be

Table 3.13.1

State	Average SAT Score	State	Average SAT Score
AK	911	MT	986
AL	1011	NE	1025
AZ	939	NV	913
AR	935	NH	924
CA	895	NJ	893
CO	969	NM	1003
CT	898	NY	888
DE	892	NC	860
DC	849	ND	1056
FL	879	OH	966
GA	844	OK	1019
HI	881	OR	927
ID	969	PA	879
IL	1024	RI	882
IN	876	SC	838
IA	1080	SD	1031
KS	1044	TN	1023
KY	997	TX	886
LA	1011	UT	1067
ME	883	VT	899
MD	908	VA	893
MA	901	WA	922
MI	1009	WV	921
MN	1057	WI	1044
MS	1013	WY	980
MO	1017		

competitive at Ivy League–type schools. The number 1023, then, is the average of *something* (in this case, an elite subset of all Tennessee students), but it does not correspond to the center of the SAT distribution for *all* Tennessee students.

The moral here is that analyzing data effectively requires that we look beyond the obvious. What we have learned in Chapter 3 about random variables and probability distributions and expected values will be helpful only if we take the time to learn about the context and the idiosyncrasies of the phenomenon being studied. To do otherwise is likely to lead to conclusions that are, at best, superficial and, at worst, incorrect.

Appendix 3.A.1 Minitab Applications

Numerous software packages are available for performing a variety of probability and statistical calculations. Among the first to be developed, and one that continues to be very popular, is Minitab. Beginning here, we will include at the ends of certain chapters a short discussion of Minitab solutions to some of the problems that were discussed in the chapter. What other software packages can do and the ways their outputs are formatted are likely to be quite similar.

Contained in Minitab are subroutines that can do some of the more important pdf and cdf computations described in Sections 3.3 and 3.4. In the case of binomial random variables, for instance, the statements

```
MTB  > pdf k;
SUBC > binomial n p.
```

and

```
MTB  > cdf k;
SUBC > binomial n p.
```

will calculate $\binom{n}{k}p^k(1-p)^{n-k}$ and $\sum_{r=0}^{k}\binom{n}{r}p^r(1-p)^{n-r}$, respectively. Figure 3.A.1.1 shows the Minitab program for doing the cdf calculation [$= P(X \leq 15)$] asked for in part (a) of Example 3.2.2.

The commands `pdf k` and `cdf k` can be run on many of the probability models most likely to be encountered in real-world problems. Those on the list that we have already seen are the binomial, Poisson, normal, uniform, and exponential distributions.

Figure 3.A.1.1

```
MTB > cdf 15;
SUBC > binomial 30 0.60.
```
Cumulative Distribution Function
```
Binomial with n = 30 and p = 0.600000
        x    P(X <= x)
    15.00     0.1754
```

For discrete random variables, the cdf can be printed out in its entirety (that is, for every integer) by deleting the argument k and using the command `MTB < cdf;`. Typical is the output in Figure 3.A.1.2, corresponding to the cdf for a binomial random variable with $n = 4$ and $p = \frac{1}{6}$.

Figure 3.A.1.2

```
MTB > cdf;
SUBC > binomial 4 0.167.
```
Cumulative Distribution Function
```
Binomial with n = 4 and p =0.167000
         x    P( X <= x)
         0        0.4815
         1        0.8676
         2        0.9837
         3        0.9992
         4        1.0000
```

Figure 3.A.1.3

```
MTB > invcdf 0.60;
SUBC > exponential 1.
```
Inverse Cumulative Distribution Function
```
Exponential with mean = 1.00000
   P(X <= x)  x
    0.6000    0.9163
```

Also available is an *inverse cdf* command, which in the case of a continuous random variable Y and a specified probability p identifies the value y having the property that $P(Y \leq y) = F_Y(Y) = p$. For example, if $p = 0.60$ and Y is an exponential random variable with pdf $f_Y(y) = e^{-y}$, $y > 0$, the value $y = 0.9163$ has the property that $P(Y \leq 0.9163) = F_Y(0.9163) = 0.60$. That is,

$$F_Y(0.9163) = \int_0^{0.9163} e^{-y}\, dy = 0.60$$

With Minitab the number 0.9163 is found by using the command `MTB>invcdf 0.60` (see Figure 3.A.1.3).

Chapter 4

SPECIAL DISTRIBUTIONS

Although he maintained lifelong literary and artistic interests, Quetelet's mathematical talents led him to a doctorate from the University of Ghent and from there to a college teaching position in Brussels. In 1833 he was appointed astronomer at the Brussels Royal Observatory after having been largely responsible for its founding. His work with the Belgian census marked the beginning of his pioneering efforts in what today would be called mathematical sociology. Quetelet was well known throughout Europe in scientific and literary circles: At the time of his death he was a member of more than one hundred learned societies.

—Lambert Adolphe Jacques Quetelet (1796–1874)

4.1 Introduction

To "qualify" as a probability model, a function defined over a sample space S needs to satisfy only two criteria: (1) It must be nonnegative for all outcomes in S, and (2) it must sum or integrate to 1. That means, for example, that $f_Y(y) = \frac{y}{4} + \frac{7y^3}{2}$, $0 \le y \le 1$, can be considered a pdf because $f_Y(y) \ge 0$ for all $0 \le y \le 1$ and $\int_0^1 \left(\frac{y}{4} + \frac{7y^3}{2}\right) dy = 1$.

It certainly does not follow, though, that every $f_Y(y)$ and $p_X(k)$ that satisfy these two criteria would actually be used as probability models. A pdf has practical significance only if it does, indeed, model the probabilistic behavior of real-world phenomena. In point of fact, only a handful of functions do [and $f_Y(y) = \frac{y}{4} + \frac{7y^3}{2}$, $0 \le y \le 1$, is not one of them!].

Whether a probability function—say, $f_Y(y)$—adequately models a given phenomenon ultimately depends on whether the physical factors that influence the value of Y parallel the mathematical assumptions implicit in $f_Y(y)$. Surprisingly, many measurements (i.e., random variables) that seem to be very different are actually the consequence of the same set of assumptions (and will, therefore, be modeled by the same pdf). That said, it makes sense to single out these "real-world" pdfs and investigate their properties in more detail. This, of course, is not an idea we are seeing for the first time—recall the attention given to the binomial and hypergeometric distributions in Section 3.2.

Chapter 4 continues in the spirit of Section 3.2 by examining five other widely used models. Three of the five are discrete; the other two are continuous. One of the continuous pdfs is the normal (or Gaussian) distribution, which, by far, is the most important of all probability models. As we will see, the normal "curve" figures prominently in every chapter from this point on.

Examples play a major role in Chapter 4. The only way to appreciate fully the generality of a probability model is to look at some of its specific applications. Thus, included in this chapter are case studies ranging from the discovery of alpha-particle radiation to an early ESP experiment to an analysis of volcanic eruptions to counting bug parts in peanut butter.

4.2 The Poisson Distribution

The binomial distribution problems that appeared in Section 3.2 all had relatively small values for n, so evaluating $p_X(k) = P(X = k) = \binom{n}{k} p^k(1-p)^{n-k}$ was not particularly difficult. But suppose n were 1000 and k, 500. Evaluating $p_X(500)$ would be a formidable task for many handheld calculators, even today. Two hundred years ago, the prospect of doing cumbersome binomial calculations *by hand* was a catalyst for mathematicians to develop some easy-to-use approximations. One of the first such approximations was the *Poisson limit,* which eventually gave rise to the *Poisson distribution*. Both are described in Section 4.2.

Simeon Denis Poisson (1781–1840) was an eminent French mathematician and physicist, an academic administrator of some note, and, according to an 1826 letter from the mathematician Abel to a friend, a man who knew "how to behave with a great deal of dignity." One of Poisson's many interests was the application of probability to the law, and in 1837 he wrote *Recherches sur la Probabilite de Jugements*. Included in the latter is a limit for $p_X(k) = \binom{n}{k} p^k(1-p)^{n-k}$ that holds when n approaches ∞, p approaches 0, and np remains constant. In practice, Poisson's limit is used to approximate hard-to-calculate binomial probabilities where the values of n and p reflect the conditions of the limit—that is, when n is large and p is small.

The Poisson Limit

Deriving an asymptotic expression for the binomial probability model is a straightforward exercise in calculus, given that np is to remain fixed as n increases.

Theorem 4.2.1 *Suppose X is a binomial random variable, where*

$$P(X = k) = p_X(k) = \binom{n}{k} p^k(1-p)^{n-k}, \quad k = 0, 1, \ldots, n$$

If $n \to \infty$ and $p \to 0$ in such a way that $\lambda = np$ remains constant, then

$$\lim_{\substack{n\to\infty \\ p\to 0 \\ np = \text{const.}}} P(X = k) = \lim_{\substack{n\to\infty \\ p\to 0 \\ np = \text{const.}}} \binom{n}{k} p^k(1-p)^{n-k} = \frac{e^{-np}(np)^k}{k!}$$

Proof We begin by rewriting the binomial probability in terms of λ:

$$\lim_{n\to\infty}\binom{n}{k}p^k(1-p)^{n-k} = \lim_{n\to\infty}\binom{n}{k}\left(\frac{\lambda}{n}\right)^k\left(1-\frac{\lambda}{n}\right)^{n-k}$$

$$= \lim_{n\to\infty}\frac{n!}{k!(n-k)!}\lambda^k\left(\frac{1}{n^k}\right)\left(1-\frac{\lambda}{n}\right)^{-k}\left(1-\frac{\lambda}{n}\right)^n$$

$$= \frac{\lambda^k}{k!}\lim_{n\to\infty}\frac{n!}{(n-k)!}\frac{1}{(n-\lambda)^k}\left(1-\frac{\lambda}{n}\right)^n$$

But since $[1-(\lambda/n)]^n \to e^{-\lambda}$ as $n\to\infty$, we need show only that

$$\frac{n!}{(n-k)!(n-\lambda)^k}\to 1$$

to prove the theorem. However, note that

$$\frac{n!}{(n-k)!(n-\lambda)^k} = \frac{n(n-1)\cdots(n-k+1)}{(n-\lambda)(n-\lambda)\cdots(n-\lambda)}$$

a quantity that, indeed, tends to 1 as $n\to\infty$ (since λ remains constant). □

Example 4.2.1

Theorem 4.2.1 is an *asymptotic* result. Left unanswered is the question of the relevance of the Poisson limit for *finite* n and p. That is, how large does n have to be and how small does p have to be before $e^{-np}(np)^k/k!$ becomes a good approximation to the binomial probability, $p_X(k)$?

Since "good approximation" is undefined, there is no way to answer that question in any completely specific way. Tables 4.2.1 and 4.2.2, though, offer a partial solution by comparing the closeness of the approximation for two particular sets of values for n and p.

In both cases $\lambda = np$ is equal to 1, but in the former, n is set equal to 5—in the latter, to 100. We see in Table 4.2.1 ($n = 5$) that for some k the agreement between the binomial probability and Poisson's limit is not very good. If n is as large as 100, though (Table 4.2.2), the agreement is remarkably good for all k.

Table 4.2.1 Binomial Probabilities and Poisson Limits; $n = 5$ and $p = \frac{1}{5}$ ($\lambda = 1$)

k	$\binom{5}{k}(0.2)^k(0.8)^{5-k}$	$\frac{e^{-1}(1)^k}{k!}$
0	0.328	0.368
1	0.410	0.368
2	0.205	0.184
3	0.051	0.061
4	0.006	0.015
5	0.000	0.003
6+	0	0.001
	1.000	1.000

Table 4.2.2 Binomial Probabilities and Poisson Limits; $n = 100$ and $p = \frac{1}{100}$ $(\lambda = 1)$

k	$\binom{100}{k}(0.01)^k(0.99)^{100-k}$	$\frac{e^{-1}(1)^k}{k!}$
0	0.366032	0.367879
1	0.369730	0.367879
2	0.184865	0.183940
3	0.060999	0.061313
4	0.014942	0.015328
5	0.002898	0.003066
6	0.000463	0.000511
7	0.000063	0.000073
8	0.000007	0.000009
9	0.000001	0.000001
10	0.000000	0.000000
	1.000000	0.999999

■

Example 4.2.2

According to the IRS, 137.8 million individual tax returns were filed in 2008. Out of that total, 1.4 million taxpayers, or 1.0%, had the good fortune of being audited. Not everyone had the same chance of getting caught in the IRS's headlights: millionaires had the considerably higher audit rate of 5.6% (and that number might even go up a bit more if the feds find out about your bank accounts in the Caymans and your vacation home in Rio). Criminal investigations were initiated against 3749 of all those audited, and 1735 of that group were eventually convicted of tax fraud and sent to jail.

Suppose your hometown has 65,000 taxpayers, whose income profile and proclivity for tax evasion are similar to those of citizens of the United States as a whole, and suppose the IRS enforcement efforts remain much the same in the foreseeable future. What is the probability that at least three of your neighbors will be house guests of Uncle Sam next year?

Let X denote the number of your neighbors who will be incarcerated. Note that X is a binomial random variable based on a very large n $(= 65{,}000)$ and a very small p $(= 1735/137{,}800{,}000 = 0.0000126)$, so Poisson's limit is clearly applicable (and helpful). Here,

$$P(\text{At least three neighbors go to jail}) = P(X \geq 3)$$

$$= 1 - P(X \leq 2)$$

$$= 1 - \sum_{k=0}^{2} \binom{65{,}000}{k} (0.0000126)^k (0.9999874)^{65{,}000-k}$$

$$\doteq 1 - \sum_{k=0}^{2} e^{-0.819} \frac{(0.819)^k}{k!} = 0.050$$

where $\lambda = np = 65{,}000(0.0000126) = 0.819$.

■

Case Study 4.2.1

Leukemia is a rare form of cancer whose cause and mode of transmission remain largely unknown. While evidence abounds that excessive exposure to radiation can increase a person's risk of contracting the disease, it is at the same time true that most cases occur among persons whose history contains no such overexposure. A related issue, one maybe even more basic than the causality question, concerns the *spread* of the disease. It is safe to say that the prevailing medical opinion is that most forms of leukemia are not contagious—still, the hypothesis persists that some forms of the disease, particularly the childhood variety, may be. What continues to fuel this speculation are the discoveries of so-called "leukemia clusters," aggregations in time and space of unusually large numbers of cases.

To date, one of the most frequently cited leukemia clusters in the medical literature occurred during the late 1950s and early 1960s in Niles, Illinois, a suburb of Chicago (75). In the $5\frac{1}{3}$-year period from 1956 to the first four months of 1961, physicians in Niles reported a total of eight cases of leukemia among children less than fifteen years of age. The number at risk (that is, the number of residents in that age range) was 7076. To assess the likelihood of that many cases occurring in such a small population, it is necessary to look first at the leukemia incidence in neighboring towns. For all of Cook County, excluding Niles, there were 1,152,695 children less than fifteen years of age—and among those, 286 diagnosed cases of leukemia. That gives an average $5\frac{1}{3}$-year leukemia rate of 24.8 cases per 100,000:

$$\frac{286 \text{ cases for } 5\frac{1}{3} \text{ years}}{1{,}152{,}695 \text{ children}} \times \frac{100{,}000}{100{,}000} = 24.8 \text{ cases}/100{,}000 \text{ children in } 5\tfrac{1}{3} \text{ years}$$

Now, imagine the 7076 children in Niles to be a series of $n = 7076$ (independent) Bernoulli trials, each having a probability of $p = 24.8/100{,}000 = 0.000248$ of contracting leukemia. The question then becomes, given an n of 7076 and a p of 0.000248, how likely is it that eight "successes" would occur? (The expected number, of course, would be $7076 \times 0.000248 = 1.75$.) Actually, for reasons that will be elaborated on in Chapter 6, it will prove more meaningful to consider the related event, eight *or more* cases occurring in a $5\frac{1}{3}$-year span. If the probability associated with the latter is very small, it could be argued that leukemia did not occur randomly in Niles and that, perhaps, contagion was a factor.

Using the binomial distribution, we can express the probability of eight or more cases as

$$P(8 \text{ or more cases}) = \sum_{k=8}^{7076} \binom{7076}{k} (0.000248)^k (0.999752)^{7076-k} \qquad (4.2.1)$$

Much of the computational unpleasantness implicit in Equation 4.2.1 can be avoided by appealing to Theorem 4.2.1. Given that $np = 7076 \times 0.000248 = 1.75$,

(Continued on next page)

(Case Study 4.2.1 continued)

$$
\begin{aligned}
P(X \geq 8) &= 1 - P(X \leq 7) \\
&\doteq 1 - \sum_{k=0}^{7} \frac{e^{-1.75}(1.75)^k}{k!} \\
&= 1 - 0.99953 \\
&= 0.00047
\end{aligned}
$$

How close can we expect 0.00047 to be to the "true" binomial sum? Very close. Considering the accuracy of the Poisson limit when n is as small as one hundred (recall Table 4.2.2), we should feel very confident here, where n is 7076.

Interpreting the 0.00047 probability is not nearly as easy as assessing its accuracy. The fact that the probability is so very small tends to denigrate the hypothesis that leukemia in Niles occurred at random. On the other hand, rare events, such as clusters, *do* happen by chance. The basic difficulty of putting the probability associated with a given cluster into any meaningful perspective is not knowing in how many similar communities leukemia did *not* exhibit a tendency to cluster. That there is no obvious way to do this is one reason the leukemia controversy is still with us.

About the Data Publication of the Niles cluster led to a number of research efforts on the part of biostatisticians to find quantitative methods capable of detecting clustering in space and time for diseases having low epidemicity. Several techniques were ultimately put forth, but the inherent "noise" in the data—variations in population densities, ethnicities, risk factors, and medical practices—often proved impossible to overcome.

Questions

4.2.1. If a typist averages one misspelling in every 3250 words, what are the chances that a 6000-word report is free of all such errors? Answer the question two ways—first, by using an exact binomial analysis, and second, by using a Poisson approximation. Does the similarity (or dissimilarity) of the two answers surprise you? Explain.

4.2.2. A medical study recently documented that 905 mistakes were made among the 289,411 prescriptions written during one year at a large metropolitan teaching hospital. Suppose a patient is admitted with a condition serious enough to warrant 10 different prescriptions. Approximate the probability that at least one will contain an error.

4.2.3. Five hundred people are attending the first annual "I was Hit by Lighting" Club. Approximate the probability that at most one of the five hundred was born on Poisson's birthday.

4.2.4. A chromosome mutation linked with colorblindness is known to occur, on the average, once in every ten thousand births.

(a) Approximate the probability that exactly three of the next twenty thousand babies born will have the mutation.

(b) How many babies out of the next twenty thousand would have to be born with the mutation to convince you that the "one in ten thousand" estimate is too low? [*Hint:* Calculate $P(X \geq k) = 1 - P(X \leq k-1)$ for various k. (Recall Case Study 4.2.1.)]

4.2.5. Suppose that 1% of all items in a supermarket are not priced properly. A customer buys ten items. What is the probability that she will be delayed by the cashier because one or more of her items require a price check?

Calculate both a binomial answer and a Poisson answer. Is the binomial model "exact" in this case? Explain.

4.2.6. A newly formed life insurance company has underwritten term policies on 120 women between the ages of forty and forty-four. Suppose that each woman has a 1/150 probability of dying during the next calendar year, and that each death requires the company to pay out \$50,000 in benefits. Approximate the probability that the company will have to pay at least \$150,000 in benefits next year.

4.2.7. According to an airline industry report (178), roughly 1 piece of luggage out of every 200 that are checked is lost. Suppose that a frequent-flying businesswoman will be checking 120 bags over the course of the next year. Approximate the probability that she will lose 2 of more pieces of luggage.

4.2.8. Electromagnetic fields generated by power transmission lines are suspected by some researchers to be a cause of cancer. Especially at risk would be telephone linemen because of their frequent proximity to high-voltage wires. According to one study, two cases of a rare form of cancer were detected among a group of 9500 linemen (174). In the general population, the incidence of that particular condition is on the order of one in a million. What would you conclude? (*Hint:* Recall the approach taken in Case Study 4.2.1.)

4.2.9. Astronomers estimate that as many as one hundred billion stars in the Milky Way galaxy are encircled by planets. If so, we may have a plethora of cosmic neighbors. Let p denote the probability that any such solar system contains intelligent life. How small can p be and still give a fifty-fifty chance that we are not alone?

The Poisson Distribution

The real significance of Poisson's limit theorem went unrecognized for more than fifty years. For most of the latter part of the nineteenth century, Theorem 4.2.1 was taken strictly at face value: It provides a convenient approximation for $p_X(k)$ when X is binomial, n is large, and p is small. But then in 1898 a German professor, Ladislaus von Bortkiewicz, published a monograph entitled *Das Gesetz der Kleinen Zahlen (The Law of Small Numbers)* that would quickly transform Poisson's "limit" into Poisson's "distribution."

What is best remembered about Bortkiewicz's monograph is the curious set of data described in Question 4.2.10. The measurements recorded were the numbers of Prussian cavalry soldiers who had been kicked to death by their horses. In analyzing those figures, Bortkiewicz was able to show that the function $e^{-\lambda}\lambda^k/k!$ is a useful probability model in its own right, even when (1) no explicit binomial random variable is present and (2) values for n and p are unavailable. Other researchers were quick to follow Bortkiewicz's lead, and a steady stream of Poisson distribution applications began showing up in technical journals. Today the function $p_X(k) = e^{-\lambda}\lambda^k/k!$ is universally recognized as being among the three or four most important data models in all of statistics.

Theorem 4.2.2 *The random variable X is said to have a Poisson distribution if*

$$p_X(k) = P(X = k) = \frac{e^{-\lambda}\lambda^k}{k!}, \quad k = 0, 1, 2, \ldots$$

where λ is a positive constant. Also, for any Poisson random variable, $E(X) = \lambda$ and $\text{Var}(X) = \lambda$.

Proof To show that $p_X(k)$ qualifies as a probability function, note, first of all, that $p_X(k) \geq 0$ for all nonnegative integers k. Also, $p_X(k)$ sums to 1:

$$\sum_{k=0}^{\infty} p_X(k) = \sum_{k=0}^{\infty} \frac{e^{-\lambda}\lambda^k}{k!} = e^{-\lambda} \sum_{k=0}^{\infty} \frac{\lambda^k}{k!} = e^{-\lambda} \cdot e^{\lambda} = 1$$

since $\sum_{k=0}^{\infty} \frac{\lambda^k}{k!}$ is the Taylor series expansion of e^{λ}. Verifying that $E(X) = \lambda$ and $\text{Var}(X) = \lambda$ has already been done in Example 3.12.9, using moment-generating functions. □

Fitting the Poisson Distribution to Data

Poisson data invariably refer to the numbers of times a certain event occurs during each of a series of "units" (often *time* or *space*). For example, X might be the weekly number of traffic accidents reported at a given intersection. If such records are kept for an entire year, the resulting data would be the sample $k_1, k_2, \ldots, k_{52}$, where each k_i is a nonnegative integer.

Whether or not a set of k_i's can be viewed as Poisson data depends on whether the proportions of 0's, 1's, 2's, and so on, *in the sample* are numerically similar to the probabilities that $X = 0, 1, 2$, and so on, as predicted by $p_X(k) = e^{-\lambda}\lambda^k/k!$. The next two case studies show data sets where the variability in the observed k_i's *is* consistent with the probabilities predicted by the Poisson distribution. Notice in each case that the λ in $p_X(k)$ is replaced by the sample mean of the k_i's—that is, by $\bar{k} = (1/n)\sum_{c=1}^{n} k_i$. Why these phenomena are described by the Poisson distribution will be discussed later in this section; why λ is replaced by $\bar{k}$ will be explained in Chapter 5.

Case Study 4.2.2

Among the early research projects investigating the nature of radiation was a 1910 study of α-particle emission by Ernest Rutherford and Hans Geiger (152). For each of 2608 eighth-minute intervals, the two physicists recorded the number of α particles emitted from a polonium source (as detected by what would eventually be called a Geiger counter). The numbers and proportions of times that k such particles were detected in a given eighth-minute ($k = 0, 1, 2, \ldots$) are detailed in the first three columns of Table 4.2.3. Two α particles, for example, were detected in each of 383 eighth-minute intervals, meaning that $X = 2$ was the observation recorded 15% ($= 383/2608 \times 100$) of the time.

Table 4.2.3

No. Detected, k	Frequency	Proportion	$p_X(k) = e^{-3.87}(3.87)^k/k!$
0	57	0.02	0.02
1	203	0.08	0.08
2	383	0.15	0.16
3	525	0.20	0.20
4	532	0.20	0.20
5	408	0.16	0.15
6	273	0.10	0.10
7	139	0.05	0.05
8	45	0.02	0.03
9	27	0.01	0.01
10	10	0.00	0.00
11+	6	0.00	0.00
	2608	1.0	1.0

(Continued on next page)

To see whether a probability function of the form $p_X(k) = e^{-\lambda}\lambda^k/k!$ can adequately model the observed proportions in the third column, we first need to replace λ with the sample's average value for X. Suppose the six observations comprising the "11+" category are each assigned the value 11. Then

$$\bar{k} = \frac{57(0) + 203(1) + 383(2) + \cdots + 6(11)}{2608} = \frac{10{,}092}{2608}$$
$$= 3.87$$

and the presumed model is $p_X(k) = e^{-3.87}(3.87)^k/k!$, $k = 0, 1, 2, \ldots$. Notice how closely the entries in the fourth column [i.e., $p_X(0), p_X(1), p_X(2), \ldots$] agree with the sample proportions appearing in the third column. The conclusion here is inescapable: The phenomenon of radiation can be modeled very effectively by the Poisson distribution.

About the Data The most obvious (and frequent) application of the Poisson/radioactivity relationship is to use the former to describe and predict the behavior of the latter. But the relationship is also routinely used in reverse. Workers responsible for inspecting areas where radioactive contamination is a potential hazard need to know that their monitoring equipment is functioning properly. How do they do that? A standard safety procedure before entering what might be a life-threatening "hot zone" is to take a series of readings on a known radioactive source (much like the Rutherford/Geiger experiment itself). If the resulting set of counts does not follow a Poisson distribution, the meter is assumed to be broken and must be repaired or replaced.

Case Study 4.2.3

In the 432 years from 1500 to 1931, war broke out somewhere in the world a total of 299 times. By definition, a military action was a war if it either was legally declared, involved over fifty thousand troops, or resulted in significant boundary realignments. To achieve greater uniformity from war to war, major confrontations were split into smaller "subwars": World War I, for example, was treated as five separate conflicts (143).

Let X denote the number of wars starting in a given year. The first two columns in Table 4.2.4 show the distribution of X for the 432-year period in question. Here the average number of wars beginning in a given year was 0.69:

$$\bar{k} = \frac{0(223) + 1(142) + 2(48) + 3(15) + 4(4)}{432} = 0.69$$

The last two columns in Table 4.2.4 compare the observed proportions of years for which $X = k$ with the proposed Poisson model

$$p_X(k) = e^{-0.69}\frac{(0.69)^k}{k!}, \quad k = 0, 1, 2, \ldots$$

(Continued on next page)

(Case Study 4.2.3 continued)

Table 4.2.4

Number of Wars, k	Frequency	Proportion	$p_X(k) = e^{-0.69}\dfrac{(0.69)^k}{k!}$
0	223	0.52	0.50
1	142	0.33	0.35
2	48	0.11	0.12
3	15	0.03	0.03
4+	4	0.01	0.00
	432	1.00	1.00

Clearly, there is a very close agreement between the two—the number of wars beginning in a given year can be considered a Poisson random variable.

The Poisson Model: The Law of Small Numbers

Given that the expression $e^{-\lambda}\lambda^k/k!$ models phenomena as diverse as α-radiation and outbreak of war raises an obvious question: *Why* is that same $p_X(k)$ describing such different random variables? The answer is that the underlying physical conditions that produce those two sets of measurements are actually much the same, despite how superficially different the resulting data may seem to be. Both phenomena are examples of a set of mathematical assumptions known as the *Poisson model*. Any measurements that are derived from conditions that mirror those assumptions will necessarily vary in accordance with the Poisson distribution.

Suppose a series of events is occurring during a time interval of length T. Imagine dividing T into n nonoverlapping subintervals, each of length $\frac{T}{n}$, where n is large (see Figure 4.2.1). Furthermore, suppose that

Figure 4.2.1

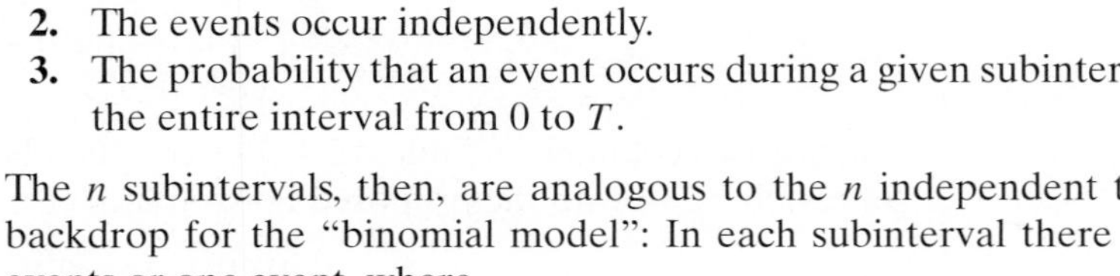

1. The probability that two or more events occur in any given subinterval is essentially 0.
2. The events occur independently.
3. The probability that an event occurs during a given subinterval is constant over the entire interval from 0 to T.

The n subintervals, then, are analogous to the n independent trials that form the backdrop for the "binomial model": In each subinterval there will be either zero events or one event, where

$$p_n = P(\text{Event occurs in a given subinterval})$$

remains constant from subinterval to subinterval.

Let the random variable X denote the total number of events occurring during time T, and let λ denote the *rate* at which events occur (e.g., λ might be expressed as 2.5 events per minute). Then

$$E(X) = \lambda T = np_n \quad \text{(why?)}$$

which implies that $p_n = \frac{\lambda T}{n}$. From Theorem 4.2.1, then,

$$\begin{aligned} p_X(k) = P(X = k) &= \binom{n}{k}\left(\frac{\lambda T}{n}\right)^k \left(1 - \frac{\lambda T}{n}\right)^{n-k} \\ &\doteq \frac{e^{-n(\lambda T/n)}\left[n(\lambda T/n)\right]^k}{k!} \\ &= \frac{e^{-\lambda T}(\lambda T)^k}{k!} \end{aligned} \tag{4.2.2}$$

Now we can see more clearly why Poisson's "limit," as given in Theorem 4.2.1, is so important. The three Poisson model assumptions are so unexceptional that they apply to countless real-world phenomena. Each time they do, the pdf $p_X(k) = e^{-\lambda T}(\lambda T)^k/k!$ finds another application.

Example 4.2.3

It is not surprising that the number of α particles emitted by a radioactive source in a given unit of time follows a Poisson distribution. Nuclear physicists have known for a long time that the phenomenon of radioactivity obeys the same assumptions that define the Poisson model. Each is a poster child for the other. Case Study 4.2.3, on the other hand, is a different matter altogether. It is not so obvious why the number of wars starting in a given year should have a Poisson distribution. Reconciling the data in Table 4.2.4 with the "picture" of the Poisson model in Figure 4.2.1 raises a number of questions that never came up in connection with radioactivity.

Imagine recording the data summarized in Table 4.2.4. For each year, new wars would appear as "occurrences" on a grid of cells, similar to the one pictured in Figure 4.2.2 for 1776. Civil wars would be entered along the diagonal and wars between two countries, above the diagonal. Each cell would contain either a 0 (no war) or a 1 (war). The year 1776 saw the onset of only one major conflict, the Revolutionary War between the United States and Britain. If the random variable

$$X_i = \text{number of outbreaks of war in year } i, \; i = 1500, 1501, \ldots, 1931$$

then $X_{1776} = 1$.

What do we know, in general, about the random variable X_i? If each cell in the grid is thought of as a "trial," then X_i is clearly the number of "successes" in those n trials. Does that make X_i a binomial random variable? Not necessarily. According to Theorem 3.2.1, X_i qualifies as a binomial random variable only if the trials are independent and the probability of success is the same from trial to trial.

At first glance, the independence assumption would seem to be problematic. There is no denying that some wars are linked to others. The timing of the French Revolution, for example, is widely thought to have been influenced by the success of the American Revolution. Does that make the two wars dependent? In a historical sense, yes; in a statistical sense, no. The French Revolution began in 1789, thirteen years after the onset of the American Revolution. The random variable X_{1776}, though, focuses only on wars starting in 1776, so linkages that are years apart do not compromise the binomial's independence assumption.

Not all wars identified in Case Study 4.2.3, though, can claim to be independent in the statistical sense. The last entry in Column 2 of Table 4.2.4 shows that four

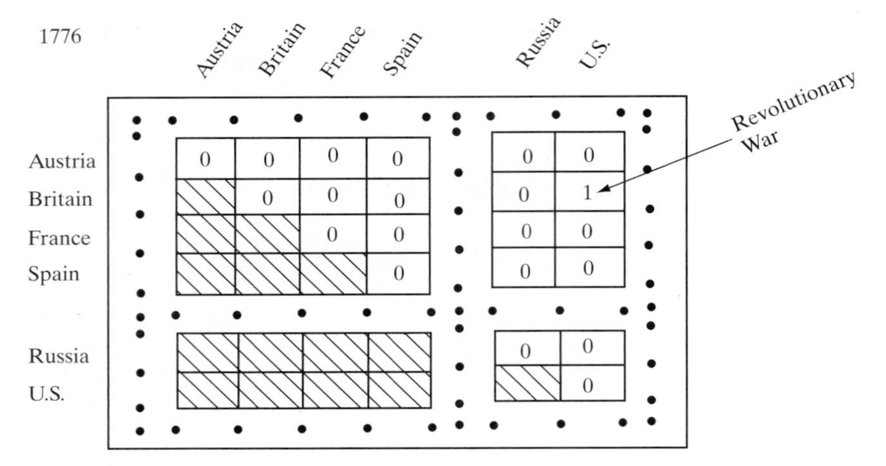

Figure 4.2.2

or more wars erupted on four separate occasions; the Poisson model (Column 4) predicted that no years would experience that many new wars. Most likely, those four years had a decided excess of new wars because of political alliances that led to a cascade of new wars being declared simultaneously. Those wars definitely violated the binomial assumption of independent trials, but they accounted for only a very small fraction of the entire data set.

The other binomial assumption—that each trial has the same probability of success—holds fairly well. For the vast majority of years and the vast majority of countries, the probabilities of new wars will be very small and most likely similar. For almost every year, then, X_i can be considered a binomial random variable based on a very large n and a very small p. That being the case, it follows by Theorem 4.2.1 that each $X_i, i = 1500, 1501, \ldots, 1931$ can be approximated by a Poisson distribution.

One other assumption needs to be addressed. Knowing that $X_{1500}, X_{1501}, \ldots, X_{1931}$—individually—are Poisson random variables does not guarantee that the distribution of all 432 X_i's will have a Poisson distribution. Only if the X_i's are independent observations having basically the same Poisson distribution—that is, the same value for λ—will their overall distribution be Poisson. But Table 4.2.4 *does* have a Poisson distribution, implying that the set of X_i's does, in fact, behave like a random sample. Along with that sweeping conclusion, though, comes the realization that, as a species, our levels of belligerence at the national level [that is, the 432 values for $\lambda = E(X_i)$] have remained basically the same for the past five hundred years. Whether that should be viewed as a reason for celebration or a cause for alarm is a question best left to historians, not statisticians. ■

Calculating Poisson Probabilities

Three formulas have appeared in connection with the Poisson distribution:

1. $p_X(k) \doteq e^{-np}\frac{(np)^k}{k!}$
2. $p_X(k) = e^{-\lambda}\frac{\lambda^k}{k!}$
3. $p_X(k) = e^{-\lambda T}\frac{(\lambda T)^k}{k!}$

The first is the approximating Poisson limit, where the $p_X(k)$ on the left-hand side refers to the probability that a *binomial* random variable (with parameters n and p)

is equal to k. Formulas (2) and (3) are sometimes confused because both presume to give the probability that a Poisson random variable equals k. Why are they different?

Actually, all three formulas are the same in the sense that the right-hand sides of each could be written as

4. $e^{-E(X)} \frac{[E(X)]^k}{k!}$

In formula (1), X is binomial, so $E(X) = np$. In formula (2), which comes from Theorem 4.2.2, λ is defined to be $E(X)$. Formula (3) covers all those situations where the units of X and λ are not consistent, in which case $E(X) \neq \lambda$. However, λ can always be multiplied by an appropriate constant T to make λT equal to $E(X)$.

For example, suppose a certain radioisotope is known to emit α particles at the rate of $\lambda = 1.5$ emissions/second. For whatever reason, though, the experimenter defines the Poisson random variable X to be the number of emissions counted in a given *minute*. Then $T = 60$ seconds and

$$\begin{aligned} E(X) &= 1.5 \text{ emissions/second} \times 60 \text{ seconds} \\ &= \lambda T = 90 \text{ emissions} \end{aligned}$$

Example 4.2.4

Entomologists estimate that an average person consumes almost a pound of bug parts each year (173). There are that many insect eggs, larvae, and miscellaneous body pieces in the foods we eat and the liquids we drink. The Food and Drug Administration (FDA) sets a Food Defect Action Level (FDAL) for each product: Bug-part concentrations below the FDAL are considered acceptable. The legal limit for peanut butter, for example, is thirty insect fragments per hundred grams. Suppose the crackers you just bought from a vending machine are spread with twenty grams of peanut butter. What are the chances that your snack will include at least five crunchy critters?

Let X denote the number of bug parts in twenty grams of peanut butter. Assuming the worst, suppose the contamination level equals the FDA limit—that is, thirty fragments per hundred grams (or 0.30 fragment/g). Notice that T in this case is twenty grams, making $E(X) = 6.0$:

$$\frac{0.30 \text{ fragment}}{\text{g}} \times 20 \text{ g} = 6.0 \text{ fragments}$$

It follows, then, that the probability that your snack contains five or more bug parts is a disgusting *0.71*:

$$\begin{aligned} P(X \geq 5) = 1 - P(X \leq 4) &= 1 - \sum_{k=0}^{4} \frac{e^{-6.0}(6.0)^k}{k!} \\ &= 1 - 0.29 \\ &= 0.71 \end{aligned}$$

Bon appetit! ■

Questions

4.2.10. During the latter part of the nineteenth century, Prussian officials gathered information relating to the hazards that horses posed to cavalry soldiers. A total of ten cavalry corps were monitored over a period of twenty years. Recorded for each year and each corps was X, the annual number of fatalities due to kicks. Summarized in the following table are the two hundred values recorded for X (12). Show that these data can be modeled by a Poisson pdf. Follow the procedure illustrated in Case Studies 4.2.2 and 4.2.3.

No. of Deaths, k	Observed Number of Corps-Years in Which k Fatalities Occurred
0	109
1	65
2	22
3	3
4	1
	200

4.2.11. A random sample of 356 seniors enrolled at the University of West Florida was categorized according to X, the number of times they had changed majors (110). Based on the summary of that information shown in the following table, would you conclude that X can be treated as a Poisson random variable?

Number of Major Changes	Frequency
0	237
1	90
2	22
3	7

4.2.12. Midwestern Skies books ten commuter flights each week. Passenger totals are much the same from week to week, as are the numbers of pieces of luggage that are checked. Listed in the following table are the numbers of bags that were lost during each of the first forty weeks in 2009. Do these figures support the presumption that the number of bags lost by Midwestern during a typical week is a Poisson random variable?

Week	Bags Lost	Week	Bags Lost	Week	Bags Lost
1	1	14	2	27	1
2	0	15	1	28	2
3	0	16	3	29	0
4	3	17	0	30	0
5	4	18	2	31	1
6	1	19	5	32	3
7	0	20	2	33	1
8	2	21	1	34	2
9	0	22	1	35	0
10	2	23	1	36	1
11	3	24	2	37	4
12	1	25	1	38	2
13	2	26	3	39	1
				40	0

4.2.13. In 1893, New Zealand became the first country to permit women to vote. Scattered over the ensuing 113 years, various countries joined the movement to grant this right to women. The table below (121) shows how many countries took this step in a given year. Do these data seem to follow a Poisson distribution?

Yearly Number of Countries Granting Women the Vote	Frequency
0	82
1	25
2	4
3	0
4	2

4.2.14. The following are the daily numbers of death notices for women over the age of eighty that appeared in the *London Times* over a three-year period (74).

Number of Deaths	Observed Frequency
0	162
1	267
2	271
3	185
4	111
5	61
6	27
7	8
8	3
9	1
	1096

(a) Does the Poisson pdf provide a good description of the variability pattern evident in these data?

(b) If your answer to part (a) is "no," which of the Poisson model assumptions do you think might not be holding?

4.2.15. A certain species of European mite is capable of damaging the bark on orange trees. The following are the results of inspections done on one hundred saplings chosen at random from a large orchard. The measurement recorded, X, is the number of mite infestations found on the trunk of each tree. Is it reasonable to assume that X is a Poisson random variable? If not, which of the Poisson model assumptions is likely not to be true?

No. of Infestations, k	No. of Trees
0	55
1	20
2	21
3	1
4	1
5	1
6	0
7	1

4.2.16. A tool and die press that stamps out cams used in small gasoline engines tends to break down once every five hours. The machine can be repaired and put back on line quickly, but each such incident costs \$50. What is the probability that maintenance expenses for the press will be no more than \$100 on a typical eight-hour workday?

4.2.17. In a new fiber-optic communication system, transmission errors occur at the rate of 1.5 per ten seconds. What is the probability that more than two errors will occur during the next half-minute?

4.2.18. Assume that the number of hits, X, that a baseball team makes in a nine-inning game has a Poisson distribution. If the probability that a team makes zero hits is $\frac{1}{3}$, what are their chances of getting two or more hits?

4.2.19. Flaws in metal sheeting produced by a high-temperature roller occur at the rate of one per ten square feet. What is the probability that three or more flaws will appear in a five-by-eight-foot panel?

4.2.20. Suppose a radioactive source is metered for two hours, during which time the total number of alpha particles counted is 482. What is the probability that exactly three particles will be counted in the next two minutes? Answer the question two ways—first, by defining X to be the number of particles counted in two minutes, and second, by defining X to be the number of particles counted in one minute.

4.2.21. Suppose that on-the-job injuries in a textile mill occur at the rate of 0.1 per day.

(a) What is the probability that two accidents will occur during the next (five-day) workweek?

(b) Is the probability that four accidents will occur over the next two workweeks the square of your answer to part (a)? Explain.

4.2.22. Find $P(X = 4)$ if the random variable X has a Poisson distribution such that $P(X = 1) = P(X = 2)$.

4.2.23. Let X be a Poisson random variable with parameter λ. Show that the probability that X is even is $\frac{1}{2}(1 + e^{-2\lambda})$.

4.2.24. Let X and Y be independent Poisson random variables with parameters λ and μ, respectively. Example 3.12.10 established that $X + Y$ is also Poisson with parameter $\lambda + \mu$. Prove that same result using Theorem 3.8.3.

4.2.25. If X_1 is a Poisson random variable for which $E(X_1) = \lambda$ and if the conditional pdf of X_2 given that $X_1 = x_1$ is binomial with parameters x_1 and p, show that the marginal pdf of X_2 is Poisson with $E(X_2) = \lambda p$.

Intervals Between Events: The Poisson/Exponential Relationship

Situations sometimes arise where the time interval between consecutively occurring events is an important random variable. Imagine being responsible for the maintenance on a network of computers. Clearly, the number of technicians you would need to employ in order to be capable of responding to service calls in a timely fashion would be a function of the "waiting time" from one breakdown to another.

Figure 4.2.3 shows the relationship between the random variables X and Y, where X denotes the number of occurrences in a unit of time and Y denotes the interval between consecutive occurrences. Pictured are six intervals: $X = 0$ on one occasion, $X = 1$ on three occasions, $X = 2$ once, and $X = 3$ once. Resulting from those eight occurrences are seven measurements on the random variable Y. Obviously, the pdf for Y will depend on the pdf for X. One particularly important special case of that dependence is the Poisson/exponential relationship outlined in Theorem 4.2.3.

Figure 4.2.3

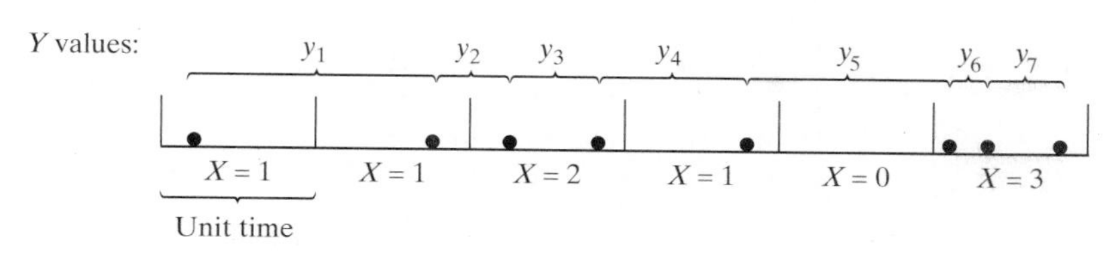

Theorem 4.2.3 *Suppose a series of events satisfying the Poisson model are occurring at the rate of λ per unit time. Let the random variable Y denote the interval between consecutive events. Then Y has the exponential distribution*

$$f_Y(y) = \lambda e^{-\lambda y}, \quad y > 0$$

Proof Suppose an event has occurred at time a. Consider the interval that extends from a to $a+y$. Since the (Poisson) events are occurring at the rate of λ per unit time, the probability that no outcomes will occur in the interval $(a, a+y)$ is $\frac{e^{-\lambda y}(\lambda y)^0}{0!} = e^{-\lambda y}$. Define the random variable Y to denote the interval between consecutive occurrences. Notice that there will be no occurrences in the interval $(a, a+y)$ only if $Y > y$. Therefore,

$$P(Y > y) = e^{-\lambda y}$$

or, equivalently,

$$F_Y(y) = P(Y \leq y) = 1 - P(Y > y) = 1 - e^{-\lambda y}$$

Let $f_Y(y)$ be the (unknown) pdf for Y. It must be true that

$$P(Y \leq y) = \int_0^y f_Y(t)\,dt$$

Taking derivatives of the two expressions for $F_Y(y)$ gives

$$\frac{d}{dy}\int_0^y f_Y(t)\,dt = \frac{d}{dy}(1 - e^{-\lambda y})$$

which implies that

$$f_Y(y) = \lambda e^{-\lambda y}, \quad y > 0$$

□

Case Study 4.2.4

Over "short" geological periods, a volcano's eruptions are believed to be Poisson events—that is, they are thought to occur independently and at a constant rate. If so, the pdf describing the intervals between eruptions should have the form $f_Y(y) = \lambda e^{-\lambda y}$. Collected for the purpose of testing that presumption are the data in Table 4.2.5, showing the intervals (in months) that elapsed between thirty-seven consecutive eruptions of Mauna Loa, a fourteen-thousand-foot volcano in Hawaii (106). During the period covered—1832 to 1950—eruptions were occurring at the rate of $\lambda = 0.027$ per month (or once every 3.1 years). Is the variability in these thirty-six y_i's consistent with the statement of Theorem 4.2.3?

Table 4.2.5

126	73	3	6	37	23
73	23	2	65	94	51
26	21	6	68	16	20
6	18	6	41	40	18
41	11	12	38	77	61
26	3	38	50	91	12

To answer that question requires that the data be reduced to a density-scaled histogram and superimposed on a graph of the predicted exponential pdf

(Continued on next page)

(recall Case Study 3.4.1). Table 4.2.6 details the construction of the histogram. Notice in Figure 4.2.4 that the shape of that histogram *is* entirely consistent with the theoretical model—$f_Y(y) = 0.027e^{-0.027y}$—stated in Theorem 4.2.3.

Table 4.2.6

Interval (mos), y	Frequency	Density
$0 \le y < 20$	13	0.0181
$20 \le y < 40$	9	0.0125
$40 \le y < 60$	5	0.0069
$60 \le y < 80$	6	0.0083
$80 \le y < 100$	2	0.0028
$100 \le y < 120$	0	0.0000
$120 \le y < 140$	1	0.0014
	36	

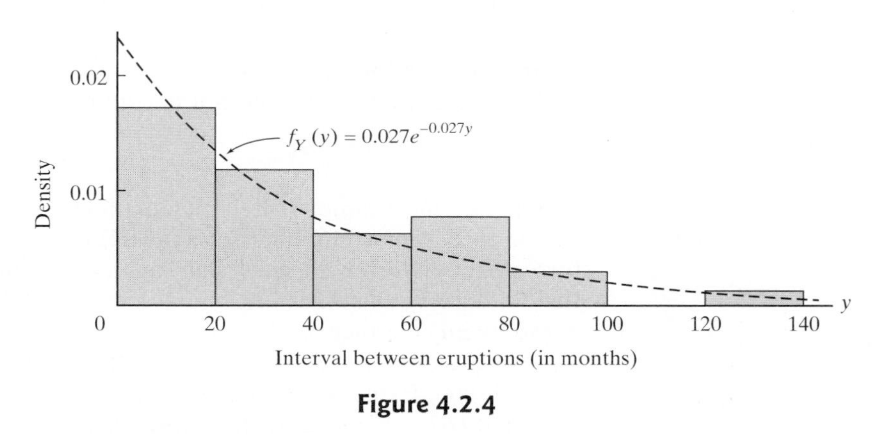

Figure 4.2.4

About the Data Among pessimists, a favorite saying is "Bad things come in threes." Optimists, not to be outdone, claim that "Good things come in threes." Are they right? In a sense, yes, but not because of fate, bad karma, or good luck. Bad things (and good things and so-so things) seem to come in threes because of (1) our intuition's inability to understand randomness and (2) the Poisson/exponential relationship. Case Study 4.2.4—specifically, the shape of the exponential pdf pictured in Figure 4.2.4—illustrates the statistics behind the superstition.

Random events, such as volcanic eruptions, do not occur at equally spaced intervals. Nor do the intervals between consecutive occurrences follow some sort of symmetric distribution, where the most common separations are close to the average separations. Quite the contrary. The Poisson/exponential relationship guarantees that the distribution of interval lengths between consecutive occurrences will be sharply skewed [look again at $f_Y(y)$], implying that the most common separation lengths will be the *shortest* ones.

Suppose that bad things are, in fact, happening to us randomly in time. Our intuitions unconsciously get a sense of the rate at which those bad things are occurring. If they happen at the rate of, say, twelve bad things per year, we mistakenly think they should come one month apart. But that is simply not the way random events behave, as Theorem 4.2.3 clearly shows.

Look at the entries in Table 4.2.5. The average of those thirty-six (randomly occurring) eruption separations was 37.7 months, yet seven of the separations were extremely short (less than or equal to six months). If two of those extremely short separations happened to occur consecutively, it would be tempting (but wrong) to conclude that the eruptions (since they came so close together) were "occurring in threes" for some supernatural reason.

Using the combinatorial techniques discussed in Section 2.6, we can calculate the probability that two extremely short intervals would occur consecutively. Think of the thirty-six intervals as being either "normal" or "extremely short." There are twenty-nine in the first group and seven in the second. Using the method described in Example 2.6.21, the probability that two extremely short separations would occur consecutively at least once is 61%, which hardly qualifies as a rare event:

$$P(\text{Two extremely short separations occur consecutively at least once})$$
$$= \frac{\binom{30}{6}\cdot\binom{6}{1}+\binom{30}{5}\cdot\binom{5}{2}+\binom{30}{4}\cdot\binom{4}{3}}{\binom{36}{29}} = 0.61$$

So, despite what our intuitions might tell us, the phenomenon of bad things coming in threes is neither mysterious nor uncommon or unexpected.

Example 4.2.5

Among the most famous of all meteor showers are the Perseids, which occur each year in early August. In some areas the frequency of visible Perseids can be as high as forty per hour. Given that such sightings are Poisson events, calculate the probability that an observer who has just seen a meteor will have to wait at least five minutes before seeing another one.

Let the random variable Y denote the interval (in minutes) between consecutive sightings. Expressed in the units of Y, the *forty-per-hour* rate of visible Perseids becomes *0.67 per minute*. A straightforward integration, then, shows that the probability is *0.035* that an observer will have to wait five minutes or more to see another meteor:

$$\begin{aligned} P(Y > 5) &= \int_5^\infty 0.67e^{-0.67y}\,dy \\ &= \int_{3.35}^\infty e^{-u}\,du \quad (\text{where } u = 0.67y) \\ &= -e^{-u}\Big|_{3.35}^\infty = e^{-3.35} \\ &= 0.035 \end{aligned}$$

■

Questions

4.2.26. Suppose that commercial airplane crashes in a certain country occur at the rate of 2.5 per year.

(a) Is it reasonable to assume that such crashes are Poisson events? Explain.
(b) What is the probability that four or more crashes will occur next year?
(c) What is the probability that the next two crashes will occur within three months of one another?

4.2.27. Records show that deaths occur at the rate of 0.1 per day among patients residing in a large nursing home. If someone dies today, what are the chances that a week or more will elapse before another death occurs?

4.2.28. Suppose that Y_1 and Y_2 are independent exponential random variables, each having pdf $f_Y(y) = \lambda e^{-\lambda y}$, $y > 0$. If $Y = Y_1 + Y_2$, it can be shown that

$$f_{Y_1+Y_2}(y) = \lambda^2 y e^{-\lambda y}, \quad y > 0$$

Recall Case Study 4.2.4. What is the probability that the next three eruptions of Mauna Loa will be less than forty months apart?

4.2.29. Fifty spotlights have just been installed in an outdoor security system. According to the manufacturer's specifications, these particular lights are expected to burn out at the rate of 1.1 per one hundred hours. What is the expected number of bulbs that will fail to last for at least seventy-five hours?

4.2.30. Suppose you want to invent a new superstition that "Bad things come in fours." Using the data given in Case Study 4.2.4 and the type of analysis described on p. 238, calculate the probability that your superstition would appear to be true.

4.3 The Normal Distribution

The Poisson limit described in Section 4.2 was not the only, or the first, approximation developed for the purpose of facilitating the calculation of binomial probabilities. Early in the eighteenth century, Abraham DeMoivre proved that areas under the curve $f_z(z) = \frac{1}{\sqrt{2\pi}} e^{-z^2/2}$, $-\infty < z < \infty$, can be used to estimate $P\left[a \le \frac{X - n\left(\frac{1}{2}\right)}{\sqrt{n\left(\frac{1}{2}\right)\left(\frac{1}{2}\right)}} \le b\right]$, where X is a binomial random variable with a large n and $p = \frac{1}{2}$.

Figure 4.3.1 illustrates the central idea in DeMoivre's discovery. Pictured is a probability histogram of the binomial distribution with $n = 20$ and $p = \frac{1}{2}$. Superimposed over the histogram is the function $f_Y(y) = \frac{1}{\sqrt{2\pi} \cdot \sqrt{5}} e^{-\frac{1}{2}\frac{(y-10)^2}{5}}$. Notice how closely the area under the curve approximates the area of the bar, even for this relatively small value of n. The French mathematician Pierre-Simon Laplace generalized DeMoivre's original idea to binomial approximations for arbitrary p and brought this theorem to the full attention of the mathematical community by including it in his influential 1812 book, *Theorie Analytique des Probabilities.*

Figure 4.3.1

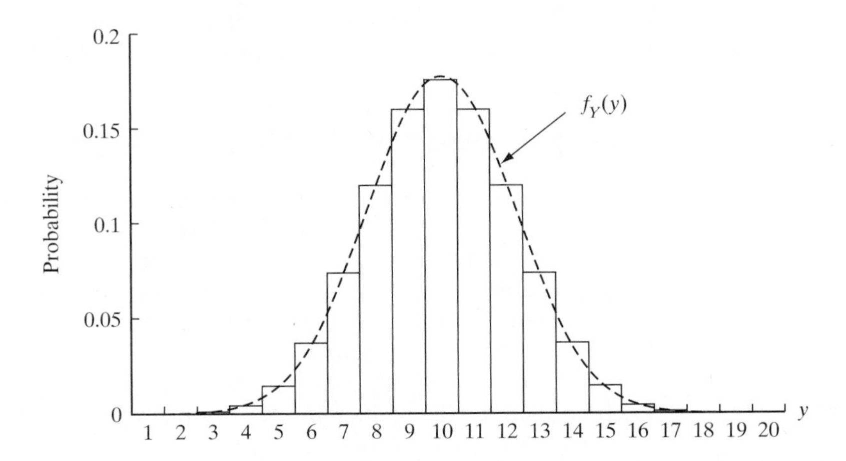

Theorem 4.3.1 *Let X be a binomial random variable defined on n independent trials for which $p = P(\text{success})$. For any numbers a and b,*

$$\lim_{n \to \infty} P\left[a \le \frac{X - np}{\sqrt{np(1-p)}} \le b\right] = \frac{1}{\sqrt{2\pi}} \int_a^b e^{-z^2/2}\, dz$$

Proof One of the ways to verify Theorem 4.3.1 is to show that the limit of the moment-generating function for $\frac{X-np}{\sqrt{np(1-p)}}$ as $n \to \infty$ is $e^{t^2/2}$ and that $e^{t^2/2}$ is also the value of $\int_{-\infty}^{\infty} e^{tz} \cdot \frac{1}{\sqrt{2\pi}} e^{-z^2/2} dz$. By Theorem 3.12.2, then, the limiting pdf of $Z = \frac{X-np}{\sqrt{np(1-p)}}$ is the function $f_Z(z) = \frac{1}{\sqrt{2\pi}} e^{-z^2/2}$, $-\infty < z < \infty$. See Appendix 4.A.2 for the proof of a more general result. □

Comment We saw in Section 4.2 that Poisson's *limit* is actually a special case of Poisson's *distribution*, $p_X(k) = \frac{e^{-\lambda}\lambda^k}{k!}$, $k = 0, 1, 2, \ldots$. Similarly, the DeMoivre-Laplace limit is a pdf in its own right. Justifying that assertion, of course, requires proving that $f_Z(z) = \frac{1}{\sqrt{2\pi}} e^{-z^2/2}$ integrates to 1 for $-\infty < z < \infty$.

Curiously, there is no algebraic or trigonometric substitution that can be used to demonstrate that the area under $f_Z(z)$ is 1. However, by using polar coordinates, we can verify a necessary and sufficient alternative—namely, that the *square* of $\int_{-\infty}^{\infty} \frac{1}{\sqrt{2\pi}} e^{-z^2/2} dz$ equals 1.

To begin, note that

$$\frac{1}{\sqrt{2\pi}} \int_{-\infty}^{\infty} e^{-x^2/2} dx \cdot \frac{1}{\sqrt{2\pi}} \int_{-\infty}^{\infty} e^{-y^2/2} dy = \frac{1}{2\pi} \int_{-\infty}^{\infty} \int_{-\infty}^{\infty} e^{-\frac{1}{2}(x^2+y^2)} dx\, dy$$

Let $x = r \cos\theta$ and $y = r \sin\theta$, so $dx\, dy = r\, dr\, d\theta$. Then

$$\begin{aligned} \frac{1}{2\pi} \int_{-\infty}^{\infty} \int_{-\infty}^{\infty} e^{-\frac{1}{2}(x^2+y^2)} dx\, dy &= \frac{1}{2\pi} \int_0^{2\pi} \int_0^{\infty} e^{-r^2/2} r\, dr\, d\theta \\ &= \frac{1}{2\pi} \int_0^{\infty} r e^{-r^2/2} dr \cdot \int_0^{2\pi} d\theta \\ &= 1 \end{aligned}$$

Comment The function $f_Z(z) = \frac{1}{\sqrt{2\pi}} e^{-z^2/2}$ is referred to as the *standard normal* (or *Gaussian*) *curve*. By convention, any random variable whose probabilistic behavior is described by a standard normal curve is denoted by Z (rather than X, Y, or W). Since $M_Z(t) = e^{t^2/2}$, it follows readily that $E(Z) = 0$ and $\text{Var}(Z) = 1$.

Finding Areas Under the Standard Normal Curve

In order to use Theorem 4.3.1, we need to be able to find the area under the graph of $f_Z(z)$ above an arbitrary interval $[a, b]$. In practice, such values are obtained in one of two ways—either by using a *normal table,* a copy of which appears at the back of every statistics book, or by running a computer software package. Typically, both approaches give the *cdf,* $F_Z(z) = P(Z \le z)$, associated with Z (and from the cdf we can deduce the desired area).

Table 4.3.1 shows a portion of the normal table that appears in Appendix A.1. Each row under the Z heading represents a number along the horizontal axis of $f_Z(z)$ rounded off to the nearest tenth; Columns 0 through 9 allow that number to be written to the hundredths place. Entries in the body of the table are areas under the graph of $f_Z(z)$ *to the left* of the number indicated by the entry's row and column. For example, the number listed at the intersection of the "1.1" row and the "4" column is 0.8729, which means that the area under $f_Z(z)$ from $-\infty$ to 1.14 is 0.8729. That is,

$$\int_{-\infty}^{1.14} \frac{1}{\sqrt{2\pi}} e^{-z^2/2} dz = 0.8729 = P(-\infty < Z \le 1.14) = F_Z(1.14)$$

Table 4.3.1

Z	0	1	2	3	4	5	6	7	8	9
−3.	0.0013	0.0010	0.0007	0.0005	0.0003	0.0002	0.0002	0.0001	0.0001	0.0000
⋮						⋮				
−0.4	0.3446	0.3409	0.3372	0.3336	0.3300	0.3264	0.3228	0.3192	0.3156	0.3121
−0.3	0.3821	0.3783	0.3745	0.3707	0.3669	0.3632	0.3594	0.3557	0.3520	0.3483
−0.2	0.4207	0.4168	0.4129	0.4090	0.4052	0.4013	0.3974	0.3936	0.3897	0.3859
−0.1	0.4602	0.4562	0.4522	0.4483	0.4443	0.4404	0.4364	0.4325	0.4286	0.4247
−0.0	0.5000	0.4960	0.4920	0.4880	0.4840	0.4801	0.4761	0.4721	0.4681	0.4641
0.0	0.5000	0.5040	0.5080	0.5120	0.5160	0.5199	0.5239	0.5279	0.5319	0.5359
0.1	0.5398	0.5438	0.5478	0.5517	0.5557	0.5596	0.5636	0.5675	0.5714	0.5753
0.2	0.5793	0.5832	0.5871	0.5910	0.5948	0.5987	0.6026	0.6064	0.6103	0.6141
0.3	0.6179	0.6217	0.6255	0.6293	0.6331	0.6368	0.6406	0.6443	0.6480	0.6517
0.4	0.6554	0.6591	0.6628	0.6664	0.6700	0.6736	0.6772	0.6808	0.6844	0.6879
0.5	0.6915	0.6950	0.6985	0.7019	0.7054	0.7088	0.7123	0.7157	0.7190	0.7224
0.6	0.7257	0.7291	0.7324	0.7357	0.7389	0.7422	0.7454	0.7486	0.7517	0.7549
0.7	0.7580	0.7611	0.7642	0.7673	0.7703	0.7734	0.7764	0.7794	0.7823	0.7852
0.8	0.7881	0.7910	0.7939	0.7967	0.7995	0.8023	0.8051	0.8078	0.8106	0.8133
0.9	0.8159	0.8186	0.8212	0.8238	0.8264	0.8289	0.8315	0.8340	0.8365	0.8389
1.0	0.8413	0.8438	0.8461	0.8485	0.8508	0.8531	0.8554	0.8577	0.8599	0.8621
1.1	0.8643	0.8665	0.8686	0.8708	0.8729	0.8749	0.8770	0.8790	0.8810	0.8830
1.2	0.8849	0.8869	0.8888	0.8907	0.8925	0.8944	0.8962	0.8980	0.8997	0.9015
1.3	0.9032	0.9049	0.9066	0.9082	0.9099	0.9115	0.9131	0.9147	0.9162	0.9177
1.4	0.9192	0.9207	0.9222	0.9236	0.9251	0.9265	0.9278	0.9292	0.9306	0.9319
⋮						⋮				
3.	0.9987	0.9990	0.9993	0.9995	0.9997	0.9998	0.9998	0.9999	0.9999	1.0000

(see Figure 4.3.2).

Figure 4.3.2

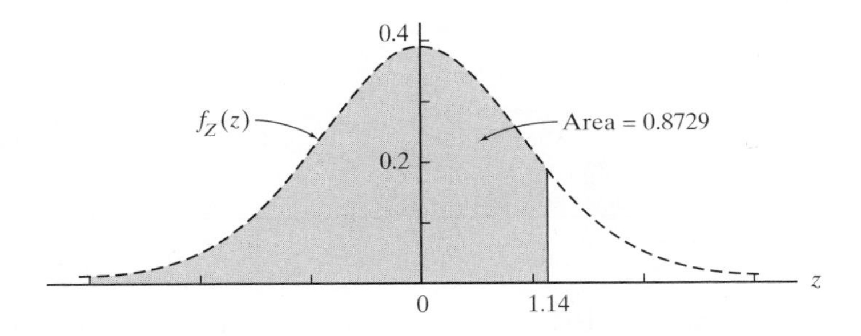

Areas under $f_z(z)$ *to the right of a number* or *between two numbers* can also be calculated from the information given in normal tables. Since the total area under $f_Z(z)$ is 1,

$$\begin{aligned} P(b < Z < +\infty) &= \text{area under } f_Z(z) \text{ to the right of } b \\ &= 1 - \text{area under } f_Z(z) \text{ to the left of } b \\ &= 1 - P(-\infty < Z \leq b) \\ &= 1 - F_Z(b) \end{aligned}$$

Similarly, the area under $f_z(z)$ *between* two numbers a and b is necessarily the area under $f_Z(z)$ to the left of b *minus* the area under $f_Z(z)$ to the left of a:

$$\begin{aligned} P(a \le Z \le b) &= \text{area under } f_Z(z) \text{ between } a \text{ and } b \\ &= \text{area under } f_Z(z) \text{ to the left of } b - \text{area under } f_Z(z) \text{ to the left of } a \\ &= P(-\infty < Z \le b) - P(-\infty < Z < a) \\ &= F_Z(b) - F_Z(a) \end{aligned}$$

The Continuity Correction

Figure 4.3.3 illustrates the underlying "geometry" implicit in the DeMoivre-Laplace Theorem. Pictured there is a continuous curve, $f(y)$, approximating a histogram, where we can presume that the areas of the rectangles are representing the probabilities associated with a discrete random variable X. Clearly, $\int_a^b f(y)\,dy$ is numerically similar to $P(a \le X \le b)$, but the diagram suggests that the approximation would be even better if the integral extended from $a-0.5$ to $b+0.5$, which would then include the cross-hatched areas. That is, a refinement of the technique of using areas under continuous curves to estimate probabilities of discrete random variables would be to write

$$P(a \le X \le b) \doteq \int_{a-0.5}^{b+0.5} f(y)\,dy$$

The substitution of $a-0.5$ for a and $b+0.5$ for b is called the *continuity correction*. Applying the latter to the DeMoivre-Laplace approximation leads to a slightly different statement for Theorem 4.3.1: If X is a binomial random variable with parameters n and p,

$$P(a \le X \le b) = F_Z\left[\frac{b+0.5-np}{\sqrt{np(1-p)}}\right] - F_Z\left[\frac{a-0.5-np}{\sqrt{np(1-p)}}\right]$$

Figure 4.3.3

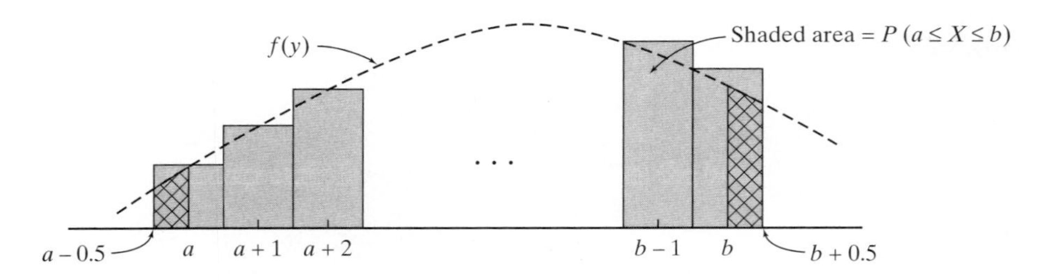

Comment Even with the continuity correction refinement, normal curve approximations can be inadequate if n is too small, especially when p is close to 0 or to 1. As a rule of thumb, the DeMoivre-Laplace limit should be used only if the magnitudes of n and p are such that $n > 9\frac{p}{1-p}$ and $n > 9\frac{1-p}{p}$.

Example 4.3.1

Boeing 757s flying certain routes are configured to have 168 economy-class seats. Experience has shown that only 90% of all ticket holders on those flights will actually show up in time to board the plane. Knowing that, suppose an airline sells 178 tickets for the 168 seats. What is the probability that not everyone who arrives at the gate on time can be accommodated?

Let the random variable X denote the number of would-be passengers who show up for a flight. Since travelers are sometimes with their families, not every ticket holder constitutes an independent event. Still, we can get a useful approximation to the probability that the flight is overbooked by assuming that X is binomial with $n = 178$ and $p = 0.9$. What we are looking for is $P(169 \leq X \leq 178)$, the probability that more ticket holders show up than there are seats on the plane. According to Theorem 4.3.1 (and using the continuity correction),

$$
\begin{aligned}
P(\text{Flight is overbooked}) &= P(169 \leq X \leq 178) \\
&= P\left[\frac{169 - 0.5 - np}{\sqrt{np(1-p)}} \leq \frac{X - np}{\sqrt{np(1-p)}} \leq \frac{178 + 0.5 - np}{\sqrt{np(1-p)}}\right] \\
&= P\left[\frac{168.5 - 178(0.9)}{\sqrt{178(0.9)(0.1)}} \leq \frac{X - 178(0.9)}{\sqrt{178(0.9)(0.1)}} \leq \frac{178.5 - 178(0.9)}{\sqrt{178(0.9)(0.1)}}\right] \\
&\doteq P(2.07 \leq Z \leq 4.57) = F_z(4.57) - F_z(2.07)
\end{aligned}
$$

From Appendix A.1, $F_Z(4.57) = P(Z \leq 4.57)$ is equal to 1, for all practical purposes, and the area under $f_Z(z)$ to the left of 2.07 is 0.9808. Therefore,

$$
\begin{aligned}
P(\text{Flight is overbooked}) &= 1.0000 - 0.9808 \\
&= 0.0192
\end{aligned}
$$

implying that the chances are about one in fifty that not every ticket holder will have a seat. ■

Case Study 4.3.1

Research in extrasensory perception has ranged from the slightly unconventional to the downright bizarre. Toward the latter part of the nineteenth century and even well into the twentieth century, much of what was done involved spiritualists and mediums. But beginning around 1910, experimenters moved out of the seance parlors and into the laboratory, where they began setting up controlled studies that could be analyzed statistically. In 1938, Pratt and Woodruff, working out of Duke University, did an experiment that became a prototype for an entire generation of ESP research (71).

The investigator and a subject sat at opposite ends of a table. Between them was a screen with a large gap at the bottom. Five blank cards, visible to both participants, were placed side by side on the table beneath the screen. On the subject's side of the screen one of the standard ESP symbols (see Figure 4.3.4) was hung over each of the blank cards.

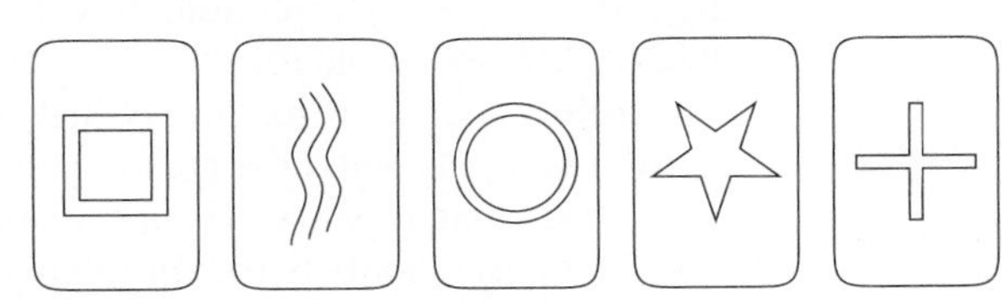

Figure 4.3.4

(Continued on next page)

(Case Study 4.3.1 continued)

The experimenter shuffled a deck of ESP cards, picked up the top one, and concentrated on it. The subject tried to guess its identity: If he thought it was a *circle,* he would point to the blank card on the table that was beneath the circle card hanging on his side of the screen. The procedure was then repeated. Altogether, a total of thirty-two subjects, all students, took part in the experiment. They made a total of sixty thousand guesses—and were correct 12,489 times.

With five denominations involved, the probability of a subject's making a correct identification just by chance was $\frac{1}{5}$. Assuming a binomial model, the expected number of correct guesses would be $60{,}000 \times \frac{1}{5}$, or 12,000. The question is, how "near" to 12,000 is 12,489? Should we write off the observed excess of 489 as nothing more than luck, or can we conclude that ESP has been demonstrated?

To effect a resolution between the conflicting "luck" and "ESP" hypotheses, we need to compute the probability of the subjects' getting 12,489 or more correct answers *under the presumption that* $p = \frac{1}{5}$. Only if that probability is very small can 12,489 be construed as evidence in support of ESP.

Let the random variable X denote the number of correct responses in sixty thousand tries. Then

$$P(X \geq 12{,}489) = \sum_{k=12{,}489}^{60{,}000} \binom{60{,}000}{k} \left(\frac{1}{5}\right)^k \left(\frac{4}{5}\right)^{60{,}000-k} \tag{4.3.1}$$

At this point the DeMoivre-Laplace limit theorem becomes a welcome alternative to computing the 47,512 binomial probabilities implicit in Equation 4.3.1. First we apply the continuity correction and rewrite $P(X \geq 12{,}489)$ as $P(X \geq 12{,}488.5)$. Then

$$\begin{aligned}
P(X \geq 12{,}489) &= P\left[\frac{X - np}{\sqrt{np(1-p)}} \geq \frac{12{,}488.5 - 60{,}000(1/5)}{\sqrt{60{,}000(1/5)(4/5)}}\right] \\
&= P\left[\frac{X - np}{\sqrt{np(1-p)}} \geq 4.99\right] \\
&\doteq \frac{1}{\sqrt{2\pi}} \int_{4.99}^{\infty} e^{-z^2/2}\, dz \\
&= 0.0000003
\end{aligned}$$

this last value being obtained from a more extensive version of Table A.1 in the Appendix.

Here, the fact that $P(X \geq 12{,}489)$ is so extremely small makes the "luck" hypothesis $\left(p = \frac{1}{5}\right)$ untenable. It would appear that something other than chance had to be responsible for the occurrence of so many correct guesses. Still, it does not follow that ESP has necessarily been demonstrated. Flaws in the experimental setup as well as errors in reporting the scores could have inadvertently produced what appears to be a statistically significant result. Suffice it to say that a great many scientists remain highly skeptical of ESP research in general and of the Pratt-Woodruff experiment in particular. [For a more thorough critique of the data we have just described, see (43).]

About the Data This is a good set of data for illustrating why we need formal mathematical methods for interpreting data. As we have seen on other occasions, our intuitions, when left unsupported by probability calculations, can often be deceived. A typical first reaction to the Pratt-Woodruff results is to dismiss as inconsequential the 489 additional correct answers. To many, it seems entirely believable that sixty thousand guesses could produce, by chance, an extra 489 correct responses. Only after making the $P(X \geq 12{,}489)$ computation do we see the utter implausibility of that conclusion. What statistics is doing here is what we would like it to do in general—rule out hypotheses that are not supported by the data and point us in the direction of inferences that are more likely to be true.

Questions

4.3.1. Use Appendix Table A.1 to evaluate the following integrals. In each case, draw a diagram of $f_Z(z)$ and shade the area that corresponds to the integral.

(a) $\int_{-0.44}^{1.33} \frac{1}{\sqrt{2\pi}} e^{-z^2/2}\, dz$
(b) $\int_{-\infty}^{0.94} \frac{1}{\sqrt{2\pi}} e^{-z^2/2}\, dz$
(c) $\int_{-1.48}^{\infty} \frac{1}{\sqrt{2\pi}} e^{-z^2/2}\, dz$
(d) $\int_{-\infty}^{-4.32} \frac{1}{\sqrt{2\pi}} e^{-z^2/2}\, dz$

4.3.2. Let Z be a standard normal random variable. Use Appendix Table A.1 to find the numerical value for each of the following probabilities. Show each of your answers as an area under $f_Z(z)$.

(a) $P(0 \leq Z \leq 2.07)$
(b) $P(-0.64 \leq Z < -0.11)$
(c) $P(Z > -1.06)$
(d) $P(Z < -2.33)$
(e) $P(Z \geq 4.61)$

4.3.3.

(a) Let $0 < a < b$. Which number is larger?

$$\int_a^b \frac{1}{\sqrt{2\pi}} e^{-z^2/2}\, dz \quad \text{or} \quad \int_{-b}^{-a} \frac{1}{\sqrt{2\pi}} e^{-z^2/2}\, dz$$

(b) Let $a > 0$. Which number is larger?

$$\int_a^{a+1} \frac{1}{\sqrt{2\pi}} e^{-z^2/2}\, dz \quad \text{or} \quad \int_{a-1/2}^{a+1/2} \frac{1}{\sqrt{2\pi}} e^{-z^2/2}\, dz$$

4.3.4.

(a) Evaluate $\int_0^{1.24} e^{-z^2/2}\, dz$.
(b) Evaluate $\int_{-\infty}^{\infty} 6e^{-z^2/2}\, dz$.

4.3.5. Assume that the random variable Z is described by a standard normal curve $f_Z(z)$. For what values of z are the following statements true?

(a) $P(Z \leq z) = 0.33$
(b) $P(Z \geq z) = 0.2236$
(c) $P(-1.00 \leq Z \leq z) = 0.5004$
(d) $P(-z < Z < z) = 0.80$
(e) $P(z \leq Z \leq 2.03) = 0.15$

4.3.6. Let z_α denote the value of Z for which $P(Z \geq z_\alpha) = \alpha$. By definition, the *interquartile range, Q*, for the standard normal curve is the difference

$$Q = z_{.25} - z_{.75}$$

Find Q.

4.3.7. Oak Hill has 74,806 registered automobiles. A city ordinance requires each to display a bumper decal showing that the owner paid an annual wheel tax of \$50. By law, new decals need to be purchased during the month of the owner's birthday. This year's budget assumes that at least \$306,000 in decal revenue will be collected in November. What is the probability that the wheel taxes reported in that month will be less than anticipated and produce a budget shortfall?

4.3.8. Hertz Brothers, a small, family-owned radio manufacturer, produces electronic components domestically but subcontracts the cabinets to a foreign supplier. Although inexpensive, the foreign supplier has a quality-control program that leaves much to be desired. On the average, only 80% of the standard 1600-unit shipment that Hertz receives is usable. Currently, Hertz has back orders for 1260 radios but storage space for no more than 1310 cabinets. What are the chances that the number of usable units in Hertz's latest shipment will be large enough to allow Hertz to fill all the orders already on hand, yet small enough to avoid causing any inventory problems?

4.3.9. Fifty-five percent of the registered voters in Sheridanville favor their incumbent mayor in her bid for re-election. If four hundred voters go to the polls, approximate the probability that

(a) the race ends in a tie.
(b) the challenger scores an upset victory.

4.3.10. State Tech's basketball team, the Fighting Logarithms, have a 70% foul-shooting percentage.

(a) Write a formula for the exact probability that out of their next one hundred free throws, they will make between seventy-five and eighty, inclusive.
(b) Approximate the probability asked for in part (a).

4.3.11. A random sample of 747 obituaries published recently in Salt Lake City newspapers revealed that 344 (or 46%) of the decedents died in the three-month period following their birthdays (123). Assess the statistical significance of that finding by approximating the probability that 46% or more would die in that particular interval if deaths occurred randomly throughout the year. What would you conclude on the basis of your answer?

4.3.12. There is a theory embraced by certain parapsychologists that hypnosis can enhance a person's ESP ability. To test that hypothesis, an experiment was set up with fifteen hypnotized subjects (21). Each was asked to make 100 guesses using the same sort of ESP cards and protocol that were described in Case Study 4.3.1. A total of 326 correct identifications were made. Can it be argued on the basis of those results that hypnosis does have an effect on a person's ESP ability? Explain.

4.3.13. If $p_X(k) = \binom{10}{k}(0.7)^k(0.3)^{10-k}$, $k = 0, 1, \ldots, 10$, is it appropriate to approximate $P(4 \leq X \leq 8)$ by computing the following?

$$P\left[\frac{3.5 - 10(0.7)}{\sqrt{10(0.7)(0.3)}} \leq Z \leq \frac{8.5 - 10(0.7)}{\sqrt{10(0.7)(0.3)}}\right]$$

Explain.

4.3.14. A sell-out crowd of 42,200 is expected at Cleveland's Jacobs Field for next Tuesday's game against the Baltimore Orioles, the last before a long road trip. The ballpark's concession manager is trying to decide how much food to have on hand. Looking at records from games played earlier in the season, she knows that, on the average, 38% of all those in attendance will buy a hot dog. How large an order should she place if she wants to have no more that a 20% chance of demand exceeding supply?

Central Limit Theorem

It was pointed out in Example 3.9.3 that every binomial random variable X can be written as the sum of n independent Bernoulli random variables $X_1, X_2, \ldots, X_n$, where

$$X_i = \begin{cases} 1 & \text{with probability } p \\ 0 & \text{with probability } 1-p \end{cases}$$

But if $X = X_1 + X_2 + \cdots + X_n$, Theorem 4.3.1 can be reexpressed as

$$\lim_{n\to\infty} P\left[a \leq \frac{X_1 + X_2 + \cdots + X_n - np}{\sqrt{np(1-p)}} \leq b\right] = \frac{1}{\sqrt{2\pi}} \int_a^b e^{-z^2/2}\,dz \tag{4.3.2}$$

Implicit in Equation 4.3.2 is an obvious question: Does the DeMoivre-Laplace limit apply to sums of other types of random variables as well? Remarkably, the answer is "yes." Efforts to extend Equation 4.3.2 have continued for more than 150 years. Russian probabilists—A. M. Lyapunov, in particular—made many of the key advances. In 1920, George Polya gave these new generalizations a name that has been associated with the result ever since: He called it the *central limit theorem* (136).

Theorem 4.3.2 *(Central Limit Theorem) Let $W_1, W_2, \ldots$ be an infinite sequence of independent random variables, each with the same distribution. Suppose that the mean μ and the variance σ^2 of $f_W(w)$ are both finite. For any numbers a and b,*

$$\lim_{n\to\infty} P\left(a \le \frac{W_1 + \cdots + W_n - n\mu}{\sqrt{n}\sigma} \le b\right) = \frac{1}{\sqrt{2\pi}} \int_a^b e^{-z^2/2}\, dz$$

Proof See Appendix 4.A.2. □

Comment The central limit theorem is often stated in terms of the *average of* W_1, $W_2, \ldots,$ and W_n, rather than their sum. Since

$$E\left[\frac{1}{n}(W_1 + \cdots + W_n)\right] = E(\overline{W}) = \mu \quad \text{and} \quad \text{Var}\left[\frac{1}{n}(W_1 + \cdots + W_n)\right] = \sigma^2/n,$$

Theorem 4.3.2 can be stated in the equivalent form

$$\lim_{n\to\infty} P\left(a \le \frac{\overline{W} - \mu}{\sigma/\sqrt{n}} \le b\right) = \frac{1}{\sqrt{2\pi}} \int_a^b e^{-z^2/2}\, dz$$

We will use both formulations, the choice depending on which is more convenient for the problem at hand.

Example 4.3.2 The top of Table 4.3.2 shows a Minitab simulation where forty random samples of size 5 were drawn from a uniform pdf defined over the interval [0, 1]. Each row corresponds to a different sample. The sum of the five numbers appearing in a given sample is denoted "y" and is listed in column C6. For this particular uniform pdf, $\mu = \frac{1}{2}$ and $\sigma^2 = \frac{1}{12}$ (recall Question 3.6.4), so

$$\frac{W_1 + \cdots + W_n - n\mu}{\sqrt{n}\sigma} = \frac{Y - \frac{5}{2}}{\sqrt{\frac{5}{12}}}$$

Table 4.3.2

	C1 y1	C2 y2	C3 y3	C4 y4	C5 y5	C6 y	C7 Z ratio
1	0.556099	0.646873	0.354373	0.673821	0.233126	2.46429	−0.05532
2	0.497846	0.588979	0.272095	0.956614	0.819901	3.13544	0.98441
3	0.284027	0.209458	0.414743	0.614309	0.439456	1.96199	−0.83348
4	0.599286	0.667891	0.194460	0.839481	0.694474	2.99559	0.76777
5	0.280689	0.692159	0.036593	0.728826	0.314434	2.05270	−0.69295
6	0.462741	0.349264	0.471254	0.613070	0.489125	2.38545	−0.17745
7	0.556940	0.246789	0.719907	0.711414	0.918221	3.15327	1.01204
8	0.102855	0.679119	0.559210	0.014393	0.518450	1.87403	−0.96975
9	0.642859	0.004636	0.728131	0.299165	0.801093	2.47588	−0.03736
10	0.017770	0.568188	0.416351	0.908079	0.075108	1.98550	−0.79707
11	0.331291	0.410705	0.118571	0.979254	0.242582	2.08240	−0.64694
12	0.355047	0.961126	0.920597	0.575467	0.585492	3.39773	1.39076
13	0.626197	0.304754	0.530345	0.933018	0.675899	3.07021	0.88337
14	0.211714	0.404505	0.045544	0.213012	0.520614	1.39539	−1.71125
15	0.535199	0.130715	0.603642	0.333023	0.405782	2.00836	−0.76164
16	0.810374	0.153955	0.082226	0.827269	0.897901	2.77172	0.42095

Table 4.3.2 (*continued*)

	C1 y1	C2 y2	C3 y3	C4 y4	C5 y5	C6 y	C7 Z ratio
17	0.687550	0.185393	0.620878	0.013395	0.819712	2.32693	−0.26812
18	0.424193	0.529199	0.201554	0.157073	0.090455	1.40248	−1.70028
19	0.397373	0.143507	0.973991	0.234845	0.681147	2.43086	−0.10711
20	0.413788	0.653468	0.017335	0.556255	0.900568	2.54141	0.06416
21	0.602607	0.094162	0.247676	0.638875	0.653910	2.23723	−0.40708
22	0.963678	0.375850	0.909377	0.307358	0.828882	3.38515	1.37126
23	0.967499	0.868809	0.940770	0.405564	0.814348	3.99699	2.31913
24	0.439913	0.446679	0.075227	0.983295	0.554581	2.49970	−0.00047
25	0.215774	0.407494	0.002307	0.971140	0.437144	2.03386	−0.72214
26	0.108881	0.271860	0.972351	0.604762	0.210347	2.16820	−0.51402
27	0.337798	0.173911	0.309916	0.300208	0.666831	1.78866	−1.10200
28	0.635017	0.187311	0.365419	0.831417	0.463567	2.48273	−0.02675
29	0.563097	0.065293	0.841320	0.518055	0.685137	2.67290	0.26786
30	0.687242	0.544286	0.980337	0.649507	0.077364	2.93874	0.67969
31	0.784501	0.745614	0.459559	0.565875	0.529171	3.08472	0.90584
32	0.505460	0.355340	0.163285	0.352540	0.896521	2.27315	−0.35144
33	0.336992	0.734869	0.824409	0.321047	0.682283	2.89960	0.61906
34	0.784279	0.194038	0.323756	0.430020	0.459238	2.19133	−0.47819
35	0.548008	0.788351	0.831117	0.200790	0.823102	3.19137	1.07106
36	0.096383	0.844281	0.680927	0.656946	0.050867	2.32940	−0.26429
37	0.161502	0.972933	0.038113	0.515530	0.553788	2.24187	−0.39990
38	0.677552	0.232181	0.307234	0.588927	0.365403	2.17130	−0.50922
39	0.470454	0.267230	0.652802	0.633286	0.410964	2.43474	−0.10111
40	0.104377	0.819950	0.047036	0.189226	0.399502	1.56009	−1.45610

At the bottom of Table 4.3.2 is a density-scaled histogram of the forty "Z ratios," $\frac{y-5/2}{\sqrt{5/12}}$ (as listed in column C7). Notice the close agreement between the distribution of those ratios and $f_Z(z)$: What we see there is entirely consistent with the statement of Theorem 4.3.2.

Comment Theorem 4.3.2 is an asymptotic result, yet it can provide surprisingly good approximations *even when n is very small*. Example 4.3.2 is a typical case in point: The uniform pdf over [0, 1] looks nothing like a bell-shaped curve, yet random samples as small as $n = 5$ yield sums that behave probabilistically much like the theoretical limit.

In general, samples from symmetric pdfs will produce sums that "converge" quickly to the theoretical limit. On the other hand, if the underlying pdf is sharply skewed—for example, $f_Y(y) = 10e^{-10y}$, $y > 0$—it would take a larger n to achieve the level of agreement present in Figure 4.3.2. ■

Example 4.3.3

A random sample of size $n = 15$ is drawn from the pdf $f_Y(y) = 3(1-y)^2$, $0 \le y \le 1$. Let $\bar{Y} = \left(\frac{1}{15}\right)\sum_{i=1}^{15} Y_i$. Use the central limit theorem to approximate $P\left(\frac{1}{8} \le \bar{Y} \le \frac{3}{8}\right)$.

Note, first of all, that

$$E(Y) = \int_0^1 y \cdot 3(1-y)^2\, dy = \frac{1}{4}$$

and

$$\sigma^2 = \text{Var}(Y) = E(Y^2) - \mu^2 = \int_0^1 y^2 \cdot 3(1-y)^2\, dy - \left(\frac{1}{4}\right)^2 = \frac{3}{80}$$

According, then, to the central limit theorem formulation that appears in the comment on p. 247, the probability that $\bar{Y}$ will lie between $\frac{1}{8}$ and $\frac{3}{8}$ is approximately *0.99*:

$$P\left(\frac{1}{8} \le \bar{Y} \le \frac{3}{8}\right) = P\left(\frac{\frac{1}{8} - \frac{1}{4}}{\sqrt{\frac{3}{80}}/\sqrt{15}} \le \frac{\bar{Y} - \frac{1}{4}}{\sqrt{\frac{3}{80}}/\sqrt{15}} \le \frac{\frac{3}{8} - \frac{1}{4}}{\sqrt{\frac{3}{80}}/\sqrt{15}}\right)$$
$$= P(-2.50 \le Z \le 2.50)$$
$$= 0.9876$$

■

Example 4.3.4

In preparing next quarter's budget, the accountant for a small business has one hundred different expenditures to account for. Her predecessor listed each entry to the penny, but doing so grossly overstates the precision of the process. As a more truthful alternative, she intends to record each budget allocation to the nearest \$100. What is the probability that her total estimated budget will end up differing from the actual cost by more than \$500? Assume that $Y_1, Y_2, \ldots, Y_{100}$, the rounding errors she makes on the one hundred items, are independent and uniformly distributed over the interval [−\$50, +\$50].

Let

$$S_{100} = Y_1 + Y_2 + \cdots + Y_{100}$$
$$= \text{total rounding error}$$

What the accountant wants to estimate is $P(|S_{100}| > \$500)$. By the distribution assumption made for each Y_i,

$$E(Y_i) = 0, \quad i = 1, 2, \ldots, 100$$

and

$$\text{Var}(Y_i) = E\left(Y_i^2\right) = \int_{-50}^{50} \frac{1}{100} y^2 \, dy$$
$$= \frac{2500}{3}$$

Therefore,

$$E(S_{100}) = E(Y_1 + Y_2 + \cdots + Y_{100}) = 0$$

and

$$\text{Var}(S_{100}) = \text{Var}(Y_1 + Y_2 + \cdots + Y_{100}) = 100\left(\frac{2500}{3}\right)$$
$$= \frac{250{,}000}{3}$$

Applying Theorem 4.3.2, then, shows that her strategy has roughly an *8%* chance of being in error by more than \$500:

$$P(|S_{100}| > \$500) = 1 - P(-500 \le S_{100} \le 500)$$
$$= 1 - P\left(\frac{-500 - 0}{500/\sqrt{3}} \le \frac{S_{100} - 0}{500/\sqrt{3}} \le \frac{500 - 0}{500/\sqrt{3}}\right)$$
$$= 1 - P(-1.73 < Z < 1.73)$$
$$= 0.0836$$

■

Questions

4.3.15. A fair coin is tossed two hundred times. Let $X_i = 1$ if the ith toss comes up heads and $X_i = 0$ otherwise, $i = 1, 2, \ldots, 200$; $X = \sum_{i=1}^{200} X_i$. Calculate the central limit theorem approximation for $P(|X - E(X)| \le 5)$. How does this differ from the DeMoivre-Laplace approximation?

4.3.16. Suppose that one hundred fair dice are tossed. Estimate the probability that the sum of the faces showing exceeds 370. Include a continuity correction in your analysis.

4.3.17. Let X be the amount won or lost in betting \$5 on red in roulette. Then $p_x(5) = \frac{18}{38}$ and $p_x(-5) = \frac{20}{38}$. If a gambler bets on red one hundred times, use the central limit theorem to estimate the probability that those wagers result in less than \$50 in losses.

4.3.18. If $X_1, X_2, \ldots, X_n$ are independent Poisson random variables with parameters $\lambda_1, \lambda_2, \ldots, \lambda_n$, respectively, and if $X = X_1 + X_2 + \cdots + X_n$, then X is a Poisson random variable with parameter $\lambda = \sum_{i=1}^{n} \lambda_i$ (recall Example 3.12.10). What specific form does the ratio in Theorem 4.3.2 take if the X_i's are Poisson random variables?

4.3.19. An electronics firm receives, on the average, fifty orders per week for a particular silicon chip. If the company has sixty chips on hand, use the central limit theorem to approximate the probability that they will be unable to fill all their orders for the upcoming week. Assume that weekly demands follow a Poisson distribution. (*Hint:* See Question 4.3.18.)

4.3.20. Considerable controversy has arisen over the possible aftereffects of a nuclear weapons test conducted in Nevada in 1957. Included as part of the test were some three thousand military and civilian "observers." Now, more than fifty years later, eight cases of leukemia have been diagnosed among those three thousand. The expected number of cases, based on the demographic characteristics of the observers, was three. Assess the statistical significance of those findings. Calculate both an exact answer using the Poisson distribution as well as an approximation based on the central limit theorem.

The Normal Curve as a Model for Individual Measurements

Because of the central limit theorem, we know that sums (or averages) of virtually any set of random variables, when suitably scaled, have distributions that can be approximated by a standard normal curve. Perhaps even more surprising is the fact that many *individual* measurements, when suitably scaled, also have a standard normal distribution. Why should the latter be true? What do single observations have in common with samples of size n?

Astronomers in the early nineteenth century were among the first to understand the connection. Imagine looking through a telescope for the purpose of determining the location of a star. Conceptually, the data point, Y, eventually recorded is the sum of two components: (1) the star's *true* location μ^* (which remains unknown) and (2) measurement error. By definition, measurement error is the net effect of all those factors that cause the random variable Y to have a value different from μ^*. Typically, these effects will be additive, in which case the random variable can be written as a sum,

$$Y = \mu^* + W_1 + W_2 + \cdots + W_t \tag{4.3.3}$$

where W_1, for example, might represent the effect of atmospheric irregularities, W_2 the effect of seismic vibrations, W_3 the effect of parallax distortions, and so on.

If Equation 4.3.3 is a valid representation of the random variable Y, then it would follow that the central limit theorem applies to the *individual* Y_i's. Moreover, if

$$E(Y) = E(\mu^* + W_1 + W_2 + \cdots + W_t) = \mu$$

and

$$\text{Var}(Y) = \text{Var}(\mu^* + W_1 + W_2 + \cdots + W_t) = \sigma^2$$

the ratio in Theorem 4.3.2 takes the form $\frac{Y-\mu}{\sigma}$. Furthermore, t is likely to be very large, so the approximation implied by the central limit theorem is essentially an equality—that is, *we take the pdf of* $\frac{Y-\mu}{\sigma}$ *to be* $f_Z(z)$.

Finding an actual formula for $f_Y(y)$, then, becomes an exercise in applying Theorem 3.8.2. Given that $\frac{Y-\mu}{\sigma} = Z$,

$$Y = \mu + \sigma Z$$

and

$$\begin{aligned} f_Y(y) &= \frac{1}{\sigma} f_Z\left(\frac{y-\mu}{\sigma}\right) \\ &= \frac{1}{\sqrt{2\pi}\sigma} e^{-\frac{1}{2}\left(\frac{y-\mu}{\sigma}\right)^2}, \quad -\infty < y < \infty \end{aligned}$$

Definition 4.3.1. A random variable Y is said to be normally distributed with mean μ and variance σ^2 if

$$f_Y(y) = \frac{1}{\sqrt{2\pi}\sigma} e^{-\frac{1}{2}\left(\frac{y-\mu}{\sigma}\right)^2}, \quad -\infty < y < \infty$$

The symbol $Y \sim N(\mu, \sigma^2)$ will sometimes be used to denote the fact that Y has a normal distribution with mean μ and variance σ^2.

Comment Areas under an "arbitrary" normal distribution, $f_Y(y)$, are calculated by finding the equivalent area under the standard normal distribution, $f_Z(z)$:

$$P(a \leq Y \leq b) = P\left(\frac{a-\mu}{\sigma} \leq \frac{Y-\mu}{\sigma} \leq \frac{b-\mu}{\sigma}\right) = P\left(\frac{a-\mu}{\sigma} \leq Z \leq \frac{b-\mu}{\sigma}\right)$$

The ratio $\frac{Y-\mu}{\sigma}$ is often referred to as either a Z *transformation* or a Z *score*.

Example 4.3.5

In most states a motorist is legally drunk, or driving under the influence (DUI), if his or her blood alcohol concentration, Y, is 0.08% or higher. When a suspected DUI offender is pulled over, police often request a sobriety test. Although the breath analyzers used for that purpose are remarkably precise, the machines do exhibit a certain amount of measurement error. Because of that variability, the possibility exists that a driver's *true* blood alcohol concentration may be *under* 0.08% even if the analyzer gives a reading *over* 0.08%.

Experience has shown that repeated breath analyzer measurements taken from the same person produce a distribution of responses that can be described by a normal pdf with μ equal to the person's true blood alcohol concentration and σ equal to 0.004%. Suppose a driver is stopped at a roadblock on his way home from a party. Having celebrated a bit more than he should have, he has a true blood alcohol concentration of 0.075%, just barely under the legal limit. If he takes the breath analyzer test, what are the chances that he will be incorrectly booked on a DUI charge?

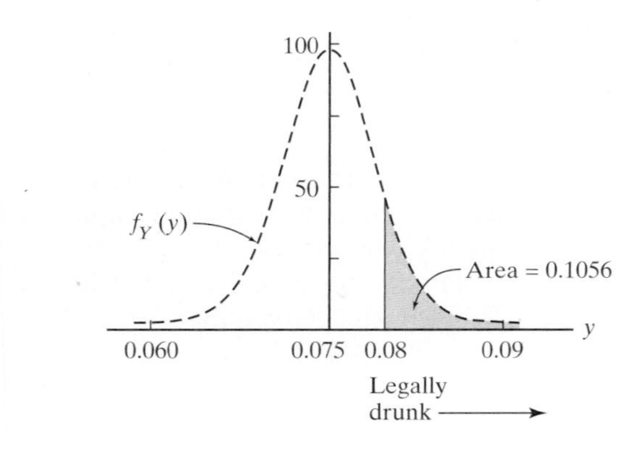

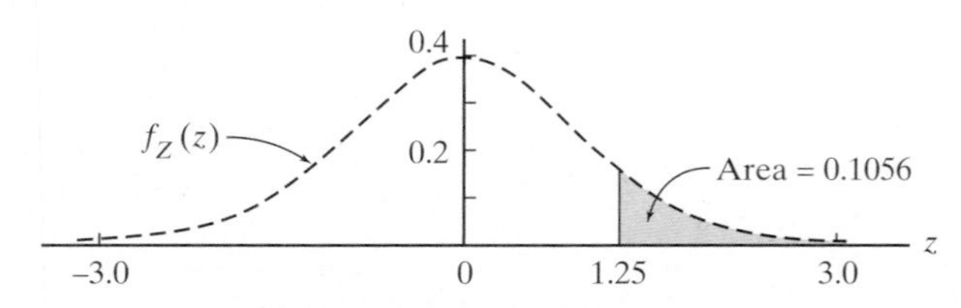

Figure 4.3.5

Since a DUI arrest occurs when $Y \geq 0.08\%$, we need to find $P(Y \geq 0.08)$ when $\mu = 0.075$ and $\sigma = 0.004$ (the percentage is irrelevant to any probability calculation and can be ignored). An application of the Z transformation shows that the driver has almost an *11%* chance of being falsely accused:

$$P(Y \geq 0.08) = P\left(\frac{Y - 0.075}{0.004} \geq \frac{0.080 - 0.075}{0.004}\right)$$
$$= P(Z \geq 1.25) = 1 - P(Z < 1.25)$$
$$= 1 - 0.8944 = 0.1056$$

Figure 4.3.5 shows $f_Y(y)$, $f_Z(z)$, and the two areas that are equal. ■

Case Study 4.3.2

For his many notable achievements, Sir Francis Galton (1822–1911) is much admired by scientists and statisticians, but not so much by criminals (at least not criminals who know something about history). What should rankle the incarcerated set is the fact that Galton did groundbreaking work in using fingerprints for identification purposes. Late in the nineteenth century, he showed that all fingerprints could be classified into three generic types—the whorl, the loop, and the arch (see Figure 4.3.6). A few years later, Sir Edward Richard Henry, who would eventually become Commissioner of Scotland Yard, refined Galton's system to include eight generic types. The Henry system, as it came to be known, was quickly adopted by law enforcement agencies worldwide and ultimately became the foundation for the first AFIS (Automated Fingerprint Identification System) databases introduced in the 1990s.

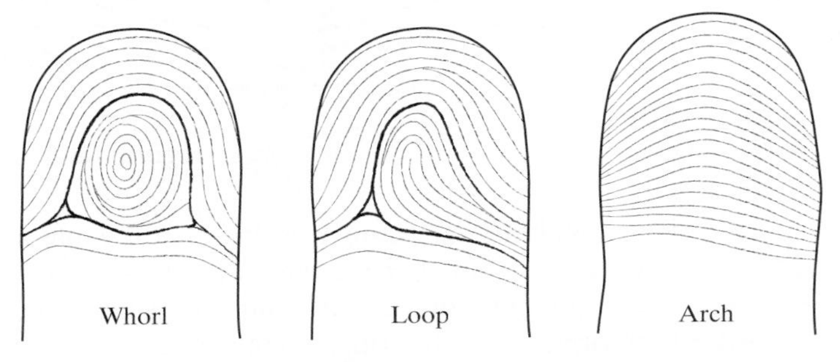

Figure 4.3.6

There are many characteristics besides the three proposed by Galton and the eight proposed by Henry that can be used to distinguish one fingerprint from another. Among the most objective of these is the ridge count. In the loop pattern, there is a point—the triradius—where the three opposing ridge systems come together. A straight line drawn from the triradius to the center of the loop will cross a certain number of ridges; in Figure 4.3.7, that number is eleven. Adding the numbers of ridge crossings for each finger yields a sum known as the *ridge count*.

Consider the following scenario. Police are investigating the murder of a pedestrian in a heavily populated urban area that is thought to have been a gang-related, drive-by shooting, perhaps as part of an initiation ritual. No eyewitnesses have come forth, but an unregistered gun was found nearby that the

(Continued on next page)

(Case Study 4.3.2 continued)

ballistics lab has confirmed was the murder weapon. Lifted from the gun was a partially smudged set of latent fingerprints. None of the features typically used for identification purposes was recognizable except for the ridge count, which appeared to be at least 270. The police have arrested a young man who lives in the area, is known to belong to a local gang, has no verifiable alibi for the night of the shooting, and has a ridge count of 275. His trial is about to begin.

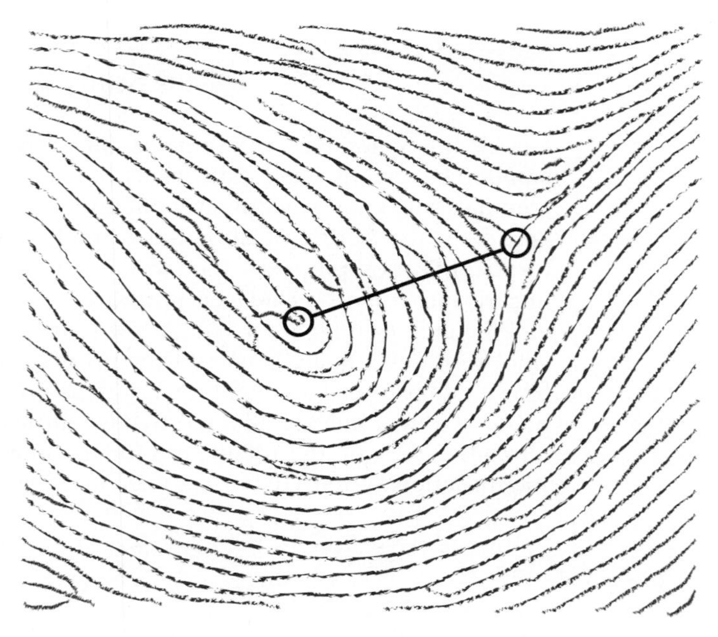

Figure 4.3.7

Neither the state nor the defense has a strong case. Both sides have no choice but to base their arguments on the statistical implications of the defendant's ridge count. And both sides have access to the same background information—that ridge counts for males are normally distributed with a mean (μ) of 146 and a standard deviation (σ) of 52.

The state's case

Clearly, the defendant has an unusually high ridge count. The strength of the prosecutor's case hinges on *how* unusual. According to the Z transformation given on p. 252 (together with a continuity correction), the probability of a ridge count, Y, being at least 275 is 0.0068:

$$P(Y \geq 275) \doteq P\left(\frac{Y-146}{52} \geq \frac{274.5-146}{52}\right) = P(Z \geq 2.47) = 0.0068$$

This is great news for the prosecutor: jurors will most likely interpret a probability that small as being very strong evidence against the defendant.

(Continued on next page)

The defense's case

The defense must necessarily try to establish "reasonable doubt" by showing that the probability is fairly high that someone other than the defendant could have committed the murder. To make that argument requires an application of conditional probability as it pertains to the binomial distribution.

Suppose n male gang members were riding around on the night of the shooting and could conceivably have committed the crime, and let X denote the number of those individuals who have ridge counts of at least 270. Then

$$P(X=k)=\binom{n}{k}p^k(1-p)^{n-k}$$

where

$$p=P(Y\geq 270)\doteq P\left(\frac{Y-146}{52}\geq\frac{269.5-146}{52}\right)=P(Z\geq 2.38)=0.0087$$

Also,

$$P(X=1)=\binom{n}{1}p^1(1-p)^{n-1}=np(1-p)^{n-1}$$

$$P(X\geq 1)=1-P(X=0)=1-(1-p)^n$$

and

$$P(X\geq 2)=1-(1-p)^n-np(1-p)^{n-1}$$

Therefore,

$$\begin{aligned}P(X\geq 2)|P(X\geq 1)&=\frac{P(X\geq 2)}{P(X\geq 1)}\\&=\frac{1-(1-p)^n-np(1-p)^{n-1}}{1-(1-p)^n}\\&=P(\text{at least two persons have ridge counts}\\&\qquad\geq 270|\text{ at least one person has a ridge count}\geq 270)\\&=P(\text{at least one other person besides the defendant}\\&\qquad\text{could have committed the murder})\end{aligned}$$

How large was n on the night of the shooting? There is no way to know, but it could have been sizeable, given the amount of gang activity found in many metropolitan areas. Table 4.3.3 lists the values of $P(X\geq 2|X\geq 1)$ calculated for

Table 4.3.3

n	$P(X\geq 2\|X\geq 1)$
25	0.10
50	0.20
100	0.37
150	0.51
200	0.63

various values of n ranging from 25 to 200. For example, if $n = 50$ gang members including the defendant were riding around on the night of the murder, there is a 20% chance that at least one other individual besides the defendant has a ridge count of at least 270 (and might be the shooter).

Imagine yourself on the jury. Which is the more persuasive statistical analysis, the state's calculation that $P(Y \geq 275) = 0.0068$, or the defense's tabulation of $P(X \geq 2 | X \geq 1)$? Would your verdict be "guilty" or "not guilty"?

About the Data Given that astrologers, psychics, and Tarot card readers still abound, it should come as no surprise that fingerprint patterns gave rise to their own form of fortune-telling (known more elegantly as *dactylomancy*). According to those who believe in such things, a person "having whorls on all fingers is restless, vacillating, doubting, sensitive, clever, eager for action, and inclined to crime." Needless to say, of course, "A mixture of loops and whorls signifies a neutral character, a person who is kind, obedient, truthful, but often undecided and impatient" (32).

Example 4.3.6 Mensa (from the Latin word for "mind") is an international society devoted to intellectual pursuits. Any person who has an IQ in the upper 2% of the general population is eligible to join. What is the *lowest* IQ that will qualify a person for membership? Assume that IQs are normally distributed with $\mu = 100$ and $\sigma = 16$.

Let the random variable Y denote a person's IQ, and let the constant y_L be the lowest IQ that qualifies someone to be a card-carrying Mensan. The two are related by a probability equation:

$$P(Y \geq y_L) = 0.02$$

or, equivalently,

$$P(Y < y_L) = 1 - 0.02 = 0.98 \tag{4.3.4}$$

(see Figure 4.3.8).

Applying the Z transformation to Equation 4.3.4 gives

$$P(Y < y_L) = P\left(\frac{Y - 100}{16} < \frac{y_L - 100}{16}\right) = P\left(Z < \frac{y_L - 100}{16}\right) = 0.98$$

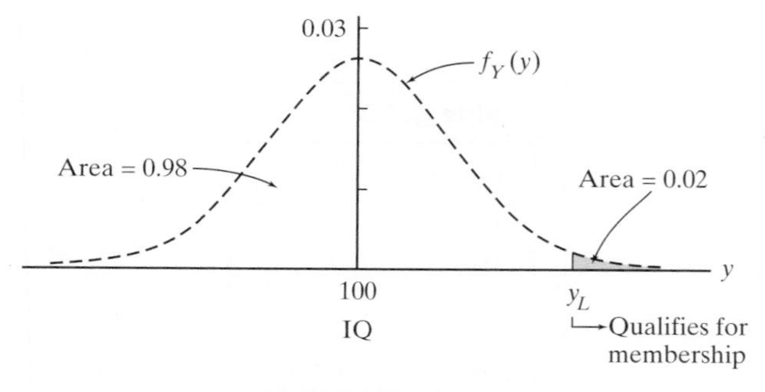

Figure 4.3.8

From the standard normal table in Appendix Table A.1, though,

$$P(Z < 2.05) = 0.9798 \doteq 0.98$$

Since $\frac{y_L - 100}{16}$ and 2.05 are both cutting off the same area of 0.02 under $f_Z(z)$, they must be equal, which implies that *133* is the lowest acceptable IQ for Mensa:

$$y_L = 100 + 16(2.05) = 133$$

■

Example 4.3.7 Suppose a random variable Y has the moment-generating function $M_Y(t) = e^{3t+8t^2}$. Calculate $P(-1 \le Y \le 9)$.

To begin, notice that $M_Y(t)$ has the same *form* as the moment-generating function for a normal random variable. That is,

$$e^{3t+8t^2} = e^{\mu t + (\sigma^2 t^2)/2}$$

where $\mu = 3$ and $\sigma^2 = 16$ (recall Example 3.12.4). To evaluate $P(-1 \le Y \le 9)$, then, requires an application of the Z transformation:

$$\begin{aligned} P(-1 \le Y \le 9) = P\left(\frac{-1-3}{4} \le \frac{Y-3}{4} \le \frac{9-3}{4}\right) &= P(-1.00 \le Z \le 1.50) \\ &= 0.9332 - 0.1587 \\ &= 0.7745 \end{aligned}$$

■

Theorem 4.3.3 *Let Y_1 be a normally distributed random variable with mean μ_1 and variance σ_1^2, and let Y_2 be a normally distributed random variable with mean μ_2 and variance σ_2^2. Define $Y = Y_1 + Y_2$. If Y_1 and Y_2 are independent, Y is normally distributed with mean $\mu_1 + \mu_2$ and variance $\sigma_1^2 + \sigma_2^2$.*

Proof Let $M_{Y_i}(t)$ denote the moment-generating function for Y_i, $i = 1, 2$, and let $M_Y(t)$ be the moment-generating function for Y. Since $Y = Y_1 + Y_2$, and the Y_i's are independent,

$$\begin{aligned} M_Y(t) &= M_{Y_1}(t) \cdot M_{Y_2}(t) \\ &= e^{\mu_1 t + \left(\sigma_1^2 t^2\right)/2} \cdot e^{\mu_2 t + \left(\sigma_2^2 t^2\right)/2} \qquad \text{(See Example 3.12.4)} \\ &= e^{(\mu_1+\mu_2)t + \left(\sigma_1^2+\sigma_2^2\right)t^2/2} \end{aligned}$$

We recognize the latter, though, to be the moment-generating function for a normal random variable with mean $\mu_1 + \mu_2$ and variance $\sigma_1^2 + \sigma_2^2$. The result follows by virtue of the uniqueness property stated in Theorem 3.12.2. □

Corollary 4.3.1 *Let $Y_1, Y_2, \ldots, Y_n$ be a random sample of size n from a normal distribution with mean μ and variance σ^2. Then the sample mean, $\bar{Y} = \frac{1}{n}\sum_{i=1}^{n} Y_i$, is also normally distributed with mean μ but with variance equal to σ^2/n (which implies that $\frac{\bar{Y}-\mu}{\sigma/\sqrt{n}}$ is a standard normal random variable, Z).* ◄

Corollary 4.3.2 *Let $Y_1, Y_2, \ldots, Y_n$ be any set of independent normal random variables with means $\mu_1, \mu_2, \ldots, \mu_n$ and variances $\sigma_1^2, \sigma_2^2, \ldots, \sigma_n^2$, respectively. Let $a_1, a_2, \ldots, a_n$ be any set of constants. Then $Y = a_1Y_1 + a_2Y_2 + \cdots + a_nY_n$ is normally distributed with mean $\mu = \sum_{i=1}^{n} a_i\mu_i$ and variance $\sigma^2 = \sum_{i=1}^{n} a_i^2\sigma_i^2$.*

Example 4.3.8 The elevator in the athletic dorm at Swampwater Tech has a maximum capacity of twenty-four hundred pounds. Suppose that ten football players get on at the twentieth floor. If the weights of Tech's players are normally distributed with a mean of two hundred twenty pounds and a standard deviation of twenty pounds, what is the probability that there will be ten fewer Muskrats at tomorrow's practice?

Let the random variables $Y_1, Y_2, \ldots, Y_{10}$ denote the weights of the ten players. At issue is the probability that $Y = \sum_{i=1}^{10} Y_i$ exceeds twenty-four hundred pounds. But

$$P\left(\sum_{i=1}^{10} Y_i > 2400\right) = P\left(\frac{1}{10}\sum_{i=1}^{10} Y_i > \frac{1}{10} \cdot 2400\right) = P(\bar{Y} > 240.0)$$

A Z transformation can be applied to the latter expression using the corollary on p. 257:

$$P(\bar{Y} > 240.0) = P\left(\frac{\bar{Y} - 220}{20/\sqrt{10}} > \frac{240.0 - 220}{20/\sqrt{10}}\right) = P(Z > 3.16)$$
$$= 0.0008$$

Clearly, the chances of a Muskrat splat are minimal. (How much would the probability change if eleven players squeezed onto the elevator?)

Questions

4.3.21. Econo-Tire is planning an advertising campaign for its newest product, an inexpensive radial. Preliminary road tests conducted by the firm's quality-control department have suggested that the lifetimes of these tires will be normally distributed with an average of thirty thousand miles and a standard deviation of five thousand miles. The marketing division would like to run a commercial that makes the claim that at least nine out of ten drivers will get at least twenty-five thousand miles on a set of Econo-Tires. Based on the road test data, is the company justified in making that assertion?

4.3.22. A large computer chip manufacturing plant under construction in Westbank is expected to result in an additional fourteen hundred children in the county's public school system once the permament workforce arrives. Any child with an IQ under 80 or over 135 will require individualized instruction that will cost the city an additional $1750 per year. How much money should Westbank anticipate spending next year to meet the needs of its new special ed students? Assume that IQ scores are normally distributed with a mean (μ) of 100 and a standard deviation (σ) of 16.

4.3.23. Records for the past several years show that the amount of money collected daily by a prominent televangelist is normally distributed with a mean (μ) of $20,000 and a standard deviation (σ) of $5000. What are the chances that tomorrow's donations will exceed $30,000?

4.3.24. The following letter was written to a well-known dispenser of advice to the lovelorn (171):

> Dear Abby: You wrote in your column that a woman is pregnant for 266 days. Who said so? I carried my baby for ten months and five days, and there is no doubt about it because I know the exact date my baby was conceived. My husband is in the Navy and it couldn't have possibly been conceived any other time because I saw him only once for an hour, and I didn't see him again until the day before the baby was born.
>
> I don't drink or run around, and there is no way this baby isn't his, so please print a retraction about the 266-day carrying time because otherwise I am in a lot of trouble.
>
> San Diego Reader

Whether or not San Diego Reader is telling the truth is a judgment that lies beyond the scope of any statistical analysis, but quantifying the plausibility of her story does not. According to the collective experience of generations of pediatricians, pregnancy durations, Y, tend to be normally distributed with $\mu = 266$ days and $\sigma = 16$ days. Do a probability calculation that addresses San Diego Reader's credibility. What would you conclude?

4.3.25. A criminologist has developed a questionnaire for predicting whether a teenager will become a delinquent. Scores on the questionnaire can range from 0 to 100, with higher values reflecting a presumably greater criminal tendency. As a rule of thumb, the criminologist decides to classify a teenager as a potential delinquent if his or her score exceeds 75. The questionnaire has already been tested on a large sample of teenagers, both delinquent and nondelinquent. Among those considered nondelinquent, scores were normally distributed with a mean (μ) of 60 and a standard deviation (σ) of 10. Among those considered delinquent, scores were normally distributed with a mean of 80 and a standard deviation of 5.

(a) What proportion of the time will the criminologist misclassify a nondelinquent as a delinquent? A delinquent as a nondelinquent?

(b) On the same set of axes, draw the normal curves that represent the distributions of scores made by delinquents and nondelinquents. Shade the two areas that correspond to the probabilities asked for in part (a).

4.3.26. The cross-sectional area of plastic tubing for use in pulmonary resuscitators is normally distributed with $\mu = 12.5$ mm^2 and $\sigma = 0.2$ mm^2. When the area is less than 12.0 mm^2 or greater than 13.0 mm^2, the tube does not fit properly. If the tubes are shipped in boxes of one thousand, how many wrong-sized tubes per box can doctors expect to find?

4.3.27. At State University, the average score of the entering class on the verbal portion of the SAT is 565, with a standard deviation of 75. Marian scored a 660. How many of State's other 4250 freshmen did better? Assume that the scores are normally distributed.

4.3.28. A college professor teaches Chemistry 101 each fall to a large class of freshmen. For tests, she uses standardized exams that she knows from past experience produce bell-shaped grade distributions with a mean of 70 and a standard deviation of 12. Her philosophy of grading is to impose standards that will yield, in the long run, 20% A's, 26% B's, 38% C's, 12% D's, and 4% F's. Where should the cutoff be between the A's and the B's? Between the B's and the C's?

4.3.29. Suppose the random variable Y can be described by a normal curve with $\mu = 40$. For what value of σ is

$$P(20 \leq Y \leq 60) = 0.50$$

4.3.30. It is estimated that 80% of all eighteen-year-old women have weights ranging from 103.5 to 144.5 lb. Assuming the weight distribution can be adequately modeled by a normal curve and that 103.5 and 144.5 are equidistant from the average weight μ, calculate σ.

4.3.31. Recall the breath analyzer problem described in Example 4.3.5. Suppose the driver's blood alcohol concentration is actually 0.09% rather than 0.075%. What is the probability that the breath analyzer will make an error in his favor and indicate that he is *not* legally drunk? Suppose the police offer the driver a choice—either take the sobriety test once or take it twice and average the readings. Which option should a "0.075%" driver take? Which option should a "0.09%" driver take? Explain.

4.3.32. If a random variable Y is normally distributed with mean μ and standard deviation σ, the Z ratio $\frac{Y-\mu}{\sigma}$ is often referred to as a *normed* score: It indicates the magnitude of y *relative to the distribution from which it came.* "Norming" is sometimes used as an affirmative-action mechanism in hiring decisions. Suppose a cosmetics company is seeking a new sales manager. The aptitude test they have traditionally given for that position shows a distinct gender bias: Scores for men are normally distributed with $\mu = 62.0$ and $\sigma = 7.6$, while scores for women are normally distributed with $\mu = 76.3$ and $\sigma = 10.8$. Laura and Michael are the two candidates vying for the position: Laura has scored 92 on the test and Michael 75. If the company agrees to norm the scores for gender bias, whom should they hire?

4.3.33. The IQs of nine randomly selected people are recorded. Let $\overline{Y}$ denote their average. Assuming the distribution from which the Y_i's were drawn is normal with a mean of 100 and a standard deviation of 16, what is the probability that $\overline{Y}$ will exceed 103? What is the probability that any arbitrary Y_i will exceed 103? What is the probability that exactly three of the Y_i's will exceed 103?

4.3.34. Let $Y_1, Y_2, \ldots, Y_n$ be a random sample from a normal distribution where the mean is 2 and the variance is 4. How large must n be in order that

$$P(1.9 \leq \overline{Y} \leq 2.1) \geq 0.99$$

4.3.35. A circuit contains three resistors wired in series. Each is rated at 6 ohms. Suppose, however, that the true resistance of each one is a normally distributed random variable with a mean of 6 ohms and a standard deviation of 0.3 ohm. What is the probability that the combined resistance will exceed 19 ohms? How "precise" would the manufacturing process have to be to make the probability less than 0.005 that the combined resistance of the circuit would exceed 19 ohms?

4.3.36. The cylinders and pistons for a certain internal combustion engine are manufactured by a process that gives a normal distribution of cylinder diameters with a mean of 41.5 cm and a standard deviation of 0.4 cm. Similarly, the distribution of piston diameters is normal with a mean of 40.5 cm and a standard deviation of 0.3 cm. If the piston diameter is greater than the cylinder diameter, the former can be reworked until the two "fit." What proportion of cylinder-piston pairs will need to be reworked?

4.3.37. Use moment-generating functions to prove the two corollaries to Theorem 4.3.3.

4.3.38. Let $Y_1, Y_2, \ldots, Y_9$ be a random sample of size 9 from a normal distribution where $\mu = 2$ and $\sigma = 2$. Let $Y_1^*, Y_2^*, \ldots, Y_9^*$ be an independent random sample from a normal distribution having $\mu = 1$ and $\sigma = 1$. Find $P(\bar{Y} \geq \bar{Y}^*)$.

4.4 The Geometric Distribution

Consider a series of independent trials, each having one of two possible outcomes, success or failure. Let $p = P(\text{Trial ends in success})$. Define the random variable X to be the trial *at which the first success occurs*. Figure 4.4.1 suggests a formula for the pdf of X:

$$\begin{aligned}
p_X(k) = P(X = k) &= P(\text{First success occurs on } k\text{th trial}) \\
&= P(\text{First } k-1 \text{ trials end in failure and } k\text{th trial ends in success}) \\
&= P(\text{First } k-1 \text{ trials end in failure}) \cdot P(k\text{th trial ends in success}) \\
&= (1-p)^{k-1}p, \quad k = 1, 2, \ldots \qquad (4.4.1)
\end{aligned}$$

We call the probability model in Equation 4.4.1 a *geometric distribution* (with parameter p).

Figure 4.4.1

F	F	$\cdots$	F	S
1	2		$k-1$	k

Comment Even without its association with independent trials and Figure 4.4.1, the function $p_X(k) = (1-p)^{k-1}p$, $k = 1, 2, \ldots$ qualifies as a discrete pdf because (1) $p_X(k) \geq 0$ for all k and (2) $\sum_{\text{all } k} p_X(k) = 1$:

$$\begin{aligned}
\sum_{k=1}^{\infty} (1-p)^{k-1}p &= p\sum_{j=0}^{\infty}(1-p)^j \\
&= p \cdot \left[\frac{1}{1-(1-p)}\right] \\
&= 1
\end{aligned}$$

Example 4.4.1 A pair of fair dice are tossed until a sum of 7 appears for the first time. What is the probability that more than four rolls will be required for that to happen?

Each throw of the dice here is an independent trial for which

$$p = P(\text{sum} = 7) = \frac{6}{36} = \frac{1}{6}$$

Let X denote the roll at which the first sum of 7 appears. Clearly, X has the structure of a geometric random variable, and

$$P(X > 4) = 1 - P(X \leq 4) = 1 - \sum_{k=1}^{4} \left(\frac{5}{6}\right)^{k-1} \left(\frac{1}{6}\right)$$
$$= 1 - \frac{671}{1296}$$
$$= 0.48$$

■

Theorem 4.4.1 *Let X have a geometric distribution with $p_X(k) = (1-p)^{k-1}p$, $k = 1, 2, \ldots$. Then*

1. $M_X(t) = \frac{pe^t}{1-(1-p)e^t}$
2. $E(X) = \frac{1}{p}$
3. $\text{Var}(X) = \frac{1-p}{p^2}$

Proof See Examples 3.12.1 and 3.12.5 for derivations of $M_X(t)$ and $E(X)$. The formula for $\text{Var}(X)$ is left as an exercise. □

Example 4.4.2 A grocery store is sponsoring a sales promotion where the cashiers give away one of the letters A, E, L, S, U, or V for each purchase. If a customer collects all six (spelling *VALUES*), he or she gets \$10 worth of groceries free. What is the expected number of trips to the store a customer needs to make in order to get a complete set? Assume the different letters are given away randomly.

Let X_i denote the number of purchases necessary to get the ith different letter, $i = 1, 2, \ldots, 6$, and let X denote the number of purchases necessary to qualify for the \$10. Then $X = X_1 + X_2 + \cdots + X_6$ (see Figure 4.4.2). Clearly, X_1 equals 1 with probability 1, so $E(X_1) = 1$. Having received the first letter, the chances of getting a different one are $\frac{5}{6}$ for each subsequent trip to the store. Therefore,

$$f_{X_2}(k) = P(X_2 = k) = \left(\frac{1}{6}\right)^{k-1} \frac{5}{6}, \quad k = 1, 2, \ldots$$

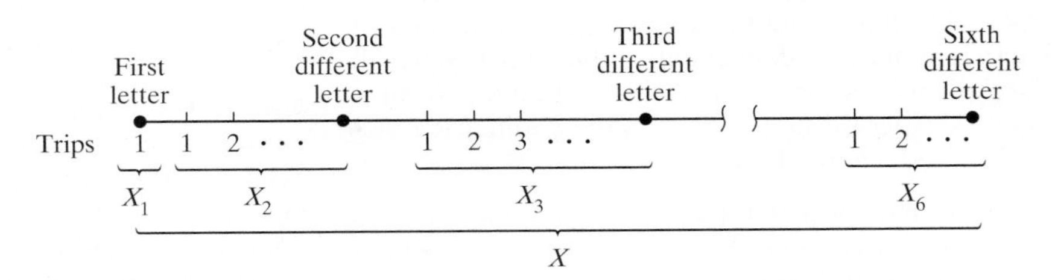

Figure 4.4.2

That is, X_2 is a geometric random variable with parameter $p = \frac{5}{6}$. By Theorem 4.4.1, $E(X_2) = \frac{6}{5}$. Similarly, the chances of getting a *third* different letter are $\frac{4}{6}$ (for each purchase), so

$$f_{X_3}(k) = P(X_3 = k) = \left(\frac{2}{6}\right)^{k-1} \left(\frac{4}{6}\right), \quad k = 1, 2, \ldots$$

and $E(X_3) = \frac{6}{4}$. Continuing in this fashion, we can find the remaining $E(X_i)$'s. It follows that a customer will have to make *14.7* trips to the store, on the average, to collect a complete set of six letters:

$$E(X) = \sum_{i=1}^{6} E(X_i)$$
$$= 1 + \frac{6}{5} + \frac{6}{4} + \frac{6}{3} + \frac{6}{2} + \frac{6}{1}$$
$$= 14.7$$

Questions

4.4.1. Because of her past convictions for mail fraud and forgery, Jody has a 30% chance each year of having her tax returns audited. What is the probability that she will escape detection for at least three years? Assume that she exaggerates, distorts, misrepresents, lies, and cheats every year.

4.4.2. A teenager is trying to get a driver's license. Write out the formula for the pdf $p_x(k)$, where the random variable X is the number of tries that he needs to pass the road test. Assume that his probability of passing the exam on any given attempt is 0.10. On the average, how many attempts is he likely to require before he gets his license?

4.4.3. Is the following set of data likely to have come from the geometric pdf $p_X(k) = \left(\frac{3}{4}\right)^{k-1} \cdot \left(\frac{1}{4}\right)$, $k = 1, 2, \ldots$? Explain.

2	8	1	2	2	5	1	2	8	3
5	4	2	4	7	2	2	8	4	7
2	6	2	3	5	1	3	3	2	5
4	2	2	3	6	3	6	4	9	3
3	7	5	1	3	4	3	4	6	2

4.4.4. Recently married, a young couple plans to continue having children until they have their first girl. Suppose the probability that a child is a girl is $\frac{1}{2}$, the outcome of each birth is an independent event, and the birth at which the first girl appears has a geometric distribution. What is the couple's expected family size? Is the geometric pdf a reasonable model here? Discuss.

4.4.5. Show that the cdf for a geometric random variable is given by $F_X(t) = P(X \leq t) = 1 - (1-p)^{[t]}$, where $[t]$ denotes the greatest integer in t, $t \geq 0$.

4.4.6. Suppose three fair dice are tossed repeatedly. Let the random variable X denote the roll on which a sum of 4 appears for the first time. Use the expression for $F_x(t)$ given in Question 4.4.5 to evaluate $P(65 \leq X \leq 75)$.

4.4.7. Let Y be an exponential random variable, where $f_Y(y) = \lambda e^{-\lambda y}$, $0 \leq y$. For any positive integer n, show that $P(n \leq Y \leq n+1) = e^{-\lambda n}(1 - e^{-\lambda})$. Note that if $p = 1 - e^{-\lambda}$, the "discrete" version of the exponential pdf is the geometric pdf.

4.4.8. Sometimes the geometric random variable is defined to be the number of trials, *X, preceding* the first success. Write down the corresponding pdf and derive the moment-generating function for X two ways—(1) by evaluating $E(e^{tX})$ directly and (2) by using Theorem 3.12.3.

4.4.9. Differentiate the moment-generating function for a geometric random variable and verify the expressions given for $E(X)$ and $\text{Var}(X)$ in Theorem 4.4.1.

4.4.10. Suppose that the random variables X_1 and X_2 have mgfs $M_{X_1}(t) = \frac{\frac{1}{2}e^t}{1-\left(1-\frac{1}{2}\right)e^t}$ and $M_{X_2}(t) = \frac{\frac{1}{4}e^t}{1-\left(1-\frac{1}{4}t\right)e^t}$, respectively. Let $X = X_1 + X_2$. Does X have a geometric distribution? Assume that X_1 and X_2 are independent.

4.4.11. The *factorial moment-generating function* for any random variable W is the expected value of t^w. Moreover $\frac{d^r}{dt^r} E(t^W)\mid_{t=1} = E[W(W-1)\cdots(W-r+1)]$. Find the factorial moment-generating function for a geometric random variable and use it to verify the expected value and variance formulas given in Theorem 4.4.1.

4.5 The Negative Binomial Distribution

The geometric distribution introduced in Section 4.4 can be generalized in a very straightforward fashion. Imagine waiting for the rth (instead of the first) success in a series of independent trials, where each trial has a probability of p of ending in success (see Figure 4.5.1).

Figure 4.5.1

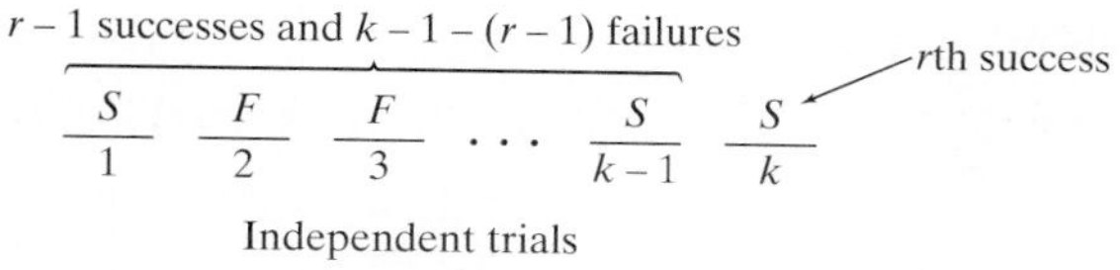

$r-1$ successes and $k-1-(r-1)$ failures

$$\underset{1}{\underline{S}} \quad \underset{2}{\underline{F}} \quad \underset{3}{\underline{F}} \quad \cdots \quad \underset{k-1}{\underline{S}} \quad \underset{k}{\underline{S}}$$

rth success

Independent trials

Let the random variable X denote the trial at which the rth success occurs. Then

$$\begin{aligned} p_X(k) = P(X=k) &= P(r\text{th success occurs on } k\text{th trial}) \\ &= P(r-1 \text{ successes occur in first } k-1 \text{ trials and} \\ &\qquad \text{success occurs on } k\text{th trial}) \\ &= P(r-1 \text{ successes occur in first } k-1 \text{ trials}) \\ &\quad \cdot P(\text{Success occurs on } k\text{th trial}) \\ &= \binom{k-1}{r-1} p^{r-1}(1-p^{k-1-(r-1)}) \cdot p \\ &= \binom{k-1}{r-1} p^r(1-p)^{k-r},\ k=r, r+1, \ldots \end{aligned} \tag{4.5.1}$$

Any random variable whose pdf has the form given in Equation 4.5.1 is said to have a *negative binomial distribution* (with parameter p).

Comment Two equivalent formulations of the negative binomial structure are widely used. Sometimes X is defined to be the number of trials *preceding* the rth success; other times, X is taken to be the number of trials in *excess of r* that are necessary to achieve the rth success. The underlying probability structure is the same, however X is defined. We will primarily use Equation 4.5.1; properties of the other two definitions for X will be covered in the exercises.

Theorem 4.5.1 *Let X have a negative binomial distribution with* $p_X(k) = \binom{k-1}{r-1} p^r.(1-p)^{k-r}$, $k=r$, $r+1, \ldots$. *Then*

1. $M_X(t) = \left[\frac{pe^t}{1-(1-p)e^t}\right]^r$
2. $E(X) = \frac{r}{p}$
3. $\text{Var}(X) = \frac{r(1-p)}{p^2}$

Proof All of these results follow immediately from the fact that X can be written as the sum of r independent geometric random variables, $X_1, X_2, \ldots, X_r$, each with parameter p. That is,

$$\begin{aligned} X &= \text{total number of trials to achieve } r\text{th success} \\ &= \text{number of trials to achieve 1st success} \\ &\quad + \text{number of additional trials to achieve 2nd success} + \cdots \\ &\quad + \text{number of additional trials to achieve } r\text{th success} \\ &= X_1 + X_2 + \cdots + X_r \end{aligned}$$

where

$$p_{X_i}(k) = (1-p)^{k-1}p, \quad k=1,2,\ldots, \quad i=1,2,\ldots,r$$

Therefore,

$$M_X(t) = M_{X_1}(t)M_{X_2}(t)\ldots M_{X_r}(t)$$
$$= \left[\frac{pe^t}{1-(1-p)e^t}\right]^r$$

Also, from Theorem 4.4.1,

$$E(X) = E(X_1) + E(X_2) + \cdots + E(X_r)$$
$$= \frac{1}{p} + \frac{1}{p} + \cdots + \frac{1}{p}$$
$$= \frac{r}{p}$$

and

$$\text{Var}(X) = \text{Var}(X_1) + \text{Var}(X_2) + \cdots + \text{Var}(X_r)$$
$$= \frac{1-p}{p^2} + \frac{1-p}{p^2} + \cdots + \frac{1-p}{p^2}$$
$$= \frac{r(1-p)}{p^2}$$

□

Example 4.5.1

The California Mellows are a semipro baseball team. Eschewing all forms of violence, the laid-back Mellow batters never swing at a pitch, and should they be fortunate enough to reach base on a walk, they never try to steal. On the average, how many runs will the Mellows score in a nine-inning road game, assuming the opposing pitcher has a 50% probability of throwing a strike on any given pitch (83)?

The solution to this problem illustrates very nicely the interplay between the physical constraints imposed by a question (in this case, the rules of baseball) and the mathematical characteristics of the underlying probability model. The negative binomial distribution appears *twice* in this analysis, along with several of the properties associated with expected values and linear combinations.

To begin, we calculate the probability of a Mellow batter striking out. Let the random variable X denote the number of pitches necessary for that to happen. Clearly, $X = 3, 4, 5,$ or 6 (why can X not be larger than 6?), and

$$p_X(k) = P(X = k) = P(\text{2 strikes are called in the first } k-1 \text{ pitches and the } k\text{th pitch is the 3rd strike})$$
$$= \binom{k-1}{2}\left(\frac{1}{2}\right)^3\left(\frac{1}{2}\right)^{k-3}, \quad k = 3, 4, 5, 6$$

Therefore,

$$P(\text{Batter strikes out}) = \sum_{k=3}^{6} p_X(k) = \left(\frac{1}{2}\right)^3 + \binom{3}{2}\left(\frac{1}{2}\right)^4 + \binom{4}{2}\left(\frac{1}{2}\right)^5 + \binom{5}{2}\left(\frac{1}{2}\right)^6$$
$$= \frac{21}{32}$$

Now, let the random variable W denote the number of walks the Mellows get in a given inning. In order for W to take on the value w, exactly two of the first $w+2$ batters must strike out, as must the $(w+3)$rd (see Figure 4.5.2). The pdf for W, then, is a negative binomial with $p = P(\text{Batter strikes out}) = \frac{21}{32}$:

$$p_W(w) = P(W = w) = \binom{w+2}{2}\left(\frac{21}{32}\right)^3\left(\frac{11}{32}\right)^w, \quad w = 0, 1, 2, \ldots$$

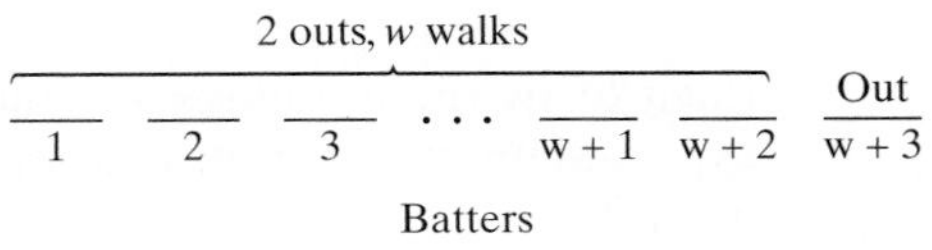

Figure 4.5.2

In order for a run to score, the pitcher must walk a Mellows batter with the bases loaded. Let the random variable R denote the total number of runs walked in during a given inning. Then

$$R = \begin{cases} 0 & \text{if } w \leq 3 \\ w-3 & \text{if } w > 3 \end{cases}$$

and

$$\begin{aligned} E(R) = &= \sum_{w=4}^{\infty}(w-3)\binom{w+2}{2}\left(\frac{21}{32}\right)^3\left(\frac{11}{32}\right)^w \\ &= \sum_{w=0}^{\infty}(w-3)\cdot P(W=w) - \sum_{w=0}^{3}(w-3)\cdot P(W=w) \\ &= E(W) - 3 + \sum_{w=0}^{3}(3-w)\cdot\binom{w+2}{2}\left(\frac{21}{32}\right)^3\left(\frac{11}{32}\right)^w \end{aligned} \tag{4.5.2}$$

To evaluate $E(W)$ using the statement of Theorem 4.5.1 requires a linear transformation to rescale W to the format of Equation 4.5.1. Let

$T = W + 3 =$ total number of Mellow batters appearing in a given inning

Then

$$p_T(t) = p_W(t-3) = \binom{t-1}{2}\left(\frac{21}{32}\right)^3\left(\frac{11}{32}\right)^{t-3}, \quad t = 3, 4, \ldots$$

which we recognize as a negative binomial pdf with $r = 3$ and $p = \frac{21}{32}$. Therefore,

$$E(T) = \frac{3}{21/32} = \frac{32}{7}$$

which makes $E(W) = E(T) - 3 = \frac{32}{7} - 3 = \frac{11}{7}$.

From Equation 4.5.2, then, the expected number of runs scored by the Mellows in a given inning is *0.202:*

$$E(R) = \frac{11}{7} - 3 + 3 \cdot \binom{2}{2}\left(\frac{21}{32}\right)^3\left(\frac{11}{32}\right)^0 + 2 \cdot \binom{3}{2}\left(\frac{21}{32}\right)^3\left(\frac{11}{32}\right)^1$$
$$+ 1 \cdot \binom{4}{2}\left(\frac{21}{32}\right)^3\left(\frac{11}{32}\right)^2$$
$$= 0.202$$

Each of the nine innings, of course, would have the same value for $E(R)$, so the expected number of runs in a *game* is the sum $0.202 + 0.202 + \cdots + 0.202 = 9(0.202)$, or *1.82*. ■

Case Study 4.5.1

Natural phenomena that are particularly complicated for whatever reasons may be impossible to describe with any single, easy-to-work-with probability model. An effective Plan B in those situations is to break the phenomenon down into simpler components and simulate the contributions of each of those components by using randomly generated observations. These are called Monte Carlo analyses, an example of which is described in detail in Section 4.7.

The fundamental requirement of any simulation technique is the ability to generate random observations from specified pdfs. In practice, this is done using computers because the number of observations needed is huge. In principle, though, the same, simple procedure can be used, by hand, to generate random observations from any discrete pdf.

Recall Example 4.5.1 and the random variable W, where W is the number of walks the Mellow batters are issued in a given inning. It was shown that $p_W(w)$ is the particular negative binomial pdf,

$$p_W(w) = P(W = w) = \binom{w+2}{w}\left(\frac{21}{32}\right)^3\left(\frac{11}{32}\right)^w, \quad w = 0, 1, 2, \ldots$$

Suppose a record is kept of the numbers of walks the Mellow batters receive in each of the next one hundred innings the team plays. What might that record look like?

The answer is, the record will look like a random sample of size 100 drawn from $p_W(w)$. Table 4.5.1 illustrates a procedure for generating such a sample. The first two columns show $p_W(w)$ for the nine values of w likely to occur (0 through 8). The third column parcels out the one hundred digits 00 through 99 into nine intervals whose lengths correspond to the values of $p_W(w)$.

There are twenty-nine two-digit numbers, for example, in the interval 28 to 56, with each of those numbers having the same probability of 0.01. Any random two-digit number that falls anywhere in that interval will then be mapped into the value $w = 1$ (which will happen, in the long run, 29% of the time).

Tables of random digits are typically presented in blocks of twenty-five (see Figure 4.5.3).

(Continued on next page)

Table 4.5.1

w	$p_W(w)$	Random Number Range
0	0.28	00–27
1	0.29	28–56
2	0.20	57–76
3	0.11	77–87
4	0.06	88–93
5	0.03	94–96
6	0.01	97
7	0.01	98
8+	0.01	99

23107	15053	39098
65402	70659	84864
75528	18738	05624
85830	56869	15227
13300	08158	48968
75604	22878	02011
01188	17564	85393
71585	83287	97265
23495	57484	61680
51851	27186	16656

Figure 4.5.3

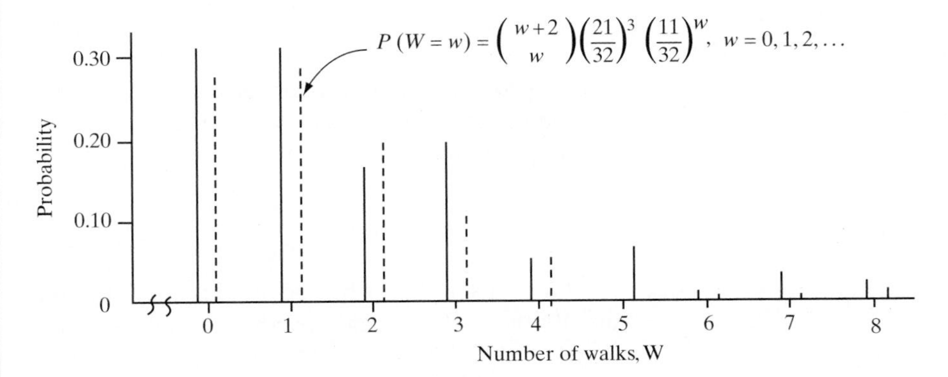

Figure 4.5.4

For the particular block circled, the first two columns,

$$22 \quad 17 \quad 83 \quad 57 \quad 27$$

would correspond to the negative binomial values

$$0 \quad 0 \quad 3 \quad 2 \quad 0$$

(Continued on next page)

(Case Study 4.5.1 continued)

Figure 4.5.4 shows the results of using a table of random digits and Table 4.5.1 to generate a sample of one hundred random observations from $p_W(w)$. The agreement is not perfect (as it shouldn't be), but certainly very good (as it should be).

About the Data Random number generators for continuous pdfs use random digits in ways that are much different from the strategy illustrated in Table 4.5.1 and much different from each other. The standard normal pdf and the exponential pdf are two cases in point.

Let $U_1, U_2, \ldots$ be a set of random observations drawn from the uniform pdf defined over the interval [0, 1]. Standard normal observations are generated by appealing to the central limit theorem. Since each U_i has $E(U_i) = 1/2$ and Var(U_i) = 1/12, it follows that

$$E\left(\sum_{i=1}^{k} U_i\right) = k/2$$

and

$$\text{Var}\left(\sum_{i=1}^{k} U_i\right) = k/12$$

and by the central limit theorem,

$$\frac{\sum_{i=1}^{k} U_i - k/2}{\sqrt{k/12}} \doteq Z$$

The approximation improves as k increases, but a particularly convenient (and sufficiently large) value is $k = 12$. The formula for generating a standard normal observation, then, reduces to

$$Z = \sum_{i=1}^{12} U_i - 6$$

Once a set of Z_i's has been calculated, random observations from any normal distribution can be easily produced. Suppose the objective is to generate a set of Y_i's that would be a random sample from a normal distribution having mean μ and variance σ^2. Since

$$\frac{Y - \mu}{\sigma} = Z$$

or, equivalently,

$$Y = \mu + \sigma Z$$

it follows that the random sample from $f_Y(y)$ would be

$$Y_i = \mu + \sigma Z_i, i = 1, 2, \ldots$$

By way of contrast, all that is needed to generate random observations from the exponential pdf, $f_Y(y) = \lambda e^{-\lambda y}$, $y \geq 0$, is a simple transformation. If U_i, $i = 1, 2, \ldots,$

is a set of uniform random variables as defined earlier, then $Y_i = -(1/\lambda) \ln U_i$, $i = 1, 2, \ldots$, will be the desired set of exponential observations. Why that should be so is an exercise in differentiating the cdf of Y. By definition,

$$F_Y(y) = P(Y \leq y) = P(\ln U > -\lambda y) = P(U > e^{-\lambda y})$$

$$= \int_{e^{-\lambda y}}^{1} 1\, du = 1 - e^{-\lambda y}$$

which implies that

$$f_Y(y) = F_Y'(y) = \lambda e^{-\lambda y}, \ y \geq 0$$

Questions

4.5.1. A door-to-door encyclopedia salesperson is required to document five in-home visits each day. Suppose that she has a 30% chance of being invited into any given home, with each address representing an independent trial. What is the probability that she requires fewer than eight houses to achieve her fifth success?

4.5.2. An underground military installation is fortified to the extent that it can withstand up to three direct hits from air-to-surface missiles and still function. Suppose an enemy aircraft is armed with missiles, each having a 30% chance of scoring a direct hit. What is the probability that the installation will be destroyed with the seventh missile fired?

4.5.3. Darryl's statistics homework last night was to flip a fair coin and record the toss, X, when heads appeared for the second time. The experiment was to be repeated a total of one hundred times. The following are the one hundred values for X that Darryl turned in this morning. Do you think that he actually did the assignment? Explain.

3	7	3	2	9	3	4	3	3	2
7	3	8	4	3	3	3	4	3	3
4	3	2	2	4	5	2	2	2	4
2	5	6	4	2	6	2	8	3	2
8	2	3	2	4	3	2	6	3	3
3	2	5	3	6	4	5	6	5	6
3	5	2	7	2	10	4	3	2	2
4	2	4	5	5	5	6	2	4	3
3	4	4	6	3	4	2	5	5	2
5	7	5	3	2	7	4	4	4	3

4.5.4. When a machine is improperly adjusted, it has probability 0.15 of producing a defective item. Each day, the machine is run until three defective items are produced. When this occurs, it is stopped and checked for adjustment. What is the probability that an improperly adjusted machine will produce five or more items before being stopped? What is the average number of items an improperly adjusted machine will produce before being stopped?

4.5.5. For a negative binomial random variable whose pdf is given by Equation 4.5.1, find $E(X)$ directly by evaluating $\sum_{k=r}^{\infty} k \binom{k-1}{r-1} p^r (1-p)^{k-r}$. (*Hint:* Reduce the sum to one involving negative binomial probabilities with parameters $r+1$ and p.)

4.5.6. Let the random variable X denote the number of trials in *excess of* r that are required to achieve the rth success in a series of independent trials, where p is the probability of success at any given trial. Show that

$$p_X(k) = \binom{k+r-1}{k} p^r (1-p)^k, \quad k = 0, 1, 2, \ldots$$

[*Note:* This particular formula for $p_X(k)$ is often used in place of Equation 4.5.1 as the definition of the pdf for a negative binomial random variable.]

4.5.7. Calculate the mean, variance, and moment-generating function for a negative binomial random variable X whose pdf is given by the expression

$$p_X(k) = \binom{k+r-1}{k} p^r (1-p)^k, \quad k = 0, 1, 2, \ldots$$

(See Question 4.5.6.)

4.5.8. Let X_1, X_2, and X_3 be three independent negative binomial random variables with pdfs

$$p_{X_i}(k) = \binom{k-1}{2} \left(\frac{4}{5}\right)^3 \left(\frac{1}{5}\right)^{k-3}, \quad k = 3, 4, 5, \ldots$$

for $i = 1, 2, 3$. Define $X = X_1 + X_2 + X_3$. Find $P(10 \leq X \leq 12)$. (*Hint:* Use the moment-generating functions of X_1, X_2, and X_3 to deduce the pdf of X.)

4.5.9. Differentiate the moment-generating function $M_X(t) = \left[\frac{pe^t}{1-(1-p)e^t}\right]^r$ to verify the formula given in Theorem 4.5.1 for $E(X)$.

4.5.10. Suppose that $X_1, X_2, \ldots, X_k$ are independent negative binomial random variables with parameters r_1 and p, r_2 and $p, \ldots,$ and r_k and p, respectively. Let $X = X_1 + X_2 + \cdots + X_k$. Find $M_X(t)$, $p_X(t)$, $E(X)$, and $\text{Var}(X)$.

4.6 The Gamma Distribution

Suppose a series of independent events are occurring at the constant rate of λ per unit time. If the random variable Y denotes the interval between consecutive occurrences, we know from Theorem 4.2.3 that $f_Y(y) = \lambda e^{-\lambda y}$, $y > 0$. Equivalently, Y can be interpreted as the "waiting time" for the first occurrence. This section generalizes the Poisson/exponential relationship and focuses on the interval, or waiting time, required for the rth event to occur (see Figure 4.6.1).

Figure 4.6.1

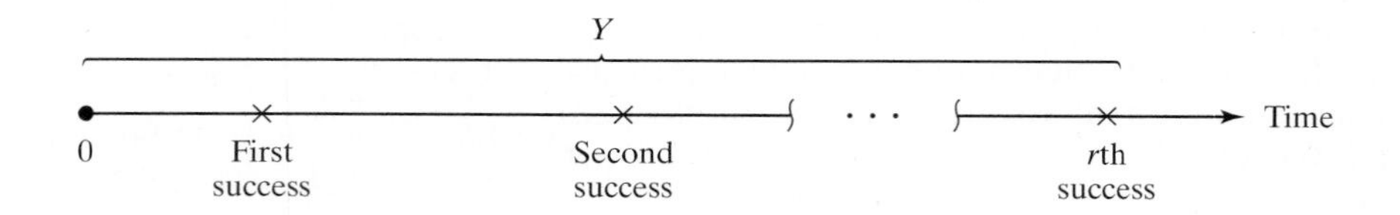

Theorem 4.6.1 *Suppose that Poisson events are occurring at the constant rate of λ per unit time. Let the random variable Y denote the waiting time for the rth event. Then Y has pdf $f_Y(y)$, where*

$$f_Y(y) = \frac{\lambda^r}{(r-1)!} y^{r-1} e^{-\lambda y}, \quad y > 0$$

Proof We will establish the formula for $f_Y(y)$ by deriving and differentiating its cdf, $F_Y(y)$. Let Y denote the waiting time to the rth occurrence. Then

$$\begin{aligned} F_Y(y) = P(Y \leq y) &= 1 - P(Y > y) \\ &= 1 - P(\text{Fewer than } r \text{ events occur in } [0, y]) \\ &= 1 - \sum_{k=0}^{r-1} e^{-\lambda y} \frac{(\lambda y)^k}{k!} \end{aligned}$$

since the number of events that occur in the interval $[0, y]$ is a Poisson random variable with parameter λy.

From Theorem 3.4.1,

$$\begin{aligned} f_Y(y) = F'_Y(y) &= \frac{d}{dy}\left[1 - \sum_{k=0}^{r-1} e^{-\lambda y} \frac{(\lambda y)^k}{k!}\right] \\ &= \sum_{k=0}^{r-1} \lambda e^{-\lambda y} \frac{(\lambda y)^k}{k!} - \sum_{k=1}^{r-1} \lambda e^{-\lambda y} \frac{(\lambda y)^{k-1}}{(k-1)!} \\ &= \sum_{k=0}^{r-1} \lambda e^{-\lambda y} \frac{(\lambda y)^k}{k!} - \sum_{k=0}^{r-2} \lambda e^{-\lambda y} \frac{(\lambda y)^k}{k!} \\ &= \frac{\lambda^r}{(r-1)!} y^{r-1} e^{-\lambda y}, \quad y > 0 \end{aligned}$$

□

Example 4.6.1 Engineers designing the next generation of space shuttles plan to include two fuel pumps—one active, the other in reserve. If the primary pump malfunctions, the second will automatically be brought on line.

Suppose a typical mission is expected to require that fuel be pumped for at most fifty hours. According to the manufacturer's specifications, pumps are expected to fail once every one hundred hours (so $\lambda = 0.01$). What are the chances that such a fuel pump system would not remain functioning for the full fifty hours?

Let the random variable Y denote the time that will elapse before the second pump breaks down. According to Theorem 4.6.1, the pdf for Y has parameters $r = 2$ and $\lambda = 0.01$, and we can write

$$f_Y(y) = \frac{(0.01)^2}{1!} ye^{-0.01y}, \quad y > 0$$

Therefore,

$$\begin{aligned} P(\text{System fails to last for fifty hours}) &= \int_0^{50} 0.0001 ye^{-0.01y}\, dy \\ &= \int_0^{0.50} ue^{-u}\, du \end{aligned}$$

where $u = 0.01y$. The probability, then, that the primary pump and its backup would not remain operable for the targeted fifty hours is *0.09*:

$$\begin{aligned} \int_0^{0.50} ue^{-u}\, du &= (-u-1)e^{-u}\Big|_{\mu=0}^{0.50} \\ &= 0.09 \end{aligned}$$

■

Generalizing the Waiting Time Distribution

By virtue of Theorem 4.6.1, $\int_0^\infty y^{r-1}e^{-\lambda y}\, dy$ converges for any integer $r > 0$. But the convergence also holds for any *real number* $r > 0$, because for any such r there will be an integer $t > r$ and $\int_0^\infty y^{r-1}e^{-\lambda y}\, dy \le \int_0^\infty y^{t-1}e^{-\lambda y}\, dy < \infty$. The finiteness of $\int_0^\infty y^{r-1}e^{-\lambda y}\, dy$ justifies the consideration of a related definite integral, one that was first studied by Euler, but named by Legendre.

Definition 4.6.1. For any real number $r > 0$, the *gamma function of r* is denoted $\Gamma(r)$, where

$$\Gamma(r) = \int_0^\infty y^{r-1}e^{-y}\, dy$$

Theorem 4.6.2 *Let $\Gamma(r) = \int_0^\infty y^{r-1}e^{-y}\, dy$ for any real number $r > 0$. Then*

1. $\Gamma(1) = 1$
2. $\Gamma(r) = (r-1)\Gamma(r-1)$
3. *If r is an integer, then $\Gamma(r) = (r-1)!$*

Proof

1. $\Gamma(1) = \int_0^\infty y^{1-1} e^{-y}\, dy = \int_0^\infty e^{-y}\, dy = 1$
2. Integrate the gamma function by parts. Let $u = y^{r-1}$ and $dv = e^{-y}$. Then

$$\int_0^\infty y^{r-1} e^{-y}\, dy = -y^{r-1} e^{-y}\Big|_0^\infty + \int_0^\infty (r-1) y^{r-2} e^{-y}\, dy$$
$$= (r-1)\int_0^\infty y^{r-2} e^{-y}\, dy = (r-1)\Gamma(r-1)$$

3. Use part (2) as the basis for an induction argument. The details will be left as an exercise. □

Definition 4.6.2. Given real numbers $r > 0$ and $\lambda > 0$, the random variable Y is said to have the *gamma pdf* with parameters r and λ if

$$f_Y(y) = \frac{\lambda^r}{\Gamma(r)} y^{r-1} e^{-\lambda y}, \quad y > 0$$

Comment To justify Definition 4.6.2 requires a proof that $f_Y(y)$ integrates to 1. Let $u = \lambda y$. Then

$$\int_0^\infty \frac{\lambda^r}{\Gamma(r)} y^{r-1} e^{-\lambda y} dy = \frac{\lambda^r}{\Gamma(r)} \int_0^\infty \left(\frac{u}{\lambda}\right)^{r-1} e^{-u} \frac{1}{\lambda} du$$
$$= \frac{1}{\Gamma(r)} \int_0^\infty u^{r-1} e^{-u} du = \frac{1}{\Gamma(r)} \Gamma(r) = 1$$

Theorem 4.6.3 *Suppose that Y has a gamma pdf with parameters r and λ. Then*

1. $E(Y) = r/\lambda$
2. $Var(Y) = r/\lambda^2$

Proof

1. $$E(Y) = \int_0^\infty y \frac{\lambda^r}{\Gamma(r)} y^{r-1} e^{-\lambda y}\, dy = \frac{\lambda^r}{\Gamma(r)} \int_0^\infty y^r e^{-\lambda y}\, dy$$
$$= \frac{\lambda^r}{\Gamma(r)} \frac{\Gamma(r+1)}{\lambda^{r+1}} \int_0^\infty \frac{\lambda^{r+1}}{\Gamma(r+1)} y^r e^{-\lambda y}\, dy$$
$$= \frac{\lambda^r}{\Gamma(r)} \frac{r\Gamma(r)}{\lambda^{r+1}} (1) = r/\lambda$$
2. A calculation similar to the integration carried out in part (1) shows that $E(Y^2) = r(r+1)/\lambda^2$. Then

$$\text{Var}(Y) = E(Y^2) - [E(Y)]^2$$
$$= r(r+1)/\lambda^2 - (r/\lambda)^2$$
$$= r/\lambda^2$$

□

Sums of Gamma Random Variables

We have already seen that certain random variables satisfy an additive property that "reproduces" the pdf—the sum of two independent binomial random variables with the same p, for example, is binomial (recall Example 3.8.2). Similarly, the sum of two independent Poissons is Poisson and the sum of two independent normals is normal. That said, most random variables are *not* additive. The sum of two independent uniforms is not uniform; the sum of two independent exponentials is not exponential; and so on. Gamma random variables belong to the short list making up the first category.

Theorem 4.6.4 *Suppose U has the gamma pdf with parameters r and λ, V has the gamma pdf with parameters s and λ, and U and V are independent. Then $U+V$ has a gamma pdf with parameters $r+s$ and λ.*

Proof The pdf of the sum is the convolution integral

$$\begin{aligned} f_{U+V}(t) &= \int_{-\infty}^{\infty} f_U(u) f_V(t-u)\,du \\ &= \int_0^t \frac{\lambda^r}{\Gamma(r)} u^{r-1} e^{-\lambda u} \frac{\lambda^s}{\Gamma(s)} (t-u)^{s-1} e^{-\lambda(t-u)}\,du \\ &= e^{-\lambda t} \frac{\lambda^{r+s}}{\Gamma(r)\Gamma(s)} \int_0^t u^{r-1}(t-u)^{s-1}\,du \end{aligned}$$

Make the substitution $v = u/t$. Then the integral becomes

$$t^{r-1} t^{s-1} t \int_0^t v^{r-1}(1-v)^{s-1}\,dv = t^{r+s-1} \int_0^1 v^{r-1}(1-v)^{s-1}\,dv$$

and

$$f_{U+V}(t) = \lambda^{r+s} t^{r+s-1} e^{-\lambda t} \left[\frac{1}{\Gamma(r)\Gamma(s)} \int_0^1 v^{r-1}(1-v)^{s-1}\,dv \right] \qquad (4.6.1)$$

The numerical value of the constant in brackets in Equation 4.6.1 is not immediately obvious, but the factors in front of the brackets correspond to the functional part of a gamma pdf with parameters $r+s$ and λ. It follows, then, that $f_{U+V}(t)$ must be that particular gamma pdf. It also follows that the constant in brackets must equal $1/\Gamma(r+s)$ (to comply with Definition 4.6.2), so, as a "bonus" identity, Equation 4.6.1 implies that

$$\int_0^1 v^{r-1}(1-v)^{s-1}\,dv = \frac{\Gamma(r)\Gamma(s)}{\Gamma(r+s)}$$

□

Theorem 4.6.5 *If Y has a gamma pdf with parameters r and λ, then $M_Y(t) = (1 - t/\lambda)^{-r}$.*

Proof

$$\begin{aligned} M_Y(t) = E(e^{tY}) &= \int_0^{\infty} e^{ty} \frac{\lambda^r}{\Gamma(r)} y^{r-1} e^{-\lambda y}\,dy = \frac{\lambda^r}{\Gamma(r)} \int_0^{\infty} y^{r-1} e^{-(\lambda - t)y}\,dy \\ &= \frac{\lambda^r}{\Gamma(r)} \frac{\Gamma(r)}{(\lambda-t)^r} \int_0^{\infty} \frac{(\lambda-t)^r}{\Gamma(r)} y^r e^{-(\lambda-t)y}\,dy \\ &= \frac{\lambda^r}{(\lambda-t)^r}(1) = (1 - t/\lambda)^{-r} \end{aligned}$$

□

Questions

4.6.1. An Arctic weather station has three electronic wind gauges. Only one is used at any given time. The lifetime of each gauge is exponentially distributed with a mean of one thousand hours. What is the pdf of Y, the random variable measuring the time until the last gauge wears out?

4.6.2. A service contact on a new university computer system provides twenty-four free repair calls from a technician. Suppose the technician is required, on the average, three times a month. What is the average time it will take for the service contract to be fulfilled?

4.6.3. Suppose a set of measurements $Y_1, Y_2, \ldots, Y_{100}$ is taken from a gamma pdf for which $E(Y) = 1.5$ and $\text{Var}(Y) = 0.75$. How many Y_i's would you expect to find in the interval [1.0, 2.5]?

4.6.4. Demonstrate that λ plays the role of a scale parameter by showing that if Y is gamma with parameters r and λ, then λY is gamma with parameters r and 1.

4.6.5. Show that a gamma pdf has the unique mode $\frac{r-1}{\lambda}$; that is, show that the function $f_Y(y) = \frac{\lambda^r}{\Gamma(r)} y^{r-1} e^{-\lambda y}$ takes its maximum value at $y_{\text{mode}} = \frac{r-1}{\lambda}$ and at no other point.

4.6.6. Prove that $\Gamma\left(\frac{1}{2}\right) = \sqrt{\pi}$. [*Hint:* Consider $E(Z^2)$, where Z is a standard normal random variable.]

4.6.7. Show that $\Gamma\left(\frac{7}{2}\right) = \frac{15}{8}\sqrt{\pi}$.

4.6.8. If the random variable Y has the gamma pdf with integer parameter r and arbitrary $\lambda > 0$, show that

$$E(Y^m) = \frac{(m+r-1)!}{(r-1)!\lambda^m}$$

[*Hint:* Use the fact that $\int_0^\infty y^{r-1} e^{-y}\, dy = (r-1)!$ when r is a positive integer.]

4.6.9. Differentiate the gamma moment-generating function to verify the formulas for $E(Y)$ and $\text{Var}(Y)$ given in Theorem 4.6.3.

4.6.10. Differentiate the gamma moment-generating function to show that the formula for $E(Y^m)$ given in Question 4.6.8 holds for arbitrary $r > 0$.

4.7 Taking a Second Look at Statistics (Monte Carlo Simulations)

Calculating probabilities associated with (1) single random variables and (2) functions of sets of random variables has been the overarching theme of Chapters 3 and 4. Facilitating those computations has been a variety of transformations, summation properties, and mathematical relationships linking one pdf with another. Collectively, these results are enormously effective. Sometimes, though, the intrinsic complexity of a random variable overwhelms our ability to model its probabilistic behavior in any formal or precise way. An alternative in those situations is to use a computer to draw random samples from one or more distributions that model portions of the random variable's behavior. If a large enough number of such samples is generated, a histogram (or density-scaled histogram) can be constructed that will accurately reflect the random variable's true (but unknown) distribution. Sampling "experiments" of this sort are known as *Monte Carlo studies*.

Real-life situations where a Monte Carlo analysis could be helpful are not difficult to imagine. Suppose, for instance, you just bought a state-of-the-art, high-definition, plasma screen television. In addition to the pricey initial cost, an optional warranty is available that covers all repairs made during the first two years. According to an independent laboratory's reliability study, this particular television is likely to require 0.75 service call per year, on the average. Moreover, the costs of service calls are expected to be normally distributed with a mean (μ) of \$100 and a standard deviation (σ) of \$20. If the warranty sells for \$200, should you buy it?

Like any insurance policy, a warranty may or may not be a good investment, depending on what events unfold, and when. Here the relevant random variable is W, the total amount spent on repair calls during the first two years. For any particular customer, the value of W will depend on (1) the number of repairs needed in the first two years and (2) the cost of each repair. Although we have reliability and cost assumptions that address (1) and (2), the two-year limit on the warranty introduces a complexity that goes beyond what we have learned in Chapters 3 and 4. What remains is the option of using random samples to simulate the repair costs that might accrue during those first two years.

Note, first, that it would not be unreasonable to assume that the service calls are Poisson events (occurring at the rate of 0.75 per year). If that were the case, Theorem 4.2.3 implies that the interval, Y, between successive repair calls would have an exponential distribution with pdf

$$f_Y(y) = 0.75e^{-0.75y}, \quad y > 0$$

(see Figure 4.7.1). Moreover, if the random variable C denotes the cost associated with a particular maintenance call, then,

$$f_C(c) = \frac{1}{\sqrt{2\pi}(20)} e^{-\left(\frac{1}{2}\right)[(c-100)/20]^2}, \quad -\infty < c < \infty$$

(see Figure 4.7.2).

Figure 4.7.1

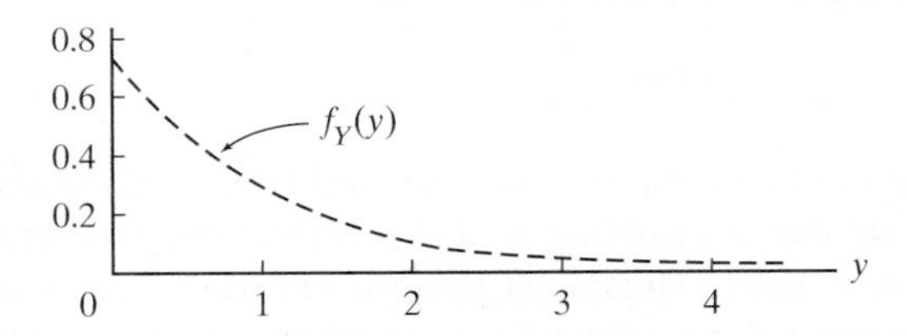

Figure 4.7.2

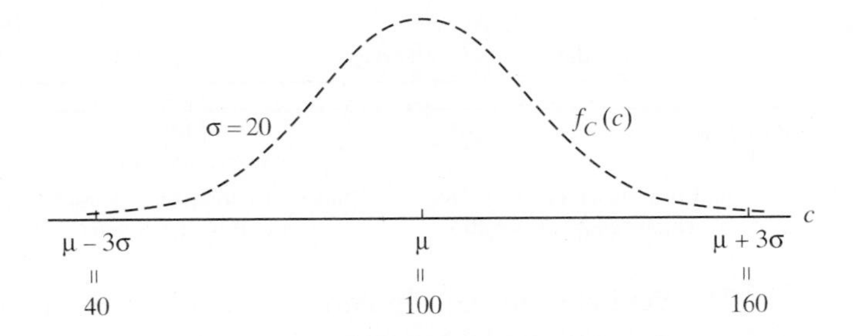

Now, with the pdfs for Y and C fully specified, we can use the computer to generate representative repair cost scenarios. We begin by generating a random sample (of size 1) from the pdf, $f_Y(y) = 0.75e^{-0.75y}$. Either of two equivalent Minitab procedures can be followed:

Session Window Method		**Menu-Driven Method**
Click on EDITOR, then on ENABLE COMMANDS (this activates the Session Window). Then type MTB > random 1 c1; SUBC > exponential 1.33. MTB > print c1	OR	Click on CALC, then on RANDOM DATA, then on EXPONENTIAL. Type 1 in "Number of rows" box; type 1.33 in "Scale" box; type c1 in "Store" box. Click on OK. The generated exponential deviate appears in the upper left hand corner of the WORKSHEET.

```
Data Display

c1
      1.15988
```

(*Note:* For both methods, Minitab uses $1/\lambda$ as the exponential parameter. Here, $1/\lambda = 1/0.75 = 1.33$.)

As shown in Figure 4.7.3, the number generated was 1.15988 yrs (corresponding to a first repair call occurring *423* days ($= 1.15988 \times 365$) after the purchase of the TV).

Figure 4.7.3

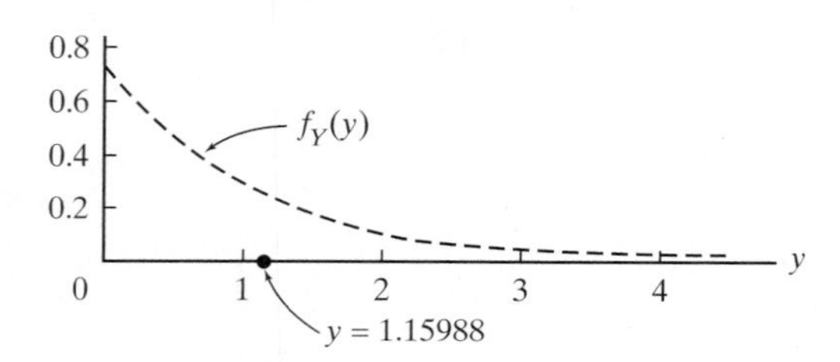

Applying the same syntax a second time yielded the random sample 0.284931 year ($=$ *104* days); applying it still a third time produced the observation 1.46394 years ($=$ *534* days). These last two observations taken on $f_Y(y)$ correspond to the second repair call occurring 104 days after the first, and the third occurring 534 days after the second (see Figure 4.7.4). Since the warranty does not extend past the first 730 days, the third repair would not be covered.

Figure 4.7.4

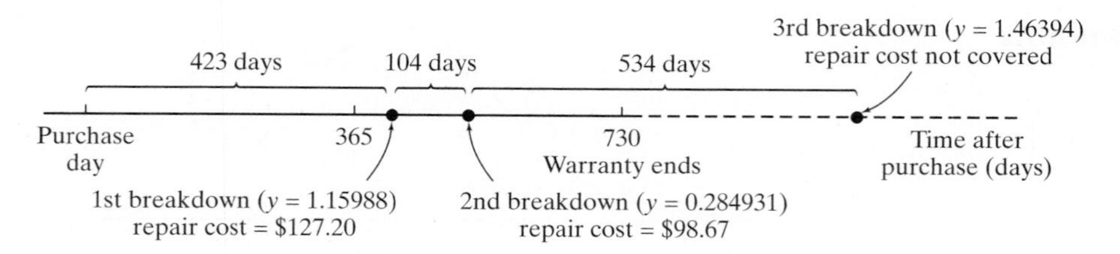

The next step in the simulation would be to generate two observations from $f_C(c)$ that would model the costs of the two repairs that occurred during the warranty period. The session-window syntax for simulating each repair cost would be the statements

```
MTB  > random 1 c1;
SUBC > normal 100 20.
MTB  > print c1
```

```
MTB  > random 1 c1;
SUBC > exponential 1.33.
MTB  > print c1
c1
    1.15988
```

```
MTB  > random 1 c1;
SUBC > normal 100 20.
MTB  > print c1
c1
    127.199
```

```
MTB  > random 1 c1;
SUBC > exponential 1.33.
MTB  > print c1
c1
    0.284931
```

```
MTB  > random 1 c1;
SUBC > normal 100 20.
MTB  > print c1
c1
    98.6673
```

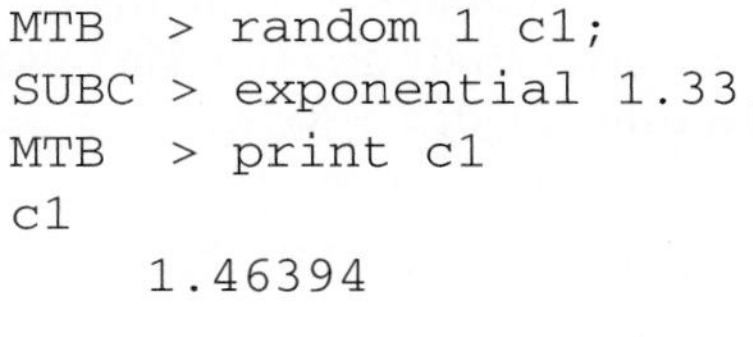

```
MTB  > random 1 c1;
SUBC > exponential 1.33.
MTB  > print c1
c1
    1.46394
```

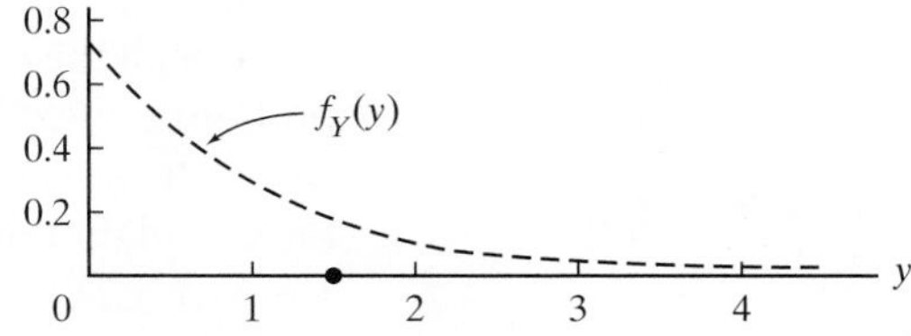

Figure 4.7.5

Running those commands twice produced c-values of 127.199 and 98.6673 (see Figure 4.7.5), corresponding to repair bills of \$127.20 and \$98.67, meaning that a total of \$225.87 (= \$127.20 + \$98.67) would have been spent on maintenance during the first two years. In that case, the \$200 warranty would have been a good investment.

The final step in the Monte Carlo analysis is to repeat many times the sampling process that led to Figure 4.7.5—that is, to generate a series of y_i's whose sum (in days) is less than or equal to 730, and for each y_i in that sample, to generate a corresponding cost, c_i. The sum of those c_i's becomes a simulated value of the maintenance-cost random variable, W.

Figure 4.7.6

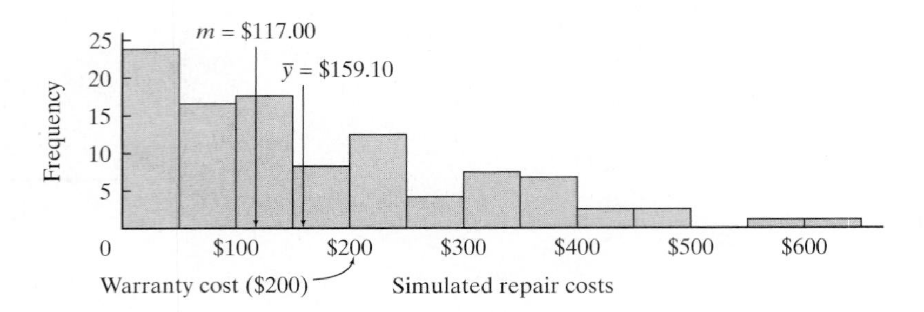

The histogram in Figure 4.7.6 shows the distribution of repair costs incurred in *one hundred simulated two-year periods*, one being the sequence of events detailed in Figure 4.7.5. There is much that it tells us. First of all (and not surprisingly), the warranty costs more than either the median repair bill (= \$117.00) or the mean repair bill (= \$159.10).

The customer, in other words, will tend to lose money on the optional protection, and the company will tend to make money. On the other hand, a full 33% of the simulated two-year breakdown scenarios led to repair bills in excess of \$200, including 6% that were more than twice the cost of the warranty. At the other extreme, 24% of the samples produced no maintenance problems whatsoever; for those customers, the \$200 spent up front is totally wasted!

So, should you buy the warranty? Yes, if you feel the need to have a financial cushion to offset the (small) probability of experiencing exceptionally bad luck; no, if you can afford to absorb an occasional big loss.

Appendix 4.A.1 Minitab Applications

Examples at the end of Chapter 3 and earlier in this chapter illustrated the use of Minitab's PDF, CDF, and INVCDF commands on the binomial, exponential, and normal distributions. Altogether, those same commands can be applied to more than twenty of the probability distributions most frequently encountered, including the Poisson, geometric, negative binomial, and gamma pdfs featured in Chapter 4.

Recall the leukemia cluster study described in Case Study 4.2.1. The data's interpretation hinged on the value of $P(X \geq 8)$, where X was a Poisson random variable with pdf, $p_X(k) = e^{-1.75}\frac{(1.75)^k}{k!}$, $k = 0, 1, 2, \ldots$. The printout in Figure 4.A.1.1 shows the calculation of $P(X \geq 8)$ using the CDF command and the fact that $P(X \geq 8) = 1 - P(X \leq 7)$.

Figure 4.A.1.1

```
MTB > cdf 7;
SUBC > poisson 1.75.

Cumulative Distribution Function
Poisson with mean = 1.75
x     P(X <= x)
7        0.999532

MTB > let k1 = 1 - 0.999532
MTB > print k1

Data Display

k1          0.000468000
```

Areas under normal curves between points a and b are calculated by subtracting $F_Y(a)$ from $F_Y(b)$, just as we did in Section 4.3 (recall the Comment after Definition 4.3.1). There is no need, however, to reexpress the probability as an area under the *standard* normal curve. Figure 4.A.1.2 shows the Minitab calculation for the probability that the random variable Y lies between 48 and 51, where Y is normally distributed with $\mu = 50$ and $\sigma = 4$. According to the computer,

$$\begin{aligned} P(48 < Y < 51) &= F_Y(51) - F_Y(48) \\ &= 0.598706 - 0.308538 \\ &= 0.290168 \end{aligned}$$

Figure 4.A.1.2

```
MTB > cdf 51;
SUBC> normal 50 4.

Cumulative Distribution Function
Normal with mean = 50 and standard deviation = 4
  x       P( X <= x)
  51         0.598706
MTB > cdf 48;
SUBC> normal 50 4.

Cumulative Distribution Function
Normal with mean = 50.0000 and standard deviation = 4.00000
  x       P( X <= x)
  48         0.308538
MTB > let k1 = 0.598706-0.308538
MTB > print k1
Data Display
k1          0.290168
```

On several occasions in Chapter 4 we made use of Minitab's RANDOM command, a subroutine that generates samples from a specific pdf. Simulations of that sort can be very helpful in illustrating a variety of statistical concepts. Shown in Figure 4.A.1.3, for example, is the syntax for generating a random sample of size 50 from a binomial pdf having $n = 60$ and $p = 0.40$. And calculated for each of those fifty observations is its Z ratio, given by

$$Z\text{-ratio} = \frac{X - E(X)}{\sqrt{\text{Var}(X)}} = \frac{X - 60(0.40)}{\sqrt{60(0.40)(0.60)}} = \frac{X - 24}{\sqrt{14.4}}$$

[By the DeMoivre-Laplace theorem, of course, the distribution of those ratios should have a shape much like the standard normal pdf, $f_Z(z)$.]

Figure 4.A.1.3

```
MTB > random 50 c1;
SUBC> binomial 60 0.40.
MRB > print c1
Data Display
C1
     27  29  23  22  21  21  22  26  26  20  26  25  27
     32  22  27  22  20  19  19  21  23  28  23  27  29
     13  24  22  25  25  20  25  26  15  24  17  28  21
     16  24  22  25  25  21  23  23  20  25  30
MTB > let c2 = (c1 - 24)/sqrt(14.4)
MTB > name c2 'Z-ratio'
MTB > print c2
Data Display
Z-ratio
  0.79057   1.31762 -0.26352 -0.52705 -0.79057 -0.79057 -0.52705
  0.52705   0.52705 -1.05409  0.52705  0.26352  0.79057  2.10819
 -0.52705   0.79057 -0.52705 -1.05409 -1.31762 -1.31762 -0.79057
 -0.26352   1.05409 -0.26352  0.79057  1.31762 -2.89875  0.00000
 -0.52705   0.26352  0.26352 -1.05409  0.26352  0.52705 -2.37171
  0.00000  -1.84466  1.05409 -0.79057 -2.10819  0.00000 -0.52705
  0.26352   0.26352 -0.79057 -0.26352 -0.26352 -1.05409  0.26352
  1.58114
```

Appendix 4.A.2 A Proof of the Central Limit Theorem

Proving Theorem 4.3.2 in its full generality is beyond the level of this text. However, we can establish a slightly weaker version of the result by assuming that the moment-generating function of each W_i exists.

Lemma *Let $W_1, W_2, \ldots$ be a set of random variables such that $\lim_{n\to\infty} M_{W_n}(t) = M_W(t)$ for all t in some interval about 0. Then $\lim_{n\to\infty} F_{W_n}(w) = F_W(w)$ for all $-\infty < w < \infty$.*

To prove the central limit theorem using moment-generating functions requires showing that

$$\lim_{n\to\infty} M_{(W_1+\cdots+W_n-n\mu)/(\sqrt{n}\sigma)}(t) = M_Z(t) = e^{t^2/2}$$

For notational simplicity, let

$$\frac{W_1+\cdots+W_n-n\mu}{\sqrt{n}\sigma} = \frac{S_1+\cdots+S_n}{\sqrt{n}}$$

where $S_i = (W_i - \mu)/\sigma$. Notice that $E(S_i) = 0$ and $\text{Var}(S_i) = 1$. Moreover, from Theorem 3.12.3,

$$M_{(S_1+\cdots+S_n)/\sqrt{n}}(t) = \left[M\left(\frac{t}{\sqrt{n}}\right)\right]^n$$

where $M(t)$ denotes the moment-generating function common to each of the S_i's.

By virtue of the way the S_i's are defined, $M(0) = 1$, $M^{(1)}(0) = E(S_i) = 0$, and $M^{(2)}(0) = \text{Var}(S_i) = 1$. Applying Taylor's theorem, then, to $M(t)$, we can write

$$M(t) = 1 + M^{(1)}(0)t + \frac{1}{2}M^{(2)}(r)t^2 = 1 + \frac{1}{2}t^2M^{(2)}(r)$$

for some number r, $|r| < |t|$. Thus

$$\begin{aligned}\lim_{n\to\infty}\left[M\left(\frac{t}{\sqrt{n}}\right)\right]^n &= \lim_{n\to\infty}\left[1+\frac{t^2}{2n}M^{(2)}(s)\right]^n, \quad |s| < \frac{|t|}{\sqrt{n}} \\ &= \exp \lim_{n\to\infty} n \ln\left[1+\frac{t^2}{2n}M^{(2)}(s)\right] \\ &= \exp \lim_{n\to\infty} \frac{t^2}{2}\cdot M^{(2)}(s)\cdot \frac{\ln\left[1+\frac{t^2}{2n}M^{(2)}(s)\right] - \ln(1)}{\frac{t^2}{2n}M^{(2)}(s)}\end{aligned}$$

The existence of $M(t)$ implies the existence of all its derivatives. In particular, $M^{(3)}(t)$ exists, so $M^{(2)}(t)$ is continuous. Therefore, $\lim_{t\to 0} M^{(2)}(t) = M^{(2)}(0) = 1$. Since $|s| < |t|/\sqrt{n}$, $s \to 0$ as $n \to \infty$, so

$$\lim_{n\to\infty} M^{(2)}(s) = M^{(2)}(0) = 1$$

Also, as $n \to \infty$, the quantity $(t^2/2n)M^{(2)}(s) \to 0 \cdot 1 = 0$, so it plays the role of "Δx" in the definition of the derivative. Hence we obtain

$$\lim_{n\to\infty}\left[M\left(\frac{t}{\sqrt{n}}\right)\right]^n = \exp\frac{t^2}{2}\cdot 1 \cdot \ln^{(1)}(1) = e^{(1/2)t^2}$$

Since this last expression is the moment-generating function for a standard normal random variable, the theorem is proved.

Chapter 5

ESTIMATION

A towering figure in the development of both applied and mathematical statistics, Fisher had formal training in mathematics and theoretical physics, graduating from Cambridge in 1912. After a brief career as a teacher, he accepted a post in 1919 as statistician at the Rothamsted Experimental Station. There, the day-to-day problems he encountered in collecting and interpreting agricultural data led directly to much of his most important work in the theory of estimation and experimental design. Fisher was also a prominent geneticist and devoted considerable time to the development of a quantitative argument that would support Darwin's theory of natural selection. He returned to academia in 1933, succeeding Karl Pearson as the Galton Professor of Eugenics at the University of London. Fisher was knighted in 1952.

—Ronald Aylmer Fisher (1890–1962)

5.1 Introduction

The ability of probability functions to describe, or *model*, experimental data was demonstrated in numerous examples in Chapter 4. In Section 4.2, for example, the Poisson distribution was shown to predict very well the number of alpha emissions from a radioactive source as well as the number of wars starting in a given year. In Section 4.3 another probability model, the normal curve, was applied to phenomena as diverse as breath analyzer readings and IQ scores. Other models illustrated in Chapter 4 included the exponential, negative binomial, and gamma distributions.

All of these probability functions, of course, are actually *families* of models in the sense that each includes one or more *parameters*. The Poisson model, for instance, is indexed by the occurrence rate, λ. Changing λ changes the probabilities associated with $p_X(k)$ [see Figure 5.1.1, which compares $p_X(k) = e^{-\lambda}\lambda^k/k!, k = 0, 1, 2, \ldots,$ for $\lambda = 1$ and $\lambda = 4$]. Similarly, the binomial model is defined in terms of the success probability p; the normal distribution, by the two parameters μ and σ.

Before any of these models can be applied, values need to be assigned to their parameters. Typically, this is done by taking a random sample (of n observations) and using those measurements to *estimate* the unknown parameter(s).

Figure 5.1.1

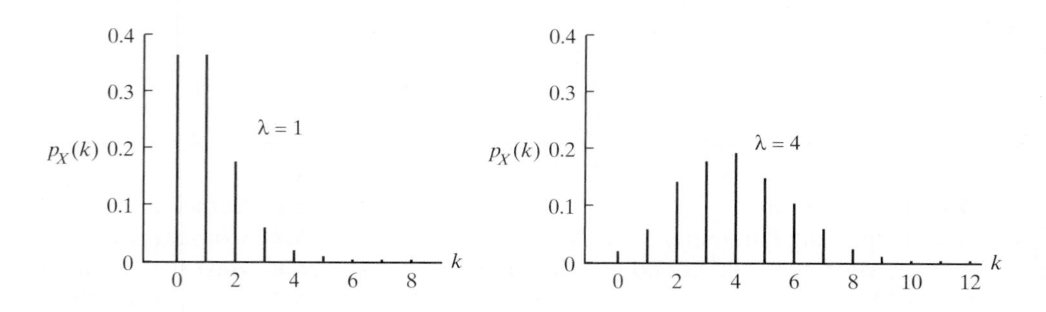

Example 5.1.1

Imagine being handed a coin whose probability, p, of coming up heads is unknown. Your assignment is to toss the coin three times and use the resulting sequence of H's and T's to suggest a value for p. Suppose the sequence of three tosses turns out to be HHT. Based on those outcomes, what can be reasonably inferred about p?

Start by defining the random variable X to be the number of heads on a given toss. Then

$$X = \begin{cases} 1 & \text{if a toss comes up heads} \\ 0 & \text{if a toss comes up tails} \end{cases}$$

and the theoretical probability model for X is the function

$$p_X(k) = p^k(1-p)^{1-k} = \begin{cases} p & \text{for } k = 1 \\ 1-p & \text{for } k = 0 \end{cases}$$

Expressed in terms of X, the sequence HHT corresponds to a sample of size $n = 3$, where $X_1 = 1$, $X_2 = 1$, and $X_3 = 0$.

Since the X_i's are independent random variables, the probability associated with the sample is $p^2(1-p)$:

$$P(X_1 = 1 \cap X_2 = 1 \cap X_3 = 0) = P(X_1 = 1) \cdot P(X_2 = 1) \cdot P(X_3 = 0) = p^2(1-p)$$

Knowing that our objective is to identify a plausible value (i.e., an "estimate") for p, it could be argued that a reasonable choice for that parameter would be the value that maximizes the probability of the sample. Figure 5.1.2 shows $P(X_1 = 1, X_2 = 1, X_3 = 0)$ *as a function of* p. By inspection, we see that the value that maximizes the probability of HHT is $p = \frac{2}{3}$.

More generally, suppose we toss the coin n times and record a set of outcomes $X_1 = k_1$, $X_2 = k_2, \ldots,$ and $X_n = k_n$. Then

$$\begin{aligned} P(X_1 = k_1, X_2 = k_2, \ldots, X_n = k_n) &= p^{k_1}(1-p)^{1-k_1} \ldots p^{k_n}(1-p)^{1-k_n} \\ &= p^{\sum_{i=1}^{n} k_i}(1-p)^{n-\sum_{i=1}^{n} k_i} \end{aligned}$$

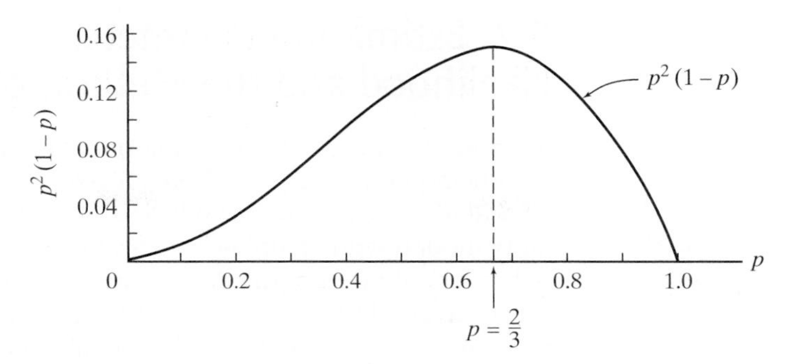

Figure 5.1.2

The value of p that maximizes $P(X_1 = k_1, \ldots, X_n = k_n)$ is, of course, the value for which the derivative of $p^{\sum_{i=1}^{n} k_i}(1-p)^{n-\sum_{i=1}^{n} k_i}$ *with respect to* p is 0. But

$$d/dp\left[p^{\sum_{i=1}^{n} k_i}(1-p)^{n-\sum_{i=1}^{n} k_i}\right] = \sum_{i=1}^{n} k_i\left[p^{\sum_{i=1}^{n} k_i - 1}(1-p)^{n-\sum_{i=1}^{n} k_i}\right] + \left[\sum_{i=1}^{n} k_i - n\right] p^{\sum_{i=1}^{n} k_i}(1-p)^{n-\sum_{i=1}^{n} k_i - 1} \tag{5.1.1}$$

If the derivative is set equal to zero, Equation 5.1.1 reduces to

$$\sum_{i=1}^{n} k_i(1-p) + \left(\sum_{i=1}^{n} k_i - n\right)p = 0$$

Solving for p identifies

$$\left(\frac{1}{n}\right)\sum_{i=1}^{n} k_i$$

as the value of the parameter that is most consistent with the n observations $k_1, k_2, \ldots, k_n$.

Comment Any function of a random sample whose objective is to approximate a parameter is called a *statistic*, or an *estimator*. If θ is the parameter being approximated, its estimator will be denoted $\hat{\theta}$. When an estimator is evaluated (by substituting the actual measurements recorded), the resulting number is called an *estimate*. In Example 5.1.1, the function $\left(\frac{1}{n}\right)\sum_{i=1}^{n} X_i$ is an estimator for p; the value $\frac{2}{3}$ that is calculated when the $n = 3$ observations are $X_1 = 1$, $X_2 = 1$, and $X_3 = 0$ is an estimate of p. More specifically, $\left(\frac{1}{n}\right)\sum_{i=1}^{n} X_i$ is a *maximum likelihood estimator* (for p) and $\frac{2}{3}\left[= \left(\frac{1}{n}\right)\sum_{i=1}^{n} k_i = \left(\frac{1}{3}\right)(2)\right]$ is a *maximum likelihood estimate* (for p). ■

In this chapter, we look at some of the practical, as well as the mathematical, issues involved in the problem of estimating parameters. How is the functional form of an estimator determined? What statistical properties does a given estimator have? What properties would we *like* an estimator to have? As we answer these questions, our focus will begin to shift away from the study of probability and toward the study of statistics.

5.2 Estimating Parameters: The Method of Maximum Likelihood and the Method of Moments

Suppose $Y_1, Y_2, \ldots, Y_n$ is a random sample from a continuous pdf $f_Y(y)$ whose unknown parameter is θ. [*Note*: To emphasize that our focus is on the parameter, we will identify continuous pdfs in this chapter as $f_Y(y;\theta)$; similarly, discrete probability models with an unknown parameter θ will be denoted $p_X(k;\theta)$]. The question is, how should we use the data to approximate θ?

In Example 5.1.1, we saw that the parameter p in the discrete probability model $p_X(k;p) = p^k(1-p)^{1-k}$, $k = 0, 1$ could reasonably be estimated by the function $\left(\frac{1}{n}\right)\sum_{i=1}^{n} k_i$, based on the random sample $X_1 = k_1, X_2 = k_2, \ldots, X_n = k_n$. How would the form of the estimate change if the data came from, say, an exponential distribution? Or a Poisson distribution?

In this section we introduce two techniques for finding estimates—the method of maximum likelihood and the method of moments. Others are available, but these are the two that are the most widely used. Often, but not always, they give the same answer.

The Method of Maximum Likelihood

The basic idea behind maximum likelihood estimation is the rationale that was appealed to in Example 5.1.1. That is, it seems plausible to choose as the estimate for θ the value of the parameter that maximizes the "likelihood" of the sample. The latter is measured by a *likelihood function*, which is simply the product of the underlying pdf evaluated for each of the data points. In Example 5.1.1, the likelihood function for the sample HHT (i.e., for $X_1 = 1$, $X_2 = 1$, and $X_3 = 0$) is the product $p^2(1-p)$.

Definition 5.2.1. Let $k_1, k_2, \ldots, k_n$ be a random sample of size n from the discrete pdf $p_X(k;\theta)$, where θ is an unknown parameter. The *likelihood function*, $L(\theta)$, is the product of the pdf evaluated at the n k_i's. That is,

$$L(\theta) = \prod_{i=1}^{n} p_X(k_i;\theta)$$

If $y_1, y_2, \ldots, y_n$ is a random sample of size n from a continuous pdf, $f_Y(y;\theta)$, where θ is an unknown parameter, the likelihood function is written

$$L(\theta) = \prod_{i=1}^{n} f_Y(y_i;\theta)$$

Comment Joint pdfs and likelihood functions look the same, but the two are interpreted differently. A joint pdf defined for a set of n random variables is a multivariate function of the values of those n random variables, either $k_1, k_2, \ldots, k_n$ or $y_1, y_2, \ldots, y_n$. By contrast, L is a function of θ; it should not be considered a function of either the k_i's or the y_i's.

Definition 5.2.2. Let $L(\theta) = \prod_{i=1}^{n} p_X(k_i; \theta)$ and $L(\theta) = \prod_{i=1}^{n} f_Y(y_i; \theta)$ be the likelihood functions corresponding to random samples $k_1, k_2, \ldots, k_n$ and $y_1, y_2, \ldots, y_n$ drawn from the discrete pdf $p_X(k; \theta)$ and continuous pdf $f_Y(y; \theta)$, respectively, where θ is an unknown parameter. In each case, let θ_e be a value of the parameter such that $L(\theta_e) \geq L(\theta)$ for all possible values of θ. Then θ_e is called a *maximum likelihood estimate* for θ.

Applying the Method of Maximum Likelihood

We will see in Example 5.2.1 and many subsequent examples that finding the θ_e that maximizes a likelihood function is often an application of the calculus. Specifically, we solve the equation $\frac{d}{d\theta} L(\theta) = 0$ for θ. In some cases, a more tractable equation results by setting the derivative of $\ln L(\theta)$ equal to 0. Since $\ln L(\theta)$ increases with $L(\theta)$, the same θ_e that maximizes $\ln L(\theta)$ also maximizes $L(\theta)$.

Example 5.2.1

In Case Study 4.2.2, which discussed modeling α-particle emissions, the mean of the data $\bar{k}$ was used as the parameter λ of the Poisson distribution. This choice seems reasonable, since λ is the mean of the pdf.

In this example, the choice of the sample mean as an estimate of the parameter λ of the Poisson distribution will be justified via the method of maximum likelihood, using a small data set to introduce the technique. So, suppose that $X_1 = 3$, $X_2 = 5$, $X_3 = 4$, and $X_4 = 2$ is a set of four independent observations representing the Poisson probability model,

$$p_X(k;\ \lambda) = e^{-\lambda} \frac{\lambda^k}{k!},\ k = 0, 1, 2, \ldots$$

Find the maximum likelihood for λ.

According to Definition 5.2.1,

$$L(\lambda) = e^{-\lambda} \frac{\lambda^3}{3!} \cdot e^{-\lambda} \frac{\lambda^5}{5!} \cdot e^{-\lambda} \frac{\lambda^4}{4!} \cdot e^{-\lambda} \frac{\lambda^2}{2!} = e^{-4\lambda} \lambda^{14} \frac{1}{3!5!4!2!}$$

Then $\ln L(\lambda) = -4\lambda + 14 \ln \lambda - \ln(3!5!4!2!)$. Differentiating $\ln L(\lambda)$ with respect to λ gives

$$\frac{d \ln L(\lambda)}{d\lambda} = -4 + \frac{14}{\lambda}$$

To find the λ that maximizes $L(\lambda)$, we set the derivative equal to zero. Here $-4 + \frac{14}{\lambda} = 0$ implies that $4\lambda = 14$, and the solution to this equation is $\lambda = \frac{14}{4} = 3.5$.

Notice that the second derivative of $L(\lambda)$ is $-\frac{14}{\lambda^2}$, which is negative for all λ. Thus, $\lambda = 3.5$ is indeed a true maximum of the likelihood function, as well as the only one. (Following the notation introduced in Definition 5.2.2, the number 3.5 is called the *maximum likelihood estimate for* λ, and we would write $\lambda_e = 3.5$.) ■

Comment There is a better way to answer the question posed in Example 5.2.1. Rather than evaluate—and differentiate—the likelihood function for a particular sample observed (in this case, the four observations 3, 5, 4, and 2), we can get a more informative answer by considering the more general problem of taking a random sample of *size n* from $p_X(k;\ \lambda) = e^{-\lambda}\frac{\lambda^k}{k!}$ and using the outcomes—$X_1 = k_1$, $X_2 = k_2, \ldots, X_n = k_n$—to find a *formula* for the maximum likelihood estimate.

For the Poisson pdf, the likelihood function based on such a sample would be written

$$L(\lambda) = \prod_{i=1}^{n} e^{-\lambda}\frac{\lambda^{k_i}}{k_i!} = e^{-n\lambda}\lambda^{\sum_{i=1}^{n} k_i}\frac{1}{\prod_{i=1}^{n} k_i!}$$

As was the case in Example 5.2.1, $\ln L(\lambda)$ is easier to work with than $L(\lambda)$. Here,

$$\ln L(\lambda) = -n\lambda + \left(\sum_{i=1}^{n} k_i\right)\ln\lambda - \ln\prod_{i=1}^{n} k_i!$$

and

$$\frac{d\ln L(\lambda)}{d\lambda} = -n + \frac{\sum_{i=1}^{n} k_i}{\lambda}$$

Setting the derivative equal to 0 gives

$$-n + \frac{\sum_{i=1}^{n} k_i}{\lambda} = 0$$

which implies that $\lambda_e = \frac{\sum_{i=1}^{n} k_i}{n} = \bar{k}$.

Reassuringly, for the particular example used in Example 5.2.1—$n = 4$ and $\sum_{i=1}^{4} k_i = 14$—the formula just derived reduces to the maximum likelihood estimate of $14/4 = 3.5$ that we found at the outset.

The general result of $\bar{k}$ also justifies the choice of parameter estimate made in Case Study 4.2.2.

Comment Implicit in Example 5.2.1 and the remarks following it is the important distinction between a *maximum likelihood estimate* and a *maximum likelihood estimator*. The first is a number or an expression representing a number; the second is a random variable.

Both the number 3.5 and the formula $\frac{1}{n}\sum_{i=1}^{n} k_i$ are maximum likelihood *estimates* for λ and would be denoted λ_e because both are considered numerical constants.

If, on the other hand, we imagine the measurements *before they are recorded*—that is, as the random variables $X_1, X_2, \ldots, X_n$—then the estimate formula $\frac{1}{n}\sum_{i=1}^{n} k_i$ is more properly written as the random variable $\frac{1}{n}\sum_{i=1}^{n} X_i = \overline{X}$.

This last expression is the maximum likelihood *estimator* for λ and would be denoted $\hat{\lambda}$. Maximum likelihood estimators such as $\hat{\lambda}$ have pdfs, expected values, and variances, whereas maximum likelihood *estimates* such as λ_e have none of these statistical properties.

Example 5.2.2

Suppose an isolated weather-reporting station has an electronic device whose time to failure is given by the exponential model

$$f_Y(y;\ \theta) = \frac{1}{\theta}e^{-y/\theta}, \quad 0 \le y < \infty; 0 < \theta < \infty$$

The station also has a spare device, so the time until this instrument is not available is the sum of these two exponential pdfs, which is

$$f_Y(y;\ \theta) = \frac{1}{\theta^2}ye^{-y/\theta}, \quad 0 \le y < \infty;\ 0 < \theta < \infty$$

Five data points have been collected—9.2, 5.6, 18.4, 12.1, and 10.7. Find the maximum likelihood estimate for θ.

Following the advice given in the Comment on p. 285, we begin by deriving a general formula for θ_e—that is, by assuming that the data are the n observations $y_1, y_2, \ldots, y_n$. The likelihood function then becomes

$$L(\theta) = \prod_{i=1}^{n} \frac{1}{\theta^2} y_i e^{-y_i/\theta}$$

$$= \theta^{-2n}\left(\prod_{i=1}^{n} y_i\right) e^{-(1/\theta)\sum\limits_{i=1}^{n} y_i}$$

and

$$\ln L(\theta) = -2n \ln\theta + \ln\left(\prod_{i=1}^{n} y_i\right) - \frac{1}{\theta}\sum_{i=1}^{n} y_i$$

Setting the derivative of $\ln L(\theta)$ equal to 0 gives

$$\frac{d \ln L(\theta)}{d\theta} = \frac{-2n}{\theta} + \frac{1}{\theta^2}\sum_{i=1}^{n} y_i = 0$$

which implies that

$$\theta_e = \frac{1}{2n}\sum_{i=1}^{n} y_i$$

The final step is to evaluate numerically the formula for θ_e. Substituting the actual $n = 5$ sample values recorded gives $\sum\limits_{i=1}^{5} y_i = 9.2 + 5.6 + 18.4 + 12.1 + 10.7 = 56.0$, so

$$\theta_e = \frac{1}{2(5)}(56.0) = 5.6$$

■

Using Order Statistics as Maximum Likelihood Estimates

Situations exist for which the equations $\frac{dL(\theta)}{d\theta} = 0$ or $\frac{d \ln L(\theta)}{d\theta} = 0$ are not meaningful and neither will yield a solution for θ_e. These occur when the range of the pdf from which the data are drawn is a function of the parameter being estimated. [This happens, for instance, when the sample of y_i's come from the uniform pdf, $f_Y(y;\theta) = 1/\theta$, $0 \le y \le \theta$.] The maximum likelihood estimates in these cases will be an order statistic, typically either $y_{\min}$ or $y_{\max}$.

Example 5.2.3 Suppose $y_1, y_2, \ldots, y_n$ is a set of measurements representing an exponential pdf with $\lambda = 1$ but with an unknown "threshold" parameter, θ. That is,

$$f_Y(y;\theta) = e^{-(y-\theta)}, \quad y \geq \theta; \quad \theta > 0$$

(see Figure 5.2.1). Find the maximum likelihood estimate for θ.

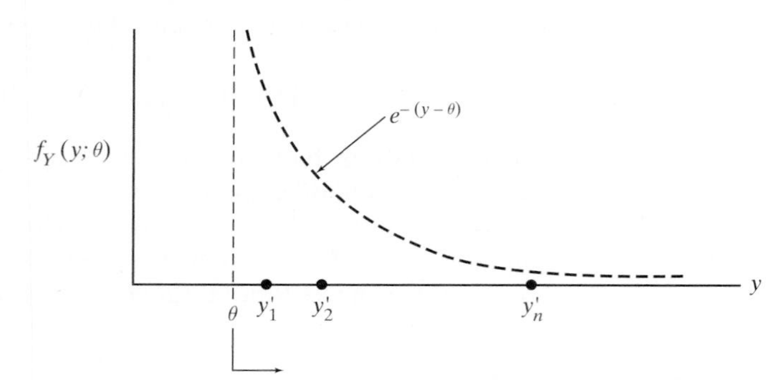

Figure 5.2.1

Proceeding in the usual fashion, we start by deriving an expression for the likelihood function:

$$L(\theta) = \prod_{i=1}^{n} e^{-(y_i-\theta)}$$

$$= e^{-\sum_{i=1}^{n} y_i + n\theta}$$

Here, finding θ_e by solving the equation $\frac{d \ln L(\theta)}{d\theta} = 0$ will not work because $\frac{d \ln L(\theta)}{d\theta} = \frac{d}{d\theta}\left(-\sum_{i=1}^{n} y_i + n\theta\right) = n$. Instead, we need to look at the likelihood function directly.

Notice that $L(\theta) = e^{-\sum_{i=1}^{n} y_i + n\theta}$ is maximized when the exponent of e is maximized. But for given $y_1, y_2, \ldots, y_n$ (and n), making $-\sum_{i=1}^{n} y_i + n\theta$ as large as possible requires that θ be as large as possible. Figure 5.2.1 shows how large θ can be: It can be moved to the right *only as far as the smallest order statistic*. Any value of θ larger than $y_{\min}$ would violate the condition on $f_Y(y;\theta)$ that $y \geq \theta$. Therefore, $\theta_e = y_{\min}$. ■

Case Study 5.2.1

Each evening, the media report various averages and indices that are presented as portraying the state of the stock market. But do they? Are these numbers conveying any really useful information? Some financial analysts would say "no," arguing that speculative markets tend to rise and fall randomly, as though some hidden roulette wheel were spinning out the figures.

One way to test this theory is to model the up-and-down behavior of the markets as a geometric random variable. If this model were to fit, we would be able to argue that the market doesn't use yesterday's history to "decide"

(Continued on next page)

whether to rise or fall the next day, nor does this history change the probability p of a rise or $1 - p$ of a fall the following day.

So, suppose that on a given Day 0 the market rose and the following Day 1 it fell. Let the geometric random variable X represent the number of days the market falls (failures) before it rises again (a success). For example, if on Day 2 the market rises, then $X = 1$. In that case $p_X(1) = p$. If the market declines on Days 2, 3, and 4, and then rises on Day 5, $X = 4$ and $p_X(4) = (1 - p)^3 p$.

This model can be examined by comparing the theoretical distribution for $pX(k)$ to what is observed in a speculative market. However, to do so, the parameter p must be estimated. The maximum likelihood estimate will prove a good choice. Suppose a random sample from the geometric distribution, $k_1, k_2, \ldots, k_n$, is given. Then

$$L(p) = \prod_{i=1}^{n} p_X(k_i) = \prod_{i=1}^{n} (1 - p)^{k_i - 1} p = (1 - p)^{\sum_{i=1}^{n} k_i - n} p^n$$

and

$$\ln L(p) = \ln\left[(1 - p)^{\sum_{i=1}^{n} k_i - n} p^n\right] = \left(\sum_{i=1}^{n} k_i - n\right) \ln(1 - p) + n \ln p$$

Setting the derivative of $\ln L(p)$ equal to 0 gives the equation

$$-\frac{\sum_{i=1}^{n} k_i - n}{1 - p} + \frac{n}{p} = 0$$

or, equivalently,

$$\left(n - \sum_{i=1}^{n} k_i\right) p + n(1 - p) = 0$$

Solving this equation gives $p_e = n / \sum_{i=1}^{n} k_i = 1/\bar{k}$.

Now, turning to a data set to compare to the geometric model, we employ the widely used closing Dow Jones average for the years 2006 and 2007. The first column gives the value of k, the argument of the random variable X. Column 2 presents the number of times $X = k$ in the data set.

Table 5.2.1

k	Observed Frequency	Expected Frequency
1	72	74.14
2	35	31.20
3	11	13.13
4	6	5.52
5	2	2.32
6	2	1.69

Source: finance.yahoo.com/of/hp.s=%SEDJI.

(Continued on next page)

(Case Study 5.2.1 continued)

Note that the Observed Frequency column totals 128, which is n in the formula above for p_e. From the table, we obtain

$$\sum_{i=1}^{n} k_i = 1(72) + 2(35) + 3(11) + 4(6) + 5(2) + 6(2) = 221$$

Then $p_e = 128/221 = 0.5792$. Using this value, the estimated probability of, for example, $p_X(2) = (1 - 0.5792)(0.5792) = 0.2437$. If the model gives the probability of $k = 2$ to be 0.2437, then it seems reasonable to expect to see $n(0.2437) = 128(0.2437) = 31.20$ occurrences of $X = 2$. This is the second entry in the Expected Frequency column of the table. The other expected values are calculated similarly, except for the value corresponding to $k = 6$. In that case, we fill in whatever value makes the expected frequencies sum to $n = 128$.

The close agreement between the Observed and Expected Frequency columns argues for the validity of the geometric model, using the maximum likelihood estimate. This suggests that the stock market doesn't remember yesterday.

Finding Maximum Likelihood Estimates When More Than One Parameter Is Unknown

If a family of probability models is indexed by two or more unknown parameters—say, $\theta_1, \theta_2, \ldots, \theta_k$—finding maximum likelihood estimates for the θ_i's requires the solution of a set of k simultaneous equations. If $k = 2$, for example, we would typically need to solve the system

$$\frac{\partial \ln L(\theta_1, \theta_2)}{\partial \theta_1} = 0$$

$$\frac{\partial \ln L(\theta_1, \theta_2)}{\partial \theta_2} = 0$$

Example 5.2.4

Suppose a random sample of size n is drawn from the two-parameter normal pdf

$$f_Y(y; \mu, \sigma^2) = \frac{1}{\sqrt{2\pi}\sqrt{\sigma^2}} e^{-\frac{1}{2}\frac{(y-\mu)^2}{\sigma^2}} \qquad -\infty < y < \infty; -\infty < \mu < \infty; \sigma^2 > 0$$

Use the method of maximum likelihood to find formulas for μ_e and σ_e^2.

We start by finding $L(\mu, \sigma^2)$ and $\ln L(\mu, \sigma^2)$:

$$L(\mu, \sigma^2) = \prod_{i=1}^{n} \frac{1}{\sqrt{2\pi}\sigma} e^{\frac{1}{2}\frac{(y_i-\mu)}{\sigma^2}}$$

$$= (2\pi\sigma^2)^{-n/2} e^{\frac{1}{2}\frac{(y_i-\mu)}{\sigma^2}}$$

and

$$\ln L(\mu, \sigma^2) = -\frac{n}{2}\ln(2\pi\sigma^2) - \frac{1}{2}\frac{1}{\sigma^2}\sum_{i=1}^{n}(y_i - \mu)^2$$

Moreover,

$$\frac{\partial \ln L(\mu, \sigma^2)}{\partial \mu} = \frac{1}{\sigma^2} \sum_{i=1}^{n} (y_i - \mu)$$

and

$$\frac{\partial \ln L(\mu, \sigma^2)}{\partial \sigma^2} = -\frac{n}{2} \cdot \frac{1}{\sigma^2} + \frac{1}{2} \left(\frac{1}{\sigma^2}\right)^2 \sum_{i=1}^{n} (y_i - \mu)^2$$

Setting the two derivatives equal to zero gives the equations

$$\sum_{i=1}^{n} (y_i - \mu) = 0 \tag{5.2.1}$$

and

$$-n\sigma^2 + \sum_{i=1}^{n} (y_i - \mu)^2 = 0 \tag{5.2.2}$$

Equation 5.2.1 simplifies to

$$\sum_{i=1}^{n} y_i = n\mu$$

which implies that $\mu_e = \frac{1}{n} \sum_{i=1}^{n} y_i = \overline{y}$. Substituting μ_e, then, into Equation 5.2.2 gives

$$-n\sigma^2 + \sum_{i=1}^{n} (y_i - \overline{y})^2 = 0$$

or

$$\sigma_e^2 = \frac{1}{n} \sum_{i=1}^{n} (y_i - \overline{y})^2$$

Comment The method of maximum likelihood has a long history: Daniel Bernoulli was using it as early as 1777 (130). It was Ronald Fisher, though, in the early years of the twentieth century, who first studied the mathematical properties of likelihood estimation in any detail, and the procedure is often credited to him. ■

Questions

5.2.1. A random sample of size 8—$X_1 = 1$, $X_2 = 0$, $X_3 = 1$, $X_4 = 1$, $X_5 = 0$, $X_6 = 1$, $X_7 = 1$, and $X_8 = 0$—is taken from the probability function

$$p_X(k; \theta) = \theta^k (1 - \theta)^{1-k}, \quad k = 0, 1; \quad 0 < \theta < 1$$

Find the maximum likelihood estimate for θ.

5.2.2. The number of red chips and white chips in an urn is unknown, but the *proportion*, p, of reds is either $\frac{1}{3}$ or $\frac{1}{2}$. A sample of size 5, drawn with replacement, yields the sequence red, white, white, red, and white. What is the maximum likelihood estimate for p?

5.2.3. Use the sample $Y_1 = 8.2$, $Y_2 = 9.1$, $Y_3 = 10.6$, and $Y_4 = 4.9$ to calculate the maximum likelihood estimate for λ in the exponential pdf

$$f_Y(y; \lambda) = \lambda e^{-\lambda y}, \quad y \geq 0$$

5.2.4. Suppose a random sample of size n is drawn from the probability model

$$p_X(k; \theta) = \frac{\theta^{2k} e^{-\theta^2}}{k!}, \quad k = 0, 1, 2, \ldots$$

Find a formula for the maximum likelihood estimator, $\hat{\theta}$.

5.2.5. Given that $Y_1 = 2.3$, $Y_2 = 1.9$, and $Y_3 = 4.6$ is a random sample from

$$f_Y(y;\theta) = \frac{y^3 e^{-y/\theta}}{6\theta^4}, \quad y \geq 0$$

calculate the maximum likelihood estimate for θ.

5.2.6. Use the method of maximum likelihood to estimate θ in the pdf

$$f_Y(y;\theta) = \frac{\theta}{2\sqrt{y}} e^{-\theta\sqrt{y}}, \quad y \geq 0$$

Evaluate θ_e for the following random sample of size 4: $Y_1 = 6.2$, $Y_2 = 7.0$, $Y_3 = 2.5$, and $Y_4 = 4.2$.

5.2.7. An engineer is creating a project scheduling program and recognizes that the tasks making up the project are not always completed on time. However, the completion proportion tends to be fairly high. To reflect this condition, he uses the pdf

$$f_Y(y;\theta) = \theta y^{\theta-1}, \quad 0 \leq y \leq 1, \quad \text{and} \quad 0 < \theta$$

where y is the proportion of the task completed. Suppose that in his previous project, the proportions of tasks completed were 0.77, 0.82, 0.92, 0.94, and 0.98. Estimate θ.

5.2.8. The following data show the number of occupants in passenger cars observed during one hour at a busy intersection in Los Angeles (69). Suppose it can be assumed that these data follow a geometric distribution, $p_X(k;p) = (1-p)^{k-1}p$, $k = 1, 2, \ldots$. Estimate p and compare the observed and expected frequencies for each value of X.

Number of Occupants	Frequency
1	678
2	227
3	56
4	28
5	8
6+	14
	1011

5.2.9. For the Major League Baseball seasons from 1950 through 2008, there were fifty-nine nine-inning games in which one of the teams did not manage to get a hit. The data in the table give the number of no-hitters *per season* over this period. Assume that the data follow a Poisson distribution,

$$p_X(k;\lambda) = e^{-\lambda}\frac{\lambda^k}{k!}, k = 0, 1, 2, \ldots$$

(a) Estimate λ and compare the observed and expected frequencies.

(b) Does the agreement (or lack of agreement) in part (a) come as a surprise? Explain.

No. of No-Hitters	Frequency
0	6
1	19
2	12
3	13
4+	9

Source: en.wikipedia.org/wiki/List_of_Major_League_Baseball_no-hitlers.

5.2.10. **(a)** Based on the random sample $Y_1 = 6.3$, $Y_2 = 1.8$, $Y_3 = 14.2$, and $Y_4 = 7.6$, use the method of maximum likelihood to estimate the parameter θ in the uniform pdf

$$f_Y(y;\theta) = \frac{1}{\theta}, \quad 0 \leq y \leq \theta$$

(b) Suppose the random sample in part (a) represents the two-parameter uniform pdf

$$f_Y(y;\theta_1,\theta_2) = \frac{1}{\theta_2 - \theta_1}, \quad \theta_1 \leq y \leq \theta_2$$

Find the maximum likelihood estimates for θ_1 and θ_2.

5.2.11. Find the maximum likelihood estimate for θ in the pdf

$$f_Y(y;\theta) = \frac{2y}{1-\theta^2}, \quad \theta \leq y \leq 1$$

if a random sample of size 6 yielded the measurements 0.70, 0.63, 0.92, 0.86, 0.43, and 0.21.

5.2.12. A random sample of size n is taken from the pdf

$$f_Y(y;\theta) = \frac{2y}{\theta^2}, \quad 0 \leq y \leq \theta$$

Find an expression for $\hat{\theta}$, the maximum likelihood estimator for θ.

5.2.13. If the random variable Y denotes an individual's income, Pareto's law claims that $P(Y \geq y) = \left(\frac{k}{y}\right)^{\theta}$, where k is the entire population's minimum income. It follows that $F_Y(y) = 1 - \left(\frac{k}{y}\right)^{\theta}$, and, by differentiation,

$$f_Y(y;\theta) = \theta k^{\theta}\left(\frac{1}{y}\right)^{\theta+1}, \quad y \geq k; \quad \theta \geq 1$$

Assume k is known. Find the maximum likelihood estimator for θ if income information has been collected on a random sample of 25 individuals.

5.2.14. The exponential pdf is a measure of lifetimes of devices that do not age (see Question 3.11.11). However, the exponential pdf is a special case of the *Weibull distribution*, which measures time to failure of devices where the probability of failure increases as time does. A Weibull random variable Y has pdf $f_Y(y;\alpha,\beta) = \alpha\beta y^{\beta-1}e^{-\alpha y^{\beta}}$, $0 \leq y$, $0 < \alpha$, $0 < \beta$.

(a) Find the maximum likelihood estimator for α assuming that β is known.

(b) Suppose α and β are both unknown. Write down the equations that would be solved simultaneously to find the maximum likelihood estimators of α and β.

5.2.15. Suppose a random sample of size n is drawn from a normal pdf where the mean μ is known but the variance σ^2 is unknown. Use the method of maximum likelihood to find a formula for $\hat{\sigma}^2$. Compare your answer to the maximum likelihood estimator found in Example 5.2.4.

The Method of Moments

A second procedure for estimating parameters is the *method of moments*. Proposed near the turn of the twentieth century by the great British statistician, Karl Pearson, the method of moments is often more tractable than the method of maximum likelihood in situations where the underlying probability model has multiple parameters.

Suppose that Y is a continuous random variable and that its pdf is a function of s unknown parameters, $\theta_1, \theta_2, \ldots, \theta_s$. The first s moments of Y, if they exist, are given by the integrals

$$E(Y^j) = \int_{-\infty}^{\infty} y^j \cdot f_Y(y; \theta_1, \theta_2, \ldots, \theta_s)\, dy, \quad j = 1, 2, \ldots, s$$

In general, each $E(Y^j)$ will be a different function of the s parameters. That is,

$$E(Y^1) = g_1(\theta_1, \theta_2, \ldots, \theta_s)$$
$$E(Y^2) = g_2(\theta_1, \theta_2, \ldots, \theta_s)$$
$$\vdots$$
$$E(Y^s) = g_s(\theta_1, \theta_2, \ldots, \theta_s)$$

Corresponding to each *theoretical* moment, $E(Y^j)$, is a *sample* moment, $\frac{1}{n}\sum_{i=1}^{n} y_i^j$. Intuitively, the jth sample moment is an approximation to the jth theoretical moment. Setting the two equal *for each* j produces a system of s simultaneous equations, the solutions to which are the desired set of estimates, $\theta_{1e}, \theta_{2e}, \ldots,$ and θ_{se}.

Definition 5.2.3. Let $y_1, y_2, \ldots, y_n$ be a random sample from the continuous pdf $f_Y(y; \theta_1, \theta_2, \ldots, \theta_s)$. The *method of moments* estimates, $\theta_{1e}, \theta_{2e}, \ldots,$ and θ_{se}, for the model's unknown parameters are the solutions of the s simultaneous equations

$$\int_{-\infty}^{\infty} y\, f_Y(y; \theta_1, \theta_2, \ldots, \theta_s)\, dy = \left(\frac{1}{n}\right)\sum_{i=1}^{n} y_i$$
$$\int_{-\infty}^{\infty} y^2 f_Y(y; \theta_1, \theta_2, \ldots, \theta_s)\, dy = \left(\frac{1}{n}\right)\sum_{i=1}^{n} y_i^2$$
$$\vdots \qquad\qquad \vdots$$
$$\int_{-\infty}^{\infty} y^s f_Y(y; \theta_1, \theta_2, \ldots, \theta_s)\, dy = \left(\frac{1}{n}\right)\sum_{i=1}^{n} y_i^s$$

If the underlying random variable is discrete with pdf $p_X(k;\theta_1,\theta_2,\ldots,\theta_s)$, the method of moments estimates are the solutions of the system of equations,

$$\sum_{\text{all }k} k^j p_X(k;\theta_1,\theta_2,\ldots,\theta_s) = \left(\frac{1}{n}\right)\sum_{\text{all }k} k^j, \quad j=1,2,\ldots,s$$

Example 5.2.5

Suppose that $Y_1 = 0.42$, $Y_2 = 0.10$, $Y_3 = 0.65$, and $Y_4 = 0.23$ is a random sample of size 4 from the pdf

$$f_Y(y;\theta) = \theta y^{\theta-1}, \quad 0 \le y \le 1$$

Find the method of moments estimate for θ.

Taking the same approach that we followed in finding maximum likelihood estimates, we will derive a general expression for the method of moments estimate before making any use of the four data points. Notice that only one equation needs to be solved because the pdf is indexed by just a single parameter.

The first theoretical moment of Y is $\frac{\theta}{\theta+1}$:

$$\begin{aligned} E(Y) &= \int_0^1 y\cdot\theta y^{\theta-1}\,dy \\ &= \theta\cdot\frac{y^{\theta+1}}{\theta+1}\bigg|_0^1 \\ &= \frac{\theta}{\theta+1} \end{aligned}$$

Setting $E(Y)$ equal to $\frac{1}{n}\sum_{i=1}^{n} y_i (=\overline{y})$, the first sample moment, gives

$$\frac{\theta}{\theta+1} = \overline{y}$$

which implies that the method of moments estimate for θ is

$$\theta_e = \frac{\overline{y}}{1-\overline{y}}$$

Here, $\overline{y} = \frac{1}{4}(0.42+0.10+0.65+0.23) = 0.35$, so

$$\theta_e = \frac{0.35}{1-0.35} = 0.54$$

■

The gamma distribution, $f_Y(y;\ r,\lambda) = \frac{\lambda^r}{\Gamma(r)} y^{r-1}e^{-\lambda y}$, $y \ge 0$, often provides a good model for data that are inherently not symmetric, as the rainfall example below will show. Deriving maximum likelihood estimators for r and λ, though, is difficult because $\Gamma(r)$ does not have a closed-form derivative. However, the method of moments estimators are not hard to find.

From Theorem 4.6.3, $E(Y) = \frac{r}{\lambda}$ and $\text{Var}(Y) = \frac{r}{\lambda^2}$. Recall that

$$E(Y^2) = \text{Var}(Y) + [E(Y)]^2,$$

so for the gamma distribution,

$$E(Y^2) = \frac{r}{\lambda^2} + \left(\frac{r}{\lambda}\right)^2 = \frac{r(r+1)}{\lambda^2}$$

To find the method of moments estimators, form the two equations

$$\frac{1}{n}\sum_{i=1}^{n} y_i = \frac{r}{\lambda} \text{ and } \frac{1}{n}\sum_{i=1}^{n} y_i^2 = \frac{r(r+1)}{\lambda^2}$$

From the first equation, $r = \frac{\lambda}{n}\sum_{i=1}^{n} y_i$.
Substituting that value into the second equation gives

$$\frac{1}{n}\sum_{i=1}^{n} y_i^2 = \frac{\frac{\lambda}{n}\sum_{i=1}^{n} y_i\left(\frac{\lambda}{n}\sum_{i=1}^{n} y_i + 1\right)}{\lambda^2}$$

The solution of that equation for λ gives its method of moments estimate:

$$\lambda_e = \frac{\sum_{i=1}^{n} y_i}{\sum_{i=1}^{n} y_i^2 - \frac{1}{n}\left(\sum_{i=1}^{n} y_i\right)^2}$$

and then

$$r_e = \frac{\lambda_e}{n}\sum_{i=1}^{n} y_i = \bar{y}\lambda_e$$

Case Study 5.2.2

In the western United States, the supply of water to support daily living, agriculture, and industry is a matter of serious concern. For that reason, the U.S. Department of Agriculture has established a network of stations to record precipitation. One such site is in California just south of Lake Tahoe, with the inviting name of Heavenly Valley. Columns 1 and 2 of Table 5.2.2 below give the monthly rainfall in inches for the 294 months in which there was some measurable precipitation, over a period of twenty-eight years.

Table 5.2.2

Inches Rainfall	Observed Frequency	Expected Frequency
0–1	87	80.54
1–2	58	57.19
2–3	42	41.62
3–4	23	30.44
4–5	20	22.31
5–6	14	16.38
6–7	10	12.03
7–8	13	8.84
8–9	9	6.51
9–10	4	4.79
>10	14	13.35

Source: www.wcc.nrcs.usda.gov.

(Continued on next page)

(Case Study 5.2.2 continued)

These data are clearly not symmetric, which suggests that a gamma distribution might provide a good fit. The original, ungrouped data set provides the necessary sums for estimating r and λ:

$$\sum_{i=1}^{294} y_i = 942.0 \text{ and } \sum_{i=1}^{294} y_i^2 = 6117.82$$

Then $\lambda_e = \frac{942.0}{6117.82 - \frac{1}{294}(942.0)^2} = 0.3039$ and $r_e = \frac{942.0}{294}(0.3039) = 0.9737$.

Integrating the gamma pdf over the rainfall interval limits and multiplying by $n(= 294)$ gives the expected frequencies in the third column. The second entry in that column, for example, is given by

$$294 \cdot \int_1^2 \frac{0.3039^{0.9737}}{\Gamma(0.9737)} y^{0.9737-1} e^{-0.3039y} \, dy = 57.19$$

The above quantity and the others in the third column were calculated using the Minitab routine

```
MTB  > cdf c1;
SUBC > gamma 0.9737 1/0.3039
```

Clearly, the agreement between observed and expected frequencies is quite good. A visual approach to examining the fit between data and model is presented in Figure 5.2.2, where the estimated gamma curve is superimposed on the data's density-scaled histogram.

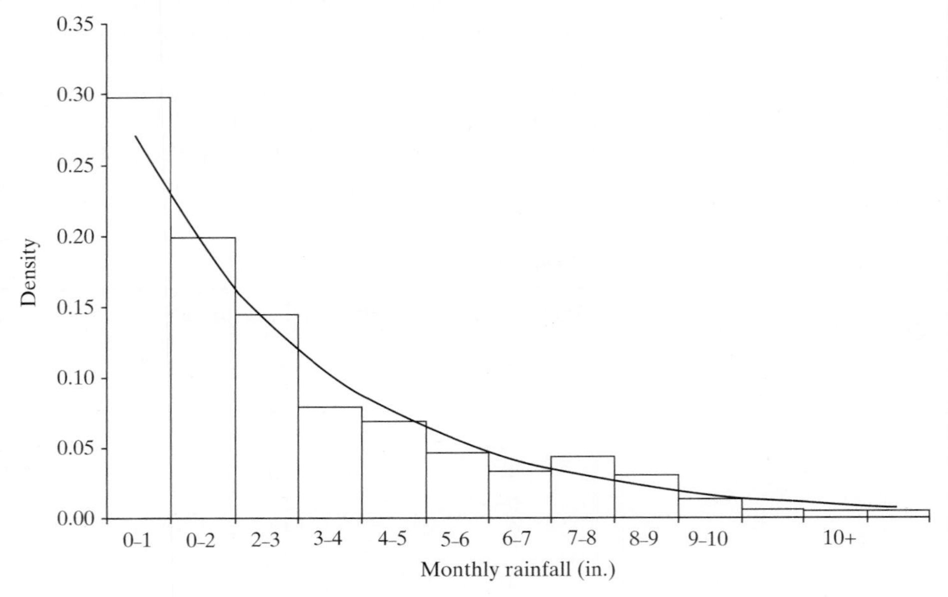

Figure 5.2.2

The adequacy of the approximation here would come as no surprise to a meteorologist. The gamma distribution is frequently used to describe the variation in precipitation levels.

Questions

5.2.16. Let $y_1, y_2, \ldots, y_n$ be a random sample of size n from the pdf $f_Y(y;\theta) = \frac{2y}{\theta^2}, 0 \le y \le \theta$. Find a formula for the method of moments estimate for θ. Compare the values of the method of moments estimate and the maximum likelihood estimate if a random sample of size 5 consists of the numbers 17, 92, 46, 39, and 56 (recall Question 5.2.12).

5.2.17. Use the method of moments to estimate θ in the pdf

$$f_Y(y;\theta) = (\theta^2+\theta)y^{\theta-1}(1-y), \quad 0 \le y \le 1$$

Assume that a random sample of size n has been collected.

5.2.18. A criminologist is searching through FBI files to document the prevalence of a rare double-whorl fingerprint. Among six consecutive sets of 100,000 prints scanned by a computer, the numbers of persons having the abnormality are 3, 0, 3, 4, 2, and 1, respectively. Assume that double whorls are Poisson events. Use the method of moments to estimate their occurrence rate, λ. How would your answer change if λ were estimated using the method of maximum likelihood?

5.2.19. Find the method of moments estimate for λ if a random sample of size n is taken from the exponential pdf, $f_Y(y;\lambda) = \lambda e^{-\lambda y}, y \ge 0$.

5.2.20. Suppose that $Y_1 = 8.3$, $Y_2 = 4.9$, $Y_3 = 2.6$, and $Y_4 = 6.5$ is a random sample of size 4 from the two-parameter uniform pdf,

$$f_Y(y;\theta_1,\theta_2) = \frac{1}{2\theta_2}, \quad \theta_1 - \theta_2 \le y \le \theta_1 + \theta_2$$

Use the method of moments to calculate θ_{1e} and θ_{2e}.

5.2.21. Find a formula for the method of moments estimate for the parameter θ in the Pareto pdf,

$$f_Y(y;\theta) = \theta k^{\theta}\left(\frac{1}{y}\right)^{\theta+1}, \quad y \ge k; \quad \theta \ge 1$$

Assume that k is known and that the data consist of a random sample of size n. Compare your answer to the maximum likelihood estimator found in Question 5.2.13.

5.2.22. Calculate the method of moments estimate for the parameter θ in the probability function

$$p_X(k;\theta) = \theta^k(1-\theta)^{1-k}, \quad k = 0, 1$$

if a sample of size 5 is the set of numbers 0, 0, 1, 0, 1.

5.2.23. Find the method of moments estimates for μ and σ^2, based on a random sample of size n drawn from a normal pdf, where $\mu = E(Y)$ and $\sigma^2 = \text{Var}(Y)$. Compare your answers with the maximum likelihood estimates derived in Example 5.2.4.

5.2.24. Use the method of moments to derive formulas for estimating the parameters r and p in the negative binomial pdf,

$$p_X(k;r,p) = \binom{k-1}{r-1} p^r(1-p)^{k-r}, \quad k = r, r+1, \ldots$$

5.2.25. Bird songs can be characterized by the number of clusters of "syllables" that are strung together in rapid succession. If the last cluster is defined as a "success," it may be reasonable to treat the *number* of clusters in a song as a geometric random variable. Does the model $p_X(k) = (1-p)^{k-1}p, k = 1, 2, \ldots$, adequately describe the following distribution of 250 song lengths (100)? Begin by finding the method of moments estimate for p. Then calculate the set of "expected" frequencies.

No. of Clusters/Song	Frequency
1	132
2	52
3	34
4	9
5	7
6	5
7	5
8	6
	250

5.2.26. Let $y_1, y_2, \ldots, y_n$ be a random sample from the continuous pdf $f_Y(y;\ \theta_1, \theta_2)$. Let $\hat{\sigma}^2 = \frac{1}{n}\sum_{i=1}^{n}(y_i - \overline{y})^2$. Show that the solutions of the equations

$$E(Y) = \overline{y} \text{ and } \text{Var}(Y) = \hat{\sigma}^2$$

for θ_1 and θ_2 give the same results as using the equations in Definition 5.2.3.

5.3 Interval Estimation

Point estimates, no matter how they are determined, share the same fundamental weakness: They provide no indication of their inherent precision. We know, for instance, that $\hat{\lambda} = \overline{X}$ is both the maximum likelihood and the method of moments estimator for the Poisson parameter, λ. But suppose a sample of size 6 is taken from the probability model $p_X(k) = e^{-\lambda}\lambda^k/k!$ and we find that $\lambda_e = 6.8$. Does it follow that the true λ is likely to be close to λ_e—say, in the interval from 6.7 to 6.9—or is the estimation process so imprecise that λ might actually be as small as 1.0, or as large as 12.0? Unfortunately, point estimates, by themselves, do not allow us to make those kinds of extrapolations. Any such statements require that the *variation* of the estimator be taken into account.

The usual way to quantify the amount of uncertainty in an estimator is to construct a *confidence interval*. In principle, confidence intervals are ranges of numbers that have a high probability of "containing" the unknown parameter as an interior point. By looking at the *width* of a confidence interval, we can get a good sense of the estimator's precision.

Example 5.3.1

Suppose that 6.5, 9.2, 9.9, and 12.4 constitute a random sample of size 4 from the pdf

$$f_Y(y;\mu) = \frac{1}{\sqrt{2\pi}(0.8)} e^{-\frac{1}{2}\left(\frac{y-\mu}{0.8}\right)^2}, \quad -\infty < y < \infty$$

That is, the four y_i's come from a normal distribution where σ is equal to 0.8, but the mean, μ, is unknown. What values of μ are believable in light of the four data points?

To answer that question requires that we keep the distinction between *estimates* and *estimators* clearly in mind. First of all, we know from Example 5.2.4 that the maximum likelihood estimate for μ is $\mu_e = \bar{y} = \left(\frac{1}{n}\right)\sum_{i=1}^{n} y_i = \left(\frac{1}{4}\right)(38.0) = 9.5$. We also know something very specific about the probabilistic behavior of the maximum likelihood estimator, $\bar{Y}$: According to the corollary to Theorem 4.3.3, $\frac{\bar{Y}-\mu}{\sigma/\sqrt{n}} = \frac{\bar{Y}-\mu}{0.8/\sqrt{4}}$ has a standard normal pdf, $f_Z(z)$. The probability, then, that $\frac{\bar{Y}-\mu}{0.8/\sqrt{4}}$ will fall between two specified values can be deduced from Table A.1 in the Appendix. For example,

$$P(-1.96 \le Z \le 1.96) = 0.95 = P\left(-1.96 \le \frac{\bar{Y}-\mu}{0.8/\sqrt{4}} \le 1.96\right) \quad (5.3.1)$$

(see Figure 5.3.1).

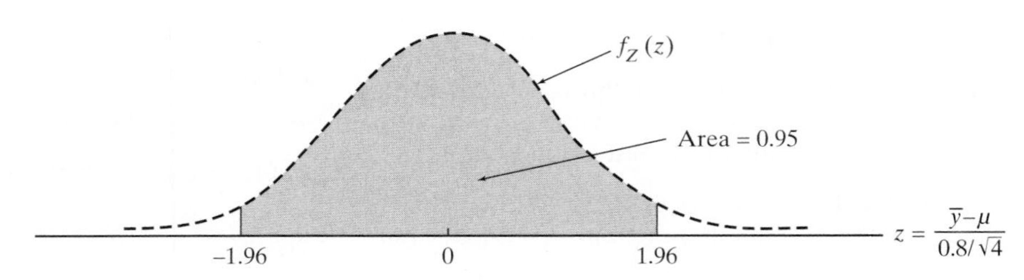

Figure 5.3.1

"Inverting" probability statements of the sort illustrated in Equation 5.3.1 is the mechanism by which we can identify a set of parameter values compatible with the sample data. If

$$P\left(-1.96 \le \frac{\bar{Y}-\mu}{0.8/\sqrt{4}} \le 1.96\right) = 0.95$$

then

$$P\left(\overline{Y} - 1.96\frac{0.8}{\sqrt{4}} \le \mu \le \overline{Y} + 1.96\frac{0.8}{\sqrt{4}}\right) = 0.95$$

which implies that the random interval

$$\left(\overline{Y} - 1.96\frac{0.8}{\sqrt{4}}, \overline{Y} + 1.96\frac{0.8}{\sqrt{4}}\right)$$

has a 95% chance of containing μ as an interior point.

After substituting for $\overline{Y}$, the random interval in this case reduces to

$$\left(9.50 - 1.96\frac{0.8}{\sqrt{4}}, 9.50 + 1.96\frac{0.8}{\sqrt{4}}\right) = (8.72, 10.28)$$

We call (8.72, 10.28) a *95% confidence interval for* μ. In the long run, 95% of the intervals constructed in this fashion will contain the unknown μ; the remaining 5% will lie either entirely to the left of μ or entirely to the right. For a given set of data, of course, we have no way of knowing whether the calculated $\left(\overline{y} - 1.96 \cdot \frac{0.8}{\sqrt{4}}, \overline{y} + 1.96 \cdot \frac{0.8}{\sqrt{4}}\right)$ is one of the 95% that contains μ or one of the 5% that does not.

Figure 5.3.2 illustrates graphically the statistical implications associated with the random interval $\left(\overline{Y} - 1.96\frac{0.8}{\sqrt{4}}, \overline{Y} + 1.96\frac{0.8}{\sqrt{4}}\right)$. For every different $\overline{y}$, the interval will have a different location. While there is no way to know whether or not a given interval—in particular, the one the experimenter has just calculated—will include the unknown μ, we do have the reassurance that *in the long run*, 95% of all such intervals will.

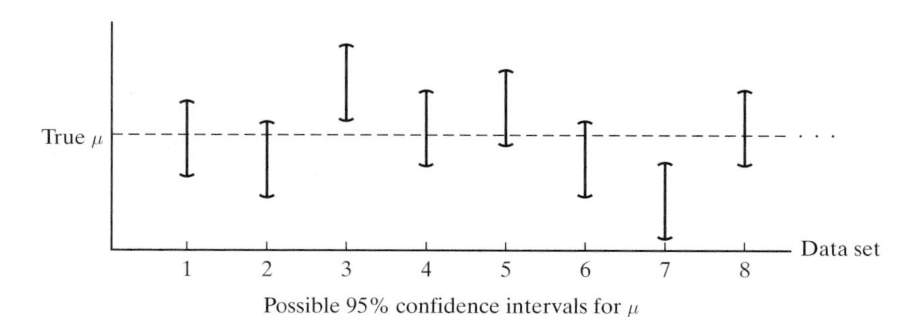

Figure 5.3.2

Comment The behavior of confidence intervals can be modeled nicely by using a computer's random number generator. The output in Table 5.3.1 is a case in point. Fifty simulations of the confidence interval described in Example 5.3.1 are displayed. That is, fifty samples, each of size $n = 4$, were drawn from the normal pdf

$$f_Y(y; \mu) = \frac{1}{\sqrt{2\pi}(0.8)} e^{-\frac{1}{2}\left(\frac{y-\mu}{0.8}\right)^2}, \quad -\infty < y < \infty$$

using Minitab's RANDOM command. (To fully specify the model—and to know the value that each confidence interval was seeking to contain—the true μ was assumed to equal ten). For each sample of size $n = 4$, the lower and upper limits of the corresponding 95% confidence interval were calculated, using the formulas

Table 5.3.1

```
MTB  > random 50 c1-c4;
SUBC > normal 10 0.8.
MTB  > rmean c1-c4 c5
MTB  > let c6 = c5 - 1.96*(0.8)/sqrt(4)
MTB  > let c7 = c5 + 1.96*(0.8)/sqrt(4)
MTB  > name c6 'Low.Lim.' c7 'Upp.Lim.'
MTB  > print c6 c7
```

Data Display

Row	Low.Lim.	Upp.Lim.	Contains μ = 10?
1	8.7596	10.3276	Yes
2	8.8763	10.4443	Yes
3	8.8337	10.4017	Yes
4	9.5800	11.1480	Yes
5	8.5106	10.0786	Yes
6	9.6946	11.2626	Yes
7	8.7079	10.2759	Yes
8	10.0014	11.5694	NO
9	9.3408	10.9088	Yes
10	9.5428	11.1108	Yes
11	8.4650	10.0330	Yes
12	9.6346	11.2026	Yes
13	9.2076	10.7756	Yes
14	9.2517	10.8197	Yes
15	8.7568	10.3248	Yes
16	9.8439	11.4119	Yes
17	9.3297	10.8977	Yes
18	9.5685	11.1365	Yes
19	8.9728	10.5408	Yes
20	8.5775	10.1455	Yes
21	9.3979	10.9659	Yes
22	9.2115	10.7795	Yes
23	9.6277	11.1957	Yes
24	9.4252	10.9932	Yes
25	9.6868	11.2548	Yes
26	8.8779	10.4459	Yes
27	9.1570	10.7250	Yes
28	9.3277	10.8957	Yes
29	9.1606	10.7286	Yes
30	8.8919	10.4599	Yes
31	9.3838	10.9518	Yes
32	8.7575	10.3255	Yes
33	10.4602	12.0282	NO
34	8.9437	10.5117	Yes
35	9.0049	10.5729	Yes
36	9.0148	10.5828	Yes
37	8.8110	10.3790	Yes
38	9.1981	10.7661	Yes
39	9.0042	10.5722	Yes
40	9.7019	11.2699	Yes
41	9.2167	10.7847	Yes
42	8.3901	9.9581	NO
43	8.6337	10.2017	Yes
44	9.4606	11.0286	Yes
45	9.3278	10.8958	Yes
46	8.5843	10.1523	Yes
47	9.0541	10.6221	Yes
48	9.2042	10.7722	Yes
49	9.2710	10.8390	Yes
50	9.5697	11.1377	Yes

47 of the 50 95% confidence intervals contain the true $\mu(= 10)$

$$\text{Low.Lim.} = \bar{y} - 1.96\frac{0.8}{\sqrt{4}}$$

$$\text{Upp.Lim.} = \bar{y} + 1.96\frac{0.8}{\sqrt{4}}$$

As the last column in the DATA DISPLAY indicates, only three of the fifty confidence intervals fail to contain $\mu = 10$: Samples eight and thirty-three yield intervals that lie entirely to the right of the parameter, while sample forty-two produces a range of values that lies entirely to the left. The remaining forty-seven intervals, though, or *94%* $\left(= \frac{47}{50} \times 100\right)$, *do* contain the true value of μ as an interior point.

Case Study 5.3.1

In the eighth century B.C., the Etruscan civilization was the most advanced in all of Italy. Its art forms and political innovations were destined to leave indelible marks on the entire Western world. Originally located along the western coast between the Arno and Tiber Rivers (the region now known as Tuscany), it spread quickly across the Apennines and eventually overran much of Italy. But as quickly as it came, it faded. Militarily it was to prove no match for the burgeoning Roman legions, and by the dawn of Christianity it was all but gone.

No written history from the Etruscan empire has ever been found, and to this day its origins remain shrouded in mystery. Were the Etruscans native Italians, or were they immigrants? And if they were immigrants, where did they come from? Much of what *is* known has come from anthropometric studies—that is, investigations that use body measurements to determine racial characteristics and ethnic origins.

A case in point is the set of data given in Table 5.3.2, showing the sizes of eighty-four Etruscan skulls unearthed in various archaeological digs throughout Italy (6). The sample mean, $\bar{y}$, of those measurements is 143.8 mm. Researchers believe that skull widths of present-day Italian males are normally distributed with a mean (μ) of 132.4 mm and a standard deviation (σ) of 6.0 mm. What does

Table 5.3.2

Maximum Head Breadths (mm) of 84 Etruscan Males

141	148	132	138	154	142	150
146	155	158	150	140	147	148
144	150	149	145	149	158	143
141	144	144	126	140	144	142
141	140	145	135	147	146	141
136	140	146	142	137	148	154
137	139	143	140	131	143	141
149	148	135	148	152	143	144
141	143	147	146	150	132	142
142	143	153	149	146	149	138
142	149	142	137	134	144	146
147	140	142	140	137	152	145

(Continued on next page)

(Case Study 5.3.1 continued)

the difference between $\overline{y} = 143.8$ and $\mu = 132.4$ imply about the likelihood that Etruscans and Italians share the same ethnic origin?

One way to answer that question is to construct a 95% confidence interval for the true mean of the population represented by the eighty-four y_i's in Table 5.3.2. If that confidence interval fails to contain $\mu = 132.4$, it could be argued that the Etruscans were not the forebears of modern Italians. (Of course, it would also be necessary to factor in whatever evolutionary trends in skull sizes have occurred for *Homo sapiens*, in general, over the past three thousand years.)

It follows from the discussion in Example 5.3.1 that the endpoints for a 95% confidence interval for μ are given by the general formula

$$\left(\overline{y} - 1.96 \cdot \frac{\sigma}{\sqrt{n}}, \overline{y} + 1.96 \cdot \frac{\sigma}{\sqrt{n}}\right)$$

Here, that expression reduces to

$$\left(143.8 - 1.96 \cdot \frac{6.0}{\sqrt{84}}, 143.8 + 1.96 \cdot \frac{6.0}{\sqrt{84}}\right) = (142.5 \text{ mm}, 145.1 \text{ mm})$$

Since the value $\mu = 132.4$ is *not* contained in the 95% confidence interval (or even close to being contained), we would conclude that a sample mean of 143.8 (based on a sample of size 84) is not likely to have come from a normal population where $\mu = 132.4$ (and $\sigma = 6.0$). It would appear, in other words, that Italians are not the direct descendants of Etruscans.

Comment Random intervals can be constructed to have whatever "confidence" we choose. Suppose $z_{\alpha/2}$ is defined to be the value for which $P(Z \geq z_{\alpha/2}) = \alpha/2$. If $\alpha = 0.05$, for example, $z_{\alpha/2} = z_{.025} = 1.96$. A *$100(1-\alpha)$% confidence interval for μ*, then, is the range of numbers

$$\left(\overline{y} - z_{\alpha/2} \cdot \frac{\sigma}{\sqrt{n}}, \overline{y} + z_{\alpha/2} \cdot \frac{\sigma}{\sqrt{n}}\right)$$

In practice, α is typically set at either 0.10, 0.05, or 0.01, although in some fields *50%* confidence intervals are frequently used.

Confidence Intervals for the Binomial Parameter, *p*

Perhaps the most frequently encountered applications of confidence intervals are those involving the binomial parameter, p. Opinion surveys are often the context: When polls are released, it has become standard practice to issue a disclaimer by saying that the findings have a certain *margin of error*. As we will see later in this section, margins of error are related to 95% confidence intervals.

The inversion technique followed in Example 5.3.1 can be applied to large-sample binomial random variables as well. We know from Theorem 4.3.1 that

$$(X - np)/\sqrt{np(1-p)} = (X/n - p)/\sqrt{p(1-p)/n}$$

has approximately a standard normal distribution when X is binomial and n is large. It is also true that the pdf describing

$$\frac{X/n - p}{\sqrt{\frac{(X/n)(1-X/n)}{n}}}$$

can be approximated by $f_Z(z)$, a result that seems plausible given that $\frac{X}{n}$ is the maximum likelihood estimator for p.

Therefore,

$$P\left[-z_{\alpha/2} \le \frac{X/n - p}{\sqrt{\frac{(X/n)(1-X/n)}{n}}} \le z_{\alpha/2}\right] \doteq 1-\alpha \tag{5.3.2}$$

Rewriting Equation 5.3.2 by isolating p in the center of the inequalities leads to the formula given in Theorem 5.3.1.

Theorem 5.3.1 *Let k be the number of successes in n independent trials, where n is large and $p = P(success)$ is unknown. An approximate $100(1-\alpha)\%$ confidence interval for p is the set of numbers*

$$\left[\frac{k}{n} - z_{\alpha/2}\sqrt{\frac{(k/n)(1-k/n)}{n}},\ \frac{k}{n} + z_{\alpha/2}\sqrt{\frac{(k/n)(1-k/n)}{n}}\right]$$

Case Study 5.3.2

A majority of Americans have favored increased fuel efficiency for automobiles. Some do not, primarily because of concern over increased costs, or from general opposition to government mandates. The public's intensity about the issue tends to fluctuate with the price of gasoline. In the summer of 2008, when the national average of prices for regular unleaded gasoline exceeded \$4 per gallon, fuel efficiency became part of the political landscape.

How much the public does favor increased fuel efficiency has been the subject of numerous polls. A Gallup telephone poll of 1012 adults (18 and over) in March of 2009 reported that 810 favored the setting of higher fuel-efficiency standards for automobiles.

Given that $n = 1012$ and $k = 810$, the "believable" values for p, the probability that an adult does favor efficiency, according to Theorem 5.3.1, are the proportions from *0.776* to *0.825*:

$$\left[\frac{810}{1012} - 1.96\sqrt{\frac{(810/1012)(1-810/1012)}{1012}},\ \frac{810}{1012} + 1.96\sqrt{\frac{(810/1012)(1-810/1012)}{1012}}\right]$$

$$= (0.776, 0.825)$$

If the true proportion of Americans, in other words, who support increased fuel efficiency is less than 0.776 or greater than 0.825, it would be unlikely that a *sample proportion* (based on 1012 responses) would be the observed $810/1012 = 0.800$.

Source: http://www.gallup.com/poll/118543/Americans-Green-Light-Higher-Fuel-Efficiency-Standards.aspx.

Comment We call (0.776, 0.825) a 95% confidence interval for p, but it does not follow that p has a 95% chance of lying between 0.776 and 0.825. The parameter p is a constant, so it falls between 0.776 and 0.825 either 0% of the time or 100% of the time. The "95%" refers to the *procedure* by which the interval is constructed, not to any particular interval. This, of course, is entirely analogous to the interpretation given earlier to 95% confidence intervals for μ.

Comment Robert Frost was certainly more familiar with iambic pentameter than he was with estimated parameters, but in 1942 he wrote a couplet that sounds very much like a poet's perception of a confidence interval (98):

We dance round in a ring and suppose,
But the Secret sits in the middle and knows.

Example 5.3.2

Central to every statistical software package is a random number generator. Two or three simple commands are typically all that are required to output a sample of size n representing any of the standard probability models. But how can we be certain that numbers purporting to be random observations from, say, a normal distribution with $\mu = 50$ and $\sigma = 10$ actually *do* represent that particular pdf?

The answer is, we cannot; however, a number of "tests" are available to check whether the simulated measurements appear to be random *with respect to a given criterion*. One such procedure is the *median test*.

Suppose $y_1, y_2, \ldots, y_n$ denote measurements presumed to have come from a continuous pdf $f_Y(y)$. Let k denote the number of y_i's that are less than the median of $f_Y(y)$. If the sample *is* random, we would expect the difference between $\frac{k}{n}$ and $\frac{1}{2}$ to be small. More specifically, a 95% confidence interval based on $\frac{k}{n}$ should contain the value 0.5.

Listed in Table 5.3.3 is a set of sixty y_i's generated by Minitab to represent the exponential pdf, $f_Y(y) = e^{-y}$, $y \geq 0$. Does this sample pass the median test?

The median here is $m = 0.69315$:

$$\int_0^m e^{-y}\,dy = -e^{-y}\Big|_0^m = 1 - e^{-m} = 0.5$$

which implies that $m = -\ln(0.5) = 0.69315$. Notice that of the sixty entries in Table 5.3.3, a total of $k = 26$ (those marked with an asterisk, *) fall to the left of the median. For these particular y_i's, then, $\frac{k}{n} = \frac{26}{60} = 0.433$.

Table 5.3.3

0.00940*	0.75095	2.32466	0.66715*	3.38765	3.01784	0.05509*
0.93661	1.39603	0.50795*	0.11041*	2.89577	1.20041	1.44422
0.46474*	0.48272*	0.48223*	3.59149	1.38016	0.41382*	0.31684*
0.58175*	0.86681	0.55491*	0.07451*	1.88641	2.40564	1.07111
5.05936	0.04804*	0.07498*	1.52084	1.06972	0.62928*	0.09433*
1.83196	1.91987	1.92874	1.93181	0.78811	2.16919	1.16045
0.81223	1.84549	1.20752	0.11387*	0.38966*	0.42250*	0.77279
1.31728	0.81077	0.59111*	0.36793*	0.16938*	2.41135	0.21528*
0.54938*	0.73217	0.52019*	0.73169			

$*$ number ≤ 0.69315 [= median of $f_Y(y) = e^{-y}$, $y > 0$]

Let p denote the (unknown) probability that a random observation produced by Minitab's generator will lie to the left of the pdf's median. Based on these sixty observations, the 95% confidence interval for p is the range of numbers extending from *0.308* to *0.558:*

$$\left(\frac{26}{30} - 1.96\sqrt{\frac{(26/60)(1-26/60)}{60}},\ \frac{26}{60} + 1.96\sqrt{\frac{(26/60)(1-26/60)}{60}}\right) = (0.308, 0.558)$$

The fact that the value $p = 0.50$ is contained in the confidence interval implies that these data *do* pass the median test. It is entirely believable, in other words, that a bona fide exponential random sample of size 60 would have twenty-six observations falling below the pdf's median, and thirty-four above. ■

Margin of Error

In the popular press, estimates for p (i.e., values of $\frac{k}{n}$) are typically accompanied by a *margin of error*, as opposed to a confidence interval. The two are related: A margin of error is half the maximum width of a 95% confidence interval. (The number actually quoted is usually expressed as a percentage.)

Let w denote the width of a 95% confidence interval for p. From Theorem 5.3.1,

$$\begin{aligned} w &= \frac{k}{n} + 1.96\sqrt{\frac{(k/n)(1-k/n)}{n}} - \left[\frac{k}{n} - 1.96\sqrt{\frac{(k/n)(1-k/n)}{n}}\right] \\ &= 3.92\sqrt{\frac{(k/n)(1-k/n)}{n}} \end{aligned}$$

Notice that for fixed n, w is a function of the product $\left(\frac{k}{n}\right)\left(1-\frac{k}{n}\right)$. But given that $0 \le \frac{k}{n} \le 1$, the largest value that $\left(\frac{k}{n}\right)\left(1-\frac{k}{n}\right)$ can achieve is $\frac{1}{2} \cdot \frac{1}{2}$, or $\frac{1}{4}$ (see Question 5.3.18). Therefore,

$$\max w = 3.92\sqrt{\frac{1}{4n}}$$

Definition 5.3.1. The *margin of error* associated with an estimate $\frac{k}{n}$, where k is the number of successes in n independent trials, is $100d\%$, where

$$d = \frac{1.96}{2\sqrt{n}}$$

Example 5.3.3

In the mid-term elections of 2006, the political winds were shifting. One of the key races for control of the Senate was in Virginia, where challenger Jim Webb and incumbent George Allen were in a very tight race. Just a week before the election, the Associated Press reported on a CNN poll based on telephone interviews of 597 registered voters who identified themselves as likely to vote. Webb was the choice of 299 of those surveyed. The article went on to state, "Because Webb's edge is equal to the margin of error of plus or minus 4 percentage points, it means that he can be considered slightly ahead."

Is the margin of error in fact 4%? Applying Definition 5.3.1 (with $n = 597$) shows that the margin of error associated with the poll's result, using a 95% confidence interval, is indeed 4%:

$$\frac{1.96}{2\sqrt{597}} = 0.040$$

Notice that the margin of error has nothing to do with the actual survey results. Had the percentage of respondents preferring Webb been 25%, 75%, or any other number, the margin of error, by definition, would have been the same.

The more important question is whether these results have any real meaning in what was clearly to be a close election.

Source: http://archive.newsmax.com/archives/ic/2006/10/31/72811.shtml?s=ic. ■

About the Data Example 5.3.3 shows how the use of the margin of error has been badly handled by the media. The faulty interpretations are particularly prevalent in the context of political polls, especially since media reports of polls fail to give the confidence level, which is always taken to be 95%. Another issue is whether the confidence intervals provided are in fact useful. In Example 5.3.3, the 95% confidence interval has margin of error 4% and is

$$(0.501 - 0.040, 0.501 + 0.040) = (0.461, 0.541)$$

However, such a margin of error yields a confidence interval that is too wide to provide any meaningful information. The campaign had had media attention for months. Even a less-than-astute political observer would have been quite certain that the proportion of people voting for Webb would be between 0.461 and 0.541. As it turned out, the race was as close as predicted, and Webb won by a margin of just over seven thousand votes out of more than two million cast.

Even when political races are not as close as the Webb–Allen race, persistent misinterpretations abound. Here is what happens. A poll (based on a sample of n voters) is conducted, showing, for example, that 52% of the respondents intend to support Candidate A and 48%, Candidate B. Moreover, the corresponding margin of error, based on the sample of size n, is (correctly) reported to be, say, 5%. What often comes next is a statement that the race is a "statistical tie" or a "statistical dead heat" because the difference between the two percentages, $52\% - 48\% = 4\%$, is *within* the 5% margin of error. Is that statement true? No. Is it even close to being true? No.

If the observed difference in the percentages supporting Candidate A and Candidate B is 4% and the margin of error is 5%, then the widest possible 95% confidence interval for p, the true difference between the two percentages ($p =$ Candidate A's true % – Candidate B's true %) would be

$$(4\% - 5\%, 4\% + 5\%) = (-1\%, 9\%)$$

The latter implies that we should not rule out the possibility that the true value for p could be as small as -1% (in which case Candidate B would win a tight race) or as large as $+9\%$ (in which case Candidate A would win in a landslide). The serious mistake in the "statistical tie" terminology is the implication that all the possible values from -1% to $+9\%$ are equally likely. That is simply not true. For every confidence interval, parameter values near the center are much more plausible than those near either the left-hand or right-hand endpoints. Here, a 4% lead for Candidate A in a poll that has a 5% margin of error is not a "tie"—quite the contrary, it would more properly be interpreted as almost a guarantee that Candidate A will win.

Misinterpretations aside, there is yet a more fundamental problem in using the margin of error as a measure of the day-to-day or week-to-week variation in political polls. By definition, the margin of error refers to *sampling* variation—that is, it reflects the extent to which the estimator $\hat{p} = \frac{X}{n}$ varies if repeated samples of size n are drawn from the *same* population. Consecutive political polls, though, do not represent the same population. Between one poll and the next, a variety of scenarios can transpire that can fundamentally change the opinions of the voting population—one candidate may give an especially good speech or make an embarrassing gaffe, a scandal can emerge that seriously damages someone's reputation, or a world event comes to pass that for one reason or another reflects more negatively on one candidate than the other. Although all of these possibilities have the potential to influence the value of $\frac{X}{n}$ much more than sampling variability can, none of them is included in the margin of error.

Choosing Sample Sizes

Related to confidence intervals and margins of error is an important experimental design question. Suppose a researcher wishes to estimate the binomial parameter p based on results from a series of n independent trials, *but n has yet to be determined.* Larger values of n will, of course, yield estimates having greater precision, but more observations also demand greater expenditures of time and money. How can those two concerns best be reconciled?

If the experimenter can articulate the minimal degree of precision that would be considered acceptable, a Z transformation can be used to calculate the smallest (i.e., the cheapest) sample size capable of achieving that objective. For example, suppose we want $\frac{X}{n}$ to have at least a $100(1-\alpha)\%$ probability of lying within a distance d of p. The problem is solved, then, if we can find the smallest n for which

$$P\left(-d \leq \frac{X}{n} - p \leq d\right) = 1 - \alpha \tag{5.3.3}$$

Theorem 5.3.2 *Let $\frac{X}{n}$ be the estimator for the parameter p in a binomial distribution. In order for $\frac{X}{n}$ to have at least a $100(1-\alpha)\%$ probability of being within a distance d of p, the sample size should be no smaller than*

$$n = \frac{z_{\alpha/2}^2}{4d^2}$$

where $z_{\alpha/2}$ is the value for which $P(Z \geq z_{\alpha/2}) = \alpha/2$.

Proof Start by dividing the terms in the probability portion of Equation 5.3.3 by the standard deviation of $\frac{X}{n}$ to form an approximate Z ratio:

$$P\left(-d \leq \frac{X}{n} - p \leq d\right) = P\left[\frac{-d}{\sqrt{p(1-p)/n}} \leq \frac{X/n - p}{\sqrt{p(1-p)/n}} \leq \frac{d}{\sqrt{p(1-p)/n}}\right]$$

$$\doteq P\left[\frac{-d}{\sqrt{p(1-p)/n}} \leq Z \leq \frac{d}{\sqrt{p(1-p)/n}}\right] = 1 - \alpha$$

But $P(-z_{\alpha/2} \le Z \le z_{\alpha/2}) = 1 - \alpha$, so

$$\frac{d}{\sqrt{p(1-p)/n}} = z_{\alpha/2}$$

which implies that

$$n = \frac{z_{\alpha/2}^2 p(1-p)}{d^2} \tag{5.3.4}$$

Equation 5.3.4 is not an acceptable final answer, though, because the right-hand side is a function of p, the unknown parameter. But $p(1-p) \le \frac{1}{4}$ for $0 \le p \le 1$, so the sample size

$$n = \frac{z_{\alpha/2}^2}{4d^2}$$

would necessarily cause $\frac{X}{n}$ to satisfy Equation 5.3.3, regardless of the actual value of p. (Notice the connection between the statements of Theorem 5.3.2 and Definition 5.3.1.) □

Example 5.3.4

A public health survey is being planned in a large metropolitan area for the purpose of estimating the proportion of children, ages zero to fourteen, who are lacking adequate polio immunization. Organizers of the project would like the sample proportion of inadequately immunized children, $\frac{X}{n}$, to have at least a 98% probability of being within 0.05 of the true proportion, p. How large should the sample be?

Here $100(1-\alpha) = 98$, so $\alpha = 0.02$ and $z_{\alpha/2} = 2.33$. By Theorem 5.3.2, then, the smallest acceptable sample size is *543*:

$$n = \frac{(2.33)^2}{4(0.05)^2} = 543$$

Comment Occasionally, there may be reason to believe that p is necessarily less than some number r_1, where $r_1 < \frac{1}{2}$, or greater than some number r_2, where $r_2 > \frac{1}{2}$. If so, the factors $p(1-p)$ in Equation 5.3.4 can be replaced by either $r_1(1-r_1)$ or $r_2(1-r_2)$, and the sample size required to estimate p with a specified precision will be reduced, perhaps by a considerable amount.

Suppose, for example, that previous immunization studies suggest that no more than 20% of children between the ages of zero and fourteen are inadequately immunized. The smallest sample size, then, for which

$$P\left(-0.05 \le \frac{X}{n} - p \le 0.05\right) = 0.98$$

is *348*, an n that represents almost a 36% reduction $\left(= \frac{543-348}{543} \times 100\right)$ from the original 543:

$$n = \frac{(2.33)^2}{(0.05)^2}(0.20)(0.80) = 348$$

Comment Theorems 5.3.1 and 5.3.2 are both based on the assumption that the X in $\frac{X}{n}$ varies according to a binomial model. What we learned in Section 3.3, though, seems to contradict that assumption: Samples used in opinion surveys are invariably drawn *without replacement*, in which case X is hypergeometric, not binomial. The consequences of that particular "error," however, are easily corrected and frequently negligible.

It can be shown mathematically that the expected value of $\frac{X}{n}$ is the same regardless of whether X is binomial or hypergeometric; its variance, though, is different. If X is binomial,

$$\text{Var}\left(\frac{X}{n}\right) = \frac{p(1-p)}{n}$$

If X is hypergeometric,

$$\text{Var}\left(\frac{X}{n}\right) = \frac{p(1-p)}{n}\left(\frac{N-n}{N-1}\right)$$

where N is the total number of subjects in the population.

Since $\frac{N-n}{N-1} < 1$, the actual variance of $\frac{X}{n}$ is somewhat smaller than the (binomial) variance we have been assuming, $\frac{p(1-p)}{n}$. The ratio $\frac{N-n}{N-1}$ is called the *finite correction factor.* If N is much larger than n, which is typically the case, then the magnitude of $\frac{N-n}{N-1}$ will be so close to 1 that the variance of $\frac{X}{n}$ is equal to $\frac{p(1-p)}{n}$ for all practical purposes. Thus the "binomial" assumption in those situations is more than adequate. Only when the sample is a sizeable fraction of the population do we need to include the finite correction factor in any calculations that involve the variance of $\frac{X}{n}$. ■

Questions

5.3.1. A commonly used IQ test is scaled to have a mean of 100 and a standard deviation of $\sigma = 15$. A school counselor was curious about the average IQ of the students in her school and took a random sample of fifty students' IQ scores. The average of these was $\overline{y} = 107.9$. Find a 95% confidence interval for the student IQ in the school.

5.3.2. The production of a nationally marketed detergent results in certain workers receiving prolonged exposures to a *Bacillus subtilis* enzyme. Nineteen workers were tested to determine the effects of those exposures, if any, on various respiratory functions. One such function, air-flow rate, is measured by computing the ratio of a person's forced expiratory volume (FEV_1) to his or her vital capacity (VC). (Vital capacity is the maximum volume of air a person can exhale after taking as deep a breath as possible; FEV_1 is the maximum volume of air a person can exhale in one second.) In persons with no lung dysfunction, the "norm" for FEV_1/VC ratios is 0.80. Based on the following data (164), is it believable that exposure to the *Bacillus subtilis* enzyme has no effect on the FEV_1/VC ratio? Answer the question by constructing a 95% confidence interval. Assume that FEV_1/VC ratios are normally distributed with $\sigma = 0.09$.

Subject	FEV_1/VC	Subject	FEV_1/VC
RH	0.61	WS	0.78
RB	0.70	RV	0.84
MB	0.63	EN	0.83
DM	0.76	WD	0.82
WB	0.67	FR	0.74
RB	0.72	PD	0.85
BF	0.64	EB	0.73
JT	0.82	PC	0.85
PS	0.88	RW	0.87
RB	0.82		

5.3.3. Mercury pollution is widely recognized as a serious ecological problem. Much of the mercury released into the environment originates as a byproduct of coal burning and other industrial processes. It does not become dangerous until it falls into large bodies of water, where microorganisms convert it to methylmercury (CH_3^{203}), an organic form that is particularly toxic. Fish are the intermediaries: They ingest and absorb the methylmercury and are then eaten by humans. Men and women, however, may not metabolize CH_3^{203} at the same rate. In one study investigating that issue, six women were given a known amount of protein-bound methylmercury. Shown in the following table are the half-lives of the methylmercury in their systems (114). For men, the average CH_3^{203} half-life is believed to be eighty days. Assume that for both genders, CH_3^{203} half-lives are normally distributed with a standard deviation (σ) of eight days. Construct a 95% confidence interval for the true female CH_3^{203} half-life. Based on these data, is it believable that males and females metabolize methylmercury at the same rate? Explain.

Females	CH_3^{203} Half-Life
AE	52
EH	69
LJ	73
AN	88
KR	87
LU	56

5.3.4. A physician who has a group of thirty-eight female patients aged 18 to 24 on a special diet wishes to estimate the effect of the diet on total serum cholesterol. For this group, their average serum cholesterol is 188.4 (measured in mg/100mL). Because of a large-scale government study, the physician is willing to assume that the total serum cholesterol measurements are normally distributed with standard deviation of $\sigma = 40.7$. Find a 95% confidence interval of the mean serum cholesterol of patients on the special diet. Does the diet seem to have any effect on their serum cholesterol, given that the national average for women aged 18 to 24 is 192.0?

5.3.5. Suppose a sample of size n is to be drawn from a normal distribution where σ is known to be 14.3. How large does n have to be to guarantee that the length of the 95% confidence interval for μ will be less than 3.06?

5.3.6. What "confidence" would be associated with each of the following intervals? Assume that the random variable Y is normally distributed and that σ is known.

(a) $\left(\overline{y} - 1.64 \cdot \frac{\sigma}{\sqrt{n}}, \overline{y} + 2.33 \cdot \frac{\sigma}{\sqrt{n}}\right)$

(b) $\left(-\infty, \overline{y} + 2.58 \cdot \frac{\sigma}{\sqrt{n}}\right)$

(c) $\left(\overline{y} - 1.64 \cdot \frac{\sigma}{\sqrt{n}}, \overline{y}\right)$

5.3.7. Five independent samples, each of size n, are to be drawn from a normal distribution where σ is known. For each sample, the interval $\left(\overline{y} - 0.96 \cdot \frac{\sigma}{\sqrt{n}}, \overline{y} + 1.06 \cdot \frac{\sigma}{\sqrt{n}}\right)$ will be constructed. What is the probability that at least four of the intervals will contain the unknown μ?

5.3.8. Suppose that $y_1, y_2, \ldots, y_n$ is a random sample of size n from a normal distribution where σ is known. Depending on how the tail-area probabilities are split up, an infinite number of random intervals having a 95% probability of containing μ can be constructed. What is unique about the particular interval $\left(\overline{y} - 1.96 \cdot \frac{\sigma}{\sqrt{n}}, \overline{y} + 1.96 \cdot \frac{\sigma}{\sqrt{n}}\right)$?

5.3.9. If the standard deviation (σ) associated with the pdf that produced the following sample is 3.6, would it be correct to claim that

$$\left(2.61 - 1.96 \cdot \frac{3.6}{\sqrt{20}}, 2.61 + 1.96 \cdot \frac{3.6}{\sqrt{20}}\right) = (1.03, 4.19)$$

is a 95% confidence interval for μ? Explain.

2.5	0.1	0.2	1.3
3.2	0.1	0.1	1.4
0.5	0.2	0.4	11.2
0.4	7.4	1.8	2.1
0.3	8.6	0.3	10.1

5.3.10. In 1927, the year he hit sixty home runs, Babe Ruth batted .356, having collected 192 hits in 540 official at-bats (140). Based on his performance that season, construct a 95% confidence interval for Ruth's probability of getting a hit in a future at-bat.

5.3.11. To buy a thirty-second commercial break during the telecast of Super Bowl XXIX cost approximately \$1,000,000. Not surprisingly, potential sponsors wanted to know how many people might be watching. In a survey of 1015 potential viewers, 281 said they expected to see less than a quarter of the advertisements aired during the game. Define the relevant parameter and estimate it using a 90% confidence interval.

5.3.12. During one of the first "beer wars" in the early 1980s, a taste test between Schlitz and Budweiser was the focus of a nationally broadcast TV commercial. One hundred people agreed to drink from two unmarked mugs and indicate which of the two beers they liked better; fifty-four said, "Bud." Construct and interpret the corresponding 95% confidence interval for p, the true proportion of beer drinkers who prefered Budweiser to Schlitz. How would Budweiser and Schlitz executives each have put these results in the best possible light for their respective companies?

5.3.13. The Pew Research Center did a survey of 2253 adults and discovered that 63% of them had broadband Internet connections in their homes. The survey report noted that this figure represented a "significant jump" from the similar figure of 54% from two years earlier. One way to define "significant jump" is to show that the earlier number does not lie in the 95% confidence interval. Was the increase significant by this definition?

Source: http://www.pewinternet.org/Reports/2009/10-Home-Broadband-Adoption-2009.aspx.

5.3.14. If (0.57, 0.63) is a 50% confidence interval for p, what does $\frac{k}{n}$ equal and how many observations were taken?

5.3.15. Suppose a coin is to be tossed n times for the purpose of estimating p, where $p = P(\text{heads})$. How large must n be to guarantee that the length of the 99% confidence interval for p will be less than 0.02?

5.3.16. On the morning of November 9, 1994—the day after the electoral landslide that had returned Republicans to power in both branches of Congress—several key races were still in doubt. The most prominent was the Washington contest involving Democrat Tom Foley, the reigning speaker of the house. An Associated Press story showed how narrow the margin had become (120):

> With 99 percent of precincts reporting, Foley trailed Republican challenger George Nethercutt by just 2,174 votes, or 50.6 percent to 49.4 percent. About 14,000 absentee ballots remained uncounted, making the race too close to call.

Let $p = P(\text{Absentee voter prefers Foley})$. How small could p have been and still have given Foley a 20% chance of overcoming Nethercutt's lead and winning the election?

5.3.17. Which of the following two intervals has the greater probability of containing the binomial parameter p?

$$\left[\frac{X}{n} - 0.67\sqrt{\frac{(X/n)(1-X/n)}{n}},\ \frac{X}{n} + 0.67\sqrt{\frac{(X/n)(1-X/n)}{n}}\right]$$

$$\text{or} \quad \left(\frac{X}{n}, \infty\right)$$

5.3.18. Examine the first two derivatives of the function $g(p) = p(1-p)$ to verify the claim on p. 305 that $p(1-p) \leq \frac{1}{4}$ for $0 < p < 1$.

5.3.19. The financial crisis of 2008 highlighted the issue of excessive compensation for business CEOs. In a Gallup poll in the summer of 2009, 998 adults were asked, "Do you favor or oppose the federal government taking steps to limit the pay of executives at major companies?", with 59% responding in favor. The report of the poll noted a margin of error of ± 3 percentage points. Verify the margin of error and construct a 95% confidence interval.

Source: http://www.gallup.com/poll/120872/Americans-Favor-Gov-Action-Limit-Executive-Pay.aspx.

5.3.20. Viral infections contracted early during a woman's pregnancy can be very harmful to the fetus. One study found a total of 86 deaths and birth defects among 202 pregnancies complicated by a first-trimester German measles infection (45). Is it believable that the true proportion of abnormal births under similar circumstances could be as high as 50%? Answer the question by calculating the margin of error for the sample proportion, 86/202.

5.3.21. Rewrite Definition 5.3.1 to cover the case where a finite correction factor needs to be included (i.e., situations where the sample size n is not negligible relative to the population size N).

5.3.22. A public health official is planning for the supply of influenza vaccine needed for the upcoming flu season. She took a poll of 350 local citizens and found that only 126 said they would be vaccinated.

(a) Find the 90% confidence interval for the true proportion of people who plan to get the vaccine.
(b) Find the confidence interval, including the finite correction factor, assuming the town's population is 3000.

5.3.23. Given that n observations will produce a binomial parameter estimator, $\frac{X}{n}$, having a margin of error equal to 0.06, how many observations are required for the proportion to have a margin of error half that size?

5.3.24. Given that a political poll shows that 52% of the sample favors Candidate A, whereas 48% would vote for Candidate B, and given that the margin of error associated with the survey is 0.05, does it make sense to claim that the two candidates are tied? Explain.

5.3.25. Assume that the binomial parameter p is to be estimated with the function $\frac{X}{n}$, where X is the number of successes in n independent trials. Which demands the larger sample size: requiring that $\frac{X}{n}$ have a 96% probability of being within 0.05 of p, or requiring that $\frac{X}{n}$ have a 92% probability of being within 0.04 of p?

5.3.26. Suppose that p is to be estimated by $\frac{X}{n}$ and we are willing to assume that the true p will not be greater than 0.4. What is the smallest n for which $\frac{X}{n}$ will have a 99% probability of being within 0.05 of p?

5.3.27. Let p denote the true proportion of college students who support the movement to colorize classic films. Let the random variable X denote the number of students (out of n) who prefer colorized versions to black and white. What is the smallest sample size for which the probability is 80% that the difference between $\frac{X}{n}$ and p is less than 0.02?

5.3.28. University officials are planning to audit 1586 new appointments to estimate the proportion p who have been incorrectly processed by the payroll department.

(a) How large does the sample size need to be in order for $\frac{X}{n}$, the sample proportion, to have an 85% chance of lying within 0.03 of p?
(b) Past audits suggest that p will not be larger than 0.10. Using that information, recalculate the sample size asked for in part (a).

5.4 Properties of Estimators

The method of maximum likelihood and the method of moments described in Section 5.2 both use very reasonable criteria to identify estimators for unknown parameters, yet the two do not always yield the same answer. For example, given that $Y_1, Y_2, \ldots, Y_n$ is a random sample from the pdf $f_Y(y;\theta) = \frac{2y}{\theta^2}, 0 \le y \le \theta$, the maximum likelihood estimator for θ is $\hat{\theta} = Y_{\max}$ while the method of moments estimator is $\hat{\theta} = \frac{3}{2}\overline{Y}$. (See Questions 5.2.12 and 5.2.15.) Implicit in those two formulas is an obvious question—which should we use?

More generally, the fact that parameters have multiple estimators (actually, an infinite number of $\hat{\theta}$'s can be found for any given θ) requires that we investigate the statistical properties associated with the estimation process. What qualities should a "good" estimator have? Is it possible to find a "best" $\hat{\theta}$? These and other questions relating to the theory of estimation will be addressed in the next several sections.

To understand the *mathematics* of estimation, we must first keep in mind that every estimator is a function of a set of random variables—that is, $\hat{\theta} = h(Y_1, Y_2, \ldots, Y_n)$. As such, any $\hat{\theta}$, itself, is a random variable: It has a pdf, an expected value, and a variance, all three of which play key roles in evaluating its capabilities.

We will denote the pdf of an estimator (at some point u) with the symbol $f_{\hat{\theta}}(u)$ or $p_{\hat{\theta}}(u)$, depending on whether $\hat{\theta}$ is a continuous or a discrete random variable. Probability calculations involving θ will reduce to integrals of $f_{\hat{\theta}}(u)$ (if $\hat{\theta}$ is continuous) or sums of $p_{\hat{\theta}}(u)$ (if $\hat{\theta}$ is discrete).

Example 5.4.1

a. Suppose a coin, for which $p = P(\text{heads})$ is unknown, is to be tossed ten times for the purpose of estimating p with the function $\hat{p} = \frac{X}{10}$, where X is the observed number of heads. *If* $p = 0.60$, what is the probability that $\left|\frac{X}{10} - 0.60\right| \le 0.10$? That is, what are the chances that the estimator will fall within 0.10 of the true value of the parameter? Here $\hat{p}$ is discrete—the only values $\frac{X}{10}$ can take on are $\frac{0}{10}, \frac{1}{10}, \ldots, \frac{10}{10}$. Moreover, when $p = 0.60$,

$$p_{\hat{p}}\left(\frac{k}{10}\right) = P\left(\hat{p} = \frac{k}{10}\right) = P(X = k) = \binom{10}{k}(0.60)^k(0.40)^{10-k}, \quad k = 0, 1, \ldots, 10$$

Therefore,

$$\begin{aligned} P\left(\left|\frac{X}{10} - 0.60\right| \le 0.10\right) &= P\left(0.60 - 0.10 \le \frac{X}{10} \le 0.60 + 0.10\right) \\ &= P(5 \le X \le 7) \\ &= \sum_{k=5}^{7}\binom{10}{k}(0.60)^k(0.40)^{10-k} \\ &= 0.6665 \end{aligned}$$

b. How likely is the estimator $\frac{X}{n}$ to lie within 0.10 of p if the coin in part (a) is tossed *one hundred* times? Given that n is so large, a Z transformation can be

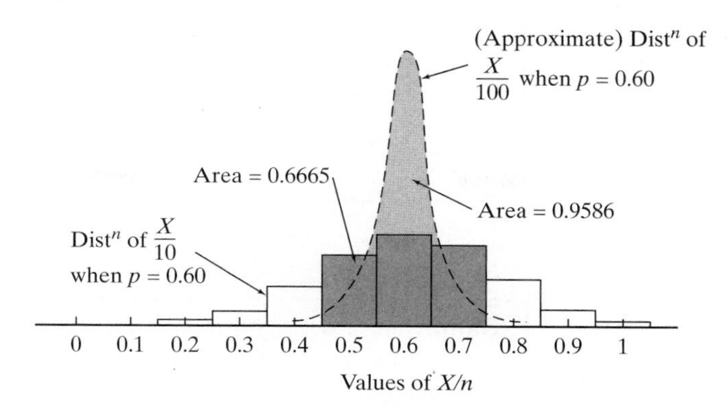

Figure 5.4.1

used to approximate the variation in $\frac{X}{100}$. Since $E\left(\frac{X}{n}\right) = p$ and $\text{Var}\left(\frac{X}{n}\right) = p(1-p)/n$, we can write

$$P\left(\left|\frac{X}{100} - 0.60\right| \leq 0.10\right) = P\left(0.50 \leq \frac{X}{100} \leq 0.70\right)$$

$$= P\left[\frac{0.50 - 0.60}{\sqrt{\frac{(0.60)(0.40)}{100}}} \leq \frac{X/100 - 0.60}{\sqrt{\frac{(0.60)(0.40)}{100}}} \leq \frac{0.70 - 0.60}{\sqrt{\frac{(0.60)(0.40)}{100}}}\right]$$

$$\doteq P(-2.04 \leq Z \leq 2.04)$$

$$= 0.9586$$

Figure 5.4.1 shows the two probabilities just calculated as areas under the probability functions describing $\frac{X}{10}$ and $\frac{X}{100}$. As we would expect, the larger sample size produces a more precise estimator—with $n = 10$, $\frac{X}{10}$ has only a 67% chance of lying in the range from 0.50 to 0.70; for $n = 100$, though, the probability of $\frac{X}{100}$ falling within 0.10 of the true $p\,(= 0.60)$ increases to 96%.

Are the additional ninety observations worth the gain in precision that we see in Figure 5.4.1? Maybe yes and maybe no. In general, the answer to that sort of question depends on two factors: (1) the cost of taking additional measurements, and (2) the cost of making bad decisions or inappropriate inferences because of inaccurate estimates. In practice, both costs—especially the latter—can be very difficult to quantify. ■

Unbiasedness

Because they are random variables, estimators will take on different values from sample to sample. Typically, some samples will yield θ_e's that underestimate θ while others will lead to θ_e's that are numerically too large. Intuitively, we would like the underestimates to somehow "balance out" the overestimates—that is, $\hat{\theta}$ should not systematically err in any one particular direction.

Figure 5.4.2 shows the pdfs for two estimators, $\hat{\theta}_1$ and $\hat{\theta}_2$. Common sense tells us that $\hat{\theta}_1$ is the better of the two because $f_{\hat{\theta}_1}(u)$ is centered with respect to the true θ; $\hat{\theta}_2$, on the other hand, will tend to give estimates that are too large because the bulk of $f_{\hat{\theta}_2}(u)$ lies to the right of the true θ.

Figure 5.4.2

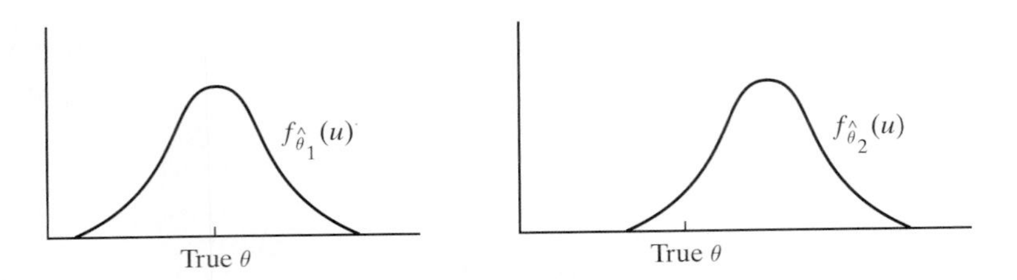

Definition 5.4.1. Suppose that $Y_1, Y_2, \ldots, Y_n$ is a random sample from the continuous pdf $f_Y(y;\theta)$, where θ is an unknown parameter. An estimator $\hat{\theta}\,[= h(Y_1, Y_2, \ldots, Y_n)]$ is said to be *unbiased* (for θ) if $E(\hat{\theta}) = \theta$ for all θ. [The same concept and terminology apply if the data consist of a random sample $X_1, X_2, \ldots, X_n$ drawn from a discrete pdf $p_X(k;\theta)$].

Example 5.4.2

It was mentioned at the outset of this section that $\hat{\theta}_1 = \frac{3}{2}\overline{Y}$ and $\hat{\theta}_2 = Y_{\max}$ are two estimators for θ in the pdf $f_Y(y;\,\theta) = \frac{2y}{\theta^2}, 0 \le y \le \theta$. Are either or both unbiased?

First we need $E(Y)$, which is $\int_0^\theta y \cdot \frac{2y}{\theta^2} dy = \frac{2}{3}\theta$. Then using the properties of expected values, we can show that $\hat{\theta}_1$ is unbiased for all θ:

$$E(\hat{\theta}_1) = E\left(\frac{3}{2}\overline{Y}\right) = \frac{3}{2}E(\overline{Y}) = \frac{3}{2}E(Y) = \frac{3}{2}\cdot\frac{2}{3}\theta = \theta$$

The maximum likelihood estimator, on the other hand, is obviously biased—since $Y_{\max}$ is necessarily less than or equal to θ, its pdf will not be centered with respect to θ, and $E(Y_{\max})$ will be *less than* θ. The exact factor by which $Y_{\max}$ tends to underestimate θ is readily calculated. Recall from Theorem 3.10.1 that

$$f_{Y_{\max}}(y) = nF_Y(y)^{n-1} f_Y(y)$$

The cdf for Y is

$$F_Y(y) = \int_0^y \frac{2t}{\theta^2} dt = \frac{y^2}{\theta^2}$$

Then

$$f_{Y_{\max}}(y) = n\left(\frac{y^2}{\theta^2}\right)^{n-1}\frac{2y}{\theta^2} = \frac{2n}{\theta^{2n}}y^{2n-1}, 0 \le y \le \theta$$

Therefore,

$$E(Y_{\max}) = \int_0^\theta y \cdot \frac{2n}{\theta^{2n}}y^{2n-1}dy = \frac{2n}{\theta^{2n}}\int_0^\theta y^{2n}dy = \frac{2n}{\theta^{2n}}\cdot\frac{\theta^{2n+1}}{2n+1} = \frac{2n}{2n+1}\theta$$

$\lim\limits_{n\to\infty} \frac{2n}{2n+1}\theta = \theta$. Intuitively, this decrease in the bias makes sense because $f_{\hat{\theta}_2}$ becomes increasingly concentrated around θ as n grows. ■

Comment For any finite n, we can construct an estimator based on $Y_{\max}$ that *is* unbiased. Let $\hat{\theta}_3 = \frac{2n+1}{2n}\cdot Y_{\max}$. Then

$$E(\hat{\theta}_3) = E\left(\frac{2n+1}{2n}\cdot Y_{\max}\right) = \frac{2n+1}{2n}E(Y_{\max}) = \frac{2n+1}{2n}\cdot\frac{2n}{2n+1}\theta = \theta$$

Example 5.4.3

Let $X_1, X_2, \ldots, X_n$ be a random sample from a discrete pdf $p_X(k;\theta)$, where $\theta = E(X)$ is an unknown parameter. Consider the estimator

$$\hat{\theta} = \sum_{i=1}^{n} a_i X_i$$

where the a_i's are constants. For what values of $a_1, a_2, \ldots, a_n$ will $\hat{\theta}$ be unbiased?

By assumption, $\theta = E(X)$, so

$$\begin{aligned} E(\hat{\theta}) &= E\left(\sum_{i=1}^{n} a_i X_i\right) \\ &= \sum_{i=1}^{n} a_i E(X_i) = \sum_{i=1}^{n} a_i \theta \\ &= \theta \sum_{i=1}^{n} a_i \end{aligned}$$

Clearly, $\hat{\theta}$ will be unbiased for any set of a_i's for which $\sum_{i=1}^{n} a_i = 1$. ■

Example 5.4.4

Given a random sample $Y_1, Y_2, \ldots, Y_n$ from a normal distribution whose parameters μ and σ^2 are both unknown, the maximum likelihood estimator for σ^2 is

$$\hat{\sigma}^2 = \frac{1}{n} \sum_{i=1}^{n} (Y_i - \overline{Y})^2$$

(recall Example 5.2.4). Is $\hat{\sigma}^2$ unbiased for σ^2? If not, what function of $\hat{\sigma}^2$ does have an expected value equal to σ^2?

Notice, first, from Theorem 3.6.1 that for any random variable Y, $\text{Var}(Y) = E(Y^2) - [E(Y)]^2$. Also, from Section 3.9, for any average, $\overline{Y}$, of a sample of n random variables, $Y_1, Y_2, \ldots, Y_n$, $E(\overline{Y}) = E(Y_i)$ and $\text{Var}(\overline{Y}) = (1/n)\text{Var}(Y_i)$. Using those results, we can write

$$\begin{aligned} E(\hat{\sigma}^2) &= E\left[\frac{1}{n} \sum_{i=1}^{n} (Y_i - \overline{Y})^2\right] \\ &= E\left[\frac{1}{n} \sum_{i=1}^{n} \left(Y_i^2 - 2Y_i\overline{Y} + \overline{Y}^2\right)\right] \\ &= E\left[\frac{1}{n} \left(\sum_{i=1}^{n} Y_i^2 - n\overline{Y}^2\right)\right] \\ &= \frac{1}{n}\left[\sum_{i=1}^{n} E\left(Y_i^2\right) - nE(\overline{Y}^2)\right] \\ &= \frac{1}{n}\left[\sum_{i=1}^{n} (\sigma^2 + \mu^2) - n(\frac{\sigma^2}{n} + \mu^2)\right] \\ &= \frac{n-1}{n}\sigma^2 \end{aligned}$$

Since the latter is not equal to σ^2, $\hat{\sigma}^2$ is biased.

To "unbias" the maximum likelihood estimator in this case, we need simply multiply $\hat{\sigma}^2$ by $\frac{n}{n-1}$. By convention, the unbiased version of the maximum likelihood estimator for σ^2 in a normal distribution is denoted S^2 and is referred to as the *sample variance*:

$$S^2 = \text{sample variance} = \frac{n}{n-1} \cdot \frac{1}{n} \sum_{i=1}^{n} (Y_i - \overline{Y})^2$$

$$= \frac{1}{n-1} \sum_{i=1}^{n} (Y_i - \overline{Y})^2$$

Comment The square root of the sample variance is called the *sample standard deviation*:

$$S = \text{sample standard deviation} = \sqrt{\frac{1}{n-1} \sum_{i=1}^{n} (Y_i - \overline{Y})^2}$$

In practice, S is the most commonly used estimator for σ *even though* $E(S) \neq \sigma$ [despite the fact that $E(S^2) = \sigma^2$]. ■

Questions

5.4.1. Two chips are drawn without replacement from an urn containing five chips, numbered 1 through 5. The average of the two drawn is to be used as an estimator, $\hat{\theta}$, for the true average of all the chips ($\theta = 3$). Calculate $P(|\hat{\theta} - 3| > 1.0)$.

5.4.2. Suppose a random sample of size $n = 6$ is drawn from the uniform pdf $f_Y(y;\theta) = 1/\theta, 0 \le y \le \theta$, for the purpose of using $\hat{\theta} = Y_{\max}$ to estimate θ.

(a) Calculate the probability that $\hat{\theta}$ falls within 0.2 of θ given that the parameter's true value is 3.0.
(b) Calculate the probability of the event asked for in part (a), assuming the sample size is 3 instead of 6.

5.4.3. Five hundred adults are asked whether they favor a bipartisan campaign finance reform bill. If the true proportion of the electorate in favor of the legislation is 52%, what are the chances that fewer than half of those in the sample support the proposal? Use a Z transformation to approximate the answer.

5.4.4. A sample of size $n = 16$ is drawn from a normal distribution where $\sigma = 10$ but μ is unknown. If $\mu = 20$, what is the probability that the estimator $\hat{\mu} = \overline{Y}$ will lie between 19.0 and 21.0?

5.4.5. Suppose $X_1, X_2, \ldots, X_n$ is a random sample of size n drawn from a Poisson pdf where λ is an unknown parameter. Show that $\hat{\lambda} = \overline{X}$ is unbiased for λ. For what type of parameter, in general, will the sample mean necessarily be an unbiased estimator? (*Hint:* The answer is implicit in the derivation showing that $\overline{X}$ is unbiased for the Poisson λ.)

5.4.6. Let $Y_{\min}$ be the smallest order statistic in a random sample of size n drawn from the uniform pdf, $f_Y(y;\theta) = 1/\theta, 0 \le y \le \theta$. Find an unbiased estimator for θ based on $Y_{\min}$.

5.4.7. Let Y be the random variable described in Example 5.2.3, where $f_Y(y,\theta) = e^{-(y-\theta)}$, $y \ge \theta, \theta > 0$. Show that $Y_{\min} - \frac{1}{n}$ is an unbiased estimator of θ.

5.4.8. Suppose that 14, 10, 18, and 21 constitute a random sample of size 4 drawn from a uniform pdf defined over the interval $[0,\theta]$, where θ is unknown. Find an unbiased estimator for θ based on Y_3', the third order statistic. What numerical value does the estimator have for these particular observations? Is it possible that we would know that an estimate for θ based on Y_3' was incorrect, even if we had no idea what the true value of θ might be? Explain.

5.4.9. A random sample of size 2, Y_1 and Y_2, is drawn from the pdf

$$f_Y(y;\theta) = 2y\theta^2, \quad 0 < y < \frac{1}{\theta}$$

What must c equal if the statistic $c(Y_1 + 2Y_2)$ is to be an unbiased estimator for $\frac{1}{\theta}$?

5.4.10. A sample of size 1 is drawn from the uniform pdf defined over the interval $[0,\theta]$. Find an unbiased estimator for θ^2. (*Hint:* Is $\hat{\theta} = Y^2$ unbiased?)

5.4.11. Suppose that W is an unbiased estimator for θ. Can W^2 be an unbiased estimator for θ^2?

5.4.12. We showed in Example 5.4.4 that $\hat{\sigma}^2 = \frac{1}{n}\sum_{i=1}^{n}(Y_i - \overline{Y})^2$ is biased for σ^2. Suppose μ is known and does not have to be estimated by $\overline{Y}$. Show that $\hat{\sigma}^2 = \frac{1}{n}\sum_{i=1}^{n}(Y_i - \mu)^2$ is unbiased for σ^2.

5.4.13. As an alternative to imposing unbiasedness, an estimator's distribution can be "centered" by requiring that its median be equal to the unknown parameter θ. If it is, $\hat{\theta}$ is said to be *median unbiased*. Let $Y_1, Y_2, \ldots, Y_n$ be a random sample of size n from the uniform pdf, $f_Y(y;\theta) = 1/\theta, 0 \leq y \leq \theta$. For arbitrary n, is $\hat{\theta} = \frac{n+1}{n} \cdot Y_{\max}$ median unbiased? Is it median unbiased for any value of n?

5.4.14. Let $Y_1, Y_2, \ldots, Y_n$ be a random sample of size n from the pdf $f_Y(y;\theta) = \frac{1}{\theta}e^{-y/\theta}, y > 0$. Let $\hat{\theta} = n \cdot Y_{\min}$. Is $\hat{\theta}$ unbiased for θ? Is $\hat{\theta} = \frac{1}{n}\sum_{i=1}^{n} Y_i$ unbiased for θ?

5.4.15. An estimator $\hat{\theta}_n = h(W_1, \ldots, W_n)$ is said to be *asymptotically unbiased for* θ if $\lim_{n\to\infty} E(\hat{\theta}_n) = \theta$. Suppose W is a random variable with $E(W) = \mu$ and with variance σ^2. Show that $\overline{W}^2$ is an asymptotically unbiased estimator for μ^2.

5.4.16. Is the maximum likelihood estimator for σ^2 in a normal pdf, where both μ and σ^2 are unknown, asymptotically unbiased?

Efficiency

As we have seen, unknown parameters can have a multiplicity of unbiased estimators. For samples drawn from the uniform pdf, $f_Y(y;\theta) = 1/\theta, 0 \leq y \leq \theta$, for example, both $\hat{\theta} = \frac{n+1}{n} \cdot Y_{\max}$ and $\hat{\theta} = \frac{2}{n}\sum_{i=1}^{n} Y_i$ have expected values equal to θ. Does it matter which we choose?

Yes. Unbiasedness is not the only property we would like an estimator to have; also important is its *precision*. Figure 5.4.3 shows the pdfs associated with two hypothetical estimators, $\hat{\theta}_1$ and $\hat{\theta}_2$. Both are unbiased for θ, but $\hat{\theta}_2$ is clearly the better of the two because of its smaller variance. For any value r,

$$P(\theta - r \leq \hat{\theta}_2 \leq \theta + r) > P(\theta - r \leq \hat{\theta}_1 \leq \theta + r)$$

That is, $\hat{\theta}_2$ has a greater chance of being within a distance r of the unknown θ than does $\hat{\theta}_1$.

Definition 5.4.2. Let $\hat{\theta}_1$ and $\hat{\theta}_2$ be two unbiased estimators for a parameter θ. If

$$\text{Var}(\hat{\theta}_1) < \text{Var}(\hat{\theta}_2)$$

we say that $\hat{\theta}_1$ is *more efficient* than $\hat{\theta}_2$. Also, the *relative efficiency of* $\hat{\theta}_1$ *with respect to* $\hat{\theta}_2$ is the ratio $\text{Var}(\hat{\theta}_2)/\text{Var}(\hat{\theta}_1)$.

Figure 5.4.3

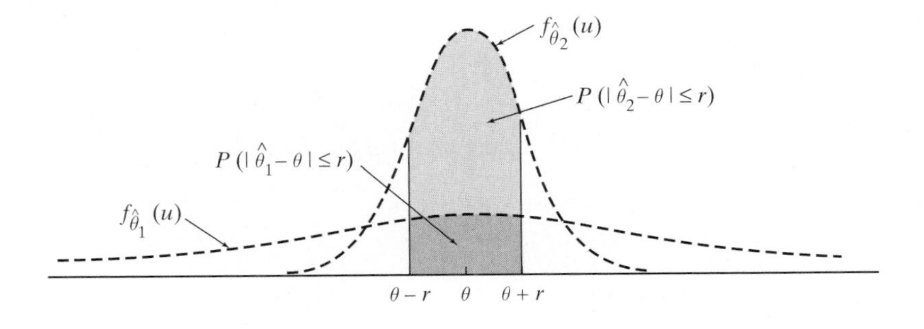

Example 5.4.5 Let Y_1, Y_2, and Y_3 be a random sample from a normal distribution where both μ and σ are unknown. Which of the following is a more efficient estimator for μ?

$$\hat{\mu}_1 = \frac{1}{4}Y_1 + \frac{1}{2}Y_2 + \frac{1}{4}Y_3$$

or

$$\hat{\mu}_2 = \frac{1}{3}Y_1 + \frac{1}{3}Y_2 + \frac{1}{3}Y_3$$

Notice, first, that both $\hat{\mu}_1$ and $\hat{\mu}_2$ are unbiased for μ:

$$\begin{aligned} E(\hat{\mu}_1) &= E\left(\frac{1}{4}Y_1 + \frac{1}{2}Y_2 + \frac{1}{4}Y_3\right) \\ &= \frac{1}{4}E(Y_1) + \frac{1}{2}E(Y_2) + \frac{1}{4}E(Y_3) \\ &= \frac{1}{4}\mu + \frac{1}{2}\mu + \frac{1}{4}\mu \\ &= \mu \end{aligned}$$

and

$$\begin{aligned} E(\hat{\mu}_2) &= E\left(\frac{1}{3}Y_1 + \frac{1}{3}Y_2 + \frac{1}{3}Y_3\right) \\ &= \frac{1}{3}E(Y_1) + \frac{1}{3}E(Y_2) + \frac{1}{3}E(Y_3) \\ &= \frac{1}{3}\mu + \frac{1}{3}\mu + \frac{1}{3}\mu \\ &= \mu \end{aligned}$$

But $\text{Var}(\hat{\mu}_2) < \text{Var}(\hat{\mu}_1)$ so $\hat{\mu}_2$ is the more efficient of the two:

$$\begin{aligned} \text{Var}(\hat{\mu}_1) &= \text{Var}\left(\frac{1}{4}Y_1 + \frac{1}{2}Y_2 + \frac{1}{4}Y_3\right) \\ &= \frac{1}{16}\text{Var}(Y_1) + \frac{1}{4}\text{Var}(Y_2) + \frac{1}{16}\text{Var}(Y_3) \\ &= \frac{3\sigma^2}{8} \\ \text{Var}(\hat{\mu}_2) &= \text{Var}\left(\frac{1}{3}Y_1 + \frac{1}{3}Y_2 + \frac{1}{3}Y_3\right) \\ &= \frac{1}{9}\text{Var}(Y_1) + \frac{1}{9}\text{Var}(Y_2) + \frac{1}{9}\text{Var}(Y_3) \\ &= \frac{3\sigma^2}{9} \end{aligned}$$

(The relative efficiency of $\hat{\mu}_2$ to $\hat{\mu}_1$ is

$$\frac{3\sigma^2}{8} \Big/ \frac{3\sigma^2}{9}$$

or 1.125.) ■

Example 5.4.6 Let $Y_1, \ldots, Y_n$ be a random sample from the pdf $f_Y(y;\theta) = \frac{2y}{\theta^2}$, $0 \le y \le \theta$. We know from Example 5.4.2 that $\hat{\theta}_1 = \frac{3}{2}\overline{Y}$ and $\hat{\theta}_2 = \frac{2n+1}{2n}Y_{max}$ are both unbiased for θ. Which estimator is more efficient?

First, let us calculate the variance of $\hat{\theta}_1 = \frac{3}{2}\overline{Y}$. To do so, we need the variance of Y. To that end, note that

$$E(Y^2) = \int_0^\theta y^2 \cdot \frac{2y}{\theta^2}dy = \frac{2}{\theta^2}\int_0^\theta y^3 dy = \frac{2}{\theta^2}\cdot\frac{\theta^4}{4} = \frac{1}{2}\theta^2$$

and

$$\text{Var}(Y) = E(Y^2) - E(Y)^2 = \frac{1}{2}\theta^2 - \left(\frac{2}{3}\theta\right)^2 = \frac{\theta^2}{18}$$

Then

$$\text{Var}(\hat{\theta}_1) = \text{Var}\left(\frac{3}{2}\overline{Y}\right) = \frac{9}{4}\text{Var}(\overline{Y}) = \frac{9}{4}\frac{\text{Var}(Y)}{n} = \frac{9}{4n}\cdot\frac{\theta^2}{18} = \frac{\theta^2}{8n}$$

To address the variance of $\hat{\theta}_2 = \frac{2n+1}{2n}Y_{max}$, we start with finding the variance of Y_{max}. Recall that its pdf is

$$nF_Y(y)^{n-1}f_Y(y) = \frac{2n}{\theta^{2n}}y^{2n-1}, 0 \le y \le \theta$$

From that expression, we obtain

$$E(Y_{max}^2) = \int_0^\theta y^2 \cdot \frac{2n}{\theta^{2n}}y^{2n-1}dy = \frac{2n}{\theta^{2n}}\int_0^\theta y^{2n+1}dy = \frac{2n}{\theta^{2n}}\cdot\frac{\theta^{2n+2}}{2n+2} = \frac{n}{n+1}\theta^2$$

and then

$$\text{Var}(Y_{max}) = E(Y_{max}^2) - E(Y_{max})^2 = \frac{n}{n+1}\theta^2 - \left(\frac{2n}{2n+1}\theta\right)^2 = \frac{n}{(n+1)(2n+1)^2}\theta^2$$

Finally,

$$\text{Var}(\hat{\theta}_2) = \text{Var}\left(\frac{2n+1}{2n}Y_{max}\right) = \frac{(2n+1)^2}{4n^2}\text{Var}(Y_{max}) = \frac{(2n+1)^2}{4n^2}\cdot\frac{n}{(n+1)(2n+1)^2}\theta^2$$

$$= \frac{1}{4n(n+1)}\theta^2$$

Note that $\text{Var}(\hat{\theta}_2) = \frac{1}{4n(n+1)}\theta^2 < \frac{1}{8n}\theta^2 = \text{Var}(\hat{\theta}_1)$ for $n > 1$, so we say that $\hat{\theta}_2$ is *more efficient* than $\hat{\theta}_1$. The *relative efficiency* of $\hat{\theta}_2$ with respect to $\hat{\theta}_1$ is the ratio of their variances:

$$\frac{\text{Var}(\hat{\theta}_1)}{\text{Var}(\hat{\theta}_2)} = \frac{1}{8n}\theta^2 \div \frac{1}{4n(n+1)}\theta^2 = \frac{4n(n+1)}{8n} = \frac{(n+1)}{2}$$

■

Questions

5.4.17. Let $X_1, X_2, \ldots, X_n$ denote the outcomes of a series of n independent trials, where

$$X_i = \begin{cases} 1 & \text{with probability } p \\ 0 & \text{with probability } 1-p \end{cases}$$

for $i = 1, 2, \ldots, n$. Let $X = X_1 + X_2 + \cdots + X_n$.

(a) Show that $\hat{p}_1 = X_1$ and $\hat{p}_2 = \frac{X}{n}$ are unbiased estimators for p.

(b) Intuitively, $\hat{p}_2$ is a better estimator than $\hat{p}_1$ because $\hat{p}_1$ fails to include any of the information about the parameter contained in trials 2 through n. Verify that speculation by comparing the variances of $\hat{p}_1$ and $\hat{p}_2$.

5.4.18. Suppose that $n = 5$ observations are taken from the uniform pdf, $f_Y(y;\theta) = 1/\theta, 0 \le y \le \theta$, where θ is unknown. Two unbiased estimators for θ are

$$\hat{\theta}_1 = \frac{6}{5} \cdot Y_{\text{max}} \quad \text{and} \quad \hat{\theta}_2 = 6 \cdot Y_{\text{min}}$$

Which estimator would be better to use? [*Hint:* What must be true of $\text{Var}(Y_{\text{max}})$ and $\text{Var}(Y_{\text{min}})$ given that $f_Y(y;\theta)$ is symmetric?] Does your answer as to which estimator is better make sense on intuitive grounds? Explain.

5.4.19. Let $Y_1, Y_2, \ldots, Y_n$ be a random sample of size n from the pdf $f_Y(y;\theta) = \frac{1}{\theta}e^{-y/\theta}, y > 0$.

(a) Show that $\hat{\theta}_1 = Y_1, \hat{\theta}_2 = \overline{Y}$, and $\hat{\theta}_3 = n \cdot Y_{\text{min}}$ are all unbiased estimators for θ.
(b) Find the variances of $\hat{\theta}_1, \hat{\theta}_2$, and $\hat{\theta}_3$.
(c) Calculate the relative efficiencies of $\hat{\theta}_1$ to $\hat{\theta}_3$ and $\hat{\theta}_2$ to $\hat{\theta}_3$.

5.4.20. Given a random sample of size n from a Poisson distribution, $\hat{\lambda}_1 = X_1$ and $\hat{\lambda}_2 = \overline{X}$ are two unbiased estimators for λ. Calculate the relative efficiency of $\hat{\lambda}_1$ to $\hat{\lambda}_2$.

5.4.21. If $Y_1, Y_2, \ldots, Y_n$ are random observations from a uniform pdf over $[0, \theta]$, both $\hat{\theta}_1 = \left(\frac{n+1}{n}\right) \cdot Y_{\text{max}}$ and $\hat{\theta}_2 = (n+1) \cdot Y_{\text{min}}$ are unbiased estimators for θ. Show that $\text{Var}(\hat{\theta}_2)/\text{Var}(\hat{\theta}_1) = n^2$.

5.4.22. Suppose that W_1 is a random variable with mean μ and variance σ_1^2 and W_2 is a random variable with mean μ and variance σ_2^2. From Example 5.4.3, we know that $cW_1 + (1-c)W_2$ is an unbiased estimator of μ for any constant $c > 0$. If W_1 and W_2 are independent, for what value of c is the estimator $cW_1 + (1-c)W_2$ most efficient?

5.5 Minimum-Variance Estimators: The Cramér-Rao Lower Bound

Given two estimators, $\hat{\theta}_1$ and $\hat{\theta}_2$, each unbiased for the parameter θ, we know from Section 5.4 which is "better"—the one with the smaller variance. But nothing in that section speaks to the more fundamental question of how good $\hat{\theta}_1$ and $\hat{\theta}_2$ are *relative to the infinitely many other unbiased estimators for* θ. Is there a $\hat{\theta}_3$, for example, that has a smaller variance than either $\hat{\theta}_1$ or $\hat{\theta}_2$ has? Can we identify the unbiased estimator having the *smallest* variance? Addressing those concerns is one of the most elegant, yet practical, theorems in all of mathematical statistics, a result known as the *Cramér-Rao lower bound*.

Suppose a random sample of size n is taken from, say, a continuous probability distribution $f_Y(y;\theta)$, where θ is an unknown parameter. Associated with $f_Y(y;\theta)$ is a theoretical limit below which the variance of any unbiased estimator for θ cannot fall. That limit is the Cramér-Rao lower bound. If the variance of a given $\hat{\theta}$ is *equal* to the Cramér-Rao lower bound, we know that estimator is *optimal* in the sense that no unbiased $\hat{\theta}$ can estimate θ with greater precision.

Theorem 5.5.1 *(Cramér-Rao Inequality.) Let $f_Y(y;\theta)$ be a continuous pdf with continuous first-order and second-order derivatives. Also, suppose that the set of y values, where $f_Y(y;\theta) \neq 0$, does not depend on θ.*

Let $Y_1, Y_2, \ldots, Y_n$ be a random sample from $f_Y(y;\theta)$, and let $\hat{\theta} = h(Y_1, Y_2, \ldots, Y_n)$ be any unbiased estimator of θ. Then

$$\text{Var}(\hat{\theta}) \ge \left\{ nE\left[\left(\frac{\partial \ln f_Y(Y;\theta)}{\partial\theta}\right)^2\right]\right\}^{-1} = \left\{-nE\left[\frac{\partial^2 \ln f_Y(Y;\theta)}{\partial\theta^2}\right]\right\}^{-1}$$

[A similar statement holds if the n observations come from a discrete pdf, $p_X(k;\theta)$].

Proof See (93). □

Example 5.5.1 Suppose the random variables $X_1, X_2, \ldots, X_n$ denote the number of successes (0 or 1) in each of n independent trials, where $p = P$(Success occurs at any given trial) is an unknown parameter. Then

$$p_{X_i}(k; p) = p^k(1-p)^{1-k}, \quad k = 0, 1; \quad 0 < p < 1$$

Let $X = X_1 + X_2 + \cdots + X_n$ = total number of successes and define $\hat{p} = \frac{X}{n}$. Clearly, $\hat{p}$ is unbiased for p $\left[E(\hat{p}) = E\left(\frac{X}{n}\right) = \frac{E(X)}{n} = \frac{np}{n} = p\right]$. How does $\text{Var}(\hat{p})$ compare with the Cramér-Rao lower bound for $p_{X_i}(k; p)$?

Note, first, that

$$\text{Var}(\hat{p}) = \text{Var}\left(\frac{X}{n}\right) = \frac{1}{n^2}\text{Var}(X) = \frac{1}{n^2}np(1-p) = \frac{p(1-p)}{n}$$

(since X is a binomial random variable). To evaluate, say, the *second* form of the Cramér-Rao lower bound, we begin by writing

$$\ln p_{X_i}(X_i; p) = X_i \ln p + (1 - X_i)\ln(1-p)$$

Moreover,

$$\frac{\partial \ln p_{X_i}(X_i; p)}{\partial p} = \frac{X_i}{p} - \frac{1 - X_i}{1 - p}$$

and

$$\frac{\partial^2 \ln p_{X_i}(X_i; p)}{\partial p^2} = -\frac{X_i}{p^2} - \frac{1 - X_i}{(1-p)^2}$$

Taking the expected value of the second derivative gives

$$E\left[\frac{\partial^2 \ln p_{X_i}(X_i; p)}{\partial p^2}\right] = -\frac{p}{p^2} - \frac{(1-p)}{(1-p)^2} = -\frac{1}{p(1-p)}$$

The Cramér-Rao lower bound, then, reduces to

$$\frac{1}{-n\left[-\frac{1}{p(1-p)}\right]} = \frac{p(1-p)}{n}$$

which *equals* the variance of $\hat{p} = \frac{X}{n}$. It follows that $\frac{X}{n}$ is the preferred statistic for estimating the binomial parameter p: No unbiased estimator can possibly be more precise.

Definition 5.5.1. Let Θ denote the set of all estimators $\hat{\theta} = h(Y_1, Y_2, \ldots, Y_n)$ that are unbiased for the parameter θ in the continuous pdf $f_Y(y; \theta)$. We say that $\hat{\theta}^*$ is a *best* (or *minimum-variance*) estimator if $\hat{\theta}^* \in \Theta$ and

$$\text{Var}(\hat{\theta}^*) \le \text{Var}(\hat{\theta}) \quad \text{for all} \quad \hat{\theta} \in \Theta$$

[Similar terminology applies if Θ is the set of all unbiased estimators for the parameter Θ in a discrete pdf, $p_X(k; \theta)$].

Related to the notion of a *best estimator* is the concept of *efficiency*. The connection is spelled out in Definition 5.5.2 for the case where $\hat{\theta}$ is based on data coming from a continuous pdf $f_Y(y; \theta)$. The same terminology applies if the data are a set of X_i's from a discrete pdf $p_X(k; \theta)$.

Definition 5.5.2. Let $Y_1, Y_2, \ldots, Y_n$ be a random sample of size n drawn from the continuous pdf $f_Y(y;\theta)$. Let $\hat{\theta} = h(Y_1, Y_2, \ldots, Y_n)$ be an unbiased estimator for θ.

a. The unbiased estimator $\hat{\theta}$ is said to be *efficient* if the variance of $\hat{\theta}$ equals the Cramér-Rao lower bound associated with $f_Y(y;\theta)$.
b. The *efficiency* of an unbiased estimator $\hat{\theta}$ is the ratio of the Cramér-Rao lower bound for $f_Y(y;\theta)$ to the variance of $\hat{\theta}$.

Comment The designations "efficient" and "best" are not synonymous. If the variance of an unbiased estimator is equal to the Cramér-Rao lower bound, then that estimator by definition is a best estimator. The converse, though, is not always true. There are situations for which the variances of *no* unbiased estimators achieve the Cramér-Rao lower bound. None of those, then, is *efficient*, but one (or more) could still be termed *best*. For the independent trials described in Example 5.5.1, $\hat{p} = \frac{X}{n}$ is both efficient *and* best. ■

Example 5.5.2

If $Y_1, Y_2, \ldots, Y_n$ is a random sample from $f_Y(y;\theta) = 2y/\theta^2, 0 \le y \le \theta, \hat{\theta} = \frac{3}{2}\overline{Y}$ is an unbiased estimator for θ (see Example 5.4.2). Show that the variance of $\hat{\theta}$ is *less than* the Cramér-Rao lower bound for $f_Y(y;\theta)$.

From Example 5.4.6, we know that

$$\operatorname{Var}(\hat{\theta}) = \frac{\theta^2}{8n}$$

To calculate the Cramér-Rao lower bound for $f_Y(y;\theta)$, we first note that

$$\ln f_Y(Y;\theta) = \ln(2Y\theta^{-2}) = \ln 2Y - 2\ln\theta$$

and

$$\frac{\partial \ln f_Y(Y;\theta)}{\partial\theta} = \frac{-2}{\theta}$$

Therefore,

$$E\left[\left[\frac{\partial \ln f_Y(Y;\theta)}{\partial\theta}\right]^2\right] = E\left(\frac{4}{\theta^2}\right) = \int_0^\theta \frac{4}{\theta^2}\cdot\frac{2y}{\theta^2}\,dy$$

$$= \frac{4}{\theta^2}$$

and

$$\left\{nE\left[\left(\frac{\partial \ln f_Y(Y;\theta)}{\partial\theta}\right)^2\right]\right\}^{-1} = \frac{\theta^2}{4n}$$

Is the variance of $\hat{\theta}$ less than the Cramér-Rao lower bound? Yes, $\frac{\theta^2}{8n} < \frac{\theta^2}{4n}$. Is the statement of Theorem 5.5.1 contradicted? No, because the theorem does not apply in this situation: The set of y's where $f_Y(y;\theta) \neq 0$ is a function of θ, a condition that violates one of the Cramér-Rao assumptions. ■

Questions

5.5.1. Let $Y_1, Y_2, \ldots, Y_n$ be a random sample from $f_Y(y;\theta) = \frac{1}{\theta}e^{-y/\theta}$, $y > 0$. Compare the Cramér-Rao lower bound for $f_Y(y;\theta)$ to the variance of the maximum likelihood estimator for θ, $\hat{\theta} = \frac{1}{n}\sum_{i=1}^{n} Y_i$. Is $\overline{Y}$ a best estimator for θ?

5.5.2. Let $X_1, X_2, \ldots, X_n$ be a random sample of size n from the Poisson distribution, $p_X(k;\lambda) = \frac{e^{-\lambda}\lambda^k}{k!}$, $k = 0, 1, \ldots$. Show that $\hat{\lambda} = \frac{1}{n}\sum_{i=1}^{n} X_i$ is an efficient estimator for λ.

5.5.3. Suppose a random sample of size n is taken from a normal distribution with mean μ and variance σ^2, where σ^2 is known. Compare the Cramér-Rao lower bound for $f_Y(y;\mu)$ with the variance of $\hat{\mu} = \overline{Y} = \frac{1}{n}\sum_{i=1}^{n} Y_i$. Is $\overline{Y}$ an efficient estimator for μ?

5.5.4. Let $Y_1, Y_2, \ldots, Y_n$ be a random sample from the uniform pdf $f_Y(y;\theta) = 1/\theta$, $0 \le y \le \theta$. Compare the Cramér-Rao lower bound for $f_Y(y;\theta)$ with the variance of the unbiased estimator $\hat{\theta} = \frac{n+1}{n} \cdot Y_{\max}$. Discuss.

5.5.5. Let X have the pdf $f_X(k;\theta) = \frac{(\theta-1)^{k-1}}{\theta^k}$, $k = 1, 2, 3, \ldots$, $\theta > 1$, which is geometric ($p = 1/\theta$). For this pdf $E(X) = \theta$ and $\text{Var}(X) = \theta(\theta - 1)$ (see Theorem 4.4.1). Is the statistic $\overline{X}$ efficient?

5.5.6. Let $Y_1, Y_2, \ldots, Y_n$ be a random sample of size n from the pdf

$$f_Y(y;\theta) = \frac{1}{(r-1)!\theta^r} y^{r-1} e^{-y/\theta}, \quad y > 0$$

(a) Show that $\hat{\theta} = \frac{1}{r}\overline{Y}$ is an unbiased estimator for θ.

(b) Show that $\hat{\theta} = \frac{1}{r}\overline{Y}$ is a minimum-variance estimator for θ.

5.5.7. Prove the equivalence of the two forms given for the Cramér-Rao lower bound in Theorem 5.5.1. [*Hint*: Differentiate the equation $\int_{-\infty}^{\infty} f_Y(y)\,dy = 1$ with respect to θ and deduce that $\int_{-\infty}^{\infty} \frac{\partial \ln f_Y(y)}{\partial \theta} f_Y(y)\,dy = 1$. Then differentiate again with respect to θ.]

5.6 Sufficient Estimators

Statisticians have proven to be quite diligent (and creative) in articulating properties that good estimators should exhibit. Sections 5.4 and 5.5, for example, introduced the notions of an estimator being unbiased and having minimum variance; Section 5.7 will explain what it means for an estimator to be "consistent." All of those properties are easy to motivate, and they impose conditions on the probabilistic behavior of $\hat{\theta}$ that make eminently good sense. In this section, we look at a deeper property of estimators, one that is not so intuitive but has some particularly important theoretical implications.

Whether or not an estimator is *sufficient* refers to the amount of "information" it contains about the unknown parameter. Estimates, of course, are calculated using values obtained from random samples [drawn from either $p_X(k;\theta)$ or $f_Y(y;\theta)$]. If everything that we can possibly know from the data about θ is encapsulated in the estimate θ_e, then the corresponding estimator $\hat{\theta}$ is said to be *sufficient*. A comparison of two estimators, one sufficient and the other not, should help clarify the concept.

An Estimator That Is Sufficient

Suppose that a random sample of size n—$X_1 = k_1$, $X_2 = k_2, \ldots, X_n = k_n$—is taken from the Bernoulli pdf,

$$p_X(k;p) = p^k(1-p)^{1-k}, \quad k = 0, 1$$

where p is an unknown parameter. We know from Example 5.1.1 that the maximum likelihood estimator for p is

$$\hat{p} = \left(\frac{1}{n}\right) \sum_{i=1}^{n} X_i$$

[and the maximum likelihood estimate is $p_e = \left(\frac{1}{n}\right) \sum_{i=1}^{n} k_i$]. To show that $\hat{p}$ is a sufficient estimator for p requires that we calculate the conditional probability that $X_1 = k_1, \ldots, X_n = k_n$ given that $\hat{p} = p_e$.

Generalizing the Comment following Example 3.11.3, we can write

$$\begin{aligned} P(X_1 = k_1, \ldots, X_n = k_n \mid \hat{p} = p_e) &= \frac{P(X_1 = k_1, \ldots, X_n = k_n \cap \hat{p} = p_e)}{P(\hat{p} = p_e)} \\ &= \frac{P(X_1 = k_1, \ldots, X_n = k_n)}{P(\hat{p} = p_e)} \end{aligned}$$

But

$$\begin{aligned} P(X_1 = k_1, \ldots, X_n = k_n) &= p^{k_1}(1-p)^{1-k_1} \cdots p^{k_n}(1-p)^{1-k_n} \\ &= p^{\sum_{i=1}^{n} k_i}(1-p)^{n - \sum_{i=1}^{n} k_i} \\ &= p^{np_e}(1-p)^{n-np_e} \end{aligned}$$

and

$$P(\hat{p} = p_e) = P\left(\sum_{i=1}^{n} X_i = np_e\right) = \binom{n}{np_e} p^{np_e}(1-p)^{n-np_e}$$

since $\sum_{i=1}^{n} X_i$ has a binomial distribution with parameters n and p (recall Example 3.9.3). Therefore,

$$P(X_1 = k_1, \ldots, X_n = k_n \mid \hat{p} = p_e) = \frac{p^{np_e}(1-p)^{n-np_e}}{\binom{n}{np_e} p^{np_e}(1-p)^{n-np_e}} = \frac{1}{\binom{n}{np_e}} \tag{5.6.1}$$

Notice that $P(X_1 = k_1, \ldots, X_n = k_n \mid \hat{p} = p_e)$ *is not a function of* p. That is precisely the condition that makes $\hat{p} = \left(\frac{1}{n}\right) \sum_{i=1}^{n} X_i$ a *sufficient* estimator. Equation 5.6.1 says, in effect, that everything the data can possibly tell us about the parameter p is contained in the estimate p_e. Remember that, initially, the joint pdf of the sample, $P(X_1 = k_1, \ldots, X_n = k_n)$, is a function of the k_i's *and* p. What we have just shown, though, is that if that probability is conditioned on the value of this particular estimate—that is, on $\hat{p} = p_e$—then p is eliminated and the probability of the sample is completely determined [in this case, it equals $\binom{n}{np_e}^{-1}$, where $\binom{n}{np_e}$ is the number of ways to arrange the 0's and 1's in a sample of size n for which $\hat{p} = p_e$].

If we had used some other estimator—say, $\hat{p}^*$—and if $P(X_1 = k_1, \ldots, X_n = k_n \mid \hat{p}^* = p_e^*)$ had remained a function of p, the conclusion would be that the information in p_e^* was not "sufficient" to eliminate the parameter p from the conditional probability. A simple example of such a $\hat{p}^*$ would be $\hat{p}^* = X_1$. Then p_e^* would be k_1 and the conditional probability of $X_1 = k_1, \ldots, X_n = k_n$ given that $\hat{p}^* = p_e^*$ would remain a function of p:

$$P(X_1 = k_1, \ldots, X_n = k_n \mid \hat{p}^* = k_1) = \frac{p^{\sum_{i=1}^{n} k_i}(1-p)^{n - \sum_{i=1}^{n} k_i}}{p^{k_1}(1-p)^{1-k_1}} = p^{\sum_{i=2}^{n} k_i}(1-p)^{n-1-\sum_{i=2}^{n} k_i}$$

Comment Some of the dice problems we did in Section 2.4 have aspects that parallel to some extent the notion of an estimator being sufficient. Suppose, for example, we roll a pair of fair dice without being allowed to view the outcome. Our objective is to calculate the probability that the sum showing is an even number. If we had no other information, the answer would be $\frac{1}{2}$. Suppose, though, that two people *do* see the outcome—which was, in fact, a sum of 7—and each is allowed to characterize the outcome without providing us with the exact sum that occurred. Person A tells us that "the sum was less than or equal to 7"; Person B says that "the sum was an odd number."

Whose information is more helpful? Person B's. The conditional probability of the sum being even given that the sum is less than or equal to 7 is $\frac{9}{21}$, which still leaves our initial question largely unanswered:

$$P(\text{Sum is even} \mid \text{sum} \leq 7) = \frac{P(2) + P(4) + P(6)}{P(2) + P(3) + P(4) + P(5) + P(6) + P(7)}$$

$$= \frac{\frac{1}{36} + \frac{3}{36} + \frac{5}{36}}{\frac{1}{36} + \frac{2}{36} + \frac{3}{36} + \frac{4}{36} + \frac{5}{36} + \frac{6}{36}}$$

$$= \frac{9}{21}$$

In contrast, Person B utilized the data in a way that definitely answered the original question:

$$P(\text{Sum is even} \mid \text{Sum is odd}) = 0$$

In a sense, B's information was "sufficient"; A's information was not.

An Estimator That Is Not Sufficient

Suppose a random sample of size n—$Y_1, Y_2, \ldots, Y_n$—is drawn from the pdf $f_Y(y;\theta) = \frac{2y}{\theta^2}$, $0 \leq y \leq \theta$, where θ is an unknown parameter. Recall that the method of moments estimator is

$$\hat{\theta} = \frac{3}{2}\overline{Y} = \frac{3}{2n}\sum_{i=1}^{n} Y_i$$

This statistic is not sufficient because all the information in the data that pertains to the parameter θ is not necessarily contained in the numerical value θ_e.

If $\hat{\theta}$ were a sufficient statistic, then any two random samples of size n *having the same value for* θ_e should yield exactly the same information about θ. However, a simple numerical example shows this not to be the case. Consider two random samples of size 3—$y_1 = 3$, $y_2 = 4$, $y_3 = 5$ and $y_1 = 1$, $y_2 = 3$, $y_3 = 8$. In both cases,

$$\theta_e = \frac{3}{2}\overline{y} = \frac{3}{2 \cdot 3}\sum_{i=1}^{3} y_i = 6$$

Do both samples, though, convey the same information about the possible value of θ? No. Based on the first sample, the true θ could, in fact, be equal to 4. On the other hand, the second sample rules out the possibility that θ is 4 because one of the observations ($y_3 = 8$) is larger than 4, but according to the definition of the pdf, all Y_i's must be less than θ.

A Formal Definition

Suppose that $X_1 = k_1, \ldots, X_n = k_n$ is a random sample of size n from the discrete pdf $p_X(k;\theta)$, where θ is an unknown parameter. Conceptually, $\hat{\theta}$ is a sufficient statistic for θ if

$$P(X_1 = k_1, \ldots, X_n = k_n \mid \hat{\theta} = \theta_e) = \frac{P(X_1 = k_1, \ldots, X_n = k_n \cap \hat{\theta} = \theta_e)}{P(\hat{\theta} = \theta_e)}$$

$$= \frac{\prod_{i=1}^{n} p_X(k_i;\theta)}{p_{\hat{\theta}}(\theta_e;\theta)} = b(k_1, \ldots, k_n) \tag{5.6.2}$$

where $p_{\hat{\theta}}(\theta_e;\theta)$ is the pdf of the statistic evaluated at the point $\hat{\theta} = \theta_e$ and $b(k_1, \ldots, k_n)$ is a constant independent of θ. Equivalently, the condition that qualifies a statistic as being sufficient can be expressed by cross-multiplying Equation 5.6.2.

Definition 5.6.1. Let $X_1 = k_1, \ldots, X_n = k_n$ be a random sample of size n from $p_X(k;\theta)$. The statistic $\hat{\theta} = h(X_1, \ldots, X_n)$ is *sufficient* for θ if the likelihood function, $L(\theta)$, factors into the product of the pdf for $\hat{\theta}$ and a constant that does not involve θ—that is, if

$$L(\theta) = \prod_{i=1}^{n} p_X(k_i;\theta) = p_{\hat{\theta}}(\theta_e;\theta)b(k_1, \ldots, k_n)$$

A similar statement holds if the data consist of a random sample $Y_1 = y_1, \ldots, Y_n = y_n$ drawn from a continuous pdf $f_Y(y;\theta)$.

Comment If $\hat{\theta}$ is sufficient for θ, then any one-to-one function of $\hat{\theta}$ is also a sufficient statistic for θ. As a case in point, we showed on p. 324 that

$$\hat{p} = \left(\frac{1}{n}\right)\sum_{i=1}^{n} X_i$$

is a sufficient statistic for the parameter p in a Bernoulli pdf. It is also true, then, that

$$\hat{p}^* = n\hat{p} = \sum_{i=1}^{n} X_i$$

is sufficient for p.

Example 5.6.1 Let $X_1 = k_1, \ldots, X_n = k_n$ be a random sample of size n from the Poisson pdf, $p_X(k;\lambda) = e^{-\lambda}\lambda^k/k!$, $k = 0, 1, 2, \ldots$. Show that

$$\hat{\lambda} = \sum_{i=1}^{n} X_i$$

is a sufficient statistic for λ.

From Example 3.12.10, we know that $\hat{\lambda}$, being a sum of n independent Poisson random variables, each with parameter λ, is itself a Poisson random variable with parameter $n\lambda$. By Definition 5.6.1, then, $\hat{\lambda}$ is a sufficient statistic for λ if the sample's likelihood function factors into a product of the pdf for $\hat{\lambda}$ times a constant that is independent of λ.

But

$$L(\lambda) = \prod_{i=1}^{n} e^{-\lambda}\lambda^{k_i}/k_i! = e^{-n\lambda}\lambda^{\sum_{i=1}^{n} k_i} \Big/ \prod_{i=1}^{n} k_i!$$

$$= \frac{e^{-n\lambda} n^{\sum_{i=1}^{n} k_i} \lambda^{\sum_{i=1}^{n} k_i} \left(\sum_{i=1}^{n} k_i\right)!}{\left(\sum_{i=1}^{n} k_i\right)! \prod_{i=1}^{n} k_i! n^{\sum_{i=1}^{n} k_i}}$$

$$= \frac{e^{-n\lambda}(n\lambda)^{\sum_{i=1}^{n} k_i}}{\left(\sum_{i=1}^{n} k_i\right)!} \cdot \frac{\left(\sum_{i=1}^{n} k_i\right)!}{\prod_{i=1}^{n} k_i! n^{\sum_{i=1}^{n} k_i}}$$

$$= p_{\hat{\lambda}}(\lambda_e; \lambda) \cdot b(k_1, \ldots, k_n) \tag{5.6.3}$$

proving that $\hat{\lambda} = \sum_{i=1}^{n} X_i$ is a sufficient statistic for λ.

Comment The factorization in Equation 5.6.3 implies that $\hat{\lambda} = \sum_{i=1}^{n} X_i$ is a sufficient statistic for λ. It is not, however, an unbiased estimator for λ:

$$E(\hat{\lambda}) = \sum_{i=1}^{n} E(X_i) = \sum_{i=1}^{n} \lambda = n\lambda$$

Constructing an unbiased estimator *based* on the sufficient statistic, though, is a simple matter. Let

$$\hat{\lambda}^* = \frac{1}{n}\hat{\lambda} = \frac{1}{n}\sum_{i=1}^{n} X_i$$

Then $E(\hat{\lambda}^*) = \frac{1}{n}E(\hat{\lambda}) = \frac{1}{n}n\lambda = \lambda$, so $\hat{\lambda}^*$ is unbiased for λ. Moreover, $\hat{\lambda}^*$ is a one-to-one function of $\hat{\lambda}$, so, by the Comment on p. 326, $\hat{\lambda}^*$ is, itself, a sufficient estimator for λ. ■

A Second Factorization Criterion

Using Definition 5.6.1 to verify that a statistic is sufficient requires that the pdf $p_{\hat{\theta}}[h(k_1, \ldots, k_n); \theta]$ or $f_{\hat{\theta}}[h(y_1, \ldots, y_n); \theta]$ be explicitly identified as one of the two factors whose product equals the likelihood function. If $\hat{\theta}$ is complicated, though, finding its pdf may be prohibitively difficult. The next theorem gives an alternative factorization criterion for establishing that a statistic is sufficient. It does not require that the pdf for $\hat{\theta}$ be known.

Theorem 5.6.1 *Let $X_1 = k_1, \ldots, X_n = k_n$ be a random sample of size n from the discrete pdf $p_X(k; \theta)$. The statistic $\hat{\theta} = h(X_1, \ldots, X_n)$ is sufficient for θ if and only if there are functions $g[h(k_1, \ldots, k_n); \theta]$ and $b(k_1, \ldots, k_n)$ such that*

$$L(\theta) = g[h(k_1, \ldots, k_n); \theta] \cdot b(k_1, \ldots, k_n) \tag{5.6.4}$$

where the function $b(k_1, \ldots, k_n)$ does not involve the parameter θ. A similar statement holds in the continuous case.

Proof First, suppose that $\hat{\theta}$ is sufficient for θ. Then the factorization criterion of Definition 5.6.1 includes Equation 5.6.4 as a special case.

Now, assume that Equation 5.6.4 holds. The theorem will be proved if it can be shown that $g[b(k_1, \ldots, k_n); \dot{\theta}]$ can always be "converted" to include the pdf of $\hat{\theta}$ (at which point Definition 5.6.1 would apply). Let c be some value of the function $b(k_1, \ldots, k_n)$ and let A be the set of samples of size n that constitute the inverse image of c—that is, $A = h^{-1}(c)$. Then

$$p_{\hat{\theta}}(c; \theta) = \sum_{(k_1, k_2, \ldots, k_n) \varepsilon A} p_{X_1, X_2, \ldots, X_n}(k_1, k_2, \ldots, k_n) = \sum_{(k_1, k_2, \ldots, k_n) \varepsilon A} \prod_{i=1}^{n} p_{X_i}(k_i)$$

$$= \sum_{(k_1, k_2, \ldots, k_n) \varepsilon A} g(c; \theta) \cdot b(k_1, k_2, \ldots, k_n) = g(c; \theta) \cdot \sum_{(k_1, k_2, \ldots, k_n) \varepsilon A} b(k_1, k_2, \ldots, k_n)$$

Since we are interested only in points where $p_{\hat{\theta}}(c; \theta) \neq 0$, we can assume that $\sum_{(k_1, k_2, \ldots, k_n) \varepsilon A} b(k_1, k_2, \ldots, k_n) \neq 0$. Therefore,

$$g(c; \theta) = p_{\hat{\theta}}(c; \theta) \cdot \frac{1}{\sum_{(k_1, k_2, \ldots, k_n) \varepsilon A} b(k_1, k_2, \ldots, k_n)} \tag{5.6.5}$$

Substituting the right-hand side of Equation 5.6.5 into Equation 5.6.4 shows that $\hat{\theta}$ qualifies as a sufficient statistic for θ. A similar argument can be made if the data consist of a random sample $Y_1 = y_1, \ldots, Y_n = y_n$ drawn from a continuous pdf $f_Y(y; \theta)$. See (200) for more details. □

Example 5.6.2

Suppose $Y_1, \ldots, Y_n$ is a random sample from $f_Y(y; \theta) = \frac{2y}{\theta^2}, 0 \leq y \leq \theta$. We know from Question 5.2.12 that the maximum likelihood estimator for θ is $\hat{\theta} = Y_{\max}$. Is $Y_{\max}$ also sufficient for θ?

Since the set of Y values where $f_Y(y; \theta) \neq 0$ depends on θ, the likelihood function must be written in a way to include that restriction. The device achieving that goal is called an *indicator function*. We define the function $I_{[0,\theta]}(y)$ by

$$I_{[0,\theta]}(y) = \begin{cases} 1 & 0 \leq y \leq \theta \\ 0 & \text{otherwise} \end{cases}$$

Then we can write $f_Y(y; \theta) = \frac{2y}{\theta^2} \cdot I_{[0,\theta]}(y)$ for all y.

The likelihood function is

$$L(\theta) = \prod_{i=1}^{n} \frac{2y_i}{\theta^2} \cdot I_{[0,\theta]}(y_i) = \left(\prod_{i=1}^{n} 2y_i\right)\left(\frac{1}{\theta^{2n}}\right) \prod_{i=1}^{n} I_{[0,\theta]}(y_i)$$

But the critical fact is that

$$\prod_{i=1}^{n} I_{[0,\theta]}(y_i) = I_{[0,\theta]}(y_{\max})$$

Thus the likelihood function decomposes in such a way that the factor involving θ contains only the y_i's through $y_{\max}$:

$$L(\theta) = \left(\prod_{i=1}^{n} 2y_i\right)\left(\frac{1}{\theta^{2n}}\right)\prod_{i=1}^{n} I_{[0,\theta]}(y_i) = \left[\frac{I_{[0,\theta]}(y_{\max})}{\theta^{2n}}\right]\cdot\left(\prod_{i=1}^{n} 2y_i\right)$$

This decomposition meets the criterion of Theorem 5.6.1, and $Y_{\max}$ is sufficient for θ. (Why doesn't this argument work for $Y_{\min}$?) ■

Sufficiency as It Relates to Other Properties of Estimators

This chapter has constructed a rather elaborate facade of mathematical properties and procedures associated with estimators. We have asked whether $\hat{\theta}$ is unbiased, efficient, and/or sufficient. How we *find* $\hat{\theta}$ has also come under scrutiny—some estimators have been derived using the method of maximum likelihood; others have come from the method of moments. Not all of these aspects of estimators and estimation, though, are entirely disjoint—some are related and interconnected in a variety of ways.

Suppose, for example, that a sufficient estimator $\hat{\theta}_S$ exists for a parameter θ, and suppose that $\hat{\theta}_M$ is the maximum likelihood estimator for that same θ. If, for a given sample, $\hat{\theta}_S = \theta_e$, we know from Theorem 5.6.1 that

$$L(\theta) = g(\theta_e;\theta)\cdot b(k_1,\ldots,k_n)$$

Since the maximum likelihood estimate, by definition, maximizes $L(\theta)$, it must also maximize $g(\theta_e;\theta)$. But any θ that maximizes $g(\theta_e;\theta)$ will necessarily be a function of θ_e. It follows, then, that maximum likelihood estimators are necessarily functions of sufficient estimators—that is, $\hat{\theta}_M = f(\hat{\theta}_S)$ (which is the primary theoretical justification for why maximum likelihood estimators are preferred to method of moments estimators).

Sufficient estimators also play a critical role in the search for efficient estimators—that is, unbiased estimators whose variance equals the Cramér-Rao lower bound. There will be an infinite number of unbiased estimators for any unknown parameter in any pdf. That said, there may be a subset of those unbiased estimators that are functions of sufficient estimators. If so, it can be proved [see (93)] that the variance of every unbiased estimator based on a sufficient estimator will necessarily be less than the variance of every unbiased estimator that is not a function of a sufficient estimator. It follows, then, that to find an efficient estimator for θ, we can restrict our attention to functions of sufficient estimators for θ.

Questions

5.6.1. Let $X_1, X_2, \ldots, X_n$ be a random sample of size n from the geometric distribution, $p_X(k;p) = (1-p)^{k-1}p$, $k = 1, 2, \ldots$. Show that $\hat{p} = \sum_{i=1}^{n} X_i$ is sufficient for p.

5.6.2. Let X_1, X_2, and X_3 be a set of three independent Bernoulli random variables with unknown parameter $p = P(X_i = 1)$. It was shown on p. 324 that $\hat{p} = X_1 + X_2 + X_3$ is sufficient for p. Show that the linear combination $\hat{p}^* = X_1 + 2X_2 + 3X_3$ is *not* sufficient for p.

5.6.3. If $\hat{\theta}$ is sufficient for θ, show that any one-to-one function of $\hat{\theta}$ is also sufficient for θ.

5.6.4. Show that $\hat{\sigma}^2 = \sum_{i=1}^{n} Y_i^2$ is sufficient for σ^2 if $Y_1, Y_2, \ldots, Y_n$ is a random sample from a normal pdf with $\mu = 0$.

5.6.5. Let $Y_1, Y_2, \ldots, Y_n$ be a random sample of size n from the pdf of Question 5.5.6,

$$f_Y(y;\theta) = \frac{1}{(r-1)!\theta^r} y^{r-1} e^{-y/\theta}, \quad 0 \le y$$

for positive parameter θ and r a known positive integer. Find a sufficient statistic for θ.

5.6.6. Let $Y_1, Y_2, \ldots, Y_n$ be a random sample of size n from the pdf $f_Y(y;\theta) = \theta y^{\theta-1}, 0 \le y \le 1$. Use Theorem 5.6.1 to show that $W = \prod_{i=1}^{n} Y_i$ is a sufficient statistic for θ. Is the maximum likelihood estimator of θ a function of W?

5.6.7. Suppose a random sample of size n is drawn from the pdf

$$f_Y(y;\theta) = e^{-(y-\theta)}, \quad \theta \le y$$

(a) Show that $\hat{\theta} = Y_{\min}$ is sufficient for the threshold parameter θ.
(b) Show that $Y_{\max}$ is not sufficient for θ.

5.6.8. Suppose a random sample of size n is drawn from the pdf

$$f_Y(y;\theta) = \frac{1}{\theta}, \quad 0 \le y \le \theta$$

Find a sufficient statistic for θ.

5.6.9. A probability model $g_W(w;\theta)$ is said to be expressed in *exponential form* if it can be written as

$$g_W(w;\theta) = e^{K(w)p(\theta)+S(w)+q(\theta)}$$

where the range of W is independent of θ. Show that $\hat{\theta} = \sum_{i=1}^{n} K(W_i)$ is sufficient for θ.

5.6.10. Write the pdf $f_Y(y;\lambda) = \lambda e^{-\lambda y}$, $y > 0$, in exponential form and deduce a sufficient statistic for λ (see Question 5.6.9). Assume that the data consist of a random sample of size n.

5.6.11. Let $Y_1, Y_2, \ldots, Y_n$ be a random sample from a Pareto pdf,

$$f_Y(y;\theta) = \theta/(1+y)^{\theta+1}, \quad 0 \le y \le \infty; \quad 0 < \theta < \infty$$

Write $f_Y(y;\theta)$ in exponential form and deduce a sufficient statistic for θ (see Question 5.6.9).

5.7 Consistency

The properties of estimators that we have examined thus far—for instance, unbiasedness and sufficiency—have assumed that the data consist of a *fixed* sample size n. It sometimes makes sense, though, to consider the *asymptotic* behavior of estimators: We may find, for example, that an estimator possesses a desirable property *in the limit* that it fails to exhibit for any finite n.

Recall Example 5.4.4, which focused on the maximum likelihood estimator for σ^2 in a sample of size n drawn from a normal pdf [that is, on $\hat{\sigma}^2 = \frac{1}{n}\sum_{i=1}^{n}(Y_i - \overline{Y})^2$]. For any finite n, $\hat{\sigma}^2$ is biased:

$$E\left[\frac{1}{n}\sum_{i=1}^{n}(Y_i - \overline{Y})^2\right] = \frac{n-1}{n}\sigma^2 \ne \sigma^2$$

As n goes to infinity, though, the limit of $E(\hat{\sigma}^2)$ does equal σ^2, and we say that $\hat{\sigma}^2$ is *asymptotically unbiased*.

Introduced in this section is a second asymptotic characteristic of an estimator, a property known as *consistency*. Unlike asymptotic unbiasedness, consistency refers to the shape of the pdf for $\hat{\theta}_n$ and how that shape changes as a function of n. (To emphasize the fact that the estimator for a parameter is now being viewed as a *sequence* of estimators, we will write $\hat{\theta}_n$ instead of $\hat{\theta}$.)

Definition 5.7.1. An estimator $\hat{\theta}_n = h(W_1, W_2, \ldots, W_n)$ is said to be *consistent* for θ if it converges in probability to θ—that is, if for all $\varepsilon > 0$,

$$\lim_{n\to\infty} P(|\hat{\theta}_n - \theta| < \varepsilon) = 1$$

Comment To solve certain kinds of sample-size problems, it can be helpful to think of Definition 5.7.1 in an epsilon/delta context; that is, $\hat{\theta}_n$ is consistent for θ if for all $\varepsilon > 0$ and $\delta > 0$, there exists an $n(\varepsilon, \delta)$ such that

$$P(|\hat{\theta}_n - \theta| < \varepsilon) > 1 - \delta \quad \text{for} \quad n > n(\varepsilon, \delta)$$

Example 5.7.1 Let $Y_1, Y_2, \ldots, Y_n$ be a random sample from the uniform pdf

$$f_Y(y; \theta) = \frac{1}{\theta}, \quad 0 \le y \le \theta$$

and let $\hat{\theta}_n = Y_{\max}$. We already know that $Y_{\max}$ is biased for θ, but is it consistent?

Recall from Question 5.4.2 that

$$f_{Y_{\max}}(y) = \frac{ny^{n-1}}{\theta^n}, \quad 0 \le y \le \theta$$

Therefore,

$$P(|\hat{\theta}_n - \theta| < \varepsilon) = P(\theta - \varepsilon < \hat{\theta}_n < \theta) = \int_{\theta-\varepsilon}^{\theta} \frac{ny^{n-1}}{\theta^n}\, dy = \left.\frac{y^n}{\theta^n}\right|_{\theta-\varepsilon}^{\theta}$$

$$= 1 - \left(\frac{\theta - \varepsilon}{\theta}\right)^n$$

Since $[(\theta - \varepsilon)/\theta] < 1$, it follows that $[(\theta - \varepsilon)/\theta]^n \to 0$ as $n \to \infty$. Therefore, $\lim_{n\to\infty} P(|\hat{\theta}_n - \theta| < \varepsilon) = 1$, proving that $\hat{\theta}_n = Y_{\max}$ is consistent for θ.

Figure 5.7.1 illustrates the convergence of $\hat{\theta}_n$. As n increases, the shape of $f_{Y_{\max}}(y)$ changes in such a way that the pdf becomes increasingly concentrated in an ε-neighborhood of θ. For any $n > n(\varepsilon, \delta)$, $P(|\hat{\theta}_n - \theta| < \varepsilon) > 1 - \delta$.

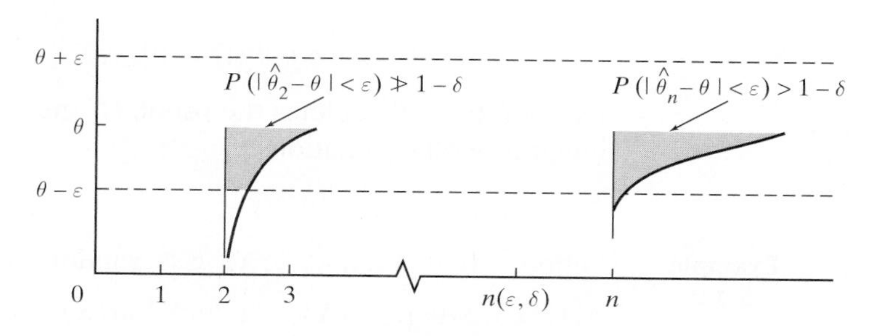

Figure 5.7.1

If θ, ε, and δ are specified, we can calculate $n(\varepsilon, \delta)$, the smallest sample size that will enable $\hat{\theta}_n$ to achieve a given precision. For example, suppose $\theta = 4$. How large a sample is required to give $\hat{\theta}_n$ an 80% chance of lying within 0.10 of θ?

In the terminology of the Comment on p. 331, $\varepsilon = 0.10$, $\delta = 0.20$, and

$$P(|\hat{\theta} - 4| < 0.10) = 1 - \left(\frac{4 - 0.10}{4}\right)^n \ge 1 - 0.20$$

Therefore,

$$(0.975)^{n(\varepsilon,\delta)} = 0.20$$

which implies that $n(\varepsilon, \delta) = 64$. ■

A useful result for establishing consistency is *Chebyshev's inequality*, which appears here as Theorem 5.7.1. More generally, the latter serves as an upper bound for the probability that any random variable lies outside an ε-neighborhood of its mean.

Theorem 5.7.1 *(Chebyshev's inequality.) Let W be any random variable with mean μ and variance σ^2. For any $\varepsilon > 0$,*

$$P(|W - \mu| < \varepsilon) \geq 1 - \frac{\sigma^2}{\varepsilon^2}$$

or, equivalently,

$$P(|W - \mu| \geq \varepsilon) \leq \frac{\sigma^2}{\varepsilon^2}$$

Proof In the continuous case,

$$\begin{aligned}\text{Var}(Y) &= \int_{-\infty}^{\infty} (y-\mu)^2 f_Y(y)\,dy \\ &= \int_{-\infty}^{\mu-\varepsilon} (y-\mu)^2 f_Y(y)\,dy + \int_{\mu-\varepsilon}^{\mu+\varepsilon} (y-\mu)^2 f_Y(y)\,dy + \int_{\mu+\varepsilon}^{\infty} (y-\mu)^2 f_Y(y)\,dy\end{aligned}$$

Omitting the nonnegative middle integral gives an inequality:

$$\begin{aligned}\text{Var}(Y) &\geq \int_{-\infty}^{\mu-\varepsilon} (y-\mu)^2 f_Y(y)\,dy + \int_{\mu+\varepsilon}^{\infty} (y-\mu)^2 f_Y(y)\,dy \\ &\geq \int_{|y-\mu|\geq\varepsilon} (y-\mu)^2 f_Y(y)\,dy \\ &\geq \int_{|y-\mu|\geq\varepsilon} \varepsilon^2 f_Y(y)\,dy \\ &= \varepsilon^2 P(|Y-\mu| \geq \varepsilon)\end{aligned}$$

Division by ε^2 completes the proof. (If the random variable is discrete, replace the integrals with summations.) □

Example 5.7.2 Suppose that $X_1, X_2, \ldots, X_n$ is a random sample of size n from a discrete pdf $p_X(k; \mu)$, where $E(X) = \mu$ and $\text{Var}(X) = \sigma^2 < \infty$. Let $\hat{\mu}_n = \left(\frac{1}{n}\right)\sum_{i=1}^{n} X_i$. Is $\hat{\mu}_n$ a consistent estimator for μ?

According to Chebyshev's inequality,

$$P(|\hat{\mu}_n - \mu| < \varepsilon) > 1 - \frac{\text{Var}(\hat{\mu}_n)}{\varepsilon^2}$$

But $\text{Var}(\hat{\mu}_n) = \text{Var}\left(\frac{1}{n}\sum_{i=1}^{n} X_i\right) = \frac{1}{n^2}\sum_{i=1}^{n}\text{Var}(X_i) = (1/n^2)\cdot n\sigma^2 = \sigma^2/n$, so

$$P(|\hat{\mu}_n - \mu| < \varepsilon) > 1 - \frac{\sigma^2}{n\varepsilon^2}$$

For any ε, δ, and σ^2, an n can be found that makes $\frac{\sigma^2}{n\varepsilon^2} < \delta$. Therefore, $\lim_{n\to\infty} P(|\hat{\mu}_n - \mu| < \varepsilon) = 1$ (i.e., $\hat{\mu}_n$ is consistent for μ).

Comment The fact that the sample mean, $\hat{\mu}_n$, is necessarily a consistent estimator for the true mean μ, no matter what pdf the data come from, is often referred to as the *weak law of large numbers*. It was first proved by Chebyshev in 1866. ■

Comment We saw in Section 5.6 that one of the theoretical reasons that justifies using the method of maximum likelihood to identify good estimators is the fact that maximum likelihood estimators are necessarily functions of sufficient statistics. As an additional rationale for seeking maximum likelihood estimators, it can be proved under very general conditions that maximum likelihood estimators are also consistent (see 93).

Questions

5.7.1. How large a sample must be taken from a normal pdf where $E(Y) = 18$ in order to guarantee that $\hat{\mu}_n = \overline{Y}_n = \frac{1}{n}\sum_{i=1}^{n} Y_i$ has a 90% probability of lying somewhere in the interval [16, 20]? Assume that $\sigma = 5.0$.

5.7.2. Let $Y_1, Y_2, \ldots, Y_n$ be a random sample of size n from a normal pdf having $\mu = 0$. Show that $S_n^2 = \frac{1}{n}\sum_{i=1}^{n} Y_i^2$ is a consistent estimator for $\sigma^2 = \text{Var}(Y)$.

5.7.3. Suppose $Y_1, Y_2, \ldots, Y_n$ is a random sample from the exponential pdf, $f_Y(y; \lambda) = \lambda e^{-\lambda y}$, $y > 0$.

(a) Show that $\hat{\lambda}_n = Y_1$ is not consistent for λ.

(b) Show that $\hat{\lambda}_n = \sum_{i=1}^{n} Y_i$ is not consistent for λ.

5.7.4. An estimator $\hat{\theta}_n$ is said to be *squared-error consistent* for θ if $\lim_{n\to\infty} E[(\hat{\theta}_n - \theta)^2] = 0$.

(a) Show that any squared-error consistent $\hat{\theta}_n$ is asymptotically unbiased (see Question 5.4.15).

(b) Show that any squared-error consistent $\hat{\theta}_n$ is consistent in the sense of Definition 5.7.1.

5.7.5. Suppose $\hat{\theta}_n = Y_{\max}$ is to be used as an estimator for the parameter θ in the uniform pdf, $f_Y(y; \theta) = 1/\theta, 0 \le y \le \theta$. Show that $\hat{\theta}_n$ is squared-error consistent (see Question 5.7.4).

5.7.6. If $2n + 1$ random observations are drawn from a continuous and symmetric pdf with mean μ and if $f_Y(\mu; \mu) \neq 0$, then the *sample median*, Y'_{n+1}, is unbiased for μ, and $\text{Var}(Y'_{n+1}) \doteq 1/(8[f_Y(\mu; \mu)]^2 n)$ [see (54)]. Show that $\hat{\mu}_n = Y'_{n+1}$ is consistent for μ.

5.8 Bayesian Estimation

Bayesian analysis is a set of statistical techniques based on inverse probabilities calculated from Bayes' Theorem (recall Section 2.4). In particular, Bayesian statistics provide formal methods for incorporating prior knowledge into the estimation of unknown parameters.

An interesting example of a Bayesian solution to an unusual estimation problem occurred some years ago in the search for a missing nuclear submarine. In the spring of 1968, the USS *Scorpion* was on maneuvers with the Sixth Fleet in Mediterranean waters. In May, she was ordered to proceed to her homeport of Norfolk, Virginia. The last message from the *Scorpion* was received on May 21, and indicated her position to be about fifty miles south of the Azores, a group of islands eight hundred miles off the coast of Portugal. Navy officials decided that the sub had sunk somewhere along the eastern coast of the United States. A massive search was mounted, but to no avail, and the *Scorpion's* fate remained a mystery.

Enter John Craven, a Navy expert in deep-water exploration, who believed the *Scorpion* had not been found because it had never reached the eastern seaboard and was still somewhere near the Azores. In setting up a search strategy, Craven divided the area near the Azores into a grid of n squares, and solicited the advice of a group of veteran submarine commanders on the chances of the *Scorpion* having been lost in each of those regions. Combining their opinions resulted in a set of probabilities, $P(A_1), P(A_2), \ldots, P(A_n)$, that the sub had sunk in areas $1, 2, \ldots, n$, respectively.

Now, suppose $P(A_k)$ was the largest of the $P(A_i)$'s. Then area k would be the first region searched. Let B_k be the event that the *Scorpion* would be found if it had sunk in area k and area k was searched. Assume that the sub was *not* found. From Theorem 2.4.2,

$$P(A_k \mid B_k^C) = \frac{P(B_k^C \mid A_k)P(A_k)}{P(B_k^C \mid A_k)P(A_k) + P(B_k^C \mid A_k^C)P(A_k^C)}$$

becomes an updated $P(A_k)$—call it $P^*(A_k)$. The remaining $P(A_i)$'s, $i \neq k$, can then be normalized to form the revised probabilities $P^*(A_i), i \neq k$, where $\sum_{i=1}^{n} P^*(A_i) = 1$. If $P^*(A_j)$ was the largest of the $P^*(A_i)$'s, then area j would be searched next. If the sub was not found there, a third set of probabilities, $P^{**}(A_1), P^{**}(A_2), \ldots, P^{**}(A_n)$, would be calculated in the same fashion, and the search would continue.

In October of 1968, the USS *Scorpion* was, indeed, found near the Azores; all ninety-nine men aboard had perished. *Why* it sunk has never been disclosed. One theory has suggested that one of its torpedoes accidentally exploded; Cold War conspiracy advocates think it may have been sunk while spying on a group of Soviet subs. What *is* known is that the strategy of using Bayes' Theorem to update the location probabilities of where the *Scorpion* might have sunk proved to be successful.

Prior Distributions and Posterior Distributions

Conceptually, a major difference between Bayesian analysis and non-Bayesian analysis is the assumptions associated with unknown parameters. In a non-Bayesian analysis (which would include all the statistical methodology in this book except the present section), unknown parameters are viewed as constants; in a Bayesian analysis, parameters are treated as random variables, meaning they have a pdf.

At the outset in a Bayesian analysis, the pdf assigned to the parameter may be based on little or no information and is referred to as the *prior distribution*. As soon as some data are collected, it becomes possible—via Bayes' Theorem—to revise and refine the pdf ascribed to the parameter. Any such updated pdf is referred to as a *posterior distribution*. In the search for the USS *Scorpion*, the unknown parameters were the probabilities of finding the sub in each of the grid areas surrounding the Azores. The prior distribution on those parameters were the probabilities $P(A_1), P(A_2), \ldots, P(A_n)$. Each time an area was searched and the sub not found, a posterior distribution was calculated—the first was the set of probabilities $P^*(A_1), P^*(A_2), \ldots, P^*(A_n)$; the second was the set of probabilities $P^{**}(A_1), P^{**}(A_2), \ldots, P^{**}(A_n)$; and so on.

Example 5.8.1

Suppose a retailer is interested in modeling the number of calls arriving at a phone bank in a five-minute interval. Section 4.2 established that the Poisson distribution would be the pdf to choose. But what value should be assigned to the Poisson's parameter, λ?

If the rate of calls was constant over a twenty-four-hour period, an estimate λ_e for λ could be calculated by dividing the total number of calls received during a full day by 288, the latter being the number of five-minute intervals in a twenty-four-hour period. If the random variable X, then, denotes the number of calls received during a random five-minute interval, the estimated probability that $X = k$ would be $p_X(k) = e^{-\lambda_e}\frac{\lambda_e^k}{k!}$, $k = 0, 1, 2, \ldots$.

In reality, though, the incoming call rate is not likely to remain constant over an entire twenty-four-hour period. Suppose, in fact, that an examination of telephone logs for the past several months suggests that λ equals 10 about three-quarters of the time, and it equals 8 about one-quarter of the time. Described in Bayesian terminology, the rate parameter is a random variable Λ, and the (discrete) prior distribution for Λ is defined by two probabilities:

$$p_\Lambda(8) = P(\Lambda = 8) = 0.25$$

and

$$p_\Lambda(10) = P(\Lambda = 10) = 0.75$$

Now, suppose certain facets of the retailer's operation have recently changed (different products to sell, different amounts of advertising, etc.). Those changes may very well affect the distribution associated with the call rate. Updating the prior distribution for Λ requires (a) some data and (b) an application of Bayes' Theorem. Being both frugal and statistically challenged, the retailer decides to construct a posterior distribution for Λ on the basis of a single observation. To that end, a five-minute interval is preselected at random and the corresponding value for X is found to be 7. How should $p_\Lambda(8)$ and $p_\Lambda(10)$ be revised?

Using Bayes' Theorem,

$$P(\Lambda = 10 \mid X = 7) = \frac{P(X = 7 \mid \Lambda = 10)P(\Lambda = 10)}{P(X = 7 \mid \Lambda = 8)P(\Lambda = 8) + P(X = 7 \mid \Lambda = 10)P(\Lambda = 10)}$$

$$= \frac{e^{-10}\frac{10^7}{7!}(0.75)}{\left(e^{-8}\frac{8^7}{7!}\right)(0.25) + e^{-10}\frac{10^7}{7!}(0.75)}$$

$$= \frac{(0.090)(0.75)}{(0.140)(0.25) + (0.090)(0.75)} = 0.659$$

which implies that

$$P(\Lambda = 8 \mid X = 7) = 1 - 0.659 = 0.341$$

Notice that the posterior distribution for Λ has changed in a way that makes sense intuitively. Initially, $P(\Lambda = 8)$ was 0.25. Since the data point, $x = 7$, is more consistent with $\Lambda = 8$ than with $\Lambda = 10$, the posterior pdf has *increased* the probability that $\Lambda = 8$ (from 0.25 to 0.341) and *decreased* the probability that $\Lambda = 10$ (from 0.75 to 0.659). ■

Definition 5.8.1. Let W be a statistic dependent on a parameter θ. Call its pdf $f_W(w \mid \theta)$. Assume that θ is the value of a random variable Θ, whose prior distribution is denoted $p_\Theta(\theta)$, if Θ is discrete, and $f_\Theta(\theta)$, if Θ is continuous. The *posterior distribution* of Θ, given that $W = w$, is the quotient

$$g_\theta(\theta \mid W = w) = \begin{cases} \dfrac{p_W(w|\theta)f_\Theta(\theta)}{\int_{-\infty}^{\infty} p_W(w|\theta)f_\Theta(\theta)\,d\theta} & \text{if } W \text{ is discrete} \\[2ex] \dfrac{f_W(w|\theta)f_\Theta(\theta)}{\int_{-\infty}^{\infty} f_W(w|\theta)f_\theta(\theta)\,d\theta} & \text{if } W \text{ is continuous} \end{cases}$$

[*Note:* If Θ is discrete, call its pdf $p_\theta(\theta)$ and replace the integrations with summations.]

Comment Definition 5.8.1 can be used to construct a posterior distribution even if no information is available on which to base a prior distribution. In such cases, the uniform pdf is substituted for either $p_\Theta(\theta)$ or $f_\Theta(\theta)$ and referred to as a *noninformative prior*.

Example 5.8.2

Max, a video game pirate (and Bayesian), is trying to decide how many illegal copies of *Zombie Beach Party* to have on hand for the upcoming holiday season. To get a rough idea of what the demand might be, he talks with n potential customers and finds that $X = k$ would buy a copy for a present (or for themselves). The obvious choice for a probability model for X, of course, would be the binomial pdf. Given n potential customers, the probability that k would actually buy one of Max's illegal copies is the familiar

$$p_X(k \mid \theta) = \binom{n}{k} \theta^k (1-\theta)^{n-k}, \quad k = 0, 1, \ldots, n$$

where the maximum likelihood estimate for θ is given by $\theta_e = \frac{k}{n}$.

It may very well be the case, though, that Max has some additional insight about the value of θ on the basis of similar video games that he illegally marketed in previous years. Suppose he suspects, for example, that the percentage of potential customers who will buy *Zombie Beach Party* is likely to be between 3% and 4% and probably will not exceed 7%. A reasonable prior distribution for Θ, then, would be a pdf mostly concentrated over the interval 0 to 0.07 with a mean or median in the 0.035 range.

One such probability model whose shape would comply with the restraints that Max is imposing is the *beta pdf*. Written with Θ as the random variable, the (two-parameter) beta pdf is given by

$$f_\Theta(\theta) = \frac{\Gamma(r+s)}{\Gamma(r)\Gamma(s)} \theta^{r-1} (1-\theta)^{s-1}, \quad 0 \le \theta \le 1$$

The beta distribution with $r = 2$ and $s = 4$ is pictured in Figure 5.8.1. By choosing different values for r and s, $f_\Theta(\theta)$ can be skewed more sharply to the right or to the left, and the bulk of the distribution can be concentrated close to zero or close to one. The question is, if an appropriate beta pdf is used as a *prior* distribution for Θ, and if a random sample of k potential customers (out of n) said they would buy the video game, what would be a reasonable *posterior* distribution for Θ?

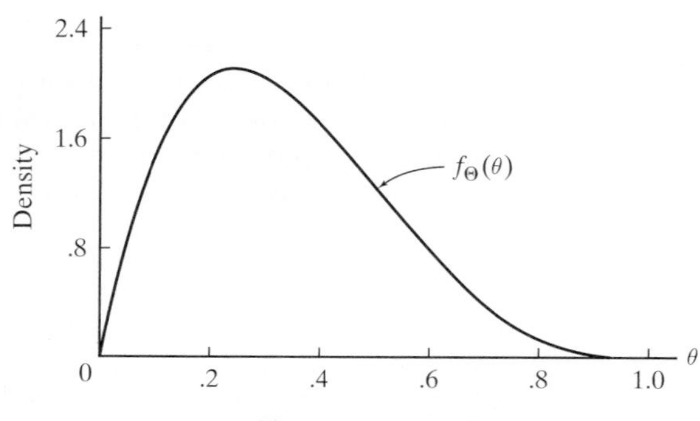

Figure 5.8.1

From Definition 5.8.1 for the case where $W\ (= X)$ is discrete and Θ is continuous,

$$g_\Theta(\theta \mid X = k) = \frac{p_X(k \mid \theta) f_\Theta(\theta)}{\int_{-\infty}^{\infty} p_X(k \mid \theta) f_\Theta(\theta)\, d\theta}$$

Substituting into the numerator gives

$$\begin{aligned} p_X(k \mid \theta) f_\Theta(\theta) &= \binom{n}{k} \theta^k (1-\theta)^{n-k} \frac{\Gamma(r+s)}{\Gamma(r)\Gamma(s)} \theta^{r-1}(1-\theta)^{s-1} \\ &= \binom{n}{k} \frac{\Gamma(r+s)}{\Gamma(r)\Gamma(s)} \theta^{k+r-1}(1-\theta)^{n-k+s-1} \end{aligned}$$

so

$$\begin{aligned} g_\Theta(\theta \mid X = k) &= \frac{\binom{n}{k} \frac{\Gamma(r+s)}{\Gamma(r)\Gamma(s)} \theta^{k+r-1}(1-\theta)^{n-k+s-1}}{\int_0^1 \binom{n}{k} \frac{\Gamma(r+s)}{\Gamma(r)\Gamma(s)} \theta^{k+r-1}(1-\theta)^{n-k+s-1}\, d\theta} \\ &= \left[\frac{\binom{n}{k} \frac{\Gamma(r+s)}{\Gamma(r)\Gamma(s)}}{\int_0^1 \binom{n}{k} \frac{\Gamma(r+s)}{\Gamma(r)\Gamma(s)} \theta^{k+r-1}(1-\theta)^{n-k+s-1}\, d\theta} \right] \theta^{k+r-1}(1-\theta)^{n-k+s-1} \end{aligned}$$

Notice that if the parameters r and s in the beta pdf were relabeled $k+r$ and $n-k+s$, respectively, the equation for $f_\Theta(\theta)$ would be

$$f_\Theta(\theta) = \frac{\Gamma(n+r+s)}{\Gamma(k+r)\Gamma(n-k+s)} \theta^{k+r-1}(1-\theta)^{n-k+s-1}$$

But those same exponents for θ and $(1-\theta)$ appear outside the brackets in the expression for $g_\Theta(\theta \mid X = k)$. Since there can be only one $f_\Theta(\theta)$ whose variable factors are $\theta^{k+r-1}(1-\theta)^{n-k+s-1}$, it follows that $g_\Theta(\theta \mid X = k)$ is a beta pdf with parameters $k+r$ and $n-k+s$.

The final step in the construction of a posterior distribution for Θ is to choose values for r and s that would produce a (prior) beta distribution having the configuration described on p. 336—that is, with a mean or median at 0.035 and the bulk of the distribution between 0 and 0.07. It can be shown [see (92)] that the expected value of a beta pdf is $r/(r+s)$. Setting 0.035, then, equal to that quotient implies that

$$s \doteq 28r$$

By trial and error with a calculator that can integrate a beta pdf, the values $r = 4$ and $s = 28(4) = 102$ are found to yield an $f_\Theta(\theta)$ having almost all of its area to the left of 0.07. Substituting those values for r and s into $g_\Theta(\theta \mid X = k)$ gives the completed posterior distribution:

$$\begin{aligned} g_\Theta(\theta \mid X = k) &= \frac{\Gamma(n+106)}{\Gamma(k+4)\Gamma(n-k+102)} \theta^{k+4-1}(1-\theta)^{n-k+102-1} \\ &= \frac{(n+105)!}{(k+3)!(n-k+101)!} \theta^{k+3}(1-\theta)^{n-k+101} \end{aligned}$$

■

Example 5.8.3

Certain prior distributions "fit" especially well with certain parameters in the sense that the resulting posterior distributions are easy to work with. Example 5.8.2 was a case in point—assigning a beta prior distribution to the unknown parameter in a binomial pdf led to a beta posterior distribution. A similar relationship holds if a gamma pdf is used as the prior distribution for the parameter in a Poisson model.

Suppose $X_1, X_2, \ldots, X_n$ denotes a random sample from the Poisson pdf, $p_X(k \mid \theta) = e^{-\theta}\theta^k/k!$, $k = 0, 1, \ldots$. Let $W = \sum_{i=1}^{n} X_i$. By Example 3.12.10, W has a Poisson distribution with parameter $n\theta$—that is, $p_W(w \mid \theta) = e^{-n\theta}(n\theta)^w/w!$, $w = 0, 1, 2, \ldots$.

Let the gamma pdf,

$$f_\Theta(\theta) = \frac{\mu^s}{\Gamma(s)}\theta^{s-1}e^{-\mu\theta}, \quad 0 < \theta < \infty$$

be the prior distribution assigned to Θ. Then

$$g_\Theta(\theta \mid W = w) = \frac{p_W(w \mid \theta) f_\Theta(\theta)}{\int_\theta p_W(w \mid \theta) f_\Theta(\theta)\, d\theta}$$

where

$$\begin{aligned} p_W(w \mid \theta) f_\Theta(\theta) &= e^{-n\theta}\frac{(n\theta)^w}{w!}\frac{\mu^s}{\Gamma(s)}\theta^{s-1}e^{-\mu\theta} \\ &= \frac{n^w}{w!}\frac{\mu^s}{\Gamma(s)}\theta^{w+s-1}e^{-(\mu+n)\theta} \end{aligned}$$

Now, using the same argument that simplified the calculation of the posterior distribution in Example 5.8.2, we can write

$$g_\Theta(\theta \mid W = w) = \left[\frac{\frac{n^w}{w!}\frac{\mu^s}{\Gamma(s)}}{\int_\theta p_W(w \mid \theta) f_\Theta(\theta)\, d\theta}\right]\theta^{w+s-1}e^{-(\mu+n)\theta}$$

But the only pdf having the factors $\theta^{w+s-1}e^{-(\mu+n)\theta}$ is the gamma distribution with parameters $w + s$ and $\mu + n$. It follows, then, that

$$g_\Theta(\theta \mid W = w) = \frac{(\mu + n)^{w+s}}{\Gamma(w + s)}\theta^{w+s-1}e^{-(\mu+n)\theta}$$

■

Case Study 5.8.1

Predicting the annual number of hurricanes that will hit the U.S. mainland is a problem receiving a great deal of public attention, given the disastrous summer of 2004, when four major hurricanes struck Florida causing billions of dollars of damage and several mass evacuations. For all the reasons discussed in Section 4.2, the obvious pdf for modeling the number of hurricanes reaching the mainland is the Poisson, where the unknown parameter θ would be the expected number in a given year.

Table 5.8.1 shows the numbers of hurricanes that actually did come ashore for three fifty-year periods. Use that information to construct a posterior distribution for θ. Assume that the prior distribution is a gamma pdf.

(Continued on next page)

Table 5.8.1

Years	Number of Hurricanes
1851–1900	88
1901–1950	92
1951–2000	72

Not surprisingly, meteorologists consider the data from the earliest period, 1851 to 1900 to be the least reliable. Those eighty-eight hurricanes, then, will be used to formulate the prior distribution. Let

$$f_\Theta(\theta) = \frac{\mu^s}{\Gamma(s)}\theta^{s-1}e^{-\mu\theta}, \quad 0 < \theta < \infty$$

Recall from Theorem 4.6.3 that for a gamma pdf, $E(\Theta) = s/\mu$. For the years from 1851 to 1900, though, the sample average number of hurricanes per year was $\frac{88}{50}$. Setting the latter equal to $E(\Theta)$ allows $s = 88$ and $\mu = 50$ to be assigned to the gamma's parameters. That is, we can take the prior distribution to be

$$f_\Theta(\theta) = \frac{50^{88}}{\Gamma(88)}\theta^{88-1}e^{-50\theta}$$

Also, the posterior distribution given at the end of Example 5.8.3 becomes

$$g_\Theta(\theta \mid W = w) = \frac{(50+n)^{w+88}}{\Gamma(w+88)}\theta^{w+87}e^{-(50+n)\theta}$$

The data, then, to incorporate into the posterior distribution would be the fact that $w = 92 + 72 = 164$ hurricanes occurred over the most recent $n = 100$ years included in the database. Therefore,

$$g_\Theta(\theta \mid W = w) = \frac{(50+100)^{164+88}}{\Gamma(164+88)}\theta^{164+87}e^{-(50+100)\theta} = \frac{(150)^{252}}{\Gamma(252)}\theta^{251}e^{-150\theta}$$

Example 5.8.4

In the examples seen thus far, the joint pdf $g_{W,\theta}(w, \theta) = p_W(w \mid \theta) f_\theta(\theta)$ of a statistic W and a parameter Θ [with a prior distribution $f_\theta(\theta)$] was the starting point in finding the posterior distribution of Θ. For some applications, though, the objective is not to derive $g_\theta(\theta \mid W = w)$, but, rather, to find the *marginal pdf of W*.

For instance, suppose a sample of size $n = 1$ is drawn from a Poisson pdf, $p_W(w \mid \theta) = e^{-\theta}\theta^w/w!$, $w = 0, 1, \ldots$, where the prior distribution is the gamma pdf, $f_\theta(\theta) = \frac{\mu^s}{\Gamma(s)}\theta^{s-1}e^{-\mu\theta}$. According to Example 5.8.3,

$$g_{W,\theta}(w, \theta) = p_W(w \mid \theta) f_\theta(\theta) = \frac{1}{w!}\frac{\mu^s}{\Gamma(s)}\theta^{w+s-1}e^{-(\mu+1)\theta}$$

What is the corresponding *marginal pdf of W*—that is, $p_W(w)$?

Recall Theorem 3.7.2. Integrating the joint pdf of W and Θ over θ gives

$$\begin{aligned} p_W(w) &= \int_0^\infty g_{W,\theta}(w, \theta)\, d\theta \\ &= \int_0^\infty \frac{1}{w!}\frac{\mu^s}{\Gamma(s)}\theta^{w+s-1}e^{-(\mu+1)\theta}\, d\theta \end{aligned}$$

$$= \frac{1}{w!}\frac{\mu^s}{\Gamma(s)}\int_0^\infty \theta^{w+s-1}e^{-(\mu+1)\theta}\,d\theta$$

$$= \frac{1}{w!}\frac{\mu^s}{\Gamma(s)}\frac{\Gamma(w+s)}{(\mu+1)^{w+s}}$$

$$= \frac{\Gamma(w+s)}{w!\Gamma(s)}\left(\frac{\mu}{\mu+1}\right)^s\left(\frac{1}{\mu+1}\right)^w$$

But $\frac{\Gamma(w+s)}{w!\Gamma(s)} = \binom{w+s-1}{w}$. Finally, let $p = \mu/(\mu+1)$, so $1-p = 1/(\mu+1)$, and the marginal pdf reduces to a *negative binomial distribution* with parameters s and p:

$$p_W(w) = \binom{w+s-1}{w} p^s(1-p)^w$$

(see Question 4.5.6). ■

Case Study 5.8.2

Psychologists use a special coordination test to study a person's likelihood of making manual errors. For any given person, the number of such errors made on the test is known to follow a Poisson distribution with some particular value for the rate parameter, θ. But as we all know (from watching the clumsy people around us who spill things and get in our way), θ varies considerably from person to person. Suppose, in fact, that variability in Θ can be described by a gamma pdf. If so, the marginal pdf of the number of errors made by a individual should have a negative binomial distribution (according to Example 5.8.4).

Columns 1 and 2 of Table 5.8.2 show the number of errors made on the coordination test by a sample of 504 subjects—82 made zero errors, 57 made one error, and so on. To know whether those responses can be adequately modeled

Table 5.8.2

Number of Errors, w	Observed Frequency	Negative Binomial Predicted Frequency
0	82	79.2
1	57	57.1
2	46	46.3
3	39	38.9
4	33	33.3
5	28	28.8
6	25	25.1
7	22	22.0
8	19	19.3
9	17	17.0
10	15	15.0
11	13	13.3
12	12	11.8
13	10	10.4

(Continued on next page)

Table 5.8.2 *(continued)*

Number of Errors, w	Observed Frequency	Negative Binomial Predicted Frequency
14	9	9.3
15	8	8.3
16	7	7.3
17	6	6.5
18	6	5.8
19	5	5.2
20	5	4.6
21	4	4.1
22	4	3.7
23	3	3.3
24	3	2.9
25	3	2.6
26	2	2.4
27	2	2.1
28	2	1.9
29	2	1.7
30	2	1.5
≥ 31	13	13.1
Total	504	504.0

by a negative binomial distribution requires that the parameters p and s be estimated. To that end, it should be noted that the maximum likelihood estimate for p in a negative binomial is $ns/\sum_{i=1}^{n} w_i$. Expected frequencies, then, can be calculated by choosing a value for s and solving for p. By trial and error, the entries shown in Column 3 were based on a negative binomial pdf for which $s = 0.8$ and $p = (504)(0.8)/3821 = 0.106$. Clearly, the model fits exceptionally well, which supports the analysis carried out in Example 5.8.4.

Bayesian Estimation

Fundamental to the philosophy of Bayesian analysis is the notion that all relevant information about an unknown parameter, θ, is encoded in the parameter's posterior distribution, $g_\Theta(\theta \mid W = w)$. Given that premise, an obvious question arises: How can $g_\Theta(\theta \mid W = w)$ be used to calculate an appropriate *point estimator*, $\hat{\theta}$? One approach, similar to using the likelihood function to find a maximum likelihood estimator, is to differentiate the posterior distribution, in which case the value for which $dg_\Theta(\theta \mid W = w)/d\theta = 0$—that is, the *mode*—becomes $\hat{\theta}$.

For theoretical reasons, though, a method much preferred by Bayesians is to use some key ideas from *decision theory* as a framework for identifying a reasonable $\hat{\theta}$. In particular, Bayesian estimates are chosen to minimize the *risk* associated with $\hat{\theta}$, where the risk is the expected value of the *loss* incurred by the error in the estimate. Presumably, as $\hat{\theta} - \theta$ gets further away from 0—that is, as the estimation error gets larger—the loss associated with $\hat{\theta}$ will increase.

Definition 5.8.2. Let $\hat{\theta}$ be an estimator for θ based on a statistic W. The *loss function* associated with $\hat{\theta}$ is denoted $L(\hat{\theta}, \theta)$, where $L(\hat{\theta}, \theta) \geq 0$ and $L(\theta, \theta) = 0$.

Example 5.8.5 It is typically the case that quantifying in any precise way the consequences, economic or otherwise, of $\hat{\theta}$ not being equal to θ is all but impossible. The "generic" loss functions defined in those situations are chosen primarily for their mathematical convenience. Two of the most frequently used are $L(\hat{\theta}, \theta) = |\hat{\theta} - \theta|$ and $L(\hat{\theta}, \theta) = (\hat{\theta} - \theta)^2$. Sometimes, though, the context in which a parameter is being estimated *does* allow for a loss function to be defined in a very specific and relevant way.

Consider the inventory dilemma faced by Max, the Bayesian video game pirate whose illegal activities were described in Example 5.8.2. The unknown parameter in question was θ, the proportion of his n potential customers who would purchase a copy of *Zombie Beach Party*. Suppose that Max decides—for whatever reasons—to estimate θ with $\hat{\theta}$. As a consequence, it would follow that he should have $n\,\hat{\theta}$ copies of the video game available. That said, what would be the corresponding loss funciton?

Here, the implications of $\hat{\theta}$ not being equal to θ are readily quantifiable. If $\hat{\theta} < \theta$, then $n(\theta - \hat{\theta})$ sales will be lost (at a cost of, say, \$$c$ per video). On the other hand, if $\hat{\theta} > \theta$, there will be $n(\hat{\theta} - \theta)$ unsold videos, each of which will incur a storage cost of, say, \$$d$ per unit. The loss function that applies to Max's situation, then, is clearly defined:

$$L(\hat{\theta}, \theta) = \begin{cases} \$cn(\theta - \hat{\theta}) & \text{if } \hat{\theta} < \theta \\ \$dn(\hat{\theta} - \theta) & \text{if } \hat{\theta} > \theta \end{cases}$$

■

Definition 5.8.3. Let $L(\hat{\theta}, \theta)$ be the loss function associated with an estimate of the parameter θ. Let $g_\theta(\theta \mid W = w)$ be the posterior distribution of the random variable Θ. Then the *risk* associated with $\hat{\theta}$ is the expected value of the loss function with respect to the posterior distribution of θ.

$$\text{risk} = \begin{cases} \int_\theta L(\hat{\theta}, \theta) g_\Theta(\theta \mid W = w)\, d\theta & \text{if } \Theta \text{ is continuous} \\ \sum_{\text{all } \theta} L(\hat{\theta}, \theta) g_\Theta(\theta \mid W = w) & \text{if } \Theta \text{ is discrete} \end{cases}$$

Using the Risk Function to Find $\hat{\theta}$

Given that the risk function represents the *expected loss* associated with the estimator $\hat{\theta}$, it makes sense to look for the $\hat{\theta}$ that *minimizes* the risk. Any $\hat{\theta}$ that achieves that objective is said to be a *Bayes estimate.* In general, finding the Bayes estimate requires solving the equation $d(\text{risk})/d\hat{\theta} = 0$. For two of the most frequently used loss functions, $L(\hat{\theta}, \theta) = |\hat{\theta} - \theta|$ and $L(\hat{\theta}, \theta) = (\hat{\theta} - \theta)^2$, though, there is a much easier way to calculate $\hat{\theta}$.

Theorem 5.8.1 *Let $g_\theta(\theta \mid W = w)$ be the posterior distribution for the unknown parameter θ.*

a. *If the loss function associated with $\hat{\theta}$ is $L(\hat{\theta}, \theta) = |\hat{\theta} - \theta|$, then the Bayes estimate for θ is the* median *of $g_\Theta(\theta \mid W = w)$.*

b. *If the loss function associated with $\hat{\theta}$ is $L(\hat{\theta}, \theta) = (\hat{\theta} - \theta)^2$, then the Bayes estimate for θ is the mean of $g_\Theta(\theta \mid W = w)$.*

Proof

a. The proof follows from a general result for the expected value of a random variable. The fact that the pdf in the expectation here is a posterior distribution is irrelevant. The derivation will be given for a continuous random variable (having a finite expected value); the proof for the discrete case is similar.

Let $f_W(w)$ be the pdf for the random variable W, where the median of W is m. Then

$$\begin{aligned} E(|W-m|) &= \int_{-\infty}^{\infty} |w-m| f_W(w)\,dw \\ &= \int_{-\infty}^{m} (m-w) f_W(w)\,dw + \int_{m}^{\infty} (w-m) f_W(w)\,dw \\ &= m\int_{-\infty}^{m} f_W(w)\,dw - \int_{-\infty}^{m} w f_W(w)\,dw \\ &\quad + \int_{m}^{\infty} w f_W(w)\,dw - m\int_{m}^{\infty} f_W(w)\,dw \end{aligned}$$

The first and last integrals are equal by definition of the median so,

$$E(|W-m|) = -\int_{-\infty}^{m} w f_W(w)\,dw + \int_{m}^{\infty} w f_W(w)\,dw$$

Now, suppose $m \geq 0$ (the proof for negative m is similar). Splitting the first integral into two parts gives

$$E(|W-m|) = -\int_{-\infty}^{0} w f_W(w)\,dw - \int_{0}^{m} w f_W(w)\,dw + \int_{m}^{\infty} w f_W(w)\,dw$$

Notice that the middle integral is positive, so changing its negative sign to a plus implies that

$$\begin{aligned} E(|W-m|) &\leq -\int_{-\infty}^{0} w f_W(w)\,dw + \int_{0}^{m} w f_W(w)\,dw + \int_{m}^{\infty} w f_W(w)\,dw \\ &\leq \int_{-\infty}^{0} -w f_W(w)\,dw + \int_{0}^{\infty} w f_W(w)\,dw \end{aligned}$$

Therefore,

$$E(|W-m|) \leq E(|W|) \tag{5.8.1}$$

Finally, suppose b is any constant. Then

$$\frac{1}{2} = P(W \leq m) = P(W-b \leq m-b),$$

showing that $m-b$ is the median of the random variable $W-b$. Applying Equation 5.8.1 to the variable $W-b$, we can write

$$E(|W-m|) = E[|(W-b)-(m-b)|] \leq E(|W-b|)$$

which implies that the *median* of $g_\Theta(\theta \mid W=w)$ is the Bayes estimate for θ when $L(\hat{\theta},\theta) = |\hat{\theta}-\theta|$.

b. Let W be any random variable whose mean is μ and whose variance is finite, and let b be any constant. Then

$$\begin{aligned} E[(W-b)^2] &= E[(W-\mu)+(\mu-b)]^2 \\ &= E[(W-\mu)^2]+2(\mu-b)E(W-\mu)+(\mu-b)^2 \\ &= \text{Var}(W)+0+(\mu-b)^2 \end{aligned}$$

implying that $E[(W-b)]^2$ is minimized when $b=\mu$. It follows that the Bayes estimate for θ, given a quadratic loss function, is the *mean* of the posterior distribution. □

Example 5.8.6 Recall Example 5.8.3, where the parameter in a Poisson distribution was assumed to have a gamma prior distribution. For a random sample of size n, where $W=\sum_{i=1}^{n} X_i$,

$$p_W(w|\theta)=e^{-n\theta}(n\theta)^w/w!, \quad w=0,1,2,\ldots$$

$$f_\Theta(\theta)=\frac{\mu^s}{\Gamma(s)}\theta^{s-1}e^{-\mu\theta}$$

which resulted in the posterior distribution being a gamma pdf with parameters $w+s$ and $\mu+n$.

Suppose the loss function associated with $\hat{\theta}$ is quadratic, $L(\hat{\theta},\theta)=(\hat{\theta}-\theta)^2$. By part (b) of Theorem 5.8.1, the Bayes estimate for θ is the mean of the posterior distribution. From Theorem 4.6.3, though, the mean of $g_\Theta(\theta \mid W=w)$ is $(w+s)/(\mu+n)$.

Notice that

$$\frac{w+s}{\mu+n}=\frac{n}{\mu+n}\left(\frac{w}{n}\right)+\frac{\mu}{\mu+n}\left(\frac{s}{\mu}\right)$$

which shows that the Bayes estimate is a weighted average of $\frac{w}{n}$, the maximum likelihood estimate for θ and $\frac{s}{\mu}$, the mean of the prior distribution. Moreover, as n gets large, the Bayes estimate converges to the maximum likelihood estimate. ■

Questions

5.8.1. Suppose that X is a geometric random variable, where $p_X(k|\theta)=(1-\theta)^{k-1}\theta, k=1,2,\ldots$. Assume that the prior distribution for θ is the beta pdf with parameters r and s. Find the posterior distribution for θ.

5.8.2. Find the squared-error loss $[L(\hat{\theta},\theta)=(\hat{\theta}-\theta)^2]$ Bayes estimate for θ in Example 5.8.2 and express it as a weighted average of the maximum likelihood estimate for θ and the mean of the prior pdf.

5.8.3. Suppose the binomial pdf described in Example 5.8.2 refers to the number of votes a candidate might receive in a poll conducted before the general election. Moreover, suppose a beta prior distribution has been assigned to θ, and every indicator suggests the election will be close. The pollster, then, has good reason for concentrating the bulk of the prior distribution around the value $\theta=\frac{1}{2}$. Setting the two beta parameters r and s both equal to 135 will accomplish that objective (in the event $r=s=135$, the probability of θ being between 0.45 and 0.55 is approximately 0.90).

(a) Find the corresponding posterior distribution.

(b) Find the squared-error loss Bayes estimate for θ and express it as a weighted average of the maximum likelihood estimate for θ and the mean of the prior pdf.

5.8.4. What is the squared-error loss Bayes estimate for the parameter θ in a binomial pdf, where θ has a uniform distribution—that is, a noninformative prior? (Recall that a uniform prior is a beta pdf for which $r=s=1$.)

5.8.5. In Questions 5.8.2–5.8.4, is the Bayes estimate unbiased? Is it asymptotically unbiased?

5.8.6. Suppose that Y is a gamma random variable with parameters r and θ and the prior is also gamma with parameters s and μ. Show that the posterior pdf is gamma with parameters $r+s$ and $y+\mu$.

5.8.7. Let $Y_1, Y_2, \ldots, Y_n$ be a random sample from a gamma pdf with parameters r and θ, where the prior distribution assigned to θ is the gamma pdf with parameters s and μ. Let $W = Y_1 + Y_2 + \cdots + Y_n$. Find the posterior pdf for θ.

5.8.8. Find the squared-error loss Bayes estimate for θ in Question 5.8.7.

5.8.9. Consider, again, the scenario described in Example 5.8.2—a binomial random variable X has parameters n and θ, where the latter has a beta prior with integer parameters r and s. Integrate the joint pdf $p_X(k\,|\,\theta) f_\Theta(\theta)$ with respect to θ to show that the marginal pdf of X is given by

$$p_X(k) = \frac{\binom{k+r-1}{k}\binom{n-k+s-1}{n-k}}{\binom{n+r+s-1}{n}}, \quad k = 0, 1, \ldots, n$$

5.9 Taking a Second Look at Statistics (Beyond Classical Estimation)

The theory of estimation presented in this chapter can properly be called *classical*. It is a legacy of the late nineteenth and early twentieth centuries, culminating in the work of R.A. Fisher, especially his foundational paper published in 1922 (47).

This chapter covers the historical, yet still vibrant, theory and technique of estimation. This material is the basis for many of the modern advances in statistics. And, these approaches still provide useful methods for estimating parameters and building models.

But statistics, like every other branch of knowledge, progresses. As is the case for most sciences, the computer has dramatically changed the landscape. Classical problems—such as finding maximum likelihood estimators—that were difficult, if not impossible, to solve in Fisher's day can now be attacked through computer approximations.

However, modern computers not only give new methods for old problems, but they also provide new avenues of approach. One such set of new methods goes under the general name of *resampling*. One part of resampling is known as *bootstrapping*. This technique is useful when classical inference is impossible.

A general explication of bootstrapping is not possible in this section, but an example of its application to estimating the *standard* error should provide a sense of the idea.

The *standard error* of an estimator $\hat{\theta}$ is just its standard deviation; that is, $\sqrt{\text{Var}(\hat{\theta})}$. The standard error, or an approximation of it, is an essential part of the construction of confidence intervals. For the normal case, $\overline{Y}$ is the basis of the confidence interval, and its standard error is $\sigma/\sqrt{n}$. If X is a binomial random variable with parameters n and p, then the standard error $\sqrt{\frac{p(1-p)}{n}}$ is readily approximated by $\sqrt{\frac{\frac{k}{n}(1-\frac{k}{n})}{n}}$, where k is the observed number of successes.

In general, though, estimating the standard error may not be so straightforward. As a case in point, consider the gamma pdf with $r = 2$ and unknown parameter θ, $f_Y(y;\theta) = \frac{1}{\theta^2} y e^{-y/\theta}$. Recall from Example 5.2.2 that the maximum likelihood estimator for θ is $\frac{1}{2}\overline{Y}$. Then its variance is

$$\text{Var}\left(\frac{1}{2}\overline{Y}\right) = \frac{1}{4}\text{Var}(\overline{Y}) = \frac{1}{4}\frac{\text{Var}(\overline{Y})}{n} = \frac{1}{4n}2\theta^2 = \frac{\theta^2}{2n}$$

and the standard error is the square root of the variance, or $\frac{\theta}{\sqrt{2n}}$.

To understand the technique of the bootstrapping estimate of the standard error in this case, let us consider a numerical example, given in a series of steps.

Step 1. Bootstrapping begins with a random sample from the pdf of interest. If we let $n = 15$, Table 5.9.1 is the given sample from $f_Y(y;\, \theta) = \frac{1}{\theta^2} y e^{-y/\theta}$:

Table 5.9.1				
30.987	9.949	26.720	9.651	29.137
47.653	33.250	4.933	17.923	2.400
7.580	9.941	16.624	28.514	10.693

Step 2. The sum of the entries in the table is 285.955, so the maximum likelihood estimate of the parameter θ is

$$\theta_e = \frac{1}{2}\overline{y} = \frac{1}{2} \cdot \frac{1}{15}(285.955) = 9.5318$$

Step 3. Then using the estimate $\theta_e = 9.5318$ for θ, two hundred random samples from the pdf $f_Y(y;\, 9.5318) = \frac{1}{(9.5318)^2} y e^{-y/9.5318}$ are generated. How this is done using Minitab will be discussed in Appendix 5.A.1.

It suffices here to note that samples appear as an array of numbers with fifteen columns and two hundred rows. Each row represents a random sample of size 15 from the indicated gamma pdf.

Table 5.9.2														
19.445	10.867	6.183	3.517	20.388	51.501	14.735	52.809	11.244	59.533	15.135	15.579	14.354	22.670	2
11.808	4.380	12.44	9.208	9.222	2.674	63.703	36.037	46.190	22.793	23.329	40.706	23.872	40.909	4
.														
.	(Additional 197 rows)													
.														
7.536	4.693	7.452	22.606	11.512	2.136	2.718	25.778	16.023	27.405	18.801	65.723	0.853	7.536	4

Step 4. Use each row of Table 5.9.2 to obtain $\theta_e = \frac{1}{2}\overline{y}$, the estimate of the unknown parameter θ. For the first row in Table 5.9.2, we obtain $\theta_e = 11.2873$, and for the second, $\theta_e = 11.6986$.

Step 5. From Step 4, two hundred estimates of θ result. Calculate the sample standard deviation of these two hundred numbers, which gives the value 1.83491. This is the bootstrap estimate of the standard error.

The value of θ that generated the original sample in Table 5.9.1 was 10. Thus, the actual standard error is

$$\frac{\theta}{\sqrt{2n}} = \frac{10}{\sqrt{2 \cdot 15}} = 1.82574$$

The bootstrap estimate of 1.83491 is quite close to the actual value.

Appendix 5.A.1 Minitab Applications

Because of their ability to generate random observations from many of the standard probability distributions, computers can be very effective in illustrating estimation properties and procedures. We have also seen in Sections 4.7 and 5.9 that computers are essential tools for new estimation techniques.

The meaning of confidence intervals can also be nicely demonstrated using Minitab's RANDOM command. Deriving formulas for confidence intervals is straightforward, but calling attention to their variability from sample to sample is best accomplished using a Monte Carlo analysis. Example 5.3.1 is a case in point. The fifty simulated 95% confidence intervals displayed in Table 5.3.1 reinforce the interpretation that should be accorded to any particular evaluation of the formula $\left(\bar{y} - 1.96\frac{0.8}{\sqrt{4}}, \bar{y} + 1.96\frac{0.8}{\sqrt{4}}\right)$.

The distributions of estimators—and some of their important properties—can also be easily examined using the computer. Recall the serial number analysis described in Case Study 1.2.2. If the production numbers to be estimated are large, then the assumption that the captured serial numbers represent a random sample from a *discrete* uniform pdf can reasonably be replaced by the assumption that the captured serial numbers represent a random sample from the (easier-to-work-with) *continuous* uniform pdf, defined over the interval $[0, \theta]$. Two unbiased estimators for θ, then, would be

$$\hat{\theta}_1 = (2/n)\sum_{i=1}^{n} Y_i$$

and

$$\hat{\theta}_2 = [(n+1)/n]Y_{\max}$$

Question 5.4.18 gave a special case of the more general result that

$$\text{Var}(\hat{\theta}_2) = \theta^2/[n(n+2)] < \text{Var}(\hat{\theta}_1) = \theta^2/3n$$

But suppose the complexity of two unbiased estimators precluded the calculation of their variances. How would we decide which to use? Probably the simplest solution would be to simulate each one's distribution and compare their sample standard deviations.

Figures 5.A.1.1 and 5.A.1.2 illustrate that technique on the two estimators

$$\hat{\theta}_1 = (2/n)\sum_{i=1}^{n} Y_i \quad \text{and} \quad \hat{\theta}_2 = [(n+1)/n]Y_{\max}$$

for the uniform parameter θ. Suppose that $n = 5$ serial numbers have been "captured" and the true value for θ is 3400. Figure 5.A.1.1 shows the Minitab syntax for generating two hundred samples of size 5 from $f_Y(y;\theta) = 1/3400$, $0 \leq y \leq 3400$, and calculating $\hat{\theta}_1$. The DESCRIBE command shows that the average of the θ_e's is 3383.8 and the sample standard deviation of the two hundred estimates is 913.2.

In contrast, Figure 5.A.1.2 details a similar simulation (two hundred samples, each of size 5) for the estimator $\hat{\theta}_2$. The accompanying DESCRIBE output lends support to the claim that $\hat{\theta}_2$ is the better estimator—it shows the average θ_e to be closer to the true value of 3400 than the average θ_e calculated from $\hat{\theta}_1$ (3398.4 versus 3383.8) and its sample standard deviation is smaller than the sample standard deviation of the θ_es from $\hat{\theta}_1$ (563.9 versus 913.2).

Figure 5.A.1.1

```
MTB  > random 200 c1-c5;
SUBC > uniform 0 3400.
MTB  > rmean c1-c5 c6
MTB  > let c7 = 2*c6
MTB  > histogram c7;
SUBC > start 2800;
SUBC > increment 200.
Histogram of C7    N = 200
48 Obs. below the first class
Midpoint   Count
   2800      12   ************
   3000      12   ************
   3200      19   *******************
   3400      13   *************
   3600      22   **********************
   3800      17   *****************
   4000      11   ***********
   4200      14   **************
   4400       8   ********
   4600      10   **********
   4800       3   ***
   5000       6   ******
   5200       3   ***
   5400       2   **
MTB > describe c7
           N     MEAN   MEDIAN   TRMEAN    STDEV   SEMEAN
C7       200   3383.8   3418.3   3388.6    913.2     64.6
         MIN      MAX       Q1       Q3
C7     997.0   5462.9   2718.0   4002.1
```

Figure 5.A.1.2

```
MTB  > random 200 c1-c5;
SUBC > uniform 0 3400.
MTB  > rmaximum c1-c5 c6
MTB  > let c7 = (6/5)*c6
MTB  > histogram c7;
SUBC > start 2800;
SUBC > increment 200.
Histogram of C7    N = 200
32 Obs. below the first class
Midpoint   Count
   2800       8   ********
   3000      10   **********
   3200      17   *****************
   3400      22   **********************
   3600      36   ************************************
   3800      37   *************************************
   4000      38   **************************************
MTB > describe c7
           N     MEAN   MEDIAN   TRMEAN    STDEV   SEMEAN
C7       200   3398.4   3604.6   3437.1    563.9     39.9
         MIN      MAX       Q1       Q3
C7    1513.9   4077.4   3093.2   3847.9
```

The sample necessary for the bootstrapping example in Section 5.9 was generated by a similar set of commands:

MTB > random 200 c1-c15;

SUBC > gamma 2 10.

Given the array in Table 5.9.2, the estimate of the parameter from each row sample was obtained by

MTB > rmean c1-c15 c16

MTB > let c17 = .5 * c16

Finally, the bootstrap estimate was the standard deviation of the numbers in Column 17 given by

MTB > stdev c17

with the resulting printout

Standard deviation of C17 = 1.83491

Chapter 6

HYPOTHESIS TESTING

As a young man, Laplace went to Paris to seek his fortune as a mathematician, disregarding his father's wishes that he enter the clergy. He soon became a protégé of d'Alembert and at the age of twenty-four was elected to the Academy of Sciences. Laplace was recognized as one of the leading figures of that group for his work in physics, celestial mechanics, and pure mathematics. He also enjoyed some political prestige, and his friend, Napoleon Bonaparte, made him Minister of the Interior for a brief period. With the restoration of the Bourbon monarchy, Laplace renounced Napoleon for Louis XVIII, who later made him a marquis.

—Pierre-Simon, Marquis de Laplace (1749–1827)

6.1 Introduction

Inferences, as we saw in Chapter 5, often reduce to numerical estimates of parameters, in the form of either single points or confidence intervals. But not always. In many experimental situations, the conclusion to be drawn is *not* numerical and is more aptly phrased as a choice between two conflicting theories, or *hypotheses*. A court psychiatrist, for example, may be called upon to pronounce an accused murderer either "sane" or "insane"; the FDA must decide whether a new flu vaccine is "effective" or "ineffective"; a geneticist concludes that the inheritance of eye color in a certain strain of *Drosophila melanogaster* either "does" or "does not" follow classical Mendelian principles. In this chapter we examine the statistical methodology and the attendant consequences involved in making decisions of this sort.

The process of dichotomizing the possible conclusions of an experiment and then using the theory of probability to choose one option over the other is known as *hypothesis testing*. The two competing propositions are called the *null hypothesis* (written H_0) and the *alternative hypothesis* (written H_1). How we go about choosing between H_0 and H_1 is conceptually similar to the way a jury deliberates in a court trial. The null hypothesis is analogous to the defendant: Just as the latter is presumed innocent until "proven" guilty, so is the null hypothesis "accepted" unless the data argue overwhelmingly to the contrary. Mathematically, choosing between H_0 and H_1 is an exercise in applying courtroom protocol to situations where the "evidence" consists of measurements made on random variables.

Chapter 6 focuses on basic principles—in particular, on the probabilistic structure that underlies the decision-making process. Most of the important specific applications of hypothesis testing will be taken up later, beginning in Chapter 7.

6.2 The Decision Rule

Imagine an automobile company looking for additives that might increase gas mileage. As a pilot study, they send thirty cars fueled with a new additive on a road trip from Boston to Los Angeles. Without the additive, those same cars are known to average 25.0 mpg with a standard deviation (σ) of 2.4 mpg.

Suppose it turns out that the thirty cars average $\overline{y} = 26.3$ mpg *with* the additive. What should the company conclude? If the additive *is* effective but the position is taken that the increase from 25.0 to 26.3 is due solely to chance, the company will mistakenly pass up a potentially lucrative product. On the other hand, if the additive is *not* effective but the firm interprets the mileage increase as "proof" that the additive works, time and money will ultimately be wasted developing a product that has no intrinsic value.

In practice, researchers would assess the increase from 25.0 mpg to 26.3 mpg by framing the company's choices in the context of the courtroom analogy mentioned in Section 6.1. Here, the null hypothesis, which is typically a statement reflecting the status quo, would be the assertion that the additive has no effect; the alternative hypothesis would claim that the additive does work. By agreement, we give H_0 (like the defendant) the benefit of the doubt. If the road trip average, then, is "close" to 25.0 in some probabilistic sense still to be determined, we must conclude that the new additive has not demonstrated its superiority. The problem is that whether 26.3 mpg qualifies as being "close" to 25.0 mpg is not immediately obvious.

At this point, rephrasing the question in random variable terminology will prove helpful. Let $y_1, y_2, \ldots, y_{30}$ denote the mileages recorded by each of the cars during the cross-country test run. We will assume that the y_i's are normally distributed with an unknown mean μ. Furthermore, suppose that prior experience with road tests of this type suggests that σ will equal *2.4*.[1] That is,

$$f_Y(y; \mu) = \frac{1}{\sqrt{2\pi}(2.4)} e^{-\frac{1}{2}\left(\frac{y-\mu}{2.4}\right)^2}, \quad -\infty < y < \infty$$

The two competing hypotheses, then, can be expressed as statements about μ. In effect, we are *testing*

$$H_0\colon \mu = 25.0 \quad \text{(Additive is } \textit{not} \text{ effective)}$$

versus

$$H_1\colon \mu > 25.0 \quad \text{(Additive } \textit{is} \text{ effective)}$$

Values of the sample mean, $\overline{y}$, less than or equal to 25.0 are certainly not grounds for rejecting the null hypothesis; averages a bit larger than 25.0 would also lead to that conclusion (because of the commitment to give H_0 the benefit of the doubt). On the other hand, we would probably view a cross-country average of, say, 35.0 mpg as

---

[1] In practice, the value of σ usually needs to be estimated; we will return to that more frequently encountered scenario in Chapter 7.

exceptionally strong evidence *against* the null hypothesis, and our decision would be "reject H_0." In effect, somewhere between 25.0 and 35.0 there is a point—call it $\bar{y}^*$—where for all practical purposes the credibility of H_0 ends (see Figure 6.2.1).

Figure 6.2.1

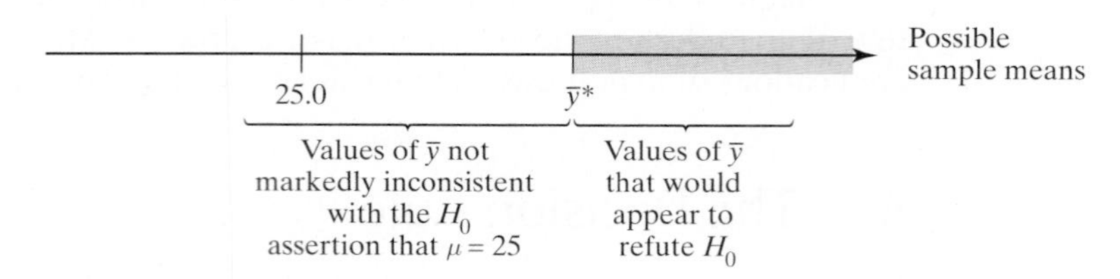

Finding an appropriate numerical value for $\bar{y}^*$ is accomplished by combining the courtroom analogy with what we know about the probabilistic behavior of $\bar{Y}$. Suppose, for the sake of argument, we set $\bar{y}^*$ equal to 25.25—that is, we would reject H_0 if $\bar{y} \geq 25.25$. Is that a good decision rule? No. If 25.25 defined "close," then H_0 would be rejected 28% of the time *even if H_0 were true*:

$$\begin{aligned} P(\text{We reject } H_0 \mid H_0 \text{ is true}) &= P(\bar{Y} \geq 25.25 \mid \mu = 25.0) \\ &= P\left(\frac{\bar{Y} - 25.0}{2.4/\sqrt{30}} \geq \frac{25.25 - 25.0}{2.4/\sqrt{30}}\right) \\ &= P(Z \geq 0.57) \\ &= 0.2843 \end{aligned}$$

(see Figure 6.2.2). Common sense, though, tells us that 28% is an inappropriately large probability for making this kind of incorrect inference. No jury, for example, would convict a defendant knowing it had a 28% chance of sending an innocent person to jail.

Figure 6.2.2

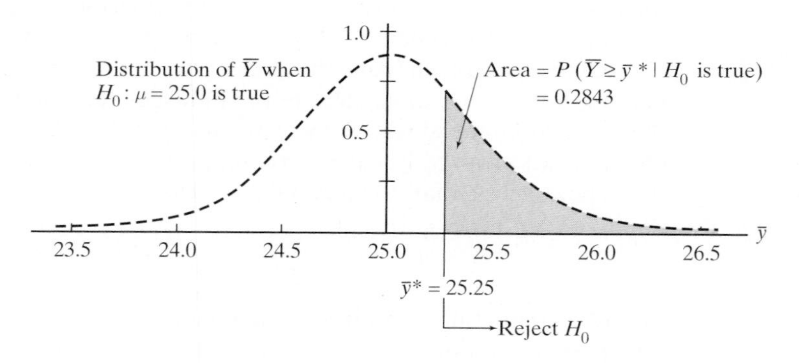

Clearly, we need to make $\bar{y}^*$ larger. Would it be reasonable to set $\bar{y}^*$ equal to, say, *26.50*? Probably not, because setting $\bar{y}^*$ that large would err in the other direction by giving the null hypothesis *too much* benefit of the doubt. If $\bar{y}^* = 26.50$, the probability of rejecting H_0 *if H_0 were true* is only *0.0003*:

$$\begin{aligned} P(\text{We reject } H_0 \mid H_0 \text{ is true}) &= P(\bar{Y} \geq 26.50 \mid \mu = 25.0) \\ &= P\left(\frac{\bar{Y} - 25.0}{2.4/\sqrt{30}} \geq \frac{26.50 - 25.0}{2.4/\sqrt{30}}\right) \\ &= P(Z \geq 3.42) \\ &= 0.0003 \end{aligned}$$

(see Figure 6.2.3). Requiring that much evidence before rejecting H_0 would be analogous to a jury not returning a guilty verdict unless the prosecutor could produce a roomful of eyewitnesses, an obvious motive, a signed confession, and a dead body in the trunk of the defendant's car!

Figure 6.2.3

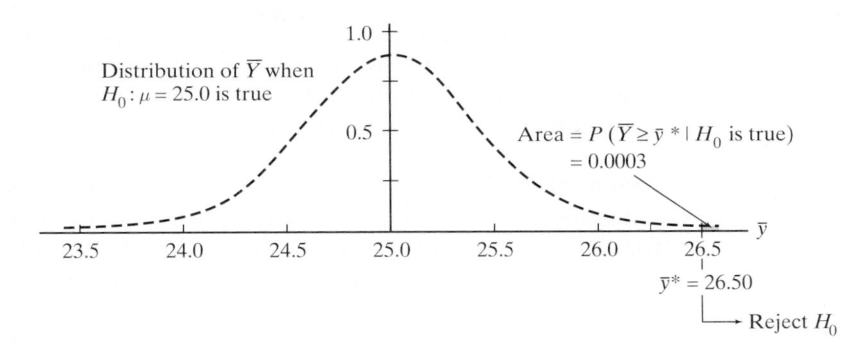

If a probability of 0.28 represents too little benefit of the doubt being accorded to H_0 and 0.0003 represents too much, what value *should* we choose for $P(\overline{Y} \geq \overline{y}^* \mid H_0 \text{ is true})$? While there is no way to answer that question definitively or mathematically, researchers who use hypothesis testing have come to the consensus that the probability of rejecting H_0 when H_0 is true should be somewhere in the neighborhood of *0.05*. Experience seems to suggest that when a 0.05 probability is used, null hypotheses are neither dismissed too capriciously nor embraced too wholeheartedly. (More will be said about this particular probability, and its consequences, in Section 6.3.)

Comment In 1768, British troops were sent to Boston to quell an outbreak of civil disturbances. Five citizens were killed in the aftermath, and several soldiers were subsequently put on trial for manslaughter. Explaining the guidelines under which a verdict was to be reached, the judge told the jury, "If upon the whole, ye are in any reasonable doubt of their guilt, ye must then, agreeable to the rule of law, declare them innocent" (177). Ever since, the expression "beyond all reasonable doubt" has been a frequently used indicator of how much evidence is needed in a jury trial to overturn a defendant's presumption of innocence. For many experimenters, choosing $\overline{y}^*$ such that

$$P(\text{We reject } H_0 \mid H_0 \text{ is true}) = 0.05$$

is comparable to a jury convicting a defendant only if the latter's guilt is established "beyond all reasonable doubt."

Suppose the 0.05 "criterion" is applied here. Finding the corresponding $\overline{y}^*$ is a calculation similar to what was done in Example 4.3.6. Given that

$$P(\overline{Y} \geq \overline{y}^* \mid H_0 \text{ is true}) = 0.05$$

it follows that

$$P\left(\frac{\overline{Y} - 25.0}{2.4/\sqrt{30}} \geq \frac{\overline{y}^* - 25.0}{2.4/\sqrt{30}}\right) = P\left(Z \geq \frac{\overline{y}^* - 25.0}{2.4/\sqrt{30}}\right) = 0.05$$

But we know from Appendix A.1 that $P(Z \geq 1.64) = 0.05$. Therefore,

$$\frac{\overline{y}^* - 25.0}{2.4/\sqrt{30}} = 1.64 \tag{6.2.1}$$

which implies that $\overline{y}^* = 25.718$.

The company's statistical strategy is now completely determined: They should reject the null hypothesis that the additive has no effect if $\overline{y} \geq 25.718$. Since the sample mean was 26.3, the appropriate decision is, indeed, to *reject* H_0. It appears that the additive *does* increase mileage.

Comment It must be remembered that rejecting H_0 does not *prove* that H_0 is false, any more than a jury's decision to convict guarantees that the defendant is guilty. The 0.05 decision rule is simply saying that *if* the *true* mean (μ) is 25.0, sample means ($\overline{y}$) as large or larger than 25.718 are expected to occur only 5% of the time. Because of that small probability, a reasonable conclusion when $\overline{y} \geq 25.718$ is that μ is *not* 25.0.

Table 6.2.1 is a computer simulation of this particular 0.05 decision rule. A total of seventy-five random samples, each of size 30, have been drawn from a normal distribution having $\mu = 25.0$ and $\sigma = 2.4$. The corresponding $\overline{y}$ for each sample is then compared with $\overline{y}^* = 25.718$. As the entries in the table indicate, five of the samples lead to the erroneous conclusion that $H_0: \mu = 25.0$ should be rejected.

Since each sample mean has a 0.05 probability of exceeding 25.718 (when μ =25.0), we would expect 75(0.05), or *3.75*, of the data sets to result in a "reject

Table 6.2.1

$\overline{y}$	≥ 25.718?	$\overline{y}$	≥ 25.718?	$\overline{y}$	≥ 25.718?
25.133	no	25.259	no	25.200	no
24.602	no	25.866	yes	25.653	no
24.587	no	25.623	no	25.198	no
24.945	no	24.550	no	24.758	no
24.761	no	24.919	no	24.842	no
24.177	no	24.770	no	25.383	no
25.306	no	25.080	no	24.793	no
25.601	no	25.307	no	24.874	no
24.121	no	24.004	no	25.513	no
25.516	no	24.772	no	24.862	no
24.547	no	24.843	no	25.034	no
24.235	no	25.771	yes	25.150	no
25.809	yes	24.233	no	24.639	no
25.719	yes	24.853	no	24.314	no
25.307	no	25.018	no	25.045	no
25.011	no	25.176	no	24.803	no
24.783	no	24.750	no	24.780	no
25.196	no	25.578	no	25.691	no
24.577	no	24.807	no	24.207	no
24.762	no	24.298	no	24.743	no
25.805	yes	24.807	no	24.618	no
24.380	no	24.346	no	25.401	no
25.224	no	25.261	no	24.958	no
24.371	no	25.062	no	25.678	no
25.033	no	25.391	no	24.795	no

H_0" conclusion. Reassuringly, the observed number of incorrect inferences (= 5) is quite close to that expected value.

> **Definition 6.2.1.** If $H_0\colon \mu = \mu_o$ is rejected using a 0.05 decision rule, the difference between $\overline{y}$ and μ_o is said to be *statistically significant.*

Expressing Decision Rules in Terms of Z Ratios

As we have seen, decision rules are statements that spell out the conditions under which a null hypothesis is to be rejected. The format of those statements, though, can vary. Depending on the context, one version may be easier to work with than another.

Recall Equation 6.2.1. Rejecting $H_0\colon \mu = 25.0$ when

$$\overline{y} \geq \overline{y}^* = 25.0 + 1.64 \cdot \frac{2.4}{\sqrt{30}} = 25.718$$

is clearly equivalent to rejecting H_0 when

$$\frac{\overline{y} - 25.0}{2.4/\sqrt{30}} \geq 1.64 \tag{6.2.2}$$

(if one rejects the null hypothesis, the other will necessarily do the same).

We know from Chapter 4 that the random variable $\frac{\overline{Y}-25.0}{2.4/\sqrt{30}}$ has a standard normal distribution (if $\mu = 25.0$). When a particular $\overline{y}$ is substituted for $\overline{Y}$ (as in Inequality 6.2.2), we call $\frac{\overline{y}-25.0}{2.4/\sqrt{30}}$ the *observed z*. Choosing between H_0 and H_1 is typically (and most conveniently) done in terms of the observed z. In Section 6.4, though, we will encounter certain questions related to hypothesis testing that are best answered by phrasing the decision rule in terms of $\overline{y}^*$.

> **Definition 6.2.2.** Any function of the observed data whose numerical value dictates whether H_0 is accepted or rejected is called a *test statistic*. The set of values for the test statistic that result in the null hypothesis being rejected is called the *critical region* and is denoted C. The particular point in C that separates the rejection region from the acceptance region is called the *critical value*.

Comment For the gas mileage example, both $\overline{y}$ and $\frac{\overline{y}-25.0}{2.4/\sqrt{30}}$ qualify as test statistics. If the sample mean is used, the associated critical region would be written

$$C = \{\overline{y};\ \overline{y} \geq 25.718\}$$

(and 25.718 is the critical value). If the decision rule is framed in terms of a Z ratio,

$$C = \left\{z;\ z = \frac{\overline{y} - 25.0}{2.4/\sqrt{30}} \geq 1.64\right\}$$

In this latter case, the critical value is 1.64.

> **Definition 6.2.3.** The probability that the test statistic lies in the critical region *when H_0 is true* is called the *level of significance* and is denoted α.

Comment In principle, the value chosen for α should reflect the consequences of making the mistake of rejecting H_0 when H_0 is true. As those consequences get more severe, the critical region C should be defined so that α gets smaller. In practice, though, efforts to quantify the costs of making incorrect inferences are arbitrary at best. In most situations, experimenters abandon any such attempts and routinely set the level of significance equal to 0.05. If another α is used, it is likely to be either 0.001, 0.01, or 0.10.

Here again, the similarity between hypothesis testing and courtroom protocol is worth keeping in mind. Just as experimenters can make α larger or smaller to reflect the consequences of mistakenly rejecting H_0 when H_0 is true, so can juries demand more or less evidence to return a conviction. For juries, any such changes are usually dictated by the severity of the possible punishment. A grand jury deciding whether or not to indict someone for fraud, for example, will inevitably require less evidence to return a conviction than will a jury impaneled for a murder trial.

One-Sided Versus Two-Sided Alternatives

In most hypothesis tests, H_0 consists of a single number, typically the value of the parameter that represents the status quo. The "25.0" in $H_0\colon \mu = 25.0$, for example, is the mileage that would be expected when the additive has no effect. If the mean of a normal distribution is the parameter being tested, our general notation for the null hypothesis will be $H_0\colon \mu = \mu_o$, where μ_o is the status quo value of μ.

Alternative hypotheses, by way of contrast, invariably embrace entire ranges of parameter values. If there is reason to believe *before any data are collected* that the parameter being tested is necessarily restricted to one particular "side" of H_0, then H_1 is defined to reflect that limitation and we say that the alternative hypothesis is *one-sided*. Two variations are possible: H_1 can be one-sided *to the left* ($H_1\colon \mu < \mu_o$) or it can be one-sided *to the right* ($H_1\colon \mu > \mu_o$). If no such a priori information is available, the alternative hypothesis needs to accommodate the possibility that the true parameter value might lie on either side of μ_0. Any such alternative is said to be *two-sided*. For testing $H_0\colon \mu = \mu_o$, the two-sided alternative is written $H_1\colon \mu \neq \mu_o$.

In the gasoline example, it was tacitly assumed that the additive either would have *no* effect (in which case $\mu = 25.0$ and H_0 would be true) or would *increase* mileage (implying that the true mean would lie somewhere "to the right" of H_0). Accordingly, we wrote the alternative hypothesis as $H_1\colon \mu > 25.0$. If we had reason to suspect, though, that the additive might interfere with the gasoline's combustibility and possibly *decrease* mileage, it would have been necessary to use a two-sided alternative ($H_1\colon \mu \neq 25.0$).

Whether the alternative hypothesis is defined to be one-sided or two-sided is important because the nature of H_1 plays a key role in determining the form of the critical region. We saw earlier that the 0.05 decision rule for testing

$$H_0\colon \mu = 25.0$$

versus

$$H_1\colon \mu > 25.0$$

calls for H_0 to be rejected if $\frac{\bar{y}-25.0}{2.4/\sqrt{30}} \geq 1.64$. That is, only if the sample mean is substantially *larger* than 25.0 will we reject H_0.

If the alternative hypothesis had been two-sided, sample means either much smaller than 25.0 *or* much larger than 25.0 would be evidence against H_0 (and in support of H_1). Moreover, the 0.05 probability associated with the critical region C would be split into two halves, with 0.025 being assigned to the left-most portion of C, and 0.025 to the right-most portion. From Appendix Table A.1, though, $P(Z \leq -1.96) = P(Z \geq 1.96) = 0.025$, so the two-sided 0.05 decision rule would call for $H_0\colon \mu = 25.0$ to be rejected if $\frac{\bar{y}-25.0}{2.4/\sqrt{30}}$ is either (1) ≤ -1.96 or (2) ≥ 1.96.

Testing H_0: $\mu = \mu_o$ (σ Known)

Let z_α be the number having the property that $P(Z \geq z_\alpha) = \alpha$. Values for z_α can be found from the standard normal cdf tabulated in Appendix A.1. If $\alpha = 0.05$, for example, $z_{.05} = 1.64$ (see Figure 6.2.4). Of course, by the symmetry of the normal curve, $-z_\alpha$ has the property that $P(Z \leq -z_\alpha) = \alpha$.

Figure 6.2.4

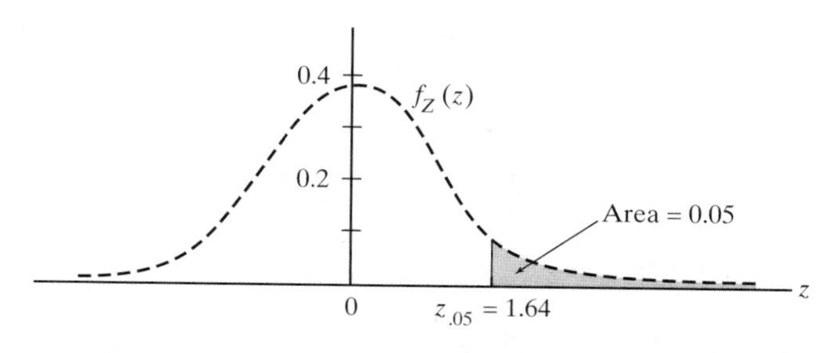

Theorem 6.2.1 *Let $y_1, y_2, \ldots, y_n$ be a random sample of size n from a normal distribution where σ is known. Let $z = \frac{\bar{y}-\mu_o}{\sigma/\sqrt{n}}$.*

a. *To test $H_0\colon \mu = \mu_o$ versus $H_1\colon \mu > \mu_o$ at the α level of significance, reject H_0 if $z \geq z_\alpha$.*
b. *To test $H_0\colon \mu = \mu_o$ versus $H_1\colon \mu < \mu_o$ at the α level of significance, reject H_0 if $z \leq -z_\alpha$.*
c. *To test $H_0\colon \mu = \mu_o$ versus $H_1\colon \mu \neq \mu_o$ at the α level of significance, reject H_0 if z is either* (1) $\leq -z_{\alpha/2}$ or (2) $\geq z_{\alpha/2}$. □

Example 6.2.1 As part of a "Math for the Twenty-First Century" initiative, Bayview High was chosen to participate in the evaluation of a new algebra and geometry curriculum. In the recent past, Bayview's students were considered "typical," having earned scores on standardized exams that were very consistent with national averages.

Two years ago, a cohort of eighty-six Bayview sophomores, all randomly selected, were assigned to a special set of classes that integrated algebra and geometry. According to test results that have just been released, those students averaged 502 on the SAT-I math exam; nationwide, seniors averaged 494 with a standard deviation of 124. Can it be claimed at the $\alpha = 0.05$ level of significance that the new curriculum had an effect?

To begin, we define the parameter μ to be the true average SAT-I math score that we could expect the new curriculum to produce. The obvious "status quo" value for μ is the current national average—that is, $\mu_o = 494$. The alternative hypothesis here should be two-sided because the possibility certainly exists that a revised curriculum—however well intentioned—would actually *lower* a student's achievement.

According to part (c) of Theorem 6.2.1, then, we should reject $H_0\colon \mu = 494$ in favor of $H_1\colon \mu \neq 494$ at the $\alpha = 0.05$ level of significance if the test statistic z is either (1) $\leq -z_{.025}(= -1.96)$ or (2) $\geq z_{.025}(= 1.96)$. But $\bar{y} = 502$, so

$$z = \frac{502 - 494}{124/\sqrt{86}} = 0.60$$

implying that our decision should be "Fail to reject H_0." Even though Bayview's 502 is eight points above the national average, it does not follow that the improvement was due to the new curriculum: An increase of that magnitude could easily have occurred by chance, even if the new curriculum had no effect whatsoever (see Figure 6.2.5).

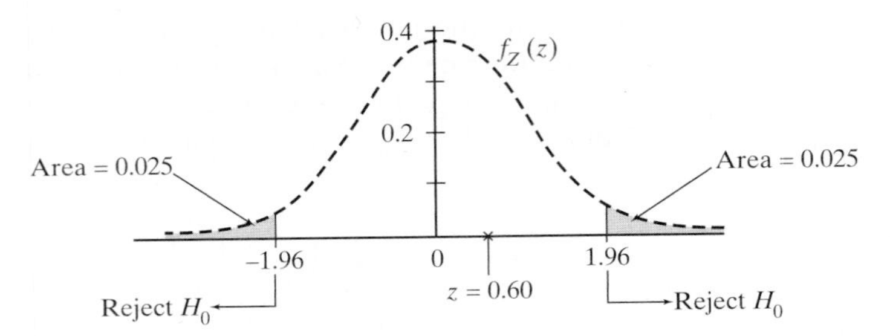

Figure 6.2.5

Comment If the null hypothesis is *not* rejected, we should phrase the conclusion as "Fail to reject H_0" rather than "Accept H_0." Those two statements may seem to be the same, but, in fact, they have very different connotations. The phrase "Accept H_0" suggests that the experimenter is concluding that H_0 is true. But that may not be the case. In a court trial, when a jury returns a verdict of "Not guilty," they are not saying that they necessarily believe that the defendant is innocent. They are simply asserting that the evidence—in their opinion—is not sufficient to overturn the presumption that the defendant is innocent. That same distinction applies to hypothesis testing. If a test statistic does not fall in the critical region (which was the case in Example 6.2.1), the proper interpretation is to conclude that we "Fail to reject H_0."

The *P*-Value

There are two general ways to quantify the amount of evidence against H_0 that is contained in a given set of data. The first involves the *level of significance* concept introduced in Definition 6.2.3. Using that format, the experimenter selects a value for α (usually 0.05 or 0.01) *before any data are collected.* Once α is specified, a corresponding critical region can be identified. If the test statistic falls in the critical region, we reject H_0 at the α level of significance. Another strategy is to calculate a *P-value.*

Definition 6.2.4. The *P-value* associated with an observed test statistic is the probability of getting a value for that test statistic as extreme as or more extreme than what was actually observed (relative to H_1) *given that H_0 is true.*

Comment Test statistics that yield small P-values should be interpreted as evidence *against* H_0. More specifically, if the P-value calculated for a test statistic is less than or equal to α, the null hypothesis can be rejected at the α level of significance. Or, put another way, the P-value is the smallest α at which we can reject H_0.

Example 6.2.2

Recall Example 6.2.1. Given that $H_0: \mu = 494$ is being tested against $H_1: \mu \neq 494$, what P-value is associated with the calculated test statistic, $z = 0.60$, and how should it be interpreted?

If $H_0: \mu = 494$ is true, the random variable $Z = \frac{\bar{Y} - 494}{124/\sqrt{86}}$ has a standard normal pdf. Relative to the two-sided H_1, any value of Z greater than or equal to 0.60 *or* less than or equal to -0.60 qualifies as being "as extreme as or more extreme than" the observed z. Therefore, by Definition 6.2.4,

$$
\begin{aligned}
P\text{-value} &= P(Z \geq 0.60) + P(Z \leq -0.60) \\
&= 0.2743 + 0.2743 \\
&= 0.5486
\end{aligned}
$$

(see Figure 6.2.6).

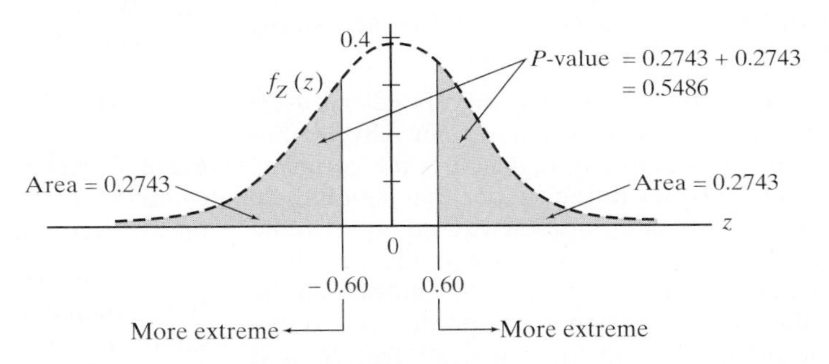

Figure 6.2.6

As noted in the preceding comment, P-values can be used as decision rules. In Example 6.2.1, 0.05 was the stated level of significance. Having determined here that the P-value associated with $z = 0.60$ is 0.5486, we know that $H_0: \mu = 494$ would *not* be rejected at the given α. Indeed, the null hypothesis would not be rejected for any value of α up to and including 0.5486.

Notice that the P-value would have been halved had H_1 been one-sided. Suppose we were confident that the new algebra and geometry classes would not *lower* a student's math SAT. The appropriate hypothesis test in that case would be $H_0: \mu = 494$ versus $H_1: \mu > 494$. Moreover, only values in the right-hand tail of $f_Z(z)$ would be considered more extreme than the observed $z = 0.60$, so

$$
P\text{-value} = P(Z \geq 0.60) = 0.2743
$$

■

Questions

6.2.1. State the decision rule that would be used to test the following hypotheses. Evaluate the appropriate test statistic and state your conclusion.

(a) $H_0: \mu = 120$ versus $H_1: \mu < 120$; $\bar{y} = 114.2, n = 25, \sigma = 18, \alpha = 0.08$

(b) $H_0: \mu = 42.9$ versus $H_1: \mu \neq 42.9$; $\overline{y} = 45.1, n = 16, \sigma = 3.2, \alpha = 0.01$

(c) $H_0: \mu = 14.2$ versus $H_1: \mu > 14.2$; $\overline{y} = 15.8, n = 9, \sigma = 4.1, \alpha = 0.13$

6.2.2. An herbalist is experimenting with juices extracted from berries and roots that may have the ability to affect the Stanford-Binet IQ scores of students afflicted with mild cases of attention deficit disorder (ADD). A random sample of twenty-two children diagnosed with the condition have been drinking Brain-Blaster daily for two months. Past experience suggests that children with ADD score an average of 95 on the IQ test with a standard deviation of 15. If the data are to be analyzed using the $\alpha = 0.06$ level of significance, what values of $\overline{y}$ would cause H_0 to be rejected? Assume that H_1 is two-sided.

6.2.3. **(a)** Suppose $H_0: \mu = \mu_o$ is rejected in favor of $H_1: \mu \neq \mu_o$ at the $\alpha = 0.05$ level of significance. Would H_0 necessarily be rejected at the $\alpha = 0.01$ level of significance?
(b) Suppose $H_0: \mu = \mu_o$ is rejected in favor of $H_1: \mu \neq \mu_o$ at the $\alpha = 0.01$ level of significance. Would H_0 necessarily be rejected at the $\alpha = 0.05$ level of significance?

6.2.4. Company records show that drivers get an average of 32,500 miles on a set of Road Hugger All-Weather radial tires. Hoping to improve that figure, the company has added a new polymer to the rubber that should help protect the tires from deterioration caused by extreme temperatures. Fifteen drivers who tested the new tires have reported getting an average of 33,800 miles. Can the company claim that the polymer has produced a statistically significant increase in tire mileage? Test $H_0: \mu = 32{,}500$ against a one-sided alternative at the $\alpha = 0.05$ level. Assume that the standard deviation (σ) of the tire mileages has not been affected by the addition of the polymer and is still 4000 miles.

6.2.5. If $H_0: \mu = \mu_o$ is rejected in favor of $H_1: \mu > \mu_o$, will it necessarily be rejected in favor of $H_1: \mu \neq \mu_o$? Assume that α remains the same.

6.2.6. A random sample of size 16 is drawn from a normal distribution having $\sigma = 6.0$ for the purpose of testing $H_0: \mu = 30$ versus $H_1: \mu \neq 30$. The experimenter chooses to define the critical region C to be the set of sample means lying in the interval (29.9, 30.1). What level of significance does the test have? Why is (29.9, 30.1) a poor choice for the critical region? What range of $\overline{y}$ values should comprise C, assuming the same α is to be used?

6.2.7. Recall the breath analyzers described in Example 4.3.5. The following are thirty blood alcohol determinations made by Analyzer GTE-10, a three-year-old unit that may be in need of recalibration. All thirty measurements were made using a test sample on which a properly adjusted machine would give a reading of 12.6%.

12.3	12.7	13.6	12.7	12.9	12.6
12.6	13.1	12.6	13.1	12.7	12.5
13.2	12.8	12.4	12.6	12.4	12.4
13.1	12.9	13.3	12.6	12.6	12.7
13.1	12.4	12.4	13.1	12.4	12.9

(a) If μ denotes the true average reading that Analyzer GTE-10 would give for a person whose blood alcohol concentration is 12.6%, test

$$H_0: \mu = 12.6$$

versus

$$H_1: \mu \neq 12.6$$

at the $\alpha = 0.05$ level of significance. Assume that $\sigma = 0.4$. Would you recommend that the machine be readjusted?

(b) What statistical assumptions are implicit in the hypothesis test done in part (a)? Is there any reason to suspect that those assumptions may not be satisfied?

6.2.8. Calculate the P-values for the hypothesis tests indicated in Question 6.2.1. Do they agree with your decisions on whether or not to reject H_0?

6.2.9. Suppose $H_0: \mu = 120$ is tested against $H_1: \mu \neq 120$. If $\sigma = 10$ and $n = 16$, what P-value is associated with the sample mean $\overline{y} = 122.3$? Under what circumstances would H_0 be rejected?

6.2.10 As a class research project, Rosaura wants to see whether the stress of final exams elevates the blood pressures of freshmen women. When they are not under any untoward duress, healthy eighteen-year-old women have systolic blood pressures that average 120 mm Hg with a standard deviation of 12 mm Hg. If Rosaura finds that the average blood pressure for the fifty women in Statistics 101 on the day of the final exam is 125.2, what should she conclude? Set up and test an appropriate hypothesis.

6.2.11. As input for a new inflation model, economists predicted that the average cost of a hypothetical "food basket" in east Tennessee in July would be \$145.75. The standard deviation (σ) of basket prices was assumed to be \$9.50, a figure that has held fairly constant over the years. To check their prediction, a sample of twenty-five baskets representing different parts of the region were checked in late July, and the average cost was \$149.75. Let $\alpha = 0.05$. Is the difference between the economists' prediction and the sample mean statistically significant?

6.3 Testing Binomial Data—H_0: $p = p_o$

Suppose a set of data—$k_1, k_2, \ldots, k_n$—represents the outcomes of n Bernoulli trials, where $k_i = 1$ or 0, depending on whether the ith trial ended in success or failure, respectively. If $p = P(i\text{th trial ends in success})$ is unknown, it may be appropriate to test the null hypothesis H_0: $p = p_o$, where p_o is some particularly relevant (or status quo) value of p. Any such procedure is called a *binomial hypothesis test* because the appropriate test statistic is the *sum* of the k_i's—call it k—and we know from Theorem 3.2.1 that the total number of successes, X, in a series of independent trials has a binomial distribution,

$$p_X(k; p) = P(X = k) = \binom{n}{k} p^k (1-p)^{n-k}, \quad k = 0, 1, 2, \ldots, n$$

Two different procedures for testing H_0: $p = p_o$ need to be considered, the distinction resting on the magnitude of n. If

$$0 < np_o - 3\sqrt{np_o(1-p_o)} < np_o + 3\sqrt{np_o(1-p_o)} < n \tag{6.3.1}$$

a "large-sample" test of H_0: $p = p_o$ is done, based on an approximate Z ratio. Otherwise, a "small-sample" decision rule is used, one where the critical region is defined in terms of the exact binomial distribution associated with the random variable X.

A Large-Sample Test for the Binomial Parameter p

Suppose the number of observations, n, making up a set of Bernoulli random variables is sufficiently large that Inequality 6.3.1 is satisfied. We know in that case from Section 4.3 that the random variable $\frac{X - np_o}{\sqrt{np_o(1-p_o)}}$ has approximately a standard normal pdf, $f_Z(z)$ if $p = p_o$. Values of $\frac{X - np_o}{\sqrt{np_o(1-p_o)}}$ close to zero, of course, would be evidence in favor of H_0: $p = p_o$ [since $E\left(\frac{X - np_o}{\sqrt{np_o(1-p_o)}}\right) = 0$ when $p = p_o$]. Conversely, the credibility of H_0: $p = p_o$ clearly diminishes as $\frac{X - np_o}{\sqrt{np_o(1-p_o)}}$ moves further and further away from zero. The large-sample test of H_0: $p = p_o$, then, takes the same basic form as the test of H_0: $\mu = \mu_o$ in Section 6.2.

Theorem 6.3.1 *Let $k_1, k_2, \ldots, k_n$ be a random sample of n Bernoulli random variables for which $0 < np_o - 3\sqrt{np_o(1-p_o)} < np_o + 3\sqrt{np_o(1-p_o)} < n$. Let $k = k_1 + k_2 + \cdots + k_n$ denote the total number of "successes" in the n trials. Define $z = \frac{k - np_o}{\sqrt{np_o(1-p_o)}}$.*

a. *To test H_0: $p = p_o$ versus H_1: $p > p_o$ at the α level of significance, reject H_0 if $z \geq z_\alpha$.*

b. *To test H_0: $p = p_o$ versus H_1: $p < p_o$ at the α level of significance, reject H_0 if $z \leq -z_\alpha$.*

c. *To test H_0: $p = p_o$ versus H_1: $p \neq p_o$ at the α level of significance, reject H_0 if z is either (1) $\leq -z_{\alpha/2}$ or (2) $\geq z_{\alpha/2}$.* □

Case Study 6.3.1

In gambling parlance, a *point spread* is a hypothetical increment added to the score of the presumably weaker of two teams playing. By intention, its magnitude should have the effect of making the game a toss-up; that is, each team should have a 50% chance of beating the spread.

In practice, setting the "line" on a game is a highly subjective endeavor, which raises the question of whether or not the Las Vegas crowd actually gets it right (113). Addressing that issue, a recent study examined the records of 124 National Football League games; it was found that in sixty-seven of the matchups (or *54%*), the favored team beat the spread. Is the difference between 54% and 50% small enough to be written off to chance, or did the study uncover convincing evidence that oddsmakers are *not* capable of accurately quantifying the competitive edge that one team holds over another?

Let $p = P(\text{Favored team beats spread})$. If p is any value other than 0.50, the bookies are assigning point spreads incorrectly. To be tested, then, are the hypotheses

$$H_0: p = 0.50$$

versus

$$H_1: p \neq 0.50$$

Suppose 0.05 is taken to be the level of significance.

In the terminology of Theorem 6.3.1, $n = 124$, $p_o = 0.50$, and

$$k_i = \begin{cases} 1 & \text{if favored team beats spread in } i\text{th game} \\ 0 & \text{if favored team does not beat spread in } i\text{th game} \end{cases}$$

for $i = 1, 2, \ldots, 124$. Therefore, the sum $k = k_1 + k_2 + \cdots + k_{124}$ denotes the total number of times the favored team beat the spread.

According to the two-sided decision rule given in part (c) of Theorem 6.3.1, the null hypothesis should be rejected if z is either less than or equal to $-1.96\,(= -z_{.05/2})$ or greater than or equal to $1.96\,(= z_{.05/2})$. But

$$z = \frac{67 - 124(0.50)}{\sqrt{124(0.50)(0.50)}} = 0.90$$

does *not* fall in the critical region, so $H_0: p = 0.50$ should not be rejected at the $\alpha = 0.05$ level of significance. The outcomes of these 124 games, in other words, are entirely consistent with the presumption that bookies know which of two teams is better, and by how much.

About the Data Here the observed z is 0.90 and H_1 is two-sided, so the P-value is 0.37:

$$P\text{-value} = P(Z \leq -0.90) + P(Z \geq 0.90) = 0.1841 + 0.1841 \doteq 0.37$$

According to the Comment following Definition 6.2.4, then, the conclusion could be written

"Fail to reject H_0 for any $\alpha < 0.37$."

Would it also be correct to summarize the data with the statement

"Reject H_0 at the $\alpha = 0.40$ level of significance"?

In theory, yes; in practice, no. For all the reasons discussed in Section 6.2, the rationale underlying hypothesis testing demands that α be kept small (and "small" usually means less than or equal to 0.10).

It is typically the experimenter's objective to reject H_0, because H_0 represents the status quo, and there is seldom a compelling reason to devote time and money to a study for the purpose of confirming what is already believed. That being the case, experimenters are always on the lookout for ways to increase their probability of rejecting H_0. There are a number of entirely appropriate actions that can be taken to accomplish that objective, several of which will be discussed in Section 6.4. However, raising α above 0.10 is not one of the appropriate actions; and raising α as high as 0.40 would absolutely never be done.

Case Study 6.3.2

There is a theory that people may tend to "postpone" their deaths until after some event that has particular meaning to them has passed (134). Birthdays, a family reunion, or the return of a loved one have all been suggested as the sorts of personal milestones that might have such an effect. National elections may be another. Studies have shown that the mortality rate in the United States drops noticeably during the Septembers and Octobers of presidential election years. If the postponement theory is to be believed, the reason for the decrease is that many of the elderly who would have died in those two months "hang on" until they see who wins.

Some years ago, a national periodical reported the findings of a study that looked at obituaries published in a Salt Lake City newspaper. Among the 747 decedents, the paper identified that only 60, or *8.0%*, had died in the three-month period preceding their birth months (123). If individuals are dying randomly with respect to their birthdays, we would expect 25% to die during any given three-month interval. What should we make, then, of the decrease from 25% to 8%? Has the study provided convincing evidence that the death months reported for the sample do not constitute a random sample of months?

Imagine the 747 deaths being divided into two categories: those that occurred in the three-month period prior to a person's birthday and those that occurred at other times during the year. Let $k_i = 1$ if the ith person belongs to the first category and $k_i = 0$, otherwise. Then $k = k_1 + k_2 + \cdots + k_{747}$ denotes the total number of deaths in the first category. The latter, of course, is the value of a binomial random variable with parameter p, where

$$p = P(\text{Person dies in three months prior to birth month})$$

If people do *not* postpone their deaths (to wait for a birthday), p should be $\frac{3}{12}$, or 0.25; if they do, p will be something *less than* 0.25. Assessing the decrease from 25% to 8%, then, is done with a one-sided binomial hypothesis test:

$$H_0: p = 0.25$$

versus

$$H_1: p < 0.25$$

(Continued on next page)

(Case Study 6.3.2 continued)

Let $\alpha = 0.05$. According to part (b) of Theorem 6.3.1, H_0 should be rejected if

$$z = \frac{k - np_o}{\sqrt{np_o(1 - p_o)}} \le -z_{.05} = -1.64$$

Substituting for k, n, and p_o, we find that the test statistic falls far to the left of the critical value:

$$z = \frac{60 - 747(0.25)}{\sqrt{747(0.25)(0.75)}} = -10.7$$

The evidence is overwhelming, therefore, that the decrease from 25% to 8% is due to something other than chance. Explanations other than the postponement theory, of course, may be wholly or partially responsible for the nonrandom distribution of deaths. Still, the data show a pattern entirely consistent with the notion that we do have some control over when we die.

About the Data A similar conclusion was reached in a study conducted among the Chinese community living in California. The "significant event" in that case was not a birthday—it was the annual Harvest Moon festival, a celebration that holds particular meaning for elderly women. Based on census data tracked over a twenty-four-year period, it was determined that fifty-one deaths among elderly Chinese women should have occurred during the week *before* the festivals, and fifty-two deaths *after* the festivals. In point of fact, thirty-three died the week before and seventy died the week after (22).

A Small-Sample Test for the Binomial Parameter p

Suppose that $k_1, k_2, \ldots, k_n$ is a random sample of Bernoulli random variables where n is too small for Inequality 6.3.1 to hold. The decision rule, then, for testing H_0: $p = p_o$ that was given in Theorem 6.3.1 would not be appropriate. Instead, the critical region is defined by using the exact binomial distribution (rather than a normal approximation).

Example 6.3.1

Suppose that $n = 19$ elderly patients are to be given an experimental drug designed to relieve arthritis pain. The standard treatment is known to be effective in 85% of similar cases. If p denotes the probability that the new drug will reduce a patient's pain, the researcher wishes to test

$$H_0: p = 0.85$$

versus

$$H_1: p \neq 0.85$$

The decision will be based on the magnitude of k, the total number in the sample for whom the durg is effective—that is, on

$$k = k_1 + k_2 + \cdots + k_{19}$$

where

$$k_i = \begin{cases} 0 & \text{if the new drug fails to relieve } i\text{th patient's pain} \\ 1 & \text{if the new drug does relieve } i\text{th patient's pain} \end{cases}$$

What should the decision rule be if the intention is to keep α somewhere near 10%? [Note that Theorem 6.3.1 does not apply here because Inequality 6.3.1 is not satisfied—specifically, $np_o + 3\sqrt{np_o(1-p_o)} = 19(0.85) + 3\sqrt{19(0.85)(0.15)} = 20.8$ is not less than $n(= 19)$.]

If the null hypothesis is true, the expected number of successes would be $np_o = 19(0.85)$. or *16.2*. It follows that values of k to the extreme right or extreme left of 16.2 should constitute the critical region.

```
MTB > pdf;
SUBC > binomial 19  0.85.
```

Probability Density Function

Binomial with n = 19 and p = 0.85

x	P(X = x)	
6	0.000000	
7	0.000002	
8	0.000018	
9	0.000123	→ $P(X \leq 13) = 0.053696$
10	0.000699	
11	0.003242	
12	0.012246	
13	0.037366	
14	0.090746	
15	0.171409	
16	0.242829	
17	0.242829	
18	0.152892	
19	0.045599	→ $P(X = 19) = 0.045599$

Figure 6.3.1

Figure 6.3.1 is a Minitab printout of $p_X(k) = \binom{19}{k}(0.85)^k(0.15)^{19-k}$. By inspection, we can see that the critical region

$$C = \{k : k \leq 13 \quad \text{or} \quad k = 19\}$$

would produce an α close to the desired 0.10 (and would keep the probabilities associated with the two sides of the rejection region roughly the same). In random variable notation,

$$\begin{aligned} P(X \in C \mid H_0 \text{ is true}) &= P(X \leq 13 \mid p = 0.85) + P(X = 19 \mid p = 0.85) \\ &= 0.053696 + 0.045599 \\ &= 0.099295 \\ &\doteq 0.10 \end{aligned}$$

■

Questions

6.3.1. Commercial fishermen working certain parts of the Atlantic Ocean sometimes find their efforts hindered by the presence of whales. Ideally, they would like to scare away the whales without frightening the fish. One of the strategies being experimented with is to transmit underwater the sounds of a killer whale. On the fifty-two occasions that technique has been tried, it worked twenty-four times (that is, the whales immediately left the area). Experience has shown, though, that 40% of all whales sighted near fishing boats leave of their own accord, probably just to get away from the noise of the boat.

(a) Let $p = P$(Whale leaves area after hearing sounds of killer whale). Test H_0: $p = 0.40$ versus H_1: $p > 0.40$ at the $\alpha = 0.05$ level of significance. Can it be argued on the basis of these data that transmitting underwater predator sounds is an effective technique for clearing fishing waters of unwanted whales?

(b) Calculate the P-value for these data. For what values of α would H_0 be rejected?

6.3.2. Efforts to find a genetic explanation for why certain people are right-handed and others left-handed have been largely unsuccessful. Reliable data are difficult to find because of environmental factors that also influence a child's "handedness." To avoid that complication, researchers often study the analogous problem of "pawedness" in animals, where both genotypes and the environment can be partially controlled. In one such experiment (27), mice were put into a cage having a feeding tube that was equally accessible from the right or the left. Each mouse was then carefully watched over a number of feedings. If it used its right paw more than half the time to activate the tube, it was defined to be "right-pawed." Observations of this sort showed that 67% of mice belonging to strain A/J are right-pawed. A similar protocol was followed on a sample of thirty-five mice belonging to strain A/HeJ. Of those thirty-five, a total of eighteen were eventually classified as right-pawed. Test whether the proportion of right-pawed mice found in the A/HeJ sample was significantly different from what was known about the A/J strain. Use a two-sided alternative and let 0.05 be the probability associated with the critical region.

6.3.3. Defeated in his most recent attempt to win a congressional seat because of a sizeable gender gap, a politician has spent the last two years speaking out in favor of women's rights issues. A newly released poll claims to have contacted a random sample of 120 of the politician's current supporters and found that 72 were men. In the election that he lost, exit polls indicated that 65% of those who voted for him were men. Using an $\alpha = 0.05$ level of significance, test the null hypothesis that the proportion of his male supporters has remained the same. Make the alternative hypothesis one-sided.

6.3.4. Suppose H_0: $p = 0.45$ is to be tested against H_1: $p > 0.45$ at the $\alpha = 0.14$ level of significance, where $p = P(i\text{th}$ trial ends in success). If the sample size is 200, what is the smallest number of successes that will cause H_0 to be rejected?

6.3.5. Recall the median test described in Example 5.3.2. Reformulate that analysis as a hypothesis test rather than a confidence interval. What P-value is associated with the outcomes listed in Table 5.3.3?

6.3.6. Among the early attempts to revisit the death postponement theory introduced in Case Study 6.3.2 was an examination of the birth dates and death dates of 348 U.S. celebrities (134). It was found that 16 of those individuals had died in the month preceding their birth month. Set up and test the appropriate H_0 against a one-sided H_1. Use the 0.05 level of significance.

6.3.7. What α levels are possible with a decision rule of the form "Reject H_0 if $k \geq k^*$" when H_0: $p = 0.5$ is to be tested against H_1: $p > 0.5$ using a random sample of size $n = 7$?

6.3.8. The following is a Minitab printout of the binomial pdf $p_X(k) = \binom{9}{k}(0.6)^k(0.4)^{9-k}$, $k = 0, 1, \ldots, 9$. Suppose H_0: $p = 0.6$ is to be tested against H_1: $p > 0.6$ and we wish the level of significance to be *exactly* 0.05. Use Theorem 2.4.1 to combine two different critical regions into a single *randomized decision rule* for which $\alpha = 0.05$.

```
MTB > pdf;
SUBC > binomial 9  0.6.
```

Probability Density Function

Binomial with n = 9 and p = 0.6

x	P(X = x)
0	0.000262
1	0.003539
2	0.021234
3	0.074318
4	0.167215
5	0.250823
6	0.250823
7	0.161243
8	0.060466
9	0.010078

6.3.9. Suppose H_0: $p = 0.75$ is to be tested against H_1: $p < 0.75$ using a random sample of size $n = 7$ and the decision rule "Reject H_0 if $k \leq 3$."

(a) What is the test's level of significance?

(b) Graph the probability that H_0 will be rejected *as a function of p*.

6.4 Type I and Type II Errors

The possibility of drawing incorrect conclusions is an inevitable byproduct of hypothesis testing. No matter what sort of mathematical facade is laid atop the decision-making process, there is no way to guarantee that what the test tells us is the truth. One kind of error—rejecting H_0 when H_0 is true—figured prominently in Section 6.3: It was argued that critical regions should be defined so as to keep the probability of making such errors small, often on the order of 0.05.

In point of fact, there are two different kinds of errors that can be committed with any hypothesis test: (1) We can reject H_0 when H_0 is true and (2) we can fail to reject H_0 when H_0 is false. These are called *Type I* and *Type II* errors, respectively. At the same time, there are two kinds of correct decisions: (1) We can fail to reject a true H_0 and (2) we can reject a false H_0. Figure 6.4.1 shows these four possible "Decision/State of nature" combinations.

Figure 6.4.1

		True State of Nature	
		H_0 is true	H_1 is true
Our Decision	Fail to reject H_0	Correct decision	Type II error
	Reject H_0	Type I error	Correct decision

Computing the Probability of Committing a Type I Error

Once an inference is made, there is no way to know whether the conclusion reached was correct. It *is* possible, though, to calculate the probability of having made an error, and the magnitude of that probability can help us better understand the "power" of the hypothesis test and its ability to distinguish between H_0 and H_1.

Recall the fuel additive example developed in Section 6.2: $H_0\colon \mu = 25.0$ was to be tested against $H_1\colon \mu > 25.0$ using a sample of size $n = 30$. The decision rule stated that H_0 should be rejected if $\bar{y}$, the average mpg with the new additive, equalled or exceeded 25.718. In that case, the probability of committing a Type I error is *0.05*:

$$\begin{aligned} P(\text{Type I error}) &= P(\text{Reject } H_0 \mid H_0 \text{ is true}) \\ &= P(\overline{Y} \geq 25.718 \mid \mu = 25.0) \\ &= P\left(\frac{\overline{Y} - 25.0}{2.4/\sqrt{30}} \geq \frac{25.718 - 25.0}{2.4/\sqrt{30}}\right) \\ &= P(Z \geq 1.64) = 0.05 \end{aligned}$$

Of course, the fact that the probability of committing a Type I error equals 0.05 should come as no surprise. In our earlier discussion of how "beyond reasonable doubt" should be interpreted numerically, we specifically chose the critical region so that the probability of the decision rule rejecting H_0 when H_0 is true *would* be 0.05.

In general, the probability of committing a Type I error is referred to as a test's *level of significance* and is denoted α (recall Definition 6.2.3). The concept is a crucial one: The level of significance is a single-number summary of the "rules" by which the decision process is being conducted. In essence, α reflects the amount of evidence the experimenter is demanding to see before abandoning the null hypothesis.

Computing the Probability of Committing a Type II Error

We just saw that calculating the probability of a Type I error is a nonproblem: There are no computations necessary, since the probability equals whatever value the experimenter sets a priori for α. A similar situation does not hold for Type

II errors. To begin with, Type II error probabilities are not specified explicitly by the experimenter; also, each hypothesis test has an infinite number of Type II error probabilities, one for each value of the parameter admissible under H_1.

As an example, suppose we want to find the probability of committing a Type II error in the gasoline experiment if the true μ (*with the additive*) were 25.750. By definition,

$$\begin{aligned} P(\text{Type II error} \mid \mu = 25.750) &= P(\text{We fail to reject } H_0 \mid \mu = 25.750) \\ &= P(\overline{Y} < 25.718 \mid \mu = 25.750) \\ &= P\left(\frac{\overline{Y} - 25.75}{2.4/\sqrt{30}} < \frac{25.718 - 25.75}{2.4/\sqrt{30}}\right) \\ &= P(Z < -0.07) = 0.4721 \end{aligned}$$

So, even if the new additive increased the fuel economy to 25.750 mpg (from 25 mpg), our decision rule would be "tricked" 47% of the time: that is, it would tell us on those occasions *not* to reject H_0.

The symbol for the probability of committing a Type II error is β. Figure 6.4.2 shows the sampling distribution of $\overline{Y}$ when $\mu = 25.0$ (i.e., when H_0 is true) and when $\mu = 25.750$ (H_1 is true); the areas corresponding to α and β are shaded.

Figure 6.4.2

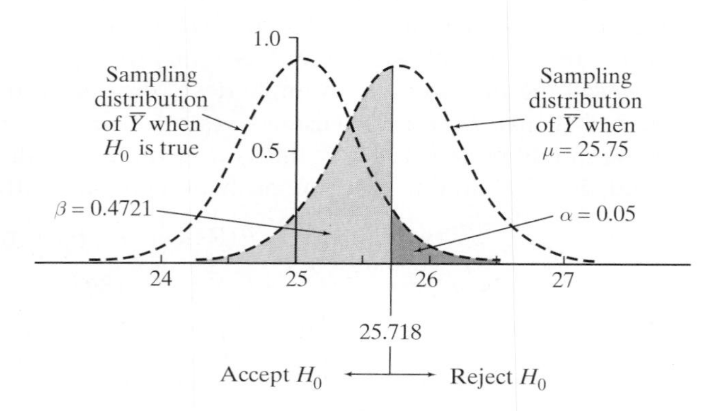

Figure 6.4.3

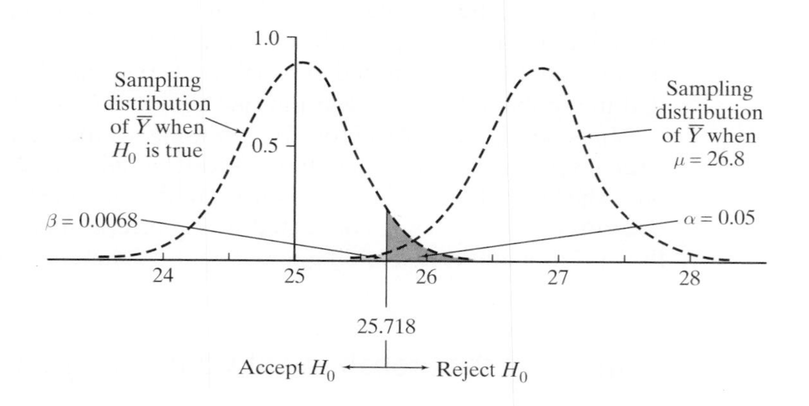

Clearly, the magnitude of β is a function of the presumed value for μ. If, for example, the gasoline additive is so effective as to raise fuel efficiency to 26.8 mpg, the probability that our decision rule would lead us to make a Type II error is a much smaller *0.0068*:

$$P(\text{Type II error} \mid \mu = 26.8) = P(\text{We fail to reject } H_0 \mid \mu = 26.8)$$
$$= P(\overline{Y} < 25.718 \mid \mu = 26.8) = P\left(\frac{\overline{Y} - 26.8}{2.4/\sqrt{30}} < \frac{25.718 - 26.8}{2.4/\sqrt{30}}\right)$$
$$= P(Z < -2.47) = 0.0068$$

(See Figure 6.4.3.)

Power Curves

If β is the probability that we fail to reject H_0 when H_1 is true, then $1-\beta$ is the probability of the complement—that we *reject* H_0 when H_1 is true. We call $1-\beta$ the *power* of the test; it represents the ability of the decision rule to "recognize" (correctly) that H_0 is false.

The alternative hypothesis H_1 usually depends on a parameter, which makes $1-\beta$ a function of that parameter. The relationship they share can be pictured by drawing a *power curve*, which is simply a graph of $1-\beta$ versus the set of all possible parameter values.

Figure 6.4.4 shows the power curve for testing

$$H_0: \mu = 25.0$$
versus
$$H_1: \mu > 25.0$$

where μ is the mean of a normal distribution with $\sigma = 2.4$, and the decision rule is "Reject H_0 if $\overline{y} \geq 25.718$." The two marked points on the curve represent the $(\mu, 1-\beta)$ pairs just determined, (25.75, 0.5297) and (26.8, 0.9932). One other point can be gotten for every power curve, without doing any calculations: When $\mu = \mu_0$ (the value specified by H_0), $1-\beta = \alpha$. Of course, as the true mean gets further and further away from the H_0 mean, the power will converge to 1.

Figure 6.4.4

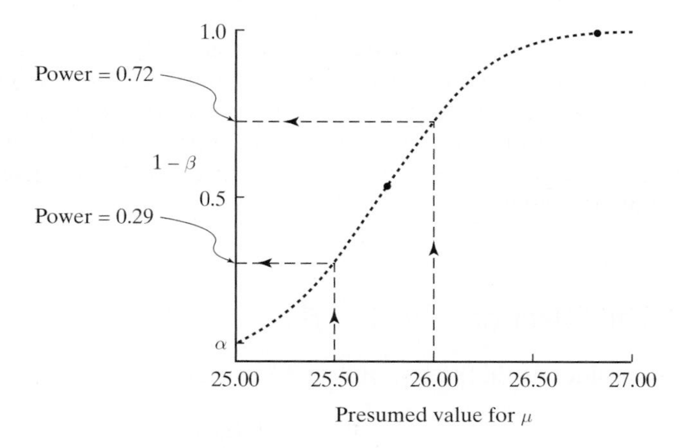

Power curves serve two different purposes. On the one hand, they completely characterize the performance that can be expected from a hypothesis test. In

Figure 6.4.5

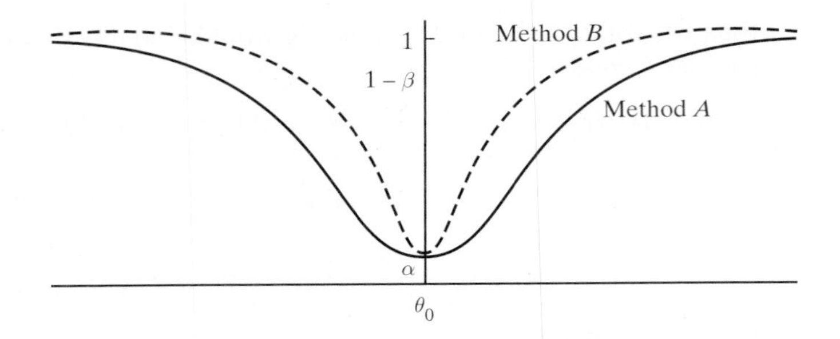

Figure 6.4.4, for example, the two arrows show that the probability of rejecting $H_0: \mu = 25$ in favor of $H_1: \mu > 25$ when $\mu = 26.0$ is approximately *0.72*. (Or, equivalently, Type II errors will be committed roughly *28%* of the time when $\mu = 26.0$.) As the true mean moves closer to μ_o (and becomes more difficult to distinguish) the power of the test understandably diminishes. If $\mu = 25.5$, for example, the graph shows that $1 - \beta$ falls to 0.29.

Power curves are also useful for *comparing* one inference procedure with another. For every conceivable hypothesis testing situation, a variety of procedures for choosing between H_0 and H_1 will be available. How do we know which to use?

The answer to that question is not always simple. Some procedures will be computationally more convenient or easier to explain than others; some will make slightly different assumptions about the pdf being sampled. Associated with each of them, though, is a power curve. If the selection of a hypothesis test is to hinge solely on its ability to distinguish H_0 from H_1, then the procedure to choose is the one having the *steepest* power curve.

Figure 6.4.5 shows the power curves for two hypothetical methods A and B, each of which is testing $H_0{:}\,\theta = \theta_o$ versus $H_1{:}\,\theta \neq \theta_o$ at the α level of significance. From the standpoint of power, Method B is clearly the better of the two—it always has a higher probability of correctly rejecting H_0 when the parameter θ is not equal to θ_o.

Factors That Influence the Power of a Test

The ability of a test procedure to reject H_0 when H_0 is false is clearly of prime importance, a fact that raises an obvious question: What can an experimenter do to influence the value of $1 - \beta$? In the case of the Z test described in Theorem 6.2.1, $1 - \beta$ is a function of α, σ, and n. By appropriately raising or lowering the values of those parameters, the power of the test against any given μ can be made to equal any desired level.

The Effect of α on $1 - \beta$

Consider again the test of

$$H_0{:}\,\mu = 25.0$$

versus

$$H_1{:}\,\mu > 25.0$$

discussed earlier in this section. In its original form, $\alpha = 0.05$, $\sigma = 2.4$, $n = 30$, and the decision rule called for H_0 to be rejected if $\bar{y} \geq 25.718$.

Figure 6.4.6 shows what happens to $1 - \beta$ (when $\mu = 25.75$) if σ, n, and μ are held constant but α is increased to 0.10. The top pair of distributions shows the configuration that appears in Figure 6.4.2; the power in this case is $1 - 0.4721$, or *0.53*. The bottom portion of the graph illustrates what happens when α is set at 0.10 instead of 0.05—the decision rule changes from "Reject H_0 if $\bar{y} \geq 25.718$" to "Reject H_0 if $\bar{y} \geq 25.561$" (see Question 6.4.2) and the power increases from 0.53 to *0.67*:

$$\begin{aligned}
1 - \beta &= P(\text{Reject } H_0 \mid H_1 \text{ is true}) \\
&= P(\bar{Y} \geq 25.561 \mid \mu = 25.75) \\
&= P\left(\frac{\bar{Y} - 25.75}{2.4/\sqrt{30}} \geq \frac{25.561 - 25.75}{2.4/\sqrt{30}}\right) \\
&= P(Z \geq -0.43) \\
&= 0.6664
\end{aligned}$$

Figure 6.4.6

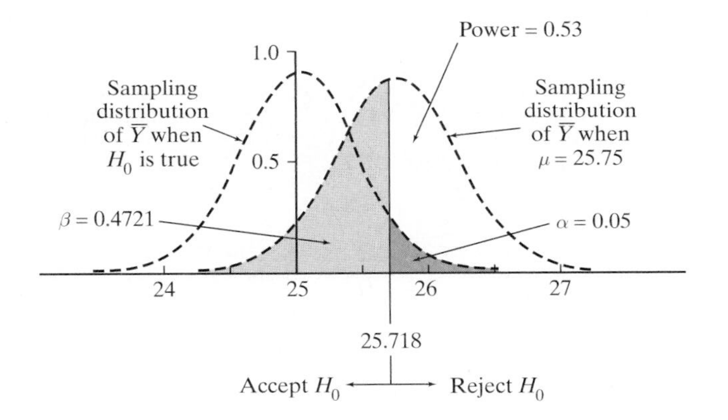

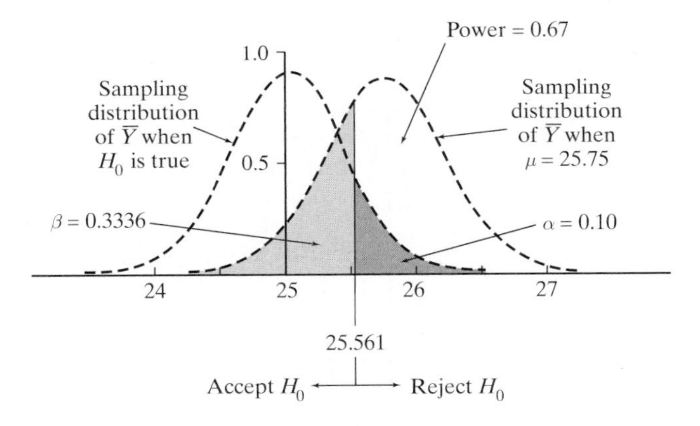

The specifics of Figure 6.4.6 accurately reflect what is true in general: *Increasing α decreases β and increases the power*. That said, it does not follow *in practice* that experimenters should manipulate α to achieve a desired $1 - \beta$. For all the reasons cited in Section 6.2, α should typically be set equal to a number somewhere in the neighborhood of 0.05. If the corresponding $1 - \beta$ against a particular μ is deemed to be inappropriate, adjustments should be made in the values of σ and/or n.

The Effects of σ and n on $1 - \beta$

Although it may not always be feasible (or even possible), *decreasing* σ will necessarily *increase* $1-\beta$. In the gasoline additive example, σ is assumed to be 2.4 mpg, the latter being a measure of the variation in gas mileages from driver to driver achieved in a cross-country road trip from Boston to Los Angeles (recall p. 351). Intuitively, the environmental differences inherent in a trip of that magnitude would be considerable. Different drivers would encounter different weather conditions and varying amounts of traffic, and would perhaps take alternate routes.

Suppose, instead, that the drivers simply did laps around a test track rather than drive on actual highways. Conditions from driver to driver would then be much more uniform and the value of σ would surely be smaller. What would be the effect on $1-\beta$ when $\mu = 25.75$ (and $\alpha = 0.05$) if σ could be reduced from 2.4 mpg to 1.2 mpg?

As Figure 6.4.7 shows, reducing σ has the effect of making the H_0 distribution of $\overline{Y}$ more concentrated around $\mu_o(=25)$ and the H_1 distribution of $\overline{Y}$ more concentrated around $\mu(=25.75)$. Substituting into Equation 6.2.1 (with 1.2 for σ in place

Figure 6.4.7

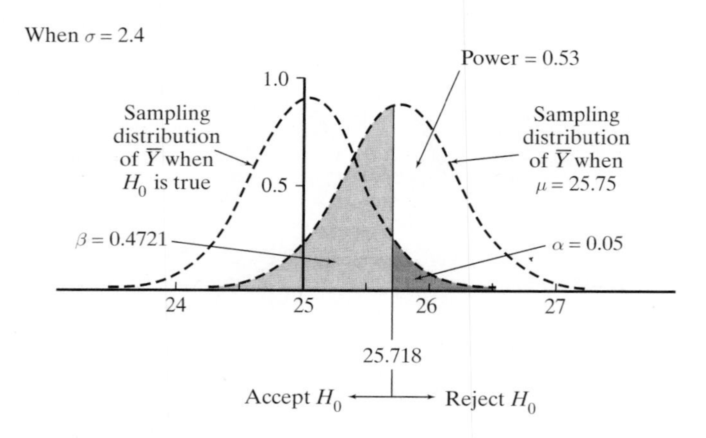

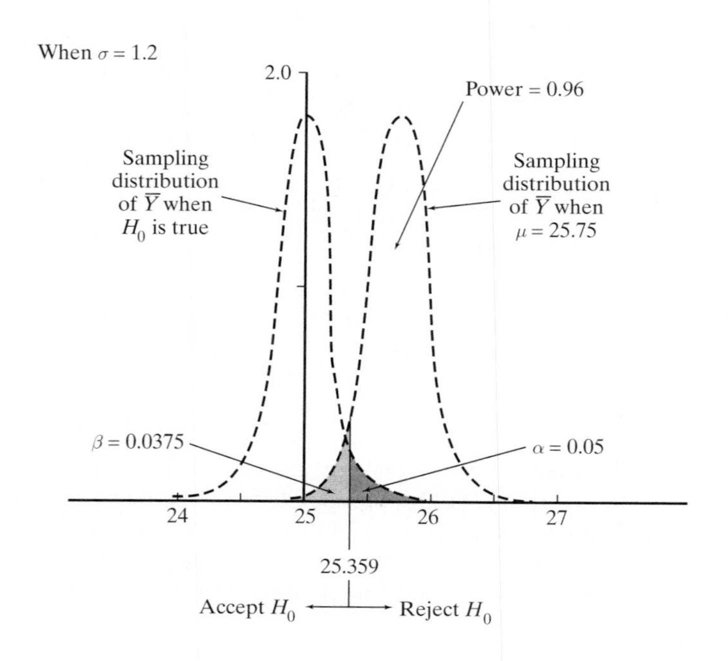

of 2.4), we find that the critical value $\overline{y}^*$ moves closer to μ_o [from 25.718 to *25.359* $\left(=25+1.64\cdot\frac{1.2}{\sqrt{30}}\right)$] and the proportion of the H_1 distribution above the rejection region (i.e., the power) *increases* from 0.53 to *0.96*:

$$\begin{aligned}1-\beta &= P(\overline{Y}\geq 25.359 \mid \mu=25.75)\\ &= P\left(Z\geq\frac{25.359-25.75}{1.2/\sqrt{30}}\right)=P(Z\geq -1.78)=0.9625\end{aligned}$$

In theory, reducing σ can be a very effective way of increasing the power of a test, as Figure 6.4.7 makes abundantly clear. In practice, though, refinements in the way data are collected that would have a substantial impact on the magnitude of σ are often either difficult to identify or prohibitively expensive. More typically, experimenters achieve the same effect by simply increasing the sample size.

Look again at the two sets of distributions in Figure 6.4.7. The increase in $1-\beta$ from 0.53 to 0.96 was accomplished by cutting the denominator of the test statistic $\left(z=\frac{\overline{y}-25}{\sigma/\sqrt{30}}\right)$ in half by reducing the standard deviation from 2.4 to 1.2. The same numerical effect would be produced if σ were left unchanged but n was increased from 30 to 120—that is, $\frac{1.2}{\sqrt{30}}=\frac{2.4}{\sqrt{120}}$. Because it can easily be increased or decreased, the sample size is the parameter that researchers almost invariably turn to as the mechanism for ensuring that a hypothesis test will have a sufficiently high power against a given alternative.

Example 6.4.1

Suppose an experimenter wishes to test

$$H_0: \mu=100$$

versus

$$H_1: \mu>100$$

at the $\alpha=0.05$ level of significance and wants $1-\beta$ to equal 0.60 when $\mu=103$. What is the smallest (i.e., cheapest) sample size that will achieve that objective? Assume that the variable being measured is normally distributed with $\sigma=14$.

Finding n, given values for $\alpha, 1-\beta, \sigma$, and μ, requires that two simultaneous equations be written for the critical value $\overline{y}^*$, one in terms of the H_0 distribution and the other in terms of the H_1 distribution. Setting the two equal will yield the minimum sample size that achieves the desired α and $1-\beta$.

Consider, first, the consequences of the level of significance being equal to 0.05. By definition,

$$\begin{aligned}\alpha &= P(\text{We reject } H_0 \mid H_0 \text{ is true})\\ &= P(\overline{Y}\geq\overline{y}^*\mid\mu=100)\\ &= P\left(\frac{\overline{Y}-100}{14/\sqrt{n}}\geq\frac{\overline{y}^*-100}{14/\sqrt{n}}\right)\\ &= P\left(Z\geq\frac{\overline{y}^*-100}{14/\sqrt{n}}\right)\\ &=0.05\end{aligned}$$

But $P(Z\geq 1.64)=0.05$, so

$$\frac{\overline{y}^*-100}{14/\sqrt{n}}=1.64$$

or, equivalently,

$$\overline{y}^* = 100 + 1.64 \cdot \frac{14}{\sqrt{n}} \tag{6.4.1}$$

Similarly,

$$1 - \beta = P(\text{We reject } H_0 \mid H_1 \text{ is true}) = P(\overline{Y} \geq \overline{y}^* \mid \mu = 103)$$

$$= P\left(\frac{\overline{Y} - 103}{14/\sqrt{n}} \geq \frac{\overline{y}^* - 103}{14/\sqrt{n}}\right) = 0.60$$

From Appendix Table A.1, though, $P(Z \geq -0.25) = 0.5987 \doteq 0.60$, so

$$\frac{\overline{y}^* - 103}{14/\sqrt{n}} = -0.25$$

which implies that

$$\overline{y}^* = 103 - 0.25 \cdot \frac{14}{\sqrt{n}} \tag{6.4.2}$$

It follows, then, from Equations 6.4.1 and 6.4.2 that

$$100 + 1.64 \cdot \frac{14}{\sqrt{n}} = 103 - 0.25 \cdot \frac{14}{\sqrt{n}}$$

Solving for n shows that a minimum of *seventy-eight* observations must be taken to guarantee that the hypothesis test will have the desired precision. ■

Decision Rules for Nonnormal Data

Our discussion of hypothesis testing thus far has been confined to inferences involving either binomial data or normal data. Decision rules for other types of probability functions are rooted in the same basic principles.

In general, to test $H_0{:}\,\theta = \theta_o$, where θ is the unknown parameter in a pdf $f_Y(y; \theta)$, we initially define the decision rule in terms of $\hat{\theta}$, where the latter is a sufficient statistic for θ. The corresponding critical region is the set of values of $\hat{\theta}$ least compatible with θ_o (but admissible under H_1) whose total probability when H_0 is true is α. In the case of testing $H_0{:}\,\mu = \mu_o$ versus $H_1{:}\,\mu > \mu_o$, for example, where the data are normally distributed, $\overline{Y}$ is a sufficient statistic for μ, and the least likely values for the sample mean that are admissible under H_1 are those for which $\overline{y} \geq \overline{y}^*$, where $P(\overline{Y} \geq \overline{y}^* \mid H_0 \text{ is true}) = \alpha$.

Example 6.4.2

A random sample of size $n = 8$ is drawn from the uniform pdf, $f_Y(y; \theta) = 1/\theta, 0 \leq y \leq \theta$, for the purpose of testing

$$H_0{:}\,\theta = 2.0$$

$$\text{versus}$$

$$H_1{:}\,\theta < 2.0$$

at the $\alpha = 0.10$ level of significance. Suppose the decision rule is to be based on Y'_8, the largest order statistic. What would be the probability of committing a Type II error when $\theta = 1.7$?

If H_0 is true, Y_8' should be close to 2.0, and values of the largest order statistic that are much *smaller* than 2.0 would be evidence in favor of $H_1: \theta < 2.0$. It follows, then, that the form of the decision rule should be

$$\text{"Reject } H_0: \theta = 2.0 \quad \text{if} \quad y_8' \leq c\text{"}$$

where $P(Y_8' \leq c \mid H_0 \text{ is true}) = 0.10$.

From Theorem 3.10.1,

$$f_{Y_8'}(y; \theta = 2) = 8\left(\frac{y}{2}\right)^7 \cdot \frac{1}{2}, \quad 0 \leq y \leq 2$$

Therefore, the constant c that appears in the $\alpha = 0.10$ decision rule must satisfy the equation

$$\int_0^c 8\left(\frac{y}{2}\right)^7 \cdot \frac{1}{2}\, dy = 0.10$$

or, equivalently,

$$\left(\frac{c}{2}\right)^8 = 0.10$$

implying that $c = 1.50$.

Now, β when $\theta = 1.7$ is, by definition, the probability that Y_8' falls in the acceptance region when $H_1: \theta = 1.7$ is true. That is,

$$\beta = P(Y_8' > 1.50 \mid \theta = 1.7) = \int_{1.50}^{1.7} 8\left(\frac{y}{1.7}\right)^7 \cdot \frac{1}{1.7}\, dy$$

$$= 1 - \left(\frac{1.5}{1.7}\right)^8 = 0.63$$

(See Figure 6.4.8.)

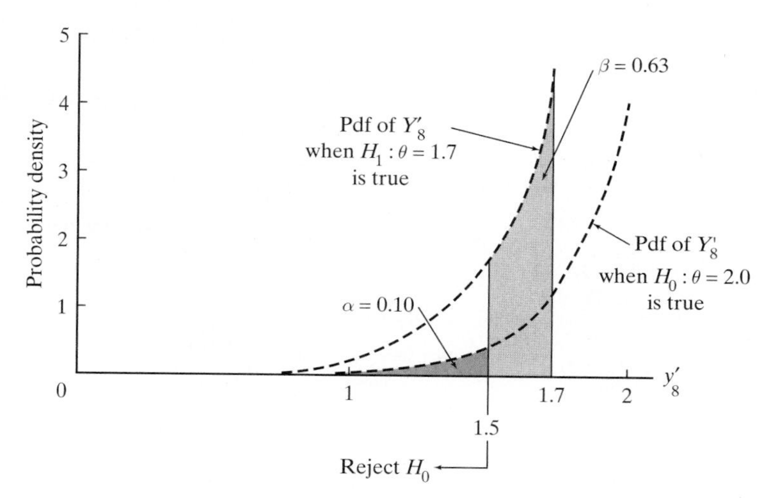

Figure 6.4.8

Example 6.4.3

Four measurements—k_1, k_2, k_3, k_4—are taken on a Poisson random variable, X, where $p_X(k; \lambda) = e^{-\lambda}\lambda^k / k!$, $k = 0, 1, 2, \ldots$, for the purpose of testing

$$H_0: \lambda = 0.8$$

versus

$$H_1: \lambda > 0.8$$

What decision rule should be used if the level of significance is to be 0.10, and what will be the power of the test when $\lambda = 1.2$?

From Example 5.6.1, we know that $\overline{X}$ is a sufficient statistic for λ; the same would be true, of course, for $\sum_{i=1}^{4} X_i$. It will be more convenient to state the decision rule in terms of the latter because we already know the probability model that describes its behavior: If X_1, X_2, X_3, X_4 are four independent Poisson random variables, each with parameter λ, then $\sum_{i=1}^{4} X_i$ has a Poisson distribution with parameter 4λ (recall Example 3.12.10).

Figure 6.4.9 is a Minitab printout of the Poisson probability function having $\lambda = 3.2$, which would be the sampling distribution of $\sum_{i=1}^{4} X_i$ when $H_0\colon \lambda = 0.8$ is true.

```
MTB > pdf;
SUBC > poisson 3.2.
```

Probability Density Function

Poisson with mean = 3.2

x	P(X = x)
0	0.040762
1	0.130439
2	0.208702
3	0.222616
4	0.178093
5	0.113979
6	0.060789
7	0.027789
8	0.011116
9	0.003952
10	0.001265
11	0.000368
12	0.000098
13	0.000024
14	0.000006
15	0.000001
16	0.000000

Critical region (x = 6 through 16): $\alpha = P(\text{Reject}\, H_0 \mid H_0 \text{ is true}) = 0.105408$

Figure 6.4.9

```
MTB > pdf;
SUBC > poisson 4.8.
```

Probability Density Function

Poisson with mean = 4.8

x	P(X = x)
0	0.008230
1	0.039503
2	0.094807
3	0.151691
4	0.182029
5	0.174748
6	0.139798
7	0.095862
8	0.057517
9	0.030676
10	0.014724
11	0.006425
12	0.002570
13	0.000949
14	0.000325
15	0.000104
16	0.000031
17	0.000009
18	0.000002
19	0.000001
20	0.000000

(x = 6 through 20): $1 - \beta = P(\text{Reject}\, H_0 \mid H_1 \text{ is true}) = 0.348993$

Figure 6.4.10

By inspection, the decision rule "Reject $H_0: \lambda = 0.8$ if $\sum_{i=1}^{4} k_i \geq 6$" gives an α close to the desired 0.10.

If H_1 is true and $\lambda = 1.2$, $\sum_{i=1}^{4} X_i$ will have a Poisson distribution with a parameter equal to 4.8. According to Figure 6.4.10, the probability that the sum of a random sample of size 4 from such a distribution would equal or exceed 6 (i.e., $1 - \beta$ when $\lambda = 1.2$) is *0.348993*. ■

Example 6.4.4

Suppose a random sample of seven observations is taken from the pdf $f_Y(y;\theta) = (\theta+1)y^{\theta}, 0 \leq y \leq 1$, to test

$$H_0: \theta = 2$$

versus

$$H_1: \theta > 2$$

As a decision rule, the experimenter plans to record X, the number of y_i's that exceed 0.9, and reject H_0 if $X \geq 4$. What proportion of the time would such a decision rule lead to a Type I error?

To evaluate $\alpha = P(\text{Reject } H_0 \mid H_0 \text{ is true})$, we first need to recognize that X is a binomial random variable where $n = 7$ and the parameter p is an area under $f_Y(y; \theta = 2)$:

$$\begin{aligned} p = P(Y \geq 0.9 \mid H_0 \text{ is true}) &= P[Y \geq 0.9 \mid f_Y(y;2) = 3y^2] \\ &= \int_{0.9}^{1} 3y^2\, dy \\ &= 0.271 \end{aligned}$$

It follows, then, that H_0 will be incorrectly rejected *9.2%* of the time:

$$\begin{aligned} \alpha = P(X \geq 4 \mid \theta = 2) &= \sum_{k=4}^{7} \binom{7}{k} (0.271)^k (0.729)^{7-k} \\ &= 0.092 \end{aligned}$$

Comment The basic notions of Type I and Type II errors first arose in a quality-control context. The pioneering work was done at the Bell Telephone Laboratories: There the terms *producer's risk* and *consumer's risk* were introduced for what we now call α and β. Eventually, these ideas were generalized by Neyman and Pearson in the 1930s and evolved into the theory of hypothesis testing as we know it today. ■

Questions

6.4.1. Recall the "Math for the Twenty-First Century" hypothesis test done in Example 6.2.1. Calculate the power of that test when the true mean is 500.

6.4.2. Carry out the details to verify the decision rule change cited on p. 371 in connection with Figure 6.4.6.

6.4.3. For the decision rule found in Question 6.2.2 to test $H_0: \mu = 95$ versus $H_1: \mu \neq 95$ at the $\alpha = 0.06$ level of significance, calculate $1 - \beta$ when $\mu = 90$.

6.4.4. Construct a power curve for the $\alpha = 0.05$ test of $H_0: \mu = 60$ versus $H_1: \mu \neq 60$ if the data consist of a random sample of size 16 from a normal distribution having $\sigma = 4$.

6.4.5. If $H_0: \mu = 240$ is tested against $H_1: \mu < 240$ at the $\alpha = 0.01$ level of significance with a random sample of twenty-five normally distributed observations, what proportion of the time will the procedure fail to recognize that μ has dropped to 220? Assume that $\sigma = 50$.

6.4.6. Suppose $n = 36$ observations are taken from a normal distribution where $\sigma = 8.0$ for the purpose of testing $H_0: \mu = 60$ versus $H_1: \mu \neq 60$ at the $\alpha = 0.07$ level of significance. The lead investigator skipped statistics class the day decision rules were being discussed and intends to reject H_0 if $\overline{y}$ falls in the region $(60 - \overline{y}^*, 60 + \overline{y}^*)$.

(a) Find $\overline{y}^*$.
(b) What is the power of the test when $\mu = 62$?
(c) What would the power of the test be when $\mu = 62$ if the critical region had been defined the correct way?

6.4.7. If $H_0: \mu = 200$ is to be tested against $H_1: \mu < 200$ at the $\alpha = 0.10$ level of significance based on a random sample of size n from a normal distribution where $\sigma = 15.0$, what is the smallest value for n that will make the power equal to at least 0.75 when $\mu = 197$?

6.4.8. Will $n = 45$ be a sufficiently large sample to test $H_0: \mu = 10$ versus $H_1: \mu \neq 10$ at the $\alpha = 0.05$ level of significance if the experimenter wants the Type II error probability to be no greater than 0.20 when $\mu = 12$? Assume that $\sigma = 4$.

6.4.9. If $H_0: \mu = 30$ is tested against $H_1: \mu > 30$ using $n = 16$ observations (normally distributed) and if $1 - \beta = 0.85$ when $\mu = 34$, what does α equal? Assume that $\sigma = 9$.

6.4.10. Suppose a sample of size 1 is taken from the pdf $f_Y(y) = (1/\lambda)e^{-y/\lambda}$, $y > 0$, for the purpose of testing

$$H_0: \lambda = 1$$
$$\text{versus}$$
$$H_1: \lambda > 1$$

The null hypothesis will be rejected if $y \geq 3.20$.

(a) Calculate the probability of committing a Type I error.
(b) Calculate the probability of committing a Type II error when $\lambda = \frac{4}{3}$.
(c) Draw a diagram that shows the α and β calculated in parts (a) and (b) as areas.

6.4.11. Polygraphs used in criminal investigations typically measure five bodily functions: (1) thoracic respiration, (2) abdominal respiration, (3) blood pressure and pulse rate, (4) muscular movement and pressure, and (5) galvanic skin response. In principle, the magnitude of these responses when the subject is asked a relevant question ("Did you murder your wife?") indicate whether he is lying or telling the truth. The procedure, of course, is not infallible, as a recent study bore out (82). Seven experienced polygraph examiners were given a set of forty records—twenty were from innocent suspects and twenty from guilty suspects. The subjects had been asked eleven questions, on the basis of which each examiner was to make an overall judgment: "Innocent" or "Guilty." The results are as follows:

		Suspect's True Status	
		Innocent	Guilty
Examiner's	"Innocent"	131	15
Decision	"Guilty"	9	125

What would be the numerical values of α and β in this context? In a judicial setting, should Type I and Type II errors carry equal weight? Explain.

6.4.12. An urn contains ten chips. An unknown number of the chips are white; the others are red. We wish to test

H_0: exactly half the chips are white

versus

H_1: more than half the chips are white

We will draw, without replacement, three chips and reject H_0 if two or more are white. Find α. Also, find β when the urn is (a) 60% white and (b) 70% white.

6.4.13. Suppose that a random sample of size 5 is drawn from a uniform pdf:

$$f_Y(y; \theta) = \begin{cases} \frac{1}{\theta}, & 0 < y < \theta \\ 0, & \text{elsewhere} \end{cases}$$

We wish to test

$$H_0: \theta = 2$$
$$\text{versus}$$
$$H_1: \theta > 2$$

by rejecting the null hypothesis if $y_{max} \geq k$. Find the value of k that makes the probability of committing a Type I error equal to 0.05.

6.4.14. A sample of size 1 is taken from the pdf

$$f_Y(y) = (\theta + 1)y^{\theta}, \quad 0 \leq y \leq 1$$

The hypothesis $H_0: \theta = 1$ is to be rejected in favor of $H_1: \theta > 1$ if $y \geq 0.90$. What is the test's level of significance?

6.4.15. A series of n Bernoulli trials is to be observed as data for testing

$$H_0: p = \tfrac{1}{2}$$
$$\text{versus}$$
$$H_1: p > \tfrac{1}{2}$$

The null hypothesis will be rejected if k, the observed number of successes, equals n. For what value of p will the probability of committing a Type II error equal 0.05?

6.4.16. Let X_1 be a binomial random variable with $n=2$ and $p_{X_1} = P(\text{success})$. Let X_2 be an independent binomial random variable with $n=4$ and $p_{X_2} = P(\text{success})$. Let $X = X_1 + X_2$. Calculate α if

$$H_0: p_{X_1} = p_{X_2} = \tfrac{1}{2}$$
versus
$$H_1: p_{X_1} = p_{X_2} > \tfrac{1}{2}$$

is to be tested by rejecting the null hypothesis when $k \geq 5$.

6.4.17. A sample of size 1 from the pdf $f_Y(y) = (1+\theta)y^\theta$, $0 \leq y \leq 1$, is to be the basis for testing

$$H_0: \theta = 1$$
versus
$$H_1: \theta < 1$$

The critical region will be the interval $y \leq \frac{1}{2}$. Find an expression for $1 - \beta$ as a function of θ.

6.4.18. An experimenter takes a sample of size 1 from the Poisson probability model, $p_X(k) = e^{-\lambda}\lambda^k/k!$, $k = 0, 1, 2, \ldots$, and wishes to test

$$H_0: \lambda = 6$$
versus
$$H_1: \lambda < 6$$

by rejecting H_0 if $k \leq 2$.

(a) Calculate the probability of committing a Type I error.

(b) Calculate the probability of committing a Type II error when $\lambda = 4$.

6.4.19. A sample of size 1 is taken from the geometric probability model, $p_X(k) = (1-p)^{k-1}p$, $k = 1, 2, 3, \ldots$, to test $H_0: p = \frac{1}{3}$ versus $H_1: p > \frac{1}{3}$. The null hypothesis is to be rejected if $k \geq 4$. What is the probability that a Type II error will be committed when $p = \frac{1}{2}$?

6.4.20. Suppose that one observation from the exponential pdf, $f_Y(y) = \lambda e^{-\lambda y}$, $y > 0$, is to be used to test $H_0: \lambda = 1$ versus $H_1: \lambda < 1$. The decision rule calls for the null hypothesis to be rejected if $y \geq \ln 10$. Find β as a function of λ.

6.4.21. A random sample of size 2 is drawn from a uniform pdf defined over the interval $[0, \theta]$. We wish to test

$$H_0: \theta = 2$$
versus
$$H_1: \theta < 2$$

by rejecting H_0 when $y_1 + y_2 \leq k$. Find the value for k that gives a level of significance of 0.05.

6.4.22. Suppose that the hypotheses of Question 6.4.21 are to be tested with a decision rule of the form "Reject $H_0: \theta = 2$ if $y_1 y_2 \leq k^*$." Find the value of k^* that gives a level of significance of 0.05 (see Theorem 3.8.5).

6.5 A Notion of Optimality: The Generalized Likelihood Ratio

In the next several chapters we will be studying some of the particular hypothesis tests that statisticians most often use in dealing with real-world problems. All of these have the same conceptual heritage—a fundamental notion known as the *generalized likelihood ratio*, or *GLR*. More than just a principle, the generalized likelihood ratio is a working criterion for actually *suggesting* test procedures.

As a first look at this important idea, we will conclude Chapter 6 with an application of the generalized likelihood ratio to the problem of testing the parameter θ in a uniform pdf. Notice the relationship here between the likelihood ratio and the definition of an "optimal" hypothesis test.

Suppose $y_1, y_2, \ldots, y_n$ is a random sample from a uniform pdf over the interval $[0, \theta]$, where θ is unknown, and our objective is to test

$$H_0: \theta = \theta_o$$
versus
$$H_1: \theta < \theta_o$$

at a specified level of significance α. What is the "best" decision rule for choosing between H_0 and H_1, and by what criterion is it considered optimal?

As a starting point in answering those questions, it will be necessary to define two parameter spaces, ω and Ω. In general, ω is the set of unknown parameter values admissible under H_0. In the case of the uniform, the only parameter is θ, and the null hypothesis restricts it to a single point:

$$\omega = \{\theta : \theta = \theta_o\}$$

The second parameter space, Ω, is the set of all possible values of all unknown parameters. Here,

$$\Omega = \{\theta : 0 < \theta \leq \theta_o\}$$

Now, recall the definition of the likelihood function, L, from Definition 5.2.1. Given a sample of size n from a uniform pdf,

$$L = L(\theta) = \prod_{i=1}^{n} f_Y(y_i; \theta) = \begin{cases} \left(\frac{1}{\theta}\right)^n, & 0 \leq y_i \leq \theta \\ 0, & \text{otherwise} \end{cases}$$

For reasons that will soon be clear, we need to maximize $L(\theta)$ twice, once under ω and again under Ω. Since θ can take on only one value—θ_o—under ω,

$$\max_{\omega} L(\theta) = L(\theta_o) = \begin{cases} \left(\frac{1}{\theta_o}\right)^n, & 0 \leq y_i \leq \theta_o \\ 0, & \text{otherwise} \end{cases}$$

Maximizing $L(\theta)$ under Ω—that is, with *no* restrictions—is accomplished by simply substituting the maximum likelihood estimate for θ into $L(\theta)$. For the uniform parameter, y_{max} is the maximum likelihood estimate (recall Question 5.2.10). Therefore,

$$\max_{\Omega} L(\theta) = \left(\frac{1}{y_{max}}\right)^n$$

For notational simplicity, we denote $\max_{\omega} L(\theta)$ and $\max_{\Omega} L(\theta)$ by $L(\omega_e)$ and $L(\Omega_e)$, respectively.

Definition 6.5.1. Let $y_1, y_2, \ldots, y_n$ be a random sample from $f_Y(y; \theta_1, \ldots, \theta_k)$. The generalized likelihood ratio, λ, is defined to be

$$\lambda = \frac{\max_{\omega} L(\theta_1, \ldots, \theta_k)}{\max_{\Omega} L(\theta_1, \ldots, \theta_k)} = \frac{L(\omega_e)}{L(\Omega_e)}$$

For the uniform distribution,

$$\lambda = \frac{(1/\theta_0)^n}{(1/y_{max})^n} = \left(\frac{y_{max}}{\theta_0}\right)^n$$

Note that, in general, λ will always be positive but never greater than 1 (why?). Furthermore, values of the likelihood ratio close to 1 suggest that the data are very compatible with H_0. That is, the observations are "explained" almost as well by the H_0 parameters as by *any* parameters [as measured by $L(\omega_e)$ and $L(\Omega_e)$]. For these values of λ we should *accept* H_0. Conversely, if $L(\omega_e)/L(\Omega_e)$ were close to 0, the data would not be very compatible with the parameter values in ω and it would make sense to *reject* H_0.

Definition 6.5.2. A generalized likelihood ratio test (GLRT) is one that rejects H_0 whenever

$$0 < \lambda \leq \lambda^*$$

where λ^* is chosen so that

$$P(0 < \Lambda \leq \lambda^* \mid H_0 \text{ is true}) = \alpha$$

(Note: In keeping with the capital letter notation introduced in Chapter 3, Λ denotes the generalized likelihood ratio expressed as a random variable.)

Let $f_\Lambda(\lambda \mid H_0)$ denote the pdf of the generalized likelihood ratio when H_0 is true. If $f_\Lambda(\lambda \mid H_0)$ were known, λ^* (and, therefore, the decision rule) could be determined by solving the equation

$$\alpha = \int_0^{\lambda^*} f_\Lambda(\lambda \mid H_0)\, d\lambda$$

(see Figure 6.5.1). In many situations, though, $f_\Lambda(\lambda \mid H_0)$ is *not* known, and it becomes necessary to show that Λ is a monotonic function of some quantity W, where the distribution of W *is* known. Once we have found such a statistic, any test based on w will be equivalent to one based on λ.

Here, a suitable W is easy to find. Note that

$$P(\Lambda \leq \lambda^* \mid H_0 \text{ is true}) = \alpha = P\left[\left(\frac{Y_{\max}}{\theta_0}\right)^n \leq \lambda^* \mid H_0 \text{ is true}\right]$$

$$= P\left(\frac{Y_{\max}}{\theta_0} \leq \sqrt[n]{\lambda^*} \mid H_0 \text{ is true}\right)$$

Figure 6.5.1

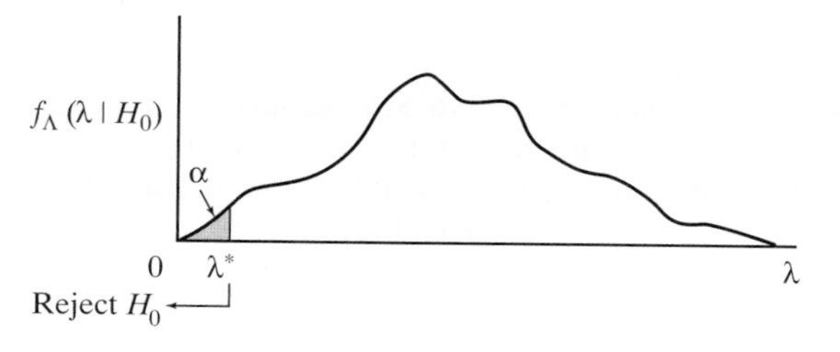

Let $W = Y_{\max}/\theta_0$ and $w^* = \sqrt[n]{\lambda^*}$. Then

$$P(\Lambda \leq \lambda^* \mid H_0 \text{ is true}) = P(W \leq w^* \mid H_0 \text{ is true}) \tag{6.5.1}$$

Here the right-hand side of Equation 6.5.1 can be evaluated from what we already know about the density function for the largest order statistic from a uniform distribution. Let $f_{Y_{\max}}(y; \theta_0)$ be the density function for $Y_{\max}$. Then

$$f_W(w; \theta_0) = \theta_0 f_{Y_{\max}}(\theta_0 w; \theta_0) \qquad \text{(recall Theorem 3.8.2)}$$

which, from Theorem 3.10.1, reduces to

$$\frac{\theta_0 n (\theta_0 w)^{n-1}}{\theta_0^n} = n w^{n-1}, \quad 0 \leq w \leq 1$$

Therefore,

$$P(W \leq w^* \mid H_0 \text{ is true}) = \int_0^{w^*} nw^{n-1}dw = (w^*)^n = \alpha$$

implying that the critical value for W is

$$w^* = \sqrt[n]{\alpha}$$

That is, the GLRT calls for H_0 to be rejected if

$$w = \frac{y_{\max}}{\theta_0} \leq \sqrt[n]{\alpha}$$

Questions

6.5.1. Let $k_1, k_2, \ldots, k_n$ be a random sample from the geometric probability function

$$p_X(k; p) = (1-p)^{k-1}p, \quad k = 1, 2, \ldots$$

Find λ, the generalized likelihood ratio for testing $H_0: p = p_0$ versus $H_1: p \neq p_0$.

6.5.2. Let $y_1, y_2, \ldots, y_{10}$ be a random sample from an exponential pdf with unknown parameter λ. Find the form of the GLRT for $H_0: \lambda = \lambda_0$ versus $H_1: \lambda \neq \lambda_0$. What integral would have to be evaluated to determine the critical value if α were equal to 0.05?

6.5.3. Let $y_1, y_2, \ldots, y_n$ be a random sample from a normal pdf with unknown mean μ and variance 1. Find the form of the GLRT for $H_0: \mu = \mu_0$ versus $H_1: \mu \neq \mu_0$.

6.5.4. In the scenario of Question 6.5.3, suppose the alternative hypothesis is $H_1: \mu = \mu_1$, for some particular value of μ_1. How does the likelihood ratio test change in this case? In what way does the critical region depend on the particular value of μ_1?

6.5.5. Let k denote the number of successes observed in a sequence of n independent Bernoulli trials, where $p = P(\text{success})$.

(a) Show that the critical region of the likelihood ratio test of $H_0: p = \frac{1}{2}$ versus $H_1: p \neq \frac{1}{2}$ can be written in the form

$$k \cdot \ln(k) + (n-k) \cdot \ln(n-k) \geq \lambda^{**}$$

(b) Use the symmetry of the graph of

$$f(k) = k \cdot \ln(k) + (n-k) \cdot \ln(n-k)$$

to show that the critical region can be written in the form

$$\left| \bar{k} - \frac{1}{2} \right| \geq c$$

where c is a constant determined by α.

6.5.6. Suppose a sufficient statistic exists for the parameter θ. Use Theorem 5.6.1 to show that the critical region of a likelihood ratio test will depend on the sufficient statistic.

6.6 Taking a Second Look at Statistics (Statistical Significance versus "Practical" Significance)

The most important concept in this chapter—the notion of *statistical significance*—is also the most problematic. Why? Because statistical significance does not always mean what it *seems* to mean. By definition, the difference between, say, $\overline{y}$ and μ_o is statistically significant if $H_0: \mu = \mu_o$ can be rejected at the $\alpha = 0.05$ level. What that implies is that a *sample mean* equal to the observed $\overline{y}$ is not likely to have come from a (normal) distribution whose *true mean* was μ_o. What it does *not* imply is that the true mean is necessarily much different than μ_o.

Recall the discussion of power curves in Section 6.4 and, in particular, the effect of n on $1 - \beta$. The example illustrating those topics involved an additive that might be able to increase a car's gas mileage. The hypotheses being tested were

$$H_0: \mu = 25.0$$

versus

$$H_1: \mu > 25.0$$

where σ was assumed to be 2.4 (mpg) and α was set at 0.05. *If* $n = 30$, the decision rule called for H_0 to be rejected when $\bar{y} \geq 25.718$ (see p. 354). Figure 6.6.1 is the test's power curve [the point $(\mu, 1 - \beta) = (25.75, 1 - 0.47)$ was calculated on p. 368].

Figure 6.6.1

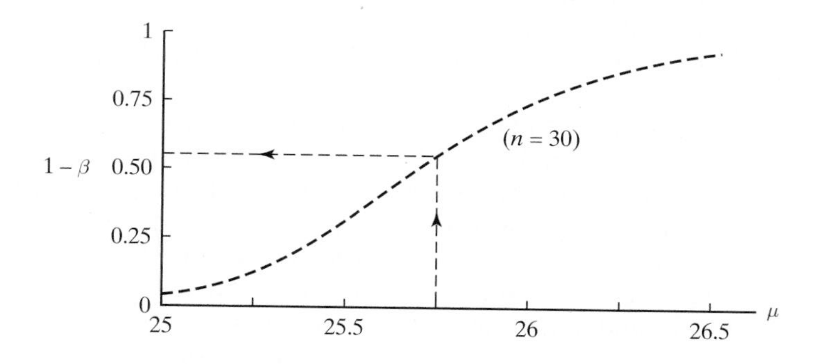

The important point was made in Section 6.4 that researchers have a variety of ways to *increase* the power of a test—that is, to decrease the probability of committing a Type II error. Experimentally, the usual way is to increase the sample size, which has the effect of reducing the overlap between the H_0 and H_1 distributions (Figure 6.4.7 pictured such a reduction when the sample size was kept fixed but σ was decreased from 2.4 to 1.2). Here, to show the effect of n on $1 - \beta$, Figure 6.6.2 superimposes the power curves for testing $H_0: \mu = 25.0$ versus $H_1: \mu > 25.0$ in the cases where $n = 30$, $n = 60$, and $n = 900$ (keeping $\alpha = 0.05$ and $\sigma = 2.4$).

Figure 6.6.2

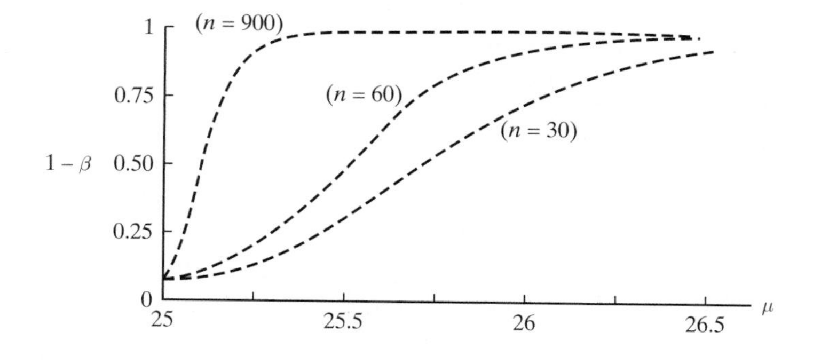

There is good news in Figure 6.6.2 and there is bad news in Figure 6.6.2. The good news—not surprisingly—is that the probability of rejecting a false hypothesis increases dramatically as n increases. If the true mean μ is 25.25, for example, the Z test will (correctly) reject $H_0: \mu = 25.0$ *14%* of the time when $n = 30$, *20%* of the time when $n = 60$, and a robust *93%* of the time when $n = 900$.

The bad news implicit in Figure 6.6.2 is that *any* false hypothesis, even one where the true μ is just "epsilon" away from μ_o, can be rejected virtually 100% of the time if a large enough sample size is used. Why is that bad? Because saying that a difference (between $\bar{y}$ and μ_o) is *statistically significant* makes it sound meaningful when, in fact, it may be totally inconsequential.

Suppose, for example, an additive could be found that would increase a car's gas mileage from 25.000 mpg to 25.001 mpg. Such a minuscule improvement would mean basically nothing to the consumer, yet if a large enough sample size were used, the probability of rejecting $H_0: \mu = 25.000$ in favor of $H_1: \mu > 25.000$ could be made arbitrarily close to 1. That is, the difference between $\overline{y}$ and 25.000 would qualify as being statistically significant even though it had no "practical significance" whatsoever.

Two lessons should be learned here, one old and one new. The new lesson is to be wary of inferences drawn from experiments or surveys based on huge sample sizes. Many statistically significant conclusions are likely to result in those situations, but some of those "reject H_0's" may be driven primarily by the sample size. Paying attention to the *magnitude* of $\overline{y} - \mu_o$ (or $\frac{k}{n} - p_o$) is often a good way to keep the conclusion of a hypothesis test in perspective.

The second lesson has been encountered before and will come up again: Analyzing data is not a simple exercise in plugging into formulas or reading computer printouts. Real-world data are seldom simple, and they cannot be adequately summarized, quantified, or interpreted with any single statistical technique. Hypothesis tests, like every other inference procedure, have strengths and weaknesses, assumptions and limitations. Being aware of what they can tell us—and how they can trick us—is the first step toward using them properly.

Chapter 7

Inferences Based on the Normal Distribution

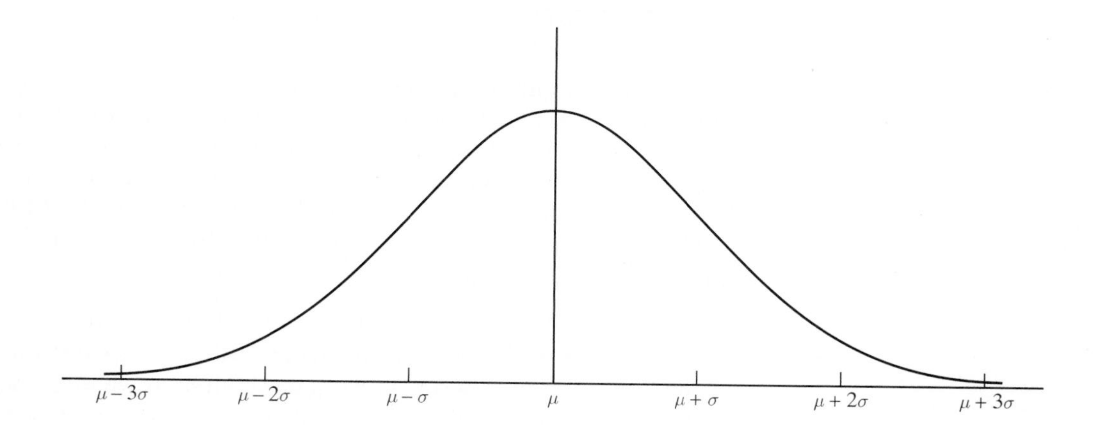

I know of scarcely anything so apt to impress the imagination as the wonderful form of cosmic order expressed by the "law of frequency of error" (the normal distribution). The law would have been personified by the Greeks and deified, if they had known of it. It reigns with serenity and in complete self effacement amidst the wildest confusion. The huger the mob, and the greater the anarchy, the more perfect is its sway. It is the supreme law of Unreason.

—Francis Galton

7.1 Introduction

Finding probability distributions to describe—and, ultimately, to predict—empirical data is one of the most important contributions a statistician can make to the research scientist. Already we have seen a number of functions playing that role.

The binomial is an obvious model for the number of correct responses in the Pratt-Woodruff ESP experiment (Case Study 4.3.1); the probability of holding a winning ticket in the Florida Lottery is given by the hypergeometric (Example 3.2.6); and applications of the Poisson have run the gamut from radioactive decay (Case Study 4.2.2) to the number of wars starting in a given year (Case Study 4.2.3). Those examples notwithstanding, by far the most widely used probability model in statistics is the *normal* (or *Gaussian*) distribution,

$$f_Y(y) = \frac{1}{\sqrt{2\pi}\,\sigma} e^{-(1/2)[(y-\mu)/\sigma]^2}, \quad -\infty < y < \infty \tag{7.1.1}$$

Some of the history surrounding the normal curve has already been discussed in Chapter 4—how it first appeared as a limiting form of the binomial, but then soon found itself used most often in nonbinomial situations. We also learned how to find areas under normal curves and did some problems involving sums and averages. Chapter 5 provided estimates of the parameters of the normal density and showed their role in fitting normal curves to data. In this chapter, we will take a second look at the properties and applications of this singularly important pdf, this time paying attention to the part it plays in estimation and hypothesis testing.

7.2 Comparing $\frac{\overline{Y}-\mu}{\sigma/\sqrt{n}}$ and $\frac{\overline{Y}-\mu}{S/\sqrt{n}}$

Suppose that a random sample of n measurements, $Y_1, Y_2, \ldots, Y_n$, is to be taken on a trait that is thought to be normally distributed, the objective being to draw an inference about the underlying pdf's true mean, μ. *If the variance σ^2 is known*, we already know how to proceed: A decision rule for testing $H_0: \mu = \mu_0$ is given in Theorem 6.2.1, and the construction of a confidence interval for μ is described in Section 5.3. As we learned, both of those procedures are based on the fact that the ratio $Z = \frac{\overline{Y}-\mu}{\sigma/\sqrt{n}}$ has a standard normal distribution, $f_Z(z)$.

In practice, though, the parameter σ^2 is seldom known, so the ratio $\frac{\overline{Y}-\mu}{\sigma/\sqrt{n}}$ cannot be calculated, even if a value for the mean—say, μ_0—is substituted for μ. Typically, the only information experimenters have about σ^2 is what can be gleaned from the Y_i's themselves. The usual estimator for the population variance, of course, is $S^2 = \frac{1}{n-1}\sum_{i=1}^{n}(Y_i - \overline{Y})^2$, the unbiased version of the maximum likelihood estimator for σ^2. The question is, what effect does replacing σ with S have on the Z ratio? Are there are probabilistic differences between $\frac{\overline{Y}-\mu}{\sigma/\sqrt{n}}$ and $\frac{\overline{Y}-\mu}{S/\sqrt{n}}$?

Historically, many early practitioners of statistics felt that replacing σ with S had, in fact, no effect on the distribution of the Z ratio. Sometimes they were right. If the sample size is very large (which was not an unusual state of affairs in many of the early applications of statistics), the estimator S is essentially a constant and for all intents and purposes equal to the true σ. Under those conditions, the ratio $\frac{\overline{Y}-\mu}{S/\sqrt{n}}$ *will* behave much like a standard normal random variable, Z. When the sample size n is small, though, replacing σ with S *does* matter, and it changes the way we draw inferences about μ.

Credit for recognizing that $\frac{\overline{Y}-\mu}{\sigma/\sqrt{n}}$ and $\frac{\overline{Y}-\mu}{S/\sqrt{n}}$ do not have the same distribution goes to William Sealy Gossett. After graduating in 1899 from Oxford with First Class degrees in Chemistry and Mathematics, Gossett took a position at Arthur Guinness, Son & Co., Ltd., a firm that brewed a thick, dark ale known as stout. Given the task of making the art of brewing more scientific, Gossett quickly realized that any experimental studies would necessarily face two obstacles. First, for a variety of economic and logistical reasons, sample sizes would invariably be small; and second, there would never be any way to know the exact value of the true variance, σ^2, associated with any set of measurements.

So, when the objective of a study was to draw an inference about μ, Gossett found himself working with the ratio $\frac{\overline{Y}-\mu}{S/\sqrt{n}}$, where n was often on the order of four or five. The more he encountered that situation, the more he became convinced that ratios of that sort are *not* adequately described by the standard normal pdf. In particular, the distribution of $\frac{\overline{Y}-\mu}{S/\sqrt{n}}$ seemed to have the same general bell-shaped configuration as $f_Z(z)$, but the tails were "thicker"—that is, ratios much smaller than zero or much greater than zero were not as rare as the standard normal pdf would predict.

Figure 7.2.1 illustrates the distinction between the distributions of $\frac{\overline{Y}-\mu}{\sigma/\sqrt{n}}$ and $\frac{\overline{Y}-\mu}{S/\sqrt{n}}$ that caught Gossett's attention. In Figure 7.2.1a, five hundred samples of size $n=4$ have been drawn from a normal distribution where the value of σ is known. For each sample, the ratio $\frac{\overline{Y}-\mu}{\sigma/\sqrt{4}}$ has been computed. Superimposed over the shaded histogram of those five hundred ratios is the standard normal curve, $f_Z(z)$. Clearly, the probabilistic behavior of the random variable $\frac{\overline{Y}-\mu}{\sigma/\sqrt{4}}$ is entirely consistent with $f_Z(z)$.

Figure 7.2.1

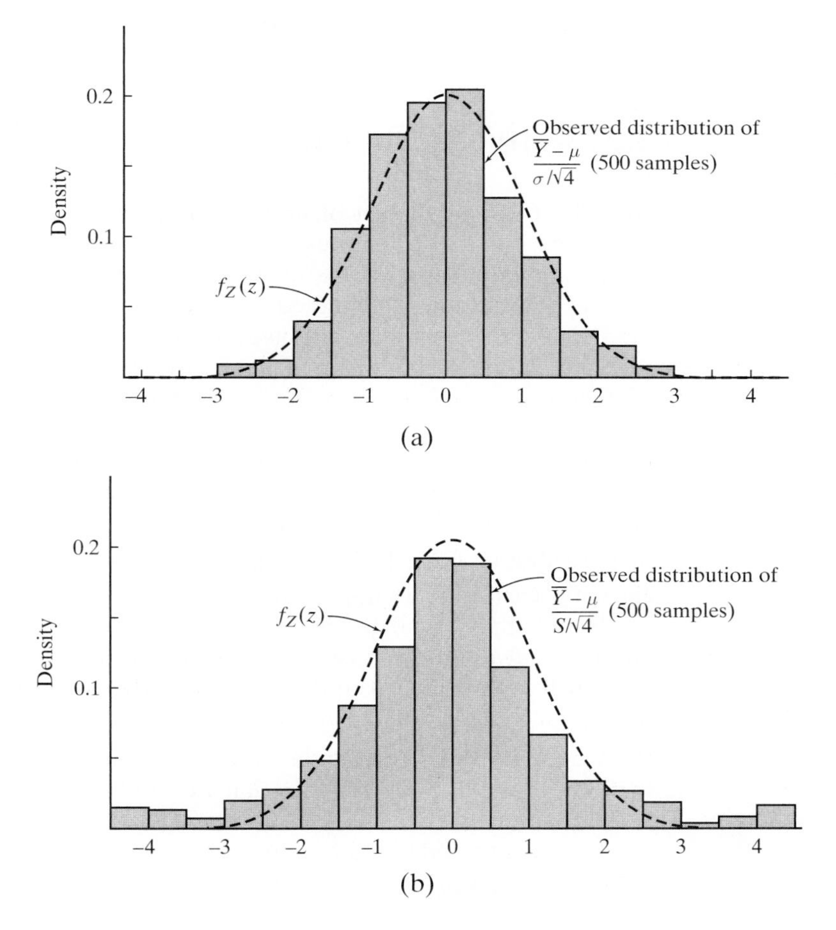

The histogram pictured in Figure 7.2.1b is also based on five hundred samples of size $n = 4$ drawn from a normal distribution. Here, though, S has been calculated for each sample, so the ratios comprising the histogram are $\frac{\overline{Y}-\mu}{S/\sqrt{4}}$ rather than $\frac{\overline{Y}-\mu}{\sigma/\sqrt{4}}$. In this case, the superimposed standard normal pdf does *not* adequately describe the histogram—specifically, it underestimates the number of ratios much less than zero as well as the number much larger than zero (which is exactly what Gossett had noted).

Gossett published a paper in 1908 entitled "The Probable Error of a Mean," in which he derived a formula for the pdf of the ratio $\frac{\overline{Y}-\mu}{S/\sqrt{n}}$. To prevent disclosure of confidential company information, Guinness prohibited its employees from publishing any papers, regardless of content. So, Gossett's work, one of the major statistical breakthroughs of the twentieth century, was published under the name "Student."

Initially, Gossett's discovery attracted very little attention. Virtually none of his contemporaries had the slightest inkling of the impact that Gossett's paper would have on modern statistics. Indeed, fourteen years after its publication, Gossett sent R.A. Fisher a tabulation of his distribution, with a note saying, "I am sending you a copy of Student's Tables as you are the only man that's ever likely to use them."

Fisher very much understood the value of Gossett's work and believed that Gossett had effected a "logical revolution." Fisher presented a rigorous mathematical derivation of Gossett's pdf in 1924, the core of which appears in Appendix 7.A.2. Fisher somewhat arbitrarily chose the letter t for the $\frac{\overline{Y}-\mu}{S/\sqrt{n}}$ statistic. Consequently, its pdf is known as the *Student t distribution*.

7.3 Deriving the Distribution of $\frac{\overline{Y}-\mu}{S/\sqrt{n}}$

Broadly speaking, the set of probability functions that statisticians have occasion to use fall into two categories. There are a dozen or so that can effectively model the *individual* measurements taken on a variety of real-world phenomena. These are the distributions we studied in Chapters 3 and 4—most notably, the normal, binomial, Poisson, exponential, hypergeometric, and uniform. There is a smaller set of probability distributions that model the behavior of *functions* based on sets of n random variables. These are called *sampling distributions*, and they are typically used for inference purposes.

The normal distribution belongs to both categories. We have seen a number of scenarios (IQ scores, for example) where the Gaussian distribution is very effective at describing the distribution of repeated measurements. At the same time, the normal distribution is used to model the probabilistic behavior of $T = \frac{\overline{Y}-\mu}{\sigma/\sqrt{n}}$. In the latter capacity, it serves as a sampling distribution.

Next to the normal distribution, the three most important sampling distributions are the *Student t distribution,* the *chi square distribution,* and the *F distribution*. All three will be introduced in this section, partly because we need the latter two to derive $f_T(t)$, the pdf for the t ratio, $T = \frac{\overline{Y}-\mu}{S/\sqrt{n}}$. So, although our primary objective in this section is to study the Student t distribution, we will in the process introduce the two other sampling distributions that we will be encountering over and over again in the chapters ahead.

Deriving the pdf for a t ratio is not a simple matter. That may come as a surprise, given that deducing the pdf for $\frac{\overline{Y}-\mu}{\sigma/\sqrt{n}}$ is quite easy (using moment-generating functions). But going from $\frac{\overline{Y}-\mu}{\sigma/\sqrt{n}}$ to $\frac{\overline{Y}-\mu}{S/\sqrt{n}}$ creates some major mathematical complications because T (unlike Z) is the ratio of *two* random variables, $\overline{Y}$ and S, both of which are functions of n random variables, $Y_1, Y_2, \ldots, Y_n$. In general—and this ratio is no exception—finding pdfs of quotients of random variables is difficult, especially when the numerator and denominator random variables have cumbersome pdfs to begin with.

As we will see in the next few pages, the derivation of $f_T(t)$ plays out in several steps. First, we show that $\sum_{j=1}^{m} Z_j^2$, where the Z_j's are independent standard normal random variables, has a gamma distribution (more specifically, a special case of the gamma distribution, called a *chi square distribution*). Then we show that $\overline{Y}$ and S^2, based on a random sample of size n from a normal distribution, are independent random variables and that $\frac{(n-1)S^2}{\sigma^2}$ has a chi square distribution. Next we derive the pdf of the ratio of two independent chi square random variables (which is called the *F distribution*). The final step in the proof is to show that $T^2 = \left(\frac{\overline{Y}-\mu}{S/\sqrt{n}}\right)^2$ can be written as the quotient of two independent chi square random variables, making it a special case of the F distribution. Knowing the latter allows us to deduce $f_T(t)$.

Theorem 7.3.1 *Let $U = \sum_{j=1}^{m} Z_j^2$, where $Z_1, Z_2, \ldots, Z_m$ are independent standard normal random variables. Then U has a gamma distribution with $r = \frac{m}{2}$ and $\lambda = \frac{1}{2}$. That is,*

$$f_U(u) = \frac{1}{2^{m/2}\Gamma\left(\frac{m}{2}\right)} u^{(m/2)-1} e^{-u/2}, \quad u \geq 0$$

Proof First take $m = 1$. For any $u \geq 0$,

$$F_{Z^2}(u) = P(Z^2 \leq u) = P\left(-\sqrt{u} \leq Z \leq \sqrt{u}\right) = 2P\left(0 \leq Z \leq \sqrt{u}\right)$$

$$= \frac{2}{\sqrt{2\pi}} \int_0^{\sqrt{u}} e^{-z^2/2}\, dz$$

Differentiating both sides of the equation for $F_{Z^2}(u)$ gives $f_{Z^2}(u)$:

$$f_{Z^2}(u) = \frac{d}{du} F_{Z^2}(u) = \frac{2}{\sqrt{2\pi}} \frac{1}{2\sqrt{u}} e^{-u/2} = \frac{1}{2^{1/2}\Gamma\left(\frac{1}{2}\right)} u^{(1/2)-1} e^{-u/2}$$

Notice that $f_U(u) = f_{Z^2}(u)$ has the form of a gamma pdf with $r = \frac{1}{2}$ and $\lambda = \frac{1}{2}$. By Theorem 4.6.4, then, the sum of m such squares has the stated gamma distribution with $r = m\left(\frac{1}{2}\right) = \frac{m}{2}$ and $\lambda = \frac{1}{2}$. □

The distribution of the sum of squares of independent standard normal random variables is sufficiently important that it gets its own name, despite the fact that it represents nothing more than a special case of the gamma distribution.

Definition 7.3.1. The pdf of $U = \sum_{j=1}^{m} Z_j^2$, where $Z_1, Z_2, \ldots, Z_m$ are independent standard normal random variables, is called the *chi square distribution with m degrees of freedom*.

The next theorem is especially critical in the derivation of $f_T(t)$. Using simple algebra, it can be shown that the square of a t ratio can be written as the quotient of two chi square random variables, one a function of $\overline{Y}$ and the other a function of S^2. By showing that $\overline{Y}$ and S^2 are independent (as Theorem 7.3.2 does), Theorem 3.8.4 can be used to find an expression for the pdf of the quotient.

Theorem 7.3.2 *Let $Y_1, Y_2, \ldots, Y_n$ be a random sample from a normal distribution with mean μ and variance σ^2. Then*

a. *S^2 and $\overline{Y}$ are independent.*

b. *$\frac{(n-1)S^2}{\sigma^2} = \frac{1}{\sigma^2}\sum_{i=1}^{n}(Y_i - \overline{Y})^2$ has a chi square distribution with $n-1$ degrees of freedom.*

Proof See Appendix 7.A.2 □

As we will see shortly, the square of a t ratio is a special case of an F random variable. The next definition and theorem summarize the properties of the F distribution that we will need to find the pdf associated with the Student t distribution.

Definition 7.3.2. Suppose that U and V are independent chi square random variables with n and m degrees of freedom, respectively. A random variable of the form $\frac{V/m}{U/n}$ is said to have an *F distribution with m and n degrees of freedom.*

Comment The F in the name of this distribution commemorates the renowned statistician Sir Ronald Fisher.

Theorem 7.3.3 *Suppose $F_{m,n} = \frac{V/m}{U/n}$ denotes an F random variable with m and n degrees of freedom. The pdf of $F_{m,n}$ has the form*

$$f_{F_{m,n}}(w) = \frac{\Gamma\left(\frac{m+n}{2}\right) m^{m/2} n^{n/2} w^{(m/2)-1}}{\Gamma\left(\frac{m}{2}\right)\Gamma\left(\frac{n}{2}\right)(n+mw)^{(m+n)/2}}, \quad w \geq 0$$

Proof We begin by finding the pdf for V/U. From Theorem 7.3.1 we know that $f_V(v) = \frac{1}{2^{m/2}\Gamma(m/2)} v^{(m/2)-1} e^{-v/2}$ and $f_U(u) = \frac{1}{2^{n/2}\Gamma(n/2)} u^{(n/2)-1} e^{-u/2}$.

From Theorem 3.8.4, we have that the pdf of $W = V/U$ is

$$f_{V/U}(w) = \int_0^\infty |u| f_U(u) f_V(uw)\, du$$

$$= \int_0^\infty u \frac{1}{2^{n/2}\Gamma(n/2)} u^{(n/2)-1} e^{-u/2} \frac{1}{2^{m/2}\Gamma(m/2)} (uw)^{(m/2)-1} e^{-uw/2}\, du$$

$$= \frac{1}{2^{(n+m)/2}\Gamma(n/2)\Gamma(m/2)} w^{(m/2)-1} \int_0^\infty u^{\frac{n+m}{2}-1} e^{-[(1+w)/2]u}\, du$$

The integrand is the variable part of a gamma density with $r = (n+m)/2$ and $\lambda = (1+w)/2$. Thus, the integral equals the inverse of the density's constant. This gives

$$f_{V/U} = \frac{1}{2^{(n+m)/2}\Gamma(n/2)\Gamma(m/2)} w^{(m/2)-1} \frac{\Gamma\left(\frac{n+m}{2}\right)}{[(1+w)/2]^{\frac{n+m}{2}}} = \frac{\Gamma\left(\frac{n+m}{2}\right)}{\Gamma(n/2)\Gamma(m/2)} \frac{w^{(m/2)-1}}{(1+w)^{\frac{n+m}{2}}}$$

The statement of the theorem, then, follows from Theorem 3.8.2:

$$f_{\frac{V/m}{U/n}}(w) = f_{\frac{n}{m}V/U}(w) = \frac{1}{n/m} f_{V/U}\left(\frac{w}{n/m}\right) = \frac{m}{n} f_{V/U}\left(\frac{m}{n}w\right)$$ □

F Tables

When graphed, an F distribution looks very much like a typical chi square distribution—values of $\frac{V/m}{U/n}$ can never be negative and the F pdf is skewed sharply to the right. Clearly, the complexity of $f_{F_{m,n}}(r)$ makes the function difficult to work with directly. Tables, though, are widely available that give various percentiles of F distributions for different values of m and n.

Figure 7.3.1 shows $f_{F_{3,5}}(r)$. In general, the symbol $F_{p,m,n}$ will be used to denote the 100 pth percentile of the F distribution with m and n degrees of freedom. Here, the 95th percentile of $f_{F_{3,5}}(r)$—that is, $F_{.95,3,5}$—is *5.41* (see Appendix Table A.4).

Figure 7.3.1

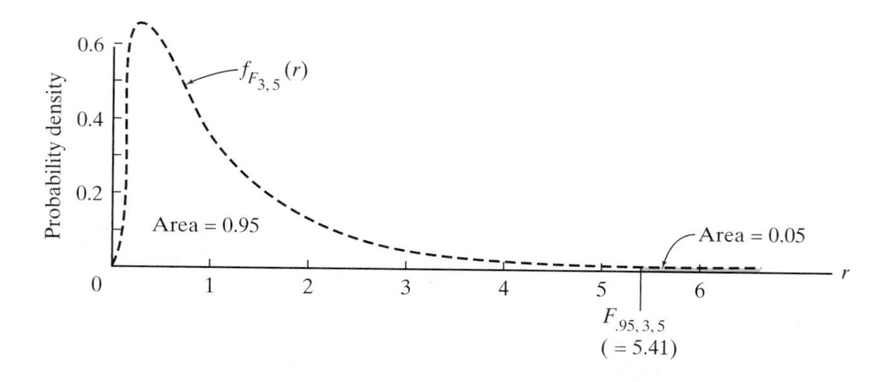

Using the F Distribution to Derive the pdf for t Ratios

Now we have all the background results necessary to find the pdf of $\frac{\overline{Y}-\mu}{S/\sqrt{n}}$. Actually, though, we can do better than that because what we have been calling the "t ratio" is just one special case of an entire family of quotients known as t ratios. Finding the pdf for that entire family will give us the probability distribution for $\frac{\overline{Y}-\mu}{S/\sqrt{n}}$ as well.

Definition 7.3.3. Let Z be a standard normal random variable and let U be a chi square random variable independent of Z with n degrees of freedom. The *Student t ratio with n degrees of freedom* is denoted T_n, where

$$T_n = \frac{Z}{\sqrt{\frac{U}{n}}}$$

Comment The term "degrees of freedom" is often abbrieviated by df.

Lemma *The pdf for T_n is symmetric: $f_{T_n}(t) = f_{T_n}(-t)$, for all t.*

Proof For convenience of notation, let $V = \sqrt{\frac{U}{n}}$. Then by Theorem 3.8.4 and the symmetry of the pdf of Z,

$$f_{T_n}(t) = \int_0^\infty v f_V(v) f_Z(tv) dv = \int_0^\infty v f_V(v) f_Z(-tv) dv = f_{T_n}(-t)$$

□

Theorem 7.3.4 *The pdf for a Student t random variable with n degrees of freedom is given by*

$$f_{T_n}(t) = \frac{\Gamma\left(\frac{n+1}{2}\right)}{\sqrt{n\pi}\,\Gamma\left(\frac{n}{2}\right)\left(1+\frac{t^2}{n}\right)^{(n+1)/2}}, \qquad -\infty < t < \infty$$

Proof Note that $T_n^2 = \frac{Z^2}{U/n}$ has an F distribution with 1 and n df. Therefore,

$$f_{T_n^2}(t) = \frac{n^{n/2}\Gamma\left(\frac{n+1}{2}\right)}{\Gamma\left(\frac{1}{2}\right)\Gamma\left(\frac{n}{2}\right)} t^{-1/2} \frac{1}{(n+t)^{(n+1)/2}}, \qquad t > 0$$

Suppose that $t > 0$. By the symmetry of $f_{T_n}(t)$,

$$\begin{aligned} F_{T_n}(t) = P(T_n \le t) &= \frac{1}{2} + P(0 \le T_n \le t) \\ &= \frac{1}{2} + \frac{1}{2}P(-t \le T_n \le t) \\ &= \frac{1}{2} + \frac{1}{2}P\left(0 \le T_n^2 \le t^2\right) \\ &= \frac{1}{2} + \frac{1}{2}F_{T_n^2}(t^2) \end{aligned}$$

Differentiating $F_{T_n}(t)$ gives the stated result:

$$\begin{aligned} f_{T_n}(t) = F'_{T_n}(t) &= t \cdot f_{T_n^2}(t^2) \\ &= t\frac{n^{n/2}\Gamma\left(\frac{n+1}{2}\right)}{\Gamma\left(\frac{1}{2}\right)\Gamma\left(\frac{n}{2}\right)}(t^2)^{-(1/2)}\frac{1}{(n+t^2)^{(n+1)/2}} \\ &= \frac{\Gamma\left(\frac{n+1}{2}\right)}{\sqrt{n\pi}\,\Gamma\left(\frac{n}{2}\right)} \cdot \frac{1}{\left[1+\left(\frac{t^2}{n}\right)\right]^{(n+1)/2}} \end{aligned}$$

□

Comment Over the years, the lowercase t has come to be the accepted symbol for the random variable of Definition 7.3.3. We will follow that convention when the context allows some flexibility. In mathematical statements about distributions, though, we will be consistent with random variable notation and denote the Student t ratio as T_n.

All that remains to be verified, then, to accomplish our original goal of finding the pdf for $\frac{\overline{Y}-\mu}{S/\sqrt{n}}$ is to show that the latter is a special case of the Student t random variable described in Definition 7.3.3. Theorem 7.3.5 provides the details. Notice that a sample of size n yields a t ratio in this case having $n-1$ degrees of freedom.

Theorem 7.3.5 *Let $Y_1, Y_2, \ldots, Y_n$ be a random sample from a normal distribution with mean μ and standard deviation σ. Then*

$$T_{n-1} = \frac{\overline{Y} - \mu}{S/\sqrt{n}}$$

has a Student t distribution with $n-1$ degrees of freedom.

Proof We can rewrite $\frac{\overline{Y}-\mu}{S/\sqrt{n}}$ in the form

$$\frac{\overline{Y} - \mu}{S/\sqrt{n}} = \frac{\frac{\overline{Y}-\mu}{\sigma/\sqrt{n}}}{\sqrt{\frac{(n-1)S^2}{\sigma^2(n-1)}}}$$

But $\frac{\overline{Y}-\mu}{\sigma/\sqrt{n}}$ is a standard normal random variable and $\frac{(n-1)S^2}{\sigma^2}$ has a chi square distribution with $n-1$ df. Moreover, Theorem 7.3.2 shows that

$$\frac{\overline{Y} - \mu}{\sigma/\sqrt{n}} \quad \text{and} \quad \frac{(n-1)S^2}{\sigma^2}$$

are independent. The statement of the theorem follows immediately, then, from Definition 7.3.3. □

$f_{T_n}(t)$ and $f_Z(Z)$: How the Two Pdfs Are Related

Despite the considerable disparity in the appearance of the formulas for $f_{T_n}(t)$ and $f_Z(z)$, Student t distributions and the standard normal distribution have much in common. Both are bell shaped, symmetric, and centered around zero. Student t curves, though, are flatter.

Figure 7.3.2 is a graph of two Student t distributions—one with 2 df and the other with 10 df. Also pictured is the standard normal pdf, $f_Z(z)$. Notice that as n increases, $f_{T_n}(t)$ becomes more and more like $f_Z(z)$.

Figure 7.3.2

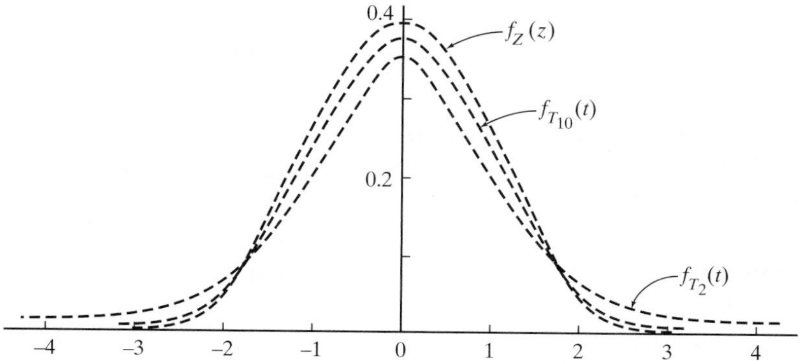

The convergence of $f_{T_n}(t)$ to $f_Z(z)$ is a consequence of two estimation properties:

1. The sample standard deviation is asymptotically unbiased for σ.
2. The standard deviation of S goes to 0 as n approaches ∞. (See Question 7.3.4.)

Therefore as n gets large, the probabilistic behavior of $\frac{\overline{Y}-\mu}{S/\sqrt{n}}$ will become increasingly similar to the distribution of $\frac{\overline{Y}-\mu}{\sigma/\sqrt{n}}$—that is, to $f_Z(z)$.

Questions

7.3.1. Show directly—without appealing to the fact that χ_n^2 is a gamma random variable—that $f_U(u)$ as stated in Definition 7.3.1 is a true probability density function.

7.3.2. Find the moment-generating function for a chi square random variable and use it to show that $E(\chi_n^2) = n$ and $\text{Var}(\chi_n^2) = 2n$.

7.3.3. Is it believable that the numbers 65, 30, and 55 are a random sample of size 3 from a normal distribution with $\mu = 50$ and $\sigma = 10$? Answer the question by using a chi square distribution. [*Hint:* Let $Z_i = (Y_i - 50)/10$ and use Theorem 7.3.1.]

7.3.4. Use the fact that $(n-1)S^2/\sigma^2$ is a chi square random variable with $n-1$ df to prove that

$$\text{Var}(S^2) = \frac{2\sigma^4}{n-1}$$

(*Hint:* Use the fact that the variance of a chi square random variable with k df is $2k$.)

7.3.5. Let $Y_1, Y_2, \ldots, Y_n$ be a random sample from a normal distribution. Use the statement of Question 7.3.4 to prove that S^2 is consistent for σ^2.

7.3.6. If Y is a chi square random variable with n degrees of freedom, the pdf of $(Y-n)/\sqrt{2n}$ converges to $f_Z(z)$ as n goes to infinity (recall Question 7.3.2). Use the asymptotic normality of $(Y-n)/\sqrt{2n}$ to approximate the fortieth percentile of a chi square random variable with 200 degrees of freedom.

7.3.7. Use Appendix Table A.4 to find

(a) $F_{.50,6,7}$
(b) $F_{.001,15,5}$
(c) $F_{.90,2,2}$

7.3.8. Let V and U be independent chi square random variables with 7 and 9 degrees of freedom, respectively. Is it more likely that $\frac{V/7}{U/9}$ will be between (1) 2.51 and 3.29 or (2) 3.29 and 4.20?

7.3.9. Use Appendix Table A.4 to find the values of x that satisfy the following equations:

(a) $P(0.109 < F_{4,6} < x) = 0.95$
(b) $P(0.427 < F_{11,7} < 1.69) = x$
(c) $P(F_{x,x} > 5.35) = 0.01$
(d) $P(0.115 < F_{3,x} < 3.29) = 0.90$
(e) $P\left(x < \frac{V/2}{U/3}\right) = 0.25$, where V is a chi square random variable with 2 df and U is an independent chi square random variable with 3 df.

7.3.10. Suppose that two independent samples of size n are drawn from a normal distribution with variance σ^2. Let S_1^2 and S_2^2 denote the two sample variances. Use the fact that $\frac{(n-1)S^2}{\sigma^2}$ has a chi square distribution with $n-1$ df to explain why

$$\lim_{\substack{n\to\infty \\ m\to\infty}} F_{m,n} = 1$$

7.3.11. If the random variable F has an F distribution with m and n degrees of freedom, show that $1/F$ has an F distribution with n and m degrees of freedom.

7.3.12. Use the result claimed in Question 7.3.11 to express percentiles of $f_{F_{n,m}}(r)$ in terms of percentiles from $f_{F_{m,n}}(r)$. That is, if we know the values a and b for which $P(a \le F_{m,n} \le b) = q$, what values of c and d will satisfy the equation $P(c \le F_{n,m} \le d) = q$? "Check" your answer with Appendix Table A.4 by comparing the values of $F_{.05,2,8}$, $F_{.95,2,8}$, $F_{.05,8,2}$, and $F_{.95,8,2}$.

7.3.13. Show that as $n \to \infty$, the pdf of a Student t random variable with n df converges to $f_Z(z)$. (*Hint:* To show that the constant term in the pdf for T_n converges to $1/\sqrt{2\pi}$, use Stirling's formula,

$$n! \doteq \sqrt{2\pi n}\, n^n e^{-n})$$

Also, recall that $\lim_{n\to\infty}\left(1+\frac{a}{n}\right)^n = e^a$.

7.3.14. Evaluate the integral

$$\int_0^\infty \frac{1}{1+x^2}\,dx$$

using the Student t distribution.

7.3.15. For a Student t random variable Y with n degrees of freedom and any positive integer k, show that $E(Y^{2k})$ exists if $2k < n$. (*Hint:* Integrals of the form

$$\int_0^\infty \frac{1}{(1+y^\alpha)^\beta}\,dy$$

are finite if $\alpha > 0$, $\beta > 0$, and $\alpha\beta > 1$.)

7.4 Drawing Inferences About μ

One of the most common of all statistical objectives is to draw inferences about the *mean* of the population being represented by a set of data. Indeed, we already took a first look at that problem in Section 6.2. If the Y_i's come from a normal distibution where σ is known, the null hypothesis $H_0 : \mu = \mu_0$ can be tested by calculating a Z ratio, $\frac{\overline{Y}-\mu}{\sigma/\sqrt{n}}$ (recall Theorem 6.2.1).

Implicit in that solution, though, is an assumption not likely to be satisfied: rarely does the experimenter actually know the value of σ. Section 7.3 dealt with precisely that scenario and derived the pdf of the ratio $T_{n-1} = \frac{\overline{Y}-\mu}{S/\sqrt{n}}$, where σ has been replaced by an estimator, S. Given T_{n-1} (which we learned has a *Student t distribution with $n-1$ degrees of freedom*), we now have the tools necessary to draw inferences about μ in the all-important case where σ is not known. Section 7.4 illustrates these various techniques and also examines the key assumption underlying the "t test" and looks at what happens when that assumption is not satisfied.

t Tables

We have already seen that doing hypothesis tests and constructing confidence intervals using $\frac{\overline{Y}-\mu}{\sigma/\sqrt{n}}$ or some other Z ratio requires that we know certain upper and/or lower percentiles from the standard normal distribution. There will be a similar need to identify appropriate "cutoffs" from Student t distributions when the inference procedure is based on $\frac{\overline{Y}-\mu}{S/\sqrt{n}}$, or some other t ratio.

Figure 7.4.1 shows a portion of the *t table* that appears in the back of every statistics book. Each row corresponds to a different Student t pdf. The column headings give the area *to the right* of the number appearing in the body of the table.

Figure 7.4.1

	α						
df	.20	.15	.10	.05	.025	.01	.005
1	1.376	1.963	3.078	6.3138	12.706	31.821	63.657
2	1.061	1.386	1.886	2.9200	4.3027	6.965	9.9248
3	0.978	1.250	1.638	2.3534	3.1825	4.541	5.8409
4	0.941	1.190	1.533	2.1318	2.7764	3.747	4.6041
5	0.920	1.156	1.476	2.0150	2.5706	3.365	4.0321
6	0.906	1.134	1.440	1.9432	2.4469	3.143	3.7074
⋮			⋮				
30	0.854	1.055	1.310	1.6973	2.0423	2.457	2.7500
∞	0.84	1.04	1.28	1.64	1.96	2.33	2.58

For example, the entry *4.541* listed in the $\alpha = .01$ column and the $df = 3$ row has the property that $P(T_3 \geq 4.541) = 0.01$.

More generally, we will use the symbol $t_{\alpha,n}$ to denote the $100(1-\alpha)$th percentile of $f_{T_n}(t)$. That is, $P(T_n \geq t_{\alpha,n}) = \alpha$ (see Figure 7.4.2). No lower percentiles of Student t curves need to be tabulated because the symmetry of $f_{T_n}(t)$ implies that $P(T_n \leq -t_{\alpha,n}) = \alpha$.

The number of different Student t pdfs summarized in a t table varies considerably. Many tables will provide cutoffs for degrees of freedom ranging only from 1 to 30; others will include df values from 1 to 50, or even from 1 to 100. The last row in any t table, though, is always labeled "∞": Those entries, of course, correspond to z_α.

Figure 7.4.2

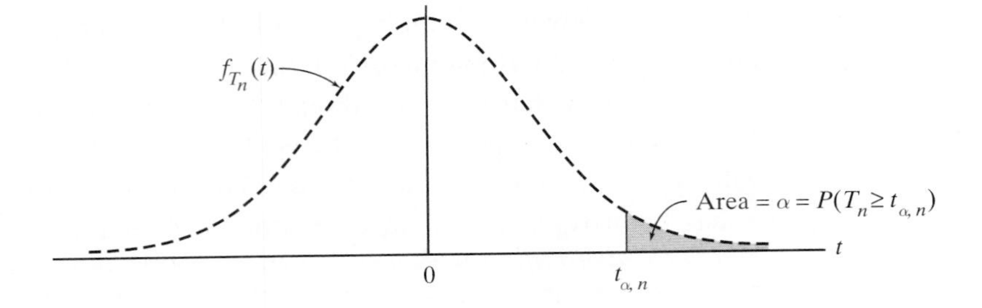

Constructing a Confidence Interval for μ

The fact that $\frac{\overline{Y}-\mu}{S/\sqrt{n}}$ has a Student t distribution with $n-1$ degrees of freedom justifies the statement that

$$P\left(-t_{\alpha/2,n-1} \leq \frac{\overline{Y}-\mu}{S/\sqrt{n}} \leq t_{\alpha/2,n-1}\right) = 1-\alpha$$

or, equivalently, that

$$P\left(\overline{Y} - t_{\alpha/2,n-1} \cdot \frac{S}{\sqrt{n}} \leq \mu \leq \overline{Y} + t_{\alpha/2,n-1} \cdot \frac{S}{\sqrt{n}}\right) = 1-\alpha \qquad (7.4.1)$$

(provided the Y_i's are a random sample from a normal distribution).

When the actual data values are then used to evaluate $\overline{Y}$ and S, the lower and upper endpoints identified in Equation 7.4.1 define a $100(1-\alpha)\%$ confidence interval for μ.

Theorem 7.4.1 *Let $y_1, y_2, \ldots, y_n$ be a random sample of size n from a normal distribution with (unknown) mean μ. A $100(1-\alpha)\%$ confidence interval for μ is the set of values*

$$\left(\overline{y} - t_{\alpha/2,n-1} \cdot \frac{s}{\sqrt{n}},\ \overline{y} + t_{\alpha/2,n-1} \cdot \frac{s}{\sqrt{n}}\right)$$

□

Case Study 7.4.1

To hunt flying insects, bats emit high-frequency sounds and then listen for their echoes. Until an insect is located, these pulses are emitted at intervals of from fifty to one hundred milliseconds. When an insect *is* detected, the pulse-to-pulse interval suddenly decreases—sometimes to as low as ten milliseconds—thus enabling the bat to pinpoint its prey's position. This raises an interesting question: How far apart are the bat and the insect when the bat first senses that the insect is there? Or, put another way, what is the effective range of a bat's echolocation system?

The technical problems that had to be overcome in measuring the bat-to-insect detection distance were far more complex than the statistical problems involved in analyzing the actual data. The procedure that finally evolved was to put a bat into an eleven-by-sixteen-foot room, along with an ample supply

(Continued on next page)

of fruit flies, and record the action with two synchronized sixteen-millimeter sound-on-film cameras. By examining the two sets of pictures frame by frame, scientists could follow the bat's flight pattern and, at the same time, monitor its pulse frequency. For each insect that was caught (65), it was therefore possible to estimate the distance between the bat and the insect at the precise moment the bat's pulse-to-pulse interval decreased (see Table 7.4.1).

Table 7.4.1

Catch Number	Detection Distance (cm)
1	62
2	52
3	68
4	23
5	34
6	45
7	27
8	42
9	83
10	56
11	40

Define μ to be a bat's true average detection distance. Use the eleven observations in Table 7.4.1 to construct a 95% confidence interval for μ.

Letting $y_1 = 62, y_2 = 52, \ldots, y_{11} = 40$, we have that

$$\sum_{i=1}^{11} y_i = 532 \quad \text{and} \quad \sum_{i=1}^{11} y_i^2 = 29{,}000$$

Therefore,

$$\overline{y} = \frac{532}{11} = 48.4 \text{ cm}$$

and

$$s = \sqrt{\frac{11(29{,}000) - (532)^2}{11(10)}} = 18.1 \text{ cm}$$

If the population from which the y_i's are being drawn is normal, the behavior of

$$\frac{\overline{Y} - \mu}{S/\sqrt{n}}$$

will be described by a Student t curve with 10 degrees of freedom. From Table A.2 in the Appendix,

$$P(-2.2281 < T_{10} < 2.2281) = 0.95$$

(Continued on next page)

(Case Study 7.4.1 continued)

Accordingly, the 95% confidence interval for μ is

$$\left[\overline{y} - 2.2281\left(\frac{s}{\sqrt{11}}\right), \overline{y} + 2.2281\left(\frac{s}{\sqrt{11}}\right)\right]$$
$$= \left[48.4 - 2.2281\left(\frac{18.1}{\sqrt{11}}\right), 48.4 + 2.2281\left(\frac{18.1}{\sqrt{11}}\right)\right]$$
$$= (36.2 \text{ cm}, 60.6 \text{ cm}).$$

Example 7.4.1

The sample mean and sample standard deviation for the random sample of size $n = 20$ given in the following list are 2.6 and 3.6, respectively. Let μ denote the true mean of the distribution being represented by these y_i's.

2.5	0.1	0.2	1.3
3.2	0.1	0.1	1.4
0.5	0.2	0.4	11.2
0.4	7.4	1.8	2.1
0.3	8.6	0.3	10.1

Is it correct to say that a 95% confidence interval for μ is the set of following values?

$$\left(\overline{y} - t_{.025,n-1} \cdot \frac{s}{\sqrt{n}}, \overline{y} + t_{.025,n-1} \cdot \frac{s}{\sqrt{n}}\right)$$
$$= \left(2.6 - 2.0930 \cdot \frac{3.6}{\sqrt{20}}, 2.6 + 2.0930 \cdot \frac{3.6}{\sqrt{20}}\right)$$
$$= (0.9, 4.3)$$

No. It *is* true that all the correct factors have been used in calculating (0.9, 4.3), but Theorem 7.4.1 does not apply in this case because the normality assumption it makes is clearly being violated. Figure 7.4.3 is a histogram of the twenty y_i's. The extreme skewness that is so evident there is not consistent with the presumption that the data's underlying pdf is a normal distribution. As a result, the pdf describing the probabilistic behavior of $\frac{\overline{Y}-\mu}{S/\sqrt{20}}$ would *not* be $f_{T_{19}}(t)$.

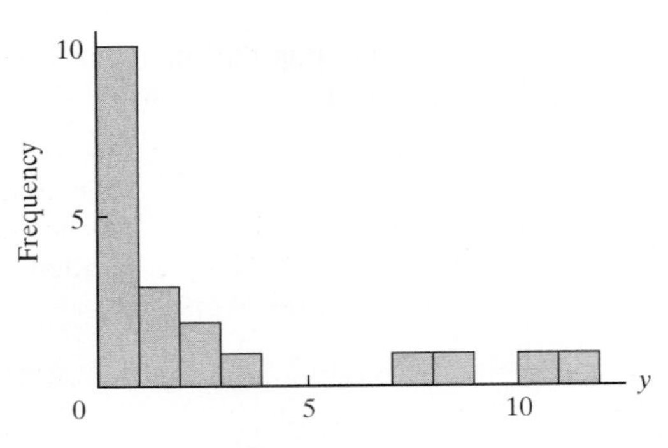

Figure 7.4.3

■

Comment To say that $\frac{\overline{Y}-\mu}{S/\sqrt{20}}$ in this situation is not *exactly* a T_{19} random variable leaves unanswered a critical question: Is the ratio *approximately* a T_{19} random variable? We will revisit the normality assumption—and what happens when that assumption is not satisfied—later in this section when we discuss a critically important property known as *robustness*.

Questions

7.4.1. Use Appendix Table A.2 to find the following probabilities:

(a) $P(T_6 \geq 1.134)$
(b) $P(T_{15} \leq 0.866)$
(c) $P(T_3 \geq -1.250)$
(d) $P(-1.055 < T_{29} < 2.462)$

7.4.2. What values of x satisfy the following equations?

(a) $P(-x \leq T_{22} \leq x) = 0.98$
(b) $P(T_{13} \geq x) = 0.85$
(c) $P(T_{26} < x) = 0.95$
(d) $P(T_2 \geq x) = 0.025$

7.4.3. Which of the following differences is larger? Explain.

$$t_{.05,n} - t_{.10,n} \quad \text{or} \quad t_{.10,n} - t_{.15,n}$$

7.4.4. A random sample of size $n = 9$ is drawn from a normal distribution with $\mu = 27.6$. Within what interval $(-a, +a)$ can we expect to find $\frac{\overline{Y}-27.6}{S/\sqrt{9}}$ 80% of the time? 90% of the time?

7.4.5. Suppose a random sample of size $n = 11$ is drawn from a normal distribution with $\mu = 15.0$. For what value of k is the following true?

$$P\left(\left|\frac{\overline{Y} - 15.0}{S/\sqrt{11}}\right| \geq k\right) = 0.05$$

7.4.6. Let $\overline{Y}$ and S denote the sample mean and sample standard deviation, respectively, based on a set of $n = 20$ measurements taken from a normal distribution with $\mu = 90.6$. Find the function $k(S)$ for which

$$P[90.6 - k(S) \leq \overline{Y} \leq 90.6 + k(S)] = 0.99$$

7.4.7. Cell phones emit radio frequency energy that is absorbed by the body when the phone is next to the ear and may be harmful. The table in the next column gives the absorption rate for a random sample of twenty cell phones. (The Federal Communication Commission sets a maximum of 1.6 watts per kilogram for the absorption rate of such energy.) Construct a 90% confidence interval for the true average cell phone absorption rate.

0.87	0.72
1.30	1.05
0.79	0.61
1.45	1.01
1.15	0.20
1.31	0.67
1.09	1.35
0.66	1.27
0.49	1.28
1.40	1.55

Source: reviews.cnet.com/cell-phone-radiation-levels/

7.4.8. The following table lists the typical cost of repairing the bumper of a moderately priced midsize car damaged by a corner collision at 3 mph. Use these observations to construct a 95% confidence interval for μ, the true average repair cost for all such automobiles with similar damage. The sample standard deviation for these data is $s = \$369.02$.

Make/Model	Repair Cost	Make/Model	Repair Cost
Hyundai Sonata	$1019	Honda Accord	$1461
Nissan Altima	$1090	Volkswagen Jetta	$1525
Mitsubishi Galant	$1109	Toyota Camry	$1670
Saturn AURA	$1235	Chevrolet Malibu	$1685
Subaru Legacy	$1275	Volkswagen Passat	$1783
Pontiac G6	$1361	Nissan Maxima	$1787
Mazda 6	$1437	Ford Fusion	$1889
Volvo S40	$1446	Chrysler Sebring	$2484

Source: www.iihs.org/ratings/bumpersbycategory.aspx?

7.4.9. Creativity, as any number of studies have shown, is very much a province of the young. Whether the focus is music, literature, science, or mathematics, an individual's best work seldom occurs late in life. Einstein, for example, made his most profound discoveries at the age of twenty-six; Newton, at the age of twenty-three. The following are twelve scientific breakthroughs dating from the middle of the sixteenth century to the early years of the twentieth century (205). All represented high-water marks in the careers of the scientists involved.

Discovery	Discoverer	Year	Age, y
Earth goes around sun	Copernicus	1543	40
Telescope, basic laws of astronomy	Galileo	1600	34
Principles of motion, gravitation, calculus	Newton	1665	23
Nature of electricity	Franklin	1746	40
Burning is uniting with oxygen	Lavoisier	1774	31
Earth evolved by gradual processes	Lyell	1830	33
Evidence for natural selection controlling evolution	Darwin	1858	49
Field equations for light	Maxwell	1864	33
Radioactivity	Curie	1896	34
Quantum theory	Planck	1901	43
Special theory of relativity, $E=mc^2$	Einstein	1905	26
Mathematical foundations for quantum theory	Schrödinger	1926	39

(a) What can be inferred from these data about the *true* average age at which scientists do their best work? Answer the question by constructing a 95% confidence interval.

(b) Before constructing a confidence interval for a set of observations extending over a long period of time, we should be convinced that the y_i's exhibit no biases or trends. If, for example, the age at which scientists made major discoveries decreased from century to century, then the parameter μ would no longer be a constant, and the confidence interval would be meaningless. Plot "date" versus "age" for these twelve discoveries. Put "date" on the abscissa. Does the variability in the y_i's appear to be random with respect to time?

7.4.10. How long does it take to fly from Atlanta to New York's LaGuardia airport? There are many components of the time elapsed, but one of the more stable measurements is the actual in-air time. For a sample of sixty-one flights between these destinations on Sundays in April, the time in minutes (y) gave the following results:

$$\sum_{i=1}^{61} y_i = 6450 \text{ and } \sum_{i=1}^{61} y_i^2 = 684,900$$

Find a 99% confidence interval for the average flight time.

Source: www.bts.gov/xml/ontimesummarystatistics/src/dstat/OntimeSummaryDepaturesData.xml.

7.4.11. In a nongeriatric population, platelet counts ranging from 140 to 440 (thousands per mm^3 of blood) are considered "normal." The following are the platelet counts recorded for twenty-four female nursing home residents (169).

Subject	Count	Subject	Count
1	125	13	180
2	170	14	180
3	250	15	280
4	270	16	240
5	144	17	270
6	184	18	220
7	176	19	110
8	100	20	176
9	220	21	280
10	200	22	176
11	170	23	188
12	160	24	176

Use the following sums:

$$\sum_{i=1}^{24} y_i = 4645 \quad \text{and} \quad \sum_{i=1}^{24} y_i^2 = 959,265$$

How does the definition of "normal" above compare with the 90% confidence interval?

7.4.12. If a normally distributed sample of size $n=16$ produces a 95% confidence interval for μ that ranges from 44.7 to 49.9, what are the values of $\bar{y}$ and s?

7.4.13. Two samples, each of size n, are taken from a normal distribution with unknown mean μ and unknown standard deviation σ. A 90% confidence interval for μ is constructed with the first sample, and a 95% confidence interval for μ is constructed with the second. Will the 95% confidence interval necessarily be longer than the 90% confidence interval? Explain.

7.4.14. Revenues reported last week from nine boutiques franchised by an international clothier averaged $59,540 with a standard deviation of $6860. Based on those figures, in what range might the company expect to find the average revenue of all of its boutiques?

7.4.15. What "confidence" is associated with each of the following random intervals? Assume that the Y_i's are normally distributed.

(a) $\left[\bar{Y} - 2.0930\left(\frac{S}{\sqrt{20}}\right), \bar{Y} + 2.0930\left(\frac{S}{\sqrt{20}}\right)\right]$

(b) $\left[\bar{Y} - 1.345\left(\frac{S}{\sqrt{15}}\right), \bar{Y} + 1.345\left(\frac{S}{\sqrt{15}}\right)\right]$

(c) $\left[\bar{Y} - 1.7056\left(\frac{S}{\sqrt{27}}\right), \bar{Y} + 2.7787\left(\frac{S}{\sqrt{27}}\right)\right]$

(d) $\left[-\infty, \bar{Y} + 1.7247\left(\frac{S}{\sqrt{21}}\right)\right]$

7.4.16. The weather station at Dismal Swamp, California, recorded monthly precipitation (y) for twenty-eight years. For these data, $\sum_{i=1}^{336} y_i = 1392.6$ and $\sum_{i=1}^{336} y_i^2 = 10,518.84$.

(a) Find the 95% confidence interval for the mean monthly precipitation.

(b) The table on the right gives a frequency ditribution for the Dismal Swamp precipitation data. Does this distribution raise questions about using Theorem 7.4.1?

Rainfall in inches	Frequency
0–1	85
1–2	38
2–3	35
3–4	41
4–5	28
5–6	24
6–7	18
7–8	16
8–9	16
9–10	5
10–11	9
11–12	21

Source: www.wcc.nrcs.usda.gov.

Testing $H_o : \mu = \mu_o$ (The One-Sample t Test)

Suppose a normally distributed random sample of size n is observed for the purpose of testing the null hypothesis that $\mu = \mu_o$. If σ is unknown—which is usually the case—the procedure we use is called a *one-sample t test*. Conceptually, the latter is much like the *Z test* of Theorem 6.2.1, except that the decision rule is defined in terms of $t = \frac{\overline{y}-\mu_o}{s/\sqrt{n}}$ rather than $z = \frac{\overline{y}-\mu_o}{\sigma/\sqrt{n}}$ [which requires that the critical values come from $f_{T_{n-1}}(t)$ rather than $f_Z(z)$].

Theorem 7.4.2 *Let $y_1, y_2, \ldots, y_n$ be a random sample of size n from a normal distribution where σ is unknown. Let $t = \frac{\overline{y}-\mu_o}{s/\sqrt{n}}$.*

a. To test $H_0 : \mu = \mu_o$ versus $H_1 : \mu > \mu_o$ at the α level of significance, reject H_0 if $t \geq t_{\alpha,n-1}$.

b. To test $H_0 : \mu = \mu_o$ versus $H_1 : \mu < \mu_o$ at the α level of significance, reject H_0 if $t \leq -t_{\alpha,n-1}$.

c. To test $H_0 : \mu = \mu_o$ versus $H_1 : \mu \neq \mu_o$ at the α level of significance, reject H_0 if t is either (1) $\leq -t_{\alpha/2,n-1}$ or (2) $\geq t_{\alpha/2,n-1}$.

Proof Appendix 7.A.3 gives the complete derivation that justifies using the procedure described in Theorem 7.4.2. In short, the test statistic $t = \frac{\overline{y}-\mu_o}{s/\sqrt{n}}$ is a monotonic function of the λ that appears in Definition 6.5.2, which makes the one-sample t test a GLRT. □

Case Study 7.4.2

Not all rectangles are created equal. Since antiquity, societies have expressed aesthetic preferences for rectangles having certain width (w) to length (l) ratios.

One "standard" calls for the width-to-length ratio to be equal to the ratio of the length to the sum of the width and the length. That is,

(Continued on next page)

(Case Study 7.4.2 continued)

$$\frac{w}{l} = \frac{l}{w+l} \tag{7.4.2}$$

Equation 7.4.2 implies that the width is $\frac{1}{2}(\sqrt{5}-1)$, or approximately 0.618, times as long as the length. The Greeks called this the golden rectangle and used it often in their architecture (see Figure 7.4.4). Many other cultures were similarly inclined. The Egyptians, for example, built their pyramids out of stones whose faces were golden rectangles. Today in our society, the golden rectangle remains an architectural and artistic standard, and even items such as driver's licenses, business cards, and picture frames often have w/l ratios close to 0.618.

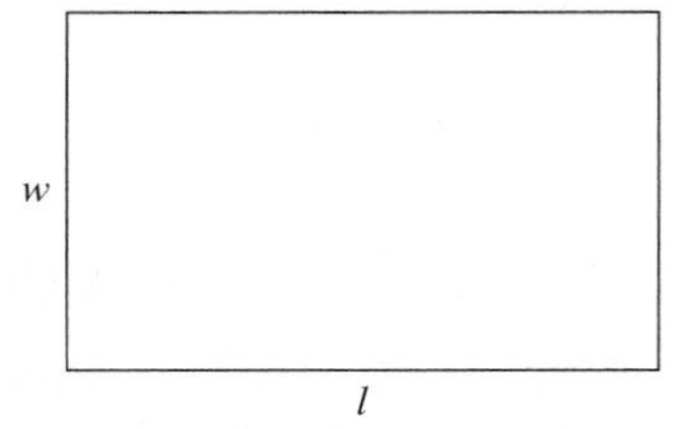

Figure 7.4.4 A golden rectangle $\left(\frac{w}{l} = \frac{l}{w+l}\right)$

The fact that many societies have embraced the golden rectangle as an aesthetic standard has two possible explanations. One, they "learned" to like it because of the profound influence that Greek writers, philosophers, and artists have had on cultures all over the world. Or two, there is something unique about human perception that predisposes a preference for the golden rectangle.

Researchers in the field of experimental aesthetics have tried to test the plausibility of those two hypotheses by seeing whether the golden rectangle is accorded any special status by societies that had no contact whatsoever with the Greeks or with their legacy. One such study (37) examined the w/l ratios of beaded rectangles sewn by the Shoshoni Indians as decorations on their blankets and clothes. Table 7.4.2 lists the ratios found for twenty such rectangles.

If, indeed, the Shoshonis also had a preference for golden rectangles, we would expect their ratios to be "close" to 0.618. The average value of the entries in Table 7.4.2, though, is *0.661*. What does that imply? Is 0.661 close enough to 0.618 to support the position that liking the golden rectangle is a human characteristic, or is 0.661 so far from 0.618 that the only prudent conclusion is that the Shoshonis did *not* agree with the aesthetics espoused by the Greeks?

Table 7.4.2 Width-to-Length Ratios of Shoshoni Rectangles

0.693	0.749	0.654	0.670
0.662	0.672	0.615	0.606
0.690	0.628	0.668	0.611
0.606	0.609	0.601	0.553
0.570	0.844	0.576	0.933

(Continued on next page)

Let μ denote the true average width-to-length ratio of Shoshoni rectangles. The hypotheses to be tested are

$$H_0 : \mu = 0.618$$

versus

$$H_1 : \mu \neq 0.618$$

For tests of this nature, the value of $\alpha = 0.05$ is often used. For that value of α and a two-sided test, the critical values, using part (c) of Theorem 7.4.2 and Appendix Table A.2, are $t_{.025,19} = 2.0930$ and $-t_{.025,19} = -2.0930$.

The data in Table 7.4.2 have $\bar{y} = 0.661$ and $s = 0.093$. Substituting these values into the t ratio gives a test statistic that lies just inside of the interval between -2.0930 and 2.0930:

$$t = \frac{\bar{y} - \mu_0}{s/\sqrt{n}} = \frac{0.661 - 0.618}{0.093/\sqrt{20}} = 2.068$$

Thus, these data do not rule out the possibility that the Shoshoni Indians also embraced the golden rectangle as an aesthetic standard.

About the Data Like π and e, the ratio w/l for golden rectangles (more commonly referred to as either *phi* or the *golden ratio*), is an irrational number with all sorts of fascinating properties and connections.

Algebraically, the solution of the equation

$$\frac{w}{l} = \frac{l}{w + l}$$

is the continued fraction

$$\frac{w}{l} = 1 + \cfrac{1}{1 + \cfrac{1}{1 + \cfrac{1}{1 + \cfrac{1}{1 + \cdots}}}}$$

Among the curiosities associated with phi is its relationship with the *Fibonacci series*. The latter, of course, is the famous sequence in which each term is the sum of its two predecessors—that is,

1 1 2 3 5 8 13 21 34 55 89 ...

Example 7.4.2

Three banks serve a metropolitan area's inner-city neighborhoods: Federal Trust, American United, and Third Union. The state banking commission is concerned that loan applications from inner-city residents are not being accorded the same consideration that comparable requests have received from individuals in rural areas. Both constituencies claim to have anecdotal evidence suggesting that the other group is being given preferential treatment.

Records show that last year these three banks approved 62% of all the home mortgage applications filed by rural residents. Listed in Table 7.4.3 are the approval rates posted over that same period by the twelve branch offices of Federal Trust

Table 7.4.3

Bank	Location	Affiliation	Percent Approved
1	3rd & Morgan	AU	59
2	Jefferson Pike	TU	65
3	East 150th & Clark	TU	69
4	Midway Mall	FT	53
5	N. Charter Highway	FT	60
6	Lewis & Abbot	AU	53
7	West 10th & Lorain	FT	58
8	Highway 70	FT	64
9	Parkway Northwest	AU	46
10	Lanier & Tower	TU	67
11	King & Tara Court	AU	51
12	Bluedot Corners	FT	59

(FT), American United (AU), and Third Union (TU) that work primarily with the inner-city community. Do these figures lend any credence to the contention that the banks are treating inner-city residents and rural residents differently? Analyze the data using an $\alpha = 0.05$ level of significance.

As a starting point, we might want to test

$$H_0 : \mu = 62$$

versus

$$H_1 : \mu \neq 62$$

where μ is the true average approval rate for all inner-city banks. Table 7.4.4 summarizes the analysis. The two critical values are $\pm t_{.025,11} = \pm 2.2010$, and the observed t ratio is $-1.66 \left(= \frac{58.667 - 62}{6.946/\sqrt{12}}\right)$, so our decision is "Fail to reject H_0."

Table 7.4.4

Banks	n	$\bar{y}$	s	t Ratio	Critical Value	Reject H_0?
All	12	58.667	6.946	−1.66	±2.2010	No

About the Data The "overall" analysis of Table 7.4.4, though, may be too simplistic. Common sense would tell us to look also at the three banks separately. What emerges, then, is an entirely different picture (see Table 7.4.5). Now we can see why both groups felt discriminated against: American United ($t = -3.63$) and Third

Table 7.4.5

Banks	n	$\bar{y}$	s	t Ratio	Critical Value	Reject H_0?
American United	4	52.25	5.38	−3.63	±3.1825	Yes
Federal Trust	5	58.80	3.96	−1.81	±2.7764	No
Third Union	3	67.00	2.00	+4.33	±4.3027	Yes

Union ($t = +4.33$) each had rates that differed significantly from 62%—*but in opposite directions!* Only Federal Trust seems to be dealing with inner-city residents and rural residents in an even-handed way. ■

Questions

7.4.17. Recall the *Bacillus subtilis* data in Question 5.3.2. Test the null hypothesis that exposure to the enzyme does not affect a worker's respiratory capacity (as measured by the FEV_1/VC ratio). Use a one-sided H_1 and let $\alpha = 0.05$. Assume that σ is not known.

7.4.18. Recall Case Study 5.3.1. Assess the credibility of the theory that Etruscans were native Italians by testing an appropriate H_0 against a two-sided H_1. Set α equal to 0.05. Use 143.8 mm and 6.0 mm for $\overline{y}$ and s, respectively, and let $\mu_o = 132.4$. Do these data appear to satisfy the distribution assumption made by the t test? Explain.

7.4.19. MBAs R Us advertises that its program increases a person's score on the GMAT by an average of forty points. As a way of checking the validity of that claim, a consumer watchdog group hired fifteen students to take both the review course and the GMAT. Prior to starting the course, the fifteen students were given a diagnostic test that predicted how well they would do on the GMAT in the absence of any special training. The following table gives each student's actual GMAT score minus his or her predicted score. Set up and carry out an appropriate hypothesis test. Use the 0.05 level of significance.

Subject	y_i = act. GMAT − pre. GMAT	y_i^2
SA	35	1225
LG	37	1369
SH	33	1089
KN	34	1156
DF	38	1444
SH	40	1600
ML	35	1225
JG	36	1296
KH	38	1444
HS	33	1089
LL	28	784
CE	34	1156
KK	47	2209
CW	42	1764
DP	46	2116

7.4.20. In addition to the Shoshoni data of Case Study 7.4.2, a set of rectangles that might tend to the golden ratio are national flags. The table below gives the width-to-length ratios for a random sample of the flags of thirty-four countries. Let μ be the width-to-length ratio for national flags. At the $\alpha = 0.01$ level, test $H_0: \mu = 0.618$ versus $H_1: \mu \neq 0.618$.

Country	Ratio Width to Height	Country	Ratio Width to Height
Afghanistan	0.500	Iceland	0.720
Albania	0.714	Iran	0.571
Algeria	0.667	Israel	0.727
Angola	0.667	Laos	0.667
Argentina	0.667	Lebanon	0.667
Bahamas	0.500	Liberia	0.526
Denmark	0.757	Macedonia	0.500
Djibouti	0.553	Mexico	0.571
Ecuador	0.500		
Egypt	0.667	Monaco	0.800
El Salvador	0.600	Namibia	0.667
		Nepal	1.250
Estonia	0.667	Romania	0.667
Ethiopia	0.500	Rwanda	0.667
Gabon	0.750	South Africa	0.667
Fiji	0.500	St. Helena	0.500
France	0.667	Sweden	0.625
Honduras	0.500	United Kingdom	0.500

Source: http://www.anyflag.com/country/costaric.php.

7.4.21. A manufacturer of pipe for laying underground electrical cables is concerned about the pipe's rate of corrosion and whether a special coating may retard that rate. As a way of measuring corrosion, the manufacturer examines a short length of pipe and records the depth of the maximum pit. The manufacturer's tests have shown that in a year's time in the particular kind of soil the manufacturer must deal with, the average depth of the maximum pit in a foot of pipe is 0.0042 inch. To see whether that average can be reduced, ten pipes are coated with a new plastic and buried in the same soil. After one year, the following maximum pit depths are recorded (in inches): 0.0039, 0.0041, 0.0038, 0.0044, 0.0040, 0.0036, 0.0034, 0.0036, 0.0046, and 0.0036. Given that the sample standard deviation for these ten measurements is 0.00383 inch, can it be concluded at the $\alpha = 0.05$ level of significance that the plastic coating is beneficial?

7.4.22. The first analysis done in Example 7.4.2 (using all $n = 12$ banks with $\bar{y} = 58.667$) failed to reject H_0: $\mu = 62$ at the $\alpha = 0.05$ level. Had μ_o been, say, 61.7 or 58.6, the same conclusion would have been reached. What do we call the entire set of μ_o's for which H_0: $\mu = \mu_o$ would *not* be rejected at the $\alpha = 0.05$ level?

Testing H_o: $\mu = \mu_o$ When the Normality Assumption Is Not Met

Every t test makes the same explicit assumption—namely, that the set of n y_i's is normally distributed. But suppose the normality assumption is *not* true. What are the consequences? Is the validity of the t test compromised?

Figure 7.4.5 addresses the first question. We know that if the normality assumption *is* true, the pdf describing the variation of the t ratio, $\frac{\bar{Y}-\mu_o}{S/\sqrt{n}}$, is $f_{T_{n-1}}(t)$. The latter, of course, provides the decision rule's critical values. If $H_0 : \mu = \mu_o$ is to be tested against $H_1 : \mu \neq \mu_o$, for example, the null hypothesis is rejected if t is either (1) $\leq -t_{\alpha/2,n-1}$ or (2) $\geq t_{\alpha/2,n-1}$ (which makes the Type I error probability equal to α).

Figure 7.4.5

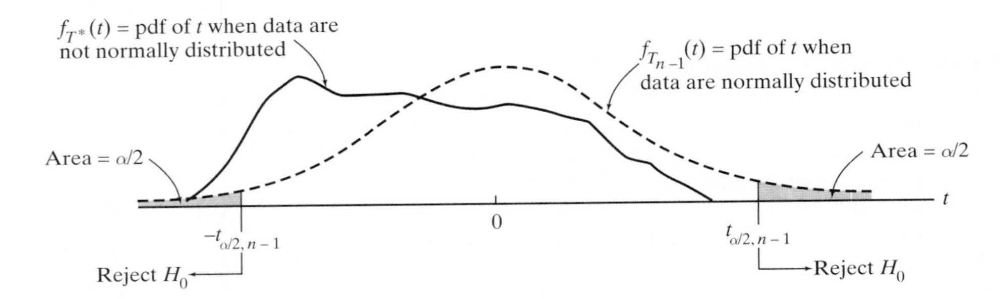

If the normality assumption is *not* true, the pdf of $\frac{\bar{Y}-\mu_o}{S/\sqrt{n}}$ will not be $f_{T_{n-1}}(t)$ and

$$P\left(\frac{\bar{Y}-\mu_o}{S/\sqrt{n}} \leq -t_{\alpha/2,n-1}\right) + P\left(\frac{\bar{Y}-\mu_o}{S/\sqrt{n}} \geq t_{\alpha/2,n-1}\right) \neq \alpha$$

In effect, violating the normality assumption creates *two* α's: The "nominal" α is the Type I error probability we specify at the outset—typically, 0.05 or 0.01. The "true" α is the actual probability that $\frac{\bar{Y}-\mu_o}{S/\sqrt{n}}$ falls in the rejection region (when H_0 is true). For the two-sided decision rule pictured in Figure 7.4.5,

$$\text{true}\,\alpha = \int_{-\infty}^{-t_{\alpha/2,n-1}} f_{T^*}(t)\,dt + \int_{t_{\alpha/2,n-1}}^{\infty} f_{T^*}(t)\,dt$$

Whether or not the validity of the t test is "compromised" by the normality assumption being violated depends on the numerical difference between the two α's. If $f_{T^*}(t)$ is, in fact, quite similar in shape and location to $f_{T_{n-1}}(t)$, then the true α will be approximately equal to the nominal α. In that case, the fact that the y_i's are not normally distributed would be essentially irrelevant. On the other hand, if $f_{T^*}(t)$ and $f_{T_{n-1}}(t)$ are dramatically different (as they appear to be in Figure 7.4.5), it would follow that the normality assumption *is* critical, and establishing the "significance" of a t ratio becomes problematic.

Unfortunately, getting an exact expression for $f_{T^*}(t)$ is essentially impossible, because the distribution depends on the pdf being sampled, and there is seldom any way of knowing precisely what that pdf might be. However, we can still meaningfully explore the sensitivity of the t ratio to violations of the normality assumption by simulating samples of size n from selected distributions and comparing the resulting histogram of t ratios to $f_{T_{n-1}}(t)$.

Figure 7.4.6 shows four such simulations, using Minitab; the first three consist of one hundred random samples of size $n = 6$. In Figure 7.4.6(a), the samples come from a uniform pdf defined over the interval [0, 1]; in Figure 7.4.6(b), the underlying pdf is the exponential with $\lambda = 1$; and in Figure 7.4.6(c), the data are coming from a Poisson pdf with $\lambda = 5$.

If the normality assumption were true, t ratios based on samples of size 6 would vary in accordance with the Student t distribution with 5 df. On pp. 407–408, $f_{T_5}(t)$ has been superimposed over the histograms of the t ratios coming from the three different pdfs. What we see there is really quite remarkable. The t ratios based on y_i's coming from a uniform pdf, for example, are behaving much the same way as t ratios would vary if the y_i's were normally distributed—that is, $f_{T^*}(t)$ in this case appears to be very similar to $f_{T_5}(t)$. The same is true for samples coming from a Poisson distribution (see Theorem 4.2.2). For both of those underlying pdfs, in other words, the true α would not be much different from the nominal α.

Figure 7.4.6(b) tells a slightly different story. When samples of size 6 are drawn from an exponential pdf, the t ratios are *not* in particularly close agreement with

Figure 7.4.6

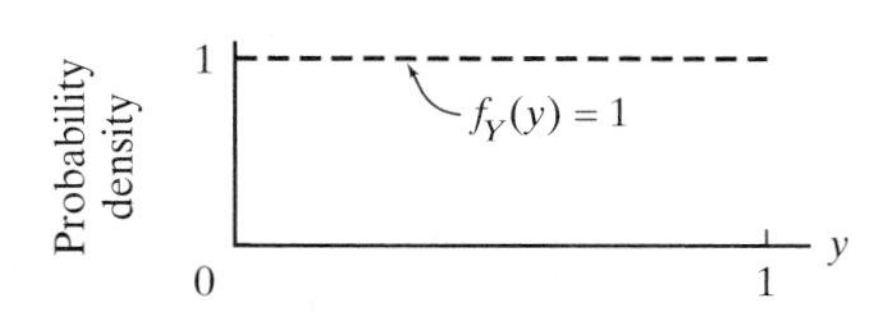


```
MTB  > random 100 c1-c6;
SUBC> uniform 0 1.
MTB  > rmean c1-c6 c7
MTB  > rstdev c1-c6 c8
MTB  > let c9 = sqrt(6)*(((c7)-0.5)/(c8))
MTB  > histogram c9
```

This command calculates $\frac{\bar{y} - \mu}{s/\sqrt{n}} = \frac{\bar{y} - 0.5}{s/\sqrt{6}}$

Sample distribution

$f_{T_5}(t)$

t ratio ($n = 6$)

Figure 7.4.6 (*Continued*) **(b)**

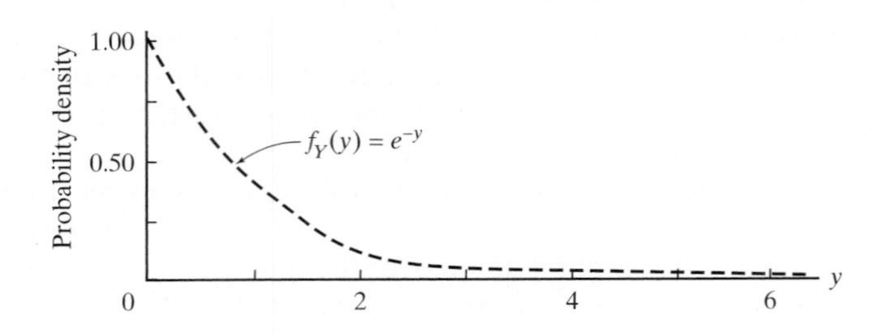


```
MTB > random¨100¨cl-c6;
SUBC> exponential¨1.
MTB > rmean¨cl-c6¨c7
MTB > rstdev¨cl-c6¨c8
MTB > let¨c9¨ = ¨sqrt(6)*(((c7)¨-¨1.0)/(c8))
MTB¨>¨histogram¨c9
```

$$\left[= \frac{\bar{y} - \mu}{s/\sqrt{6}}\right]$$

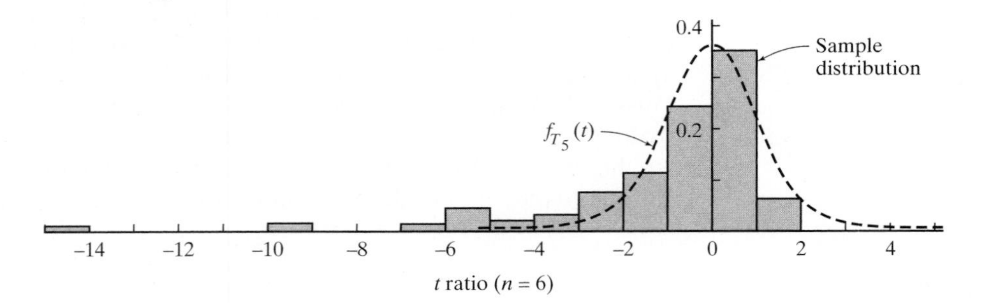

(c)

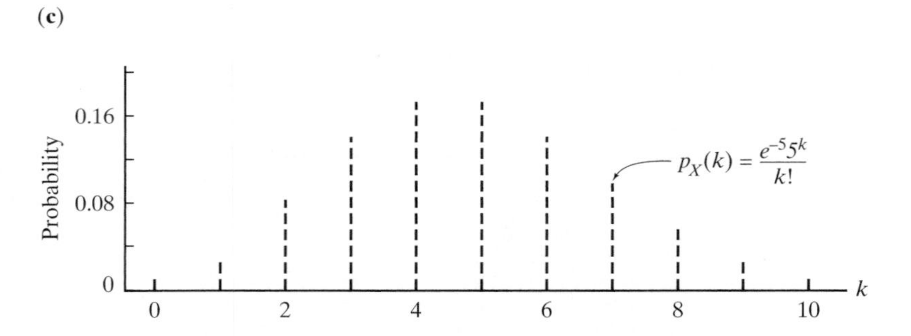


```
MTB¨>¨random¨100¨cl-c6;
SUBC¨>¨poisson¨5.
MTB¨>¨rmean¨cl-c6¨c7
MTB¨>¨rstdev¨cl-c6¨c8
MTB¨>¨let¨c9¨=¨sqrt(6)*(((c7)¨-¨5.0)/(c8))
MTB¨>¨histogram¨c9
```

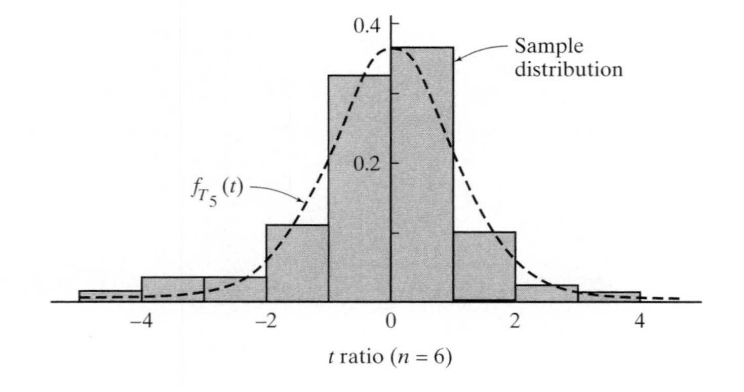

$f_{T_5}(t)$. Specifically, very negative t ratios are occurring much more often than the Student t curve would predict, while large positive t ratios are occurring less often (see Question 7.4.23). But look at Figure 7.4.6(d). When the sample size is increased to $n = 15$, the skewness so prominent in Figure 7.4.6(b) is mostly gone.

Figure 7.4.6 *(Continued)* **(d)**

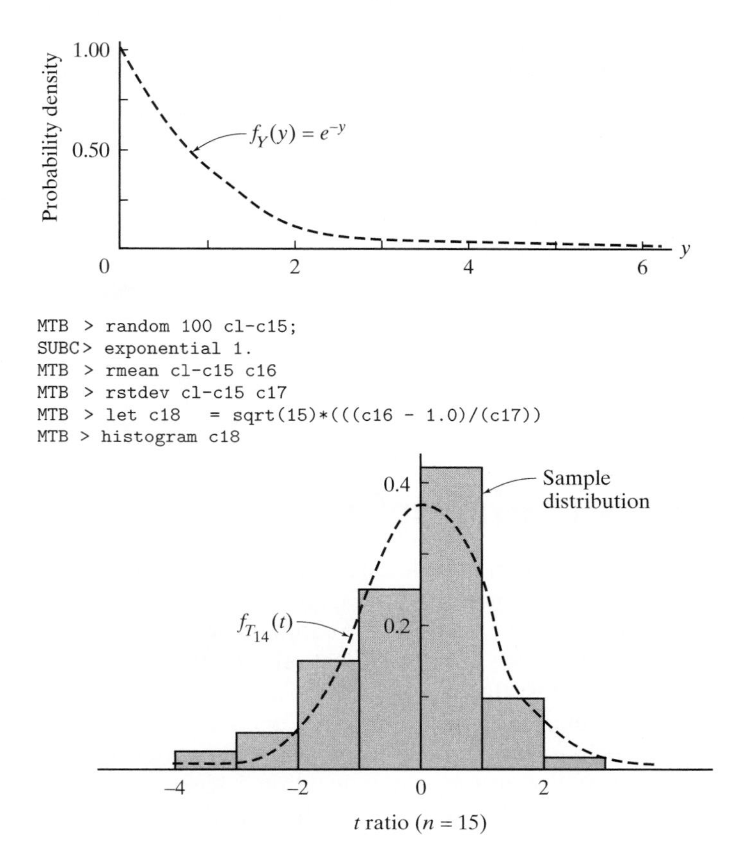

Reflected in these specific simulations are some general properties of the t ratio:

1. The distribution of $\frac{\overline{Y}-\mu}{S/\sqrt{n}}$ is relatively unaffected by the pdf of the y_i's [provided $f_Y(y)$ is not too skewed and n is not too small].
2. As n increases, the pdf of $\frac{\overline{Y}-\mu}{S/\sqrt{n}}$ becomes increasingly similar to $f_{T_{n-1}}(t)$.

In mathematical statistics, the term *robust* is used to describe a procedure that is not heavily dependent on whatever assumptions it makes. Figure 7.4.6 shows that the t *test is robust with respect to departures from normality*.

From a practical standpoint, it would be difficult to overstate the importance of the t test being robust. If the pdf of $\frac{\overline{Y}-\mu}{S/\sqrt{n}}$ varied dramatically depending on the origin of the y_i's, we would never know if the true α associated with, say, a 0.05 decision rule was anywhere near 0.05. That degree of uncertainty would make the t test virtually worthless.

Questions

7.4.23. Explain why the distribution of t ratios calculated from small samples drawn from the exponential pdf, $f_Y(y) = e^{-y}$, $y \geq 0$, will be skewed to the left [recall Figure 7.4.6(b)]. [*Hint:* What does the shape of $f_Y(y)$ imply about the possibility of each y_i being close to 0? If the entire sample did consist of y_i's close to 0, what value would the t ratio have?]

7.4.24. Suppose one hundred samples of size $n = 3$ are taken from each of the pdfs

$$(1)\ f_Y(y) = 2y, \qquad 0 \leq y \leq 1$$

and

$$(2)\ f_Y(y) = 4y^3, \qquad 0 \leq y \leq 1$$

and for each set of three observations, the ratio

$$\frac{\overline{y} - \mu}{s/\sqrt{3}}$$

is calculated, where μ is the expected value of the particular pdf being sampled. How would you expect the distributions of the two sets of ratios to be different? How would they be similar? Be as specific as possible.

7.4.25. Suppose that random samples of size n are drawn from the uniform pdf, $f_Y(y) = 1, 0 \leq y \leq 1$. For each sample, the ratio $t = \frac{\overline{y}-0.5}{s/\sqrt{n}}$ is calculated. Parts (b) and (d) of Figure 7.4.6 suggest that the pdf of t will become increasingly similar to $f_{T_{n-1}}(t)$ as n increases. To which pdf is $f_{T_{n-1}}(t)$, itself, converging as n increases?

7.4.26. On which of the following sets of data would you be reluctant to do a t test? Explain.

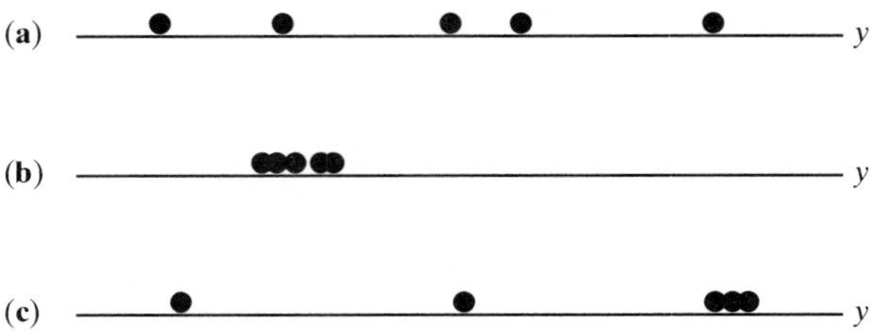

7.5 Drawing Inferences About σ^2

When random samples are drawn from a normal distribution, it is usually the case that the parameter μ is the target of the investigation. More often than not, the mean mirrors the "effect" of a treatment or condition, in which case it makes sense to apply what we learned in Section 7.4—that is, either construct a confidence interval for μ or test the hypothesis that $\mu = \mu_o$.

But exceptions are not that uncommon. Situations occur where the "precision" associated with a measurement is, itself, important—perhaps even more important than the measurement's "location." If so, we need to shift our focus to the *scale parameter,* σ^2. Two key facts that we learned earlier about the population variance will now come into play. First, an unbiased estimator for σ^2 based on its maximum likelihood estimator is the sample variance, S^2, where

$$S^2 = \frac{1}{n-1}\sum_{i=1}^{n}(Y_i - \overline{Y})^2$$

And, second, the ratio

$$\frac{(n-1)S^2}{\sigma^2} = \frac{1}{\sigma^2}\sum_{i=1}^{n}(Y_i - \overline{Y})^2$$

has a chi square distribution with $n - 1$ degrees of freedom. Putting these two pieces of information together allows us to draw inferences about σ^2—in particular, we can construct confidence intervals for σ^2 and test the hypothesis that $\sigma^2 = \sigma_o^2$.

Chi Square Tables

Just as we need a *t table* to carry out inferences about μ (when σ^2 is unknown), we need a *chi square table* to provide the cutoffs for making inferences involving σ^2. The layout of chi square tables is dictated by the fact that all chi square pdfs (unlike Z and t distributions) are skewed (see, for example, Figure 7.5.1, showing a chi square curve having 5 degrees of freedom). Because of that asymmetry, chi square tables need to provide cutoffs for both the left-hand tail and the right-hand tail of each chi square distribution.

Figure 7.5.1

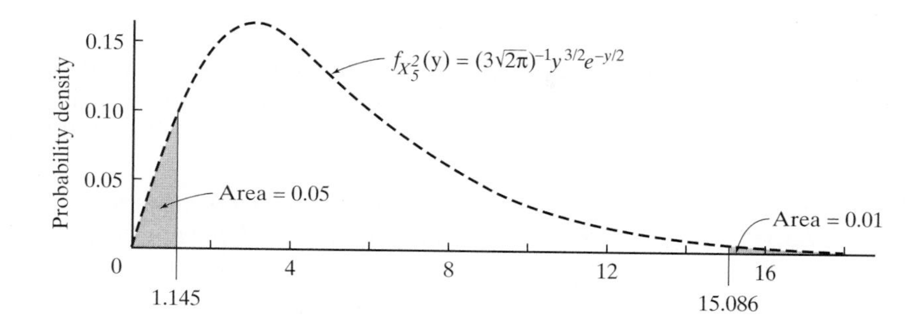

Figure 7.5.2 shows the top portion of the chi square table that appears in Appendix A.3. Successive rows refer to different chi square distributions (each having a different number of degrees of freedom). The column headings denote the areas *to the left* of the numbers listed in the body of the table.

Figure 7.5.2

	p							
df	.01	.025	.05	.10	.90	.95	.975	.99
1	0.000157	0.000982	0.00393	0.0158	2.706	3.841	5.024	6.635
2	0.0201	0.0506	0.103	0.211	4.605	5.991	7.378	9.210
3	0.115	0.216	0.352	0.584	6.251	7.815	9.348	11.345
4	0.297	0.484	0.711	1.064	7.779	9.488	11.143	13.277
5	0.554	0.831	1.145	1.610	9.236	11.070	12.832	15.086
6	0.872	1.237	1.635	2.204	10.645	12.592	14.449	16.812
7	1.239	1.690	2.167	2.833	12.017	14.067	16.013	18.475
8	1.646	2.180	2.733	3.490	13.362	15.507	17.535	20.090
9	2.088	2.700	3.325	4.168	14.684	16.919	19.023	21.666
10	2.558	3.247	3.940	4.865	15.987	18.307	20.483	23.209
11	3.053	3.816	4.575	5.578	17.275	19.675	21.920	24.725
12	3.571	4.404	5.226	6.304	18.549	21.026	23.336	26.217

We will use the symbol $\chi^2_{p,n}$ to denote the number along the horizontal axis that cuts off, to its left, an area of p under the chi square distribution with n degrees of freedom. For example, from the fifth row of the chi square table, we see the numbers *1.145* and *15.086* under the column headings *.05* and *.99*, respectively. It follows that

$$P(\chi_5^2 \le 1.145) = 0.05$$

and

$$P(\chi_5^2 \le 15.086) = 0.99$$

(see Figure 7.5.1). In terms of the $\chi^2_{p,n}$ notation, $1.145 = \chi^2_{.05,5}$ and $15.086 = \chi^2_{.99,5}$. (The area *to the right* of 15.086, of course, must be 0.01.)

Constructing Confidence Intervals for σ^2

Since $\frac{(n-1)S^2}{\sigma^2}$ has a chi square distribution with $n-1$ degrees of freedom, we can write

$$P\left[\chi^2_{\alpha/2,n-1} \le \frac{(n-1)S^2}{\sigma^2} \le \chi^2_{1-\alpha/2,n-1}\right] = 1-\alpha \tag{7.5.1}$$

If Equation 7.5.1 is then inverted to isolate σ^2 in the center of the inequalities, the two endpoints will necessarily define a $100(1-\alpha)\%$ confidence interval for the population variance. The algebraic details will be left as an exercise.

Theorem 7.5.1 *Let s^2 denote the sample variance calculated from a random sample of n observations drawn from a normal distribution with mean μ and variance σ^2. Then*

a. *a $100(1-\alpha)\%$ confidence interval for σ^2 is the set of values*

$$\left[\frac{(n-1)s^2}{\chi^2_{1-\alpha/2,n-1}}, \frac{(n-1)s^2}{\chi^2_{\alpha/2,n-1}}\right]$$

b. *a $100(1-\alpha)\%$ confidence interval for σ is the set of values*

$$\left[\sqrt{\frac{(n-1)s^2}{\chi^2_{1-\alpha/2,n-1}}}, \sqrt{\frac{(n-1)s^2}{\chi^2_{\alpha/2,n-1}}}\right]$$

□

Case Study 7.5.1

The chain of events that define the geological evolution of the Earth began hundreds of millions of years ago. Fossils play a key role in documenting the *relative* times those events occurred, but to establish an *absolute* chronology, scientists rely primarily on radioactive decay.

One of the newest dating techniques uses a rock's potassium-argon ratio. Almost all minerals contain potassium (K) as well as certain of its isotopes, including ^{40}K. The latter, though, is unstable and decays into isotopes of argon and calcium, ^{40}Ar and ^{40}Ca. By knowing the rates at which the various daughter products are formed and by measuring the amounts of ^{40}Ar and ^{40}K present in a specimen, geologists can estimate the object's age.

Critical to the interpretation of any such dates, of course, is the precision of the underlying procedure. One obvious way to estimate that precision is to use the technique on a sample of rocks known to have the same age. Whatever variation occurs, then, from rock to rock is reflecting the inherent precision (or lack of precision) of the procedure.

Table 7.5.1 lists the potassium-argon estimated ages of nineteen mineral samples, all taken from the Black Forest in southeastern Germany (111). Assume that the procedure's estimated ages are normally distributed with (unknown) mean μ and (unknown) variance σ^2. Construct a 95% confidence interval for σ.

(Continued on next page)

Table 7.5.1

Specimen	Estimated Age (millions of years)
1	249
2	254
3	243
4	268
5	253
6	269
7	287
8	241
9	273
10	306
11	303
12	280
13	260
14	256
15	278
16	344
17	304
18	283
19	310

Here

$$\sum_{i=1}^{19} y_i = 5261$$

$$\sum_{i=1}^{19} y_i^2 = 1{,}469{,}945$$

so the sample variance is *733.4*:

$$s^2 = \frac{19(1{,}469{,}945) - (5261)^2}{19(18)} = 733.4$$

Since $n = 19$, the critical values appearing in the left-hand and right-hand limits of the σ confidence interval come from the chi square pdf with *18* df. According to Appendix Table A.3,

$$P\left(8.23 < \chi^2_{18} < 31.53\right) = 0.95$$

so the 95% confidence interval for the potassium-argon method's precision is the set of values

$$\left[\sqrt{\frac{(19-1)(733.4)}{31.53}}, \sqrt{\frac{(19-1)(733.4)}{8.23}}\right] = (20.5 \text{ million years}, 40.0 \text{ million years})$$

Example 7.5.1 The width of a confidence interval for σ^2 is a function of both n and S^2:

$$\begin{aligned}\text{Width} &= \text{upper limit} - \text{lower limit}\\ &= \frac{(n-1)S^2}{\chi^2_{\alpha/2,n-1}} - \frac{(n-1)S^2}{\chi^2_{1-\alpha/2,n-1}}\\ &= (n-1)S^2\left(\frac{1}{\chi^2_{\alpha/2,n-1}} - \frac{1}{\chi^2_{1-\alpha/2,n-1}}\right)\end{aligned} \tag{7.5.2}$$

As n gets larger, the interval will tend to get narrower because the unknown σ^2 is being estimated more precisely. What is the smallest number of observations that will guarantee that the average width of a 95% confidence interval for σ^2 is no greater than σ^2?

Since S^2 is an unbiased estimator for σ^2, Equation 7.5.2 implies that the expected width of a 95% confidence interval for the variance is the expression

$$E(\text{width}) = (n-1)\sigma^2\left(\frac{1}{\chi^2_{.025,n-1}} - \frac{1}{\chi^2_{.975,n-1}}\right)$$

Clearly, then, for the expected width to be less than or equal to σ^2, n must be chosen so that

$$(n-1)\left(\frac{1}{\chi^2_{.025,n-1}} - \frac{1}{\chi^2_{.975,n-1}}\right) \leq 1$$

Trial and error can be used to identify the desired n. The first three columns in Table 7.5.2 come from the chi square distribution in Appendix Table A.3. As the computation in the last column indicates, $n = 39$ is the smallest sample size that will yield 95% confidence intervals for σ^2 whose average width is less than σ^2.

Table 7.5.2

n	$\chi^2_{.025,n-1}$	$\chi^2_{.975,n-1}$	$(n-1)\left(\frac{1}{\chi^2_{.025,n-1}} - \frac{1}{\chi^2_{.975,n-1}}\right)$
15	5.629	26.119	1.95
20	8.907	32.852	1.55
30	16.047	45.722	1.17
38	22.106	55.668	1.01
39	22.878	56.895	0.99

■

Testing H_0: $\sigma^2 = \sigma_o^2$

The generalized likelihood ratio criterion introduced in Section 6.5 can be used to set up hypothesis tests for σ^2. The complete derivation appears in Appendix 7.A.4. Theorem 7.5.2 states the resulting decision rule. Playing a key role—just as it did in the construction of confidence intervals for σ^2—is the chi square ratio from Theorem 7.3.2.

Theorem 7.5.2 *Let s^2 denote the sample variance calculated from a random sample of n observations drawn from a normal distribution with mean μ and variance σ^2. Let $\chi^2 = (n-1)s^2/\sigma_o^2$.*

a. To test $H_0: \sigma^2 = \sigma_o^2$ versus $H_1: \sigma^2 > \sigma_o^2$ at the α level of significance, reject H_0 if $\chi^2 \geq \chi^2_{1-\alpha,n-1}$.

b. To test $H_0: \sigma^2 = \sigma_o^2$ versus $H_1: \sigma^2 < \sigma_o^2$ at the α level of significance, reject H_0 if $\chi^2 \leq \chi^2_{\alpha,n-1}$.

c. To test $H_0: \sigma^2 = \sigma_o^2$ versus $H_1: \sigma^2 \neq \sigma_o^2$ at the α level of significance, reject H_0 if χ^2 is either (1) $\leq \chi^2_{\alpha/2,n-1}$ or (2) $\geq \chi^2_{1-\alpha/2,n-1}$. □

Case Study 7.5.2

Mutual funds are investment vehicles consisting of a portfolio of various types of investments. If such an investment is to meet annual spending needs, the owner of shares in the fund is interested in the average of the annual returns of the fund. Investors are also concerned with the volatility of the annual returns, measured by the variance or standard deviation. One common method of evaluating a mutual fund is to compare it to a benchmark, the Lipper Average being one of these. This index number is the average of returns from a universe of mutual funds.

The Global Rock Fund is a typical mutual fund, with heavy investments in international funds. It claimed to best the Lipper Average in terms of volatility over the period from 1989 through 2007. Its returns are given in the table below.

Year	Investment Return %	Year	Investment Return %
1989	15.32	1999	27.43
1990	1.62	2000	8.57
1991	28.43	2001	1.88
1992	11.91	2002	−7.96
1993	20.71	2003	35.98
1994	−2.15	2004	14.27
1995	23.29	2005	10.33
1996	15.96	2006	15.94
1997	11.12	2007	16.71
1998	0.37		

The standard deviation for these returns is 11.28%, while the corresponding figure for the Lipper Average is 11.67%. Now, clearly, the Global Rock Fund has a smaller standard deviation than the Lipper Average, but is this small difference due just to random variation? The hypothesis test is meant to answer such questions.

Let σ^2 denote the variance of the population represented by the return percentages shown in the table above. To judge whether the observed standard deviation less than 11.67 is significant requires that we test

(Continued on next page)

(Case Study 7.5.2 continued)

$$H_0 : \sigma^2 = (11.67)^2$$

versus

$$H_1 : \sigma^2 < (11.67)^2$$

Let $\alpha = 0.05$. With $n = 19$, the critical value for the chi square ratio [from part (b) of Theorem 7.5.2] is $\chi^2_{1-\alpha,n-1} = \chi^2_{.05,18} = 9.390$ (see Figure 7.5.3). But

$$\chi^2 = \frac{(n-1)s^2}{\sigma_0^2} = \frac{(19-1)(11.28)^2}{(11.67)^2} = 16.82$$

so our decision is clear: Do not reject H_0.

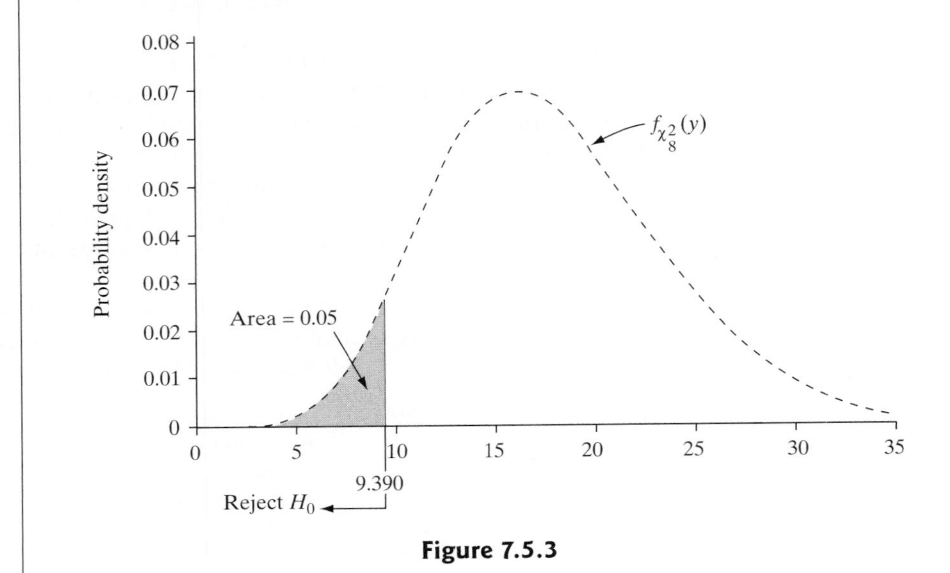

Figure 7.5.3

Questions

7.5.1. Use Appendix Table A.3 to find the following cutoffs and indicate their location on the graph of the appropriate chi square distribution.

(a) $\chi^2_{.95,14}$
(b) $\chi^2_{.90,2}$
(c) $\chi^2_{.025,9}$

7.5.2. Evaluate the following probabilities:

(a) $P(\chi^2_{17} \geq 8.672)$
(b) $P(\chi^2_6 < 10.645)$
(c) $P(9.591 \leq \chi^2_{20} \leq 34.170)$
(d) $P(\chi^2_2 < 9.210)$

7.5.3. Find the value y that satisfies each of the following equations:

(a) $P(\chi^2_9 \geq y) = 0.99$
(b) $P(\chi^2_{15} \leq y) = 0.05$
(c) $P(9.542 \leq \chi^2_{22} \leq y) = 0.09$
(d) $P(y \leq \chi^2_{31} \leq 48.232) = 0.95$

7.5.4. For what value of n is each of the following statements true?

(a) $P(\chi^2_n \geq 5.009) = 0.975$
(b) $P(27.204 \leq \chi^2_n \leq 30.144) = 0.05$
(c) $P(\chi^2_n \leq 19.281) = 0.05$
(d) $P(10.085 \leq \chi^2_n \leq 24.769) = 0.80$

7.5.5. For df values beyond the range of Appendix Table A.3, chi square cutoffs can be approximated by using a formula based on cutoffs from the standard normal pdf, $f_Z(z)$. Define $\chi^2_{p,n}$ and z^*_p so that $P(\chi^2_n \le \chi^2_{p,n}) = p$ and $P(Z \le z^*_p) = p$, respectively. Then

$$\chi^2_{p,n} \doteq n\left(1 - \frac{2}{9n} + z^*_p\sqrt{\frac{2}{9n}}\right)^3$$

Approximate the 95th percentile of the chi square distribution with 200 df. That is, find the value of y for which

$$P(\chi^2_{200} \le y) \doteq 0.95$$

7.5.6. Let $Y_1, Y_2, \ldots, Y_n$ be a random sample of size n from a normal distribution having mean μ and variance σ^2. What is the smallest value of n for which the following is true?

$$P\left(\frac{S^2}{\sigma^2} < 2\right) \ge 0.95$$

(*Hint:* Use a trial-and-error method.)

7.5.7. Start with the fact that $(n-1)S^2/\sigma^2$ has a chi square distribution with $n-1$ df (if the Y_i's are normally distributed) and derive the confidence interval formulas given in Theorem 7.5.1.

7.5.8. A random sample of size $n = 19$ is drawn from a normal distribution for which $\sigma^2 = 12.0$. In what range are we likely to find the sample variance, s^2? Answer the question by finding two numbers a and b such that

$$P(a \le S^2 \le b) = 0.95$$

7.5.9. How long sporting events last is quite variable. This variability can cause problems for TV broadcasters, since the amount of commercials and commentator blather varies with the length of the event. As an example of this variability, the table below gives the lengths for a random sample of middle-round contests at the 2008 Wimbledon Championships in women's tennis.

Match	Length (minutes)
Cirstea-Kuznetsova	73
Srebotnik-Meusburger	76
De Los Rios-V. Williams	59
Kanepi-Mauresmo	104
Garbin-Szavay	114
Bondarenko-Lisicki	106
Vaidisova-Bremond	79
Groenefeld-Moore	74
Govortsova-Sugiyama	142
Zheng-Jankovic	129
Perebiynis-Bammer	95
Bondarenko-V. Williams	56
Coin-Mauresmo	84
Petrova-Pennetta	142
Wozniacki-Jankovic	106
Groenefeld-Safina	75

Source: 2008.usopen.org/en_US/scores/cmatch/index.html?promo=t.

(a) Assume that match lengths are normally distributed. Use Theorem 7.5.1 to construct a 95% confidence interval for the standard deviation of match lengths.

(b) Use these same data to construct two *one-sided* 95% confidence intervals for σ.

7.5.10. How much interest certificates of deposit (CDs) pay varies by financial institution and also by length of the investment. A large sample of national one-year CD offerings in 2009 showed an average interest rate of 1.84 and a standard deviation $\sigma = 0.262$. A five-year CD ties up an investor's money, so it usually pays a higher rate of interest. However, higher rates might cause more variability. The table lists the five-year CD rate offerings from $n = 10$ banks in the northeast United States. Find a 95% confidence interval for the standard deviation of 5-year CD rates. Do these data suggest that interest rates for five-year CDs are more variable than those for one-year certificates?

Bank	Interest Rate (%)
Domestic Bank	2.21
Stonebridge Bank	2.47
Waterfield Bank	2.81
NOVA Bank	2.81
American Bank	2.96
Metropolitan National Bank	3.00
AIG Bank	3.35
iGObanking.com	3.44
Discover Bank	3.44
Intervest National Bank	3.49

Source: Company reports.

7.5.11. In Case Study 7.5.1, the 95% confidence interval was constructed for σ rather than for σ^2. In practice, is an experimenter more likely to focus on the standard deviation or on the variance, or do you think that both formulas in Theorem 7.5.1 are likely to be used equally often? Explain.

7.5.12. **(a)** Use the asymptotic normality of chi square random variables (see Question 7.3.6) to derive large-sample confidence interval formulas for σ and σ^2.

(b) Use your answer to part (a) to construct an approximate 95% confidence interval for the standard deviation of estimated potassium-argon ages based on the 19 y_i's in Table 7.5.1. How does this confidence interval compare with the one in Case Study 7.5.1?

7.5.13. If a 90% confidence interval for σ^2 is reported to be $(51.47, 261.90)$, what is the value of the sample standard deviation?

7.5.14. Let $Y_1, Y_2, \ldots, Y_n$ be a random sample of size n from the pdf

$$f_Y(y) = \left(\frac{1}{\theta}\right) e^{-y/\theta}, \quad y > 0; \quad \theta > 0$$

(a) Use moment-generating functions to show that the ratio $2n\overline{Y}/\theta$ has a chi square distribution with $2n$ df.

(b) Use the result in part (a) to derive a $100(1-\alpha)\%$ confidence interval for θ.

7.5.15. Another method for dating rocks was used before the advent of the potassium-argon method described in Case Study 7.5.1. Because of a mineral's lead content, it was capable of yielding estimates for this same time period with a standard deviation of 30.4 million years. The potassium-argon method in Case Study 7.5.1 had a smaller sample standard deviation of $\sqrt{733.4} = 27.1$ million years. Is this "proof" that the potassium-argon method is more precise? Using the data in Table 7.5.1, test at the 0.05 level whether the potassium-argon method has a smaller standard deviation than the older procedure using lead.

7.5.16. When working properly, the amounts of cement that a filling machine puts into 25-kg bags have a standard deviation (σ) of 1.0 kg. In the next column are the weights recorded for thirty bags selected at random from a day's production. Test H_0: $\sigma^2 = 1$ versus H_1: $\sigma^2 > 1$ using the $\alpha = 0.05$ level of significance. Assume that the weights are normally distributed.

26.18	24.22	24.22
25.30	26.48	24.49
25.18	23.97	25.68
24.54	25.83	26.01
25.14	25.05	25.50
25.44	26.24	25.84
24.49	25.46	26.09
25.01	25.01	25.21
25.12	24.71	26.04
25.67	25.27	25.23

Use the following sums:

$$\sum_{i=1}^{30} y_i = 758.62 \text{ and } \sum_{i=1}^{30} y_i^2 = 19{,}195.7938$$

7.5.17. A stock analyst claims to have devised a mathematical technique for selecting high-quality mutual funds and promises that a client's portfolio will have higher average ten-year annualized returns and lower volatility; that is, a smaller standard deviation. After ten years, one of the analyst's twenty-four-stock portfolios showed an average ten-year annualized return of 11.50% and a standard deviation of 10.17%. The benchmarks for the type of funds considered are a mean of 10.10% and a standard deviation of 15.67%.

(a) Let μ be the mean for a twenty-four-stock portfolio selected by the analyst's method. Test at the 0.05 level that the portfolio beat the benchmark; that is, test H_0: $\mu = 10.1$ versus H_1: $\mu > 10.1$.

(b) Let σ be the standard deviation for a twenty-four-stock portfolio selected by the analyst's method. Test at the 0.05 level that the portfolio beat the benchmark; that is, test H_0: $\sigma = 15.67$ versus H_1: $\sigma < 15.67$.

7.6 Taking a Second Look at Statistics (Type II Error)

For data that are normal, *and when the variance σ^2 is known*, both Type I errors and Type II errors can be determined, staying within the family of normal distributions. (See Example 6.4.1, for instance.) As the material in this chapter shows, the situation changes radically when σ^2 is not known. With the development of the Student t distribution, tests of a given level of significance α can be constructed. But what is the Type II error of such a test?

To answer this question, let us first recall the form of the test statistic and critical region testing, for example,

$$H_0: \mu = \mu_0 \text{ versus } H_1: \mu > \mu_0$$

The null hypothesis is rejected if

$$\frac{\overline{Y} - \mu_0}{S/\sqrt{n}} \geq t_{\alpha,n-1}$$

The probability of the Type II error, β, of the test at some value $\mu_1 > \mu_0$ is

$$P\left(\frac{\overline{Y} - \mu_0}{S/\sqrt{n}} < t_{\alpha,n-1}\right)$$

However, since μ_0 is not the mean of $\overline{Y}$ under H_1, the distribution of $\frac{\overline{Y}-\mu_0}{S/\sqrt{n}}$ is *not* Student t. Indeed, a new distribution is called for.

The following algebraic manipulations help to place the needed density into a recognizable form.

$$\frac{\overline{Y} - \mu_0}{S/\sqrt{n}} = \frac{\overline{Y} - \mu_1 + (\mu_1 - \mu_0)}{S/\sqrt{n}} = \frac{\frac{\overline{Y}-\mu_1}{\sigma} + \frac{(\mu_1-\mu_0)}{\sigma}}{\frac{S/\sqrt{n}}{\sigma}} = \frac{\frac{\overline{Y}-\mu_1}{\sigma/\sqrt{n}} + \frac{(\mu_1-\mu_0)}{\sigma/\sqrt{n}}}{S/\sigma}$$

$$= \frac{\frac{\overline{Y}-\mu_1}{\sigma/\sqrt{n}} + \frac{(\mu_1-\mu_0)}{\sigma/\sqrt{n}}}{\sqrt{\frac{(n-1)S^2/\sigma^2}{n-1}}} = \frac{\frac{\overline{Y}-\mu_1}{\sigma/\sqrt{n}} + \delta}{\sqrt{\frac{(n-1)S^2/\sigma^2}{n-1}}} = \frac{Z + \delta}{\sqrt{\frac{U}{n-1}}}$$

where $Z = \frac{\overline{Y}-\mu_1}{\sigma/\sqrt{n}}$ is normal, $U = \frac{(n-1)S^2}{\sigma^2}$ is a chi square variable with $n - 1$ degrees of freedom, and $\delta = \frac{(\mu_1-\mu_0)}{\sigma/\sqrt{n}}$ is an (unknown) constant. Note that the random variable $\frac{Z+\delta}{\sqrt{\frac{U}{n-1}}}$ differs from the Student t with $n - 1$ degrees of freedom $\frac{Z}{\sqrt{\frac{U}{n-1}}}$ only because of the additive term δ in the numerator. But adding δ changes the nature of the pdf significantly.

An expression of the form $\frac{Z+\delta}{\sqrt{\frac{U}{n-1}}}$ is said to have a *noncentral t distribution with $n - 1$ degrees of freedom and noncentrality parameter δ*.

The probability density function for a noncentral t variable is now well known (97). Even though there are computer approximations to the distribution, not knowing σ^2 means that δ is also unknown. One approach often taken is to specify the difference between the true mean and the hypothesized mean *as a given proportion of σ*. That is, the Type II error is given as a function of $\frac{\mu_1-\mu_0}{\sigma}$ rather than μ_1. In some cases, this quantity can be approximated by $\frac{\mu_1-\mu_0}{s}$.

The following numerical example will help to clarify these ideas.

Example 7.6.1

Suppose we wish to test $H_0 : \mu = \mu_0$ versus $H_1 : \mu > \mu_0$ at the $\alpha = 0.05$ level of significance. Let $n = 20$. In this case the test is to reject H_0 if the test statistic $\frac{\overline{y}-\mu_0}{s/\sqrt{n}}$ is greater than $t_{.05,19} = 1.7291$. What will be the Type II error if the mean has shifted by 0.5 standard deviation to the right of μ_0?

Saying that the mean has shifted by 0.5 standard deviation to the right of μ_0 is equivalent to setting $\frac{\mu_1-\mu_0}{\sigma} = 0.5$. In that case, the noncentrality parameter is $\delta = \frac{\mu_1-\mu_0}{\sigma/\sqrt{n}} = (0.5) \cdot \sqrt{20} = 2.236$.

The probability of a Type II error is

$$P(T_{19,2.236} \leq 1.7291)$$

where $T_{19,2.236}$ is a noncentral t variable with 19 degrees of freedom and noncentrality parameter 2.236.

To calculate this quantity, we need the cdf of $T_{19,2.236}$. Fortunately, many statistical software programs have this function. The Minitab commands for calculating the desired probability are

```
MTB > CDF 1.7291;
SUBC > T 19 2.236
```

with output

Cumulative Distribution Function

Student's t distribution with 19 DF and noncentrality parameter 2.236

```
     x   P(X <= x)
1.7291   0.304828
```

The sought-after Type II error to three decimal places is 0.305.

Simulations

As we have seen, with enough distribution theory, the tools for finding Type II errors for the Student t test exist. Also, there are noncentral chi square and F distributions.

However, the assumption that the underlying data are normally distributed is necessary for such results. In the case of Type I errors, we have seen that the t test is somewhat robust with regard to the data deviating from normality. (See Section 7.4.) In the case of the noncentral t, dealing with departures from normality presents significant analytical challenges. But the empirical approach of using simulations can bypass such difficulties and still give meaningful results.

To start, consider a simulation of the problem presented in Example 7.6.1. Suppose the data have a normal distribution with $\mu_0 = 5$ and $\sigma = 3$. The sample size is $n = 20$. Suppose we want to find the Type I error when the true $\delta = 2.236$. For the given $\sigma = 3$, this is equivalent to

$$2.236 = \frac{\mu_1 - \mu_0}{\sigma/\sqrt{n}} = \frac{\mu_1 - 5}{3/\sqrt{20}}$$

or $\mu_1 = 6.5$.

A Type II error occurs if the test statistic is less than 1.7291. In this case, H_0 would be accepted when rejection is the proper decision.

Using Minitab, two hundred samples of size 20 from the normal distribution with $\mu = 6.5$ and $\sigma^2 = 9$ are generated: Minitab produces a 200×20 array. For each row of the array, the test statistic $\frac{\bar{y}-5}{s/\sqrt{20}}$ is calculated and placed in Column 21. If this value is less than 1.7291, a 1 is placed in that row of Column 22; otherwise a 0 goes there. The sum of the entries in Column 22 gives the observed number of Type II errors. Based on the computed value of the Type II error, 0.305, for the assumed value of δ, this observed number should be approximately $200(0.305) = 61$.

The Minitab simulation gave sixty-four observed Type II errors—a very close figure to what was expected.

The robustness for Type II errors can lead to analytical thickets. However, simulation can again shed some light on Type II errors in some cases. As an example, suppose the data are not normal, but gamma with $r = 4.694$ and $\lambda = 0.722$. Even though the distribution is skewed, these values make the mean $\mu = 6.5$ and the variance $\sigma^2 = 9$, as in the normal case above. Again relying on Minitab to give two hundred random samples of size 20, the observed number of Type II errors is sixty, so the test has some robustness for Type II errors in that case. Even though the data are not normal, the key statistic in the analysis, $\bar{y}$, will be approximately normal by the central limit theorem.

If the distribution of the underlying data is unknown or extremely skewed, nonparametric tests, like the ones covered in Chapter 14 and in (28) are advised.

Appendix 7.A.1 Minitab Applications

Many statistical procedures, including several featured in this chapter, require that the sample mean and sample standard deviation be calculated. Minitab's DESCRIBE command gives $\bar{y}$ and s, along with several other useful numerical characteristics of a sample. Figure 7.A.1.1 shows the DESCRIBE input and output for the twenty observations cited in Example 7.4.1.

Figure 7.A.1.1

```
MTB  > set c1
DATA > 2.5 3.2 0.5 0.4 0.3 0.1 0.1 0.2 7.4 8.6 0.2 0.1
DATA > 0.4 1.8 0.3 1.3 1.4 11.2 2.1 10.1
DATA > end
MTB  > describe c1
```

Descriptive Statistics: C1

Variable	N	N*	Mean	SE Mean	StDev	Minimum	Q1	Median	Q3	Maximum
C1	20	0	2.610	0.809	3.617	0.100	0.225	0.900	3.025	11.200

Here,

N = sample size
N* = number of observations missing from c1 (that is, the number of "interior" blanks)
Mean = sample mean = $\bar{y}$
SE Mean = standard error of the mean = $\frac{s}{\sqrt{n}}$
StDev = sample standard deviation = s
Minimum = smallest observation
Q1 = first quartile = 25th percentile
Median = middle observation (in terms of magnitude), or average of the middle two if n is even
Q3 = third quartile = 75th percentile
Maximum = largest observation

Describing Samples Using Minitab Windows

1. Enter data under C1 in the WORKSHEET. Click on STAT, then on BASIC STATISTICS, then on DISPLAY DESCRIPTIVE STATISTICS.
2. Type C1 in VARIABLES box; click on OK.

Percentiles of chi square, t, and F distributions can be obtained using the INVCDF command introduced in Appendix 3.A.1. Figure 7.A.1.2 shows the syntax for printing out $\chi^2_{.95,6}(=12.5916)$ and $F_{.01,4,7}(=0.0667746)$.

Figure 7.A.1.2

```
MTB  > invcdf 0.95;
SUBC > chisq 6.
```

Inverse Cumulative Distribution Function

```
Chi-Square with 6 DF

P(X <= x)          x
     0.95    12.5916

MTB > invcdf 0.01;
SUBC> f 4 7.
```

Inverse Cumulative Distribution Function

```
F distribution with 4 DF in numerator and 7 DF in denominator

P(X <= x)            x
     0.01    0.0667746
```

To find Student t cutoffs, the $t_{\alpha,n-1}$ notation needs to be expressed as a percentile. We have defined $t_{.10,13}$, for example, to be the value for which

$$P(T_{13} \geq t_{.10,13}) = 0.10$$

In the terminology of the INVCDF command, though, $t_{.10,13}(= 1.35017)$ is the ninetieth percentile of the $f_{T_{13}}(t)$ pdf (see Figure 7.A.1.3).

Figure 7.A.1.3

```
MTB > invcdf 0.90;
SUBC> t 13.
```

Inverse Cumulative Distribution Function

```
Student's t distribution with 13 DF

P(X <= x)          x
      0.9    1.35017
```

The Minitab command for constructing a confidence interval for μ (Theorem 7.4.1) is "TINTERVAL X Y," where X denotes the desired value for the confidence coefficient $1 - \alpha$ and Y is the column where the data are stored. Figure 7.A.1.4 shows the TINTERVAL command applied to the bat data from Case Study 7.4.1; $1 - \alpha$ is taken to be 0.95.

Figure 7.A.1.4

```
MTB  > set  c1
DATA > 62  52  68  23  34  45  27  42  83  56  40
DATA > end
MTB  > tinterval 0.95 c1
```

One-Sample T: C1

```
Variable  N     Mean    StDev   SE Mean        95%      CI
C1       11    48.36    18.08      5.45    (36.21, 60.51)
```

Constructing Confidence Intervals Using Minitab Windows

1. Enter data under C1 in the WORKSHEET.
2. Click on STAT, then on BASIC STATISTICS, then on 1-SAMPLE T.
3. Enter C1 in the SAMPLES IN COLUMNS box, click on OPTIONS, and enter the value of $100(1 - \alpha)$ in the CONFIDENCE LEVEL box.
4. Click on OK. Click on OK.

Figure 7.A.1.5 shows the input and output for doing a t test on the approval data given in Table 7.4.3. The basic command is "TTEST X Y," where X is the value of μ_o and Y is the column where the data are stored. If no other punctuation is used,

Figure 7.A.1.5

```
MTB  > set c1
DATA > 59 65 69 53 60 53 58 64 46 67 51 59
DATA > end
MTB  > ttest 62 c1
```

One-Sample T: C1

```
Test of mu = 62 vs not = 62

Variable  N     Mean   StDev  SE Mean             95% CI          T       P
C1       12    58.66    6.95     2.01    (54.25,63.08)      -1.66   0.125
```

the program automatically takes H_1 to be two-sided. If a one-sided test *to the right* is desired, we write

```
MTB > ttest X Y;
SUBC > alternative +1.
```

For a one-sided test *to the left,* the subcommand becomes "alternative −1".

Notice that no value for α is entered, and that the conclusion is not phrased as either "Accept H_0" or "Reject H_0." Rather, the analysis ends with the calculation of the data's *P-value*.

Here,

$$\begin{aligned} P\text{-value} &= P(T_{11} \leq -1.66) + P(T_{11} \geq 1.66) \\ &= 0.0626 + 0.0626 \\ &= 0.125 \end{aligned}$$

(recall Definition 6.2.4). Since the P-value exceeds the intended $\alpha(= 0.05)$, the conclusion is "Fail to reject H_0."

Testing $H_0: \mu = \mu_o$ Using Minitab Windows

1. Enter data under C1 in the WORKSHEET.
2. Click on STAT, then on BASIC STATISTICS, then on 1-SAMPLE T.
3. Type C1 in SAMPLES IN COLUMNS box; click on PERFORM HYPOTHESIS TEST and enter the value of μ_o. Click on OPTIONS, then choose NOT EQUAL.
4. Click on OK; then click on OK.

Appendix 7.A.2 Some Distribution Results for $\overline{Y}$ and S^2

Theorem 7.A.2.1 *Let $Y_1, Y_2, \ldots, Y_n$ be a random sample of size n from a normal distribution with mean μ and variance σ^2. Define*

$$\overline{Y} = \frac{1}{n}\sum_{i=1}^{n} Y_i \quad \text{and} \quad S^2 = \frac{1}{n-1}\sum_{i=1}^{n}(Y_i - \overline{Y})^2$$

Then

a. *$\overline{Y}$ and S^2 are independent.*

b. *$\frac{(n-1)S^2}{\sigma^2}$ has a chi square distribution with $n-1$ degrees of freedom.*

Proof The proof of this theorem relies on certain linear algebra techniques as well as a change-of-variables formula for multiple integrals. Definition 7.A.2.1 and the Lemma that follows review the necessary background results. For further details, see (44) or (213). □

Definition 7.A.2.1.

a. A matrix A is said to be *orthogonal* if $AA^T = I$.
b. Let β be any n-dimensional vector over the real numbers. That is, $\beta = (c_1, c_2, \ldots, c_n)$, where each c_j is a real number. The *length* of β will be defined as

$$\| \beta \| = \left(c_1^2 + \cdots + c_n^2\right)^{1/2}$$

(Note that $\| \beta \|^2 = \beta\beta^T$.)

Lemma

a. *A matrix A is orthogonal if and only if*

$$\| A\beta \| = \| \beta \| \quad \textit{for each } \beta$$

b. *If a matrix A is orthogonal, then det* $A = 1$.
c. *Let g be a one-to-one continuous mapping on a subset, D, of n-space. Then*

$$\int_{g(D)} f(x_1, \ldots, x_n)\, dx_1 \cdots dx_n = \int_D f[g(y_1, \ldots, y_n)] \text{ det } J(g)\, dy_1 \cdots dy_n$$

where $J(g)$ *is the Jacobian of the transformation.*

Set $X_i = (Y_i - \mu)/\sigma$ for $i = 1, 2, \ldots, n$. Then all the X_i's are $N(0, 1)$. Let A be an $n \times n$ orthogonal matrix whose last row is $(\frac{1}{\sqrt{n}}, \frac{1}{\sqrt{n}}, \ldots, \frac{1}{\sqrt{n}})$. Let $\vec{X} = (X_1, \ldots, X_n)^T$ and define $\vec{Z} = (Z_1, Z_2, \ldots, Z_n)^T$ by the transformation $\vec{Z} = A\vec{X}$. [Note that $Z_n = (\frac{1}{\sqrt{n}})X_1 + \cdots + (\frac{1}{\sqrt{n}})X_n = \sqrt{n}\,\overline{X}$.]

For any set D,

$$\begin{aligned} P(\vec{Z} \in D) &= P(A\vec{X} \in D) = P(\vec{X} \in A^{-1}D) \\ &= \int_{A^{-1}D} f_{X_1,\ldots,X_n}(x_1, \ldots, x_n)\, dx_1 \cdots dx_n \\ &= \int_D f_{X_1,\ldots,X_n}[g(\vec{z})] \text{ det } J(g)\, dz_1 \cdots dz_n \\ &= \int_D f_{X_1,\ldots,X_n}(A^{-1}\vec{z}) \cdot 1 \cdot dz_1 \cdots dz_n \end{aligned}$$

where $g(\vec{z}) = A^{-1}\vec{z}$. But A^{-1} is orthogonal, so setting $(x_1, \ldots, x_n)^T = A^{-1}z$, we have that

$$x_1^2 + \cdots + x_n^2 = z_1^2 + \cdots + z_n^2$$

Thus

$$\begin{aligned} f_{X_1,\ldots,X_n}(\vec{x}) &= (2\pi)^{-n/2} e^{-(1/2)\left(x_1^2 + \cdots + x_n^2\right)} \\ &= (2\pi)^{-n/2} e^{-(1/2)\left(z_1^2 + \cdots + z_n^2\right)} \end{aligned}$$

From this we conclude that

$$P(\vec{Z} \in D) = \int_D (2\pi)^{-n/2} e^{-(n/2)\left(z_1^2 + \cdots + z_n^2\right)} dz_1 \cdots dz_n$$

implying that the Z_j's are independent standard normals.

Finally,

$$\sum_{j=1}^{n} Z_j^2 = \sum_{j=1}^{n-1} Z_j^2 + n\overline{X}^2 = \sum_{j=1}^{n} X_j^2 = \sum_{j=1}^{n} (X_j - \overline{X})^2 + n\overline{X}^2$$

Therefore,

$$\sum_{j=1}^{n-1} Z_j^2 = \sum_{j=1}^{n} (X_j - \overline{X})^2$$

and $\overline{X}^2$ (and thus $\overline{X}$) is independent of $\sum_{j=1}^{n} (X_i - \overline{X})^2$, so the conclusion follows for standard normal variables. Also, since $\overline{Y} = \sigma\overline{X} + \mu$ and $\sum_{i=1}^{n} (Y_i - \overline{Y})^2 = \sigma^2 \sum_{i=1}^{n} (X_i - \overline{X})^2$, the conclusion follows for $N(\mu, \sigma^2)$ variables.

Comment As part of the proof just presented, we established a version of *Fisher's lemma:*

> Let $X_1, X_2, \ldots, X_n$ be independent standard normal random variables and let A be an orthogonal matrix. Define $(Z_1, \ldots, Z_n)^T = A(X_1, \ldots, X_n)^T$. Then the Z_i's are independent standard normal random variables.

Appendix 7.A.3 A Proof that the One-Sample t Test is a GLRT

Theorem 7.A.3.1 *The one-sample t test, as outlined in Theorem 7.4.2, is a GLRT.*

Proof Consider the test of $H_0: \mu = \mu_o$ versus $H_1: \mu \neq \mu_o$. The two parameter spaces restricted to H_0 and $H_0 \cup H_1$—that is, ω and Ω, respectively—are given by

$$\omega = \{(\mu, \sigma^2): \mu = \mu_0; \quad 0 \leq \sigma^2 < \infty\}$$

and

$$\Omega = \{(\mu, \sigma^2): -\infty < \mu < \infty; \quad 0 \leq \sigma^2 < \infty\}$$

Without elaborating the details (see Example 5.2.4 for a very similar problem), it can be readily shown that, under ω,

$$\mu_e = \mu_0 \quad \text{and} \quad \sigma_e^2 = \frac{1}{n}\sum_{i=1}^{n} (y_i - \mu_0)^2$$

Under Ω,

$$\mu_e = \overline{y} \quad \text{and} \quad \sigma_e^2 = \frac{1}{n}\sum_{i=1}^{n} (y_i - \overline{y})^2$$

Therefore, since

$$L(\mu, \sigma^2) = \left(\frac{1}{\sqrt{2\pi}\sigma}\right)^n \exp\left[-\frac{1}{2}\sum_{i=1}^{n}\left(\frac{y_i - \mu}{\sigma}\right)^2\right]$$

direct substitution gives

$$L(\omega_e) = \left[\frac{\sqrt{n}}{\sqrt{2\pi}\sqrt{\sum_{i=1}^{n} (y_i - \mu_0)^2}}\right]^n e^{-n/2}$$

$$= \left[\frac{ne^{-1}}{2\pi \sum_{i=1}^{n} (y_i - \mu_0)^2} \right]^{n/2}$$

and

$$L(\Omega_e) = \left[\frac{ne^{-1}}{2\pi \sum_{i=1}^{n} (y_i - \overline{y})^2} \right]^{n/2}$$

From $L(\omega_e)$ and $L(\Omega_e)$ we get the likelihood ratio:

$$\lambda = \frac{L(\omega_e)}{L(\Omega_e)} = \left[\frac{\sum_{i=1}^{n} (y_i - \overline{y})^2}{\sum_{i=1}^{n} (y_i - \mu_0)^2} \right]^{n/2}, \quad 0 < \lambda \leq 1$$

As is often the case, it will prove to be more convenient to base a test on a monotonic function of λ, rather than on λ itself. We begin by rewriting the ratio's denominator:

$$\sum_{i=1}^{n} (y_i - \mu_0)^2 = \sum_{i=1}^{n} [(y_i - \overline{y}) + (\overline{y} - \mu_0)]^2$$

$$= \sum_{i=1}^{n} (y_i - \overline{y})^2 + n(\overline{y} - \mu_0)^2$$

Therefore,

$$\lambda = \left[1 + \frac{n(\overline{y} - \mu_0)^2}{\sum_{i=1}^{n} (y_i - \overline{y})^2} \right]^{-n/2}$$

$$= \left(1 + \frac{t^2}{n-1} \right)^{-n/2}$$

where

$$t = \frac{\overline{y} - \mu_0}{s/\sqrt{n}}$$

Observe that as t^2 increases, λ decreases. This implies that the original GLRT—which, by definition, would have rejected H_0 for any λ that was too small, say, less than λ^*—is equivalent to a test that rejects H_0 whenever t^2 is too large. But t is an observation of the random variable

$$T = \frac{\overline{Y} - \mu_0}{S/\sqrt{n}} \quad (= T_{n-1} \text{ by Theorem 7.3.5})$$

Thus "too large" translates numerically into $t_{\alpha/2,n-1}$:

$$0 < \lambda \leq \lambda^* \Leftrightarrow t^2 \geq (t_{\alpha/2,n-1})^2$$

But

$$t^2 \geq (t_{\alpha/2,n-1})^2 \Leftrightarrow t \leq -t_{\alpha/2,n-1} \quad \text{or} \quad t \geq t_{\alpha/2,n-1}$$

and the theorem is proved. □

Appendix 7.A.4 A Proof of Theorem 7.5.2

We begin by considering the test of $H_0: \sigma^2 = \sigma_o^2$ against a two-sided H_1. The relevant parameter spaces are

$$\omega = \{(\mu, \sigma^2): -\infty < \mu < \infty, \quad \sigma^2 = \sigma_0^2\}$$

and

$$\Omega = \{(\mu, \sigma^2): -\infty < \mu < \infty, \quad 0 \leq \sigma^2\}$$

In both, the maximum likelihood estimate for μ is $\overline{y}$. In ω, the maximum likelihood estimate for σ^2 is simply σ_0^2; in Ω, $\sigma_e^2 = (1/n) \sum_{i=1}^{n} (y_i - \overline{y})^2$ (see Example 5.4.4). Therefore, the two likelihood functions, maximized over ω and over Ω, are

$$L(\omega_e) = \left(\frac{1}{2\pi\sigma_0^2}\right)^{n/2} \exp\left[-\frac{1}{2}\sum_{i=1}^{n}\left(\frac{y_i - \overline{y}}{\sigma_0}\right)^2\right]$$

and

$$L(\Omega_e) = \left[\frac{n}{2\pi \sum_{i=1}^{n} (y_i - \overline{y})^2}\right]^{n/2} \exp\left\{-\frac{n}{2}\sum_{i=1}^{n}\left[\frac{y_i - \overline{y}}{\sqrt{\sum_{i=1}^{n} (y_i - \overline{y})^2}}\right]^2\right\}$$

$$= \left[\frac{n}{2\pi \sum_{i=1}^{n} (y_i - \overline{y})^2}\right]^{n/2} e^{-n/2}$$

It follows that the generalized likelihood ratio is given by

$$\lambda = \frac{L(\omega_e)}{L(\Omega_e)}$$

$$= \left[\frac{\sum_{i=1}^{n} (y_i - \overline{y})^2}{n\sigma_0^2}\right]^{n/2} \cdot \exp\left[-\frac{1}{2}\sum_{i=1}^{n}\left(\frac{y_i - \overline{y}}{\sigma_0}\right)^2 + \frac{n}{2}\right]$$

$$= \left(\frac{\sigma_e^2}{\sigma_0^2}\right)^{n/2} \cdot e^{-(n/2)\left(\sigma_e^2/\sigma_0^2\right) + n/2}$$

We need to know the behavior of λ, considered as a function of (σ_e^2/σ_0^2). For simplicity, let $x = (\sigma_e^2/\sigma_0^2)$. Then $\lambda = x^{n/2} e^{-(n/2)x + n/2}$ and the inequality $\lambda \leq \lambda^*$ is equivalent to $xe^{-x} \leq e^{-1}(\lambda^*)^{2/n}$. The right-hand side is again an arbitrary constant, say, k^*. Figure 7.A.4.1 is a graph of $y = xe^{-x}$. Notice that the values of $x = (\sigma_e^2/\sigma_0^2)$ for which $xe^{-x} \leq k^*$, and equivalently $\lambda \leq \lambda^*$, fall into two regions, one for values of σ_e^2/σ_0^2 close to 0 and the other for values of σ_e^2/σ_0^2 much larger than 1. According to the likelihood ratio principle, we should reject H_0 for any $\lambda \leq \lambda^*$, where $P(\Lambda \leq \lambda^* | H_0) = \alpha$. But λ^* determines (via k^*) numbers a and b, so the critical region is $C = \{(\sigma_e^2/\sigma_0^2) : (\sigma_e^2/\sigma_0^2) \leq a \text{ or } (\sigma_e^2/\sigma_0^2) \geq b\}$.

Figure 7.A.4.1

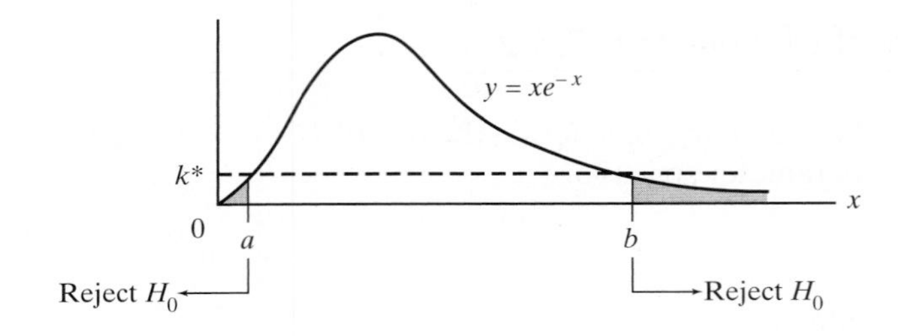

Comment At this point it is necessary to make a slight approximation. Just because $P(\Lambda \leq \lambda^* | H_0) = \alpha$, it does not follow that

$$P\left[\frac{\left(\frac{1}{n}\right)\sum_{i=1}^{n}(Y_i - \overline{Y})^2}{\sigma_0^2} \leq a\right] = \frac{\alpha}{2} = P\left[\frac{\left(\frac{1}{n}\right)\sum_{i=1}^{n}(Y_i - \overline{Y})^2}{\sigma_0^2} \geq b\right]$$

and, in fact, the two tails of the critical regions will *not* have exactly the same probability. Nevertheless, the two are numerically close enough that we will not substantially compromise the likelihood ratio criterion by setting each one equal to $\alpha/2$.

Note that

$$P\left[\frac{\left(\frac{1}{n}\right)\sum_{i=1}^{n}(Y_i - \overline{Y})^2}{\sigma_0^2} \leq a\right] = P\left[\frac{\sum_{i=1}^{n}(Y_i - \overline{Y})^2}{\sigma_0^2} \leq na\right]$$

$$= P\left[\frac{(n-1)S^2}{\sigma_0^2} \leq na\right]$$

$$= P\left(\chi^2_{n-1} \leq na\right)$$

and, similarly,

$$P\left[\frac{\left(\frac{1}{n}\right)\sum_{i=1}^{n}(Y_i - \overline{Y})^2}{\sigma_0^2} \geq b\right] = P\left(\chi^2_{n-1} \geq nb\right)$$

Thus we will choose as critical values $\chi^2_{\alpha/2,n-1}$ and $\chi^2_{1-\alpha/2,n-1}$ and reject H_0 if either

$$\frac{(n-1)s^2}{\sigma_0^2} \leq \chi^2_{\alpha/2,n-1}$$

or

$$\frac{(n-1)s^2}{\sigma_0^2} \geq \chi^2_{1-\alpha/2,n-1}$$

(see Figure 7.A.4.2).

Figure 7.A.4.2

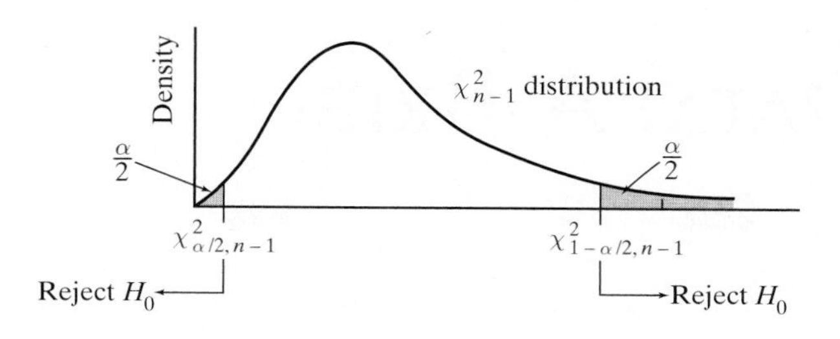

Comment One-sided tests for dispersion are set up in a similar fashion. In the case of

$$H_0: \sigma^2 = \sigma_0^2$$

versus

$$H_1: \sigma^2 < \sigma_0^2$$

H_0 is rejected if

$$\frac{(n-1)s^2}{\sigma_0^2} \leq \chi^2_{\alpha,n-1}$$

For

$$H_0: \sigma^2 = \sigma_0^2$$

versus

$$H_1: \sigma^2 > \sigma_0^2$$

H_0 is rejected if

$$\frac{(n-1)s^2}{\sigma_0^2} \geq \chi^2_{1-\alpha,n-1}$$

Chapter 8

Types of Data: A Brief Overview

The practice of statistics is typically conducted on two distinct levels. Analyzing data requires first and foremost an understanding of random variables. Which pdfs are modeling the observations? What parameters are involved, and how should they be estimated? Broader issues, though, need to be addressed as well. How is the entire set of measurements configured? Which factors are being investigated; in what ways are they related? Altogether, seven different types of data are profiled in Chapter 8. Collectively, they represent a sizeable fraction of the "experimental designs" that many researchers are likely to encounter.

8.1 Introduction

Chapters 6 and 7 have introduced the basic principles of statistical inference. The typical objective in that material was either to construct a confidence interval or to test the credibility of a null hypothesis. A variety of formulas and decision rules were derived to accommodate distinctions in the nature of the data and the parameter being investigated. It should not go unnoticed, though, that every set of data in those two chapters, despite their superficial differences, shares a critically important common denominator—each represents the exact same *experimental design*.

A working knowledge of statistics requires that the subject be pursued at two different levels. On one level, attention needs to be paid to the mathematical properties inherent *in the individual measurements*. These are what might be thought of as the "micro" structure of statistics. What is the pdf of the Y_i's? Do we know $E(Y_i)$ or $\text{Var}(Y_i)$? Are the Y_i's independent?

Viewed collectively, though, every set of measurements also has a certain overall structure, or *design*. It will be those "macro" features that we focus on in this chapter. A number of issues need to be addressed. How is one design different from another? Under what circumstances is a given design desirable? Or undesirable? How does the design of an experiment influence the analysis of that experiment?

The answers to some of these questions will need to be deferred until each design is taken up individually and in detail later in the text. For now our objective is much more limited—Chapter 8 is meant to be a brief introduction to some of the important ideas involved in the classification of data. What we learn here will serve as a backdrop and a frame of reference for the multiplicity of statistical procedures derived in Chapters 9 through 14.

Definitions

To describe an experimental design, and to distinguish one design from another, requires that we understand several key definitions.

Factors and Factor Levels The word *factor* is used to denote any treatment or therapy "applied to" the subjects being measured or any relevant feature (age, sex, ethnicity, etc.) "characteristic" of those subjects. Different versions, extents, or aspects of a factor are referred to as *levels*.

Case Study 8.1.1

Generations of athletes have been cautioned that cigarette smoking impedes performance. One measure of the truth of that warning is the effect of smoking on heart rate. In one study (73), six nonsmokers, six light smokers, six moderate smokers, and six heavy smokers each engaged in sustained physical exercise. Table 8.1.1 lists their heart rates after they had rested for three minutes.

Table 8.1.1 Heart Rates

	Nonsmokers	Light Smokers	Moderate Smokers	Heavy Smokers
	69	55	66	91
	52	60	81	72
	71	78	70	81
	58	58	77	67
	59	62	57	95
	65	66	79	84
Averages:	62.3	63.2	71.7	81.7

The single factor in this experiment is smoking, and its levels are the four different column headings in Table 8.1.1. A more elaborate study addressing this same concern about smoking could easily be designed to incorporate three factors. Common sense tells us that the harmful effects of smoking may not be the same for men as they are for women, and they may be more (or less) pronounced in senior citizens than they are in young adults. As a factor, gender would have two levels, male and female, and age could easily have

(Continued on next page)

(Case Study 8.1.1 continued)

at least three—for example, 18–34, 35–64, and 65+. If all three factors were included, the format of the data table would look like Figure 8.1.1.

		Nonsmokers		Light Smokers		Moderate Smokers		Heavy Smokers	
		M	F	M	F	M	F	M	F
	18–34								
Age	35–64								
	65+								

Figure 8.1.1

Blocks Sometimes subjects or environments share certain characteristics that affect the way that levels of a factor respond, yet those characteristics are of no intrinsic interest to the experimenter. Any such set of conditions or subjects is called a *block*.

Case Study 8.1.2

Table 8.1.2 summarizes the results of a rodent-control experiment that was carried out in Milwaukee, Wisconsin, over a period of ten weeks. The study's single factor was *rat poison flavor*, and it had four levels—plain, butter-vanilla, roast beef, and bread.

Table 8.1.2 Bait-Acceptance Percentages

Survey Number	Plain	Butter-Vanilla	Roast Beef	Bread
1	13.8	11.7	14.0	12.6
2	12.9	16.7	15.5	13.8
3	25.9	29.8	27.8	25.0
4	18.0	23.1	23.0	16.9
5	15.2	20.2	19.0	13.7

Eight hundred baits of each flavor were placed around garbage-storage areas. After two weeks, the percentages of baits taken were recorded. For the next two weeks, another set of 3200 baits were placed at a different set of locations, and the same protocol was followed. Altogether, five two-week "surveys" were completed (85).

Clearly, each survey created a unique experimental environment. Baits were placed at different locations, weather conditions would not be the same, and the availability of other sources of food might change. For those reasons

(Continued on next page)

and maybe others, Survey 3, for example, yielded percentages noticeably higher than those of Surveys 1 and 2. The experimenters' sole objective, though, was to compare the four flavors—which did the rodents prefer? The fact that the survey "environments" were not identical was both anticipated and irrelevant. The five different surveys, then, qualify as *blocks*.

About the Data To an applied statistician, the data in Table 8.1.2 would be classified as a *complete block* experiment, because the entire set of factor levels was compared within each block. Sometimes physical limitations prevent that from being possible, and only subsets of factor levels can appear in a given block. Experiments of that sort are referred to as *incomplete block* designs. Not surprisingly, they are much more difficult to analyze.

Independent and Dependent Observations Whatever the context, measurements collected for the purpose of comparing two or more factor levels are necessarily either *dependent* or *independent*. Two or more observations are *dependent* if they share a particular commonality relevant to what is being measured. If there is no such linkage, the observations are *independent*.

An example of dependent data is the acceptance percentages recorded in Table 8.1.2. The 13.8, for example, shown in the upper-left-hand corner is measuring both the rodents' preference for plain baits and also the environmental conditions that prevailed in Survey 1; similarly, the observation immediately to its right, 11.7, measures the rodents' preference for the butter-vanilla flavor and *the same survey environmental conditions*. By definition, then, 13.8 and 11.7 are *dependent* measurements because their values have the commonality of sharing the same conditions of Survey 1. Taken together, then, the data in Table 8.1.2 are five sets of dependent observations, each set being a sample of size 4.

By way of contrast, the observations in Table 8.1.1 are *independent*. The 69 and 55 in the first row, for example, have nothing exceptional in common—they are simply measuring the effects of two different factor levels applied to two different people. Would the first two entries in the first column, 69 and 52, be considered dependent? No. Simply sharing the same factor level does not make observations dependent.

For reasons that will be examined in detail in later chapters, factor levels can often be compared much more efficiently with dependent observations than with independent observations. Fortunately, dependent observations come about quite naturally in a number of different ways. Measurements made on twins, siblings, or littermates are automatically dependent because of the subjects' shared genetic structure (and, of course, repeated measurements taken on the same individual are dependent). In agricultural experiments, crops grown in the same general location are dependent because they share similar soil quality, drainage, and weather conditions. Industrial measurements taken with the same piece of equipment or by the same operator are likewise dependent. And, of course, time and place (like the surveys in Table 8.1.2) are often used to induce shared conditions. Those are some of the "standard" ways to make observations dependent. Over the years, experimenters have become very adept at finding clever, "nonstandard" ways as well.

Similar and Dissimilar Units Units also play a role in a data set's macrostructure. Two measurements are said to be *similar* if their units are the same and *dissimilar* otherwise. Comparing the effects of different factor levels is the typical objective when the units in a set of data are all the same. This was the situation in both Case Studies 8.1.1 and 8.1.2. Dissimilar measurements are analyzed by quantifying their relationship.

Quantitative Measurements and Qualitative Measurements Measurements are considered *quantitative* if their possible values are numerical. The heart rates in Table 8.1.1 and the bait-acceptance percentages in Table 8.1.2 are two examples. *Qualitative* measurements have "values" that are either categories, characteristics, or conditions.

Case Study 8.1.3

Certain viral infections contracted during pregnancy—particularly early in the first trimester—can cause birth defects. By far the most dangerous of these are Rubella infections, also known as German measles. Table 8.1.3 (45) summarizes the history of 578 pregnancies, each complicated by a Rubella infection either "early" (first trimester) or "late" (second and third trimesters).

Table 8.1.3

		When Infection Occurred	
		Early	Late
Outcome	Abnormal birth	59	27
	Normal birth	143	349
	% of abnormal births	29.2	7.2

Despite all the numbers displayed in Table 8.1.3, these are not quantitative measurements. What we are seeing is a *summary* of *qualitative* measurements. When the data were originally recorded, they would have looked like Figure 8.1.2. The qualitative time variable had two values (early or late), as did the qualitative outcome variable (normal or abnormal).

Patient no.	Name	Time of Infection	Birth outcome
1	ML	Early	Abnormal
2	JG	Late	Normal
3	DF	Early	Normal
.	.	.	.
.	.	.	.
.	.	.	.
578	CW	Early	Abnormal

Figure 8.1.2

Possible Designs

The definitions just cited can give rise to a sizeable number of different experimental designs, far more than can be covered in this text. Still, the number of designs *that are widely used* is fairly small. Much of the data likely to be encountered fall into one of the following seven formats:

One-sample data
Two-sample data
k-sample data
Paired data
Randomized block data
Regression data
Categorical data

The heart rates listed in Table 8.1.1, for example, qualify as *k-sample data*; the rodent bait acceptances in Table 8.1.2 are *randomized block data*; and the Rubella/pregnancy outcomes in Table 8.1.3 are *categorical data*.

In Section 8.2, each design will be profiled, illustrated, and reduced to a mathematical model. Special attention will be given to each design's objectives—that is, for what type of inference is it likely to be used?

8.2 Classifying Data

The answers to no more than four questions are needed to classify a set of data as one of the seven basic models listed in the preceding section:

1. Are the observations quantitative or qualitative?
2. Are the units similar or dissimilar?
3. How many factor levels are involved?
4. Are the observations dependent or independent?

In Section 8.2, we use these four questions as the starting point in distinguishing one experimental design from another.

One-Sample Data

The simplest of all experimental designs, the *one-sample data* design, consists of a single random sample of size n. Necessarily, the n observations reflect one particular set of conditions or one specific factor. During presidential election years, a familiar example (probably too familiar...) is the political poll. A random sample of n voters, all representing the same demographic group, are asked whether they intend to vote for Candidate X—1 for yes, 0 for no. Recorded, then, are the outcomes of n Bernoulli trials, where the unknown parameter p is the true proportion of voters in that particular demographic constituency who intend to support Candidate X.

Other discrete random variables can also appear as one-sample data. Recall Case Study 4.2.3, describing the outbreaks of war from 1500 to 1931. Those 432 observations were shown to follow a Poisson distribution. In practice, though, one-sample data will more typically consist of measurements on a *continuous* random variable. In Case Study 4.2.4, the sample of thirty-six intervals between consecutive eruptions of Mauna Loa had a distribution entirely consistent with an exponential random variable.

All these examples notwithstanding, by far the most frequently encountered set of assumptions associated with one-sample data is that the Y_i's are a random sample of size n from a normal distribution with unknown mean μ and unknown standard deviation σ. Possible inference procedures would be either hypothesis tests or confidence intervals for μ and/or σ, whichever would be appropriate for the experimenter's objectives.

In describing experimental designs, the assumptions given for a set of measurements are often written in the form of a *model equation*, which, by definition, expresses the value of an aribitrary Y_i as the sum of fixed and variable components. For one-sample data, the usual model equation is

$$Y_i = \mu + \varepsilon_i, \quad i = 1, 2, \ldots, n$$

where ε_i is a normally distributed random variable with mean 0 and standard deviation σ.

Case Study 8.2.1

Inventions, whether simple or complex, can take a long time to become marketable. Minute Rice, for example, was developed in 1931 but appeared for the first time on grocery shelves in 1949, some eighteen years later. Listed in Table 8.2.1 are the conception dates and realization dates for seventeen familiar products (197). Computed for each and shown in the last column is the product's development time, y. In the case of Minute Rice, $y = 18$ $(= 1949 - 1931)$.

Table 8.2.1

Invention	Conception Date	Realization Date	Development Time (years)
Automatic transmission	1930	1946	16
Ballpoint pen	1938	1945	7
Filter cigarettes	1953	1955	2
Frozen foods	1908	1923	15
Helicopter	1904	1941	37
Instant coffee	1934	1956	22
Minute Rice	1931	1949	18
Nylon	1927	1939	12
Photography	1782	1838	56
Radar	1904	1939	35
Roll-on deodorant	1948	1955	7
Telegraph	1820	1838	18
Television	1884	1947	63
Transistor	1940	1956	16
VCR	1950	1956	6
Xerox copying	1935	1950	15
Zipper	1883	1913	30
			Average 22.2

About the Data In addition to exhibiting one-sample data, Table 8.2.1 is typical of the "fun list" format that appears so often in the print media. These are entertainment data more so than serious scientific research. Here, for example, the average development time is 22.2 years. Would it make sense to use that average as part of a formal inference procedure? Not really. If it could be assumed that these seventeen inventions were in some sense a random sample of all possible inventions, then using 22.2 years to draw an inference about the "true" average development time would be legitimate. But the arbitrariness of the inventions included in Table 8.2.1 makes that assumption highly questionable at best. Data like these are meant to be enjoyed and to inform, not to be analyzed.

Two-Sample Data

Two-sample data consist of two independent random samples of sizes m and n, each having quantitative, similar unit measurements. Each sample is associated with a different factor level. Sometimes the two samples are sequences of Bernoulli trials, in which case the measurements are 0's and 1's. Given that scenario, the data's two parameters are the unknown "success" probabilities p_X and p_Y, and the usual inference procedure would be to test $H_0: p_X = p_Y$.

Much more often, the two samples are normally distributed with possibly different means and possibly different standard deviations. If $X_1, X_2, \ldots, X_n$ denotes the first sample and $Y_1, Y_2, \ldots, Y_m$ the second, the usual model equation assumptions would be written

$$X_i = \mu_X + \varepsilon_i, \quad i = 1, 2, \ldots, n$$

$$Y_j = \mu_Y + \varepsilon'_j, \quad j = 1, 2, \ldots, m$$

where ε_i is normally distributed with mean 0 and standard deviation σ_X, and ε^1_j is normally distributed with mean 0 and standard deviation σ_Y.

With two-sample data, inference procedures are more likely to be hypothesis tests than confidence intervals. A *two-sample t test* is used to assess the credibility of $H_0 : \mu_X = \mu_Y$; an *F test* is used when the objective is to choose between H_0: $\sigma_X = \sigma_Y$ and, say, H_1: $\sigma_X \neq \sigma_Y$. Both procedures will be described in Chapter 9.

To experimenters, two-sample data address what is sometimes a serious flaw with one-sample data. The usual one-sample hypothesis test, H_0: $\mu = \mu_0$, makes the tacit assumption that the Y_i's (whose true mean is μ) were collected under the same conditions that gave rise to the "standard" value μ_0, against which μ is being tested. There may be no way to know whether that assumption is true, or even remotely true. The two-sample format, on the other hand, lets the experimenter control the conditions (and subjects) under which *both* sets of measurements are taken. Doing so heightens the chances that the true means are being compared in a fair and equitable way.

Case Study 8.2.2

Forensic scientists sometimes have difficulty identifying the sex of a murder victim whose body is discovered badly decomposed. Often, dental structure can provide useful clues because female teeth and male teeth have different physical

(Continued on next page)

(Case Study 8.2.2 continued)

and chemical characteristics. The extent to which X-rays can penetrate tooth enamel, for instance, is not the same for the two sexes.

Table 8.2.2 lists the enamel spectropenetration gradients for eight male teeth and eight female teeth (57). These measurements have all the characteristics of the two-sample format: the data are *quantitative*, the units are *similar*, *two* factor levels (male and female) are involved, and the observations are *independent*.

Table 8.2.2 Enamel Spectropenetration Gradients

	Male	Female
	4.9	4.8
	5.4	5.3
	5.0	3.7
	5.5	4.1
	5.4	5.6
	6.6	4.0
	6.3	3.6
	4.3	5.0
Averages:	5.4	4.5

The sample averages are 5.4 for the male teeth and 4.5 for the female teeth. According to the *two-sample t test* introduced in Chapter 9, the difference between those two sample means is, indeed, statistically significant.

About the Data In analyzing these data, the assumption would be made that the male gradients (X_i's) and the female gradients (Y_j's) are normally distributed. How do we know if that assumption is correct? We don't. For large data sets—sample sizes of 30 or more—the assumption that observations are normally distributed can be investigated using a *goodness-of-fit test*, the details of which are presented in Chapter 10. For small samples like those in Table 8.2.2, the best that we can do is to plot the data along a horizontal line and see if the spacing is consistent with the shape of a normal curve. That is, does the pattern show signs of symmetry and is the bulk of the data near the center of the range?

Figure 8.2.1

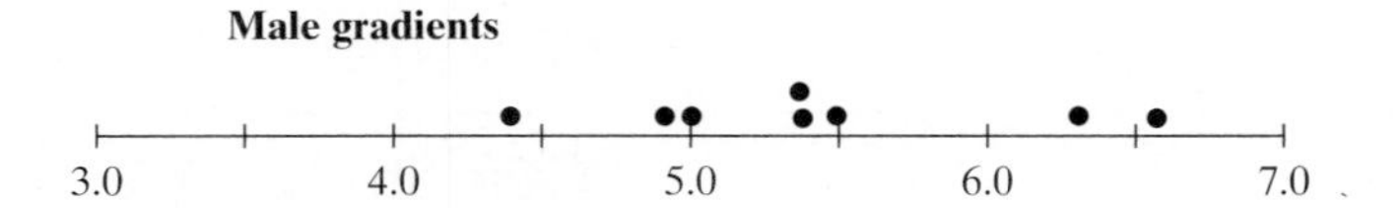

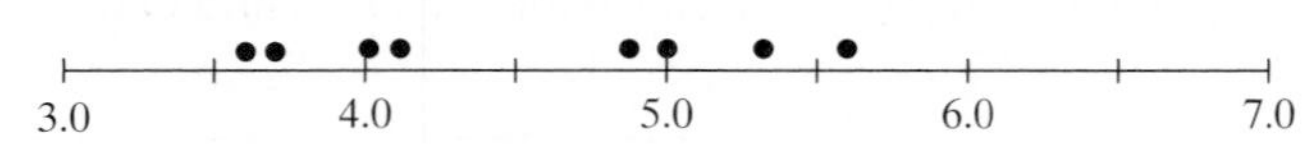

Figure 8.2.1 shows two such graphs for the gradients listed in Table 8.2.2. By the criteria just mentioned, there is nothing about either sample that would be inconsistent with the assumption that both the X_i's and the Y_j's are normally distributed.

k-Sample Data

When more than two factor levels are being compared, and when the observations are quantitative, have similar units, and are independent, the measurements are said to be *k-sample data*. Although their assumptions are comparable, two-sample data and k-sample data are treated as distinct experimental designs because the ways they are analyzed are totally different. The t test format that figures so prominently in the interpretation of one-sample and two-sample data cannot be extended to accommodate k-sample data. A more powerful technique, the *analysis of variance*, is needed and will be the sole topic of Chapters 12 and 13.

Their multiplicity of factor levels also requires that k-sample data be identified using double-subscript notation. The ith observation appearing in the jth factor level will be denoted Y_{ij}, so the model equations take the form

$$Y_{ij} = \mu_j + \varepsilon_{ij}, i = 1, 2, \ldots n_j, j = 1, 2, \ldots, k$$

where n_j denotes the sample size associated with the jth factor level ($n_1 + n_2 + \cdots + n_k = n$), and ε_{ij} is a normally distributed random variable with mean 0 and the same standard deviation σ for all i and j.

The first step in analyzing k-sample data is to test H_0: $\mu_1 = \mu_2 = \cdots = \mu_k$. Procedures are also available for testing *subhypotheses* involving certain factor levels irrespective of all the others—in effect, fine-tuning the focus of the inferences.

Case Study 8.2.3

Many studies have been undertaken to document the directional changes over time in the Earth's magnetic field. One approach compared the 1669, 1780, and 1865 eruptions of Mount Etna. For each seismic event, the magnetic field in the resulting molten lava aligned itself with the Earth's magnetic field as it prevailed at that time. When the lava cooled and hardened, the magnetic field was "captured" and its direction remained fixed. Table 8.2.3 lists the declinations of the magnetic field measured in three blocks of lava, randomly sampled from each of those three eruptions (170).

Table 8.2.3 Declination of Magnetic Field

	In 1669	In 1780	In 1865
	57.8	57.9	52.7
	60.2	55.2	53.0
	60.3	54.8	49.4
Averages:	59.4	56.0	51.7

About the Data Every factor in every experiment is said to be either a *fixed effect* or a *random effect*—a fixed effect if the factor's levels have been preselected by the experimenter, a random effect otherwise. Here "time" would be considered a random effect because its three levels—1669, 1780, and 1865—were not preselected. They were simply the times when the volcano erupted. Whether a factor is fixed or random does not affect the analysis of the experimental designs considered in this text. For more complicated, multifactor designs, though, the distinction is critical and often dictates the way an analysis proceeds.

Paired Data

In the two-sample and k-sample designs, factor levels are compared using *independent* samples. An alternative is to use *dependent* samples by grouping the subjects into n *blocks*. If only two factor levels are involved, the blocks are referred to as *pairs*, which gives the design its name.

The responses to factor levels X and Y in the *ith pair* are recorded as X_i and Y_i, respectively. Whatever contributions to those values are due to the conditions prevailing in Pair i will be denoted P_i. The model equations, then, can be written

$$X_i = \mu_X + P_i + \varepsilon_i, \quad i = 1, 2, \ldots, n$$

and

$$Y_i = \mu_Y + P_i + \varepsilon_i', \quad i = 1, 2, \ldots, n$$

where ε_i and ε_i' are independent normally distributed random variables with mean 0 and the same standard deviation σ. The fact that P_i is the same for both X_i and Y_i is what makes the samples dependent.

The statistical objective of two-sample data and paired data is often the same. Both use t tests to focus on the null hypothesis that the true means (μ_X and μ_Y) associated with the two factor levels are equal. A paired-data analysis, though, tests H_0: $\mu_X = \mu_Y$ by defining $\mu_D = \mu_X - \mu_Y$ and testing H_0: $\mu_0 = 0$. In effect, a *paired t test* is a *one-sample t test* done on the set of within-pair differences, $d_i = x_i - y_i$, $i = 1, 2, \ldots, n$.

Some of the more common ways to form paired data have already been mentioned on p. 433. A not-so-common application of one of those ways—time and place—is described in Case Study 8.2.4.

Case Study 8.2.4

There are many factors that predispose bees to sting (other than sheer orneriness...). A person wearing dark clothing, for instance, is more likely to get stung than someone wearing white. And someone whose movements are quick and jerky runs a higher risk than does a person who moves more slowly. Still another factor—one particularly important to apiarists—is whether or not the person has just been stung by other bees.

The influence of prior stings was simulated in an experiment by dangling eight cotton balls wrapped in muslin up and down in front of the entrance to

(Continued on next page)

a hive (53). Four of the balls had just been exposed to a swarm of angry bees and were filled with stings; the other four were "fresh." After a specified length of time, the number of new stings in each of the balls was counted. The entire procedure was repeated eight more times (see Table 8.2.4).

Table 8.2.4

Trial	Cotton Balls Previously Stung	Fresh Cotton Balls	Difference
1	27	33	−6
2	9	9	0
3	33	21	12
4	33	15	18
5	4	6	−2
6	21	16	5
7	20	19	1
8	33	15	18
9	70	10	60
		Average:	11.8

The last column in Table 8.2.4 gives the nine within-pair differences. The average of those differences is 11.8. The issue to be resolved—and what we need the paired t test to tell us—is whether the difference between 11.8 and 0 is statistically significant.

About the Data When two factor levels are to be compared, experimenters often have the choice of using either the two-sample format or the paired-data format. That would be the case here. If the experimenter dangled previously stung cotton balls in front of the hive on, say, nine occasions and did the same with fresh cotton balls on nine *other* occasions, the two samples would be independent, and the data set would qualify as a two-sample design.

Neither design is always better than the other, for a number of reasons detailed in Chapter 13. Sometimes, which is likely to be more effective is not obvious. The situation described in Case Study 8.2.4, though, is not one of those times! In general, the paired-data format is superior when excessive heterogeneity in the experimental environment or among the subjects is present. Is that the case here? Definitely. Bees have a well-deserved reputation for erratic, Jekyll-and-Hyde-type behavior. All sorts of transient factors might conceivably influence their responses to balls dangling in front of their hive. The two-sample format would allow all of that trial-to-trial variability *within* the factor levels to obscure the difference *between* the factor levels. That would be a very serious drawback to using a two-sample design in this particular context. In contrast, by targeting the within-pair *differences*, the paired-data design effectively eliminates the component P_i that appears in the model equations:

$$X_i - Y_i = \mu_X + P_i + \varepsilon_i - (\mu_Y + P_i + \varepsilon_i') = \mu_X - \mu_Y + \varepsilon_i - \varepsilon_i'$$

In short, the choice of an experimental design here is a no-brainer. The researchers who conducted this study did exactly what they should have done.

Randomized Block Data

Randomized block data have the same basic structure as paired data—quantitative measurements, similar units, and dependent samples; the only difference is that more than two factor levels are involved in randomized block data. Those additional factor levels, though, add a degree of complexity that the paired t test is unable to handle. Like k-sample data, randomized block data require the analysis of variance for their interpretation.

Suppose the data set consists of k factor levels, all of which are applied in each of b blocks. The model equation for Y_{ij}, the observation appearing in the ith block and receiving the jth factor level, then becomes

$$Y_{ij} = \mu_j + B_i + \varepsilon_{ij}, \quad i = 1, 2, \ldots, b; \quad j = 1, 2, \ldots, k$$

where μ_j is the true average response associated with the jth factor level, B_i is the portion of the value of Y_{ij} that can be attributed to the net effect of all the conditions that characterize Block i, and ε_{ij} is a normally distributed random variable with mean 0 and the same standard deviation σ for all i and j.

Case Study 8.2.5

Table 8.2.5 summarizes the results of a randomized block experiment set up to investigate the possible effects of "blood doping," a controversial procedure whereby athletes are injected with additional red blood cells for the purpose of enhancing their performance (15). Six runners were the subjects. Each was timed in three ten thousand-meter races: once after receiving extra red blood cells, once after being injected with a placebo, and once after receiving no treatment whatsoever. Listed are their times (in minutes) to complete the race.

Table 8.2.5

Subject	No Injection	Placebo	Blood Doping
1	34.03	34.53	33.03
2	32.85	32.70	31.55
3	33.50	33.62	32.33
4	32.52	31.23	31.20
5	34.15	32.85	32.80
6	33.77	33.05	33.07

Clearly, the times in a given row are dependent—all three reflect to some extent the inherent speed of the subject, regardless of which factor level might also be operative. Documenting differences from subject to subject, though,

(Continued on next page)

would not be the objective for doing this sort of study. If μ_1, μ_2, and μ_3 denote the true average times characteristic of the *no injection, placebo*, and *blood doping* factor levels, respectively, the experimenter's first priority would be to test H_0: $\mu_1 = \mu_2 = \mu_3$.

About the Data The name *randomized block* derives from one of the properties that such data supposedly have—namely, that the factor levels *within each block* have been applied in a random order. To do otherwise—that is, to take the measurements in any sort of systematic fashion (however well intentioned)—is to create the opportunity for the observations to become *biased*. If that worst-case scenario should happen, the data are worthless because there is no way to separate the "factor effect" from the "bias effect" (and, of course, there is no way to know for certain whether the data were biased in the first place).

For the same reasons, two-sample data and k-sample data should be *completely randomized*, which means that the entire set of measurements should be taken in a random order. Figure 8.2.2 shows an acceptable measurement sequence for the performance times in Table 8.2.5 and the magnetic field declinations in Table 8.2.3.

Figure 8.2.2

Subject	No Injection	Placebo	Blood Doping
1	2	3	1
2	1	2	3
3	2	1	3
4	2	1	3
5	3	2	1
6	3	1	2

In 1669	In 1780	In 1865
3	8	5
4	1	9
7	6	2

Regression Data

All the experimental designs introduced up to this point share the property that their measurements have the same units. Moreover, each has had the same basic objective: to quantify or to compare the effects of one or more factor levels. In contrast, *regression data* typically consist of measurements with dissimilar units, and the objective with them is to study the functional relationship between the variables.

Regression data often have the form (x_i, Y_i), $i = 1, 2, \ldots, n$, where x_i is the value of an independent variable (typically preselected by the experimenter) and Y_i is a dependent random variable (usually having units different from those of x_i). A particularly important special case is the *simple linear model*,

$$Y_i = \beta_0 + \beta_1 x_i + \varepsilon_i, \; i = 1, 2, \ldots, n$$

where ε_i is assumed to be normally distributed with mean 0 and standard deviation σ. Here $E(Y_i) = \beta_0 + \beta_1 x_i$, but more generally, $E(Y_i)$ can be any function $g(x_i)$—for example,

$$E(Y_i) = \beta_0 x_i^{\beta_1} \text{ or } E(Y_i) = \beta_0 e^{\beta_1 x_i}$$

The details will not be presented in this text, but the simple linear model can be extended to include k independent variables. The result is a *multiple linear regression* model,

$$Y_i = \beta_0 + \beta_1 x_{1i} + \beta_2 x_{2i} + \cdots + \beta_k x_{ki} + \varepsilon_i,\ i = 1, 2, \ldots, n$$

An important special case of the regression model occurs when the x_i's are not preselected by the experimenter. Suppose, for example, that the relationship between height and weight is to be studied in adult males. One way to collect a set of relevant data would be to choose a random sample of n adult males and record each subject's height and weight. Neither variable in that case would be preselected or controlled by the experimenter: the height, X_i, and the weight, Y_i, of the ith subject are both random variables, and the measurements in that case—(X_1, Y_1), (X_2, Y_2), ..., (X_n, Y_n)—are said to be *correlation data*. The usual assumption invoked for correlation data is that the (X, Y)'s are jointly distributed according to a bivariate normal distribution (see Figure 8.2.3).

Figure 8.2.3

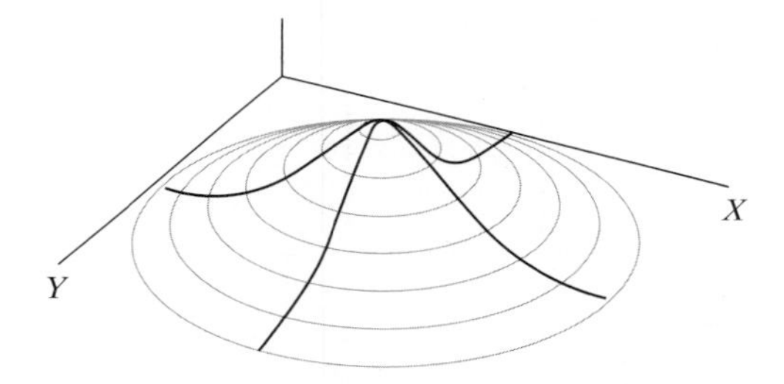

The implications of the independent variable being either preselected (x_i) or random (X_i) will be explored at length in Chapter 11. Suffice it to say that if the objective is to summarize the relationship between the two variables with a straight line, as it is in Figure 8.2.4, it makes absolutely no difference whether the data have the form (x_i, Y_i) or (X_i, Y_i)—the resulting equation will be the same.

Case Study 8.2.6

One of the most startling and profound scientific revelations of the twentieth century was the evidence, discovered in 1929 by the American astronomer Edwin Hubble, that the universe is expanding. Hubble's research shattered forever the ancient belief that the heavens are basically in a state of cosmic equilibrium: quite the contrary, galaxies are receding from each other at mind-bending velocities (the cluster Hydra, for example, is moving away from other clusters at the rate of 38.0 thousand miles/sec).

(Continued on next page)

If y is a galaxy's recession velocity (relative to that of any other galaxy) and x is its distance (from that other galaxy), Hubble's law states that

$$y = Hx$$

where H is known as Hubble's constant. Table 8.2.6 summarizes his findings—listed are distance and velocity determinations made for eleven galactic clusters (23).

Table 8.2.6

Cluster	Distance, x (millions of light-years)	Velocity, y (thousands of miles/sec)
Virgo	22	0.75
Pegasus	68	2.4
Perseus	108	3.2
Coma Berenices	137	4.7
Ursa Major No. 1	255	9.3
Leo	315	12.0
Corona Borealis	390	13.4
Gemini	405	14.4
Bootes	685	24.5
Ursa Major No. 2	700	26.0
Hydra	1100	38.0

For these data, the value H is estimated to be 0.03544 (using a technique covered in Chapter 11). Figure 8.2.4 shows that

$$y = 0.03544x$$

fits the data exceptionally well.

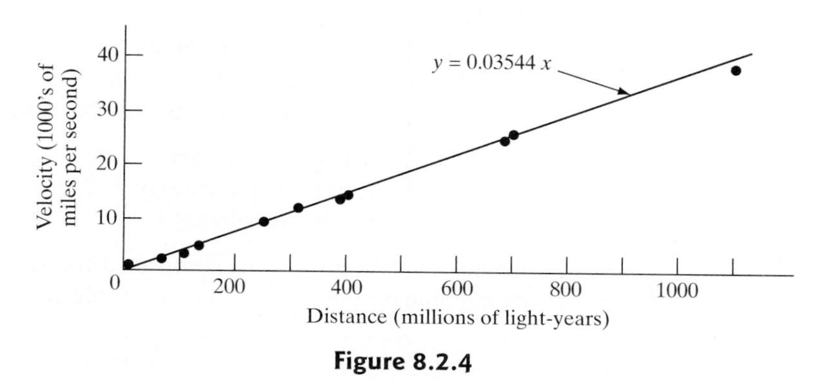

Figure 8.2.4

About the Data Techniques for measuring interstellar distances have been greatly refined since the 1920s when Hubble reported the data in Table 8.2.6. The most recent estimates yield a value for Hubble's constant about a third as large as the slope shown in Figure 8.2.4. That particular adjustment is critical because the reciprocal of Hubble's constant can be used to calculate the age of the universe, or, at the very least, the time elapsed since the Big Bang [see (96)]. Based on the revised data, the Big Bang occurred some fifteen billion years ago, a number that agrees with estimates found using other methods.

Comment Look again at the graph of Hubble's data in Figure 8.2.4. Which is the appropriate description of the eleven (distance, velocity) measurements—are they (x_i, Y_i)'s or (X_i, Y_i)'s? The answer is not obvious. At first glance, these would appear to be correlation data—distance (X) and velocity (Y) measurements having been made jointly on a random sample of eleven galactic clusters. Arguing against that conclusion is the spacing of the points. With correlation data, the bulk of the X measurements lie near the center of their range, which is not the case here. Perhaps the reason for the unusual spacing is a set of constraints imposed by the other-worldly nature of the data, or maybe it suggests that Hubble, for whatever reasons, preselected the clusters because of their distances.

Categorical Data

Suppose two *qualitative, dissimilar* variables are observed on each of n subjects, where the first variable has R possible values and the second variable, C possible values. We call such measurements *categorical data*.

The number of times each value of one variable occurs with each value of the other variable is typically displayed in a *contingency table*, which necessarily has R rows and C columns. Whether the two variables are *independent* is the question that an experimenter can use categorical data to answer.

Case Study 8.2.7

Is there a relationship between a physician's malpractice history (X) and his or her specialty (Y)? Three "values" of X were looked at, as well as three "values" of Y (29):

$$X = \begin{cases} \text{no malpractice claims} \\ \text{one or more claims ending in damages awarded} \\ \text{one or more claims filed but none requiring compensation} \end{cases}$$

$$Y = \begin{cases} \text{orthopedic surgery} \\ \text{obstetrics-gynecology} \\ \text{internal medicine} \end{cases}$$

A total of 1942 physicians comprised the sample. The resulting (X, Y) values are summarized in the contingency table shown in Figure 8.2.5.

	Orth. Surg.	Ob-Gyn	Int. Med.	Totals
No claims	147	349	709	1205
At least one claim lost	106	149	62	317
No claims lost	156	149	115	420
Totals:	409	647	886	1942

Figure 8.2.5

(Continued on next page)

The hypotheses to be tested in any categorical-data problem always have the same form:

$$H_0: X \text{ and } Y \text{ are independent}$$

versus

$$H_1: X \text{ and } Y \text{ are dependent}$$

The formal procedure for choosing between H_0 and H_1 is a *chi square test*, which will be covered in Chapter 10. A quick look at these data leaves no doubt that H_0 would be overwhelmingly rejected. If X and Y are independent, the probability that a physician receives, say, *no claims*, should be the same for all three specialties. The sample proportions of *no claims*, though, are dramatically different from specialty to specialty:

for Orthopedic surgery—147/409 = 35.9%

for Ob-Gyn—349/647 = 53.9%

for Internal medicine—706/886 = 80.0%

Clearly, the variables X and Y are dependent.

About the Data The categorical-data format "overlaps" the two-sample data format for one particular type of measurement. Consider the simplest version of categorical data, where both X and Y have only two values. Call the two values of X "success" and "failure," and the two values of Y "Level 1" and "Level 2." Given a sample of n such observations, the corresponding contingency table would look like Figure 8.2.6.

Figure 8.2.6

			Y		
			Level 1	Level 2	Totals
	Success		a	b	a + b
X	Failure		c	d	c + d
		Totals:	a + c	b + d	n = a + b + c + d

Notice that the "a" and "c" in Column 1 are another way of expressing the numbers of 1's and 0's, respectively, in a sequence of a + c Bernoulli trials. Similarly, the "b" and "d" in Column 2 tally up the 1's and 0's, respectively, in a second set of b + d Bernoulli trials. To say that X and Y are dependent (in the categorical-data sense) is to say that the difference between a/(a + c) and b/(b + d) is statistically significant (in the two-sample data sense). The two data models answer their respective questions with different statistical tests, but the two procedures (a chi square test and a Z test) are equivalent—one will reject H_0 if and only if the other rejects H_0.

A Flowchart for Classifying Data

Differentiating the seven data formats just discussed are the answers to the four questions cited at the beginning of this section: Are the data qualitative or quantitative? Are the units similar or dissimilar? How many factor levels are involved? Are the samples dependent or independent? The flowchart pictured in Figure 8.2.7 shows the sequence of responses that leads to each of the seven models.

Figure 8.2.7

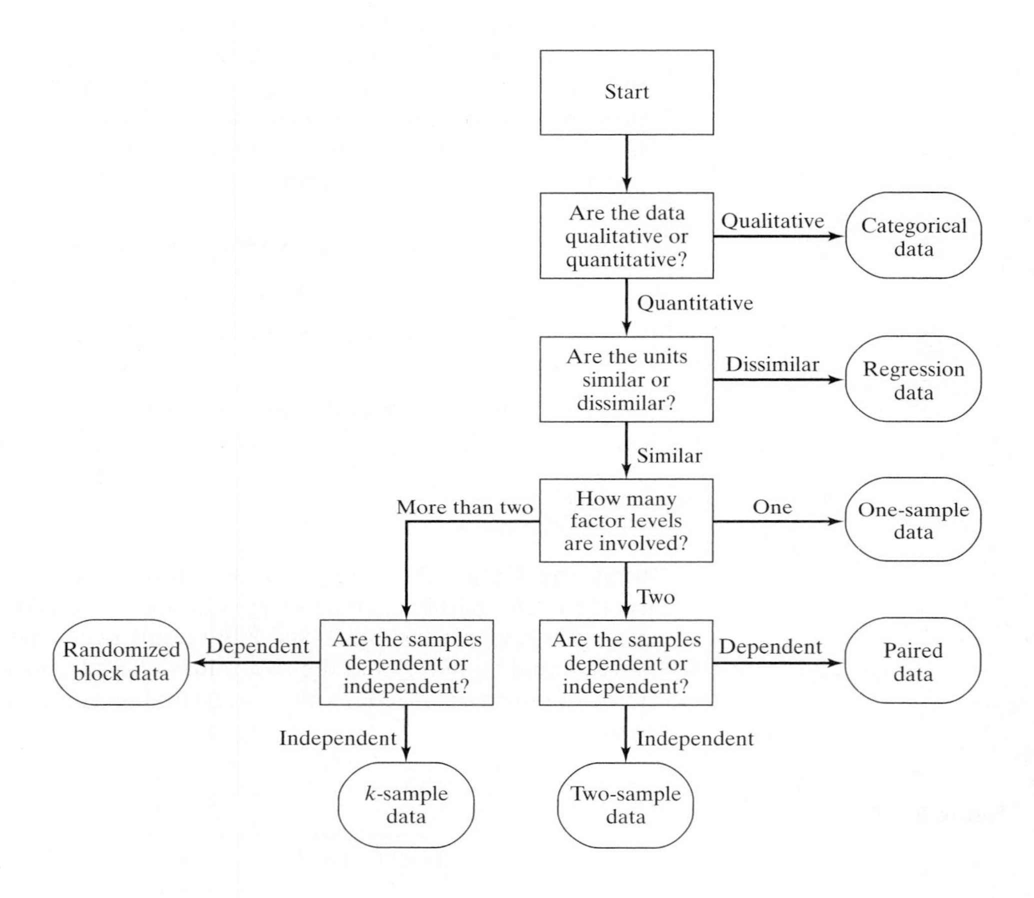

Example 8.2.1

The federal Community Reinvestment Act of 1977 was enacted out of concern that banks were reluctant to make loans in low- and moderate-income areas, even when applicants seemed otherwise acceptable. The figures in Table 8.2.7 show one particular bank's credit penetration in ten low-income census tracts (A through J) and ten high-income census tracts (K through T). To which of the seven models do these data belong?

Note, first, that the measurements (1) are quantitative and (2) have similar units. *Low-income* and *High-income* correspond to *two* treatment levels, and the two samples are clearly independent (the *4.6* recorded in tract A, for example, has nothing specific in common with the *11.6* recorded in tract K). From the flowchart, then, the answers *quantitative/similar/two/independent* imply that these are *two-sample data*.

Table 8.2.7

Low-Income Census Tract	Percent of Households with Credit	High-Income Census Tract	Percent of Households with Credit
A	4.6	K	11.6
B	6.6	L	8.5
C	3.3	M	8.2
D	9.8	N	15.1
E	6.9	O	12.6
F	11.0	P	11.3
G	6.0	Q	9.1
H	4.6	R	4.2
I	4.2	S	6.4
J	5.1	T	5.9

■

Example 8.2.2

Individuals looking at the vertical lines in Figure 8.2.8 will tend to perceive the right one as shorter, even though the two are equal. Moreover, the perceived difference in those lengths—what psychologists call the "strength" of the illusion—has been shown to be a function of age.

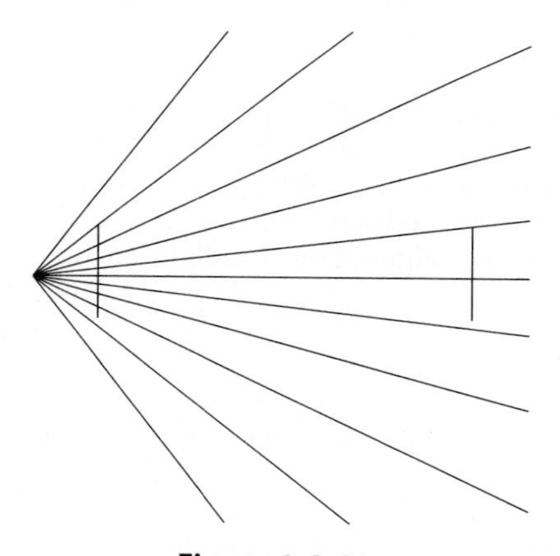

Figure 8.2.8

A study was done to see whether individuals who are hypnotized and regressed to different ages also perceive the illusion differently. Table 8.2.8 shows the illusion strengths measured for eight subjects while they were (1) awake, (2) regressed to age nine, and (3) regressed to age five (137). Which of the seven experimental designs do these data represent?

Look again at the sequence of questions posed by the flowchart in Figure 8.2.7:

1. Are the data qualitative or quantitative? *Quantitative*
2. Are the units similar or dissimilar? *Similar*
3. How many factor levels are involved? *More than two*
4. Are the observations dependent or independent? *Dependent*

Table 8.2.8

Subject	(1) Awake	(2) Regressed to Age 9	(3) Regressed to Age 5
1	0.81	0.69	0.56
2	0.44	0.31	0.44
3	0.44	0.44	0.44
4	0.56	0.44	0.44
5	0.19	0.19	0.31
6	0.94	0.44	0.69
7	0.44	0.44	0.44
8	0.06	0.19	0.19

According to the flowchart, then, these measurements qualify as *randomized block data.* ■

Questions

For Questions 8.2.1–8.2.12 use the flowchart in Figure 8.2.7 to identify the experimental designs represented. In each case, answer whichever of the questions on p. 435 are necessary to make the determination.

8.2.1. Kepler's Third Law states that "the squares of the periods of the planets are proportional to the cubes of their mean distance from the Sun." Listed below are the periods of revolution (x), the mean distances from the sun (y), and the values x^2/y^3 for the eight planets in the solar system (3).

Planet	x_i (years)	y_i (astronomical units)	x_i^2/y_i^3
Mercury	0.241	0.387	1.002
Venus	0.615	0.723	1.001
Earth	1.000	1.000	1.000
Mars	1.881	1.524	1.000
Jupiter	11.86	5.203	0.999
Saturn	29.46	9.54	1.000
Uranus	84.01	19.18	1.000
Neptune	164.8	30.06	1.000

8.2.2. Mandatory helmet laws for motorcycle riders remain a controversial issue. Some states have a "limited" ordinance that applies only to younger riders; others have a "comprehensive" statute requiring all riders to wear helmets. Listed in the next column are the deaths per ten thousand registered motorcycles reported by states having each type of legislation (184).

Limited Helmet Law			Comprehensive Helmet Law		
6.8	7.0	9.1	7.1	4.8	7.0
10.6	4.1	0.5	11.2	5.0	6.8
9.6	5.7	6.7	17.9	8.1	7.3
9.1	7.6	6.4	11.3	5.5	12.9
5.2	3.0	4.7	8.5	11.7	3.7
13.2	6.7	15.0	9.3	4.0	5.2
6.9	7.3	4.7	5.4	7.0	6.9
8.1	4.2	4.8	10.5	9.3	8.6

8.2.3. *Aedes aegypti* is the scientific name of the mosquito that transmits yellow fever. Although no longer a major health problem in the Western world, yellow fever was perhaps the most devastating communicable disease in the United States for almost two hundred years. To see how long it takes the *Aedes* mosquito to complete a feeding, five young females were allowed to bite an exposed human forearm without the threat of being swatted. The resulting blood-sucking times (in seconds) are summarized below (89).

Mosquito	Bite Duration (sec)
1	176.0
2	202.9
3	315.0
4	374.6
5	352.5

8.2.4. Male cockroaches can be very antagonistic toward other male cockroaches. Encounters may be fleeting or quite spirited, the latter often resulting in missing antennae and broken wings. A study was done to see whether cockroach density has any effect on the frequency of serious altercations. Ten groups of four male cockroaches (*Byrsotria fumigata*) were each subjected to three levels of density: high, intermediate, and low. The following are the numbers of "serious" encounters per minute that were observed (14).

Group	High	Intermediate	Low
1	0.30	0.11	0.12
2	0.20	0.24	0.28
3	0.17	0.13	0.20
4	0.25	0.36	0.15
5	0.27	0.20	0.31
6	0.19	0.12	0.16
7	0.27	0.19	0.20
8	0.23	0.08	0.17
9	0.37	0.18	0.18
10	0.29	0.20	0.20
Averages:	0.25	0.18	0.20

8.2.5. Luxury suites, many costing more than $100,000 to rent, have become big-budget status symbols in new sports arenas. Below are the numbers of suites (x) and their projected revenues (y) for nine of the country's newest facilities (196).

Arena	Number of Suites, x	Projected Revenues (in millions), y
Palace (Detroit)	180	$11.0
Orlando Arena	26	1.4
Bradley Center (Milwaukee)	68	3.0
America West (Phoenix)	88	6.0
Charlotte Coliseum	12	0.9
Target Center (Minneapolis)	67	4.0
Salt Lake City Arena	56	3.5
Miami Arena	18	1.4
ARCO Arena (Sacramento)	30	2.7

8.2.6. Depth perception is a life-or-death ability for lambs inhabiting rugged mountain terrain. How quickly a lamb develops that faculty may depend on the amount of time it spends with its ewe. Thirteen sets of lamb littermates were the subjects of an experiment that addressed that question (99). One member of each litter was left with its mother; the other was removed immediately after birth. Once every hour, the lambs were placed on a simulated cliff, part of which included a platform of glass. If a lamb placed its feet on the glass, it "failed" the test, since that would have been equivalent to walking off the cliff. Below are the trial numbers when the lambs first learned not to walk on the glass—that is, when they first developed depth perception.

Number of Trials to Learn Depth Perception		
Group	Mothered, x_i	Unmothered, y_i
1	2	3
2	3	11
3	5	10
4	3	5
5	2	5
6	1	4
7	1	2
8	5	7
9	3	5
10	1	4
11	7	8
12	3	12
13	5	7

8.2.7. To see whether teachers' expectations for students can become self-fulfilling prophecies, fifteen first graders were given a standard IQ test. The childrens' teachers, though, were told it was a special test for predicting whether a child would show sudden spurts of intellectual growth in the near future (see 147). Researchers divided the children into three groups of sizes 6, 5, and 4 *at random*, but they informed the teachers that, according to the test, the children in group I would not demonstrate any pronounced intellectual growth for the next year, those in group II would develop at a moderate rate, and those in group III could be expected to make exceptional progress. A year later, the same fifteen children were again given a standard IQ test. Below are the differences in the two scores for each child (second test – first test).

Changes in IQ (second test – first test)		
Group I	Group II	Group III
3	10	20
2	4	9
6	11	18
10	14	19
10	3	
5		

8.2.8. Among young drivers, roughly a third of all fatal automobile accidents are speed-related; by age 60 that proportion drops to about one-tenth. Listed below are a recent year's percentages of speed-related fatalities for ages ranging from 16 to 72 (189).

Age	Percent Speed-Related Fatalities
16	37
17	32
18	33
19	34
20	33
22	31
24	28
27	26
32	23
42	16
52	13
57	10
62	9
72	7

8.2.9. Gorillas are not the solitary creatures that they are often made out to be: they live in groups whose average size is about 16, which usually includes three adult males, six adult females, and seven "youngsters." Listed below are the sizes of ten groups of mountain gorillas observed in the volcanic highlands of the Albert National Park in the Congo (157).

Group	No. of Gorillas
1	8
2	19
3	5
4	24
5	11
6	20
7	18
8	21
9	27
10	16

8.2.10. Roughly 360,000 bankruptcies were filed in U.S. Federal Court during 1981; by 1990 the annual number was more than twice that figure. The following are the numbers of business failures reported year by year through the 1980s (175).

Year	Bankruptcies Filed
1981	360,329
1982	367,866
1983	374,734
1984	344,275
1985	364,536
1986	477,856
1987	561,274
1988	594,567
1989	642,993
1990	726,484

8.2.11. The diversity of bird species in a given area is related to plant diversity, as measured by variation in foliage heights as well as the variety of flora. Below are indices measured on those two traits for thirteen desert-type habitats (109).

Area	Plant Cover Diversity, x_i	Bird Species Diversity, y_i
1	0.90	1.80
2	0.76	1.36
3	1.67	2.92
4	1.44	2.61
5	0.20	0.42
6	0.16	0.49
7	1.12	1.90
8	1.04	2.38
9	0.48	1.24
10	1.33	2.80
11	1.10	2.41
12	1.56	2.80
13	1.15	2.16

8.2.12. Male toads often have trouble distinguishing between other male toads and female toads, a state of affairs that can lead to awkward moments during mating season. When male toad A inadvertently makes inappropriate romantic overtures to male toad B, the latter emits a short call known as a release chirp. Below are the lengths of the release chirps measured for fifteen male toads innocently caught up in misadventures of the heart (17).

Toad	Length of Release Chirp (sec)
1	0.11
2	0.06
3	0.06
4	0.06
5	0.11
6	0.08
7	0.08
8	0.10
9	0.06
10	0.06
11	0.15
12	0.16
13	0.11
14	0.10
15	0.07

For Questions 8.2.13–8.2.32 identify the experimental design (one-sample, two-sample, etc.) that each set of data represents.

8.2.13. A pharmaceutical company is testing two new drugs designed to improve the blood-clotting ability of hemophiliacs. Six subjects volunteering for the study are randomly divided into two groups of size 3. The first group is given drug A; the second group, drug B. The response variable in each case is the subject's prothrombin time, a number that reflects the time it takes for a clot to form. The results (in seconds) for group A are 32.6, 46.7, and 81.2; for group B, 25.9, 33.6, and 35.1.

8.2.14. Investment firms financing the construction of new shopping centers pay close attention to the amount of retail floor space already available. Listed below are population and floor space figures for five southern cities.

City	Population, x	Retail Floor Space (in million square meters), y
1	400,000	3,450
2	150,000	1,825
3	1,250,000	7,480
4	2,975,000	14,260
5	760,000	5,290

8.2.15. Nine political writers were asked to assess the United States' culpability in murders committed by revolutionary groups financed by the CIA. Scores were assigned using a scale of 0 to 100. Three of the writers were native Americans living in the United States, three were native Americans living abroad, and three were foreign nationals.

Americans in U.S.	Americans Abroad	Foreign Nationals
45	65	75
45	50	90
40	55	85

8.2.16. To see whether low-priced homes are easier to sell than moderately priced homes, a national realty company collected the following information on the lengths of times homes were on the market before being sold.

	Number of Days on Market	
City	Low-Priced	Moderately Priced
Buffalo	55	70
Charlotte	40	30
Newark	70	110

8.2.17. The following is a breakdown of what 120 college freshmen intend to do next summer.

	Work	School	Play
Male	22	14	19
Female	14	31	20

8.2.18. An efficiency study was done on the delivery of first-class mail originating from the four cities listed in the following table. Recorded for each city was the average length of time (in days) that it took a letter to reach a destination in that same city. Samples were taken on two occasions, Sept. 1, 2001 and Sept. 1, 2004.

City	Sept. 1, 2001	Sept. 1, 2004
Wooster	1.8	1.7
Midland	2.0	2.0
Beaumont	2.2	2.5
Manchester	1.9	1.7

8.2.19. Two methods (A and B) are available for removing dangerous heavy metals from public water supplies. Eight water samples collected from various parts of the United States were used to compare the two methods. Four were treated with Method A and four were treated with Method B. After the processes were completed, each sample was rated for purity on a scale of 1 to 100.

Method A	Method B
88.6	81.4
92.1	84.6
90.7	91.4
93.6	78.6

8.2.20. Out of 120 senior citizens polled, 65 favored a complete overhaul of the health care system while 55 preferred more modest changes. When the same choice was put to 85 first-time voters, 40 said they were in favor of major reform while 45 opted for minor revisions.

8.2.21. To illustrate the complexity and arbitrariness of IRS regulations, a tax-reform lobbying group has sent the same five clients to each of two professional tax preparers. The following are the estimated tax liabilities quoted by each of the preparers.

Client	Preparer A	Preparer B
GS	$31,281	$26,850
MB	14,256	13,958
AA	26,197	25,520
DP	8,283	9,107
SB	47,825	43,192

8.2.22. The production of a certain organic chemical requires ammonium chloride. The manufacturer can obtain the ammonium chloride in one of three forms: powdered, moderately ground, and coarse. To see if the consistency of the NH_4Cl is itself a factor that needs to be considered, the manufacturer decides to run the reaction seven times with each form of ammonium chloride. The following are the resulting yields (in pounds).

Powdered NH_4Cl	Moderately Ground NH_4Cl	Coarse NH_4Cl
146	150	141
152	144	138
149	148	142
161	155	146
158	154	139
154	148	137
149	150	145

8.2.23. An investigation was conducted of 107 fatal poisonings of children. Each death was caused by one of three drugs. In each instance it was determined how the child received the fatal overdose. Responsibility for the 107 accidents was assessed according to the following breakdown.

	Drug A	Drug B	Drug C
Child Responsible	10	10	18
Parent Responsible	10	14	10
Another Person Responsible	4	18	13

8.2.24. As part of an affirmative-action litigation, records were produced showing the average salaries earned by white, black, and Hispanic workers in a large manufacturing plant. Three different departments were selected at random for the comparison. The entries shown are average annual salaries, in thousands of dollars.

	White	Black	Hispanic
Department 1	40.2	39.8	39.9
Department 2	40.6	39.0	39.2
Department 3	39.7	40.0	38.4

8.2.25. In Eastern Europe a study was done on fifty people bitten by rabid animals. Twenty victims were given the standard Pasteur treatment, while the other thirty were given the Pasteur treatment in addition to one or more doses of antirabies gamma globulin. Nine of those given the standard treatment survived; twenty survived in the gamma globulin group.

8.2.26. To see if any geographical pricing differences exist, the cost of a basic-cable TV package was determined for a random sample of six cities, three in the southeast and three in the northwest. Monthly charges for the southeastern cities were $13.20, $11.55, and $16.75; residents in the three northwestern cities paid $14.80, $17.65, and $19.20.

8.2.27. A public relations firm hired by a would-be presidential candidate has conducted a poll to see whether their client faces a gender gap. Out of 800 men interviewed, 325 strongly supported the candidate, 151 were strongly opposed, and 324 were undecided. Among the 750 women included in the sample, 258 were strong supporters, 241 were strong opponents, and 251 were undecided.

8.2.28. As part of a review of its rate structure, an automobile insurance company has compiled the following data on claims filed by five male policyholders and five female policyholders.

Client (male)	Claims Filed in 2004	Client (female)	Claims Filed in 2004
MK	$2750	SB	0
JM	0	ML	0
AK	0	MS	0
KT	$1500	BM	$2150
JT	0	LL	0

8.2.29. A company claims to have produced a blended gasoline that can improve a car's fuel consumption. They decide to compare their product with the leading gas currently on the market. Three different cars were used for the test: a Porsche, a Buick, and a VW. The Porsche got 13.6 mpg with the new gas and 12.2 mpg with the "standard" gas; the Buick got 18.7 mpg with the new gas and 18.5 with the standard; the figures for the VW were 34.5 and 32.6, respectively.

8.2.30. In a survey conducted by State University's Learning Center, a sample of three freshmen said they studied 6, 4, and 10 hours, respectively, over the weekend. The same question was posed to three sophomores, who reported study times of 4, 5, and 7 hours. For three juniors, the responses were 2, 8, and 6 hours.

8.2.31. A consumer advocacy group, investigating the prices of steel-belted radial tires produced by three major manufacturers, collects the following data.

Year	Company A	Company B	Company C
1995	\$62.00	\$68.00	\$65.00
2000	\$70.00	\$72.00	\$69.00
2005	\$78.00	\$75.00	\$75.00

8.2.32. A small fourth-grade class is randomly split into two groups. Each group is taught fractions using a different method. After three weeks, both groups are given the same 100-point test. The scores of students in the first group are 82, 86, 91, 72, and 68; the scores reported for the second group are 76, 63, 80, 72, and 67.

8.3 Taking a Second Look at Statistics (Samples Are Not "Valid"!)

Designing an experiment invariably requires that two fundamental issues be resolved. First and foremost is the choice of the design itself. Based on the type of data available and the objectives to be addressed, what overall "structure" should the experiment have? Seven of the most frequently occurring answers to that question are the seven models profiled in this chapter, ranging from the simplicity of the one-sample design to the complexity of the randomized block design.

As soon as a design has been chosen, a second question immediately follows: How large should the sample size (or sample sizes) be? It is precisely that question, though, that leads to a very common sampling misconception. There is a widely held belief (even by many experienced experimenters, who should know better) that some samples are "valid" (presumably because of their size), while others are not. Every consulting statistician could probably retire to Hawaii at an early age if he or she got a dollar for every time an experimenter posed the following sort of question: "I intend to compare Treatment X and Treatment Y using the two-sample format. My plan is to take twenty measurements on each of the two treatments. Will those be valid samples?"

The sentiment behind such a question is entirely understandable: the researcher is asking whether two samples of size 20 will be "adequate" (in some sense) for addressing the objectives of the experiment. Unfortunately, the word "valid" is meaningless in this context. There is no such thing as a valid sample because the word "valid" has no statistical definition.

To be sure, we have already learned how to calculate the smallest values of n that will achieve certain objectives, typically expressed in terms of the precision of an estimator or the power of a hypothesis test. Recall Theorem 5.3.2. To guarantee that the estimator X/n for the binomial parameter p has at least a $100(1-\alpha)\%$ chance of lying within a distance d of p requires that n be as least as large as $z_{\alpha/2}^2/4d^2$.

Suppose, for example, that we want a sample size capable of guaranteeing that X/n will have an $80\%[=100(1-\alpha)\%]$ chance of being within $0.05\,(=d)$ of p. By Theorem 5.3.2,

$$n \geq \frac{(1.28)^2}{4(0.05)^2} = 164$$

On the other hand, that sample of $n = 164$ would not be large enough to guarantee that X/n has, say, a *95%* chance of being within *0.03* of p. To meet these latter requirements, n would have to be as least as large as *1068* $[=(1.96)^2/4(0.03)^2]$.

Therein lies the problem. Sample sizes that can satisfy one set of specifications will not necessarily be capable of satisfying another. There is no "one size fits all" value for n that qualifies a sample as being "adequate" or "sufficient" or "valid."

In a broader sense, the phrase "valid sample" is much like the expression "statistical tie" discussed in Section 5.3. Both are widely used, and each is a well-intentioned attempt to simplify an important statistical concept. Unfortunately, both also share the dubious distinction of being mathematical nonsense.

Chapter 9

TWO-SAMPLE INFERENCES

After earning an Oxford degree in mathematics and chemistry, Gosset began working in 1899 for Messrs. Guinness, a Dublin brewery. Fluctuations in materials and temperature and the necessarily small-scale experiments inherent in brewing convinced him of the necessity for a new, small-sample theory of statistics. Writing under the pseudonym "Student," he published work with the t ratio that was destined to become a cornerstone of modern statistical methodology.

—William Sealy Gosset ("Student") (1876–1937)

9.1 Introduction

The simplicity of the one-sample model makes it the logical starting point for any discussion of statistical inference, but it also limits its applicability to the real world. Very few experiments involve just a single treatment or a single set of conditions. On the contrary, researchers almost invariably design experiments to compare responses to *several* treatment levels—or, at the very least, to compare a single treatment with a control.

In this chapter we examine the simplest of these multilevel designs, *two-sample inferences*. Structurally, two-sample inferences always fall into one of two different formats: Either two (presumably) different treatment levels are applied to two independent sets of similar subjects or the same treatment is applied to two (presumably) different kinds of subjects. Comparing the effectiveness of germicide A relative to that of germicide B by measuring the zones of inhibition each one produces in two sets of similarly cultured Petri dishes would be an example of the first type. On the other hand, examining the bones of sixty-year-old men and sixty-year-old women, all lifelong residents of the same city, to see whether both sexes absorb environmental strontium-90 at the same rate would be an example of the second type.

Inference in two-sample problems usually reduces to a comparison of *location* parameters. We might assume, for example, that the population of responses associated with, say, treatment X is normally distributed with mean μ_X and standard deviation σ_X while the Y distribution is normal with mean μ_Y and standard deviation σ_Y. Comparing location parameters, then, reduces to testing H_0: $\mu_X = \mu_Y$. As always, the alternative may be either one-sided, H_1: $\mu_X < \mu_Y$ or H_1: $\mu_X > \mu_Y$, or two-sided, H_1: $\mu_X \neq \mu_Y$. (If the data are binomial, the location parameters are p_X and p_Y, the true "success" probabilities for treatments X and Y, and the null hypothesis takes the form H_0: $p_X = p_Y$.)

Sometimes, although much less frequently, it becomes more relevant to compare the *variabilities* of two treatments, rather than their locations. A food company, for example, trying to decide which of two types of machines to buy for filling cereal boxes would naturally be concerned about the *average* weights of the boxes filled by each type, but they would also want to know something about the *variabilities* of the weights. Obviously, a machine that produces high proportions of "underfills" and "overfills" would be a distinct liability. In a situation of this sort, the appropriate null hypothesis is H_0: $\sigma_X^2 = \sigma_Y^2$.

For comparing the means of two normal populations when $\sigma_X = \sigma_Y$, the standard procedure is the *two-sample t test*. As described in Section 9.2, this is a relatively straightforward extension of Chapter 7's one-sample t test. If $\sigma_X \neq \sigma_Y$, an approximate t test is used. For comparing variances, though, it will be necessary to introduce a completely new test—this one based on the F distribution of Section 7.3. The binomial version of the two-sample problem, testing H_0: $p_X = p_Y$, is taken up in Section 9.4.

It was mentioned in connection with one-sample problems that certain inferences, for various reasons, are more aptly phrased in terms of confidence intervals rather than hypothesis tests. The same is true of two-sample problems. In Section 9.5, confidence intervals are constructed for the location *difference* of two populations, $\mu_X - \mu_Y$ (or $p_X - p_Y$), and the variability *quotient*, σ_X^2/σ_Y^2.

9.2 Testing H_0: $\mu_X = \mu_Y$

We will suppose that the data for a given experiment consist of two independent random samples, $X_1, X_2, \ldots, X_n$ and $Y_1, Y_2, \ldots, Y_m$, representing either of the models referred to in Section 9.1. Furthermore, the two populations from which the X's and Y's are drawn will be presumed normal. Let μ_X and μ_Y denote their means. Our objective is to derive a procedure for testing H_0: $\mu_X = \mu_Y$.

As it turns out, the precise form of the test we are looking for depends on the variances of the X and Y populations. If it can be assumed that σ_X^2 and σ_Y^2 are equal, it is a relatively straightforward task to produce the GLRT for H_0: $\mu_X = \mu_Y$. (This is, in fact, what we will do in Theorem 9.2.2.) But if the variances of the two populations are *not* equal, the problem becomes much more complex. This second case, known as the Behrens-Fisher problem, is more than seventy-five years old and remains one of the more famous "unsolved" problems in statistics. What headway investigators *have* made has been confined to approximate solutions. These will be discussed later in this section. For what follows next, it can be assumed that $\sigma_X^2 = \sigma_Y^2$.

For the one-sample test $\mu = \mu_0$, the GLRT was shown to be a function of a special case of the t ratio introduced in Definition 7.3.3 (recall Theorem 7.3.5). We begin this section with a theorem that gives still another special case of Definition 7.3.3.

Theorem 9.2.1 *Let $X_1, X_2, \ldots, X_n$ be a random sample of size n from a normal distribution with mean μ_X and standard deviation σ and let $Y_1, Y_2, \ldots, Y_m$ be an independent random sample of size m from a normal distribution with mean μ_Y and standard deviation σ.*

Let S_X^2 and S_Y^2 be the two corresponding sample variances, and S_p^2 the pooled variance, where

$$S_p^2 = \frac{(n-1)S_X^2 + (m-1)S_Y^2}{n+m-2} = \frac{\sum_{i=1}^{n}(X_i - \overline{X})^2 + \sum_{i=1}^{m}(Y_i - \overline{Y})^2}{n+m-2}$$

Then

$$T_{n+m-2} = \frac{\overline{X} - \overline{Y} - (\mu_X - \mu_Y)}{S_p\sqrt{\frac{1}{n} + \frac{1}{m}}}$$

has a Student t distribution with $n + m - 2$ degrees of freedom.

Proof The method of proof here is very similar to what was used for Theorem 7.3.5. Note that an equivalent formulation of T_{n+m-2} is

$$T_{n+m-2} = \frac{\frac{\overline{X}-\overline{Y}-(\mu_X-\mu_Y)}{\sigma\sqrt{\frac{1}{n}+\frac{1}{m}}}}{\sqrt{S_p^2/\sigma^2}}$$

$$= \frac{\frac{\overline{X}-\overline{Y}-(\mu_X-\mu_Y)}{\sigma\sqrt{\frac{1}{n}+\frac{1}{m}}}}{\sqrt{\frac{1}{n+m-2}\left[\sum_{i=1}^{n}\left(\frac{X_i-\overline{X}}{\sigma}\right)^2 + \sum_{i=1}^{m}\left(\frac{Y_i-\overline{Y}}{\sigma}\right)^2\right]}}$$

But $E(\overline{X} - \overline{Y}) = \mu_X - \mu_Y$ and $\text{Var}(\overline{X} - \overline{Y}) = \sigma^2/n + \sigma^2/m$, so the numerator of the ratio has a standard normal distribution, $f_Z(z)$.

In the denominator,

$$\sum_{i=1}^{n}\left(\frac{X_i - \overline{X}}{\sigma}\right)^2 = \frac{(n-1)S_X^2}{\sigma^2}$$

and

$$\sum_{i=1}^{m}\left(\frac{Y_i - \overline{Y}}{\sigma}\right)^2 = \frac{(m-1)S_Y^2}{\sigma^2}$$

are independent χ^2 random variables with $n - 1$ and $m - 1$ df, respectively, so

$$\sum_{i=1}^{n}\left(\frac{X_i - \overline{X}}{\sigma}\right)^2 + \sum_{i=1}^{m}\left(\frac{Y_i - \overline{Y}}{\sigma}\right)^2$$

has a χ^2 distribution with $n + m - 2$ df (recall Theorem 7.3.1 and Theorem 4.6.4). Also, by Appendix 7.A.2, the numerator and denominator are independent. It follows from Definition 7.3.3, then, that

$$\frac{\overline{X} - \overline{Y} - (\mu_X - \mu_Y)}{S_p\sqrt{\frac{1}{n} + \frac{1}{m}}}$$

has a Student t distribution with $n + m - 2$ df. □

Theorem 9.2.2 *Let $x_1, x_2, \ldots, x_n$ and $y_1, y_2, \ldots, y_m$ be independent random samples from normal distributions with means μ_X and μ_Y, respectively, and with the same standard deviation σ. Let*

$$t = \frac{\overline{x} - \overline{y}}{s_p \sqrt{\frac{1}{n} + \frac{1}{m}}}$$

a. *To test H_0: $\mu_X = \mu_Y$ versus H_1: $\mu_X > \mu_Y$ at the α level of significance, reject H_0 if $t \geq t_{\alpha,n+m-2}$.*

b. *To test H_0: $\mu_X = \mu_Y$ versus H_1: $\mu_X < \mu_Y$ at the α level of significance, reject H_0 if $t \leq -t_{\alpha,n+m-2}$.*

c. *To test H_0: $\mu_X = \mu_Y$ versus H_1: $\mu_X \neq \mu_Y$ at the α level of significance, reject H_0 if t is either (1) $\leq -t_{\alpha/2,n+m-2}$ or (2) $\geq t_{\alpha/2,n+m-2}$.*

Proof See Appendix 9.A.1. □

Case Study 9.2.1

The mystery surrounding the nature of Mark Twain's participation in the Civil War was discussed (but not resolved) in Case Study 1.2.2. Recall that historians are still unclear as to whether the creator of *Huckleberry Finn* and *Tom Sawyer* was a civilian or a combatant in the early 1860s and whether his sympathies lay with the North or with the South.

A tantalizing clue that might shed some light on the matter is a set of ten war-related essays written by one Quintus Curtius Snodgrass, who claimed to be in the Louisiana militia, although no records documenting his service have ever been found. If Snodgrass was just a pen name Twain used, as some suspect, then these essays are basically a diary of Twain's activities during the war, and the mystery is solved. If Quintus Curtius Snodgrass was *not* a pen name, these essays are just a red herring, and all questions about Twain's military activities remain unanswered.

Assessing the likelihood that Twain and Snodgrass were one and the same would be the job of a "forensic statistician." Authors have characteristic word-length profiles that effectively serve as verbal fingerprints (much like incriminating evidence left at a crime scene). If Authors A and B tend to use, say, three-letter words with significantly different frequencies, a reasonable inference would be that A and B are different people.

Table 9.2.1 shows the proportions of three-letter words in each of the ten Snodgrass essays and in eight essays known to have been written by Mark Twain. If x_i denotes the ith Twain proportion, $i = 1, 2, \ldots, 8$, and y_i denotes the ith Snodgrass proportion, $i = 1, 2, \ldots, 10$, then

$$\sum_{i=1}^{8} x_i = 1.855 \text{ so } \overline{x} = 1.855/8 = 0.2319$$

(Continued on next page)

Table 9.2.1 Proportion of Three-Letter Words

Twain	Proportion	QCS	Proportion
Sergeant Fathom letter	0.225	Letter I	0.209
Madame Caprell letter	0.262	Letter II	0.205
Mark Twain letters in		Letter III	0.196
Territorial Enterprise		Letter IV	0.210
First letter	0.217	Letter V	0.202
Second letter	0.240	Letter VI	0.207
Third letter	0.230	Letter VII	0.224
Fourth letter	0.229	Letter VIII	0.223
First *Innocents Abroad* letter		Letter IX	0.220
First half	0.235	Letter X	0.201
Second half	0.217		

and

$$\sum_{i=1}^{10} y_i = 2.097 \text{ so } \bar{y} = 2.097/10 = 0.2097$$

The question to be answered is whether the difference between 0.2319 and 0.2097 is statistically significant.

Let μ_X and μ_Y denote the true average proportions of three-letter words that Twain and Snodgrass, respectively, tended to use. Our objective is to test

$$H_0 : \mu_X = \mu_Y$$

versus

$$H_1 : \mu_X \neq \mu_Y$$

Since

$$\sum_{i=1}^{8} x_i^2 = 0.4316 \quad \text{and} \quad \sum_{i=1}^{10} y_i^2 = 0.4406$$

the two sample variances are

$$s_X^2 = \frac{8(0.4316) - (1.855)^2}{8(7)}$$
$$= 0.0002103$$

and

$$s_Y^2 = \frac{10(0.4406) - (2.097)^2}{10(9)}$$
$$= 0.0000955$$

(Continued on next page)

(Case Study 9.2.1 continued)

Combined, they give a pooled standard deviation of 0.0121:

$$s_p = \sqrt{\frac{\sum_{i=1}^{8}(x_i - 0.2319)^2 + \sum_{i=1}^{10}(y_i - 0.2097)^2}{n+m-2}}$$

$$= \sqrt{\frac{(n-1)s_X^2 + (m-1)s_Y^2}{n+m-2}}$$

$$= \sqrt{\frac{7(0.0002103) + 9(0.0000955)}{8+10-2}}$$

$$= \sqrt{0.0001457}$$

$$= 0.0121$$

According to Theorem 9.2.1, if H_0: $\mu_X = \mu_Y$ is true, the sampling distribution of

$$T = \frac{\overline{X} - \overline{Y}}{S_p\sqrt{\frac{1}{8} + \frac{1}{10}}}$$

is described by a Student t curve with 16 $(= 8 + 10 - 2)$ degrees of freedom.

Suppose we let $\alpha = 0.01$. By part (c) of Theorem 9.2.2, H_0 should be rejected in favor of a two-sided H_1 if either (1) $t \leq -t_{\alpha/2,n+m-2} = -t_{.005,16} = -2.9208$ or (2) $t \geq t_{\alpha/2,n+m-2} = t_{.005,16} = 2.9208$ (see Figure 9.2.1). But

$$t = \frac{0.2319 - 0.2097}{0.0121\sqrt{\frac{1}{8} + \frac{1}{10}}}$$

$$= 3.88$$

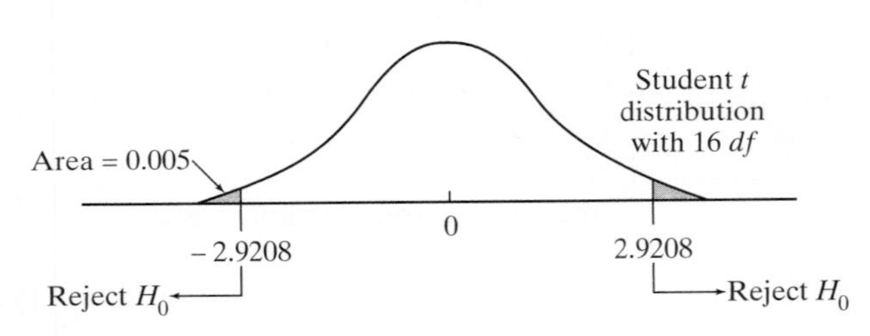

Figure 9.2.1

a value falling considerably to the right of $t_{.005,16}$. Therefore, we should *reject* H_0—it appears that Twain and Snodgrass were not the same person. So, unfortunately, nothing that Twain did can be inferred from anything that Snodgrass wrote.

About the Data The X_i's and Y_i's in Table 9.2.1, being proportions, are necessarily *not* normally distributed random variables with the same variance, so the basic conditions of Theorem 9.2.2 are not met. Fortunately, the consequences of violated assumptions on the probabilistic behavior of T_{n+m-2} are frequently minimal. The robustness property of the *one-sample t ratio* that we investigated in Chapter 7 also holds true for the *two-sample t ratio*.

Case Study 9.2.2

Dislike your statistics instructor? Retaliation time will come at the end of the semester, when you pepper the student course evaluation form with 1's. Were you pleased? Then send a signal with a load of 5's. Either way, students' evaluations of their instructors do matter. These instruments are commonly used for promotion, tenure, and merit raise decisions.

Studies of student course evaluations show that they do have value. They tend to show reliability and consistency. Yet questions remain as to the ability of these questionnaires to identify good teachers and courses.

A veteran instructor of developmental psychology decided to do a study (201) on how a single changed factor might affect his students' course evaluations. He had attended a workshop extolling the virtue of an enthusiastic style in the classroom—more hand gestures, increased voice pitch variability, and the like. The vehicle for the study was the large-lecture undergraduate developmental psychology course he had taught in the fall semester. He set about to teach the spring-semester offering in the same way, with the exception of a more enthusiastic style.

The professor fully understood the difficulty of controlling for the many variables. He selected the spring class to have the same demographics as the one in the fall. He used the same textbook, syllabus, and tests. He listened to audiotapes of the fall lectures and reproduced them as closely as possible, covering the same topics in the same order.

The first step in examining the effect of enthusiasm on course evaluations is to establish that students have, in fact, perceived an increase in enthusiasm. Table 9.2.2 summarizes the ratings the instructor received on the "enthusiasm" question for the two semesters. Unless the increase in sample means (2.14 to 4.21) is statistically significant, there is no point in trying to compare fall and spring responses to other questions.

Table 9.2.2

Fall, x_i	Spring, y_i
$n = 229$	$m = 243$
$\overline{x} = 2.14$	$\overline{y} = 4.21$
$s_X = 0.94$	$s_Y = 0.83$

Let μ_X and μ_Y denote the true means associated with the two different teaching styles. There is no reason to think that increased enthusiasm on the part of the instructor would *decrease* the students' perception of enthusiasm, so it can be argued here that H_1 should be one-sided. That is, we want to test

$$H_0: \mu_X = \mu_Y$$

versus

$$H_1: \mu_X < \mu_Y$$

(Continued on next page)

(Case Study 9.2.2 continued)

Let $\alpha = 0.05$.

Since $n = 229$ and $m = 243$, the t statistic has $229 + 243 - 2 = 470$ degrees of freedom. Thus, the decision rule calls for the rejection of H_0 if

$$t = \frac{\bar{x} - \bar{y}}{s_P\sqrt{\frac{1}{229} + \frac{1}{243}}} \leq -t_{\alpha,n+m-2} = -t_{.05,470}$$

A glance at Table A.2 in the Appendix shows that for any value $n > 100$, z_α is a good approximation of $t_{\alpha,n}$. That is, $-t_{.05,470} \doteq -z_{.05} = -1.64$.

The pooled standard deviation for these data is *0.885*:

$$s_P = \sqrt{\frac{228(0.94)^2 + 242(0.83)^2}{229 + 243 - 2}} = 0.885$$

Therefore,

$$t = \frac{2.14 - 4.21}{0.885\sqrt{\frac{1}{229} + \frac{1}{243}}} = -25.42$$

and our conclusion is a resounding rejection of H_0—the increased enthusiasm was, indeed, noticed.

The real question of interest is whether the change in enthusiasm produced a *perceived* change in some other aspect of teaching that we know did *not* change. For example, the instructor did not become more knowledgeable about the material over the course of the two semesters. The student ratings, though, disagree.

Table 9.2.3 shows the instructor's fall and spring ratings on the "knowledgeable" question. Is the increase from $\bar{x} = 3.61$ to $\bar{y} = 4.05$ statistically significant? Yes. For these data, $s_P = 0.898$ and

$$t = \frac{3.61 - 4.05}{0.898\sqrt{\frac{1}{229} + \frac{1}{243}}} = -5.33$$

which falls far to the left of the 0.05 critical value $(= -1.64)$.

What we can glean from these data is both reassuring yet a bit disturbing. Table 9.2.2 appears to confirm the widely held belief that enthusiasm is an important factor in effective teaching. Table 9.2.3, on the other hand, strikes a more cautionary note. It speaks to another widely held belief—that student evaluations can sometimes be difficult to interpret. Questions that purport to be measuring one trait may, in fact, be reflecting something entirely different.

Table 9.2.3

Fall, x_i	Spring, y_i
$n = 229$	$m = 243$
$\bar{x} = 3.61$	$\bar{y} = 4.05$
$s_X = 0.84$	$s_Y = 0.95$

About the Data The five-choice responses in student evaluation forms are very common in survey questionnaires. Such questions are known as Likert items, named after the psychologist Rensis Likert. The item typically asks the respondent to choose his or her level of agreement with a statement, for example, "The instructor shows concern for students." The choices start with "strongly disagree," which is scored with a "1," and go up to a "5" for "strongly agree." The statistic for a given question in a survey is the *average* value taken over all responses.

Is a t test an appropriate way to analyze data of this sort? Maybe, but the nature of the responses raises some serious concerns. First of all, the fact that students talk with each other about their instructors suggests that not all the sample values will be independent. More importantly, the five-point Likert scale hardly resembles the normality assumption implicit in a Student t analysis. For many practitioners—but not all—the robustness of the t test would be enough to justify the analysis described in Case Study 9.2.2.

The Behrens-Fisher Problem

Finding a statistic with known density for testing the equality of two means from normally distributed random samples when the standard deviations of the samples are not equal is known as the *Behrens-Fisher problem*. No exact solution is known, but a widely used approximation is based on the test statistic

$$W = \frac{\overline{X} - \overline{Y} - (\mu_X - \mu_Y)}{\sqrt{\frac{S_X^2}{n} + \frac{S_Y^2}{m}}}$$

where, as usual, $\overline{X}$ and $\overline{Y}$ are the sample means, and S_X^2 and S_Y^2 are the unbiased estimators of the variance. B. L. Welch, a faculty member at University College, London, in a 1938 Biometrika article showed that W is approximately distributed as a Student t random variable with degrees of freedom given by the nonintuitive expression

$$\frac{\left(\frac{\sigma_1^2}{n_1} + \frac{\sigma_2^2}{n_2}\right)^2}{\frac{\sigma_1^4}{n_1^2(n_1-1)} + \frac{\sigma_2^4}{n_2^2(n_2-1)}}$$

To understand Welch's approximation, it helps to rewrite the random variable W as

$$W = \frac{\overline{X} - \overline{Y} - (\mu_X - \mu_Y)}{\sqrt{\frac{S_X^2}{n} + \frac{S_Y^2}{m}}} = \frac{\overline{X} - \overline{Y} - (\mu_X - \mu_Y)}{\sqrt{\frac{\sigma_X^2}{n} + \frac{\sigma_Y^2}{m}}} \div \frac{\sqrt{\frac{S_X^2}{n} + \frac{S_Y^2}{m}}}{\sqrt{\frac{\sigma_X^2}{n} + \frac{\sigma_Y^2}{m}}}$$

In this form, the numerator is a standard normal variable. Suppose there is a chi square random variable V with ν degrees of freedom such that the square of the denominator is equal to V/ν. Then the expression would indeed be a Student t variable with ν degrees of freedom. However, in general, the denominator will not have exactly that distribution. The strategy, then, is to find an *approximate* equality for

$$\frac{\frac{S_X^2}{n} + \frac{S_Y^2}{m}}{\frac{\sigma_X^2}{n} + \frac{\sigma_Y^2}{m}} = \frac{V}{\nu}$$

or, equivalently,

$$\frac{S_X^2}{n} + \frac{S_Y^2}{m} = \left(\frac{\sigma_X^2}{n} + \frac{\sigma_Y^2}{m}\right)\frac{V}{\nu}$$

At issue is the value of ν. The method of moments (recall Section 5.2) suggests a solution. If the means and variances of both sides are equated, it can be shown that

$$\nu = \frac{\left(\frac{\sigma_X^2}{n} + \frac{\sigma_Y^2}{m}\right)^2}{\frac{\sigma_X^4}{n^2(n-1)} + \frac{\sigma_Y^4}{m^2(m-1)}}$$

Moreover, the expression for ν depends only on the *ratio* of the variances, $\theta = \frac{\sigma_X^2}{\sigma_Y^2}$. To see why, divide the numerator and denominator by σ_Y^4. Then

$$\frac{\left(\frac{1}{n}\frac{\sigma_X^2}{\sigma_Y^2} + \frac{1}{m}\right)^2}{\frac{1}{n^2(n-1)}\left(\frac{\sigma_X^2}{\sigma_Y^2}\right)^2 + \frac{1}{m^2(m-1)}} = \frac{\left(\frac{1}{n}\theta + \frac{1}{m}\right)^2}{\frac{1}{n^2(n-1)}\theta^2 + \frac{1}{m^2(m-1)}}$$

and multiplying numerator and denominator by n^2 gives the somewhat more appealing form

$$\nu = \frac{\left(\theta + \frac{n}{m}\right)^2}{\frac{1}{(n-1)}\theta^2 + \frac{1}{(m-1)}\left(\frac{n}{m}\right)^2}$$

Of course, the main application of this theory occurs when σ_X^2 and σ_Y^2 are unknown and θ must thus be estimated, the obvious choice being $\theta = \frac{s_X^2}{s_Y^2}$.

This leads us to the following theorem for testing the equality of means when the variances cannot be assumed equal.

Theorem 9.2.3 *Let $X_1, X_2, \ldots, X_n$ and $Y_1, Y_2, \ldots, Y_m$ be independent random samples from normal distributions with means μ_X and μ_Y, and standard deviations σ_X and σ_Y, respectively. Let*

$$W = \frac{\overline{X} - \overline{Y} - (\mu_X - \mu_Y)}{\sqrt{\frac{S_X^2}{n} + \frac{S_Y^2}{m}}}$$

Using $\hat{\theta} = \frac{s_X^2}{s_Y^2}$, take ν to be the expression $\frac{\left(\hat{\theta} + \frac{n}{m}\right)^2}{\frac{1}{(n-1)}\hat{\theta}^2 + \frac{1}{(m-1)}\left(\frac{n}{m}\right)^2}$, rounded to the nearest integer. Then W has approximately a Student t distribution with ν degrees of freedom.

Case Study 9.2.3

Does size matter? While a successful company's large number of sales should mean bigger profits, does it yield greater *profitability*? *Forbes* magazine periodically rates the top two hundred small companies (52), and for each gives the profitability as measured by the five-year percentage return on equity. Using data from the *Forbes* article, Table 9.2.4 gives the return on equity for the twelve companies with the largest number of sales (ranging from \$679 million to \$738

(Continued on next page)

million) and for the twelve companies with the smallest number of sales (ranging from \$25 million to \$66 million). Based on these data, can we say that the return on equity differs between the two types of companies?

Table 9.2.4

Large-Sales Companies	Return on Equity (%)	Small-Sales Companies	Return on Equity (%)
Deckers Outdoor	21	NVE	21
Jos. A. Bank Clothiers	23	Hi-Shear Technology	21
National Instruments	13	Bovie Medical	14
Dolby Laboratories	22	Rocky Mountain Chocolate Factory	31
Quest Software	7	Rochester Medical	19
Green Mountain Coffee Roasters	17	Anika Therapeutics	19
Lufkin Industries	19	Nathan's Famous	11
Red Hat	11	Somanetics	29
Matrix Service	2	Bolt Technology	20
DXP Enterprises	30	Energy Recovery	27
Franklin Electric	15	Transcend Services	27
LSB Industries	43	IEC Electronics	24

Let μ_X and μ_Y be the respective average returns on equity. The indicated test of hypotheses is

$$H_0 : \mu_X = \mu_Y$$

versus

$$H_1 : \mu_X \neq \mu_Y$$

For the data in the table, $\overline{x} = 18.6$, $\overline{y} = 21.9$, $s_X^2 = 115.9929$, and $s_Y^2 = 35.7604$. The test statistic is

$$w = \frac{\overline{x} - \overline{y} - (\mu_X - \mu_Y)}{\sqrt{\frac{s_X^2}{n} + \frac{s_Y^2}{m}}} = \frac{18.6 - 21.9}{\sqrt{\frac{115.9929}{12} + \frac{35.7604}{12}}} = -0.928$$

Also,

$$\hat{\theta} = \frac{s_X^2}{s_Y^2} = \frac{115.9929}{35.7604} = 3.244$$

so

$$\frac{\left(3.244 + \frac{12}{12}\right)^2}{\frac{1}{11}(3.244)^2 + \frac{1}{11}\left(\frac{12}{12}\right)^2} = 17.2$$

which implies that $\nu = 17$.

We should reject H_0 at the $\alpha = 0.05$ level of significance if $w > t_{0.025,17} = 2.1098$ or $w < -t_{0.025,17} = -2.1098$. Here, $w = -0.928$ falls in between the two critical values, so the difference between $\overline{x}$ and $\overline{y}$ is not statistically significant.

Comment It occasionally happens that an experimenter wants to test H_0: $\mu_X = \mu_Y$ and *knows* the values of σ_X^2 and σ_Y^2. For those situations, the t test of Theorem 9.2.2 is inappropriate. If the n X_i's and m Y_i's are normally distributed, it follows from the corollary to Theorem 4.3.3 that

$$Z = \frac{\overline{X} - \overline{Y} - (\mu_X - \mu_Y)}{\sqrt{\frac{\sigma_X^2}{n} + \frac{\sigma_Y^2}{m}}} \tag{9.2.1}$$

has a standard normal distribution. Any such test of H_0: $\mu_X = \mu_Y$, then, should be based on an observed Z ratio rather than an observed t ratio.

If the degrees of freedom for a t test exceed 100, then the test statistic of Equation 9.2.1 is used, but it is treated as a Z ratio. In either the test of Theorem 9.2.2 or 9.2.3, if the degrees of freedom exceed 100, the statistic of Theorem 9.2.3 is used with the z tables.

Questions

9.2.1. Some states that operate a lottery believe that restricting the use of lottery profits to supporting education makes the lottery more profitable. Other states permit general use of the lottery income. The profitability of the lottery for a group of states in each category is given below.

State Lottery Profits

For Education		For General Use	
State	% Profit	State	% Profit
New Mexico	24	Massachusetts	21
Idaho	25	Maine	22
Kentucky	28	Iowa	24
South Carolina	28	Colorado	27
Georgia	28	Indiana	27
Missouri	29	Dist. Columbia	28
Ohio	29	Connecticut	29
Tennessee	31	Pennsylvania	32
Florida	31	Maryland	32
California	35		
North Carolina	35		
New Jersey	35		

Source: *New York Times*, National Section, October 7, 2007, p. 14.

Test at the $\alpha = 0.01$ level whether the mean profit of states using the lottery for education is higher than that of states permitting general use. Assume that the variances of the two random variables are equal.

9.2.2. As the United States has struggled with the growing obesity of its citizens, diets have become big business. Among the many competing regimens for those seeking weight reduction are the Atkins and Zone diets. In a comparison of these two diets for one-year weight loss, a study (59) found that seventy-seven subjects on the Atkins diet had an average weight loss of $\overline{x} = -4.7$ kg and a sample standard deviation of $s_X = 7.05$ kg. Similar figures for the seventy-nine people on the Zone diet were $\overline{y} = -1.6$ kg and $s_Y = 5.36$ kg. Is the greater reduction with the Atkins diet statistically significant? Test for $\alpha = 0.05$.

9.2.3. A medical researcher believes that women typically have lower serum cholesterol than men. To test this hypothesis, he took a sample of 476 men between the ages of nineteen and forty-four and found their mean serum cholesterol to be 189.0 mg/dl with a sample standard deviation of 34.2. A group of 592 women in the same age range averaged 177.2 mg/dl and had a sample standard deviation of 33.3. Is the lower average for the women statistically significant? Set $\alpha = 0.05$.

9.2.4. In the academic year 2004–05, 1126 high school freshmen took the SAT Reasoning Test. On the Critical Reasoning portion, this group had a mean score of 491 with a standard deviation of 119. The following year, 5042 sophomores (none of them in the 2004–05 freshmen group) scored an average of 498, with a standard deviation of 129. Is the higher average score for the sophomores a result of such factors as additional schooling and increased maturity or simply a random effect? Test at the $\alpha = 0.05$ level of significance.

Source: College Board SAT, Total Group Profile Report, 2008.

9.2.5. The University of Missouri–St. Louis gave a validation test to entering students who had taken calculus in high school. The group of ninety-three students receiving no college credit had a mean score of 4.17 on the validation test with a sample standard deviation of 3.70. For the twenty-eight students who received credit from a high school dual-enrollment class, the mean score was 4.61 with a sample standard deviation of 4.28. Is there a significant difference in these means at the $\alpha = 0.01$ level?

Source: MAA Focus, December 2008, p. 19.

9.2.6. Ring Lardner was one of this country's most popular writers during the 1920s and 1930s. He was also a chronic alcoholic who died prematurely at the age of forty-eight. The following table lists the life spans of some of Lardner's contemporaries (36). Those in the sample on the left were all problem drinkers; they died, on the average, at age sixty-five. The twelve (sober) writers on the right tended to live a full ten years longer. Can it be argued that an increase of that magnitude is statistically significant? Test an appropriate null hypothesis against a one-sided H_1. Use the 0.05 level of significance. (*Note*: The pooled sample standard deviation for these two samples is *13.9*.)

Authors Noted for Alchohol Abuse		*Authors Not Noted for Alchohol Abuse*	
Name	Age at Death	Name	Age at Death
Ring Lardner	48	Carl Van Doren	65
Sinclair Lewis	66	Ezra Pound	87
Raymond Chandler	71	Randolph Bourne	32
Eugene O'Neill	65	Van Wyck Brooks	77
Robert Benchley	56	Samuel Eliot Morrison	89
J.P. Marquand	67	John Crowe Ransom	86
Dashiell Hammett	67	T.S. Eliot	77
e.e. cummings	70	Conrad Aiken	84
Edmund Wilson	77	Ben Ames Williams	64
Average:	65.2	Henry Miller	88
		Archibald MacLeish	90
		James Thurber	67
		Average:	75.5

9.2.7. Poverty Point is the name given to a number of widely scattered archaeological sites throughout Louisiana, Mississippi, and Arkansas. These are the remains of a society thought to have flourished during the period from 1700 to 500 B.C. Among their characteristic artifacts are ornaments that were fashioned out of clay and then baked. The following table shows the dates (in years B.C.) associated with four of these baked clay ornaments found in two different Poverty Point sites, Terral Lewis and Jaketown (86). The averages for the two samples are 1133.0 and 1013.5, respectively. Is it believable that these two settlements developed the technology to manufacture baked clay ornaments at the same time? Set up and test an appropriate H_0 against a two-sided H_1 at the $\alpha = 0.05$ level of significance. For these data $s_x = 266.9$ and $s_y = 224.3$.

Terral Lewis Estimates, x_i	Jaketown Estimates, y_i
1492	1346
1169	942
883	908
988	858

9.2.8. A major source of "mercury poisoning" comes from the ingestion of methylmercury (CH_3^{203}), which is found in contaminated fish (recall Question 5.3.3). Among the questions pursued by medical investigators trying to understand the nature of this particular health problem is whether methylmercury is equally hazardous to men and women. The following (114) are the half-lives of methylmercury in the systems of six women and nine men who volunteered for a study where each subject was given an oral administration of CH_3^{203}. Is there evidence here that women metabolize methylmercury at a different rate than men do? Do an appropriate two-sample t test at the $\alpha = 0.01$ level of significance. The two sample standard deviations for these data are $s_X = 15.1$ and $s_Y = 8.1$.

Methylmercury (CH_3^{203}) Half-Lives (in Days)	
Females, x_i	Males, y_i
52	72
69	88
73	87
88	74
87	78
56	70
	78
	93
	74

9.2.9. Lipton, a company primarily known for tea, considered using coupons to stimulate sales of its packaged dinner entrees. The company was particularly interested whether there was a diffences in the effect of coupons on singles versus married couples. A poll of consumers asked them to respond to the question "Do you use coupons regularly?" by a numerical scale, where 1 stands for agree strongly, 2 for agree, 3 for neutral, 4 for disagree, and 5 for disagree strongly. The results of the poll are given in the following table (19).

Use Coupons Regularly	
Single (X)	Married (Y)
$n = 31$	$n = 57$
$\bar{x} = 3.10$	$\bar{y} = 2.43$
$s_X = 1.469$	$s_Y = 1.350$

Is the observed difference significant at the $\alpha = 0.05$ level?

9.2.10. A company markets two brands of latex paint—regular and a more expensive brand that claims to dry an hour faster. A consumer magazine decides to test this claim by painting ten panels with each product. The average drying time of the regular brand is 2.1 hours with a sample standard deviation of 12 minutes. The fast-drying version has an average of 1.6 hours with a sample standard deviation of 16 minutes. Test the null hypothesis that the more expensive brand dries an hour quicker. Use a one-sided H_1. Let $\alpha = 0.05$.

9.2.11. **(a)** Suppose H_0: $\mu_X = \mu_Y$ is to be tested against H_1: $\mu_X \neq \mu_Y$. The two sample sizes are 6 and 11. If $s_p = 15.3$, what is the smallest value for $|\overline{x} - \overline{y}|$ that will result in H_0 being rejected at the $\alpha = 0.01$ level of significance?
(b) What is the smallest value for $\overline{x} - \overline{y}$ that will lead to the rejection of H_0: $\mu_X = \mu_Y$ in favor of H_1: $\mu_X > \mu_Y$ if $\alpha = 0.05$, $s_P = 214.9$, $n = 13$, and $m = 8$?

9.2.12. Suppose that H_0: $\mu_X = \mu_Y$ is being tested against H_1: $\mu_X \neq \mu_Y$, where σ_X^2 and σ_Y^2 are known to be 17.6 and 22.9, respectively. If $n = 10$, $m = 20$, $\overline{x} = 81.6$, and $\overline{y} = 79.9$, what P-value would be associated with the observed Z ratio?

9.2.13. An executive has two routes that she can take to and from work each day. The first is by interstate; the second requires driving through town. On the average it takes her 33 minutes to get to work by the interstate and 35 minutes by going through town. The standard deviations for the two routes are 6 and 5 minutes, respectively. Assume the distributions of the times for the two routes are approximately normally distributed.

(a) What is the probability that on a given day, driving through town would be the quicker of her choices?
(b) What is the probability that driving through town for an entire week (ten trips) would yield a lower average time than taking the interstate for the entire week?

9.2.14. Prove that the Z ratio given in Equation 9.2.1 has a standard normal distribution.

9.2.15. If $X_1, X_2, \ldots, X_n$ and $Y_1, Y_2, \ldots, Y_m$ are independent random samples from normal distributions with the same σ^2, prove that their pooled sample variance, s_p^2, is an unbiased estimator for σ^2.

9.2.16. Let $X_1, X_2, \ldots, X_n$ and $Y_1, Y_2, \ldots, Y_m$ be independent random samples drawn from normal distributions with means μ_X and μ_Y, respectively, and with the same known variance σ^2.Use the generalized likelihood ratio criterion to derive a test procedure for choosing between H_0: $\mu_X = \mu_Y$ and H_1: $\mu_X \neq \mu_Y$.

9.2.17. A person exposed to an infectious agent, either by contact or by vaccination, normally develops antibodies to that agent. Presumably, the severity of an infection is related to the number of antibodies produced. The degree of antibody response is indicated by saying that the person's blood serum has a certain *titer*, with higher titers indicating greater concentrations of antibodies. The following table gives the titers of twenty-two persons involved in a tularemia epidemic in Vermont (18). Eleven were quite ill; the other eleven were asymptomatic. Use an approximate t ratio to test H_0: $\mu_X = \mu_Y$ against a one-sided H_1 at the 0.05 level of significance.

The sample standard deviations for the "Severely Ill" and "Asymptomatic" groups are 428 and 183, respectively.

Severely Ill		*Asymptomatic*	
Subject	Titer	Subject	Titer
1	640	12	10
2	80	13	320
3	1280	14	320
4	160	15	320
5	640	16	320
6	640	17	80
7	1280	18	160
8	640	19	10
9	160	20	640
10	320	21	160
11	160	22	320

9.2.18. For the approximate two-sample t test described in Question 9.2.17, it will be true that

$$v < n + m - 2$$

Why is that a disadvantage for the approximate test? That is, why is it better to use the Theorem 9.2.1 version of the t test if, in fact, $\sigma_X^2 = \sigma_Y^2$?

9.2.19. The two-sample data described in Question 8.2.2 would be analyzed by testing H_0: $\mu_X = \mu_Y$, where μ_X and μ_Y denote the true average motorcycle-related fatality rates for states having "limited" and "comprehensive" helmet laws, respectively.

(a) Should the t test for H_0: $\mu_X = \mu_Y$ follow the format of Theorem 9.2.2 or the approximation given in Theorem 9.2.3? Explain.
(b) Is there anything unusual about these data? Explain.

9.2.20. Some financial analysts believe that the election of a Republican president is good for the stock market. To test this claim, one study (155) recorded the ten-year growth in Standard & Poor's index following each election of a new president. The results are given in the table below.

Democrats		Republicans	
Winner	S&P Growth	Winner	S&P Growth
Roosevelt '36	22.4	Eisenhower '52	45.7
Roosevelt '40	24.0	Eisenhower '56	28.6
Roosevelt '44	38.0	Nixon '68	14.2
Truman '48	45.7	Nixon '72	18.8
Kennedy '60	21.2	Reagan '80	50.3
Johnson '64	17.9	Reagan '84	40.1
Carter '76	38.2	Bush '88	52.4
Clinton '92	33.7		
Clinton '96	23.8		

Is the higher average for the Republicans statistically significant? Test at the 0.01 level. Do not assume the variances are equal.

9.3 Testing H_0: $\sigma_X^2 = \sigma_Y^2$—The F Test

Although by far the majority of two-sample problems are set up to detect possible shifts in location parameters, situations sometimes arise where it is equally important—perhaps even more important—to compare variability parameters. Two machines on an assembly line, for example, may be producing items whose *average* dimensions (μ_X and μ_Y) of some sort—say, thickness—are not significantly different but whose variabilities (as measured by σ_X^2 and σ_Y^2) are. This becomes a critical piece of information if the increased variability results in an unacceptable proportion of items from one of the machines falling outside the engineering specifications (see Figure 9.3.1).

Figure 9.3.1 Variability of machine outputs.

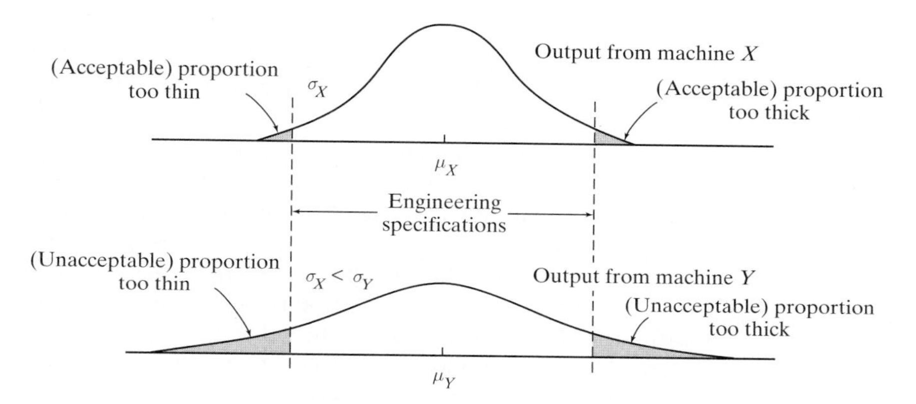

In this section we will examine the generalized likelihood ratio test of H_0: $\sigma_X^2 = \sigma_Y^2$ versus H_1: $\sigma_X^2 \neq \sigma_Y^2$. The data will consist of two independent random samples of sizes n and m: The first—$x_1, x_2, \ldots, x_n$—is assumed to have come from a normal distribution having mean μ_X and variance σ_X^2; the second—$y_1, y_2, \ldots, y_m$—from a normal distribution having mean μ_Y and variance σ_Y^2. (All four parameters are assumed to be unknown.) Theorem 9.3.1 gives the test procedure that will be used. The proof will not be given, but it follows the same basic pattern we have seen in other GLRTs; the important step is showing that the likelihood ratio is a monotonic function of the F random variable described in Definition 7.3.2.

Comment Tests of H_0: $\sigma_X^2 = \sigma_Y^2$ arise in another, more routine context. Recall that the procedure for testing the equality of μ_X and μ_Y depends on whether or not the two population variances are equal. This implies that a test of H_0: $\sigma_X^2 = \sigma_Y^2$ should precede every test of H_0: $\mu_X = \mu_Y$. If the former is accepted, the t test on μ_X and μ_Y is done according to Theorem 9.2.2; but if H_0: $\sigma_X^2 = \sigma_Y^2$ is rejected, Theorem 9.2.2 is not entirely appropriate. A frequently used alternative in that case is the approximate t test described in Theorem 9.2.3.

Theorem 9.3.1 *Let $x_1, x_2, \ldots, x_n$ and $y_1, y_2, \ldots, y_m$ be independent random samples from normal distributions with means μ_X and μ_Y and standard deviations σ_X and σ_Y, respectively.*

a. *To test H_0: $\sigma_X^2 = \sigma_Y^2$ versus H_1: $\sigma_X^2 > \sigma_Y^2$ at the α level of significance, reject H_0 if $s_Y^2/s_X^2 \leq F_{\alpha, m-1, n-1}$.*

b. *To test* H_0: $\sigma_X^2 = \sigma_Y^2$ *versus* H_1: $\sigma_X^2 < \sigma_Y^2$ *at the* α *level of significance, reject* H_0 *if* $s_Y^2/s_X^2 \geq F_{1-\alpha,m-1,n-1}$.

c. *To test* H_0: $\sigma_X^2 = \sigma_Y^2$ *versus* H_1: $\sigma_X^2 \neq \sigma_Y^2$ *at the* α *level of significance, reject* H_0 *if* s_Y^2/s_X^2 *is either* (1) $\leq F_{\alpha/2,m-1,n-1}$ *or* (2) $\geq F_{1-\alpha/2,m-1,n-1}$.

Comment The GLRT described in Theorem 9.3.1 is *approximate* for the same sort of reason the GLRT for H_0: $\sigma^2 = \sigma_0^2$ is approximate (see Theorem 7.5.2). The distribution of the test statistic, S_Y^2/S_X^2, is not symmetric, and the two ranges of variance ratios yielding λ's less than or equal to λ^* (i.e., the left tail and right tail of the critical region) have slightly different areas. For the sake of convenience, though, it is customary to choose the two critical values so that each cuts off the same area, $\alpha/2$.

Case Study 9.3.1

Electroencephalograms are records showing fluctuations of electrical activity in the brain. Among the several different kinds of brain waves produced, the dominant ones are usually *alpha* waves. These have a characteristic frequency of anywhere from eight to thirteen cycles per second.

The objective of the experiment described in this example was to see whether sensory deprivation over an extended period of time has any effect on the alpha-wave pattern. The subjects were twenty inmates in a Canadian prison who were randomly split into two equal-sized groups. Members of one group were placed in solitary confinement; those in the other group were allowed to remain in their own cells. Seven days later, alpha-wave frequencies were measured for all twenty subjects (60), as shown in Table 9.3.1.

Table 9.3.1 Alpha-Wave Frequencies (CPS)

Nonconfined, x_i	Solitary Confinement, y_i
10.7	9.6
10.7	10.4
10.4	9.7
10.9	10.3
10.5	9.2
10.3	9.3
9.6	9.9
11.1	9.5
11.2	9.0
10.4	10.9

Judging from Figure 9.3.2, there was an apparent *decrease* in alpha-wave frequency for persons in solitary confinement. There also appears to have been an *increase* in the variability for that group. We will use the F test to determine whether the observed difference in variability ($s_X^2 = 0.21$ versus $s_Y^2 = 0.36$) is statistically significant.

(Continued on next page)

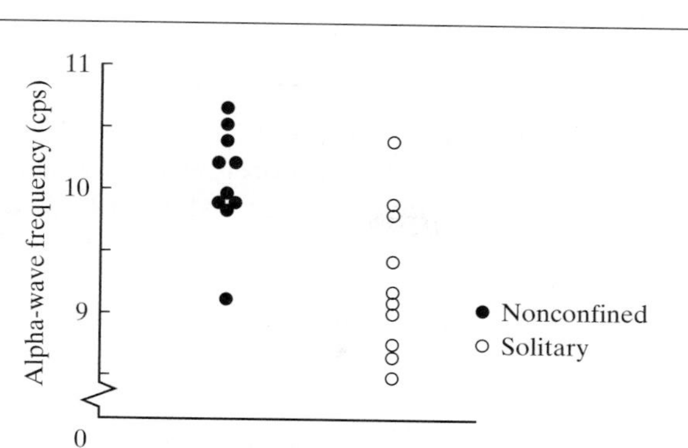

Figure 9.3.2 Alpha-wave frequencies (cps).

Let σ_X^2 and σ_Y^2 denote the true variances of alpha-wave frequencies for nonconfined and solitary-confined prisoners, respectively. The hypotheses to be tested are

$$H_0: \sigma_X^2 = \sigma_Y^2$$

versus

$$H_1: \sigma_X^2 \neq \sigma_Y^2$$

Let $\alpha = 0.05$ be the level of significance. Given that

$$\sum_{i=1}^{10} x_i = 105.8 \qquad \sum_{i=1}^{10} x_i^2 = 1121.26$$

$$\sum_{i=1}^{10} y_i = 97.8 \qquad \sum_{i=1}^{10} y_i^2 = 959.70$$

the sample variances become

$$s_X^2 = \frac{10(1121.26) - (105.8)^2}{10(9)} = 0.21$$

and

$$s_Y^2 = \frac{10(959.70) - (97.8)^2}{10(9)} = 0.36$$

Dividing the sample variances gives an observed F ratio of *1.71*:

$$F = \frac{s_Y^2}{s_X^2} = \frac{0.36}{0.21} = 1.71$$

Both n and m are ten, so we would expect S_Y^2/S_X^2 to behave like an F random variable with nine and nine degrees of freedom (assuming H_0: $\sigma_X^2 = \sigma_Y^2$ is true). From Table A.4 in the Appendix, we see that the values cutting off areas of 0.025 in either tail of that distribution are 0.248 and 4.03 (see Figure 9.3.3).

Since the observed F ratio falls between the two critical values, our decision is to *fail to reject* H_0—a ratio of sample variances equal to 1.71 does not rule out

(Continued on next page)

(Case Study 9.3.1 continued)

the possibility that the two true variances are equal. (In light of the Comment preceding Theorem 9.3.1, it would now be appropriate to test H_0: $\mu_X = \mu_Y$ using the two-sample t test described in Section 9.2.)

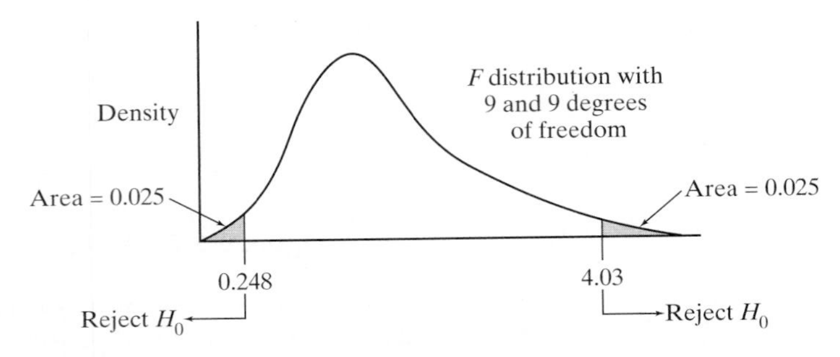

Figure 9.3.3 Distribution of S_Y^2/S_X^2 when H_0 is true.

Questions

9.3.1. Case Study 9.2.3 was offered as an example of testing means when the variances are not assumed equal. Was this a correct assumption about the variances? Test at the 0.05 level of significance.

9.3.2. Two popular forms of mortgage are the thirty-year fixed-rate mortgage, where the borrower has thirty years to repay the loan at a constant rate, and the adjustable-rate mortgage (ARM), one version of which is for five years with the possibility of yearly changes in the interest rate. Since the ARM offers less certainty, its rates are usually lower than those of fixed-rate mortgages. However, such vehicles should show more variability in rates. Test this hypothesis at the 0.10 level of significance using the following samples of mortgage offerings for a loan of $160,000 (the borrower needs $200,000, but must pay $40,000 up front).

$160,000 Mortgage Rates	
30-Year Fixed	ARM
5.500	3.875
5.500	5.125
5.250	5.000
5.125	4.750
5.875	4.375
5.625	
5.250	
4.875	

9.3.3. Among the standard personality inventories used by psychologists is the thematic apperception test (TAT) in which a subject is shown a series of pictures and is asked to make up a story about each one. Interpreted properly, the content of the stories can provide valuable insights into the subject's mental well-being. The following data show the TAT results for 40 women, 20 of whom were the mothers of normal children and 20 the mothers of schizophrenic children. In each case the subject was shown the same set of 10 pictures. The figures recorded were the numbers of stories (out of 10) that revealed a *positive* parent–child relationship, one where the mother was clearly capable of interacting with her child in a flexible, open-minded way (199).

TAT Scores									
Mothers of Normal Children					Mothers of Schizophrenic Children				
8	4	6	3	1	2	1	1	3	2
4	4	6	4	2	7	2	1	3	1
2	1	1	4	3	0	2	4	2	3
3	2	6	3	4	3	0	1	2	2

(a) Test $H_0: \sigma_X^2 = \sigma_Y^2$ versus $H_1: \sigma_X^2 \neq \sigma_Y^2$, where σ_X^2 and σ_Y^2 are the variances of the scores of mothers of normal children and scores of mothers of schizophrenic children, respectively. Let $\alpha = 0.05$.

(b) If $H_0: \sigma_X^2 = \sigma_Y^2$ is accepted in part (a), test $H_0: \mu_X = \mu_Y$ versus $H_1: \mu_X \neq \mu_Y$. Set α equal to 0.05.

9.3.4. In a study designed to investigate the effects of a strong magnetic field on the early development of mice

(7), 10 cages, each containing three 30-day-old albino female mice, were subjected for a period of 12 days to a magnetic field having an average strength of 80 Oe/cm. Thirty other mice, housed in 10 similar cages, were not put in the magnetic field and served as controls. Listed in the table are the weight gains, in grams, for each of the 20 sets of mice.

In Magnetic Field		*Not in Magnetic Field*	
Cage	Weight Gain (g)	Cage	Weight Gain (g)
1	22.8	11	23.5
2	10.2	12	31.0
3	20.8	13	19.5
4	27.0	14	26.2
5	19.2	15	26.5
6	9.0	16	25.2
7	14.2	17	24.5
8	19.8	18	23.8
9	14.5	19	27.8
10	14.8	20	22.0

Test whether the variances of the two sets of weight gains are significantly different. Let $\alpha = 0.05$. For the mice in the magnetic field, $s_X = 5.67$; for the other mice, $s_Y = 3.18$.

9.3.5. Raynaud's syndrome is characterized by the sudden impairment of blood circulation in the fingers, a condition that results in discoloration and heat loss. The magnitude of the problem is evidenced in the following data, where twenty subjects (ten "normals" and ten with Raynaud's syndrome) immersed their right forefingers in water kept at 19°C. The heat output (in cal/cm²/minute) of the forefinger was then measured with a calorimeter (105).

Normal Subjects		*Subjects with Raynaud's Syndrome*	
Patient	Heat Output (cal/cm²/min)	Patient	Heat Output (cal/cm²/min)
W.K.	2.43	R.A.	0.81
M.N.	1.83	R.M.	0.70
S.A.	2.43	F.M.	0.74
Z.K.	2.70	K.A.	0.36
J.H.	1.88	H.M.	0.75
J.G.	1.96	S.M.	0.56
G.K.	1.53	R.M.	0.65
A.S.	2.08	G.E.	0.87
T.E.	1.85	B.W.	0.40
L.F.	2.44	N.E.	0.31
	$\bar{x} = 2.11$		$\bar{y} = 0.62$
	$s_X = 0.37$		$s_Y = 0.20$

Test that the heat-output variances for normal subjects and those with Raynaud's syndrome are the same. Use a two-sided alternative and the 0.05 level of significance.

9.3.6. The bitter, eight-month baseball strike that ended the 1994 season so abruptly was expected to have substantial repercussions at the box office when the 1995 season finally got under way. It did. By the end of the first week of play, American League teams were playing to 12.8% fewer fans than the year before; National League teams fared even worse—their attendance was down 15.1% (190). Based on the team-by-team attendance figures given below, would it be appropriate to use the pooled two-sample t test of Theorem 9.2.2 to assess the statistical significance of the difference between those two means?

American League		*National League*	
Team	Change	Team	Change
Baltimore	–2%	Atlanta	–49%
Boston	+16	Chicago	–4
California	+7	Cincinnati	–18
Chicago	–27	Colorado	–27
Cleveland	No home games	Florida	–15
Detroit	–22	Houston	–16
Kansas City	–20	Los Angeles	–10
Milwaukee	–30	Montreal	–1
Minnesota	–8	New York	+34
New York	–2	Philadelphia	–9
Oakland	No home games	Pittsburgh	–28
Seattle	–3	San Diego	–10
Texas	–39	San Francisco	–45
Toronto	–24	St. Louis	–14
Average:	–12.8%	Average:	–15.1%

9.3.7. For the data in Question 9.2.8, the sample variances for the methylmercury half-lives are 227.77 for the females and 65.25 for the males. Does the magnitude of that difference invalidate using Theorem 9.2.2 to test H_0: $\mu_X = \mu_Y$? Explain.

9.3.8. Crosstown busing to compensate for de facto segregation was begun on a fairly large scale in Nashville during the 1960s. Progress was made, but critics argued that too many racial imbalances were left unaddressed. Among the data cited in the early 1970s are the following figures, showing the percentages of African-American students enrolled in a random sample of eighteen public schools (165). Nine of the schools were located in predominantly African-American neighborhoods; the other nine, in predominantly white neighborhoods. Which version of the two-sample t test, Theorem 9.2.2 or the Behrens–Fisher approximation given in Theorem 9.2.3, would be more appropriate for deciding whether the difference between 35.9% and 19.7% is statistically significant? Justify your answer.

Schools in African-American Neighborhoods	Schools in White Neighborhoods
36%	21%
28	14
41	11
32	30
46	29
39	6
24	18
32	25
45	23
Average: 35.9%	Average: 19.7%

9.3.9. Show that the generalized likelihood ratio for testing H_0: $\sigma_X^2 = \sigma_Y^2$ versus H_1: $\sigma_X^2 \neq \sigma_Y^2$ as described in Theorem 9.3.1 is given by

$$\lambda = \frac{L(\omega_e)}{L(\Omega_e)} = \frac{(m+n)^{(n+m)/2}}{n^{n/2}m^{m/2}} \frac{\left[\sum_{i=1}^{n}(x_i-\bar{x})^2\right]^{n/2}\left[\sum_{j=1}^{m}(y_j-\bar{y})^2\right]^{m/2}}{\left[\sum_{i=1}^{n}(x_i-\bar{x})^2+\sum_{j=1}^{m}(y_j-\bar{y})^2\right]^{(m+n)/2}}$$

9.3.10. Let $X_1, X_2, \ldots, X_n$ and $Y_1, Y_2, \ldots, Y_m$ be independent random samples from normal distributions with means μ_X and μ_Y and standard deviations σ_X and σ_Y, respectively, where μ_X and μ_Y are known. Derive the GLRT for H_0: $\sigma_X^2 = \sigma_Y^2$ versus H_1: $\sigma_X^2 > \sigma_Y^2$.

9.4 Binomial Data: Testing H_0: $p_X = p_Y$

Up to this point, the data considered in this chapter have been independent random samples of sizes n and m drawn from two *continuous* distributions—in fact, from two *normal* distributions. Other scenarios, of course, are quite possible. The X's and Y's might represent continuous random variables but have density functions other than the normal. Or they might be *discrete*. In this section we consider the most common example of this latter type: situations where the two sets of data are *binomial*.

Applying the Generalized Likelihood Ratio Criterion

Suppose that n Bernoulli trials related to treatment X have resulted in x successes, and that m (independent) Bernoulli trials related to treatment Y have yielded y successes. We wish to test whether p_X and p_Y, the *true* probabilities of success for treatment X and treatment Y, are equal:

$$H_0: p_X = p_Y (= p)$$

versus

$$H_1: p_X \neq p_Y$$

Let α be the level of significance.

Following the notation used for GLRTs, the two parameter spaces here are

$$\omega = \{(p_X, p_Y): 0 \leq p_X = p_Y \leq 1\}$$

and

$$\Omega = \{(p_X, p_Y): 0 \leq p_X \leq 1, 0 \leq p_Y \leq 1\}$$

Furthermore, the likelihood function can be written

$$L = p_X^x(1-p_X)^{n-x} \cdot p_Y^y(1-p_Y)^{m-y}$$

Setting the derivative of ln L with respect to $p(= p_X = p_Y)$ equal to 0 and solving for p gives a not-too-surprising result—namely,

$$p_e = \frac{x+y}{n+m}$$

That is, the maximum likelihood estimate for p under H_0 is the pooled success proportion. Similarly, solving $\partial \ln L / \partial p_X = 0$ and $\partial \ln L / \partial p_Y = 0$ gives the two original sample proportions as the unrestricted maximum likelihood estimates, for p_X and p_Y:

$$p_{X_e} = \frac{x}{n}, \quad p_{Y_e} = \frac{y}{m}$$

Putting p_e, p_{X_e}, and p_{Y_e} back into L gives the generalized likelihood ratio:

$$\lambda = \frac{L(\omega_e)}{L(\Omega_e)} = \frac{\left[(x+y)/(n+m)\right]^{x+y}\left[1-(x+y)/(n+m)\right]^{n+m-x-y}}{(x/n)^x\left[1-(x/n)\right]^{n-x}(y/m)^y\left[1-(y/m)\right]^{m-y}} \tag{9.4.1}$$

Equation 9.4.1 is such a difficult function to work with that it is necessary to find an approximation to the usual generalized likelihood ratio test. There are several available. It can be shown, for example, that $-2 \ln \lambda$ for this problem has an asymptotic χ^2 distribution with 1 degree of freedom (200). Thus, an approximate two-sided, $\alpha = 0.05$ test is to reject H_0 if $-2 \ln \lambda \geq 3.84$.

Another approach, and the one most often used, is to appeal to the central limit theorem and make the observation that

$$\frac{\frac{X}{n} - \frac{Y}{m} - E\left(\frac{X}{n} - \frac{Y}{m}\right)}{\sqrt{\text{Var}\left(\frac{X}{n} - \frac{Y}{m}\right)}}$$

has an approximate standard normal distribution. Under H_0, of course,

$$E\left(\frac{X}{n} - \frac{Y}{m}\right) = 0$$

and

$$\text{Var}\left(\frac{X}{n} - \frac{Y}{m}\right) = \frac{p(1-p)}{n} + \frac{p(1-p)}{m}$$
$$= \frac{(n+m)p(1-p)}{nm}$$

If p is now replaced by $\frac{x+y}{n+m}$, its maximum likelihood estimate under ω, we get the statement of Theorem 9.4.1.

Theorem 9.4.1 *Let x and y denote the numbers of successes observed in two independent sets of n and m Bernoulli trials, respectively, where p_X and p_Y are the true success probabilities associated with each set of trials. Let $p_e = \frac{x+y}{n+m}$ and define*

$$z = \frac{\frac{x}{n} - \frac{y}{m}}{\sqrt{\frac{p_e(1-p_e)}{n} + \frac{p_e(1-p_e)}{m}}}$$

a. *To test H_0: $p_X = p_Y$ versus H_1: $p_X > p_Y$ at the α level of significance, reject H_0 if $z \geq z_\alpha$.*

b. *To test H_0: $p_X = p_Y$ versus H_1: $p_X < p_Y$ at the α level of significance, reject H_0 if $z \leq -z_\alpha$.*

c. *To test H_0: $p_X = p_Y$ versus H_1: $p_X \neq p_Y$ at the α level of significance, reject H_0 if z is either (1) $\leq -z_{\alpha/2}$ or (2) $\geq z_{\alpha/2}$.*

Comment The utility of Theorem 9.4.1 actually extends beyond the scope we have just described. Any continuous variable can always be dichotomized and "transformed" into a Bernoulli variable. For example, blood pressure can be recorded in terms of "mm Hg," a continuous variable, or simply as "normal" or "abnormal," a Bernoulli variable. The next two case studies illustrate these two sources of binomial data. In the first, the measurements begin and end as Bernoulli variables; in the second, the initial measurement of "number of nightmares per month" is dichotomized into "often" and "seldom."

Case Study 9.4.1

Until almost the end of the nineteenth century, the mortality associated with surgical operations—even minor ones—was extremely high. The major problem was infection. The germ theory as a model for disease transmission was still unknown, so there was no concept of sterilization. As a result, many patients died from postoperative complications.

The major breakthrough that was so desperately needed finally came when Joseph Lister, a British physician, began reading about some of the work done by Louis Pasteur. In a series of classic experiments, Pasteur had succeeded in demonstrating the role that yeasts and bacteria play in fermentation. Lister conjectured that human infections might have a similar organic origin. To test his theory he began using carbolic acid as an operating-room disinfectant. He performed forty amputations with the aid of carbolic acid, and thirty-four patients survived. He also did thirty-five amputations *without* carbolic acid, and nineteen patients survived. While it seems clear that carbolic acid did improve survival rates, a test of statistical significance helps to rule out a difference due to chance (202).

Let p_X be the true probability of survival with carbolic acid, and let p_Y denote the true survival probability without the antiseptic. The hypotheses to be tested are

$$H_0: \ p_X = p_Y (= p)$$

versus

$$H_1: \ p_X > p_Y$$

Take $\alpha = 0.01$.

If H_0 is true, the pooled estimate of p would be the overall survival rate. That is,

$$p_e = \frac{34 + 19}{40 + 35} = \frac{53}{75} = 0.707$$

The sample proportions for survival with and without carbolic acid are $34/40 = 0.850$ and $19/35 = 0.543$, respectively. According to Theorem 9.4.1, then, the test statistic is

$$z = \frac{0.850 - 0.543}{\sqrt{\frac{(0.707)(0.293)}{40} + \frac{(0.707)(0.293)}{35}}} = 2.92$$

Since z exceeds the $\alpha = 0.01$ critical value ($z_{.01} = 2.33$), we should reject the null hypothesis and conclude that the use of carbolic acid saves lives.

About the Data In spite of this study and a growing body of similar evidence, the theory of antiseptic surgery was not immediately accepted in Lister's native England. Continental European surgeons, though, understood the value of Lister's work and in 1875 presented him with a humanitarian award.

Case Study 9.4.2

Over the years, numerous studies have sought to characterize the nightmare sufferer. Out of these has emerged the stereotype of someone with high anxiety, low ego strength, feelings of inadequacy, and poorer-than-average physical health. What is not so well known, though, is whether men fall into this pattern with the same frequency as women. To this end, a clinical survey (77) looked at nightmare frequencies for a sample of 160 men and 192 women. Each subject was asked whether he (or she) experienced nightmares "often" (at least once a month) or "seldom" (less than once a month). The percentages of men and women saying "often" were 34.4% and 31.3%, respectively (see Table 9.4.1). Is the difference between those two percentages statistically significant?

Table 9.4.1 Frequency of Nightmares

	Men	Women	Total
Nightmares often	55	60	115
Nightmares seldom	105	132	237
Totals	160	192	
% often:	34.4	31.3	

Let p_M and p_W denote the true proportions of men having nightmares often and women having nightmares often, respectively. The hypotheses to be tested are

$$H_0: p_M = p_W$$

versus

$$H_1: p_M \neq p_W$$

Let $\alpha = 0.05$. Then $\pm z_{.025} = \pm 1.96$ become the two critical values. Moreover, $p_e = \frac{55+60}{160+192} = 0.327$, so

$$z = \frac{0.344 - 0.313}{\sqrt{\frac{(0.327)(0.673)}{160} + \frac{(0.327)(0.673)}{192}}}$$
$$= 0.62$$

The conclusion, then, is clear: We fail to reject the null hypothesis—these data provide no convincing evidence that the frequency of nightmares is different for men than for women.

About the Data The results of every statistical study are intended to be generalized—from the subjects measured to a broader population that the sample might reasonably be expected to represent. Obviously, then, knowing something about the subjects is essential if a set of data is to be interpreted (and extrapolated) properly. Table 9.4.1 is a cautionary case in point. The 352 individuals interviewed were not the typical sort of subjects solicited for a university research project. They were all institutionalized mental patients.

Questions

9.4.1. The phenomenon of handedness has been extensively studied in human populations. The percentages of adults who are right-handed, left-handed, and ambidextrous are well documented. What is not so well known is that a similar phenomenon is present in lower animals. Dogs, for example, can be either right-pawed or left-pawed. Suppose that in a random sample of 200 beagles, it is found that 55 are left-pawed and that in a random sample of 200 collies, 40 are left-pawed. Can we conclude that the difference in the two sample proportions of left-pawed dogs is statistically significant for $\alpha = 0.05$?

9.4.2. In a study designed to see whether a controlled diet could retard the process of arteriosclerosis, a total of 846 randomly chosen persons were followed over an eight-year period. Half were instructed to eat only certain foods; the other half could eat whatever they wanted. At the end of eight years, 66 persons in the diet group were found to have died of either myocardial infarction or cerebral infarction, as compared to 93 deaths of a similar nature in the control group (203). Do the appropriate analysis. Let $\alpha = 0.05$.

9.4.3. Water witching, the practice of using the movements of a forked twig to locate underground water (or minerals), dates back over 400 years. Its first detailed description appears in Agricola's *De re Metallica,* published in 1556. That water witching works remains a belief widely held among rural people in Europe and throughout the Americas. [In 1960 the number of "active" water witches in the United States was estimated to be more than 20,000 (193).] Reliable evidence supporting or refuting water witching is hard to find. Personal accounts of isolated successes or failures tend to be strongly biased by the attitude of the observer. Of all the wells dug in Fence Lake, New Mexico, 29 "witched" wells and 32 "nonwitched" wells were sunk. Of the "witched" wells, 24 were successful. For the "nonwitched" wells, there were 27 successes. What would you conclude?

9.4.4. If flying saucers are a genuine phenomenon, it would follow that the nature of sightings (that is, their physical characteristics) would be similar in different parts of the world. A prominent UFO investigator compiled a listing of 91 sightings reported in Spain and 1117 reported elsewhere. Among the information recorded was whether the saucer was on the ground or hovering. His data are summarized in the following table (87). Let p_S and p_{NS} denote the true probabilities of "Saucer on ground" in Spain and not in Spain, respectively. Test H_0: $p_S = p_{NS}$ against a two-sided H_1. Let $\alpha = 0.01$.

	In Spain	Not in Spain
Saucer on ground	53	705
Saucer hovering	38	412

9.4.5. In some criminal cases, the judge and the defendant's lawyer will enter into a plea bargain, where the accused pleads guilty to a lesser charge. The proportion of time this happens is called the *mitigation rate*. A Florida Corrections Department study showed that Escambia County had the state's fourth highest rate, 61.7% (1033 out of 1675 cases). Concerned that the guilty were not getting appropriate sentences, the state attorney put in new policies to limit the number of plea bargains. A follow-up study (133) showed that the mitigation rate dropped to 52.1% (344 out of 660 cases). Is it fair to conclude that the drop was due to the new policies, or can the decline be written off to chance? Test at the $\alpha = 0.01$ level.

9.4.6. Suppose H_0: $p_X = p_Y$ is being tested against H_1: $p_X \neq p_Y$ on the basis of two independent sets of one hundred Bernoulli trials. If x, the number of successes in the first set, is sixty and y, the number of successes in the second set, is forty-eight, what P-value would be associated with the data?

9.4.7. A total of 8605 students are enrolled full-time at State University this semester, 4134 of whom are women. Of the 6001 students who live on campus, 2915 are women. Can it be argued that the difference in the proportion of men and women living on campus is statistically significant? Carry out an appropriate analysis. Let $\alpha = 0.05$.

9.4.8. The kittiwake is a seagull whose mating behavior is basically monogamous. Normally, the birds separate for several months after the completion of one breeding season and reunite at the beginning of the next. Whether or not the birds actually do reunite, though, may be affected by the success of their "relationship" the season before. A total of 769 kittiwake pair-bonds were studied (30) over the course of two breeding seasons; of those 769, some 609 successfully bred during the first season; the remaining 160 were unsuccessful. The following season, 175 of the previously successful pair-bonds "divorced," as did 100 of the 160 whose prior relationship left something to be desired.

Can we conclude that the difference in the two divorce rates (29% and 63%) is statistically significant?

	Breeding in Previous Year	
	Successful	Unsuccessful
Number divorced	175	100
Number not divorced	434	60
Total	609	160
Percent divorced	29	63

9.4.9. A utility infielder for a National League club batted .260 last season in three hundred trips to the plate. This year he hit .250 in two hundred at-bats. The owners are trying to cut his pay for next year on the grounds that his output has deteriorated. The player argues, though, that his performances the last two seasons have not been significantly different, so his salary should not be reduced. Who is right?

9.4.10. Compute $-2 \ln \lambda$ (see Equation 9.4.1) for the nightmare data of Case Study 9.4.2, and use it to test the hypothesis that $p_X = p_Y$. Let $\alpha = 0.01$.

9.5 Confidence Intervals for the Two-Sample Problem

Two-sample data lend themselves nicely to the hypothesis testing format because a meaningful H_0 can always be defined (which is not the case for every set of *one-sample* data). The same inferences, though, can just as easily be phrased in terms of confidence intervals. Simple inversions similar to the derivation of Equation 7.4.1 will yield confidence intervals for $\mu_X - \mu_Y$, σ_X^2/σ_Y^2, and $p_X - p_Y$.

Theorem 9.5.1 *Let $x_1, x_2, \ldots, x_n$ and $y_1, y_2, \ldots, y_m$ be independent random samples drawn from normal distributions with means μ_X and μ_Y, respectively, and with the same standard deviation, σ. Let s_p denote the data's pooled standard deviation. A $100(1-\alpha)\%$ confidence interval for $\mu_X - \mu_Y$ is given by*

$$\left(\bar{x} - \bar{y} - t_{\alpha/2,n+m-2} \cdot s_p\sqrt{\frac{1}{n} + \frac{1}{m}},\ \bar{x} - \bar{y} + t_{\alpha/2,n+m-2} \cdot s_p\sqrt{\frac{1}{n} + \frac{1}{m}}\right)$$

Proof We know from Theorem 9.2.1 that

$$\frac{\overline{X} - \overline{Y} - (\mu_X - \mu_Y)}{S_p\sqrt{\frac{1}{n} + \frac{1}{m}}}$$

has a Student t distribution with $n + m - 2$ df. Therefore,

$$P\left[-t_{\alpha/2,n+m-2} \leq \frac{\overline{X} - \overline{Y} - (\mu_X - \mu_Y)}{S_p\sqrt{\frac{1}{n} + \frac{1}{m}}} \leq t_{\alpha/2,n+m-2}\right] = 1 - \alpha \tag{9.5.1}$$

Rewriting Equation 9.5.1 by isolating $\mu_X - \mu_Y$ in the center of the inequalities gives the endpoints stated in the theorem. □

Case Study 9.5.1

Case Study 8.2.2 made the claim that X-rays penetrate the tooth enamel of men and women differently, a fact that allows dental structure to help identify the sex of badly decomposed bodies. In this case study, the statistical analysis for

(Continued on next page)

(Case Study 9.5.1 continued)

that assertion is provided. Moreover, the resulting confidence interval gives an estimate of the difference in the mean enamel spectropenetration gradients for the two sexes.

Listed in Table 9.5.1 (and Table 8.2.2) are the gradients for eight female teeth and eight male teeth (57). These numbers are measures of the rate of change in the amount of X-ray penetration through a 500-micron section of tooth enamel at a wavelength of 600 nm as opposed to 400 nm.

Table 9.5.1 Enamel Spectropenetration Gradients

Male, x_i	Female, y_i
4.9	4.8
5.4	5.3
5.0	3.7
5.5	4.1
5.4	5.6
6.6	4.0
6.3	3.6
4.3	5.0

Let μ_X and μ_Y be the population means of the spectropenetration gradients associated with male teeth and with female teeth, respectively. Note that

$$\sum_{i=1}^{8} x_i = 43.4 \quad \text{and} \quad \sum_{i=1}^{8} x_i^2 = 239.32$$

from which

$$\bar{x} = \frac{43.4}{8} = 5.4$$

and

$$s_X^2 = \frac{8(239.32) - (43.4)^2}{8(7)} = 0.55$$

Similarly,

$$\sum_{i=1}^{8} y_i = 36.1 \quad \text{and} \quad \sum_{i=1}^{8} y_i^2 = 166.95$$

so that

$$\bar{y} = \frac{36.1}{8} = 4.5$$

and

$$s_Y^2 = \frac{8(166.95) - (36.1)^2}{8(7)} = 0.58$$

Therefore, the pooled standard deviation is equal to 0.75:

$$s_P = \sqrt{\frac{7(0.55) + 7(0.58)}{8 + 8 - 2}} = \sqrt{0.565} = 0.75$$

(Continued on next page)

We know that the ratio

$$\frac{\overline{X}-\overline{Y}-(\mu_X-\mu_Y)}{S_p\sqrt{\frac{1}{8}+\frac{1}{8}}}$$

will be approximated by a Student t curve with 14 degrees of freedom. Since $t_{.025,14}=2.1448$, the 95% confidence interval for $\mu_X-\mu_Y$ is given by

$$\left(\bar{x}-\bar{y}-2.1448\, s_p\sqrt{\frac{1}{8}+\frac{1}{8}},\ \bar{x}-\bar{y}+2.1448\, s_p\sqrt{\frac{1}{8}+\frac{1}{8}}\right)$$

$$=\left[5.4-4.5-2.1448(0.75)\sqrt{0.25},\ 5.4-4.5+2.1448(0.75)\sqrt{0.25}\right]$$

$$=(0.1,\ 1.7)$$

Comment Here the 95% confidence interval does not include the value 0. This means that had we tested

$$H_0\colon \mu_X=\mu_Y$$

versus

$$H_1\colon \mu_X\neq\mu_Y$$

at the $\alpha=0.05$ level of significance, H_0 would have been rejected.

Comment For the scenario of Theorem of 9.5.1, if the variances are not equal, then an approximate $100(1-\alpha)$% confidence interval is given by

$$\left(\bar{x}-\bar{y}-t_{\alpha/2,\nu}\sqrt{\frac{s_X^2}{n}+\frac{s_Y^2}{m}},\ \bar{x}-\bar{y}+t_{\alpha/2,\nu}\sqrt{\frac{s_X^2}{n}+\frac{s_Y^2}{m}}\right)$$

where $\nu=\dfrac{\left(\hat{\theta}+\frac{n}{m}\right)^2}{\frac{1}{(n-1)}\hat{\theta}^2+\frac{1}{(m-1)}\left(\frac{n}{m}\right)^2}$ for $\hat{\theta}=\frac{s_X^2}{s_Y^2}$.

If the degrees or freedom exceed 100, then the form above is used, with $z_{\alpha/2}$ replacing $t_{\alpha/2,\nu}$.

Theorem 9.5.2 *Let $x_1, x_2, \ldots, x_n$ and $y_1, y_2, \ldots, y_m$ be independent random samples drawn from normal distributions with standard deviations σ_X and σ_Y, respectively. A $100(1-\alpha)$% confidence interval for the variance ratio, σ_X^2/σ_Y^2, is given by*

$$\left(\frac{s_X^2}{s_Y^2}F_{\alpha/2,m-1,n-1},\ \frac{s_X^2}{s_Y^2}F_{1-\alpha/2,m-1,n-1}\right)$$

Proof Start with the fact that $\frac{S_Y^2/\sigma_Y^2}{S_X^2/\sigma_X^2}$ has an F distribution with $m-1$ and $n-1$ df, and follow the strategy used in the proof of Theorem 9.5.1—that is, isolate σ_X^2/σ_Y^2 in the center of the analogous inequalities. □

Case Study 9.5.2

The easiest way to measure the movement, or flow, of a glacier is with a camera. First a set of reference points is marked off at various sites near the glacier's edge. Then these points, along with the glacier, are photographed from an airplane. The problem is this: How long should the time interval be between photographs? If too *short* a period has elapsed, the glacier will not have moved very far and the errors associated with the photographic technique will be relatively large. If too *long* a period has elapsed, parts of the glacier might be deformed by the surrounding terrain, an eventuality that could introduce substantial variability into the point-to-point velocity estimates.

Two sets of flow rates for the Antarctic's Hoseason Glacier have been calculated (115), one based on photographs taken *three* years apart, the other, *five* years apart (see Table 9.5.2). On the basis of other considerations, it can be assumed that the "true" flow rate was constant for the eight years in question.

Table 9.5.2 Flow Rates Estimated for the Hoseason Glacier (Meters per Day)

Three-Year Span, x_i	Five-Year Span, y_i
0.73	0.72
0.76	0.74
0.75	0.74
0.77	0.72
0.73	0.72
0.75	
0.74	

The objective here is to assess the relative variabilities associated with the three- and five-year time periods. One way to do this—assuming the data to be normal—is to construct, say, a 95% confidence interval for the variance ratio. If that interval does not contain the value 1, we infer that the two time periods lead to flow rate estimates of significantly different precision.

From Table 9.5.2,

$$\sum_{i=1}^{7} x_i = 5.23 \quad \text{and} \quad \sum_{i=1}^{7} x_i^2 = 3.9089$$

so that

$$s_X^2 = \frac{7(3.9089) - (5.23)^2}{7(6)} = 0.000224$$

Similarly,

$$\sum_{i=1}^{5} y_i = 3.64 \quad \text{and} \quad \sum_{i=1}^{5} y_i^2 = 2.6504$$

(Continued on next page)

making

$$s_Y^2 = \frac{5(2.6504) - (3.64)^2}{5(4)} = 0.000120$$

The two critical values come from Table A.4 in the Appendix:

$$F_{.025,4,6} = 0.109 \quad \text{and} \quad F_{.975,4,6} = 6.23$$

Substituting, then, into the statement of Theorem 9.5.2 gives (0.203, 11.629) as a 95% confidence interval for σ_X^2/σ_Y^2:

$$\left(\frac{0.000224}{0.000120}0.109, \frac{0.000224}{0.000120}6.23\right) = (0.203, 11.629)$$

Thus, although the three-year data have a larger *sample* variance than the five-year data, no conclusions can be drawn about the *true* variances being different, because the ratio $\sigma_X^2/\sigma_Y^2 = 1$ is contained in the confidence interval.

Theorem 9.5.3 *Let x and y denote the numbers of successes observed in two independent sets of n and m Bernoulli trials, respectively. If p_X and p_Y denote the true success probabilities, an approximate $100(1-\alpha)\%$ confidence interval for $p_X - p_Y$ is given by*

$$\left[\frac{x}{n} - \frac{y}{m} - z_{\alpha/2}\sqrt{\frac{\left(\frac{x}{n}\right)\left(1-\frac{x}{n}\right)}{n} + \frac{\left(\frac{y}{m}\right)\left(1-\frac{y}{m}\right)}{m}},\right.$$
$$\left.\frac{x}{n} - \frac{y}{m} + z_{\alpha/2}\sqrt{\frac{\left(\frac{x}{n}\right)\left(1-\frac{x}{n}\right)}{n} + \frac{\left(\frac{y}{m}\right)\left(1-\frac{y}{m}\right)}{m}}\right]$$

Proof See Question 9.5.11. □

Case Study 9.5.3

If a hospital patient's heart stops, an emergency message, *code blue*, is called. A team rushes to the bedside and attempts to revive the patient. A study (131) suggests that patients are better off not suffering cardiac arrest after 11 P.M., the so-called *graveyard shift*. The study lasted seven years and used non–emergency room data from over five hundred hospitals. During the day and early evening hours, 58,593 cardiac arrests occurred and 11,604 patients survived to leave the hospital. For the 11 P.M. shift, of the 28,155 heart stoppages, 4139 patients lived to be discharged.

Let p_X (estimated by $11{,}604/58{,}593 = 0.198$) be the true probability of survival during the earlier hours. Let p_Y denote the true survival probability for the graveyard shift (estimated by $4139/28{,}155 = 0.147$). To construct a 95% confidence interval for $p_X - p_Y$, take $z_{\alpha/2} = 1.96$. Then Theorem 9.5.3 gives the lower limit of the confidence interval as

$$0.198 - 0.147 - 1.96\sqrt{\frac{(0.198)(0.802)}{58{,}593} + \frac{(0.147)(0.853)}{28{,}155}} = 0.0458$$

(Continued on next page)

(Case Study 9.5.3 continued)

and the upper limit as

$$0.198 - 0.147 + 1.96\sqrt{\frac{(0.198)(0.802)}{58{,}593} + \frac{(0.147)(0.853)}{28{,}155}} = 0.0562$$

so the 95% confidence interval is (0.0458, 0.0562).

Since $p_X - p_Y = 0$ is not included in the interval (which lies entirely to the *right* of 0), we can conclude that survival rates are worse during the graveyard shift.

Questions

9.5.1 In 1965 a silver shortage in the United States prompted Congress to authorize the minting of silverless dimes and quarters. They also recommended that the silver content of half-dollars be reduced from 90% to 40%. Historically, fluctuations in the amount of rare metals found in coins are not uncommon (76). The following data may be a case in point. Listed are the silver percentages found in samples of a Byzantine coin minted on two separate occasions during the reign of Manuel I (1143–1180). Construct a 90% confidence interval for $\mu_X - \mu_Y$, the true average difference in the coin's silver content (= "early" − "late"). What does the interval imply about the outcome of testing H_0: $\mu_X = \mu_Y$? For these data $s_X = 0.54$ and $s_Y = 0.36$.

Early Coinage, x_i (% Ag)	Late Coinage, y_i (% Ag)
5.9	5.3
6.8	5.6
6.4	5.5
7.0	5.1
6.6	6.2
7.7	5.8
7.2	5.8
6.9	
6.2	
Average: 6.7	Average: 5.6

9.5.2 Male fiddler crabs solicit attention from the opposite sex by standing in front of their burrows and waving their claws at the females who walk by. If a female likes what she sees, she pays the male a brief visit in his burrow. If everything goes well and the crustacean chemistry clicks, she will stay a little longer and mate. In what may be a ploy to lessen the risk of spending the night alone, some of the males build elaborate mud domes over their burrows. Do the following data (215) suggest that a male's time spent waving to females is influenced by whether his burrow has a dome? Answer the question by constructing and interpreting a 95% confidence interval for $\mu_X - \mu_Y$. Use the value $s_p = 11.2$.

% of Time Spent Waving to Females	
Males with Domes, x_i	Males without Domes, y_i
100.0	76.4
58.6	84.2
93.5	96.5
83.6	88.8
84.1	85.3
	79.1
	83.6

9.5.3 Construct two 99% confidence intervals for $\mu_X - \mu_Y$ using the data of Case Study 9.2.3, first assuming the variances are equal, and then assuming they are not.

9.5.4 Carry out the details to complete the proof of Theorem 9.5.1.

9.5.5 Suppose that $X_1, X_2, \ldots, X_n$ and $Y_1, Y_2, \ldots, Y_m$ are independent random samples from normal distributions with means μ_X and μ_Y and *known* standard deviations σ_X and σ_Y, respectively. Derive a $100(1-\alpha)\%$ confidence interval for $\mu_X - \mu_Y$.

9.5.6 Construct a 95% confidence interval for σ_X^2/σ_Y^2 based on the data in Case Study 9.2.1. The hypothesis test referred to tacitly assumed that the variances were equal. Does that agree with your confidence interval? Explain.

9.5.7 One of the parameters used in evaluating myocardial function is the end diastolic volume (EDV). The following table shows EDVs recorded for eight persons considered to have normal cardiac function and for six with constrictive pericarditis (192). Would it be correct to use Theorem 9.2.2 to test H_0: $\mu_X = \mu_Y$? Answer the question by constructing a 95% confidence interval for σ_X^2/σ_Y^2.

Normal, x_i	Constrictive Pericarditis, y_i
62	24
60	56
78	42
62	74
49	44
67	28
80	
48	

9.5.8 Complete the proof of Theorem 9.5.2.

9.5.9 Flonase is a nasal spray for diminishing nasal allergic symptoms. In clinical trials for side effects, 782 sufferers from allergic rhinitis were given a daily dose of 200 mcg of Flonase. Of this group, 126 reported headaches. A group of 758 subjects were given a placebo, and 111 of them reported headaches. Find a 95% confidence interval for the difference in proportion of headaches for the two groups. Does the confidence interval suggest a statistically significant difference in the frequency of headaches for Flonase users?
Source: http://www.drugs.com/sfx/flonase-side-effects.html.

9.5.10 Construct an 80% confidence interval for the difference $p_M - p_W$ in the nightmare frequency data summarized in Case Study 9.4.2.

9.5.11 If p_X and p_Y denote the true success probabilities associated with two sets of n and m independent Bernoulli trials, respectively, the ratio

$$\frac{\frac{X}{n} - \frac{Y}{m} - (p_X - p_Y)}{\sqrt{\frac{(X/n)(1-X/n)}{n} + \frac{(Y/m)(1-Y/m)}{m}}}$$

has approximately a standard normal distribution. Use that fact to prove Theorem 9.5.3.

9.5.12 Suicide rates in the United States tend to be much higher for men than for women, at all ages. That pattern may not extend to all professions, though. Death certificates obtained for the 3637 members of the American Chemical Society who died over a twenty-year period revealed that 106 of the 3522 male deaths were suicides, as compared to 13 of the 115 female deaths (101). Construct a 95% confidence interval for the difference in suicide rates. What would you conclude?

9.6 Taking a Second Look at Statistics (Choosing Samples)

Choosing sample *sizes* is a topic that invariably receives extensive coverage whenever applied statistics and experimental design are discussed. For good reason. Whatever the context, the number of observations making up a data set figures prominently in the ability of those data to address any and all of the questions raised by the experimenter. As sample sizes get larger, we know that estimators become more precise and hypothesis tests get better at distinguishing between H_0 and H_1. Larger sample sizes, of course, are also more expensive. The trade-off between how many observations researchers can *afford* to take and how many they would *like* to take is a choice that has to be made early on in the design of any experiment. If the sample sizes ultimately decided upon are too small, there is a risk that the objectives of the study will not be fully achieved—parameters may be estimated with insufficient precision and hypothesis tests may reach incorrect conclusions.

That said, choosing sample sizes is often not as critical to the success of an experiment as choosing sample *subjects*. In a two-sample design, for example, how should we decide which particular subjects to assign to treatment X and which to treatment Y? If the subjects comprising a sample are somehow "biased" with respect to the measurement being recorded, the integrity of the conclusions is irretrievably compromised. There are no statistical techniques for "correcting" inferences based on measurements that were biased in some unknown way. It is also true that biases can be very subtle, yet still have a pronounced effect on the final measurements. That being the case, it is incumbent on researchers to take every possible precaution at the outset to prevent inappropriate assignments of subjects to treatments.

For example, suppose for your Senior Project you plan to study whether a new synthetic testosterone can affect the behavior of female rats. Your intention is to set up a two-sample design where ten rats will be given weekly injections of the new testosterone compound and another ten rats will serve as a control group, receiving weekly injections of a placebo. At the end of eight weeks, all twenty rats will be put in a large community cage, and the behavior of each one will be closely monitored for signs of aggression.

Last week you placed an order for twenty female *Rattus norvegicus* from the local Rats 'R Us franchise. They arrived today, all housed in one large cage. Your plan is to remove ten of the twenty "at random," and then put those ten in a similarly large cage. The ten removed will be receiving the testosterone injections; the ten remaining in the original cage will constitute the control group. The question is, *which* ten should be removed?

The obvious answer—reach in and pull out ten—is very much the wrong answer! Why? Because the samples formed in such a way might very well be biased if, for example, you (understandably) tended to avoid grabbing the rats that looked like they might bite. If that were the case, the ones you drew out would be biased, by virtue of being more passive than the ones left behind. Since the measurements ultimately to be taken deal with aggression, biasing the samples in that particular way would be a fatal flaw. Whether the total sample size was twenty or twenty thousand, the results would be worthless.

In general, relying on our intuitive sense of the word "random" to allocate subjects to different treatments is risky, to say the least. The correct approach would be to number the rats from 1 to 20 and then use a random number table or a computer's random number generator to identify the ten to be removed. Figure 9.6.1 shows the Minitab syntax for choosing a random sample of ten numbers from the integers 1 through 20. According to this particular run of the SAMPLE routine, the ten rats to be removed for the testosterone injections are (in order) numbers 1, 5, 8, 9, 10, 14, 15, 18, 19 and 20.

Figure 9.6.1

```
MTB   > set c1
DATA  > 1:20
DATA  > end
MTB   > sample 10 c1 c2
MTB   > print c2
```

Data Display

```
C2  18  1  20  19  9  10  8  15  14  5
```

There is a moral here. Designing, carrying out, and analyzing an experiment is an exercise that draws on a variety of scientific, computational, and statistical skills, some of which may be quite sophisticated. No matter how well those complex issues are attended to, though, the enterprise will fail if the simplest and most basic aspects of the experiment—such as assigning subjects to treatments—are not carefully scrutinized and properly done. The Devil, as the saying goes, is in the details.

Appendix 9.A.1 A Derivation of the Two-Sample *t* Test (A Proof of Theorem 9.2.2)

To begin, we note that both the restricted and unrestricted parameter spaces, ω and Ω, are three dimensional:

$$\omega = \{(\mu_X, \mu_Y, \sigma): -\infty < \mu_X = \mu_Y < \infty, 0 < \sigma < \infty\}$$

and

$$\Omega = \{(\mu_X, \mu_Y, \sigma): -\infty < \mu_X < \infty, -\infty < \mu_Y < \infty, 0 < \sigma < \infty\}$$

Since the X's and Y's are independent (and normal),

$$L(\omega) = \prod_{i=1}^{n} f_X(x_i) \prod_{j=1}^{m} f_Y(y_j)$$

$$= \left(\frac{1}{\sqrt{2\pi}\sigma}\right)^{n+m} \exp\left\{-\frac{1}{2\sigma^2}\left[\sum_{i=1}^{n}(x_i-\mu)^2 + \sum_{j=1}^{m}(y_i-\mu)^2\right]\right\} \quad (9.A.1.1)$$

where $\mu = \mu_X = \mu_Y$. If we take $\ln L(\omega)$ and solve $\partial \ln L(\omega)/\partial\mu = 0$ and $\partial \ln L(\omega)/\partial\sigma^2 = 0$ simultaneously, the solutions will be the restricted maximum likelihood estimates:

$$\mu_{\omega_e} = \frac{\sum_{i=1}^{n} x_i + \sum_{j=1}^{m} y_j}{n+m} \quad (9.A.1.2)$$

and

$$\sigma^2_{\omega_e} = \frac{\sum_{i=1}^{n}(x_i-\mu_e)^2 + \sum_{j=1}^{m}(y_j-\mu_e)^2}{n+m} \quad (9.A.1.3)$$

Substituting Equations 9.A.1.2 and 9.A.1.3 into Equation 9.A.1.1 gives the numerator of the generalized likelihood ratio:

$$L(\omega_e) = \left(\frac{e^{-1}}{2\pi\sigma^2_{\omega_e}}\right)^{(n+m)/2}$$

Similarly, the likelihood function unrestricted by the null hypothesis is

$$L(\Omega) = \left(\frac{1}{\sqrt{2\pi}\sigma}\right)^{n+m} \exp\left\{-\frac{1}{2\sigma^2}\left[\sum_{i=1}^{n}(x_i-\mu_X)^2 + \sum_{j=1}^{m}(y_j-\mu_Y)^2\right]\right\} \quad (9.A.1.4)$$

Here, solving

$$\frac{\partial \ln L(\Omega)}{\partial \mu_X} = 0 \qquad \frac{\partial \ln L(\Omega)}{\partial \mu_Y} = 0 \qquad \frac{\partial \ln L(\Omega)}{\partial \sigma^2} = 0$$

gives

$$\mu_{X_e} = \bar{x} \quad \mu_{Y_e} = \bar{y}$$

$$\sigma^2_{\Omega_e} = \frac{\sum_{i=1}^{n} (x_i - \bar{x})^2 + \sum_{j=1}^{m} (y_j - \bar{y})^2}{n+m}$$

If these estimates are substituted into Equation 9.A.1.4, the maximum value for $L(\Omega)$ simplifies to

$$L(\Omega_e) = \left(e^{-1}/2\pi\sigma^2_{\Omega_e}\right)^{(n+m)/2}$$

It follows, then, that the generalized likelihood ratio, λ, is equal to

$$\lambda = \frac{L(\omega_e)}{L(\Omega_e)} = \left(\frac{\sigma^2_{\Omega_e}}{\sigma^2_{\omega_e}}\right)^{(n+m)/2}$$

or, equivalently,

$$\lambda^{2/(n+m)} = \frac{\sum_{i=1}^{n} (x_i - \bar{x})^2 + \sum_{j=1}^{m} (y_j - \bar{y})^2}{\sum_{i=1}^{n} \left[x_i - \left(\frac{n\bar{x}+m\bar{y}}{n+m}\right)\right]^2 + \sum_{j=1}^{m} \left[y_j - \left(\frac{n\bar{x}+m\bar{y}}{n+m}\right)\right]^2}$$

Using the identity

$$\sum_{i=1}^{n} \left(x_i - \frac{n\bar{x}+m\bar{y}}{n+m}\right)^2 = \sum_{i=1}^{n} (x_i - \bar{x})^2 + \frac{m^2 n}{(n+m)^2}(\bar{x} - \bar{y})^2$$

we can write $\lambda^{2/(n+m)}$ as

$$\lambda^{2/(n+m)} = \frac{\sum_{i=1}^{n} (x_i - \bar{x})^2 + \sum_{j=1}^{m} (y_j - \bar{y})^2}{\sum_{i=1}^{n} (x_i - \bar{x})^2 + \sum_{j=1}^{m} (y_j - \bar{y})^2 + \frac{nm}{n+m}(\bar{x} - \bar{y})^2}$$

$$= \frac{1}{1 + \frac{(\bar{x}-\bar{y})^2}{\left[\sum_{i=1}^{n} (x_i - \bar{x})^2 + \sum_{j=1}^{m} (y_j - \bar{y})^2\right]\left(\frac{1}{n} + \frac{1}{m}\right)}}$$

$$= \frac{n+m-2}{n+m-2+\frac{(\bar{x}-\bar{y})^2}{s_p^2[(1/n)+(1/m)]}}$$

where s_p^2 is the pooled variance:

$$s_p^2 = \frac{1}{n+m-2}\left[\sum_{i=1}^{n} (x_i - \bar{x})^2 + \sum_{j=1}^{m} (y_j - \bar{y})^2\right]$$

Therefore, in terms of the observed t ratio, $\lambda^{2/(n+m)}$ simplifies to

$$\lambda^{2/(n+m)} = \frac{n+m-2}{n+m-2+t^2} \tag{9.A.1.5}$$

At this point the proof is almost complete. The generalized likelihood ratio criterion, rejecting H_0: $\mu_X = \mu_Y$ when $0 < \lambda \leq \lambda^*$, is clearly equivalent to rejecting the null hypothesis when $0 < \lambda^{2/(n+m)} \leq \lambda^{**}$. But both of these, from Equation 9.A.1.5, are the same as rejecting H_0 when t^2 is too large. Thus the decision rule in terms of t^2 is

$$\text{Reject } H_0\text{: } \mu_X = \mu_Y \text{ in favor of } H_1\text{: } \mu_X \neq \mu_Y \text{ if } t^2 \geq t^{*2}$$

Or, phrasing this in still another way, we should reject H_0 if either $t \geq t^*$ or $t \leq -t^*$, where

$$P(-t^* < T < t^* \mid H_0\text{: } \mu_X = \mu_Y \text{ is true}) = 1 - \alpha$$

By Theorem 9.2.1, though, T has a Student t distribution with $n + m - 2$ df, which makes $\pm t^* = \pm t_{\alpha/2,n+m-2}$, and the theorem is proved.

Appendix 9.A.2 Minitab Applications

Minitab has a simple command—TWOSAMPLE C1 C2—for doing a two-sample t test on a set of x_i's and y_i's stored in columns C1 and C2, respectively. The same command automatically constructs a 95% confidence interval for $\mu_X - \mu_Y$.

Figure 9.A.2.1

```
MTB   > set c1
DATA  > 0.225 0.262 0.217 0.240 0.230 0.229 0.235 0.217
DATA  > end
MTB   > set c2
DATA  > 0.209 0.205 0.196 0.210 0.202 0.207 0.224 0.223
DATA  > 0.220 0.201
DATA  > end
MTB   > name c1 'X' c2 'Y'
MTB   > twosample c1 c2;
SUBC  > pooled.
```

Two-Sample T-Test and CI: X, Y

```
Two-sample T for X vs Y

    N     Mean      StDev     SE Mean
X   8    0.2319    0.0146     0.0051
Y  10    0.20970   0.00966    0.0031

Difference = mu (X) - mu (Y)
Estimate for difference: 0.02217
95% CI for difference: (0.01005, 0.03430)
T-Test of difference = 0 (vs not =): T-Value = 3.88 P-Value = 0.001 DF = 16
Both use Pooled StDev = 0.0121
```

Figure 9.A.2.1 shows the syntax for analyzing the Quintus Curtius Snodgrass data in Table 9.2.1. Notice that a subcommand is included. If we write

```
MTB > twosample c1 c2
```

Minitab will assume the two population variances are not equal, and it will perform the approximate t test described in Theorem 9.2.3. If the intention is to assume that $\sigma_X^2 = \sigma_Y^2$ (and do the t test as described in Theorem 9.2.1), the proper syntax is

```
MTB  > twosample c1 c2;
SUBC > pooled.
```

As is typical, Minitab associates the test statistic with a P-value rather than an "Accept H_0" or "Reject H_0" conclusion. Here, $P = 0.001$, which is consistent with the decision reached in Case Study 9.2.1 to "reject H_0 at the $\alpha = 0.01$ level of significance." Figure 9.A.2.2 shows the "unpooled" analysis of these same data. The conclusion is the same, although the P-value has almost tripled, because both the test statistic and its degrees of freedom have decreased (recall Question 9.2.18).

Figure 9.A.2.2

```
MTB   > set c1
DATA  > 0.225 0.262 0.217 0.240 0.230 0.229 0.235 0.217
DATA  > end
MTB   > set c2
DATA  > 0.209 0.205 0.196 0.210 0.202 0.207 0.224 0.223 0.220 0.201
DATA  > end
MTB   > name c1 'X' c2 'Y'
MTB   > twosample c1 c2
```

Two-Sample T-Test and CI: X, Y

```
Two-sample T for X vs Y

    N     Mean     StDev    SE Mean
X   8   0.2319    0.0146     0.0051
Y  10  0.20970   0.00966     0.0031

Difference = mu (X) - mu (Y)
Estimate for difference: 0.02217
95% CI for difference: (0.00900, 0.03535)
T-Test of difference = 0 (vs not =): T-Value = 3.70 P-Value = 0.003 DF = 11
```

Testing $H_0: \mu_X = \mu_Y$ Using Minitab Windows

1. Enter the two samples in C1 and C2, respectively.
2. Click on STAT, then on BASIC STATISTICS, then on 2-SAMPLE t.
3. Click on SAMPLES IN DIFFERENT COLUMNS, and type C1 in FIRST box and C2 in SECOND box.
4. Click on ASSUME EQUAL VARIANCES (if a pooled t test is desired).
5. Click on OPTIONS.
6. Enter value for $100(1-\alpha)$ in CONFIDENCE LEVEL box.
7. Click on NOT EQUAL; then click on whichever H_1 is desired.
8. Click on OK; click on remaining OK.

Chapter 10

Goodness-of-Fit Tests

Called by some the founder of twentieth-century statistics, Pearson received his university education at Cambridge, concentrating on physics, philosophy, and law. He was called to the bar in 1881 but never practiced. In 1911 Pearson resigned his chair of applied mathematics and mechanics at University College, London, and became the first Galton Professor of Eugenics, as was Galton's wish. Together with Weldon, Pearson founded the prestigious journal Biometrika *and served as its principal editor from 1901 until his death.*

—Karl Pearson (1857–1936)

10.1 Introduction

The give-and-take between the mathematics of probability and the empiricism of statistics should be, by now, a comfortably familiar theme. Time and time again we have seen repeated measurements, no matter their source, exhibiting a regularity of pattern that can be well approximated by one or more of the handful of probability functions introduced in Chapter 4. Until now, all the inferences resulting from this interfacing have been parameter specific, a fact to which the many hypothesis tests about means, variances, and binomial proportions paraded forth in Chapters 6, 7, and 9 bear ample testimony. Still, there are other situations where the basic *form of* $p_X(k)$ or $f_Y(y)$, rather than the value of its parameters, is the most important question at issue. These situations are the focus of Chapter 10.

A geneticist, for example, might want to know whether the inheritance of a certain set of traits follows the same set of ratios as those prescribed by Mendelian theory. The objective of a psychologist, on the other hand, might be to confirm or refute a newly proposed model for cognitive serial learning. Probably the most habitual users of inference procedures directed at the entire pdf, though, are statisticians themselves: As a prelude to doing any sort of hypothesis test or confidence interval, an attempt should be made, sample size permitting, to verify that the data are, indeed, representative of whatever distribution that procedure presumes. Usually, this will mean testing to see whether a set of y_i's might conceivably represent a *normal* distribution.

In general, any procedure that seeks to determine whether a set of data could reasonably have originated from some given probability distribution, or *class* of probability distributions, is called a *goodness-of-fit* test. The principle behind the particular goodness-of-fit test we will look at is very straightforward: First the observed data are grouped, more or less arbitrarily, into k classes; then each class's "expected" occupancy is calculated on the basis of the presumed model. If it should happen that the set of observed and expected frequencies shows considerably more disagreement than sampling variability would predict, our conclusion will be that the supposed $p_X(k)$ or $f_Y(y)$ was incorrect.

In practice, goodness-of-fit tests have several variants, depending on the specificity of the null hypothesis. Section 10.3 describes the approach to take when both the form of the presumed data model and the values of its parameters are known. More typically, we know the form of $p_X(k)$ or $f_Y(y)$, but their parameters need to be estimated; these are taken up in Section 10.4.

A somewhat different application of goodness-of-fit testing is the focus of Section 10.5. There, the null hypothesis is that two random variables are *independent.* In more than a few fields of endeavor, tests for independence are among the most frequently used of all inference procedures.

10.2 The Multinomial Distribution

Their diversity notwithstanding, most goodness-of-fit tests are based on essentially the same statistic, one that has an asymptotic chi square distribution. The underlying structure of that statistic, though, derives from the *multinomial distribution,* a direct extension of the familiar *binomial.* In this section we define the multinomial and state those of its properties that relate to goodness-of-fit testing.

Given a series of n independent Bernoulli trials, each with success probability p, we know that the pdf for X, the total number of successes, is

$$P(X=k)=p_X(k)=\binom{n}{k}p^k(1-p)^{n-k}, \quad k=0,1,\ldots,n \tag{10.2.1}$$

One of the obvious ways to generalize Equation 10.2.1 is to consider situations in which at each trial, one of t outcomes can occur, rather than just one of two. That is, we will assume that each trial will result in one of the outcomes $r_1, r_2, \ldots, r_t$, where $p(r_i)=p_i$, $i=1,2,\ldots,t$ (see Figure 10.2.1). It follows, of course, that $\sum_{i=1}^{t} p_i = 1$.

Figure 10.2.1

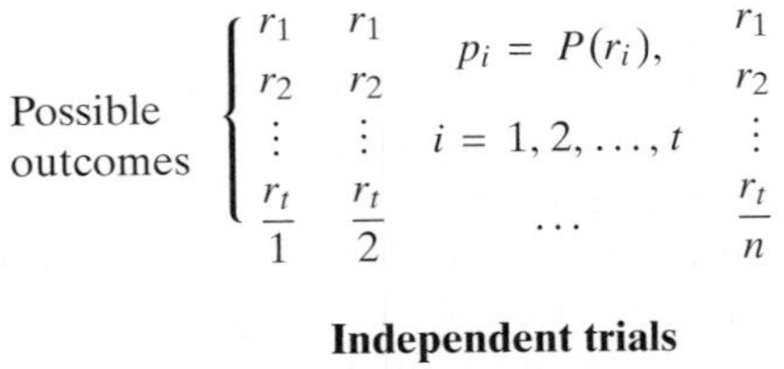

$$\text{Possible outcomes}\left\{\begin{array}{ccccc} r_1 & r_1 & & & r_1 \\ r_2 & r_2 & p_i = P(r_i), & & r_2 \\ \vdots & \vdots & i=1,2,\ldots,t & & \vdots \\ r_t & r_t & & & r_t \\ \hline 1 & 2 & \cdots & & n \end{array}\right.$$

Independent trials

In the binomial model, the two possible outcomes are denoted s and f, where $P(s)=p$ and $P(f)=1-p$. Moreover, the outcomes of the n trials can be nicely summarized with a single random variable X, where X denotes the number of successes. In the more general multinomial model, we will need a random variable to count the number of times that *each* of the r_i's occurs. To that end, we define

$$X_i = \text{number of times } r_i \text{ occurs}, \quad i = 1, 2, \ldots, t$$

For a given set of n trials, $X_1 = k_1, X_2 = k_2, \ldots, X_t = k_t$ and $\sum_{i=1}^{t} k_i = n$.

Theorem 10.2.1 *Let X_i denote the number of times that the outcome r_i occurs, $i = 1, 2, \ldots, t$, in a series of n independent trials, where $p_i = P(r_i)$. Then the vector $(X_1, X_2, \ldots, X_t)$ has a multinomial distribution and*

$$p_{X_1, X_2, \ldots, X_t}(k_1, k_2, \ldots, k_t) = P(X_1 = k_1, X_2 = k_2, \ldots, X_t = k_t)$$
$$= \frac{n!}{k_1!\, k_2! \cdots k_t!} p_1^{k_1} p_2^{k_2} \cdots p_t^{k_t},$$
$$k_i = 0, 1, \ldots, n; \quad i = 1, 2, \ldots, t; \quad \sum_{i=1}^{t} k_i = n$$

Proof Any particular sequence of $k_1\, r_1$'s, $k_2\, r_2$'s, …, and $k_t\, r_t$'s has probability $p_1^{k_1} p_2^{k_2} \ldots p_t^{k_t}$. Moreover, the total number of outcome sequences that will generate the values $(k_1, k_2, \ldots, k_t)$ is the number of ways to permute n objects, k_1 of one type, k_2 of a second type, …, and k_t of a tth type. By Theorem 2.6.2 that number is $n!/k_1!k_2! \ldots k_t!$, and the statement of the theorem follows. □

Depending on the context, the r_i's associated with the n trials in Figure 10.2.1 can be either single numerical values (or categories) or ranges of numerical values (or categories). Example 10.2.1 illustrates the first type; Example 10.2.2, the second. The only requirements imposed on the r_i's are (1) they must span all of the outcomes possible at a given trial and (2) they must be mutually exclusive.

Example 10.2.1 Suppose a loaded die is tossed twelve times, where

$$p_i = P(\text{Face } i \text{ appears}) = ci, \quad i = 1, 2, \ldots, 6$$

What is the probability that each face will appear exactly twice?

Note that

$$\sum_{i=1}^{6} p_i = 1 = \sum_{i=1}^{6} ci = c \cdot \frac{6(6+1)}{2}$$

which implies that $c = \frac{1}{21}$ (and $p_i = i/21$). In the terminology of Theorem 10.2.1, the possible outcomes at each trial are the $t = 6$ faces, 1 $(= r_1)$ through 6 $(= r_6)$, and X_i is the number of times face i occurs, $i = 1, 2, \ldots, 6$.

The question is asking for the probability of the vector

$$(X_1, X_2, X_3, X_4, X_5, X_6) = (2, 2, 2, 2, 2, 2)$$

According to Theorem 10.2.1,

$$P(X_1 = 2, X_2 = 2, \ldots, X_6 = 2) = \frac{12!}{2!\,2! \cdots 2!} \left(\frac{1}{21}\right)^2 \left(\frac{2}{21}\right)^2 \cdots \left(\frac{6}{21}\right)^2$$
$$= 0.0005$$

■

Example 10.2.2 Five observations are drawn at random from the pdf

$$f_Y(y) = 6y(1-y), \quad 0 \leq y \leq 1$$

What is the probability that one of the observations lies in the interval [0, 0.25), none in the interval [0.25, 0.50), three in the interval [0.50, 0.75), and one in the interval [0.75, 1.00]?

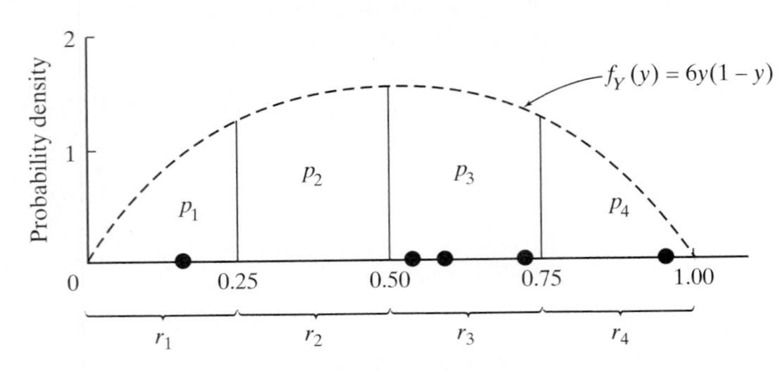

Figure 10.2.2

Figure 10.2.2 shows the pdf being sampled, together with the *ranges* r_1, r_2, r_3, and r_4, and the intended disposition of the five data points. The p_i's of Theorem 10.2.1 are now areas. Integrating $f_Y(y)$ from 0 to 0.25, for example, gives:

$$\begin{aligned} p_1 &= \int_0^{0.25} 6y(1-y)\,dy \\ &= 3y^2 \Big|_0^{0.25} - 2y^3 \Big|_0^{0.25} \\ &= \frac{5}{32} \end{aligned}$$

By symmetry, $p_4 = \frac{5}{32}$. Moreover, since the area under $f_Y(y)$ equals 1,

$$p_2 = p_3 = \frac{1}{2}\left(1 - \frac{10}{32}\right) = \frac{11}{32}$$

Let X_i denote the number of observations that fall into the ith range, $i = 1, 2, 3, 4$. The probability associated with the multinomial vector (1, 0, 3, 1), then, is *0.0198*:

$$\begin{aligned} P(X_1 = 1, X_2 = 0, X_3 = 3, X_4 = 1) &= \frac{5!}{1!\,0!\,3!\,1!}\left(\frac{5}{32}\right)^1\left(\frac{11}{32}\right)^0\left(\frac{11}{32}\right)^3\left(\frac{5}{32}\right)^1 \\ &= 0.0198 \end{aligned}$$

■

A Multinomial/Binomial Relationship

Since the multinomial pdf is conceptually a straightforward generalization of the binomial pdf, it should come as no surprise that each X_i in a multinomial vector is, itself, a binomial random variable.

Theorem 10.2.2 *Suppose the vector $(X_1, X_2, \ldots, X_t)$ is a multinomial random variable with parameters n, $p_1, p_2, \ldots$, and p_t. Then the marginal distribution of $X_i, i = 1, 2, \ldots, t$, is the binomial pdf with parameters n and p_i.*

Proof To deduce the pdf for X_i we need simply to dichotomize the possible outcomes at each of the trials into "r_i" and "not r_i." Then X_i becomes, in effect, the number of "successes" in n independent Bernoulli trials, where the probability of success at any given trial is p_i. By Theorem 3.2.1, it follows that X_i is a binomial random variable with parameters n and p_i. □

Comment Theorem 10.2.2 gives the pdf for any given X_i in a multinomial vector. Since that pdf is the binomial, we also know that the mean and variance of each X_i are $E(X_i) = np_i$ and $\text{Var}(X_i) = np_i(1 - p_i)$, respectively.

Example 10.2.3 A physics professor has just given an exam to fifty students enrolled in a thermodynamics class. From past experience, she has reason to believe that the scores will be normally distributed with $\mu = 80.0$ and $\sigma = 5.0$. Students scoring ninety or above will receive A's, between eighty and eighty-nine, B's, and so on. What are the expected values and variances for the numbers of students receiving each of the five letter grades?

Let Y denote the score a student earns on the exam, and let r_1, r_2, r_3, r_4, and r_5 denote the ranges corresponding to the letter grades A, B, C, D, and F, respectively. Then

$$\begin{aligned} p_1 &= P(\text{Student earns an A}) \\ &= P(90 \le Y \le 100) \\ &= P\left(\frac{90-80}{5} \le \frac{Y-80}{5} \le \frac{100-80}{5}\right) \\ &= P(2.00 \le Z \le 4.00) \\ &= 0.0228 \end{aligned}$$

If X_1 is the number of A's that are earned,

$$E(X_1) = np_1 = 50(0.0228) = 1.14$$

and

$$\text{Var}(X_1) = np_1(1 - p_1) = 50(0.0228)(0.9772) = 1.11$$

Table 10.2.1 lists the means and variances for all the X_i's. Each is an illustration of the Comment following Theorem 10.2.2.

Table 10.2.1

Score	Grade	p_i	$E(X_i)$	$\text{Var}(X_i)$
$90 \le Y \le 100$	A	0.0228	1.14	1.11
$80 \le Y < 90$	B	0.4772	23.86	12.47
$70 \le Y < 80$	C	0.4772	23.86	12.47
$60 \le Y < 70$	D	0.0228	1.14	1.11
$Y < 60$	F	0.0000	0.00	0.00

■

Questions

10.2.1 The Advanced Placement Program allows high school students to enroll in special classes in which a subject is studied at the college level. Proficiency is measured by a national examination. Universities typically grant course credit for a sufficiently strong performance. The possible scores are 1, 2, 3, 4, and 5, with 5 being the highest. The following table gives the probabilities associated with the scores recently made on the U.S. history test (1):

Score	Probability
1	0.116
2	0.325
3	0.236
4	0.211
5	0.112

Suppose six students from a class take the test. What is the probability they earn three 5's, two 4's, and a 3?

10.2.2 In Mendel's classical experiments with peas, he produced hybrids in such a way that the probabilities of observing the different phenotypes listed below were $\frac{9}{16}$, $\frac{3}{16}$, $\frac{3}{16}$, and $\frac{1}{16}$, respectively. Suppose that four such hybrid plants were selected at random. What is the probability that each of the four phenotypes would be represented?

Type	Probability
Round and yellow	9/16
Round and green	3/16
Angular and yellow	3/16
Angular and green	1/16

10.2.3 In classifying hypertension, three categories are used: individuals whose systolic blood pressures are less than 140, those with blood pressures between 140 and 160, and those with blood pressures over 160. For males between the ages of eighteen and twenty-four, systolic blood pressures are normally distributed with a mean equal to 124 and a standard deviation equal to 13.7. Suppose a random sample of ten individuals from that particular demographic group are examined. What is the probability that six of the blood pressures will be in the first group, three in the second, and one in the third?

10.2.4 An army enlistment officer categorizes potential recruits by IQ into three groups—class I: < 90, class II: 90–110, and class III: > 110. Given that the IQs in the population from which the recruits are drawn are normally distributed with $\mu = 100$ and $\sigma = 16$, calculate the probability that of seven enlistees, two will belong to class I, four to class II, and one to class III.

10.2.5 A disgruntled Anchorage bush pilot, upset because his gasoline credit card was cancelled, fires six air-to-surface missiles at the Alaskan pipeline. If a missile lands anywhere within twenty yards of the pipeline, major structural damage will be sustained. Assume that the probability function reflecting the pilot's expertise as a bombardier is the expression

$$f_Y(y) = \begin{cases} \dfrac{60+y}{3600}, & -60 < y < 0 \\ \dfrac{60-y}{3600}, & 0 \le y < 60 \\ 0, & \text{elsewhere} \end{cases}$$

where y denotes the perpendicular distance (in yards) from the pipeline to the point of impact. What is the probability that two of the missiles will land within twenty yards to the left of the pipeline and four will land within twenty yards to the right?

10.2.6 Based on his performance so far this season, a baseball player has the following probabilities associated with each official at-bat:

Outcome	Probability
Out	.713
Single	.270
Double	.010
Triple	.002
Home run	.005

If he has five official at-bats in tomorrow's game, what are the chances he makes two outs and hits two singles and a double?

10.2.7 Suppose that a random sample of fifty observations are taken from the pdf

$$f_Y(y) = 3y^2, \quad 0 \le y \le 1$$

Let X_i be the number of observations in the interval [0, 1/4), X_2 the number in [1/4, 2/4), X_3 the number in [2/4, 3/4), and X_4 the number in [3/4, 1].

(a) Write a formula for $f_{X_1,X_2,X_3,X_4}(3, 7, 15, 25)$.

(b) Find $\text{Var}(X_3)$.

10.2.8 Let the vector of random variables (X_1, X_2, X_3) have the trinomial pdf with parameters n, p_1, p_2, and $p_3 = 1 - p_1 - p_2$. That is,

$$P(X_1 = k_1, X_2 = k_2, X_3 = k_3) = \frac{n!}{k_1!k_2!k_3!} p_1^{k_1} p_2^{k_2} p_3^{k_3},$$

$$k_i = 0, 1, \ldots, n; \quad i = 1, 2, 3; \quad k_1 + k_2 + k_3 = n$$

By definition, the moment-generating function for (X_1, X_2, X_3) is given by

$$M_{X_1,X_2,X_3}(t_1, t_2, t_3) = E(e^{t_1X_1+t_2X_2+t_3X_3})$$

Show that

$$M_{X_1,X_2,X_3}(t_1, t_2, t_3) = (p_1e^{t_1} + p_2e^{t_2} + p_3e^{t_3})^n$$

10.2.9 If $M_{X_1,X_2,X_3}(t_1, t_2, t_3)$ is the moment-generating function for (X_1, X_2, X_3), then $M_{X_1,X_2,X_3}(t_1, 0, 0)$, $M_{X_1,X_2,X_3}(0, t_2, 0)$, and $M_{X_1,X_2,X_3}(0, 0, t_3)$ are the moment-generating functions for the marginal pdfs of X_1, X_2, and X_3, respectively. Use this fact, together with the result of Question 10.2.8, to verify the statement of Theorem 10.2.2.

10.2.10 Let $(k_1, k_2, \ldots, k_t)$ be the vector of sample observations representing a multinomial random variable with parameters n, p_1, p_2, ..., and p_t. Show that the maximum likelihood estimate for p_i is k_i/n, $i = 1, 2, \ldots, t$.

10.3 Goodness-of-Fit Tests: All Parameters Known

The simplest version of a goodness-of-fit test arises when an experimenter is able to specify completely the probability model from which the sample data are alleged to have come. It might be supposed, for example, that a set of y_i's is being generated by an exponential pdf with parameter equal to 6.3, or by a normal distribution with $\mu = 500$ and $\sigma = 100$. For continuous pdfs such as those, the hypotheses to be tested will be written

$$H_0: f_Y(y) = f_o(y)$$

versus

$$H_1: f_Y(y) \neq f_o(y)$$

where $f_Y(y)$ and $f_o(y)$ are the true and presumed pdfs, respectively. For a typical discrete model, the null hypothesis would be written H_0: $p_X(k) = p_o(k)$. It is not uncommon, though, for discrete random variables to be characterized simply by a set of probabilities associated with the t r_i's defined in Section 10.2, rather than by an equation. Then the hypotheses to be tested take the form

$$H_0: p_1 = p_{1_o}, p_2 = p_{2_o}, \ldots, p_t = p_{t_o}$$

versus

$$H_1: p_i \neq p_{i_o} \text{ for at least one } i$$

The first procedure for testing goodness-of-fit hypotheses was proposed by Karl Pearson in 1900. Couched in the language of the multinomial, the prototype of Pearson's method requires that (1) the n observations be grouped into t classes and (2) the presumed model be completely specified. Theorem 10.3.1 defines Pearson's test statistic and gives the decision rule for choosing between H_0 and H_1. In effect, H_0 is rejected if there is too much disagreement between the *actual* values for the multinomial X_i's and the *expected* values of those same X_i's.

Theorem 10.3.1 *Let $r_1, r_2, \ldots, r_t$ be the set of possible outcomes (or ranges of outcomes) associated with each of n independent trials, where $P(r_i) = p_i$, $i = 1, 2, \ldots, t$. Let X_i = number of times r_i occurs, $i = 1, 2, \ldots, t$. Then*

a. *The random variable*

$$D = \sum_{i=1}^{t} \frac{(X_i - np_i)^2}{np_i}$$

has approximately a χ^2 distribution with $t - 1$ degrees of freedom. For the approximation to be adequate, the t classes should be defined so that $np_i \geq 5$, for all i.

b. Let $k_1, k_2, \ldots, k_t$ be the observed *frequencies for the outcomes $r_1, r_2, \ldots, r_t$, respectively, and let $np_{1_o}, np_{2_o}, \ldots, np_{t_o}$ be the corresponding* expected *frequencies based on the null hypothesis. At the α level of significance, H_0: $f_Y(y) = f_o(y)$ [or H_0: $p_X(k) = p_o(k)$ or H_0: $p_1 = p_{1_o}, p_2 = p_{2_o}, \ldots, p_t = p_{t_o}$] is rejected if*

$$d = \sum_{i=1}^{t} \frac{(k_i - np_{i_o})^2}{np_{i_o}} \geq \chi^2_{1-\alpha, t-1}$$

(where $np_{i_o} \geq 5$ for all i).

Proof A formal proof of part (a) lies beyond the scope of this text, but the direction it takes can be illustrated for the simple case where $t = 2$. Under that scenario,

$$\begin{aligned} D &= \frac{(X_1 - np_1)^2}{np_1} + \frac{(X_2 - np_2)^2}{np_2} \\ &= \frac{(X_1 - np_1)^2}{np_1} + \frac{[n - X_1 - n(1 - p_1)]^2}{n(1 - p_1)} \\ &= \frac{(X_1 - np_1)^2(1 - p_1) + (-X_1 + np_1)^2 p_1}{np_1(1 - p_1)} \\ &= \frac{(X_1 - np_1)^2}{np_1(1 - p_1)} \end{aligned}$$

From Theorem 10.2.2, $E(X_1) = np_1$ and $\text{Var}(X_1) = np_1(1 - p_1)$, implying that D can be written

$$D = \left[\frac{X_1 - E(X_1)}{\sqrt{\text{Var}(X_1)}} \right]^2$$

By Theorem 4.3.1, then, D is the square of a variable that is asymptotically a standard normal, and the statement of part (a) follows (for $k = 2$) from Definition 7.3.1. [Proving the general statement is accomplished by showing that the limit of the moment-generating function for D—as n goes to ∞—is the moment-generating function for a χ^2_{t-1} random variable. See (63).] □

Comment Although Pearson formulated his statistic before any general theories of hypothesis testing had been developed, it can be shown that a decision rule based on D is asymptotically equivalent to the generalized likelihood ratio test of H_0: $p_1 = p_{1_o}, p_2 = p_{2_o}, \ldots, p_t = p_{t_o}$.

Case Study 10.3.1

Inhabiting many tropical waters is a small (<1 mm) crustacean, *Ceriodaphnia cornuta*, that occurs in two distinct morphological forms: One has a series of "horns" protruding from its exoskeleton, while the other is more rounded (see Figure 10.3.1). Are these two variants equally likely to end up as fish food, or do their predators have a preference (211)?

(Continued on next page)

Figure 10.3.1 Forms of *C. cornuta*.

A large number of *C. cornuta* were introduced into a holding tank in a three-to-one ratio—three of the unhorned variety were added for every one with horns. Also present in the tank was a natural predator of *C. cornuta,* a small (6-cm) fish, *Melaniris chagresi*. After approximately one hour, long enough for the predator to have completed its feeding, the fish was sacrificed and the contents of its stomach examined. Among the forty-four crustacean casualties, the unhorned-to-horned ratio was forty to four. What do these body counts imply?

Here, the two natural classes for the response variable are "unhorned" and "horned," and under the null hypothesis that morphology has no effect on survival, it would follow that the probability of either form's being eaten should be proportional to the numbers of each kind available. If $p_1 = P$ (Unhorned *C. cornuta* is eaten) and $p_2 = P$ (Horned *C. cornuta* is eaten), the experimenter's objective reduces to a test of

$$H_0 : p_1 = \frac{3}{4}, \quad p_2 = \frac{1}{4}$$

versus

$$H_1 : p_1 \neq \frac{3}{4}, \quad p_2 \neq \frac{1}{4}$$

Let $\alpha = 0.05$.

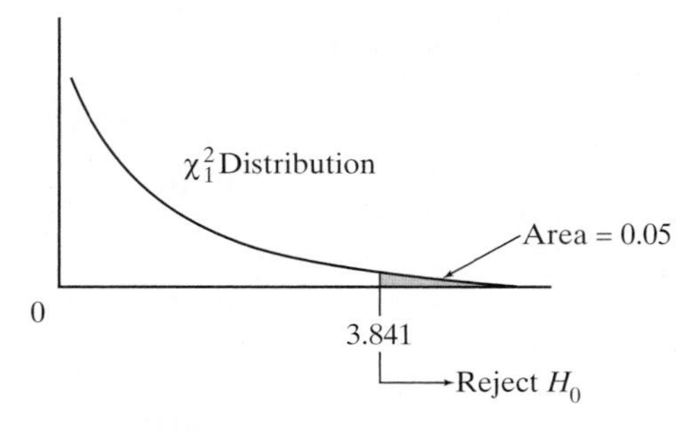

Figure 10.3.2 χ_1^2 distribution.

Since $t = 2$, the behavior of D will be approximated by a χ_1^2 distribution, for which the 0.05 critical value is 3.841 (see Figure 10.3.2). Substituting the values for the k_i's and np_{i_o}'s into the test statistic gives a d of 5.93:

(Continued on next page)

(Case Study 10.3.1 continued)

$$d = \frac{\left[40 - 44\left(\frac{3}{4}\right)\right]^2}{44\left(\frac{3}{4}\right)} + \frac{\left[4 - 44\left(\frac{1}{4}\right)\right]^2}{44\left(\frac{1}{4}\right)}$$

$$= 5.93$$

Our conclusion, then, is to *reject* H_0—it would appear that morphology *does* have an effect on *C. cornuta*'s chances of being eaten.

About the Data Rejecting H_0 in this case does not actually imply what the analysis seems to suggest. The calculation of d shows that more of the unhorned *cornuta* (= 40) were eaten than the null hypothesis had predicted (= 33), and vice versa for the horned *cornuta* (four eaten as opposed to the eleven predicted). But the presence or absence of horns was, in fact, irrelevant! A series of follow-up experiments analyzed in much the same way clearly indicated that the reason the unhorned *cornuta* were snacked on more often was their enlarged eyespot, which made them more visible—and sadly (for them) more edible.

Case Study 10.3.2

Once upon a time, when there were no computers (and calculations were actually done using pencil and paper!), log tables were used to facilitate lengthy multiplications. In the early 1930s, Frank Benford, a physicist, reexamined the claim made many years earlier by Simon Newcomb that the first several pages in library logarithm books are dirtier than the last several pages (recall Example 3.3.3). Why should students and researchers have more reason to look up logarithms beginning with 1 or 2, rather than 8 or 9? Benford began looking closely at a variety of data sets, including molecular weights of chemicals, surface areas of rivers, and baseball statistics.

Table 10.3.1

Digit, i	$\log_{10}(i+1) - \log_{10}(i)$
1	0.301
2	0.176
3	0.125
4	0.097
5	0.079
6	0.067
7	0.058
8	0.051
9	0.046

What he confirmed to his surprise was the fact that the first nonzero digits in these various numbers are *not* equally likely to be 1's, 2's, . . . , and 9's, contrary

(Continued on next page)

to what our intuition would almost certainly suggest. For reasons discussed in (78), the probability that the first nonzero digit is i tends to be

$$p_i = \log_{10}(i+1) - \log_{10}(i), \quad i = 1, 2, \ldots, 9 \tag{10.3.1}$$

These latter probabilities are now known as *Benford's law* (see Table 10.3.1).

One particularly intriguing application of Benford's law occurs in auditing, where eagle-eyed examiners are ever on the lookout for budgets whose numbers have been fabricated to cover up falsified records. Bookkeepers are not likely to be aware of Equation 10.3.1 and would tend to "make up" entries in such a way that each first digit from 1 to 9 would occur roughly the same percentage of the time. Let p_i denote the probability that the first nonzero digit in a set of data is i, $i = 1, 2, \ldots, 9$. A goodness-of-fit test to identify possible instances of "creative" accounting would define the null hypothesis to be $H_0: p_1 = p_{1_o}, p_2 = p_{2_o}, \cdots, p_9 = p_{9_o}$, where the Benford law probabilities become the p_{i_o}'s.

An example of such a test is summarized in Table 10.3.2. The values in Column 2 are a breakdown of the 355 first digits appearing in the 1997–98 operating budget for the University of West Florida (110). The corresponding expected frequencies based on Benford's law are listed in Column 4, and the goodness-of-fit test statistic, d, is the sum of the entries in Column 5:

$$d = \frac{[111 - 355 \cdot (0.301)]^2}{355 \cdot (0.301)} + \cdots + \frac{[20 - 355 \cdot (0.046)]^2}{355 \cdot (0.046)}$$

$$= 2.49$$

Table 10.3.2

Digit	Observed, k_i	Benford p_{i_o}	Expected $(= 355 \cdot p_{i_o})$	$(k_i - 355p_{i_o})^2/355p_{i_o}$
1	111	0.301	106.9	0.16
2	60	0.176	62.5	0.10
3	46	0.125	44.4	0.06
4	29	0.097	34.4	0.86
5	26	0.079	28.0	0.15
6	22	0.067	23.8	0.13
7	21	0.058	20.6	0.01
8	20	0.051	18.1	0.20
9	20	0.046	16.3	0.82
	355	1.000	355.0	2.49

Here, with $t = 9$ classes, the critical value for the hypothesis test comes from the chi square distribution with 8 df. If α is set equal to 0.05, $\chi^2_{.95,8} = 15.507$, so our conclusion is "fail to reject H_0."

About the Data There is no denying that Benford's law is extremely counterintuitive. On everyone's credibility scale, it would lie somewhere to the right of ridiculous. That said, *why* Benford's law holds for so many different phenomena has a surprisingly simple explanation.

Suppose a random variable Y takes on values that range over several orders of magnitude—say, from 10 to 1,000,000. Also, suppose the pdf of Y *when plotted on a log base 10 scale* tapers off slowly, to the extent that, for example,

$$P(100 \leq Y \leq 1000) \doteq P(1000 \leq Y \leq 10,000) \doteq P(10,000 \leq Y \leq 100,000)$$

That is,

$$P(2 \leq \log Y \leq 3) \doteq P(3 \leq \log Y \leq 4) \doteq P(4 \leq \log Y \leq 5) \tag{10.3.2}$$

which implies that $\log Y$ has approximately a uniform distribution.

Now, consider the log cycle for Y values ranging from, say, 1000 up to but not including 10,000. Table 10.3.3 shows the log interval associated with each of the possible first digits, 1 through 9. In the last column are the *widths* of each of the nine associated log intervals.

Table 10.3.3

Values of Y	Associated Logs	Width of Log Interval
$1000 \leq Y \leq 1999+$	$3.00000 \leq \log Y \leq 3.30103$	0.30103
$2000 \leq Y \leq 2999+$	$3.30103 \leq \log Y \leq 3.47712$	0.17609
$3000 \leq Y \leq 3999+$	$3.47712 \leq \log Y \leq 3.60206$	0.12494
$4000 \leq Y \leq 4999+$	$3.60206 \leq \log Y \leq 3.69897$	0.09691
$5000 \leq Y \leq 5999+$	$3.69897 \leq \log Y \leq 3.77815$	0.07918
$6000 \leq Y \leq 6999+$	$3.77815 \leq \log Y \leq 3.84510$	0.06695
$7000 \leq Y \leq 7999+$	$3.84510 \leq \log Y \leq 3.90309$	0.05799
$8000 \leq Y \leq 8999+$	$3.90309 \leq \log Y \leq 3.95424$	0.05115
$9000 \leq Y \leq 9999+$	$3.95424 \leq \log Y \leq 4.00000$	0.04576

By the earlier assumption, $\log Y$ has approximately a uniform distribution over much of the range of Y. It follows that *for a given log cycle* (in this case, $3 \leq \log Y \leq 4$),

$$P(a \leq \log Y \leq b) \doteq b - a$$

Therefore, if a value is chosen at random from the interval ($1000 \leq Y \leq 10,000$), the probability that its first digit will be 1 is the width of the interval of logs associated with numbers in the range $1000 \leq Y < 2000$—that is, $3.30103 - 3.00000 = 0.30103$. Applying that same argument to each of the possible first digits, 1 through 9, gives the entries listed in the third column of Table 10.3.3.

The interval widths just described, of course, are the same for every log cycle. It follows, then, that if a random sample from $f_Y(y)$ is taken over its entire range, roughly 30% of the y_i's will have a first digit of 1, roughly 18% will have a first digit of 2, and so on. The entries in the third column are, in fact, Benford's law:

$$P(\text{First digit is } i) = \log_{10}(i+1) - \log_{10}(i),\ i = 1, 2, \ldots, 9$$

One question still remains: Are there any frequently encountered probability functions that satisfy the assumptions imposed earlier on $f_Y(y)$? The answer is "yes." There is an entire family of pdfs known as *power models* that have the extremely long tail necessary for Benford's law to be applicable. Perhaps the most familiar member of that family is the *Pareto distribution*, where

$$f_Y(y) = ay^{-a-1};\ a > 0,\ 1 \leq y < \infty$$

Originally developed as a model for wealth allocation among members of a population (recall Question 5.2.14), Pareto's distribution has been shown more recently to describe phenomena as diverse as meteorite size, areas burned by forest fires, population sizes of human settlements, monetary value of oil reserves, and lengths of jobs assigned to supercomputers. Figure 10.3.3 shows two examples of Pareto pdfs.

Figure 10.3.3

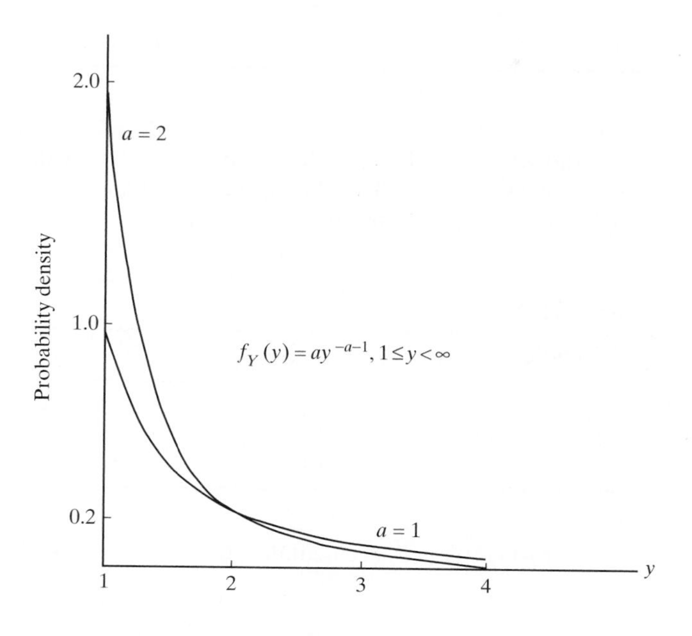

Example 10.3.1

A new statistics software package claims to be able to generate random samples from any continuous pdf. Asked to produce forty observations representing the pdf $f_Y(y) = 6y(1-y), 0 \le y \le 1$, it printed out the numbers displayed in Table 10.3.4. Are these forty y_i's a believable random sample from $f_Y(y)$? Do an appropriate goodness-of-fit test using the $\alpha = 0.05$ level of significance.

Table 10.3.4

0.18	0.06	0.27	0.58	0.98
0.55	0.24	0.58	0.97	0.36
0.48	0.11	0.59	0.15	0.53
0.29	0.46	0.21	0.39	0.89
0.34	0.09	0.64	0.52	0.64
0.71	0.56	0.48	0.44	0.40
0.80	0.83	0.02	0.10	0.51
0.43	0.14	0.74	0.75	0.22

To apply Theorem 10.3.1 to a continuous pdf requires that the data first be reduced to a set of classes. Table 10.3.5 shows one possible grouping. The p_{i_o}'s in Column 3 are the areas under $f_Y(y)$ above each of the five classes. For example,

$$p_{1_o} = \int_0^{0.20} 6y(1-y)\,dy = 0.104$$

Table 10.3.5

Class	Observed Frequency, k_i	P_{i_o}	$40p_{i_o}$
$0 \le y < 0.20$	8	0.104	4.16
$0.20 \le y < 0.40$	8	0.248	9.92
$0.40 \le y < 0.60$	14	0.296	11.84
$0.60 \le y < 0.80$	5	0.248	9.92
$0.80 \le y < 1.00$	5	0.104	4.16

Column 4 shows the expected frequencies for each of the classes. Notice that $40p_{1_o}$ and $40p_{5_o}$ are both less than 5 and fail to satisfy the "$np_i \ge 5$" restriction cited in part (a) of Theorem 10.3.1. That violation can be easily corrected, though—we need simply to combine the first two classes and the last two classes (see Table 10.3.6).

Table 10.3.6

Class	Observed Frequency, k_i	P_{i_o}	$40p_{i_o}$
$0 \le y < 0.40$	16	0.352	14.08
$0.40 \le y < 0.60$	14	0.296	11.84
$0.60 \le y \le 1.00$	10	0.352	14.08

The test statistic d, is calculated from the entries in Table 10.3.6:

$$d = \frac{(16 - 14.08)^2}{14.08} + \frac{(14 - 11.84)^2}{11.84} + \frac{(10 - 14.08)^2}{14.08}$$
$$= 1.84$$

Since the number of classes ultimately being used is three, the number of degrees of freedom associated with d is 2, and we should reject the null hypothesis that the forty y_i's are a random sample from $f_Y(y) = 6y(1-y),\ 0 \le y \le 1$ if $d \ge \chi^2_{0.95,2}$. But the latter is 5.991, so—based on these data—there is no compelling reason to doubt the advertised claim. ■

The Goodness-of-Fit Decision Rule—An Exception

The fact that the decision rule given in part (b) of Theorem 10.3.1 is *one-sided to the right* seems perfectly reasonable—simple logic tells us that the goodness-of-fit null hypothesis should be rejected if d is large, *but not if d is small.* After all, small values of d will occur only if the observed frequencies are matching up very well with the predicted frequencies, and it seems that it would never make sense to reject H_0 if that should happen. Not so. There is one specific scenario in which the appropriate goodness-of-fit test is *one-sided to the left.*

Human nature being what it is, researchers have been known (shame on them) to massage, embellish, and otherwise falsify their data. Moreover, in their overzealous efforts to support whatever theory they claim is true, they often make a second mistake of fabricating data that are too good—that is, that fit their model too closely. How can that be detected? By calculating the goodness-of-fit statistic and seeing if it falls *less than* $\chi^2_{\alpha,t-1}$, where α would be set equal to, say, 0.05 or 0.01.

Case Study 10.3.3

Gregor Mendel (1822–1884) was an Austrian monk and a scientist ahead of his time. In 1866 he wrote "Experiments in Plant Hybridization," which summarized his exhaustive studies on the way inherited traits in garden peas are passed from generation to generation. It was a landmark piece of work in which he correctly deduced the basic laws of genetics without knowing anything about genes, chromosomes, or molecular biology. But for reasons not entirely clear, no one paid any attention and his findings were virtually ignored for the next thirty-five years.

Early in the twentieth century, Mendel's work was rediscovered and quickly revolutionized the cultivation of plants and the breeding of domestic animals. With his posthumous fame, though, came some blistering criticism. No less an authority than Ronald A. Fisher voiced the opinion that Mendel's results in that 1866 paper were too good to be true—the data had to have been falsified.

Table 10.3.7 summarizes one of the data sets that attracted Fisher's attention (112). Two traits of garden peas were being studied—their *shape* (round or angular) and their *color* (yellow or green). If "round" and "yellow" are dominant and if the alleles controlling those two traits separate independently, then (according to Mendel) dihybrid crosses should produce four possible phenotypes, with probabilities 9/16, 3/16, 3/16, and 1/16, respectively.

Table 10.3.7

Phenotype	Obs. Freq.	Mendel's Model	Exp. Freq.
(round, yellow)	315	9/16	312.75
(round, green)	108	3/16	104.25
(angular, yellow)	101	3/16	104.25
(angular, green)	32	1/16	34.75

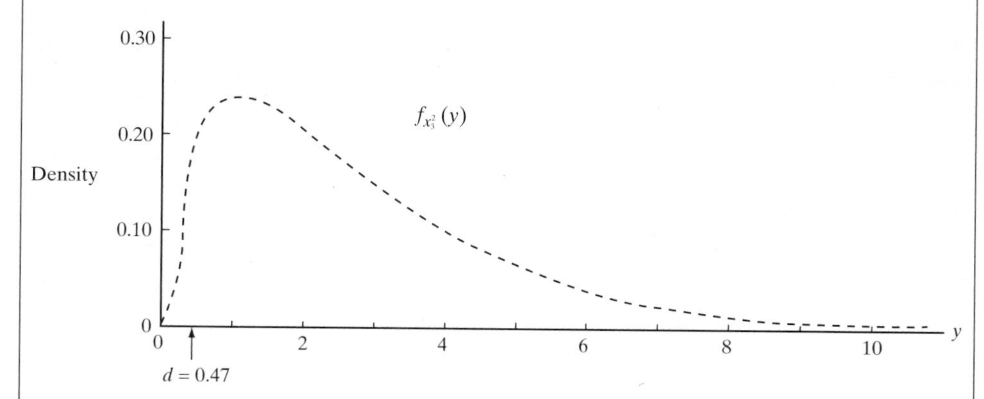

Figure 10.3.4

Notice how closely the observed frequencies approximate the expected frequencies. The goodness-of-fit statistic from Theorem 10.3.1 (with $4-1=3$ df) is equal to *0.47*:

(Continued on next page)

(Case Study 10.3.3 continued)

$$d = \frac{(315 - 312.75)^2}{312.75} + \frac{(108 - 104.25)^2}{104.25} + \frac{(101 - 104.25)^2}{104.25} + \frac{(32 - 34.75)^2}{34.75} = 0.47$$

Figure 10.3.4 shows that the value of $d = 0.47$ does look suspiciously small. By itself, it does not rise to the level of a "smoking gun," but Mendel's critics had similar issues with other portions of his data as well.

About the Data Almost seventy-five years have passed since Fisher raised his concerns about the legitimacy of Mendel's data, but there is still no broad consensus on whether or not portions of the data were falsified. And if they were, who was responsible? Mendel, of course, would be the logical suspect, but some would-be cold case detectives think the gardener did it! What actually happened back in 1866 may never be known, because many of Mendel's original notes and records have been lost or destroyed.

Questions

10.3.1 Verify the following identity concerning the statistic of Theorem 10.3.1. Note that the right-hand side is more convenient for calculations.

$$\sum_{i=1}^{t} \frac{(X_i - np_i)^2}{np_i} = \sum_{i=1}^{t} \frac{X_i^2}{np_i} - n$$

10.3.2 One hundred unordered samples of size 2 are drawn without replacement from an urn containing six red chips and four white chips. Test the adequacy of the hypergeometric model if zero whites were obtained 35 times; one white, 55 times; and two whites, 10 times. Use the 0.10 decision rule.

10.3.3 Consider again the previous question. Suppose, however, that we do not know whether the samples had been drawn with or without replacement. Test whether sampling *with* replacement is a reasonable model.

10.3.4 Show that the common belief in the propensity of babies to choose an inconvenient hour for birth has a basis in observation. A maternity hospital reported that out of one year's total of 2650 births, some 494 occurred between midnight and 4 A.M. (168). Use the goodness-of-fit test to show that the data are not what we would expect if births are assumed to occur uniformly in all time periods. Let $\alpha = 0.05$.

10.3.5 Analyze the data in the previous problem using the techniques of Section 6.3. What is the relationship between the two test statistics?

10.3.6 A number of reports in the medical literature suggest that the season of birth and the incidence of schizophrenia may be related, with a higher proportion of schizophrenics being born during the early months of the year. A study (72) following up on this hypothesis looked at 5139 persons born in England or Wales during the years 1921–1955 who were admitted to a psychiatric ward with a diagnosis of schizophrenia. Of these 5139, 1383 were born in the first quarter of the year. Based on census figures in the two countries, the expected number of persons, out of a random 5139, who would be born in the first quarter is 1292.1. Do an appropriate χ^2 test with $\alpha = 0.05$.

10.3.7 In a move that shocked candy traditionalists, the M&M/Mars Company recently replaced the tan M&M's with blue ones. More than ten million people had voted in an election to select the new color. On learning of the change, one concerned consumer counted the number of each color appearing in three pounds of M&M's (55). His tally, shown in the following table, suggests that not all the colors appear equally often—blues, in particular, are decidedly less common than browns. According to an M&M/Mars spokesperson, there are actually three frequencies associated with the six colors: 30% of M&M's are brown, yellow and red each account for 20%, and orange, blue, and green each occur 10% of the time. Test at the $\alpha = 0.05$ level of significance the hypothesis that the consumer's data are consistent with the company's stated intentions.

Color	Number
Brown	455
Yellow	343
Red	318
Orange	152
Blue	130
Green	129

10.3.8 The following table lists World Series lengths for the fifty years from 1926 to 1975. Test at the 0.10 level whether these data are compatible with the model that each World Series game is an independent Bernoulli trial with $p = P(\text{AL wins}) = P(\text{NL wins}) = \frac{1}{2}$.

Number of Games	Number of Years
4	9
5	11
6	8
7	22

10.3.9 Records kept at an eastern racetrack showed the following distribution of winners as a function of their starting-post position. All 144 races were run with a full field of eight horses.

Starting Post	1	2	3	4	5	6	7	8
Number of Winners	32	21	19	20	16	11	14	11

Test an appropriate goodness-of-fit hypothesis. Let $\alpha = 0.05$.

10.3.10 It was noted in Question 4.3.24 that the mean (μ) and standard deviation (σ) of pregnancy durations are 266 days and 16 days, respectively. Accepting those as the true parameter values, test whether the additional assumption that pregnancy durations are normally distributed is supported by the following list of seventy pregnancy durations reported by County General Hospital. Let $\alpha = 0.10$ be the level of significance. Use "$220 \le y < 230$," "$230 \le y < 240$," and so on, as the classes.

251	264	234	283	226	244	269	241	276	274
263	243	254	276	241	232	260	248	284	253
265	235	259	279	256	256	254	256	250	269
240	261	263	262	259	230	268	284	259	261
268	268	264	271	263	259	294	259	263	278
267	293	247	244	250	266	286	263	274	253
281	286	266	249	255	233	245	266	265	264

10.3.11 In the past, defendants convicted of grand theft auto served Y years in prison, where the pdf describing the variation in Y had the form

$$f_Y(y) = \frac{1}{9}y^2, \quad 0 < y \le 3$$

Recent judicial reforms, though, may have impacted the punishment meted out for this particular crime. A review of 50 individuals convicted of grand theft auto five years ago showed that 8 served less than one year in jail, 16 served between one and two years, and 26 served between two and three years. Are these data consistent with $f_Y(y)$? Do an appropriate hypothesis test using the $\alpha = 0.05$ level of significance.

10.4 Goodness-of-Fit Tests: Parameters Unknown

More common than the sort of problems described in Section 10.3 are situations where the experimenter has reason to believe that the response variable follows some particular *family* of pdfs—say, the normal or the Poisson—but has little or no prior information to suggest what values should be assigned to the model's parameters. In cases such as these, we will carry out the goodness-of-fit test by first estimating all unknown parameters, preferably with the method of maximum likelihood. The appropriate test statistic, denoted d_1, is a modified version of Pearson's d:

$$d_1 = \sum_{i=1}^{t} \frac{(k_i - n\hat{p}_{i_o})^2}{n\hat{p}_{i_o}}$$

Here, the factors $\hat{p}_{1_o}, \hat{p}_{2_o}, \ldots, \hat{p}_{t_o}$ denote the *estimated* probabilities associated with the outcomes $r_1, r_2, \ldots, r_t$.

For example, suppose $n = 100$ observations are taken from a distribution hypothesized to be an exponential pdf, $f_o(y) = \lambda e^{-\lambda y}, y \ge 0$, and suppose that r_1 is defined to be the interval from 0 to 1.5. *If* the numerical value of λ is known—say, $\lambda = 0.4$—then the probability associated with r_1 would be denoted p_{1_o}, where

$$p_{1_o} = \int_0^{1.5} 0.4e^{-0.4y}\,dy = 0.45$$

On the other hand, suppose λ is not known but $\sum_{i=1}^{100} y_i = 200$. Since the maximum likelihood estimate for λ in this case is

$$\lambda_e = n \bigg/ \sum_{i=1}^{100} y_i = \frac{100}{200} = 0.50$$

(recall Question 5.2.3), the *estimated* null hypothesis exponential model is $f_o(y) = 0.50e^{-0.50y}$, $y \geq 0$, and the corresponding *estimated* probability associated with r_1 is denoted $\hat{p}_{1_o}$, where

$$\hat{p}_{1_o} = \int_0^{1.5} f_o(y; \lambda_e)\, dy = \int_0^{1.5} \lambda_e e^{-\lambda_e y}\, dy = \int_0^{1.5} 0.5e^{-0.5y}\, dy = 0.53$$

So, whereas d compares the observed frequencies of the r_i's with their expected frequencies, d_1 compares the observed frequencies of the r_i's with their estimated expected frequencies.

We pay a price for having to rely on the data to fill in details about the presumed model: Each estimated parameter reduces by 1 the number of degrees of freedom associated with the χ^2 distribution approximating the sampling distribution of D_1. And, as we have seen in other hypothesis testing situations, as the number of degrees of freedom associated with the test statistic decreases, so does the power of the test.

Theorem 10.4.1 *Suppose that a random sample of n observations is taken from $f_Y(y)$ [or $p_X(k)$], a pdf having s unknown parameters. Let $r_1, r_2, \ldots, r_t$ be a set of mutually exclusive ranges (or outcomes) associated with each of the n observations. Let $\hat{p}_i =$ estimated probability of $r_i, i = 1, 2, \ldots, t$ (as calculated from $f_Y(y)$ [or $p_X(k)$] after the pdfs s unknown parameters have been replaced by their maximum likelihood estimates). Let X_i denote the number of times that r_i occurs, $i = 1, 2, \ldots, t$. Then*

a. *the random variable*

$$D_1 = \sum_{i=1}^{t} \frac{(X_i - n\hat{p}_i)^2}{n\hat{p}_i}$$

has approximately a χ^2 distribution with $t - 1 - s$ degrees of freedom. For the approximation to be fully adequate, the r_i's should be defined so that $n\hat{p}_i \geq 5$ for all i.

b. *to test H_0: $f_Y(y) = f_o(y)$ [or H_0: $p_X(k) = p_o(k)$] at the α level of significance, calculate*

$$d_1 = \sum_{i=1}^{t} \frac{(k_i - n\hat{p}_{i_o})^2}{n\hat{p}_{i_o}}$$

where $k_1, k_2, \ldots, k_t$ are the observed *frequencies of $r_1, r_2, \ldots, r_t$, respectively, and $n\hat{p}_{1_o}, n\hat{p}_{2_o}, \ldots, n\hat{p}_{t_o}$ are the corresponding estimated* expected *frequencies based on the null hypothesis. If*

$$d_1 \geq \chi^2_{1-\alpha, t-1-s}$$

H_0 should be rejected. (The r_i's should be defined so that $n\hat{p}_{i_o} \geq 5$ for all i.)

Case Study 10.4.1

Despite the fact that batters occasionally go on lengthy hitting streaks (and slumps), there is reason to believe that the number of hits a baseball player gets in a game behaves much like a binomial random variable. Data demonstrating that claim have come from a study (132) of National League box scores from Opening Day through mid-July in 1996. Players had exactly four official at-bats a total of 4096 times during that period. The resulting distribution of their hits is summarized in Table 10.4.1. Are these numbers consistent with the hypothesis that the number of hits a player gets in four at-bats is binomially distributed?

Table 10.4.1

	Number of Hits, i	Obs. Freq., k_i	Estimated Exp. Freq., $n\hat{p}_{i_o}$
r_i's	0	1280	1289.1
	1	1717	1728.0
	2	915	868.6
	3	167	194.0
	4	17	16.3

Here the five possible outcomes associated with each four-at-bat game would be the number of hits a player makes, so $r_1 = 0, r_2 = 1, \ldots, r_5 = 4$. The presumption to be tested is that the probabilities of those r_i's are given by the binomial distribution—that is,

$$P(\text{Player gets } i \text{ hits in four at-bats}) = \binom{4}{i} p^i (1-p)^{4-i}, \quad i = 0, 1, 2, 3, 4$$

where $p = P(\text{Player gets a hit on a given at-bat})$.

In this case, p qualifies as an unknown parameter and needs to be estimated before the goodness-of-fit analysis can go any further. Recall from Example 5.1.1 that the maximum likelihood estimate for p is the ratio of the total number of successes divided by the total number of trials. With successes being "hits" and trials being "at-bats," it follows that

$$p_e = \frac{1280(0) + 1717(1) + 915(2) + 167(3) + 17(4)}{4096(4)} = \frac{4116}{16{,}384} = 0.251$$

The precise null hypothesis being tested, then, can be written

$$H_0\text{: } P(\text{Player gets } i \text{ hits}) = \binom{4}{i} (0.251)^i (0.749)^{4-i}, \quad i = 0, 1, 2, 3, 4$$

The third column in Table 10.4.1 shows the estimated expected frequencies based on the estimated H_0 pdf. For example,

$$\begin{aligned} n\hat{p}_{1_o} &= \text{estimated expected frequency for } r_1 \\ &= \text{estimated number of times players would get 0 hits} \\ &= 4096 \cdot \binom{4}{0} (0.251)^0 (0.749)^4 \\ &= 1289.1 \end{aligned}$$

(Continued on next page)

(Case Study 10.4.1 continued)

Corresponding to 1289.1, of course, is the entry in the first row of Column 2 in Table 10.4.1, listing the *observed* number of times players got zero hits (= 1280).

If we elect to test the null hypothesis at the $\alpha = 0.05$ level of significance, then by Theorem 10.4.1 H_0 should be rejected if

$$d_1 \geq \chi^2_{0.95,5-1-1} = 7.815$$

Here the degrees of freedom associated with the test statistic would be $t - 1 - s = 5 - 1 - 1 = 3$ because $s = 1$ df is lost as a result of p having been replaced by its maximum likelihood estimate.

Putting the entries from the last two columns of Table 10.4.1 into the formula for d_1 gives

$$\begin{aligned} d_1 &= \frac{(1280 - 1289.1)^2}{1289.1} + \frac{(1717 - 1728.0)^2}{1728.0} + \frac{(915 - 868.6)^2}{868.6} \\ &\quad + \frac{(167 - 194.0)^2}{194.0} + \frac{(17 - 16.3)^2}{16.3} \\ &= 6.401 \end{aligned}$$

Our conclusion, then, is to *fail to reject* H_0—the data summarized in Table 10.4.1 do not rule out the possibility that the numbers of hits players get in four-at-bat games follow a binomial distribution.

About the Data The fact that the binomial pdf is not ruled out as a model for the number of hits a player gets in a game is perhaps a little surprising in light of the fact that some of its assumptions are clearly not being satisfied. The parameter p, for example, is presumed to be constant over the entire set of trials. That is certainly not true for the data in Table 10.4.1. Not only does the "true" value of p obviously vary from player to player, it varies from at-bat to at-bat for the same player if different pitchers are used during the course of a game. Also in question is whether each at-bat qualifies as a truly independent event. As a game progresses, Major League players (hitters and pitchers alike) surely rehash what happened on previous at-bats and try to make adjustments accordingly. To borrow a term we used earlier in connection with hypothesis tests, it would appear that the binomial model is somewhat "robust" with respect to departures from its two most basic assumptions.

Case Study 10.4.2

The Poisson probability function often models rare events that occur over time, which suggests that it may prove useful in describing actuarial phenomena. Table 10.4.2 raises one such possibility—listed are the daily numbers of death notices for women over the age of eighty that appeared in the London *Times* over a three-year period (74). Is it believable that these fatalities are occurring in a pattern consistent with a Poisson pdf?

(Continued on next page)

Table 10.4.2

Number of Deaths, i	Obs. Freq., k_i	Est. Exp. Freq., $n\hat{p}_{i_o}$
0	162	126.8
1	267	273.5
2	271	294.9
3	185	212.1
4	111	114.3
5	61	49.3
6	27	17.8
7	8	5.5
8	3	1.4
9	1	0.3
10+	0	0.1
	1096	1096

To claim that a Poisson pdf can model these data is to say that

$$P(i \text{ women over the age of eighty die on a given day}) = e^{-\lambda}\lambda^i/i!, \quad i = 0, 1, 2, \ldots$$

where λ is the expected number of such fatalities on a given day. Other than what the data may suggest, there is no obvious numerical value to assign to λ at the outset. However, from Chapter 5, we know that the maximum likelihood estimate for the parameter in a Poisson pdf is the sample average rate at which the events occurred—that is, the total number of occurrences divided by the total number of time periods covered. Here, that quotient comes to *2.157*:

$$\begin{aligned}\lambda_e &= \frac{\text{total number of fatalities}}{\text{total number of days}} \\ &= \frac{0(162) + 1(267) + 2(271) + \cdots + 9(1)}{1096} \\ &= 2.157\end{aligned}$$

The estimated expected frequencies, then, are calculated by multiplying 1096 times $e^{-2.157}(2.157)^i/i!, i = 0, 1, 2, \ldots$. The third column in Table 10.4.2 lists the entire set of $n\hat{p}_{i_o}$'s. [*Note*: Whenever the model being fitted has an infinite number of possible outcomes (as is the case with the Poisson), the last expected frequency is calculated by subtracting the sum of all the others from n. This guarantees that the sum of the observed frequencies is equal to the sum of the estimated expected frequencies.] Applied to these data, that proviso implies that

$$\text{estimated expected frequency for "10+"} = 1096 - 126.8 - 273.5 - \cdots - 0.3 = 0.1$$

One final modification needs to be made before the test statistic, d_1, can be calculated. Recall that each estimated expected frequency should be at least 5 in order for the χ^2 approximation to the pdf of D_1 to be adequate. The last three

(Continued on next page)

(Case Study 10.4.2 continued)

classes in Table 10.4.2, though, all have very small values for $n\hat{p}_{i_o}$ (1.4, 0.3, and 0.1). To comply with the "$n\hat{p}_{i_o} \geq 5$" requirement, we need to pool the last four rows into a "7+" category, which would have an observed frequency of *12* $(=0+1+3+8)$ and an estimated expected frequency of *7.3* $(=0.1+0.3+1.4+5.5)$ (see Table 10.4.3).

Table 10.4.3

	Number of Deaths, i	Obs. Freq., k_i	Est. Exp. Freq., $n\hat{p}_{i_o}$
$r_1, r_2, \ldots, r_8$	0	162	126.8
	1	267	273.5
	2	271	294.9
	3	185	212.1
	4	111	114.3
	5	61	49.3
	6	27	17.8
	7+	12	7.3
		1096	1096

Based on the observed and estimated expected frequencies for the eight r_i's identified in Table 10.4.3, the test statistic, d_1, equals *25.98*:

$$d_1 = \frac{(162-126.8)^2}{126.8} + \frac{(267-273.5)^2}{273.5} + \cdots + \frac{(12-7.3)^2}{7.3}$$

$$= 25.98$$

With eight classes and one estimated parameter, the number of degrees of freedom associated with d_1 is 6 $(=8-1-1)$. To test

H_0: $P(i \text{ women over eighty die on a given day}) = e^{-2.157}(2.157)^i/i!, \quad i = 0, 1, 2, \ldots$

at the $\alpha = 0.05$ level of significance, we should reject H_0 if

$$d_1 \geq \chi^2_{0.95,6}$$

But the 95th percentile of the χ^2_6 distribution is *12.592,* which lies well to the left of d_1, so our conclusion is to *reject* H_0—there is too much disagreement between the observed and estimated expected frequencies in Table 10.4.3 to be consistent with the hypothesis that the data's underlying probability model is a Poisson pdf.

About the Data A row-by-row comparison of the entries in Table 10.4.3 shows a pronounced excess of days having zero fatalities and also an excess of days having large numbers of fatalities (five, six, or seven plus). One possible explanation for those disparities would be that the Poisson assumption that λ remains constant over the entire time covered is not satisfied. Events such as flu epidemics, for example, might cause λ to vary considerably from month to month and contribute to the data's "disconnect" from the Poisson model.

Case Study 10.4.3

Listed in Table 10.4.4 are the times (in days) that it takes each of the fifty states, the District of Columbia, and Puerto Rico to process a Social Security disability claim (185). Can these fifty-two measurements be considered a random sample from a normal distribution? Test the appropriate hypothesis at the $\alpha = 0.05$ level of significance.

Table 10.4.4

State	Time	State	Time	State	Time
Alabama	67.4	Louisiana	86.0	Oklahoma	104.2
Alaska	81.8	Maine	51.3	Oregon	101.6
Arizona	106.5	Maryland	74.1	Pennsylvania	60.1
Arkansas	53.5	Massachusetts	77.8	Puerto Rico	108.3
California	122.8	Michigan	75.1	Rhode Island	84.8
Colorado	71.6	Minnesota	56.5	S. Carolina	70.8
Connecticut	71.4	Mississippi	63.2	S. Dakota	47.3
Delaware	73.1	Missouri	57.1	Tennessee	72.4
D.C.	100.5	Montana	62.2	Texas	72.5
Florida	63.9	Nebraska	70.2	Utah	81.1
Georgia	74.6	Nevada	113.2	Vermont	92.5
Hawaii	115.8	New Hampshire	76.4	Virginia	46.2
Idaho	47.9	New Jersey	109.6	Washington	76.0
Illinois	68.1	New Mexico	74.1	W. Virginia	78.8
Indiana	55.3	New York	86.2	Wisconsin	66.7
Iowa	61.2	N. Carolina	59.5	Wyoming	45.6
Kansas	78.9	N. Dakota	53.9		
Kentucky	61.1	Ohio	69.8		

Shown in Column 1 of Table 10.4.5 is an initial breakdown of the range of Y into nine intervals. Notice that the first and last intervals are open-ended to reflect the fact that the presumed underlying normal distribution is defined for the entire real line.

Table 10.4.5

Interval	Obs. Freq., k_i	$\hat{p}_{i_0}$	Est. Exp. Freq.
$y < 50.0$	4	0.0968	5.03
$50.0 \le y < 60.0$	7	0.1209	6.29
$60.0 \le y < 70.0$	10	0.1797	9.34
$70.0 \le y < 80.0$	16	0.2052	10.67
$80.0 \le y < 90.0$	5	0.1797	9.34
$90.0 \le y < 100.0$	1	0.1209	6.29
$100.0 \le y < 110.0$	6	0.0616	3.20
$110.0 \le y < 120.0$	2	0.0253	1.31
$y \ge 120.0$	1	0.0099	0.51

(Continued on next page)

(Case Study 10.4.3 continued)

In a typical test of normality—and these data are no exception—both parameters, μ and σ, need to be estimated before any expected frequencies can be calculated. Here, using the formulas for the sample mean and sample standard deviation given in Chapter 5,

$$\mu_e = \bar{y} = 75.0 \text{ days}$$

and

$$\sigma_e = s = 19.3 \text{ days}$$

The estimated probability, $\hat{p}_{i_0}$, associated with the ith interval is calculated by using $\bar{y}$ and s to define an approximate Z transformation. For example,

$$\hat{p}_{3_0} = P(60.0 \leq Y < 70.0) \doteq P\left(\frac{60.0 - 75.0}{19.3} \leq Z < \frac{70.0 - 75.0}{19.3}\right)$$
$$\doteq P(-0.78 \leq Z < -0.26) = 0.1797$$

The estimated expected frequencies, then, are the products $52 \cdot \hat{p}_{i_0}$, for $i = 1, 2, \ldots, 9$. For the interval $60 \leq y < 70.0$,

$$n \cdot \hat{p}_{3_0} = 52(0.1797) = 9.34$$

Notice that the three bottom-most subintervals in Column 4 of Table 10.4.5 have estimated expected frequencies less than 5, which violates the condition imposed in Theorem 10.4.1. Collapsing those three into a single interval yields a revised set of data on which the goodness-of-fit statistic can be calculated (see Table 10.4.6).

Table 10.4.6

Interval	Obs. Freq. k_i	$\hat{p}_{i_0}$	Est. Exp. Freq.
$y < 50.0$	4	0.0968	5.03
$50.0 \leq y < 60.0$	7	0.1209	6.29
$60.0 \leq y < 70.0$	10	0.1797	9.34
$70.0 \leq y < 80.0$	16	0.2052	10.67
$80.0 \leq y < 90.0$	5	0.1797	9.34
$90.0 \leq y < 100.0$	1	0.1209	6.29
$y \geq 100.0$	9	0.0968	5.03
	52	1	52.0

According to Theorem 10.4.1, the assumption that Y is a normally distributed random variable should be rejected at the $\alpha = 0.05$ level of significance if

$$d_1 \geq \chi^2_{0.95,7-1-2} = \chi^2_{0.95,4} = 9.488$$

since the revised data grouped into seven classes and two parameters in $f_Y(y)$ have been estimated. But

$$d_1 = \frac{(4 - 5.03)^2}{5.03} + \frac{(7 - 6.29)^2}{6.29} + \cdots + \frac{(9 - 5.03)^2}{5.03} = 12.59$$

so the conclusion is to reject the normality assumption.

About the Data These data raise two obvious questions: (1) What effect does the conclusion of the goodness-of-fit test have on the legitimacy of other analyses that might be done—for example, the construction of a confidence interval for μ? and (2) What might account for the distribution of processing times *not* being normal?

The answer to the first question is easy—*none*. It is true that the derivation of the formula for, say, a confidence interval for μ assumes that the data are normally distributed (recall Theorem 7.4.1). In this case, though, mitigating circumstances make that assumption not so critical. The sample size is large ($n = 52$); the degree of nonnormality is not egregious (had α been set equal to 0.01, H_0 would not have been rejected); and, as the discussion on pp. 406–410 pointed out, procedures involving the Student t distribution are very robust with respect to departures from normality.

The second question is more problematic. The second column in Table 10.4.5 shows that the data are clearly skewed to the right, and there is even a suggestion that the fifty-two observations might represent a mixture of two distributions, each having a different mean. The nine states representing the highest processing times appear to have nothing in common in terms of size, location, or demographics. So why is the right-hand tail of the distribution so different from the left-hand tail? Perhaps the states with the longest waiting times have smaller staffs (relative to their workloads) or they use less up-to-date equipment or follow different procedures. Another possibility—and one that can always be a factor when data are coming from different sources—is that not every state is defining or measuring "processing time" in the same way. From a public policy standpoint, researching the second question is obviously more important than simply doing a goodness-of-fit test to answer the first.

Questions

10.4.1 A public policy polling group is investigating whether people living in the same household tend to make independent political choices. They select two hundred homes where exactly three voters live. The residents are asked separately for their opinion ("yes" or "no") on a city charter amendment. If their opinions are formed independently, the number saying "yes" should be binomially distributed. Do an appropriate goodness-of-fit test on the data below. Let $\alpha = 0.05$.

No. Saying "yes"	Frequency
0	30
1	56
2	73
3	41

10.4.2 From 1837 to 1932, the U.S. Supreme Court had forty-eight vacancies. The table in the next column shows the number of years in which exactly k of the vacancies occurred (185). At the $\alpha = 0.01$ level of significance, test the hypothesis that these data can be described by a Poisson pdf.

Number of Vacancies	Number of Years
0	59
1	27
2	9
3	1
4+	0

10.4.3 As a way of studying the spread of a plant disease known as creeping rot, a field of cabbage plants was divided into 270 *quadrats*, each quadrat containing the same number of plants. The following table lists the numbers of plants per quadrat showing signs of creeping rot infestation.

Number of Infected Plants/Quadrat	Number of Quadrats
0	38
1	57
2	68
3	47
4	23

Number of Infected Plants/Quadrat	Number of Quadrats
5	9
6	10
7	7
8	3
9	4
10	2
11	1
12	1
13+	0

Can the number of plants infected with creeping rot per quadrat be described by a Poisson pdf? Let $\alpha = 0.05$. What might be a physical reason for the Poisson not being appropriate in this situation? Which assumption of the Poisson appears to be violated?

10.4.4 Carry out the details for a goodness-of-fit test on the horse kick data of Question 4.2.10. Use the 0.01 level of significance.

10.4.5 In rotogravure, a method of printing by rolling paper over engraved, chrome-plated cylinders, the printed paper can be flawed by undesirable lines called *bands*. Bands occur when grooves form on the cylinder's surface. When this happens, the presses must be stopped, and the cylinders repolished or replated. The following table gives the number of workdays a printing firm experienced between successive banding shutdowns (39). Fit these data with an exponential model and perform the appropriate goodness-of-fit test at the 0.05 level of significance.

Workdays Between Shutdowns	Number Observed
0–1	130
1–2	41
2–3	25
3–4	8
4–5	2
5–6	3
6–7	1
7–8	1

10.4.6 Do a goodness-of-fit test for normality on the SAT data in Table 3.13.1. Take the sample mean and sample standard deviation to be 949.4 and 68.4, respectively.

10.4.7 A sociologist is studying various aspects of the personal lives of preeminent nineteenth-century scholars. A total of 120 subjects in her sample had families consisting of two children. The distribution of the number of boys in those families is summarized in the following table. Can it be concluded that the number of boys in two-child families of preeminent scholars is binomially distributed? Let $\alpha = 0.05$.

Number of boys	0	1	2
Number of families	24	64	32

10.4.8 In theory, Monte Carlo studies rely on computers to generate large sets of random numbers. Particularly important are random variables representing the uniform pdf defined over the unit interval, $f_Y(y) = 1, 0 \leq y \leq 1$. In practice, though, computers typically generate *pseudorandom numbers*, the latter being values produced systematically by sophisticated algorithms that presumably mimic "true" random variables. Below are one hundred pseudorandom numbers from a uniform pdf. Set up and test the appropriate goodness-of-fit hypothesis. Let $\alpha = 0.05$.

.216	.673	.130	.587	.044	.501	.958	.415	.872	.329
.786	.243	.700	.157	.614	.071	.528	.985	.442	.899
.356	.813	.270	.727	.184	.641	.098	.555	.012	.469
.926	.383	.840	.297	.754	.211	.668	.125	.582	.039
.496	.953	.410	.867	.324	.781	.238	.695	.152	.609
.066	.523	.980	.437	.894	.351	.808	.265	.722	.179
.636	.093	.550	.007	.464	.921	.378	.835	.292	.749
.206	.663	.120	.577	.034	.491	.948	.405	.862	.319
.776	.233	.690	.147	.604	.061	.518	.975	.432	.889
.346	.803	.260	.717	.174	.631	.088	.545	.002	.459

10.4.9 Because it satisfies all the assumptions implicit in the Poisson model, radioactive decay should be described by a probability function of the form $p_X(k) = e^{-\lambda}\lambda^k/k!, k = 0, 1, 2, \ldots,$ where the random variable X denotes the number of particles emitted (or counted) during a given time interval. Does that hold true for the Rutherford and Geiger data given in Case Study 4.2.2? Set up and carry out an appropriate analysis.

10.4.10 Carry out the details to test whether the suffrage data described in Question 4.2.13 follow a Poisson model.

10.4.11 Is the following set of data likely to have come from the geometric pdf, $p_X(k) = (1 - p)^{k-1}p$, $k = 1, 2, \ldots$?

2	8	1	2	2	5	1	2	8	3
5	4	2	4	7	2	2	8	4	7
2	6	2	3	5	1	3	3	2	5
4	2	2	3	6	3	6	4	9	3
3	7	5	1	3	4	3	4	6	2

10.4.12 To raise money for a new rectory, the members of a church hold a raffle. A total of n tickets are sold (numbered 1 through n), out of which a total of fifty winners are to be drawn presumably at random. The following are the fifty lucky numbers. Set up a goodness-of-fit test that focuses on the randomness of the draw. Use the 0.05 level of significance.

108	110	21	6	44
89	68	50	13	63
84	64	69	92	12
46	78	113	104	105
9	115	58	2	20
19	96	28	72	81
32	75	3	49	86
94	61	35	31	56
17	100	102	114	76
106	112	80	59	73

10.5 Contingency Tables

Hypothesis tests, as we have seen, take several fundamentally different forms. Those covered in Chapters 6, 7, and 9 focus on parameters of pdfs—the one-sample, two-sided t test, for example, reduces to a choice between H_0: $\mu = \mu_o$ and H_1: $\mu \neq \mu_o$. Earlier in this chapter, the pdf itself was the issue, and the goodness-of-fit tests in Sections 10.3 and 10.4 dealt with null hypotheses of the form H_0: $f_Y(y) = f_o(y)$.

A third (and final) category of hypothesis tests remains. These apply to situations where the *independence* of two random variables is being questioned. Examples are commonplace. Are the incidence rates of cancer related to mental health? Do a politician's approval ratings depend on the gender of the respondents? Are trends in juvenile delinquency linked to the increasing violence in video games? In this section, we will modify the goodness-of-fit statistic D_1 in such a way that it can distinguish between events that are independent and events that are dependent.

Testing for Independence: A Special Case

A simple example is the best way to motivate the changes that need to be made to the structure of D_1 to make it capable of testing for independence. The key is Definition 2.5.1.

Suppose A is some trait (or random variable) that has two mutually exclusive categories, A_1 and A_2, and suppose that B is a second trait (or random variable) that also has two mutually exclusive categories, B_1 and B_2. To say that *A is independent of B* is to say that the likelihoods of A_1 or A_2 occurring are not influenced by B_1 or B_2. More specifically, four separate conditional probability equations must hold if A and B are to be independent:

$$P(A_1 \mid B_1) = P(A_1) \qquad P(A_1 \mid B_2) = P(A_1)$$
$$P(A_2 \mid B_1) = P(A_2) \qquad P(A_2 \mid B_2) = P(A_2) \tag{10.5.1}$$

By Definition 2.4.1, $P(A_i|B_j) = \frac{P(A_i \cap B_j)}{P(B_j)}$, for all i and j, so the conditions specified in Equation 10.5.1 are equivalent to

$$P(A_1 \cap B_1) = P(A_1)P(B_1) \qquad P(A_1 \cap B_2) = P(A_1)P(B_2)$$
$$P(A_2 \cap B_1) = P(A_2)P(B_1) \qquad P(A_2 \cap B_2) = P(A_2)P(B_2) \tag{10.5.2}$$

Now, suppose a random sample of n observations is taken, and n_{ij} is defined to be the number of observations belonging to A_i and B_j (so $n = n_{11} + n_{12} + n_{21} + n_{22}$). If

Table 10.5.1

		Trait B		
		B_1	B_2	Row Totals
Trait A	A_1	n_{11}	n_{12}	R_1
	A_2	n_{21}	n_{22}	R_2
Column totals:		C_1	C_2	n

we imagine the two categories of A and the two categories of B defining a matrix with two rows and two columns, the four *observed frequencies* can be displayed in the *contingency table* pictured in Table 10.5.1.

If A and B are independent, the probability statements in Equation 10.5.2 would be true, and (by virtue of Theorem 10.2.2), the *expected frequencies* for the four combinations of A_i and B_j would be the entries shown in Table 10.5.2.

Table 10.5.2

		Trait B		
		B_1	B_2	Row Totals
Trait A	A_1	$nP(A_1)P(B_1)$	$nP(A_1)P(B_2)$	R_1
	A_2	$nP(A_2)P(B_1)$	$nP(A_2)P(B_2)$	R_2
Column totals:		C_1	C_2	n

Although $P(A_1)$, $P(A_2)$, $P(B_1)$, and $P(B_2)$ are unknown, they all have obvious estimates—namely, the sample proportion of the time that each occurs. That is,

$$\hat{P}(A_1) = \frac{R_1}{n} \qquad \hat{P}(B_1) = \frac{C_1}{n}$$

$$\hat{P}(A_2) = \frac{R_2}{n} \qquad \hat{P}(B_2) = \frac{C_2}{n} \tag{10.5.3}$$

Table 10.5.3, then, shows the *estimated expected frequencies* (corresponding to n_{11}, n_{12}, n_{21}, and n_{22}) based on the assumption that A and B are independent.

Table 10.5.3

		Trait B	
		B_1	B_2
Trait A	A_1	R_1C_1/n	R_1C_2/n
	A_2	R_2C_1/n	R_2C_2/n

If traits A and B are independent, the observed frequencies in Table 10.5.1 should agree fairly well with the estimated expected frequencies in Table 10.5.3 *because the latter were calculated under the presumption that A and B are independent.* The analog of the test statistic d_1, then, would be the sum d_2, where

$$d_2 = \frac{\left(n_{11} - \frac{R_1C_1}{n}\right)^2}{\frac{R_1C_1}{n}} + \frac{\left(n_{12} - \frac{R_1C_2}{n}\right)^2}{\frac{R_1C_2}{n}} + \frac{\left(n_{21} - \frac{R_2C_1}{n}\right)^2}{\frac{R_2C_1}{n}} + \frac{\left(n_{22} - \frac{R_2C_2}{n}\right)^2}{\frac{R_2C_2}{n}}$$

In the event that d_2 is "large," meaning that one or more of the observed frequencies is substantially different from the corresponding estimated expected frequency, H_0: A and B are independent should be rejected. (In this simple case where both A and B have only two categories, D_2 has approximately a χ_1^2 pdf when H_0 is true, so if α were set at 0.05, H_0 would be rejected if $d_2 \geq \chi^2_{0.95,1} = 3.841$.)

Testing for Independence: The General Case

Suppose n observations are taken on a sample space S partitioned by the set of events $A_1, A_2, \ldots, A_r$ and also partitioned by the set of events $B_1, B_2, \ldots, B_c$. That is,

$$A_i \cap A_j = \emptyset \quad \text{for all } i \neq j \qquad \text{and} \qquad \bigcup_{i=1}^{r} A_i = S$$

and

$$B_i \cap B_j = \emptyset \quad \text{for all } i \neq j \qquad \text{and} \qquad \bigcup_{j=1}^{c} B_j = S$$

Let the random variables $X_{ij}, i = 1, 2, \ldots, r, j = 1, 2, \ldots, c$, denote the number of observations that belong to $A_i \cap B_j$. Our objective is to test whether the A_i's are independent of the B_j's.

Table 10.5.4 shows the two sets of events defining the rows and columns of an $r \times c$ matrix; the k_{ij}'s that appear in the body of the table are the observed values of the X_{ij}'s (recall Table 10.5.1).

Table 10.5.4

	B_1	B_2	$\cdots$	B_c	Row Totals
A_1	k_{11}	k_{12}		k_{1c}	R_1
A_2	k_{21}	k_{22}		k_{2c}	R_2
$\vdots$		$\vdots$	$\cdots$	$\vdots$	$\vdots$
A_r	k_{r1}	k_{r2}		k_{rc}	R_r
Column totals	C_1	C_2		C_c	n

[*Note*: In the terminology of Section 10.2, the X_{ij}'s are a set of *rc multinomial* random variables. Moreover, each individual X_{ij} is a *binomial* random variable with parameters n and p_{ij}, where $p_{ij} = P(A_i \cap B_j)$.]

Let $p_i = P(A_i), i = 1, 2, \ldots, r$, and let $q_j = P(B_j), j = 1, 2, \ldots, c$, so

$$\sum_{i=1}^{r} p_i = 1 = \sum_{j=1}^{c} q_j$$

Invariably, the p_i's and q_j's will be unknown, but their maximum likelihood estimates are simply the corresponding row and column sample proportions:

$$\hat{p}_1 = R_1/n, \quad \hat{p}_2 = R_2/n, \ldots, \hat{p}_r = R_r/n$$
$$\hat{q}_1 = C_1/n, \quad \hat{q}_2 = C_2/n, \ldots, \hat{q}_c = C_c/n$$

(recall Equation 10.5.3).

If the A_i's and B_j's are independent, then

$$P(A_i \cap B_j) = P(A_i)P(B_j) = p_i q_j$$

and the expected frequency corresponding to k_{ij} would be $np_iq_j, i = 1, 2, \ldots, r$; $j = 1, 2, \ldots, c$ (recall the Comment following Theorem 10.2.2). Also, the *estimated expected frequency* for $A_i \cap B_j$ would be

$$n\hat{p}_i\hat{q}_j = n \cdot R_i/n \cdot C_j/n = R_iC_j/n \tag{10.5.4}$$

(recall Table 10.5.3).

So, for each of the rc row-and-column combinations pictured in Table 10.5.4, we have an observed frequency (k_{ij}) and an estimated expected frequency (R_iC_j/n) based on the null hypothesis that the A_i's are independent of the B_j's. The test statistic that would be analogous to d_1, then, would be the double sum d_2, where

$$d_2 = \sum_{i=1}^{r}\sum_{j=1}^{c} \frac{(k_{ij} - n\hat{p}_i\hat{q}_j)^2}{n\hat{p}_i\hat{q}_j}$$

Large values of d_2 would be considered evidence *against* the independence assumption.

Theorem 10.5.1 *Suppose that n observations are taken on a sample space partitioned by the events $A_1, A_2, \ldots, A_r$ and also by the events $B_1, B_2, \ldots, B_c$. Let $p_i = P(A_i)$ $q_j = P(B_j)$, and $p_{ij} = P(A_i \cap B_j)$, $i = 1, 2, \ldots, r$; $j = 1, 2, \ldots, c$. Let X_{ij} denote the number of observations belonging to the intersection $A_i \cap B_j$. Then*

a. *the random variable*

$$D_2 = \sum_{i=1}^{r}\sum_{j=1}^{c} \frac{(X_{ij} - np_{ij})^2}{np_{ij}}$$

has approximately a χ^2 distribution with rc − 1 degrees of freedom (provided $np_{ij} \geq 5$ for all i and j).

b. *to test H_0: the A_i's are independent of the B_j's, calculate the test statistic*

$$d_2 = \sum_{i=1}^{r}\sum_{j=1}^{c} \frac{(k_{ij} - n\hat{p}_i\hat{q}_j)^2}{n\hat{p}_i\hat{q}_j}$$

where k_{ij} is the number of observations in the sample that belong to $A_i \cap B_j$, $i = 1, 2, \ldots, r$; $j = 1, 2, \ldots, c$ and $\hat{p}_i$ and $\hat{q}_j$ are the maximum likelihood estimates for p_i and q_j, respectively. The null hypothesis should be rejected at the α level of significance if

$$d_2 \geq \chi^2_{1-\alpha,(r-1)(c-1)}$$

(Analogous to the condition stipulated for all other goodness-of-fit tests, it will be assumed that $n\hat{p}_i\hat{q}_j \geq 5$ for all i and j.)

Comment In general, the number of degrees of freedom associated with a goodness-of-fit statistic is given by the formula

df = number of classes − 1 − number of estimated parameters

(recall Theorem 10.4.1). For the double sum that defines d_2,

$$\text{number of classes} = rc$$
$$\text{number of estimated parameters} = r - 1 + c - 1$$

(because once $r - 1$ of the p_i's are estimated, the one that remains is predetermined by the fact that $\sum_{i=1}^{r} p_i = 1$; similarly, only $c - 1$ of the q_j's need to be estimated). But

$$rc - 1 - (r - 1) - (c - 1) = (r - 1)(c - 1)$$

Comment The χ^2 distribution with $(r - 1)$ $(c - 1)$ degrees of freedom provides an adequate approximation to the distribution of d_2 only if $n\hat{p}_i\hat{q}_j \geq 5$ for all i and j. If one or more cells in a contingency table have estimated expected frequencies that are substantially less than 5, the table should be "collapsed" and the rows and/or columns redefined.

Case Study 10.5.1

Gene Siskel and Roger Ebert were popular movie critics for a syndicated television show. Viewers of the program were entertained by the frequent flare-ups of acerbic disagreement between the two. They were immediately recognizable to a large audience of movie goers by their rating system of "thumbs up" for good films, "thumbs down" for bad ones, and an occasional "sideways" for those in between.

Table 10.5.5 summarizes their evaluations of 160 movies (2). Do these numbers suggest that Siskel and Ebert had completely different aesthetics—in which case their ratings would be independent—or do they demonstrate that the two shared considerable common ground, despite their many on-the-air verbal jabs?

Table 10.5.5

		Ebert Ratings			
		Down	Sideways	Up	Total
Siskel Ratings	Down	24	8	13	45
	Sideways	8	13	11	32
	Up	10	9	64	83
	Total	42	30	88	160

Using Equation 10.5.4, we can calculate the estimated expected number of times that both reviewers would say "thumbs down" if, in fact, their ratings were independent:

$$\hat{E}(X_{11}) = \frac{R_1 \cdot C_1}{n} = \frac{(45)(42)}{160}$$
$$= 11.8$$

(Continued on next page)

(Case Study 10.5.1 continued)

Table 10.5.6 displays the entire set of estimated expected frequencies, all calculated the same way.

Table 10.5.6

		Ebert Ratings			
		Down	Sideways	Up	Total
	Down	24 (11.8)	8 (8.4)	13 (24.8)	45
Siskel Ratings	Sideways	8 (8.4)	13 (6.0)	11 (17.6)	32
	Up	10 (21.8)	9 (15.6)	64 (45.6)	83
	Total	42	30	88	160

Now, suppose we wish to test

H_0: Siskel ratings and Ebert ratings were independent

versus

H_1: Siskel ratings and Ebert ratings were dependent

at the $\alpha = 0.01$ level of significance. With $r = 3$ and $c = 3$, the number of degrees of freedom associated with the test statistic is $(3 - 1)(3 - 1) = 4$, and H_0 should be rejected if

$$d_2 \geq \chi^2_{0.99,4} = 13.277$$

But

$$d_2 = \frac{(24 - 11.8)^2}{11.8} + \frac{(8 - 8.4)^2}{8.4} + \cdots + \frac{(64 - 45.6)^2}{45.6}$$
$$= 45.37$$

so the evidence is overwhelming that Siskel and Ebert's judgments were not independent.

"Reducing" Continuous Data to Contingency Tables

Most applications of contingency tables begin with *qualitative* data, Case Study 10.5.1 being a typical case in point. Sometimes, though, contingency tables can provide a particularly convenient format for testing the independence of two random variables that initially appear as *quantitative* data. If those x and y measurements are each *reduced* to being either "high" or "low," for example, the original x_i's and y_i's become frequencies in a 2×2 contingency table (and can be used to test H_0: X and Y are independent).

Case Study 10.5.2

Sociologists have speculated that feelings of alienation may be a major factor contributing to an individual's risk of committing suicide. If so, cities with more transient populations should have higher suicide rates than urban areas where neighborhoods are more stable. Listed in Table 10.5.7 is the "mobility index" (y) and the "suicide rate" (x) for each of twenty-five U.S. cities (210). (*Note*: The mobility index was defined in such a way that smaller values of y correspond to higher levels of transiency.) Do these data support the sociologists' suspicion?

Table 10.5.7

City	Suicides per 100,000, x_i	Mobility Index, y_i	City	Suicides per 100,000, x_i	Mobility Index, y_i
New York	19.3	54.3	Washington	22.5	37.1
Chicago	17.0	51.5	Minneapolis	23.8	56.3
Philadelphia	17.5	64.6	New Orleans	17.2	82.9
Detroit	16.5	42.5	Cincinnati	23.9	62.2
Los Angeles	23.8	20.3	Newark	21.4	51.9
Cleveland	20.1	52.2	Kansas City	24.5	49.4
St. Louis	24.8	62.4	Seattle	31.7	30.7
Baltimore	18.0	72.0	Indianapolis	21.0	66.1
Boston	14.8	59.4	Rochester	17.2	68.0
Pittsburgh	14.9	70.0	Jersey City	10.1	56.5
San Francisco	40.0	43.8	Louisville	16.6	78.7
Milwaukee	19.3	66.2	Portland	29.3	33.2
Buffalo	13.8	67.6			

To reduce these data to a 2×2 contingency table, we redefine each x_i as being either "$\geq \bar{x}$" or "$< \bar{x}$" and each y_i as being either "$\geq \bar{y}$" or "$< \bar{y}$." Here,

$$\bar{x} = \frac{19.3 + 17.0 + \cdots + 29.3}{25} = 20.8$$

and

$$\bar{y} = \frac{54.3 + 51.5 + \cdots + 33.2}{25} = 56.0$$

so the twenty-five (x_i, y_i)'s produce the 2×2 contingency table shown in Table 10.5.8.

Table 10.5.8

		Mobility Index	
		Low (<56.0)	High (≥56.0)
Suicide Rate	High (≥20.8)	7	4
	Low (<20.8)	3	11

(*Continued on next page*)

(Case Study 10.5.2 continued)

If X and Y are independent, the four estimated expected frequencies associated with the contingency table (and calculated from Equation 10.5.4) are the entries appearing in Table 10.5.9.

Table 10.5.9

		Mobility Index	
		Low (<56.0)	High ($\geq$56.0)
Suicide	High ($\geq$20.8)	4.4*	6.6
Rate	Low (<20.8)	5.6	8.4

*$\hat{E}(X_{11}) = 4.4$ does not quite satisfy the "$n\hat{p}_i\hat{q}_j \geq 5$" restriction stated in Theorem 10.5.1, but 4.4 is close enough to 5 to maintain the integrity of the χ^2 approximation.

Substituting into the test statistic from Theorem 10.5.1 gives

$$d_2 = \frac{(7-4.4)^2}{4.4} + \frac{(4-6.6)^2}{6.6} + \frac{(3-5.6)^2}{5.6} + \frac{(11-8.4)^2}{8.4}$$
$$= 4.57$$

With $(r-1)(c-1) = (2-1)(2-1) = 1$ df, the $\alpha = 0.05$ critical value associated with d_2 is $\chi^2_{0.95,1} = 3.841$. The appropriate conclusion, then, is to *reject* H_0 since $d_2 \geq 3.841$—the sociologists' suspicion that suicide rates and transiency in urban areas are dependent *is* borne out by the data.

Case Study 10.5.3

Beginning in 1647, witchcraft accusations, trials, and executions were an on-again, off-again phenomenon in the New England colonies. Occasionally, Puritan angst over matters Satanic would flare up like an epidemic, and entire communities would become convinced that many of their neighbors were in league with the Devil. The most famous of these paranoid outbreaks were the Salem witch trials that occurred in 1692 and 1693.

Altogether, a total of 185 adults (and children) were accused of witchcraft in Salem during those two years, 141 females and 44 males. Fourteen of the women (9.9%) were eventually hanged; a similar fate befell five (11.4%) of the men (90). Is the difference between 9.9% and 11.4% statistically significant?

Recall the discussion on p. 447. Testing whether the difference between two independent binomial proportions is statistically significant is equivalent to testing whether the two factors represented in a 2×2 contingency table are independent—that is, $H_0: p_X = p_Y$ will be rejected if and only if we reject H_0: X and Y are independent.

Table 10.5.10 shows the Salem data presented as a 2×2 contingency table. In parentheses are the expected frequencies calculated under the

(Continued on next page)

null hypothesis that the chances of an accused witch being executed were independent of gender.

Table 10.5.10

	Accused Witches		
	Female	Male	
Executed	14 (14.5)	5 (4.5)	19
Not Executed	127 (126.5)	39 (39.5)	166
			185
Totals	141	44	

By part (b) of Theorem 10.5.1, $d_2 = 0.08$ (with 1 df). If the independence hypothesis is to be tested at the $\alpha = 0.05$ level of significance, the appropriate critical value is

$$\chi^2_{1-\alpha,(r-1)(c-1)} = \chi^2_{0.95,(2-1)(2-1)} = \chi^2_{0.95,1} = 3.84$$

so the conclusion is "fail to reject H_0."

If these data were viewed as two independent sets of Bernoulli trials of sizes 141 and 44, respectively, the appropriate test statistic would be the observed Z ratio described in Theorem 9.4.1. With $x = 14$, $n = 141$, $y = 5$, $m = 44$, $z = -0.28$, and $\alpha = 0.05$, critical values would be ± 1.96 (see Question 10.5.10), implying that the difference between 9.9% and 11.4% is not statistically significant.

Questions

10.5.1 Market researchers often gather information by telephone, but calling only listed numbers may badly skew the responses, if listed and unlisted households are fundamentally different with respect to the question being asked. The following is the slightly modified summary of a survey done by Pacific Bell to see whether homeownership is related to telephone listing (142). At the $\alpha = 0.05$ and $\alpha = 0.10$ levels of significance, test whether those two "conditions" are independent.

	Listed	Unlisted
Own	628	146
Rent	172	54

10.5.2 Many factors influence a company's decision to relocate to another site. The state of Florida, hoping to attract such relocations, sponsored a study (50) on how different companies view various factors. One part of the study compared the importance of a high-quality workforce to manufacturing firms and to nonmanufacturing firms. At the $\alpha = 0.05$ level of significance, do the following data suggest that the importance of a high-quality workforce is not viewed the same by all types of businesses?

		Manufacturing	Other
Importance	Extremely or somewhat	168	73
	Not very	42	26

10.5.3 A total of 1154 girls attending a public high school were given a questionnaire that measured how much each had exhibited delinquent behavior (124). From an analysis of the results, the researchers categorized 111 of the girls as "delinquent." The following is a cross-classification of the delinquents and the nondelinquents according to their birth order. At the $\alpha = 0.01$ level of significance, is there evidence here to support the contention that birth order and delinquency are related?

	Delinquent	Not Delinquent
Oldest	24	450
In Between	29	312
Youngest	35	211
Only Child	23	70

10.5.4 Recall the rubella/birth defect study described in Case Study 8.1.3. At the $\alpha = 0.01$ level of significance, can it be concluded that the risk of an abnormal birth is affected by *when* a rubella infection is contracted during pregnancy?

10.5.5 Research has suggested that regular use of aspirin or other nonsteroidal anti-inflammatory drugs (NSAIDs) may be effective in reducing the risk of breast cancer. In one recent study (179), 1442 women with breast cancer were asked whether they had used aspirin regularly one year prior to their diagnosis; 301 said "yes." Among a matched control group of 1420 women without breast cancer, 345 reported that they were regular aspirin users. What would you conclude? Set up and test an appropriate hypothesis. Let 0.05 be the level of significance.

10.5.6 High blood pressure is known to be one of the major contributors to coronary heart disease. A study was done to see whether or not there is a significant relationship between the blood pressures of children and those of their fathers (88). If such a relationship did exist, it might be possible to use one group to screen for high-risk individuals in the other group. The subjects were 92 eleventh graders, 47 males and 45 females, and their fathers. Blood pressures for both the children and the fathers were categorized as belonging to either the lower, middle, or upper third of their respective distributions. Test whether or not the blood pressures of children can be considered to be independent of the blood pressures of their fathers. Let $\alpha = 0.05$.

		Child's Blood Pressure		
		Lower Third	Middle Third	Upper Third
Father's blood pressure	Lower third	14	11	8
	Middle third	11	11	9
	Upper third	6	10	12

10.5.7 The following data were collected as part of a study to see whether a mouse's early upbringing has any effect on its aggressiveness later in life (84). A total of 307 mice were divided into two groups shortly after birth. Each of the 167 mice in the first group was raised by its natural mother; the remaining 140 in the second group were raised by "foster" mice. When each mouse was three months old, it was put into a small cage with another mouse it had not seen before. The two were then watched for a predetermined period of time (six minutes) to see whether they would start fighting. Set up and carry out an appropriate χ^2 test. Let $\alpha = 0.05$.

	Natural Mother	Foster Mother
Number fighting	27	47
Number not fighting	140	93
	167	140

10.5.8 The Hopwood Decision resulted from a 1996 U.S. Fifth Circuit Court of Appeals case that greatly limited Texas universities' affirmative-action programs for admission of minority students. As a consequence, minority enrollment dropped significantly. One solution proposed was to accept all students in the top 10% of their graduating class. The success of such a plan in achieving diversity would hinge on the enrollment rates for the different racial groups. The following are the average numbers of freshmen in the top 10% of their classes admitted and enrolled, by race, at UT-Austin for the years 1990–1996. Are the enrollment rates dependent on the racial groups? Do the appropriate analysis using the $\alpha = 0.05$ level of significance.

	Admitted	Enrolled
White	2592	1481
African-American	159	78
Hispanic	800	375
Asian	667	399

10.5.9 Portfolio turnover expresses the past year's trading activity as a percentage of an account's average assets. The following table summarizes the performances of one hundred mutual funds cross-classified according to portfolio turnover and annual return. Test the independence assumption. Let $\alpha = 0.05$.

		Annual Return	
		$\leq$10%	>10%
Portfolio Return	$\geq$100%	11	10
	<100%	55	24

10.5.10 (a) For the witchcraft data described in Case Study 10.5.3, verify that $z = -0.28$.
(b) Notice that $(-0.28)^2 = 0.08$ and $(\pm 1.96)^2 = 3.84$. Why should those equalities be true?

10.6 Taking a Second Look at Statistics (Outliers)

This chapter has explored important questions related to the "pedigree" of a set of data. Given the measurements $y_1, y_2, \ldots, y_n$, for example, is it believable that they represent a random sample from some particular pdf, $f_o(y)$? Or, given a set of bivariate observations, $(x_1, y_1), (x_2, y_2), \ldots, (x_n, y_n)$, representing the random variables X and Y, is it believable that X and Y are independent?

In practice, experimenters sometimes encounter a slightly different sort of pedigree question, one that focuses on individual measurements rather than on entire data sets. For example, suppose a laboratory experiment has yielded the twenty observations listed in Table 10.6.1. Grouped into classes of width 10, the data have the frequency distribution shown in Table 10.6.2. The question is, what (if anything) should be done with the measurement $y = 127.6$ that lies considerably to the right of the rest of the data? Is it simply the largest observation in the sample (in which case it should be kept), or does its separation from the bulk of the distribution reflect some sort of fundamental measurement error (in which case it should be discarded)?

Table 10.6.1

73.5	45.6	51.2	15.6	49.2
55.7	24.8	127.6	49.7	53.8
91.6	82.9	78.4	58.4	67.9
44.3	62.4	37.4	30.8	59.6

Table 10.6.2

Observation	Frequency
$10.0 \le y < 20.0$	1
$20.0 \le y < 30.0$	1
$30.0 \le y < 40.0$	2
$40.0 \le y < 50.0$	4
$50.0 \le y < 60.0$	5
$60.0 \le y < 70.0$	2
$70.0 \le y < 80.0$	2
$80.0 \le y < 90.0$	1
$90.0 \le y < 100.0$	1
$100.0 \le y < 110.0$	0
$110.0 \le y < 120.0$	0
$120.0 \le y < 130.0$	1

While there is no way to answer that question with any certainty, there are test procedures that can shed some light on the likelihood (subject to certain assumptions) of an "outlier" being a sample from the same pdf that generated all the other observations. One such procedure, due to Dixon (38), assumes that the observations are coming from a normal distribution and is based on either the ratio

$$r_{01} = \frac{y'_n - y'_{n-1}}{y'_n - y'_1} \qquad \text{or} \qquad r_{10} = \frac{y'_2 - y'_1}{y'_n - y'_1}$$

where y'_i is the ith order statistic in the sample of size n. If the potential outlier is the *largest* observation in the sample, the test statistic is r_{01}; if the potential outlier is the smallest observation, the test statistic is r_{10}.

Table 10.6.3 gives upper percentage points for the distribution of r_{01} (and r_{10}) as a function of the sample size n. For the data in Table 10.6.1, $n = 20$ and the largest observation is the measurement in question, so the test statistic is

$$r_{01} = \frac{y'_{20} - y'_{19}}{y'_{20} - y'_1}$$

From Table 10.6.1, $y'_1 = 15.6$, $y'_{19} = 91.6$, and $y'_{20} = 127.6$, so

$$r_{01} = \frac{127.6 - 91.6}{127.6 - 15.6} = 0.32$$

According to Table 10.6.3, the P-value associated with the outcome $r_{01} = 0.32$ is between 0.05 and 0.02, since the 95th percentile of the r_{01} distribution when $n = 20$

Table 10.6.3

PERCENTAGE POINTS OF THE DISTRIBUTION OF r_{10}

n \ $1\text{-}\alpha$	.80	.90	.95	.98	.99	.995
3	.781	.886	.941	.976	.988	.994
4	.560	.679	.765	.846	.889	.926
5	.451	.557	.642	.729	.780	.821
6	.386	.482	.560	.644	.698	.740
7	.344	.434	.507	.586	.637	.680
8	.314	.399	.468	.543	.590	.634
9	.290	.370	.437	.510	.555	.598
10	.273	.349	.412	.483	.527	.568
11	.259	.332	.392	.460	.502	.542
12	.247	.318	.376	.441	.482	.522
13	.237	.305	.361	.425	.465	.503
14	.228	.294	.349	.411	.450	.488
15	.220	.285	.338	.399	.438	.475
16	.213	.277	.329	.388	.426	.463
17	.207	.269	.320	.379	.416	.452
18	.202	.263	.313	.370	.407	.442
19	.197	.258	.306	.363	.398	.433
20	.193	.252	.300	.356	.391	.425
21	.189	.247	.295	.350	.384	.418
22	.185	.242	.290	.344	.378	.411
23	.182	.238	.285	.338	.372	.404
24	.179	.234	.281	.333	.367	.399
25	.176	.230	.277	.329	.362	.393
26	.173	.227	.273	.324	.357	.388
27	.171	.224	.269	.320	.353	.384
28	.168	.220	.266	.316	.349	.380
29	.166	.218	.263	.312	.345	.376
30	.164	.215	.260	.309	.341	.372

Source: Dunn, Olive Jean and Clark, Virginia A. *Applied Statistics: Analysis of Variance and Regression*. New York: John Wiley & Sons, 1974, p. 374.

is 0.300, and the 98th percentile is 0.356. So, should y'_{20} be discarded? Probably not, unless there is reason to believe that its large value was the result of a measurement error. The distribution of r_{01} makes it clear that a value of 127.6 in this case is not dramatically out of line.

A word of caution—the simplicity of testing for outliers does not mean the procedure should be used capriciously. More than a few experimenters have woefully regretted discarding suspicious observations under the guise of "cleaning up" their data. Sometimes the observations *not* fitting the presumed model constitute the most important information in a data set, because they may be the first and only clues that the presumed model is, in fact, incorrect.

Appendix 10.A.1 Minitab Applications

The Minitab command CHISQUARE, followed by the columns in which the observed frequencies have been entered, performs the χ^2 test for independence described in Theorem 10.5.1. Figure 10.A.1.1 shows the input and output for the Minitab analysis of the data in Case Study 10.5.1. In addition to the estimated expected frequencies and the value of the test statistic, the CHISQUARE routine also indicates the number of degrees of freedom associated with d_2 and its P-value. Here the P-value is so small there is no question that the null hypothesis of independence should be rejected.

```
MTB   > set c1
DATA  > 24 8 10
DATA  > end
MTB   > set c2
DATA  > 8 13 9
DATA  > end
MTB   > set c3
DATA  > 13 11 64
DATA  > end
MTB   > chisquare c1-c3
```

Figure 10.A.1.1

Chi Square Test: C1, C2, C3

```
Expected counts are printed below observed counts
Chi square contributions are printed below expected counts
            C1      C2      C3                 Total
1           24       8      13                    45
         11.81    8.44   24.75
        12.574   0.023   5.578

2            8      13      11                    32
          8.40    6.00   17.60
         0.019   8.167   2.475

3           10       9      64                    83
         21.79   15.56   45.65
         6.377   2.767   7.376

Total       42      30      88                   160

Chi-Sq = 45.357, DF = 4, P-Value = 0.000
```

Testing for Independence Using Minitab Windows

1. Enter each column of observed frequencies in a separate column.
2. Click on STAT, then on TABLES, then on CHISQUARE TEST.
3. Enter the columns containing the data, and click on OK.

Chapter 11

REGRESSION

Galton had earned a Cambridge mathematics degree and completed two years of medical school when his father died, leaving him with a substantial inheritance. Free to travel, he became an explorer of some note, but when The Origin of Species *was published in 1859, his interests began to shift from geography to statistics and anthropology (Charles Darwin was his cousin). It was Galton's work on fingerprints that made possible their use in human identification. He was knighted in 1909.*

—Francis Galton (1822–1911)

11.1 Introduction

High on the list of problems that experimenters most frequently need to deal with is the determination of the relationships that exist among the various components of a complex system. If those relationships are sufficiently understood, there is a good possibility that the system's output can be effectively modeled, maybe even controlled.

Consider, for example, the formidable problem of relating the incidence of cancer to its many contributing causes—diet, genetic makeup, pollution, and cigarette smoking, to name only a few. Or think of the Wall Street financier trying to anticipate trends in stock prices by tracking market indices and corporate performances, as well as the overall economic climate. In those situations, a host of variables are involved, and the analysis becomes very intricate. Fortunately, many of the fundamental ideas associated with the study of relationships can be nicely illustrated when only *two* variables are involved. This two-variable model will be the focus of Chapter 11.

Section 11.2 gives a computational technique for determining the "best" equation describing a set of points $(x_1, y_1), (x_2, y_2), \ldots$, and (x_n, y_n), where *best* is defined geometrically. Section 11.3 adds a probability distribution to the y-variable, which allows for a variety of inference procedures to be developed. The consequences of both measurements being random variables is the topic of Section 11.4. Then Section 11.5 takes up a special case of Section 11.4, where the variability in X and Y is described by the *bivariate normal pdf*.

11.2 The Method of Least Squares

We begin our study of the relationship between two variables by asking a simple geometry question. Given a set of n points—(x_1, y_1), (x_2, y_2), ..., (x_n, y_n)—and a positive integer m, which polynomial of degree m is "closest" to the given points?

Suppose that the desired polynomial, $p(x)$, is written

$$p(x) = a + \sum_{i=1}^{m} b_i x^i$$

where $a, b_1, \ldots, b_m$ are to be determined. The *method of least squares* answers the question by finding the coefficient values that minimize the sum of the squares of the vertical distances from the data points to the presumed polynomial. That is, the polynomial $p(x)$ that we will call "best" is the one whose coefficients minimize the function L, where

$$L = \sum_{i=1}^{n} [y_i - p(x_i)]^2$$

Theorem 11.2.1 summarizes the method of least squares as it applies to the important special case where $p(x)$ is a *linear* polynomial. (*Note*: To simplify notation, the linear polynomial $y = a + b_1 x^1$ will be written $y = a + bx$.)

Theorem 11.2.1 *Given n points (x_1, y_1), (x_2, y_2), ..., (x_n, y_n), the straight line $y = a + bx$ minimizing*

$$L = \sum_{i=1}^{n} [y_i - (a + bx_i)]^2$$

has slope

$$b = \frac{n \sum_{i=1}^{n} x_i y_i - \left(\sum_{i=1}^{n} x_i\right)\left(\sum_{i=1}^{n} y_i\right)}{n\left(\sum_{i=1}^{n} x_i^2\right) - \left(\sum_{i=1}^{n} x_i\right)^2}$$

and y-intercept

$$a = \frac{\sum_{i=1}^{n} y_i - b \sum_{i=1}^{n} x_i}{n} = \bar{y} - b\bar{x}$$

Proof The proof is accomplished by the familiar calculus technique of taking the partial derivatives of L with respect to a and b, setting the resulting expressions equal to 0, and solving. By the first step we get

$$\frac{\partial L}{\partial b} = \sum_{i=1}^{n} (-2)x_i[y_i - (a + bx_i)]$$

and

$$\frac{\partial L}{\partial a} = \sum_{i=1}^{n} (-2)[y_i - (a + bx_i)]$$

Setting the right-hand sides of $\partial L/\partial a$ and $\partial L/\partial b$ equal to 0 and simplifying yields the two equations

$$na + \left(\sum_{i=1}^{n} x_i\right) b = \sum_{i=1}^{n} y_i$$

and

$$\left(\sum_{i=1}^{n} x_i\right) a + \left(\sum_{i=1}^{n} x_i^2\right) b = \sum_{i=1}^{n} x_i y_i$$

An application of Cramer's rule gives the solution for b stated in the theorem. The expression for a follows immediately. □

Case Study 11.2.1

A manufacturer of air conditioning units is having assembly problems due to the failure of a connecting rod to meet finished-weight specifications. Too many rods are being completely tooled, then rejected as overweight. To reduce that cost, the company's quality-control department wants to quantify the relationship between the weight of the finished rod, y, and that of the rough casting, x (139). Castings likely to produce rods that are too heavy can then be discarded before undergoing the final (and costly) tooling process.

As a first step in examining the xy-relationship, twenty-five (x_i, y_i) pairs are measured (see Table 11.2.1). Graphed, the points suggest that the weight of the finished rod is linearly related to the weight of the rough casting (see Figure 11.2.1). Use Theorem 11.2.1 to find the best straight line approximating the xy-relationship.

Table 11.2.1

Rod Number	Rough Weight, x	Finished Weight, y	Rod Number	Rough Weight, x	Finished Weight, y
1	2.745	2.080	14	2.635	1.990
2	2.700	2.045	15	2.630	1.990
3	2.690	2.050	16	2.625	1.995
4	2.680	2.005	17	2.625	1.985
5	2.675	2.035	18	2.620	1.970
6	2.670	2.035	19	2.615	1.985
7	2.665	2.020	20	2.615	1.990
8	2.660	2.005	21	2.615	1.995
9	2.655	2.010	22	2.610	1.990
10	2.655	2.000	23	2.590	1.975
11	2.650	2.000	24	2.590	1.995
12	2.650	2.005	25	2.565	1.955
13	2.645	2.015			

From Table 11.2.1, we find that

$$\sum_{i=1}^{25} x_i = 66.075 \qquad \sum_{i=1}^{25} x_i^2 = 174.672925$$

$$\sum_{i=1}^{25} y_i = 50.12 \qquad \sum_{i=1}^{25} y_i^2 = 100.49865$$

$$\sum_{i=1}^{25} x_i y_i = 132.490725$$

(Continued on next page)

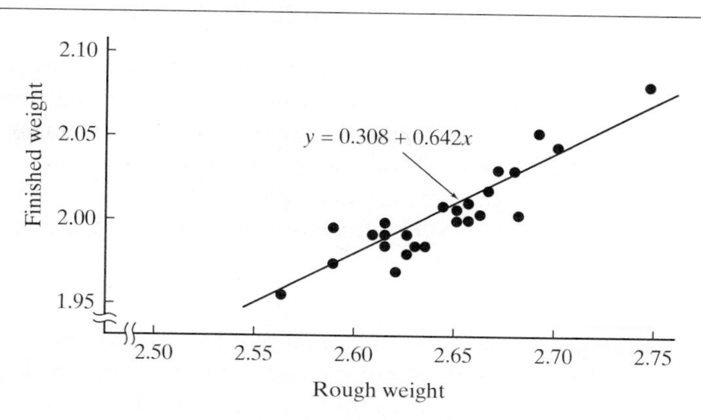

Figure 11.2.1

Therefore,

$$b = \frac{25(132.490725) - (66.075)(50.12)}{25(174.672925) - (66.075)^2} = 0.642$$

and

$$a = \frac{50.12 - 0.642(66.075)}{25} = 0.308$$

making the least squares line

$$y = 0.308 + 0.642x$$

The manufacturer is now in a position to make some informed policy decisions. If the weight of a rough casting is, say, 2.71 oz., the least squares line predicts that its finished weight will be 2.05 oz.:

$$\text{estimated weight} = a + b(2.71) = 0.308 + 0.642(2.71) = 2.05$$

In the event that finished weights of 2.05 oz. are considered to be too heavy, rough castings weighing 2.71 oz. (or more) should be discarded.

Residuals

The difference between an observed y_i and the value of the least squares line when $x = x_i$ is called the ith *residual*. Its magnitude reflects the failure of the least squares line to "model" that particular point.

Definition 11.2.1. Let a and b be the least squares coefficients associated with the sample $(x_1, y_1), (x_2, y_2), \ldots, (x_n, y_n)$. For any value of x, the quantity $\hat{y} = a + bx$ is known as the *predicted value* of y. For any given $i, i = 1, 2, \ldots, n$, the difference $y_i - \hat{y}_i = y_i - (a + bx_i)$ is called the ith *residual*. A graph of $y_i - \hat{y}_i$ versus x_i, for all i, is called a *residual plot*.

Interpreting Residual Plots

Applied statisticians find residual plots to be very helpful in assessing the appropriateness of fitting a straight line through a given set of n points. If the relationship between x and y *is* linear, the corresponding residual plot typically shows no patterns, cycles, trends, or outliers. For nonlinear relationships, though, residual plots often take on dramatically nonrandom appearances that can very effectively highlight and illuminate the underlying association between x and y.

Example 11.2.1 Make the residual plot for the data in Case Study 11.2.1. What does its appearance imply about the suitability of fitting those points with a straight line?

We begin by calculating the residuals for each of the twenty-five data points. The first observation recorded, for example, was $(x_1, y_1) = (2.745, 2.080)$. The corresponding predicted value, $\hat{y}_1$, is 2.070:

$$\hat{y}_1 = 0.308 + 0.642(2.745)$$
$$= 2.070$$

The first residual, then, is $y_1 - \hat{y}_1 = 2.080 - 2.070$, or 0.010. The complete set of residuals appears in the last column of Table 11.2.2.

Table 11.2.2

x_i	y_i	$\hat{y}_i$	$y_i - \hat{y}_i$
2.745	2.080	2.070	0.010
2.700	2.045	2.041	0.004
2.690	2.050	2.035	0.015
2.680	2.005	2.029	−0.024
2.675	2.035	2.025	0.010
2.670	2.035	2.022	0.013
2.665	2.020	2.019	0.001
2.660	2.005	2.016	−0.011
2.655	2.010	2.013	−0.003
2.655	2.000	2.013	−0.013
2.650	2.000	2.009	−0.009
2.650	2.005	2.009	−0.004
2.645	2.015	2.006	0.009
2.635	1.990	2.000	−0.010
2.630	1.990	1.996	−0.006
2.625	1.995	1.993	0.002
2.625	1.985	1.993	−0.008
2.620	1.970	1.990	−0.020
2.615	1.985	1.987	−0.002
2.615	1.990	1.987	0.003
2.615	1.995	1.987	0.008
2.610	1.990	1.984	0.006
2.590	1.975	1.971	0.004
2.590	1.995	1.971	0.024
2.565	1.955	1.955	0.000

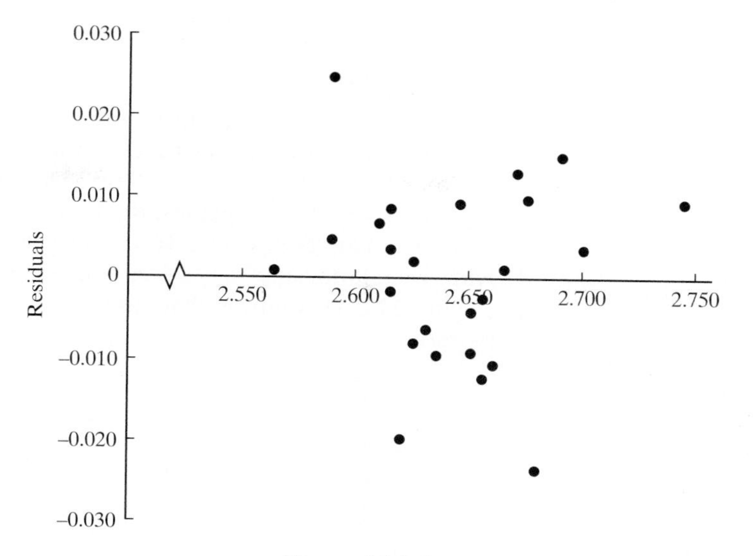

Figure 11.2.2

Figure 11.2.2 shows the residual plot generated by fitting the least squares straight line, $y = 0.308 + 0.642x$, to the twenty-five (x_i, y_i)'s. To an applied statistician, there is nothing here that would raise any serious doubts about using a straight line to describe the xy-relationship—the points appear to be randomly scattered and exhibit no obvious anomalies or patterns. ■

Case Study 11.2.2

Table 11.2.3 lists Social Security expenditures for five-year intervals from 1965 to 2005. During that period, payouts rose from \$19.2 billion to \$529.9 billion. Substituting these nine (x_i, y_i)'s into the formulas in Theorem 11.2.1 gives

$$y = -38.0 + 12.9x$$

Table 11.2.3

Year	Years after 1965, x	Social Security Expenditures (\$ billions), y
1965	0	19.2
1970	5	33.1
1975	10	69.2
1980	15	123.6
1985	20	190.6
1990	25	253.1
1995	30	339.8
2000	35	415.1
2005	40	529.9

Source: www.socialsecurity.gov/history/trustfunds.html.

(Continued on next page)

(Case Study 11.2.2 continued)

as the least squares straight line describing the xy-relationship. Based on the data from 1965 to 2005, is it reasonable to predict that Social Security costs in the year 2010 (when $x = 45$) will be \$543 billion [$= -38.0 + 12.9(45)$]?

Not at all. At first glance, the least squares line does appear to fit the data quite well (see Figure 11.2.3). A closer look, though, suggests that the underlying xy-relationship may be curvilinear rather than linear. The residual plot (Figure 11.2.4) confirms that suspicion—there we see a distinctly nonrandom pattern.

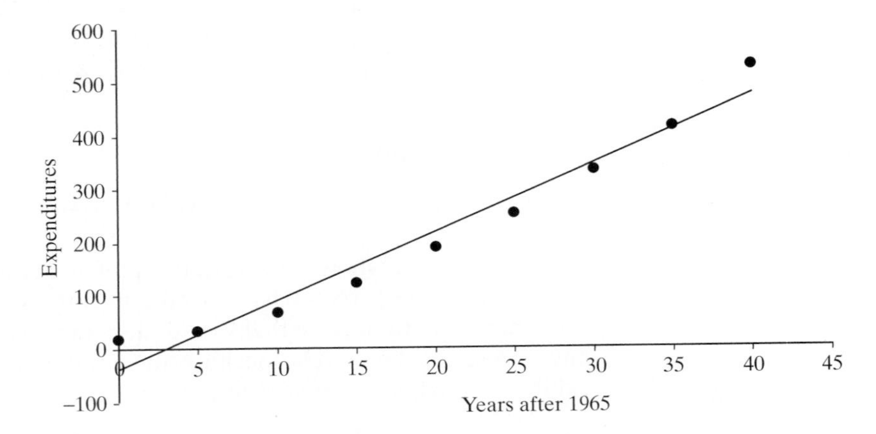

Figure 11.2.3

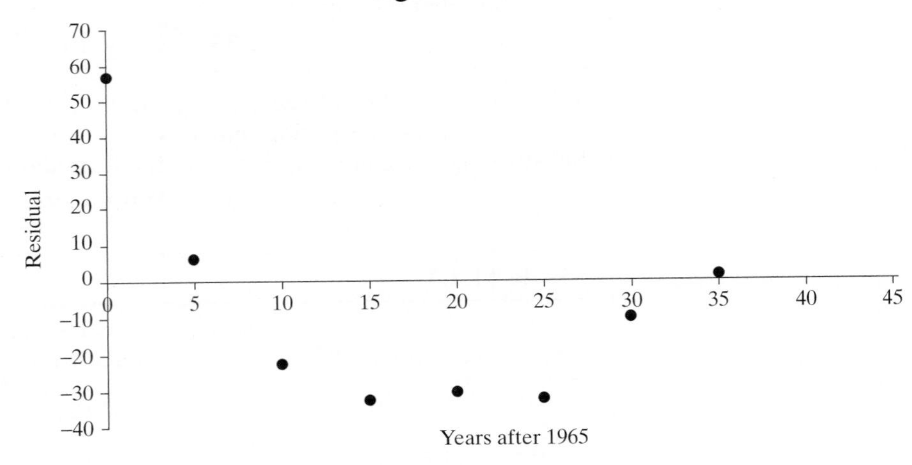

Figure 11.2.4

Clearly, extrapolating these data would be foolish. The figure for the next year, 2006, of \$555 billion already exceeded the linear projection of \$543 billion, leading economists to predict rapidly accelerating expenditures in the future.

Comment For the data in Table 11.2.3, the suggestion that the xy-relationship may be curvilinear is certainly present in Figure 11.2.3, but the residual plot makes the case much more emphatically. In point of fact, that will often be the case, which is why residual plots are such a valuable diagnostic tool—departures from randomness that may be only hinted at in an xy-plot will be exaggerated and highlighted in the corresponding residual plot.

Case Study 11.2.3

A new, presumably simpler laboratory procedure has been proposed for recovering calcium oxide (CaO) from solutions that contain magnesium. Critics of the method argue that the results are too dependent on the person who performs the analysis. To demonstrate their concern, they arrange for the procedure to be run on ten samples, each containing a known amount of CaO. Nine of the ten tests are done by Chemist A; the other is run by Chemist B. Based on the results summarized in Table 11.2.4, does their criticism seem justified?

Table 11.2.4

Chemist	CaO Present (in mg), x	CaO Recovered (in mg), y
A	4.0	3.7
A	8.0	7.8
A	12.5	12.1
A	16.0	15.6
A	20.0	19.8
A	25.0	24.5
B	31.0	31.1
A	36.0	35.5
A	40.0	39.4
A	40.0	39.5

Figure 11.2.5 shows the scatterplot of y versus x. The linear function appears to fit all ten points exceptionally well, which would suggest that the critics' concerns are unwarranted. But look at the residual plot (Figure 11.2.6). The

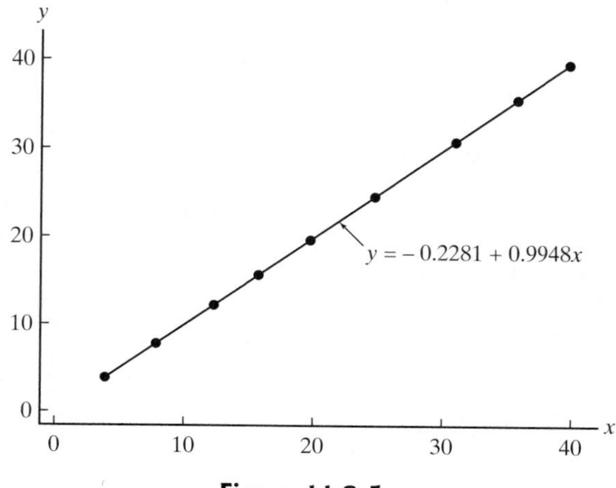

Figure 11.2.5

(Continued on next page)

(Case Study 11.2.3 continued)

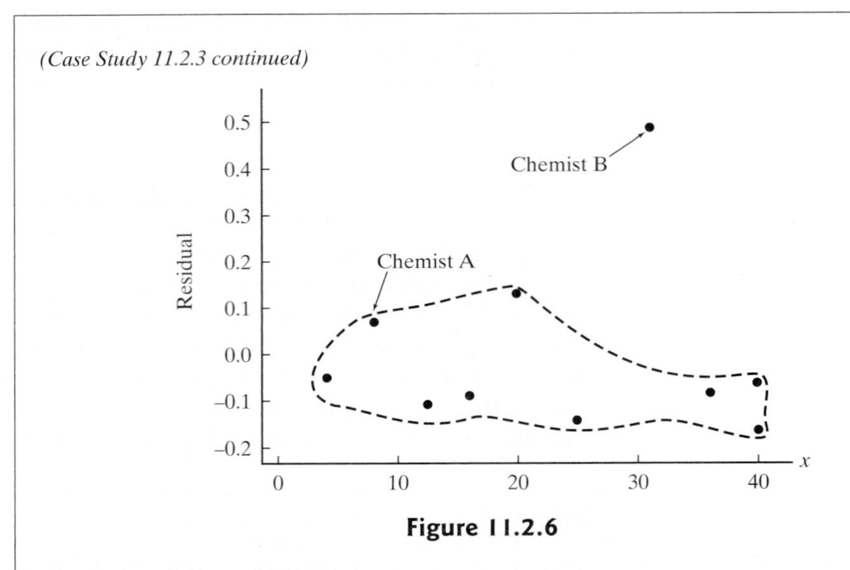

Figure 11.2.6

latter shows one point located noticeably further away from zero than any of the others, and that point corresponds to the one measurement attributed to Chemist B. So, while the scatterplot has failed to identify anything unusual about the data, the residual plot has focused on precisely the question the data set out to answer.

Does the appearance of the residual plot—specifically, the separation between the Chemist B data point and the nine Chemist A data points—"prove" that the output from the new procedure is dependent on the analyst? No, but it does speak to the magnitude of the disparity and, in so doing, provides the critics with at least a partial answer to their original question.

Questions

11.2.1. Crickets make their chirping sound by sliding one wing cover very rapidly back and forth over the other. Biologists have long been aware that there is a linear relationship between *temperature* and the *frequency* with which a cricket chirps, although the slope and y-intercept of the relationship vary from species to species. The following table lists fifteen frequency-temperature observations recorded for the striped ground cricket, *Nemobius fasciatus fasciatus* (135). Plot these data and find the equation of the least squares line, $y = a + bx$. Suppose a cricket of this species is observed to chirp eighteen times per second. What would be the estimated temperature?

For the data in the table, the sums needed are:

$$\sum_{i=1}^{15} x_i = 249.8 \qquad \sum_{i=1}^{15} x_i^2 = 4{,}200.56$$

$$\sum_{i=1}^{15} y_i = 1{,}200.6 \qquad \sum_{i=1}^{15} x_i y_i = 20{,}127.47$$

Observation Number	Chirps per Second, x	Temperature, y (°F)
1	20.0	88.6
2	16.0	71.6
3	19.8	93.3
4	18.4	84.3
5	17.1	80.6
6	15.5	75.2
7	14.7	69.7
8	17.1	82.0
9	15.4	69.4
10	16.2	83.3
11	15.0	79.6
12	17.2	82.6
13	16.0	80.6
14	17.0	83.5
15	14.4	76.3

11.2.2. The aging of whisky in charred oak barrels brings about a number of chemical changes that enhance its taste and darken its color. The following table shows the change in a whisky's proof as a function of the number of years it is stored (159).

Age, x (years)	Proof, y
0	104.6
0.5	104.1
1	104.4
2	105.0
3	106.0
4	106.8
5	107.7
6	108.7
7	110.6
8	112.1

(*Note*: The proof initially decreases because of dilution by moisture in the staves of the barrels.) Graph these data and draw in the least squares line.

11.2.3. As water temperature increases, sodium nitrate ($NaNO_3$) becomes more soluble. The following table (103) gives the number of parts of sodium nitrate that dissolve in one hundred parts of water.

Temperature (degrees Celsius), x	Parts Dissolved, y
0	66.7
4	71.0
10	76.3
15	80.6
21	85.7
29	92.9
36	99.4
51	113.6
68	125.1

Calculate the residuals, $y_1 - \hat{y}_1, \ldots, y_9 - \hat{y}_9$, and draw the residual plot. Does it suggest that fitting a straight line through these data would be appropriate? Use the following sums:

$$\sum_{i=1}^{9} x_i = 234 \qquad \sum_{i=1}^{9} y_i = 811.3$$

$$\sum_{i=1}^{9} x_i^2 = 10{,}144 \qquad \sum_{i=1}^{9} x_i y_i = 24{,}628.6$$

11.2.4. What, if anything, is unusual about the following residual plots?

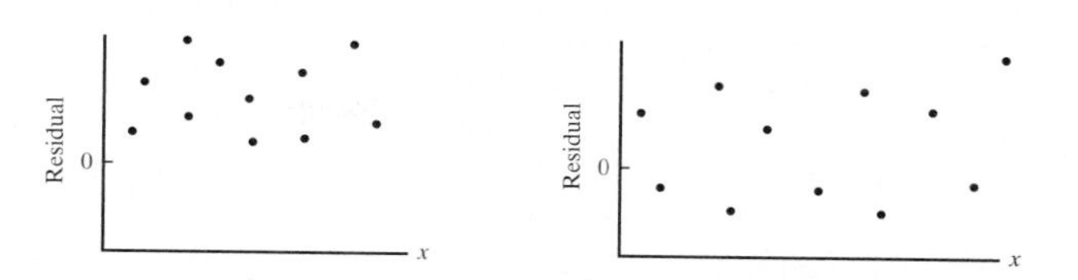

11.2.5. The following is the residual plot that results from fitting the equation $y = 6.0 + 2.0x$ to a set of $n = 10$ points. What, if anything, would be wrong with predicting that y will equal 30.0 when $x = 12$?

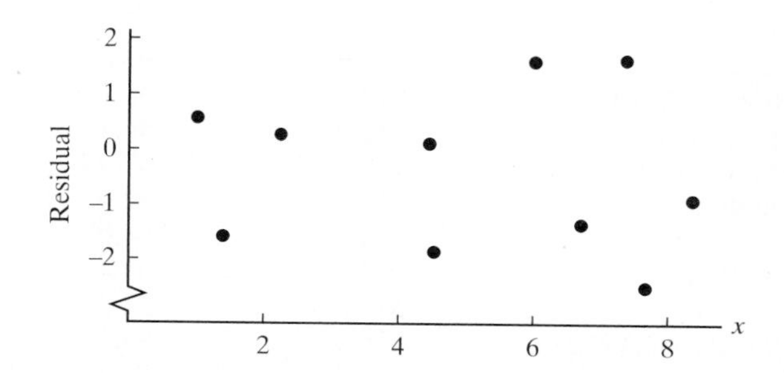

11.2.6. Would the following residual plot produced by fitting a least squares straight line to a set of $n = 13$ points cause you to doubt that the underlying xy-relationship is linear? Explain.

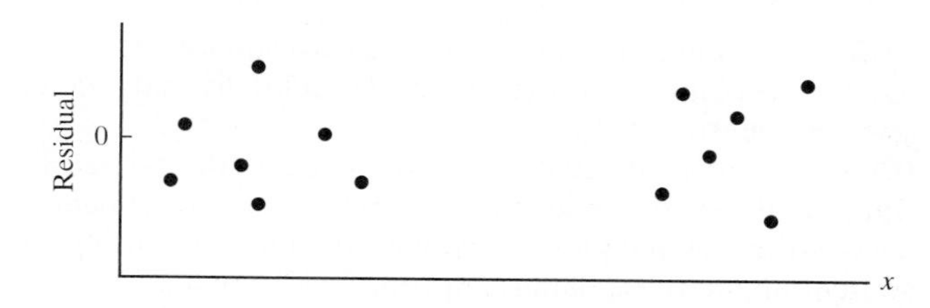

11.2.7. The relationship between school funding and student performance continues to be a hotly debated political and philosophical issue. Typical of the data available are the following figures, showing the per-pupil expenditures and graduation rate for 26 randomly chosen districts in Massachusetts.

Graph the data and superimpose the least squares line, $y = a + bx$. What would you conclude about the xy-relationship? Use the following sums:

$$\sum_{i=1}^{26} x_i = 360 \qquad \sum_{i=1}^{26} y_i = 2{,}256.6$$

$$\sum_{i=1}^{26} x_i^2 = 5{,}365.08 \qquad \sum_{i=1}^{26} x_i y_i = 31{,}402$$

District	Spending per Pupil (in 1000s), x	Graduation Rate
Dighton-Rehoboth	\$10.0	88.7
Duxbury	\$10.2	93.2
Tyngsborough	\$10.2	95.1
Lynnfield	\$10.3	94.0
Southwick-Tolland	\$10.3	88.3
Clinton	\$10.8	89.9
Athol-Royalston	\$11.0	67.7
Tantasqua	\$11.0	90.2
Ayer	\$11.2	95.5
Adams-Cheshire	\$11.6	75.2
Danvers	\$12.1	84.6
Lee	\$12.3	85.0
Needham	\$12.6	94.8
New Bedford	\$12.7	56.1
Springfield	\$12.9	54.4
Manchester Essex	\$13.0	97.9
Dedham	\$13.9	83.0
Lexington	\$14.5	94.0
Chatham	\$14.7	91.4
Newton	\$15.5	94.2
Blackstone Valley	\$16.4	97.2
Concord Carlisle	\$17.5	94.4
Pathfinder	\$18.1	78.6
Nantucket	\$20.8	87.6
Essex	\$22.4	93.3
Provincetown	\$24.0	92.3

Source: profiles.doe.mass.edu/state–report/ppx.aspx.

11.2.8. (a) Find the equation of the least squares straight line for the plant cover diversity/bird species diversity data given in Question 8.2.11.
(b) Make the residual plot associated with the least squares fit asked for in part (a). Based on the appearance of the residual plot, would you conclude that fitting a straight line to these data is appropriate? Explain.

11.2.9. A nuclear plant was established in Hanford, Washington, in 1943. Over the years, a significant amount of strontium 90 and cesium 137 leaked into the Columbia River. In a study to determine how much this radioactivity caused serious medical problems for those who lived along the river, public health officials created an index of radioactive exposure for nine Oregon counties in the vicinity of the river. As a covariate, cancer mortality was determined for each of the counties (40). The results are given in the table in the next column. For the nine (x_i, y_i)'s in the table,

$$\sum_{i=1}^{9} x_i = 41.56 \qquad \sum_{i=1}^{9} x_i^2 = 289.4222$$

$$\sum_{i=1}^{9} y_i = 1{,}416.1 \qquad \sum_{i=1}^{9} x_i y_i = 7{,}439.37$$

County	Index of Exposure	Cancer Mortality per 100,000
Umatilla	2.49	147.1
Morrow	2.57	130.1
Gilliam	3.41	129.9
Sherman	1.25	113.5
Wasco	1.62	137.5
Hood River	3.83	162.3
Portland	11.64	207.5
Columbia	6.41	177.9
Clatsop	8.34	210.3

Find the least squares straight line for these points. Also, construct the corresponding residual plot. Does it seem reasonable to conclude that x and y are linearly related?

11.2.10. Would you have any reservations about fitting the following data with a straight line? Explain.

x	y
3	20
7	37
5	29
1	10
10	59
12	69
6	39
11	58
8	47
9	48
2	18
4	29

11.2.11. When two closely related species are crossed, the progeny will tend to have physical traits that lie somewhere between those of the two parents. Whether a similar mixing occurs with behavioral traits was the focus of an experiment where the subjects were mallard and pintail ducks (162). A total of eleven males were studied; all were second-generation crosses. A rating scale was devised that measured the extent to which the plumage of each of the ducks resembled the plumage of the first generation's parents. A score of 0 indicated that the hybrid had the same appearance (phenotype) as a pure mallard; a score of 20 meant that the hybrid looked like a pintail. Similarly, certain behavioral traits were quantified and a second scale was constructed that ranged from 0 (completely mallard-like) to 15 (completely pintail-like). Use Theorem 11.2.1 and the following data to summarize the relationship between the plumage and behavioral indices. Does a linear model seem adequate?

Male	Plumage Index, x	Behavioral Index, y
R	7	3
S	13	10
D	14	11
F	6	5
W	14	15
K	15	15
U	4	7
O	8	10
V	7	4
J	9	9
L	14	11

11.2.12. Verify that the coefficients a and b of the least squares straight line are solutions of the matrix equation

$$\begin{pmatrix} n & \sum_{i=1}^{n} x_i \\ \sum_{i=1}^{n} x_i & \sum_{i=1}^{n} x_i^2 \end{pmatrix} \begin{pmatrix} a \\ b \end{pmatrix} = \begin{pmatrix} \sum_{i=1}^{n} y_i \\ \sum_{i=1}^{n} x_i y_i \end{pmatrix}$$

11.2.13. Prove that a least squares straight line must necessarily pass through the point $(\bar{x}, \bar{y})$.

11.2.14. In some regression situations, there are *a priori* reasons for assuming that the xy-relationship being approximated passes through the origin. If so, the equation to be fit to the (x_i, y_i)'s has the form $y = bx$. Use the least squares criterion to show that the "best" slope in that case is given by

$$b = \frac{\sum_{i=1}^{n} x_i y_i}{\sum_{i=1}^{n} x_i^2}$$

11.2.15. One of the most startling scientific discoveries of the twentieth century was the announcement in 1929 by the American astronomer Edwin Hubble that the universe is expanding. If v is a galaxy's recession velocity (relative to that of any other galaxy) and d is its distance (from that same galaxy), Hubble's law states that

$$v = Hd$$

where H is known as Hubble's constant. (To cosmologists, Hubble's constant is a critically important number—its reciprocal, after being properly scaled, is an estimate of the age of the universe.) The following are distance and velocity measurements made on eleven galactic clusters (23). Use the formula cited in Question 11.2.14 to estimate Hubble's constant.

Cluster	Distance (millions of light-years)	Velocity (thousands of miles/sec)
Virgo	22	0.75
Pegasus	68	2.4
Perseus	108	3.2
Coma Berenices	137	4.7
Ursa Major No. 1	255	9.3
Leo	315	12.0
Corona Borealis	390	13.4
Gemini	405	14.4
Bootes	685	24.5
Ursa Major No. 2	700	26.0
Hydra	1100	38.0

11.2.16. Given a set of n linearly related points, $(x_1, y_1), (x_2, y_2), \ldots,$ and (x_n, y_n), use the least squares criterion to find formulas for

(a) a if the slope of the xy-relationship is known to be b^*.

(b) b if the y-intercept of the xy-relationship is known to be a^*.

11.2.17. Among the problems faced by job seekers wanting to reenter the workforce, eroded skills and outdated backgrounds are two of the most difficult to overcome. Knowing that, employers are often wary of hiring individuals who have spent lengthy periods of time away from the job. The following table shows the percentages of hospitals willing to rehire medical technicians who have been away from that career for x years (145). It can be argued that the fitted line should necessarily have a y-intercept of 100 because no employer would refuse to hire someone (due to outdated skills) whose career had not been interrupted at all—that is, applicants for whom $x = 0$. Under that assumption, use the result from Question 11.2.16 to fit these data with the model $y = 100 + bx$.

Years of Inactivity, x	Percent of Hospitals Willing to Hire, y
0.5	100
1.5	94
4	75
8	44
13	28
18	17

11.2.18. A graph of the luxury suite data in Question 8.2.5 suggests that the xy-relationship is linear. Moreover, it makes sense to constrain the fitted line to go through the origin, since $x = 0$ suites will necessarily produce $y = 0$ revenue.

(a) Find the equation of the least squares line, $y = bx$. (*Hint*: Recall Question 11.2.14.)

(b) How much revenue would 120 suites be expected to generate?

11.2.19. Set up (but do not solve) the equations necessary to determine the least squares estimates for the trigonometric model,

$$y = a + bx + c \sin x$$

Assume that the data consist of the random sample (x_1, y_1), (x_2, y_2), ..., and (x_n, y_n).

Nonlinear Models

Obviously, not all xy-relationships can be adequately described by straight lines. Curvilinear relationships of all sorts can be found in every field of endeavor. Many of these nonlinear models, though, can still be fit using Theorem 11.2.1, provided the data have been initially "linearized" by a suitable transformation.

Exponential Regression Suppose the relationship between two variables is best described by an exponential function of the form

$$y = ae^{bx} \tag{11.2.1}$$

Depending on the value of b, Equation 11.2.1 will look like one of the graphs pictured in Figure 11.2.7. Those curvilinear shapes notwithstanding, though, there is a *linear* model also related to Equation 11.2.1.

Figure 11.2.7

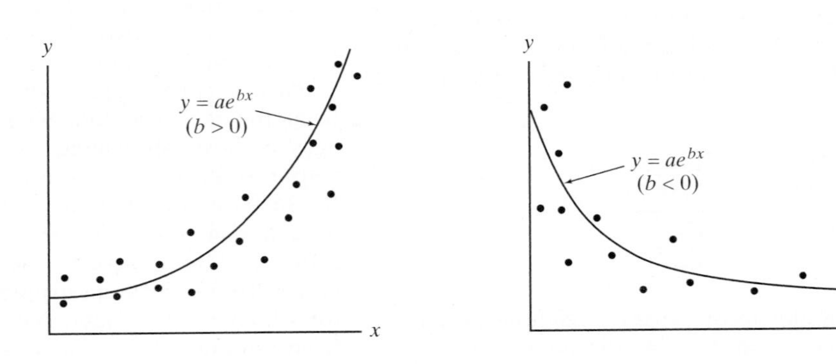

If $y = ae^{bx}$, it is necessarily true that

$$\ln y = \ln a + bx \tag{11.2.2}$$

which implies that *ln y and x have a linear relationship*. That being the case, the formulas of Theorem 11.2.1 *applied to x and ln y* should yield the slope and y-intercept of Equation 11.2.2.

Specifically,

$$b = \frac{n \sum_{i=1}^{n} x_i \ln y_i - \left(\sum_{i=1}^{n} x_i\right)\left(\sum_{i=1}^{n} \ln y_i\right)}{n \sum_{i=1}^{n} x_i^2 - \left(\sum_{i=1}^{n} x_i\right)^2}$$

and

$$\ln a = \frac{\sum_{i=1}^{n} \ln y_i - b \sum_{i=1}^{n} x_i}{n}$$

Comment Transformations that induce linearity often require that the slope and/or y-intercept of the transformed model be transformed "back" to the original model. Here, for example, Theorem 11.2.1 leads to a formula for $\ln a$, which means that the constant a appearing in the original exponential model is evaluated by calculating $e^{\ln a}$.

Case Study 11.2.4

Beginning in the 1970s, computers have steadily decreased in size as they have grown in power. The ability to have more computing potential in a four-pound laptop than in a mainframe of the 1970s is a result of engineers squeezing more and more transistors onto silicon chips. The rate at which this miniaturization occurs is known as Moore's law, after Gordon Moore, one of the founders of Intel Corporation. His prediction, first articulated in 1965, was that the number of transistors per chip would double every eighteen months.

Table 11.2.5 lists some of the growth benchmarks—namely, the number of transistors per chip—associated with the Intel chips marketed over the twenty-year period from 1975 through 1995. Based on these figures, is it believable that chip capacity is, in fact, doubling at a fixed rate (meaning that Equation 11.2.1 applies)? And if so, how close is the actual doubling time to Moore's prediction of eighteen months?

A plot of y versus x shows that their relationship is certainly not linear (see Figure 11.2.8). The scatterplot more closely resembles the graph of $y = ae^{bx}$ when $b > 0$, as shown in Figure 11.2.7.

Table 11.2.5

Chip	Year	Years after 1975, x	Transistors per Chip, y
8080	1975	0	4,500
8086	1978	3	29,000
80286	1982	7	90,000
80386	1985	10	229,000
80486	1989	14	1,200,000
Pentium	1993	18	3,100,000
Pentium Pro	1995	20	5,500,000

Source: en.wikipedia.org/wiki/Transistor—count.

Table 11.2.6 shows the calculation of the sums required to evaluate the formulas for b and $\ln a$. Here the slope and the y-intercept

(Continued on next page)

(Case Study 11.2.4 continued)

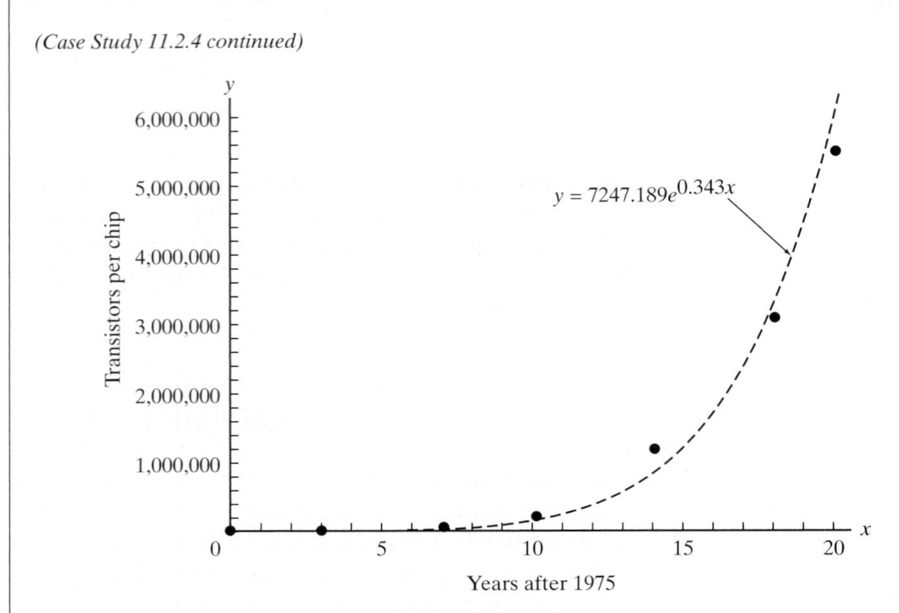

Figure 11.2.8

Table 11.2.6

Years after 1975, x_i	x_i^2	Transistors per Chip, y_i	$\ln y_i$	$x_i \cdot \ln y_i$
0	0	4,500	8.41183	0
3	9	29,000	10.27505	30.82515
7	49	90,000	11.40756	79.85292
10	100	229,000	12.34148	123.41480
14	196	1,200,000	13.99783	195.96962
18	324	3,100,000	14.94691	269.04438
20	400	5,500,000	15.52026	310.40520
72	1078		86.90093	1009.51207

of the linearized model (Equation 11.2.2) are 0.342810 and 8.888369, respectively:

$$b = \frac{7(1009.51207) - 72(86.90093)}{7(1078) - (72)^2}$$

$$= 0.342810$$

and

$$\ln a = \frac{86.9093 - (0.342810)(72)}{7}$$

$$= 8.888369$$

Therefore,

$$a = e^{\ln a} = e^{8.888369} = 7247.189$$

(Continued on next page)

which implies that the best-fitting exponential model describing Intel's technological advances in chip design has the equation

$$y = 7247.189e^{0.343x}$$

(see Figure 11.2.8).

To compare Equation 11.2.1 to Moore's "eighteen-month doubling time" prediction requires that we write $y = 7247.189e^{0.343x}$ in the form $y = 7247.189(2)^x$. But

$$e^{0.343} = 2^{0.495}$$

so another way to express the fitted curve would be

$$y = 7247.189(2^{0.495x}) \qquad (11.2.3)$$

In Equation 11.2.3, though, y doubles when $2^{0.495x} = 2$, or, equivalently, when $0.495x = 1$, which implies that *2.0 years* is the empirically determined technology doubling time, a pace not too much slower than Moore's prediction of eighteen months.

About the Data In April of 2005, Gordon Moore pronounced his law dead. He said, "It can't continue forever. The nature of exponentials is that you push them out and eventually disaster happens." If by "disaster" he meant that technology often makes a quantum leap, moving well beyond what an extrapolated law could predict, he was quite correct. Indeed, he could have made this declaration in 2003. By that year, the Itanium 2 featured 220,000,000 transistors on a chip, whereas the model of the case study predicts the number to be only

$$y = 7247.189e^{0.343(28)} = 107,432,032$$

(In the equation, $x = 2003 - 1975 = 28$.)

Logarithmic Regression Another frequently encountered curvilinear model that can be easily linearized is the equation

$$y = ax^b \qquad (11.2.4)$$

Taking the common log of both sides of Equation 11.2.4 gives

$$\log y = \log a + b \log x$$

which implies that *log y is linear with log x*. Therefore,

$$b = \frac{n\sum_{i=1}^{n} \log x_i \cdot \log y_i - \left(\sum_{i=1}^{n} \log x_i\right)\left(\sum_{i=1}^{n} \log y_i\right)}{n\sum_{i=1}^{n} (\log x_i)^2 - \left(\sum_{i=1}^{n} \log x_i\right)^2}$$

and

$$\log a = \frac{\sum_{i=1}^{n} \log y_i - b\sum_{i=1}^{n} \log x_i}{n}$$

Regressions of this type have slower growth rates than exponential models and are particularly useful in describing biological and engineering phenomena.

Case Study 11.2.5

Among mammals, the relationship between the age at which an animal develops locomotion and the age at which it first begins to play has been widely studied. Table 11.2.7 lists "onset" times for locomotion and for play in eleven different species (41). Graphed, the data show a pattern that suggests that $y = ax^b$ would be a good function for modeling the xy-relationship (see Figure 11.2.9).

Table 11.2.7

Species	Locomotion Begins, x (days)	Play Begins, y (days)
Homo sapiens	360	90
Gorilla gorilla	165	105
Felis catus	21	21
Canis familiaris	23	26
Rattus norvegicus	11	14
Turdus merula	18	28
Macaca mulatta	18	21
Pan troglodytes	150	105
Saimiri sciurens	45	68
Cercocebus alb.	45	75
Tamiasciureus hud.	18	46

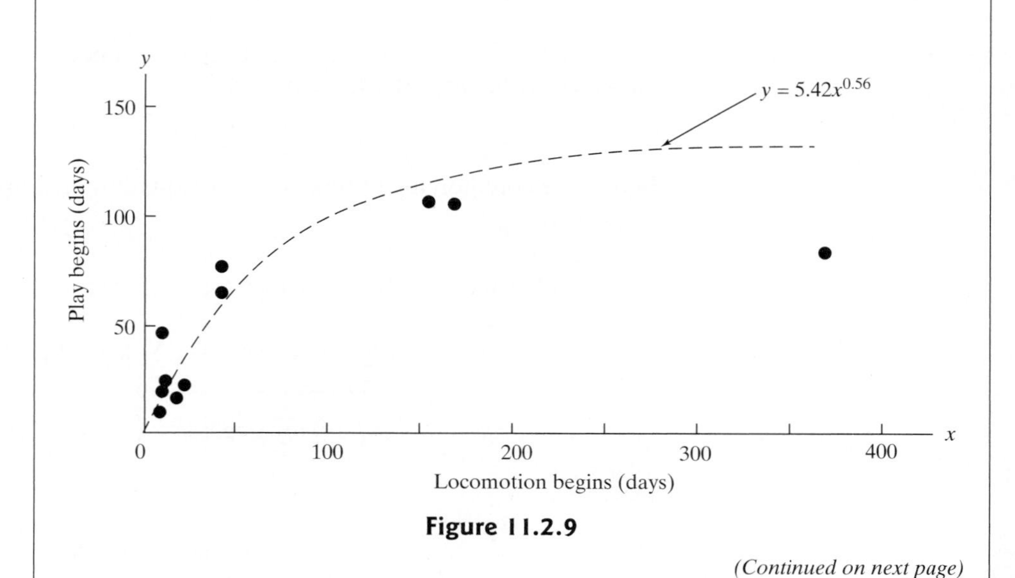

Figure 11.2.9

(Continued on next page)

Table 11.2.8

x_i	$\log x_i$	y_i	$\log y_i$	$(\log x_i)^2$	$\log x_i \log y_i$
360	2.55630	90	1.95424	6.53467	4.99562
165	2.21748	105	2.02119	4.91722	4.48195
21	1.32222	21	1.32222	1.74827	1.74827
23	1.36173	26	1.41497	1.85431	1.92681
11	1.04139	14	1.14613	1.08449	1.19357
18	1.25527	28	1.44716	1.57570	1.81658
18	1.25527	21	1.32222	1.57570	1.65974
150	2.17609	105	2.02119	4.73537	4.39829
45	1.65321	68	1.83251	2.73310	3.02952
45	1.65321	75	1.87506	2.73310	3.09987
18	1.25527	46	1.66276	1.57570	2.08721
	17.74744		18.01965	31.06763	30.43743

The sums and sums of squares necessary to find a and b are calculated in Table 11.2.8. Substituting into the formulas on p. 547 for the slope and y-intercept of the linearized model gives

$$b = \frac{11(30.43743) - (17.74744)(18.01965)}{11(31.06763) - (17.74744)^2}$$

$$= 0.56$$

and

$$\log a = \frac{18.01965 - (0.56)(17.74744)}{11}$$

$$= 0.73364$$

Therefore, $a = 10^{0.73364} = 5.42$, and the equation describing the xy-relationship is $y = 5.42x^{0.56}$ (see Figure 11.2.9).

Logistic Regression *Growth* is a fundamental characteristic of organisms, institutions, and ideas. In biology, it might refer to the change in size of a *Drosophila* population; in economics, to the proliferation of global markets; in political science, to the gradual acceptance of tax reform. Prominent among the many growth models capable of describing situations of this sort is the logistic equation

$$y = \frac{L}{1 + e^{a+bx}} \tag{11.2.5}$$

where a, b, and L are constants. For different values of a and b, Equation 11.2.5 generates a variety of S-chaped curves.

To linearize Equation 11.2.5, we start with its reciprocal:

$$\frac{1}{y} = \frac{1 + e^{a+bx}}{L}$$

Therefore,

$$\frac{L}{y} = 1 + e^{a+bx}$$

and

$$\frac{L - y}{y} = e^{a+bx}$$

Equivalently,

$$\ln\left(\frac{L - y}{y}\right) = a + bx$$

which implies that $\ln\left(\frac{L - y}{y}\right)$ *is linear with* x.

Comment The parameter L is interpreted as the limit to which y is converging as x increases. In practice, L is often estimated simply by plotting the data and "eye-balling" the y-asymptote.

Case Study 11.2.6

Biological organisms often exhibit exponential growth. However, in some cases, that rapid rate of growth cannot be sustained. Such factors as lack of nutrition to support a large population or the buildup of toxins limit the rate of growth. In such cases the curve begins concave up, inflects at some point, and becomes concave down and asymptotic to a limit.

A now-classical experiment provides data with the above characteristics. Carlson (20) measured the amount of biomass of brewer's yeast (*Saccharomyces Cerevisiae*) at one-hour intervals. Table 11.2.9 shows the results.

Table 11.2.9

Hour	Yeast Count	Hour	Yeast Count
0	9.6	9	441.0
1	18.3	10	513.3
2	29.0	11	559.7
3	47.2	12	594.8
4	71.1	13	629.4
5	119.1	14	640.8
6	174.6	15	651.1
7	257.3	16	655.9
8	350.7	17	659.6

The scatterplot for these eighteen data points has a definite S-shaped appearance (see Figure 11.2.10), which makes Equation 11.2.5 a good candidate for modeling the xy-relationship. The limit to which the population is converging appears to be about *700*. Quantify the population/time relationship by fitting a logistic equation to these data. Let $L = 700$.

(Continued on next page)

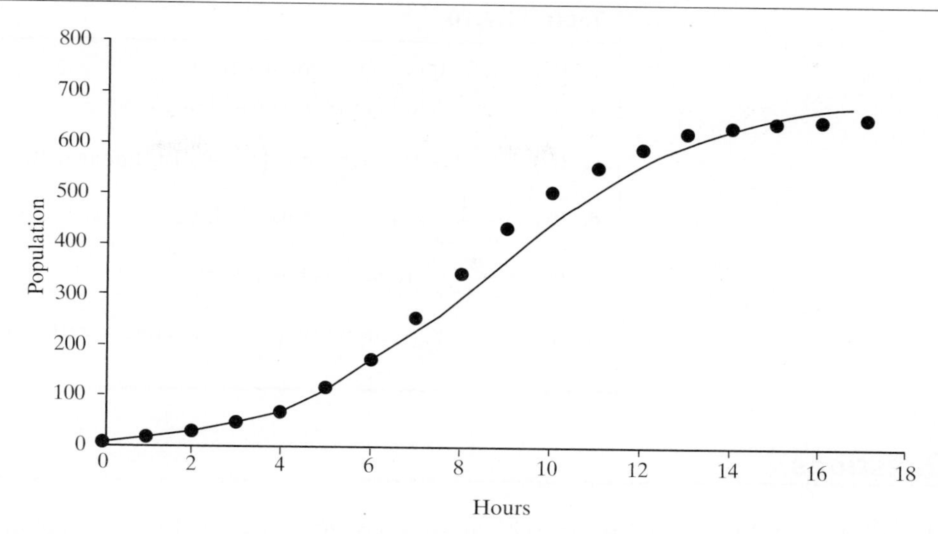

Figure 11.2.10

The form of the linearized version of Equation 11.2.5 requires that we find the following sums:

$$\sum_{i=1}^{18} x_i = 153, \quad \sum_{i=1}^{18} \ln\left(\frac{700 - y_i}{y_i}\right) = 1.75603, \quad \sum_{i=1}^{18} x_i^2 = 1785, \text{ and}$$

$$\sum_{i=1}^{18} x_i \cdot \ln\left(\frac{700 - y_i}{y_i}\right) = -197.40071$$

Substituting $\ln\left(\frac{700-y_i}{y_i}\right)$ for y_i into the formulas for a and b in Theorem 11.2.1 gives

$$b = \frac{18(-197.40071) - (153)(1.75603)}{18(1785) - (153)^2} = -0.4382$$

and

$$a = \frac{1.75603 - (-0.4382)(153)}{18} = 3.822$$

so the best-fitting logistic curve has equation

$$y = \frac{700}{1 + e^{3.822 - 0.4382x}}$$

Other Curvilinear Models While the exponential, logarithmic, and logistic equations are three of the most common curvilinear models, there are several others that deserve mention as well. Table 11.2.10 lists a total of six nonlinear equations, including the three already described. Along with each is the particular transformation that reduces the equation to a linear form. Proofs for parts (d), (e), and (f) will be left as exercises.

Table 11.2.10

a. If $y = ae^{bx}$, then $\ln y$ is linear with x.
b. If $y = ax^b$, then $\log y$ is linear with $\log x$.
c. If $y = L/(1 + e^{a+bx})$, then $\ln\left(\frac{L-y}{y}\right)$ is linear with x.
d. If $y = \frac{1}{a+bx}$, then $\frac{1}{y}$ is linear with x.
e. If $y = \frac{x}{a+bx}$, then $\frac{1}{y}$ is linear with $\frac{1}{x}$.
f. If $y = 1 - e^{-x^b/a}$, then $\ln \ln\left(\frac{1}{1-y}\right)$ is linear with $\ln x$.

Questions

11.2.20. Radioactive gold (^{195}Au-aurothiomalate) has an affinity for inflamed tissues and is sometimes used as a tracer to diagnose arthritis. The data in the following table (62) come from an experiment investigating the length of time and the concentrations that ^{195}Au-aurothiomalate is retained in a person's blood. Listed are the serum gold concentrations found in ten blood samples taken from patients given an initial dose of 50 mg. Follow-up readings were made at various times, ranging from one to seven days after injection. In each case, the retention is expressed as a percentage of the patient's day-zero serum gold concentration.

Days after Injection, x	Serum Gold % Concentration, y
1	94.5
1	86.4
2	71.0
2	80.5
2	81.4
3	67.4
5	49.3
6	46.8
6	42.3
7	36.6

(a) Fit an exponential curve to these data.
(b) Estimate the half-life of ^{195}Au-aurothiomalate; that is, how long does it take for half the gold to disappear from a person's blood?

If x denotes days after injection and y denotes serum gold % concentration, then $\sum_{i=1}^{10} x_i = 35$, $\sum_{i=1}^{10} x_i^2 = 169$, $\sum_{i=1}^{10} \ln\, y_i = 41.35720$, and $\sum_{i=1}^{10} x_i \ln\, y_i = 137.97415$.

11.2.21. The growth of the federal debt is one of the characteristic features of the U.S. economy. The rapidity of the increases from 1996 to 2006, as shown in the table below, suggests an exponential model.

Year	Years after 1995, x	Gross Federal Debt (in \$ trillions), y
1996	1	5.181
1997	2	5.396
1998	3	5.478
1999	4	5.606
2000	5	5.629
2001	6	5.770
2002	7	6.198
2003	8	6.760
2004	9	7.355
2005	10	7.905
2006	11	8.451

Source: whitehouse.gov/omb/budget/fy2008/pdf/hist.pdf.

(a) Find the best-fitting exponential curve, using the method of least squares together with an appropriate linearizing transformation. Use the sums: $\sum_{i=1}^{20} \ln\, y_i = 20.16825$ and $\sum_{i=1}^{20} x_i \cdot \ln\, y_i = 126.33786$.
(b) The official Office of Management and Budget prediction for 2007 was $9 trillion. Compare this figure to the projection using the model from part (a).
(c) Even though the model of part (a) is considered "good" by a criterion to be given in Section 11.4 (*r squared*), plot the residuals and consider what they say about the exponential fit.

11.2.22. Used cars are often sold wholesale at auctions, and from these sales, retail sales prices are recommended.

The following table gives the recommended prices in 2009 for a four-door manual transmission Toyota Corolla based on the age of the car.

Age (in years), x	Suggested retail price, y
1	$14,680
2	12,150
3	11,215
4	10,180
5	9,230
6	8,455
7	7,730
8	6,825
9	6,135
10	5,620

Source: www.bb.com.

(a) Fit these data with a model of the form $y = ae^{bx}$. Graph the (x_i, y_i)'s and superimpose the least squares exponential curve.

(b) What would you predict the retail price of an eleven-year-old Toyota Corolla to be?

(c) The price of a new Corolla in 2009 was $16,200. Is that figure consistent with the widely held belief that a new car depreciates substantially the moment it is purchased? Explain.

11.2.23. The stock market showed steady and significant growth during the period from 1981 to 2000. This growth was reflected in the Dow Jones Industrial Average. The table gives the Dow Jones average (rounded to the nearest whole number) for the opening of the stock market in January for the years 1981 to 2000.

Years after 1981, x	Dow Jones Industrial Average, y
0	947
1	871
2	1,076
3	1,221
4	1,287
5	1,571
6	2,158
7	1,958
8	2,342
9	2,591
10	2,736
11	3,223
12	3,310
13	3,978
14	3,844
15	5,395
16	6,813
17	7,907
18	9,359
19	10,941

Source: finance.yahoo.com/of/hp?s=%5EDJI.

Use the fact that $\sum_{i=1}^{20} \ln y_i = 158.58560$ and $\sum_{i=1}^{20} x_i \cdot \ln y_i = 1591.99387$ to fit the data with an exponential model.

11.2.24. Suppose a set of n (x_i, y_i)'s are measured on a phenomenon whose theoretical xy-relationship is of the form $y = ae^{bx}$.

(a) Show that $\frac{dy}{dx} = by$ implies that $y = ae^{bx}$.

(b) On what kind of graph paper would the (x_i, y_i)'s show a linear relationship?

11.2.25. In 1959, the Ise Bay typhoon devastated parts of Japan. For seven metropolitan areas in the storm's path, the following table gives the number of homes damaged as a function of peak wind gust (118). Show that a function of the form $y = ax^b$ provides a good model for the data.

City	Peak Wind Gust (hundred mph), x	Numbers of Damaged Homes (in thousands), y
A	0.98	25.000
B	0.74	0.950
C	1.12	200.000
D	1.34	150.000
E	0.87	0.940
F	0.65	0.090
G	1.39	260.000

Use the following sums:

$$\sum_{i=1}^{7} \log x_i = -0.067772 \qquad \sum_{i=1}^{7} \log y_i = 7.1951$$

$$\sum_{i=1}^{7} (\log x_i)^2 = 0.0948679 \qquad \sum_{i=1}^{7} (\log x_i)(\log y_i) = 0.92314$$

11.2.26. Studies have shown that certain ants in a colony are assigned foraging duties, which require them to come and go from the colony on a regular basis. Furthermore, if y is the colony size and x is the number of ants that forage, the relationship between y and x has the form $y = ax^b$, where a and b vary from species to species. Once the parameter values have been estimated for a particular kind of ant, biologists can count the (relatively small) number of ants that forage and then use the regression equation to estimate the (much larger) number of ants living in the colony. The table on p. 554 gives the results of a "calibration" study done on the red wood ant (*Formica polyctena*): Listed are the actual colony sizes, y, and the foraging sizes, x, recorded for fifteen of their colonies (94).

(a) Find a and b using the sums below.
(b) If the number of red wood ants seen foraging is 2500, what would be a reasonable estimate for the size of the colony from which they came?

$$\sum_{i=1}^{15} \log x_i = 41.77441 \qquad \sum_{i=1}^{15} \log y_i = 52.79857$$

$$\sum_{i=1}^{15} (\log x_i)^2 = 126.60450 \qquad \sum_{i=1}^{15} \log x_i \cdot \log y_i = 156.03811$$

Foraging Size, x	Colony Size, y
45	280
74	222
118	288
70	601
220	1,205
823	2,769
647	2,828
446	3,229
765	3,762
338	7,551
611	8,834
4,119	12,584
850	12,605
11,600	34,661
64,512	139,043

11.2.27. Over the years, many efforts have been made to demonstrate that the human brain is appreciably different in structure from the brains of lower-order primates. In point of fact, such differences in gross anatomy are disconcertingly difficult to discern. The following are the average areas of the striate cortex (x) and the prestriate cortex (y) found for humans and for three species of chimpanzees (129).

Primate	Area: Striate Cortex, x (mm^2)	Area: Prestriate Cortex, y (mm^2)
Homo	2613	7838
Pongo	1876	2864
Cercopithecus	933	1334
Galago	78.9	40.8

Plot the data and superimpose the least squares curve, $y = ax^b$.

11.2.28. Years of experience buying and selling commercial real estate have convinced many investors that the value of land zoned for business (y) is inversely related to its distance (x) from the center of town—that is, $y = a + b \cdot \frac{1}{x}$. If that suspicion is correct, what should be the appraised value of a piece of property located $\frac{1}{4}$ mile from the town square, based on the sales listed below?

Land Parcel	Distance from Center of City (in thousand feet), x	Value (in thousands), y
H1	1.00	\$20.5
B6	0.50	42.7
Q4	0.25	80.4
L4	2.00	10.5
T7	4.00	6.1
D9	6.00	6.0
E4	10.00	3.5

11.2.29. Verify the claims made in parts (d), (e), and (f) of Table 11.2.10—that is, prove that the transformations cited will linearize the original models.

11.2.30. During the 1960s, when the Cold War was fueling an arms race between the Soviet Union and the United States, the number of American intercontinental ballistic missiles (ICBMs) rose from 18 to 1054 (9). Moreover, the sizes of the ICBM stockpile during that decade had an S-shaped pattern, suggesting that the logistic curve would provide a good model. Graph the following data, and approximate the xy-relationship with the function $y = \frac{L}{1 + e^{a+bx}}$. Assume that $L = 1055$.

Years	Years after 1959, x	Number of ICBMs, y
1960	1	18
1961	2	63
1962	3	294
1963	4	424
1964	5	834
1965	6	854
1966	7	904
1967	8	1054
1968	9	1054
1969	10	1054

11.2.31. The following table shows a portion of the results from a clinical trial investigating the effectiveness of a monoamine oxidase inhibitor as a treatment for depression (207). The relationship between y, the percentage of subjects showing improvement, and x, the patient's age, appears to be S-shaped. Graph the data and superimpose a graph of the least squares curve $y = \frac{L}{1 + e^{a+bx}}$. Take L to be 60.

Age Group	Age Mid-Point, x	% Improved, y	$\ln\left(\frac{60-y}{y}\right)$
[28, 32)	30	11	1.49393
[32, 36)	34	14	1.18958
[36, 40)	38	19	0.76913
[40, 44)	42	32	−0.13353
[44, 48)	46	42	−0.84730
[48, 52)	50	48	−1.38629
[52, 56)	54	50	−1.60944
[56, 60)	58	52	−1.87180

11.3 The Linear Model

Section 11.2 views the problem of "curve fitting" from a purely geometrical perspective. The observed (x_i, y_i)'s are assumed to be nothing more than points in the xy-plane, devoid of any statistical properties. It is more realistic, though, to think of each y as the value recorded for a random variable Y, meaning that a *distribution* of possible y-values is associated with every value of x.

Consider, for example, the connecting rod weights analyzed in Case Study 11.2.1. The first rod listed in Table 11.2.1 had an initial weight of $x = 2.745$ oz. and, after the tooling process was completed, a finished weight of $y = 2.080$ oz. It does not follow from that one observation, of course, that an initial weight of 2.745 oz. necessarily leads to a finished weight of 2.080 oz. Common sense tells us that the tooling process will not always have exactly the same effect, even on rods having the same initial weight. Associated with each x, then, there *will* be a range of possible y-values. The symbol $f_{Y|x}(y)$ is used to denote the pdfs of these "conditional" distributions.

Definition 11.3.1. Let $f_{Y|x}(y)$ denote the pdf of the random variable Y for a given value x, and let $E(Y \mid x)$ denote the expected value associated with $f_{Y|x}(y)$. The function

$$y = E(Y \mid x)$$

is called the *regression curve of Y on x*.

Example 11.3.1

Suppose that corresponding to each value of x in the interval $0 \le x \le 1$ is a distribution of y-values having the pdf

$$f_{Y|x}(y) = \frac{x + y}{x + \frac{1}{2}}, \quad 0 \le y \le 1; \quad 0 \le x \le 1$$

Find and graph the regression curve of Y on x.

Notice, first of all, that for any x between 0 and 1, $f_{Y|x}(y)$ does qualify as a pdf:

1. $f_{Y|x}(y) \ge 0, \quad$ for $\quad 0 \le y \le 1 \quad$ and any $\quad 0 \le x \le 1$
2. $\int_0^1 f_{Y|x}(y)\, dy = \int_0^1 \left(\frac{x + y}{x + \frac{1}{2}}\right) dy = 1$

Moreover,

$$E(Y \mid x) = \int_0^1 y \cdot f_{Y|x}(y)\, dy = \int_0^1 y \cdot \frac{x+y}{x+\frac{1}{2}}\, dy$$

$$= \left[\frac{xy^2}{2\left(x+\frac{1}{2}\right)} + \frac{y^3}{3\left(x+\frac{1}{2}\right)} \right]\Bigg|_0^1$$

$$= \frac{3x+2}{6x+3}, \quad 0 \le x \le 1$$

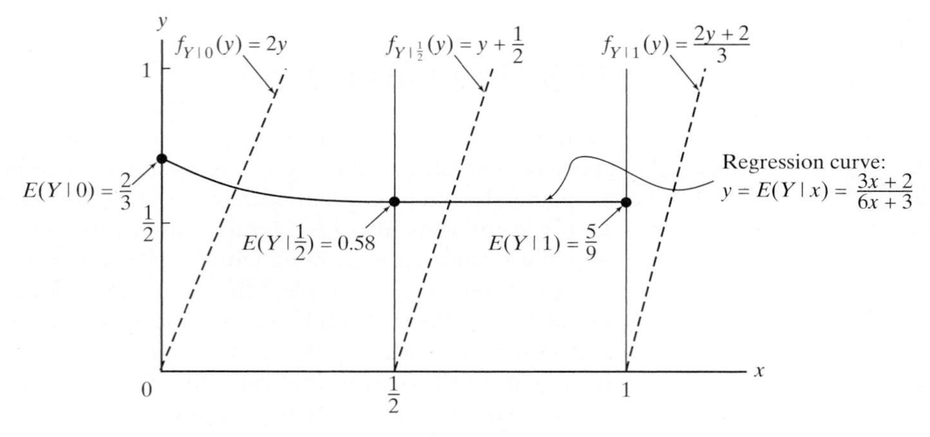

Figure 11.3.1

Figure 11.3.1 shows the regression curve, $y = E(Y \mid x) = \dfrac{3x+2}{6x+3}$, together with three of the conditional distributions—$f_{Y|0}(y) = 2y$, $f_{Y|\frac{1}{2}}(y) = y + \frac{1}{2}$, and $f_{y|1}(y) = \dfrac{2y+2}{3}$. The $f_{Y|x}(y)$'s, of course, should be visualized as coming out of the plane of the paper. ■

A Special Case

Definition 11.3.1 introduces the notion of a regression curve in the most general of contexts. In practice, there is one special case of the function $y = E(Y \mid x)$ that is particularly important. Known as the *simple linear model,* it makes four assumptions:

1. $f_{Y|x}(y)$ is a normal pdf for all x.
2. The standard deviation, σ, associated with $f_{Y|x}(y)$ is the same for all x.
3. The means of all the conditional Y distributions are collinear—that is,

$$y = E(Y \mid x) = \beta_0 + \beta_1 x$$

4. All of the conditional distributions represent independent random variables. (See Figure 11.3.2.)

Figure 11.3.2

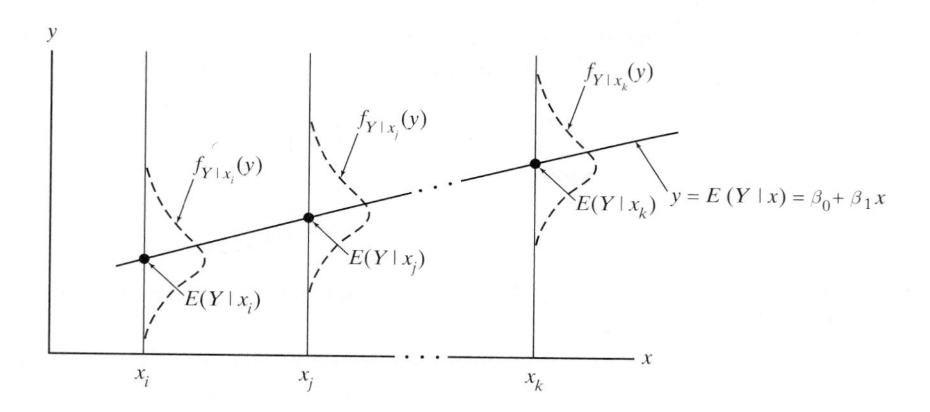

Estimating the Linear Model Parameters

Implicit in the simple linear model are three parameters—β_0, β_1, and σ^2. Typically, all three will be unknown and need to be estimated. Since the model assumes a probability structure for the Y-variable, estimates can be obtained using the method of maximum likelihood, as opposed to the method of least squares that we saw in Section 11.2. (Maximum likelihood estimates are preferable to least squares estimates because the former have probability distributions that can be used to set up hypothesis tests and confidence intervals.)

Comment It would be entirely consistent with the notation used previously to denote the sample in Theorem 11.3.1 as $(x_1, y_1), (x_2, y_2), \ldots,$ and (x_n, y_n). To emphasize the important distinction, though, between the (lack of) assumptions on the y_i's made in Section 11.2 and the conditional pdfs $f_{Y|x}(y)$ introduced in Definition 11.3.1, we will use random variable notation to write linear model data as $(x_1, Y_1), (x_2, Y_2), \ldots,$ and (x_n, Y_n).

Theorem 11.3.1 *Let $(x_1, Y_1), (x_2, Y_2), \ldots,$ and (x_n, Y_n) be a set of points satisfying the simple linear model, $E(Y|x) = \beta_0 + \beta_1 x$. The maximum likelihood estimators for β_0, β_1, and σ^2 are given by*

$$\hat{\beta}_1 = \frac{n\sum_{i=1}^{n} x_i Y_i - \left(\sum_{i=1}^{n} x_i\right)\left(\sum_{i=1}^{n} Y_i\right)}{n\left(\sum_{i=1}^{n} x_i^2\right) - \left(\sum_{i=1}^{n} x_i\right)^2}$$

$$\hat{\beta}_0 = \bar{Y} - \hat{\beta}_1 \bar{x}$$

and

$$\hat{\sigma}^2 = \frac{1}{n}\sum_{i=1}^{n}(Y_i - \hat{Y}_i)^2$$

where $\hat{Y}_i = \hat{\beta}_0 + \hat{\beta}_1 x_i$, $i = 1, \ldots, n$.

Proof Since each Y_i is assumed to be normally distributed with mean equal to $\beta_0 + \beta_1 x_i$ and variance equal to σ^2, the sample's likelihood function, L, is the product

$$L = \prod_{i=1}^{n} f_{Y|x_i}(y_i)$$

$$= \prod_{i=1}^{n} \frac{1}{\sqrt{2\pi}\sigma} e^{-\frac{1}{2}\left(\frac{y_i - \beta_0 - \beta_1 x_i}{\sigma}\right)^2}$$

The maximum of L occurs when the partial derivatives with respect to β_0, β_1, and σ^2 all vanish.

It will be easier, computationally, to differentiate $-2\ln L$, and the latter will be minimized for the same parameter values that maximize L. Here,

$$-2\ln L = n \cdot \ln(2\pi) + n \cdot \ln(\sigma^2) + \frac{1}{\sigma^2}\sum_{i=1}^{n}(y_i - \beta_0 - \beta_1 x_i)^2$$

Setting the three partial derivatives equal to 0 gives

$$\frac{\partial(-2\ln L)}{\partial \beta_0} = \frac{2}{\sigma^2}\sum_{i=1}^{n}(y_i - \beta_0 - \beta_1 x_i)(-1) = 0$$

$$\frac{\partial(-2\ln L)}{\partial \beta_1} = \frac{2}{\sigma^2}\sum_{i=1}^{n}(y_i - \beta_0 - \beta_1 x_i)(-x_i) = 0$$

$$\frac{\partial(-2\ln L)}{\partial \sigma^2} = \frac{n}{\sigma^2} - \frac{2}{(\sigma^2)^2}\sum_{i=1}^{n}(y_i - \beta_0 - \beta_1 x_i)^2 = 0$$

The first two equations depend only on β_0 and β_1, and the resulting solutions for $\hat{\beta}_0$ and $\hat{\beta}_1$ have the same forms that are given in the statement of the theorem. Substituting the solutions from the first two equations into the third gives the expression for $\hat{\sigma}^2$. □

Comment Note the similarity in the formulas for the maximum likelihood estimators and the least squares estimates for $\hat{\beta}_0$ and $\hat{\beta}_1$. The least squares estimates, of course, are numbers, while the maximum likelihood estimators are random variables.

Up to this point, random variables have been denoted with uppercase letters and their values with lowercase letters. In this section, boldface $\hat{\boldsymbol{\beta}}_0$ and $\hat{\boldsymbol{\beta}}_1$ will represent the maximum likelihood *random variables,* and plain-text $\hat{\beta}_0$ and $\hat{\beta}_1$ will refer to specific values taken on by those random variables.

Properties of Linear Model Estimators

By virtue of the assumptions that define the simple linear model, we know that the estimators $\hat{\boldsymbol{\beta}}_0$, $\hat{\boldsymbol{\beta}}_1$, and $\hat{\boldsymbol{\sigma}}^2$ are random variables. Before those estimators can be used to set up inference procedures, though, we need to establish their basic statistical properties—specifically, their means, variances, and pdfs.

Theorem 11.3.2 *Let $(x_1, Y_1), (x_2, Y_2), \ldots,$ and (x_n, Y_n) be a set of points satisfying the simple linear model, $E(Y \mid x) = \beta_0 + \beta_1 x$. Let $\hat{\boldsymbol{\beta}}_0$, $\hat{\boldsymbol{\beta}}_1$, and $\hat{\boldsymbol{\sigma}}^2$ be the maximum likelihood estimators for β_0, β_1, and σ^2, respectively. Then*

a. $\hat{\beta}_0$ *and* $\hat{\beta}_1$ *are both normally distributed.*

b. $\hat{\beta}_0$ *and* $\hat{\beta}_1$ *are both unbiased:* $E(\hat{\beta}_0) = \beta_0$ *and* $E(\hat{\beta}_1) = \beta_1$.

c. $\text{Var}(\hat{\beta}_1) = \dfrac{\sigma^2}{\sum_{i=1}^{n}(x_i - \bar{x})^2}$

d. $\text{Var}(\hat{\beta}_0) = \dfrac{\sigma^2 \sum_{i=1}^{n} x_i^2}{n \sum_{i=1}^{n}(x_i - \bar{x})^2} = \sigma^2 \left[\dfrac{1}{n} + \dfrac{\bar{x}^2}{\sum_{i=1}^{n}(x_i - \bar{x})^2} \right]$

Proof We will prove the statements for $\hat{\beta}_1$; the results for $\hat{\beta}_0$ follow similarly.

The equation for the estimator $\hat{\beta}_1$ given in Theorem 11.3.1 is the simplest form that solves the likelihood equations (and the least squares equations as well). It is also convenient for computation. However, two other expressions for $\hat{\beta}_1$ are useful for theoretical results.

To begin, take the version of $\hat{\beta}_1$ from Theorem 11.3.1:

$$\hat{\beta}_1 = \frac{n \sum_{i=1}^{n} x_i Y_i - \left(\sum_{i=1}^{n} x_i \right)\left(\sum_{i=1}^{n} Y_i \right)}{n \left(\sum_{i=1}^{n} x_i^2 \right) - \left(\sum_{i=1}^{n} x_i \right)^2}$$

Dividing numerator and denominator by n gives

$$\hat{\beta}_1 = \frac{\sum_{i=1}^{n} x_i Y_i - \frac{1}{n}\left(\sum_{i=1}^{n} x_i \right)\left(\sum_{i=1}^{n} Y_i \right)}{\left(\sum_{i=1}^{n} x_i^2 \right) - \frac{1}{n}\left(\sum_{i=1}^{n} x_i \right)^2}$$

$$= \frac{\sum_{i=1}^{n} x_i Y_i - \bar{x}\left(\sum_{i=1}^{n} Y_i \right)}{\left(\sum_{i=1}^{n} x_i^2 \right) - n\bar{x}^2}$$

$$= \frac{\sum_{i=1}^{n} (x_i - \bar{x}) Y_i}{\left(\sum_{i=1}^{n} x_i^2 \right) - n\bar{x}^2} \tag{11.3.1}$$

Equation 11.3.1 expresses $\hat{\beta}_1$ as a linear combination of independent normal variables, so by the second corollary to Theorem 4.3.3, it is itself normal, proving part (a).

To see that $\hat{\beta}_1$ is unbiased, note that

$$E(\hat{\beta}_1) = \frac{\sum_{i=1}^{n} (x_i - \bar{x}) E(Y_i)}{\left(\sum_{i=1}^{n} x_i^2 \right) - n\bar{x}^2} = \frac{\sum_{i=1}^{n} (x_i - \bar{x})(\beta_0 + \beta_1 x_i)}{\left(\sum_{i=1}^{n} x_i^2 \right) - n\bar{x}^2} = \frac{\beta_0 \sum_{i=1}^{n} (x_i - \bar{x}) + \beta_1 \sum_{i=1}^{n} (x_i - \bar{x}) x_i}{\left(\sum_{i=1}^{n} x_i^2 \right) - n\bar{x}^2}$$

$$= \frac{0 + \beta_1 \sum_{i=1}^{n} (x_i - \bar{x}) x_i}{\left(\sum_{i=1}^{n} x_i^2 \right) - n\bar{x}^2} = \frac{\beta_1 \left(\sum_{i=1}^{n} x_i^2 - n\bar{x}^2 \right)}{\left(\sum_{i=1}^{n} x_i^2 \right) - n\bar{x}^2} = \beta_1$$

To find $\text{Var}(\hat{\beta}_1)$, rewrite the denominator of Equation 11.3.1 in the form

$$\left(\sum_{i=1}^{n} x_i^2\right) - n\bar{x}^2 = \sum_{i=1}^{n}(x_i^2 - 2x_i\bar{x} + \bar{x}^2) = \sum_{i=1}^{n}(x_i - \bar{x})^2$$

which makes

$$\beta_1 = \frac{\sum_{i=1}^{n}(x_i - \bar{x})Y_i}{\sum_{i=1}^{n}(x_i - \bar{x})^2} \tag{11.3.2}$$

Using Equation 11.3.2, Theorem 3.6.2, and the second corollary to Theorem 3.9.5 gives

$$\begin{aligned}
\text{Var}(\hat{\beta}_1) &= \text{Var}\left[\frac{1}{\sum_{i=1}^{n}(x_i - \bar{x})^2}\sum_{i=1}^{n}(x_i - \bar{x})Y_i\right] \\
&= \frac{1}{\left[\sum_{i=1}^{n}(x_i - \bar{x})^2\right]^2}\sum_{i=1}^{n}(x_i - \bar{x})^2\sigma^2 \\
&= \frac{\sigma^2}{\sum_{i=1}^{n}(x_i - \bar{x})^2}
\end{aligned}$$

□

Theorem 11.3.3 *Let $(x_1, Y_1), (x_2, Y_2), \ldots, (x_n, Y_n)$ satisfy the assumptions of the simple linear model. Then*

a. *$\hat{\beta}_1$, $\bar{Y}$, and $\hat{\sigma}^2$ are mutually independent.*

b. *$\dfrac{n\hat{\sigma}^2}{\sigma^2}$ has a chi square distribution with $n-2$ degrees of freedom.*

Proof See Appendix 11.A.2. □

Corollary *Let $\hat{\sigma}^2$ be the maximum likelihood estimator for σ^2 in a simple linear model. Then $\dfrac{n}{n-2} \cdot \hat{\sigma}^2$ is an unbiased estimator for σ^2.*

Proof Recall that the expected value of a χ_k^2 distribution is k (see Theorems 4.6.3 and 7.3.1). Therefore,

$$\begin{aligned}
E\left(\frac{n}{n-2} \cdot \hat{\sigma}^2\right) &= \frac{\sigma^2}{n-2} E\left(\frac{n\hat{\sigma}^2}{\sigma^2}\right) \\
&= \frac{\sigma^2}{n-2} \cdot (n-2) \qquad \text{[by part (b) of Theorem 11.3.3]} \\
&= \sigma^2
\end{aligned}$$

◄

Corollary *The random variables $\hat{Y}$ and $\hat{\sigma}^2$ are independent.* ◄

Estimating σ^2

We know that the (biased) maximum likelihood estimator for σ^2 in a simple linear model is

$$\hat{\sigma}^2 = \frac{1}{n}\sum_{i=1}^{n}(Y_i - \hat{\beta}_0 - \hat{\beta}_1 x_i)^2$$

The unbiased estimator for σ^2 based on $\hat{\sigma}^2$ is denoted S^2, where

$$S^2 = \frac{n}{n-2}\hat{\sigma}^2 = \frac{1}{n-2}\sum_{i=1}^{n}(Y_i - \hat{\beta}_0 - \hat{\beta}_1 x_i)^2$$

Statistical software packages—including Minitab—typically print out s, rather than $\hat{\sigma}$, in summarizing the calculations associated with linear model data. To accommodate that convention, we will use s^2 rather than $\hat{\sigma}^2$ in writing the formulas for the test statistics and confidence intervals that arise in connection with the simple linear model.

Comment Calculating $\sum_{i=1}^{n}(y_i - \hat{\beta}_0 - \hat{\beta}_1 x_i)^2 = \sum_{i=1}^{n}(y_i - \hat{y}_i)^2$ can be cumbersome. Three (algebraically equivalent) computing formulas are available that may be easier to use, depending on the data:

$$\sum_{i=1}^{n}(y_i - \hat{y}_i)^2 = \sum_{i=1}^{n}(y_i - \bar{y})^2 - \hat{\beta}_1^2\sum_{i=1}^{n}(x_i - \bar{x})^2 \tag{11.3.3}$$

$$\sum_{i=1}^{n}(y_i - \hat{y}_i)^2 = \sum_{i=1}^{n}y_i^2 - \frac{1}{n}\sum_{i=1}^{n}y_i - \frac{\left[\sum_{i=1}^{n}x_i y_i - \frac{1}{n}\left(\sum_{i=1}^{n}x_i\right)\left(\sum_{i=1}^{n}y_i\right)\right]^2}{\sum_{i=1}^{n}x_i^2 - \frac{1}{n}\sum_{i=1}^{n}x_i} \tag{11.3.4}$$

$$\sum_{i=1}^{n}(y_i - \hat{y}_i)^2 = \sum_{i=1}^{n}y_i^2 - \hat{\beta}_0\sum_{i=1}^{n}y_i - \hat{\beta}_1\sum_{i=1}^{n}x_i y_i \tag{11.3.5}$$

Drawing Inferences about β_1

Hypothesis tests and confidence intervals for β_1 can be carried out by defining a t statistic based on the properties that appear in Theorems 11.3.2 and 11.3.3.

Theorem 11.3.4 *Let $(x_1, Y_1), (x_2, Y_2), \ldots,$ and (x_n, Y_n) be a set of points that satisfy the assumptions of the simple linear model, and let $S^2 = \frac{1}{n-2}\sum_{i=1}^{n}(Y_i - \hat{\beta}_0 - \hat{\beta}_1 x_i)^2$. Then*

$$T_{n-2} = \frac{\hat{\beta}_1 - \beta_1}{S\Big/\sqrt{\sum_{i=1}^{n}(x_i - \bar{x})^2}}$$

has a Student t distribution with $n-2$ degrees of freedom.

Proof We know from Theorem 11.3.2 that

$$Z = \frac{\hat{\beta}_1 - \beta_1}{\sigma \Big/ \sqrt{\sum_{i=1}^{n} (x_i - \bar{x})^2}}$$

has a standard normal pdf. Furthermore, $\frac{n\hat{\sigma}^2}{\sigma^2} = \frac{(n-2)S^2}{\sigma^2}$ has a χ^2 pdf with $n-2$ degrees of freedom, and, by Theorem 11.3.3, Z and $\frac{(n-2)S^2}{\sigma^2}$ are independent. From Definition 7.3.3, then, it follows that

$$Z \Big/ \sqrt{\frac{(n-2)S^2}{\sigma^2} \Big/ (n-2)} = \frac{\hat{\beta}_1 - \beta_1}{S \Big/ \sqrt{\sum_{i=1}^{n} (x_i - \bar{x})^2}}$$

has a Student t distribution with $n-2$ degrees of freedom. □

Theorem 11.3.5 *Let $(x_1, Y_1), (x_2, Y_2), \ldots,$ and (x_n, Y_n) be a set of points that satisfy the assumptions of the simple linear model. Let*

$$t = \frac{\hat{\beta}_1 - \beta_{1_o}}{s \Big/ \sqrt{\sum_{i=1}^{n} (x_i - \bar{x})^2}}$$

a. *To test $H_0{:}\,\beta_1 = \beta_{1_o}$ versus $H_1{:}\,\beta_1 > \beta_{1_o}$ at the α level of significance, reject H_0 if $t \geq t_{\alpha,n-2}$.*

b. *To test $H_0{:}\,\beta_1 = \beta_{1_o}$ versus $H_1{:}\,\beta_1 < \beta_{1_o}$ at the α level of significance, reject H_0 if $t \leq -t_{\alpha,n-2}$.*

c. *To test $H_0{:}\,\beta_1 = \beta_{1_o}$ versus $H_1{:}\,\beta_1 \neq \beta_{1_o}$ at the α level of significance, reject H_0 if t is either (1) $\leq -t_{\alpha/2,n-2}$ or (2) $\geq t_{\alpha/2,n-2}$.*

Proof The decision rule given here is, in fact, a GLRT. A formal proof proceeds along the lines followed in Appendix 7.A.4. We will omit the details. □

Comment A particularly common application of Theorem 11.3.5 is to test $H_0{:}\,\beta_1 = 0$. If the null hypothesis that the slope is zero is rejected, it can be concluded (at the α level of significance) that $E(Y)$ changes with x. Conversely, if $H_0{:}\,\beta_1 = 0$ is *not* rejected, the data have not ruled out the possibility that variation in Y is unaffected by x.

Case Study 11.3.1

By late 1971, all cigarette packs had to be labeled with the words, "Warning: The Surgeon General Has Determined That Smoking Is Dangerous To Your Health." The case against smoking rested heavily on statistical, rather than laboratory, evidence. Extensive surveys of smokers and nonsmokers had revealed the former to have much higher risks of dying from a variety of causes, including heart disease.

(Continued on next page)

Typical of that research are the data in Table 11.3.1, showing the annual cigarette consumption, x, and the corresponding mortality rate, Y, due to coronary heart disease (CHD) for twenty-one countries (116). Do these data support the suspicion that smoking contributes to CHD mortality? Test $H_0: \beta_1 = 0$ versus $H_1: \beta_1 > 0$ at the $\alpha = 0.05$ level of significance.

Table 11.3.1

Country	Cigarette Consumption per Adult per Year, x	CHD Mortality per 100,000 (ages 35–64), y
United States	3900	256.9
Canada	3350	211.6
Australia	3220	238.1
New Zealand	3220	211.8
United Kingdom	2790	194.1
Switzerland	2780	124.5
Ireland	2770	187.3
Iceland	2290	110.5
Finland	2160	233.1
West Germany	1890	150.3
Netherlands	1810	124.7
Greece	1800	41.2
Austria	1770	182.1
Belgium	1700	118.1
Mexico	1680	31.9
Italy	1510	114.3
Denmark	1500	144.9
France	1410	59.7
Sweden	1270	126.9
Spain	1200	43.9
Norway	1090	136.3

From Table 11.3.1,

$$\sum_{i=1}^{21} x_i = 45,110 \qquad \sum_{i=1}^{21} y_i = 3,042.2$$

$$\sum_{i=1}^{21} x_i^2 = 109,957,100 \qquad \sum_{i=1}^{21} y_i^2 = 529,321.58$$

$$\sum_{i=1}^{21} x_i y_i = 7,319,602$$

and it follows that

$$\hat{\beta}_1 = \frac{n \sum_{i=1}^{n} x_i y_i - \left(\sum_{i=1}^{n} x_i\right)\left(\sum_{i=1}^{n} y_i\right)}{n\left(\sum_{i=1}^{n} x_i^2\right) - \left(\sum_{i=1}^{n} x_i\right)^2}$$

$$= \frac{21(7,319,602) - (45,110)(3,042.2)}{21(109,957,100) - (45,110)^2} = 0.0601$$

(Continued on next page)

(Case Study 11.3.1 continued)

and

$$\hat{\beta}_0 = \frac{\sum_{i=1}^{n} y_i - \hat{\beta}_1 \sum_{i=1}^{n} x_i}{n}$$

$$= \frac{3{,}042.2 - 0.0601(45{,}110)}{21} = 15.771$$

The two other quantities needed for the test statistic are

$$\sum_{i=1}^{n}(x_i - \bar{x})^2 = \sum_{i=1}^{n} x_i^2 - \left(\frac{1}{n}\right)\left(\sum_{i=1}^{n} x_i\right)^2$$

$$= 109{,}957{,}100 - \left(\frac{1}{21}\right)(45{,}100)^2 = 13{,}056{,}523.81$$

so $\sqrt{\sum_{i=1}^{n}(x_i - \bar{x})^2} = \sqrt{13{,}056{,}523.81} = 3{,}613.38$.

From Equation 11.3.5,

$$s^2 = \frac{1}{21-2}\left(\sum_{i=1}^{21} y_i^2 - \hat{\beta}_0 \sum_{i=1}^{21} y_i - \hat{\beta}_1 \sum_{i=1}^{21} x_i y_i\right)$$

$$= \frac{1}{19}[529{,}321.58 - (15.766)(3{,}042.2) - (0.0601)(7{,}319{,}602)] = 2{,}181.588$$

and $s = \sqrt{2{,}181.588} = 46.707$

To test

$$H_0\colon \beta_1 = 0$$

versus

$$H_0\colon \beta_1 > 0$$

at the $\alpha = 0.05$ level of significance, we should reject the null hypothesis if $t \geq t_{.05,19} = 1.7291$. But

$$t = \frac{\hat{\beta}_1 - \beta_{1_o}}{s \Big/ \sqrt{\sum_{i=1}^{n}(x_i - \bar{x})^2}} = \frac{0.0601 - 0}{46.707/3{,}613.38}$$

$$= 4.65$$

so our conclusion is clear-cut—reject H_0. It would appear that the level of CHD mortality in a country *is* affected by its citizens' smoking habits. More specifically, as the number of people who smoke increases, so will the number who die of coronary heart disease.

Theorem 11.3.6 *Let $(x_1, Y_1), (x_2, Y_2), \ldots,$ and (x_n, Y_n) be a set of points that satisfy the assumptions of the simple linear model, and let $s^2 = \frac{1}{n-2}\sum_{i=1}^{n}(y_i - \hat{\beta}_0 - \hat{\beta}_1 x_i)^2$. Then*

$$\left[\hat{\beta}_1 - t_{\alpha/2,n-2} \cdot \frac{s}{\sqrt{\sum_{i=1}^{n}(x_i - \bar{x})^2}},\ \hat{\beta}_1 + t_{\alpha/2,n-2} \cdot \frac{s}{\sqrt{\sum_{i=1}^{n}(x_i - \bar{x})^2}}\right]$$

is a $100(1-\alpha)\%$ *confidence interval for* β_1.

Proof Let T_{n-2} denote a Student t random variable with $n-2$ degrees of freedom, in which case

$$P(-t_{\alpha/2,n-2} \leq T_{n-2} \leq t_{\alpha/2,n-2}) = 1 - \alpha$$

Substitute the expression for T_{n-2} given in Theorem 11.3.4 and isolate β_1 in the center of the inequalities. The resulting endpoints will be the expressions appearing in the statement of the theorem. □

Case Study 11.3.2

For many firms, the cost of sales is a linear function of net revenue. This seems to be the case for Starbucks, now a staple of the coffee-drinking public. Prior to 1971, Americans drinking coffee outside of their homes had little choice but a weak, watery brew often kept for hours on a hotplate, giving a burned, bitter taste. In 1971, a company opened a coffee shop in Seattle's famous Pike Place Market to serve robust and fresh coffee. The shop was named after a character in Herman Melville's *Moby Dick*, and it signified the import of coffee across the seas. By 2007, the chain had grown to over fifteen thousand outlets.

Table 11.3.2 shows Starbucks' annual net revenue (x) and the cost of operating the stores (y) primarily responsible for generating that revenue. Graphed, the xy-relationship is described very well by the line $y = 18.57 + 0.41x$, where 18.57 and 0.41 are the values of $\hat{\beta}_0$ and $\hat{\beta}_1$ calculated from the formulas in Theorem 11.3.1 (see Figure 11.3.3).

Table 11.3.2

Year	Net Revenue (in $ millions), x	Cost of Sales (in $ millions), y
1999	1687	748
2000	2178	962
2001	2649	1113
2002	3289	1350
2003	4076	1686
2004	5294	2199
2005	6369	2605
2006	7787	3179
2007	9411	3999

Source: Company reports.

(Continued on next page)

(Case Study 11.3.2 continued)

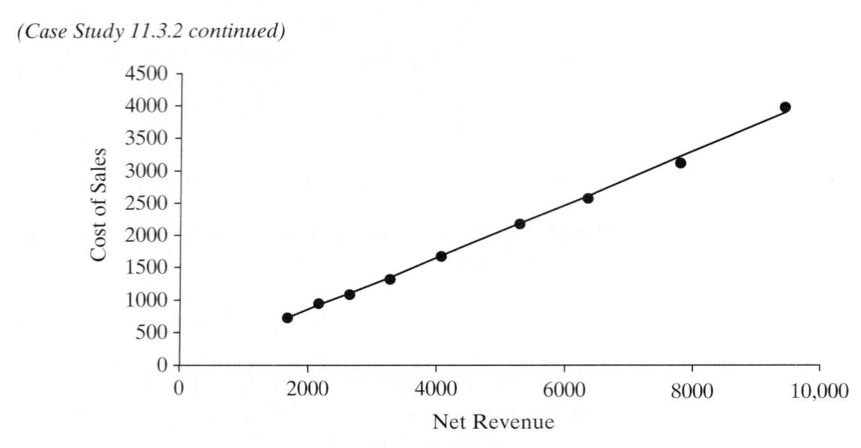

Figure 11.3.3

The true slope in this situation—β_1—is particularly important from the company's perspective because it represents the amount that costs are likely to increase when revenues go up by one unit. That said, it makes sense to construct, say, a 95% confidence interval for β_1 based on the observed $\hat{\beta}_1$.

Here,

$$\sum_{i=1}^{9}(x_i-\bar{x})^2 = 56,865,526.89$$

$$\text{so} \quad \sqrt{\sum_{i=1}^{9}(x_i-\bar{x})^2} = \sqrt{56,865,526.89} = 7540.92$$

and from Equation 11.3.5,

$$s^2 = \frac{1}{9-2}\left[\sum_{i=1}^{9} y_i^2 - \hat{\beta}_0\sum_{i=1}^{9} y_i - \hat{\beta}_1\sum_{i=1}^{9} x_i y_i\right]$$

$$= \frac{1}{7}[45,108,481 - (18.57)(17,841) - (0.41)(108,239,948)] = 2535.01$$

so $s = \sqrt{2535.01} = 50.35$.

Using $t_{.025,7} = 2.3646$, the expression given in Theorem 11.3.6 reduces to

$$\left(0.41 - 2.3646\cdot\frac{50.35}{7540.92},\ 0.41 + 2.3646\cdot\frac{50.35}{7540.92}\right) = (\$0.394,\ \$0.426)$$

Judging from these data, then, the company can anticipate that costs will rise somewhere between thirty-nine and forty-three cents for every one-dollar increase in revenues.

About the Data The predictive value of the regression equation in Case Study 11.3.2 depends on a continuing healthy economic climate after the years 1999–2007, the period for which the data were generated. In the case of Starbucks, a serious economic downturn began in 2008, and in the summer of that year, Starbucks announced plans to close six hundred stores. An equation based on 1999–2007 data might still be useful, but a more prudent strategy would be to revisit the equation in light of what happened in 2008 and 2009, when consumers' discretionary expenses were curtailed.

Drawing Inferences about $\boldsymbol{\beta}_0$

In practice, the value of β_0 is not likely to be as important as the value of β_1. Slopes often quantify particularly important aspects of xy-relationships, which was true, for example, in Case Study 11.3.2. Nevertheless, hypothesis tests and confidence intervals for β_0 can be easily derived from the results given in Theorems 11.3.2 and 11.3.3.

The GLRT procedure for assessing the credibility of $H_0: \beta_0 = \beta_{0_o}$ is based on a Student t random variable with $n-2$ degrees of freedom:

$$T_{n-2} = \frac{(\hat{\boldsymbol{\beta}}_0 - \beta_{0_o})\sqrt{n}\sqrt{\sum_{i=i}^{n}(x_i-\bar{x})^2}}{S\sqrt{\sum_{i=1}^{n}x_i^2}} = \frac{\hat{\boldsymbol{\beta}}_0 - \beta_{0_o}}{\sqrt{\widehat{\text{Var}}(\hat{\boldsymbol{\beta}}_0)}} \tag{11.3.6}$$

"Inverting" Equation 11.3.6 (recall the proof of Theorem 11.3.6) yields

$$\left[\hat{\beta}_0 - t_{\alpha/2,n-2}\cdot\frac{s\sqrt{\sum_{i=1}^{n}x_i^2}}{\sqrt{n}\sqrt{\sum_{i=1}^{n}(x_i-\bar{x})^2}},\ \hat{\beta}_0 + t_{\alpha/2,n-2}\cdot\frac{s\sqrt{\sum_{i=1}^{n}x_i^2}}{\sqrt{n}\sqrt{\sum_{i=1}^{n}(x_i-\bar{x})^2}}\right]$$

as the formula for a $100(1-\alpha)\%$ confidence interval for β_0.

Drawing Inferences about $\boldsymbol{\sigma}^2$

Since $(n-2)S^2/\sigma^2$ has a χ^2 pdf with $n-2$ df (if the n observations satisfy the stipulations implicit in the simple linear model), it follows that

$$P\left[\chi^2_{\alpha/2,n-2} \le \frac{(n-2)S^2}{\sigma^2} \le \chi^2_{1-\alpha/2,n-2}\right] = 1-\alpha$$

Equivalently,

$$P\left[\frac{(n-2)S^2}{\chi^2_{1-a/2,n-2}} \le \sigma^2 \le \frac{(n-2)S^2}{\chi^2_{\alpha/2,n-2}}\right] = 1-\alpha$$

in which case

$$\left[\frac{(n-2)s^2}{\chi^2_{1-\alpha/2,n-2}}, \frac{(n-2)s^2}{\chi^2_{\alpha/2,n-2}}\right]$$

becomes the *100(1 − α)% confidence interval for* σ^2 (recall Theorem 7.5.1). Testing $H_0: \sigma^2 = \sigma_o^2$ is done by calculating the ratio

$$\chi^2 = \frac{(n-2)s^2}{\sigma_o^2}$$

which has a χ^2 distribution with $n-2$ df when the null hypothesis is true. Except for the degrees of freedom ($n-2$ rather than $n-1$), the appropriate decision rules for one-sided and two-sided H_1's are similar to those given in Theorem 7.5.2.

Questions

11.3.1. Insect flight ability can be measured in a laboratory by attaching the insect to a nearly frictionless rotating arm with a thin wire. The "tethered" insect then flies in circles until exhausted. The nonstop distance flown can easily be calculated from the number of revolutions made by the arm. The following are measurements of this sort made on *Culex tarsalis* mosquitos of four different ages. The response variable is the average distance flown until exhaustion for forty females of the species (150).

Age, x (weeks)	Distance Flown, y (thousand meters)
1	12.6
2	11.6
3	6.8
4	9.2

Fit a straight line to these data and test that the slope is zero. Use a two-sided alternative and the 0.05 level of significance.

11.3.2. The best straight line through the Massachusetts funding/graduation rate data described in Question 11.2.7 has the equation $y = 81.088 + 0.412x$, where $s = 11.78848$.

(a) Construct a 95% confidence interval for β_1.
(b) What does your answer to part (a) imply about the outcome of testing H_0: $\beta_1 = 0$ versus H_1: $\beta_1 \neq 0$ at the $\alpha = 0.05$ level of significance?
(c) Graph the data and superimpose the regression line. How would you summarize these data, and their implications, to a meeting of the state School Board?

11.3.3. Based on the data in Question 11.2.1, the relationship between y, the ambient temperature, and x, the frequency of a cricket's chirping, is given by $y = 25.2 + 3.29x$, where $s = 3.83$. At the $\alpha = 0.01$ level of significance, can the hypothesis that chirping frequency is not related to temperature be rejected?

11.3.4. Suppose an experimenter intends to do a regression analysis by taking a total of $2n$ data points, where the x_i's are restricted to the interval [0, 5]. If the xy-relationship is assumed to be linear and if the objective is to estimate the slope with the greatest possible precision, what values should be assigned to the x_i's?

11.3.5. Suppose a total of $n = 9$ measurements are to be taken on a simple linear model, where the x_i's will be set equal to $1, 2, \ldots,$ and 9. If the variance associated with the xy-relationship is known to be 45.0, what is the probability that the estimated slope will be within 1.5 units of the true slope?

11.3.6. Prove the useful computing formula (Equation 11.3.5) that

$$\sum_{i=1}^{n}(y_i - \hat{\beta}_0 - \hat{\beta}_1 x_i)^2 = \sum_{i=1}^{n} y_i^2 - \hat{\beta}_0 \sum_{i=1}^{n} y_i - \hat{\beta}_1 \sum_{i=1}^{n} x_i y_i$$

11.3.7. The sodium nitrate ($NaNO_3$) solubility data in Question 11.2.3 is described nicely by the regression line $y = 67.508 + 0.871x$, where $s = 0.959$. Construct a 90% confidence interval for the y-intercept, β_0.

11.3.8. Set up and carry out an appropriate hypothesis test for the Hanford radioactive contamination data given in Question 11.2.9. Let $\alpha = 0.05$. Justify your choice of H_0 and H_1. What do you conclude?

11.3.9. Test H_0: $\beta_1 = 0$ versus H_1: $\beta_1 \neq 0$ for the plumage index/behavioral index data given in Question 11.2.11 Let $\alpha = 0.05$. Use the fact that $y = 0.61 + 0.84x$ is the best straight line describing the xy-relationship.

11.3.10. Let (x_1, Y_1), $(x_2, Y_2), \ldots,$ and (x_n, Y_n) be a set of points satisfying the assumptions of the simple linear model. Prove that

$$E(\bar{Y}) = \beta_0 + \beta_1 \bar{x}$$

11.3.11. Derive a formula for a 95% confidence interval for β_0 if n (x_i, Y_i)'s are taken on a simple linear model where σ is known.

11.3.12. Which, if any, of the assumptions of the simple linear model appear to be violated in the following scatterplot? Which, if any, appear to be satisfied? Which, if any, cannot be assessed by looking at the scatterplot?

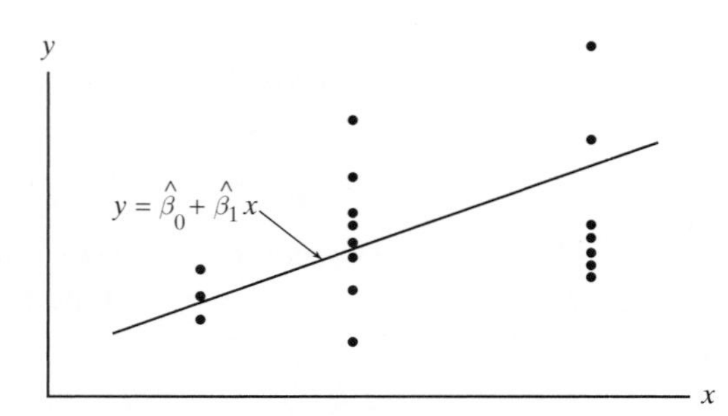

11.3.13. State the decision rule and the conclusion if H_0: $\sigma^2 = 12.6$ is to be tested against H_1: $\sigma^2 \neq 12.6$ where $n = 24$, $s^2 = 18.2$, and $\alpha = 0.05$.

11.3.14. Construct a 90% confidence interval for σ^2 in the cigarette-consumption/CHD mortality data given in Case Study 11.3.1.

11.3.15. Recall Kepler's Third Law data given in Question 8.2.1. The estimated regression line describing the xy-relationship has the equation $y = 2.27 + 0.16x$, where $s = 2.31$. Construct a 90% confidence interval for σ^2.

Drawing Inferences about $E(Y \mid x)$

In Case Study 11.3.1, the random variable Y represents the CHD mortality resulting from x cigarette consumption. A public health official would certainly want to have some idea of the range of mortality likely to be encountered in a country where x is, say, 4200.

Intuition tells us that a reasonable point estimator for $E(Y \mid x)$ is the height of the regression line at x—that is, $\hat{Y} = \hat{\beta}_0 + \hat{\beta}_1 x$. By Theorem 11.3.2, the latter is unbiased:

$$E(\hat{Y}) = E(\hat{\beta}_0 + \hat{\beta}_1 x) = E(\hat{\beta}_0) + xE(\hat{\beta}_1) = \beta_0 + \beta_1 x$$

Of course, to use $\hat{Y}$ in any inference procedure requires that we know its variance. But

$$\begin{aligned}
\text{Var}(\hat{Y}) &= \text{Var}(\hat{\beta}_0 + \hat{\beta}_1 x) = \text{Var}(\bar{Y} - \hat{\beta}_1 \bar{x} + \hat{\beta}_1 x) \\
&= \text{Var}[\bar{Y} + \hat{\beta}_1 (x - \bar{x})] \\
&= \text{Var}(\bar{Y}) + (x - \bar{x})^2 \text{Var}(\hat{\beta}_1) \quad \text{(why?)} \\
&= \frac{1}{n}\sigma^2 + \frac{(x - \bar{x})^2}{\sum_{i=1}^{n} (x_i - \bar{x})^2}\sigma^2 \\
&= \sigma^2 \left[\frac{1}{n} + \frac{(x - \bar{x})^2}{\sum_{i=1}^{n} (x_i - \bar{x})^2} \right]
\end{aligned}$$

An application of Definition 7.3.3, then, allows us to construct a Student t random variable based on $\hat{Y}$. Specifically,

$$T_{n-2} = \frac{\hat{Y} - (\beta_0 + \beta_1 x)}{\sigma \sqrt{\frac{1}{n} + \frac{(x-\bar{x})^2}{\sum_{i=1}^{n} (x_i - \bar{x})^2}}} \Bigg/ \sqrt{\frac{(n-2)S^2}{\frac{\sigma^2}{n-2}}} = \frac{\hat{Y} - (\beta_0 + \beta_1 x)}{S \sqrt{\frac{1}{n} + \frac{(x-\bar{x})^2}{\sum_{i=1}^{n} (x_i - \bar{x})^2}}}$$

has a Student t distribution with $n - 2$ degrees of freedom. Isolating $\beta_0 + \beta_1 x = E(Y \mid x)$ in the center of the inequalities $P(-t_{\alpha/2,n-2} \le T_{n-2} \le t_{\alpha/2,n-2}) = 1 - \alpha$ produces a $100(1 - \alpha)\%$ confidence interval for $E(Y \mid x)$.

Theorem 11.3.7 *Let $(x_1, Y_1), (x_2, Y_2), \ldots,$ and (x_n, Y_n) be a set of points that satisfy the assumptions of the simple linear model. A $100(1 - \alpha)\%$ confidence interval for $E(Y \mid x) = \beta_0 + \beta_1 x$ is given by $(\hat{y} - w, \hat{y} + w)$, where*

$$w = t_{\alpha/2,n-2} \cdot s\sqrt{\frac{1}{n} + \frac{(x-\bar{x})^2}{\sum_{i=1}^{n}(x_i-\bar{x})^2}}$$

and $\hat{y} = \hat{\beta}_0 + \hat{\beta}_1 x$.

Example 11.3.2 Look again at Case Study 11.3.1. Suppose a country's public health officials estimate cigarette consumption to be 4200 per adult per year. If that were the case, what CHD mortality would they expect? Answer the question by constructing a 95% confidence interval for $E(Y|4200)$.

Here, $n = 21$, $t_{.025,19} = 2.0930$, $\sum_{i=1}^{21}(x_i - \bar{x})^2 = 13{,}056{,}523.81$, $s = 46.707$, $\hat{\beta}_0 = 15.7661$, $\hat{\beta}_1 = 0.0601$, and $\bar{x} = 2148.095$. From Theorem 11.3.7, then,

$$\hat{y} = 15.7661 + 0.0601(4200) = 268.1861$$

$$w = 2.0930(46.707)\sqrt{\frac{1}{21} + \frac{(4200-2148.095)^2}{13{,}056{,}523.81}} = 59.4714$$

and the 95% confidence interval for $E(Y|4200)$ is

$$(268.1861 - 59.4714,\ 268.1861 - 59.4714)$$

which rounded to two decimal places is

$$(208.71, 327.66)$$

Comment Notice from the formula in Theorem 11.3.7 that the width of a confidence interval for $E(Y \mid x)$ increases as the value of x becomes more extreme. That is, we are better able to predict the location of the regression line for an x-value close to $\bar{x}$ than we are for x-values that are either very small or very large.

Figure 11.3.4 shows the dependence of w on x for the data from Case Study 11.3.1. The lower and upper limits for the 95% confidence interval for $E(Y \mid x)$ have been calculated for all x. Pictured is the dotted curve (or *95% confidence band*) connecting those endpoints. The width of the band is smallest when $x = 2148.1$ $(= \bar{x})$.

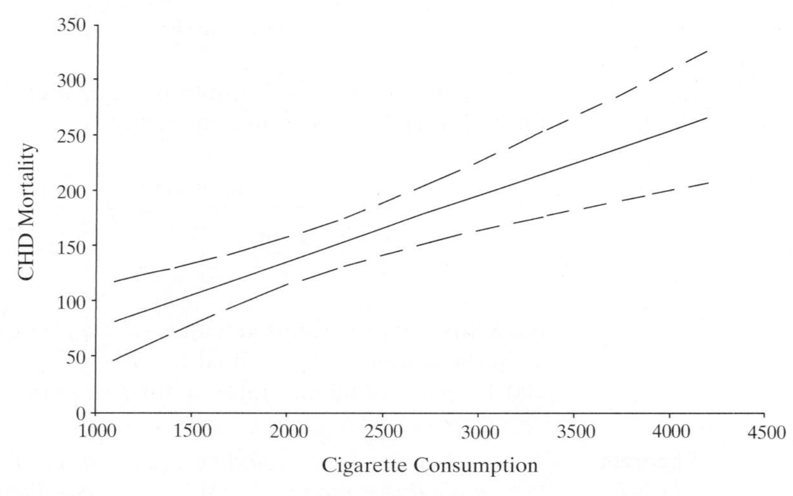

Figure 11.3.4

Drawing Inferences about Future Observations

A variation on Theorem 11.3.7 is the determination of a range of numbers that would have a high probability of including the value Y of a *single future observation* to be recorded at some given level of x. In Case Study 11.3.1, public health officials might want to predict the *actual* (not the *average*) CHD mortality that would occur if cigarette consumption is x.

Let $(x_1, Y_1), (x_2, Y_2), \ldots, (x_n, Y_n)$ be a set of n points that satisfy the assumptions of the simple linear model, and let (x, Y) be a hypothetical future observation, where Y is independent of the n Y_i's. A *prediction interval* is a range of numbers that contains Y with a specified probability.

Consider the difference $\hat{Y} - Y$. Clearly,

$$E(\hat{Y} - Y) = E(\hat{Y}) - E(Y) = (\beta_0 + \beta_1 x) - (\beta_0 + \beta_1 x) = 0$$

and

$$\begin{aligned} \text{Var}(\hat{Y} - Y) &= \text{Var}(\hat{Y}) + \text{Var}(Y) \\ &= \sigma^2 \left[\frac{1}{n} + \frac{(x - \bar{x})^2}{\sum_{i=1}^{n} (x_i - \bar{x})^2} \right] + \sigma^2 \\ &= \sigma^2 \left[1 + \frac{1}{n} + \frac{(x - \bar{x})^2}{\sum_{i=1}^{n} (x_i - \bar{x})^2} \right] \end{aligned}$$

Following exactly the same steps that were taken in the derivation of Theorem 11.3.7, a Student t random variable with $n - 2$ degrees of freedom can be constructed from $\hat{Y} - Y$ (using Definition 7.3.3). Inverting the equation $P(-t_{\alpha/2,n-2} \leq T_{n-2} \leq t_{\alpha/2,n-2}) = 1 - \alpha$ will then yield the prediction interval $(\hat{y} - w, \hat{y} + w)$ given in Theorem 11.3.8.

Theorem 11.3.8 *Let $(x_1, Y_1), (x_2, Y_2), \ldots,$ and (x_n, Y_n) be a set of n points that satisfy the assumptions of the simple linear model. A $100(1 - \alpha)\%$ prediction interval for Y at the fixed value x is given by $(\hat{y} - w, \hat{y} + w)$, where*

$$w = t_{\alpha/2,n-2} \cdot s \sqrt{1 + \frac{1}{n} + \frac{(x - \bar{x})^2}{\sum_{i=1}^{n} (x_i - \bar{x})^2}}$$

and $\hat{y} = \hat{\beta}_0 + \hat{\beta}_1 x$. □

Example 11.3.3 Based on the data in Case Study 11.3.1, we calculated in Example 11.3.2 that a 95% confidence interval for $E(Y|4200)$ is $(208.71, 327.66)$. How does that compare to the corresponding 95% prediction interval for Y?

When $x = 4200$, $\hat{y} = 268.1861$ for both intervals. From Theorem 11.3.8, the width of the 95% prediction interval for Y is:

$$w = 2.0930(46.707)\sqrt{1 + \frac{1}{21} + \frac{(4200 - 2148.095)^2}{13{,}056{,}523.81}} = 114.4725$$

The 95% prediction interval, then, is

$$(268.1861 - 114.4725, 268.1861 + 114.4725)$$

which rounded to two decimal places is

$$(153.76, 382.61)$$

which makes it 92% wider than the 95% confidence interval for $E(Y|4200)$. ■

Testing the Equality of Two Slopes

We saw in Chapter 9 that the comparison of two treatments or two conditions often leads to a hypothesis test that the mean of one is equal to the mean of the other. Similarly, the comparison of two linear xy-relationships often requires that we test $H_0: \beta_1 = \beta_1^*$, where β_1 and β_1^* are the true slopes associated with the two regressions.

If the data points taken on the two regressions are all independent, a two-sample t test can be set up based on the properties in Theorems 11.3.2 and 11.3.3. Theorem 11.3.9 identifies the appropriate test statistic and summarizes the GLRT decision rule. Details of the proof will be omitted.

Theorem 11.3.9 *Let $(x_1, Y_1), (x_2, Y_2), \ldots, (x_n, Y_n)$ and $(x_1^*, Y_1^*), (x_2^*, Y_2^*), \ldots, (x_m^*, Y_m^*)$ be two independent sets of points, each satisfying the assumptions of the simple linear model—that is, $E(Y \mid x) = \beta_0 + \beta_1 x$ and $E(Y^* \mid x^*) = \beta_0^* + \beta_1^* x^*$.*

a. Let

$$T = \frac{\hat{\beta}_1 - \hat{\beta}_1^* - (\beta_1 - \beta_1^*)}{S\sqrt{\dfrac{1}{\sum_{i=1}^{n}(x_i - \bar{x})^2} + \dfrac{1}{\sum_{i=1}^{m}(x_i^* - \bar{x}^*)^2}}}$$

where

$$S = \sqrt{\frac{\sum_{i=1}^{n}[Y_i - (\hat{\beta}_0 + \hat{\beta}_1 x_i)]^2 + \sum_{i=1}^{m}[Y_i^* - (\hat{\beta}_0^* + \hat{\beta}_1^* x_i)]^2}{n + m - 4}}$$

Then T has a Student t distribution with $n + m - 4$ degrees of freedom.

b. To test $H_0: \beta_1 = \beta_1^*$ versus $H_1: \beta_1 \neq \beta_1^*$ at the α level of significance, reject H_0 if t is either (1) $\leq -t_{\alpha/2,n+m-4}$ or (2) $\geq t_{\alpha/2,n+m-4}$, where

$$t = \frac{\hat{\beta}_1 - \hat{\beta}_1^*}{s\sqrt{\dfrac{1}{\sum_{i=1}^{n}(x_i - \bar{x})^2} + \dfrac{1}{\sum_{i=1}^{m}(x_i^* - \bar{x}^*)^2}}}$$

(One-sided tests are defined in the usual way by replacing $\pm t_{\alpha/2,n+m-4}$ with either $t_{\alpha,n+m-4}$ or $-t_{\alpha,n+m-4}$.)

Example 11.3.4 Genetic variability is thought to be a key factor in the survival of a species, the idea being that "diverse" populations should have a better chance of coping with changing environments. Table 11.3.3 summarizes the results of a study designed to test that hypothesis experimentally [data slightly modified from (4)]. Two populations

Table 11.3.3

Date	Day no., $x(=x^*)$	Strain A pop^n, y	Strain B pop^n, y^*
Feb 2	0	100	100
May 13	100	250	203
Aug 21	200	304	214
Nov 29	300	403	295
Mar 8	400	446	330
Jun 16	500	482	324

of fruit flies (*Drosophila serrata*)—one that was cross-bred (Strain A) and the other, in-bred (Strain B)—were put into sealed containers where food and space were kept to a minimum. Recorded every hundred days were the numbers of *Drosophila* alive in each population.

Figure 11.3.5 shows a graph of the two sets of population figures. For both strains, growth was approximately linear over the period covered. Strain A, though, with an estimated slope of 0.74, increased at a faster rate than did Strain B, where the estimated slope was 0.45. The question is, do we have enough evidence here to reject the null hypothesis that the two true slopes are equal? Is the difference between 0.74 and 0.45, in other words, statistically significant?

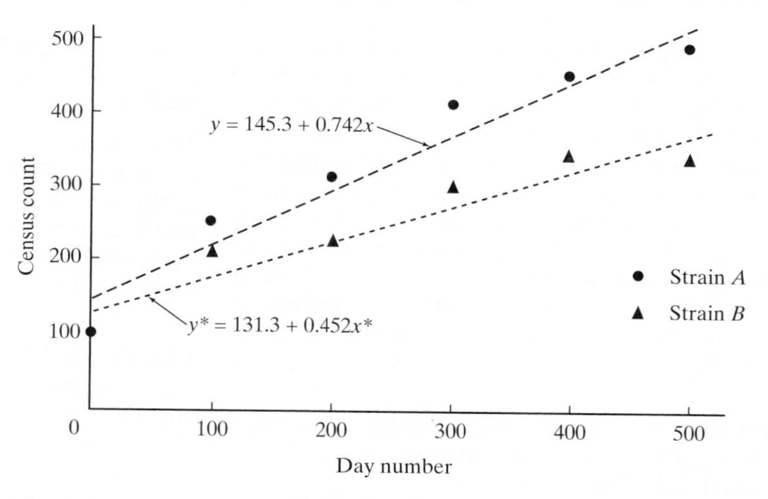

Figure 11.3.5

Let $\alpha = 0.05$ and let $(x_i, y_i), i = 1, 2, \ldots, 6$, and $(x_i^*, y_i^*), i = 1, 2, \ldots, 6$, denote the times and population sizes for Strain A and Strain B, respectively. Our objective is to test $H_0 : \beta_1 = \beta_1^*$ versus $H_1 : \beta_1 > \beta_1^*$. Rejecting H_0, of course, would support the contention that genetic variability benefits a species' chances of survival.

From Table 11.3.3, $\bar{x} = \bar{x}^* = 250$ and

$$\sum_{i=1}^{6} (x_i - \bar{x})^2 = \sum_{i=1}^{6} (x_i^* - \bar{x}^*)^2 = 175{,}000$$

Also,

$$\sum_{i=1}^{6} [y_i - (145.3 - 0.742x_i)]^2 = 5512.14$$

and

$$\sum_{i=1}^{6} [y_i^* - (131.3 + 0.452x_i^*)]^2 = 3960.14$$

so

$$s = \sqrt{\frac{5512.14 + 3960.14}{6 + 6 - 4}} = 34.41$$

Since H_1 is one-sided to the right, we should reject H_0 if $t \geq t_{.05,8} = 1.8595$. But

$$t = \frac{0.742 - 0.452}{34.41\sqrt{\frac{1}{175{,}000} + \frac{1}{175{,}000}}}$$
$$= 2.50$$

These data, then, *do* support the theory that genetically mixed populations have a better chance of survival in hostile environments. ■

Questions

11.3.16. Regression techniques can be very useful in situations where one variable—say, y—is difficult to measure but x is not. Once such an xy-relationship has been "calibrated," based on a set of (x_i, y_i)'s, future values of Y can be easily estimated using $\hat{\beta}_0 + \hat{\beta}_1 x$. Determining the volume of an irregularly shaped object, for example, is often difficult, but weighing that object is likely to be easy. The following table shows the weights (in kilograms) and the volumes (in cubic decimeters) of eighteen children between the ages of five and eight (13). The estimated regression line has the equation $y = -0.104 + 0.988x$, where $s = 0.202$.

(a) Construct a 95% confidence interval for $E(Y|14.0)$.
(b) Construct a 95% prediction interval for the volume of a child weighing 14.0 kilograms.

Weight, x	Volume, y	Weight, x	Volume, y
17.1	16.7	15.8	15.2
10.5	10.4	15.1	14.8
13.8	13.5	12.1	11.9
15.7	15.7	18.4	18.3
11.9	11.6	17.1	16.7
10.4	10.2	16.7	16.6
15.0	14.5	16.5	15.9
16.0	15.8	15.1	15.1
17.8	17.6	15.1	14.5

11.3.17. Construct a 95% confidence interval for $E(Y\,|\,2.750)$ using the connecting rod data given in Case Study 11.2.1.

11.3.18. For the CHD mortality data of Case Study 11.3.1, construct a 99% confidence interval for the expected death rate in a country where the cigarette consumption is 2500 per adult per year. Is a public health official more likely to be interested in a 99% confidence interval for $E(Y\,|\,2500)$ or a 99% prediction interval for Y when $x = 2500$?

11.3.19. The Master of Business Administration (M.B.A.) degree typically prepares its possessors for a high-salaried position, most often in business or industry. So, a reasonable measure of the effectiveness of an M.B.A. program is the median salary of its graduates five years after graduation. The table gives the tuition paid and the median five-year-out salary for graduates of sixteen highly ranked private M.B.A. programs.

University	Tuition ($ thousands)	Median Salary ($ thousands)
Wake Forest	71	108
Emory	81	121
SMU	81	122
Georgetown	83	147
USC	86	155
Vanderbilt	87	128

NYU	89	170
Cornell	92	168
Yale	93	160
Duke	93	148
Dartmouth	94	205
Northwestern	96	165
MIT	96	190
Chicago	97	210
Carnegie Mellon	98	145
Columbia	99	182

Source: www.forbes.com/lists/2009/95/best-business-schools-09_Best-Business-Schools.

Find the 95% confidence interval for $E(Y|102)$. Harvard's tuition during this time period was \$102,000. Does the interval include the Harvard graduate median salary of \$215,000?

11.3.20. In the radioactive exposure example in Question 11.2.9, find the 95% confidence interval for $E(Y|9.00)$ and the prediction interval for the value 9.00.

11.3.21. Attorneys representing a group of male buyers employed by Flirty Fashions are filing a reverse discrimination suit against the female-owned company. Central to their case are the following data, showing the relationship between years of service and annual salary for the firm's fourteen buyers, six of whom are men. The plaintiffs claim that the difference in slopes (0.606 for men versus 1.07 for women) is prima facie evidence that the company's salary policies discriminate against men. As the lawyer for Flirty Fashions, how would you respond? Use the following sums:

$$\sum_{i=1}^{6}(y_i - 21.3 - 0.606x_i)^2 = 5.983$$

and

$$\sum_{i=1}^{8}(y_i^* - 23.2 - 1.07x_i^*)^2 = 13.804$$

Also, $\sum_{i=1}^{6}(x_i - \bar{x})^2 = 31.33$ and $\sum_{i=1}^{8}(x_i^* - \bar{x}^*)^2 = 46$.

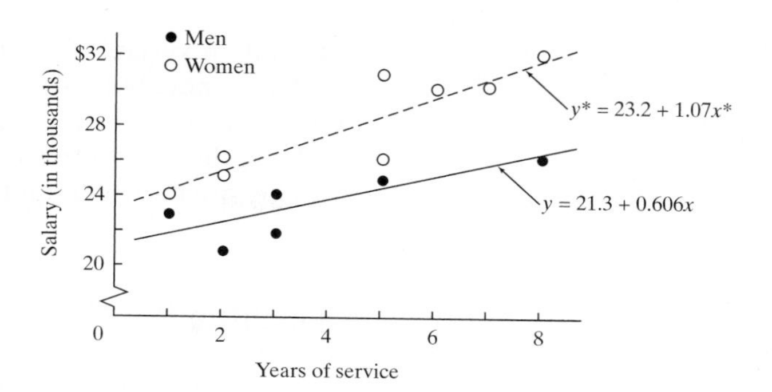

11.3.22. Polls taken during a city's last two administrations (one Democratic, one Republican) suggested that public support of the two mayors fell off linearly with years in office. Can it be concluded from the following data that the rates at which the two administrations lost favor were significantly different? Let $\alpha = 0.05$. (*Note*: $y = 69.3077 - 3.4615x$ with an estimated standard deviation of 0.9058 and $y^* = 59.9407 - 2.7373x^*$ with an estimated standard deviation of 1.2368.)

Democratic Mayor		Republican Mayor	
Years after Taking Office, x	Percent in Support, y	Years after Taking Office, x^*	Percent in Support, y^*
2	63	1	58
3	58	2	55
5	52	4	47
7	46	6	43
8	41	7	41
		8	39

11.3.23. Prove that the variance of $\hat{Y}$ can also be written

$$\text{Var}(\hat{Y}) = \frac{\sigma^2 \sum_{i=1}^{n}(x_i - x)^2}{n\sum_{i=1}^{n}(x_i - \bar{x})^2}$$

11.3.24. Show that

$$\sum_{i=1}^{n}(Y_i - \bar{Y})^2 = \sum_{i=1}^{n}(Y_i - \hat{Y}_i)^2 + \sum_{i=1}^{n}(\hat{Y}_i - \bar{Y})^2$$

for any set of points $(x_i, Y_i), i = 1, 2, \ldots, n$.

11.4 Covariance and Correlation

Our discussion of xy-relationships in Chapter 11 began with the simplest possible setup from a statistical standpoint—the case where the (x_i, y_i)'s are just numbers and have no probabilistic structure whatsoever. Then we examined the more complicated (and more "inference-friendly") scenario where x_i is a constant but Y_i

is a random variable. Introduced in this section is the next level of complexity—problems where both X_i and Y_i are assumed to be random variables. [Measurements of the form (x_i, y_i) or (x_i, Y_i) are typically referred to as *regression data*; observations satisfying the assumptions made in this section—that is, measurements of the form (X_i, Y_i)—are more commonly referred to as *correlation data*.]

Measuring the Dependence Between Two Random Variables

Given a pair of random variables, it makes sense to inquire how one varies *with respect to the other*. If X increases, for example, does Y also tend to increase? And if so, how strong is the dependence between the two?

The first step in addressing such questions was taken in Section 3.9 with the definition of covariance. In that section, its role was primarily as a tool for finding the variance of a sum of random variables. Here, it will serve as the basis for measuring the relationship between X and Y.

The Correlation Coefficient

The covariance of X and Y necessarily reflects the *units* of both random variables, which can make it difficult to interpret. In applied settings, it helps to have a *dimensionless* measure of dependency so that one xy-relationship can be compared to another. Dividing Cov(X, Y) by $\sigma_X\sigma_Y$ accomplishes not only that objective but also scales the quotient to be a number between -1 and $+1$.

Definition 11.4.1. Let X and Y be any two random variables. The *correlation coefficient of X and Y*, denoted $\rho(X, Y)$, is given by

$$\rho(X, Y) = \frac{\text{Cov}(X, Y)}{\sigma_X\sigma_Y} = \text{Cov}(X^*, Y^*)$$

where $X^* = (X - \mu_X)/\sigma_X$ and $Y^* = (Y - \mu_Y)/\sigma_Y$.

Theorem 11.4.1 *For any two random variables X and Y,*

a. $|\rho(X, Y)| \leq 1$.

b. $|\rho(X, Y)| = 1$ *if and only if* $Y = aX + b$ *for some constants a and b (except possibly on a set of probability zero).*

Proof Following the notation of Definition 11.4.1, let X^* and Y^* denote the standardized transformations of X and Y. Then

$$\begin{aligned} 0 \leq \text{Var}(X^* \pm Y^*) &= \text{Var}(X^*) \pm 2\,\text{Cov}(X^*, Y^*) + \text{Var}(Y^*) \\ &= 1 \pm 2\rho(X, Y) + 1 \\ &= 2[1 \pm \rho(X, Y)] \end{aligned}$$

But $1 \pm \rho(X, Y) \geq 0$ implies that $|\rho(X, Y)| \leq 1$, and part (a) of the theorem is proved.

Next, suppose that $\rho(X, Y) = 1$. Then $\text{Var}(X^* - Y^*) = 0$; however, a random variable with zero variance is constant, except possibly on a set of probability zero. From the constancy of $X^* - Y^*$, it readily follows that Y is a linear function of X. The case for $\rho(X, Y) = -1$ is similar.

The converse of part (b) is left as an exercise. □

Questions

11.4.1. Let X and Y have the joint pdf

$$f_{X,Y}(x, y) = \begin{cases} \frac{x+2y}{22}, & \text{for } (x, y) = (1, 1), (1, 3), (2, 1), (2, 3) \\ 0, & \text{elsewhere} \end{cases}$$

Find Cov(X, Y) and $\rho(X, Y)$.

11.4.2. Suppose that X and Y have the joint pdf

$$f_{X,Y}(x, y) = x + y, \quad 0 < x < 1, 0 < y < 1$$

Find $\rho(X, Y)$.

11.4.3. If the random variables X and Y have the joint pdf

$$f_{X,Y}(x, y) = \begin{cases} 8xy, & 0 \le y \le x \le 1 \\ 0, & \text{otherwise} \end{cases}$$

show that Cov$(X, Y) = \frac{8}{450}$. Calculate $\rho(X, Y)$.

11.4.4. Suppose that X and Y are discrete random variables with the joint pdf

(x, y)	$f_{X,Y}(x, y)$
(1, 2)	$\frac{1}{2}$
(1, 3)	$\frac{1}{4}$
(2, 1)	$\frac{1}{8}$
(2, 4)	$\frac{1}{8}$

Find the correlation coefficient between $X$and Y.

11.4.5. Prove that $\rho(a + bX, c + dY) = \rho(X, Y)$ for constants a, b, c, and d where b and d are positive. Note that this result allows for a change of scale to one convenient for computation.

11.4.6. Let the random variable X take on the values $1, 2, \ldots, n$, each with probability $1/n$. Define Y to be X^2. Find $\rho(X, Y)$ and $\lim_{n\to\infty} \rho(X, Y)$.

11.4.7. (a) For random variables X and Y, show that

$$\text{Cov}(X + Y, X - Y) = \text{Var}(X) - \text{Var}(Y)$$

(b) Suppose that Cov$(X, Y) = 0$. Prove that

$$\rho(X + Y, X - Y) = \frac{\text{Var}(X) - \text{Var}(Y)}{\text{Var}(X) + \text{Var}(Y)}$$

Estimating $\rho(X, Y)$: The Sample Correlation Coefficient

We conclude this section with an estimation problem. Suppose the correlation coefficient between X and Y is unknown, but we have some relevant information about its value in the form of n measurements $(X_1, Y_1), (X_2, Y_2), \ldots$, and (X_n, Y_n). How can we use those data to estimate $\rho(X, Y)$?

Since the correlation coefficient can be written in terms of various theoretical moments,

$$\rho(X, Y) = \frac{E(XY) - E(X)E(Y)}{\sqrt{\text{Var}(X)}\sqrt{\text{Var}(Y)}}$$

it would seem reasonable to estimate each component of $\rho(X, Y)$ with its corresponding *sample* moment. That is, let $\bar{X}$ and $\bar{Y}$ approximate $E(X)$ and $E(Y)$, replace $E(XY)$ with

$$\frac{1}{n}\sum_{i=1}^{n} X_i Y_i$$

and substitute

$$\frac{1}{n}\sum_{i=1}^{n}(X_i - \bar{X})^2 \quad \text{and} \quad \frac{1}{n}\sum_{i=1}^{n}(Y_i - \bar{Y})^2$$

for Var(X) and Var(Y), respectively.

We define the *sample correlation coefficient,* then, to be the ratio

$$R = \frac{\frac{1}{n}\sum_{i=1}^{n} X_i Y_i - \bar{X}\,\bar{Y}}{\sqrt{\frac{1}{n}\sum_{i=1}^{n}(X_i - \bar{X})^2}\sqrt{\frac{1}{n}\sum_{i=1}^{n}(Y_i - \bar{Y})^2}} \tag{11.4.1}$$

or, equivalently,

$$R = \frac{n\sum_{i=1}^{n} X_i Y_i - \left(\sum_{i=1}^{n} X_i\right)\left(\sum_{i=1}^{n} Y_i\right)}{\sqrt{n\sum_{i=1}^{n} X_i^2 - \left(\sum_{i=1}^{n} X_i\right)^2}\sqrt{n\sum_{i=1}^{n} Y_i^2 - \left(\sum_{i=1}^{n} Y_i\right)^2}} \tag{11.4.2}$$

(Sometimes R is referred to as the *Pearson product-moment correlation coefficient,* in honor of the eminent British statistician Karl Pearson.)

Questions

11.4.8. Derive Equation 11.4.2 from Equation 11.4.1.

11.4.9. Let $(x_1, y_1), (x_2, y_2), \ldots, (x_n, y_n)$ be a set of measurements whose sample correlation coefficient is r. Show that

$$r = \hat{\beta}_1 \cdot \frac{\sqrt{n\sum_{i=1}^{n} x_i^2 - \left(\sum_{i=1}^{n} x_i\right)^2}}{\sqrt{n\sum_{i=1}^{n} y_i^2 - \left(\sum_{i=1}^{n} y_i\right)^2}}$$

where $\hat{\beta}_1$ is the maximum likelihood estimate for the slope.

Interpreting R

The properties cited for $\rho(X, Y)$ in Theorem 11.4.1 are not sufficient to provide a useful interpretation of R. What does it mean, for example, to say that the sample correlation coefficient is 0.73, or 0.55, or −0.24? One way to answer such a question focuses on the *square* of R, rather than on R itself.

We know from Equation 11.3.3 that

$$\sum_{i=1}^{n}(y_i - \hat{\beta}_0 - \hat{\beta}_1 x_i)^2 = \sum_{i=1}^{n}(y_i - \bar{y})^2 - \hat{\beta}_1^2\sum_{i=1}^{n}(x_i - \bar{x})^2$$

Using the relationship between $\hat{\beta}_1$ and r in Question 11.4.9—together with the fact that $\sum_{i=1}^{n}(x_i - \bar{x})^2 = \sum_{i=1}^{n} x_i^2 - \left(\sum_{i=1}^{n} x_i\right)^2 \Big/ n$—we can write

$$\sum_{i=1}^{n}(y_i - \hat{\beta}_0 - \hat{\beta}_1 x_i)^2 = \sum_{i=1}^{n}(y_i - \bar{y})^2 - r^2 \cdot \frac{\sum_{i=1}^{n}(y_i - \bar{y})^2}{\sum_{i=1}^{n}(x_i - \bar{x})^2} \cdot \sum_{i=1}^{n}(x_i - \bar{x})^2$$

which reduces to

$$r^2 = \frac{\sum_{i=1}^{n}(y_i - \bar{y})^2 - \sum_{i=1}^{n}(y_i - \hat{\beta}_0 - \hat{\beta}_1 x_i)^2}{\sum_{i=1}^{n}(y_i - \bar{y})^2} \tag{11.4.3}$$

Equation 11.4.3 has a nice, simple interpretation. Notice that

1. $\sum_{i=1}^{n} (y_i - \bar{y})^2$ represents the *total variability* in the dependent variable—that is, the extent to which the y_i's are not all the same.
2. $\sum_{i=1}^{n} (y_i - \hat{\beta}_0 - \hat{\beta}_1 x_i)^2$ represents the variation in the y_i's *not explained* (or accounted for) by the linear regression with x.
3. $\sum_{i=1}^{n} (y_i - \bar{y})^2 - \sum_{i=1}^{n} (y_i - \hat{\beta}_0 - \hat{\beta}_1 x_i)^2$ represents the variation in the y_i's that *is explained* by the linear regression with x.

Therefore, r^2 is the *proportion of the total variation in the y_i's that can be attributed to the linear relationship with x*. So, if $r = 0.60$, we can say that *36%* of the variation in Y is explained by the linear regression with X (and that 64% is associated with other factors).

Comment The quantity r^2 is sometimes called the *coefficient of determination*.

Case Study 11.4.1

The Scholastic Aptitude Test (SAT) is widely used by colleges and universities to help choose their incoming classes. It was never designed to measure the quality of education provided by secondary schools, but critics and supporters alike seem increasingly intent on forcing it into that role. The problem is that average SAT scores associated with schools or districts or states reflect a variety of factors, some of which have little or nothing to do with the quality of instruction that students are receiving.

Table 11.4.1 shows one testing period's average SAT scores (Y), by state, as a function of participation rate (X), where the SAT score is the sum of the Critical Reading, Math, and Writing subtest scores. As Figure 11.4.1 suggests, there appears to be a strong dependency between the two measurements—as a state's participation rate goes down, its average SAT score goes up. In South Dakota, for example, only 3% of the students eligible to take the test actually did; in New York, the participation rate was a dramatically larger 84%. The average SAT score in New York was 1473; in South Dakota the average score of 1766 was 20% higher.

A good way to quantify the overall relationship between test scores and participation rates is to calculate the data's sample correlation coefficient, r.

From Table 11.4.1, we can calculate the sums necessary to evaluate Equation 11.4.2:

$$\sum_{i=1}^{51} x_i = 1{,}891 \qquad \sum_{i=1}^{51} y_i = 81{,}396$$

$$\sum_{i=1}^{51} x_i^2 = 114{,}983 \qquad \sum_{i=1}^{51} y_i^2 = 130{,}597{,}738$$

$$\sum_{i=1}^{51} x_i y_i = 2{,}863{,}056$$

(Continued on next page)

(Case Study 11.4.1 continued)

Table 11.4.1

State	Participation Rate, x	Average SAT Score, y	State	Participation Rate, x	Average SAT Score, y
AL	8%	1676	MT	24%	1612
AK	45%	1533	NE	5%	1733
AZ	26%	1538	NV	40%	1482
AR	5%	1701	NH	74%	1555
CA	48%	1512	NJ	76%	1504
CO	21%	1687	NM	12%	1645
CT	83%	1535	NY	84%	1473
DE	70%	1487	NC	63%	1489
DC	84%	1390	ND	3%	1766
FL	54%	1474	OH	24%	1599
GA	70%	1466	OK	6%	1701
HI	58%	1453	OR	53%	1552
ID	18%	1597	PA	71%	1478
IL	7%	1762	RI	66%	1486
IN	62%	1485	SC	61%	1461
IA	3%	1797	SD	3%	1766
KS	7%	1733	TN	11%	1707
KY	8%	1692	TX	50%	1473
LA	7%	1688	UT	6%	1661
ME	87%	1396	VT	64%	1549
MD	69%	1498	VA	68%	1522
MA	83%	1552	WA	52%	1568
MI	6%	1751	WV	19%	1511
MN	8%	1784	WI	5%	1768
MS	3%	1696	WY	6%	1677
MO	5%	1775			

Source: professionals.collegeboard.com/profdownload/cbs-08-Page-3-Table-3.pdf.

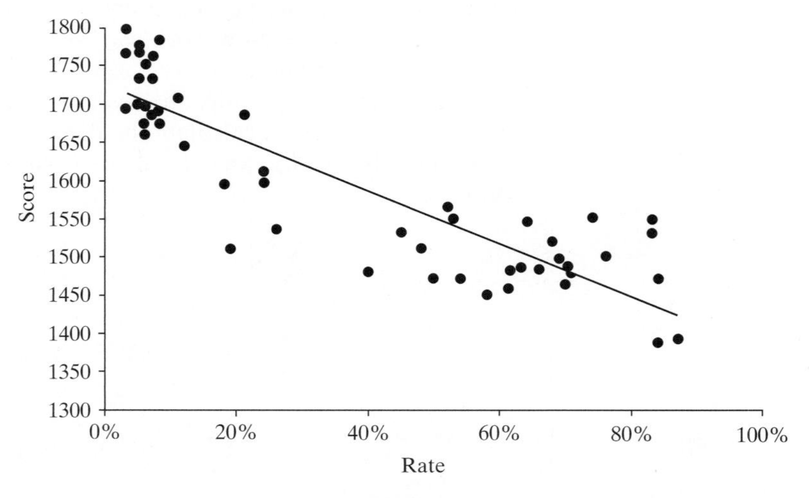

Figure 11.4.1

(Continued on next page)

Substituting the sums into the formula for r, then, shows that the sample coefficient is -0.881:

$$r = \frac{51(2{,}863{,}056) - (1{,}891)(81{,}396)}{\sqrt{51(114{,}983) - (1891)^2}\sqrt{51(130{,}597{,}738) - (81{,}396)^2}}$$
$$= -0.881$$

Since $r^2 = (-0.881)^2 = 0.776$, we can say that 77.6% of the variability in SAT scores from state to state can be attributed to the linear relationship between test scores and participation rates.

About the Data The magnitude of r^2 for these data should be a clear warning that comparing average SATs at face value from state to state or school system to school system is largely meaningless. It would make more sense to examine the residuals associated with $y = \hat{\beta}_0 + \hat{\beta}_1 x$. States with particularly large positive values for $y - \hat{y}$ may be doing something that other states might be well advised to copy.

Questions

11.4.10. In Case Study 11.3.1, how much of the variability in CHD mortality is explained by cigarette consumption?

11.4.11. Some baseball fans believe that the number of home runs a team hits is markedly affected by the altitude of the club's home park. The rationale is that the air is thinner at the higher altitudes, and balls would be expected to travel farther. The following table shows the altitudes (X) of American League ballparks and the number of home runs (Y) that each team hit during a recent season (172). Calculate the sample correlation coeffient, r, using the sums below. What would you conclude?

$$\sum_{i=1}^{12} x_i = 4936 \qquad \sum_{i=1}^{12} y_i = 1175$$
$$\sum_{i=1}^{12} x_i^2 = 3{,}071{,}116 \qquad \sum_{i=1}^{12} y_i^2 = 123{,}349$$
$$\sum_{i=1}^{12} x_i y_i = 480{,}565$$

Club	Altitude, x	Number of Home Runs, y
Cleveland	660	138
Milwaukee	635	81
Detroit	585	135
New York	55	90
Boston	21	120
Baltimore	20	84
Minnesota	815	106
Kansas City	750	57
Chicago	595	109
Texas	435	74
California	340	61
Oakland	25	120

11.4.12. The following table shows U.S. corn supplies (in millions of bushels) and corn prices (dollars per bushel rounded to the nearest $0.10) for the years 1999 through 2008. Calculate the sample correlation coefficient, r. The sums for the data in the table are:

$$\sum_{i=1}^{10} x_i = 123.1 \qquad \sum_{i=1}^{10} y_i = 25.80$$
$$\sum_{i=1}^{10} x_i^2 = 1529.63 \qquad \sum_{i=1}^{10} y_i^2 = 74.00$$
$$\sum_{i=1}^{10} x_i y_i = 325.08$$

Year	Supply, x	Price, y
1999	11.2	$1.70
2000	11.8	1.80
2001	11.5	2.00
2002	10.6	2.40
2003	11.2	2.50
2004	12.8	2.10
2005	13.1	2.00
2006	12.6	3.00
2007	14.5	4.20
2008	13.8	4.10

Source: USDA WASDE report 1.12.10, www.agmanager.info.

11.4.13. The extent to which stress is a contributing factor to the severity of chronic illnesses was the focus of the study summarized in the following table (208). Seventeen conditions were compared on a Seriousness of Illness Rating Scale (SIRS). Patients with each of these conditions were asked to fill out a Schedule of Recent Experience (SRE) questionnaire. Higher scores on the SRE reflect presumably greater levels of stress. How much of the variation in the SIRS values can be attributed to the linear regression with SRE?

Admitting Diagnosis	Average SRE, x	SIRS, y
Dandruff	26	21
Varicose veins	130	173
Psoriasis	317	174
Eczema	231	204
Anemia	325	312
Hyperthyroidism	816	393
Gallstones	563	454
Arthritis	312	468
Peptic ulcer	603	500
High blood pressure	405	520
Diabetes	599	621
Emphysema	357	636
Alcoholism	688	688
Cirrhosis	443	733
Schizophrenia	609	776
Heart failure	772	824
Cancer	777	1020

Use the following sums:

$$\sum_{i=1}^{17} x_i = 7{,}973 \qquad \sum_{i=1}^{17} y_i = 8{,}517$$

$$\sum_{i=1}^{17} x_i^2 = 4{,}611{,}291 \qquad \sum_{i=1}^{17} y_i^2 = 5{,}421{,}917$$

$$\sum_{i=1}^{17} x_i y_i = 4{,}759{,}470$$

11.4.14. Among the many strategies that investors use to try to predict trends in the stock market is the "early warning" system, which is based on the premise that what the market does in the first week in January is indicative of what it will do over the next twelve months. Listed in the following table for the eighteen years from 1991 through 2008 are x, the percentage change in the Dow Jones Industrial Average for the first week in January, and y, the percentage change for the entire year. Quantify the strength of the linear relationship between X and Y. Use the following sums:

$$\sum_{i=1}^{18} x_i = -0.9 \qquad \sum_{i=1}^{18} y_i^2 = 160.2$$

$$\sum_{i=1}^{18} x_i^2 = 92.63 \qquad \sum_{i=1}^{18} y_i^2 = 6437.68$$

$$\sum_{i=1}^{18} x_i y_i = 221.37$$

Year	% Change for First Week in January, x	% Change for Year, y
1991	−2.6	24.8
1992	−0.1	1.6
1993	−1.5	15.7
1994	1.7	2.1
1995	0.9	33.5
1996	1.2	28.2
1997	2.4	21.7
1998	−5.1	21.1
1999	4.8	25.5
2000	0.2	−6.2
2001	−1.2	−6.0
2002	−2.7	−16.2
2003	2.1	21.0
2004	0.5	1.9
2005	−1.7	−0.6
2006	2.2	16.3
2007	−0.5	7.2
2008	−1.5	−31.4

Source: finance.yahoo.com/q/hp?s=%5EDJI.

11.5 The Bivariate Normal Distribution

The singular importance of the normal distribution in univariate inference procedures should, by now, be abundantly clear. In dealing with problems that involve *two* random variables—for example, the calculation of $\rho(X, Y)$—it should come as no surprise that the most frequently encountered *joint* pdf, $f_{X,Y}(x, y)$, is a bivariate version of the normal curve. Our objectives in this section are twofold: (1) to deduce the form of the bivariate normal from basic principles and (2) to identify the particular properties of that pdf that pertain to the problem of assessing the nature of the dependence between X and Y.

Generalizing the Univariate Normal pdf

At this point, we know many things about the *univariate* normal pdf,

$$f_Y(y) = \frac{1}{\sqrt{2\pi}\,\sigma} e^{-\frac{1}{2}\left(\frac{y-\mu}{\sigma}\right)^2}, \quad -\infty < y < \infty$$

Sections upon sections have been devoted to estimating and testing its parameters, studying its transformations, and learning about its role as an approximation to the distribution of sums and averages. What has not been discussed is the generalization of $f_Y(y)$ itself, to a *bivariate, trivariate,* or *multivariate* pdf.

Given the mathematical complexities inherent in the univariate normal pdf, it should come as no surprise that its extension to higher dimensions is not a simple matter. In the bivariate case, for example, which is the only generalization we will consider, $f_{X,Y}(x, y)$ has *five* different parameters and its functional form is decidely unpleasant.

We will begin by "constructing" a bivariate normal pdf, $f_{X,Y}(x, y)$, using properties suggested by what we already know holds true for the univariate normal, $f_Y(y)$. As a first condition to impose, it seems reasonable to require that the marginal pdfs associated with $f_{X,Y}(x, y)$ be univariate normal densities. It will be sufficient to consider the case where the two marginals are *standard* normals.

If X and Y are *independent* standard normal random variables,

$$f_{X,Y}(x, y) = \frac{1}{2\pi} e^{-\frac{1}{2}(x^2+y^2)}, \quad \begin{array}{l} -\infty < x < \infty \\ -\infty < y < \infty \end{array} \tag{11.5.1}$$

Notice that the simplest extension of $f_{X,Y}(x, y)$ in Equation 11.5.1 is to replace $-\frac{1}{2}(x^2 + y^2)$ with $-\frac{1}{2}c(x^2 + uxy + y^2)$, or, equivalently, with $-\frac{1}{2}c(x^2 - 2vxy + y^2)$, where c and v are constants. The desired joint pdf, then, would have the general form

$$f_{X,Y}(x, y) = Ke^{-\frac{1}{2}c(x^2-2vxy+y^2)} \tag{11.5.2}$$

where K is the constant that makes the double integral of $f_{X,Y}(x, y)$ from $-\infty$ to ∞ equal to 1.

Now, what must be true of K, c, and v if the marginal pdfs based on $f_{X,Y}(x, y)$ are to be standard normals? Note, first, that completing the square in the exponent makes

$$\begin{aligned} x^2 - 2vxy + y^2 &= x^2 - v^2x^2 + (y^2 - 2vxy + v^2x^2) \\ &= (1 - v^2)x^2 + (y - vx)^2 \end{aligned}$$

so

$$f_{X,Y}(x, y) = Ke^{-\frac{1}{2}c(1-v^2)x^2} e^{-\frac{1}{2}c(y-vx)^2}$$

The exponents, though, must be negative, which implies that $1 - v^2 > 0$, or, equivalently, $|v| < 1$.

To find K, we start by calculating

$$\begin{aligned}
\int_{-\infty}^{\infty}\int_{-\infty}^{\infty} & e^{-(1/2)c(1-v^2)x^2} \cdot e^{-(1/2)c(y-vx)^2}\,dy\,dx \\
&= \int_{-\infty}^{\infty} e^{-(1/2)c(1-v^2)x^2}\left[\int_{-\infty}^{\infty} e^{-(1/2)c(y-vx)^2}dy\right]dx \\
&= \int_{-\infty}^{\infty} e^{-(1/2)c(1-v^2)x^2}\left(\frac{\sqrt{2\pi}}{\sqrt{c}}\right)dx \\
&= \frac{\sqrt{2\pi}}{\sqrt{c}}\,\frac{\sqrt{2\pi}}{\sqrt{c}\sqrt{1-v^2}} \\
&= \frac{2\pi}{c\sqrt{1-v^2}}
\end{aligned}$$

It follows that

$$K = \frac{c\sqrt{1-v^2}}{2\pi}$$

The constant c can be any positive value, but a convenient choice proves to be $c = 1/(1-v^2)$. Substituting K and c, then, into Equation 11.5.2 gives

$$\begin{aligned}
f_{X,Y}(x,y) &= \frac{1}{2\pi\sqrt{1-v^2}}e^{-(1/2)[1/(1-v^2)](x^2-2vxy+y^2)} \\
&= \frac{1}{2\pi\sqrt{1-v^2}}e^{-x^2}\cdot e^{-(1/2)[1/(1-v^2)](y-vx)^2}
\end{aligned} \tag{11.5.3}$$

Recall that our choice of the form of $f_{X,Y}(x,y)$ was predicated on a wish for the marginal pdfs to be normal. A simple integration shows that to be the case:

$$\begin{aligned}
f_X(x) &= \int_{-\infty}^{\infty} f_{X,Y}(x,y)\,dy \\
&= \frac{1}{2\pi\sqrt{1-v^2}}e^{-(1/2)x^2}\int_{-\infty}^{\infty} e^{-(1/2)[1/(1-v^2)](y-vx)^2}\,dy \\
&= \frac{1}{2\pi\sqrt{1-v^2}}e^{-(1/2)x^2}\cdot\sqrt{2\pi}\sqrt{1-v^2} \\
&= \frac{1}{\sqrt{2\pi}}e^{-(1/2)x^2}
\end{aligned}$$

Since $f_{X,Y}(x,y)$ is symmetric in x and y, $f_Y(y)$ is also the standard normal.

The constant v is actually the correlation coefficient between X and Y. Since $E(X) = E(Y) = 0$ and $\sigma_X = \sigma_Y = 1$,

$$\begin{aligned}
\rho(X,Y) = E(XY) &= \int_{-\infty}^{\infty}\int_{-\infty}^{\infty} xy\, f_{X,Y}(x,y)\,dx\,dy \\
&= \frac{1}{\sqrt{2\pi}}\int_{-\infty}^{\infty} xe^{-(1/2)x^2}\left[\frac{1}{\sqrt{2\pi}\sqrt{1-v^2}}\int_{-\infty}^{\infty} ye^{-(1/2)[1/(1-v^2)](y-vx)^2}dy\right]dx \\
&= \frac{1}{\sqrt{2\pi}}\int_{-\infty}^{\infty} xe^{-(1/2)x^2}\cdot vx\,dx \quad \text{(why?)} \\
&= v\frac{1}{\sqrt{2\pi}}\int_{-\infty}^{\infty} x^2e^{-(1/2)x^2}dx = v\,\mathrm{Var}(X) = v
\end{aligned}$$

Finally, we can replace x with $(x - \mu_X)/\sigma_X$ and y with $(y - \mu_Y)/\sigma_Y$. Doing so requires that the original pdf be multiplied by the derivative of both the X-transformation and the Y-transformation—that is, by $\frac{1}{\sigma_X \sigma_Y}$ [see (102)].

Definition 11.5.1. Let X and Y be random variables with joint pdf

$$f_{X,Y}(x, y) = \frac{1}{2\pi\sigma_X\sigma_Y\sqrt{1-\rho^2}}$$

$$\cdot \exp\left\{-\frac{1}{2}\left(\frac{1}{1-\rho^2}\right)\left[\frac{(x-\mu_X)^2}{\sigma_X^2} - 2\rho\frac{x-\mu_X}{\sigma_X}\cdot\frac{y-\mu_Y}{\sigma_Y} + \frac{(y-\mu_Y)^2}{\sigma_Y^2}\right]\right\}$$

for all x and y. Then X and Y are said to have the *bivariate normal distribution* (with parameters $\mu_X, \sigma_X^2, \mu_Y, \sigma_Y^2$, and ρ).

Comment For bivariate normal densities, $\rho(X, Y) = 0$ implies that X and Y are independent, a result not true in general.

Properties of the Bivariate Normal Distribution

Francis Galton, the renowned British biologist and scientist, perhaps more than any other person was responsible for launching *regression analysis* as a worthwhile field of statistical inquiry. Galton was a redoubtable data analyst whose keen insight enabled him to intuit much of the basic mathematical structure that we now associate with correlation and regression.

One of his more famous endeavors (58) was an examination of the relationship between parents' heights (X) and their adult children's heights (Y). Those particular variables have a bivariate normal distribution, the mathematical properties of which Galton knew nothing. Just by looking at cross-tabulations of X and Y, though, Galton postulated that (1) the marginal distributions of X and Y are both normal, (2) $E(Y \mid x)$ is a linear function of x, and (3) $\text{Var}(Y \mid x)$ is constant with x. As Theorem 11.5.1 shows, all of his empirically based deductions proved to be true.

Theorem 11.5.1 *Suppose that X and Y are random variables having the bivariate normal distribution given in Definition 11.5.1. Then*

a. *$f_X(x)$ is a normal pdf with mean μ_X and variance σ_X^2; $f_Y(y)$ is a normal pdf with mean μ_Y and variance σ_Y^2.*
b. *ρ is the correlation coefficient between X and Y.*
c. $E(Y \mid x) = \mu_Y + \frac{\rho\sigma_Y}{\sigma_X}(x - \mu_X)$.
d. $\text{Var}(Y \mid x) = (1 - \rho^2)\sigma_Y^2$.

Proof We have already established (a) and (b). Properties (c) and (d) will be examined for the special case $\mu_X = \mu_Y = 0$ and $\sigma_x = \sigma_y = 1$. The extension to arbitrary μ_X, μ_Y, σ_X, and σ_Y is straightforward.

First, note that

$$f_{Y|x}(y) = \frac{f_{X,Y}(x, y)}{f_X(x)}$$

$$= \frac{\frac{1}{2\pi\sqrt{1-\rho^2}} e^{-(1/2)x^2} e^{-(1/2)[1/(1-\rho^2)](y-\rho x)^2}}{\frac{1}{\sqrt{2\pi}} e^{-(1/2)x^2}}$$

$$= \frac{1}{\sqrt{2\pi}\sqrt{1-\rho^2}} e^{-(1/2)[1/(1-\rho^2)](y-\rho x)^2} \tag{11.5.4}$$

By inspection, we see that Equation 11.5.4 is the pdf of a normal random variable with mean ρx and variance $1-\rho^2$. Therefore, $E(Y \mid x) = \rho x$ and $\text{Var}(Y \mid x) = 1-\rho^2$. Replacing y with $(y-\mu_Y)/\sigma_Y$ and x with $(x-\mu_X)/\sigma_X$ gives the desired results. □

Comment The term *regression line* derives from a consequence of part (c) of Theorem 11.5.1. Suppose we make the simplifying assumption that $\mu_X = \mu_Y = \mu$ and $\sigma_X = \sigma_Y$. Then part (c) reduces to

$$E(Y \mid x) - \mu = \rho(X, Y)(x - \mu)$$

But recall that $|\rho(X, Y)| \leq 1$—and, in this case, $0 < \rho(X, Y) < 1$. Here, the positive sign of $\rho(X, Y)$ tells us that, on the average, tall parents have tall children. However, $\rho(X, Y) < 1$ means (again, *on the average*) that the children's heights are closer to the mean than are the parents'. Galton called this phenomenon "regression to mediocrity."

Questions

11.5.1. Suppose that X and Y have a bivariate normal pdf with $\mu_X = 3$, $\mu_Y = 6$, $\sigma_X^2 = 4$, $\sigma_Y^2 = 10$, and $\rho = \frac{1}{2}$. Find $P(5 < Y < 6\frac{1}{2})$ and $P(5 < Y < 6\frac{1}{2} \mid x = 2)$.

11.5.2. Suppose that X and Y have a bivariate normal distribution with $\text{Var}(X) = \text{Var}(Y)$.

(a) Show that X and $Y - \rho X$ are independent.
(b) Show that $X + Y$ and $X - Y$ are independent. [*Hint*: See Question 11.4.7(a).]

11.5.3. Suppose that X and Y have a bivariate normal distribution.

(a) Prove that $X + Y$ has a normal distribution when X and Y are standard normal random variables.
(b) Find $E(cX + dY)$ and $\text{Var}(cX + dY)$ in terms of $\mu_X, \mu_Y, \sigma_X, \sigma_Y$, and $\rho(X, Y)$, where X and Y are arbitrary normal random variables.

11.5.4. Suppose that the random variables X and Y have a bivariate normal pdf with $\mu_X = 56$, $\mu_Y = 11$, $\sigma_X^2 = 1.2$, $\sigma_Y^2 = 2.6$, and $\rho = 0.6$. Compute $P(10 < Y < 10.5 \mid x = 55)$. Suppose that $n = 4$ values were to be observed with x fixed at 55. Find $P(10.5 < \bar{Y} < 11 \mid x = 55)$.

11.5.5. If the joint pdf of the random variables X and Y is

$$f_{X,Y}(x, y) = ke^{-(2/3)[(1/4)x^2 - (1/2)xy + y^2]}$$

find $E(X)$, $E(Y)$, $\text{Var}(X)$, $\text{Var}(Y)$, $\rho(X, Y)$, and k.

11.5.6. Give conditions on $a > 0$, $b > 0$, and u so that

$$f_{X,Y}(x, y) = ke^{-(ax^2 - 2uxy + by^2)}$$

is the bivariate normal density of random variables X and Y each having expected value 0. Also, find $\text{Var}(X)$, $\text{Var}(Y)$, and $\rho(X, Y)$.

Estimating Parameters in the Bivariate Normal pdf

The five parameters in $f_{X,Y}(x, y)$ can be estimated in the usual way with the method of maximum likelihood. Given a random sample of size n from $f_{X,Y}(x, y)$—(x_1, y_1),

$(x_2, y_2), \ldots, (x_n, y_n)$—we define $L = \prod_{i=1}^{n} f_{X,Y}(x_i, y_i)$ and take the derivatives of ln L with respect to each of the parameters. Solved simultaneously, the resulting five equations (each derivative set equal to 0) yield the maximum likelihood estimators given in Theorem 11.5.2. Details of the derivation will be left as an exercise.

Theorem 11.5.2 *Given that $f_{X,Y}(x, y)$ is a bivariate normal pdf, the maximum likelihood estimators for μ_X, μ_Y, σ_X^2, σ_Y^2, and ρ, assuming that all five are unknown, are $\bar{X}$, $\bar{Y}$, $\left(\frac{1}{n}\right)\sum_{i=1}^{n}(X_i - \bar{X})^2$, $\left(\frac{1}{n}\right)\sum_{i=1}^{n}(Y_i - \bar{Y})^2$, and R, respectively.* □

Testing H_0: $\rho = 0$

If X and Y have a bivariate normal distribution, testing whether the two variables are independent is equivalent to testing whether their correlation coefficient, ρ, equals 0 (recall the Comment following Definition 11.5.1). Two different procedures are widely used for testing H_0: $\rho = 0$. One is an exact test based on the T_{n-2} random variable given in Theorem 11.5.3; the other is an approximate test based on the standard normal distribution.

Theorem 11.5.3 *Let $(X_1, Y_1), (X_2, Y_2), \ldots, (X_n, Y_n)$ be a random sample of size n drawn from a bivariate normal distribution, and let R be the sample correlation coefficient. Under the null hypothesis that $\rho = 0$, the statistic*

$$T_{n-2} = \frac{\sqrt{n-2}\,R}{\sqrt{1-R^2}}$$

has a Student t distribution with $n-2$ degrees of freedom.

Proof See (49). □

Example 11.5.1 Table 11.5.1 gives the mean temperature for twenty successive days in April and the average daily butterfat content in the milk of ten cows (138). Can we conclude that temperature and butterfat content have a nonzero correlation?

Let ρ denote the true correlation coefficient between X and Y. The hypotheses to be tested are

$$H_0: \rho = 0$$

versus

$$H_1: \rho \neq 0$$

Let $\alpha = 0.05$. Given that $n = 20$, the statistic

$$t = \frac{\sqrt{n-2} \cdot r}{\sqrt{1-r^2}}$$

follows a Student t distribution with 18 df (if H_0: $\rho = 0$ is true). That being the case, the null hypothesis will be rejected if t is either (1) $\leq -2.1009\,(= -t_{0.025,18})$ or (2) $\geq +2.1009 (= t_{0.025,18})$.

Table 11.5.1

Date	Temperature, x	Percent Butterfat, y
April 3	64	4.65
4	65	4.58
5	65	4.67
6	64	4.60
7	61	4.83
8	55	4.55
9	39	5.14
10	41	4.71
11	46	4.69
12	59	4.65
13	56	4.36
14	56	4.82
15	62	4.65
16	37	4.66
17	37	4.95
18	45	4.60
19	57	4.68
20	58	4.65
21	60	4.60
22	55	4.46

For the data in Table 11.5.1,

$$\sum_{i=1}^{20} x_i = 1{,}082 \qquad \sum_{i=1}^{20} y_i = 93.5$$

$$\sum_{i=1}^{20} x_i^2 = 60{,}304 \qquad \sum_{i=1}^{20} y_i^2 = 437.6406$$

$$\sum_{i=1}^{20} x_i y_i = 5{,}044.5$$

so

$$r = \frac{20(5{,}044.5) - (1{,}082)(93.5)}{\sqrt{20(60{,}304) - (1{,}082)^2}\sqrt{20(437.6406) - (93.5)^2}} = -0.453$$

Therefore,

$$t = \frac{\sqrt{n-2}\cdot r}{\sqrt{1-r^2}} = \frac{\sqrt{18}(-0.453)}{\sqrt{1-(-0.453)^2}} = -2.156$$

and our conclusion is *reject* H_0—it would appear that temperature and butterfat content are not independent.

Comment An alternate approach to testing H_0: $\rho = 0$ was given by Fisher (46). He showed that the statistic

$$\frac{1}{2}\ln\frac{1+R}{1-R}$$

is asymptotically normal with mean $\frac{1}{2}\ln[(1+\rho)/(1-\rho)]$ and variance approximately $1/(n-3)$. Fisher's formulation makes it relatively easy to determine the power of a correlation test—a computation that would be much more difficult if the inference had to be based on $\sqrt{n-2}\,R/\sqrt{1-R^2}$.

■

Questions

11.5.7. What would the conclusion be for the test of Example 11.5.1 if $\alpha = 0.01$?

11.5.8. In a study of heart disease (73), the weight (in pounds) and the blood cholesterol (in mg/dl) of fourteen men without a history of coronary incidents were recorded. At the $\alpha = 0.05$ level, can we conclude from these data that the two variables are independent?

Subject	Weight, x	Cholesterol, y
1	168	135
2	175	403
3	173	294
4	158	312
5	154	311
6	214	222
7	176	302
8	262	269
9	181	311
10	143	286
11	140	403
12	187	244
13	163	353
14	164	252

The data in the table give the following sums:

$$\sum_{i=1}^{14} x_i = 2,458 \qquad \sum_{i=1}^{14} y_i = 4,097$$

$$\sum_{i=1}^{14} x_i^2 = 444,118 \qquad \sum_{i=1}^{14} y_i^2 = 1,262,559$$

$$\sum_{i=1}^{14} x_i y_i = 710,499$$

11.5.9. Recall the baseball data in Question 11.4.11. Test whether home run frequency and home park altitude are independent. Let $\alpha = 0.05$.

11.5.10. Test H_0: $\rho = 0$ versus H_1: $\rho \neq 0$ for the SRE/SIRS data described in Question 11.4.13. Let 0.01 be the level of significance.

11.5.11. The National Collegiate Athletic Association has had a long-standing concern about the graduation rate of athletes. Under the urging of the Association, some prominent athletic programs increased the funds for tutoring athletes. The table below gives the amount spent (in millions of dollars) and the resulting percentage of athletes graduating in 2007. Test $H_0: \rho = 0$ versus $H_1: \rho > 0$ at the 0.10 level of significance.

University	Money Spent on Athletes Tutoring, x	Graduation Rate 2007, y
Minnesota	1.61	72
Kansas	1.61	70
Florida	1.67	87
LSU	1.74	69
Georgia	1.77	70
Tennessee	1.83	78
Kentucky	1.86	73
Ohio St.	1.89	78
Texas	1.90	72
Oklahoma	2.45	69

Source: Pensacola News Journal (Florida), December 21, 2008.

11.6 Taking a Second Look at Statistics (How *Not* to Interpret the Sample Correlation Coefficient)

Of all the "numbers" that statisticians and experimenters routinely compute, the correlation coefficient is one of the most frequently *misinterpreted*. Two errors in particular are common. First, there is a tendency to assume, either implicitly or explicitly, that a high sample correlation coefficient implies causality. It does not. Even if the linear relationship between x and y is perfect—that is, even if $r = -1$ or $r = +1$—we cannot conclude that X *causes* Y (or that Y *causes* X). The sample correlation coefficient is simply a measure of the strength of a linear relationship. *Why* the xy-relationship exists in the first place is a different question altogether.

George Bernard Shaw (an unlikely contributor to a mathematics text!) described elegantly the fallacy of using statistical relationships to infer underlying causality. Commenting on the "correlations" that exist between lifestyle and health, he wrote in *The Doctor's Dilemma* (163):

> It is easy to prove that the wearing of tall hats and the carrying of umbrellas enlarges the chest, prolongs life, and confers comparative immunity from disease; for the statistics show that the classes which use these articles are bigger, healthier, and live longer than the class which never dreams of possessing such things. It does not take much perspicacity to see that what really makes this difference is not the tall hat and the umbrella, but the wealth and nourishment of which they are evidence, and that a gold watch or membership of a club in Pall Mall might be proved in the same way to have the like sovereign virtues. A university degree, a daily bath, the owning of thirty pairs of trousers, a knowledge of Wagner's music, a pew in church, anything, in short, that implies more means and better nurture than the mass of laborers enjoy, can be statistically palmed off as a magic-spell conferring all sorts of privileges.

Examples of "spurious" correlations similar to those cited by Shaw are disturbingly commonplace. Between 1875 and 1920, for example, the correlation between the annual birthrate in Great Britain and the annual production of pig iron in the United States was an almost "perfect" -0.98. High correlations have also been found between salaries of Presbyterian ministers in Massachusetts and the price of rum in Havana and between the academic achievement of U.S. schoolchildren and the number of miles they live from the Canadian border. All too often, what looks like a cause is not a cause at all, but simply the effect of one or more factors that were not even measured. Researchers need to be very careful not to read more into the value of r than the number legitimately implies.

The second error frequently made when interpreting sample correlation coefficients is to forget that r measures the strength of a *linear* relationship. It says nothing about the strength of a *curvilinear* relationship. Computing r for the points shown in Figure 11.6.1, for example, is totally inappropriate. The (x_i, y_i) values in that scatterplot are clearly related but not in a linear way. Quoting the value of r would be misleading.

Figure 11.6.1

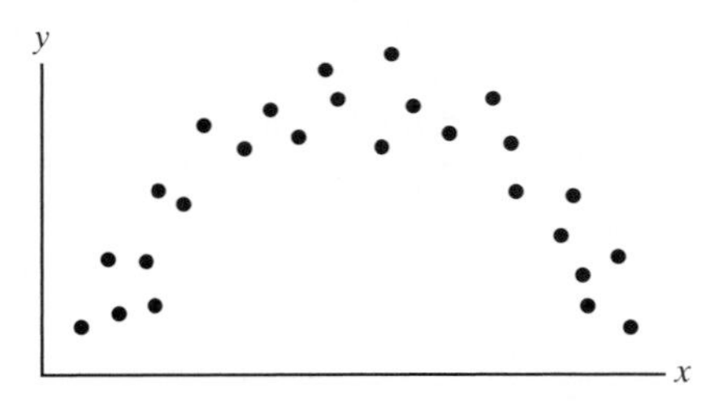

The lesson to be learned from Figure 11.6.1 is clear—*always graph the data!* No correlation coefficient should ever be calculated (much less interpreted) without first plotting the (x_i, y_i)'s to make certain that the underlying relationship is linear. Digital cameras have probably rendered photographs useless as evidence in a court of law, but for a statistician, a picture is still worth a thousand words.

Appendix 11.A.1 Minitab Applications

If a set of x_i's has been entered in Column C1 and the associated y_i's in Column C2, the Minitab command

```
MTB > regress c2 1 c1
```

will compute the estimated regression line, $y = \hat{\beta}_0 + \hat{\beta}_1 x$, and provide the calculations for testing H_0: $\beta_1 = 0$ and H_0: $\beta_0 = 0$. Also printed out automatically will be r^2 and s, the square root of the unbiased estimate for σ^2 in the simple linear model. Subcommands are available for plotting the data, calculating and graphing the residuals, and constructing confidence intervals and prediction intervals.

Figure 11.A.1.1 is the printout of the REGRESS command applied to the Sales versus Revenue data described in Case Study 11.3.2. Included is a listing of the residuals (in Column C3).

The entries in the "SE Coef" column are based on parts (c) and (d) of Theorem 11.3.2. The value *0.006677*, for example, is the estimated standard deviation of the estimated slope. That is,

$$0.006677 = \sqrt{\frac{s^2}{\sum_{i=1}^{9}(x_i - \bar{x})^2}}$$

where $s = 50.3489$ (as listed on the printout). The last entry in the "T" column is the value of T_{n-2} from Theorem 11.3.4 when $\beta_1 = 0$. That is,

$$61.93 = \frac{0.413520 - 0}{0.006677}$$

As we have seen in earlier chapters, the "conclusions" of hypothesis tests performed by computer software packages are invariably couched in terms of P-values. Here, for example, the test of H_0: $\beta_0 = 0$ versus H_1: $\beta_0 \neq 0$ yields an observed t ratio of 0.52, for which the P-value is *0.621*. Since the latter is so large, we would fail to reject H_0: $\beta_0 = 0$ at any reasonable level of α.

Figure 11.A.1.1

```
MTB  > set c1
DATA > 1687 2178 2649 3289 4076 5294 6369 7787 9411
DATA > end
MTB  > set c1
DATA > 748 962 1113 1350 1686 2199 2605 3179 3999
DATA > end
MTB  > regress c2 1 c1;
SUBC > residuals c3.

Regression Analysis: C2 versus C1

The regression equation is
C2 = 18.6 + 0.414 C1

Predictor       Coef    SE Coef      T      P
Constant       18.57      35.87   0.52  0.621
C1          0.413520   0.006677  61.93  0.000

S = 50.3489 R-Sq = 99.8% R-Sq(adj) = 99.8%

MTB > print c1 c2 c3

Data Display

Row    C1    C2        C3
1    1687   748   31.8218
2    2178   962   42.7834
3    2649  1113   -0.9845
4    3289  1350  -28.6373
5    4076  1686  -18.0775
6    5294  2199   -8.7449
7    6369  2605  -47.2789
8    7787  3179  -59.6502
9    9411  3999   88.7933
```

If SUBC > predict "x" is appended to the "regress c2 1 c1" command, Minitab will print out the 95% confidence interval for $E(Y\,|\,x)$ and the 95% prediction interval for Y at the point x. Figure 11.A.1.2 shows the input and output that provide these computations.

Figure 11.A.1.2

```
MTB  > set c1
DATA > 1687 2178 2649 3289 4076 5294 6369 7787 9411
DATA > end
MTB  > set c1
DATA > 748 962 1113 1350 1686 2199 2605 3179 3999
DATA > end
MTB  > regress c2 1 c1;
SUBC > predict 9700.

Predicted Values for New Observations

New
Obs     Fit   SE Fit        95% CI               95% PI
  1  4029.7     37.1   (3942.1, 4117.4)   (3881.9, 4177.6)
```

Doing Linear Regression Using Minitab Windows

1. Enter the x_i's in C1 and the y_i's in C2.
2. Click on STAT, then on REGRESSION, then on second REGRESSION.
3. Type C2 in RESPONSE box. Then click on PREDICTOR box and type C1.
4. Click on OK.
5. To display the line, click on STAT, then on REGRESSION, then on FITTED LINE PLOT.
6. Type C2 in RESPONSE box and C1 in PREDICTOR box.
7. Click on LINEAR; then click on OK.

Appendix 11.A.2 A Proof of Theorem 11.3.3

The strategy for the proof is to express $n\hat{\sigma}^2$ in terms of the squares of normal random variables and then apply Fisher's Lemma (see Appendix 7.A.2). The random variables to be used are $\hat{\beta}_1 - \beta_1$, $W_i = Y_i - \beta_0 - \beta_1 x_i$, $i = 1, \ldots, n$, and $\bar{W} = \frac{1}{n}\sum_{i=1}^{n} W_i = \bar{Y} - \beta_0 - \beta_1\bar{x}$. Note that

$$W_i - \bar{W} = (Y_i - \bar{Y}) - \beta_1(x_i - \bar{x})$$

or, equivalently,

$$Y_i - \bar{Y} = (W_i - \bar{W}) + \beta_1(x_i - \bar{x})$$

Next, we express $\hat{\beta}_1 - \beta_1$ as a linear combination of the W_i's. The argument begins by using Equation 11.3.1 to express $\hat{\beta}_1$:

$$\hat{\beta}_1 - \beta_1 = \frac{\sum_{i=1}^{n}(x_i - \bar{x})(Y_i - \bar{Y})}{\sum_{i=1}^{n}(x_i - \bar{x})^2} - \beta_1$$

$$= \frac{\sum_{i=1}^{n}(x_i - \bar{x})(Y_i - \bar{Y}) - \beta_1\sum_{i=1}^{n}(x_i - \bar{x})^2}{\sum_{i=1}^{n}(x_i - \bar{x})^2}$$

$$= \frac{\sum_{i=1}^{n}(x_i - \bar{x})[(W_i - \bar{W}) + \beta_1(x_i - \bar{x})] - \beta_1\sum_{i=1}^{n}(x_i - \bar{x})^2}{\sum_{i=1}^{n}(x_i - \bar{x})^2}$$

$$= \frac{\sum_{i=1}^{n}(x_i - \bar{x})(W_i - \bar{W})}{\sum_{i=1}^{n}(x_i - \bar{x})^2} \tag{11.A.2.1}$$

Recall from Equation 11.3.3 that

$$n\hat{\sigma}^2 = \sum_{i=1}^{n}(Y_i - \bar{Y})^2 - \hat{\beta}_1^2\sum_{i=1}^{n}(x_i - \bar{x})^2 \tag{11.A.2.2}$$

We need to express Equation 11.A.2.2 in terms of the W_i's—that is,

$$n\hat{\sigma}^2 = \sum_{i=1}^{n}[(W_i - \bar{W}) + \beta_1(x_i - \bar{x})]^2 - \hat{\beta}_1^2\sum_{i=1}^{n}(x_i - \bar{x})^2$$

$$= \sum_{i=1}^{n}(W_i - \bar{W})^2 + 2\beta_1\sum_{i=1}^{n}(x_i - \bar{x})(W_i - \bar{W}) + \beta_1^2\sum_{i=1}^{n}(x_i - \bar{x})^2$$

$$- \hat{\beta}_1^2\sum_{i=1}^{n}(x_i - \bar{x})^2 \tag{11.A.2.3}$$

From Equation 11.A.2.1, we can write

$$\sum_{i=1}^{n}(x_i - \bar{x})(W_i - \bar{W}) = (\hat{\beta}_1 - \beta_1)\sum_{i=1}^{n}(x_i - \bar{x})^2$$

Substituting the right-hand side of the preceding expression for $\sum_{i=1}^{n}(x_i - \bar{x})(W_i - \bar{W})$ in Equation 11.A.2.3 gives

$$n\hat{\sigma}^2 = \sum_{i=1}^{n}(W_i - \bar{W})^2 + 2\beta_1(\hat{\beta}_1 - \beta_1)\sum_{i=1}^{n}(x_i - \bar{x})^2$$

$$+ \beta_1^2\sum_{i=1}^{n}(x_i - \bar{x})^2 - \hat{\beta}_1^2\sum_{i=1}^{n}(x_i - \bar{x})^2$$

$$= \sum_{i=1}^{n}(W_i - \bar{W})^2 + \sum_{i=1}^{n}(x_i - \bar{x})^2[2\beta_1(\hat{\beta}_1 - \beta_1) + \beta_1^2 - \hat{\beta}_1^2]$$

$$= \sum_{i=1}^{n}(W_i - \bar{W})^2 - \sum_{i=1}^{n}(x_i - \bar{x})^2[\hat{\beta}_1^2 - 2\hat{\beta}_1\beta_1 + \beta_1^2]$$

$$= \sum_{i=1}^{n}(W_i - \bar{W})^2 - \sum_{i=1}^{n}(x_i - \bar{x})^2(\hat{\beta}_1 - \beta_1)^2$$

$$= \sum_{i=1}^{n}W_i^2 - n\bar{W}^2 - \sum_{i=1}^{n}(x_i - \bar{x})^2(\hat{\beta}_1 - \beta_1)^2$$

Now, choose an orthogonal matrix, $\mathbf{M}$, whose first two rows are

$$\frac{x_1 - \bar{x}}{\sqrt{\sum_{i=1}^{n}(x_i - \bar{x})^2}} \cdots \frac{x_n - \bar{x}}{\sqrt{\sum_{i=1}^{n}(x_i - \bar{x})^2}}$$

and

$$\frac{1}{\sqrt{n}} \cdots \frac{1}{\sqrt{n}}$$

Define the random variables $Z_1, \ldots, Z_n$ through the transformation

$$\begin{pmatrix} Z_1 \\ \vdots \\ Z_n \end{pmatrix} = \mathbf{M} \begin{pmatrix} W_1 \\ \vdots \\ W_n \end{pmatrix}$$

By Fisher's Lemma, the Z_i's are independent, normal random variables with mean zero and variance σ^2, and

$$\sum_{i=1}^{n} Z_i^2 = \sum_{i=1}^{n} W_i^2$$

Also, by Equation 11.A.2.1 and the choice of the first row of $\mathbf{M}$,

$$Z_1^2 = \sum_{i=1}^{n}(x_i - \bar{x})^2(\hat{\beta}_1 - \beta_1)^2$$

and, by the selection of the second row of $\mathbf{M}$,

$$Z_2^2 = n\overline{W}^2$$

Thus,

$$n\hat{\sigma}^2 = \sum_{i=1}^{n} W_i^2 - Z_1^2 - Z_2^2 = \sum_{i=3}^{n} Z_i^2$$

From this follows the independence of $n\hat{\sigma}^2$, $\hat{\beta}_1$, and $\bar{Y}$.

Finally, notice that

$$\frac{n\hat{\sigma}^2}{\sigma^2} = \sum_{i=3}^{n}\left(\frac{Z_i}{\sigma}\right)^2$$

The fact that the sum has a chi square distribution with $n - 2$ degrees of freedom proves the last part of the theorem.

The Analysis of Variance

Chapter 12

"No aphorism is more frequently repeated in connection with field trials, than that we must ask Nature few questions or, ideally, one question, at a time. The writer is convinced that this view is wholly mistaken. Nature, he suggests, will best respond to a logical and carefully thought-out questionnaire; indeed, if we ask her a single question, she will often refuse to answer until some other topic has been discussed."

—Ronald A. Fisher

12.1 Introduction

In this chapter we take up an important extension of the two-sample location problem introduced in Chapter 9. The *completely randomized one-factor design* is a conceptually similar k-sample location problem, but one that requires a substantially different sort of analysis than its prototype. Here, the appropriate test statistic turns out to be a ratio of variance estimates, the sampling behavior of which is described by an F distribution rather than a Student t. The name attached to this procedure, in deference to the form of its test statistic, is the *analysis of variance* (or ANOVA for short). A very flexible method, the analysis of variance is applied to many other experimental designs as well, a particularly important one being the *randomized block design* covered in Chapter 13.

Comment Credit for much of the early development of the analysis of variance goes to Sir Ronald A. Fisher. Shortly after the end of World War I, Fisher resigned a public school teaching position that he was none too happy with and accepted a post at the Rothamsted Agricultural Experiment Station, a facility heavily involved in agricultural research. There he found himself entangled in problems where differences in the response variable (crop yields, for example) were constantly in danger of being obscured by the high level of uncontrollable heterogeneity in the experimental environment (different soil qualities, drainage gradients, and so on). Quickly seeing that traditional techniques were hopelessly inadequate under these conditions, Fisher set out to look for alternatives and in just a few years succeeded in fashioning an entirely new statistical methodology, a panoply of data-collecting principles and mathematical tools that is today known as *experimental design*. The centerpiece of Fisher's creation—what makes it all work—is the analysis of variance.

Suppose an experimenter wishes to compare the average effects elicited by k different levels of some given factor, where k is greater than or equal to 2. The factor, for example, might be "stop-smoking" therapies and the levels, three specific methods. Or the factor might be crowdedness as it relates to aggression in captive monkeys, with the levels being five different monkey-per-square-foot densities in five separate enclosures. Still another example might be an engineering study comparing the effectiveness of four kinds of catalytic converters in reducing the concentrations of harmful emissions in automobile exhaust. Whatever the circumstances, data from a completely randomized one-factor design will consist of k independent random samples of sizes $n_1, n_2, \ldots,$ and n_k, the total sample size being denoted $n \left(= \sum_{j=1}^{k} n_j\right)$. We will let Y_{ij} represent the ith observation recorded for the jth level. Table 12.1.1 shows some additional terminology. (*Note:* To simplify notation in the next two chapters, data will always be written as random variables—that is, as Y_{ij} rather than y_{ij}.)

The dot notation of Table 12.1.1 is standard in analysis of variance problems. The presence of a dot in lieu of a subscript indicates that particular subscript has been summed over. Thus the response *total* for the jth sample is written

$$T_{.j} = \sum_{i=1}^{n_j} Y_{ij} \quad (= Y_{1j} + Y_{2j} + \cdots + Y_{n_j j})$$

and the corresponding sample *mean* becomes $\overline{Y}_{.j}$, where

$$\overline{Y}_{.j} = \frac{1}{n_j} \sum_{i=1}^{n_j} Y_{ij} = \frac{T_{.j}}{n_j}$$

Table 12.1.1

	Treatment Level			
	1	2	$\cdots$	k
	Y_{11}	Y_{12}		Y_{1k}
	Y_{21}	Y_{22}		
	$\vdots$	$\vdots$	$\cdots$	$\vdots$
	$Y_{n_1 1}$	$Y_{n_2 2}$		$Y_{n_k k}$
Sample sizes:	n_1	n_2	$\cdots$	n_k
Sample totals:	$T_{.1}$	$T_{.2}$		$T_{.k}$
Sample means:	$\overline{Y}_{.1}$	$\overline{Y}_{.2}$		$\overline{Y}_{.k}$
True means:	μ_1	μ_2		μ_k

By the same convention, $T_{..}$ and $\overline{Y}_{..}$ will denote the overall total and overall mean, respectively:

$$T_{..} = \sum_{j=1}^{k}\sum_{i=1}^{n_j} Y_{ij} = \sum_{j=1}^{k} T_{.j}$$

$$\overline{Y}_{..} = \frac{1}{n}\sum_{j=1}^{k}\sum_{i=1}^{n_j} Y_{ij} = \frac{1}{n}\sum_{j=1}^{k} n_j\overline{Y}_{.j} = \frac{1}{n}\sum_{j=1}^{k} T_{.j}$$

Appearing at the bottom of Table 12.1.1 are a set of *true means*, $\mu_1, \mu_2, \ldots, \mu_k$. Each μ_j is an unknown location parameter reflecting the true average response characteristic of level j. Often our objective will be to test the equality of the μ_j's—that is,

$$H_0\colon \mu_1 = \mu_2 = \ldots = \mu_k$$

versus

$$H_1\colon \text{not all the } \mu_j\text{'s are equal}$$

In the next several sections we will propose a variance-ratio statistic for testing H_0, investigate its sampling behavior under both H_0 and H_1, and introduce a set of computing formulas to simplify its evaluation. We will also explore the possibility of testing *subhypotheses* about the μ_j's—for example, $H_0\colon \mu_i = \mu_j$ (irrespective of the other μ_j's) or $H_0\colon \mu_3 = (\mu_4 + \mu_5)/2$.

12.2 The *F* Test

To derive a procedure for testing $H_0\colon \mu_1 = \mu_2 = \ldots = \mu_k$, we could once again invoke the generalized likelihood ratio criterion, compute $\lambda = L(\omega_e)/L(\Omega_e)$, and begin the search for a monotonic function of λ having a known distribution. But since we have already seen several examples of formal GLRT calculations in Chapters 7 and 9, the benefits of doing another would be marginal. Deducing the test statistic on intuitive grounds will be more instructive.

The data structure for a completely randomized one-factor design was outlined in Section 12.1. To that basic setup we now add a *distribution* assumption: The Y_{ij}'s will be presumed to be independent and normally distributed with mean μ_j, $j = 1, 2, \ldots, k$, and variance σ^2 (constant for all j)—that is,

$$f_{Y_{ij}}(y) = \frac{1}{\sqrt{2\pi}\,\sigma} e^{-\frac{1}{2}\left(\frac{y-\mu_j}{\sigma}\right)^2}, \qquad -\infty < y < \infty$$

In analysis of variance problems—as was true in regression problems—distribution assumptions are usually expressed in terms of *model equations*. In the latter, the response variable is represented as the sum of one or more fixed components and one or more random components. Here, one possible model equation would be

$$Y_{ij} = \mu_j + \varepsilon_{ij}$$

where ε_{ij} denotes the "noise" associated with Y_{ij}—that is, the amount by which Y_{ij} differs from its expected value. Of course, from the distribution assumption on Y_{ij}, it follows that ε_{ij} is also normal with variance σ^2, but with mean zero.

We will denote the overall average effect associated with the n observations in the sample by the symbol μ, where $\mu = \frac{1}{n}\sum_{j=1}^{k} n_j\mu_j$. If H_0 is true, of course, μ is the value that each of the μ_j's equals.

Sums of Squares

To find an appropriate test statistic, we begin by estimating each of the μ_j's. For each j, $Y_{1j}, Y_{2j}, \ldots, Y_{n_j j}$ is a random sample from a normal distribution. By Example 5.2.4, the maximum likelihood estimator of μ_j is $\overline{Y}_{.j}$. Then $\frac{1}{n}\sum_{j=1}^{k} n_j\overline{Y}_{.j} = \overline{Y}_{..}$ is the obvious choice to estimate μ. It follows that

$$SSTR = \sum_{j=1}^{k}\sum_{i=1}^{n_j}\left(\overline{Y}_{.j} - \overline{Y}_{..}\right)^2 = \sum_{j=1}^{k} n_j\left(\overline{Y}_{.j} - \overline{Y}_{..}\right)^2$$

which is called the *treatment sum of squares,* estimates the variation among the μ_j's. [If all the μ_j's were equal, the $\overline{Y}_{.j}$'s would be similar (to $\overline{Y}_{..}$) and $SSTR$ would be small.]

Analyzing the behavior of $SSTR$ requires an expression relating the $\overline{Y}_{.j}$'s and $\overline{Y}_{..}$ to the parameter μ. But

$$\begin{aligned}
SSTR &= \sum_{j=1}^{k} n_j\left(\overline{Y}_{.j} - \overline{Y}_{..}\right)^2 = \sum_{j=1}^{k} n_j\left[\left(\overline{Y}_{.j} - \mu\right) - \left(\overline{Y}_{..} - \mu\right)\right]^2 \\
&= \sum_{j=1}^{k} n_j\left[\left(\overline{Y}_{.j} - \mu\right)^2 + \left(\overline{Y}_{..} - \mu\right)^2 - 2\left(\overline{Y}_{.j} - \mu\right)\left(\overline{Y}_{..} - \mu\right)\right] \\
&= \sum_{j=1}^{k} n_j\left(\overline{Y}_{.j} - \mu\right)^2 + \sum_{j=1}^{k} n_j\left(\overline{Y}_{..} - \mu\right)^2 - 2\left(\overline{Y}_{..} - \mu\right)\sum_{j=1}^{k} n_j\left(\overline{Y}_{.j} - \mu\right) \\
&= \sum_{j=1}^{k} n_j\left(\overline{Y}_{.j} - \mu\right)^2 + n\left(\overline{Y}_{..} - \mu\right)^2 - 2\left(\overline{Y}_{..} - \mu\right)n\left(\overline{Y}_{..} - \mu\right) \\
&= \sum_{j=1}^{k} n_j\left(\overline{Y}_{.j} - \mu\right)^2 - n\left(\overline{Y}_{..} - \mu\right)^2 \qquad (12.2.1)
\end{aligned}$$

Now, with Equation 12.2.1 as background, Theorem 12.2.1 states the connection we are looking for—that the expected value of $SSTR$ increases as the differences among the μ_j's increase.

Theorem 12.2.1 *Let SSTR be the treatment sum of squares defined for k independent random samples of sizes $n_1, n_2, \ldots,$ and n_k. Then*

$$E(SSTR) = (k-1)\sigma^2 + \sum_{j=1}^{k} n_j(\mu_j - \mu)^2$$

Proof From Equation 12.2.1,

$$E(SSTR) = \sum_{j=1}^{k} n_j E\left[\left(\overline{Y}_{.j} - \mu\right)^2\right] - nE\left[\left(\overline{Y}_{..} - \mu\right)^2\right]$$

Since μ is the mean of $\overline{Y}_{..}$, then $E\left[\left(\overline{Y}_{..} - \mu\right)^2\right] = \sigma^2/n$. Also,

$$E\left[\left(\overline{Y}_{.j} - \mu\right)^2\right] = \text{Var}\left(\overline{Y}_{.j} - \mu\right) + \left[E\left(\overline{Y}_{.j} - \mu\right)\right]^2$$

by Theorem 3.6.1. But Theorem 3.6.2 implies that

$$\text{Var}\left(\overline{Y}_{.j} - \mu\right) = \text{Var}\left(\overline{Y}_{.j}\right) = \sigma^2/n_j$$

So, $E\left[\left(\overline{Y}_{.j} - \mu\right)^2\right] = \sigma^2/n_j + \left(\mu_j - \mu\right)^2$. Substituting these equalities into the expression for $E(SSTR)$ yields

$$E(SSTR) = \sum_{j=1}^{k} n_j \sigma^2/n_j + \sum_{j=1}^{k} n_j(\mu_j - \mu)^2 - n(\sigma^2/n)$$

or

$$E(SSTR) = (k-1)\sigma^2 + \sum_{j=1}^{k} n_j(\mu_j - \mu)^2$$

□

Testing H_0: $\mu_1 = \mu_2 = \ldots = \mu_k$ When σ^2 Is Known

Theorem 12.2.1 suggests that $SSTR$ can be the basis for a test of the null hypothesis that the treatment level means are all equal. When the μ_j's *are* the same, $E(SSTR) = (k-1)\sigma^2$. If the true means are not all equal, $E(SSTR)$ will be larger than $(k-1)\sigma^2$. It follows that we should reject H_0 if $SSTR$ is "significantly large." Of course, to determine the exact location of the rejection region for a given α, we need to know the pdf of $SSTR$, or some function of $SSTR$, when H_0 is true.

Theorem 12.2.2 *When H_0: $\mu_1 = \mu_2 = \ldots = \mu_k$ is true, $SSTR/\sigma^2$ has a chi square distribution with $k-1$ degrees of freedom.*

Proof The theorem can be proved directly at this point by an application of Fisher's Lemma, similar to the approaches taken in Appendices 7.A.2 and 11.A.2. Rather than repeat those arguments, we will give a moment-generating function derivation in Appendix 12.A.2. □

If α, then, is the level of significance, *and if σ^2 is known,* we should reject H_0: $\mu_1 = \mu_2 = \ldots = \mu_k$ in favor of H_1: Not all the μ_j's are equal if $SSTR/\sigma^2 \geq \chi^2_{1-\alpha,k-1}$. In practice, though, comparing a set of μ_j's is seldom that easy because σ^2 is rarely known. Almost invariably, σ^2 needs to be estimated; doing so changes both the nature and the distribution of the test statistic.

Testing H_0: $\mu_1 = \mu_2 = \ldots = \mu_k$ When σ^2 Is Unknown

We know that each of the k samples can provide an independent, unbiased estimate for σ^2 (recall Example 5.4.4 and see the following discussion). Using the notation of Table 12.1.1, the *jth sample variance* is written

$$S_j^2 = \frac{1}{n_j - 1} \sum_{i=1}^{n_j} \left(Y_{ij} - \overline{Y}_{.j}\right)^2$$

Multiplying each S_j^2 by $n_j - 1$ and summing over j gives the numerator of the obvious "pooled" estimator for σ^2 (recall the way S_p^2 was defined in the two-sample t test). We call this quantity the *error sum of squares*, or SSE:

$$SSE = \sum_{j=1}^{k} (n_j - 1)S_j^2 = \sum_{j=1}^{k} \sum_{i=1}^{n_j} \left(Y_{ij} - \overline{Y}_{.j}\right)^2$$

Theorem 12.2.3 *Whether or not H_0: $\mu_1 = \mu_2 = \ldots = \mu_k$ is true,*

1. *SSE/σ^2 has a chi square distribution with $n - k$ degrees of freedom.*
2. *SSE and $SSTR$ are independent.*

Proof By Theorem 7.3.2, $(n_j - 1)S_j^2/\sigma^2$ has a chi square distribution with $n_j - 1$ degrees of freedom. By the addition property, then, of the chi square distribution, SSE/σ^2 is a chi square random variable with $\sum_{j=1}^{k} (n_j - 1) = n - k$ degrees of freedom.

Each S_j^2 is independent of $\overline{Y}_{.i}$ for $i \neq j$ because the underlying samples are independent. Also, each S_j^2 is independent of $\overline{Y}_{.j}$ by Theorem 7.3.2. Therefore, SSE and $SSTR$ are independent. □

If we ignore the treatments and consider the data as one sample, then the variation about the parameter μ can be estimated by the double sum $\sum_{j=1}^{k} \sum_{i=1}^{n_j} (Y_{ij} - \overline{Y}_{..})^2$. This quantity is known as the *total sum of squares* and denoted $SSTOT$.

Theorem 12.2.4 *If n observations are divided into k samples of sizes $n_1, n_2, \ldots,$ and n_k,*

$$SSTOT = SSTR + SSE$$

Proof

$$SSTOT = \sum_{j=1}^{k} \sum_{i=1}^{n_j} \left(Y_{ij} - \overline{Y}_{..}\right)^2 = \sum_{j=1}^{k} \sum_{i=1}^{n_j} \left[\left(\overline{Y}_{.j} - \overline{Y}_{..}\right) + \left(Y_{ij} - \overline{Y}_{.j}\right)\right]^2 \qquad (12.2.2)$$

Expanding the right-hand side of Equation 12.2.2 gives

$$\sum_{j=1}^{k} \sum_{i=1}^{n_j} \left(\overline{Y}_{.j} - \overline{Y}_{..}\right)^2 + \sum_{j=1}^{k} \sum_{i=1}^{n_j} \left(Y_{ij} - \overline{Y}_{.j}\right)^2$$

since the cross-product term vanishes:

$$\sum_{j=1}^{k}\sum_{i=1}^{n_j}(\overline{Y}_{.j}-\overline{Y}_{..})(Y_{ij}-\overline{Y}_{.j})=\sum_{j=1}^{k}(\overline{Y}_{.j}-\overline{Y}_{..})\sum_{i=1}^{n_j}(Y_{ij}-\overline{Y}_{.j})$$

$$=\sum_{j=1}^{k}(\overline{Y}_{.j}-\overline{Y}_{..})(0)=0$$

Therefore,

$$\sum_{j=1}^{k}\sum_{i=1}^{n_j}(Y_{ij}-\overline{Y}_{..})^2=\sum_{j=1}^{k}\sum_{i=1}^{n_j}(\overline{Y}_{.j}-\overline{Y}_{..})^2+\sum_{j=1}^{k}\sum_{i=1}^{n_j}(Y_{ij}-\overline{Y}_{.j})^2$$

That is, $SSTOT = SSTR + SSE$. □

Theorem 12.2.5 *Suppose that each observation in a set of k independent random samples is normally distributed with the same variance, σ^2. Let $\mu_1, \mu_2, \ldots,$ and μ_k be the true means associated with the k samples. Then*

a. If H_0: $\mu_1 = \mu_2 = \ldots = \mu_k$ is true,

$$F = \frac{SSTR/(k-1)}{SSE/(n-k)}$$

has an F distribution with $k-1$ and $n-k$ degrees of freedom.

b. At the α level of significance, H_0: $\mu_1 = \mu_2 = \ldots = \mu_k$ should be rejected if $F \geq F_{1-\alpha,k-1,n-k}$.

Proof By Theorem 12.2.3, $SSTR$ and SSE are independent. We also know that $SSTR/\sigma^2$ and SSE/σ^2 are chi square random variables. Part (a), then, follows from the definition of the F distribution.

To justify the location of the critical region cited in part (b), we need to examine the behavior of the proposed test statistic when H_1 is true. From Theorem 12.2.1, we know the expected value of the numerator of F:

$$E[SSTR/(k-1)] = \sigma^2 + \frac{1}{k-1}\sum_{j=1}^{k}(\mu_j - \mu)^2 \qquad (12.2.3)$$

Moreover, from Theorem 12.2.3 it follows that the expected value of the denominator of the test statistic—that is, $E[SSE/(n-k)]$—is σ^2, regardless of which hypothesis is true.

Now, if H_0 is true, the expected values of both the numerator and the denominator of F will be σ^2, so the ratio is likely to be *close to 1*. If H_1 is true, though, the expected value of $SSTR/(k-1)$ will be greater than the expected value of $SSE/(n-k)$, implying that the observed F ratio will tend to be *larger than 1*. The critical region, therefore, should be in the *right-hand* tail of the $F_{k-1,n-k}$ distribution. That is, we should reject H_0: $\mu_1 = \mu_2 = \ldots = \mu_k$ if $F = \dfrac{SSTR/(k-1)}{SSE/(n-k)} \geq F_{1-\alpha,k-1,n-k}$. □

ANOVA Tables

Computations for carrying out analyses of variance are typically presented in the form of *ANOVA tables*. Highly structured, these tables are especially helpful in identifying the various test statistics that arise in connection with complicated experimental designs. Figure 12.2.1 shows the format of the ANOVA table for testing $H_0: \mu_1 = \mu_2 = \ldots = \mu_k$.

The rows in any ANOVA table correspond to the *sources of variation* singled out in an observation's model equation. More specifically, the last row always refers to the data's *total variation* (as measured by *SSTOT*); the preceding rows correspond to the variations whose sum yields the total variation. For this particular experimental design, the three rows are reflecting the fact that

$$SSTR + SSE = SSTOT$$

Figure 12.2.1

Source	df	*SS*	*MS*	*F*	*P*
Treatment	$k-1$	*SSTR*	*MSTR*	$\frac{MSTR}{MSE}$	$P(F_{k-1,n-k} \geq \text{observed}F)$
Error	$n-k$	*SSE*	*MSE*		
Total	$n-1$	*SSTOT*			

Next to each "source" is the number of *degrees of freedom* (df) associated with its sum of squares. Note that the df for *total* is the sum of the degrees of freedom for *treatments* and *error* $(n-1 = k-1+n-k)$.

The *SS* column lists the *sum of squares* associated with each source of variation—here, either *SSTR, SSE*, or *SSTOT*. The *MS*, or *mean square,* column is derived by dividing each sum of squares by its degrees of freedom. The *mean square for treatments,* then, is given by

$$MSTR = \frac{SSTR}{k-1}$$

and the *mean square for error* becomes

$$MSE = \frac{SSE}{n-k}$$

No entry is listed as being the mean square for total.

The entry in the top row of the F column is the value of the test statistic:

$$F = \frac{MSTR}{MSE} = \frac{SSTR/(k-1)}{SSE/(n-k)}$$

The final entry, also in the top row, is the P-value associated with the observed F. If $P < \alpha$, of course, we can reject $H_0: \mu_1 = \mu_2 = \ldots = \mu_k$ at the α level of significance.

Case Study 12.2.1

Generations of athletes have been cautioned that cigarette smoking retards performance. One measure of the truth of that warning is the effect of smoking on heart rate. In one study (73) examining that impact, six each of nonsmokers, light smokers, moderate smokers, and heavy smokers undertook sustained physical exercise. Their heart rates were measured after resting for three

(Continued on next page)

minutes. The results appear in Table 12.2.1. Are the differences among the $\overline{Y}_{.j}$'s statistically significant? That is, if μ_1, μ_2, μ_3, and μ_4 denote the *true* average heart rates for the four groups of smokers, can we reject H_0: $\mu_1 = \mu_2 = \mu_3 = \mu_4$?

Table 12.2.1

	Nonsmokers	Light Smokers	Moderate Smokers	Heavy Smokers
	69	55	66	91
	52	60	81	72
	71	78	70	81
	58	58	77	67
	59	62	57	95
	65	66	79	84
$T_{.j}$	374	379	430	490
$\overline{Y}_{.j}$	62.3	63.2	71.7	81.7

Let $\alpha = 0.05$. For these data, $k = 4$ and $n = 24$, so H_0: $\mu_1 = \mu_2 = \mu_3 = \mu_4$ should be rejected if

$$F = \frac{SSTR/(4-1)}{SSE/(24-4)} \geq F_{1-0.05,4-1,24-4} = F_{.95,3,20} = 3.10$$

(see Figure 12.2.2).

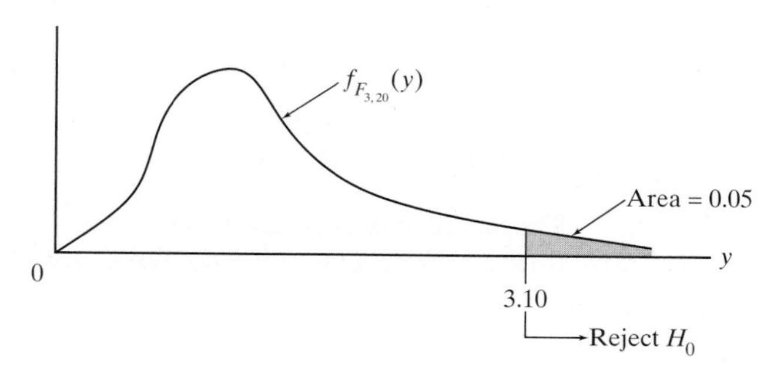

Figure 12.2.2

The overall sample mean, $\overline{Y}_{..}$, is given by

$$\overline{Y}_{..} = \frac{1}{n}\sum_{j=1}^{k} T_{.j} = \frac{374 + 379 + 430 + 490}{24}$$

$$= 69.7$$

Therefore,

$$SSTR = \sum_{j=1}^{4} n_j\left(\overline{Y}_{.j} - \overline{Y}_{..}\right)^2 = 6[(62.3 - 69.7)^2 + \cdots + (81.7 - 69.7)^2]$$

$$= 1464.125$$

(Continued on next page)

(Case Study 12.2.1 continued)

Similarly,

$$SSE = \sum_{j=1}^{4}\sum_{i=1}^{6}\left(Y_{ij} - \overline{Y}_{.j}\right)^2 = [(69 - 62.3)^2 + \cdots + (65 - 62.3)^2] + \cdots + [(91 - 81.7)^2 + \cdots + (84 - 81.7)^2]$$

$$= 1594.833$$

The observed test statistic, then, equals *6.12*:

$$F = \frac{1464.125/(4-1)}{1594.833/(24-4)} = 6.12$$

Since $6.12 > F_{.95,3,20} = 3.10$, H_0: $\mu_1 = \mu_2 = \mu_3 = \mu_4$ should be rejected. These data support the contention that smoking influences a person's heart rate.

Figure 12.2.3 shows the analysis of these data summarized in the ANOVA table format. Notice that the small P-value $(= 0.004)$ is consistent with the conclusion that H_0 should be rejected.

Source	df	*SS*	*MS*	*F*	*P*
Treatment	3	1464.125	488.04	6.12	0.004
Error	20	1594.833	79.74		
Total	23	3058.958			

Figure 12.2.3

Computing Formulas

There are easier ways to compute an F statistic than by using the "defining" formulas for $SSTR$ and SSE. Let $C = T_{..}^2/n$. Then

$$SSTOT = \sum_{j=1}^{k}\sum_{i=1}^{n_j} Y_{ij}^2 - C \tag{12.2.4}$$

$$SSTR = \sum_{j=1}^{k} \frac{T_{.j}^2}{n_j} - C \tag{12.2.5}$$

and, from Theorem 12.2.4,

$$SSE = SSTOT - SSTR$$

(The proofs of Equations 12.2.4 and 12.2.5 are left as exercises.)

Example 12.2.1

For the data in Table 12.2.1,

$$C = T_{..}^2/n = (374 + 379 + 430 + 490)^2/24 = 116{,}622.04$$

and

$$\sum_{j=1}^{4}\sum_{i=1}^{6} Y_{ij}^2 = (69)^2 + (52)^2 + \cdots + (84)^2 = 119{,}681$$

in which case

$$SSTOT = \sum_{j=1}^{4}\sum_{i=1}^{6} Y_{ij}^2 - C = 119{,}681 - 116{,}622.04 = 3058.96$$

Also,

$$SSTR = \sum_{j=1}^{4} T_{.j}^2 - C = (374)^2/6 + (379)^2/6 + (430)^2/6 + (490)^2/6 - 116{,}622.04$$
$$= 1464.13$$

so

$$SSE = SSTOT - SSTR = 3058.96 - 1464.13 = 1594.83$$

Notice that these sums of squares have the same numerical values that were found earlier in Case Study 12.2.1 using the original formulas for *SSTOT*, *SSTR*, and *SSE*. ■

Questions

12.2.1. The following are the gas mileages recorded during a series of road tests with four new models of Japanese luxury sedans. Test the null hypothesis that all four models, on the average, give the same mileage. Let $\alpha = 0.05$. Will the conclusion change if $\alpha = 0.10$?

Model			
A	B	C	D
22	28	29	23
26	24	32	24
	29	28	

12.2.2. Mount Etna erupted in 1669, 1780, and 1865. When molten lava hardens, it retains the direction of the Earth's magnetic field. Three blocks of lava were examined from each of these eruptions and the declination of the magnetic field in the block was measured (170). The results are given in the following table. Do these data suggest that the direction of the Earth's magnetic field shifted over the time period spanned by the eruptions? Let $\alpha = 0.05$.

1669	1780	1865
57.8	57.9	52.7
60.2	55.2	53.0
60.3	54.8	49.4

12.2.3. An indicator of the value of a stock relative to its earnings is its *price-earnings ratio*: the average of a given year's high and low selling prices divided by its annual earnings. The following table provides the price-earnings ratios for a sample of thirty stocks, ten each from the financial, industrial, and utility sectors of the New York Stock Exchange. Test at the 0.01 level that the true mean price-earnings ratios for the three market sectors are the same. Use the computing formulas on p. 604 to find *SSTR* and *SSE*. Use the ANOVA table format to summarize the computations; omit the *P*-value column.

Financial	Industrial	Utility
7.1	26.2	14.0
9.9	12.4	15.5
8.8	15.2	11.9
8.8	28.6	10.9
20.6	10.3	14.3
7.9	9.7	11.0
18.8	12.5	9.7
17.7	16.7	10.8
15.2	19.7	16.0
6.6	24.8	11.3

12.2.4. Each of five varieties of corn are planted in three plots in a large field. The respective yields, in bushels per acre, are in the following table.

Variety 1	Variety 2	Variety 3	Variety 4	Variety 5
46.2	49.2	60.3	48.9	52.5
51.9	58.6	58.7	51.4	54.0
48.7	57.4	60.4	44.6	49.3

Test whether the differences among the average yields are statistically significant. Show the ANOVA table. Let 0.05 be the level of significance.

12.2.5. Three pottery shards from four widely scattered and now-extinct Native American tribes have been collected by a museum. Archaeologists were asked to estimate the age of the shards. Based on the results shown in the following table, is it conceivable that the four tribes were contemporaries of one another? Let $\alpha = 0.01$.

Estimated Ages of Shards (years)			
Lakeside	Deep Gorge	Willow Ridge	Azalea Hill
1200	850	1800	950
800	900	1450	1200
950	1100	1150	1150

12.2.6. Recall the teachers' expectation data described in Question 8.2.7. Let μ_j denote the true average IQ change associated with group j, $j =$ I, II, or III. Test $H_0\colon \mu_I = \mu_{II} = \mu_{III}$ versus H_1: not all μ_j's are equal. Let $\alpha = 0.05$.

12.2.7. Fill in the entries missing from the following ANOVA table.

Source	df	*SS*	*MS*	*F*
Treatment	4			6.40
Error			10.60	
Total		377.36		

12.2.8. Do the following data appear to violate the assumptions underlying the analysis of variance? Explain.

Treatment			
A	B	C	D
16	4	26	8
17	12	22	9
16	2	23	11
17	26	24	8

12.2.9. Prove Equations 12.2.4 and 12.2.5.

12.2.10. Use Fisher's Lemma to prove Theorem 12.2.2.

Comparing the Two-Sample *t* Test with the Analysis of Variance

The analysis of variance was introduced in Section 12.1 as a k-sample *extension* of the two-sample test. The two procedures overlap, though, when k is equal to 2. An obvious question arises: Which procedure is better for testing $H_0\colon \mu_X = \mu_Y$? The answer, as Example 12.2.2 shows, is "neither." The two test procedures are entirely equivalent: If one rejects H_0, so will the other.

Example 12.2.2

Suppose that $X_1, X_2, \ldots, X_n$ and $Y_1, Y_2, \ldots, Y_m$ are two sets of independent, normally distributed random variables with the same variance, σ^2. Let μ_X and μ_Y denote their respective means. Show that the two-sample t test and the analysis of variance are equivalent for testing $H_0\colon \mu_X = \mu_Y$.

If H_0 were tested using the analysis of variance, the observed F ratio would be

$$F = \frac{SSTR/(k-1)}{SSE/(n+m-k)} = \frac{SSTR}{SSE/(n+m-2)} \tag{12.2.6}$$

and it would have 1 and $n+m-2$ degrees of freedom. The null hypothesis would be rejected if $F \geq F_{1-\alpha,1,n+m-2}$.

To compare the ANOVA decision rule with a two-sample t test requires that $SSTR$ and SSE be expressed in the "$\overline{X}$ and $\overline{Y}$" notation of t ratios. First, note that

$$\begin{aligned} SSTR &= n_1\left(\overline{Y}_{.1} - \overline{Y}_{..}\right)^2 + n_2\left(\overline{Y}_{.2} - \overline{Y}_{..}\right)^2 \\ &= n\left(\overline{X} - \overline{Y}_{..}\right)^2 + m\left(\overline{Y} - \overline{Y}_{..}\right)^2 \end{aligned}$$

In this case, $\overline{Y}_{..} = \frac{1}{n+m}(n\overline{X} + m\overline{Y})$, so

$$\begin{aligned} SSTR &= n\left[\overline{X} - \frac{1}{n+m}(n\overline{X} + m\overline{Y})\right]^2 + m\left[\overline{Y} - \frac{1}{n+m}(n\overline{X} + m\overline{Y})\right]^2 \\ &= n\left[\frac{m(\overline{X} - \overline{Y})}{n+m}\right]^2 + m\left[\frac{n(\overline{X} - \overline{Y})}{n+m}\right]^2 \\ &= \left[\frac{nm^2}{(n+m)^2} + \frac{mn^2}{(n+m)^2}\right](\overline{X} - \overline{Y})^2 \\ &= \frac{nm}{n+m}(\overline{X} - \overline{Y})^2 \end{aligned}$$

Also,

$$\begin{aligned} SSE &= (n_1 - 1)S_1^2 + (n_2 - 1)S_2^2 \\ &= (n-1)S_X^2 + (m-1)S_Y^2 \\ &= (n+m-2)S_P^2 \end{aligned}$$

Substituting these expressions for $SSTR$ and SSE into the F statistic of Equation 12.2.6 yields

$$F = \frac{\frac{nm}{n+m}(\overline{X} - \overline{Y})^2}{\frac{(n+m-2)S_P^2}{(n+m-2)}} = \frac{\frac{nm}{n+m}(\overline{X} - \overline{Y})^2}{S_P^2} = \frac{(\overline{X} - \overline{Y})^2}{S_P^2\left(\frac{1}{n} + \frac{1}{m}\right)} \tag{12.2.7}$$

Notice that the right-hand expression in Equation 12.2.7 is the square of the two-sample t statistic described in Theorem 9.2.2. Moreover,

$$\begin{aligned} \alpha &= P(T \leq -t_{\alpha/2,n+m-2} \text{ or } T \geq t_{\alpha/2,n+m-2}) = P(T^2 \geq t^2_{\alpha/2,n+m-2}) \\ &= P(F_{1,n+m-2} \geq t^2_{\alpha/2,n+m-2}) \end{aligned}$$

But the unique value c such that $P(F_{1,n+m-2} \geq c) = \alpha$ is $c = F_{1-\alpha,1,n+m-2}$, so $F_{1-\alpha,1,n+m-2} = t^2_{\alpha/2,n+m-2}$. Thus,

$$\alpha = P(T \leq -t_{\alpha/2,n+m-2} \text{ or } T \geq t_{\alpha/2,n+m-2}) = P(F \geq F_{1-\alpha,1,n+m-2})$$

It follows that if one test statistic rejects H_0 at the α level of significance, so will the other. ■

Questions

12.2.11. Verify the conclusion of Example 12.2.2 by doing a t test and an analysis of variance on the data of Question 9.2.8. Show that the observed F ratio is the square of the observed t ratio and that the F critical value is the square of the t critical value.

12.2.12. Do an analysis of variance on the Mark Twain–Quintus Curtius Snodgrass data of Case Study 9.2.1. Verify that the observed F ratio is the square of the observed t ratio.

12.2.13. Do an analysis of variance and a pooled two-sample t test on the motorcycle data given in Question 8.2.2. How are the observed F ratio and observed t ratio related? How are the two critical values related? Assume that $\alpha = 0.05$.

12.3 Multiple Comparisons: Tukey's Method

The suspicion that smoking affects heart rates was borne out by the analysis done in Case Study 12.2.1. In retrospect, the fact that $H_0: \mu_1 = \mu_2 = \mu_3 = \mu_4$ was rejected is not surprising, given the sizeable range in the $\overline{Y}_{.j}$'s (from 62.3 for nonsmokers to 81.7 for heavy smokers). But not all the treatment groups were far apart: The heart rates for nonsmokers and light smokers were fairly close—62.3 versus 63.2. That raises an obvious question: Is there some way to follow up an initial test of $H_0: \mu_1 = \mu_2 = \ldots = \mu_k$ by looking at *subhypotheses*—that is, can we test hypotheses that involve fewer than the full set of population means (for example, $H_0: \mu_1 = \mu_2$)?

The answer is "yes," but the solution is not as simple as it might appear at first glance. In particular, it would be inappropriate to do a series of standard two-sample t tests on different pairs of means—for example, applying Theorem 9.2.1 to μ_1 versus μ_2, then to μ_2 versus μ_3, and so on. If each of those tests was done at a certain level of significance α, the probability that *at least one* Type I error would be committed would be much larger than α. That being the case, the "nominal" value for α misrepresents the collective precision of the inferences.

Suppose, for example, we did ten independent tests of the form $H_0: \mu_i = \mu_j$ versus $H_1: \mu_i \neq \mu_j$, each at level $\alpha = 0.05$, on a large set of population means. Even though the probability of making a Type I error on any given test is only 0.05, the chances of incorrectly rejecting a true H_0 with at least one of the ten t tests increases dramatically to *0.40*:

$$
\begin{aligned}
P(\text{at least one Type I error}) &= 1 - P(\text{no Type I errors}) \\
&= 1 - (0.95)^{10} \\
&= 0.40
\end{aligned}
$$

Addressing that concern, mathematical statisticians have paid a good deal of attention to the so-called *multiple comparison problem*. Many different procedures, operating under various sets of assumptions, have been developed. All have the objective of keeping the probability of committing at least one Type I error small, even when the number of tests performed is large (or even infinite). In this section, we develop one of the earliest of these techniques, a still widely used method due to John Tukey.

A Background Result: The Studentized Range Distribution

The simplest multiple comparison problem is to test the equality of all *pairs* of individual means—that is, to test with one procedure $H_0: \mu_i = \mu_j$ versus $H_1: \mu_i \neq \mu_j$ *for all $i \neq j$*. In Tukey's method, these tests are performed using confidence intervals for $\mu_i - \mu_j$. The derivation depends on knowing the probabilistic behavior of the ratio R/S, where R is the range of a set of normally distributed random variables, and S is an estimator for their true standard deviation.

Definition 12.3.1. Let $W_1, W_2, \ldots,$ and W_k be a set of k independent, normally distributed random variables with mean μ and variance σ^2, and let R denote their range:

$$R = \max_i W_i - \min_i W_i$$

Suppose S^2 is based on a chi square random variable with v degrees of freedom, independent of the W_i's, where $E(S^2) = \sigma^2$. The *studentized range*, $Q_{k,v}$, is the ratio

$$Q_{k,v} = \frac{R}{S}$$

Table A.5 in the Appendix gives values of $Q_{\alpha,k,v}$, the $100(1-\alpha)$th percentile of $Q_{k,v}$, for $\alpha = 0.05$ and 0.01, and for various values of k and v. For example, if $k = 4$ and $v = 8$, $Q_{.05,4,8} = 4.53$, meaning that $P\left(\frac{R}{S} \geq 4.53\right) = 0.05$, where R is the range of four normally distributed random variables, whose true standard deviation, σ, is being estimated by a sample standard deviation, S, having 8 degrees of freedom (see Figure 12.3.1). (*Note:* For the applications of the studentized range in this chapter, S^2 will always be MSE and v will be $n - k$.)

Figure 12.3.1

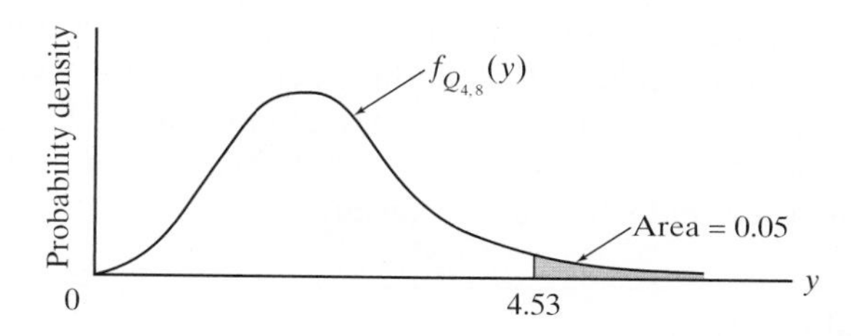

Theorem 12.3.1 *Let $\overline{Y}_{.j}$, $j = 1, 2, \ldots, k$ be the k sample means in a completely randomized one-factor design. Let $n_j = r$ be the common sample size, and let μ_j be the true means, $j = 1, 2, \ldots, k$. The probability is $1 - \alpha$ that all $\binom{k}{2}$ differences $\mu_i - \mu_j$ will simultaneously satisfy the inequalities*

$$\overline{Y}_{.i} - \overline{Y}_{.j} - D\sqrt{MSE} < \mu_i - \mu_j < \overline{Y}_{.i} - \overline{Y}_{.j} + D\sqrt{MSE}$$

where $D = Q_{\alpha,k,rk-k}/\sqrt{r}$. If, for a given i and j, zero is not contained in the preceding inequality, $H_0\colon \mu_i = \mu_j$ can be rejected in favor of $H_1\colon \mu_i \neq \mu_j$, at the α level of significance.

Proof Let $W_t = \overline{Y}_{.t} - \mu_t$. Then W_t is normally distributed with mean zero and variance σ^2/r. Let max W_t and min W_t denote the maximum and minimum values, respectively, for W_t, where t ranges from 1 to k.

Take MSE/r to be the estimator for σ^2/r. From the definition of the studentized range, $\dfrac{\max W_t - \min W_t}{\sqrt{\dfrac{MSE}{r}}}$ has a $Q_{k,rk-k}$ pdf, which implies that

$$P\left(\frac{\max W_t - \min W_t}{\sqrt{\frac{MSE}{r}}} < Q_{\alpha,k,rk-k}\right) = 1 - \alpha$$

or, equivalently,

$$P\left(\max W_t - \min W_t < D\sqrt{MSE}\right) = 1 - \alpha \tag{12.3.1}$$

where $D = Q_{\alpha,k,rk-k}/\sqrt{r}$.

Now, if Equation 12.3.1 is true, it must also be true that

$$P(|W_i - W_j| < D\sqrt{MSE}) = 1 - \alpha \quad \text{for } all \ i \text{ and } j \qquad (12.3.2)$$

Rewriting Equation 12.3.2 gives

$$P(-D\sqrt{MSE} < W_i - W_j < D\sqrt{MSE}) = 1 - \alpha \quad \text{for } all \ i \text{ and } j \qquad (12.3.3)$$

Recall that $W_t = \overline{Y}_{.t} - \mu_t$. Substituting the latter for W_i and W_j into Equation 12.3.3 yields the statement of the theorem:

$$P(\overline{Y}_{.i} - \overline{Y}_{.j} - D\sqrt{MSE} < \mu_i - \mu_j < \overline{Y}_{.i} - \overline{Y}_{.j} + D\sqrt{MSE}) = 1 - \alpha$$

for all i and j. □

Case Study 12.3.1

A certain fraction of antibiotics injected into the bloodstream are "bound" to serum proteins. This phenomenon bears directly on the effectiveness of the medication, because the binding decreases the systemic uptake of the drug. Table 12.3.1 lists the binding percentages in bovine serum measured for five widely prescribed antibiotics (214). Which antibiotics have similar binding properties, and which are different?

Table 12.3.1

	Penicillin G	Tetra-cycline	Strepto-mycin	Erythro-mycin	Chloram-phenicol
	29.6	27.3	5.8	21.6	29.2
	24.3	32.6	6.2	17.4	32.8
	28.5	30.8	11.0	18.3	25.0
	32.0	34.8	8.3	19.0	24.2
$T_{.j}$	114.4	125.5	31.3	76.3	111.2
$\overline{Y}_{.j}$	28.6	31.4	7.8	19.1	27.8

To answer that question requires that we make all $\binom{5}{2} = 10$ pairwise comparisons of μ_i versus μ_j. First, MSE must be computed. From the entries in Table 12.3.1,

$$SSE = \sum_{j=1}^{5}\sum_{i=1}^{4}(Y_{ij} - \overline{Y}_{.j})^2 = 135.83$$

so $MSE = 135.83/(20 - 5) = 9.06$. Let $\alpha = 0.05$. Since $n - k = 20 - 5 = 15$, the appropriate cutoff from the studentized range distribution is $Q_{.05,5,15} = 4.37$. Therefore, $D = 4.37/\sqrt{4} = 2.185$ and $D\sqrt{MSE} = 6.58$.

For each different pairwise subhypothesis test, $H_0\colon \mu_i = \mu_j$ versus $H_1\colon \mu_i \neq \mu_j$, Table 12.3.2 lists the value of $\overline{Y}_{.i} - \overline{Y}_{.j}$, together with the corresponding 95% Tukey confidence interval for $\mu_i - \mu_j$ calculated from Theorem 12.3.1. As the last column indicates, seven of the subhypotheses are rejected (those whose Tukey intervals do not contain zero) and three are not rejected.

(Continued on next page)

Table 12.3.2

Pairwise Difference	$\overline{Y}_{.i} - \overline{Y}_{.j}$	Tukey Interval	Conclusion
$\mu_1 - \mu_2$	−2.8	(−9.38, 3.78)	NS
$\mu_1 - \mu_3$	20.8	(14.22, 27.38)	Reject
$\mu_1 - \mu_4$	9.5	(2.92, 16.08)	Reject
$\mu_1 - \mu_5$	0.8	(−5.78, 7.38)	NS
$\mu_2 - \mu_3$	23.6	(17.02, 30.18)	Reject
$\mu_2 - \mu_4$	12.3	(5.72, 18.88)	Reject
$\mu_2 - \mu_5$	3.6	(−2.98, 10.18)	NS
$\mu_3 - \mu_4$	−11.3	(−17.88, −4.72)	Reject
$\mu_3 - \mu_5$	−20.0	(−26.58, −13.42)	Reject
$\mu_4 - \mu_5$	−8.7	(−15.28, −2.12)	Reject

Questions

12.3.1. Use Tukey's method to make all the pairwise comparisons for the heart rate data of Case Study 12.2.1 at the 0.05 level of significance.

12.3.2. Construct 95% Tukey intervals for the three pairwise differences, $\mu_i - \mu_j$, for the data of Question 12.2.3.

12.3.3. Intravenous infusion fluids produced by three different pharmaceutical companies (Cutter, Abbott, and McGaw) were tested for their concentrations of particulate contaminants. Six samples were inspected from each company. The figures listed in the table are, for each sample, the number of particles per liter greater than five microns in diameter (183).

Number of Contaminant Particles		
Cutter	Abbott	McGaw
255	105	577
264	288	515
342	98	214
331	275	413
234	221	401
217	240	260

Do the analysis of variance to test $H_0\colon \mu_C = \mu_A = \mu_M$ and then test each of the three pairwise subhypotheses by constructing 95% Tukey intervals.

12.3.4. Construct 95% Tukey intervals for all ten pairwise differences, $\mu_i - \mu_j$, for the data of Question 12.2.4. Summarize the results by plotting the five sample averages on a horizontal axis and drawing straight lines under varieties whose average yields are not significantly different.

12.3.5. Construct 95% Tukey confidence intervals for the three pairwise differences associated with the murder culpability scores described in Question 8.2.15. Which differences are statistically significant?

12.3.6. If 95% Tukey confidence intervals tell us to reject $H_0\colon \mu_1 = \mu_2$ and $H_0\colon \mu_1 = \mu_3$, will we necessarily reject $H_0\colon \mu_2 = \mu_3$?

12.3.7. The width of a Tukey confidence interval is

$$2\sqrt{MSE}\,Q_{\alpha,k,n-k}\Big/\sqrt{\frac{n}{k}}$$

If k increases, but $\frac{n}{k}$ and MSE stay the same, will the Tukey intervals get shorter or longer? Justify your answer intuitively.

12.4 Testing Subhypotheses with Contrasts

There are two general ways to test a subhypothesis, the choice depending, strangely enough, on *when* H_0 can be fully specified. If a researcher wishes to do an experiment first, and then let the results suggest a suitable subhypothesis, the appropriate analysis is any of the various multiple comparison techniques—for example, the Tukey method of Section 12.3.

If, on the other hand, physical considerations, economic factors, past experience, or any other factors suggest a particular subhypothesis *before any data are taken,* H_0 can best be tested using a *contrast*. The advantage of the latter is that tests based on contrasts have greater power than the analogous tests based on a multiple comparison procedure would have.

Definition 12.4.1. Let $\mu_1, \mu_2, \ldots, \mu_k$ denote the true means of k factor levels being sampled. A linear combination, C, of the μ_j's is said to be a *contrast* if the sum of its coefficients is 0. That is, C is a contrast if $C = \sum_{j=1}^{k} c_j \mu_j$, where the c_j's are constants such that $\sum_{j=1}^{k} c_j = 0$.

Contrasts have a direct connection with hypothesis tests. Suppose a set of data consists of five treatment levels, and we wish to test the subhypothesis $H_0\colon \mu_1 = \mu_2$. The latter could also be written $H_0\colon \mu_1 - \mu_2 = 0$, which is actually a statement about a contrast—specifically, the contrast C, where

$$C = \mu_1 - \mu_2 = (1)\mu_1 + (-1)\mu_2 + (0)\mu_3 + (0)\mu_4 + (0)\mu_5$$

Or, suppose in Case Study 12.3.1, there was a good pharmacological reason for comparing the average level of serum binding for the first two antibiotics to the average level for the last three. Written as a subhypothesis, the statement of no difference would be

$$H_0\colon \frac{\mu_1 + \mu_2}{2} = \frac{\mu_3 + \mu_4 + \mu_5}{3}$$

As a contrast, it becomes

$$C = \frac{1}{2}\mu_1 + \frac{1}{2}\mu_2 - \frac{1}{3}\mu_3 - \frac{1}{3}\mu_4 - \frac{1}{3}\mu_5$$

In both these cases, the numerical value of the contrast will be 0 if H_0 is true. This suggests that the choice between H_0 and H_1 can be accomplished by first estimating C and then determining, via a significance test, whether that estimate is too far from 0.

We begin by considering some of the mathematical properties of contrasts and their estimates. Since $\overline{Y}_{.j}$ is always an unbiased estimator for μ_j, it seems reasonable to estimate C, a linear combination of population means, with $\hat{C}$, a linear combination of *sample* means:

$$\hat{C} = \sum_{j=1}^{k} c_j \overline{Y}_{.j}$$

(The coefficients appearing in $\hat{C}$, of course, are the same as those that defined C.) It follows that

$$E(\hat{C}) = \sum_{j=1}^{k} c_j E(\overline{Y}_{.j}) = C$$

and

$$\text{Var}(\hat{C}) = \sum_{j=1}^{k} c_j^2 \text{Var}\left(\overline{Y}_{.j}\right) = \sigma^2 \sum_{j=1}^{k} \frac{c_j^2}{n_j}$$

Comment Replacing the unknown error variance, σ^2, by its estimate from the ANOVA table—*MSE*—gives a formula for the estimated variance of the estimated contrast:

$$S_{\hat{C}}^2 = MSE \sum_{j=1}^{k} \frac{c_j^2}{n_j}$$

The sampling behavior of $\hat{C}$ is easily derived. By Theorem 4.3.3, the normality of the Y_{ij}'s ensures that $\hat{C}$ is also normal, and by the usual Z transformation, the ratio

$$\frac{\hat{C} - E(\hat{C})}{\sqrt{\text{Var}(\hat{C})}} = \frac{\hat{C} - C}{\sqrt{\text{Var}(\hat{C})}}$$

is a *standard* normal. Therefore,

$$\left[\frac{\hat{C} - C}{\sqrt{\text{Var}(\hat{C})}}\right]^2$$

is a chi square random variable with 1 degree of freedom. Of course, if H_0: $\mu_1 = \mu_2 = \ldots = \mu_k$ is true, C is 0, and the ratio reduces to

$$\frac{\hat{C}^2}{\sigma^2 \sum_{j=1}^{k} \frac{c_j^2}{n_j}}$$

One additional property of contrasts is worth noting because of its connection to the treatment sum of squares in the analysis of variance. Two contrasts

$$C_1 = \sum_{j=1}^{k} c_{1j}\mu_j \quad \text{and} \quad C_2 = \sum_{j=1}^{k} c_{2j}\mu_j$$

are said to be *orthogonal* if

$$\sum_{j=1}^{k} \frac{c_{1j}c_{2j}}{n_j} = 0$$

Similarly, a set of q contrasts, $\{C_i\}_{i=1}^{q}$, are said to be *mutually orthogonal* if

$$\sum_{j=1}^{k} \frac{c_{sj}c_{tj}}{n_j} = 0 \quad \text{for all } s \neq t$$

(The same definitions apply to *estimated* contrasts.)

Definition 12.4.2 and Theorems 12.4.1 and 12.4.2, both stated here without proof, summarize the relationship between contrasts and the analysis of variance. In short, the treatment sum of squares can be partitioned into $k - 1$ "contrast" sums of squares, provided the contrasts are mutually orthogonal.

Definition 12.4.2. Let $C_i = \sum_{j=1}^{k} c_{ij}\mu_j$ be any contrast. The sum of squares associated with C_i is given by

$$SS_{C_i} = \frac{\hat{C}_i^2}{\sum_{j=1}^{k} \frac{c_{ij}^2}{n_j}}$$

where $\hat{C}_i = \sum_{j=1}^{k} c_{ij}\overline{Y}_{.j}$.

Theorem 12.4.1 *Let $\left\{C_i = \sum_{j=1}^{k} c_{ij}\mu_j\right\}_{i=1}^{k-1}$ be a set of $k-1$ mutually orthogonal contrasts. Let $\left\{\hat{C}_i = \sum_{j=1}^{k} c_{ij}\overline{Y}_{.j}\right\}_{i=1}^{k-1}$ be their estimators. Then*

$$\begin{aligned} SSTR &= \sum_{j=1}^{k}\sum_{i=1}^{n_j}\left(\overline{Y}_{.j} - \overline{Y}_{..}\right)^2 \\ &= SS_{C_1} + SS_{C_2} + \cdots + SS_{C_{k-1}} \end{aligned}$$

□

Theorem 12.4.2 *Let C be a contrast having the same coefficients as the subhypothesis $H_0\colon c_1\mu_1 + c_2\mu_2 + \cdots + c_k\mu_k = 0$, where $\sum_{j=1}^{k} c_j = 0$. Let $n = \sum_{j=1}^{k} n_j$ be the total sample size. Then*

a. $F = \dfrac{SS_C/1}{SSE/(n-k)}$ *has an F distribution with 1 and $n-k$ degrees of freedom.*

b. $H_0\colon c_1\mu_1 + c_2\mu_2 + \cdots + c_k\mu_k = 0$ *should be rejected at the α level of significance if $F \geq F_{1-\alpha,1,n-k}$.* □

Comment Theorem 12.4.1 is not meant to imply that only mutually orthogonal contrasts can, or should, be tested. It is simply a statement of a partitioning relationship that exists between $SSTR$ and the sum of squares for mutually orthogonal C_i's. In any given experiment, the contrasts that should be singled out are those the experimenter has some prior reason to test.

Case Study 12.4.1

As a rule, infants are not able to walk by themselves until they are almost fourteen months old. One study, however, investigated the possibility of reducing that time through the use of special "walking" exercises (212). A total of twenty-three infants were included in the experiment—all were one-week-old white males. They were randomly divided into four groups, and for seven weeks each group followed a different training program. Group A received special

(Continued on next page)

walking and placing exercises for twelve minutes each day. Group B also had daily twelve-minute exercise periods but was not given the special walking and placing exercises. Groups C and D received no special instruction. The progress of groups A, B, and C was checked every week; the progress of group D was checked only once, at the end of the study.

After seven weeks the formal training ended and the parents were told they could continue with whatever procedure they desired. Table 12.4.1 lists the ages (in months) at which each of the twenty-three children first walked alone. Table 12.4.2 shows the analysis of variance computations. Based on 3 and 19 degrees of freedom, the $\alpha = 0.05$ critical value is 3.13, so $H_0: \mu_A = \mu_B = \mu_C = \mu_D$ is not rejected.

Table 12.4.1 Age When Infants First Walked Alone (Months)

	Group A	Group B	Group C	Group D
	9.00	11.00	11.50	13.25
	9.50	10.00	12.00	11.50
	9.75	10.00	9.00	12.00
	10.00	11.75	11.50	13.50
	13.00	10.50	13.25	11.50
	9.50	15.00	13.00	
$T_{.j}$	60.75	68.25	70.25	61.75
$\overline{Y}_{.j}$	10.12	11.38	11.71	12.35

Table 12.4.2 ANOVA Computations

Source	df	*SS*	*MS*	F
Exercises	3	14.77	4.92	2.14
Error	19	43.70	2.30	
Total	22	58.47		

At this point the analysis could end with the overall H_0 not being rejected. We will continue with the subhypothesis procedures, however, to illustrate the application of Theorem 12.4.2.

Recall that groups A and B spent equal amounts of time exercising but followed different regimens. Consequently, a test of $H_0: \mu_A = \mu_B$ versus $H_1: \mu_A \neq \mu_B$ would be an obvious way to assess the effectiveness of the special walking and placing exercises. The associated contrast would be $C_1 = \mu_A - \mu_B$. Similarly, a test of $H_0: \mu_C = \mu_D$ (using $C_2 = \mu_C - \mu_D$) would provide an evaluation of the psychological effect of periodic progress checks.

From Definition 12.4.2 and the data in Table 12.4.1,

$$SS_{C_1} = \frac{\left[1\left(\frac{60.75}{6}\right) - 1\left(\frac{68.25}{6}\right)\right]^2}{\frac{1^2}{6} + \frac{(-1)^2}{6}} = 4.68$$

(Continued on next page)

(Case Study 12.4.1 continued)

and

$$SS_{C_2} = \frac{\left[1\left(\frac{70.25}{6}\right) - 1\left(\frac{61.75}{5}\right)\right]^2}{\frac{1^2}{6} + \frac{(-1)^2}{5}} = 1.12$$

Dividing these sums of squares by the mean square for error ($=2.30$) gives F ratios of $4.68/2.30 = 2.03$ and $1.12/2.30 = 0.49$, neither of which is significant at the $\alpha = 0.05$ level ($F_{.95,1,19} = 4.38$) (see Table 12.4.3).

Table 12.4.3 Subhypothesis Computations

Subhypothesis	Contrast	SS	F
$H_0: \mu_A = \mu_B$	$C_1 = \mu_A - \mu_B$	4.68	2.03
$H_0: \mu_C = \mu_D$	$C_2 = \mu_C - \mu_D$	1.12	0.49

Questions

12.4.1. The cathode warm-up time (in seconds) was determined for three different types of X-ray tubes using fifteen observations of each type. The results are listed in the following table.

Warm-Up Times (sec)					
Tube Type					
A		*B*		*C*	
19	27	20	24	16	14
23	31	20	25	26	18
26	25	32	29	15	19
18	22	27	31	18	21
20	23	40	24	19	17
20	27	24	25	17	19
18	29	22	32	19	18
35		18		18	

Do an analysis of variance on these data and test the hypothesis that the three tube types require the same average warm-up time. Include a pair of orthogonal contrasts in your ANOVA table. Define one of the contrasts so it tests $H_0: \mu_A = \mu_C$. What does the other contrast test? Check to see that the sums of squares associated with your two contrasts verify the statement of Theorem 12.4.1.

12.4.2. Test the hypothesis that the average of the true yields for the first three varieties of corn described in Question 12.2.4 is the same as the average for the last two. Let $\alpha = 0.05$.

12.4.3. In Case Study 12.2.1 test the hypothesis that the average of the heart rates for light and moderate smokers is the same as that for heavy smokers. Let the level of significance be 0.05.

12.4.4. Large companies have the option of limiting their growth, but does doing so lead to higher profitability? The table below gives the profitability for a sample of twenty-one top-ranked companies, where profitability is expressed in terms of annual profit as a percentage of total company assets. The firms are divided into three groups by size of assets—$50 billion or less, between $51 and $100 billion, and over $100 billion. Test the hypothesis that small- and medium-size companies are as profitable as large companies. Let $\alpha = 0.10$.

Size of Assets (billions of $)		
$50 or Less	Between $51 and $100	Greater than $100
7.2	11.3	14.8
6.5	5.6	11.3
5.7	5.3	9.2
4.4	5.3	4.8
3.4	10.4	3.9
3.4	6.2	10.2
7.8	5.3	7.3

(Note: SSE = 147.17429)

12.4.5. Verify that $C_3 = \frac{11}{12}\mu_A + \frac{11}{12}\mu_B - \mu_C - \frac{5}{6}\mu_D$ is orthogonal to the C_1 and C_2 of Case Study 12.4.1. Find SS_{C_3} and illustrate the statement of Theorem 12.4.1.

12.4.6. For many years sodium nitrite has been used as a curing agent for bacon, and until recently it was thought to be perfectly harmless. But now it appears that during frying, sodium nitrite induces the formation of nitrosopyrrolidine (NPy), a substance suspected of being a carcinogen. In one study focusing on this problem, measurements were made of the amount of NPy (in ppb) recovered after the frying of three slices of four commercially available brands of bacon (161). Do the analysis of variance for the data in the table and partition the treatment sum of squares into a complete set of three mutually orthogonal contrasts. Let the first contrast test $H_0: \mu_A = \mu_B$ and the second, $H_0: (\mu_A + \mu_B)/2 = (\mu_C + \mu_D)/2$. Do all tests at the 0.05 level of significance.

NPy Recovered from Bacon (ppb)

Brand			
A	*B*	*C*	*D*
20	75	15	25
40	25	30	30
18	21	21	31

12.5 Data Transformations

The three assumptions required by the analysis of variance have already been mentioned: the Y_{ij}'s must be independent, normally distributed, and have the same variance for all j. In practice, these three are not equally difficult to satisfy, nor do their violations have the same consequences for the F test.

Independence is certainly a critical property for the Y_{ij}'s to have, but randomizing the order in which observations are taken (relative to the different treatment levels) tends to eliminate systematic bias—and achieve independence—quite effectively. Normality is a much more difficult property to induce or even to verify (recall Section 10.4). Fortunately, violations of that particular assumption, unless extreme, do not seriously compromise the probabilistic integrity of the analysis of variance (like the t test, the F test is *robust* against departures from normality).

If the final assumption is violated, though, and the Y_{ij}'s do *not* all have the same variance, the effect on certain inference procedures—for example, the construction of confidence intervals for individual means—can be more unsettling. However, it is possible in some situations to "stabilize" the level-to-level variances by a suitable *data transformation*.

Suppose that Y_{ij} has pdf $f_Y(y_{ij}; \mu_j)$, $i = 1, 2, \ldots, n_j$; $j = 1, 2, \ldots, k$, and a known function g exists for which $\text{Var}(Y_{ij}) = g(\mu_j)$. We wish to find a transformation, A, that, when applied to the Y_{ij}'s, will generate a new set of variables having a constant variance—that is, $A(Y_{ij}) = W_{ij}$, where $\text{Var}(W_{ij}) = c_1^2$, a constant.

By Taylor's theorem,

$$W_{ij} \doteq A(\mu_j) + (Y_{ij} - \mu_j)A'(\mu_j)$$

Of course, $E(W_{ij}) = A(\mu_j)$, since $E(Y_{ij} - \mu_j) = 0$. Also,

$$\begin{aligned}
\text{Var}(W_{ij}) &= E[W_{ij} - E(W_{ij})]^2 \\
&= E[(Y_{ij} - \mu_j)A'(\mu_j)]^2 \\
&= [A'(\mu_j)]^2 \text{Var}(Y_{ij}) = [A'(\mu_j)]^2 g(\mu_j)
\end{aligned}$$

Solving for $A'(\mu_j)$ gives

$$A'(\mu_j) = \frac{\sqrt{\text{Var}(W_{ij})}}{\sqrt{g(\mu_i)}} = \frac{c_1}{\sqrt{g(\mu_j)}}$$

For Y_{ij} in the neighborhood of μ_j, it follows that

$$A(Y_{ij}) = c_1 \int \frac{1}{\sqrt{g(y_{ij})}}\, dy_{ij} + c_2 \tag{12.5.1}$$

Example 12.5.1

Suppose the Y_{ij}'s are Poisson random variables with mean μ_j, $j = 1, 2, \ldots, k$, so

$$f_Y(y_{ij}; \mu_j) = \frac{e^{-\mu_j}\mu_j^{y_{ij}}}{y_{ij}!}$$

In this case, the variance is *equal* to the mean (recall Theorem 4.2.2):

$$\text{Var}(Y_{ij}) = E(Y_{ij}) = \mu_j = g(\mu_j)$$

By Equation 12.5.1, then,

$$A(Y_{ij}) = c_1 \int \frac{1}{\sqrt{y_{ij}}}\, dy_{ij} + c_2 = 2c_1\sqrt{y_{ij}} + c_2$$

or, letting $c_1 = \frac{1}{2}$ and $c_2 = 0$ to make the transformation as simple as possible,

$$A(Y_{ij}) = \sqrt{Y_{ij}} \tag{12.5.2}$$

Equation 12.5.2 implies that if the data are known in advance to be Poisson, each of the observations should be replaced by its *square root* before we proceed with the analysis of variance. ∎

Example 12.5.2

Suppose each Y_{ij} is a binomial random variable with pdf

$$f_Y(y_{ij}; n, p_j) = \binom{n}{y_{ij}} p_j^{y_{ij}}(1 - p_j)^{n - y_{ij}}$$

Here, $E(Y_{ij}) = np_j = \mu_j$, which implies that

$$\text{Var}(Y_{ij}) = np_j(1 - p_j) = \mu_j\left(1 - \frac{\mu_j}{n}\right) = g(\mu_j)$$

It follows that the variance-stabilizing transformation for this type of data is the *inverse sine*:

$$\begin{aligned} A(Y_{ij}) &= c_1 \int \frac{1}{\sqrt{y_{ij}(1 - y_{ij}/n)}}\, dy_{ij} + c_2 \\ &= c_1 2\sqrt{n}\arcsin\left(\frac{Y_{ij}}{n}\right)^{1/2} + c_2 \end{aligned}$$

or, what is equivalent,

$$A(Y_{ij}) = \arcsin\left(\frac{Y_{ij}}{n}\right)^{1/2}$$

∎

Questions

12.5.1. A commercial film processor is experimenting with two kinds of fully automatic color developers. Six sheets of exposed film are put through each developer. The number of flaws on each negative visible with the naked eye is then counted.

Number of Visible Flaws	
Developer A	Developer B
1	8
4	6
5	4
6	9
3	11
7	10

Assume the number of flaws on a given negative is a Poisson random variable. Make an appropriate data transformation and do the indicated analysis of variance.

12.5.2. An experimenter wants to do an analysis of variance on a set of data involving five treatment groups, each with three replicates. She has computed $\overline{Y}_{.j}$ and S_j for each group and gotten the results listed in the following table.

	Treatment Group				
	1	2	3	4	5
$\overline{Y}_{.j}$	9.0	4.0	16.0	9.0	1.0
S_j	3.0	2.0	4.0	3.0	1.0

What should the experimenter do before computing the various sums of squares necessary to carry out the F test? Be as quantitative as possible.

12.5.3. Three air-to-surface missile launchers are tested for their accuracy. The same gun crew fires four rounds with each launcher, each round consisting of twenty missiles. A "hit" is scored if the missile lands within ten yards of the target. The following table gives the number of hits registered in each round.

Number of Hits per Round		
Launcher A	Launcher B	Launcher C
13	15	9
11	16	11
10	18	10
14	17	8

Compare the accuracy of these three launchers by using the analysis of variance after making a suitable data transformation. Let $\alpha = 0.05$.

12.6 Taking a Second Look at Statistics (Putting the Subject of Statistics Together—The Contributions of Ronald A. Fisher)

"The time has come," the Walrus said
"To talk of many things:
Of shoes—and ships—and sealing wax
Of cabbages—and kings.
And why the sea is boiling hot
And whether pigs have wings."
Lewis Carroll

Statistics, as we know it today, is very much a product of the twentieth century. To be sure, its roots are centuries old. The Frenchmen Blaise Pascal and Pierre Fermat did their protean work on probability in 1654. At about that same time, John Graunt was studying Bills of Mortality in England and demonstrating a remarkable flair for teasing out patterns and trends. Still, as the twentieth century dawned, there was no real *subject* of statistics. There were bits and pieces of probability theory, and there were more than a few extremely capable observers of random phenomena—Francis Galton and Adolphe Quetelet being among the most prominent—but there was nothing resembling any general principles or formal methodology.

Perhaps the most serious "gap" at the turn of the century was the almost total lack of information about sampling distributions. No one knew, for example, the pdfs that described quantities such as $\frac{\overline{Y}-\mu_0}{S/\sqrt{n}}$, $\frac{(n-1)S^2}{\sigma^2}$, $\frac{\overline{X}-\overline{Y}}{S_p\sqrt{\frac{1}{n}+\frac{1}{m}}}$, or $\frac{S_Y^2}{S_X^2}$. These, of course, turned up as test statistics in Chapters 6, 7, and 9. Not knowing their pdfs meant that no inferences other than point estimates could be made about the parameters of normal distributions. Moreover, there was very little known about point estimates and, more generally, about the mathematical properties that should be associated with the estimation process.

Two individuals who figured very prominently in the early efforts to put statistics on a solid mathematical footing were Karl Pearson and W.S. Gossett (who published under the pseudonym "Student"). In 1900, Pearson deduced the distribution of the goodness-of-fit statistic, which appeared in Chapter 10. And Gossett, in 1908, came up with the pdf for $\left(\frac{\overline{Y}-\mu_o}{S/\sqrt{n}}\right)$—that is, the t distribution. It was a third person, though, Ronald A. Fisher, who stood tallest among his peers. He not only did much of the early work in deriving sampling distributions and exploring the mathematical properties of estimation, he also created the critically important area of applied statistics known as *experimental design*.

Born in 1890 in a suburb of London, Fisher was mathematically precocious and particularly adept at visualizing complicated problems in his head, a talent that some believe he developed to compensate for his congenitally poor eyesight. He graduated with distinction from Cambridge in 1912, where his specialties were physics and optics. During his time there, he also developed what would become a lifelong interest in genetics. He was particularly intrigued with the possibility of finding a mathematical justification for Darwin's theory of evolution. (Almost two decades later, he published a book on the subject, *The Genetical Theory of Natural Selection*.)

In 1915, he derived the distribution of the sample correlation coefficient in a paper that is often thought to mark the beginning of the modern theory of sampling distributions. After teaching high school physics for several years (a job that did not seem to suit him especially well), he accepted a position as a statistician at the Rothamsted Agricultural Station. There he absolutely flourished as he immersed himself in the pursuit of both applied and mathematical statistics. Among his accomplishments was a seminal paper published in 1921, "Mathematical Foundations of Theoretical Statistics," which provided the framework for generations of future research.

The work at Rothamsted brought him face-to-face with the very difficult problem of drawing inferences from field trials where biases of various sorts (different soil qualities, uneven drainage gradients, etc.) were the rule rather than the exception. The strategies he devised for dealing with heterogeneous environments eventually coalesced into what is now referred to as *experimental design*. Guided by his twin principles of replication and randomization, he revolutionized the protocol for setting up and conducting experiments. The mathematical techniques that supported his ideas on experimental design became known, of course, as the *analysis of variance*. In 1925, Fisher published *Statistical Methods for Research Workers*, a classic text whose many subsequent editions helped countless scientists become more sophisticated in the ways of analyzing data. A decade later he wrote *The Design of Experiments*, a second highly acclaimed guide for researchers.

Fisher was knighted in 1952, ten years before he died in Adelaide, Australia, at the age of seventy-two (48).

Appendix 12.A.1 Minitab Applications

The Minitab command for doing the F test of Theorem 12.2.5 is

```
MTB > aovoneway c1-ck
```

where the Y_{ij}'s from the k samples have been entered in columns c1 through ck. The output appears in the ANOVA table format of Figure 12.2.1.

Displayed in Figure 12.A.1.1 are the input and output for analyzing the heart rate data described in Case Study 12.2.1. The program also prints out 95% confidence intervals for each μ_j—that is,

$$\left(\overline{Y}_{\cdot j} - t_{.025,n_j-1} \cdot \frac{S}{\sqrt{n_j}}, \overline{Y}_{\cdot j} + t_{.025,n_j-1} \cdot \frac{S}{\sqrt{n_j}}\right)$$

where S is the pooled standard deviation calculated from all k samples.

Figure 12.A.1.1

```
MTB  > set c1
DATA > 69 52 71 58 59 65
DATA > end
MTB  > set c2
DATA > 55 60 78 58 62 66
DATA > end
MTB  > set c3
DATA > 66 81 70 77 57 79
DATA > end
MTB  > set c4
DATA > 91 72 81 67 95 84
DATA > end
MTB  > aovoneway c1-c4
```

One-way ANOVA: C1, C2, C3, C4

```
Source   DF      SS     MS     F      P
Factor    3  1464.1  488.0  6.12  0.004
Error    20  1594.8   79.7
Total    23  3059.0

S = 8.930   R-Sq = 47.86%   R-Sq (adj) = 40.04%

                             Individual 95% CIs For Mean Based on
                             Pooled StDev
Level  N    Mean   StDev   ------+---------+---------+---------+
C1     6  62.333   7.257   (------*------)
C2     6  63.167   8.159    (------*------)
C3     6  71.667   9.158            (------*------)
C4     6  81.667  10.764                    (------*------)
                           ------+---------+---------+---------+
                                60        70        80        90

Pooled StDev = 8.930
```

Testing H_0: $\mu_1 = \ldots = \mu_k$ Using Minitab Windows

1. Enter the k samples in columns C1 through Ck, respectively.
2. Click on STAT, then on ANOVA, then on ONE-WAY (UNSTACKED).
3. Type C1-Ck in RESPONSES box, and click on OK.

Pairwise comparisons are also available in Minitab, but the Tukey method requires that the data be entered differently than how they were for the AOVONEWAY command. First, the k samples are "stacked" in a single column—say, c1. Then a second column, c2, is created whose entries identify the treatment level to which each Y_{ij} in Column 1 belongs. For example, c1 and c2 for the data

Level 1	Level 2	Level 3
4	−1	6
2	3	8

would be

$$c1 = \begin{pmatrix} 4 \\ 2 \\ -1 \\ 3 \\ 6 \\ 8 \end{pmatrix} \quad \text{and} \quad c2 = \begin{pmatrix} 1 \\ 1 \\ 2 \\ 2 \\ 3 \\ 3 \end{pmatrix}$$

The statements

```
MTB  > oneway c1 c2;
SUBC > tukey.
```

will then produce a complete set of 95% Tukey confidence intervals.

Figure 12.A.1.2 shows the Minitab input, the ANOVA table output, and the complete set of 95% Tukey confidence intervals for the serum binding data of Case Study 12.3.1. Intervals not containing 0, of course, correspond to "pairwise" null subhypotheses that should be rejected (at the $\alpha = 0.05$ level of significance). For example, the 95% Tukey confidence interval for $\mu_3 - \mu_1$ extends from −27.350 to −14.200. Since 0 is not contained in that interval, the null subhypothesis H_0: $\mu_1 = \mu_3$ should be rejected at the $\alpha = 0.05$ level of significance.

Constructing Tukey Confidence Intervals Using Minitab Windows

1. Enter entire sample in column C1, beginning with the n_1 observations in Sample 1, followed by the n_2 observations in Sample 2, and so on.
2. In column C2, enter n_1 1's, followed by n_2 2's, and so on.
3. Click on STAT, then on ANOVA, then on ONE-WAY.
4. Type C1 in RESPONSE box and C2 in FACTOR box.
5. Click on COMPARISONS, then on TUKEY'S FAMILY ERROR RATE. Enter the desired value for 100 α.
6. Double click on OK.

Figure 12.A.1.2

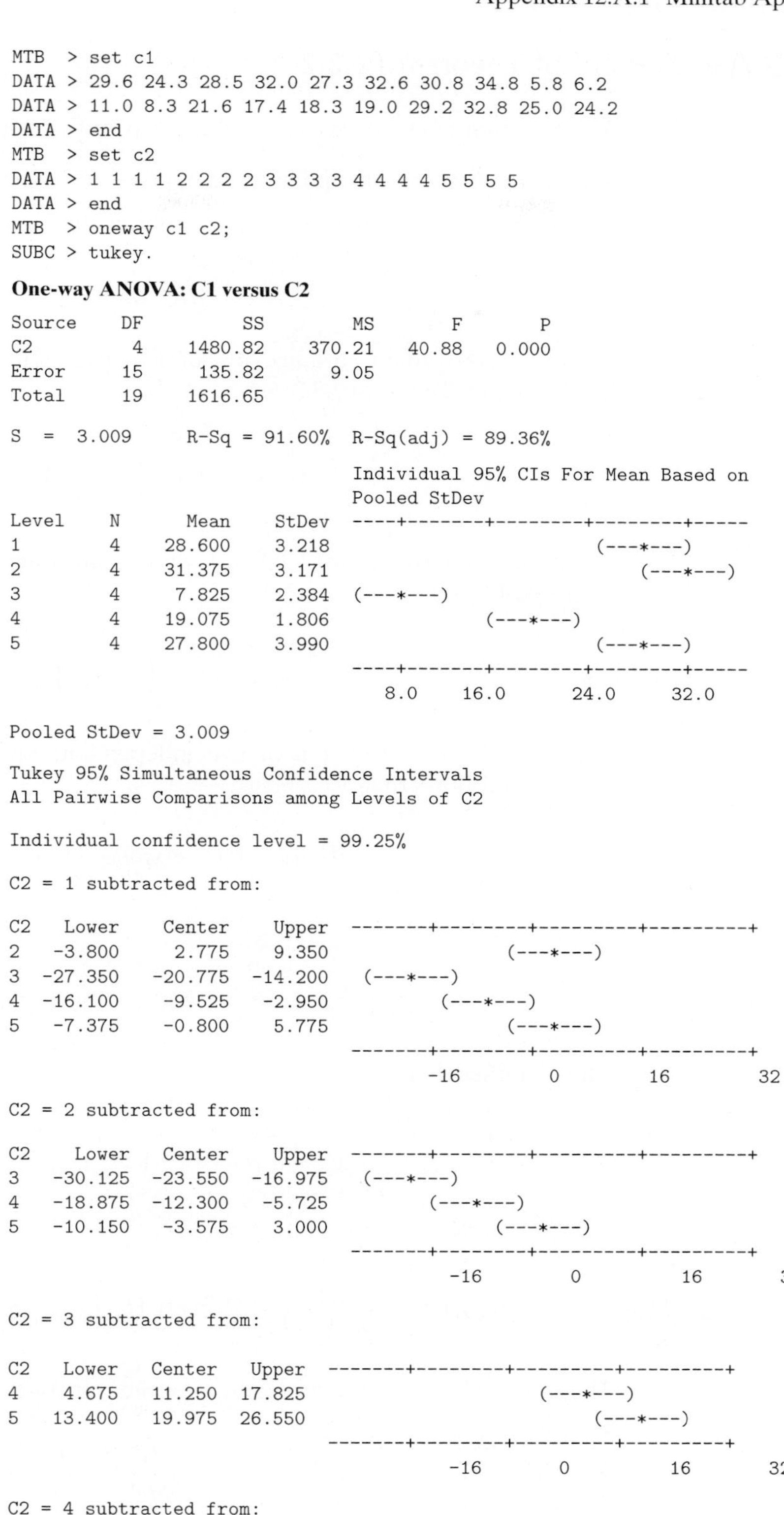

```
MTB  > set c1
DATA > 29.6 24.3 28.5 32.0 27.3 32.6 30.8 34.8 5.8 6.2
DATA > 11.0 8.3 21.6 17.4 18.3 19.0 29.2 32.8 25.0 24.2
DATA > end
MTB  > set c2
DATA > 1 1 1 1 2 2 2 2 3 3 3 3 4 4 4 4 5 5 5 5
DATA > end
MTB  > oneway c1 c2;
SUBC > tukey.
```

One-way ANOVA: C1 versus C2

```
Source  DF       SS      MS      F      P
C2       4  1480.82  370.21  40.88  0.000
Error   15   135.82    9.05
Total   19  1616.65

S  =  3.009    R-Sq = 91.60%  R-Sq(adj) = 89.36%

                              Individual 95% CIs For Mean Based on
                              Pooled StDev
Level  N    Mean   StDev  ----+--------+--------+--------+-----
1      4  28.600   3.218                       (---*---)
2      4  31.375   3.171                           (---*---)
3      4   7.825   2.384  (---*---)
4      4  19.075   1.806             (---*---)
5      4  27.800   3.990                       (---*---)
                          ----+--------+--------+--------+-----
                             8.0    16.0      24.0     32.0

Pooled StDev = 3.009

Tukey 95% Simultaneous Confidence Intervals
All Pairwise Comparisons among Levels of C2

Individual confidence level = 99.25%

C2 = 1 subtracted from:

C2   Lower   Center    Upper  -------+--------+---------+---------+
2   -3.800    2.775    9.350                (---*---)
3  -27.350  -20.775  -14.200   (---*---)
4  -16.100   -9.525   -2.950        (---*---)
5   -7.375   -0.800    5.775              (---*---)
                              -------+--------+---------+---------+
                                   -16        0        16        32

C2 = 2 subtracted from:

C2    Lower  Center    Upper  -------+--------+---------+---------+
3   -30.125 -23.550  -16.975   (---*---)
4   -18.875 -12.300   -5.725        (---*---)
5   -10.150  -3.575    3.000             (---*---)
                              -------+--------+---------+---------+
                                   -16        0        16        32

C2 = 3 subtracted from:

C2   Lower  Center   Upper  -------+--------+---------+---------+
4    4.675  11.250  17.825                     (---*---)
5   13.400  19.975  26.550                         (---*---)
                            -------+--------+---------+---------+
                                 -16        0        16        32

C2 = 4 subtracted from:

C2  Lower  Center   Upper  -------+--------+---------+---------+
5   2.150   8.725  15.300                 (---*---)
                           -------+--------+---------+---------+
                                -16        0        16        32
```

Appendix 12.A.2 A Proof of Theorem 12.2.2

To prove that $SSTR/\sigma^2$ has a chi square distribution with $k-1$ degrees of freedom, it suffices to show that the moment-generating function of $SSTR/\sigma^2$ is $\left(\frac{1}{1-2t}\right)^{(k-1)/2}$.

Note, first, that under the null hypothesis that $\mu_1 = \mu_2 = \ldots = \mu_k$,

$$SSTOT = (n-1)S^2$$

where S^2 is the sample variance of a set of n observations from a normal distribution. Therefore, by Theorem 7.3.2,

$$M_{SSTOT/\sigma^2}(t) = \left(\frac{1}{1-2t}\right)^{(n-1)/2}$$

Also, from Theorem 12.2.3, SSE/σ^2 is a chi square random variable with $n-k$ degrees of freedom, so

$$M_{SSE/\sigma^2}(t) = \left(\frac{1}{1-2t}\right)^{(n-k)/2}$$

Since $SSTOT/\sigma^2$ is the *sum* of two independent random variables, $SSTR/\sigma^2$ and SSE/σ^2, it follows that

$$M_{SSTOT/\sigma^2}(t) = M_{SSTR/\sigma^2}(t) \cdot M_{SSE/\sigma^2}(t)$$

or

$$\left(\frac{1}{1-2t}\right)^{(n-1)/2} = M_{SSTR/\sigma^2}(t) \cdot \left(\frac{1}{1-2t}\right)^{(n-k)/2}$$

which implies that

$$M_{SSTR/\sigma^2}(t) = \left(\frac{1}{1-2t}\right)^{(k-1)/2}$$

Appendix 12.A.3 The Distribution of $\frac{SSTR/(k-1)}{SSE/(n-k)}$ When H_1 Is True

Theorem 12.2.5 gives the distribution of the test statistic

$$F = \frac{SSTR/(k-1)}{SSE/(n-k)}$$

when the null hypothesis is true. To calculate either the power of the analysis of variance or the probability of committing a Type II error, though, requires that we know the pdf of the observed F *when* H_1 *is true.*

Definition 12.A.3.1. Let V_j have a normal pdf with mean μ_j and variance 1, for $j = 1, \ldots, r$, and suppose that the V_j's are independent. Then

$$V = \sum_{j=1}^{r} V_j^2$$

is said to have the *noncentral* χ^2 *distribution* with r degrees of freedom and noncentrality parameter γ, where

$$\gamma = \sum_{j=1}^{r} \mu_j^2$$

Theorem 12.A.3.1 *The moment-generating function for a noncentral χ^2 random variable, V, with r degrees of freedom and noncentrality parameter γ is given by*

$$M_V(t) = (1-2t)^{-\frac{r}{2}} e^{\frac{\gamma t}{1-2t}}, \quad t < \frac{1}{2}$$

Proof We begin by finding the moment-generating function for the special case where $r = 1$.

Let V be a normal random variable with mean μ and variance 1, and let $V = Z + \mu$, where Z is a standard normal random variable. By definition, the moment-generating function for V^2 can be written

$$M_{V^2}(t) = E\left(e^{tV^2}\right) = E\left[e^{t(Z+\mu)^2}\right]$$

$$= \frac{1}{\sqrt{2\pi}} \int_{-\infty}^{\infty} e^{t(z+\mu)^2} e^{-\frac{1}{2}z^2} dz = \frac{1}{\sqrt{2\pi}} \int_{-\infty}^{\infty} e^{t(z+\mu)^2 - \frac{1}{2}z^2} dz$$

To evaluate the integral, we first complete the square in the exponent:

$$tz^2 + 2tz\mu + t\mu^2 - \frac{1}{2}z^2$$

$$= -\frac{1}{2}[(1-2t)z^2 - 4t\mu z] + t\mu^2$$

$$= -\frac{1}{2} \cdot \frac{z^2 - \dfrac{4t\mu}{(1-2t)} z}{1/(1-2t)} + t\mu^2$$

$$= -\frac{1}{2} \cdot \frac{z^2 - \dfrac{4t\mu}{(1-2t)} z + \dfrac{4t^2\mu^2}{(1-2t)^2}}{1/(1-2t)} + t\mu^2 + \frac{\dfrac{2t^2\mu^2}{(1-2t)^2}}{1/(1-2t)}$$

$$= -\frac{1}{2} \cdot \left[\frac{z - \dfrac{2t\mu}{(1-2t)}}{1/\sqrt{1-2t}}\right]^2 + t\mu^2 + \frac{2t^2\mu^2}{(1-2t)}$$

Therefore,

$$M_{V^2}(t) = e^{\mu^2 t + \frac{2\mu^2 t^2}{(1-2t)}} \frac{1}{\sqrt{2\pi}} \int_{-\infty}^{\infty} e^{-\frac{1}{2}\cdot\left[\frac{z - \frac{2t\mu}{(1-2t)}}{1/\sqrt{1-2t}}\right]^2} dz$$

$$= (1-2t)^{-\frac{1}{2}} e^{\mu^2 \frac{t}{1-2t}}$$

The general result, where $r \neq 1$, follows from an application of Theorem 3.12.3(b). Let $V = \sum_{j=1}^{r} V_j^2$, where the V_j's are independent. Then

$$M_{\sum_{j=1}^{r} V_j^2}(t) = (1-2t)^{-\frac{r}{2}} e^{\sum_{j=1}^{r} \mu_j^2 \frac{t}{1-2t}} = (1-2t)^{-\frac{r}{2}} \cdot e^{\frac{\gamma t}{1-2t}}$$

□

Definition 12.A.3.2. Let V_1 be a noncentral χ^2 random variable with r_1 degrees of freedom and noncentrality parameter γ. Suppose V_2 is a (central) χ^2 random variable with r_2 degrees of freedom and independent of V_1. The ratio

$$\frac{V_1/r_1}{V_2/r_2}$$

is said to have a *noncentral F distribution* with r_1 and r_2 degrees of freedom and noncentrality parameter γ.

Theorem 12.A.3.2 *The ratio*

$$\frac{SSTR/(k-1)}{SSE/(n-k)}$$

has a noncentral F distribution with $k-1$ and $n-k$ degrees of freedom and noncentrality parameter $\gamma = \frac{1}{\sigma^2}\sum_{j=1}^{k} n_j(\mu_j - \mu)^2$.

Proof From Equation 12.2.1,

$$SSTR = \sum_{j=1}^{k} n_j \left(\overline{Y}_{.j} - \mu\right)^2 - n\left(\overline{Y}_{..} - \mu\right)^2$$

so

$$\frac{SSTR}{\sigma^2} = \sum_{j=1}^{k} \left(\frac{\overline{Y}_{.j} - \mu}{\sigma/\sqrt{n_j}}\right)^2 - \left(\frac{\overline{Y}_{..} - \mu}{\sigma/\sqrt{n}}\right)^2 \qquad (12.A.3.1)$$

Let $W_j = \dfrac{\overline{Y}_{.j} - \mu}{\sigma/\sqrt{n_j}}$, $j = 1, \ldots, k$. Since $E(\overline{Y}_{.j}) = \mu_j$, $E(W_j) = \sqrt{n_j}(\mu_j - \mu)/\sigma$. Also, $\text{Var}(W_j) = \dfrac{\text{Var}(\overline{Y}_{.j} - \mu)}{\sigma^2/n_j} = \dfrac{\sigma^2/n_j}{\sigma^2/n_j} = 1$. Thus, because $\overline{Y}_{.j}$ is normal, each W_j is normal with mean $\sqrt{n_j}(\mu_j - \mu)/\sigma$ and variance 1. The second component of $SSTR/\sigma^2$, $\dfrac{\overline{Y}_{..} - \mu}{\sigma/\sqrt{n}}$, is a standard normal random variable.

Now, recalling the transformation technique used in Appendix 7.A.2, choose an orthogonal matrix $\mathbf{A}$ with first row $(\sqrt{n_1/n}, \sqrt{n_2/n}, \ldots, \sqrt{n_k/n})$. Define the vector $\vec{\mathbf{V}}$ of random variables by $\vec{\mathbf{V}} = A(W_1, W_2, \ldots, W_k)^T$. First note that

$$V_1 = \sum_{j=1}^{k} \frac{\sqrt{n_j}}{\sqrt{n}} W_j = \sum_{j=1}^{k} \frac{\sqrt{n_j}}{\sqrt{n}} \frac{(\overline{Y}_{.j} - \mu)}{\sigma/\sqrt{n_j}}$$

$$= \frac{1}{\sigma\sqrt{n}} \sum_{j=1}^{k} n_j(\overline{Y}_{.j} - \mu) = \frac{1}{\sigma\sqrt{n}} \left[\sum_{j=1}^{k} n_j \overline{Y}_{.j} - \left(\sum_{j=1}^{k} n_j \right) \mu \right]$$

$$= \frac{1}{\sigma\sqrt{n}} (n\overline{Y}_{..} - n\mu) = \frac{\overline{Y}_{..} - \mu}{\sigma/\sqrt{n}}$$

which gives $V_1^2 = \left(\frac{\overline{Y}_{..} - \mu}{\sigma/\sqrt{n}} \right)^2$.

Because of the orthogonality of the matrix,

$$\sum_{j=1}^{k} V_j^2 = \sum_{j=1}^{k} W_j^2 \quad \text{or} \quad \sum_{j=2}^{k} V_j^2 = \sum_{j=1}^{k} W_j^2 - V_1^2$$

But $\sum_{j=1}^{k} W_j^2 - V_1^2 = SSTR/\sigma^2$ by Equation 12.A.3.1. Moreover, each V_j is a normal random variable for which $V_j = \sum_{i=1}^{k} a_{ji} W_i$, where the a_{ji}'s are the entries in the jth row of **A**. Therefore,

$$\text{Var}(V_j) = \sum_{i=1}^{k} \text{Var}(a_{ji} W_i) = \sum_{i=1}^{k} a_{ji}^2 \text{Var}(W_i) = \sum_{i=1}^{k} a_{ji}^2$$

since each W_j has variance 1. But the orthogonality of matrix **A** implies that $\sum_{i=1}^{k} a_{ji}^2 = 1$ for each j. So each V_j is normal with variance 1, and $\sum_{j=2}^{k} V_j^2$ has a noncentral χ^2 distribution with $k-1$ degrees of freedom.

From Question 12.A.3.4, the noncentrality parameter of $\sum_{j=2}^{k} V_j^2$ is

$$E\left(\sum_{j=2}^{k} V_j^2 \right) - (k-1) = E\left(\sum_{j=1}^{k} W_j^2 \right) - E(V_1^2) - (k-1)$$

$$= \sum_{j=1}^{k} [\text{Var}(W_j) + [E(W_j)]^2] - [\text{Var}(V_1) + E(V_1)^2] - (k-1)$$

$$= \sum_{j=1}^{k} [1 + [\sqrt{n_j}(\mu_j - \mu)/\sigma]^2] - (1+0) - (k-1)$$

$$= \frac{1}{\sigma^2} \sum_{j=1}^{k} n_j(\mu_j - \mu)^2$$

Therefore, since SSE/σ^2 has a χ^2 distribution with $n-k$ degrees of freedom, it follows immediately from Definition 12.A.3.2 that, when H_1 is true,

$$F = \frac{\frac{SSTR}{k-1}}{\frac{SSE}{n-k}}$$

has a noncentral F distribution with $k-1$ and $n-k$ degrees of freedom and noncentrality parameter $\gamma = \frac{1}{\sigma^2}\sum_{j=1}^{k} n_j(\mu_j - \mu)^2$. □

Comment As H_1 gets farther from H_0, as measured by γ, the noncentral F will shift more and more to the right of the central F. Accordingly, the power of the F test will increase. That is,

$$P(F \geq F_{1-\alpha,k-1,n-k}) \rightarrow 1 \quad \text{as } \gamma \rightarrow \infty$$

The pdf for the noncentral F is not very tractable, but its integral has been evaluated by numerical approximation. This has allowed the power function for the F test to be tabulated [see, for instance, (108)].

Questions

12.A.3.1. Suppose an experimenter has taken three independent measurements on each of five treatment levels and intends to use the analysis of variance to test

$$H_0: \mu_1 = \mu_2 = \mu_3 = \mu_4 = \mu_5 \,(=0)$$

versus

$$H_1: \text{not all the } \mu_j\text{'s are equal}$$

Two of the possible alternatives in H_1 are

$$H_1^*: \mu_1 = -1,\ \mu_2 = 2,\ \mu_3 = 0,\ \mu_4 = 1,\ \mu_5 = -2$$

and

$$H_1^{**}: \mu_1 = -3,\ \mu_2 = 2,\ \mu_3 = 1,\ \mu_4 = 0,\ \mu_5 = 0$$

Against which alternative will the F test have the greater power? Explain.

12.A.3.2. In the scenario of the previous question, is $H_1: \mu_1 = 2,\ \mu_2 = 1,\ \mu_3 = 1,\ \mu_4 = -3,\ \mu_5 = 0$ an "admissible" alternative hypothesis?

12.A.3.3. If the random variable V has a noncentral χ^2distribution with r degrees of freedom and noncentrality parameter γ, use its moment-generating function to find $E(V)$.

12.A.3.4. If the random variable V has a noncentral χ^2 distribution with r degrees of freedom and noncentrality parameter γ, show that $\gamma = E(V) - r$.

12.A.3.5. Suppose $V_1, V_2, \ldots, V_n$ are independent noncentral χ^2 random variables having $r_1, r_2, \ldots, r_n$ degrees of freedom, respectively, and with noncentrality parameters $\gamma_1, \gamma_2, \ldots, \gamma_n$. Find the distribution of $V = V_1 + V_2 + \cdots + V_n$.

Chapter 13

Randomized Block Designs

"... when I first came to study statistical methods, nothing was further from my thoughts, or from those of my contemporaries, than that the art of experimental design would ever come to be, as it now surely is, an integral part of the subject."
—Ronald A. Fisher, 1947

13.1 Introduction

In any experiment, reducing the magnitude of the experimental error is a highly desirable objective: The smaller σ^2 is, the better will be our chances of rejecting a false null hypothesis. Basically, there are two ways to reduce experimental error. The nonstatistical approach is simply to refine the experimental technique—use better equipment, minimize subject error, and so on. The statistical method, which can often produce results much more dramatic, is to collect the data in "blocks," in what is referred to as a *randomized block design*.

Historically, it was Fisher who first advanced the notion of blocking. He saw it initially as a statistical defense against the obfuscating effects of soil heterogeneity in agricultural experiments. Suppose, for example, a researcher wishes to compare the yields of four different varieties of corn. Figure 13.1.1(a) shows the simplest experimental layout: Variety A is planted in the leftmost portion of the field, variety B is planted next to A, and so on. Even to a city slicker, though, the statistical hazards in using this design should be obvious. Suppose, for example, there was a soil *gradient* in the field, with the best soil being in the westernmost part (where variety A was planted). Then if variety A achieved the highest yield, we would not know whether to attribute its success to its inherent quality or to its location (or to some combination of both).

A more sensible approach is pictured in Figure 13.1.1(b). There the field is divided into a number of smaller "blocks," each block being still further parceled into four "plots." All four varieties are planted in each block, one to a plot, with the plot assignments being chosen at random. Notice that the geographical contiguity of the four plots within a given block ensures that the environmental conditions from

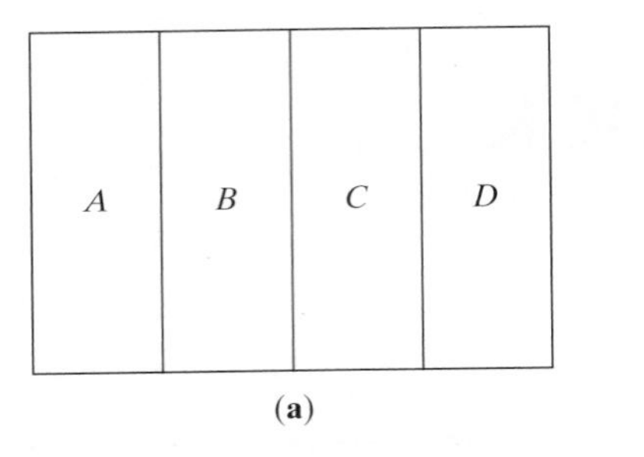

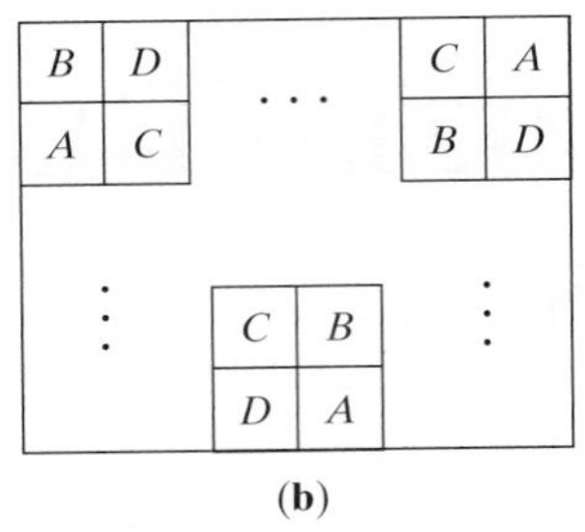

Figure 13.1.1 Two different experimental designs.

plot to plot will be relatively uniform and will not lead to any biasing of the observed yields. What the analysis of variance will then do is "pool" from block to block the *within*-block information concerning the treatment differences while bypassing the *between*-block differences—that is, the heterogeneity in the experimental environment. As a result, the treatment comparisons can be made with greater precision. Analytically, where the total sum of squares was partitioned into *two* components in a completely randomized one-factor design, it will be split into *three* separate sums in a randomized block design: one for treatments, another for blocks, and a third for experimental error.

It did not take long for scientists to realize that the benefits of blocking could be extended well beyond the confines of agricultural experimentation. In medical research, blocks are often made up of subjects of the same age, sex, and overall physical condition. A common practice in animal studies is to form blocks out of littermates. Industrial experiments often require that "time" be a blocking criterion: Measurements taken by personnel on the day shift might be considered one block and those taken by the night shift a second block. In some sense the ultimate form of blocking, although one not always physically possible, is to apply the entire set of treatment levels to each subject, thus making each subject its own block.

Section 13.2 begins with a development of the analysis of variance for the randomized block design, where k treatment levels are administered within each of b blocks. The observations within a given block, of course, are *dependent*. As was the case in Chapter 12, the hypotheses to be tested are $H_0: \mu_1 = \mu_2 = \ldots = \mu_k$ versus H_1: not all the μ_j's are equal. The section concludes with a pair of case studies that illustrate the blocking concept in two very different settings.

We saw in the previous chapter that when $k = 2$, and the samples are independent, the F test is equivalent to a two-sample t test. A similar duality exists here. When $k = 2$ treatment levels are compared within b blocks, $H_0: \mu_1 = \mu_2$ can be tested using either the analysis of variance or a *paired t test*. The latter is described in Section 13.3.

13.2 The *F* Test for a Randomized Block Design

Superficially, the structure of randomized block data looks much like the format we encountered in Chapter 12—associated with each of k treatment levels is a sample of measurements. Here, though, each column has exactly the same number of observations—that is, $n_j = b$ for all j, so the data set is necessarily a $b \times k$ matrix (see Table 13.2.1).

On the other hand, from a statistical standpoint randomized block data are fundamentally different from k-sample data (recall the discussion in Section 8.2). In Chapter 12, the k samples were *independent*. Here, the observations within a given row (which corresponds to a *block*) are *dependent*, since each reflects to some extent the conditions inherent in that block. That distinction causes the analysis of variance to proceed differently.

Table 13.2.1

		Treatment Level				Block Totals	Block Means	True Block Effects
		1	2		k			
Blocks	1	Y_{11}	Y_{12}	$\cdots$	Y_{1k}	$T_{1.}$	$\overline{Y}_{1.}$	β_1
	2	Y_{21}	Y_{22}		Y_{2k}	$T_{2.}$	$\overline{Y}_{2.}$	β_2
	$\vdots$	$\vdots$			$\vdots$	$\vdots$	$\vdots$	$\vdots$
	b	Y_{b1}	Y_{b2}		Y_{bk}	$T_{b.}$	$\overline{Y}_{b.}$	β_b
Sample totals		$T_{.1}$	$T_{.2}$		$T_{.k}$	$T_{..}$		
Sample means		$\overline{Y}_{.1}$	$\overline{Y}_{.2}$	$\cdots$	$\overline{Y}_{.k}$		$\overline{Y}_{..}$	
True means		μ_1	μ_2		μ_k			

Our objective is to test $H_0\colon \mu_1 = \mu_2 = \ldots = \mu_k$, the same as it was in Chapter 12. But here the mathematical model associated with Y_{ij} has an additional term, representing the effect of the ith block. If each "block effect," β_i, is assumed to be additive, we can write

$$Y_{ij} = \mu_j + \beta_i + \varepsilon_{ij}$$

where ε_{ij} is normally distributed with mean zero and variance σ^2, for $i = 1, 2, \ldots, b$ and $j = 1, 2, \ldots, k$. As before, we will let μ denote the overall average treatment effect associated with the bk observations—that is, $\mu = \frac{1}{k}\sum_{j=1}^{k} \mu_j$.

The basic approach followed in Chapter 12 can still be taken here, but SSE needs to be recalculated, because the "error" in a set of randomized block measurements will reflect both the block effect *and* the random error. To separate the two requires that we first estimate the set of block effects, $\beta_1, \beta_2, \ldots,$ and β_b.

Let $\overline{Y}_{i.} = \frac{1}{k}\sum_{j=1}^{k} Y_{ij}$ denote the sample average of the k observations in the ith block. Suppose the data contained no random error—that is, $\varepsilon_{ij} = 0$ for all i and j. Then

$$\overline{Y}_{i.} = \frac{1}{k}\sum_{j=1}^{k}(\mu_j + \beta_i) = \left(\frac{1}{k}\sum_{j=1}^{k}\mu_j\right) + \frac{1}{k}k\beta_i = \mu + \beta_i$$

If $\overline{Y}_{..}$ is substituted for μ, the estimate for β_i becomes $\overline{Y}_{i.} - \overline{Y}_{..}$.

Now, adding and subtracting $\overline{Y}_{i.} - \overline{Y}_{..}$ in the expression for SSE from Chapter 12 gives

$$\begin{aligned}
\sum_{i=1}^{b}\sum_{j=1}^{k}(Y_{ij} - \overline{Y}_{.j})^2 &= \sum_{i=1}^{b}\sum_{j=1}^{k}\left[(Y_{ij} - \overline{Y}_{.j}) + (\overline{Y}_{i.} - \overline{Y}_{..}) - (\overline{Y}_{i.} - \overline{Y}_{..})\right]^2 \\
&= \sum_{i=1}^{b}\sum_{j=1}^{k}\left[(\overline{Y}_{i.} - \overline{Y}_{..}) + (Y_{ij} - \overline{Y}_{.j} - \overline{Y}_{i.} + \overline{Y}_{..})\right]^2 \\
&= \sum_{i=1}^{b}\sum_{j=1}^{k}(\overline{Y}_{i.} - \overline{Y}_{..})^2 + \sum_{i=1}^{b}\sum_{j=1}^{k}(Y_{ij} - \overline{Y}_{.j} - \overline{Y}_{i.} + \overline{Y}_{..})^2 \\
&\quad + 2\sum_{i=1}^{b}\sum_{j=1}^{k}(\overline{Y}_{i.} - \overline{Y}_{..})(Y_{ij} - \overline{Y}_{.j} - \overline{Y}_{i.} + \overline{Y}_{..})
\end{aligned}$$

Notice that the cross-product term can be written

$$2\sum_{i=1}^{b}(\overline{Y}_{i.}-\overline{Y}_{..})\sum_{j=1}^{k}(Y_{ij}-\overline{Y}_{.j}-\overline{Y}_{i.}+\overline{Y}_{..})$$

But

$$\sum_{j=1}^{k}(Y_{ij}-\overline{Y}_{.j}-\overline{Y}_{i.}+\overline{Y}_{..})=k\overline{Y}_{i.}-k\overline{Y}_{i.}-\sum_{j=1}^{k}(\overline{Y}_{.j}-\overline{Y}_{..})=0$$

so

$$\sum_{i=1}^{b}\sum_{j=1}^{k}(Y_{ij}-\overline{Y}_{.j})^2=\sum_{i=1}^{b}\sum_{j=1}^{k}(\overline{Y}_{i.}-\overline{Y}_{..})^2+\sum_{i=1}^{b}\sum_{j=1}^{k}(Y_{ij}-\overline{Y}_{.j}-\overline{Y}_{i.}+\overline{Y}_{..})^2 \quad (13.2.1)$$

Equation 13.2.1 is a key result. It shows that the "old" sum of squares for error from Chapter 12—$\sum_{i=1}^{b}\sum_{j=1}^{k}(Y_{ij}-\overline{Y}_{.j})^2$—can be partitioned into the sum of two other sums of squares. The first, $\sum_{i=1}^{b}\sum_{j=1}^{k}(\overline{Y}_{i.}-\overline{Y}_{..})^2$, is called the *block sum of squares* and denoted SSB. The second is the "new" sum of squares measuring random error. That is, for randomized block data,

$$SSE=\sum_{i=1}^{b}\sum_{j=1}^{k}(Y_{ij}-\overline{Y}_{.j}-\overline{Y}_{i.}+\overline{Y}_{..})^2$$

The other sums of squares from Chapter 12 remain the same in the context of the randomized block design. Specifically,

$$SSTOT=\text{total sum of squares}=\sum_{i=1}^{b}\sum_{j=1}^{k}(Y_{ij}-\overline{Y}_{..})^2$$

and

$$SSTR=\text{treatment sum of squares}=\sum_{i=1}^{b}\sum_{j=1}^{k}(\overline{Y}_{.j}-\overline{Y}_{..})^2$$

Theorem 13.2.1 *Suppose that k treatment levels are measured over a set of b blocks. Then*

a. *$SSTOT=SSTR+SSB+SSE$.*
b. *$SSTR$, SSB, and SSE are independent random variables.*

Proof The independence of the three terms that combine to give $SSTOT$ can be established using the same approach that was taken in Chapter 12. The details will be omitted. □

Theorem 13.2.2 *Suppose that k treatment levels, with means $\mu_1, \mu_2, \ldots, \mu_k$, are measured over a set of b blocks, where the block effects are $\beta_1, \beta_2, \ldots, \beta_b$. Then*

a. *When H_0: $\mu_1=\mu_2=\ldots=\mu_k$ is true, $SSTR/\sigma^2$ has a chi square distribution with $k-1$ degrees of freedom.*

b. *When H_0: $\beta_1 = \beta_2 = \ldots = \beta_b$ is true, SSB/σ^2 has a chi square distribution with $b-1$ degrees of freedom.*

c. *Regardless of whether the μ_j's and/or the β_i's are equal, SSE/σ^2 has a chi square distribution with $(b-1)(k-1)$ degrees of freedom.*

Proof The proofs are similar to those for Theorems 12.2.2 and 12.2.3. □

Theorem 13.2.3 *Suppose that k treatment levels with means $\mu_1, \mu_2, \ldots, \mu_k$ are measured over a set of b blocks. Then*

a. *If H_0: $\mu_1 = \mu_2 = \ldots = \mu_k$ is true,*

$$F = \frac{SSTR/(k-1)}{SSE/(b-1)(k-1)}$$

has an F distribution with $k-1$ and $(b-1)(k-1)$ degrees of freedom.

b. *At the α level of significance, H_0: $\mu_1 = \mu_2 = \ldots = \mu_k$ should be rejected if $F \geq F_{1-\alpha,k-1,(b-1)(k-1)}$.* □

Theorem 13.2.4 *Suppose that k treatment levels are measured over a set of b blocks, where the block effects are $\beta_1, \beta_2, \ldots,$ and β_b. Then*

a. *If H_0: $\beta_1 = \beta_2 = \ldots = \beta_b$ is true,*

$$F = \frac{SSB/(b-1)}{SSE/(b-1)(k-1)}$$

has an F distribution with $b-1$ and $(b-1)(k-1)$ degrees of freedom.

b. *At the α level of significance, H_0: $\beta_1 = \beta_2 = \ldots = \beta_b$ should be rejected if $F \geq F_{1-\alpha,b-1,(b-1)(k-1)}$.* □

Table 13.2.2 shows the ANOVA table entries for a randomized block analysis. Notice that two F ratios are calculated, one for the treatment effect and one for the block effect.

Table 13.2.2

Source	df	SS	MS	F	P
Treatments	$k-1$	$SSTR$	$SSTR/(k-1)$	$\frac{SSTR/(k-1)}{SSE/(b-1)(k-1)}$	$P[F_{k-1,(b-1)(k-1)} \geq \text{obs. } F]$
Blocks	$b-1$	SSB	$SSB/(b-1)$	$\frac{SSB/(b-1)}{SSE/(b-1)(k-1)}$	$P[F_{b-1,(b-1)(k-1)} \geq \text{obs. } F]$
Error	$(b-1)(k-1)$	SSE	$SSE/(b-1)(k-1)$		
Total	$n-1$	$SSTOT$			

Computing Formulas

Let $C = T_{..}^2/bk$. Then

$$SSTR = \sum_{j=1}^{k} \frac{T_{.j}^2}{b} - C \tag{13.2.2}$$

$$SSB = \sum_{i=1}^{b} \frac{T_{i.}^2}{k} - C \tag{13.2.3}$$

$$SSTOT = \sum_{i=1}^{b} \sum_{j=1}^{k} Y_{ij}^2 - C \tag{13.2.4}$$

and, by Theorem 13.2.1,

$$SSE = SSTOT - SSTR - SSB$$

Equations 13.2.2, 13.2.3, and 13.2.4 are considerably easier to evaluate than their counterparts on p. 632. The proofs will be left as exercises.

Case Study 13.2.1

Acrophobia is a fear of heights. It can be treated in a number of different ways. Using contact desensitization, a therapist demonstrates some task that would be difficult for someone with acrophobia to do, such as looking over a ledge or standing on a ladder. Then he guides the subject through the very same maneuver, always keeping in physical contact. Another method of treatment is demonstration participation. Here the therapist tries to *talk* the subject through the task; no physical contact is made. A third technique, live modeling, requires the subject simply to *watch* the task being done—he does not attempt it himself.

These three techniques were compared in a study involving fifteen volunteers, all of whom had a history of severe acrophobia (144). It was realized at the outset, though, that the affliction was much more incapacitating in some subjects than in others, and that this heterogeneity might compromise the therapy comparison. Accordingly, the experiment began with each subject being given the Height Avoidance Test (HAT), a series of forty-four tasks related to ladder climbing. A subject received a "point" for each task successfully completed. On the basis of their final scores, the fifteen volunteers were divided into five blocks (A, B, C, D, and E), each of size 3. The subjects in Block A had the *lowest* scores (that is, the most severe acrophobia), those in Block B the second lowest scores, and so on.

Each of the three therapies was then assigned at random to one of the three subjects in each block. When the counseling sessions were over, the subjects retook the HAT. Table 13.2.3 lists the *changes* in their scores (score after therapy – score before therapy). Test the hypothesis that the therapies are equally effective. Let $\alpha = 0.01$.

(Continued on next page)

Table 13.2.3 HAT Score Changes

	Therapy			
Block	Contact Desensitization	Demonstration Participation	Live Modeling	$T_{i.}$
A	8	2	−2	8
B	11	1	0	12
C	9	12	6	27
D	16	11	2	29
E	24	19	11	54
$T_{.j}$	68	45	17	130

Since $C = (130)^2/15 = 1126.7$ and $\sum_{i=1}^{5}\sum_{j=1}^{3} Y_{ij}^2 = 1894$, it follows that

$$SSTOT = 1894 - 1126.7 = 767.3$$

$$SSB = \frac{(8)^2}{3} + \cdots + \frac{(54)^2}{3} - 1126.7 = 438.0$$

$$SSTR = \frac{(68)^2}{5} + \frac{(45)^2}{5} + \frac{(17)^2}{5} - 1126.7 = 260.9$$

giving an error sum of squares of:

$$SSE = 767.3 - 438.0 - 260.9 = 68.4$$

The analysis of variance is summarized in Table 13.2.4. Since the calculated value of the F statistic, 15.260, exceeds $F_{.99,2,8} = 8.65$, $H_0: \mu_1 = \mu_2 = \mu_3$ can be rejected at the 0.01 level. In fact, the P-value of 0.0019 indicates that H_0 can be rejected for α as small as 0.0019.

Table 13.2.4

Source	df	*SS*	*MS*	*F*	*P*
Therapies	2	260.93	130.47	15.260	0.0019
Blocks	4	438.00	109.50	12.807	0.0015
Error	8	68.40	8.55		
Total	14	767.33			

The small P-value for "Blocks" (=0.0015) implies that $H_0: \beta_1 = \beta_2 = \ldots = \beta_5$ would also be rejected. Of course, that should come as no surprise: The blocks were intentionally set up to be as different as possible. Indeed, if $F = \dfrac{SSB/(b-1)}{SSE/(b-1)(k-1)}$ had *not* been large, we would have questioned the validity of using HAT scores to measure the severity of acrophobia.

Comment Using a randomized block design instead of a one-way design is a trade-off. The blocks result in SSE being reduced (recall Equation 13.2.1), and that increases the probability of rejecting H_0 when H_0 is false, *provided everything else* *associated with the test is kept the same.* But everything else is not kept the same: The degrees of freedom associated with "error" in the randomized block analysis $[= (b-1)(k-1)]$ are *fewer* than the degrees of freedom associated with "error" in the one-way analysis $[= k(b-1)]$. That difference is an advantage of the one-way analysis because the power of any hypothesis test diminishes as the number of degrees of freedom associated with its test statistic decreases.

Ultimately, which design is preferable in a given situation depends on the magnitude of SSB. If SSB were "large," the advantage of a much smaller SSE would more than compensate for the reduction in the degrees of freedom for "error," and the randomized block design would be a better choice than the one-way design. On the other hand, if the block effects were essentially all the same (in which case SSB would be small), then SSE for the randomized block design would not be much smaller than the SSE for the one-way design. In that case, the degrees of freedom for "error" becomes the key issue, and the one-way design would be considered preferable.

If an experiment has already been done as a randomized block design, an F test of $H_0: \beta_1 = \beta_2 = \ldots = \beta_b$ provides some guidance as to how the treatments might best be compared if a follow-up study were to be done. If the F test of $H_0: \beta_1 = \beta_2 = \ldots = \beta_b$ rejects H_0, the decision to use the randomized block design was the right one (especially if the P-value for Blocks is very small). If $H_0: \beta_1 = \beta_2 = \ldots = \beta_b$ is not rejected, future experiments comparing those same treatments should probably either (1) utilize a different blocking criterion or (2) be set up as a one-way design.

Case Study 13.2.2

Rat poison is normally made by mixing its active chemical ingredients with ordinary cornmeal. In many urban areas, though, rats can find food that they prefer to cornmeal, so the poison is left untouched. One solution is to make the cornmeal more palatable by adding food supplements such as peanut butter or meat. Doing that is effective, but the cost is high and the supplements spoil quickly.

In Milwaukee, a study was carried out to see whether *artificial* food supplements might be a workable compromise (85). For five two-week periods, thirty-two hundred baits were placed around garbage-storage areas—eight hundred consisted of plain cornmeal; a second eight hundred had cornmeal mixed with artificial butter-vanilla flavoring; a third eight hundred contained cornmeal mixed with artificial roast beef flavoring; and the remaining eight hundred were cornmeal mixed with artificial bread flavoring.

Table 13.2.5 lists, for each survey, the percentage of each type of bait that was eaten. Do the rats show any preferences for the different flavors? Were the

Table 13.2.5

Survey Number	Plain	Butter-Vanilla	Roast Beef	Bread
1	13.8	11.7	14.0	12.6
2	12.9	16.7	15.5	13.8
3	25.9	29.8	27.8	25.0
4	18.0	23.1	23.0	16.9
5	15.2	20.2	19.0	13.7

(Continued on next page)

blocks—in this case, the surveys—helpful in reducing the error sum of squares? If a follow-up study were to be done comparing these same baits, should a completely randomized design or a randomized block design be used?

All of these questions are answered by the F ratios shown in Table 13.2.6. The P-values for $H_0: \mu_1 = \mu_2 = \mu_3 = \mu_4$ $(= 0.0042)$ and $H_0: \beta_1 = \beta_2 = \beta_3 = \beta_4 = \beta_5$ $(= 0.0000)$ are both extremely small, so both null hypotheses would be rejected. Moreover, the fact that SSB is so very large indicates that considerable variation exists from survey to survey, irrespective of the baits. It follows that any future studies should be set up in a similar fashion—that is, using a randomized block design.

Table 13.2.6

Source	df	SS	MS	F	P
Flavors	3	56.38	18.79	7.58	0.0042
Surveys	4	495.32	123.83	49.93	0.0000
Error	12	29.76	2.48		
Total	19	581.46			

Tukey Comparisons for Randomized Block Data

The Tukey pairwise comparison technique of Section 12.3 can also be applied to a randomized block design. The definition of D is slightly different, since the associated studentized range is no longer $Q_{k,rk-k}$ but rather $Q_{k,(b-1)(k-1)}$, a change reflecting the number of degrees of freedom available for MSE in estimating σ^2.

Theorem 13.2.5 *Let $\overline{Y}_{.j}$, $j = 1, 2, \ldots, k$, be the sample means in a $b \times k$ randomized block design. Let μ_j be the true treatment means, $j = 1, 2, \ldots, k$. The probability is $1 - \alpha$ that all $\binom{k}{2}$ pairwise subhypotheses $H_0: \mu_s = \mu_t$ will simultaneously satisfy the inequalities*

$$\overline{Y}_{.s} - \overline{Y}_{.t} - D\sqrt{MSE} < \mu_s - \mu_t < \overline{Y}_{.s} - \overline{Y}_{.t} + D\sqrt{MSE}$$

where $D = Q_{\alpha,k,(b-1)(k-1)}/\sqrt{b}$. If, for a given s and t, zero is not contained in the preceding inequality, $H_0: \mu_s = \mu_t$ can be rejected in favor of $H_1: \mu_s \neq \mu_t$ at the α level of significance.

Example 13.2.1 Recall the comparison of the three acrophobia therapies in Case Study 13.2.1. The F test in Table 13.2.4 showed that $H_0: \mu_1 = \mu_2 = \mu_3$ can be rejected at the $\alpha = 0.05$ (or even 0.005) level of significance. Should all three therapies, though, be considered different, or is one of them simply different from the other two?

That question can be answered by constructing the set of 95% Tukey confidence intervals for the three pairwise comparisons. Here,

$$D = \frac{Q_{.05,3,8}}{\sqrt{5}} = \frac{4.04}{2.24} = 1.81$$

and the radius of the Tukey intervals is

$$D\sqrt{MSE} = 1.81\sqrt{8.55} = 5.3$$

Table 13.2.7 summarizes the calculations called for in Theorem 13.2.5.

Now we have a much better picture of the relative values of these three therapies. Based on the Tukey intervals, the difference in the means for contact desensitization (μ_1) and demonstration participation (μ_2) is *not* statistically significant. However, the increases in both the contact desensitization mean (μ_1) and the demonstration participation mean (μ_2) relative to the live modeling mean (μ_3) *are* statistically significant.

Table 13.2.7

Pairwise Difference	$\overline{Y}_{.s} - \overline{Y}_{.t}$	Tukey Interval	Conclusion
$\mu_1 - \mu_2$	4.6	$(-0.7, 9.9)$	Not significant
$\mu_1 - \mu_3$	10.2	$(4.9, 15.5)$	Reject
$\mu_2 - \mu_3$	5.6	$(0.3, 10.9)$	Reject

Contrasts for Randomized Block Data

The techniques we learned in Section 12.4 for testing contrasts can be readily adapted to randomized block designs. If C is the contrast associated with the null hypothesis, the appropriate test statistic is

$$F = \frac{SS_C/1}{SSE/(b-1)(k-1)}$$

where F has 1 and $(b-1)(k-1)$ degrees of freedom and SSE is the error sum of squares defined for randomized block data.

Case Study 13.2.3

In folklore, the full moon is often portrayed as something sinister, a kind of evil force possessing the power to control our behavior. Over the centuries, many prominent writers and philosophers have shared this belief (126). Milton, in *Paradise Lost*, refers to

> Demoniac frenzy, moping melancholy
> And moon-struck madness.

And Othello, after the murder of Desdemona, laments:

> It is the very error of the moon,
> She comes more near the earth than she was wont
> And makes men mad.

On a more scholarly level, Sir William Blackstone, the renowned eighteenth-century English barrister, defined a "lunatic" as

(Continued on next page)

one who hath... lost the use of his reason and who hath lucid intervals, sometimes enjoying his senses and sometimes not, and that frequently depending upon changes of the moon.

The possibility of lunar phases influencing human affairs is a theory not without supporters among the scientific community. Studies by reputable medical researchers have attempted to link the "Transylvania effect," as it has come to be known, with higher suicide rates, pyromania, and even epilepsy.

The relationship between lunar cycles and mental breakdowns has also been studied. Table 13.2.8 shows the admission rates to the emergency room of a Virginia mental health clinic *before, during*, and *after* the twelve full moons from August 1971 to July 1972 (11). Here, "time," as expressed in months, is acting as the blocking variable.

Table 13.2.8

	Admission Rates (patients/day)			
Month	(1) Before Full Moon	(2) During Full Moon	(3) After Full Moon	$\overline{Y}_{i.}$
Aug.	6.4	5.0	5.8	5.73
Sept.	7.1	13.0	9.2	9.77
Oct.	6.5	14.0	7.9	9.47
Nov.	8.6	12.0	7.7	9.43
Dec.	8.1	6.0	11.0	8.37
Jan.	10.4	9.0	12.9	10.77
Feb.	11.5	13.0	13.5	12.67
Mar.	13.8	16.0	13.1	14.30
Apr.	15.4	25.0	15.8	18.73
May	15.7	13.0	13.3	14.00
June	11.7	14.0	12.8	12.83
July	15.8	20.0	14.5	16.77
$\overline{Y}_{.j}$	10.92	13.33	11.46	

Table 13.2.9 summarizes the ANOVA calculations. For 2 and 22 degrees of freedom, the 0.05 critical value for the *lunar cycle* effect is 3.44, which is greater than the observed F ($=3.22$). Therefore, we would fail to reject H_0: $\mu_1 = \mu_2 = \mu_3$, and the conclusion would be that a lunar effect has *not* been demonstrated.

Table 13.2.9

Source	df	*SS*	*MS*	*F*
Lunar cycles	2	38.59	19.30	3.22
Months	11	451.08	41.01	
Error	22	132.08	6.00	
Total	35	621.75		

(Continued on next page)

(Case Study 13.2.3 continued)

Testing the overall $H_0: \mu_1 = \mu_2 = \mu_3$ is not the only appropriate way to analyze these data, though. An *a priori* subhypothesis is clearly suggested by the circumstances of the problem—specifically, it would make sense to test whether the admission rate during the full moon is different from the *average* rate during the rest of the month. The null subhypothesis corresponding to such a test would be $H_0: \mu_2 = (\mu_1 + \mu_3)/2$.

Following the procedure outlined in Section 12.4, the contrast associated with H_0 is

$$C = -\frac{1}{2}\mu_1 + \mu_2 - \frac{1}{2}\mu_3$$

and its estimate is

$$\hat{C} = -\frac{1}{2}(10.92) + 1(13.3) - \frac{1}{2}(11.46)$$
$$= 2.11$$

From Definition 12.4.2, the sum of squares associated with C is:

$$SS_C = \frac{(2.11)^2}{\frac{1/4}{12} + \frac{1}{12} + \frac{1/4}{12}} = 35.62$$

Dividing SS_C by the mean square for error gives an F ratio of *5.93* (with 1 and 22 degrees of freedom):

$$\frac{35.62/1}{132.08/22} = 5.93$$

For $\alpha = 0.05$, though, $F_{.95,1,22} = 4.30$. Therefore, contrary to our acceptance of H_0: $\mu_1 = \mu_2 = \mu_3$, we would *reject* $H_0: \mu_2 = (\mu_1 + \mu_3)/2$ and conclude that the Transylvania effect *does* exist.

Comment It is always more than a little disconcerting when two statistical techniques applied to the same data lead to opposite conclusions. That such apparent contradictions occur, though, should not be unexpected. Different methods of analysis simply utilize the data in different ways. Disagreements from time to time are inevitable.

Questions

13.2.1. In recent years a number of research projects in extrasensory perception have examined the possibility that hypnosis may be helpful in bringing out ESP in persons who did not think they had any. The obvious way to test such a hypothesis is with a self-paired design: the ESP ability of a subject when he is awake is compared to his ability when hypnotized. In one study of this sort, fifteen college students were each asked to guess the identity of 200 Zener cards (see Case Study 4.3.1). The same "sender"—that is, the person concentrating on the card—was used for each trial. For 100 of the trials both the student and the sender were awake; for the other 100 both were hypnotized. If chance were the only factor involved, the expected number of correct identifications in each set of 100 trials would be 20. The observed average numbers of correct guesses for subjects awake and subjects hypnotized were 18.9 and 21.7, respectively (21). Use the analysis of variance to determine whether that difference is statistically significant at the 0.05 level.

Number of Correct Responses (out of 100) in ESP Experiment

Student	Sender and Student in Waking State	Sender and Student in Hypnotic State
1	18	25
2	19	20
3	16	26
4	21	26
5	16	20
6	20	23
7	20	14
8	14	18
9	11	18
10	22	20
11	19	22
12	29	27
13	16	19
14	27	27
15	15	21
$\bar{Y}_{.j}$	18.9	21.7

13.2.2. The following table shows the audience shares of the three major networks' evening news broadcasts in four major cities as reported by Arbitron. Test at the $\alpha = 0.10$ level of significance the null hypothesis that viewing levels for news are the same for ABC, CBS, and NBC.

City	ABC	CBS	NBC
A	19.7	16.1	18.2
B	18.6	15.8	17.9
C	19.1	14.6	15.3
D	17.9	17.1	18.0

13.2.3. A paint manufacturer is experimenting with an additive that might make the paint less chalky. To ensure that the additive does not affect the tint, a quality-control engineer takes a sample from each of seven batches of Osage Orange. Each sample is split in half, and the additive is put into one of the two. Both samples are examined with a spectroscope, with the output read in standardized lumen units. If the tint were exactly correct, the reading would be 1.00. Test that the mean spectroscope readings are the same for the two versions of Osage Orange. Let $\alpha = 0.05$.

Batch	Without Additive	With Additive
1	1.10	1.06
2	1.05	1.02
3	1.08	1.17
4	0.98	1.21
5	1.01	1.01
6	0.96	1.23
7	1.02	1.19

13.2.4. The number of new building permits can be a good indicator of the strength of a region's economic growth. The following table gives percentage increases over a four-year period for three geographical areas. Analyze the data. Let $\alpha = 0.05$. What are your conclusions?

Year	Eastern	North Central	Southwest
2000	1.1	0.1	0.9
2001	1.3	0.8	1.0
2002	2.9	1.1	1.4
2003	3.5	1.3	1.5

13.2.5. Analyze the Transylvania effect data in Case Study 13.2.3 by calculating 95% Tukey confidence intervals for the pairwise differences among the admission rates for the three different phases of the moon. How do your conclusions agree with (or differ from) those already discussed on p. 640? Let $Q_{.05,3,22} = 3.56$.

13.2.6. The table below gives a stock fund's quarterly returns for the years 2003 to 2007. Are the results affected by the quarter of the year? Is the variability in the return from year to year statistically significant? State your conclusions using the $\alpha = 0.05$ level of significance.

	Quarter			
Year	First	Second	Third	Fourth
2003	−5.29	8.62	5.23	6.44
2004	4.96	1.06	−0.25	6.32
2005	0.11	0.58	5.46	3.01
2006	5.30	0.82	4.81	6.54
2007	1.71	5.41	−1.92	−4.78

13.2.7. Find the 95% Tukey intervals for the data of Question 13.2.2, and use them to test the three pairwise comparisons of ABC, CBS, and NBC.

13.2.8. A comparison was made of the efficiency of four different unit-dose injection systems. A group of pharmacists and nurses were the "blocks." For each system, they were to remove the unit from its outer package, assemble it, and simulate an injection. In addition to the standard system of using a disposable syringe and needle to draw the medication from a vial, the other systems tested were Vari-Ject (CIBA Pharmaceutical), Unimatic (Squibb), and

Tubex (Wyeth). Listed in the following table are the average times (in seconds) needed to implement each of the systems (149).

Average Times (sec) for Implementing Injection Systems

Subject	Standard	Vari-Ject	Unimatic	Tubex	$T_{i.}$
1	35.6	17.3	24.4	25.0	102.3
2	31.3	16.4	22.4	26.0	96.1
3	36.2	18.1	22.8	25.3	102.4
4	31.1	17.8	21.0	24.0	93.9
5	39.4	18.8	23.3	24.2	105.7
6	34.7	17.0	21.8	26.2	99.7
7	34.1	14.5	23.0	24.0	95.6
8	36.5	17.9	24.1	20.9	99.4
9	32.2	14.6	23.5	23.5	93.8
10	40.7	16.4	31.3	36.9	125.3
$T_{.j}$	351.8	168.8	237.6	256.0	1014.2

(a) Test the equality of the means at the 0.05 level.
(b) Use Tukey's method to test all six pairwise differences of the four μ_j's. Let $\alpha = 0.05$.

(*Note:* $SSTOT = 2056.10$, $SSTR = 1709.60$, and $SSB = 193.53$; $Q_{.05,4,27} = 4.34$.)

13.2.9. Heart rates were monitored (10) for six tree shrews (*Tupaia glis*) during three different stages of sleep: LSWS (light slow-wave sleep), DSWS (deep slow-wave sleep), and REM (rapid-eye-movement sleep).

Heart Rates (beats/5 seconds)

Tree Shrew	LSWS	DSWS	REM
1	14.1	11.7	15.7
2	26.0	21.1	21.5
3	20.9	19.7	18.3
4	19.0	18.2	17.0
5	26.1	23.2	22.5
6	20.5	20.7	18.9

(a) Do the analysis of variance to test the equality of the heart rates during these three phases of sleep. Let $\alpha = 0.05$.
(b) Because of the marked physiological difference between REM sleep and LSWS and DSWS sleep, it was decided before the data were collected to test the REM rate against the average of the other two. Test the appropriate subhypothesis with a contrast. Use the 0.05 level of significance. Also, find a second contrast orthogonal to the first and verify that the sum of the sum of squares for the two contrasts equals *SSTR*.

13.2.10. Refer to the rat poison data of Case Study 13.2.2. Partition the treatment sum of squares into three orthogonal contrasts. Let one contrast test the hypothesis that the true acceptance percentage for the plain cornmeal is equal to the true acceptance percentage for the cornmeal with artificial roast beef flavoring. Let a second contrast compare the effectiveness of the "butter-vanilla" and "bread" baits. What does the third contrast test? Do all testing at the $\alpha = 0.10$ level of significance.

13.2.11. Prove the computing formulas given in Equations 13.2.2, 13.2.3, and 13.2.4.

13.2.12. Differentiate the function

$$L = \sum_{i=1}^{b} \sum_{j=1}^{k} (y_{ij} - \beta_i - \mu_j)^2$$

with respect to all *bk* parameters and calculate the least squares estimates for the β_i's and μ_j's.

13.2.13. True or false:

(a) $\sum_{i=1}^{b} \overline{Y}_{i.} = \sum_{j=1}^{k} \overline{Y}_{.j}$

(b) Either *MSTR* or *MSB* or both are greater than or equal to *MSE*.

13.2.14. For a set of randomized block data comparing k treatments within b blocks, find

(a) $E(SSB)$
(b) $E(SSE)$

13.3 The Paired *t* Test

The randomized block design is typically used when three or more treatment levels are to be compared within a set of b blocks. If an experiment involves b blocks but only *two* treatment levels, the ANOVA described in Table 13.2.2 can still be used, but a computationally simpler (and equivalent) approach is to do a *paired t test*. The latter has the additional advantage in that it shows more clearly *how* the use of blocks can facilitate the comparison of treatments.

By definition, a "pair" represents a more or less constant set of conditions under which one measurement can be made on Treatment X and one measurement on Treatment Y. Paired data, then, consist of measurements taken on Treatment X and Treatment Y *within each of b pairs*. In effect, the paired t test pools the treatment response differences *within each pair* from pair to pair.

Recall Figure 8.2.4. The two observations recorded for the ith pair can be written

$$X_i = \mu_X + B_i + \varepsilon_i$$

and

$$Y_i = \mu_Y + B_i + \varepsilon_i'$$

where

1. μ_X and μ_Y are the true means associated with Treatment X and Treatment Y, respectively

and

2. B_i is the *effect*—that is, the numerical contribution to the measurement—resulting from the conditions defining Pair i (B_i may be positive, negative, or zero).

For the purposes of this section, it will also be assumed that ε_i and ε_i' are independent, normally distributed random variables, each having mean zero, but with variances σ_X^2 and σ_Y^2, respectively.

Notice that B_i "disappears" when the two measurements *within a pair* are subtracted:

$$\begin{aligned} D_i &= \mu_X + B_i + \varepsilon_i - (\mu_Y + B_i + \varepsilon_i') \\ &= \mu_X - \mu_Y + \varepsilon_i - \varepsilon_i' \end{aligned} \tag{13.3.1}$$

Moreover, it follows that

1. $E(D_i) = \mu_D = \mu_X - \mu_Y$
2. $\text{Var}(D_i) = \sigma_D^2 = \sigma_X^2 + \sigma_Y^2$

and

3. D_i is normally distributed

Equation 13.3.1 is the key to understanding how the paired-data design works. Suppose an experimenter recognizes that taking a measurement on Treatment X (under the conditions present in, say, Pair i) will result in a possibly sizeable B_i being included in the observation. Since the actual magnitude of B_i is unknown, its presence complicates the interpretation of what the observed measurement is telling us about the effect of Treatment X. What the experimenter should do in that situation is take a measurement on Treatment Y *under the same conditions that prevailed for the measurement on Treatment* X. That measurement, then, will also include the component B_i, but if the two observations are subtracted, Equation 13.3.1 shows that the resulting difference (1) will be free of B_i and (2) will be an estimate for $\mu_D = \mu_X - \mu_Y$. In effect, the paired-data design allows for the comparison of Treatment X and Treatment Y to be made unencumbered by whatever differences might exist in the experimental environment. (A more specific example of this important idea will be described at length in Section 13.4.)

Since $\mu_D = \mu_X - \mu_Y$, testing H_0: $\mu_D = 0$ is equivalent to testing H_0: $\mu_X = \mu_Y$. The procedure for doing the former is known as a *paired t test*. The statistic for testing

$H_0\colon \mu_D = 0$ is a special case of Theorem 7.3.5. If $D_i = X_i - Y_i$, $i = 1, 2, \ldots, b$, is a set of within-pair treatment differences, where $\overline{D}$ and S_D denote the sample mean and sample standard deviation of the D_i's, respectively, then

$$\frac{\overline{D} - \mu_D}{S_D/\sqrt{b}}$$

will have a Student t distribution with $b - 1$ degrees of freedom.

Theorem 13.3.1 *Let $d_1, d_2, \ldots, d_b$ be a random sample of within-pair treatment differences from a normal distribution whose mean is μ_D. Let $\overline{d}$ and s_D denote the sample mean and sample standard deviation of the d_i's, and define $t = \overline{d}/(s_D/\sqrt{b})$.*

a. *To test $H_0\colon \mu_D = 0$ versus $H_1\colon \mu_D < 0$ at the α level of significance, reject H_0 if $t \le -t_{\alpha,b-1}$.*

b. *To test $H_0\colon \mu_D = 0$ versus $H_1\colon \mu_D > 0$ at the α level of significance, reject H_0 if $t \ge t_{\alpha,b-1}$.*

c. *To test $H_0\colon \mu_D = 0$ versus $H_1\colon \mu_D \ne 0$ at the α level of significance, reject H_0 if t is either (1) $\le -t_{\alpha/2,b-1}$ or (2) $\ge t_{\alpha/2,b-1}$.* □

Case Study 13.3.1

Prior to the 1968 appearance of Kenneth Cooper's book entitled *Aerobics*, the word did not appear in *Webster's Dictionary*. Now the term is commonly understood to refer to sustained exercises intended to strengthen the heart and lungs. The actual benefits of such physical activities, as well as their possible detrimental effects, have spawned a great deal of research in human physiology as it relates to exercise.

One such study (73) concerned changes in the blood, specifically in hemoglobin levels before and after a prolonged brisk walk. Hemoglobin helps red blood cells transport oxygen to tissues and then remove carbon dioxide. Given the stress that exercise places on the need for that particular exchange, it is not unreasonable to suspect that aerobics might alter the blood's hemoglobin levels.

Ten athletes had their hemoglobin levels measured (in g/dl) prior to embarking on a sixty-kilometer walk. After they finished, their levels were measured again (see Table 13.3.1). Set up and test an appropriate H_0 and H_1.

If μ_X and μ_Y denote the true average hemoglobin levels *before* and *after* walking, respectively, and if $\mu_D = \mu_X - \mu_Y$, then the hypotheses to be tested are

$$H_0\colon \mu_D = 0$$

versus

$$H_1\colon \mu_D \ne 0$$

Let 0.05 be the level of significance.

From Table 13.3.1,

$$\sum_{i=1}^{10} d_i = 4.7 \quad \text{and} \quad \sum_{i=1}^{10} d_i^2 = 8.17$$

(Continued on next page)

Table 13.3.1

Subject	Before Walk, x_i	After Walk, y_i	$d_i = x_i - y_i$
A	14.6	13.8	0.8
B	17.3	15.4	1.9
C	10.9	11.3	−0.4
D	12.8	11.6	1.2
E	16.6	16.4	0.2
F	12.2	12.6	−0.4
G	11.2	11.8	−0.6
H	15.4	15.0	0.4
I	14.8	14.4	0.4
J	16.2	15.0	1.2

Therefore,

$$\overline{d} = \frac{1}{10}(4.7) = 0.47$$

and

$$s_D^2 = \frac{10(8.17) - (4.7)^2}{10(9)} = 0.662$$

Since $n = 10$, the critical values for the test statistic will be the 2.5th and 97.5th percentiles of the Student t distribution with 9 degrees of freedom: $\pm t_{.025,9} = \pm 2.2622$. The appropriate decision rule from Theorem 13.3.1, then, is

$$\text{Reject } H_0\colon \mu_D = 0 \text{ if } \frac{\overline{d}}{s_D/\sqrt{10}} \text{ is either } \begin{cases} \leq -2.2622 \\ \text{or} \\ \geq 2.2622 \end{cases}$$

In this case the t ratio is

$$\frac{0.47}{\sqrt{0.662}/\sqrt{10}} = 1.827$$

and our conclusion is to *fail to reject* H_0: The difference between $\overline{d}(=0.47)$ and the H_0 value for $\mu_D(=0)$ is not statistically significant.

Case Study 13.3.2

Some rental car agencies promise to offer lower-cost rentals. Among them are the aptly named Budget and Thrifty. But is Thrifty really thriftier? Table 13.3.2 shows the rates charged by these two companies for a midsize sedan rented midweek with a month's notice at each of eleven major airports. According to the d_i's listed in the last column, $\overline{d} = \$6.29$ (= average Budget rate – average Thrifty rate). The parameter of interest here is μ_D, the true average difference between the Budget and Thrifty rates. One question to be answered is whether

(Continued on next page)

(Case Study 13.3.2 continued)

Table 13.3.2

Airport	Budget, x_i	Thrifty, y_i	$d_i = x_i - y_i$
Atlanta	93.74	88.54	5.20
Baltimore	129.75	125.00	4.75
Charlotte	100.33	99.03	1.30
Chicago	111.04	104.14	6.90
Dallas–Ft. Worth	167.15	162.08	5.07
Denver	149.56	141.41	8.15
Los Angeles	124.26	122.99	1.27
Miami	108.57	102.51	6.06
New Orleans	118.62	117.44	1.18
Seattle	129.81	121.76	8.05
St. Louis	98.58	77.32	21.26

Source: www.expedia.com

the sample mean of $\overline{d} = \$6.29$ is sufficiently positive to overturn the presumption that $\mu_D = 0$.

Notice first of all that these x_i's and y_i's are dependent: the \$93.74 and \$88.54 in first row, for example, are lower than most of the rates at other airports and may reflect lower operating costs or less demand in Atlanta. That is, included in the \$93.74 and \$88.54 is the B_1 referenced in Equation 13.3.1. Similarly, a portion of the \$129.75 and \$125.00 is the B_2 for Baltimore; and so on.

A confidence interval for μ_D will provide an estimate of the savings associated with renting Thrifty midsize sedans and also give us a way of testing the two-sided hypothesis that $\mu_D = 0$.

Theorem 7.4.1 applies to the d_i's, so the form of the $100(1-\alpha)\%$ confidence interval is

$$\left(\overline{d} - t_{\alpha/2,b-1}\frac{s}{\sqrt{n}}, \overline{d} + t_{\alpha/2,b-1}\frac{s}{\sqrt{n}}\right)$$

The average of the figures in the last column is $\overline{d} = \$6.29$ and the sample standard deviation is $s = \$5.59$. For $\alpha = 0.05$, the Student t value is

$$-t_{.025,b-1} = -t_{.025,10} = 2.2281$$

so the 95% confidence interval reduces to

$$\left(6.29 - 2.2281\frac{5.59}{\sqrt{11}},\ 6.29 + 2.2281\frac{5.59}{\sqrt{11}}\right) = (\$2.53,\ \$10.05)$$

Moreover, since 0 is not in the confidence interval, we can reject the null hypothesis that $\mu_D = 0$.

About the Data The difference in rental costs in St. Louis is clearly an "outlier" and possibly results from a Thrifty promotion of some kind. The distortion that such a deviant quantity introduces suggests that a better strategy would be to compare *average* rental costs over an extended period of time.

Criteria for Pairing

The Comment following Case Study 13.2.1 discusses the issues an experimenter needs to consider in deciding whether the comparison of k treatment levels should be done as a k-sample design or a randomized block design. When *two* treatment levels are to be compared, similar questions need to be addressed:

1. Should the comparison be done with *independent* samples (and the two-sample t test) or with *dependent* samples (and the paired t test)?
2. If the paired-data model is the experimental design chosen, what criterion should be used to define the pairs?

The pros and cons of using dependent samples will be discussed in Section 13.4. Here we want to focus on some of the ways pairs are defined in real-world applications. A representative sampling of blocking criteria in general is reflected in the five Case Studies appearing earlier in Chapter 13.

The ultimate pairing criterion is to use each subject twice. Then the experimenter can be confident that whatever "contribution" a subject makes to the numerical value for Treatment X is exactly the same as the contribution made to the numerical value for Treatment Y. Over the years, "before and after" studies of this sort have become very popular with researchers. The aerobics/hemoglobin data described in Case Study 13.3.1 are a typical example.

Not every experimental protocol, though, lends itself to the possibility of testing both treatments on each subject. Suppose the objective of a study is to compare two methods of teaching fractions to third graders. Once a subject is exposed to one method (and learns *something* about fractions), assessing the effectiveness of the second treatment would be problematic. Clearly, such a study needs to be done with pairs of *two* (similar) subjects, one being taught with Method X, and the other with Method Y. Defining what "similar" means in this case could be done in a variety of ways. The closest approximation to the "before and after" format would be to use *twins* as subjects. If the number of twins available, though, was insufficient, "similar" could be defined in terms of IQ scores or previous math grades.

Another widely used strategy for creating dependent observations is to pair up measurements taken close together in time and/or space. The rationale, of course, is that measurements sharing a variety of environmental characteristics will be inflated (or deflated) by similar amounts for both Treatment X and Treatment Y. The data in Tables 13.2.5, 13.2.8, and 13.3.2 are all cases in point.

Probably the most challenging scenarios faced by experimenters are situations where there are no obvious pairing criteria of the sort just described. Rather, some sort of "pre-test" needs to be derived that would serve as a mechanism for identifying subjects likely to respond in similar ways to the two treatments. Recall, for example, the blocks defined in Case Study 13.2.1. There, the Height Avoidance Test (HAT) was used as a way of categorizing the severity of a subject's initial level of acrophobia. By defining blocks to be subjects with similar HAT scores, a set of relatively homogeneous experimental environments were created (Blocks A through E) within which all the competing therapies could be compared.

It would be difficult to overestimate the importance of choosing the blocking and pairing criteria carefully whenever the randomized block and paired-data designs are being used. Whatever can be done to minimize the additional variation in the measurements due to specific environmental effects will allow the treatments to be compared with that much more precision.

The Equivalence of the Paired t Test and the Randomized Block ANOVA When $k = 2$

Example 12.2.2 showed that the analysis of variance done on a set of k-sample data when $k=2$ is equivalent to a (pooled) two-sample t test of H_0: $\mu_X = \mu_Y$ against a two-sided alternative. Although the numerical values of the observed t and observed F will be different, as will be the locations of the two critical regions, the final inference will necessarily be the same. A similar equivalence holds for the paired t test and the randomized block ANOVA (when $k=2$).

Recall Case Study 13.3.1. Analyzed with a paired t test, H_0: $\mu_D = 0$ should be rejected in favor of H_1: $\mu_D \neq 0$ at the $\alpha = 0.05$ level of significance if

$$t \leq -t_{\alpha/2,b-1} = -t_{0.025,9} = -2.2622 \quad \text{or if} \quad t \geq t_{\alpha/2,b-1} = 2.2622$$

But $t = \dfrac{\overline{d}}{s_D/\sqrt{10}} = 1.83$, so the conclusion is "fail to reject H_0."

Table 13.3 shows the Minitab input and output for doing the analysis of variance on those same observations. The observed F ratio for "Treatments" is 3.34, and the corresponding $\alpha = 0.05$ critical value is $F_{0.95,1,9} = 5.12$.

Table 13.3.3

```
MTB  > set c1
DATA > 14.6 17.3 10.9 12.8 16.6 12.2 11.2 15.4 14.8 16.2
DATA > 13.8 15.4 11.3 11.6 16.4 12.6 11.8 15.0 14.4 15.0
DATA > end
MTB  > set c2
DATA > 1 1 1 1 1 1 1 1 1 1 2 2 2 2 2 2 2 2 2 2
DATA > end
MTB  > set c3
DATA > 1 2 3 4 5 6 7 8 9 10 1 2 3 4 5 6 7 8 9 10
DATA > end
MTB  > name c1 'Hemglb' c2 'Activ' c3 'Blocks'
MTB  > twoway c1 c2 c3
```

Two-way ANOVA: Hemglb versus Activ, Blocks

Source	DF	SS	MS	F	P
Activ	1	1.1045	1.10450	3.34	0.101
Blocks	9	73.2405	8.13783	24.57	0.000
Error	9	2.9805	0.33117		
Total	19	77.3255			

Notice that (1) the observed F is the square of the observed t and (2) the F critical value is the square of the t critical value:

$$3.34 = (1.827)^2 \quad \text{and} \quad 5.12 = (2.2622)^2$$

It follows, then, that the paired t test will reject the null hypothesis that $\mu_D = 0$ if and only if the randomized block ANOVA rejects the null hypothesis that the two Treatment means (μ_X and μ_Y) are equal.

Questions

13.3.1. Case Study 7.5.2 compared the volatility of Global Rock Funds' return on investments to that of the benchmark Lipper fund. But can it be said that the returns themselves beat the benchmark? The table below gives the annual returns of the Global Rock Fund for the years 1989 to 2007 and the corresponding Lipper averages. Test the hypothesis that $\mu_D > 0$ for these data at the 0.05 level of significance.

	Investment return %			Investment return %	
Year	Global Rock, x	Lipper Avg., y	Year	Global Rock, x	Lipper Avg., y
1989	15.32	14.76	1999	27.43	34.44
1990	1.62	−1.91	2000	8.57	1.13
1991	28.43	20.67	2001	1.88	−3.24
1992	11.91	6.18	2002	−7.96	−8.11
1993	20.71	22.97	2003	35.98	32.57
1994	−2.15	−2.44	2004	14.27	15.37
1995	23.29	20.26	2005	10.33	11.25
1996	15.96	14.79	2006	15.94	12.70
1997	11.12	14.27	2007	16.71	9.65
1998	0.37	6.25			

Note that $\sum_{i=1}^{19} d_i = 28.17$ $\sum_{i=1}^{19} d_i^2 = 370.8197$

13.3.2. Recall the depth perception data described in Question 8.2.6. Use a paired t test with $\alpha = 0.05$ to compare the numbers of trials needed to learn depth perception for Mothered and Unmothered lambs.

13.3.3. Blood coagulates as a result of a complex sequence of chemical reactions. The protein thrombin triggers the clotting of blood under the influence of another protein called prothrombin. One measure of a person's blood clotting ability is expressed in prothrombin time, which is defined to be the interval between the initiation of the thrombin-prothrombin reaction and the formation of the clot. One study (209) looked at the effect of aspirin on prothrombin time. The preceding table gives, for each of twelve subjects, the prothrombin time (in seconds) *before* and *three hours after* taking two aspirin tablets (650 mg). Test the hypothesis that aspirin influences prothrombin times. Perform the test at both the $\alpha = 0.05$ and $\alpha = 0.01$ levels.

Subject	Before Aspirin, x	After Aspirin, y
1	12.3	12.0
2	12.0	12.3
3	12.0	12.5
4	13.0	12.0
5	13.0	13.0
6	12.5	12.5
7	11.3	10.3
8	11.8	11.3
9	11.5	11.5
10	11.0	11.5
11	11.0	11.0
12	11.3	11.5

13.3.4. Use a paired t test to analyze the hypnosis/ESP data given in Question 13.2.1. Let $\alpha = 0.05$.

13.3.5. Perform the hypothesis test indicated in Question 13.2.3 at the 0.05 level using a paired t test. Compare the square of the observed t with the observed F. Do the same for the critical values associated with the two procedures. What would you conclude?

13.3.6. Let $D_1, D_2, \ldots, D_b$ be the within-block differences as defined in this section. Assume that the D_i's are normal with mean μ_D and variance σ_D^2, for $i = 1, 2, \ldots, b$. Derive a formula for a $100(1 - \alpha)\%$ confidence interval for μ_D. Apply this formula to the data of Case Study 13.3.1 and construct a 95% confidence interval for the true average hemoglobin difference ("before walk" – "after walk").

13.3.7. Construct a 95% confidence interval for μ_D in the prothrombin time data described in Question 13.3.3. See Question 13.3.6.

13.3.8. Show that the paired t test is equivalent to the F test in a randomized block design when the number of treatment levels is two. (*Hint:* Consider the distribution of $T^2 = b\overline{D}^2/S_D^2$.)

13.4 Taking a Second Look at Statistics (Choosing between a Two-Sample *t* Test and a Paired *t* Test)

Suppose that the means μ_X and μ_Y associated with two treatments X and Y are to be compared. Theoretically, two "design" options are available:

1. test $H_0: \mu_X = \mu_Y$ with *independent samples* (using Theorem 9.2.2 or Theorem 9.2.3) or
2. test $H_0: \mu_D = 0$ with *dependent samples* (using Theorem 13.3.1).

Does it make a difference which design is used? Yes. Which one is better? That depends on the nature of the subjects, and how likely they are to respond to the treatments—neither design is always superior to the other.

The two hypothetical examples described in this section illustrate the pros and cons of each approach. In the first case, the paired-data model is clearly preferable; in the second case, μ_X and μ_Y should be compared using a two-sample format.

Example 13.4.1

Comparing two weight loss plans

Suppose that Treatment X and Treatment Y are two diet regimens. A comparison of the two is to be done by looking at the weight losses recorded by subjects who have been using one of the two diets for a period of three months. Ten people have volunteered to be subjects. Table 13.4.1 gives the gender, age, height, and initial weight for each of the ten.

Table 13.4.1

Subject	Gender	Age	Height	Weight (in pounds)
HM	M	65	5′8″	204
HW	F	41	5′4″	165
JC	M	23	6′0″	260
AF	F	63	5′3″	207
DR	F	59	5′2″	192
WT	M	22	6′2″	253
SW	F	19	5′1″	178
LT	F	38	5′5″	170
TB	M	62	5′7″	212
KS	F	23	5′3″	195

Option A: Compare Diet X and Diet Y Using Independent Samples If the two-sample design is to be used, the first step would be to divide the ten subjects *at random* into two groups of size 5. Table 13.4.2 shows one such set of independent samples.

Table 13.4.2

Diet X		Diet Y	
HW	(F, middle-aged, slightly overweight)	JC	(M, young, very overweight)
AF	(F, elderly, very overweight)	WT	(M, young, very overweight)
SW	(F, young, very overweight)	HM	(M, elderly, quite overweight)
TB	(M, elderly, quite overweight)	KS	(F, young, very overweight)
DR	(F, elderly, very overweight)	LT	(F, middle-aged, slightly overweight)

Notice that each of the two samples contains individuals who are likely to respond very differently to whichever diet they are on simply because of the huge disparities in their physical profiles. Included among the subjects representing Diet X, for example, are HW and TB; HW is a slightly overweight, middle-aged female, while TB is a quite overweight, elderly male. More than likely, their weight losses after three months will be considerably different.

If some of the subjects in Diet X lose relatively few pounds (which will probably be the case for HW) while others record sizeable reductions (which is likely to happen for AF, SW, and DR, all of whom are initially very overweight), the effect will be to inflate the numerical value of s_X^2. Similarly, the value of s_Y^2 will be inflated by the inherent differences among the subjects in Diet Y.

Now, recall the formula for the two-sample t statistic,

$$t = \frac{\overline{x} - \overline{y}}{s_p\sqrt{1/n + 1/m}}$$

If s_X^2 and s_Y^2 are large, s_p will also be large. But if s_p (in the denominator of the t ratio) is very large, the t statistic itself may be fairly small even if $\overline{x} - \overline{y}$ is substantially different from zero—that is, the considerable variation *within* the samples has the potential to "obscure" the variation *between* the samples (as measured by $\overline{x} - \overline{y}$). In effect, $H_0\colon \mu_X = \mu_Y$ might not be rejected (when it should be) only because the variation from subject to subject is so large.

Option B: Compare Diet X and Diet Y Using Dependent Samples The same differences from subject to subject that undermine the two-sample t test provide some obvious criteria for setting up a *paired t* test. Table 13.4.3 shows a grouping into five pairs of the ten subjects profiled in Table 13.4.2, where the two members of each pair are as similar as possible with respect to the amount of weight they are likely to lose: for example, Pair 2—(JC, WT)—is comprised of two very overweight, young males. In the terminology of Equation 13.3.1, the B_2 that measures the subject effect of persons fitting that description will be present in the weight losses reported by both JC and WT. When their responses are subtracted, $d_2 = x_2 - y_2$ will, in effect, be free of the subject effect and will be a more precise estimate of the intrinsic difference between the two diets. It follows that differences *between* the pairs—no matter how sizeable those differences may be—are irrelevant because the comparisons of Diet X and Diet Y (that is, the d_i's) are made *within* the pairs, and then pooled from pair to pair.

Table 13.4.3

Pair	Characteristics
(HW, LT)	Female, middle-aged, slightly overweight
(JC, WT)	Male, young, very overweight
(SW, KS)	Female, young, very overweight
(HM, TB)	Male, elderly, quite overweight
(AF, DR)	Female, elderly, very overweight

The potential benefit here of using a paired-data design should be readily apparent. Recall that the paired t statistic has the form

$$t = \frac{\overline{d}}{s_D/\sqrt{b}} = \frac{\overline{x} - \overline{y}}{s_D/\sqrt{b}} \tag{13.4.1}$$

For the reasons just cited, $s_D/\sqrt{5}$ is likely to be much smaller than the two-sample $s_p\sqrt{\frac{1}{5} + \frac{1}{5}}$, thus reducing the likelihood that the paired t test's denominator will "wash out" its numerator.

Example 13.4.2 **Comparing two eye surgery techniques**

Suppose the ten subjects profiled in Table 13.4.2 are all nearsighted and have volunteered to participate in a clinical trial comparing two laser surgery techniques. The basic plan is to use Surgery X on five of the subjects, and Surgery Y on the other five. A month later, each participant will be asked to rate (on a scale of 0 to 100) his or her satisfaction with the operation. ■

Option A: Compare Surgery X and Surgery Y Using Independent Samples

Unlike the situation encountered in the diet study, none of the information recorded on the volunteers (gender, age, height, weight) has any bearing on the measurements to be recorded here: a very overweight, young male is no more or no less likely to be satisfied with corrective eye surgery than is a slightly overweight, middle-aged female. That being the case, there is no way to group the ten subjects into five pairs in such a way that the two members of a pair are uniquely similar in terms of how they are likely to respond to the satisfaction question.

To compare Surgery X and Surgery Y, then, using the two-sample format, we would simply divide the ten subjects—at random—into two groups of size 5, and choose between $H_0: \mu_X = \mu_Y$ and $H_1: \mu_X \neq \mu_Y$ on the basis of the two-sample t statistic, which would have *8* $(= n + m - 2 = 5 + 5 - 2)$ degrees of freedom.

Option B: Compare Surgery X and Surgery Y Using Dependent Samples Given the absence of any objective criteria for linking one subject with another in any meaningful way, the pairs would have to be formed at random. Doing that would have some serious negative consequences that would definitely argue against using the paired-data format. Suppose, for example, HW was paired with LT, as was the case in the diet study. Since the information in Table 13.4.2 has nothing to do with a person's reaction to eye surgery, subtracting LT's response from HW's response would not eliminate the "subject" effect as it did in the diet study, because the "personal" contribution of LT to the observed x could be entirely different than the "personal" contribution of HW to the observed y. In general, the within-pair differences—$d_i = x_i - y_i, i = 1, 2, \ldots, 5$—would still reflect the subject effects, so the value of s_D would not be reduced (relative to s_p) as it was in the diet study.

Is a lack of reduction in the magnitude of s_D a serious problem? Yes, because the paired-data format intentionally sacrifices degrees of freedom for the express purpose of reducing s_D. If the latter does not occur, those degrees of freedom are wasted. Here, given a total of ten subjects, a two-sample t test would have *8* degrees of freedom $(= n + m - 2 = 5 + 5 - 2)$; a paired t test would have *4* degrees of freedom $(= b - 1 = 5 - 1)$. When a t test has fewer degrees of freedom, the critical values for a given level of significance move farther away from zero, which means that the test with the smaller number of degrees of freedom will have a greater probability of committing a Type II error.

Table 13.4.4 shows a comparison of the two-sided critical values for t ratios with *4* degrees of freedom and with *8* degrees of freedom for α equal to either 0.10, 0.05,

Table 13.4.4

α	$t_{\alpha/2,4}$	$t_{\alpha/2,8}$
0.10	2.1318	1.8595
0.05	2.7764	2.3060
0.01	4.6041	2.3554

or 0.01. Clearly, the same value of $\bar{x} - \bar{y}$ that would reject $H_0: \mu_X = \mu_Y$ with a t test having 8 df may not be large enough to reject $H_0: \mu_D = 0$ with a t test having 4 df.

Appendix 13.A.1 Minitab Applications

To produce the information in a randomized block ANOVA table, Minitab uses the command TWOWAY C1 C2 C3. First, the data are "stacked," treatment level over treatment level, into a single column—say, $c1$ (similar to the way the y_{ij}'s in a Tukey analysis are entered). Then two auxiliary columns must be created. The first, call it $c2$, gives the column number for each entry in $c1$. The second—say, $c3$—gives the block number (i.e., the row number) for each entry in $c1$.

Consider, again, the data in Case Study 13.2.1. Figure 13.A.1.1 is the Minitab syntax for outputting the calculations that appear in Table 13.2.4. Notice that the Windows version reverses columns C1 and C2.

Figure 13.A.1.1

```
MTB  > set c1
DATA > 8 11 9 16 24 2 1 12 11 19 -2 0 6 2 11
DATA > end
MTB  > set c2
DATA > 1 1 1 1 1 2 2 2 2 2 3 3 3 3 3
DATA > end
MTB  > set c3
DATA > 1 2 3 4 5 1 2 3 4 5 1 2 3 4 5
DATA > end
MTB  > name c1 'HAT' c2 'Therapy' c3 'Blocks'
MTB  > twoway c1 c2 c3
```

Two-way ANOVA: HAT versus Therapy, Blocks

```
Analysis of variance for HAT
Source    DF      SS       MS       F       P
Therapy    2  260.93   130.47   15.26   0.002
Blocks     4  438.00   109.50   12.81   0.001
Error      8   68.40     8.55
Total     14  767.33
```

Doing a Randomized Block Analysis of Variance Using Minitab Windows

1. Enter the entire data set in column C1, beginning with Treatment level 1, followed by Treatment level 2, and so on.
2. In column C2, enter the block number of each data point in C1; in column C3, enter the column number of each data point in C1.
3. Click on STAT, then on ANOVA, then on TWO-WAY.
4. Type C1 in RESPONSE box, C2 in ROW FACTOR box, and C3 in COLUMN FACTOR box.
5. Click on OK.

There is no special command in Minitab for doing a paired t test, but none is necessary. The appropriate P-value can be found by simply applying the (one-sample) `MTB > ttest` command to the within-pair differences (and setting μ_o

equal to 0). Figure 13.A.1.2 shows the syntax for doing the paired t test on the aerobics data described in Case Study 13.3.1.

Figure 13.A.1.2

```
MTB  > set c1
DATA > 14.6 17.3 10.9 12.8 16.6 12.2 11.2 15.4 14.8 16.2
DATA > end
MTB  > set c2
DATA > 13.8 15.4 11.3 11.6 16.4 12.6 11.8 15.0 14.4 15.0
DATA > end
MTB  > let c3 = c1 - c2
MTB  > name c3'di'
MTB  > ttest 0 c3
```

One-Sample T: di

```
Test of mu = 0 vs not = 0
Variable  N     Mean    StDev  SE Mean        95% CI            T     P
di        10    0.470    0.814  0.257      (-0.112, 1.052)     1.83  0.101
```

Chapter

14

NONPARAMETRIC STATISTICS

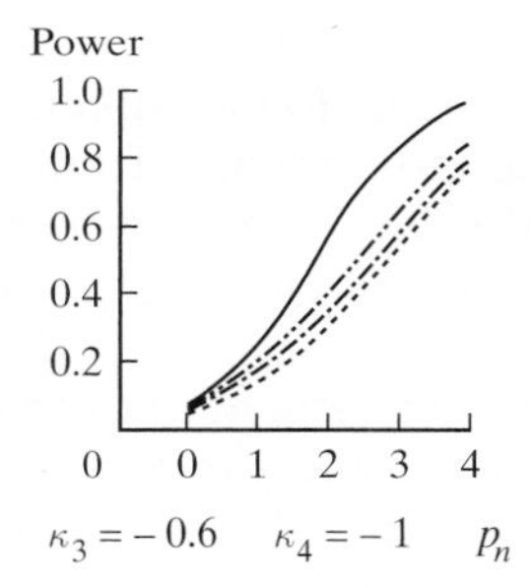

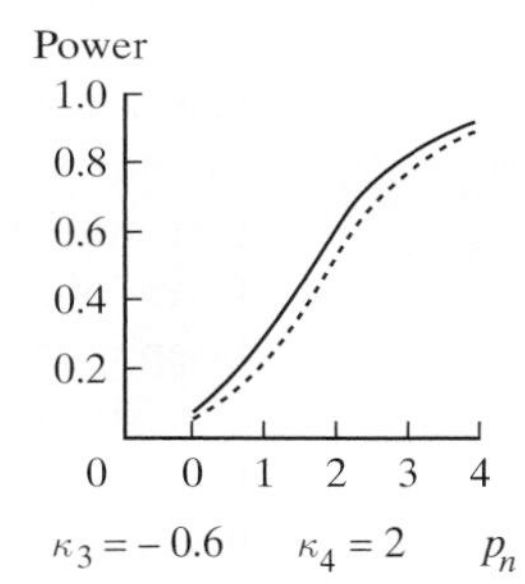

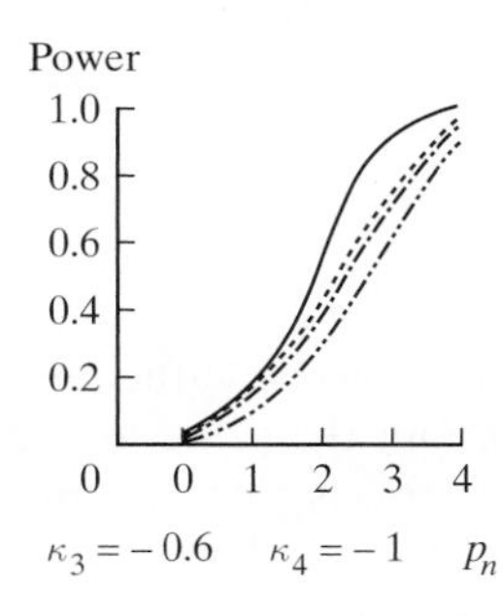

Critical to the justification of replacing a parametric test with a nonparametric test is a comparison of the power functions for the two procedures. The figures above illustrate the types of information that researchers have compiled–shown are the power functions of the one-sample t test (solid line) and the sign test (dashed lines) for three different sets of hypotheses, various degrees of nonnormality, a sample size of 10, and a level of significance of 0.05. (The parameter ρ_n measures the shift from H_0 to H_1; κ_3 and κ_4 measure the extent of nonnormality in the sampled population.)

14.1 Introduction

Behind every confidence interval and hypothesis test we have studied thus far have been very specific assumptions about the nature of the pdf that the data presumably represent. For instance, the usual Z test for proportions—$H_0: p_X = p_Y$ versus $H_1: p_X \neq p_Y$—is predicated on the assumption that the two samples consist of independent and identically distributed Bernoulli random variables. The most common assumption in data analysis, of course, is that each set of observations is a random sample from a *normal* distribution. This was the condition specified in every t test and F test that we have done.

The need to make such assumptions raises an obvious question: What changes when these assumptions are not satisfied? Certainly the statistic being calculated stays the same, as do the critical values that define the rejection region. What *does* change, of course, is the sampling distribution of the test statistic. As a result, the *actual* probability of committing, say, a Type I error will not necessarily equal the *nominal* probability of committing a Type I error. That is, if W is the test statistic with pdf $f_W(w \mid H_0)$ when H_0 is true, and C is the critical region,

$$\text{“true”}\, \alpha = \int_C f_W(w \mid H_0)\, dw$$

is not necessarily equal to the “nominal” α, because $f_W(w \mid H_0)$ is different (because of the violated assumptions) from the presumed sampling distribution of the test statistic. Moreover, there is usually no way to know the “true” functional form of $f_W(w \mid H_0)$ when the underlying assumptions about the data have not been met.

Statisticians have sought to overcome the problem implicit in not knowing the true $f_W(w \mid H_0)$ in two very different ways. One approach is the idea of *robustness*, a concept that was introduced in Section 7.4. The Monte Carlo simulations illustrated in Figure 7.4.6, for example, show that even though a set of Y_i's deviates from normality, the distribution of the t ratio,

$$t = \frac{\overline{Y} - \mu_o}{s/\sqrt{n}}$$

is likely to be sufficiently close to $f_{T_{n-1}}(t)$ that the true α, for all practical purposes, is about the same as the nominal α. The one-sample t test, in other words, is often not seriously compromised when normality fails to hold.

A second way of dealing with the additional uncertainty introduced by violated assumptions is to use test statistics whose pdfs remain the same regardless of how the population sampled may change. Inference procedures having this sort of latitude are said to be *nonparametric* or, more appropriately, *distribution free*.

The number of nonparametric procedures proposed since the early 1940s has been enormous and continues to grow. It is not the intention of Chapter 14 to survey this multiplicity of techniques in any comprehensive fashion. Instead, the objective here is to introduce some of the basic methodology of nonparametric statistics in the context of problems whose “parametric” solutions have already been discussed. Included in that list will be nonparametric treatments of the paired-data problem, the one-sample location problem, and both of the analysis of variance models covered in Chapters 12 and 13.

14.2 The Sign Test

Probably the simplest—and most general—of all nonparametric procedures is the *sign test*. Among its many applications, testing the null hypothesis that the *median* of a distribution is equal to some specific value is perhaps its most important.

By definition, the median, $\tilde{\mu}$, of a continuous pdf $f_Y(y)$ is the value for which $P(Y \leq \tilde{\mu}) = P(Y \geq \tilde{\mu}) = \frac{1}{2}$. Suppose a random sample of size n is taken from $f_Y(y)$. If the null hypothesis H_0: $\tilde{\mu} = \tilde{\mu}_0$ is true, the number of observations, X, exceeding $\tilde{\mu}_0$ is a binomial random variable with $p = P(Y_i \geq \tilde{\mu}_0) = \frac{1}{2}$. Moreover, $E(X) = n/2$, $\text{Var}(X) = n \cdot \frac{1}{2} \cdot \frac{1}{2} = n/4$, and $\frac{X-n/2}{\sqrt{n/4}}$ would have approximately a standard normal distribution (by virtue of the DeMoivre-Laplace theorem), provided n is sufficiently large. Intuitively, values of X too much larger or too much smaller than $n/2$ would be evidence that $\tilde{\mu} \neq \tilde{\mu}_0$.

Theorem 14.2.1 *Let $y_1, y_2, \ldots, y_n$ be a random sample of size n from any continuous distribution having median $\tilde{\mu}$, where $n \geq 10$. Let k denote the number of y_i's greater than $\tilde{\mu}_0$, and let $z = \frac{k-n/2}{\sqrt{n/4}}$.*

a. *To test H_0: $\tilde{\mu} = \tilde{\mu}_0$ versus H_1: $\tilde{\mu} > \tilde{\mu}_0$ at the α level of significance, reject H_0 if $z \geq z_\alpha$.*

b. *To test H_0: $\tilde{\mu} = \tilde{\mu}_0$ versus H_1: $\tilde{\mu} < \tilde{\mu}_0$ at the α level of significance, reject H_0 if $z \leq -z_\alpha$.*

c. *To test H_0: $\tilde{\mu} = \tilde{\mu}_0$ versus H_1: $\tilde{\mu} \neq \tilde{\mu}_0$ at the α level of significance, reject H_0 if z is either (1) $\leq -z_{\alpha/2}$ or (2) $\geq z_{\alpha/2}$.* □

Comment Sign tests are designed to draw inferences about *medians*. If the underlying pdf being sampled, though, is *symmetric*, the median is the same as the mean, so concluding that $\tilde{\mu} \neq \tilde{\mu}_0$ is equivalent to concluding that $\mu \neq \tilde{\mu}_0$.

Case Study 14.2.1

Synovial fluid is the clear, viscid secretion that lubricates joints and tendons. Researchers have found that certain ailments can be diagnosed on the basis of a person's synovial fluid hydrogen-ion concentration (pH). In healthy adults, the median pH for synovial fluid is 7.39. Listed in Table 14.2.1 are the pH values measured from fluids drawn from the knees of forty-three patients with arthritis (181). Does it follow from these data that synovial fluid pH can be useful in diagnosing arthritis?

Let $\tilde{\mu}$ denote the median synovial fluid pH for adults suffering from arthritis. Testing

$$H_0\text{: } \tilde{\mu} = 7.39$$

versus

$$H_1\text{: } \tilde{\mu} \neq 7.39$$

then becomes a way of quantifying the potential usefulness of synovial fluid pH as a way of diagnosing arthritis.

By inspection, a total of $k = 4$ of the $n = 43$ y_i's exceed $\tilde{\mu}_0 = 7.39$. Let $\alpha = 0.01$. The test statistic is

$$z = \frac{4 - 43/2}{\sqrt{43/4}} = -5.34$$

(Continued on next page)

(Case Study 14.2.1 continued)

Table 14.2.1

Subject	Synovial Fluid pH	Subject	Synovial Fluid pH
HW	7.02	BG	7.34
AD	7.35	GL	7.22
TK	7.32	BP	7.32
EP	7.33	NK	7.40
AF	7.15	LL	6.99
LW	7.26	KC	7.10
LT	7.25	FA	7.30
DR	7.35	ML	7.21
VU	7.38	CK	7.33
SP	7.20	LW	7.28
MM	7.31	ES	7.35
DF	7.24	DD	7.24
LM	7.34	SL	7.36
AW	7.32	RM	7.09
BB	7.34	AL	7.32
TL	7.14	BV	6.95
PM	7.20	WR	7.35
JG	7.41	HT	7.36
DH	7.77	ND	6.60
ER	7.12	SJ	7.29
DP	7.45	BA	7.31
FF	7.28		

which lies well past the left-tail critical value $(= -z_{\alpha/2} = -z_{0.005} = -2.58)$. It follows that H_0: $\tilde{\mu} = 7.39$ should be rejected, a conclusion suggesting that arthritis should be added to the list of ailments that can be detected by the pH of a person's synovial fluid.

A Small-Sample Sign Test

If $n < 10$, the decision rules given in Theorem 14.2.1 for testing $H_0\colon \tilde{\mu} = \tilde{\mu}_0$ are inappropriate because the normal approximation is not entirely adequate. Instead, decision rules need to be determined using the exact binomial distribution.

Case Study 14.2.2

Instant coffee can be formulated several different ways—freeze-drying and spray-drying being two of the most common. From a health standpoint, the most important difference from method to method is the amount of caffeine that is left as a residue. It has been shown that the median amount of caffeine left by the freeze-drying method is 3.55 grams per 100 grams of dry matter.

(Continued on next page)

Listed in Table 14.2.2 are the caffeine residues recorded for eight brands of coffee produced by the spray-dried method (182).

Table 14.2.2

Brand	Caffeine Residue (gms/100 gms dry weight)
A	4.8
B	4.0
C	3.8
D	4.3
E	3.9
F	4.6
G	3.1
H	3.7

If $\tilde{\mu}$ denotes the median caffeine residue characteristic of the spray-dried method, we compare the two methods by testing

$$H_0\colon \tilde{\mu} = 3.55$$

versus

$$H_1\colon \tilde{\mu} \neq 3.55$$

By inspection, $k = 7$ of the $n = 8$ spray-dried brands left caffeine residues in excess of $\tilde{\mu}_0 = 3.55$. Given the discrete nature of the binomial distribution, simple decision rules yielding specific α values are not likely to exist, so small-sample tests of this sort are best couched in terms of P-values. Figure 14.2.1 shows Minitab's printout of the binomial pdf when $n = 8$ and $p = \frac{1}{2}$. Since H_1 here is two-sided, the P-value associated with $k = 7$ is the probability that the corresponding binomial random variable would be greater than or equal to 7 *plus* the probability that it would be less than or equal to 1. That is,

$$\begin{aligned} P\text{-value} &= P(X \geq 7) + P(X \leq 1) \\ &= P(X = 7) + P(X = 8) + P(X = 0) + P(X = 1) \\ &= 0.031250 + 0.003906 + 0.003906 + 0.031250 \\ &= 0.070 \end{aligned}$$

The null hypothesis, then, can be rejected for any $\alpha \geq 0.07$.

```
MTB   > pdf;
SUBC  > binomial 8  0.5.

Probability Density Function

Binomial with n=8 and p=0.5

  x     P(X=x)
  0     0.003906
  1     0.031250
  2     0.109375
  3     0.218750
  4     0.273438
  5     0.218750
  6     0.109375
  7     0.031250
  8     0.003906
```

Figure 14.2.1

Using the Sign Test for Paired Data

Suppose a set of paired data—$(x_1, y_1), (x_2, y_2), \ldots, (x_b, y_b)$—has been collected, and the within-pair differences—$d_i = x_i - y_i$ $i = 1, 2, \ldots, b$—have been calculated (recall Theorem 13.3.1). The sign test becomes a viable alternative to the paired t test if there is reason to believe that the d_i's do not represent a random sample from a *normal* distribution. Let

$$p = P(X_i > Y_i), \quad i = 1, 2, \ldots, b$$

The null hypothesis that the x_i's and y_i's are representing distributions with the same median is equivalent to the null hypothesis H_0: $p = \frac{1}{2}$.

In the analysis of paired data, the generality of the sign test becomes especially apparent. The distribution of X_i need not be the same as the distribution of Y_i, nor do the distributions of X_i and X_j or Y_i and Y_j need to be the same. Furthermore, none of the distributions has to be symmetric, and they could all have different variances. The only underlying assumption is that X and Y have continuous pdfs. The null hypothesis, of course, adds the restriction that the median of the distributions *within each pair* be equal.

Let U denote the number of (x_i, y_i) pairs for which $d_i = x_i - y_i > 0$. The statistic appropriate for testing H_0: $p = \frac{1}{2}$ is either an approximate Z ratio, $\frac{U - b/2}{\sqrt{b/4}}$, or the value of U itself, which has a binomial distribution with parameters b and $\frac{1}{2}$ (when the null hypothesis is true). As before, the normal approximation is adequate if $b \geq 10$.

Case Study 14.2.3

One reason frequently cited for the mental deterioration often seen in the very elderly is the reduction in cerebral blood flow that accompanies the aging process. Addressing that concern, a study was done (5) in a nursing home to see whether cyclandelate, a drug that widens blood vessels, might be able to stimulate cerebral circulation and retard the onset of dementia.

The drug was given to eleven subjects on a daily basis. To measure its physiological effect, radioactive tracers were used to determine each subject's mean circulation time (MCT) at the start of the experiment and four months later, when the regimen was discontinued. [The MCT is the length of time (in sec) it takes blood to travel from the carotid artery to the jugular vein.] Table 14.2.3 summarizes the results.

If cyclandelate has no effect on cerebral circulation, $p = P(X_i > Y_i) = \frac{1}{2}$. Moreover, it seems reasonable here to discount the possibility that the drug might be harmful, which means that a one-sided alternative is warranted. To be tested, then, is

$$H_0\text{: } p = \frac{1}{2}$$

versus

$$H_1\text{: } p > \frac{1}{2}$$

where H_1 is one-sided *to the right* because increased cerebral circulation would result in the MCT being *reduced*, which would produce more patients for whom x_i was larger than y_i.

(Continued on next page)

Table 14.2.3

Subject	Before, x_i	After, y_i	$x_i > y_i$?	
J.B.	15	13	yes	
M.B.	12	8	yes	
A.B.	12	12.5	no	
M.B.	14	12	yes	
J.L.	13	12	yes	
S.M.	13	12.5	yes	$u = 9$
M.M.	13	12.5	yes	
S.McA.	12	14	no	
A.McL.	12.5	12	yes	
F.S.	12	11	yes	
P.W.	12.5	10	yes	

As Table 14.2.3 indicates, the number of subjects showing improvement in their MCTs was $u = 9$ (as opposed to the H_0 expected value of 5.5). Let $\alpha = 0.05$. Since $n = 11$, the normal approximation is adequate, and H_0 should be rejected if

$$\frac{u - b/2}{\sqrt{b/4}} \geq z_\alpha = z_{0.05} = 1.64$$

But

$$\frac{u - b/2}{\sqrt{b/4}} = \frac{9 - \frac{11}{2}}{\sqrt{\frac{11}{4}}} = 2.11$$

so the evidence here is fairly convincing that cyclandelate *does* speed up cerebral blood flow.

Questions

14.2.1. Recall the data in Question 8.2.9 giving the sizes of 10 gorilla groups studied in the Congo. Is it believable that the true median size, $\tilde{\mu}$, of all such groups is 9? Answer the question by finding the P-value associated with the null hypothesis H_0: $\tilde{\mu} = 9$. Assume that H_1 is two-sided. (*Note*: Tabulated on the right is the binomial pdf for the case where $n = 10$ and $p = \frac{1}{2}$.)

```
MTB  > pdf;
SUBC > binomial 10  0.5.
```

Probability Density Function

```
Binomial with n = 10 and p = 0.5

 x   P( X = x )
 0     0.000977
 1     0.009766
 2     0.043945
 3     0.117188
 4     0.205078
 5     0.246094
 6     0.205078
 7     0.117188
 8     0.043945
 9     0.009766
10     0.000977
```

14.2.2. Test H_0: $\tilde{\mu} = 0.12$ versus H_1: $\tilde{\mu} < 0.12$ for the release chirp data given in Question 8.2.12. Compare the P-value associated with the large-sample test described in Theorem 14.2.1 with the exact P-value based on the binomial distribution.

14.2.3. Below are $n = 50$ observations generated by Minitab's RANDOM command that are presumably a random sample from the exponential pdf, $f_Y(y) = e^{-y}$, $y \geq 0$. Use Theorem 14.2.1 to test whether the difference between the sample median for these y_i's ($= 0.604$) and the true median of $f_Y(y)$ is statistically significant. Let $\alpha = 0.05$.

```
0.27187 0.46495 0.19368 0.80433 1.25450 0.62962 1.88300
1.31951 2.53918 1.21187 0.95834 0.49017 0.87230 0.88571
1.41717 1.75994 0.60280 2.19654 0.00594 4.11127 0.24130
0.16473 0.08178 1.01424 0.60511 0.87973 0.06127 0.24758
0.54407 0.05267 0.75210 0.13538 0.42956 0.02261 1.20378
1.09271 1.88705 0.17500 0.50194 0.52122 0.02915 0.27348
0.08916 0.72997 0.37185 0.06500 1.47721 4.02733 0.64003
0.05603
```

14.2.4. Let $Y_1, Y_2, \ldots, Y_{22}$ be a random sample of normally distributed random variables with an unknown mean μ and a known variance of 6.0. We wish to test

$$H_0: \mu = 10$$
versus
$$H_1: \mu > 10$$

Construct a large-sample sign test having a Type I error probability of 0.05. What will the power of the test be if $\mu = 11$?

14.2.5. Suppose that $n = 7$ paired observations, (X_i, Y_i), are recorded, $i = 1, 2, \ldots, 7$. Let $p = P(Y_i > X_i)$. Write out the entire probability distribution for Y_+, the number of positive differences among the set of $Y_i - X_i$'s, $i = 1, 2, \ldots, 7$, assuming that $p = \frac{1}{2}$. What α levels are possible for testing H_0: $p = \frac{1}{2}$ versus H_1: $p > \frac{1}{2}$?

14.2.6. Analyze the Shoshoni rectangle data (Case Study 7.4.2) with a sign test. Let $\alpha = 0.05$.

14.2.7. Recall the FEV$_1$/VC data described in Question 5.3.2. Test H_0: $\tilde{\mu} = 0.80$ versus H_0: $\tilde{\mu} < 0.80$ using a sign test. Compare this conclusion with that of a t test of H_0: $\mu = 0.80$ versus H_1: $\mu < 0.80$. Let $\alpha = 0.10$. Assume that σ is unknown.

14.2.8. Do a sign test on the ESP data in Question 13.2.1. Define H_1 to be one-sided, and let $\alpha = 0.05$.

14.2.9. In a marketing research test, twenty-eight adult males were asked to shave one side of their face with one brand of razor blade and the other side with a second brand. They were to use the blades for seven days and then decide which was giving the smoother shave. Suppose that nineteen of the subjects preferred blade A. Use a sign test to determine whether it can be claimed, at the 0.05 level, that the difference in preferences is statistically significant.

14.2.10. Suppose that a random sample of size 36, $Y_1, Y_2, \ldots, Y_{36}$, is drawn from a uniform pdf defined over the interval $(0, \theta)$, where θ is unknown. Set up a large-sample sign test for deciding whether or not the 25th percentile of the Y-distribution is equal to 6. Let $\alpha = 0.05$. With what probability will your procedure commit a Type II error if 7 is the true 25th percentile?

14.2.11. Use a small-sample sign test to analyze the aerobics data given in Case Study 13.3.1. Use the binominal distribution displayed in Question 14.2.1. Let $\alpha = 0.05$. Does your conclusion agree with the inference drawn from the paired t test?

14.3 Wilcoxon Tests

Although the sign test is a bona fide nonparametric procedure, its extreme simplicity makes it somewhat atypical. The *Wilcoxon signed rank test* introduced in this section is more representative of nonparametric procedures as a whole. Like the sign test, it can be adapted to several different data structures. It can be used, for instance, as a one-sample test for location, where it becomes an alternative to the t test. It can also be applied to paired data, and with only minor modifications it can become a two-sample test for location and a two-sample test for dispersion (provided the two populations have equal locations).

Testing H_0: $\mu = \mu_o$

Let $y_1, y_2, \ldots, y_n$ be a set of independent observations drawn from the pdfs $f_{Y_1}(y), f_{Y_2}(y), \ldots, f_{Y_n}(y)$, respectively, all of which are continuous and symmetric (but not necessarily the same). Let μ denote the (common) mean of the $f_{Y_i}(y)$'s. We wish to test

$$H_0\colon \mu = \mu_0$$
versus
$$H_1\colon \mu \neq \mu_0$$

where μ_0 is some prespecified value for μ.

For an application of this sort, the signed rank test is based on the magnitudes, and directions, of the deviations of the y_i's from μ_0. Let $|y_1 - \mu_0|, |y_2 - \mu_0|, \ldots, |y_n - \mu_0|$ be the set of absolute deviations of the y_i's from μ_0. These can be ordered from smallest to largest, and we can define r_i to be the *rank* of $|y_i - \mu_0|$, where the smallest absolute deviation is assigned a rank of 1, the second smallest a rank of 2, and so on, up to n. If two or more observations are tied, each is assigned the *average* of the ranks they would have otherwise received.

Associated with each r_i will be a sign indicator, z_i, where

$$z_i = \begin{cases} 0 & \text{if } y_i - \mu_0 < 0 \\ 1 & \text{if } y_i - \mu_0 > 0 \end{cases}$$

The Wilcoxon signed rank statistic, w, is defined to be the linear combination

$$w = \sum_{i=1}^{n} r_i z_i$$

That is, w is the sum of the ranks associated with the positive deviations (from μ_0). If H_0 is true, the sum of the ranks of the positive deviations should be roughly the same as the sum of the ranks of the negative deviations.

To illustrate this terminology, consider the case where $n = 3$ and $y_1 = 6.0$, $y_2 = 4.9$, and $y_3 = 11.2$. Suppose the objective is to test

$$H_0\colon \mu = 10.0$$
versus
$$H_1\colon \mu \neq 10.0$$

Note that $|y_1 - \mu_0| = 4.0$, $|y_2 - \mu_0| = 5.1$, and $|y_3 - \mu_0| = 1.2$. Since $1.2 < 4.0 < 5.1$, it follows that $r_1 = 2$, $r_2 = 3$, and $r_3 = 1$. Also, $z_1 = 0$, $z_2 = 0$, and $z_3 = 1$. Combining the r_i's and the z_i's we have that

$$\begin{aligned} w &= \sum_{i=1}^{n} r_i z_i \\ &= (0)(2) + (0)(3) + (1)(1) \\ &= 1 \end{aligned}$$

Comment Notice that w is based on the *ranks* of the deviations from μ_0 and not on the deviations themselves. For this example, the value of w would remain unchanged if y_2 were 4.9, 3.6, or $-10{,}000$. In each case, r_2 would be 3 and z_2 would be 0. If the test statistic *did* depend on the magnitude of the deviations, it would have been necessary to specify a particular distribution for $f_Y(y)$, and the resulting procedure would no longer be nonparametric.

Theorem 14.3.1 *Let $y_1, y_2, \ldots, y_n$ be a set of independent observations drawn, respectively, from the continuous and symmetric (but not necessarily identical) pdfs $f_{Y_i}(y)$, $i = 1, 2, \ldots, n$. Suppose that each of the $f_{Y_i}(y)$'s has the same mean μ. If H_0: $\mu = \mu_0$ is true, the pdf of the data's signed rank statistic, $p_W(w)$, is given by*

$$p_W(w) = P(W = w) = \left(\frac{1}{2^n}\right) \cdot c(w)$$

where c(w) is the coefficient of e^{wt} in the expansion of

$$\prod_{i=1}^{n} (1 + e^{it})$$

Proof The statement and proof of Theorem 14.3.1 are typical of many nonparametric results. Closed-form expressions for sampling distributions are seldom possible: The combinatorial nature of nonparametric test statistics lends itself more readily to a generating function format.

To begin, note that if H_0 is true, the distribution of the signed rank statistic is equivalent to the distribution of $U = \sum_{i=1}^{n} U_i$, where

$$U_i = \begin{cases} 0 & \text{with probability } \frac{1}{2} \\ i & \text{with probability } \frac{1}{2} \end{cases}$$

Therefore, W and U have the same moment-generating function. Since the data are presumed to be a random sample, the U_i's are independent random variables, and from Theorem 3.12.3,

$$\begin{aligned} M_U(t) &= M_W(t) \\ &= \prod_{i=1}^{n} M_{U_i}(t) \\ &= \prod_{i=1}^{n} E(e^{U_i t}) \\ &= \prod_{i=1}^{n} \left(\frac{1}{2} e^{ot} + \frac{1}{2} e^{it}\right) \\ &= \left(\frac{1}{2^n}\right) \prod_{i=1}^{n} (1 + e^{it}) \end{aligned} \tag{14.3.1}$$

Now, consider the *structure* of $p_W(w)$, the pdf for the signed rank statistic. In the formation of w, r_1 can be prefixed by either a plus sign or zero; similarly for $r_2, r_3, \ldots$, and r_n. It follows that since each r_i can take on two different values, the total number of ways to "construct" signed rank sums is 2^n. Under H_0, of course, all of those scenarios are equally likely, so the pdf for the signed rank statistic must necessarily have the form

$$p_W(w) = P(W = w) = \frac{c(w)}{2^n} \tag{14.3.2}$$

where $c(w)$ is the number of ways to assign pluses and zeros to the first n integers so that $\sum_{i=1}^{n} r_i z_i$ has the value w.

The conclusion of Theorem 14.3.1 follows immediately by comparing the form of $p_W(w)$ to Equation 14.3.1 and to the general expression for a moment-generating function. By definition,

$$M_W(t) = E\left(e^{Wt}\right) = \sum_{w=1}^{n(n+1)/2} e^{wt} p_W(w)$$

but from Equations 14.3.1 and 14.3.2 we can write

$$\sum_{w=1}^{n(n+1)/2} e^{wt} p_W(w) = \left(\frac{1}{2^n}\right) \prod_{i=1}^{n} \left(1 + e^{it}\right) = \sum_{w=1}^{n(n+1)/2} e^{wt} \cdot \frac{c(w)}{2^n}$$

It follows that $c(w)$ must be the coefficient of e^{wt} in the expansion of $\prod_{i=1}^{n}(1+e^{it})$, and the theorem is proved. □

Calculating $p_W(w)$

A numerical example will help clarify the statement of Theorem 14.3.1. Suppose $n = 4$. By Equation 14.3.1, the moment-generating function for the signed rank statistic is the product

$$\begin{aligned} M_W(t) &= \left(\frac{1+e^t}{2}\right)\left(\frac{1+e^{2t}}{2}\right)\left(\frac{1+e^{3t}}{2}\right)\left(\frac{1+e^{4t}}{2}\right) \\ &= \left(\frac{1}{16}\right)\{1 + e^t + e^{2t} + 2e^{3t} + 2e^{4t} + 2e^{5t} + 2e^{6t} + 2e^{7t} + e^{8t} + e^{9t} + e^{10t}\} \end{aligned}$$

Thus, the probability that W equals, say, 2 is $\frac{1}{16}$ (since the coefficient of e^{2t} is 1); the probability that W equals 7 is $\frac{2}{16}$; and so on. The first two columns of Table 14.3.1 show the complete probability distribution of W, as given by the expansion of $M_W(t)$. The last column enumerates the particular assignments of pluses and zeros that generate each possible value w.

Tables of the cdf, $F_W(w)$

Cumulative tail area probabilities,

$$P\left(W \le w_1^*\right) = \sum_{w=0}^{w_1^*} p_W(w) \quad \text{and} \quad P\left(W \ge w_2^*\right) = \sum_{w=w_2^*}^{n(n+1)/2} p_W(w)$$

are listed in Table A.6 of the Appendix for sample sizes ranging from $n = 4$ to $n = 12$. [*Note*: The smallest possible value for w is 0, and the largest possible value is the sum of the first n integers, $n(n+1)/2$.] Based on these probabilities, decision rules for testing H_0: $\mu = \mu_0$ can be easily constructed. For example, suppose $n = 7$ and we wish to test

$$\begin{aligned} &H_0\colon \mu = \mu_0 \\ &\text{versus} \\ &H_1\colon \mu \ne \mu_0 \end{aligned}$$

at the $\alpha = 0.05$ level of significance. The critical region would be the set of w values less than or equal to 2 or greater than or equal to 26—that is, $C = \{w\colon w \le 2 \text{ or } w \ge 26\}$. That particular choice of C follows by inspection of Table A.6, because

$$\sum_{w \in C} p_W(w) = 0.023 + 0.023 \doteq 0.05$$

Table 14.3.1 Probability Distribution of W

w	$p_W(w) = P(W = w)$	r_i: 1	2	3	4
0	$\frac{1}{16}$	0	0	0	0
1	$\frac{1}{16}$	+	0	0	0
2	$\frac{1}{16}$	0	+	0	0
3	$\frac{2}{16}$	+	+	0	0
		0	0	+	0
4	$\frac{2}{16}$	+	0	+	0
		0	0	0	+
5	$\frac{2}{16}$	+	0	0	+
		0	+	+	0
6	$\frac{2}{16}$	+	+	+	0
		0	+	0	+
7	$\frac{2}{16}$	+	+	0	+
		0	0	+	+
8	$\frac{1}{16}$	+	0	+	+
9	$\frac{1}{16}$	0	+	+	+
10	$\frac{1}{16}$	+	+	+	+
	1				

Case Study 14.3.1

Swell sharks (*Cephaloscyllium ventriosum*) are small, reef-dwelling sharks that inhabit the California coastal waters south of Monterey Bay. There is a second population of these fish living nearby in the vicinity of Catalina Island, but it has been hypothesized that the two populations never mix. In between Santa Catalina and the mainland is a deep basin, which, according to the "separation" hypothesis, is an inpenetrable barrier for these particular fish (66).

One way to test this theory would be to compare the morphology of sharks caught in the two regions. If there were no mixing, we would expect a certain number of differences to have evolved. Table 14.3.2 lists the total length (TL), the height of the first dorsal fin (HDI), and the ratio TL/HDI for ten male swell sharks caught near Santa Catalina.

It has been estimated on the basis of past data that the true average TL/HDI ratio for male swell sharks caught *off the coast* is 14.60. Is that figure consistent

(Continued on next page)

with the data of Table 14.3.2? In more formal terms, if μ denotes the true mean TL/HDI ratio for the Santa Catalina population, can we reject $H_0: \mu = 14.60$, and thereby lend support to the separation theory?

Table 14.3.3 gives the values of TL/HDI $(= y_i)$, $y_i - 14.60$, $|y_i - 14.60|$, r_i, z_i, and $r_i z_i$ for the ten Santa Catalina sharks. Recall that when two or more numbers being ranked are equal, each is assigned the *average* of the ranks they would otherwise have received; here, $|y_6 - 14.60|$ and $|y_{10} - 14.60|$ are both competing for ranks for 4 and 5, so each is assigned a rank of *4.5* $[= (4+5)/2]$.

Table 14.3.2 Measurements Made on Ten Sharks Caught Near Santa Catalina

Total Length (mm)	Height of First Dorsal Fin (mm)	TL/HDI
906	68	13.32
875	67	13.06
771	55	14.02
700	59	11.86
869	64	13.58
895	65	13.77
662	49	13.51
750	52	14.42
794	55	14.44
787	51	15.43

Table 14.3.3 Computations for Wilcoxon Signed Rank Test

TL/HDI $(= y_i)$	$y_i - 14.60$	$\lvert y_i - 14.60 \rvert$	r_i	z_i	$r_i z_i$
13.32	−1.28	1.28	8	0	0
13.06	−1.54	1.54	9	0	0
14.02	−0.58	0.58	3	0	0
11.86	−2.74	2.74	10	0	0
13.58	−1.02	1.02	6	0	0
13.77	−0.83	0.83	4.5	0	0
13.51	−1.09	1.09	7	0	0
14.42	−0.18	0.18	2	0	0
14.44	−0.16	0.16	1	0	0
15.43	+0.83	0.83	4.5	1	4.5

Summing the last column of Table 14.3.3, we see that $w = 4.5$. According to Table A.6 in the Appendix, the $\alpha = 0.05$ decision rule for testing

$$H_0: \mu = 14.60$$
versus
$$H_1: \mu \neq 14.60$$

requires that H_0 be rejected if w is either less than or equal to 8 or greater than or equal to 47. (Why is the alternative hypothesis two-sided here?) (*Note*: The *exact* level of significance associated with $C = \{w: w \leq 8 \text{ or } w \geq 47\}$ is $0.024 + 0.024 = 0.048$.) Thus we should *reject* H_0, since the observed w was less than 8. These particular data, then, would support the separation hypothesis.

About the Data If data came equipped with alarm bells, the measurements in Table 14.3.3 would be ringing up a storm. The cause for concern is the fact that the y_i's being analyzed are the *quotients* of random variables (TL/HDI). A quotient can be difficult to interpret. If its value is unusually large, for example, does that imply that the numerator is unusually large or that the denominator is unusually small, or both? And what does an "average" value for a quotient imply?

Also troublesome is the fact that distributions of quotients sometimes violate critical assumptions that we typically take for granted. Here, for example, both TL and HDI might conceivably be normally distributed. *If* they were independent *standard* normal random variables (the simplest possible case), their quotient $Q = \text{TL}/\text{HDI}$ would have a *Cauchy distribution* with pdf

$$f_Q(q) = \frac{1}{\pi(1+q^2)}, \quad -\infty < q < \infty$$

Although harmless looking, $f_Q(q)$ has some highly undesirable properties: neither its mean nor its variance is finite. Moreover, it does not obey the central limit theorem—the *average* of a random sample from a Cauchy distribution,

$$\overline{Q} = \frac{1}{n}(Q_1 + Q_2 + \cdots + Q_n)$$

has the same distribution as any single observation, Q_i [see (92)]. Making matters worse, the data in Table 14.3.3 do not even represent the simplest case of a quotient of normal random variables—here the means and variances of both TL and HDI are unknown, and the two random variables may not be independent.

For all these reasons, using a nonparametric procedure on these data is clearly indicated, and the Wilcoxon signed rank test is a good choice (because the assumptions of continuity and symmetry are likely to be satisfied). The broader lesson, though, for experimenters to learn from this example is to think twice—maybe three times—before taking data in the form of quotients.

Questions

14.3.1. The average energy expenditures for eight elderly women were estimated on the basis of information received from a battery-powered heart rate monitor that each subject wore. Two overall averages were calculated for each woman, one for the summer months and one for the winter months (154), as shown in the following table. Let μ_D denote the location difference between the summer and winter energy expenditure populations. Compute $y_i - x_i, i = 1, 2, \ldots, 8$, and use the Wilcoxon signed rank procedure to test

$$H_0: \mu_D = 0$$
versus
$$H_1: \mu_D \neq 0$$

Let $\alpha = 0.15$.

Average Daily Energy Expenditures (kcal)

Subject	Summer, x_i	Winter, y_i
1	1458	1424
2	1353	1501
3	2209	1495
4	1804	1739
5	1912	2031
6	1366	934
7	1598	1401
8	1406	1339

14.3.2. Use the expansion of

$$\prod_{i=1}^{n}(1+e^{it})$$

to find the pdf of W when $n = 5$. What α levels are available for testing $H_0: \tilde{\mu} = \tilde{\mu}_0$ versus $H_1: \tilde{\mu} > \tilde{\mu}_0$?

A Large-Sample Wilcoxon Signed Rank Test

The usefulness of Table A.6 in the Appendix for testing $H_0: \mu = \mu_0$ is limited to sample sizes less than or equal to 12. For larger n, an approximate signed rank test can be constructed, using $E(W)$ and $\text{Var}(W)$ to define an approximate Z ratio.

Theorem 14.3.2 *When $H_0: \mu = \mu_0$ is true, the mean and variance of the Wilcoxon signed rank statistic, W, are given by*

$$E(W) = \frac{n(n+1)}{4}$$

and

$$\text{Var}(W) = \frac{n(n+1)(2n+1)}{24}$$

Also, for $n > 12$, the distribution of

$$\frac{W - [n(n+1)]/4}{\sqrt{[n(n+1)(2n+1)]/24}}$$

can be adequately approximated by the standard normal pdf, $f_Z(z)$.

Proof We will derive $E(W)$ and $\text{Var}(W)$; for a proof of the asymptotic normality, see (80). Recall that W has the same distribution as $U = \sum_{i=1}^{n} U_i$, where

$$U_i = \begin{cases} 0 & \text{with probability } \frac{1}{2} \\ i & \text{with probability } \frac{1}{2} \end{cases}$$

Therefore,

$$\begin{aligned} E(W) &= E\left(\sum_{i=1}^{n} U_i\right) = \sum_{i=1}^{n} E(U_i) \\ &= \sum_{i=1}^{n} \left(0 \cdot \frac{1}{2} + i \cdot \frac{1}{2}\right) = \sum_{i=1}^{n} \frac{i}{2} \\ &= \frac{n(n+1)}{4} \end{aligned}$$

Similarly,

$$\text{Var}(W) = \text{Var}(U) = \sum_{i=1}^{n} \text{Var}(U_i)$$

since the U_i's are independent. But

$$\text{Var}(U_i) = E(U_i^2) - [E(U_i)]^2$$
$$= \frac{i^2}{2} - \left(\frac{i}{2}\right)^2 = \frac{i^2}{4}$$

making

$$\text{Var}(W) = \sum_{i=1}^{n} \frac{i^2}{4} = \left(\frac{1}{4}\right)\left[\frac{n(n+1)(2n+1)}{6}\right]$$
$$= \frac{n(n+1)(2n+1)}{24}$$

□

Theorem 14.3.3 *Let w be the signed rank statistic based on n independent observations, each drawn from a continuous and symmetric pdf, where $n > 12$. Let*

$$z = \frac{w - [n(n+1)]/4}{\sqrt{[n(n+1)(2n+1)]/24}}$$

a. *To test $H_0: \mu = \mu_0$ versus $H_1: \mu > \mu_0$ at the α level of significance, reject H_0 if $z \geq z_\alpha$.*

b. *To test $H_0: \mu = \mu_0$ versus $H_1: \mu < \mu_0$ at the α level of significance, reject H_0 if $z \leq -z_\alpha$.*

c. *To test $H_0: \mu = \mu_0$ versus $H_1: \mu \neq \mu_0$ at the α level of significance, reject H_0 if z is either (1) $\leq -z_{\alpha/2}$ or (2) $\geq z_{\alpha/2}$.* □

Case Study 14.3.2

Cyclazocine, along with methadone, are two of the drugs widely used in the treatment of heroin addiction. Some years ago, a study was done (141) to evaluate the effectiveness of the former in reducing a person's psychological dependence on heroin. The subjects were fourteen males, all chronic addicts. Each was asked a battery of questions that compared his feelings when he was using heroin to his feelings when he was "clean." The resultant Q-scores ranged from a possible minimum of 11 to a possible maximum of 55, as shown in Table 14.3.4. (From the way the questions were worded, higher scores represented *less* psychological dependence.)

The shape of the histogram for these data suggests that a normality assumption may not be warranted—the weaker assumption of symmetry is more believable. That said, a case can be made for using a signed rank test on these data, rather than a one-sample t test.

(Continued on next page)

Table 14.3.4 Q-Scores of Heroin Addicts after Cyclazocine Therapy

51	43
53	45
43	27
36	21
55	26
55	22
39	43

The mean score for addicts *not* given cyclazocine is known from past experience to be 28. Can we conclude on the basis of the data in Table 14.3.4 that cyclazocine is an effective treatment?

Since high Q-scores represent *less* dependence on heroin (and assuming cyclazocine would not tend to worsen an addict's condition), the alternative hypothesis should be one-sided *to the right*. That is, we want to test

$$H_0: \mu = 28$$

versus

$$H_1: \mu > 28$$

Let α be 0.05.

Table 14.3.5 details the computations showing that the signed rank statistic, w—that is, the sum of the $r_i z_i$ column—equals *95.0*. Since $n = 14$, $E(W) = [14(14+1)]/4 = 52.5$ and $\text{Var}(W) = [14(14+1)(28+1)]/24 = 253.75$, so the approximate Z ratio is

$$z = \frac{95.0 - 52.5}{\sqrt{253.75}} = 2.67$$

Table 14.3.5 Computations to Find w

Q-Score, y_i	$(y_i - 28)$	$\|y_i - 28\|$	r_i	z_i	$r_i z_i$
51	+23	23	11	1	11
53	+25	25	12	1	12
43	+15	15	8	1	8
36	+8	8	5	1	5
55	+27	27	13.5	1	13.5
55	+27	27	13.5	1	13.5
39	+11	11	6	1	6
43	+15	15	8	1	8
45	+17	17	10	1	10
27	−1	1	1	0	0
21	−7	7	4	0	0
26	−2	2	2	0	0
22	−6	6	3	0	0
43	+15	15	8	1	8
					95.0

(Continued on next page)

(Case Study 14.3.2 continued)

The latter considerably exceeds the one-sided 0.05 critical value identified in part (a) of Theorem 14.3.3 ($= z_{0.05} = 1.64$), so the appropriate conclusion is to *reject* H_0—it would appear that cyclazocine therapy *is* helpful in reducing heroin dependence.

Testing $H_0 : \mu_D = 0$ (Paired Data)

A Wilcoxon signed rank test can also be used on paired data to test $H_0: \mu_D = 0$, where $\mu_D = \mu_X - \mu_Y$ (recall Section 13.3). Suppose that responses to two treatment levels (X and Y) are recorded within each of n pairs. Let $d_i = x_i - y_i$ be the response difference recorded for Treatment X and Treatment Y *within* the ith pair, and let r_i be the rank of $|x_i - y_i|$ in the set $|x_1 - y_1|, |x_2 - y_2|, \ldots, |x_n - y_n|$. Define

$$z_i = \begin{cases} 1 & \text{if } x_i - y_i > 0 \\ 0 & \text{if } x_i - y_i < 0 \end{cases}$$

and let $w = \sum_{i=1}^{n} r_i z_i$.

If $n < 12$, critical values for testing $H_0: \mu_D = 0$ are gotten from Table A.6 in the Appendix in exactly the same way that decision rules were determined for using the signed rank test on $H_0: \mu = \mu_0$. If $n > 12$, an approximate Z test for $H_0: \mu_D = 0$ can be carried out using the formulas given in Theorem 14.3.2.

Case Study 14.3.3

Until recently, all evaluations of college courses and instructors have been done in class using questionnaires that were filled out in pencil. But as administrators well know, tabulating those results and typing the students' written comments (to preserve anonymity) take up a considerable amount of secretarial time. To expedite that process, some schools have considered doing evaluations online. Not all faculty support such a change, though, because of their suspicion that online evaluations might result in lower ratings (which, in turn, would affect their chances for reappointment, tenure, or promotion).

To investigate the merits of that concern, one university (104) did a pilot study where a small number of instructors had their courses evaluated online. Those same teachers had taught the same course the previous year and had been evaluated the usual way in class. Table 14.3.6 shows a portion of the results. The numbers listed are the responses on a 1- to 5-point scale ("5" being the best) to the question "Overall Rating of the Instructor." Here, x_i and y_i denote the ith instructor's ratings "in-class" and "online," respectively.

To test $H_0: \mu_D = 0$ versus $H_1: \mu_D \neq 0$, where $\mu_D = \mu_X - \mu_Y$, at the $\alpha = 0.05$ level of significance requires that H_0 be rejected if the approximate Z ratio in Theorem 14.3.2 is either (1) ≤ -1.96 or (2) $\geq +1.96$. But

(Continued on next page)

$$z = \frac{w - [n(n+1)/4]}{\sqrt{[n(n+1)(2n+1)]/24}} = \frac{70 - [15(16)/4]}{\sqrt{[15(16)(31)]/24}} = 0.57$$

so the appropriate conclusion is to "fail to reject H_0." The results in Table 14.3.6 are entirely consistent, in other words, with the hypothesis that the mode of evaluation—in-class or online—has no bearing on an instructor's rating.

Table 14.3.6

Obs. #	Instr.	In-Class, x_i	Online, y_i	$\lvert x_i - y_i \rvert$	r_i	z_i	$r_i z_i$
1	EF	4.67	4.36	0.31	7	1	7
2	LC	3.50	3.64	0.14	3	0	0
3	AM	3.50	4.00	0.50	11	0	0
4	CH	3.88	3.26	0.62	12	1	12
5	DW	3.94	4.06	0.12	2	0	0
6	CA	4.88	4.58	0.30	6	1	6
7	MP	4.00	3.52	0.48	10	1	10
8	CP	4.40	3.66	0.74	13	1	13
9	RR	4.41	4.43	0.02	1	0	0
10	TB	4.11	4.28	0.17	4	0	0
11	GS	3.45	4.25	0.80	15	0	0
12	HT	4.29	4.00	0.29	5	1	5
13	DW	4.25	5.00	0.75	14	0	0
14	FE	4.18	3.85	0.33	8	1	8
15	WD	4.65	4.18	0.47	9	1	9
							$w = 70$

About the Data Theoretically, the fact that all of the in-class evaluations were done first poses some problems for the interpretation of the ratings in Table 14.3.6. If instructors tend to receive higher (or lower) ratings on successive attempts to teach the same course, then the differences $x_i - y_i$ would be biased by a time effect. However, when instructors have already taught a course several times (which was true for the faculty included in Table 14.3.6), experience has shown that trends in future attempts are not what tend to happen—instead, ratings go up and down, seemingly at random.

Testing $H_0: \mu_X = \mu_Y$ (The Wilcoxon Rank Sum Test)

Another redefinition of the statistic $w = \sum_i r_i z_i$ allows ranks to be used as a way of testing the two-sample hypothesis, $H_0: \mu_X = \mu_Y$, where μ_X and μ_Y are the means of two continuous distributions, $f_X(x)$ and $f_Y(y)$. It will be assumed that $f_X(x)$ and $f_Y(y)$ have the same shape and the same standard deviation, but they may differ with respect to location—that is, $Y = X - c$, for some constant c. When those restrictions are met, the *Wilcoxon rank sum test* can appropriately be used as a nonparametric alternative to the pooled two-sample t test.

Let $x_1, x_2, \ldots, x_n$ and $y_{n+1}, y_{n+2}, \ldots, y_{n+m}$ be two independent random samples of sizes n and m from $f_X(x)$ and $f_Y(y)$, respectively. Define r_i to be the rank of the ith observation in the combined sample (so r_i ranges from 1 for the smallest observation to $n+m$ for the largest observation).
Let

$$z_i = \begin{cases} 1 & \text{if the } i\text{th observation came from } f_X(x) \\ 0 & \text{if the } i\text{th observation came from } f_Y(y) \end{cases}$$

and define

$$w' = \sum_{i=1}^{n+m} r_i z_i$$

Here, w' denotes the sum of the ranks in the combined sample of the n observations coming from $f_X(x)$. Clearly, w' is capable of distinguishing between H_0 and H_1. If, for example, $f_X(x)$ has shifted *to the right* of $f_Y(y)$, the sum of the ranks of the x observations would tend to be larger than if $f_X(x)$ and $f_Y(y)$ had the same location.

For small values of n and m, critical values for w' have been tabulated [see, for example, (81)]. When n and m both exceed 10, a normal approximation can be used.

Theorem 14.3.4 *Let $x_1, x_2, \ldots, x_n$ and $y_{n+1}, y_{n+2}, \ldots, y_{n+m}$ be two independent random samples from $f_X(x)$ and $f_Y(y)$, respectively, where the two pdfs are the same except for a possible shift in location. Let r_i denote the rank of the ith observation in the combined sample (where the smallest observation is assigned a rank of 1 and the largest observation, a rank of $n+m$). Let*

$$w' = \sum_{i=1}^{n+m} r_i z_i$$

where z_i is 1 if the ith observation comes from $f_X(x)$ and 0, otherwise. Then

$$E(W') = \frac{n(n+m+1)}{2}$$

$$\text{Var}(W') = \frac{nm(n+m+1)}{12}$$

and

$$\frac{W' - n(n+m+1)/2}{\sqrt{nm(n+m+1)/12}}$$

has approximately a standard normal pdf if $n > 10$ and $m > 10$.

Proof See (102). □

Case Study 14.3.4

In Major League Baseball, American League teams have the option of using a "designated hitter" to bat for a particular position player, typically the pitcher. In the National League, no such substitutions are allowed, and every player must bat for himself (or be removed from the game). As a result, batting and base-running strategies employed by National League managers are much different than those used by their American League counterparts. What is not

(Continued on next page)

so obvious is whether those differences in *how* games are played have any demonstrable effect on *how long* it takes games to be played.

Table 14.3.7. shows the average home-game completion time (in minutes) reported by the twenty-six Major League teams for the 1992 season. The American League average was 173.5 minutes; the National League average, 165.8 minutes. Is the difference between those two averages statistically significant?

The entry at the bottom of the last column is the sum of the ranks of the American League times—that is, $w' = \sum_{i=1}^{26} r_i z_i = 240.5$. Since the American League and National League had $n = 14$ and $m = 12$ teams, respectively, in 1992, the formulas in Theorem 14.3.3 give

$$E(W') = \frac{14(14+12+1)}{2} = 189$$

and

$$\text{Var}(W') = \frac{14 \cdot 12(14+12+1)}{12} = 378$$

Table 14.3.7

Obs. #	Team	Time (min)	r_i	z_i	$r_i z_i$
1	Baltimore	177	21	1	21
2	Boston	177	21	1	21
3	California	165	7.5	1	7.5
4	Chicago (AL)	172	14.5	1	14.5
5	Cleveland	172	14.5	1	14.5
6	Detroit	179	24.5	1	24.5
7	Kansas City	163	5	1	5
8	Milwaukee	175	18	1	18
9	Minnesota	166	9.5	1	9.5
10	New York (AL)	182	26	1	26
11	Oakland	177	21	1	21
12	Seattle	168	12.5	1	12.5
13	Texas	179	24.5	1	24.5
14	Toronto	177	21	1	21
15	Atlanta	166	9.5	0	0
16	Chicago (NL)	154	1	0	0
17	Cincinnati	159	2	0	0
18	Houston	168	12.5	0	0
19	Los Angeles	174	16.5	0	0
20	Montreal	174	16.5	0	0
21	New York (NL)	177	21	0	0
22	Philadelphia	167	11	0	0
23	Pittsburgh	165	7.5	0	0
24	San Diego	161	3.5	0	0
25	San Francisco	164	6	0	0
26	St. Louis	161	3.5	0	0
					$w' = 240.5$

(Continued on next page)

(Case Study 14.3.4 continued)

The approximate Z statistic, then, is:

$$z = \frac{w' - E(W')}{\sqrt{\text{Var}(W')}} = \frac{240.5 - 189}{\sqrt{378}} = 2.65$$

At the $\alpha = 0.05$ level, the critical values for testing $H_0: \mu_X = \mu_Y$ versus $H_1: \mu_X \neq \mu_Y$ would be ± 1.96. The conclusion, then, is to *reject* H_0—the difference between 173.5 and 165.8 *is* statistically significant.

(*Note*: When two or more observations are tied, they are each assigned the average of the ranks they would have received had they been slightly different. There were five observations that equaled 177, and they were competing for the ranks 19, 20, 21, 22, and 23. Each, then, received the corresponding average value of 21.)

Questions

14.3.3. Two manufacturing processes are available for annealing a certain kind of copper tubing, the primary difference being in the temperature required. The critical response variable is the resulting tensile strength. To compare the methods, fifteen pieces of tubing were broken into pairs. One piece from each pair was randomly selected to be annealed at a moderate temperature, the other piece at a high temperature. The resulting tensile strengths (in tons/sq in.) are listed in the following table. Analyze these data with a Wilcoxon signed rank test. Use a two-sided alternative. Let $\alpha = 0.05$.

Tensile Strengths (tons/sq in.)

Pair	Moderate Temperature	High Temperature
1	16.5	16.9
2	17.6	17.2
3	16.9	17.0
4	15.8	16.1
5	18.4	18.2
6	17.5	17.7
7	17.6	17.9
8	16.1	16.0
9	16.8	17.3
10	15.8	16.1
11	16.8	16.5
12	17.3	17.6
13	18.1	18.4
14	17.9	17.2
15	16.4	16.5

14.3.4. To measure the effect on coordination associated with mild intoxication, thirteen subjects were each given 15.7 mL of ethyl alcohol per square meter of body surface area and asked to write a certain phrase as many times as they could in the space of one minute (119). The number of correctly written letters was then counted and scaled, with a scale value of 0 representing the score a subject not under the influence of alcohol would be expected to achieve. Negative scores indicate *decreased* writing speeds; positive scores, *increased* writing speeds.

Subject	Score	Subject	Score
1	−6	8	0
2	10	9	−7
3	9	10	5
4	−8	11	−9
5	−6	12	−10
6	−2	13	−2
7	20		

Use the signed rank test to determine whether the level of alcohol provided in this study had any effect on writing speed. Let $\alpha = 0.05$. Omit Subject 8 from your calculations.

14.3.5. Test $H_0: \tilde{\mu} = 0.80$ versus $H_1: \tilde{\mu} < 0.80$ for the FEV_1/VC ratio data of Question 5.3.2 using a Wilcoxon signed rank test. Let $\alpha = 0.10$. Compare this test to the sign test of Question 14.2.7.

14.3.6. Do a Wilcoxon signed rank test on the hemoglobin data summarized in Case Study 13.3.1. Let α

be 0.05. Compare your conclusion with the outcome of the sign test done in Question 14.2.11.

14.3.7. Suppose that the population being sampled is symmetric and we wish to test $H_0: \tilde{\mu} = \tilde{\mu}_0$. Both the sign test and the signed rank test would be valid. Which procedure, if either, would you expect to have greater power? Why?

14.3.8. Use a signed rank test to analyze the depth perception data given in Question 8.2.6. Let $\alpha = 0.05$.

14.3.9. Recall Question 9.2.6. Compare the ages at death for authors noted for alcohol abuse with the ages at death for authors *not* noted for alcohol abuse using a Wilcoxon rank sum test. Let $\alpha = 0.05$.

14.3.10. Use a large-sample Wilcoxon rank sum test to analyze the alpha wave data summarized in Table 9.3.1. Let $\alpha = 0.05$.

14.4 The Kruskal-Wallis Test

The next two sections of this chapter discuss the nonparametric counterparts for the two analysis of variance models introduced in Chapters 12 and 13. Neither of these procedures, the *Kruskal-Wallis test* and the *Friedman test,* will be derived. We will simply state the procedures and illustrate them with examples.

First, we consider the *k-sample problem.* Suppose that $k(\geq 2)$ independent random samples of sizes $n_1, n_2, \ldots, n_k$ are drawn, representing k continuous populations having the same shape but possibly different locations: $f_{Y_1}(y - c_1) = f_{Y_2}(y - c_2) = \ldots = f_{Y_k}(y - c_k)$, for constants $c_1, c_2, \ldots, c_k$. The objective is to test whether the locations of the $f_{Y_j}(y)$'s, $j = 1, 2, \ldots, k$, might all be the same—that is,

$$H_0: \mu_1 = \mu_2 = \ldots = \mu_k$$

versus

$$H_1: \text{not all the } \mu_j\text{'s are equal}$$

The Kruskal-Wallis procedure for testing H_0 is really quite simple, involving considerably fewer computations than the analysis of variance. The first step is to rank the entire set of $n = \sum_{j=1}^{k} n_j$ observations from smallest to largest. Then the rank sum, $R_{.j}$, is calculated for each sample. Table 14.4.1 shows the notation that will be used: It follows the same conventions as the dot notation of Chapter 12. The only difference is the addition of R_{ij}, the symbol for the rank corresponding to Y_{ij}.

The Kruskal-Wallis statistic, B, is defined as

$$B = \frac{12}{n(n+1)} \sum_{j=1}^{k} \frac{R_{.j}^2}{n_j} - 3(n+1)$$

Table 14.4.1 Notation for Kruskal-Wallis Procedure

	Treatment Level			
	1	2	$\cdots$	k
	$Y_{11}(R_{11})$	$Y_{12}(R_{12})$		$Y_{1k}(R_{1k})$
	$Y_{21}(R_{21})$			
	$\vdots$	$\vdots$	$\cdots$	$\vdots$
	$Y_{n_1 1}(R_{n_1 1})$	$Y_{n_2 2}(R_{n_2 2})$		$Y_{n_k k}(R_{n_k k})$
Totals	$R_{.1}$	$R_{.2}$		$R_{.k}$

Notice how B resembles the computing formula for $SSTR$ in the analysis of variance. Here $\sum_{j=1}^{k}(R_{.j}^2/n_j)$, and thus B, get larger and larger as the differences between the population locations increase. [Recall that a similar explanation was given for $SSTR$ and $\sum_{j=1}^{k}(T_{.j}^2/n_j)$.]

Theorem 14.4.1 *Suppose $n_1, n_2, \ldots, n_k$ independent observations are taken from the pdfs $f_{Y_1}(y)$, $f_{Y_2}(y)$, $\ldots$, $f_{Y_k}(y)$, respectively, where the $f_{Y_i}(y)$'s are all continuous and have the same shape. Let μ_i be the mean of $f_{Y_i}(y)$, $i = 1, 2, \ldots, k$, and let $R_{.1}, R_{.2}, \ldots, R_{.k}$ denote the rank sums associated with each of the k samples. If H_0: $\mu_1 = \mu_2 = \ldots = \mu_k$ is true,*

$$B = \frac{12}{n(n+1)}\sum_{j=1}^{k}\frac{R_{.j}^2}{n_j} - 3(n+1)$$

has approximately a χ_{k-1}^2 distribution and H_0 should be rejected at the α level of significance if $b > \chi_{1-\alpha,k-1}^2$. □

Case Study 14.4.1

On December 1, 1969, a lottery was held in Selective Service headquarters in Washington, D.C., to determine the draft status of all nineteen-year-old males. It was the first time such a procedure had been used since World War II. Priorities were established according to a person's birthday. Each of the 366 possible birth dates was written on a slip of paper and put into a small capsule. The capsules were then put into a large bowl, mixed, and drawn out one by one. By agreement, persons whose birthday corresponded to the first capsule drawn would have the highest draft priority; those whose birthday corresponded to the second capsule drawn, the second highest priority, and so on. Table 14.4.2 shows the order in which the 366 birthdates were drawn (160). The first date was September 14 (=001); the last, June 8 (=366).

We can think of the observed sequence of draft priorities as ranks from 1 to 366. If the lottery was random, the distributions of those ranks for each of the months should have been approximately equal. If the lottery was *not* random, we would expect to see certain months having a preponderance of high ranks and other months, a preponderance of low ranks.

Look at the rank totals at the bottom of Table 14.4.2. The differences from month to month are surprisingly large, ranging from a high of 7000 for March to a low of 3768 for December. Even more unexpected is the *pattern* in the variation (see Figure 14.4.1). Are the rank totals listed in Table 14.4.2 and the rank averages pictured in Figure 14.4.1 consistent with the hypothesis that the lottery was random?

Substituting the $R_{.j}$'s into the formula for B gives

$$b = \frac{12}{366(367)}\left[\frac{(6236)^2}{31} + \cdots + \frac{(3768)^2}{31}\right] - 3(367)$$

$$= 25.95$$

By Theorem 14.4.1, B has approximately a chi square distribution with 11 degrees of freedom (when H_0: $\mu_{\text{Jan}} = \mu_{\text{Feb}} = \ldots = \mu_{\text{Dec}}$ is true).

(Continued on next page)

Table 14.4.2 1969 Draft Lottery, Highest Priority (001) to Lowest Priority (366)

Date	Jan.	Feb.	Mar.	Apr.	May	June	July	Aug.	Sept.	Oct.	Nov.	Dec.
1	305	086	108	032	330	249	093	111	225	359	019	129
2	159	144	029	271	298	228	350	045	161	125	034	328
3	251	297	267	083	040	301	115	261	049	244	348	157
4	215	210	275	081	276	020	279	145	232	202	266	165
5	101	214	293	269	364	028	188	054	082	024	310	056
6	224	347	139	253	155	110	327	114	006	087	076	010
7	306	091	122	147	035	085	050	168	008	234	051	012
8	199	181	213	312	321	366	013	048	184	283	097	105
9	194	338	317	219	197	335	277	106	263	342	080	043
10	325	216	323	218	065	206	284	021	071	220	282	041
11	329	150	136	014	037	134	248	324	158	237	046	039
12	221	068	300	346	133	272	015	142	242	072	066	314
13	318	152	259	124	295	069	042	307	175	138	126	163
14	238	004	354	231	178	356	331	198	001	294	127	026
15	017	089	169	273	130	180	322	102	113	171	131	320
16	121	212	166	148	055	274	120	044	207	254	107	096
17	235	189	033	260	112	073	098	154	255	288	143	304
18	140	292	332	090	278	341	190	141	246	005	146	128
19	058	025	200	336	075	104	227	311	177	241	203	240
20	280	302	239	345	183	360	187	344	063	192	185	135
21	186	363	334	062	250	060	027	291	204	243	156	070
22	337	290	265	316	326	247	153	339	160	117	009	053
23	118	057	256	252	319	109	172	116	119	201	182	162
24	059	236	258	002	031	358	023	036	195	196	230	095
25	052	179	343	351	361	137	067	286	149	176	132	084
26	092	365	170	340	357	022	303	245	018	007	309	173
27	355	205	268	074	296	064	289	352	233	264	047	078
28	077	299	223	262	308	222	088	167	257	094	281	123
29	349	285	362	191	226	353	270	061	151	229	099	016
30	164		217	208	103	209	287	333	315	038	174	003
31	211		030		313		193	011		079		100
Totals:	6236	5886	7000	6110	6447	5872	5628	5377	4719	5656	4462	3768

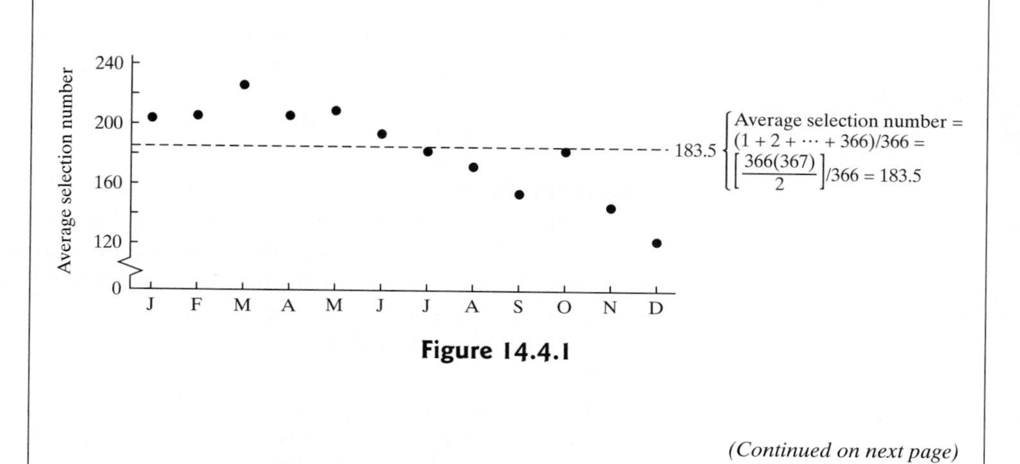

Figure 14.4.1

(Continued on next page)

(Case Study 14.4.1 continued)

Let $\alpha = 0.01$. Then H_0 should be rejected if $b \geq \chi^2_{.99,11} = 24.725$. But b *does* exceed that cutoff, implying that the lottery was *not* random.

An even more resounding rejection of the randomness hypothesis can be gotten by dividing the twelve months into two half-years—the first, January through June; the second, July through December. Then the hypotheses to be tested are

$$H_0: \mu_1 = \mu_2$$

versus

$$H_1: \mu_1 \neq \mu_2$$

Table 14.4.3, derived from Table 14.4.2, gives the new rank sums $R_{.1}$ and $R_{.2}$, associated with the two half-years. Substituting those values into the Kruskal-Wallis statistic shows that the new b (with 1 degree of freedom) is *16.85*:

$$b = \frac{12}{366(367)}\left[\frac{(37{,}551)^2}{182} + \frac{(29{,}610)^2}{184}\right] - 3(367)$$

$$= 16.85$$

Table 14.4.3 Summary of 1969 Draft Lottery by Six-Month Periods

	Jan.–June (1)	July–Dec. (2)
$R_{.j}$	37,551	29,610
n_j	182	184

The significance of 16.85 can be gauged by recalling the moments of a chi square random variable. If B has a chi square pdf with 1 degree of freedom, then $E(B) = 1$ and $\text{Var}(B) = 2$ (see Question 7.3.2). It follows, then, that the observed b is more than *11* standard deviations away from its mean:

$$\frac{16.85 - 1}{\sqrt{2}} = 11.2$$

Analyzed this way, there can be little doubt that the lottery was not random!

About the Data Needless to say, the way the 1969 draft lottery turned out was a huge embarrassment for the Selective Service Administration and a public relations nightmare. Many individuals, both inside and outside the government, argued that a "do over" was the only fair resolution. Unfortunately, any course of action would have inevitably angered a sizeable number of people, so the decision was made to stay with the original lottery, flawed though it was.

A believable explanation for *why* the selections were so nonrandom is that (1) the birthday capsules were put into the urn by month (January capsules first, February capsules second, March capsules next, and so on) and (2) the capsules were not adequately mixed before the drawings began, leaving birthdays late in the year disproportionately near the top of the urn. If (1) and (2) happened, the trend in Figure 14.4.1 would be the consequence.

What is particularly vexing about the draft lottery debacle and all the furor that it created is that setting up a "fair" lottery is so very easy. First, the birthday capsules should have been numbered from 1 to 366. Then a computer or a random number table should have been used to generate a random permutation of those numbers. That permutation would define the order in which the capsules would be put *into* the urn. If those two simple steps had been followed, the likelihood of a fiasco similar to that shown in Figure 14.4.1 would have been essentially zero.

Questions

14.4.1. Use a Kruskal-Wallis test to analyze the teacher expectation data described in Question 8.2.7. Let $\alpha = 0.05$. What assumptions are being made?

14.4.2. Recall the fiddler crab data given in Question 9.5.3. Use the Kruskal-Wallis test to compare the times spent waving to females by the two groups of males. Let $\alpha = 0.10$.

14.4.3. Use the Kruskal-Wallis method to test at the 0.05 level that methylmercury metabolism is different for males and females in Question 9.2.8.

14.4.4. Redo the analysis of the Quintus Curtius Snodgrass/Mark Twain data in Case Study 9.2.1, this time using a nonparametric procedure.

14.4.5. Use the Kruskal-Wallis technique to test the hypothesis of Case Study 12.2.1 concerning the effect of smoking on heart rate.

14.4.6. A sample of ten 40-W light bulbs was taken from each of three manufacturing plants. The bulbs were burned until failure. The number of hours that each remained lit is listed in the following table.

Plant 1	Plant 2	Plant 3
905	1109	571
1018	1155	1346
905	835	292
886	1152	825
958	1036	676
1056	926	541
904	1029	818
856	1040	90
1070	959	2246
1006	996	104

(a) Test the hypothesis that the median lives of bulbs produced at the three plants are all the same. Use the 0.05 level of significance.

(b) Are the *mean* lives of bulbs produced at the three plants all the same? Use the analysis of variance with $\alpha = 0.05$.

(c) Change the observation "2246" in the third column to "1500" and redo part (a). How does this change affect the hypothesis test?

(d) Change the observation "2246" in the third column to "1500" and redo part (b). How does this change affect the hypothesis test?

14.4.7. The production of a certain organic chemical requires the addition of ammonium chloride. The manufacturer can conveniently obtain the ammonium chloride in any one of three forms—powdered, moderately ground, and coarse. To see what effect, if any, the quality of the NH_4Cl has, the manufacturer decides to run the reaction seven times with each form of ammonium chloride. The resulting yields (in pounds) are listed in the following table. Compare the yields with a Kruskal-Wallis test. Let $\alpha = 0.05$.

Organic Chemical Yields (lb)

Powdered NH_4Cl	Moderately Ground NH_4Cl	Coarse NH_4Cl
146	150	141
152	144	138
149	148	142
161	155	146
158	154	139
149	150	145
154	148	137

14.4.8. Show that the Kruskal-Wallis statistic, B, as defined in Theorem 14.4.1 can also be written

$$B = \sum_{j=1}^{k} \left(\frac{n - n_j}{n}\right) Z_j^2$$

where

$$Z_j = \frac{\dfrac{R_{.j}}{n_j} - \dfrac{n+1}{2}}{\sqrt{\dfrac{(n+1)(n-n_j)}{12n_j}}}$$

14.5 The Friedman Test

The nonparametric analog of the analysis of variance for a randomized block design is *Friedman's test,* a procedure based on within-block *ranks*. Its form is similar to that of the Kruskal-Wallis statistic, and, like its predecessor, it has approximately a χ^2 distribution when H_0 is true.

Theorem 14.5.1 *Suppose $k(\geq 2)$ treatments are ranked independently within b blocks. Let $r_{.j}$, $j = 1, 2, \ldots, k$, be the rank sum of the jth treatment. The null hypothesis that the population medians of the k treatments are all equal is rejected at the α level of significance (approximately) if*

$$g = \frac{12}{bk(k+1)}\sum_{j=1}^{k} r_{.j}^2 - 3b(k+1) \geq \chi^2_{1-\alpha,k-1}$$ □

Case Study 14.5.1

Baseball rules allow a batter considerable leeway in how he is permitted to run from home plate to second base. Two of the possibilities are the narrow-angle and the wide-angle paths diagrammed in Figure 14.5.1. As a means of comparing the two, time trials were held involving twenty-two players (206). Each player ran both paths. Recorded for each runner was the time it took to go from a point 35 feet from home plate to a point 15 feet from second base. Based on those times, ranks (1 and 2) were assigned to each path for each player (see Table 14.5.1).

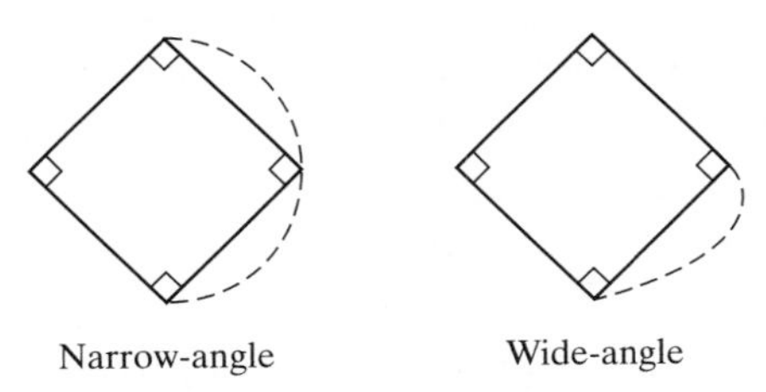

Figure 14.5.1 Batter's path from home plate to second base.

If $\tilde{\mu}_1$ and $\tilde{\mu}_2$ denote the true median rounding times associated with the narrow-angle and wide-angle paths, respectively, the hypotheses to be tested are

$$H_0: \tilde{\mu}_1 = \tilde{\mu}_2$$
versus
$$H_1: \tilde{\mu}_1 \neq \tilde{\mu}_2$$

Let $\alpha = 0.05$. By Theorem 14.5.1, the Friedman statistic (under H_0) will have approximately a χ_1^2 distribution, and the decision rule will be

$$\text{Reject } H_0 \text{ if } g \geq 3.84$$

(Continued on next page)

Table 14.5.1 Times (sec) Required to Round First Base

Player	Narrow-Angle	Rank	Wide-Angle	Rank
1	5.50	1	5.55	2
2	5.70	1	5.75	2
3	5.60	2	5.50	1
4	5.50	2	5.40	1
5	5.85	2	5.70	1
6	5.55	1	5.60	2
7	5.40	2	5.35	1
8	5.50	2	5.35	1
9	5.15	2	5.00	1
10	5.80	2	5.70	1
11	5.20	2	5.10	1
12	5.55	2	5.45	1
13	5.35	1	5.45	2
14	5.00	2	4.95	1
15	5.50	2	5.40	1
16	5.55	2	5.50	1
17	5.55	2	5.35	1
18	5.50	1	5.55	2
19	5.45	2	5.25	1
20	5.60	2	5.40	1
21	5.65	2	5.55	1
22	6.30	2	6.25	1
		39		27

But

$$g = \frac{12}{22(2)(3)}[(39)^2 + (27)^2] - 3(22)(3) = 6.54$$

implying that the two paths are *not* equivalent. The wide-angle path appears to enable runners to reach second base quicker.

Questions

14.5.1. The following data come from a field trial set up to assess the effects of different amounts of potash on the breaking strength of cotton fibers (25). The experiment was done in three blocks. The five treatment levels—36, 54, 72, 108, and 144 lbs of potash per acre—were assigned randomly within each block. The variable recorded was the Pressley strength index. Compare the effects of the different levels of potash applications using Friedman's test. Let $\alpha = 0.05$.

Pressley Strength Index for Cotton Fibers

		Treatment (pounds of potash/acre)				
		36	54	72	108	144
Blocks	1	7.62	8.14	7.76	7.17	7.46
	2	8.00	8.15	7.73	7.57	7.68
	3	7.93	7.87	7.74	7.80	7.21

14.5.2. Use Friedman's test to analyze the Transylvania effect data given in Case Study 13.2.3.

14.5.3. Until its recent indictment as a possible carcinogen, cyclamate was a widely used sweetener in soft drinks. The following data show a comparison of three laboratory methods for determining the percentage of sodium cyclamate in commercially produced orange drink. All three procedures were applied to each of twelve samples (156).

Percent Sodium Cyclamate (w/w)

	Method		
Sample	Picryl Chloride	Davies	AOAC
1	0.598	0.628	0.632
2	0.614	0.628	0.630
3	0.600	0.600	0.622
4	0.580	0.612	0.584
5	0.596	0.600	0.650
6	0.592	0.628	0.606
7	0.616	0.628	0.644
8	0.614	0.644	0.644
9	0.604	0.644	0.624
10	0.608	0.612	0.619
11	0.602	0.628	0.632
12	0.614	0.644	0.616

Use Friedman's test to determine whether the differences from method to method are statistically significant. Let $\alpha = 0.05$.

14.5.4. Use Friedman's test to compare the effects of habitat density on cockroach aggression for the data given in Question 8.2.4. Let $\alpha = 0.05$. Would the conclusion be any different if the densities were compared using the analysis of variance?

14.5.5. Compare the acrophobia therapies described in Case Study 13.2.1 using the Friedman test. Let $\alpha = 0.01$. Does your conclusion agree with the inference reached using the analysis of variance?

14.5.6. Suppose that k treatments are to be applied within each of b blocks. Let $\bar{r}_{..}$ denote the average of the bk ranks and let $\bar{r}_{.j} = (1/b)r_{.j}$. Show that the Friedman statistic given in Theorem 14.5.1 can also be written

$$g = \frac{12b}{k(k+1)} \sum_{j=1}^{k} (\bar{r}_{.j} - \bar{r}_{..})^2$$

What analysis of variance expression does this resemble?

14.6 Testing for Randomness

All hypothesis tests, parametric as well as nonparametric, make the implicit assumption that the observations comprising a given sample are *random,* meaning that the value of y_i does not predispose the value of y_j. Should that not be the case, identifying the source of the nonrandomness—and doing whatever it takes to eliminate it from future observations—necessarily becomes the experimenter's first objective.

Examples of nonrandomness are not uncommon in industrial settings, where successive measurements made on a particular piece of equipment may show a *trend,* for example, if the machine is slowly slipping out of calibration. The other extreme—where measurements show a nonrandom *alternating pattern* (high value, low value, high value, low value, ...)—can occur if successive measurements are made by two different operators, whose standards or abilities are markedly different, or, perhaps, by one operator using two different machines.

A variety of tests based on *runs* of one sort or another can be used to examine the randomness of a sequence of measurements. One of the most useful is a test based on the total number of "runs up and down."

Suppose that $y_1, y_2, \ldots, y_n$ denotes a set of n time-ordered measurements. Let $\text{sgn}(y_i - y_{i-1})$ denote the algebraic sign of the difference $y_i - y_{i-1}$. (It will be assumed that the y_i's represent a continuous random variable, so the probability of y_i and y_{i-1} being equal is zero.) The n observations, then, produce an ordered arrangement of $n-1$ pluses and/or minuses representing the signs of the differences between consecutive measurements (see Figure 14.6.1).

Figure 14.6.1

$$\text{Data:}\quad \underbrace{y_1 \qquad\qquad y_2}_{\text{sgn}(y_2 - y_1)}\underbrace{\qquad\qquad y_3}_{\text{sgn}(y_3 - y_2)} \cdots \underbrace{y_{n-1} \qquad\qquad y_n}_{\text{sgn}(y_n - y_{n-1})}$$

For example, the $n = 5$ observations

$$14.2 \quad 10.6 \quad 11.2 \quad 12.1 \quad 9.3$$

generate the "sgn" sequence

$$- \quad + \quad + \quad -$$

which corresponds to an initial *run down* (that is, going from 14.2 to 10.6), followed by two *runs up,* and ending with a final *run down.*

Let W denote the total number of runs up and down, as reflected by the sequence $\text{sgn}(y_2 - y_1), \text{sgn}(y_3 - y_2), \ldots, \text{sgn}(y_n - y_{n-1})$. For the example just cited, $W = 3$. In general, if W is too large or too small, it can be concluded that the y_i's are not random. The appropriate decision rule derives from an approximate Z ratio.

Theorem 14.6.1 *Let W denote the number of runs up and down in a sequence of n observations, where $n > 2$. If the sequence is random, then*

a. $E(W) = \frac{2n-1}{3}$

b. $\text{Var}(W) = \frac{16n-29}{90}$

and

c. $\frac{W-E(W)}{\sqrt{\text{Var}(W)}} \doteq Z$, *when* $n \geq 20$.

Proof See (125) and (204). □

Case Study 14.6.1

The first widespread labor dispute in the United States occurred in 1877. Railroads were the target, and workers were idled from Pittsburgh to San Francisco. That initial confrontation may have been a long time coming, but organizers were quick to recognize what a powerful weapon a work stoppage could be—36,757 more strikes were called between 1881 and 1905!

For that twenty-five-year period, Table 14.6.1 shows the annual numbers of strikes that were called and the percentages that were deemed successful (31). By definition, a strike was considered "successful" if most or all of the workers' demands were met.

An obvious question suggested by the nature of these data is whether the workers' successes from year to year were random. One plausible hypothesis would be that the percentages of successful strikes should show a trend and tend to increase, as unions acquired more and more power. On the other hand, it could be argued that years of high success rates might tend to alternate with years of low success rates, indicating a kind of labor and management standoff. Still another hypothesis, of course, would be that the percentages show *no* patterns whatsoever and qualify as a random sequence.

The last column shows the calculation of $\text{sgn}(y_i - y_{i-1})$ for $i = 2, 3, \ldots, 25$. By inspection, the number of runs up and down in that sequence of pluses and minuses is eighteen. To test

(Continued on next page)

(Case Study 14.6.1 continued)

Table 14.6.1

Year	Number of Strikes	% Successful, y_i	$\text{sgn}(y_i - y_{i-1})$
1881	451	61	
1882	454	53	−
1883	478	58	+
1884	443	51	−
1885	645	52	+
1886	1432	34	−
1887	1436	45	+
1888	906	52	+
1889	1075	46	−
1890	1833	52	+
1891	1717	37	−
1892	1298	39	+
1893	1305	50	+
1894	1349	38	−
1895	1215	55	+
1896	1026	59	+
1897	1078	57	−
1898	1056	64	+
1899	1797	73	+
1900	1779	46	−
1901	2924	48	+
1902	3161	47	−
1903	3494	40	−
1904	2307	35	−
1905	2077	40	+

$w = 18$

H_0: The y_i's are random with respect to the number of runs up and down

versus

H_1: The y_i's are not random with respect to the number of runs up and down

at the $\alpha = 0.05$ level of significance, we should reject the null hypothesis if $\frac{w-E(W)}{\sqrt{\text{Var}(W)}}$ is either (1) $\leq -z_{\alpha/2} = -1.96$ or (2) $\geq z_{\alpha/2} = 1.96$. Given that $n = 25$,

$$E(W) = \frac{2(25) - 1}{3} = 16.3$$

and

$$\text{Var}(W) = \frac{16(25) - 29}{90} = 4.12$$

so the observed test statistic is *+0.84*:

$$z = \frac{18 - 16.3}{\sqrt{4.12}} = 0.84$$

Our conclusion, then, is to *fail to reject* H_0—it is believable, in other words, that the observed sequence of runs up and down could, in fact, have come from a sample of twenty-five random observations.

About the Data Another hypothesis suggested by these data is that the percentage of successful strikes might vary inversely with the *number* of strikes: As the latter increased, the number of "frivolous" disputes might also have increased, which could understandably lead to a lower percentage of successful resolutions. In point of fact, that explanation does appear to have some merit. A linear fit of the twenty-five observations yields the equation

$$\% \text{ successful} = 56.17 - 0.0047 \cdot \text{ number of strikes}$$

and the null hypothesis H_0: $\beta_1 = 0$ is rejected at the $\alpha = 0.05$ level of significance.

Questions

14.6.1. The data in the table examine the relationship between stock market changes (1) during the first few days in January and (2) over the course of the entire year. Included are the years from 1950 through 1986.

(a) Use Theorem 14.6.1 to test the randomness of the January changes (relative to the number of runs up and down). Let $\alpha = 0.05$.

(b) Use Theorem 14.6.1 to test the randomness of the annual changes. Let $\alpha = 0.05$.

Year	% Change for First 5 Days in Jan., x	% Change for Year, y
1950	2.0	21.8
1951	2.3	16.5
1952	0.6	11.8
1953	−0.9	−6.6
1954	0.5	45.0
1955	−1.8	26.4
1956	−2.1	2.6
1957	−0.9	−14.3
1958	2.5	38.1
1959	0.3	8.5
1960	−0.7	−3.0
1961	1.2	23.1
1962	−3.4	−11.8
1963	2.6	18.9
1964	1.3	13.0
1965	0.7	9.1
1966	0.8	−13.1
1967	3.1	20.1
1968	0.2	7.7
1969	−2.9	−11.4
1970	0.7	0.1
1971	0.0	10.8
1972	1.4	15.6
1973	1.5	−17.4
1974	−1.5	−29.7
1975	2.2	31.5
1976	4.9	19.1
1977	−2.3	−11.5
1978	−4.6	1.1
1979	2.8	12.3
1980	0.9	25.8
1981	−2.0	−9.7
1982	−2.4	14.8
1983	3.2	17.3
1984	2.4	1.4
1985	−1.9	26.3
1986	−1.6	14.6

14.6.2. Listed below for two consecutive fiscal years are the monthly numbers of passenger boardings at a Florida airport. Use Theorem 14.6.1 to test whether these twenty-four observations can be considered a random sequence, relative to the number of runs up and down. Let $\alpha = 0.05$.

Month	Passenger Boardings	Month	Passenger Boardings
July	41,388	July	44,148
Aug.	44,880	Aug.	42,038
Sept.	33,556	Sept.	35,157
Oct.	34,805	Oct.	39,568
Nov.	33,025	Nov.	34,185
Dec.	34,873	Dec.	37,604
Jan.	31,330	Jan.	28,231
Feb.	30,954	Feb.	29,109
March	32,402	March	38,080
April	38,020	April	34,184
May	42,828	May	39,842
June	41,204	June	46,727

14.6.3. On the next page is a partial statistical summary of the first twenty-four Super Bowls (33). Of particular interest to advertisers is the network share that each game garnered. Can those shares be considered a random sequence, relative to the number of runs up and down? Test the appropriate hypothesis at the $\alpha = 0.05$ level of significance.

Game, Year	Winner, Loser	Score	MVP Is QB	Network Share (network)
I	Green Bay (NFL)	35	1	79
1967	Kansas City (AFL)	10		(CBS/NBC combined)
II	Green Bay (NFL)	33	1	68
1968	Oakland (AFL)	14		(CBS)
III	NY Jets (AFL)	16	1	71
1969	Baltimore (NFL)	7		(NBC)
IV	Kansas City (AFL)	23	1	69
1970	Minnesota (NFL)	7		(CBS)
V	Baltimore (AFC)	16	0	75
1971	Dallas (NFC)	13		(NBC)
VI	Dallas (NFC)	24	1	74
1972	Miami (AFC)	3		(CBS)
VII	Miami (AFC)	14	0	72
1973	Washington (NFC)	7		(NBC)
VIII	Miami (AFC)	24	0	73
1974	Minnesota (NFC)	7		(CBS)
IX	Pittsburgh (AFC)	16	0	72
1975	Minnesota (NFC)	6		(NBC)
X	Pittsburgh (AFC)	21	0	78
1976	Dallas (NFC)	17		(CBS)
XI	Oakland (AFC)	32	0	73
1977	Minnesota (NFC)	14		(NBC)
XII	Dallas (NFC)	27	0	67
1978	Denver (AFC)	10		(CBS)
XIII	Pittsburgh (AFC)	35	1	74
1979	Dallas (NFC)	31		(NBC)
XIV	Pittsburgh (AFC)	31	1	67
1980	Los Angeles (AFC)	19		(CBS)
XV	Oakland (AFC)	27	1	63
1981	Philadelphia (NFC)	10		(NBC)
XVI	San Francisco (NFC)	26	1	73
1982	Cincinnati (AFC)	21		(CBS)
XVII	Washington (NFC)	27	0	69
1983	Miami (AFC)	17		(NBC)
XVIII	LA Raiders (AFC)	38	0	71
1984	Washington (NFC)	9		(CBS)
XIX	San Francisco (NFC)	38	1	63
1985	Miami (AFC)	16		(ABC)
XX	Chicago (NFC)	46	0	70
1986	New England (AFC)	10		(NBC)
XXI	NY Giants (NFC)	39	1	66
1987	Denver (AFC)	20		(CBS)
XXII	Washington (NFC)	42	1	62
1988	Denver (AFC)	10		(ABC)
XXIII	San Francisco (NFC)	20	0	68
1989	Cincinnati (AFC)	16		(NBC)
XXIV	San Francisco (NFC)	55	1	63
1990	Denver (AFC)	10		(CBS)

14.6.4. In the next column are the lengths (in mm) of furniture dowels recorded as part of an ongoing quality-control program. Listed are the measurements made on thirty samples (each of size 4) taken *in order* from the assembly line. Is the variation in the sample *averages* random with respect to the number of runs up and down? Do an appropriate hypothesis test at the $\alpha = 0.05$ level of significance.

Sample	y_1	y_2	y_3	y_4	$\bar{y}$
1	46.1	44.4	45.3	44.2	45.0
2	46.0	45.4	42.5	44.4	44.6
3	44.3	44.0	45.4	43.9	44.4
4	44.9	43.7	45.2	44.8	44.7
5	43.0	45.3	45.9	43.8	44.5
6	46.0	43.2	44.4	43.7	44.3
7	46.0	44.6	45.4	46.4	45.6
8	46.1	45.5	45.0	45.5	45.5
9	42.8	45.1	44.9	44.3	44.3
10	45.0	46.7	43.0	44.8	44.9
11	45.5	44.5	45.1	47.1	45.6
12	45.8	44.6	44.8	45.1	45.1
13	45.1	45.4	46.0	45.4	45.5
14	44.6	43.8	44.2	43.9	44.1
15	44.8	45.5	45.2	46.2	45.4
16	45.8	44.1	43.3	45.8	44.8
17	44.1	44.8	46.1	45.5	45.1
18	44.5	43.6	45.1	46.9	45.0
19	45.2	43.1	46.3	46.4	45.3
20	45.9	46.8	46.8	45.8	46.3
21	44.0	44.7	46.2	45.4	45.1
22	43.4	44.6	45.4	44.4	44.5
23	43.1	44.6	44.5	45.8	44.5
24	46.6	43.3	45.1	44.2	44.8
25	46.2	44.9	45.3	46.0	45.6
26	42.5	43.4	44.3	42.7	43.2
27	43.4	43.3	43.4	43.5	43.4
28	42.3	42.4	46.6	42.3	43.4
29	41.9	42.9	42.0	42.9	42.4
30	43.2	43.5	42.2	44.7	43.4

14.6.5. Listed below are forty ordered observations generated by Minitab's RANDOM command that presumably represent a normal distribution with $\mu = 5$ and $\sigma = 2$. Can the sample be considered random with respect to the number of runs up and down?

Obs. #	y_i	Obs. #	y_i	Obs. #	y_i	Obs. #	y_i
1	7.0680	11	7.6979	21	5.9828	31	5.2625
2	4.0540	12	4.4338	22	1.4614	32	5.9047
3	6.6165	13	5.6538	23	9.2655	33	4.6342
4	1.2166	14	8.0791	24	4.9281	34	5.3089
5	4.6158	15	4.7458	25	10.5561	35	5.4942
6	7.7540	16	3.5044	26	6.1738	36	6.6914
7	7.7300	17	1.3071	27	5.4895	37	1.4380
8	6.5109	18	5.7893	28	3.6629	38	8.2604
9	3.8933	19	4.5241	29	3.7223	39	5.0209
10	2.7533	20	5.3291	30	3.5211	40	0.5544

14.6.6. Sunnydale Farms markets an all-purpose fertilizer that is supposed to contain, by weight, 15% potash (K_2O). Samples were taken daily in October from three bags chosen at random as they came off the filling machine. Tabulated on the right are the K_2O percentages recorded. Calculate the *range* ($= y_{max} - y_{min}$) for each sample. Use Theorem 14.6.1 to test whether the variation in the ranges can be considered random with respect to the number of runs up and down.

Date	y_1	y_2	y_3	Date	y_1	y_2	y_3
10/1	16.1	14.4	15.3	10/15	16.3	13.3	15.3
10/2	16.0	16.4	13.5	10/16	17.4	13.8	14.3
10/3	14.3	14.0	15.4	10/17	13.5	11.0	15.4
10/4	14.8	13.1	15.2	10/18	15.6	9.2	18.9
10/5	12.0	15.4	16.4	10/19	16.3	17.6	20.5
10/8	16.4	12.3	14.2	10/22	14.3	15.6	17.0
10/9	16.9	14.2	15.8	10/23	15.4	15.3	15.4
10/10	17.2	16.0	14.9	10/24	14.3	14.4	18.6
10/11	10.6	15.3	14.9	10/25	13.9	14.9	14.0
10/12	15.0	19.2	10.0	10/26	15.2	15.5	14.2

14.7 Taking a Second Look at Statistics (Comparing Parametric and Nonparametric Procedures)

Virtually every *parametric* hypothesis test an experimenter might consider doing has one or more *nonparametric* analogs. Using two independent samples to compare the locations of two distributions, for example, can be done with a two-sample t test or with a Wilcoxon signed rank test. Likewise, comparing k treatment levels using dependent samples can be accomplished with the (parametric) analysis of variance or with the (nonparametric) Friedman's test. Having alternative ways to analyze the same set of data inevitably raises the same sorts of questions that surfaced in Section 13.4—which procedure should be used in a given situation, and why?

The answers to those questions are rooted in the origins of the data—that is, in the pdfs generating the samples—and what those origins imply about (1) the relative *power* of the parametric and nonparametric procedures and (2) the *robustness* of the two procedures. As we have seen, parametric procedures make assumptions about the origin of the data that are much more specific than the assumptions made by nonparametric procedures. The (pooled) two-sample t test, for example, assumes that the two sets of independent observations come from *normal* distributions with the same standard deviation. The Wilcoxon signed rank test, on the other hand, makes the much weaker assumption that the observations come from *symmetric* distributions (which, of course, include normal distributions as a special case). Moreover, each observation does not have to come from the *same* symmetric distribution.

In general, if the assumptions made by a parametric test are satisfied, then that procedure will be superior to any of its nonparametric analogs in the sense that its power curve will be steeper. (Recall Figure 6.4.5—if the normality assumption is met, the parametric procedure will have a power curve similar to that for Method B; the nonparametric procedure would have a power curve similar to Method A's.)

If one or more of the parametric procedure's assumptions are not satisfied, the distribution of its test statistic will not be exactly what it would have been had the assumptions all been met (recall Figure 7.4.5). If the differences between the "theoretical" test statistic distribution and the "actual" test statistic distribution are considerable, the integrity of the parametric procedure is obviously compromised. Whether those two distributions will be considerably different depends on the robustness of the parametric procedure with respect to whichever assumptions are being violated.

Concluding this section is a set of Monte Carlo simulations that compare the one-way analysis of variance to a Kruskal-Wallis test. In each instance, the data consist of $n_j = 5$ observations taken on each of $k = 4$ treatment levels. Included are simulations that focus on (1) the power of the two procedures when the normality assumption is met and (2) the robustness of the two tests when neither the normality nor the symmetry assumptions are satisfied. Each simulation is based on one hundred replications, and the twenty observations generated for each replication (by the RANDOM command) were analyzed twice, once using the analysis of variance and again using the Kruskal-Wallis test (see Appendix 14.A.1 for an example of the Minitab syntax).

Figure 14.7.1 shows the distribution of the one hundred observed F ratios when all the H_0 assumptions made by the analysis of variance are satisfied—that is, five observations were taken on each of four treatment levels, where all twenty observations were normally distributed with the same mean and the same standard deviation. Given that $n_j = 5, k = 4$, and $n = 20$, there would be 3 df for Treatments and 16 df for Error (recall Figure 12.2.1). Superimposed over the histogram is the pdf for an $F_{3,16}$ random variable. Clearly, the agreement between the F curve and the histogram is excellent.

Figure 14.7.1

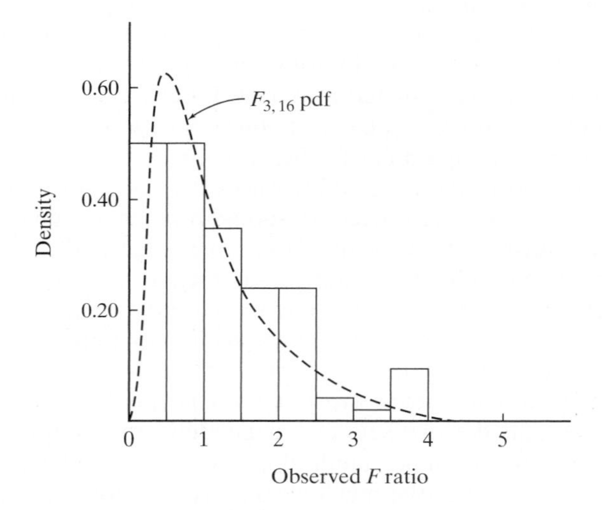

Figure 14.7.2 is the analogous "H_0" distribution for the Kruskal-Wallis test. The one hundred data sets analyzed were the same that gave rise to Figure 14.7.1. Superimposed is the χ_3^2 pdf. As predicted by Theorem 14.4.1, the distribution of observed b values is approximated very nicely by the chi square curve with 3 $(= k - 1)$ df.

One of the advantages of nonparametric procedures is that violations of their assumptions tend to have relatively mild repercussions on the distributions of their test statistics. Figure 14.7.3 is a case in point. Shown there is a histogram of Kruskal-Wallis values calculated from one hundred data sets where each of the twenty observations ($n_j = 5$ and $k = 4$) came from an *exponential* pdf with $\lambda = 1$—that is, from $f_Y(y) = e^{-y}, y > 0$. The latter is a sharply skewed pdf that violates the symmetry assumption underlying the Kruskal-Wallis test. The actual distribution of b values, though, does not appear to be much different from the values produced in Figure 14.7.2, where all the assumptions of the Kruskal-Wallis test were met.

A similar insensitivity to the data's underlying pdf is not entirely shared by the F test. Figure 14.7.4 summarizes the results of applying the analysis of variance to

Figure 14.7.2

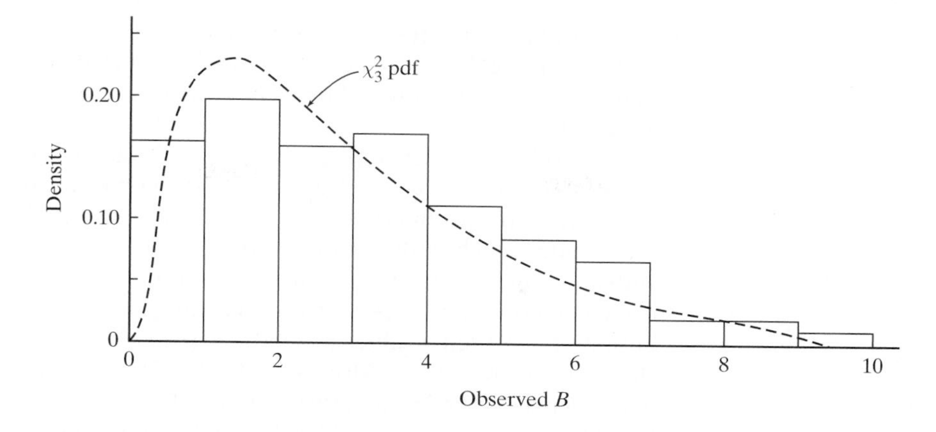

Figure 14.7.3

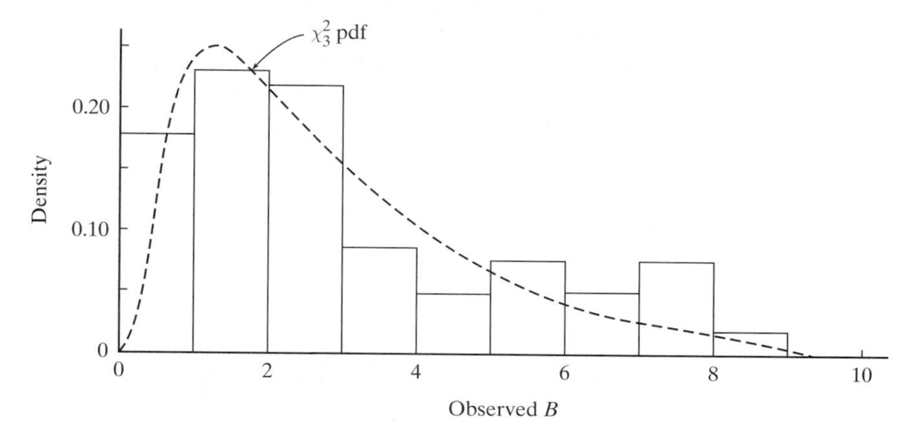

the same set of one hundred replications that produced Figure 14.7.3. Notice that a handful of the data sets yielded F ratios much larger than the $F_{3,16}$ curve would have predicted. Recall that a similar skewness was observed when the t test was applied to exponential data where n was small (see Figure 7.4.6b).

Figure 14.7.4

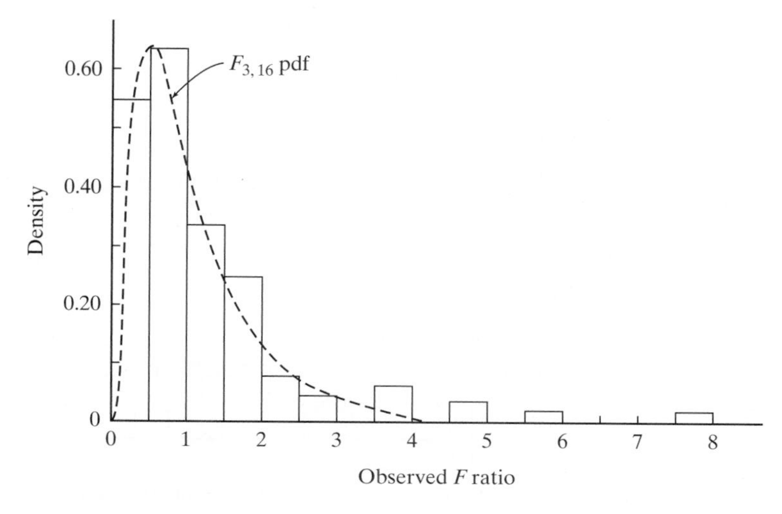

Having weaker assumptions and being less sensitive to violations of those assumptions are definite advantages that nonparametric procedures often have over their parametric counterparts. But that broader range of applicability does not come without a price: Nonparametric hypothesis tests will make Type II errors more often than will parametric procedures *when the assumptions of the parametric procedures are satisfied.*

Consider, for example, the two Monte Carlo simulations pictured in Figures 14.7.5 and 14.7.6. The former shows the results of applying the Kruskal-Wallis test to one hundred sets of k-sample data, where the five measurements representing each of the first three treatment levels came from normal distributions with $\mu = 0$ and $\sigma = 1$, while the five measurements representing the fourth treatment level came from a normal distribution with $\mu = 1$ and $\sigma = 1$. As expected, the distribution of observed b values has shifted to the right, compared to the H_0 distribution shown in Figure 14.7.3. More specifically, *26%* of the one hundred data sets produced Kruskal-Wallis values in excess of $7.815\ (= \chi^2_{0.95,3})$, meaning that H_0 would have been rejected at the $\alpha = 0.05$ level of significance. [If H_0 were true, of course, the theoretical percentage of b values exceeding 7.815 would be *5%*. Only *1%* of the data sets, though, exceeded the $\alpha = 0.01$ cutoff $(= \chi^2_{0.99,3} = 11.345)$, which is the same percentage expected if H_0 were true.]

Figure 14.7.5

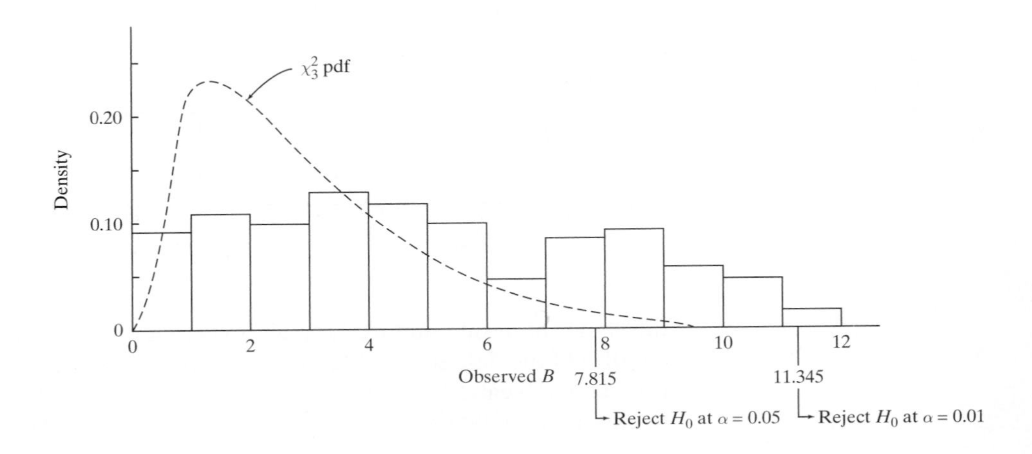

Figure 14.7.6

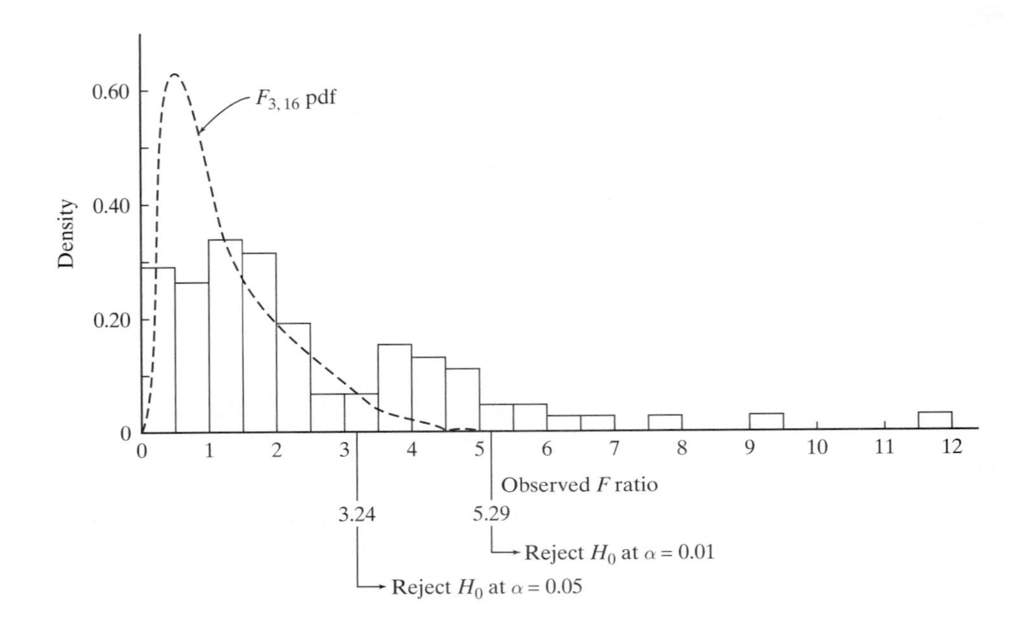

Figure 14.7.6 shows the results of doing the analysis of variance on the same one hundred data sets used for Figure 14.7.5. As was true for the Kruskal-Wallis calculations, the distribution of observed F ratios has shifted to the right (compare Figure 14.7.6 to Figure 14.7.1). What is especially noteworthy, though, is that the observed F ratios have shifted much further to the right than did the observed b values. For example, while only *1%* of the observed b values exceeded the $\alpha = 0.01$ cutoff (= 11.345), a total of *8%* of the observed F ratios exceeded their $\alpha = 0.01$ cutoff ($= F_{0.99,3,16}$).

So, is there an easy answer to the question of which type of procedure to use, parametric or nonparametric? Sometimes yes, sometimes no. If it seems reasonable to believe that all the assumptions of the parametric test are satisfied, then the parametric test should be used. For all those situations, though, where the validity of one or more of the parametric assumptions is in question, the choice becomes more problematic. If the violation of the assumptions is minimal (or if the sample sizes are fairly large), the robustness of the parametric procedures (along with their greater power) usually gives them the edge. Nonparametric tests tend to be reserved for situations where (1) sample sizes are small, *and* (2) there is reason to believe that "something" about the data is markedly inconsistent with the assumptions implicit in the available parametric procedures.

Appendix 14.A.1 Minitab Applications

The Sign Test

Figure 14.A.1.1 shows Minitab's sign test routine applied to ten paired samples—(97, 113), (106, 113), . . . , (96, 126). The basic command is

```
MTB > stest 0.0 c3;
SUBC > alternative 0.
```

where c3 contains the within-pair differences. The subcommand ALTERNATIVE 0 makes H_1 two-sided. One-sided alternative hypotheses require that ALTERNATIVE 1 (if the rejection region is to the right) or ALTERNATIVE–1 (if the rejection region is to the left) be used.

Figure 14.A.1.1

```
MTB   > set c1
DATA  > 97 106 106 95 102 111 115 104 90 96
DATA  > end
MTB   > set c2
DATA  > 113 113 101 119 111 122 121 106 110 126
DATA  > end
MTB   > let c3 = c2 — c1
MTB   > stest 0.0 c3;
SUBC  > alternative 0.
```

Sign Test for Median: C3

```
Sign test of median = 0.00000 versus not = 0.00000
          N     Below      Equal       Above          P MEDIAN
C3       10         1          0           9     0.0215  10.00
```

The Wilcoxon Signed Rank Test

The Wilcoxon signed rank statistic of Theorem 14.3.2 is calculated using the command `MTB > wtest` $\tilde{\mu}_o$ c1, where the y_i's have been entered in c1. As with the sign test, the subcommand ALTERNATIVE 0 makes H_1 two-sided. Figure 14.A.1.2 summarizes Minitab's analysis of the shark data from Case Study 14.3.1.

Figure 14.A.1.2

```
MTB  > set c1
DATA > 13.32 13.06 14.02 11.86 13.58 13.77 13.51 14.42 14.44 15.43
DATA > end
MTB  > wtest 14.6 c1;
SUBC > alternative 0.
```

Wilcoxon Signed Rank Test: C1

```
Test of median = 14.60 versus median not = 14.60
          N For   Wilcoxon          Estimated
      N   test   Statistic      P    Median
c1   10     10         4.5  0.022     13.75
```

The Kruskal-Wallis Test

Data are entered for the Kruskal-Wallis test using the stacked format seen earlier in connection with the randomized block analysis of variance in Chapter 13. The syntax, though, is different. First, the data from each treatment level are entered in a separate column. Then a *stack* command is used to transfer all those data to a single column (in this case, c5). Finally, an additional column—here, c6—is defined that identifies the treatment level represented by each data point in the stacked column.

Figure 14.A.1.3 shows the Kruskal-Wallis input and output for the heart rate data given in Case Study 12.2.1.

Figure 14.A.1.3

```
MTB   > set c1
DATA  > 69 52 71 58 59 65
DATA  > end
MTB   > set c2
DATA  > 55 60 78 58 62 66
DATA  > end
MTB   > set c3
DATA  > 66 81 70 77 57 79
DATA  > end
MTB   > set c4
DATA  > 91 72 81 67 95 84
DATA  > end
MTB   > stack c1 c2 c3 c4 c5
MTB   > set c6
DATA  > 6(1) 6(2) 6(3) 6(4)
DATA  > end
MTB   > kruskal-wallis c5 c6.
```

Kruskal-Wallis Test: C5 versus C6

```
Kruskal-Wallis Test on C5

C6         N   Median   Ave. Rank      Z

1          6   62.00          8.1  -1.77
2          6   61.00          8.3  -1.67
3          6   73.50         14.0   0.60
4          6   82.50         19.6   2.83
Overall   24                 12.5
H = 10.71  DF = 3  P = 0.013
H = 10.73  DF = 3  P = 0.013 (adjusted for ties)
```

The Friedman Test

The syntax for Friedman's test is similar to what is used for the Kruskal-Wallis procedure, except that an additional column identifying the block to which each observation belongs must be included. As before, the data from each treatment level are initially put into separate columns; then those columns are stacked. For the case of two treatment levels, the final command would be

```
MTB > friedman c3 c4 c5
```

where c3 is the stacked column of the entire data set, c4 is a column identifying the treatment level represented by each observation, and c5 is a column giving the block location of each observation.

Figure 14.A.1.4 is the Friedman analysis of the baseball data in Case Study 14.5.1. The observed test statistic is denoted S (instead of the g on p. 682).

Figure 14.A.1.4

```
MTB   > set c1
DATA  > 5.50 5.70 5.60 5.50 5.85 5.55 5.40 5.50 5.15 5.80 5.20
DATA  > 5.55 5.35 5.00 5.50 5.55 5.55 5.50 5.45 5.60 5.65 6.30
DATA  > end
MTB   > set c2
DATA  > 5.55 5.75 5.50 5.40 5.70 5.60 5.35 5.35 5.00 5.70 5.10
DATA  > 5.45 5.45 4.95 5.40 5.50 5.35 5.55 5.25 5.40 5.55 6.25
DATA  > end
MTB   > stack c1 c2 c3
MTB   > set c4
DATA  > 1 1 1 1 1 1 1 1 1 1 1 1 1 1 1 1 1 1 1 1 1 1
DATA  > 2 2 2 2 2 2 2 2 2 2 2 2 2 2 2 2 2 2 2 2 2 2
DATA  > end
MTB   > set c5
DATA  > 1 2 3 4 5 6 7 8 9 10 11 12 13 14 15 16 17 18 19 20 21 22
DATA  > 1 2 3 4 5 6 7 8 9 10 11 12 13 14 15 16 17 18 19 20 21 22
DATA  > end
MTB   > friedman c3 c4 c5.
```

Friedman Test: C3 versus C4 blocked by C5

```
S = 6.55   DF = 1   P = 0.011

                          Sum of
C4    N   Est Median       Ranks
1    22       5.5500        39.0
2    22       5.4500        27.0
Grand median = 5.5000
```

Appendix A

STATISTICAL TABLES

Table A.1 Cumulative Areas under the Standard Normal Distribution

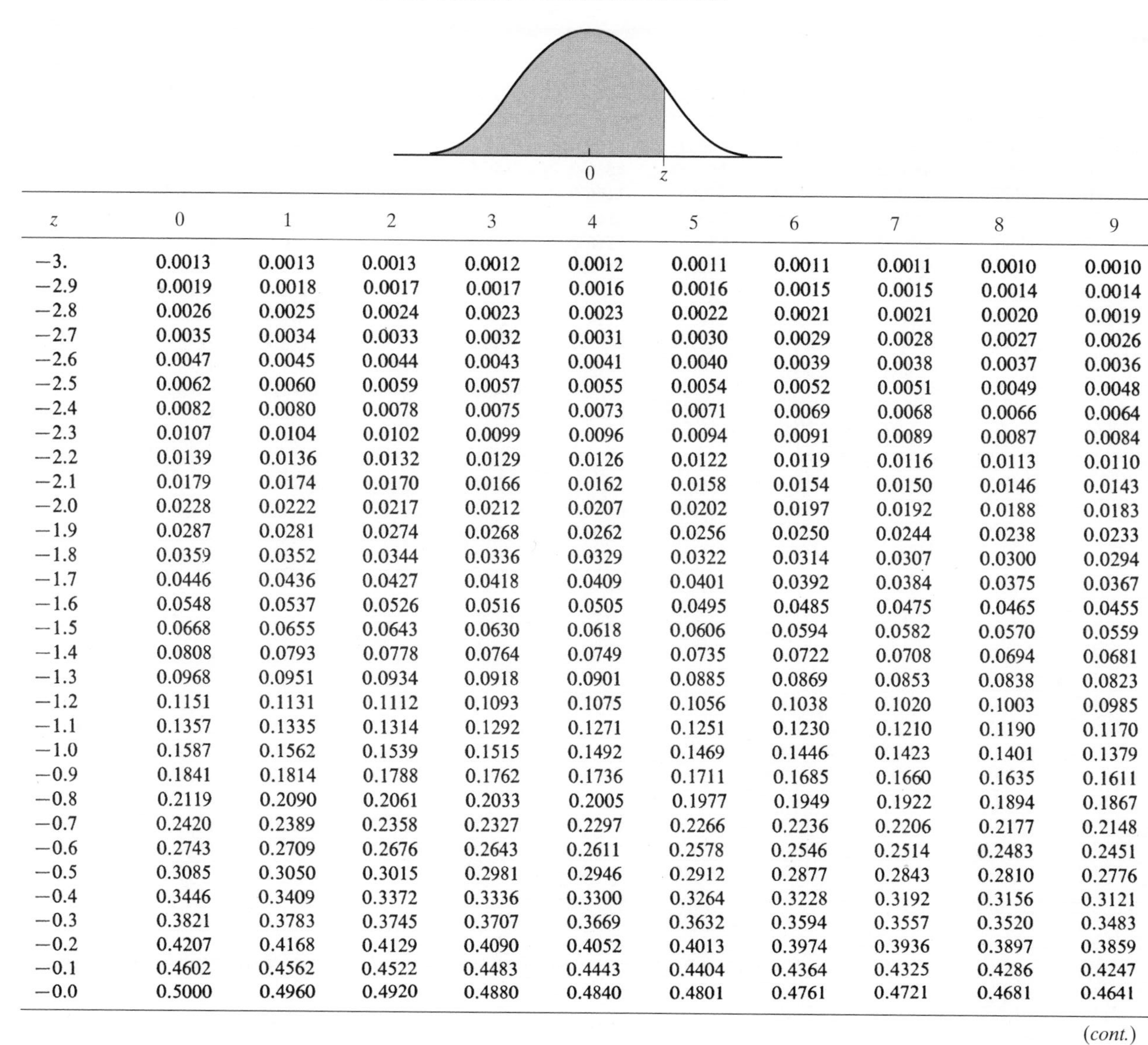

z	0	1	2	3	4	5	6	7	8	9
−3.	0.0013	0.0013	0.0013	0.0012	0.0012	0.0011	0.0011	0.0011	0.0010	0.0010
−2.9	0.0019	0.0018	0.0017	0.0017	0.0016	0.0016	0.0015	0.0015	0.0014	0.0014
−2.8	0.0026	0.0025	0.0024	0.0023	0.0023	0.0022	0.0021	0.0021	0.0020	0.0019
−2.7	0.0035	0.0034	0.0033	0.0032	0.0031	0.0030	0.0029	0.0028	0.0027	0.0026
−2.6	0.0047	0.0045	0.0044	0.0043	0.0041	0.0040	0.0039	0.0038	0.0037	0.0036
−2.5	0.0062	0.0060	0.0059	0.0057	0.0055	0.0054	0.0052	0.0051	0.0049	0.0048
−2.4	0.0082	0.0080	0.0078	0.0075	0.0073	0.0071	0.0069	0.0068	0.0066	0.0064
−2.3	0.0107	0.0104	0.0102	0.0099	0.0096	0.0094	0.0091	0.0089	0.0087	0.0084
−2.2	0.0139	0.0136	0.0132	0.0129	0.0126	0.0122	0.0119	0.0116	0.0113	0.0110
−2.1	0.0179	0.0174	0.0170	0.0166	0.0162	0.0158	0.0154	0.0150	0.0146	0.0143
−2.0	0.0228	0.0222	0.0217	0.0212	0.0207	0.0202	0.0197	0.0192	0.0188	0.0183
−1.9	0.0287	0.0281	0.0274	0.0268	0.0262	0.0256	0.0250	0.0244	0.0238	0.0233
−1.8	0.0359	0.0352	0.0344	0.0336	0.0329	0.0322	0.0314	0.0307	0.0300	0.0294
−1.7	0.0446	0.0436	0.0427	0.0418	0.0409	0.0401	0.0392	0.0384	0.0375	0.0367
−1.6	0.0548	0.0537	0.0526	0.0516	0.0505	0.0495	0.0485	0.0475	0.0465	0.0455
−1.5	0.0668	0.0655	0.0643	0.0630	0.0618	0.0606	0.0594	0.0582	0.0570	0.0559
−1.4	0.0808	0.0793	0.0778	0.0764	0.0749	0.0735	0.0722	0.0708	0.0694	0.0681
−1.3	0.0968	0.0951	0.0934	0.0918	0.0901	0.0885	0.0869	0.0853	0.0838	0.0823
−1.2	0.1151	0.1131	0.1112	0.1093	0.1075	0.1056	0.1038	0.1020	0.1003	0.0985
−1.1	0.1357	0.1335	0.1314	0.1292	0.1271	0.1251	0.1230	0.1210	0.1190	0.1170
−1.0	0.1587	0.1562	0.1539	0.1515	0.1492	0.1469	0.1446	0.1423	0.1401	0.1379
−0.9	0.1841	0.1814	0.1788	0.1762	0.1736	0.1711	0.1685	0.1660	0.1635	0.1611
−0.8	0.2119	0.2090	0.2061	0.2033	0.2005	0.1977	0.1949	0.1922	0.1894	0.1867
−0.7	0.2420	0.2389	0.2358	0.2327	0.2297	0.2266	0.2236	0.2206	0.2177	0.2148
−0.6	0.2743	0.2709	0.2676	0.2643	0.2611	0.2578	0.2546	0.2514	0.2483	0.2451
−0.5	0.3085	0.3050	0.3015	0.2981	0.2946	0.2912	0.2877	0.2843	0.2810	0.2776
−0.4	0.3446	0.3409	0.3372	0.3336	0.3300	0.3264	0.3228	0.3192	0.3156	0.3121
−0.3	0.3821	0.3783	0.3745	0.3707	0.3669	0.3632	0.3594	0.3557	0.3520	0.3483
−0.2	0.4207	0.4168	0.4129	0.4090	0.4052	0.4013	0.3974	0.3936	0.3897	0.3859
−0.1	0.4602	0.4562	0.4522	0.4483	0.4443	0.4404	0.4364	0.4325	0.4286	0.4247
−0.0	0.5000	0.4960	0.4920	0.4880	0.4840	0.4801	0.4761	0.4721	0.4681	0.4641

(cont.)

Table A.1 Cumulative Areas under the Standard Normal Distribution (*cont.*)

z	0	1	2	3	4	5	6	7	8	9
0.0	0.5000	0.5040	0.5080	0.5120	0.5160	0.5199	0.5239	0.5279	0.5319	0.5359
0.1	0.5398	0.5438	0.5478	0.5517	0.5557	0.5596	0.5636	0.5675	0.5714	0.5753
0.2	0.5793	0.5832	0.5871	0.5910	0.5948	0.5987	0.6026	0.6064	0.6103	0.6141
0.3	0.6179	0.6217	0.6255	0.6293	0.6331	0.6368	0.6406	0.6443	0.6480	0.6517
0.4	0.6554	0.6591	0.6628	0.6664	0.6700	0.6736	0.6772	0.6808	0.6844	0.6879
0.5	0.6915	0.6950	0.6985	0.7019	0.7054	0.7088	0.7123	0.7157	0.7190	0.7224
0.6	0.7257	0.7291	0.7324	0.7357	0.7389	0.7422	0.7454	0.7486	0.7517	0.7549
0.7	0.7580	0.7611	0.7642	0.7673	0.7703	0.7734	0.7764	0.7794	0.7823	0.7852
0.8	0.7881	0.7910	0.7939	0.7967	0.7995	0.8023	0.8051	0.8078	0.8106	0.8133
0.9	0.8159	0.8186	0.8212	0.8238	0.8264	0.8289	0.8315	0.8340	0.8365	0.8389
1.0	0.8413	0.8438	0.8461	0.8485	0.8508	0.8531	0.8554	0.8577	0.8599	0.8621
1.1	0.8643	0.8665	0.8686	0.8708	0.8729	0.8749	0.8770	0.8790	0.8810	0.8830
1.2	0.8849	0.8869	0.8888	0.8907	0.8925	0.8944	0.8962	0.8980	0.8997	0.9015
1.3	0.9032	0.9049	0.9066	0.9082	0.9099	0.9115	0.9131	0.9147	0.9162	0.9177
1.4	0.9192	0.9207	0.9222	0.9236	0.9251	0.9265	0.9278	0.9292	0.9306	0.9319
1.5	0.9332	0.9345	0.9357	0.9370	0.9382	0.9394	0.9406	0.9418	0.9430	0.9441
1.6	0.9452	0.9463	0.9474	0.9484	0.9495	0.9505	0.9515	0.9525	0.9535	0.9545
1.7	0.9554	0.9564	0.9573	0.9582	0.9591	0.9599	0.9608	0.9616	0.9625	0.9633
1.8	0.9641	0.9648	0.9656	0.9664	0.9671	0.9678	0.9686	0.9693	0.9700	0.9706
1.9	0.9713	0.9719	0.9726	0.9732	0.9738	0.9744	0.9750	0.9756	0.9762	0.9767
2.0	0.9772	0.9778	0.9783	0.9788	0.9793	0.9798	0.9803	0.9808	0.9812	0.9817
2.1	0.9821	0.9826	0.9830	0.9834	0.9838	0.9842	0.9846	0.9850	0.9854	0.9857
2.2	0.9861	0.9864	0.9868	0.9871	0.9874	0.9878	0.9881	0.9884	0.9887	0.9890
2.3	0.9893	0.9896	0.9898	0.9901	0.9904	0.9906	0.9909	0.9911	0.9913	0.9916
2.4	0.9918	0.9920	0.9922	0.9925	0.9927	0.9929	0.9931	0.9932	0.9934	0.9936
2.5	0.9938	0.9940	0.9941	0.9943	0.9945	0.9946	0.9948	0.9949	0.9951	0.9952
2.6	0.9953	0.9955	0.9956	0.9957	0.9959	0.9960	0.9961	0.9962	0.9963	0.9964
2.7	0.9965	0.9966	0.9967	0.9968	0.9969	0.9970	0.9971	0.9972	0.9973	0.9974
2.8	0.9974	0.9975	0.9976	0.9977	0.9977	0.9978	0.9979	0.9979	0.9980	0.9981
2.9	0.9981	0.9982	0.9982	0.9983	0.9984	0.9984	0.9985	0.9985	0.9986	0.9986
3.	0.9987	0.9987	0.9987	0.9988	0.9988	0.9989	0.9989	0.9989	0.9990	0.9990

Source: From Samuels/Witmer, *Statistics for Life Sciences*, Table 3, p. 675,

Table A.2 Upper Percentiles of Student t Distributions

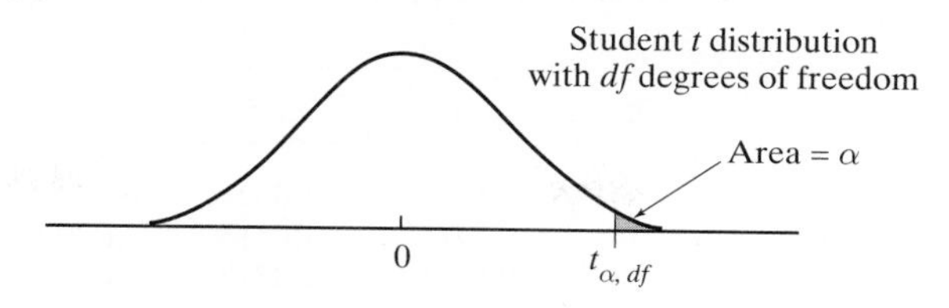

	α						
df	0.20	0.15	0.10	0.05	0.025	0.01	0.005
1	1.376	1.963	3.078	6.3138	12.706	31.821	63.657
2	1.061	1.386	1.886	2.9200	4.3027	6.965	9.9248
3	0.978	1.250	1.638	2.3534	3.1825	4.541	5.8409
4	0.941	1.190	1.533	2.1318	2.7764	3.747	4.6041
5	0.920	1.156	1.476	2.0150	2.5706	3.365	4.0321
6	0.906	1.134	1.440	1.9432	2.4469	3.143	3.7074
7	0.896	1.119	1.415	1.8946	2.3646	2.998	3.4995
8	0.889	1.108	1.397	1.8595	2.3060	2.896	3.3554
9	0.883	1.100	1.383	1.8331	2.2622	2.821	3.2498
10	0.879	1.093	1.372	1.8125	2.2281	2.764	3.1693
11	0.876	1.088	1.363	1.7959	2.2010	2.718	3.1058
12	0.873	1.083	1.356	1.7823	2.1788	2.681	3.0545
13	0.870	1.079	1.350	1.7709	2.1604	2.650	3.0123
14	0.868	1.076	1.345	1.7613	2.1448	2.624	2.9768
15	0.866	1.074	1.341	1.7530	2.1315	2.602	2.9467
16	0.865	1.071	1.337	1.7459	2.1199	2.583	2.9208
17	0.863	1.069	1.333	1.7396	2.1098	2.567	2.8982
18	0.862	1.067	1.330	1.7341	2.1009	2.552	2.8784
19	0.861	1.066	1.328	1.7291	2.0930	2.539	2.8609
20	0.860	1.064	1.325	1.7247	2.0860	2.528	2.8453
21	0.859	1.063	1.323	1.7207	2.0796	2.518	2.8314
22	0.858	1.061	1.321	1.7171	2.0739	2.508	2.8188
23	0.858	1.060	1.319	1.7139	2.0687	2.500	2.8073
24	0.857	1.059	1.318	1.7109	2.0639	2.492	2.7969
25	0.856	1.058	1.316	1.7081	2.0595	2.485	2.7874
26	0.856	1.058	1.315	1.7056	2.0555	2.479	2.7787
27	0.855	1.057	1.314	1.7033	2.0518	2.473	2.7707
28	0.855	1.056	1.313	1.7011	2.0484	2.467	2.7633
29	0.854	1.055	1.311	1.6991	2.0452	2.462	2.7564
30	0.854	1.055	1.310	1.6973	2.0423	2.457	2.7500
31	0.8535	1.0541	1.3095	1.6955	2.0395	2.453	2.7441
32	0.8531	1.0536	1.3086	1.6939	2.0370	2.449	2.7385
33	0.8527	1.0531	1.3078	1.6924	2.0345	2.445	2.7333
34	0.8524	1.0526	1.3070	1.6909	2.0323	2.441	2.7284

(cont.)

Table A.2 Upper Percentiles of Student t Distributions (*cont.*)

	α						
df	0.20	0.15	0.10	0.05	0.025	0.01	0.005
35	0.8521	1.0521	1.3062	1.6896	2.0301	2.438	2.7239
36	0.8518	1.0516	1.3055	1.6883	2.0281	2.434	2.7195
37	0.8515	1.0512	1.3049	1.6871	2.0262	2.431	2.7155
38	0.8512	1.0508	1.3042	1.6860	2.0244	2.428	2.7116
39	0.8510	1.0504	1.3037	1.6849	2.0227	2.426	2.7079
40	0.8507	1.0501	1.3031	1.6839	2.0211	2.423	2.7045
41	0.8505	1.0498	1.3026	1.6829	2.0196	2.421	2.7012
42	0.8503	1.0494	1.3020	1.6820	2.0181	2.418	2.6981
43	0.8501	1.0491	1.3016	1.6811	2.0167	2.416	2.6952
44	0.8499	1.0488	1.3011	1.6802	2.0154	2.414	2.6923
45	0.8497	1.0485	1.3007	1.6794	2.0141	2.412	2.6896
46	0.8495	1.0483	1.3002	1.6787	2.0129	2.410	2.6870
47	0.8494	1.0480	1.2998	1.6779	2.0118	2.408	2.6846
48	0.8492	1.0478	1.2994	1.6772	2.0106	2.406	2.6822
49	0.8490	1.0476	1.2991	1.6766	2.0096	2.405	2.6800
50	0.8489	1.0473	1.2987	1.6759	2.0086	2.403	2.6778
51	0.8448	1.0471	1.2984	1.6753	2.0077	2.402	2.6758
52	0.8486	1.0469	1.2981	1.6747	2.0067	2.400	2.6738
53	0.8485	1.0467	1.2978	1.6742	2.0058	2.399	2.6719
54	0.8484	1.0465	1.2975	1.6736	2.0049	2.397	2.6700
55	0.8483	1.0463	1.2972	1.6731	2.0041	2.396	2.6683
56	0.8481	1.0461	1.2969	1.6725	2.0033	2.395	2.6666
57	0.8480	1.0460	1.2967	1.6721	2.0025	2.393	2.6650
58	0.8479	1.0458	1.2964	1.6716	2.0017	2.392	2.6633
59	0.8478	1.0457	1.2962	1.6712	2.0010	2.391	2.6618
60	0.8477	1.0455	1.2959	1.6707	2.0003	2.390	2.6603
61	0.8476	1.0454	1.2957	1.6703	1.9997	2.389	2.6590
62	0.8475	1.0452	1.2954	1.6698	1.9990	2.388	2.6576
63	0.8474	1.0451	1.2952	1.6694	1.9984	2.387	2.6563
64	0.8473	1.0449	1.2950	1.6690	1.9977	2.386	2.6549
65	0.8472	1.0448	1.2948	1.6687	1.9972	2.385	2.6537
66	0.8471	1.0447	1.2945	1.6683	1.9966	2.384	2.6525
67	0.8471	1.0446	1.2944	1.6680	1.9961	2.383	2.6513
68	0.8470	1.0444	1.2942	1.6676	1.9955	2.382	2.6501
69	0.8469	1.0443	1.2940	1.6673	1.9950	2.381	2.6491
70	0.8468	1.0442	1.2938	1.6669	1.9945	2.381	2.6480
71	0.8468	1.0441	1.2936	1.6666	1.9940	2.380	2.6470
72	0.8467	1.0440	1.2934	1.6663	1.9935	2.379	2.6459
73	0.8466	1.0439	1.2933	1.6660	1.9931	2.378	2.6450
74	0.8465	1.0438	1.2931	1.6657	1.9926	2.378	2.6640
75	0.8465	1.0437	1.2930	1.6655	1.9922	2.377	2.6431
76	0.8464	1.0436	1.2928	1.6652	1.9917	2.376	2.6421
77	0.8464	1.0435	1.2927	1.6649	1.9913	2.376	2.6413
78	0.8463	1.0434	1.2925	1.6646	1.9909	2.375	2.6406
79	0.8463	1.0433	1.2924	1.6644	1.9905	2.374	2.6396

Table A.2 Upper Percentiles of Student t Distributions (*cont.*)

	α						
df	0.20	0.15	0.10	0.05	0.025	0.01	0.005
80	0.8462	1.0432	1.2922	1.6641	1.9901	2.374	2.6388
81	0.8461	1.0431	1.2921	1.6639	1.9897	2.373	2.6380
82	0.8460	1.0430	1.2920	1.6637	1.9893	2.372	2.6372
83	0.8460	1.0430	1.2919	1.6635	1.9890	2.372	2.6365
84	0.8459	1.0429	1.2917	1.6632	1.9886	2.371	2.6357
85	0.8459	1.0428	1.2916	1.6630	1.9883	2.371	2.6350
86	0.8458	1.0427	1.2915	1.6628	1.9880	2.370	2.6343
87	0.8458	1.0427	1.2914	1.6626	1.9877	2.370	2.6336
88	0.8457	1.0426	1.2913	1.6624	1.9873	2.369	2.6329
89	0.8457	1.0426	1.2912	1.6622	1.9870	2.369	2.6323
90	0.8457	1.0425	1.2910	1.6620	1.9867	2.368	2.6316
91	0.8457	1.0424	1.2909	1.6618	1.9864	2.368	2.6310
92	0.8456	1.0423	1.2908	1.6616	1.9861	2.367	2.6303
93	0.8456	1.0423	1.2907	1.6614	1.9859	2.367	2.6298
94	0.8455	1.0422	1.2906	1.6612	1.9856	2.366	2.6292
95	0.8455	1.0422	1.2905	1.6611	1.9853	2.366	2.6286
96	0.8454	1.0421	1.2904	1.6609	1.9850	2.366	2.6280
97	0.8454	1.0421	1.2904	1.6608	1.9848	2.365	2.6275
98	0.8453	1.0420	1.2903	1.6606	1.9845	2.365	2.6270
99	0.8453	1.0419	1.2902	1.6604	1.9843	2.364	2.6265
100	0.8452	1.0418	1.2901	1.6602	1.9840	2.364	2.6260
∞	0.84	1.04	1.28	1.64	1.96	2.33	2.58

Source: *Scientific Tables*, 6th ed. (Basel, Switzerland: J.R. Geigy, 1962), pp. 32–33.

Table A.3 Upper and Lower Percentiles of χ^2 Distributions

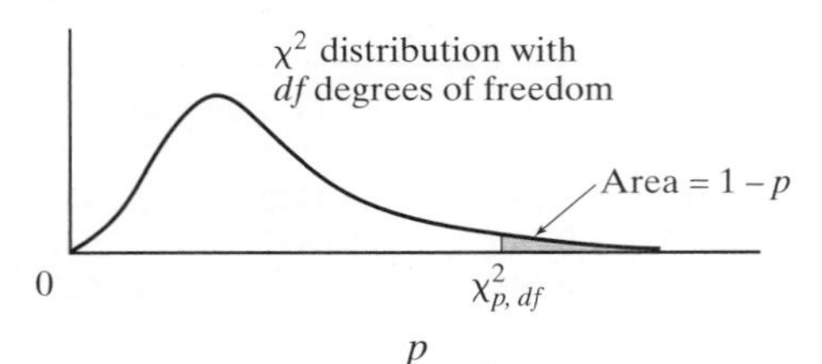

	p							
df	0.010	0.025	0.050	0.10	0.90	0.95	0.975	0.99
1	0.000157	0.000982	0.00393	0.0158	2.706	3.841	5.024	6.635
2	0.0201	0.0506	0.103	0.211	4.605	5.991	7.378	9.210
3	0.115	0.216	0.352	0.584	6.251	7.815	9.348	11.345
4	0.297	0.484	0.711	1.064	7.779	9.488	11.143	13.277
5	0.554	0.831	1.145	1.610	9.236	11.070	12.832	15.086
6	0.872	1.237	1.635	2.204	10.645	12.592	14.449	16.812
7	1.239	1.690	2.167	2.833	12.017	14.067	16.013	18.475
8	1.646	2.180	2.733	3.490	13.362	15.507	17.535	20.090
9	2.088	2.700	3.325	4.168	14.684	16.919	19.023	21.666
10	2.558	3.247	3.940	4.865	15.987	18.307	20.483	23.209
11	3.053	3.816	4.575	5.578	17.275	19.675	21.920	24.725
12	3.571	4.404	5.226	6.304	18.549	21.026	23.336	26.217
13	4.107	5.009	5.892	7.042	19.812	22.362	24.736	27.688
14	4.660	5.629	6.571	7.790	21.064	23.685	26.119	29.141
15	5.229	6.262	7.261	8.547	22.307	24.996	27.488	30.578
16	5.812	6.908	7.962	9.312	23.542	26.296	28.845	32.000
17	6.408	7.564	8.672	10.085	24.769	27.587	30.191	33.409
18	7.015	8.231	9.390	10.865	25.989	28.869	31.526	34.805
19	7.633	8.907	10.117	11.651	27.204	30.144	32.852	36.191
20	8.260	9.591	10.851	12.443	28.412	31.410	34.170	37.566
21	8.897	10.283	11.591	13.240	29.615	32.671	35.479	38.932
22	9.542	10.982	12.338	14.041	30.813	33.924	36.781	40.289
23	10.196	11.688	13.091	14.848	32.007	35.172	38.076	41.638
24	10.856	12.401	13.848	15.659	33.196	36.415	39.364	42.980
25	11.524	13.120	14.611	16.473	34.382	37.652	40.646	44.314
26	12.198	13.844	15.379	17.292	35.563	38.885	41.923	45.642
27	12.879	14.573	16.151	18.114	36.741	40.113	43.194	46.963
28	13.565	15.308	16.928	18.939	37.916	41.337	44.461	48.278
29	14.256	16.047	17.708	19.768	39.087	42.557	45.722	49.588
30	14.953	16.791	18.493	20.599	40.256	43.773	46.979	50.892
31	15.655	17.539	19.281	21.434	41.422	44.985	48.232	52.191
32	16.362	18.291	20.072	22.271	42.585	46.194	49.480	53.486
33	17.073	19.047	20.867	23.110	43.745	47.400	50.725	54.776
34	17.789	19.806	21.664	23.952	44.903	48.602	51.966	56.061

Table A.3 Upper and Lower Percentiles of χ^2 Distributions (*cont.*)

	p							
df	0.010	0.025	0.050	0.10	0.90	0.95	0.975	0.99
35	18.509	20.569	22.465	24.797	46.059	49.802	53.203	57.342
36	19.233	21.336	23.269	25.643	47.212	50.998	54.437	58.619
37	19.960	22.106	24.075	26.492	48.363	52.192	55.668	59.892
38	20.691	22.878	24.884	27.343	49.513	53.384	56.895	61.162
39	21.426	23.654	25.695	28.196	50.660	54.572	58.120	62.428
40	22.164	24.433	26.509	29.051	51.805	55.758	59.342	63.691
41	22.906	25.215	27.326	29.907	52.949	56.942	60.561	64.950
42	23.650	25.999	28.144	30.765	54.090	58.124	61.777	66.206
43	24.398	26.785	28.965	31.625	55.230	59.304	62.990	67.459
44	25.148	27.575	29.787	32.487	56.369	60.481	64.201	68.709
45	25.901	28.366	30.612	33.350	57.505	61.656	65.410	69.957
46	26.657	29.160	31.439	34.215	58.641	62.830	66.617	71.201
47	27.416	29.956	32.268	35.081	59.774	64.001	67.821	72.443
48	28.177	30.755	33.098	35.949	60.907	65.171	69.023	73.683
49	28.941	31.555	33.930	36.818	62.038	66.339	70.222	74.919
50	29.707	32.357	34.764	37.689	63.167	67.505	71.420	76.154

Source: *Scientific Tables*, 6th ed. (Basel, Switzerland: J.R. Geigy, 1962), p. 36.

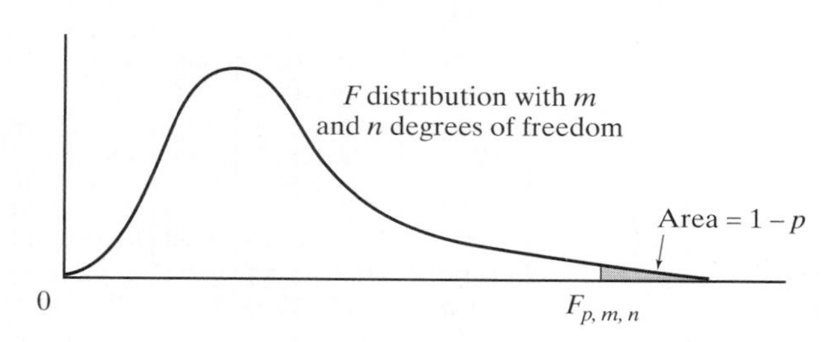

The figure above illustrates the percentiles of the F distributions shown in Table A.4. Table A.4 is used with permission from Wilfrid J. Dixon and Frank J. Massey, Jr., *Introduction to Statistical Analysis* 2nd ed. (New York: McGraw-Hill, 1957), pp. 389–404.

Table A.4 Percentiles of F Distributions

n	p \ m	1	2	3	4	5	6	7	8	9	10	11	12	p
1	.0005	$.0^{6}62$	$.0^{3}50$	$.0^{2}38$	$.0^{2}94$	.016	.022	.027	.032	.036	.039	.042	.045	.0005
	.001	$.0^{5}25$	$.0^{2}10$	$.0^{2}60$	.013	.021	.028	.034	.039	.044	.048	.051	.054	.001
	.005	$.0^{4}62$	$.0^{2}51$	.018	.032	.044	.054	.062	.068	.073	.078	.082	.085	.005
	.010	$.0^{3}25$	.010	.029	.047	.062	.073	.082	.089	.095	.100	.104	.107	.010
	.025	$.0^{2}15$	.026	.057	.082	.100	.113	.124	.132	.139	.144	.149	.153	.025
	.05	$.0^{2}62$	.054	.099	.130	.151	.167	.179	.188	.195	.201	.207	.211	.05
	.10	.025	.117	.181	.220	.246	.265	.279	.289	.298	.304	.310	.315	.10
	.25	.172	.389	.494	.553	.591	.617	.637	.650	.661	.670	.680	.684	.25
	.50	1.00	1.50	1.71	1.82	1.89	1.94	1.98	2.00	2.03	2.04	2.05	2.07	.50
	.75	5.83	7.50	8.20	8.58	8.82	8.98	9.10	9.19	9.26	9.32	9.36	9.41	.75
	.90	39.9	49.5	53.6	55.8	57.2	58.2	58.9	59.4	59.9	60.2	60.5	60.7	.90
	.95	161	200	216	225	230	234	237	239	241	242	243	244	.95
	.975	648	800	864	900	922	937	948	957	963	969	973	977	.975
	.99	405^{1}	500^{1}	540^{1}	562^{1}	576^{1}	586^{1}	593^{1}	598^{1}	602^{1}	606^{1}	608^{1}	611^{1}	.99
	.995	162^{2}	200^{2}	216^{2}	225^{2}	231^{2}	234^{2}	237^{2}	239^{2}	241^{2}	242^{2}	243^{2}	244^{2}	.995
	.999	406^{3}	500^{3}	540^{3}	562^{3}	576^{3}	586^{3}	593^{3}	598^{3}	602^{3}	606^{3}	609^{3}	611^{3}	.999
	.9995	162^{4}	200^{4}	216^{4}	225^{4}	231^{4}	234^{4}	237^{4}	239^{4}	241^{4}	242^{4}	243^{4}	244^{4}	.9995
2	.0005	$.0^{6}50$	$.0^{3}50$	$.0^{2}42$	.011	.020	.029	.037	.044	.050	.056	.061	.065	.0005
	.001	$.0^{5}20$	$.0^{2}10$	$.0^{2}68$	.016	.027	.037	.046	.054	.061	.067	.072	.077	.001
	.005	$.0^{4}50$	$.0^{2}50$	.020	.038	.055	.069	.081	.091	.099	.106	.112	.118	.005
	.01	$.0^{3}20$	.010	.032	.056	.075	.092	.105	.116	.125	.132	.139	.144	.01
	.025	$.0^{2}13$	.026	.062	.094	.119	.138	.153	.165	.175	.183	.190	.196	.025
	.05	$.0^{2}50$	.053	.105	.144	.173	.194	.211	.224	.235	.244	.251	.257	.05
	.10	.020	.111	.183	.231	.265	.289	.307	.321	.333	.342	.350	.356	.10
	.25	.133	.333	.439	.500	.540	.568	.588	.604	.616	.626	.633	.641	.25
	.50	.667	1.00	1.13	1.21	1.25	1.28	1.30	1.32	1.33	1.34	1.35	1.36	.50
	.75	2.57	3.00	3.15	3.23	3.28	3.31	3.34	3.35	3.37	3.38	3.39	3.39	.75
	.90	8.53	9.00	9.16	9.24	9.29	9.33	9.35	9.37	9.38	9.39	9.40	9.41	.90
	.95	18.5	19.0	19.2	19.2	19.3	19.3	19.4	19.4	19.4	19.4	19.4	19.4	.95
	.975	38.5	39.0	39.2	39.2	39.3	39.3	39.4	39.4	39.4	39.4	39.4	39.4	.975
	.99	98.5	99.0	99.2	99.2	99.3	99.3	99.4	99.4	99.4	99.4	99.4	99.4	.99
	.995	198	199	199	199	199	199	199	199	199	199	199	199	.995
	.999	998	999	999	999	999	999	999	999	999	999	999	999	.999
	.9995	200^{1}	200^{1}	200^{1}	200^{1}	200^{1}	200^{1}	200^{1}	200^{1}	200^{1}	200^{1}	200^{1}	200^{1}	.9995
3	.0005	$.0^{6}46$	$.0^{3}50$	$.0^{2}44$	.012	.023	.033	.043	.052	.060	.067	.074	.079	.0005
	.001	$.0^{5}19$	$.0^{2}10$	$.0^{2}71$	.018	.030	.042	.053	.063	.072	.079	.086	.093	.001
	.005	$.0^{4}46$	$.0^{2}50$	.021	.041	.060	.077	.092	.104	.115	.124	.132	.138	.005
	.01	$.0^{3}19$	.010	.034	.060	.083	.102	.118	.132	.143	.153	.161	.168	.01
	.025	$.0^{2}12$	.026	.065	.100	.129	.152	.170	.185	.197	.207	.216	.224	.025
	.05	$.0^{2}46$	.052	.108	.152	.185	.210	.230	.246	.259	.270	.279	.287	.05
	.10	.019	.109	.185	.239	.276	.304	.325	.342	.356	.367	.376	.384	.10
	.25	.122	.317	.424	.489	.531	.561	.582	.600	.613	.624	.633	.641	.25
	.50	.585	.881	1.00	1.06	1.10	1.13	1.15	1.16	1.17	1.18	1.19	1.20	.50
	.75	2.02	2.28	2.36	2.39	2.41	2.42	2.43	2.44	2.44	2.44	2.45	2.45	.75
	.90	5.54	5.46	5.39	5.34	5.31	5.28	5.27	5.25	5.24	5.23	5.22	5.22	.90
	.95	10.1	9.55	9.28	9.12	9.01	8.94	8.89	8.85	8.81	8.79	8.76	8.74	.95
	.075	17.4	16.0	15.4	15.1	14.9	14.7	14.6	14.5	14.5	14.4	14.4	14.3	.975
	.99	34.1	30.8	29.5	28.7	28.2	27.9	27.7	27.5	27.3	27.2	27.1	27.1	.99
	.995	55.6	49.8	47.5	46.2	45.4	44.8	44.4	44.1	43.9	43.7	43.5	43.4	.995
	.999	167	149	141	137	135	133	132	131	130	129	129	128	.999
	.9995	266	237	225	218	214	211	209	208	207	206	204	204	.9995

Read $.0^{3}56$ as .00056, 200^{1} as 2000, 162^{4} as 1620000, etc.

Table A.4 Percentiles of F Distributions (*cont.*)

p \ m	15	20	24	30	40	50	60	100	120	200	500	∞	p	n
.0005	.051	.058	062	.066	.069	.072	.074	.077	.078	.080	.081	.083	.0005	1
.001	.060	.067	.071	.075	.079	.082	.084	.087	.088	.089	.091	.092	.001	
.005	.093	.101	.105	.109	.113	.116	.118	.121	.122	.124	.126	.127	.005	
.01	.115	.124	.128	.132	.137	.139	.141	.145	.146	.148	.150	.151	.01	
.025	.161	.170	.175	.180	.184	.187	.189	.193	.194	.196	.198	.199	.025	
.05	.220	.230	.235	.240	.245	.248	.250	.254	.255	.257	.259	.261	.05	
.10	.325	.336	.342	.347	.353	.356	.358	.362	.364	.366	.368	.370	.10	
.25	.698	.712	.719	.727	.734	.738	.741	.747	.749	.752	.754	.756	.25	
.50	2.09	2.12	2.13	2.15	2.16	2.17	2.17	2.18	2.18	2.19	2.19	2.20	.50	
.75	9.49	9.58	9.63	9.67	9.71	9.74	9.76	9.78	9.80	9.82	9.84	9.85	.75	
.90	61.2	61.7	62.0	62.3	62.5	62.7	62.8	63.0	63.1	63.2	63.3	63.3	.90	
.95	246	248	249	250	251	252	252	253	253	254	254	254	.95	
.975	985	993	997	100^{1}	101^{1}	101^{1}	101^{1}	101^{1}	101^{1}	102^{1}	102^{1}	102^{1}	.975	
.99	616^{1}	621^{1}	623^{1}	626^{1}	629^{1}	630^{1}	631^{1}	633^{1}	634^{1}	635^{1}	636^{1}	637^{1}	.99	
.995	246^{2}	248^{2}	249^{2}	250^{2}	251^{2}	252^{2}	253^{2}	253^{2}	254^{2}	254^{2}	254^{2}	255^{2}	.995	
.999	616^{3}	621^{3}	623^{3}	626^{3}	629^{3}	630^{3}	631^{3}	633^{3}	634^{3}	635^{3}	636^{3}	637^{3}	.999	
.9995	246^{4}	248^{4}	249^{4}	250^{4}	251^{4}	252^{4}	252^{4}	253^{4}	253^{4}	253^{4}	254^{4}	254^{4}	.9995	
.0005	.076	.088	.094	.101	.108	.113	.116	.122	.124	.127	.130	.132	.0005	2
.001	.088	.100	.107	.114	.121	.126	.129	.135	.137	.140	.143	.145	.001	
.005	.130	.143	.150	.157	.165	.169	.173	.179	.181	.184	.187	.189	.005	
.01	.157	.171	.178	.186	.193	.198	.201	.207	.209	.212	.215	.217	.01	
.025	.210	.224	.232	.239	.247	.251	.255	.261	.263	.266	.269	.271	.025	
.05	.272	.286	.294	.302	.309	.314	.317	.324	.326	.329	.332	.334	.05	
.10	.371	.386	.394	.402	.410	.415	.418	.424	.426	.429	.433	.434	.10	
.25	.657	.672	.680	.689	.697	.702	.705	.711	.713	.716	.719	.721	.25	
.50	1.38	1.39	1.40	1.41	1.42	1.42	1.43	1.43	1.43	1.44	1.44	1.44	.50	
.75	3.41	3.43	3.43	3.44	3.45	3.45	3.46	3.47	3.47	3.48	3.48	3.48	.75	
.90	9.42	9.44	9.45	9.46	9.47	9.47	9.47	9.48	9.48	9.49	9.49	9.49	.90	
.95	19.4	19.4	19.5	19.5	19.5	19.5	19.5	19.5	19.5	19.5	19.5	19.5	.95	
.975	39.4	39.4	39.5	39.5	39.5	39.5	39.5	39.5	39.5	39.5	39.5	39.5	.975	
.99	99.4	99.4	99.5	99.5	99.5	99.5	99.5	99.5	99.5	99.5	99.5	99.5	.99	
.995	199	199	199	199	199	199	199	199	199	199	199	200	.995	
.999	999	999	999	999	999	999	999	999	999	999	999	999	.999	
.9995	200^{1}	200^{1}	200^{1}	200^{1}	200^{1}	200^{1}	200^{1}	200^{1}	200^{1}	200^{1}	200^{1}	200^{1}	.9995	
.0005	.093	.109	.117	.127	.136	.143	.147	.156	.158	.162	.166	.169	.0005	3
.001	.107	.123	.132	.142	.152	.158	.162	.171	.173	.177	.181	.184	.001	
.005	.154	.172	.181	.191	.201	.207	.211	.220	.222	.227	.231	.234	.005	
.01	.185	.203	.212	.222	.232	.238	.242	.251	.253	.258	.262	.264	.01	
.025	.241	.259	.269	.279	.289	.295	.299	.308	.310	.314	.318	.321	.025	
.05	.304	.323	.332	.342	.352	.358	.363	.370	.373	.377	.382	.384	.05	
.10	.402	.420	.430	.439	.449	.455	.459	.467	.469	.474	.476	.480	.10	
.25	.658	.675	.684	.693	.702	.708	.711	.719	.721	.724	.728	.730	.25	
.50	1.21	1.23	1.23	1.24	1.25	1.25	1.25	1.26	1.26	1.26	1.27	1.27	.50	
.75	2.46	2.46	2.46	2.47	2.47	2.47	2.47	2.47	2.47	2.47	2.47	2.47	.75	
.90	5.20	5.18	5.18	5.17	5.16	5.15	5.15	5.14	5.14	5.14	5.14	5.13	.90	
.95	8.70	8.66	8.63	8.62	8.59	8.58	8.57	8.55	8.55	8.54	8.53	8.53	.95	
.975	14.3	14.2	14.1	14.1	14.0	14.0	14.0	14.0	13.9	13.9	13.9	13.9	.975	
.99	26.9	26.7	26.6	26.5	26.4	26.4	26.3	26.2	26.2	26.2	26.1	26.1	.99	
.995	43.1	42.8	42.6	42.5	42.3	42.2	42.1	42.0	42.0	41.9	41.9	41.8	.995	
.999	127	126	126	125	125	125	124	124	124	124	124	123	.999	
.9995	203	201	200	199	199	198	198	197	197	197	196	196	.9995	

(*cont.*)

Table A.4 Percentiles of F Distributions (*cont.*)

n	p \ m	1	2	3	4	5	6	7	8	9	10	11	12	p
4	.0005	$.0^{6}44$	$.0^{3}50$	$.0^{2}46$	.013	.024	.036	.047	.057	.066	.075	.082	.089	.0005
	.001	$.0^{5}18$	$.0^{2}10$	$.0^{2}73$	.019	.032	.046	.058	.069	.079	.089	.097	.104	.001
	.005	$.0^{4}44$	$.0^{2}50$	.022	.043	.064	.083	.100	.114	.126	.137	.145	.153	.005
	.01	$.0^{3}18$	.010	.035	.063	.088	.109	.127	.143	.156	.167	.176	.185	.01
	.025	$.0^{2}11$	.026	.066	.104	.135	.161	.181	.198	.212	.224	.234	.243	.025
	05	$.0^{2}44$	.052	.110	.157	.193	.221	.243	.261	.275	.288	.298	.307	.05
	.10	.018	.108	.187	.243	.284	.314	.338	.356	.371	.384	.394	.403	.10
	.25	.117	.309	.418	.484	.528	.560	.583	.601	.615	.627	.637	.645	.25
	.50	.549	.828	.941	1.00	1.04	1.06	1.08	1.09	1.10	1.11	1.12	1.13	.50
	.75	1.81	2.00	2.05	2.06	2.07	2.08	2.08	2.08	2.08	2.08	2.08	2.08	.75
	.90	4.54	4.32	4.19	4.11	4.05	4.01	3.98	3.95	3.94	3.92	3.91	3.90	.90
	.95	7.71	6.94	6.59	6.39	6.26	6.16	6.09	6.04	6.00	5.96	5.94	5.91	.95
	.975	12.2	10.6	9.98	9.60	9.36	9.20	9.07	8.98	8.90	8.84	8.79	8.75	.975
	.99	21.2	18.0	16.7	16.0	15.5	15.2	15.0	14.8	14.7	14.5	14.4	14.4	.99
	.995	31.3	26.3	24.3	23.2	22.5	22.0	21.6	21.4	21.1	21.0	20.8	20.7	.995
	.999	74.1	61.2	56.2	53.4	51.7	50.5	49.7	49.0	48.5	48.0	47.7	47.4	.999
	.9995	106	87.4	80.1	76.1	73.6	71.9	70.6	69.7	68.9	68.3	67.8	67.4	.9995
5	.0005	$.0^{6}43$	$.0^{3}50$	$.0^{2}47$	.014	.025	.038	.050	.061	.070	.081	.089	.096	.0005
	.001	$.0^{5}17$	$.0^{2}10$	$.0^{2}75$	.019	.034	.048	.062	.074	.085	.095	.104	.112	.001
	.005	$.0^{4}43$	$.0^{2}50$	.022	.045	.067	.087	.105	.120	.134	.146	.156	.165	.005
	.01	$.0^{3}17$	.010	.035	.064	.091	.114	.134	.151	.165	.177	.188	.197	.01
	.025	$.0^{2}11$	.025	.067	.107	.140	.167	.189	.208	.223	.236	.248	.257	.025
	.05	$.0^{2}43$	.052	.111	.160	.198	.228	.252	.271	.287	.301	.313	.322	.05
	.10	.017	.108	.188	.247	.290	.322	.347	.367	.383	.397	.408	.418	.10
	.25	.113	.305	.415	.483	.528	.560	.584	.604	.618	.631	.641	.650	.25
	.50	.528	.799	.907	.965	1.00	1.02	1.04	1.05	1.06	1.07	1.08	1.09	.50
	.75	1.69	1.85	1.88	1.89	1.89	1.89	1.89	1.89	1.89	1.89	1.89	1.89	.75
	.90	4.06	3.78	3.62	3.52	3.45	3.40	3.37	3.34	3.32	3.30	3.28	3.27	.90
	.95	6.61	5.79	5.41	5.19	5.05	4.95	4.88	4.82	4.77	4.74	4.71	4.68	.95
	.975	10.0	8.43	7.76	7.39	7.15	6.98	6.85	6.76	6.68	6.62	6.57	6.52	.975
	.99	16.3	13.3	12.1	11.4	11.0	10.7	10.5	10.3	10.2	10.1	9.96	9.89	.99
	.995	22.8	18.3	16.5	15.6	14.9	14.5	14.2	14.0	13.8	13.6	13.5	13.4	.995
	.999	47.2	37.1	33.2	31.1	29.7	28.8	28.2	27.6	27.2	26.9	26.6	26.4	.999
	.9995	63.6	49.8	44.4	41.5	39.7	38.5	37.6	36.9	36.4	35.9	35.6	35.2	.9995
6	.0005	$.0^{6}43$	$.0^{3}50$	$.0^{2}47$	.014	.026	.039	.052	.064	.075	.085	.094	.103	.0005
	.001	$.0^{5}17$	$.0^{2}10$	$.0^{2}75$	.020	.035	.050	.064	.078	.090	.101	.111	.119	.001
	.005	$.0^{4}43$	$.0^{2}50$	.022	.045	.069	.090	.109	.126	.140	.153	.164	.174	.005
	.01	$.0^{3}17$	.010	.036	.066	.094	.118	.139	.157	.172	.186	.197	.207	.01
	.025	$.0^{2}11$	.025	.068	.109	.143	.172	.195	.215	.231	.246	.258	.268	.025
	.05	$.0^{2}43$	.052	.112	.162	.202	.233	.259	.279	.296	.311	.324	.334	.05
	.10	.017	.107	.189	.249	.294	.327	.354	.375	.392	.406	.418	.429	.10
	.25	.111	.302	.413	.481	.524	.561	.586	.606	.622	.635	.645	.654	.25
	.50	.515	.780	.886	.942	.977	1.00	1.02	1.03	1.04	1.05	1.05	1.06	.50
	.75	1.62	1.76	1.78	1.79	1.79	1.78	1.78	1.78	1.77	1.77	1.77	1.77	.75
	.90	3.78	3.46	3.29	3.18	3.11	3.05	3.01	2.98	2.96	2.94	2.92	2.90	.90
	.95	5.99	5.14	4.76	4.53	4.39	4.28	4.21	4.15	4.10	4.06	4.03	4.00	.95
	.975	8.81	7.26	6.60	6.23	5.99	5.82	5.70	5.60	5.52	5.46	5.41	5.37	.975
	.99	13.7	10.9	9.78	9.15	8.75	8.47	8.26	8.10	7.98	7.87	7.79	7.72	.99
	.995	18.6	14.5	12.9	12.0	11.5	11.1	10.8	10.6	10.4	10.2	10.1	10.0	.995
	.999	35.5	27.0	23.7	21.9	20.8	20.0	19.5	19.0	18.7	18.4	18.2	18.0	.999
	.9995	46.1	34.8	30.4	28.1	26.6	25.6	24.9	24.3	23.9	23.5	23.2	23.0	.9995

Table A.4 Percentiles of F Distributions (*cont.*)

p \ m	15	20	24	30	40	50	60	100	120	200	500	∞	p	n
.0005	.105	.125	.135	.147	.159	.166	.172	.183	.186	.191	.196	.200	.0005	**4**
.001	.121	.141	.152	.163	.176	.183	.188	.200	.202	.208	.213	.217	.001	
.005	.172	.193	.204	.216	.229	.237	.242	.253	.255	.260	.266	.269	.005	
.01	.204	.226	.237	.249	.261	.269	.274	.285	.287	.293	.298	.301	.01	
.025	.263	.284	.296	.308	.320	.327	.332	.342	.346	.351	.356	.359	.025	
.05	.327	.349	.360	.372	.384	.391	.396	.407	.409	.413	.418	.422	.05	
.10	.424	.445	.456	.467	.478	.485	.490	.500	.502	.508	.510	.514	.10	
.25	.664	.683	.692	.702	.712	.718	.722	.731	.733	.737	.740	.743	.25	
.50	1.14	1.15	1.16	1.16	1.17	1.18	1.18	1.18	1.18	1.19	1.19	1.19	.50	
.75	2.08	2.08	2.08	2.08	2.08	2.08	2.08	2.08	2.08	2.08	2.08	2.08	.75	
.90	3.87	3.84	3.83	3.82	3.80	3.80	3.79	3.78	3.78	3.77	3.76	3.76	.90	
.95	5.86	5.80	5.77	5.75	5.72	5.70	5.69	5.66	5.66	5.65	5.64	5.63	.95	
.975	8.66	8.56	8.51	8.46	8.41	8.38	8.36	8.32	8.31	8.29	8.27	8.26	.975	
.99	14.2	14.0	13.9	13.8	13.7	13.7	13.7	13.6	13.6	13.5	13.5	13.5	.99	
.995	20.4	20.2	20.0	19.9	19.8	19.7	19.6	19.5	19.5	19.4	19.4	19.3	.995	
.999	46.8	46.1	45.8	45.4	45.1	44.9	44.7	44.5	44.4	44.3	44.1	44.0	.999	
.9995	66.5	65.5	65.1	64.6	64.1	63.8	63.6	63.2	63.1	62.9	62.7	62.6	.9995	
.0005	.115	.137	.150	.163	.177	.186	.192	.205	.209	.216	.222	.226	.0005	**5**
.001	.132	.155	.167	.181	.195	.204	.210	.223	.227	.233	.239	.244	.001	
.005	.186	.210	.223	.237	.251	.260	.266	.279	.282	.288	.294	.299	.005	
.01	.219	.244	.257	.270	.285	.293	.299	.312	.315	.322	.328	.331	.01	
.025	.280	.304	.317	.330	.344	.353	.359	.370	.374	.380	.386	.390	.025	
.05	.345	.369	.382	.395	.408	.417	.422	.432	.437	.442	.448	.452	.05	
.10	.440	.463	.476	.488	.501	.508	.514	.524	.527	.532	.538	.541	.10	
.25	.669	.690	.700	.711	.722	.728	.732	.741	.743	.748	.752	.755	.25	
.50	1.10	1.11	1.12	1.12	1.13	1.13	1.14	1.14	1.14	1.15	1.15	1.15	.50	
.75	1.89	1.88	1.88	1.88	1.88	1.88	1.87	1.87	1.87	1.87	1.87	1.87	.75	
.90	3.24	3.21	3.19	3.17	3.16	3.15	3.14	3.13	3.12	3.12	3.11	3.10	.90	
.95	4.62	4.56	4.53	4.50	4.46	4.44	4.43	4.41	4.40	4.39	4.37	4.36	.95	
.975	6.43	6.33	6.28	6.23	6.18	6.14	6.12	6.08	6.07	6.05	6.03	6.02	.975	
.99	9.72	9.55	9.47	9.38	9.29	9.24	9.20	9.13	9.11	9.08	9.04	9.02	.99	
.995	13.1	12.9	12.8	12.7	12.5	12.5	12.4	12.3	12.3	12.2	12.2	12.1	.995	
.999	25.9	25.4	25.1	24.9	24.6	24.4	24.3	24.1	24.1	23.9	23.8	23.8	.999	
.9995	34.6	33.9	33.5	33.1	32.7	32.5	32.3	32.1	32.0	31.8	31.7	31.6	.9995	
.0005	.123	.148	.162	.177	.193	.203	.210	.225	.229	.236	.244	.249	.0005	**6**
.001	.141	.166	.180	.195	.211	.222	.229	.243	.247	.255	.262	.267	.001	
.005	.197	.224	.238	.253	.269	.279	.286	.301	.304	.312	.318	.324	.005	
.01	.232	.258	.273	.288	.304	.313	.321	.334	.338	.346	.352	.357	.01	
.025	.293	.320	.334	.349	.364	.375	.381	.394	.398	.405	.412	.415	.025	
.05	.358	.385	.399	.413	.428	.437	.444	.457	.460	.467	.472	.476	.05	
.10	.453	.478	.491	.505	.519	.526	.533	.546	.548	.556	.559	.564	.10	
.25	.675	.696	.707	.718	.729	.736	.741	.751	.753	.758	.762	.765	.25	
.50	1.07	1.08	1.09	1.10	1.10	1.11	1.11	1.11	1.12	1.12	1.12	1.12	.50	
.75	1.76	1.76	1.75	1.75	1.75	1.75	1.74	1.74	1.74	1.74	1.74	1.74	.75	
.90	2.87	2.84	2.82	2.80	2.78	2.77	2.76	2.75	2.74	2.73	2.73	2.72	.90	
.95	3.94	3.87	3.84	3.81	3.77	3.75	3.74	3.71	3.70	3.69	3.68	3.67	.95	
.975	5.27	5.17	5.12	5.07	5.01	4.98	4.96	4.92	4.90	4.88	4.86	4.85	.975	
.99	7.56	7.40	7.31	7.23	7.14	7.09	7.06	6.99	6.97	6.93	6.90	6.88	.99	
.995	9.81	9.59	9.47	9.36	9.24	9.17	9.12	9.03	9.00	8.95	8.91	8.88	.995	
.999	17.6	17.1	16.9	16.7	16.4	16.3	16.2	16.0	16.0	15.9	15.8	15.7	.999	
.9995	22.4	21.9	21.7	21.4	21.1	20.9	20.7	20.5	20.4	20.3	20.2	20.1	.9995	

(*cont.*)

Table A.4 Percentiles of F Distributions (*cont.*)

n	p \ m	1	2	3	4	5	6	7	8	9	10	11	12	p
7	.0005	$.0^{6}42$	$.0^{3}50$	$.0^{2}48$	.014	.027	.040	.053	.066	.078	.088	.099	.108	.0005
	.001	$.0^{5}17$	$.0^{2}10$	$.0^{2}76$	.020	.035	.051	.067	.081	.093	.105	.115	.125	.001
	.005	$.0^{4}42$	$.0^{2}50$	.023	.046	.070	.093	.113	.130	.145	.159	.171	.181	.005
	.01	$.0^{3}17$	.010	.036	.067	.096	.121	.143	.162	.178	.192	.205	.216	.01
	.025	$.0^{2}10$	.025	.068	.110	.146	.176	.200	.221	.238	.253	.266	.277	.025
	.05	$.0^{2}42$	.052	.113	.164	.205	.238	.264	.286	.304	.319	.332	.343	.05
	.10	.017	.107	.190	.251	.297	.332	.359	.381	.399	.414	.427	.438	.10
	.25	.110	.300	.412	.481	.528	.562	.588	.608	.624	.637	.649	.658	.25
	.50	.506	.767	.871	.926	.960	.983	1.00	1.01	1.02	1.03	1.04	1.04	.50
	.75	1.57	1.70	1.72	1.72	1.71	1.71	1.70	1.70	1.69	1.69	1.69	1.68	.75
	.90	3.59	3.26	3.07	2.96	2.88	2.83	2.78	2.75	2.72	2.70	2.68	2.67	.90
	.95	5.59	4.74	4.35	4.12	3.97	3.87	3.79	3.73	3.68	3.64	3.60	3.57	.95
	.975	8.07	6.54	5.89	5.52	5.29	5.12	4.99	4.90	4.82	4.76	4.71	4.67	.975
	.99	12.2	9.55	8.45	7.85	7.46	7.19	6.99	6.84	6.72	6.62	6.54	6.47	.99
	.995	16.2	12.4	10.9	10.0	9.52	9.16	8.89	8.68	8.51	8.38	8.27	8.18	.995
	.999	29.2	21.7	18.8	17.2	16.2	15.5	15.0	14.6	14.3	14.1	13.9	13.7	.999
	.9995	37.0	27.2	23.5	21.4	20.2	19.3	18.7	18.2	17.8	17.5	17.2	17.0	.9995
8	.0005	$.0^{6}42$	$.0^{3}50$	$.0^{2}48$	.014	.027	.041	.055	.068	.081	.092	.102	.112	.0005
	.001	$.0^{5}17$	$.0^{2}10$	$.0^{2}76$	.020	.036	.053	.068	.083	.096	.109	.120	.130	.001
	.005	$.0^{4}42$	$.0^{2}50$	.027	.047	.072	.095	.115	.133	.149	.164	.176	.187	.005
	.01	$.0^{3}17$	.010	.036	.068	.097	.123	.146	.166	.183	.198	.211	.222	.01
	.025	$.0^{2}10$	.025	.069	.111	.148	.179	.204	.226	.244	.259	.273	.285	.025
	.05	$.0^{2}42$	.052	.113	.166	.208	.241	.268	.291	.310	.326	.339	.351	.05
	.10	.017	.107	.190	.253	.299	.335	.363	.386	.405	.421	.435	.445	.10
	.25	.109	.298	.411	.481	.529	.563	.589	.610	.627	.640	.654	.661	.25
	.50	.499	.757	.860	.915	.948	.971	.988	1.00	1.01	1.02	1.02	1.03	.50
	.75	1.54	1.66	1.67	1.66	1.66	1.65	1.64	1.64	1.64	1.63	1.63	1.62	.75
	.90	3.46	3.11	2.92	2.81	2.73	2.67	2.62	2.59	2.56	2.54	2.52	2.50	.90
	.95	5.32	4.46	4.07	3.84	3.69	3.58	3.50	3.44	3.39	3.35	3.31	3.28	.95
	.975	7.57	6.06	5.42	5.05	4.82	4.65	4.53	4.43	4.36	4.30	4.24	4.20	.975
	.99	11.3	8.65	7.59	7.01	6.63	6.37	6.18	6.03	5.91	5.81	5.73	5.67	.99
	.995	14.7	11.0	9.60	8.81	8.30	7.95	7.69	7.50	7.34	7.21	7.10	7.01	.995
	.999	25.4	18.5	15.8	14.4	13.5	12.9	12.4	12.0	11.8	11.5	11.4	11.2	.999
	.9995	31.6	22.8	19.4	17.6	16.4	15.7	15.1	14.6	14.3	14.0	13.8	13.6	.9995
9	.0005	$.0^{6}41$	$.0^{3}50$	$.0^{2}48$	.015	.027	.042	.056	.070	.083	.094	.105	.115	.0005
	.001	$.0^{5}17$	$.0^{2}10$	$.0^{2}77$	.021	.037	.054	.070	.085	.099	.112	.123	.134	.001
	.005	$.0^{4}42$	$.0^{2}50$	.023	.047	.073	.096	.117	.136	.153	.168	.181	.192	.005
	.01	$.0^{3}17$	.010	.037	.068	.098	.125	.149	.169	.187	.202	.216	.228	.01
	.025	$.0^{2}10$	.025	.069	.112	.150	.181	.207	.230	.248	.265	.279	.291	.025
	.05	$.0^{2}40$	.052	.113	.167	.210	.244	.272	.296	.315	.331	.345	.358	.05
	.10	.017	.107	.191	.254	.302	.338	.367	.390	.410	.426	.441	.452	.10
	.25	.108	.297	.410	.480	.529	.564	.591	.612	.629	.643	.654	.664	.25
	.50	.494	.749	.852	.906	.939	.962	.978	.990	1.00	1.01	1.01	1.02	.50
	.75	1.51	1.62	1.63	1.63	1.62	1.61	1.60	1.60	1.59	1.59	1.58	1.58	.75
	.90	3.36	3.01	2.81	2.69	2.61	2.55	2.51	2.47	2.44	2.42	2.40	2.38	.90
	.95	5.12	4.26	3.86	3.63	3.48	3.37	3.29	3.23	3.18	3.14	3.10	3.07	.95
	.975	7.21	5.71	5.08	4.72	4.48	4.32	4.20	4.10	4.03	3.96	3.91	3.87	.975
	.99	10.6	8.02	6.99	6.42	6.06	5.80	5.61	5.47	5.35	5.26	5.18	5.11	.99
	.995	13.6	10.1	8.72	7.96	7.47	7.13	6.88	6.69	6.54	6.42	6.31	6.23	.995
	.999	22.9	16.4	13.9	12.6	11.7	11.1	10.7	10.4	10.1	9.89	9.71	9.57	.999
	.9995	28.0	19.9	16.8	15.1	14.1	13.3	12.8	12.4	12.1	11.8	11.6	11.4	.9995

Table A.4 Percentiles of F Distributions (*cont.*)

p \ m	15	20	24	30	40	50	60	100	120	200	500	∞	p	n
.0005	.130	.157	.172	.188	.206	.217	.225	.242	.246	.255	.263	.268	.0005	7
.001	.148	.176	.191	.208	.225	.237	.245	.261	.266	.274	.282	.288	.001	
.005	.206	.235	.251	.267	.285	.296	.304	.319	.324	.332	.340	.345	.005	
.01	.241	.270	.286	.303	.320	.331	.339	.355	.358	.366	.373	.379	.01	
.025	.304	.333	.348	.364	.381	.392	.399	.413	.418	.426	.433	.437	.025	
.05	.369	.398	.413	.428	.445	.455	.461	.476	.479	.485	.493	.498	.05	
.10	.463	.491	.504	.519	.534	.543	.550	.562	.566	.571	.578	.582	.10	
.25	.679	.702	.713	.725	.737	.745	.749	.760	.762	.767	.772	.775	.25	
.50	1.05	1.07	1.07	1.08	1.08	1.09	1.09	1.10	1.10	1.10	1.10	1.10	.50	
.75	1.68	1.67	1.67	1.66	1.66	1.66	1.65	1.65	1.65	1.65	1.65	1.65	.75	
.90	2.63	2.59	2.58	2.56	2.54	2.52	2.51	2.50	2.49	2.48	2.48	2.47	.90	
.95	3.51	3.44	3.41	3.38	3.34	3.32	3.30	3.27	3.27	3.25	3.24	3.23	.95	
.975	4.57	4.47	4.42	4.36	4.31	4.28	4.25	4.21	4.20	4.18	4.16	4.14	.975	
.99	6.31	6.16	6.07	5.99	5.91	5.86	5.82	5.75	5.74	5.70	5.67	5.65	.99	
.995	7.97	7.75	7.65	7.53	7.42	7.35	7.31	7.22	7.19	7.15	7.10	7.08	.995	
.999	13.3	12.9	12.7	12.5	12.3	12.2	12.1	11.9	11.9	11.8	11.7	11.7	.999	
.9995	16.5	16.0	15.7	15.5	15.2	15.1	15.0	14.7	14.7	14.6	14.5	14.4	.9995	
.0005	.136	.164	.181	.198	.218	.230	.239	.257	.262	.271	.281	.287	.0005	8
.001	.155	.184	.200	.218	.238	.250	.259	.277	.282	.292	.300	.306	.001	
.005	.214	.244	.261	.279	.299	.311	.319	.337	.341	.351	.358	.364	.005	
.01	.250	.281	.297	.315	.334	.346	.354	.372	.376	.385	.392	.398	.01	
.025	.313	.343	.360	.377	.395	.407	.415	.431	.435	.442	.450	.456	.025	
.05	.379	.409	.425	.441	.459	.469	.477	.493	.496	.505	.510	.516	.05	
.10	.472	.500	.515	.531	.547	.556	.563	.578	.581	.588	.595	.599	.10	
.25	.684	.707	.718	.730	.743	.751	.756	.767	.769	.775	.780	.783	.25	
.50	1.04	1.05	1.06	1.07	1.07	1.07	1.08	1.08	1.08	1.09	1.09	1.09	.50	
.75	1.62	1.61	1.60	1.60	1.59	1.59	1.59	1.58	1.58	1.58	1.58	1.58	.75	
.90	2.46	2.42	2.40	2.38	2.36	2.35	2.34	2.32	2.32	2.31	2.30	2.29	.90	
.95	3.22	3.15	3.12	3.08	3.04	3.02	3.01	2.97	2.97	2.95	2.94	2.93	.95	
.975	4.10	4.00	3.95	3.89	3.84	3.81	3.78	3.74	3.73	3.70	3.68	3.67	.975	
.99	5.52	5.36	5.28	5.20	5.12	5.07	5.03	4.96	4.95	4.91	4.88	4.86	.99	
.995	6.81	6.61	6.50	6.40	6.29	6.22	6.18	6.09	6.06	6.02	5.98	5.95	.995	
.999	10.8	10.5	10.3	10.1	9.92	9.80	9.73	9.57	9.54	9.46	9.39	9.34	.999	
.9995	13.1	12.7	12.5	12.2	12.0	11.8	11.8	11.6	11.5	11.4	11.4	11.3	.9995	
.0005	.141	.171	.188	.207	.228	.242	.251	.270	.276	.287	.297	.303	.0005	9
.001	.160	.191	.208	.228	.249	.262	.271	.291	.296	.307	.316	.323	.001	
.005	.220	.253	.271	.290	.310	.324	.332	.351	.356	.366	.376	.382	.005	
.01	.257	.289	.307	.326	.346	.358	.368	.386	.391	.400	.410	.415	.01	
.025	.320	.352	.370	.388	.408	.420	.428	.446	.450	.459	.467	.473	.025	
.05	.386	.418	.435	.452	.471	.483	.490	.508	.510	.518	.526	.532	.05	
.10	.479	.509	.525	.541	.558	.568	.575	.588	.594	.602	.610	.613	.10	
.25	.687	.711	.723	.736	.749	.757	.762	.773	.776	.782	.787	.791	.25	
.50	1.03	1.04	1.05	1.05	1.06	1.06	1.07	1.07	1.07	1.08	1.08	1.08	.50	
.75	1.57	1.56	1.56	1.55	1.55	1.54	1.54	1.53	1.53	1.53	1.53	1.53	.75	
.90	2.34	2.30	2.28	2.25	2.23	2.22	2.21	2.19	2.18	2.17	2.17	2.16	.90	
.95	3.01	2.94	2.90	2.86	2.83	2.80	2.79	2.76	2.75	2.73	2.72	2.71	.95	
.975	3.77	3.67	3.61	3.56	3.51	3.47	3.45	3.40	3.39	3.37	3.35	3.33	.975	
.99	4.96	4.81	4.73	4.65	4.57	4.52	4.48	4.42	4.40	4.36	4.33	4.31	.99	
.995	6.03	5.83	5.73	5.62	5.52	5.45	5.41	5.32	5.30	5.26	5.21	5.19	.995	
.999	9.24	8.90	8.72	8.55	8.37	8.26	8.19	8.04	8.00	7.93	7.86	7.81	.999	
.9995	11.0	10.6	10.4	10.2	9.94	9.80	9.71	9.53	9.49	9.40	9.32	9.26	.9995	

(*cont.*)

Table A.4 Percentiles of F Distributions (*cont.*)

n	p \ m	1	2	3	4	5	6	7	8	9	10	11	12	p
10	.0005	$.0^{6}41$	$.0^{3}50$	$.0^{2}49$	.015	.028	.043	.057	.071	.085	.097	.108	.119	.0005
	.001	$.0^{5}17$	$.0^{2}10$	$.0^{2}77$	.021	.037	.054	.071	.087	.101	.114	.126	.137	.001
	.005	$.0^{4}41$	$.0^{2}50$	.023	.048	.073	.098	.119	.139	.156	.171	.185	.197	.005
	.01	$.0^{3}17$	.010	.037	.069	.100	.127	.151	.172	.190	.206	.220	.233	.01
	.025	$.0^{2}10$	.025	.069	.113	.151	.183	.210	.233	.252	.269	.283	.296	.025
	.05	$.0^{2}41$	.052	.114	.168	.211	.246	.275	.299	.319	.336	.351	.363	.05
	.10	.017	.106	.191	.255	.303	.340	.370	.394	.414	.430	.444	.457	.10
	.25	.107	.296	.409	.480	.529	.565	.592	.613	.631	.645	.657	.667	.25
	.50	.490	.743	.845	.899	.932	.954	.971	.983	.992	1.00	1.01	1.01	.50
	.75	1.49	1.60	1.60	1.59	1.59	1.58	1.57	1.56	1.56	1.55	1.55	1.54	.75
	.90	3.28	2.92	2.73	2.61	2.52	2.46	2.41	2.38	2.35	2.32	2.30	2.28	.90
	.95	4.96	4.10	3.71	3.48	3.33	3.22	3.14	3.07	3.02	2.98	2.94	2.91	.95
	.975	6.94	5.46	4.83	4.47	4.24	4.07	3.95	3.85	3.78	3.72	3.66	3.62	.975
	.99	10.0	7.56	6.55	5.99	5.64	5.39	5.20	5.06	4.94	4.85	4.77	4.71	.99
	.995	12.8	9.43	8.08	7.34	6.87	6.54	6.30	6.12	5.97	5.85	5.75	5.66	.995
	.999	21.0	14.9	12.6	11.3	10.5	9.92	9.52	9.20	8.96	8.75	8.58	8.44	.999
	.9995	25.5	17.9	15.0	13.4	12.4	11.8	11.3	10.9	10.6	10.3	10.1	9.93	.9995
11	.0005	$.0^{6}41$	$.0^{3}50$	$.0^{2}49$	.015	.028	.043	.058	.072	.086	.099	.111	.121	.0005
	.001	$.0^{5}16$	$.0^{2}10$	$.0^{2}78$	.021	.038	.055	.072	.088	.103	.116	.129	.140	.001
	.005	$.0^{4}40$	$.0^{2}50$	.023	.048	.074	.099	.121	.141	.158	.174	.188	.200	.005
	.01	$.0^{3}16$	.010	.037	.069	.100	.128	.153	.175	.193	.210	.224	.237	.01
	.025	$.0^{2}10$	.025	.069	.114	.152	.185	.212	.236	.256	.273	.288	.301	.025
	.05	$.0^{2}41$	.052	.114	.168	.212	.248	.278	.302	.323	.340	.355	.368	.05
	.10	.017	.106	.192	.256	.305	.342	.373	.397	.417	.435	.448	.461	.10
	.25	.107	.295	.408	.481	.529	.565	.592	.614	.633	.645	.658	.667	.25
	.50	.486	.739	.840	.893	.926	.948	.964	.977	.986	.994	1.00	1.01	.50
	.75	1.47	1.58	1.58	1.57	1.56	1.55	1.54	1.53	1.53	1.52	1.52	1.51	.75
	.90	3.23	2.86	2.66	2.54	2.45	2.39	2.34	2.30	2.27	2.25	2.23	2.21	.90
	.95	4.84	3.98	3.59	3.36	3.20	3.09	3.01	2.95	2.90	2.85	2.82	2.79	.95
	.975	6.72	5.26	4.63	4.28	4.04	3.88	3.76	3.66	3.59	3.53	3.47	3.43	.975
	.99	9.65	7.21	6.22	5.67	5.32	5.07	4.89	4.74	4.63	4.54	4.46	4.40	.99
	.995	12.2	8.91	7.60	6.88	6.42	6.10	5.86	5.68	5.54	5.42	5.32	5.24	.995
	.999	19.7	13.8	11.6	10.3	9.58	9.05	8.66	8.35	8.12	7.92	7.76	7.62	.999
	.9995	23.6	16.4	13.6	12.2	11.2	10.6	10.1	9.76	9.48	9.24	9.04	8.88	.9995
12	.0005	$.0^{6}41$	$.0^{3}50$	$.0^{2}49$	.015	.028	.044	.058	.073	.087	.101	.113	.124	.0005
	.001	$.0^{5}16$	$.0^{2}10$	$.0^{2}78$	.021	.038	.056	.073	.089	.104	.118	.131	.143	.001
	.005	$.0^{4}39$	$.0^{2}50$	.023	.048	.075	.100	.122	.143	.161	.177	.191	.204	.005
	.01	$.0^{3}16$	.010	.037	.070	.101	.130	.155	.176	.196	.212	.227	.241	.01
	.025	$.0^{2}10$	.025	.070	.114	.153	.186	.214	.238	.259	.276	.292	.305	.025
	.05	$.0^{2}41$	.052	.114	.169	.214	.250	.280	.305	.325	.343	.358	.372	.05
	.10	.016	.106	.192	.257	.306	.344	.375	.400	.420	.438	.452	.466	.10
	.25	.106	.295	.408	.480	.530	.566	.594	.616	.633	.649	.662	.671	.25
	.50	.484	.735	.835	.888	.921	.943	.959	.972	.981	.989	.995	1.00	.50
	.75	1.46	1.56	1.56	1.55	1.54	1.53	1.52	1.51	1.51	1.50	1.50	1.49	.75
	.90	3.18	2.81	2.61	2.48	2.39	2.33	2.28	2.24	2.21	2.19	2.17	2.15	.90
	.95	4.75	3.89	3.49	3.26	3.11	3.00	2.91	2.85	2.80	2.75	2.72	2.69	.95
	.975	6.55	5.10	4.47	4.12	3.89	3.73	3.61	3.51	3.44	3.37	3.32	3.28	.975
	.99	9.33	6.93	5.95	5.41	5.06	4.82	4.64	4.50	4.39	4.30	4.22	4.16	.99
	.995	11.8	8.51	7.23	6.52	6.07	5.76	5.52	5.35	5.20	5.09	4.99	4.91	.995
	.999	18.6	13.0	10.8	9.63	8.89	8.38	8.00	7.71	7.48	7.29	7.14	7.01	.999
	.9995	22.2	15.3	12.7	11.2	10.4	9.74	9.28	8.94	8.66	8.43	8.24	8.08	.9995

Table A.4 Percentiles of F Distributions (*cont.*)

p \ m	15	20	24	30	40	50	60	100	120	200	500	∞	p	n
.0005	.145	.177	.195	.215	.238	.251	.262	.282	.288	.299	.311	.319	.0005	**10**
.001	.164	.197	.216	.236	.258	.272	.282	.303	.309	.321	.331	.338	.001	
.005	.226	.260	.279	.299	.321	.334	.344	.365	.370	.380	.391	.397	.005	
.01	.263	.297	.316	.336	.357	.370	.380	.400	.405	.415	.424	.431	.01	
.025	.327	.360	.379	.398	.419	.431	.441	.459	.464	.474	.483	.488	.025	
.05	.393	.426	.444	.462	.481	.493	.502	.518	.523	.532	.541	.546	.05	
.10	.486	.516	.532	.549	.567	.578	.586	.602	.605	.614	.621	.625	.10	
.25	.691	.714	.727	.740	.754	.762	.767	.779	.782	.788	.793	.797	.25	
.50	1.02	1.03	1.04	1.05	1.05	1.06	1.06	1.06	1.06	1.07	1.07	1.07	.50	
.75	1.53	1.52	1.52	1.51	1.51	1.50	1.50	1.49	1.49	1.49	1.48	1.48	.75	
.90	2.24	2.20	2.18	2.16	2.13	2.12	2.11	2.09	2.08	2.07	2.06	2.06	.90	
.95	2.85	2.77	2.74	2.70	2.66	2.64	2.62	2.59	2.58	2.56	2.55	2.54	.95	
.975	3.52	3.42	3.37	3.31	3.26	3.22	3.20	3.15	3.14	3.12	3.09	3.08	.975	
.99	4.56	4.41	4.33	4.25	4.17	4.12	4.08	4.01	4.00	3.96	3.93	3.91	.99	
.995	5.47	5.27	5.17	5.07	4.97	4.90	4.86	4.77	4.75	4.71	4.67	4.64	.995	
.999	8.13	7.80	7.64	7.47	7.30	7.19	7.12	6.98	6.94	6.87	6.81	6.76	.999	
.9995	9.56	9.16	8.96	8.75	8.54	8.42	8.33	8.16	8.12	8.04	7.96	7.90	.9995	
.0005	.148	.182	.201	.222	.246	.261	.271	.293	.299	.312	.324	.331	.0005	**11**
.001	.168	.202	.222	.243	.266	.282	.292	.313	.320	.332	.343	.353	.001	
.005	.231	.266	.286	.308	.330	.345	.355	.376	.382	.394	.403	.412	.005	
.01	.268	.304	.324	.344	.366	.380	.391	.412	.417	.427	.439	.444	.01	
.025	.332	.368	.386	.407	.429	.442	.450	.472	.476	.485	.495	.503	.025	
.05	.398	.433	.452	.469	.490	.503	.513	.529	.535	.543	.552	.559	.05	
.10	.490	.524	.541	.559	.578	.588	.595	.614	.617	.625	.633	.637	.10	
.25	.694	.719	.730	.744	.758	.767	.773	.780	.788	.794	.799	.803	.25	
.50	1.02	1.03	1.03	1.04	1.05	1.05	1.05	1.06	1.06	1.06	1.06	1.06	.50	
.75	1.50	1.49	1.49	1.48	1.47	1.47	1.47	1.46	1.46	1.46	1.45	1.45	.75	
.90	2.17	2.12	2.10	2.08	2.05	2.04	2.03	2.00	2.00	1.99	1.98	1.97	.90	
.95	2.72	2.65	2.61	2.57	2.53	2.51	2.49	2.46	2.45	2.43	2.42	2.40	.95	
.975	3.33	3.23	3.17	3.12	3.06	3.03	3.00	2.96	2.94	2.92	2.90	2.88	.975	
.99	4.25	4.10	4.02	3.94	3.86	3.81	3.78	3.71	3.69	3.66	3.62	3.60	.99	
.995	5.05	4.86	4.76	4.65	4.55	4.49	4.45	4.36	4.34	4.29	4.25	4.23	.995	
.999	7.32	7.01	6.85	6.68	6.52	6.41	6.35	6.21	6.17	6.10	6.04	6.00	.999	
.9995	8.52	8.14	7.94	7.75	7.55	7.43	7.35	7.18	7.14	7.06	6.98	6.93	.9995	
.0005	.152	.186	.206	.228	.253	.269	.280	.305	.311	.323	.337	.345	.0005	**12**
.001	.172	.207	.228	.250	.275	.291	.302	.326	.332	.344	.357	.365	.001	
.005	.235	.272	.292	.315	.339	.355	.365	.388	.393	.405	.417	.424	.005	
.01	.273	.310	.330	.352	.375	.391	.401	.422	.428	.441	.450	.458	.01	
.025	.337	.374	.394	.416	.437	.450	.461	.481	.487	.498	.508	.514	.025	
.05	.404	.439	.458	.478	.499	.513	.522	.541	.545	.556	.565	.571	.05	
.10	.496	.528	.546	.564	.583	.595	.604	.621	.625	.633	.641	.647	.10	
.25	.695	.721	.734	.748	.762	.771	.777	.789	.792	.799	.804	.808	.25	
.50	1.01	1.02	1.03	1.03	1.04	1.04	1.05	1.05	1.05	1.05	1.06	1.06	.50	
.75	1.48	1.47	1.46	1.45	1.45	1.44	1.44	1.43	1.43	1.43	1.42	1.42	.75	
.90	2.11	2.06	2.04	2.01	1.99	1.97	1.96	1.94	1.93	1.92	1.91	1.90	.90	
.95	2.62	2.54	2.51	2.47	2.43	2.40	2.38	2.35	2.34	2.32	2.31	2.30	.95	
.975	3.18	3.07	3.02	2.96	2.91	2.87	2.85	2.80	2.79	2.76	2.74	2.72	.975	
.99	4.01	3.86	3.78	3.70	3.62	3.57	3.54	3.47	3.45	3.41	3.38	3.36	.99	
.995	4.72	4.53	4.43	4.33	4.23	4.17	4.12	4.04	4.01	3.97	3.93	3.90	.995	
.999	6.71	6.40	6.25	6.09	5.93	5.83	5.76	5.63	5.59	5.52	5.46	5.42	.999	
.9995	7.74	7.37	7.18	7.00	6.80	6.68	6.61	6.45	6.41	6.33	6.25	6.20	.9995	

(*cont.*)

Table A.4 Percentiles of F Distributions (*cont.*)

n	p \ m	1	2	3	4	5	6	7	8	9	10	11	12	p
15	.0005	$.0^{6}41$	$.0^{3}50$	$.0^{2}49$	.015	.029	.045	.061	.076	.091	.105	.117	.129	.0005
	.001	$.0^{5}16$	$.0^{2}10$	$.0^{2}79$	.021	.039	.057	.075	.092	.108	.123	.137	.149	.001
	.005	$.0^{4}39$	$.0^{2}50$	.023	.049	.076	.102	.125	.147	.166	.183	.198	.212	.005
	.01	$.0^{3}16$	.010	.037	.070	.103	.132	.158	.181	.202	.219	.235	.249	.01
	.025	$.0^{2}10$	.025	.070	.116	.156	.190	.219	.244	.265	.284	.300	.315	.025
	.05	$.0^{2}41$	.051	.115	.170	.216	.254	.285	.311	.333	.351	.368	.382	.05
	.10	.016	.106	.192	.258	.309	.348	.380	.406	.427	.446	.461	.475	.10
	.25	.105	.293	.407	.480	.531	.568	.596	.618	.637	.652	.667	.676	.25
	.50	.478	.726	.826	.878	.911	.933	.948	.960	.970	.977	.984	.989	.50
	.75	1.43	1.52	1.52	1.51	1.49	1.48	1.47	1.46	1.46	1.45	1.44	1.44	.75
	.90	3.07	2.70	2.49	2.36	2.27	2.21	2.16	2.12	2.09	2.06	2.04	2.02	.90
	.95	4.54	3.68	3.29	3.06	2.90	2.79	2.71	2.64	2.59	2.54	2.51	2.48	.95
	.975	6.20	4.76	4.15	3.80	3.58	3.41	3.29	3.20	3.12	3.06	3.01	2.96	.975
	.99	8.68	6.36	5.42	4.89	4.56	4.32	4.14	4.00	3.89	3.80	3.73	3.67	.99
	.995	10.8	7.70	6.48	5.80	5.37	5.07	4.85	4.67	4.54	4.42	4.33	4.25	.995
	.999	16.6	11.3	9.34	8.25	7.57	7.09	6.74	6.47	6.26	6.08	5.93	5.81	.999
	.9995	19.5	13.2	10.8	9.48	8.66	8.10	7.68	7.36	7.11	6.91	6.75	6.60	.9995
20	.0005	$.0^{6}40$	$.0^{3}50$	$.0^{2}50$	.015	.029	.046	.063	.079	.094	.109	.123	.136	.0005
	.001	$.0^{5}16$	$.0^{2}10$	$.0^{2}79$	.022	.039	.058	.077	.095	.112	.128	.143	.156	.001
	.005	$.0^{4}39$	$.0^{2}50$	.023	.050	.077	.104	.129	.151	.171	.190	.206	.221	.005
	.01	$.0^{3}16$	.010	.037	.071	.105	.135	.162	.187	.208	.227	.244	.259	.01
	.025	$.0^{2}10$	.025	.071	.117	.158	.193	.224	.250	.273	.292	.310	.325	.025
	.05	$.0^{2}40$	.051	.115	.172	.219	.258	.290	.318	.340	.360	.377	.393	.05
	.10	.016	.106	.193	.260	.312	.353	.385	.412	.435	.454	.472	.485	.10
	.25	.104	.292	.407	.480	.531	.569	.598	.622	.641	.656	.671	.681	.25
	.50	.472	.718	.816	.868	.900	.922	.938	.950	.959	.966	.972	.977	.50
	.75	1.40	1.49	1.48	1.47	1.45	1.44	1.43	1.42	1.41	1.40	1.39	1.39	.75
	.90	2.97	2.59	2.38	2.25	2.16	2.09	2.04	2.00	1.96	1.94	1.91	1.89	.90
	.95	4.35	3.49	3.10	2.87	2.71	2.60	2.51	2.45	2.39	2.35	2.31	2.28	.95
	.975	5.87	4.46	3.86	3.51	3.29	3.13	3.01	2.91	2.84	2.77	2.72	2.68	.975
	.99	8.10	5.85	4.94	4.43	4.10	3.87	3.70	3.56	3.46	3.37	3.29	3.23	.99
	.995	9.94	6.99	5.82	5.17	4.76	4.47	4.26	4.09	3.96	3.85	3.76	3.68	.995
	.999	14.8	9.95	8.10	7.10	6.46	6.02	5.69	5.44	5.24	5.08	4.94	4.82	.999
	.9995	17.2	11.4	9.20	8.02	7.28	6.76	6.38	6.08	5.85	5.66	5.51	5.38	.9995
24	.0005	$.0^{6}40$	$.0^{3}50$	$.0^{2}50$	.015	.030	.046	.064	.080	.096	.112	.126	.139	.0005
	.001	$.0^{5}16$	$.0^{2}10$	$.0^{2}79$	.022	.040	.059	.079	.097	.115	.131	.146	.160	.001
	.005	$.0^{4}40$	$.0^{2}50$	.023	.050	.078	.106	.131	.154	.175	.193	.210	.226	.005
	.01	$.0^{3}16$	.010	.038	.072	.106	.137	.165	.189	.211	.231	.249	.264	.01
	.025	$.0^{2}10$	.025	.071	.117	.159	.195	.227	.253	.277	.297	.315	.331	.025
	.05	$.0^{2}40$	.051	.116	.173	.221	.260	.293	.321	.345	.365	.383	.399	.05
	.10	.016	.106	.193	.261	.313	.355	.388	.416	.439	.459	.476	.491	.10
	.25	.104	.291	.406	.480	.532	.570	.600	.623	.643	.659	.671	.684	.25
	.50	.469	.714	.812	.863	.895	.917	.932	.944	.953	.961	.967	.972	.50
	.75	1.39	1.47	1.46	1.44	1.43	1.41	1.40	1.39	1.38	1.38	1.37	1.36	.75
	.90	2.93	2.54	2.33	2.19	2.10	2.04	1.98	1.94	1.91	1.88	1.85	1.83	.90
	.95	4.26	3.40	3.01	2.78	2.62	2.51	2.42	2.36	2.30	2.25	2.21	2.18	.95
	.975	5.72	4.32	3.72	3.38	3.15	2.99	2.87	2.78	2.70	2.64	2.59	2.54	.975
	.99	7.82	5.61	4.72	4.22	3.90	3.67	3.50	3.36	3.26	3.17	3.09	3.03	.99
	.995	9.55	6.66	5.52	4.89	4.49	4.20	3.99	3.83	3.69	3.59	3.50	3.42	.995
	.999	14.0	9.34	7.55	6.59	5.98	5.55	5.23	4.99	4.80	4.64	4.50	4.39	.999
	.9995	16.2	10.6	8.52	7.39	6.68	6.18	5.82	5.54	5.31	5.13	4.98	4.85	.9995

Table A.4 Percentiles of F Distributions *(cont.)*

p \ m	15	20	24	30	40	50	60	100	120	200	500	∞	p	n
.0005	.159	.197	.220	.244	.272	.290	.303	.330	.339	.353	.368	.377	.0005	**15**
.001	.181	.219	.242	.266	.294	.313	.325	.352	.360	.375	.388	.398	.001	
.005	.246	.286	.308	.333	.360	.377	.389	.415	.422	.435	.448	.457	.005	
.01	.284	.324	.346	.370	.397	.413	.425	.450	.456	.469	.483	.490	.01	
.025	.349	.389	.410	.433	.458	.474	.485	.508	.514	.526	.538	.546	.025	
.05	.416	.454	.474	.496	.519	.535	.545	.565	.571	.581	.592	.600	.05	
.10	.507	.542	.561	.581	.602	.614	.624	.641	.647	.658	.667	.672	.10	
.25	.701	.728	.742	.757	.772	.782	.788	.802	.805	.812	.818	.822	.25	
.50	1.00	1.01	1.02	1.02	1.03	1.03	1.03	1.04	1.04	1.04	1.04	1.05	.50	
.75	1.43	1.41	1.41	1.40	1.39	1.39	1.38	1.38	1.37	1.37	1.36	1.36	.75	
.90	1.97	1.92	1.90	1.87	1.85	1.83	1.82	1.79	1.79	1.77	1.76	1.76	.90	
.95	2.40	2.33	2.39	2.25	2.20	2.18	2.16	2.12	2.11	2.10	2.08	2.07	.95	
.975	2.86	2.76	2.70	2.64	2.59	2.55	2.52	2.47	2.46	2.44	2.41	2.40	.975	
.99	3.52	3.37	3.29	3.21	3.13	3.08	3.05	2.98	2.96	2.92	2.89	2.87	.99	
.995	4.07	3.88	3.79	3.69	3.59	3.52	3.48	3.39	3.37	3.33	3.29	3.26	.995	
.999	5.54	5.25	5.10	4.95	4.80	4.70	4.64	4.51	4.47	4.41	4.35	4.31	.999	
.9995	6.27	5.93	5.75	5.58	5.40	5.29	5.21	5.06	5.02	4.94	4.87	4.83	.9995	
.0005	.169	.211	.235	.263	.295	.316	.331	.364	.375	.391	.408	.422	.0005	**20**
.001	.191	.233	.258	.286	.318	.339	.354	.386	.395	.413	.429	.441	.001	
.005	.258	.301	.327	.354	.385	.405	.419	.448	.457	.474	.490	.500	.005	
.01	.297	.340	.365	.392	.422	.441	.455	.483	.491	.508	.521	.532	.01	
.025	.363	.406	.430	.456	.484	.503	.514	.541	.548	.562	.575	.585	.025	
.05	.430	.471	.493	.518	.544	.562	.572	.595	.603	.617	.629	.637	.05	
.10	.520	.557	.578	.600	.623	.637	.648	.671	.675	.685	.694	.704	.10	
.25	.708	.736	.751	.767	.784	.794	.801	.816	.820	.827	.835	.840	.25	
.50	.989	1.00	1.01	1.01	1.02	1.02	1.02	1.03	1.03	1.03	1.03	1.03	.50	
.75	1.37	1.36	1.35	1.34	1.33	1.33	1.32	1.31	1.31	1.30	1.30	1.29	.75	
.90	1.84	1.79	1.77	1.74	1.71	1.69	1.68	1.65	1.64	1.63	1.62	1.61	.90	
.95	2.20	2.12	2.08	2.04	1.99	1.97	1.95	1.91	1.90	1.88	1.86	1.84	.95	
.975	2.57	2.46	2.41	2.35	2.29	2.25	2.22	2.17	2.16	2.13	2.10	2.09	.975	
.99	3.09	2.94	2.86	2.78	2.69	2.64	2.61	2.54	2.52	2.48	2.44	2.42	.99	
.995	3.50	3.32	3.22	3.12	3.02	2.96	2.92	2.83	2.81	2.76	2.72	2.69	.995	
.999	4.56	4.29	4.15	4.01	3.86	3.77	3.70	3.58	3.54	3.48	3.42	3.38	.999	
.9995	5.07	4.75	4.58	4.42	4.24	4.15	4.07	3.93	3.90	3.82	3.75	3.70	.9995	
.0005	.174	.218	.244	.274	.309	.331	.349	.384	.395	.416	.434	.449	.0005	**24**
.001	.196	.241	.268	.298	.332	.354	.371	.405	.417	.437	.455	.469	.001	
.005	.264	.310	.337	.367	.400	.422	.437	.469	.479	.498	.515	.527	.005	
.01	.304	.350	.376	.405	.437	.459	.473	.505	.513	.529	.546	.558	.01	
.025	.370	.415	.441	.468	.498	.518	.531	.562	.568	.585	.599	.610	.025	
.05	.437	.480	.504	.530	.558	.575	.588	.613	.622	.637	.649	.659	.05	
.10	.527	.566	.588	.611	.635	.651	.662	.685	.691	.704	.715	.723	.10	
.25	.712	.741	.757	.773	.791	.802	.809	.825	.829	.837	.844	.850	.25	
.50	.983	.994	1.00	1.01	1.01	1.02	1.02	1.02	1.02	1.02	1.03	1.03	.50	
.75	1.35	1.33	1.32	1.31	1.30	1.29	1.29	1.28	1.28	1.27	1.27	1.26	.75	
.90	1.78	1.73	1.70	1.67	1.64	1.62	1.61	1.58	1.57	1.56	1.54	1.53	.90	
.95	2.11	2.03	1.98	1.94	1.89	1.86	1.84	1.80	1.79	1.77	1.75	1.73	.95	
.975	2.44	2.33	2.27	2.21	2.15	2.11	2.08	2.02	2.01	1.98	1.95	1.94	.975	
.99	2.89	2.74	2.66	2.58	2.49	2.44	2.40	2.33	2.31	2.27	2.24	2.21	.99	
.995	3.25	3.06	2.97	2.87	2.77	2.70	2.66	2.57	2.55	2.50	2.46	2.43	.995	
.999	4.14	3.87	3.74	3.59	3.45	3.35	3.29	3.16	3.14	3.07	3.01	2.97	.999	
.9995	4.55	4.25	4.09	3.93	3.76	3.66	3.59	3.44	3.41	3.33	3.27	3.22	.9995	

(cont.)

Table A.4 Percentiles of F Distributions (*cont.*)

n	p \ m	1	2	3	4	5	6	7	8	9	10	11	12	p
30	.0005	$.0^{6}40$	$.0^{3}50$	$.0^{2}50$	.015	.030	.047	.065	.082	.098	.114	.129	.143	.0005
	.001	$.0^{5}16$	$.0^{2}10$	$.0^{2}80$	.022	.040	.060	.080	.099	.117	.134	.150	.164	.001
	.005	$.0^{4}40$	$.0^{2}50$	.024	.050	.079	.107	.133	.156	.178	.197	.215	.231	.005
	.01	$.0^{3}16$	.010	.038	.072	.107	.138	.167	.192	.215	.235	.254	.270	.01
	.025	$.0^{2}10$	.025	.071	.118	.161	.197	.229	.257	.281	.302	.321	.337	.025
	.05	$.0^{2}40$	.051	.116	.174	.222	.263	.296	.325	.349	.370	.389	.406	.05
	.10	.016	.106	.193	.262	.315	.357	.391	.420	.443	.464	.481	.497	.10
	.25	.103	.290	.406	.480	.532	.571	.601	.625	.645	.661	.676	.688	.25
	.50	.466	.709	.807	.858	.890	.912	.927	.939	.948	.955	.961	.966	.50
	.75	1.38	1.45	1.44	1.42	1.41	1.39	1.38	1.37	1.36	1.35	1.35	1.34	.75
	.90	2.88	2.49	2.28	2.14	2.05	1.98	1.93	1.88	1.85	1.82	1.79	1.77	.90
	.95	4.17	3.32	2.92	2.69	2.53	2.42	2.33	2.27	2.21	2.16	2.13	2.09	.95
	.975	5.57	4.18	3.59	3.25	3.03	2.87	2.75	2.65	2.57	2.51	2.46	2.41	.975
	.99	7.56	5.39	4.51	4.02	3.70	3.47	3.30	3.17	3.07	2.98	2.91	2.84	.99
	.995	9.18	6.35	5.24	4.62	4.23	3.95	3.74	3.58	3.45	3.34	3.25	3.18	.995
	.999	13.3	8.77	7.05	6.12	5.53	5.12	4.82	4.58	4.39	4.24	4.11	4.00	.999
	.9995	15.2	9.90	7.90	6.82	6.14	5.66	5.31	5.04	4.82	4.65	4.51	4.38	.9995
40	.0005	$.0^{6}40$	$.0^{3}50$	$.0^{2}50$	.016	.030	.048	.066	.084	.100	.117	.132	.147	.0005
	.001	$.0^{5}16$	$.0^{2}10$	$.0^{2}80$	.022	.042	.061	.081	.101	.119	.137	.153	.169	.001
	.005	$.0^{4}40$	$.0^{2}50$	.024	.051	.080	.108	.135	.159	.181	.201	.220	.237	.005
	.01	$.0^{3}16$	.010	.038	.073	.108	.140	.169	.195	.219	.240	.259	.276	.01
	.025	$.0^{3}99$	.025	.071	.119	.162	.199	.232	.260	.285	.307	.327	.344	.025
	.05	$.0^{2}40$	.051	.116	.175	.224	.265	.299	.329	.354	.376	.395	.412	.05
	.10	.016	.106	.194	.263	.317	.360	.394	.424	.448	.469	.488	.504	.10
	.25	.103	.290	.405	.480	.533	.572	.603	.627	.647	.664	.680	.691	.25
	.50	.463	.705	.802	.854	.885	.907	.922	.934	.943	.950	.956	.961	.50
	.75	1.36	1.44	1.42	1.40	1.39	1.37	1.36	1.35	1.34	1.33	1.32	1.31	.75
	.90	2.84	2.44	2.23	2.09	2.00	1.93	1.87	1.83	1.79	1.76	1.73	1.71	.90
	.95	4.08	3.23	2.84	2.61	2.45	2.34	2.25	2.18	2.12	2.08	2.04	2.00	.95
	.975	5.42	4.05	3.46	3.13	2.90	2.74	2.62	2.53	2.45	2.39	2.33	2.29	.975
	.99	7.31	5.18	4.31	3.83	3.51	3.29	3.12	2.99	2.89	2.80	2.73	2.66	.99
	.995	8.83	6.07	4.98	4.37	3.99	3.71	3.51	3.35	3.22	3.12	3.03	2.95	.995
	.999	12.6	8.25	6.60	5.70	5.13	4.73	4.44	4.21	4.02	3.87	3.75	3.64	.999
	.9995	14.4	9.25	7.33	6.30	5.64	5.19	4.85	4.59	4.38	4.21	4.07	3.95	.9995
60	.0005	$.0^{6}40$	$.0^{3}50$	$.0^{2}51$	.016	.031	.048	.067	.085	.103	.120	.136	.152	.0005
	.001	$.0^{5}16$	$.0^{2}10$	$.0^{2}80$	.022	.041	.062	.083	.103	.122	.140	.157	.174	.001
	.005	$.0^{4}40$	$.0^{2}50$	.024	.051	.081	.110	.137	.162	.185	.206	.225	.243	.005
	.01	$.0^{3}16$	.010	.038	.073	.109	.142	.172	.199	.223	.245	.265	.283	.01
	.025	$.0^{3}99$	.025	.071	.120	.163	.202	.235	.264	.290	.313	.333	.351	.025
	.05	$.0^{2}40$	.051	.116	.176	.226	.267	.303	.333	.359	.382	.402	.419	.05
	.10	.016	.106	.194	.264	.318	.362	.398	.428	.453	.475	.493	.510	.10
	.25	.102	.289	.405	.480	.534	.573	.604	.629	.650	.667	.680	.695	.25
	.50	.461	.701	.798	.849	.880	.901	.917	.928	.937	.945	.951	.956	.50
	.75	1.35	1.42	1.41	1.38	1.37	1.35	1.33	1.32	1.31	1.30	1.29	1.29	.75
	.90	2.79	2.39	2.18	2.04	1.95	1.87	1.82	1.77	1.74	1.71	1.68	1.66	.90
	.95	4.00	3.15	2.76	2.53	2.37	2.25	2.17	2.10	2.04	1.99	1.95	1.92	.95
	.975	5.29	3.93	3.34	3.01	2.79	2.63	2.51	2.41	2.33	2.27	2.22	2.17	.975
	.99	7.08	4.98	4.13	3.65	3.34	3.12	2.95	2.82	2.72	2.63	2.56	2.50	.99
	.995	8.49	5.80	4.73	4.14	3.76	3.49	3.29	3.13	3.01	2.90	2.82	2.74	.995
	.999	12.0	7.76	6.17	5.31	4.76	4.37	4.09	3.87	3.69	3.54	3.43	3.31	.999
	.9995	13.6	8.65	6.81	5.82	5.20	4.76	4.44	4.18	3.98	3.82	3.69	3.57	.9995

Table A.4 Percentiles of F Distributions (*cont.*)

p \ m	15	20	24	30	40	50	60	100	120	200	500	∞	p	n
.0005	.179	.226	.254	.287	.325	.350	.369	.410	.420	.444	.467	.483	.0005	**30**
.001	.202	.250	.278	.311	.348	.373	.391	.431	.442	.465	.488	.503	.001	
.005	.271	.320	.349	.381	.416	.441	.457	.495	.504	.524	.543	.559	.005	
.01	.311	.360	.388	.419	.454	.476	.493	.529	.538	.559	.575	.590	.01	
.025	.378	.426	.453	.482	.515	.535	.551	.585	.592	.610	.625	.639	.025	
.05	.445	.490	.516	.543	.573	.592	.606	.637	.644	.658	.676	.685	.05	
.10	.534	.575	.598	.623	.649	.667	.678	.704	.710	.725	.735	.746	.10	
.25	.716	.746	.763	.780	.798	.810	.818	.835	.839	.848	.856	.862	.25	
.50	.978	.989	.994	1.00	1.01	1.01	1.01	1.02	1.02	1.02	1.02	1.02	.50	
.75	1.32	1.30	1.29	1.28	1.27	1.26	1.26	1.25	1.24	1.24	1.23	1.23	.75	
.90	1.72	1.67	1.64	1.61	1.57	1.55	1.54	1.51	1.50	1.48	1.47	1.46	.90	
.95	2.01	1.93	1.89	1.84	1.79	1.76	1.74	1.70	1.68	1.66	1.64	1.62	.95	
.975	2.31	2.20	2.14	2.07	2.01	1.97	1.94	1.88	1.87	1.84	1.81	1.79	.975	
.99	2.70	2.55	2.47	2.39	2.30	2.25	2.21	2.13	2.11	2.07	2.03	2.01	.99	
.995	3.01	2.82	2.73	2.63	2.52	2.46	2.42	2.32	2.30	2.25	2.21	2.18	.995	
999	3.75	3.49	3.36	3.22	3.07	2.98	2.92	2.79	2.76	2.69	2.63	2.59	.999	
.9995	4.10	3.80	3.65	3.48	3.32	3.22	3.15	3.00	2.97	2.89	2.82	2.78	.9995	
.0005	.185	.236	.266	.301	.343	.373	.393	.441	.453	.480	.504	.525	.0005	**40**
.001	.209	.259	.290	.326	.367	.396	.415	.461	.473	.500	.524	.545	.001	
.005	.279	.331	.362	.396	.436	.463	.481	.524	.534	.559	581	.599	.005	
.01	.319	.371	.401	.435	.473	.498	.516	.556	.567	.592	.613	.628	.01	
.025	.387	.437	.466	.498	.533	.556	.573	.610	620	.641	.662	.674	.025	
.05	.454	.502	.529	.558	.591	.613	.627	.658	.669	.685	.704	.717	.05	
.10	.542	.585	.609	.636	.664	.683	.696	.724	.731	.747	.762	.772	.10	
.25	.720	.752	.769	.787	.806	.819	.828	.846	.851	.861	.870	.877	.25	
.50	.972	.983	.989	.994	1.00	1.00	1.01	1.01	1.01	1.01	1.02	1.02	.50	
.75	1.30	1.28	1.26	1.25	1.24	1.23	1.22	1.21	1.21	1.20	1.19	1.19	.75	
.90	1.66	1.61	1.57	1.54	1.51	1.48	1.47	1.43	1.42	1.41	1.39	1.38	90	
.95	1.92	1.84	1.79	1.74	1.69	1.66	1.64	1.59	1.58	1.55	1.53	1.51	.95	
.975	2.18	2.07	2.01	1.94	1.88	1.83	1.80	1.74	1.72	1.69	1.66	1.64	.975	
.99	2.52	2.37	2.29	2.20	2.11	2.06	2.02	1.94	1.92	1.87	1.83	1.80	.99	
.995	2.78	2.60	2.50	2.40	2.30	2.23	2.18	2.09	2.06	2.01	1.96	1.93	.995	
.999	3.40	3.15	3.01	2.87	2.73	2.64	2.57	2.44	2.41	2.34	2.28	2.23	.999	
.9995	3.68	3.39	3.24	3.08	2.92	2.82	2.74	2.60	2.57	2.49	2.41	2.37	.9995	
.0005	.192	.246	.278	.318	.365	.398	.421	.478	.493	.527	.561	.585	.0005	**60**
.001	.216	.270	.304	.343	.389	.421	.444	.497	.512	.545	.579	.602	.001	
.005	.287	.343	.376	.414	.458	.488	.510	.559	.572	.602	.633	.652	.005	
.01	.328	.383	.416	.453	.495	.524	.545	.592	.604	.633	.658	.679	.01	
.025	.396	.450	.481	.515	.555	.581	.600	.641	.654	.680	.704	.720	.025	
.05	.463	.514	.543	.575	.611	.633	.652	.690	.700	.719	.746	.759	.05	
.10	.550	.596	.622	.650	.682	.703	.717	.750	.758	.776	.793	.806	.10	
.25	.725	.758	.776	.796	.816	.830	.840	.860	.865	.877	.888	.896	.25	
.50	.967	.978	.983	.989	.994	.998	1.00	1.00	1.01	1.01	1.01	1.01	.50	
.75	1.27	1.25	1.24	1.22	1.21	1.20	1.19	1.17	1.17	1.16	1.15	1.15	.75	
.90	1.60	1.54	1.51	1.48	1.44	1.41	1.40	1.36	1.35	1.33	1.31	1.29	.90	
.95	1.84	1.75	1.70	1.65	1.59	1.56	1.53	1.48	1.47	1.44	1.41	1.39	.95	
.975	2.06	1.94	1.88	1.82	1.74	1.70	1.67	1.60	1.58	1.54	1.51	1.48	.975	
.99	2.35	2.20	2.12	2.03	1.94	1.88	1.84	1.75	1.73	1.68	1.63	1.60	.99	
.995	2.57	2.39	2.29	2.19	2.08	2.01	1.96	1.86	1.83	1.78	1.73	1.69	.995	
.999	3.08	2.83	2.69	2.56	2.41	2.31	2.25	2.11	2.09	2.01	1.93	1.89	.999	
.9995	3.30	3.02	2.87	2.71	2.55	2.45	2.38	2.23	2.19	2.11	2.03	1.98	.9995	

(*cont.*)

Table A.4 Percentiles of F Distributions (*cont.*)

n	p \ m	1	2	3	4	5	6	7	8	9	10	11	12	p
120	.0005	.0^{6}40	.0^{3}50	.0^{2}51	.016	.031	.049	.067	.087	.105	.123	.140	.156	.0005
	.001	.0^{5}16	.0^{2}10	.0^{2}81	.023	.042	.063	.084	.105	.125	.144	.162	.179	.001
	.005	.0^{4}39	.0^{2}50	.024	.051	.081	.111	.139	.165	.189	.211	.230	.249	.005
	.01	.0^{3}16	.010	.038	.074	.110	.143	.174	.202	.227	.250	.271	.290	.01
	.025	.0^{3}99	.025	.072	.120	.165	.204	.238	.268	.295	.318	.340	.359	.025
	.05	.0^{2}39	.051	.117	.177	.227	.270	.306	.337	.364	.388	.408	.427	.05
	.10	.016	.105	.194	.265	.320	.365	.401	.432	.458	.480	.500	.518	.10
	.25	.102	.288	.405	.481	.534	.574	.606	.631	.652	.670	.685	.699	.25
	.50	.458	.697	.793	.844	.875	.896	.912	.923	.932	.939	.945	.950	.50
	.75	1.34	1.40	1.39	1.37	1.35	1.33	1.31	1.30	1.29	1.28	1.27	1.26	.75
	.90	2.75	2.35	2.13	1.99	1.90	1.82	1.77	1.72	1.68	1.65	1.62	1.60	.90
	.95	3.92	3.07	2.68	2.45	2.29	2.18	2.09	2.02	1.96	1.91	1.87	1.83	.95
	.975	5.15	3.80	3.23	2.89	2.67	2.52	2.39	2.30	2.22	2.16	2.10	2.05	.975
	.99	6.85	4.79	3.95	3.48	3.17	2.96	2.79	2.66	2.56	2.47	2.40	2.34	.99
	.995	8.18	5.54	4.50	3.92	3.55	3.28	3.09	2.93	2.81	2.71	2.62	2.54	.995
	.999	11.4	7.32	5.79	4.95	4.42	4.04	3.77	3.55	3.38	3.24	3.12	3.02	.999
	.9995	12.8	8.10	6.34	5.39	4.79	4.37	4.07	3.82	3.63	3.47	3.34	3.22	.9995
∞	.0005	.0^{6}39	.0^{3}50	.0^{2}51	.016	.032	.050	.069	.088	.108	.127	.144	.161	.0005
	.001	.0^{5}16	.0^{2}10	.0^{2}81	.023	.042	.063	.085	.107	.128	.148	.167	.185	.001
	.005	.0^{4}39	.0^{2}50	.024	.052	.082	.113	.141	.168	.193	.216	.236	.256	.005
	.01	.0^{3}16	.010	.038	.074	.111	.145	.177	.206	.232	.256	.278	.298	.01
	.025	.0^{3}98	.025	.072	.121	.166	.206	.241	.272	.300	.325	.347	.367	.025
	.05	.0^{2}39	.051	.117	.178	.229	.273	.310	.342	.369	.394	.417	.436	.05
	.10	.016	.105	.195	.266	.322	.367	.405	.436	.463	.487	.508	.525	.10
	.25	.102	.288	.404	.481	.535	.576	.608	.634	.655	.674	.690	.703	.25
	.50	.455	.693	.789	.839	.870	.891	.907	.918	.927	.934	.939	.945	.50
	.75	1.32	1.39	1.37	1.35	1.33	1.31	1.29	1.28	1.27	1.25	1.24	1.24	.75
	.90	2.71	2.30	2.08	1.94	1.85	1.77	1.72	1.67	1.63	1.60	1.57	1.55	.90
	.95	3.84	3.00	2.60	2.37	2.21	2.10	2.01	1.94	1.88	1.83	1.79	1.75	.95
	.975	5.02	3.69	3.12	2.79	2.57	2.41	2.29	2.19	2.11	2.05	1.99	1.94	.975
	.99	6.63	4.61	3.78	3.32	3.02	2.80	2.64	2.51	2.41	2.32	2.25	2.18	.99
	.995	7.88	5.30	4.28	3.72	3.35	3.09	2.90	2.74	2.62	2.52	2.43	2.36	.995
	.999	10.8	6.91	5.42	4.62	4.10	3.74	3.47	3.27	3.10	2.96	2.84	2.74	.999
	.9995	12.1	7.60	5.91	5.00	4.42	4.02	3.72	3.48	3.30	3.14	3.02	2.90	.9995

Table A.4 Percentiles of F Distributions (*cont.*)

p \ m	15	20	24	30	40	50	60	100	120	200	500	∞	p	n
.0005	.199	.256	.293	.338	.390	.429	.458	.524	.543	.578	.614	.676	.0005	**120**
.001	.223	.282	.319	.363	.415	.453	.480	.542	.568	.595	.631	.691	.001	
.005	.297	.356	.393	.434	.484	.520	.545	.605	.623	.661	.702	.733	.005	
.01	.338	.397	.433	.474	.522	.556	.579	.636	.652	.688	.725	.755	.01	
.025	.406	.464	.498	.536	.580	.611	.633	.684	.698	.729	.762	.789	.025	
.05	.473	.527	.559	.594	.634	.661	.682	.727	.740	.767	.785	.819	.05	
.10	.560	.609	.636	.667	.702	.726	.742	.781	.791	.815	.838	.855	.10	
.25	.730	.765	.784	.805	.828	.843	.853	.877	.884	.897	.911	.923	.25	
.50	.961	.972	.978	.983	.989	.992	.994	1.00	1.00	1.00	1.01	1.01	.50	
.75	1.24	1.22	1.21	1.19	1.18	1.17	1.16	1.14	1.13	1.12	1.11	1.10	.75	
.90	1.55	1.48	1.45	1.41	1.37	1.34	1.32	1.27	1.26	1.24	1.21	1.19	.90	
.95	1.75	1.66	1.61	1.55	1.50	1.46	1.43	1.37	1.35	1.32	1.28	1.25	.95	
.975	1.95	1.82	1.76	1.69	1.61	1.56	1.53	1.45	1.43	1.39	1.34	1.31	.975	
.99	2.19	2.03	1.95	1.86	1.76	1.70	1.66	1.56	1.53	1.48	1.42	1.38	.99	
.995	2.37	2.19	2.09	1.98	1.87	1.80	1.75	1.64	1.61	1.54	1.48	1.43	.995	
.999	2.78	2.53	2.40	2.26	2.11	2.02	1.95	1.82	1.76	1.70	1.62	1.54	.999	
.9995	2.96	2.67	2.53	2.38	2.21	2.11	2.01	1.88	1.84	1.75	1.67	1.60	.9995	
.0005	.207	.270	.311	.360	.422	.469	.505	.599	.624	.704	.804	1.00	.0005	∞
.001	.232	.296	.338	.386	.448	.493	.527	.617	.649	.719	.819	1.00	.001	
.005	.307	.372	.412	.460	.518	.559	.592	.671	.699	.762	.843	1.00	.005	
.01	.349	.413	.452	.499	.554	.595	.625	.699	.724	.782	.858	1.00	.01	
.025	.418	.480	.517	.560	.611	.645	.675	.741	.763	.813	.878	1.00	.025	
.05	.484	.543	.577	.617	.663	.694	.720	.781	.797	.840	.896	1.00	.05	
.10	.570	.622	.652	.687	.726	.752	.774	.826	.838	.877	.919	1.00	.10	
.25	.736	.773	.793	.816	.842	.860	.872	.901	.910	.932	.957	1.00	.25	
.50	.956	.967	.972	.978	.983	.987	.989	.993	.994	.997	.999	1.00	.50	
.75	1.22	1.19	1.18	1.16	1.14	1.13	1.12	1.09	1.08	1.07	1.04	1.00	.75	
.90	1.49	1.42	1.38	1.34	1.30	1.26	1.24	1.18	1.17	1.13	1.08	1.00	.90	
.95	1.67	1.57	1.52	1.46	1.39	1.35	1.32	1.24	1.22	1.17	1.11	1.00	.95	
.975	1.83	1.71	1.64	1.57	1.48	1.43	1.39	1.30	1.27	1.21	1.13	1.00	.975	
.99	2.04	1.88	1.79	1.70	1.59	1.52	1.47	1.36	1.32	1.25	1.15	1.00	.99	
.995	2.19	2.00	1.90	1.79	1.67	1.59	1.53	1.40	1.36	1.28	1.17	1.00	.995	
.999	2.51	2.27	2.13	1.99	1.84	1.73	1.66	1.49	1.45	1.34	1.21	1.00	.999	
.9995	2.65	2.37	2.22	2.07	1.91	1.79	1.71	1.53	1.48	1.36	1.22	1.00	.9995	

Table A.5 Upper Percentiles of Studentized Range Distributions

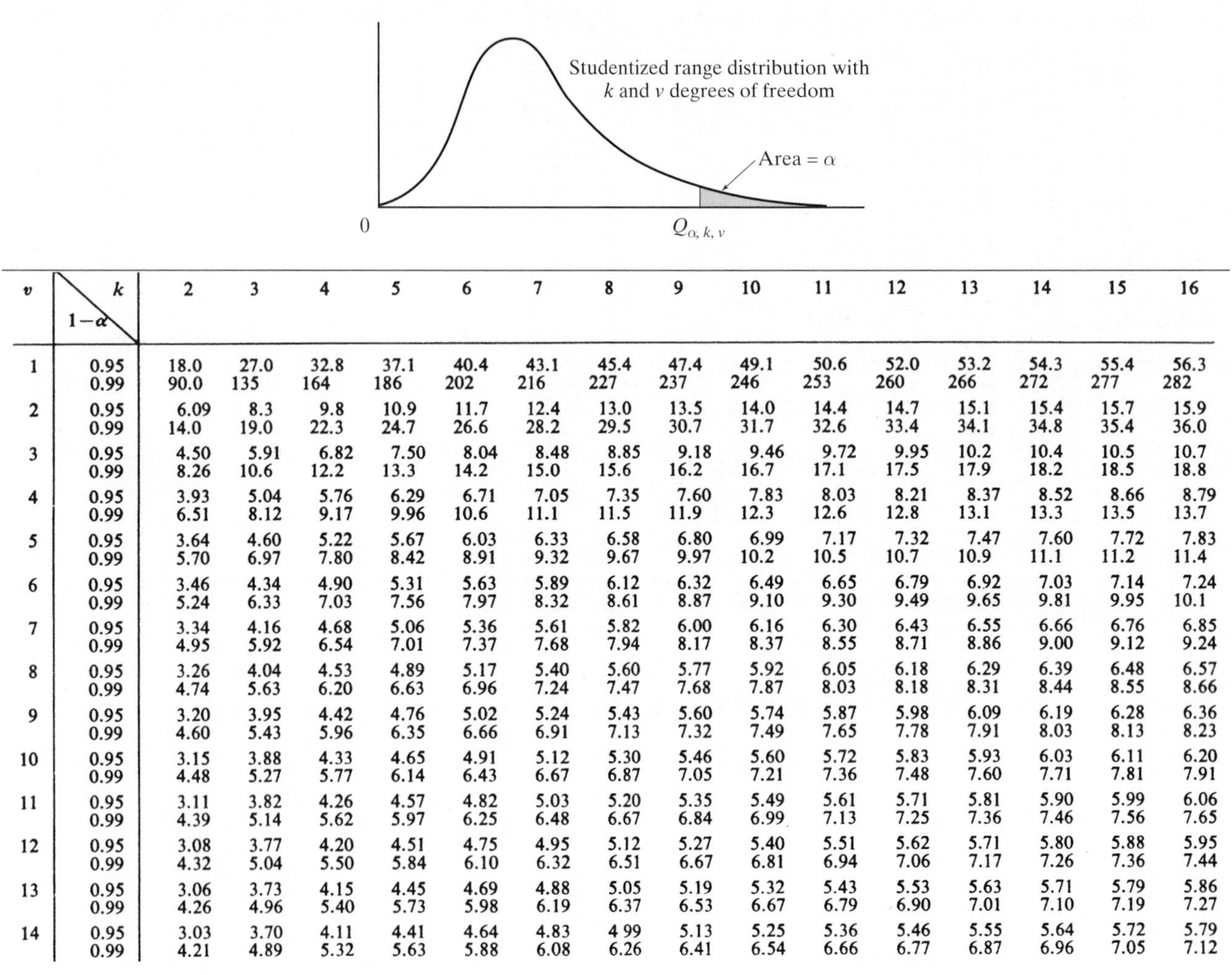

v	k / $1-\alpha$	2	3	4	5	6	7	8	9	10	11	12	13	14	15	16
1	0.95	18.0	27.0	32.8	37.1	40.4	43.1	45.4	47.4	49.1	50.6	52.0	53.2	54.3	55.4	56.3
	0.99	90.0	135	164	186	202	216	227	237	246	253	260	266	272	277	282
2	0.95	6.09	8.3	9.8	10.9	11.7	12.4	13.0	13.5	14.0	14.4	14.7	15.1	15.4	15.7	15.9
	0.99	14.0	19.0	22.3	24.7	26.6	28.2	29.5	30.7	31.7	32.6	33.4	34.1	34.8	35.4	36.0
3	0.95	4.50	5.91	6.82	7.50	8.04	8.48	8.85	9.18	9.46	9.72	9.95	10.2	10.4	10.5	10.7
	0.99	8.26	10.6	12.2	13.3	14.2	15.0	15.6	16.2	16.7	17.1	17.5	17.9	18.2	18.5	18.8
4	0.95	3.93	5.04	5.76	6.29	6.71	7.05	7.35	7.60	7.83	8.03	8.21	8.37	8.52	8.66	8.79
	0.99	6.51	8.12	9.17	9.96	10.6	11.1	11.5	11.9	12.3	12.6	12.8	13.1	13.3	13.5	13.7
5	0.95	3.64	4.60	5.22	5.67	6.03	6.33	6.58	6.80	6.99	7.17	7.32	7.47	7.60	7.72	7.83
	0.99	5.70	6.97	7.80	8.42	8.91	9.32	9.67	9.97	10.2	10.5	10.7	10.9	11.1	11.2	11.4
6	0.95	3.46	4.34	4.90	5.31	5.63	5.89	6.12	6.32	6.49	6.65	6.79	6.92	7.03	7.14	7.24
	0.99	5.24	6.33	7.03	7.56	7.97	8.32	8.61	8.87	9.10	9.30	9.49	9.65	9.81	9.95	10.1
7	0.95	3.34	4.16	4.68	5.06	5.36	5.61	5.82	6.00	6.16	6.30	6.43	6.55	6.66	6.76	6.85
	0.99	4.95	5.92	6.54	7.01	7.37	7.68	7.94	8.17	8.37	8.55	8.71	8.86	9.00	9.12	9.24
8	0.95	3.26	4.04	4.53	4.89	5.17	5.40	5.60	5.77	5.92	6.05	6.18	6.29	6.39	6.48	6.57
	0.99	4.74	5.63	6.20	6.63	6.96	7.24	7.47	7.68	7.87	8.03	8.18	8.31	8.44	8.55	8.66
9	0.95	3.20	3.95	4.42	4.76	5.02	5.24	5.43	5.60	5.74	5.87	5.98	6.09	6.19	6.28	6.36
	0.99	4.60	5.43	5.96	6.35	6.66	6.91	7.13	7.32	7.49	7.65	7.78	7.91	8.03	8.13	8.23
10	0.95	3.15	3.88	4.33	4.65	4.91	5.12	5.30	5.46	5.60	5.72	5.83	5.93	6.03	6.11	6.20
	0.99	4.48	5.27	5.77	6.14	6.43	6.67	6.87	7.05	7.21	7.36	7.48	7.60	7.71	7.81	7.91
11	0.95	3.11	3.82	4.26	4.57	4.82	5.03	5.20	5.35	5.49	5.61	5.71	5.81	5.90	5.99	6.06
	0.99	4.39	5.14	5.62	5.97	6.25	6.48	6.67	6.84	6.99	7.13	7.25	7.36	7.46	7.56	7.65
12	0.95	3.08	3.77	4.20	4.51	4.75	4.95	5.12	5.27	5.40	5.51	5.62	5.71	5.80	5.88	5.95
	0.99	4.32	5.04	5.50	5.84	6.10	6.32	6.51	6.67	6.81	6.94	7.06	7.17	7.26	7.36	7.44
13	0.95	3.06	3.73	4.15	4.45	4.69	4.88	5.05	5.19	5.32	5.43	5.53	5.63	5.71	5.79	5.86
	0.99	4.26	4.96	5.40	5.73	5.98	6.19	6.37	6.53	6.67	6.79	6.90	7.01	7.10	7.19	7.27
14	0.95	3.03	3.70	4.11	4.41	4.64	4.83	4 99	5.13	5.25	5.36	5.46	5.55	5.64	5.72	5.79
	0.99	4.21	4.89	5.32	5.63	5.88	6.08	6.26	6.41	6.54	6.66	6.77	6.87	6.96	7.05	7.12

v	k / $1-\alpha$	2	3	4	5	6	7	8	9	10	11	12	13	14	15	16
15	0.95	3.01	3.67	4.08	4.37	4.60	4.78	4.94	5.08	5.20	5.31	5.40	5.49	5.58	5.65	5.72
	0.99	4.17	4.83	5.25	5.56	5.80	5.99	6.16	6.31	6.44	6.55	6.66	6.76	6.84	6.93	7.00
16	0.95	3.00	3.65	4.05	4.33	4.56	4.74	4.90	5.03	5.15	5.26	5.35	5.44	5.52	5.59	5.66
	0.99	4.13	4.78	5.19	5.49	5.72	5.92	6.08	6.22	6.35	6.46	6.56	6.66	6.74	6.82	6.90
17	0.95	2.98	3.63	4.02	4.30	4.52	4.71	4.86	4.99	5.11	5.21	5.31	5.39	5.47	5.55	5.61
	0.99	4.10	4.74	5.14	5.43	5.66	5.85	6.01	6.15	6.27	6.38	6.48	6.57	6.66	6.73	6.80
18	0.95	2.97	3.61	4.00	4.28	4.49	4.67	4.82	4.96	5.07	5.17	5.27	5.35	5.43	5.50	5.57
	0.99	4.07	4.70	5.09	5.38	5.60	5.79	5.94	6.08	6.20	6.31	6.41	6.50	6.58	6.65	6.72
19	0.95	2.96	3.59	3.98	4.25	4.47	4.65	4.79	4.92	5.04	5.14	5.23	5.32	5.39	5.46	5.53
	0.99	4.05	4.67	5.05	5.33	5.55	5.73	5.89	6.02	6.14	6.25	6.34	6.43	6.51	6.58	6.65
20	0.95	2.95	3.58	3.96	4.23	4.45	4.62	4.77	4.90	5.01	5.11	5.20	5.28	5.36	5.43	5.49
	0.99	4.02	4.64	5.02	5.29	5.51	5.69	5.84	5.97	6.09	6.19	6.29	6.37	6.45	6.52	6.59
24	0.95	2.92	3.53	3.90	4.17	4.37	4.54	4.68	4.81	4.92	5.01	5.10	5.18	5.25	5.32	5.38
	0.99	3.96	4.54	4.91	5.17	5.37	5.54	5.69	5.81	5.92	6.02	6.11	6.19	6.26	6.33	6.39
30	0.95	2.89	3.49	3.84	4.10	4.30	4.46	4.60	4.72	4.83	4.92	5.00	5.08	5.15	5.21	5.27
	0.99	3.89	4.45	4.80	5.05	5.24	5.40	5.54	5.65	5.76	5.85	5.93	6.01	6.08	6.14	6.20
40	0.95	2.86	3.44	3.79	4.04	4.23	4.39	4.52	4.63	4.74	4.82	4.91	4.98	5.05	5.11	5.16
	0.99	3.82	4.37	4.70	4.93	5.11	5.27	5.39	5.50	5.60	5.69	5.77	5.84	5.90	5.96	6.02
60	0.95	2.83	3.40	3.74	3.98	4.16	4.31	4.44	4.55	4.65	4.73	4.81	4.88	4.94	5.00	5.06
	0.99	3.76	4.28	4.60	4.82	4.99	5.13	5.25	5.36	5.45	5.53	5.60	5.67	5.73	5.79	5.84
120	0.95	2.80	3.36	3.69	3.92	4.10	4.24	4.36	4.48	4.56	4.64	4.72	4.78	4.84	4.90	4.95
	0.99	3.70	4.20	4.50	4.71	4.87	5.01	5.12	5.21	5.30	5.38	5.44	5.51	5.56	5.61	5.66
∞	0.95	2.77	3.31	3.63	3.86	4.03	4.17	4.29	4.39	4.47	4.55	4.62	4.68	4.74	4.80	4.85
	0.99	3.64	4.12	4.40	4.60	4.76	4.88	4.99	5.08	5.16	5.23	5.29	5.35	5.40	5.45	5.49

Source: Olive Jean Dunn and Virginia A. Clark, *Applied Statistics: Analysis of Variance and Regression* (New York: Wiley, 1974), pp. 371–372. Reproduced with permission of John Wiley & Sons, Inc.

Table A.6 Upper and Lower Percentiles of the Wilcoxon Signed Rank Statistic, W

	w_1^*	w_2^*	$P(W \leq w_1^*) = P(W \geq w_2^*)$
$n = 4$	0	10	0.062
	1	9	0.125
$n = 5$	0	15	0.031
	1	14	0.062
	2	13	0.094
	3	12	0.156
$n = 6$	0	21	0.016
	1	20	0.031
	2	19	0.047
	3	18	0.078
	4	17	0.109
	5	16	0.156
$n = 7$	0	28	0.008
	1	27	0.016
	2	26	0.023
	3	25	0.039
	4	24	0.055
	5	23	0.078
	6	22	0.109
	7	21	0.148
$n = 8$	0	36	0.004
	1	35	0.008
	2	34	0.012
	3	33	0.020
	4	32	0.027
	5	31	0.039
	6	30	0.055
	7	29	0.074
	8	28	0.098
	9	27	0.125
$n = 9$	1	44	0.004
	2	43	0.006
	3	42	0.010
	4	41	0.014
	5	40	0.020
	6	39	0.027
	7	38	0.037
	8	37	0.049
	9	36	0.064
	10	35	0.082
	11	34	0.102
	12	33	0.125

Table A.6 Upper and Lower Percentiles of the Wilcoxon Signed Rank Statistic, W (*cont.*)

	w_1^*	w_2^*	$P(W \leq w_1^*) = P(W \geq w_2^*)$
$n = 10$	3	52	0.005
	4	51	0.007
	5	50	0.010
	6	49	0.014
	7	48	0.019
	8	47	0.024
	9	46	0.032
	10	45	0.042
	11	44	0.053
	12	43	0.065
	13	42	0.080
	14	41	0.097
	15	40	0.116
	16	39	0.138
$n = 11$	5	61	0.005
	6	60	0.007
	7	59	0.009
	8	58	0.012
	9	57	0.016
	10	56	0.021
	11	55	0.027
	12	54	0.034
	13	53	0.042
	14	52	0.051
	15	51	0.062
	16	50	0.074
	17	49	0.087
	18	48	0.103
	19	47	0.120
	20	46	0.139
$n = 12$	7	71	0.005
	8	70	0.006
	9	69	0.008
	10	68	0.010
	11	67	0.013
	12	66	0.017
	13	65	0.021
	14	64	0.026
	15	63	0.032
	16	62	0.039
	17	61	0.046
	18	60	0.055
	19	59	0.065
	20	58	0.076
	21	57	0.088
	22	56	0.102
	23	55	0.117
	24	54	0.133

Source: Used with permission from Wilfrid J. Dixon and Frank J. Massey, Jr., *Introduction to Statistical Analysis*, 2nd ed. (New York: McGraw-Hill, 1957), pp. 443–444.

Answers to Selected Odd-Numbered Questions

CHAPTER 2

Section 2.2

2.2.1. $S = \{(s, s, s), (s, s, f), (s, f, s), (f, s, s), (s, f, f), (f, s, f), (f, f, s), (f, f, f)\}$;
$A = \{(s, f, s), (f, s, s)\}$; $B = \{(f, f, f)\}$

2.2.3. $(1, 3, 4), (1, 3, 5), (1, 3, 6), (2, 3, 4), (2, 3, 5), (2, 3, 6)$

2.2.5. The outcome sought is $(4, 4)$.

2.2.7. $P = \{$right triangles with sides $(5, a, b)$: $a^2 + b^2 = 25\}$

2.2.9. (a) $S = \{(0, 0, 0, 0), (0, 0, 0, 1), (0, 0, 1, 0), (0, 0, 1, 1), (0, 1, 0, 0), (0, 1, 0, 1), (0, 1, 1, 0), (0, 1, 1, 1), (1, 0, 0, 0), (1, 0, 0, 1), (1, 0, 1, 0), (1, 0, 1, 1), (1, 1, 0, 0), (1, 1, 0, 1), (1, 1, 1, 0), (1, 1, 1, 1)\}$
(b) $A = \{(0, 0, 1, 1), (0, 1, 0, 1), (0, 1, 1, 0), (1, 0, 0, 1), (1, 0, 1, 0), (1, 1, 0, 0)\}$
(c) $1 + k$

2.2.11. Let p_1 and p_2 denote the two perpetrators and i_1, i_2, and i_3, the three in the lineup who are innocent. Then $S = \{(p_1, i_1), (p_1, i_2), (p_1, i_3), (p_2, i_1), (p_2, i_2), (p_2, i_3), (p_1, p_2), (i_1, i_2), (i_1, i_3), (i_2, i_3)\}$. The event A contains every outcome in S except (p_1, p_2).

2.2.13. In order for the shooter to win with a point of 9, one of the following (countably infinite) sequences of sums must be rolled: (9, 9), (9, no 7 or no 9, 9), (9, no 7 or no 9, no 7 or no 9, 9), ...

2.2.15. Let A_k be the set of chips placed in the urn at $1/2^k$ minute until midnight. For example, $A_1 = \{11, 12, \ldots, 20\}$. Then the set of chips in the urn is $\bigcup_{k=1}^{\infty}(A_k - \{k\}) = \bigcup_{k=1}^{\infty} A_k - \bigcup_{k=1}^{\infty}\{k\} = \emptyset$, since $\bigcup_{k=1}^{\infty} A_k$ is a subset of $\bigcup_{k=1}^{\infty}\{k\}$.

2.2.17. $A \cap B = \{x\colon -3 \le x \le 2\}$ and $A \cup B = \{x\colon -4 \le x \le 2\}$

2.2.19. $A = (A_{11} \cap A_{21}) \cup (A_{12} \cap A_{22})$

2.2.21. 40

2.2.23. (a) If s is a member of $A \cup (B \cap C)$, then s belongs to A or to $B \cap C$. If it is a member of A or of $B \cap C$, then it belongs to $A \cup B$ and to $A \cup C$. Thus, it is a member of $(A \cup B) \cap (A \cup C)$. Conversely, choose s in $(A \cup B) \cap (A \cup C)$. If it belongs to A, then it belongs to $A \cup (B \cap C)$. If it does not belong to A, then it must be a member of $B \cap C$. In that case it also is a member of $A \cup (B \cap C)$.

2.2.25. (a) Let s be a member of $A \cup (B \cup C)$. Then s belongs to either A or $B \cup C$ (or both). If s belongs to A, it necessarily belongs to $(A \cup B) \cup C$. If s belongs to $B \cup C$, it belongs to B or C or both, so it must belong to $(A \cup B) \cup C$. Now, suppose s belongs to $(A \cup B) \cup C$. Then it belongs to either $A \cup B$ or C or both. If it belongs to C, it must belong to $A \cup (B \cup C)$. If it belongs to $A \cup B$, it must belong to either A or B or both, so it must belong to $A \cup (B \cup C)$.
(b) The proof is similar to part (a).

2.2.27. A is a subset of B

2.2.29. (a) B and C
(b) B is a subset of A

2.2.31. 240

2.2.35. A and B are subsets of $A \cup B$.

2.2.37. 100/1200

2.2.39. 500

Section 2.3

2.3.1. 0.41

2.3.3. (a) $1 - P(A \cap B)$
(b) $P(B) - P(A \cap B)$

2.3.5. No. $P(A_1 \cup A_2 \cup A_3) = P(\text{at least one “6” appears}) = 1 - P(\text{no 6's appear}) = 1 - \left(\frac{5}{6}\right)^3 \neq \frac{1}{2}$. The A_i's are not mutually exclusive, so $P(A_1 \cup A_2 \cup A_3) \neq P(A_1) + P(A_2) + P(A_3)$.

2.3.7. By inspection, $B = (B \cap A_1) \cup (B \cap A_2) \cup \ldots \cup (B \cap A_n)$.

2.3.9. $\dfrac{3}{4}$

2.3.11. 0.30

2.3.13. 0.15

2.3.15. (a) $X^C \cap Y = \{(H, T, T, H), (T, H, H, T)\}$, so $P(X^C \cap Y) = 2/16$
(b) $X \cap Y^C = \{(H, T, T, T), (T, T, T, H), (T, H, H, H), (H, H, H, T)\}$, so $P(X \cap Y^C) = 4/16$

2.3.17. $A \cap B, (A \cap B) \cup (A \cap C), A, A \cup B, S$

Section 2.4

2.4.1. 3/10

2.4.3. If $P(A|B) = \dfrac{P(A \cap B)}{P(B)} < P(A)$, then $P(A \cap B) < P(A) \cdot P(B)$. It follows that $P(B|A) = \dfrac{P(A \cap B)}{P(A)} < \dfrac{P(A) \cdot P(B)}{P(A)} = P(B)$.

2.4.5. The answer would remain the same. Distinguishing only three family types does not make them equally likely; (girl, boy) families will occur twice as often as either (boy, boy) or (girl, girl) families.

2.4.7. 3/8

2.4.9. 5/6

2.4.11. (a) 5/100 **(b)** 70/100 **(c)** 95/100 **(d)** 75/100 **(e)** 70/95 **(f)** 25/95 **(g)** 30/35

2.4.13. 3/5

2.4.15. 1/5

2.4.17. 2/3

2.4.19. 20/55

2.4.21. 1800/360, 360; 1/360, 360

2.4.23. 1/6, 497, 400

2.4.25. 0.027

2.4.27. 0.23

2.4.29. 0.70

2.4.31. 0.02%

2.4.33. 0.645

2.4.35. No. Let B denote the event that the person calling the toss is correct. Let A_H be the event that the coin comes up Heads and let A_T be the event that the coin comes up Tails. Then $P(B) = P(B|A_H)P(A_H) + P(B|A_T)P(A_T) = (0.7)\left(\frac{1}{2}\right) + (0.3)\left(\frac{1}{2}\right) = \frac{1}{2}$.

2.4.37. 0.415

2.4.39. 0.46

2.4.41. 5/12

2.4.43. Hearthstone

2.4.45. 0.74

2.4.47. 14

2.4.49. 0.441

2.4.51. 0.64

2.4.53. 1/3

Section 2.5

2.5.1. **(a)** No, $P(A \cap B) > 0$ **(b)** No, $P(A \cap B) = 0.2 \neq 0.3 = P(A)P(B)$ **(c)** 0.8

2.5.3. 6/36

2.5.5. 0.51

2.5.7. **(a)** **(1)** 3/8 **(2)** 11/32 **(b)** **(1)** 0 **(2)** 1/4

2.5.9. 6/16

2.5.11. Equation 2.5.3:

$$P(A \cap B \cap C) = P(\{(1,3)\}) = 1/36 = (2/6)(3/6)(6/36) = P(A)P(B)P(C)$$

Equation 2.5.4:

$$P(B \cap C) = P(\{(1,3),(5,6)\}) = 2/36 \neq (3/6)(6/36) = P(B)P(C)$$

2.5.13. 11

2.5.15. $P(A \cap B \cap C) = 0$ (since the sum of two odd numbers is necessarily even) $\neq P(A) \cdot P(B) \cdot P(C) > 0$, so A, B, and C are not mutually independent. However, $P(A \cap B) = \frac{9}{36} = P(A) \cdot P(B) = \frac{3}{6} \cdot \frac{3}{6}$, $P(A \cap C) = \frac{9}{36} = P(A) \cdot P(C) = \frac{3}{6} \cdot \frac{18}{36}$, and $P(B \cap C) = \frac{9}{36} = P(B) \cdot P(C) = \frac{3}{6} \cdot \frac{18}{36}$, so A, B, and C are pairwise independent.

2.5.17. 0.56

2.5.19. Let p be the probability of having a winning game card.

Then $0.32 = P(\text{winning at least once in 5 tries})$
$= 1 - P(\text{not winning in 5 tries})$
$= 1 - (1-p)^5$, so $p = 0.074$

2.5.21. 7

2.5.23. 63/384

2.5.25. 25

2.5.27. $\frac{w}{w+r}$

2.5.29. 12

Section 2.6

2.6.1. $2 \cdot 3 \cdot 2 \cdot 2 = 24$

2.6.3. $3 \cdot 3 \cdot 5 = 45$; included are *aeu* and *cdx*

2.6.5. $9 \cdot 9 \cdot 8 = 648$; $8 \cdot 8 \cdot 5 = 320$

2.6.7. $5 \cdot 2^7 = 640$

2.6.9. $4 \cdot 14 \cdot 6 + 4 \cdot 6 \cdot 5 + 14 \cdot 6 \cdot 5 + 4 \cdot 14 \cdot 5 = 1156$

2.6.11. $2^8 - 1 = 255$; five families can be added

2.6.13. $2^8 - 1 = 255$

2.6.15. $12 \cdot 4 + 1 \cdot 3 = 51$

2.6.17. $6 \cdot 5 \cdot 4 = 120$

2.6.19. 2.645×10^{32}

2.6.21. $2 \cdot 6 \cdot 5 = 60$

2.6.23. $4 \cdot {}_{10}P_3 = 2880$

2.6.25. $6! - 1 = 719$

2.6.27. $(2!)(8!)(6) = 483,840$

2.6.29. $(13!)^4$

2.6.31. $9(8)4 = 288$

2.6.33. **(a)** $(4!)(5!) = 2880$ **(b)** $6(4!)(5!) = 17,280$ **(c)** $(4!)(5!) = 2880$ **(d)** $2(9)8(7)6(5) = 30,240$

2.6.35. $\frac{6!}{3!(1!)^3} + \frac{6!}{2!2!(1!)^2} = 180$

2.6.37. **(a)** $4!3!3! = 864$ **(b)** $3!4!3!3! = 5184$ **(c)** $10! = 3,628,800$ **(d)** $10!/4!3!3! = 4200$

2.6.39. $(2n)!/n!(2!)^n = 1 \cdot 3 \cdot 5 \ldots \cdot (2n-1)$

2.6.41. $11 \cdot 10!/3! = 6,652,800$

2.6.43. $4!/2!2! = 6$

2.6.45. $6!/3!3! = 20$

2.6.47. $\frac{1}{30} \cdot \frac{14!}{2!2!1!2!2!3!1!1!} = 30,270,240$

2.6.49. The three courses with A grade can be:
English, math, French English, math, psychology
English, math, history English, French, psychology
English, French, history English, psychology, history
math, French, psychology math, French, history
math, psychology, history French, psychology, history

2.6.51. $\binom{10}{6}\binom{15}{3} = 95,550$

2.6.53. **(a)** $\binom{9}{4} = 126$ **(b)** $\binom{5}{2}\binom{4}{2} = 60$ **(c)** $\binom{9}{4} - \binom{5}{4} - \binom{4}{4} = 120$

2.6.55. $\binom{10}{5}\Big/2 = 126$

2.6.57. $\binom{8}{4}\frac{7!}{2!4!1!} = 7350$

2.6.59. Consider the problem of selecting an unordered sample of n objects from a set of $2n$ objects, where the $2n$ have been divided into two groups, each of size n. Clearly, we could choose n from the first group and 0 from the second group, or $n-1$ from the first group and 1 from the second group, and so on. Altogether $\binom{2n}{n}$ must equal $\binom{n}{n}\binom{n}{0}+\binom{n}{n-1}\binom{n}{1}+\ldots+\binom{n}{0}\binom{n}{n}$. But $\binom{n}{j}=\binom{n}{n-j}$, $j=0,1,\ldots,n$ so $\binom{2n}{n}=\sum_{j=0}^{n}\binom{n}{j}^2$.

2.6.61. The ratio of two successive terms in the sequence is $\binom{n}{j+1}\Big/\binom{n}{j}=\frac{n-j}{j+1}$. For small j, $n-j>j+1$, implying that the terms are increasing. For $j>\frac{n-1}{2}$, though, the ratio is less than 1, meaning the terms are decreasing.

2.6.63. Using Newton's binomial expansion, the equation $(1+t)^d\cdot(1+t)^e=(1+t)^{d+e}$ can be written

$$\left(\sum_{j=0}^{d}\binom{d}{j}t^j\right)\cdot\left(\sum_{j=0}^{e}\binom{e}{j}t^j\right)=\sum_{j=0}^{d+e}\binom{d+e}{j}t^j$$

Since the exponent k can arise as $t^0\cdot t^k$, $t^1\cdot t^{k-1},\ldots,$ or $t^k\cdot t^0$, it follows that $\binom{d}{0}\binom{e}{k}+\binom{d}{1}\binom{e}{k-1}+\cdots+\binom{d}{k}\binom{e}{0}=\binom{d+e}{k}$. That is, $\binom{d+e}{k}=\sum_{j=0}^{k}\binom{d}{j}\binom{e}{k-j}$.

Section 2.7

2.7.1. 63/210

2.7.3. $1-\frac{37}{190}$

2.7.5. 10/19 (recall Bayes's rule)

2.7.7. $1/6^{n-1}$

2.7.9. $2(n!)^2/(2n)!$

2.7.11. $7!/7^7$; $1/7^6$. The assumption being made is that all possible departure patterns are equally likely, which is probably not true, since residents living on lower floors would be less inclined to wait for the elevator than would those living on the top floors.

2.7.13. $2^{10}\Big/\binom{20}{10}$

2.7.15. $\binom{k}{2}\cdot\frac{365\cdot364\cdots(365-k+2)}{(365)^k}$

2.7.17. $\binom{11}{3}\Big/\binom{47}{3}$

2.7.19. $2\Big/\binom{47}{2}$; $\left[\binom{10}{2}-2\right]\Big/\binom{4}{2}$

2.7.21. $\binom{5}{3}\binom{4}{2}^3\binom{3}{1}\binom{4}{2}\binom{2}{1}\binom{4}{1}\Big/\binom{52}{9}$

2.7.23. $\left[\binom{2}{1}\binom{2}{1}\right]^4\binom{32}{4}\Big/\binom{48}{12}$

CHAPTER 3

Section 3.2

3.2.1. 0.211

3.2.3. $(0.23)^{12}\doteq 1/45,600,000$

3.2.5. 0.0185

3.2.7 The probability that a two engine plane lands safely is 0.84. The probability that a four engine plane lands safely is 0.8208.

3.2.9. $n=6$: 0.67 $n=12$: 0.62 $n=18$: 0.60

3.2.11. The probability of two girls and two boys is 0.375. The probability of three of one sex and one of the other is 0.5.

3.2.13. 7

3.2.15. **(1)** 0.273 **(2)** 0.756

3.2.17. Expanding $[(p+(1-p)]^n$ gives $1=[(p+(1-p)]^n=\sum_{k=0}^{n}\binom{n}{k}p^k(1-p)^{n-k}$

3.2.19. 0.031

3.2.21. 64/84

3.2.23. 0.967

3.2.25. 0.964

3.2.27. 0.129

3.2.31. 53/99

3.2.33. $\frac{\binom{n_1}{r_1}\binom{n_2}{r_2}\binom{n_3}{r_3}}{\binom{N}{r}}$

3.2.35. $\frac{\binom{n_1}{k_1}\binom{n_2}{k_2}\cdots\binom{n_t}{k_t}}{\binom{N}{n}}$

Section 3.3

3.3.1. (a)

k	$p_X(k)$
2	1/10
3	2/10
4	3/10
5	4/10

(b)

k	$p_V(k)$
3	1/10
4	1/10
5	2/10
6	2/10
7	2/10
8	1/10
9	1/10

$k) = k^3/216 - (k-1)^3/216$

$(3) = 1/8;\ p_X(1) = 3/8;\ p_X(-1) = 3/8,\ p_X(-3) = 1/8$

$p_X(2k-4) = \binom{4}{k}\frac{1}{16}, \quad k = 0, 1, 2, 3, 4$

k	$p_X(k)$
0	4/10
1	3/10
2	2/10
3	1/10

3.3.11. $p_{2X+1}(k) = p_X\left(\frac{k-1}{2}\right) = \binom{4}{\frac{k-1}{2}}\left(\frac{2}{3}\right)^{\frac{k-1}{2}}\left(\frac{1}{3}\right)^{4-\frac{k-1}{2}}$, $k = 1, 3, 5, 7, 9$

3.3.13. $F_X(k) = \sum_{j=0}^{k}\binom{4}{j}\left(\frac{1}{6}\right)^j\left(\frac{5}{6}\right)^{4-j}$

3.3.15. See answer to Question 3.3.3.

Section 3.4

3.4.1. 1/16

3.4.3. 13/64

3.4.5. **(a)** 0.135 **(b)** 0.23355

3.4.7. $F_Y(y) = y^4 \quad P(Y \le 1/2) = 1/16$

3.4.9. $F_Y(y) = \begin{cases} \frac{1}{2} + y + \frac{y^2}{2}, & -1 \le y \le 0 \\ \frac{1}{2} + y - \frac{y^2}{2}, & 0 \le y \le 1 \end{cases}$

3.4.11. **(a)** 0.693 **(b)** 0.223 **(c)** 0.223 **(d)** $f_Y(y) = \frac{1}{y}$, $1 \le y \le e$

3.4.13. $f_Y(y) = \frac{1}{6}y + \frac{1}{4}y^2, \quad 0 \le y \le 2$

Section 3.5

3.5.1. -0.144668

3.5.3. \$28,200

3.5.5. \$227.58

3.5.7. 15

3.5.9. 9/4 years

3.5.11. $1/\lambda$

3.5.13. $E(X) = \sum_{k=1}^{200} k\binom{200}{k}(0.80)^k(0.20)^{200-k}$

$E(X) = np = 200(0.80) = 160$

3.5.15. \$307,421.92

3.5.17. 10/3

3.5.19. \$10.95

3.5.21. 91/36

3.5.23. 5.8125

3.5.27. **(a)** $(0.5)^{\frac{1}{\theta+1}}$ **(b)** $\frac{-1+\sqrt{5}}{2}$

3.5.29. $E(Y) = \$132$

3.5.31. \$50,000

3.5.33. Class average = 53.3, so the professor's "curve" did not work.

3.5.35. 16.33

Section 3.6

3.6.1. 12/25

3.6.3. 0.748

3.6.5. 3/80

3.6.7. 1.115

3.6.9. Johnny should pick $(a+b)/2$

3.6.11. $E(Y) = \int_0^\infty y\lambda e^{-\lambda y}dy = 1/\lambda$. $E(Y^2) = \int_0^\infty y^2\lambda e^{-\lambda y}dy = 2/\lambda^2$, using integration by parts. Thus, $\text{Var}(Y) = 2/\lambda^2 - (1/\lambda)^2 = 1/\lambda^2$.

3.6.13. $E[(X-a)^2] = \text{Var}(X) + (\mu - a)^2$ since $E(X-\mu) = 0$. This is minimized when $a = \mu$, so the minimum of $g(a) = \text{Var}(X)$.

3.6.15. 8.7°C

3.6.17. **(a)** $f_Y(y) = \frac{1}{b-a}f_U\left(\frac{y-a}{b-a}\right) = \frac{1}{b-a}$. The interval where Y is non-zero is $(b-a)(0) + a \le y \le (b-a)(1) + a$, or equivalently $a \le y \le b$
(b) $\text{Var}(Y) = \text{Var}[(b-a)U + a] = (b-a)^2\,\text{Var}(U) = (b-a)^2/12$

3.6.19. $\frac{2^r}{r+1}$; 1/7

3.6.21. 9/5

3.6.23. Let $E(X) = \mu$ and $\text{Var}(X) = \sigma^2$. Then $E(aX + b) = a\mu + b$ and $\text{Var}(aX + b) = a^2\sigma^2$. Thus, the standard deviation of $aX + b = a\sigma$ and

$$\gamma_1 = \frac{E\left[((aX+b)-(a\mu+b))^3\right]}{(a\sigma)^3} = \frac{a^3E\left[(X-\mu)^3\right]}{a^3\sigma^3} = \frac{E\left[(X-\mu)^3\right]}{\sigma^3} = \gamma_1(X)$$

The demonstration for γ_2 is similar.

3.6.25. **(a)** $c = 5$ **(b)** Highest integral moment = 4

Section 3.7

3.7.1. 1/10

3.7.3. 2

3.7.5. $P(X = x, Y = y) = \frac{\binom{3}{x}\binom{2}{y}\binom{4}{3-x-y}}{\binom{9}{3}}$, $0 \le x \le 3, 0 \le y \le 2, x + y \le 3$

3.7.7. 13/50

3.7.9. $p_z(0) = 16/36 \quad p_z(1) = 16/36 \quad p_z(2) = 4/36$

3.7.11. 1/2

3.7.13. 19/24

3.7.15. 3/4

3.7.17. $p_X(0) = 1/2 \quad p_X(1) = 1/2$
$p_Y(0) = 1/8 \quad p_X(1) = 3/8 \quad p_X(2) = 3/8 \quad p_X(3) = 1/8$

3.7.19. **(a)** $f_X(x) = 1/2, \quad 0 \le x \le 2$
$f_Y(y) = 1, \quad 0 \le y \le 1$
(b) $f_X(x) = 1/2, \quad 0 \le x \le 2$
$f_Y(y) = 3y^2, \quad 0 \le y \le 1$
(c) $f_X(x) = \frac{2}{3}(x+1), \quad 0 \le x \le 1$
$f_Y(y) = \frac{4}{3}y + \frac{1}{3}, \quad 0 \le y \le 1$
(d) $f_X(x) = x + \frac{1}{2}, \quad 0 \le x \le 1$
$f_Y(y) = y + \frac{1}{2}, \quad 0 \le y \le 1$
(e) $f_X(x) = 2x, \quad 0 \le x \le 1$
$f_Y(y) = 2y, \quad 0 \le y \le 1$
(f) $f_X(x) = xe^{-x}, \quad 0 \le x$
$f_Y(y) = ye^{-y}, \quad 0 \le y$
(g) $f_X(x) = \left(\frac{1}{x+1}\right)^2, \quad 0 \le x$
$f_Y(y) = e^{-y}, \quad 0 \le y$

3.7.21. $f_X(x) = 3 - 6x + 3x^2, \quad 0 \le x \le 1$

3.7.23. X is binomial with $n = 4$ and $p = 1/2$. Similarly, Y is binomial with $n = 4$ and $p = 1/3$.

3.7.25. **(a)** {(H, 1), (H, 2), (H, 3), (H, 4), (H, 5), (H, 6), (T, 1), (T, 2), (T, 3), (T, 4), (T, 5), (T, 6)} **(b)** 4/12

3.7.27. **(a)** $F_{X,Y}(u, v) = \frac{1}{2}uv^3, \quad 0 \le u \le 2, 0 \le v \le 1$
(b) $F_{X,Y}(u, v) = \frac{1}{3}u^2v + \frac{2}{3}uv^2, \quad 0 \le u \le 1, 0 \le v \le 1$
(c) $F_{X,Y}(u, v) = u^2v^2, \quad 0 \le u \le 1, 0 \le v \le 1$

3.7.29. $f_{X,Y}(x, y) = 1, \quad 0 \le x \le 1, 0 \le y \le 1$
The graph of $f_{X,Y}(x, y)$ is a plane of height one over the unit square.

3.7.31. 11/32

3.7.33. 0.015

3.7.35. 25/576

3.7.37. $f_{W,X}(w, x) = 4wx, \quad 0 \le w \le 1, 0 \le x \le 1$
$P(0 \le W \le 1/2, 1/2 \le X \le 1) = 3/16$

3.7.39. $f_X(x) = \lambda e^{-\lambda x}, 0 \le x$ and $f_Y(y) = \lambda e^{-\lambda y}, 0 \le y$

3.7.41. Note that $P(Y \ge 3/4) \ne 0$. Similarly $P(X \ge 3/4) \ne 0$. However, $(X \ge 3/4) \cap (Y \ge 3/4)$ is in the region where the density is 0. Thus, $P((X \ge 3/4) \cap (Y \ge 3/4))$ is zero, but the product $P(X \ge 3/4)P(Y \ge 3/4)$ is not zero.

3.7.43. 2/5

3.7.45. 1/12

3.7.47. $P(0 \le X \le 1/2, 0 \le Y \le 1/2) = 5/32 \ne (3/8)(1/2) = P(0 \le X \le 1/2)P(0 \le Y \le 1/2)$

3.7.49. Let K be the region of the plane where $f_{X,Y} \ne 0$. If K is not a rectangle with sides parallel to the coordinate axes, there exists a rectangle $A = \{(x, y) | a \le x \le b, c \le y \le d\}$ with $A \cap K = \emptyset$, but for $A_1 = \{(x, y) | a \le x \le b, \text{all } y\}$ and $A_2 = \{(x, y) | \text{all } x, c \le y \le d\}$, $A_1 \cap K \ne \emptyset$ and $A_2 \cap K \ne \emptyset$. Then $P(A) = 0$, but $P(A_1) \ne 0$ and $P(A_2) \ne 0$. But $A = A_1 \cap A_2$, so $P(A_1 \cap A_2) \ne P(A_1)P(A_2)$.

3.7.51. **(a)** 1/16 **(b)** 0.206
(c) $f_{X_1,X_2,X_3,X_4}(x_1, x_2, x_3, x_4) = 256(x_1x_2x_3x_4)^3$ where $0 \le x_1, x_2, x_3, x_4 \le 1$ **(d)** $F_{X_2,X_3}(x_2, x_3) = x_2^4x_3^4, \quad 0 \le x_2, x_3 \le 1$

Section 3.8

3.8.1. **(a)** $p_{X+Y}(w) = e^{-(\lambda+\mu)}\frac{(\lambda+\mu)^w}{w!}, \quad w = 0, 1, 2, \ldots,$ so $X + Y$ does belong to the same family.
(b) $p_{X+Y}(w) = (w-1)(1-p)^{w-2}p^2, \quad w = 2, 3, 4, \ldots$
$X + Y$ does not have the same form of pdf as X and Y, but Section 4.5 will show that they all belong to the same family—the negative binomial.

3.8.3. $f_{X+Y}(w) = \begin{cases} w & 0 \le w \le 1 \\ 2 - w & 1 \le w \le 2 \end{cases}$

3.8.5. $F_W(w) = P(W \le w) = P(Y^2 \le w) = P(Y \le \sqrt{w}) = F_Y(\sqrt{w})$
$f_W(w) = F'_W(w) = F'_Y(\sqrt{w}) = \frac{1}{2\sqrt{w}} f_Y(\sqrt{w})$

3.8.7. $3(1 - \sqrt{w}), 0 \le w \le 1$

3.8.9. **(a)** $f_W(w) = -\ln w, \quad 0 \le w \le 1$
(b) $f_W(w) = -4w \ln w, \quad 0 \le w \le 1$

3.8.11. $f_W(w) = \frac{2}{(1+w)^3}, \quad 0 \le w$

Section 3.9

3.9.1. $\frac{r(n+1)}{2}$

3.9.3. $\frac{5}{9} + \frac{11}{18} = \frac{7}{6}$

3.9.5. If and only if $\sum_{i=1}^{n} a_i = 1$

3.9.7. **(a)** $E(X_i)$ is the probability that the i-th ball drawn is red, $1 \le i \le n$. Draw the balls in order without replacement, but do not note the colors. Then look at the i-th ball *first*. The probability that it is red is surely independent of when it is drawn. Thus, all of these expected values are the same and $= r/(r+w)$.
(b) Let X be the number of red balls drawn. Then $X = \sum_{i=1}^{n} X_i$ and $E(X) = \sum_{i=1}^{n} E(X_i) = nr/(r+w)$.

3.9.9. 7.5

3.9.11. 1/8

3.9.13. 105/72

3.9.15. $E(X) = E(Y) = E(XY) = 0$. Then $\text{Cov}(X, Y) = 0$. But X and Y are functionally dependent, $Y = \sqrt{1 - X^2}$, so they are probabilistically dependent.

3.9.17. $2/\lambda^2$

3.9.19. 17/324

3.9.21. \$6750, \$373,500

3.9.23. $\sigma \leq 0.163$

Section 3.10

3.10.1. 5/16

3.10.3. 0.64

3.10.5. $P(Y_1' > m) = P(Y_1, \ldots, Y_n > m) = \left(\frac{1}{2}\right)^n$
$P(Y_n' > m) = 1 - P(Y_n' < m) = 1 - P(Y_1, \ldots, Y_n < m)$
$= 1 - P(Y_1 < m) \cdot \ldots \cdot P(Y_n < m) = 1 - \left(\frac{1}{2}\right)^n$
If $n \geq 2$, the latter probability is greater.

3.10.7. 0.200

3.10.9. $P(Y_{\min} > 20) = (1/2)^n$

3.10.11. 0.725; 0.951

3.10.15. 1/2

Section 3.11

3.11.1. $p_{Y|x}(y) = \dfrac{p_{X,Y}(x, y)}{p_X(x)} = \dfrac{x + y + xy}{3 + 5x}, \quad y = 1, 2$

3.11.3. $p_{Y|x}(y) = \dfrac{\binom{6}{y}\binom{4}{3-x-y}}{\binom{10}{3-x}}, \quad y = 0, 1, \ldots, 3 - x$

3.11.5. **(a)** $k = 1/36$
(b) $p_{Y|x}(1) = \dfrac{x+1}{3x+6}, \quad x = 1, 2, 3$

3.11.7. $p_{X,Y|z}(x, y) = \dfrac{xy + xz + yz}{9 + 12z}, \quad x = 1, 2 \quad y = 1, 2$
$z = 1, 2$

3.11.13. $f_{Y|x}(y) = \dfrac{x+y}{x+\frac{1}{2}}, \quad 0 \leq y \leq 1$

3.11.15. $f_Y(y) = \frac{1}{3}(2y + 2), \quad 0 \leq y \leq 1$

3.11.17. 2/3

3.11.19. $f_{X_1,X_2,X_3|x_4x_5}(x_1, x_2, x_3) = 8x_1x_2x_3, \quad 0 \leq x_1, x_2, x_3 \leq 1$
Note: the five random variables are independent, so the conditional pdf's are just the marginal pdf's.

Section 3.12

3.12.1. $M_X(t) = E\left(e^{tX}\right) = \sum_{k=0}^{n-1} e^{tk} p_X(k) = \sum_{k=0}^{n-1} e^{tk}\dfrac{1}{n}$
$= \dfrac{1}{n}\sum_{k=0}^{n-1}(e^t)^k = \dfrac{1 - e^{nt}}{n(1 - e^t)}$

3.12.3. $\dfrac{1}{3^{10}}(2 + e^3)^{10}$

3.12.5. **(a)** Normal with $\mu = 0$ and $\sigma^2 = 12$
(b) Exponential with $\lambda = 2$
(c) Binomial with $n = 4$ and $p = 1/2$
(d) Geometric with $p = 0.3$

3.12.7. $M_X(t) = e^{\lambda(e^t - 1)}$

3.12.9. 0

3.12.11. $M_Y^{(1)}(t) = \frac{d}{dt}e^{at+b^2t^2/2} = (a + b^2t)e^{at+b^2t^2/2}$, so $M_Y^{(1)}(0) = a$
$M_Y^{(2)}(t) = (a + b^2t)^2e^{at+b^2t^2/2} + b^2e^{at+b^2t^2/2}$, so
$M_Y^{(2)}(0) = a^2 + b^2$. Then $\text{Var}(Y) = (a^2 + b^2) - a^2 = b^2$.

3.12.13. 9

3.12.15 $E(Y) = \dfrac{a+b}{2}$

3.12.17. $M_Y(t) = \left(\dfrac{1}{1 - t/\lambda}\right)^2$

3.12.19. **(a)** True
(b) False
(c) True

3.12.21. $\overline{Y}$ is normally distributed with mean μ and variance σ^2/n.

3.12.23. **(a)** $M_W(t) = M_{3X}(t) = M_X(3t) = e^{-\lambda+\lambda e^{3t}}$. This last term is not the moment-generating function of a Poisson random variable, so W is not Poisson.
(b) $M_W(t) = M_{3X+1}(t) = e^t M_X(3t) = e^t e^{-\lambda+\lambda e^{3t}}$. This last term is not the moment-generating function of a Poisson random variable, so W is not Poisson.

CHAPTER 4

Section 4.2

4.2.1. Binomial answer: 0.158; Poisson approximation: 0.158. The agreement is not surprising because n (= 6000) is so large and p (= 1/3250) is so small.

4.2.3. 0.602

4.2.5. For both the binomial formula and the Poisson approximation, $P(X \geq 1) = 0.10$. The exact model that applies here is the hypergeometric, rather than the binomial, because $p = P$ (ith item must be checked) is a function of the previous $i - 1$ items purchased. However, the variation in p is likely to be so small that the binomial and hypergeometric distributions in this case are essentially the same.

4.2.7. 0.122

4.2.9. 6.9×10^{-12}

4.2.11. The Poisson model $p_X(k) = e^{-0.435}(0.435)^k/k!$, $k = 0, 1, \ldots$ fits the data fairly well. The expected frequencies corresponding to $k = 0, 1, 2$, and 3+ are 230.3, 100.4, 21.7, and 3.6, respectively.

4.2.13. The model $p_x(k) = e^{-0.363}\dfrac{0.363^k}{k!}$ fits the data well if we follow the usual statistical practice of collapsing low frequency categories, in this case $k = 2, 3, 4$.

Number of Countries, k	Frequency	$p_x(k)$	Expected Frequency
0	82	0.696	78.6
1	25	0.252	28.5
2+	6	0.052	5.9

The level of agreement between the observed and expected frequencies suggests that the Poisson is a good model for these data.

4.2.15. If the mites exhibit any sort of "contagion" effect, the independence assumption implicit in the Poisson model will be violated. Here, $\bar{x} = \frac{1}{100}[55(0) + 20(1) + \ldots + 1(7)] = 0.81$, but $p_X(k) = e^{-0.81}(0.81)^k/k!$, $k = 0, 1, \ldots$ does not adequately approximate the infestation distribution.

No. of Infestations, k	Frequency	Proportion	$p_X(k)$
0	55	0.55	0.4449
1	20	0.20	0.3603
2	21	0.21	0.1459
3	1	0.01	0.0394
4	1	0.01	0.0080
5	1	0.01	0.0013
6	0	0	0.0002
7+	1	0.01	0.0000
		1.00	1.00

4.2.17. 0.826

4.2.19. 0.762

4.2.21. **(a)** 0.076
(b) No. P(4 accidents occur during next two weeks) $= P(X = 4) \cdot P(X = 0) + P(X = 3) \cdot P(X = 1) + P(X = 2) \cdot P(X = 2) + P(X = 1) \cdot P(X = 3) + P(X = 0) \cdot P(X = 4)$.

4.2.23. $P(X \text{ is even}) = \sum_{k=0}^{\infty} \frac{e^{-\lambda}\lambda^{2k}}{(2k)!} = e^{-\lambda}\left\{1 + \frac{\lambda^2}{2!} + \frac{\lambda^4}{4!} + \frac{\lambda^6}{6!} + \cdots\right\}$

$$= e^{-\lambda} \cdot \cosh\lambda = e^{-\lambda}\left(\frac{e^{\lambda}+e^{-\lambda}}{2}\right) = \frac{1}{2}(1 + e^{-2\lambda}).$$

4.2.25. $P(X_2 = k) = \sum_{x_1=k}^{\infty} \binom{x_1}{k} p^k (1-p)^{x_1-k} \cdot \frac{e^{-\lambda}\lambda^{x_1}}{x_1!}$. Let $y = x_1 - k$. Then $P(X_2 = k) = \sum_{y=0}^{\infty} \binom{y+k}{k} p^k (1-p)^y \cdot \frac{e^{-\lambda}\lambda^{y+k}}{(y+k)!} = \frac{e^{-\lambda}(\lambda p)^k}{k!} \cdot \sum_{y=0}^{\infty} \frac{[\lambda(1-p)]^y}{y!} = \frac{e^{-\lambda}(\lambda p)^k}{k!} \cdot e^{\lambda(1-p)} = \frac{e^{-\lambda p}(\lambda p)^k}{k!}$.

4.2.27. 0.50

4.2.29. 28

Section 4.3

4.3.1. **(a)** 0.5782 **(b)** 0.8264 **(c)** 0.9306 **(d)** 0.0000

4.3.3. **(a)** Both are the same **(b)** $\int_{a-\frac{1}{2}}^{a+\frac{1}{2}} \frac{1}{\sqrt{2\pi}} e^{-z^2/2}\, dz$

4.3.5. **(a)** −0.44 **(b)** 0.76 **(c)** 0.41 **(d)** 1.28
(e) 0.95

4.3.7. 0.0655

4.3.9. **(a)** 0.0053 **(b)** 0.0197

4.3.11. $P(X \geq 344) \doteq P(Z \geq 13.25) = 0.0000$, which strongly discredits the hypothesis that people die randomly with respect to their birthdays.

4.3.13. The normal approximation does not apply because the needed condition $n > 9p/(1-p) = 9(0.7)/0.3 = 21$ does not hold.

4.3.15. 0.5646
For binomial data, the central limit theorem and DeMoivre-Laplace approximations differ only if the continuity correction is used in the DeMoivre-Laplace approximation.

4.3.17. 0.6808

4.3.19. 0.0694

4.3.21. No, only 84% of drivers are likely to get at least 25,000 miles on the tires.

4.3.23. 0.0228

4.3.25. **(a)** 6.68%; 15.87%

4.3.27. 434

4.3.29. 29.85

4.3.31. 0.0062. The "0.075%" driver should ask to take the test twice; the "0.09%" driver has a greater chance of not being charged by taking the test only once. As n, the number of times the test is taken, increases, the precision of the average reading increases. It is to the sober driver's advantage to have a reading as precise as possible; the opposite is true for the drunk driver.

4.3.33. 0.23

4.3.35. $\sigma = 0.22$ ohms

Section 4.4

4.4.1. 0.343

4.4.3. No, the expected frequencies ($= 50 \cdot p_X(k)$) differ considerably from the observed frequencies, especially for small values of k. The observed number of 1's, for example, is 4, while the expected number is 12.5.

4.4.5. $F_X(t) = P(X \leq t) = p\sum_{s=0}^{[t]}(1-p)^s$. But $\sum_{s=0}^{[t]}(1-p)^s = \frac{1-(1-p)^{[t]}}{1-(1-p)} = \frac{1-(1-p)^{[t]}}{p}$, and the result follows.

4.4.7. $P(n \leq Y \leq n+1) = \int_n^{n+1} \lambda e^{-\lambda y}\, dy = (1 - e^{-\lambda y})\Big|_n^{n+1} = e^{-\lambda n} - e^{-\lambda(n+1)} = e^{-\lambda n}(1 - e^{-\lambda})$. Setting $p = 1 - e^{-\lambda}$ gives $P(n \leq Y \leq n+1) = (1-p)^n p$.

Section 4.5

4.5.1. 0.029

4.5.3. Probably not. The presumed model, $p_X(k) = \binom{k-1}{1}\left(\frac{1}{2}\right)^2\left(\frac{1}{2}\right)^{k-2}$, $k = 2, 3, \ldots$ fits the data almost perfectly, as the table shows. Agreement this good is often an indication that the data have been fabricated.

k	$p_X(k)$	Obs. Freq.	Exp. Freq.
2	1/4	24	25
3	2/8	26	25
4	3/16	19	19
5	4/32	13	12
6	5/64	8	8
7	6/128	5	5
8	7/256	3	3
9	8/512	1	2
10	9/1024	1	1

4.5.5. $E(X)=\sum_{k=r}^{\infty} k\binom{k-1}{r-1}p^r(1-p)^{k-r}$

$$=\frac{r}{p}\sum_{k=r}^{\infty}\binom{k}{r}p^{r+1}(1-p)^{k-r}=\frac{r}{p}.$$

4.5.7. The given $X=Y-r$, where Y has the negative binomial pdf as described in Theorem 4.5.1. Then $E(X)=\frac{r}{p}-p=\frac{r(1-p)}{p}$, $\text{Var}(X)=\frac{r(1-p)}{p^2}$, $M_X(t)=\left[\frac{p}{1-(1-p)e^t}\right]^r$

4.5.9. $M_X^{(1)}(t)=r\left[\frac{pe^t}{1-(1-p)e^t}\right]^{r-1}\{pe^t[1-(1-p)e^t]^{-2}$ $(1-p)e^t+[1-(1-p)e^t]^{-1}pe^t\}$. When $t=0$, $M_X^{(1)}(0)=E(X)$ $=r\left[\frac{p(1-p)}{p^2}+\frac{p}{p}\right]=\frac{r}{p}$.

Section 4.6

4.6.1. $f_Y(y)=\frac{(0.001)^3}{2}y^2e^{-0.001y},\quad 0\le y$

4.6.3. If $E(Y)=\frac{r}{\lambda}=1.5$ and $\text{Var}(Y)=\frac{r}{\lambda^2}=0.75$, then $r=3$ and $\lambda=2$, which makes $f_Y(y)=4y^2e^{-2y}$, $y>0$. Then $P(1.0\le Y_i\le 2.5)=\int_{1.0}^{2.5}4y^2e^{-2y}dy=0.55$. Let X = number of Y_i's in the interval (1.0, 2.5). Since X is a binomial random variable with $n=100$ and $p=0.55$, $E(X)=np=55$.

4.6.5. Setting the first derivative of $f_Y^{(y)}$ equal to 0 gives

$$\frac{\lambda^r}{\Gamma(r)}e^{-\lambda y}\{-\lambda_y^{r-1}+(r-1)_y^{r-2}\}=0$$

which implies that $(r-1)_y^{r-2}=\lambda_y^{r-1}$, so $y=\frac{r-1}{\lambda}$ is a mode. Its uniqueness follows from the fact that the second derivative of $f_Y(y)$ is negative for all other y for which $f_Y(y)$ is defined.

$$f_{\lambda Y}(y)=\frac{1}{\lambda}f_Y(y/\lambda)=\frac{1}{\lambda}\frac{\lambda^r}{\Gamma(r)}\left(\frac{y}{\lambda}\right)^{r-1}e^{-\lambda(y/\lambda)}$$
$$=\frac{1}{\Gamma(r)}y^{r-1}e^{-y}$$

4.6.7. $\Gamma\left(\frac{7}{2}\right)=\frac{5}{2}\Gamma\left(\frac{5}{2}\right)=\frac{5}{2}\frac{3}{2}\Gamma\left(\frac{3}{2}\right)=\frac{5}{2}\frac{3}{2}\frac{1}{2}\Gamma\left(\frac{1}{2}\right)=\frac{15}{8}\Gamma\left(\frac{1}{2}\right)$ by Theorem 4.6.2 (2), and $\Gamma\left(\frac{1}{2}\right)=\sqrt{\pi}$ by Question 4.6.6.

4.6.9. Write the gamma moment-generating function in the form $M_Y(t)=(1-t/\lambda)^{-r}$. Then $M_Y^{(1)}(t)=-r(1-t/\lambda)^{-r-1}(-1/\lambda)=(r/\lambda)(1-t/\lambda)^{-r-1}$ and $M_Y^{(2)}(t)=(r/\lambda)(-r-1)(1-t/\lambda)^{-r-2}\cdot(-1/\lambda)=(r/\lambda^2)(r+1)(1-t/\lambda)^{-r-2}$. Therefore, $E(Y)=M_Y^{(1)}(0)=\frac{r}{\lambda}$ and $\text{Var}(Y)=M_Y^{(2)}(0)-\left[M_Y^{(1)}(0)\right]^2=\frac{r(r+1)}{\lambda^2}-\frac{r^2}{\lambda^2}=\frac{r}{\lambda^2}$.

CHAPTER 5

Section 5.2

5.2.1. 5/8

5.2.3. 0.122

5.2.5. 0.733

5.2.7. 8.00

5.2.9. **(a)** $\lambda=[0(6)+1(19)+2(12)+3(13)+4(9)]/59=2.00$

Number of No-hitters	Observed Frequency	Expected Frequency
0	6	8.0
1	19	16.0
2	12	16.0
3	13	10.6
4	9	8.4

(b) The agreement is reasonably good considering the number of changes in baseball over the 59 years—most notably the change in the height of the pitcher's mound.

5.2.11. $y_{\min}$

5.2.13. $\frac{25}{-25\ln k+\sum_{i=1}^{n}\ln y_i}$

5.2.15. $\theta_e=\sigma_e^2=\frac{1}{n}\sum_{i=1}^{n}(y_i-\mu)^2$

5.2.17. $\frac{2\bar{y}}{1-\bar{y}}$

5.2.19. $1/\bar{y}$

5.2.21. $\bar{y}/(\bar{y}-k)$

5.2.23. $\frac{1}{n}\sum_{i=1}^{n}y_i^2-\bar{y}^2$

5.2.25. $E(X)=1/p$ and $p_e=\frac{1}{\bar{x}}$. For the given data, $p_e=0.479$. The expected frequencies are:

X	Observed Frequency	Expected Frequency
1	132	119.8
2	52	62.4
3	34	32.5
4	9	16.9
5	7	8.8
6	5	4.6
7	5	2.4
≥ 8	6	2.6

Section 5.3

5.3.1. Confidence interval is (103.7, 112.1).

5.3.3. The confidence interval is (64.432, 77.234). Since 80 does not fall within the confidence interval, that men and women metabolize methylmercury at the same rate is not believable.

5.3.5. 336

5.3.7. 0.501

5.3.9. The interval given is correctly *calculated*. However, the data do not appear to be normal, so claiming that it is a 95% confidence interval would not be correct.

5.3.9. (0.316, 0.396)

5.3.11. (0.254, 0.300)

5.3.13. Since 0.54 does not fall in the confidence interval of (0.61, 0.65), the increase could be considered significant.

5.3.15. 16,641

5.3.17. Both intervals have confidence level approximately 50%.

5.3.19. The margin of error is correct at the 95% level. The confidence interval is (0.559, 0.621).

5.3.21. In Definition 5.3.1, substitute $d = \dfrac{1.96}{2\sqrt{n}}\sqrt{\dfrac{N-n}{N-1}}$

5.3.23. For margin of error 0.06, $n = 267$. For margin of error 0.03, $n = 1068$.

5.3.25. The first case requires $n = 421$; the second, $n = 479$.

5.3.27. 1024

Section 5.4

5.4.1. 2/10

5.4.3. 0.1841

5.4.5. (a) $E(\bar{X}) = E\left(\dfrac{1}{n}\sum_{i=1}^{n} X_i\right) = \dfrac{1}{n}\sum_{i=1}^{n} E(X_i) = \dfrac{1}{n}\sum_{i=1}^{n}\lambda = \lambda$

(b) In general, the sample mean is an unbiased estimator of the mean μ.

5.4.7. By Theorem 3.10.1, $f_{Y_{\min}}(y) = ne^{-n(y-\theta)}, \theta \le y$. Then

$$E(Y_{\min}) = \int_{\theta}^{\infty} y \cdot ne^{-(y-\theta)}dy$$

$$= \int_{0}^{\infty} (u+\theta)\cdot ne^{-u}\,du = \int_{0}^{\infty} u\cdot ne^{-u}\,du$$

$$+\theta\int_{0}^{\infty} ne^{-u}\,du = \frac{1}{n} + \theta,$$

and $E(Y_{\min} - \frac{1}{n}) = \frac{1}{n} + \theta - \frac{1}{n} = \theta$

5.4.9. 1/2

5.4.11. $E(W^2) = \mathrm{Var}(W) + E(W)^2 = \mathrm{Var}(W) + \theta^2$. Thus, W^2 is unbiased only if $\mathrm{Var}(W) = 0$, which in essence means that W is constant.

5.4.13. The median of $\hat{\theta}$ is $\dfrac{(n+1)}{n\sqrt[n]{2}}\theta$, which is unbiased only if $n = 1$.

5.4.15. $E(\bar{W}^2) = \mathrm{Var}(\bar{W}) + E(\bar{W})^2 = \dfrac{\sigma^2}{n} + \mu^2$, so $\lim_{n\to\infty} E(\bar{W}^2) = \lim_{n\to\infty}\left(\dfrac{\sigma^2}{n} + \mu^2\right) = \mu^2$

5.4.17. (a) $E(\hat{p}_1) = E(X_1) = p$. $E(\hat{p}_2) = E\left(\dfrac{X}{n}\right) = \dfrac{1}{n}E(X) = \dfrac{1}{n}np = p$, so both $\hat{p}_1$ and $\hat{p}_2$ are unbiased estimators of p.

(b) $\mathrm{Var}(\hat{p}_1) = p(1-p)$; $\mathrm{Var}(\hat{p}_2) = p(1-p)/n$, so $\mathrm{Var}(\hat{p}_2)$ is smaller by a factor of n.

5.4.19. (a) See the solution to Question 5.4.14.

(b) $\mathrm{Var}(\hat{\theta}_1) = \mathrm{Var}(Y_2) = \theta^2$, since Y_1 is a gamma variable with parameters 1 and $1/\theta$.

$$\mathrm{Var}(\hat{\theta}_2) = \mathrm{Var}(\overline{Y}) = \theta^2/n.$$

From the solution to Question 5.4.14, it follows that $nY_{\min}$ is a gamma variable with parameters 1 and θ^2/n^2, so $\mathrm{Var}(\hat{\theta}_3) = \mathrm{Var}(nY_{\min}) = \theta^2/n^2$.

(c) $\mathrm{Var}(\hat{\theta}_3)/\mathrm{Var}(\hat{\theta}_1) = ((\theta^2/n^2)/\theta^2) = 1/n^2$

$\mathrm{Var}(\hat{\theta}_3)/Var(\hat{\theta}_2) = (\theta^2/n^2)/(\theta^2/n) = 1/n$

5.4.21. $\mathrm{Var}(\hat{\theta}_1) = \mathrm{Var}\left(\dfrac{n+1}{n}Y_{\max}\right) = \dfrac{\theta^2}{n(n+2)}$

$$\mathrm{Var}(\hat{\theta}_2) = \mathrm{Var}((\mathrm{n}+1)Y_{\min}) = \frac{n\theta^2}{(n+2)}$$

$$\mathrm{Var}(\hat{\theta}_2)/\mathrm{Var}(\hat{\theta}_1) = \frac{n\theta^2}{(n+2)} \bigg/ \frac{\theta^2}{n(n+2)} = n^2$$

Section 5.5

5.5.1. The Cramer-Rao bound is θ^2/n. $\mathrm{Var}(\hat{\theta}) = \mathrm{Var}(\bar{Y}) = \mathrm{Var}(Y)/n = \theta^2/n$, so $\hat{\theta}$ is a best estimator.

5.5.3. The Cramer-Rao bound is σ^2/n. $\mathrm{Var}(\hat{\mu}) = \mathrm{Var}(\bar{Y}) = \mathrm{Var}(Y)/n = \sigma^2/n$, so $\hat{\mu}$ is an efficient estimator.

5.5.5. The Cramer-Rao bound is $\dfrac{(\theta-1)\theta}{n}$. $\mathrm{Var}(\hat{\theta}) = \mathrm{Var}(\bar{X}) = \mathrm{Var}(X)/n = \dfrac{(\theta-1)\theta}{n}$, so $\hat{\theta}$ is an efficient estimator.

5.5.7. $E\left(\dfrac{\partial^2 \ln f_W(W;\theta)}{\partial\theta^2}\right) = \displaystyle\int_{-\infty}^{\infty}\frac{\partial}{\partial\theta}\left(\frac{\partial \ln f_W(w;\theta)}{\partial\theta}\right) \times f_W(w;\theta)\,dw$

$$= \int_{-\infty}^{\infty}\frac{\partial}{\partial\theta}\left(\frac{1}{f_W(w;\theta)}\frac{\partial f_W(w;\theta)}{\partial\theta}\right) \times f_W(w;\theta)\,dw$$

$$= \int_{-\infty}^{\infty}\left[\frac{1}{f_W(w;\theta)}\frac{\partial^2 f_W(w;\theta)}{\partial\theta^2} - \frac{1}{(f_W(w;\theta))^2}\left(\frac{\partial f_W(w;\theta)}{\partial\theta}\right)^2\right] \times f_W(w;\theta)\,dw$$

$$= \int_{-\infty}^{\infty} \frac{\partial^2 f_W(w;\theta)}{\partial \theta^2} dw - \int_{-\infty}^{\infty} \frac{1}{(f_W(w;\theta))^2} \left(\frac{\partial f_W(w;\theta)}{\partial \theta}\right)^2 \times f_W(w;\theta)\, dw$$

$$= 0 - \int_{-\infty}^{\infty} \left(\frac{\partial \ln f_W(w;\theta)}{\partial \theta}\right)^2 \times f_W(w;\theta)\, dw$$

The 0 occurs because $1 = \int_{-\infty}^{\infty} f_W(w;\theta)\, dw$, so

$$0 = \frac{\partial^2 \int_{-\infty}^{\infty} f_W(w;\theta)\, dw}{\partial \theta^2} = \int_{-\infty}^{\infty} \frac{\partial^2 f_W(w;\theta)}{\partial \theta^2} dw$$

The above argument shows that

$$E\left(\frac{\partial^2 \ln f_W(W;\theta)}{\partial \theta^2}\right) = -E\left(\frac{\partial \ln f_W(W;\theta)}{\partial \theta}\right)^2$$

Multiplying both sides of the equality by n and inverting gives the desired equality.

Section 5.6

5.6.1. $\prod_{i=1}^{n} p_X(k_i;p) = \prod_{i=1}^{n}(1-p)^{k_i-1}p = (1-p)^{\left(\sum_{i=1}^{n} k_i\right)-n} p^n$.

Let $g\left(\sum_{i=1}^{n} k_i; p\right) = (1-p)^{\left(\sum_{i=1}^{n} k_i\right)-n} p^n$ and $u(k_1, k_2, \ldots, k_n) = 1$.

By Theorem 5.6.1, $\sum_{i=1}^{n} X_i$ is sufficient.

5.6.3. In the discrete case, and for a one-to-one function g, note that $P(X_1 = x_1, X_2 = x_2, \ldots, X_n = x_1 | g(\hat{\theta}) = \theta_e) = P(X_1 = x_1, X_2 = x_2, \ldots, X_n = x_1 | \hat{\theta} = g^{-1}(\theta_e))$

The right hand term does not depend on θ, because $\hat{\theta}$ is sufficient.

5.6.5. The likelihood function is $\left[\frac{1}{\theta^{rn}} e^{\frac{1}{\theta}\sum_{i=1}^{n} y_i}\right]$ $\frac{1}{[(r-1)!]^n}\left(\prod_{i=1}^{n} y_i\right)^{r-1}$ so $\sum_{i=1}^{n} Y_i$ is a sufficient statistic for θ. So also is $\frac{1}{r}\bar{Y}$. (See Question 5.6.3.)

5.6.7. (a) Write the pdf in the form $f_Y(y) = e^{-(y-\theta)} \cdot I_{[\theta,\infty]}(y)$ where $I_{[\theta,\infty]}(y)$ is the indicator function introduced in Example 5.6.2. Then the likelihood function is

$$L(\theta) = \prod_{i=1}^{n} e^{-(y_i-\theta)} \cdot I_{[\theta,\infty]}(y_i) = e^{-\sum_{i=1}^{n} y_i} e^{n\theta} \prod_{i=1}^{n} I_{[\theta,\infty]}(y_i)$$

But $\prod_{i=1}^{n} I_{[\theta,\infty]}(y_i) = I_{[\theta,\infty]}(y_{\min})$, so the likelihood function factors into

$$L(\theta) = \left(e^{-\sum_{i=1}^{n} y_i}\right)[e^{n\theta}, I_{[\theta,\infty]}(y_{\min})]$$

Thus the likelihood function decomposes in such a way that the factor involving θ only contains the y_i's through $y_{\min}$. By Theorem 5.6.1, $y_{\min}$ is sufficient.

(b) We need to show that the likelihood function given y_{max} is independent of θ. But the likelihood function is

$$\prod_{i=1}^{n} e^{-(y_i-\theta)} = \begin{cases} e^{\theta}\, e^{-\sum_{i=1}^{n} y_i} & \text{if } \theta \le y_1, y_2, \ldots y_n \\ 0 & \text{otherwise} \end{cases}$$

Regardless of the value of $y_{\max}$, the expression for the likelihood does depend on θ. If any one of the y_i, other than $y_{\max}$, is less than θ, the expression is 0. Otherwise it is non-zero.

5.6.9. $\prod_{i=1}^{n} g_W(w_i;\theta) = \left(e^{\left(\sum_{i=1}^{n} K(w_i)\right)p(\theta)+nq(\theta)}\right)\left(e^{\sum_{i=1}^{n} S(w_i)}\right)$, so $\sum_{i=1}^{n} K(W_i)$ is a sufficient statistic for θ by Theorem 5.6.1.

5.6.11. $\theta/(1+y)^{\theta+1} = e^{[\ln(1+y)](-\theta-1)+\ln\theta}$. Take $K(Y) = \ln(1+y)$, $p(\theta) = -\theta - 1$, and $q(\theta) = \ln\theta$. Then $\sum_{i=1}^{n} K(Y_i) = \sum_{i=1}^{n} \ln(1+Y_i)$ is sufficient for θ.

Section 5.7

5.7.1. 17

5.7.3. (a) $P(Y_1 > 2\lambda) = \int_{2\lambda}^{\infty} \lambda e^{-\lambda y} dy = e^{-2\lambda^2}$. Then $P(|Y_1 - \lambda| < \lambda/2) < 1 - e^{-2\lambda^2} < 1$. Thus, $\lim_{n\to\infty} P(|Y_1 - \lambda| < \lambda/2) < 1$.

(b) $P\left(\sum_{i=1}^{n} Y_i > 2\lambda\right) \ge P(Y_1 > 2\lambda) = e^{-2\lambda^2}$. The proof now proceeds along the lines of Part (a).

5.7.5 $E[(Y_{\max} - \theta)^2] = \int_0^{\theta} (y-\theta)^2 \frac{n}{\theta}\left(\frac{y}{\theta}\right)^{n-1} dy$

$$= \frac{n}{\theta^n}\int_0^{\theta} \left(y^{n+1} - 2\theta y^n + \theta^2 y^{n-1}\right) dy$$

$$= \frac{n}{\theta^n}\left(\frac{\theta^{n+2}}{n+2} - \frac{2\theta^{n+2}}{n+1} + \frac{\theta^{n+2}}{n}\right)$$

$$= \left(\frac{n}{n+2} - \frac{2n}{n+1} + 1\right)\theta^2$$

Then $\lim_{n\to\infty} E[(Y_{\max} - \theta)^2] = \lim_{n\to\infty}\left(\frac{n}{n+2} - \frac{2n}{n+1} + 1\right)\theta^2 = 0$, and the estimator is squared error consistent.

Section 5.8

5.8.1. The numerator of $g_\Theta(\theta|X=k)$ is

$$p_X(k|\theta)f_\Theta(\theta)=[(1-\theta)^{k-1}\theta]\frac{\Gamma(r+s)}{\Gamma(r)\Gamma(s)}\theta^{r-l}(1-\theta)^{s-l}$$

$$=\frac{\Gamma(r+s)}{\Gamma(r)\Gamma(s)}\theta^{r}(1-\theta)^{s+k-2}$$

The term $\theta^r(1-0)^{s+k-2}$ is the variable part of the beta distribution with parameters $r+1$ and $s+k-1$, so that is the pdf $g_\Theta(\theta|\ X=k)$.

5.8.3. (a) The posterior distribution is a beta pdf with parameters $k+135$ and $n-k+135$.

(b) The mean of the Bayes pdf given in part (a) is

$$\frac{k+135}{k+135+n-k+135}=\frac{k+135}{n+270}$$

$$=\frac{n}{n+270}\left(\frac{k}{n}\right)+\frac{270}{n+270}\left(\frac{135}{270}\right)$$

$$=\frac{n}{n+270}\left(\frac{k}{n}\right)+\frac{270}{n+270}\left(\frac{1}{2}\right)$$

5.8.5. In each case the estimator is biased, since the mean of the estimator is a weighted average of the unbiased maximum likelihood estimator and a non-zero constant. However, in each case, the weighting on the maximum likelihood estimator tends to 1 as n tends to ∞, so these estimators are asymptotically unbiased.

5.8.7. Since the sum of gamma random variables is gamma, then W is gamma with parameters nr and λ. Then $g_\Theta(\theta|X=k)$ is a gamma pdf with parameters $nr+s$ and $\sum_{i=1}^{n} y_i+\mu$.

5.8.9. $p_X(k|\theta)f_\Theta(\theta)=\binom{n}{k}\frac{\Gamma(r+s)}{\Gamma(r)\Gamma(s)}\theta^{k+r-1}(1-\theta)^{n-k+s-1}$, so

$$p_X(k|\theta)=\binom{n}{k}\frac{\Gamma(r+s)}{\Gamma(r)\Gamma(s)}\int_0^1\theta^{k+r-1}(1-\theta)^{n-k+s-1}d\theta$$

$$=\binom{n}{k}\frac{\Gamma(r+s)}{\Gamma(r)\Gamma(s)}\frac{\Gamma(k+r)\Gamma(n-k+s)}{\Gamma(n+r+s)}$$

$$=\frac{n!}{k!(n-k)!}\frac{(r+s-1)!}{(r-1)!(s-1)!}\frac{(k+r-1)!(n-k+s-1)!}{(n+r+s-1)!}$$

$$=\frac{(k+r-1)!}{k!(r-1)!}\frac{(n-k+s-1)!}{(n-k)!(s-1)!}\frac{n!(r+s-1)!}{(n+r+s-1)!}$$

$$=\binom{k+r-1}{k}\binom{n-k+s-1}{n-k}\Big/\binom{n+r+s-1}{n}$$

CHAPTER 6

Section 6.2

6.2.1. (a) Reject H_0 if $\frac{\bar{y}-120}{18/\sqrt{25}}\le-1.41$; $z=-1.61$; reject H_0.

(b) Reject H_0 if $\frac{\bar{y}-42.9}{3.2/\sqrt{16}}$ is either 1) ≤ -2.58 or 2) ≥ 2.58; $z=2.75$; reject H_0.

(c) Reject H_0 if $\frac{\bar{y}-14.2}{4.1/\sqrt{9}}\ge1.13$; $z=1.17$; reject H_0.

6.2.3. (a) No **(b)** Yes

6.2.5. No

6.2.7. (a) H_0 should be rejected if $\frac{\bar{y}-12.6}{0.4/\sqrt{30}}$ is either (1) ≤ -1.96 or (2) ≥ 1.96. But $\bar{y}=12.76$ and $z=2.19$, suggesting that the machine should be readjusted.

(b) The test assumes that the y_i's constitute a random sample from a normal distribution. Graphed, a histogram of the 30 y_i's shows a mostly bell-shaped pattern. There is no reason to suspect that the normality assumption is not being met.

6.2.9. P-value $=P(Z\le-0.92)+P(Z\ge0.92)=0.3576$; H_0 would be rejected if α had been set at any value greater than or equal to 0.3576.

6.2.11. H_0 should be rejected if $\frac{\bar{y}-145.75}{9.50/\sqrt{25}}$ is 1) ≤ -1.96 or 2) ≥ 1.96. Here, $\bar{y}=149.75$ and $z=2.10$, so the difference between \$145.75 and \$149.75 *is* statistically significant.

Section 6.3

6.3.1. (a) $z=0.91$, which is not larger than $z_{.05}$ $(=1.64)$, so H_0 would not be rejected. These data do not provide convincing evidence that transmitting predator sounds helps to reduce the number of whales in fishing waters.

(b) P-value $=P(Z\ge0.91)=0.1824$; H_0 would be rejected for any $\alpha\ge0.1814$.

6.3.3. $z=\frac{72-120(0.65)}{\sqrt{120(0.65)(0.35)}}=-1.15$, which is not less than $-z_{.05}$ $(=-1.64)$, so H_0: $p=0.65$ would not be rejected.

6.3.5. Let $p=P(Y_i\le0.69315)$. Test H_0: $p=\frac{1}{2}$ versus H_1: $p\neq\frac{1}{2}$. Given that $x=26$ and $n=60$, the P-value $=P(X\le 26)+P(X\ge34)=0.3030$.

6.3.7. Reject H_0 if $x\ge4$ gives $\alpha=0.50$; reject H_0 if $x\ge5$ gives $\alpha=0.23$; reject H_0 if $x\ge6$ gives $\alpha=0.06$; reject H_0 if $x\ge7$ gives $\alpha=0.01$.

6.3.9. (a) 0.07

Section 6.4

6.4.1. 0.0735

6.4.3. 0.3786

6.4.5. 0.6293

6.4.7. 95

6.4.9. 0.23

6.4.11. $\alpha=0.064$; $\beta=0.107$. A Type I error (convicting an innocent defendant) would be considered more serious than a Type II error (acquitting a guilty defendant).

6.4.13. 1.98

6.4.15. $\sqrt[n]{0.95}$

6.4.17. $1-\beta=\left(\frac{1}{2}\right)^{\theta+1}$

6.4.19. $\frac{7}{8}$

6.4.21. 0.63

Section 6.5

6.5.1. $\lambda = \max_{\omega} L(p)/\max_{\Omega} L(p)$, where $\max_{\omega} L(p) = p_0^n(1-p_0)^{\sum_{i=1}^{n} k_i - n}$ and $\max_{\Omega} L(p) = \left(n\Big/\sum_{i=1}^{n} k_i\right)^n \left[1-\left(n\Big/\sum_{i=1}^{n} k_i\right)\right]^{\sum_{i=1}^{n} k_i - n}$

6.5.3. $\lambda = \left\{(2\pi)^{-n/2} e^{-\frac{1}{2}\sum_{i=1}^{n}(y_i-\mu_0)^2}\right\} \Big/ \left\{(2\pi)^{-n/2} e^{-\frac{1}{2}\sum_{i=1}^{n}(y_i-\bar{y})^2}\right\}$

$= e^{-\frac{1}{2}\left((\bar{y}-\mu_0)/\left(1/\sqrt{n}\right)\right)^2}.$

Base the test on $z = (\bar{y}-\mu_0)/\left(1/\sqrt{n}\right)$.

6.5.5. (a) $\lambda = \left(\frac{1}{2}\right)^n/[(k/n)^k(1-k/n)^{n-k}] = 2^{-n}k^{-k}(n-k)^{k-n}n^n$. Rejecting H_0 when $0 < \lambda \le \lambda^*$ is equivalent to rejecting H_0 when $k\ln k + (n-k)\ln(n-k) \ge \lambda^{**}$.

(b) By inspection, $k\ln k + (n-k)\ln(n-k)$ is symmetric in k. Therefore, the left-tail and right-tail critical regions will be equidistant from $p = \frac{1}{2}$, which implies that H_0 should be rejected if $\left|k - \frac{1}{2}\right| \ge c$, where c is a function of α.

CHAPTER 7

Section 7.3

7.3.1. Clearly, $f_U(u) > 0$ for all $u > 0$. To verify that $f_U(u)$ is a pdf requires proving that $\int_0^\infty f_U(u)\,du = 1$. But $\int_0^\infty f_U(u)\,du = \frac{1}{\Gamma(n/2)}\int_0^\infty \frac{1}{2^{n/2}}u^{n/2-1}e^{-u/2}du = \frac{1}{\Gamma(n/2)}\int_0^\infty \left(\frac{u}{2}\right)^{n/2-1}e^{-u/2}(du/2) = \frac{1}{\Gamma(\frac{n}{2})} = \int_0^\infty v^{n/2-1}e^{-v}dv$, where $v = \frac{u}{2}$ and $dv = \frac{du}{2}$. By definition, $\Gamma(\frac{n}{2}) = \int_0^\infty v^{n/2-1}e^{-v}dv$. Thus, $\int_0^\infty f_U(u)dy = \frac{1}{\Gamma(n/2)}\cdot\Gamma(\frac{n}{2}) = 1$.

7.3.3. If $\mu = 50$ and $\sigma = 10$, $\sum_{i=1}^{3}\left(\frac{Y_i-50}{10}\right)^2$ should have a χ_3^2 distribution, implying that the numerical value of the sum is likely to be between, say, $\chi^2_{.025,3}(= 0.216)$ and $\chi^2_{.975,3}(= 9.348)$. Here, $\sum_{i=1}^{3}\left(\frac{Y_i-50}{10}\right)^2 = \left(\frac{65-50}{10}\right)^2 + \left(\frac{30-50}{10}\right)^2 + \left(\frac{55-50}{10}\right)^2 = 6.50$, so the data are not inconsistent with the hypothesis that the Y_i's are normally distributed with $\mu = 50$ and $\sigma = 10$.

7.3.5. Since $E(S^2) = \sigma^2$, it follows from Chebyshev's inequality that $P(|S^2-\sigma^2| < \varepsilon) > 1 - \frac{\text{Var}(S^2)}{\varepsilon^2}$. But $\text{Var}(S^2) = \frac{2\sigma^4}{n-1} \to 0$ as $n \to \infty$. Therefore, S^2 is consistent for σ^2.

7.3.7. (a) 0.983
(b) 0.132
(c) 9.00

7.3.9. (a) 6.23
(b) 0.65
(c) 9
(d) 15
(e) 2.28

7.3.11. $F = \frac{V/m}{U/n}$, where V and U are independent χ^2 variables with m and n degrees of freedom, respectively. Then $\frac{1}{F} = \frac{U/n}{V/m}$, which implies that $\frac{1}{F}$ has an F distribution with n and m degrees of freedom.

7.3.15. Let T be a Student t random variable with n degrees of freedom. Then $E(T^{2k}) = C\int_{-\infty}^{\infty} t^{2k}\frac{1}{\left(1+\frac{t^2}{n}\right)^{(n+1)/2}}dt$, where C is the product of the constants appearing in the definition of the Student t pdf. The change of variable $y = t/\sqrt{n}$ results in the integral $E(T^{2k}) = C^*\int_{-\infty}^{\infty} y^{2k}\frac{1}{(1+y^2)^{(n+1)/2}}dy$ for some constant C^*. Because of the symmetry of the integrand, $E(T^{2k})$ is finite if the integral $\int_0^\infty \frac{y^{2k}}{(1+y^2)^{(n+1)/2}}dy$ is finite. But

$$\int_0^\infty \frac{y^{2k}}{(1+y^2)^{(n+1)/2}}dy < \int_0^\infty \frac{\left(1+y^2\right)^k}{(1+y^2)^{(n+1)/2}}dy$$

$$= \int_0^\infty \frac{1}{(1+y^2)^{(n+1)/2-k}}dy = \int_0^\infty \frac{1}{(1+y^2)^{\frac{n-2k}{2}+\frac{1}{2}}}dy$$

To apply the hint, take $\alpha = 2$ and $\beta = \frac{n-2k}{2} + \frac{1}{2}$. Then $2k < n$, $\beta > 0$, and $\alpha\beta > 1$, so the integral is finite.

Section 7.4

7.4.1. (a) 0.15
(b) 0.80
(c) 0.85
(d) $0.99 - 0.15 = 0.84$

7.4.3. Both differences represent intervals associated with 5% of the area under $f_{T_n}(t)$. Because the pdf is closer to the horizontal axis the further t is away from 0, the difference $t_{.05,n} - t_{.10,n}$ is the larger of the two.

7.4.5. $k = 2.2281$

7.4.7. (0.869, 1.153)

7.4.9. (a) (30.8 yrs, 40.0 yrs)
(b) The graph of date versus age shows no obvious patterns or trends. The assumption that μ has remained constant over time is believable.

7.4.11. (175.6, 211.4)
The medical and statistical definition of "normal" differ somewhat. There are people with medically normal platelet counts who appear in the population less than 10% of the time.

7.4.13. No, because the length of a confidence interval for μ is a function of s as well as the confidence coefficient. If the sample standard deviation for the second sample was sufficiently small (relative to the sample standard deviation for the first sample), the 95% confidence interval would be shorter than the 90% confidence interval.

7.4.15. (a) 0.95
(b) 0.80
(c) 0.945
(d) 0.95

7.4.17. Obs. $t = -1.71$; $-t_{.05,18} = -1.7341$; fail to reject H_0

7.4.19. Test H_0: $\mu = 40$ vs. H_1: $\mu < 40$; obs. $t = -2.25$; $-t_{.05,14} = -1.7613$; reject H_0.

7.4.21. Test H_0: $\mu = 0.0042$ vs. H_1: $\mu < 0.0042$; obs. $t = -2.48$; $-t_{.05,9} = -1.8331$; reject H_0.

7.4.23. Because of the skewed shape of $f_Y(y)$, and if the sample size were small, it would not be unusual for all the y_i's to lie close together near 0. When that happens, $\bar{y}$ will be less than μ, s will be considerably smaller than $E(S)$, and the t ratio will be further to the left of 0 than $f_{T_{n-1}}(t)$ would predict.

7.4.25. $f_Z(z)$

Section 7.5

7.5.1. **(a)** 23.685
(b) 4.605
(c) 2.700

7.5.3. **(a)** 2.088
(b) 7.261
(c) 14.041
(d) 17.539

7.5.5. 233.9

7.5.7. $P\left(\chi^2_{\alpha/2,n-1} \leq \frac{(n-1)S^2}{\sigma^2} \leq \chi^2_{1-\alpha/2,n-1}\right) = 1 - \alpha = P\left(\frac{(n-1)S^2}{\chi^2_{1-\alpha/2,n-1}} \leq \sigma^2 \leq \frac{(n-1)S^2}{\chi^2_{\alpha/2,n-1}}\right)$, so $\left(\frac{(n-1)s^2}{\chi^2_{1-\alpha/2,n-1}}, \frac{(n-1)s^2}{\chi^2_{\alpha/2,n-1}}\right)$ is a $100(1-\alpha)\%$ confidence interval for σ^2. Taking the square root of both sides gives a $100(1-\alpha)\%$ confidence interval for σ.

7.5.9. **(a)** $(20.13, 42.17)$
(b) $(0, 39.16)$ and $(21.11, \infty)$

7.5.11. Confidence intervals for σ (as opposed to σ^2) are often preferred by experimenters because they are expressed in the same units as the data, which makes them easier to interpret.

7.5.13. $n = 10$, which implies that $\frac{9s^2}{3.325} = 261.92$, and $s = 9.8$.

7.5.15. Test H_0: $\sigma^2 = 30.4^2$ versus H_1: $\sigma^2 < 30.4^2$ The test statistic in this case is $\chi^2 = \frac{(n-1)s^2}{\sigma_0^2} = \frac{18(733.4)}{30.4^2} = 14.285$.

The critical value is $\chi^2_{\alpha,n-1} = \chi^2_{.05,18} = 9.390$. Accept the null hypothesis, so do not assume that the potassium-argon method is more precise.

7.5.17. **(a)** Test H_0: $\mu = 10.1$ versus H_1: $\mu > 10.1$
Test statistic is $\frac{\bar{y}-\mu_0}{s/\sqrt{n}} = \frac{11.5-10.1}{10.17/\sqrt{24}} = 0.674$. Critical value is $t_{\alpha,n-1} = t_{0.05,23} = 1.7139$.

Accept the null hypothesis. Do not ascribe the increase of the portfolio yield over the bench mark to the analyst's system for choosing stocks.
(b) Test H_0: $\sigma^2 = 15.67$ versus H_1: $\sigma^2 < 15.67$
Test statistic is $\chi^2 = \frac{23(10.17^2)}{15.67^2} = 9.688$. Critical value is $\chi^2_{.05,23} = 13.091$.
Reject the null hypothesis. The analyst's method of choosing stocks does seem to result in less volatility.

CHAPTER 8

Section 8.2

8.2.1. Regression data
8.2.3. One-sample data
8.2.5. Regression data
8.2.7. k-sample data
8.2.9. One-sample data
8.2.11. Regression data
8.2.13. Two-sample data
8.2.15. k-sample data
8.2.17. Categorical data
8.2.19. Two-sample data
8.2.21. Paired data
8.2.23. Categorical data
8.2.25. Categorical data
8.2.27. Categorical data
8.2.29. Paired data
8.2.31. Randomized block data

CHAPTER 9

Section 9.2

9.2.1. Since $t = 1.72 < t_{.01,19} = 2.539$, accept H_0.

9.2.3. Since $z_{.05} = 1.64 < t = 5.67$, reject H_0.

9.2.5. Since $-z_{.005} = -2.58 < t = -0.532 < z_{.005} = 2.58$, do not reject H_0.

9.2.7. Since $-t_{.025,6} = 2.4469 < t = 0.69 < t_{.025,6} = 2.4469$, accept H_0.

9.2.9. Since $t = 2.16 > t_{.025,86} = 1.9880$, reject H_0.

9.2.11. **(a)** 22.880 **(b)** 166.990

9.2.13. **(a)** 0.3974 **(b)** 0.2090

9.2.15. $E(S_X^2) = E(S_Y^2) = \sigma^2$ by Example 5.4.4.
$$E(S_P^2) = \frac{(n-1)E\left(S_X^2\right)+(m-1)E\left(S_Y^2\right)}{n+m-2} = \frac{(n-1)\sigma^2+(m-1)\sigma^2}{n+m-2} = \sigma^2$$

9.2.17. Since $t = 2.16 > t_{.05,13} = 1.7709$, reject H_0.

9.2.19. **(a)** The sample standard deviation for the first data set is approximately 3.15; for the second, 3.29. These seem close enough to permit the use of Theorem 9.2.2.
(b) Intuitively, the states with the comprehensive law should have fewer deaths. However, the average for these data is 8.1, which is larger than the average of 7.0 for the states with a more limited law.

Section 9.3

9.3.1. The observed $F = 35.7604/115.9929 = 0.308$. Since $F_{.025,11,11} = 0.288 < 0.308 < 3.47 = F_{.975,11,11}$, we can accept H_0 that the variances are equal.

9.3.3. **(a)** The critical values are $F_{.025,19,19}$ and $F_{.975,19,19}$. These values are not tabulated, but in this case, we can approximate them by $F_{.025,20,20} = 0.406$ and $F_{.975,20,20} = 2.46$. The observed $F = 2.41/3.52 = 0.685$. Since $0.406 < 0.685 < 2.46$, we can accept H_0 that the variances are equal.
(b) Since $t = 2.662 > t_{.025,38} = 2.0244$, reject H_0.

9.3.5. $F = (0.20)^2/(0.37)^2 = 0.292$. Since $0.248 = F_{.025,9,9} < 0.292 < 4.03 = F_{.975,9,9}$, accept H_0.

9.3.7. $F = 65.25/227.77 = 0.286$. Since $0.208 = F_{.025,8,5} < 0.286 < 6.76 = F_{.975,8,5}$, accept H_0. Thus, Theorem 9.2.2 is appropriate.

9.3.9. If $\sigma_X^2 = \sigma_Y^2 = \sigma^2$, the maximum likelihood estimator for σ^2 is

$$\hat{\sigma}^2 = \frac{1}{n+m}\left(\sum_{i=1}^{n}(x_i - \overline{x})^2 + \sum_{i=1}^{m}(y_i - \overline{y})^2\right).$$

Then $L(\hat{\omega}) = \left(\frac{1}{2\pi\hat{\sigma}^2}\right)^{(n+m)/2} e^{-\frac{1}{2\hat{\sigma}^2}\left(\sum_{i=1}^{n}(x_i-\overline{x})^2 + \sum_{i=1}^{m}(y_i-\overline{y})^2\right)}$

$= \left(\frac{1}{2\pi\hat{\sigma}^2}\right)^{(n+m)/2} e^{-(n+m)/2}$

If $\sigma_X^2 \neq \sigma_Y^2$ the maximum likelihood estimators for σ_X^2 and σ_Y^2 are

$$\hat{\sigma}_X^2 = \frac{1}{n}\sum_{i=1}^{n}(x_i - \overline{x})^2, \text{ and } \hat{\sigma}_Y^2 = \frac{1}{m}\sum_{i=1}^{m}(y_i - \overline{y})^2.$$

Then $L(\hat{\Omega}) = \left(\frac{1}{2\pi\hat{\sigma}_X^2}\right)^{n/2} e^{-\frac{1}{2\hat{\sigma}_X^2}\left(\sum_{i=1}^{n}(x_i-\overline{x})^2\right)} \left(\frac{1}{2\pi\hat{\sigma}_Y^2}\right)^{m/2}$

$\times e^{-\frac{1}{2\hat{\sigma}_Y^2}\left(\sum_{i=1}^{m}(y_i-\overline{y})^2\right)}$

$= \left(\frac{1}{2\pi\hat{\sigma}_X^2}\right)^{n/2} e^{-m/2} \left(\frac{1}{2\pi\hat{\sigma}_Y^2}\right)^{m/2} e^{-n/2}$

The ratio $\lambda = \frac{L(\hat{\omega})}{L(\hat{\Omega})} = \frac{(\hat{\sigma}_X^2)^{n/2}(\hat{\sigma}_Y^2)^{m/2}}{(\hat{\sigma}^2)^{(n+m)/2}}$, which equates to the expression given in the statement of the question.

Section 9.4

9.4.1. Since $-1.96 < z = 1.76 < 1.96 = z_{.025}$, accept H_0.

9.4.3. Since $-1.96 < z = -0.17 < 1.96 = z_{.025}$, accept H_0 at the 0.05 level of significance.

9.4.5. Since $z = 4.25 > 2.33 = z_{.01}$, reject H_0 at the 0.01 level of significance.

9.4.7. Since $-1.96 < z = 1.50 < 1.96 = z_{.025}$, accept H_0 at the 0.05 level of significance.

9.4.9. Since $= 0.25 < 1.64 = z_{.05}$, accept H_0. The player is right.

Section 9.5

9.5.1. $(0.71, 1.55)$. Since 0 is not in the interval, we can reject the null hypothesis that $\mu_X = \mu_Y$.

9.5.3. Equal variance confidence interval is $(-13.32, 6.72)$. Unequal variance confidence interval is $(-13.61, 7.01)$.

9.5.5. Begin with the statistic $\bar{X} - \bar{Y}$, which has $E(\bar{X} - \bar{Y}) = \mu_X - \mu_Y$ and $\text{Var}(\bar{X} - \bar{Y}) = \sigma_X^2/n + \sigma_Y^2/m$. Then $P\left(-z_{\alpha/2} \leq \frac{\bar{X}-\bar{Y}-(\mu_X-\mu_Y)}{\sqrt{\sigma_X^2/n+\sigma_Y^2/m}} \leq z_{\alpha/2}\right) = 1 - \alpha$, which implies

$$P\left(-z_{\alpha/2}\sqrt{\sigma_X^2/n + \sigma_Y^2/m} \leq \bar{X} - \bar{Y} - (\mu_X - \mu_Y) \leq z_{\alpha/2}\sqrt{\sigma_X^2/n + \sigma_Y^2/m}\right) = 1 - \alpha.$$

Solving the inequality for $\mu_X - \mu_Y$ gives

$$P\left(\bar{X} - \bar{Y} - z_{\alpha/2}\sqrt{\sigma_X^2/n + \sigma_Y^2/m} \leq \mu_X - \mu_Y \leq \bar{X} - \bar{Y} + z_{\alpha/2}\sqrt{\sigma_X^2/n + \sigma_Y^2/m}\right) = 1 - \alpha.$$

Thus the confidence interval is

$$\left(\bar{x} - \bar{y} - z_{\alpha/2}\sqrt{\sigma_X^2/n + \sigma_Y^2/m}, \bar{x} - \bar{y} + z_{\alpha/2}\sqrt{\sigma_X^2/n + \sigma_Y^2/m}\right).$$

9.5.7. $(0.06, 2.14)$. Since the confidence interval contains 1, we can accept H_0 that the variances are equal, and Theorem 9.2.1 applies.

9.5.9. $(-0.021, 0.051)$. Since the confidence interval contains 0, we can conclude that Flonase users do not suffer more headaches.

9.5.11. The approximate normal distribution implies that

$$P\left(-z_\alpha \leq \frac{\frac{X}{n} - \frac{Y}{m} - (p_X - p_Y)}{\sqrt{\frac{(X/n)(1-X/n)}{n} + \frac{(Y/m)(1-Y/m)}{m}}} \leq z_\alpha\right) = 1 - \alpha$$

or

$$P\left(-z_\alpha\sqrt{\frac{(X/n)(1-X/n)}{n} + \frac{(Y/m)(1-Y/m)}{m}} \leq \frac{X}{n} - \frac{Y}{m} - (p_X - p_Y) \leq z_\alpha\sqrt{\frac{(X/n)(1-X/n)}{n} + \frac{(Y/m)(1-Y/m)}{m}}\right)$$

$= 1 - \alpha$

which implies that

$$P\left(-\left(\frac{X}{n} - \frac{Y}{m}\right) - z_\alpha\sqrt{\frac{(X/n)(1-X/n)}{n} + \frac{(Y/m)(1-Y/m)}{m}} \leq -(p_X - p_Y) \leq -\left(\frac{X}{n} - \frac{Y}{m}\right) + z_\alpha\sqrt{\frac{(X/n)(1-X/n)}{n} + \frac{(Y/m)(1-Y/m)}{m}}\right) = 1 - \alpha$$

Multiplying the inequality by -1 yields the confidence interval.

CHAPTER 10

Section 10.2

10.2.1. 0.000886

10.2.3. 0.00265

10.2.5. 0.00649

10.2.7. (a) $\frac{50!}{3!7!15!25!}\left(\frac{1}{64}\right)^3\left(\frac{7}{64}\right)^7\left(\frac{19}{64}\right)^{15}\left(\frac{37}{64}\right)^{25}$

(b) $\text{Var}(X_3) = 50\left(\frac{19}{64}\right)\left(\frac{45}{64}\right) = 10.44$

10.2.9. Assume that $M_{X_1,X_2,X_3}(t_1, t_2, t_3) = (p_1e^{t_1} + p_2e^{t_2} + p_3e^{t_3})^n$. Then $M_{X_1,X_2,X_3}(t_1, 0, 0) = E(e^{t_1X_1}) = (p_1e^{t_1} + p_2 + p_3)^n = (1 - p_1 + p_1e^{t_1})^n$ is the mgf for X_1. But the latter has the form of the mgf for a binomial random variable with parameters n and p_1.

Section 10.3

10.3.1. $\sum_{i=1}^{t} \frac{(X_i - np_i)^2}{np_i} = \sum_{i=1}^{t} \frac{\left(X_i^2 - 2np_i X_i + n^2 p_i^2\right)}{np_i} = \sum_{i=1}^{t} \frac{X_i^2}{np_i} - 2\sum_{i=1}^{t} X_i + n\sum_{i=1}^{t} p_i = \sum_{i=1}^{t} \frac{X_i^2}{np_i} - n.$

10.3.3. If the sampling is done with replacement, the number of white chips drawn should follow a binomial distribution (with $n = 2$ and $p = 0.4$). Since the obs. $\chi^2 = 3.30 < 4.605 = \chi^2_{.90,2}$, fail to reject H_0.

10.3.5. Let $p = P$(baby is born between midnight and 4 A.M.). Test H_0: $p = 1/6$ vs. H_1: $p \neq 1/6$; obs. $z = 2.73$; reject H_0 if $\alpha = 0.05$. The obs. χ^2 in Question 10.3.4 will equal the square of the obs. z. The two tests are equivalent.

10.3.7. Obs. $\chi^2 = 12.23$ with 5 df; $\chi^2_{.95,5} = 11.070$; reject H_0.

10.3.9. Obs. $\chi^2 = 18.22$ with 7 df; $\chi^2_{.95,7} = 14.067$; reject H_0.

10.3.11. Obs. $\chi^2 = 8.10$; $\chi^2_{.95,1} = 3.841$; reject H_0.

Section 10.4

10.4.1. Obs. $\chi^2 = 11.72$ with $4 - 1 - 1 = 2$ df; $\chi^2_{.95,2} = 5.991$; reject H_0.

10.4.3. Obs. $\chi^2 = 46.75$ with $7 - 1 - 1 = 5$ df; $\chi^2_{.95,5} = 11.070$; reject H_0. The independence assumption would not hold if the infestation was contagious.

10.4.5. For the model $f_Y(y) = \lambda e^{-\lambda y}$, $\hat{\lambda} = 0.823$; obs. $\chi^2 = 4.181$ with $5 - 1 - 1 = 3$ df; $\chi^2_{.95,3} = 7.815$; fail to reject H_0.

10.4.7. Let $p = P$(child is a boy). Then $\hat{p} = 0.533$, obs. $\chi^2 = 0.62$, and we fail to reject the binomial model because $\chi^2_{.95,1} = 3.841$.

10.4.9. For the model $p_X(k) = e^{-3.87}(3.87)^k/k!$, obs. $\chi^2 = 12.9$ with $12 - 1 - 1 = 10$ df. But $\chi^2_{.95,10} = 18.307$, so we fail to reject H_0.

10.4.11. $\hat{p} = 0.26$; obs. $\chi^2 = 9.23$; $\chi^2_{.95,3} = 7.815$; reject H_0.

Section 10.5

10.5.1. Obs. $\chi^2 = 2.77$; $\chi^2_{.90,1} = 2.706$ and $\chi^2_{.95,1} = 3.841$, so H_0 is rejected at the $\alpha = 0.10$ level but not at the $\alpha = 0.05$ level.

10.5.3. Obs. $\chi^2 = 42.25$; $\chi^2_{.99,3} = 11.345$; reject H_0.

10.5.5. Obs. $\chi^2 = 4.80$; $\chi^2_{.95,1} = 3.841$; reject H_0. Regular use of aspirin appears to lessen the chances that a woman will develop breast cancer.

10.5.7. Obs. $\chi^2 = 12.61$; $\chi^2_{.95,1} = 3.841$; reject H_0.

10.5.9. Obs. $\chi^2 = 2.197$; $\chi^2_{.95,1} = 3.841$; fail to reject H_0.

CHAPTER 11

Section 11.2

11.2.1. $y = 25.23 + 3.29x$; 84.5°F

11.2.3.

x_i	$y_i - \hat{y}_i$
0	−0.81
4	0.01
10	0.09
15	0.03
21	−0.09
29	0.14
36	0.55
51	1.69
68	−1.61

A straight line appears to fit these data.

11.2.5. The value 12 is too "far" from the data observed.

11.2.7. The least squares line is $y = 88.1 + 0.412x$.

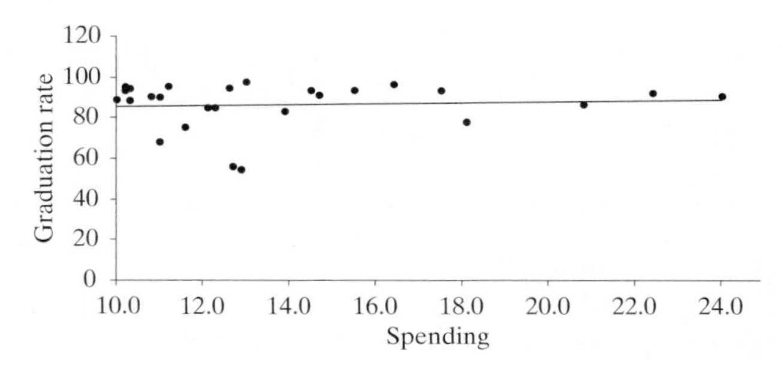

11.2.9. The least squares line is $y = 114.72 + 9.23x$.

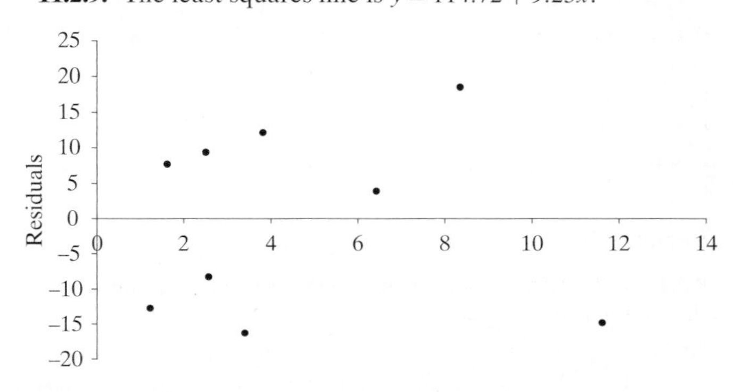

A linear fit seems reasonable.

11.2.11. The least squares line is $y = 0.61 + 0.84x$, which seems inadequate because of the large values of the residuals.

11.2.13. When $\bar{x}$ is substituted for x in the least squares equation, we obtain $y = a + b\bar{x} = \bar{y} - b\bar{x} + b\bar{x} = \bar{y}$.

11.2.15. 0.03544

11.2.17. $y = 100 - 5.19x$

11.2.19. To find the a, b, and c, solve the following set of equations.

(1) $na + \left(\sum_{i=1}^{n} x_i\right) b + \left(\sum_{i=1}^{n} \sin x_i\right) c = \sum_{i=1}^{n} y_i$

(2) $\left(\sum_{i=1}^{n} x_i\right) a + \left(\sum_{i=1}^{n} x_i^2\right) b + \left(\sum_{i=1}^{n} x_i \sin x_i\right) c = \sum_{i=1}^{n} x_i y_i$

(3) $\left(\sum_{i=1}^{n} \cos x_i\right) a + \left(\sum_{i=1}^{n} x_i \cos x_i\right) b +$

$\left(\sum_{i=1}^{n} (\cos x_i)(\sin x_i)\right) c = \sum_{i=1}^{n} y_i \cos x_i$

11.2.21. **(a)** $y = 4.6791e^{0.0484x}$ **(b)** 8.362 trillion
(c) Part (b) and the residual pattern cast doubt on the exponential model.

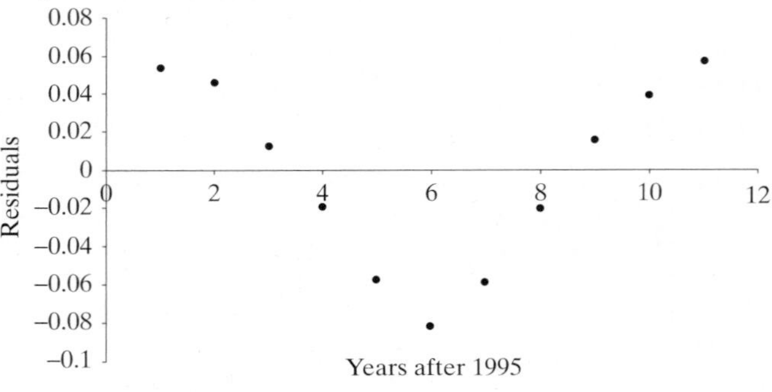

11.2.23. $y = 819.4e^{0.128x}$

11.2.25. The model is $y = 13.487x^{10.538}$.

11.2.27. The model is $y = 0.07416x^{1.43687}$.

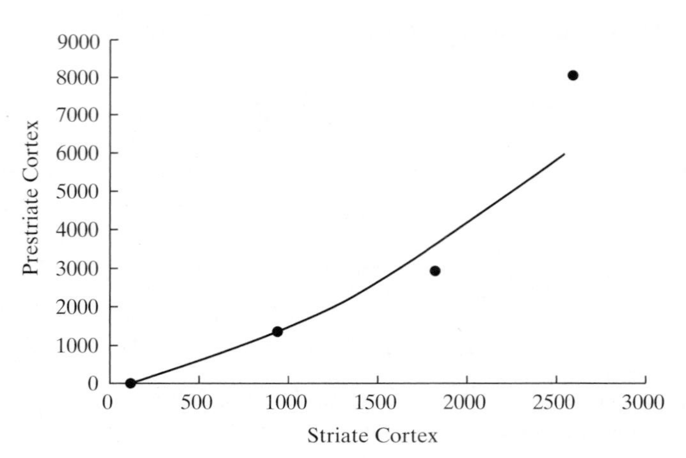

11.2.29. **(d)** If $y = \frac{1}{a+bx}$, then $\frac{1}{y} = a + bx$, and $1/y$ is linear with x.
(e) If $y = \frac{x}{a+bx}$, then $\frac{1}{y} = \frac{a+bx}{x} = b + a\frac{1}{x}$, and $1/y$ is linear with $1/x$.
(f) If $y = 1 - e^{-x^b/a}$, then $1 - y = e^{-x^b/a}$, and $\frac{1}{1-y} = e^{x^b/a}$. Taking the ln of both sides gives $\ln \frac{1}{1-y} = x^b/a$. Taking the ln again yields $\ln \ln \frac{1}{1-y} = -\ln a + b \ln x$, and $\ln \ln \frac{1}{1-y}$ is linear with $\ln x$.

11.2.31. $a = 5.55870$; $b = -0.13314$

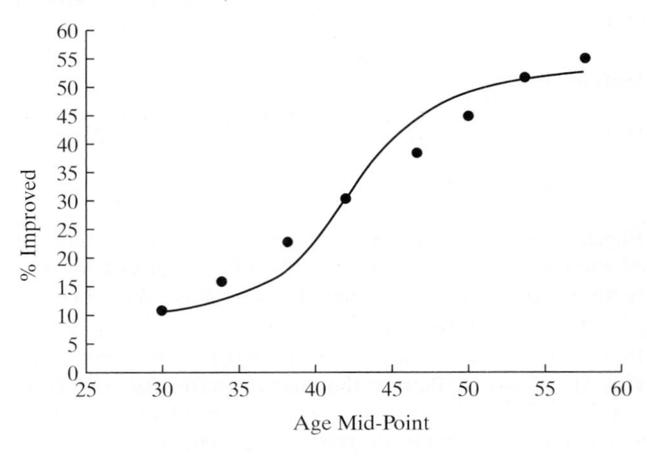

Section 11.3

11.3.1. $y = 13.8 - 1.5x$; since $-t_{.025,2} = -4.3027 < t = -1.59 < 4.3027 = t_{.025,2}$, accept H_0.

11.3.3. Since $t = 5.47 > t_{.005,13} = 3.0123$, reject H_0.

11.3.5. 0.9164

11.3.7. (66.551, 68.465)

11.3.9. Since $t = 4.38 > t_{.025,9} = 2.2622$, reject H_0.

11.3.11. By Theorem 11.3.2, $E(\hat{\beta}_0) = \beta_0$, and

$$\text{Var}(\hat{\beta}_0) = \frac{\sigma^2 \sum_{i=1}^{n} x_i}{n \sum_{i=1}^{n} (x_i - \bar{x})^2}.$$

Now $(\hat{\beta}_0 - \beta_0)/\sqrt{\text{Var}(\hat{\beta}_0)}$ is normal, so

$$P\left(-z_{\alpha/2} < (\hat{\beta}_0 - \beta_0)/\sqrt{\text{Var}(\hat{\beta}_0)} < z_{\alpha/2}\right) = 1 - \alpha.$$

Then the confidence interval is

$$\left(\hat{\beta}_0 - z_{\alpha/2}\sqrt{\text{Var}(\hat{\beta}_0)}, \quad \hat{\beta}_0 + z_{\alpha/2}\sqrt{\text{Var}(\hat{\beta}_0)}\right)$$

or

$$\left(\hat{\beta} - z_{\alpha/2}\frac{\sigma\sqrt{\sum_{i=1}^{n} x_i}}{\sqrt{n \sum_{i=1}^{n} (x_i - \bar{x})^2}}, \quad \hat{\beta}_0 + z_{\alpha/2}\frac{\sigma\sqrt{\sum_{i=1}^{n} x_i}}{\sqrt{n \sum_{i=1}^{n} (x_i - \bar{x})^2}}\right)$$

11.3.13. Reject the null hypothesis if the test statistic is $< \chi^2_{.025,22} = 10.982$ or $> \chi^2_{.975,22} = 36.781$. The observed chi square is $\frac{(n-2)s^2}{\sigma_0^2} = \frac{(24-2)(18.2)}{12.6} = 31.778$, so do not reject H_0.

11.3.15. (2.655, 17.237)

11.3.17. (2.060, 2.087)

11.3.19. The confidence interval of (173.89, 214.13) does not contain the Harvard median salary. The prediction interval of (147.40, 240.62) does.

11.3.21. The test statistic is $t = \dfrac{\hat{\beta}_1 - \hat{\beta}_1^*}{s\sqrt{\dfrac{1}{\sum_{i=1}^{6}(x_i-\bar{x})^2} + \dfrac{1}{\sum_{i=1}^{8}\left(x_i^*-\bar{x}^*\right)^2}}}$,

where $s = \sqrt{\dfrac{5.983+13.804}{6+8-4}} = 1.407$.

Then $t = \dfrac{0.606-1.07}{1.407\sqrt{\dfrac{1}{31.33}+\dfrac{1}{46}}} = -1.42$

Since the observed ratio is not less than $-t_{0.05,10} = -1.8125$, the difference in slopes can be ascribed to chance. These data do not support further investigation.

11.3.23. The form given in the text is $\text{Var}(\hat{Y}) = \sigma^2\left[\frac{1}{n} + \frac{(x-\bar{x})^2}{\sum_{i=1}^{n}(x_i-\bar{x})^2}\right]$. Putting the sum in the brackets over a least common denominator gives

$$\frac{1}{n} + \frac{(x-\bar{x})^2}{\sum_{i=1}^{n}(x_i-\bar{x})^2} = \frac{\sum_{i=1}^{n}(x_i-\bar{x})^2 + n(x-\bar{x})^2}{n\sum_{i=1}^{n}(x_i-\bar{x})^2}$$

$$= \frac{\sum_{i=1}^{n}x_i^2 - n\bar{x}^2 + n(x^2+\bar{x}^2-2x\bar{x})}{n\sum_{i=1}^{n}(x_i-\bar{x})^2}$$

$$= \frac{\sum_{i=1}^{n}x_i^2 + nx^2 - 2nx\bar{x}}{n\sum_{i=1}^{n}(x_i-\bar{x})^2}$$

$$= \frac{\sum_{i=1}^{n}x_i^2 + nx^2 - 2x\sum_{i=1}^{n}x_i}{n\sum_{i=1}^{n}(x_i-\bar{x})^2}$$

$$= \frac{\sum_{i=1}^{n}(x_i-x)^2}{n\sum_{i=1}^{n}(x_i-\bar{x})^2}.$$

Thus, $\text{Var}(\hat{Y}) = \dfrac{\sigma^2\sum_{i=1}^{n}(x_i-x)^2}{n\sum_{i=1}^{n}(x_i-\bar{x})^2}$.

Section 11.4

11.4.1. $-2/121$; $-2/15\sqrt{14}$

11.4.3. 0.492

11.4.5. $\rho(a+bX, c+dY) = \dfrac{\text{Cov}(a+bX, c+dY)}{\sqrt{\text{Var}(a+bX)\text{Var}(c+dY)}} = \dfrac{bd\,\text{Cov}(X,Y)}{\sqrt{b^2\text{Var}(X)d^2\text{Var}(Y)}}$, the equality in the numerators stemming from Question 3.9.14. Since $b > 0$, $d > 0$, this last expression is

$$\frac{bd\,\text{Cov}(X,Y)}{bd\sigma_X\sigma_Y} = \frac{\text{Cov}(X,Y)}{\sigma_X\sigma_Y} = \rho(X,Y).$$

11.4.7. (a) $\text{Cov}(X+Y, X-Y) = E[(X+Y)(X-Y)] - E(X+Y)E(X-Y) = E[X^2-Y^2] - (\mu_X+\mu_Y)(\mu_X-\mu_Y) = E(X^2) - \mu_X^2 - E(Y^2) + \mu_Y^2 = \text{Var}(X) - \text{Var}(Y)$

(b) $\rho(X+Y) = \dfrac{\text{Cov}(X+Y, X-Y)}{\sqrt{\text{Var}(X+Y)\text{Var}(X-Y)}}$. By Part (a) $\text{Cov}(X+Y, X-Y) = \text{Var}(X) - \text{Var}(Y)$.
$\text{Var}(X+Y) = \text{Var}(X) + \text{Var}(Y) = 2\text{Cov}(X,Y) = \text{Var}(X) + \text{Var}(Y) + 0$.
Similarly, $\text{Var}(X-Y) = \text{Var}(X) + \text{Var}(Y)$. Then

$$\rho(X+Y) = \frac{\text{Var}(X)-\text{Var}(Y)}{\sqrt{(\text{Var}(X)+\text{Var}(Y))(\text{Var}(X)+\text{Var}(Y))}} = \frac{\text{Var}(X)-\text{Var}(Y)}{\text{Var}(X)+\text{Var}(Y)}.$$

11.4.9. By Equation 11.4.2

$$r = \frac{n\sum_{i=1}^{n}x_iy_i - \left(\sum_{i=1}^{n}x_i\right)\left(\sum_{i=1}^{n}y_i\right)}{\sqrt{n\sum_{i=1}^{n}x_i^2 - \left(\sum_{i=1}^{n}x_i\right)^2}\sqrt{n\sum_{i=1}^{n}y_i^2 - \left(\sum_{i=1}^{n}y_i\right)^2}}$$

$$= \frac{n\sum_{i=1}^{n}x_iy_i - \left(\sum_{i=1}^{n}x_i\right)\left(\sum_{i=1}^{n}y_i\right)}{n\sum_{i=1}^{n}x_i^2 - \left(\sum_{i=1}^{n}x_i\right)^2} \times \frac{n\sum_{i=1}^{n}x_i^2 - \left(\sum_{i=1}^{n}x_i\right)^2}{\sqrt{n\sum_{i=1}^{n}x_i^2 - \left(\sum_{i=1}^{n}x_i\right)^2}\sqrt{n\sum_{i=1}^{n}y_i^2 - \left(\sum_{i=1}^{n}y_i\right)^2}}$$

$$= \hat{\beta}_1\frac{\sqrt{n\sum_{i=1}^{n}x_i^2 - \left(\sum_{i=1}^{n}x_i\right)^2}}{\sqrt{n\sum_{i=1}^{n}y_i^2 - \left(\sum_{i=1}^{n}y_i\right)^2}}.$$

11.4.11. $r = -0.030$. The data do not suggest that altitude affects home run hitting.

11.4.13. 58.1%

Section 11.5

11.5.1. 0.1891; 0.2127

11.5.3. (a) $f_{X+Y}(t) = \dfrac{1}{2\pi\sqrt{1-\rho^2}}\displaystyle\int_{-\infty}^{\infty}$

$$\exp\left\{-\frac{1}{2}\left(\frac{1}{1-\rho^2}\right)[(t-y)^2 - 2\rho(t-y)y + y^2]\right\}dy$$

The expression in the brackets can be expanded and rewritten as

$$t^2 + 2(1+\rho)y^2 - 2t(1+\rho)y = t^2 + 2(1+\rho)[y^2 - ty]$$
$$= t^2 + 2(1+\rho)\left[y^2 - ty + \frac{t^2}{4}\right] - \frac{1}{2}(1+\rho)t^2$$
$$= \frac{1-\rho}{2}t^2 + 2(1+\rho)(y - t/2)^2.$$

Placing this expression into the exponent gives

$$f_{X+Y}(t) = \frac{1}{2\pi\sqrt{1-\rho^2}} e^{-\frac{1}{2}\left(\frac{1}{1-\rho^2}\right)\frac{1-\rho}{2}t^2} \int_{-\infty}^{\infty} e^{-\frac{1}{2}\left(\frac{1}{1-\rho^2}\right)2(1+\rho)(y-t/2)^2} dy$$
$$= \frac{1}{2\pi\sqrt{1-\rho^2}} e^{-\frac{1}{2}\left(\frac{t^2}{2(1+\rho)}\right)} \int_{-\infty}^{\infty} e^{-\frac{1}{2}\left(\frac{(y-t/2)^2}{(1+\rho)/2}\right)} dy.$$

The integral is that of a normal pdf with mean $t/2$ and $\sigma^2 = (1+\rho)/2$. Thus, the integral equals $\sqrt{2\pi(1+\rho)/2} = \sqrt{\pi(1+\rho)}$. Putting this into the expression for f_{X+Y} gives

$$f_{X+Y}(t) = \frac{1}{\sqrt{2\pi}\sqrt{2(1+\rho)}} e^{-\frac{1}{2}\left(\frac{t^2}{2(1+\rho)}\right)},$$

which is the pdf of a normal variable with $\mu = 0$ and $\sigma^2 = 2(1+\rho)$.

(b) $E(X+Y) = c\mu_X + d\mu_Y$, $\text{Var}(X+Y) = c^2\sigma_X^2 + d^2\sigma_Y^2 + 2cd\sigma_X\sigma_Y\rho(X,Y)$

11.5.5. $E(X) = E(Y) = 0$; $\text{Var}(X) = 4$; $\text{Var}(Y) = 1$; $\rho(X,Y) = 1/2$; $k = 1/2\pi\sqrt{3}$

11.5.7. Since $-t_{.005,18} = -2.8784 < T_{n-2} = -2.156 < 2.8784 = t_{.005,18}$, accept H_0.

11.5.9. Since $-t_{.025,10} = -2.2281 < T_{n-2} = -0.094 < 2.2281 = t_{.025,10}$, accept H_0.

11.5.11. $r = 0.249$. $T_8 = \frac{\sqrt{8}(0.249)}{\sqrt{1-(0.249)^2}} = 0.73$
Since $T_8 = 0.73 < 1.397 = t_{.10,8}$, accept H_0.

CHAPTER 12

Section 12.2

12.2.1. Obs. $F = 3.94$ with 3 and 6 df; $F_{.95,3,6} = 4.76$ and $F_{.90,3,6} = 3.29$, so H_0 would be rejected at the $\alpha = 0.10$ level, but not at the $\alpha = 0.05$ level.

12.2.3.

Source	df	SS	MS	F
Sector	2	186.0	93.0	3.44
Error	27	728.2	27.0	
Total	29	914.2		

$F_{.99,2,27}$ does not appear in Table A.9, but $F_{.99,2,30} = 5.39 < F_{.99,2,27} < F_{.99,2,24} = 5.61$. Thus, we fail to reject H_0, since $3.44 < 5.39$

12.2.5.

Source	df	SS	MS	F	P
Tribe	3	504167	168056	3.70	0.062
Error	8	363333	45417		
Total	11	867500			

Since the P-value is greater than 0.01, we fail to reject H_0.

12.2.7.

Source	df	SS	MS	F
Treatment	4	271.36	67.84	6.40
Error	10	106.00	10.60	
Total	14	377.36		

12.2.9.

$$\text{SSTOT} = \sum_{j=1}^{k}\sum_{i=1}^{n_j}(Y_{ij} - \bar{Y}_{..})^2 = \sum_{j=1}^{k}\sum_{i=1}^{n_j}(Y_{ij}^2 - 2Y_{ij}\bar{Y}_{..} + \bar{Y}_{..}^2)$$
$$= \sum_{j=1}^{k}\sum_{i=1}^{n_j}Y_{ij}^2 - 2\bar{Y}_{..}\sum_{j=1}^{k}\sum_{i=1}^{n_j}Y_{ij} + n\bar{Y}_{..}^2$$
$$= \sum_{j=1}^{k}\sum_{i=1}^{n_j}Y_{ij}^2 - 2n\bar{Y}_{..}^2 + n\bar{Y}_{..}^2$$
$$= \sum_{j=1}^{k}\sum_{i=1}^{n_j}Y_{ij}^2 - n\bar{Y}_{..}^2 = \sum_{j=1}^{k}\sum_{i=1}^{n_j}Y_{ij}^2 - C,$$

where $C = T_{..}^2/n$. Also,

$$\text{SSTR} = \sum_{j=1}^{k}\sum_{i=1}^{n_j}(\bar{Y}_{.j}^2 - \bar{Y}_{..})^2$$
$$= \sum_{j=1}^{k} n_j(\bar{Y}_{.j}^2 - 2\bar{Y}_{.j}\bar{Y}_{..} + \bar{Y}_{..}^2)$$
$$= \sum_{j=1}^{k} T_{.j}^2/n_j - 2\bar{Y}_{..}\sum_{j=1}^{k} n_j\bar{Y}_{.j} + n\bar{Y}_{..}^2$$
$$= \sum_{j=1}^{n} T_{.j}^2/n_j - 2n\bar{Y}_{..}^2 + n\bar{Y}_{..}^2$$
$$= \sum_{j=1}^{k} T_{.j}^2/n_j - C.$$

12.2.11. Analyzed with a two-sample t test, the data in Question 9.2.8 require that $H_0\colon \mu_X = \mu_Y$ be rejected (in favor of a two-sided H_1) at the $\alpha = 0.05$ level if $|t| \geq t_{.025,6+9-2} = 2.1604$. Evaluating the test statistic gives $t = (70.83 - 79.33)/11.31\sqrt{1/6 + 1/9} = -1.43$, which implies that H_0 should not be rejected. The ANOVA table for the same data shows that $F = 2.04$. But $(-1.43)^2 = 2.04$. Moreover, H_0 would be rejected with the analysis of variance if $F \geq F_{.95,1,13} = 4.667$. But $(2.1604)^2 = 4.667$.

Source	df	SS	MS	F
Sex	1	260	260	2.04
Error	13	1661	128	
Total	14	1921		

12.2.13.

Source	df	SS	MS	F	P
Law	1	16.333	16.333	1.58	0.2150
Error	46	475.283	10.332		
Total	47	491.616			

The F critical value is 4.05.

For the pooled two-sample t test, the observed t ratio is -1.257, and the critical value is 2.0129.

Note that $(-1.257)^2 = 1.58$ (rounded to two decimal places) which is the observed F ratio. Also, $2.0129^2 = 4.05$ (rounded to two decimal places), which is the F critical value.

Section 12.3

12.3.1.

Pairwise Difference	Tukey Interval	Conclusion
$\mu_1 - \mu_2$	$(-15.27, 13.60)$	NS
$\mu_1 - \mu_3$	$(-23.77, 5.10)$	NS
$\mu_1 - \mu_4$	$(-33.77, -4.90)$	Reject
$\mu_2 - \mu_3$	$(-22.94, 5.94)$	NS
$\mu_2 - \mu_4$	$(-32.94, -4.06)$	Reject
$\mu_3 - \mu_4$	$(-24.44, 4.44)$	NS

12.3.3. Obs. $F = 5.81$ with 2 and 15 df; reject $H_0: \mu_C = \mu_A = \mu_M$ at $\alpha = 0.05$ but not at $\alpha = 0.01$.

Pairwise Difference	Tukey Interval	Conclusion
$\mu_C - \mu_A$	$(-78.9, 217.5)$	NS
$\mu_C - \mu_M$	$(-271.0, 25.4)$	NS
$\mu_A - \mu_M$	$(-340.4, -44.0)$	Reject

12.3.5.

Pairwise Difference	Tukey Interval	Conclusion
$\mu_1 - \mu_2$	$(-29.5, 2.8)$	NS
$\mu_1 - \mu_3$	$(-56.2, -23.8)$	Reject
$\mu_2 - \mu_3$	$(-42.8, -10.5)$	Reject

12.3.7. Longer. As k gets larger, the number of possible pairwise comparisons increases. To maintain the same overall probability of committing at least one Type I error, the individual intervals would need to be widened.

Section 12.4

12.4.1.

Source	df	SS	MS	F
Tube	2	510.7	255.4	11.56
Error	42	927.7	22.1	
Total	44	1438.4		

Subhypothesis	Contrast	SS	F
$H_0: \mu_A = \mu_C$	$C_1 = \mu_A - \mu_C$	264	11.95
$H_0: \mu_B = \dfrac{\mu_A + \mu_C}{2}$	$C_2 = \frac{1}{2}\mu_A - \mu_B + \frac{1}{2}\mu_C$	246.7	11.16

$H_0: \mu_A = \mu_B = \mu_C$ is strongly rejected ($F_{.99,2,42} \doteq F_{.99,2,40} = 5.18$). Theorem 12.4.1 holds true for orthogonal contrasts C_1 and $C_2 - SS_{C_1} - SS_{C_2} = 264 + 246.7 = 510.7 = SSTR$.

12.4.3. $\hat{C} = -14.25$; $SS_C = 812.25$; obs. $F = 10.19$; $F_{.95,1,20} = 4.35$; reject H_0.

12.4.5.

	μ_A	μ_B	μ_C	μ_D	$\sum_{j=1}^{4} c_j$
C_1	1	-1	0	0	0
C_2	0	0	1	-1	0
C_3	$\frac{11}{12}$	$\frac{11}{12}$	-1	$\frac{-5}{6}$	0

C_1 and C_3 are orthogonal because $\frac{1(11/12)}{6} + \frac{(-1)(11/12)}{6} = 0$; also C_2 and C_3 are orthogonal because $\frac{1(-1)}{6} + \frac{(-1)(-5/6)}{5} = 0$. $\hat{C}_3 = -2.293$ and $SS_{C_3} = 8.97$. But $SS_{C_1} + SS_{C_2} + SS_{C_3} = 4.68 + 1.12 + 8.97 = 14.77 = SSTR$.

Section 12.5

12.5.1. Replace each observation by its square root. At the $\alpha = 0.05$ level, $H_0: \mu_A = \mu_B$ is rejected. (For $\alpha = 0.01$, though, we would fail to reject H_0.)

Source	df	SS	MS	F	P
Developer	1	1.836	1.836	6.23	0.032
Error	10	2.947	0.295		
Total	11	4.783			

12.5.3. Since Y_{ij} is a binomial random variable based on $n = 20$ trials, each data point should be replaced by the arcsin of $(y_{ij}/20)^{1/2}$. Based on those transformed observations, $H_0: \mu_A = \mu_B = \mu_C$ is strongly rejected ($P < 0.001$).

Source	df	SS	MS	F	P
Launcher	2	0.30592	0.15296	22.34	0.000
Error	9	0.06163	0.00685		
Total	11	0.36755			

Appendix 12.A.3

12.A.3.1. The F test will have greater power against H_1^{**} because the latter yields a larger noncentrality parameter than does H_1^*.

12.A.3.3. $M_V(t) = (1-2t)^{-r/2} e^{\gamma t(1-2t)^{-1}}$, so $M_V^{(1)}(t) = (1-2t)^{-r/2}$

$$e^{\gamma t(1-2t)^{-1}}\left[\gamma t(-1)(1-2t)^{-2}(-2) + (1-2t)^{-1}\gamma\right] + e^{\gamma t(1-2t)^{-1}}\left(-\frac{r}{2}\right)(1-2t)^{-(r/2)-1}(-2).$$

Therefore $E(V) = M_V^{(1)}(0) = \gamma + r$.

12.A.3.5. $M_V(t) = \prod_{i=1}^{n}(1-2t)^{-r_i/2}e^{\gamma_i t/(1-2t)} = (1-2t)^{-\sum_{i=1}^{n} r_i/2} \cdot e^{\left(\sum_{i=1}^{n}\gamma_i\right)t/(1-2t)}$ which implies that V has a noncentral χ^2 distribution with $\sum_{i=1}^{n} r_i$ df and with noncentrality parameter $\sum_{i=1}^{n}\gamma_i$.

CHAPTER 13

Section 13.2

13.2.1.

Source	df	SS	MS	F	P
States	1	61.63	61.63	7.20	0.0178
Students	14	400.80	28.63	3.34	0.0155
Error	14	119.87	8.56		
Total	29	582.30			

The critical value $F_{.95,1,14}$ is approximately 4.6. Since the F statistic $= 7.20 > 4.6$, reject H_0.

13.2.3.

Source	df	SS	MS	F	P
Additive	1	0.03	0.03	4.19	0.0865
Batch	6	0.02	0.00	0.41	0.8483
Error	6	0.05	0.01		
Total	13	0.10			

Since the F statistic $= 4.19 < F_{.95,1,6} = 5.99$, accept H_0.

13.2.5. From the Table 13.2.9, we obtain $MSE = 6.00$. The radius of the Tukey interval is $D\sqrt{MSE} = (Q_{.05,3,22}/\sqrt{b})\sqrt{6.00} = (3.56/\sqrt{12})\sqrt{6.00} = 2.517$. The Tukey intervals are

Pairwise Difference	$\bar{y}_{.s} - \bar{y}_{.t}$	Tukey Interval	Conclusion
$\mu_1 - \mu_2$	−2.41	(−4.93, 0.11)	NS
$\mu_1 - \mu_3$	−0.54	(−3.06, 1.98)	NS
$\mu_2 - \mu_3$	1.87	(−0.65, 4.39)	NS

From this analysis and that of Case Study 13.2.3, we find that the significant difference occurs not for overall means testing or pairwise comparisons, but for the comparison of "during the full moon" with "not during the full moon."

13.2.7.

Pairwise Difference	$\bar{y}_{.s} - \bar{y}_{.t}$	Tukey Interval	Conclusion
$\mu_1 - \mu_2$	2.925	(0.78, 5.07)	Reject
$\mu_1 - \mu_3$	1.475	(−0.67, 3.62)	NS
$\mu_2 - \mu_3$	−1.450	(−3.60, 0.70)	NS

13.2.9. (a)

Source	df	SS	MS	F	P
Sleep stages	2	16.99	8.49	4.13	0.0493
Shrew	5	195.44	39.09	19.00	0.0001
Error	10	20.57	2.06		
Total	17	233.00			

(b) Since the observed F ratio $= 2.42 < F_{.95,1,10} = 4.96$, accept the subhypothesis. For the contrast $C_1 = -\frac{1}{2}\mu_1 - \frac{1}{2}\mu_2 + \mu_3$, $SS_{C_1} = 4.99$. For the contrast $C_2 = \mu_1 - \mu_2$, $SS_{C_2} = 12.00$. Then $SSTR = 16.99 = 4.99 + 12.00 = SS_{C_1} + SS_{C_2}$.

13.2.11. *Equation 13.2.2:*

$$SSTR = \sum_{i=1}^{b}\sum_{j=1}^{k}\left(\overline{Y}_{.j} - \overline{Y}_{..}\right)^2 = b\sum_{j=1}^{k}\left(\overline{Y}_{.j} - \overline{Y}_{..}\right)^2$$

$$= b\sum_{j=1}^{k}\left(\overline{Y}_{.j}^2 - 2\overline{Y}_{.j}\overline{Y}_{..} + \overline{Y}_{..}^2\right)$$

$$= b\sum_{j=1}^{k}\overline{Y}_{.j}^2 - 2b\overline{Y}_{..}\sum_{j=1}^{k}\overline{Y}_{.j} + bk\overline{Y}_{..}^2$$

$$= b\sum_{j=1}^{k}\frac{T_{.j}^2}{b^2} - \frac{2T_{..}^2}{bk} + \frac{T_{..}^2}{bk} = \sum_{j=1}^{k}\frac{T_{.j}^2}{b} - \frac{T_{..}^2}{bk} = \sum_{j=1}^{k}\frac{T_{.j}^2}{b} - c$$

Equation 13.2.3:

$$SSB=\sum_{i=1}^{b}\sum_{j=1}^{k}(\overline{Y}_{i.}-\overline{Y}_{..})^2=k\sum_{i=1}^{b}(\overline{Y}_{i.}-\overline{Y}_{..})^2$$

$$=k\sum_{i=1}^{b}(\overline{Y}_{i.}^2-2\overline{Y}_{i.}\overline{Y}_{..}+\overline{Y}_{..}^2)=k\sum_{i=1}^{b}\overline{Y}_{i.}^2-2k\overline{Y}_{..}\sum_{i=1}^{b}\overline{Y}_{i.}$$

$$+bk\overline{Y}_{..}^2$$

$$=k\sum_{i=1}^{b}\frac{T_{i.}^2}{k^2}-\frac{2T_{..}^2}{bk}+\frac{T_{..}^2}{bk}=\sum_{i=1}^{b}\frac{T_{i.}^2}{k}-\frac{T_{..}^2}{bk}=\sum_{i=1}^{b}\frac{T_{i.}^2}{k}-c$$

Equation 13.2.4:

$$SSTOT=\sum_{i=1}^{b}\sum_{j=1}^{k}(Y_{ij}-\overline{Y}_{..})^2=\sum_{i=1}^{b}\sum_{j=1}^{k}(Y_{ij}^2-2Y_{ij}\overline{Y}_{..}+\overline{Y}_{..}^2)$$

$$=\sum_{i=1}^{b}\sum_{j=1}^{k}Y_{ij}^2-2\overline{Y}_{..}\sum_{i=1}^{b}\sum_{j=1}^{k}Y_{ij}+bk\overline{Y}_{..}^2$$

$$=\sum_{i=1}^{b}\sum_{j=1}^{k}Y_{ij}^2-\frac{2T_{..}^2}{bk}+\frac{T_{..}^2}{bk}=\sum_{i=1}^{b}\sum_{j=1}^{k}Y_{ij}^2-c$$

13.2.13. (a) False. They are equal only when $b=k$.
(b) False. If neither treatment levels nor blocks are significant, it is possible to have F variables

$$\frac{SSTR/(k-1)}{SSE/(b-1)(k-1)} \text{ and } \frac{SSB/(b-1)}{SSE/(b-1)(k-1)} \text{ both} < 1.$$

In that case both $SSTR$ and SSB are less than SSE.

Section 13.3

13.3.1. Since $1.51 < 1.7341 = t_{.05,18}$, do not reject H_0.

13.3.3. $\alpha = 0.05$: Since $-t_{.025,11} = -2.2010 < 0.74 < 2.2010 = t_{.025,11}$, accept H_0.
$\alpha = 0.01$: Since $-t_{.005,11} = -3.1058 < 0.74 < 3.1058 = t_{.005,11}$, accept H_0.

13.3.5. Since $-t_{.025,6} = -2.4469 < -2.0481 < 2.4469 = t_{.025,6}$, accept H_0. The square of the observed Student t statistic $=$ $(-2.0481)^2 = 4.1947 =$ the observed F statistic. Also, $(t_{.025,6})^2 = (2.4469)^2 = 5.987 = F_{.95,1,6}$. Conclusion: the square of the t statistic for paired data is the randomized block design statistic for 2 treatments.

13.3.7. $(-0.21, 0.43)$

CHAPTER 14

Section 14.2

14.2.1. Here, $x = 8$ of the $n = 10$ groups were larger than the hypothesized median of 9. The P-value is $P(X \geq 8) + P(X \leq 2) = 0.000977 + 0.009766 + 0.043945 + 0.043945 + 0.009766 + 0.000977 = 2(0.054688) = 0.109376$.

14.2.3. The median of $f_Y(y)$ is 0.693. There are $x = 22$ values that exceed the hypothesized median of 0.693. The test statistic is $z = \dfrac{22 - 50/2}{\sqrt{50/4}} = -0.85$. Since $-z_{0.025} = -1.96 < -0.85 < z_{0.025} = 1.96$, do not reject H_0.

14.2.5.

y_+	$P(Y_+ = y_+)$
0	1/128
1	7/128
2	21/128
3	35/128
4	35/128
5	21/128
6	7/128
7	1/128

Possible levels for a one-sided test: 1/128, 8/128, 29/128, etc.

14.2.7. $P(Y_+ \leq 6) = 0.0835$; $P(Y_+ \leq 7) = 0.1796$. The closest test to one with $\alpha = 0.10$ is to reject H_0 if $y_+ \leq 6$. Since $y_+ = 9$, accept H_0. Since the observed t statistic $= -1.71 < -1.330 = -t_{.10,18}$, reject H_0.

14.2.9. The approximate, large-sample observed Z ratio is 1.89. Accept H_0, since $-z_{.025} = -1.96 < 1.89 < 1.96 = z_{.025}$.

14.2.11. From Table 13.3.1, the number of pairs where $x_i > y_i$ is 7. The P-value for this test is $P(U \geq 7) + P(U \leq 3) = 2(0.17186) = 0.343752$. Since the P-value exceeds $\alpha = 0.05$, do not reject the null hypothesis, which is the conclusion of Case Study 13.3.1.

Section 14.3

14.3.1. For the critical values of 7 and 29, $\alpha = 0.148$. Since $w = 9$, accept H_0.

14.3.3. The observed Z statistic has value 0.99. Since $-z_{.025} = -1.96 < 0.99 < 1.96 = z_{.025}$, accept H_0.

14.3.5. Since $w' = \dfrac{61.0 - 95}{\sqrt{617.5}} = -1.37 < -1.28 = -z_{.10}$, reject H_0. The sign test accepted H_0.

14.3.7. The signed rank test should have more power since it uses more of the information in the data.

14.3.9. A reasonable assumption is that alcohol abuse shortens life span. In that case, reject H_0 if the test statistic is less than $-z_{0.05} = -1.64$. Since the test statistic has value -1.88, reject H_0.

Section 14.4

14.4.1. Assume the data within groups are independent and that the group distributions have the same shape. Let the null hypothesis be that teachers' expectations do not matter. The Kruskal-Wallis statistics has value $b = 5.64$. Since $5.64 < 5.991 = \chi_{0.95,2}$, accept H_0.

14.4.3. Since $b = 1.68 < 3.841 = \chi^2_{.95,1}$, do not reject H_0.

14.4.5. Since $b = 10.72 > 7.815 = \chi^2_{.95,3}$, reject H_0.

14.4.7. Since $b = 12.48 > 5.991 = \chi^2_{.95,2}$, reject H_0.

Section 14.5

14.5.1. Since $g = 8.8 < 9.488 = \chi^2_{.95,4}$, accept H_0.

14.5.3. Since $g = 17.0 > 5.991 = \chi^2_{.95,2}$, reject H_0.

14.5.5. Since $g = 8.4 < 9.210 = \chi_{0.99,2}$, accept H_0. On the other hand, using the analysis of variance, the null hypothesis would be rejected at this level.

Section 14.6

14.6.1. **(a)** For these data, $w = 23$ and $z = -0.53$. Since $-z_{.025} = -1.96 < -0.53 < 1.96 = z_{.025}$, accept H_0 and assume the sequence is random.
(b) For these data, $w = 21$ and $z = -1.33$. Since $-z_{.025} = -1.96 < -1.33 < 1.96 = z_{.025}$, accept H_0 and assume the sequence is random.

14.6.3. For these data, $w = 19$ and $z = 1.68$. Since $-z_{.025} = -1.96 < 1.68 < 1.96 = z_{.025}$, accept H_0 and assume the sequence is random.

14.6.5. For these data, $w = 25$ and $z = -0.51$. Since $-z_{.025} = -1.96 < -0.51 < 1.96 = z_{.025}$, accept H_0 at the 0.05 level of significance and assume the sequence is random.

BIBLIOGRAPHY

1. Advanced Placement Program, Summary Reports. New York: The College Board, 1996.
2. Agresti, Alan, and Winner, Larry. "Evaluating Agreement and Disagreement among Movie Reviewers." *Chance*, 10, no. 2 (1997), pp. 10–14.
3. Asimov, I. *Asimov on Astronomy*. New York: Bonanza Books, 1979, p. 31.
4. Ayala, F. J. "The Mechanisms of Evolution." *Evolution, A Scientific American Book*. San Francisco: W. H. Freeman, 1978, pp. 14–27.
5. Ball, J. A. C., and Taylor, A. R. "The Effect of Cyclandelate on Mental Function and Cerebral Blood Flow in Elderly Patients," in *Research on the Cerebral Circulation*. Edited by John Stirling Meyer, Helmut Lechner, and Otto Eichhorn. Springfield, Ill.: Thomas, 1969.
6. Barnicot, N. A., and Brothwell, D. R. "The Evaluation of Metrical Data in the Comparison of Ancient and Modern Bones," in *Medical Biology and Etruscan Origins*. Edited by G. E. W. Wolstenholme and Cecilia M. O'Connor. Boston: Little, Brown and Company, 1959, pp. 131–149.
7. Barnothy, Jeno M. "Development of Young Mice," in *Biological Effects of Magnetic Fields*. Edited by Madeline F. Barnothy. New York: Plenum Press, 1964, pp. 93–99.
8. Bartle, Robert G. *The Elements of Real Analysis*, 2nd ed. New York: John Wiley & Sons, 1976.
9. Bellany, Ian. "Strategic Arms Competition and the Logistic Curve." *Survival*, 16 (1974), pp. 228–230.
10. Berger, R. J., and Walker, J. M. "A Polygraphic Study of Sleep in the Tree Shrew." *Brain, Behavior and Evolution*, 5 (1972), pp. 54–69.
11. Blackman, Sheldon, and Catalina, Don. "The Moon and the Emergency Room." *Perceptual and Motor Skills*, 37 (1973), pp. 624–626.
12. Bortkiewicz, L. *Das Gesetz der Kleinen Zahlen*. Leipzig: Teubner, 1898.
13. Boyd, Edith. "The Specific Gravity of the Human Body." *Human Biology*, 5 (1933), pp. 651–652.
14. Breed, M. D., and Byers, J. A. "The Effect of Population Density on Spacing Patterns and Behavioral Interactions in the Cockroach, *Byrsotria fumigata* (Guerin)." *Behavioral and Neural Biology*, 27 (1979), pp. 523–531.
15. Brien, A. J., and Simon, T. L. "The Effects of Red Blood Cell Infusion on 10-km Race Time." *Journal of the American Medical Association*, May 22 (1987), pp. 2761–2765.
16. Brinegar, Claude S. "Mark Twain and the Quintus Curtius Snodgrass Letters: A Statistical Test of Authorship." *Journal of the American Statistical Association*, 58 (1963), pp. 85–96.
17. Brown, L. E., and Littlejohn, M. J. "Male Release Call in the *Bufo americanus* Group," in *Evolution in the Genus Bufo*. Edited by W. F. Blair. Austin, Tx.: University of Texas Press, 1972, p. 316.
18. Buchanav, T. M., Brooks, G. F., and Brachman, P. S. "The Tularemia Skin Test." *Annals of Internal Medicine*, 74 (1971), pp. 336–343.
19. Burns, Alvin C., and Bush, Ronald F. *Marketing Research*. Englewood Cliffs, N.J.: Prentice Hall, 1995.
20. Carlson, T. "Uber Geschwindigkeit und Grosse der Hefevermehrung in Wurze." *Biochemishe Zeitschrift*, 57 (1913), pp. 313–334.
21. Casler, Lawrence. "The Effects of Hypnosis on GESP." *Journal of Parapsychology*, 28 (1964), pp. 126–134.
22. *Chronicle of Higher Education*. April 25, 1990.
23. Clason, Clyde B. *Exploring the Distant Stars*. New York: G. P. Putnam's Sons, 1958, p. 337.
24. Cochran, W. G. "Approximate Significance Levels of the Behrens–Fisher Test." *Biometrics*, 20 (1964), pp. 191–195.
25. Cochran, W. G., and Cox, Gertrude M. *Experimental Designs*, 2nd ed. New York: John Wiley & Sons, 1957, p. 108.
26. Cohen, B. "Getting Serious About Skills." *Virginia Review*, 71 (1992).
27. Collins, Robert L. "On the Inheritance of Handedness." *Journal of Heredity*, 59, no. 1 (1968).
28. Conover, W. J. *Practical Nonparametric Statistics*. New York: John Wiley & Sons, Inc., 1999.
29. Cooil, B. "Using Medical Malpractice Data to Predict the Frequency of Claims: A Study of Poisson Process Models with Random Effects." *Journal of the American Statistical Association*, 86 (1991), pp. 285–295.
30. Coulson, J. C. "The Significance of the Pair-bond in the Kittiwake," in *Parental Behavior in Birds*. Edited by Rae Silver. Stroudsburg, Pa.: Dowden, Hutchinson, & Ross, 1977, pp. 104–113.
31. Craf, John R. *Economic Development of the U.S.* New York: McGraw-Hill, 1952, pp. 368–371.
32. Cummins, Harold, and Midlo, Charles. *Finger Prints, Palms, and Soles*. Philadelphia: Blakiston Company, 1943.
33. *Dallas Morning News*. January 29, 1995.
34. David, F. N. *Games, Gods, and Gambling*. New York: Hafner, 1962, p. 16.
35. Davis, D. J. "An Analysis of Some Failure Data." *Journal of the American Statistical Association*, 47 (1952), pp. 113–150.

36. Davis, M. "Premature Mortality among Prominent American Authors Noted for Alcohol Abuse." *Drug and Alcohol Dependence*, 18 (1986), pp. 133–138.

37. Dubois, Cora, ed. *Lowie's Selected Papers in Anthropology*. Berkeley, Calif.: University of California Press, 1960, pp. 137–142.

38. Dunn, Olive Jean, and Clark, Virginia A. *Applied Statistics: Analysis of Variance and Regression*. New York: John Wiley & Sons, 1974, pp. 339–340.

39. Evans, B. Personal communication.

40. Fadelay, Robert Cunningham. "Oregon Malignancy Pattern Physiographically Related to Hanford, Washington Radioisotope Storage." *Journal of Environmental Health*, 27 (1965), pp. 883–897.

41. Fagen, Robert M. "Exercise, Play, and Physical Training in Animals," in *Perspectives in Ethology*. Edited by P. P. G. Bateson and Peter H. Klopfer. New York: Plenum Press, 1976, pp. 189–219.

42. Fairley, William B. "Evaluating the 'Small' Probability of a Catastrophic Accident from the Marine Transportation of Liquefied Natural Gas," in *Statistics and Public Policy*. Edited by William B. Fairley and Frederick Mosteller. Reading, Mass.: Addison-Wesley, 1977, pp. 331–353.

43. Feller, W. "Statistical Aspects of ESP." *Journal of Parapsychology*, 4 (1940), pp. 271–298.

44. Finkbeiner, Daniel T. *Introduction to Matrices and Linear Transformations*. San Francisco: W. H. Freeman, 1960.

45. Fishbein, Morris. *Birth Defects*. Philadelphia: Lippincott, 1962, p. 177.

46. Fisher, R. A. "On the 'Probable Error' of a Coefficient of Correlation Deduced from a Small Sample." *Metron*, 1 (1921), pp. 3–32.

47. ______. "On the Mathematical Foundations of Theoretical Statistics." *Philosophical Transactions of the Royal Society of London, Series A*, 222 (1922), pp. 309–368.

48. ______. *Contributions to Mathematical Statistics.* New York: John Wiley & Sons, 1950, pp. 265–272.

49. Fisz, Marek. *Probability Theory and Mathematical Statistics*, 3rd ed. New York: John Wiley & Sons, 1963, pp. 358–361.

50. Florida Department of Commerce. February 20, 1996.

51. *Forbes Magazine*. October 10, 1994.

52. ______. November 2, 2009.

53. Free, J. B. "The Stimuli Releasing the Stinging Response of Honeybees." *Animal Behavior*, 9 (1961), pp. 193–196.

54. Freund, John E. *Mathematical Statistics*, 2nd ed. Englewood Cliffs, N.J.: Prentice Hall, 1971, p. 226.

55. Fricker, Ronald D., Jr. "The Mysterious Case of the Blue M&M's." *Chance*, 9, no. 4 (1996), pp. 19–22.

56. Fry, Thornton C. *Probability and Its Engineering Uses*, 2nd ed. New York: Van Nostrand-Reinhold, 1965, pp. 206–209.

57. Furuhata, Tanemoto, and Yamamoto, Katsuichi. *Forensic Odontology*. Springfield, Ill.: Thomas, 1967, p. 84.

58. Galton, Francis. *Natural Inheritance*. London: Macmillan, 1908.

59. Gardner, C. D. et al. "Comparison of the Atkins, Zone, Ornish, and LEARN Diets for Change in Weight and Related Risk Factors Among Overweight Premenopausal Women." *Journal of the American Medical Association*, 297 (2007), pp. 969–977.

60. Gendreau, Paul, et al. "Changes in EEG Alpha Frequency and Evoked Response Latency During Solitary Confinement." *Journal of Abnormal Psychology*, 79 (1972), pp. 54–59.

61. Geotis, S. "Thunderstorm Water Contents and Rain Fluxes Deduced from Radar." *Journal of Applied Meteorology*, 10 (1971), p. 1234.

62. Gerber, Robert C., et al. "Kinetics of Aurothiomalate in Serum and Synovial Fluid." *Arthritis and Rheumatism*, 15 (1972), pp. 625–629.

63. Goldman, Malcomb. *Introduction to Probability and Statistics*. New York: Harcourt, Brace & World, 1970, pp. 399–403.

64. Goodman, Leo A. "Serial Number Analysis." *Journal of the American Statistical Association*, 47 (1952), pp. 622–634.

65. Griffin, Donald R., Webster, Frederick A., and Michael, Charles R. "The Echolocation of Flying Insects by Bats." *Animal Behavior*, 8 (1960), p. 148.

66. Grover, Charles A. "Population Differences in the Swell Shark *Cephaloscyllium ventriosum*." *California Fish and Game*, 58 (1972), pp. 191–197.

67. Gutenberg, B., and Richter, C. F. *Seismicity of the Earth and Associated Phenomena*. Princeton, N.J.: Princeton University Press, 1949.

68. Haggard, William H., Bilton, Thaddeus H., and Crutcher, Harold L. "Maximum Rainfall from Tropical Cyclone Systems which Cross the Appalachians." *Journal of Applied Meteorology*, 12 (1973), pp. 50–61.

69. Haight, F. A. "Group Size Distributions with Applications to Vehicle Occupancy," in *Random Counts in Physical Science, Geological Science, and Business*, vol. 3. Edited by G. P. Patil. University Park, Pa.: Pennsylvania State University Press, 1970.

70. Hankins, F. H. "Adolph Quetelet as Statistician," in *Studies in History, Economics, and Public Law*, xxxi, no. 4, New York: Longman, Green, 1908, p. 497.

71. Hansel, C. E. M. *ESP: A Scientific Evaluation*. New York: Scribner's, 1966, pp. 86–89.

72. Hare, Edward, Price, John, and Slater, Eliot. "Mental Disorders and Season of Birth: A National Sample Compared with the General Population." *British Journal of Psychiatry*, 124 (1974), pp. 81–86.

73. Hassard, Thomas H. *Understanding Biostatistics*. St. Louis, Mo.: Mosby Year Book, 1991.

74. Hasselblad, V. "Estimation of Finite Mixtures of Distributions from the Exponential Family." *Journal of the American Statistical Association*, 64 (1969), pp. 1459–1471.

75. Heath, Clark W., and Hasterlik, Robert J. "Leukemia among Children in a Suburban Community." *The American Journal of Medicine*, 34 (1963), pp. 796–812.

76. Hendy, M. F., and Charles, J. A. "The Production Techniques, Silver Content and Circulation History of the Twelfth-Century Byzantine Trachy." *Archaeometry*, 12 (1970), pp. 13–21.

77. Hersen, Michel. "Personality Characteristics of Nightmare Sufferers." *Journal of Nervous and Mental Diseases*, 153 (1971), pp. 29–31.

78. Hill, T. P. "The First Digit Phenomenon." *American Scientist*, 86 (1998), pp. 358–363.

79. Hogben, D., Pinkham, R. S., and Wilk, M. B. "The Moments of the Non-central t-distribution." *Biometrika*, 48 (1961), pp. 465–468.

80. Hogg, Robert V., McKean, Joseph W., and Craig, Allen T. *Introduction to Mathematical Statistics*, 6th ed. Upper Saddle River; N.J.: Pearson Prentice Hall, 2005.

81. Hollander, Myles, and Wolfe, Douglas A. *Nonparametric Statistical Methods.* New York: John Wiley & Sons, 1973, pp. 272–282.

82. Horvath, Frank S., and Reid, John E. "The Reliability of Polygraph Examiner Diagnosis of Truth and Deception." *Journal of Criminal Law, Criminology, and Police Science*, 62 (1971), pp. 276–281.

83. Howell, John M. "A Strange Game." *Mathematics Magazine*, 47 (1974), pp. 292–294.

84. Hudgens, Gerald A., Denenberg, Victor H., and Zarrow, M. X. "Mice Reared with Rats: Effects of Preweaning and Postweaning Social Interactions upon Behaviour." *Behaviour*, 30 (1968), pp. 259–274.

85. Hulbert, Roger H., and Krumbiegel, Edward R. "Synthetic Flavors Improve Acceptance of Anticoagulant-Type Rodenticides." *Journal of Environmental Health*, 34 (1972), pp. 402–411.

86. Huxtable, J., Aitken, M. J., and Weber, J. C. "Thermoluminescent Dating of Baked Clay Balls of the Poverty Point Culture." *Archaeometry*, 14 (1972), pp. 269–275.

87. Hyneck, Joseph Allen. *The UFO Experience: A Scientific Inquiry*. Chicago: Rognery, 1972.

88. Ibrahim, Michel A., et al. "Coronary Heart Disease: Screening by Familial Aggregation." *Archives of Environmental Health*, 16 (1968), pp. 235–240.

89. Jones, Jack Colvard, and Pilitt, Dana Richard. "Blood-feeding Behavior of Adult *Aedes Aegypti* Mosquitoes." *Biological Bulletin*, 31 (1973), pp. 127–139.

90. Karlsen, Carol F. *The Devil in the Shape of a Woman*. New York: W. W. Norton & Company, 1998, p. 51.

91. Kendall, Maurice G. "The Beginnings of a Probability Calculus," in *Studies in the History of Statistics and Probability*. Edited by E. S. Pearson and Maurice G. Kendall. Darien, Conn.: Hafner, 1970, pp. 8–11.

92. Kendall, Maurice G., and Stuart, Alan. *The Advanced Theory of Statistics*, vol. 1. New York: Hafner, 1961.

93. ______. *The Advanced Theory of Statistics*, vol. 2. New York: Hafner, 1961.

94. Kruk-DeBruin, M., Rost, Luc C. M., and Draisma, Fons G. A. M. "Estimates of the Number of Foraging Ants with the Lincoln-Index Method in Relation to the Colony Size of *Formica Polyctena*." *Journal of Animal Ecology*, 46 (1977), pp. 463–465.

95. Larsen, Richard J., and Marx, Morris L. *An Introduction to Probability and Its Applications*. Englewood Cliffs, N.J.: Prentice Hall, 1985.

96. ______. *An Introduction to Mathematical Statistics and Its Applications*, 2nd ed. Englewood Cliffs, N.J.: Prentice-Hall, 1986, pp. 452–453.

97. ______. *An Introduction to Mathematical Statistics and Its Applications*, 3rd ed. Upper Saddle River, N.J.: Prentice Hall, 2001, pp. 181–182.

98. Lathem, Edward Connery, ed. *The Poetry of Robert Frost.* New York: Holt, Rinehart and Winston, 1970.

99. Lemmon, W. B., and Patterson, G. H. "Depth Perception in Sheep: Effects of Interrupting the Mother-Neonate Bond," in *Comparative Psychology: Research in Animal Behavior.* Edited by M. R. Denny and S. Ratner. Homewood, Ill.: Dorsey Press, 1970, p. 403.

100. Lemon, Robert E., and Chatfield, Christopher. "Organization of Song in Cardinals." *Animal Behaviour*, 19 (1971), pp. 1–17.
101. Li, Frederick P. "Suicide Among Chemists." *Archives of Environmental Health*, 19 (1969), pp. 519–520.
102. Lindgren, B. W. *Statistical Theory*. New York: Macmillan, 1962.
103. Linnik, Y. V. *Method of Least Squares and Principles of the Theory of Observations*. Oxford: Pergamon Press, 1961, p. 1.
104. Longwell, William. Personal communication.
105. Lottenbach, K. "Vasomotor Tone and Vascular Response to Local Cold in Primary Raynaud's Disease." *Angiology*, 32 (1971), pp. 4–8.
106. MacDonald, G. A., and Abbott, A. T. *Volcanoes in the Sea*. Honolulu: University of Hawaii Press, 1970.
107. Maistrov, L. E. *Probability Theory—A Historical Sketch*. New York: Academic Press, 1974.
108. Mann, H. B. *Analysis and Design of Experiments*. New York: Dover, 1949.
109. Mares, M. A., et al. "The Strategies and Community Patterns of Desert Animals," in *Convergent Evolution in Warm Deserts.* Edited by G. H. Orians and O. T. Solbrig. Stroudsberg, Pa.: Dowden, Hutchinson & Ross, 1977, p. 141.
110. Marx, Morris L. Personal communication.
111. McIntyre, Donald B. "Precision and Resolution in Geochronometry," in *The Fabric of Geology*. Edited by Claude C. Albritton, Jr. Stanford, Calif.: Freeman, Cooper, and Co., 1963, pp. 112–133.
112. Mendel, J. G. "Experiments in Plant Hybridization." *Journal of the Royal Horticultural Society*, 26 (1866), pp. 1–32.
113. Merchant, L. *The National Football Lottery.* New York: Holt, Rinehart and Winston, 1973.
114. Miettinen, Jorma K. "The Accumulation and Excretion of Heavy Metals in Organisms," in *Heavy Metals in the Aquatic Environment*. Edited by P. A. Krenkel. Oxford: Pergamon Press, 1975, pp. 155–162.
115. Morgan, Peter J. "A Photogrammetric Survey of Hoseason Glacier, Kemp Coast, Antarctica." *Journal of Glaciology*, 12 (1973), pp. 113–120.
116. Mulcahy, R., McGilvray, J. W., and Hickey, N. "Cigarette Smoking Related to Geographic Variations in Coronary Heart Disease Mortality and to Expectation of Life in the Two Sexes." *American Journal of Public Health*, 60 (1970), pp. 1515–1521.
117. Munford, A. G. "A Note on the Uniformity Assumption in the Birthday Problem." *American Statistician*, 31 (1977), p. 119.
118. Nakano, T. "Natural Hazards: Report from Japan," in *Natural Hazards*. Edited by G. White. New York: Oxford University Press, 1974, pp. 231–243.
119. Nash, Harvey. *Alcohol and Caffeine*. Springfield, Ill.: Thomas, 1962, p. 96.
120. *Nashville Banner*. November 9, 1994.
121. *New York Times* (New York). May 22, 2005.
122. ______. October 7, 2007.
123. *Newsweek*. March 6, 1978.
124. Nye, Francis Iven. *Family Relationships and Delinquent Behavior*. New York: John Wiley & Sons, 1958, p. 37.
125. Olmsted, P. S. "Distribution of Sample Arrangements for Runs Up and Down." *Annals of Mathematical Statistics*, 17 (1946), pp. 24–33.
126. Olvin, J. F. "Moonlight and Nervous Disorders." *American Journal of Psychiatry*, 99 (1943), pp. 578–584.
127. Ore, O. *Cardano, The Gambling Scholar*. Princeton, N.J.: Princeton University Press, 1963, pp. 25–26.
128. Papoulis, Athanasios. *Probability, Random Variables, and Stochastic Processes*. New York: McGraw-Hill, 1965, pp. 206–207.
129. Passingham, R. E. "Anatomical Differences between the Neocortex of Man and Other Primates." *Brain, Behavior and Evolution*, 7 (1973), pp. 337–359.
130. Pearson, E. S., and Kendall, Maurice G. *Studies in the History of Statistics and Probability*. London: Griffin, 1970.
131. Peberdy, M. A., et al. "Survival from In-Hospital Cardiac Arrest During Nights and Weekends." *Journal of the American Medical Association*, 299 (2008), pp. 785–792.
132. *Pensacola News Journal* (Florida). May 25, 1997.
133. ______. September 21, 1997.
134. Phillips, David P. "Deathday and Birthday: An Unexpected Connection," in *Statistics: A Guide to the Unknown*. Edited by Judith M. Tanur, et al. San Francisco: Holden-Day, 1972.
135. Pierce, George W. *The Songs of Insects*. Cambridge, Mass.: Harvard University Press, 1949, pp. 12–21.

136. Polya, G. "Uber den Zentralen Grenzwertsatz der Wahrscheinlichkeitsrechnung und das Momenten-problem." *Mathematische Zeitschrift*, 8 (1920), pp. 171–181.

137. Porter, John W., et al. "Effect of Hypnotic Age Regression on the Magnitude of the Ponzo Illusion." *Journal of Abnormal Psychology*, 79 (1972), pp. 189–194.

138. Ragsdale, A. C., and Brody, S. *Journal of Dairy Science*, 5 (1922), p. 214.

139. Rahman, N. A. *Practical Exercises in Probability and Statistics*. New York: Hafner, 1972.

140. Reichler, Joseph L., ed. *The Baseball Encyclopedia*, 4th ed. New York: Macmillan, 1979, p. 1350.

141. Resnick, Richard B., Fink, Max, and Freedman, Alfred M. "A Cyclazocine Typology in Opiate Dependence." *American Journal of Psychiatry*, 126 (1970), pp. 1256–1260.

142. Rich, Clyde L. "Is Random Digit Dialing Really Necessary?" *Journal of Marketing Research*, 14 (1977), pp. 242–250.

143. Richardson, Lewis F. "The Distribution of Wars in Time." *Journal of the Royal Statistical Society*, 107 (1944), pp. 242–250.

144. Ritter, Brunhilde. "The Use of Contact Desensitization, Demonstration-Plus-Participation and Demonstration-Alone in the Treatment of Acrophobia." *Behaviour Research and Therapy*, 7 (1969), pp. 157–164.

145. Roberts, Charlotte A. "Retraining of Inactive Medical Technologists—Whose Responsibility?" *American Journal of Medical Technology*, 42 (1976), pp. 115–123.

146. Rohatgi, V. K. *An Introduction to Probability Theory and Mathematical Statistics*. New York: John Wiley & Sons, 1976, p. 81.

147. Rosenthal, R., and Jacobson, L. F. "Teacher Expectations for the Disadvantaged." *Scientific American*, 218 (1968), pp. 19–23.

148. Ross, Sheldon. *A First Course in Probability*, 7th ed. Upper Saddle River, N.J.: Pearson Prentice Hall, 2006, pp. 51–53.

149. Roulette, Amos. "An Assessment of Unit Dose Injectable Systems." *American Journal of Hospital Pharmacy*, 29 (1972), pp. 60–62.

150. Rowley, Wayne A. "Laboratory Flight Ability of the Mosquito *Culex Tarsalis Coq*." *Journal of Medical Entomology*, 7 (1970), pp. 713–716.

151. Roy, R. H. *The Cultures of Management*. Baltimore: Johns Hopkins University Press, 1977, p. 261.

152. Rutherford, Sir Ernest, Chadwick, James, and Ellis, C. D. *Radiations from Radioactive Substances*. London: Cambridge University Press, 1951, p. 172.

153. Sagan, Carl. *Cosmos*. New York: Random House, 1980, pp. 298–302.

154. Salvosa, Carmencita B., Payne, Philip R., and Wheeler, Erica F. "Energy Expenditure of Elderly People Living Alone or in Local Authority Homes." *American Journal of Clinical Nutrition*, 24 (1971), pp. 1467–1470.

155. Santa-Clara, P., and Valkanov, R. I. "The Presidential Puzzle: Political Cycles and the Stock Market." *Journal of Finance*, 58 (2003), pp. 1841–1872.

156. Saturley, B. A. "Colorimetric Determination of Cyclamate in Soft Drinks, Using Picryl Chloride." *Journal of the Association of Official Analytical Chemists*, 55 (1972), pp. 892–894.

157. Schaller, G. B. "The Behavior of the Mountain Gorilla," in *Primate Patterns*. Edited by P. Dolhinow. New York: Holt, Rinehart and Winston, 1972, p. 95.

158. Schell, E. D. "Samuel Pepys, Isaac Newton, and Probability." *The American Statistician*, 14 (1960), pp. 27–30.

159. Schoeneman, Robert L., Dyer, Randolph H., and Earl, Elaine M. "Analytical Profile of Straight Bourbon Whiskies." *Journal of the Association of Official Analytical Chemists*, 54 (1971), pp. 1247–1261.

160. Selective Service System. Office of the Director. Washington, D.C., 1969.

161. Sen, Nrisinha, et al. "Effect of Sodium Nitrite Concentration on the Formation of Nitrosopyrrolidine and Dimethyl Nitrosamine in Fried Bacon." *Journal of Agricultural and Food Chemistry*, 22 (1974), pp. 540–541.

162. Sharpe, Roger S., and Johnsgard, Paul A. "Inheritance of Behavioral Characters in F_2 Mallard x Pintail (*Anas Platyrhynchos L.* x *Anas Acuta L.*) Hybrids." *Behaviour*, 27 (1966), pp. 259–272.

163. Shaw, G. B. *The Doctor's Dilemma, with a Preface on Doctors*. New York: Brentano's, 1911, p. lxiv.

164. Shore, N. S., Greene, R., and Kazemi, H. "Lung Dysfunction in Workers Exposed to *Bacillus subtilis* Enzyme." *Environmental Research*, 4 (1971), pp. 512–519.

165. Stroup, Donna F. Personal communication.

166. Strutt, John William (Baron Rayleigh). "On the Resultant of a Large Number of Vibrations of the Same Pitch and of Arbitrary Phase." *Philosophical Magazine*, X (1880), pp. 73–78.

167. Sukhatme, P. V. "On Fisher and Behrens' Test of Significance for the Difference in Means of Two Normal Samples." *Sankhya*, 4 (1938), pp. 39–48.

168. Sutton, D. H. "Gestation Period." *Medical Journal of Australia*, 1 (1945), pp. 611–613.

169. Szalontai, S., and Timaffy, M. "Involutional Thrombopathy," in *Age with a Future*. Edited by P. From Hansen. Philadelphia: F. A. Davis, 1964, p. 345.
170. Tanguy, J. C. "An Archaeomagnetic Study of Mount Etna: The Magnetic Direction Recorded in Lava Flows Subsequent to the Twelfth Century." *Archaeometry*, 12, 1970, pp. 115–128.
171. *Tennessean* (Nashville). January 20, 1973.
172. ______. August 30, 1973.
173. ______. July 21, 1990.
174. ______. May 5, 1991.
175. ______. May 12, 1991.
176. ______. December 11, 1994.
177. ______. January 29, 1995.
178. ______. April 25, 1995.
179. Terry, Mary Beth, et al. "Association of Frequency and Duration of Aspirin Use and Hormone Receptor Status with Breast Cancer Risk." *Journal of the American Medical Association*, 291 (2004), pp. 2433–2436.
180. Thorndike, Frances. "Applications of Poisson's Probability Summation." *Bell System Technical Journal*, 5 (1926), pp. 604–624.
181. Treuhaft, Paul S., and McCarty, Daniel J. "Synovial Fluid pH, Lactate, Oxygen and Carbon Dioxide Partial Pressure in Various Joint Diseases." *Arthritis and Rheumatism*, 14 (1971), pp. 476–477.
182. Trugo, L. C., Macrae, R., and Dick, J. "Determination of Purine Alkaloids and Trigonelline in Instant Coffee and Other Beverages Using High Performance Liquid Chromatography." *Journal of the Science of Food and Agriculture*, 34 (1983), pp. 300–306.
183. Turco, Salvatore, and Davis, Neil. "Particulate Matter in Intravenous Infusion Fluids—Phase 3." *American Journal of Hospital Pharmacy*, 30 (1973), pp. 611–613.
184. *USA Today*. May 20, 1991.
185. ______. June 3, 1991.
186. ______. September 20, 1991.
187. ______. March 14, 1994.
188. ______. April 12, 1994.
189. ______. December 30, 1994.
190. ______. May 4, 1995.
191. Vilenkin, N. Y. *Combinatorics*. New York: Academic Press, 1971, pp. 24–26.
192. Vogel, John H. K., Horgan, John A., and Strahl, Cheryl L. "Left Ventricular Dysfunction in Chronic Constrictive Pericarditis." *Chest*, 59 (1971), pp. 484–492.
193. Vogt, E. Z., and Hyman, R. *Water Witching U.S.A.* Chicago: University of Chicago Press, 1959, p. 55.
194. Vol'kenschtein, Mikhail. *Molecules and Life*. New York: Plenum Press, 1973, pp. 301–309.
195. Walker, H. *Studies in the History of Statistical Method*. Baltimore: Williams and Wilkins, 1929.
196. *Wall Street Journal*. March 20, 1994.
197. Wallechinsky, D., Wallace, I., and Wallace, A. *The Book of Lists*. New York: Barton Books, 1978.
198. Wallis, W. A. "The Poisson Distribution and the Supreme Court." *Journal of the American Statistical Association*, 31 (1936), pp. 376–380.
199. Werner, Martha, Stabenau, James R., and Pollin, William. "Thematic Apperception Test Method for the Differentiation of Families of Schizophrenics, Delinquents, and 'Normals.' " *Journal of Abnormal Psychology*, 75 (1970), pp. 139–145.
200. Wilks, Samuel S. *Mathematical Statistics*. New York: John Wiley & Sons, 1962.
201. Williams, Wendy M., and Ceci, Stephen J. "How'm I Doing?" *Change*, 29, no. 5 (1997), pp. 12–23.
202. Winslow, Charles. *The Conquest of Epidemic Disease*. Princeton, N.J.: Princeton University Press, 1943, p. 303.
203. Wolf, Stewart, ed. *The Artery and the Process of Arteriosclerosis: Measurement and Modification*. Proceedings of an Interdisciplinary Conference on Fundamental Data on Reactions of Vascular Tissue in Man (Lindau, West Germany, April 19–25, 1970). New York: Plenum Press, 1972, p. 116.
204. Wolfowitz, J. "Asymptotic Distribution of Runs Up and Down." *Annals of Mathematical Statistics*, 15 (1944), pp. 163–172.
205. Wood, Robert M. "Giant Discoveries of Future Science." *Virginia Journal of Science*, 21 (1970), pp. 169–177.
206. Woodward, W. F. "A Comparison of Base Running Methods in Baseball." M.Sc. Thesis, Florida State University, 1970.
207. Woolson, Robert E. *Statistical Methods for the Analysis of Biomedical Data*. New York: John Wiley & Sons, 1987, p. 302.
208. Wyler, Allen R., Minoru, Masuda, and Holmes, Thomas H. "Magnitude of Life Events and Seriousness of Illness." *Psychosomatic Medicine*, 33 (1971), pp. 70–76.

209. Yochem, Donald, and Roach, Darrell. "Aspirin: Effect on Thrombus Formulation Time and Prothrombin Time of Human Subjects." *Angiology*, 22 (1971), pp. 70–76.

210. Young, P. V., and Schmid, C. *Scientific Social Surveys and Research*. Englewood Cliffs, N.J.: Prentice Hall, 1966, p. 319.

211. Zaret, Thomas M. "Predators, Invisible Prey, and the Nature of Polymorphism in the *Cladocera* (Class *Crustacea*)." *Limnology and Oceanography*, 17 (1972), pp. 171–184.

212. Zelazo, Philip R., Zelazo, Nancy Ann, and Kolb, Sarah. "'Walking' in the Newborn." *Science*, 176 (1972), pp. 314–315.

213. Zelinsky, Daniel A. *A First Course in Linear Algebra*, 2nd ed. New York: Academic Press, 1973.

214. Ziv, G., and Sulman, F. G. "Binding of Antibiotics to Bovine and Ovine Serum." *Antimicrobial Agents and Chemotherapy*, 2 (1972), pp. 206–213.

215. Zucker, N. "The Role of Hood-Building in Defining Territories and Limiting Combat in Fiddler Crabs." *Animal Behaviour*, 29 (1981), pp. 387–395.

Index